Computer Telephony ENCYCLOPEDIA

by Richard Grigonis

Published by CMP Books
An Imprint of CMP Media Inc.
12 West 21 Street
New York, NY 10010

ISBN 1-57820-045-8

For individual orders, and for information on special discounts for quantity orders, please contact:

CMP Books
6600 Silacci Way
Gilroy, CA 95020
Tel: 800-LIBRARY or 408-848-3854
Fax: 408-848-5784
Web: www.cmpbooks.com
Email: cmp@rushorder.com

Distributed to the book trade in the U.S. and Canada by
Publishers Group West
1700 Fourth St., Berkeley, CA 94710

Transferred to Digital Printing 2010

"*. . . le but d'une Encyclopédie est de rassembler les connaissances éparses sur la surface de la terre; d'en exposer le système général aux hommes avec qui nous vivons, & de le transmettre aux hommes qui viendront apres nous; afin que les travaux des siècles passés n'aient pas été des travaux inutiles pour les siècles qui succéderont; que nos neveux, devenant plus instruits, deviennent en meme temps plus vertueux & plus heureux, & que nous ne mourions pas sans avoir bien mérité du genre humain. . . .*"

Fragonard's portrait of Denis Diderot, circa 1769.

TRANSLATION:

". . . the purpose [or goal, aim] of an Encyclopedia is to assemble the knowledge [facts] scattered on the surface of the earth; to exposit its general system to the men with whom we live, & to transmit it to men who will come after us; in order that the labors of past centuries should not have been useless labors for the succeeding centuries; that our nephews, becoming better educated [literally, instructed], should at the same time become more virtuous and happier, & that we should not die without having acted meritoriously toward humankind. . . ."

> – *from the article "Encyclopédie" by Denis Diderot (1713-1784), one of the 72,998 entries in the huge 20.7 million word Encyclopédie ou Dictionnaire raisonné des sciences, des arts et des métiers (Encyclopedia or systematic dictionary of the sciences, arts and crafts), consisting of 17 volumes of text and 11 volumes of 2,569 plates published between 1751 and 1772, followed by a five volume supplement (1776-1777) and a two volume analytic index (1780).*

Acknowledgments

Upon receiving an Honorary Oscar in 1970, Orson Welles said in his acceptance speech that in order to display everyone he had worked with over the years who had contributed to his career "it would require an extreme wide-angle shot of thousands of people projected onto an immense Cinerama screen."

In my case, as time goes on, there is an ever-increasing list of people to thank for assistance and encouragement, both for their contributions to *Computer Telephony* magazine as well as the production of this encyclopedia.

First thanks should go to Harry Newton, for giving me my "big break" in publishing in 1994, sixteen years after I had given up on the idea. After Harry's departure, in 1998 Matt Kelsey, publisher of the Book Division at Miller Freeman (now CMP Media) became interested in my idea of doing a computer telephony encyclopedia – but expressed some skepticism that a single person could successfully tackle such an enormous project. Although I consequently proved it could be done, it was achieved at the cost of a tremendous amount of mental and physical strain that could have been avoided had I listened to him in the first place. Jacob wrestling with the Angel was a cinch compared to this undertaking!

Thanks goes to the many companies in the computer telephony field, their CEOs, VPs of marketing, engineers, and public relations people whom I've dealt with over the years and who have helped to bring about and sustain my seemingly perpetual condition of "information overload." Many thanks to Sharon Gacek and Sean Quigley of the Dialogic Corporation for the arduous task of collecting together information on Dialogic's analog and digital CT resource boards.

A big thank-you also goes to the writers and editors at CMP Media. Just as I was running out of time prior to publication, I was fortunate enough to be able to commission Brendan Read of *Call Center* magazine to infuse his knowledge and wit into the entries on call center outsourcing, training and telemarketing. Other authors volunteered their latest research (thanks to Tim Jones of *Teleconnect* for the key system and voice mail tables), while still others allowed me to incorporate both published and never-before-published data into the book – thanks to Bill Michael, John Jainschigg, and Ellen Muraskin of *Computer Telephony*, Elaine Rowland, Diane Boccadoro, and Andy Green of *Teleconnect*, and Joe Fleischer, Alison Ousey, Lee Hollman and Mary Lenz of *Call Center* magazine.

The final editing was entrusted to ebusiness author and consultant Janice Reynolds, whose thorough review of the manuscript unearthed and resolved a wide variety of grammatical and typographical errors.

Thanks should also go to the CMP art department in general and Robbie Alterio in particular for taking some big text files and lots of graphics and creating something that actually looks like an encyclopedia. And thanks to Christine Kern for riding herd on the editing and production process, and to Lisa Giaquinto for tracking down contact information on hundreds of companies.

And, of course, I'd like to express my sincere appreciation and gratitude to all of you out there who have comprised both *Computer Telephony* magazine's loyal readership and CT Expo's attendees. It is you who are the communications industry's true pioneers and early adopters of advanced Internet / telecom technology.

FOREWORD
Computer Telephony for the New Century

By Howard Bubb

Howard Bubb is President of Dialogic Corporation and Vice President of the Communications Products Group at Intel Corporation. He has long been recognized as one of the foremost leaders and evangelists for the computer telephony industry.

After more than a decade in the back of the telecommunications closet, computer telephony (CT) is finally poised to step out and become one of the new century's most important mainstream technologies. And it's about time.

Why is CT suddenly so hot? And why the delay? After all, CT has been commercially available since the mid-1980s. The answer is simple. In the world of mainstream computing, open systems have long been the norm. But the CT world has been made up of separate, often proprietary systems handling functions like voice processing and computer-based fax. This has made CT systems expensive and hard to manage. It's meant that to add services, customers have had to add separate systems. For instance, even though a company's voice messaging system and interactive voice response (IVR) system might use exactly the same voice processing hardware, there was no way for them to share resources and minimize hardware costs.

Today, that picture is completely changed. Next-generation CT systems are built around an open CT server that works like a PC or database server. With a standard framework, users can choose from a wide range of CT applications and technologies. Multiple applications can share media processing and network resources. Applications can simultaneously support legacy network infrastructure like PBXs and central offices, as well as next-generation Internet protocol (IP) solutions like gatekeepers and soft switches. By enabling choice and flexibility, the CT server makes it easy and affordable to build a best-in class-solution: the best applications, the best technology, the best infrastructure, and the best integration.

Tomorrow's Communication Services

In tomorrow's world, communications will be all about value-added services provided by an application-driven interactive network. With today's networks converging, there will only

be *The* Net – providing high-bandwidth packet communications from a user's pocket, palm, or portable computer to anyone, anywhere.

The convergence of the voice and data networks will enable natural voice interfaces that will make advanced communication services more accessible. CT is becoming a more responsive, application-driven interactive customer service technology that lives on the network's edge.

This new converged world will also mean expanded opportunity – for example, the hosting of services. Service providers can offer integrated applications that deliver real value to their business subscribers, who will see that communications features aren't

On Friday, September 4, 1998, *Computer Telephony* magazine officially unveiled a new product testing facility at their offices in the heart of Manhattan. Industry leader Dialogic reaffirmed its long support of *Computer Telephony* when Dialogic's then-CEO, Howard Bubb (in photo at left) arrived to handle the ribbon-cutting ceremony along with Computer Telephony's Chief Technical Editor and the laboratory's main designer, Richard "Zippy" Grigonis. Also on hand was Warren S. Meyers, the Deputy Commissioner of Network Strategies of the Department of Information Technology and Telecom for the City of New York, who read aloud a congratulatory letter from Rudolph W. Giuliani, Mayor of New York.

just for telcos anymore. Telecom hosting operates telecom applications from within the public network edge. It's a concept whose time has come. The Web has given businesses a completely new view of customer contact and employee productivity functions. Web-enabled applications have provided new ways for businesses to gain a competitive edge. Businesses will expect the same opportunities and competitive advantages with telecom services.

The market potential is huge. An estimated eight million U.S. businesses are targeted as growth companies (Source: Yankee Group). Businesses with their own infrastructure are looking to lower costs, improve operational efficiencies, accommodate growth, and meet changing needs. Large organizations are expected to increase their spending on telecom in the next two years by 56 percent (Source: Schema).

The application-driven interactive network will use open CT servers to offer users an exciting array of news services far beyond today's basic connectivity. Soon, a user's personal and business applications will all live on the same network – making it easy to interact with customers and business partners in the way that serves them best. It will be a new age in communications that changes the way all of us work and live.

Watching CT technology move so quickly into our everyday lives is exciting – especially for those of us who've been involved from the beginning. But the rapid-fire changes also make it crucial for everyone involved in the CT industry to have in-depth, up-to-date reference information in their hands. That's what this and future editions of this encyclopedia will strive to provide.

Introduction

"First, instead of 'computer,' I'd say 'computer and communication,' because unless a computer is linked to something, it really doesn't do much."

– James Burke, author, science historian and producer of such insightful television documentaries on technological history as *Connections, Connections 2,* and *The Day the Universe Changed.*

The above quote from James Burke is particularly apt, serving to evoke the most primitive underlying idea of computer telephony – that computation and communications can both be brought into a close, synergistic relationship. Burke's quote also reflects in microcosm his whole philosophy that the history of science and technology is a "web" of disparate, though linked events (he used the term "web" nearly 15 years before the World Wide Web appeared). It's across the length and breadth of this web where unlikely causal chains of serendipitous discoveries and chance occurrences form and ultimately lead to modern inventions such as the ball point pen, the combustion engine, and even the computer itself.

Indeed, Burke's unique tangled, yet slap-dash, almost picaresque version of history reminds me of the convoluted path that led me to the computer telephony industry in general and the magazine *Computer Telephony* in particular.

If I were a James Burke attempting to tell this wildly improbable tale, I'd begin by making the deliberately provocative statement that it was really the invention of word processing that was responsible for my career in computer telephony, the success of *CT* magazine, and the production of the encyclopedia you are now reading.

Ah, but we are already getting ahead of ourselves.

My first love was physics and engineering, but in the early 1970s Japan was amusing itself by sending "CARE" packages to U.S. aerospace engineers in the Pacific Northwest. While in college (what is now Rowan University in New Jersey) I began to take blocks of courses in various random areas that had no significance or connectedness other than that they caught my fancy: Cultural and physical anthropology, archaeology, art history, film theory, magazine layout and design, Indians of North America, astronomy, Magic and Religion of Primitive Tribal cultures, the medieval world, mesoamerican Pre-Columbian cultures, comparative religion, linguistics, the theory of general semantics, etc.

Watergate had just occurred, and, in the wake of the book *All the President's Men* and the movie of the same name, journalism students seemed to be budding celebrities – at least, they were getting all the girls! So I found myself graduating with a degree in Communications (the Journalism track) and a sort of minor in art history.

It was just as well, as I always had a strong interest in writing. As a child I scrawled and drew illustrations for my first book (done on the back of fanfold paper, like some enormous Mayan codex), which I proudly presented to my second grade teacher, Mrs. Johnson, and plopped it on her desk. Upon returning to my seat, I turned around to see it protruding from her wastepaper basket. I ran back up the classroom aisle, retrieved the wad of paper and plunked it down again in front of her, urging: "But you don't understand, Mrs. Johnson, this is for you to *read!*" You see, long before it became fashionable to deride American education, I already knew from an early age that educators didn't know what they were doing (or talking about) most of the time, and so I tried to set poor Mrs. Johnson straight as to what to do with the *magnum opus* I had just bestowed upon her. Mrs. Johnson to her credit thanked me for my clarification of the matter and slid the pile into her desk drawer, never to be seen again.

During my college experience I had ignored the campus newspaper but had a strong interest in the campus information and humor magazine, called *Venue*. It was there in 1977 that I hatched my first plot to start a magazine with my collegiate cohorts. One tiny problem was the $100,000 necessary to get started in those days (today one would need closer to $4 million). After everything was typeset and the artwork was produced, the tiny problem over financing became an enormous problem. In almost no time at all, the proverbial cookie had crumbled.

Undaunted, upon my graduation from college I set out like some naive Candide to make my fortune in New York City. Little did I or anyone else realize at the time that my lot was in many ways similar to that of dear old Mrs. Johnson back in elementary school: The highest unemployment rates during the 1980s were to be among journalists and teachers.

After living on a diet of fish sticks and Kool Aid for a month or so, I could very easily have lapsed into a frustrated career as an anonymous security guard. But as in the case of any would be-security guard who could type at 114 words a minute, I became the next best thing – an anonymous word processor operator! Fortunately, the system I worked with was the Wang VS100, with its 32-bit processor, 16 MB of RAM, 14 inch diameter external disk drives and dual I/O bus architecture, and which was also capable of performing data processing functions. Dr. An Wang's device opened up an interesting new (and more lucrative) world for me.

After an almost ludicrously varied career, by 1994 I had become the first Management Information Services (MIS) director in the history of Squadron, Ellenoff, Plesent & Scheinfeld, the prestigious New York law firm, perhaps best known as the attorneys representing much of the business of media baron Rupert Murdoch and his News America Corp. (I sometimes joke that my greatest claim to fame was showing Murdoch the location of the Men's room during the "glory day" of his repurchase of the *New York Post* from the overly-flamboyant parking lot tycoon, Abe Hirschfeld).

It was at this point (around September of 1994) that I saw a small advertisement in the "Want Ad" section of the *New York Times*. *Computer Telephony* magazine was looking for a new writer / editor.

I had received the first test issue of Computer Telephony a short time before, in the Fall of 1993. Its contents had spurred me to proclaim to our firm that what they needed was a "unified messaging system." As few people knew what a unified messaging system was at the time, the statement was greeted by the staff with some mumbling and a lot of blank stares.

Now I was sitting there in my office, newspaper in hand, remembering my ancient ambition of starting a magazine. But *Computer Telephony* didn't look or read like any computer or telecom magazine I had ever seen. It was designed like *Business Week*, but read more like *Spy*. I knew vaguely of *Computer Telephony* magazine's publisher and editor-in-chief, Harry Newton, who had founded *LAN* magazine in the 1980s and had sold it for millions of dollars to Miller Freeman (now CMP Media). I also knew that he published *Newton's Telecom Dictionary*, which can be found on the desk of literally every company's telecom manager as well as everyone in the telecommunications industry.

A strange impulse suddenly came over me. I commented to members of the staff that "if all else fails, I guess I could always work for this magazine - look here, why don't I send them a letter!" I decided to have a bit of mischievous fun. I fired up Microsoft Word and jokingly dashed off a letter to Harry Newton which I carefully made sure violated every tenet of good business letter writing. It read something like this:

"Dear Sir, I was recently waiting at the Port Authority Bus Terminal here in New York, ready to leave the big city for my new job tending a miniature golf course on the boardwalk of Atlantic City, when I had the urge to pick up and read a copy of my beloved *New York Times* for one last time. In it I saw your insignificant, postage stamp-size advertisement for a writer. Enclosed are some articles I wrote for *Dr. Dobb's Journal* over ten years ago. I think I may be a serial killer, but I'm sure that won't be an obstacle in getting a job with you, since I work cheap. Hope to be hearing from you."

Before mailing the letter out, I passed it around among my compatriots at the law firm and we all had a good laugh. One person said: "You're nuts. And if this guy actually responds to this letter, he's probably nuts too, so you two would get along just fine." There is, as they say, much truth in jest.

I mailed the letter with a resume and old, fading photocopies of some articles I had written years before, and then promptly forgot about the whole affair.

Several days later, while loading Windows NT 3.1 Workstation on a brand spanking new 66 MHz Treasure Chest tower PC with 16 MB of RAM, the phone rang. I picked it up. Some guy with an Australian accent was at the other end of the line.

"You haven't taken that job yet with the miniature golf course, have you?" he quizzed.

"Huh?" I answered.

"And you haven't killed anybody lately, have you?"

After a few moments of confusion, I suddenly realized that this call was about my letter!

The fellow suddenly became more serious: "Can you come down here to our offices on Twenty-First Street at five o'clock?"

"Sure, why not?" I replied, thinking that nothing was going to come of any of this, since I had no real experience in the trade publishing industry.

"All right," he said, "Just ask for Harry Newton. See you later."

Before leaving for the offices of Newton's principal company, Telecom Library, I ran a Dun & Bradstreet analysis of his organization. It appeared to be a rapidly growing company that published magazines and books, and held an annual trade exposition in Dallas, Texas.

A few hours later I found myself in a building that at the turn of the 20th century had served as a sweatshop. I was soon ushered into the office of one Harry Newton, publisher and telecom deal-maker *extraordinaire*.

Harry started off by giving me what I was soon to learn was his standard speech on why computer telephony technology was so important. It included his now-famous line: "Look at this phone, it doesn't even have a backspace key!" I dutifully noted his Mitel SX-2000 phone system.

Harry's oration soon reached a crescendo: "I'm telling you

young man, computer telephony is going to be the hottest industry ever! To work here would be your golden opportunity."

After he was finished with a spirited tirade against overly successful PBX manufacturers, stupid Nynex executives and the staid telecom industry in general, I summoned enough courage to say something: "I don't understand why on Earth you've called me in here. My only publishing experience consists of doing a collegiate humor magazine, and once when I was a temp I typed some royalty statements at Random House."

"Nonsense!" exclaimed Newton. "You're perfect for this job. You've published articles and you know computers. I can tell about these things."

Indeed he could. Part of the secret to Harry's success was a near sixth sense that allowed him to instantly size people up. It was a sort of inner divining rod that could zero in on the many talented or even brilliant people out there in the corporate world who were sequestered in minor jobs for one reason or another, never getting their "big break" because of some personal imperfection or peculiarity.

Studies going back as far as the 1960s indicate that an amazing number employees stay at the bottom of the corporate ladder or get fired simply because their employers just don't like them. Well, Harry Newton made several fortunes by hiring people he didn't like.

Like some kind of magazine industry version of movie producer Roger Corman, Harry Newton could recognize the bankable talent lurking in unknown wannabe artists, poets, stymied great American novelists, undiscovered screenwriters, frustrated stand-up comics, and a multitude of other "non-corporate animals" who continually migrate to the concrete jungle of New York City and trudge up and down its boulevards of broken dreams. Newton could look beyond the personal foibles of any would-be employee and determine in a flash whether the person had any hidden aptitude that could be exploited for a modest salary.

The word "foible," aside from meaning a minor weakness or failing of character, can also refer to "the weaker section of a sword blade, from the middle to the tip." Perhaps that's a fair description of my foible, as one of my professors in college would repeatedly (and sternly) cite to me Alexander Pope's remark that "sarcasm is a sword, not a club."

Of course, I had had 16 years to mellow (or perhaps I should say "simmer") from the time I graduated college until my fateful encounter with Mr. Newton.

That's not to say that the brilliant and charismatic though complicated Newton didn't have some considerable personal foibles of his own. Still, for all his outrageous, exasperating set of eccentricities, there were certain qualities in Newton which at times affected, and in some cases seemed well-nigh to completely overshadow all the rest. After one particularly successful Expo, for example, Harry was approached by an old acquaintance in financial need, a fellow who had taught him photography 20 years before. Harry immediately wrote out a check for $25,000 and gave it to him as a long-term loan. And when editor Rick Luhmann had his golf clubs stolen while moving to a new apartment, Harry, a long-time, near-pro tennis enthusiast, remarked that stealing someone else's sporting goods was a terrible thing to do, and promptly bought the fellow a new set of clubs.

In any case, it appeared that Newton and I had hit it off. We joked about attorneys, Bell Labs, New York, Harvard, middle age, and how many frozen TV dinners I managed to fit into my kitchen's freezer.

Shifting once more back into interview mode, Harry coyly remarked that "I don't know if we're going to hire you . . . but if we did, could you start immediately?"

"Well, I would like to walk around and see your operation," I replied. "You know, talk with some employees."

"Certainly," Harry said. He picked up the phone, punched in an extension and intoned "Hello, this is Harry. I'm sending down somebody to talk to you. He's thinking of working for us, and I want you to tell him how wonderful it is to work for me, do you understand?"

No doubt I was going to get a perfectly candid tour of Newton's little empire.

As the reader has probably determined by now, in one of those preposterous happenstances which sometimes serves Fate in good stead, I ultimately chucked my so-called computer career and took the job with the magazine. Why I did so is open to conjecture. Logically, I figured that I would get more "exposure" as a magazine writer (soon to be technical editor) than as a law firm's MIS, IS or IT Director. But illogically, it was that overly romantic, almost vanished dream of working on a magazine that "swung the ball" as it were.

I immediately learned, much to my surprise, that *CT* editor Rick Luhmann had almost single-handedly written the bimonthly throughout 1994, and the addition of myself to "the staff" was to ensure that the magazine could now become a monthly, starting in January of 1995! Needless to say, I realized that out of necessity I would soon have to become a prolific author. I went home with a copy of every book Newton's company published, immersing myself in this new world that was an amalgam of telecom and computer technologies.

But nothing could have prepared me for the whirlwind I would ride right up until the present day. In March of 1995, for example, I experienced my first "CT Expo" in Dallas, meeting many of the both hopeful start-ups and entrenched telecom players who were venturing into the computer telephony industry. The Expo's attendance had doubled over the previous year. Shortly thereafter, I was standing in Harry's office, watching him ferociously gesticulate as he exclaimed: "I tell you, we've got a tiger by the tail! The Show, the magazine, everything has doubled or tripled since last year!"

I innocently reminded Harry of his comments when we first met, of how he had already intimated that computer telephony was going to be the hottest thing since sliced bread.

"Yes, but it's *really happening!*" He was quieter now, almost in awe of the way the words sounded as he uttered them.

Harry's wildest hyperbole couldn't have adequately described the incredible success his company was to enjoy during the 1990s. By 1997, things had gotten so lucrative that Newton's partner, Gerry Friesen, convinced him that the situation was optimal for making a truly stupendous amount of money by selling the company. They sold everything, lock, stock and barrel, to Miller Freeman (now CMP Media) for over $130 million, the fourth biggest magazine deal of 1997. (In yet another bizarre twist of fate, the law firm handling the sale was none other than Squadron, Ellenoff, Plesent & Sheinfeld, my old stomping ground.) In February of 2000, a survey of circulation and advertising sales by *Folio* magazine determined that *Computer Telephony* magazine had been one of the 20 "hot start up" magazines of the 1990s.

Of course, that was then, this is now. Whereas anyone you could stop on the street knows what the Internet is, and even knows what a "call center" is, very few have probably heard of the term "computer telephony" which, ironically, links many of the underlying infrastructure technologies that make the Internet, call centers and "convergence" itself a familiar aspect of our culture.

It's enormously difficult to define computer telephony – the industry has been a moving target since the beginning. Just pick up a copy of *CT* magazine to understand what I'm referring to. First you see an article on customer relationship management, then an article on speech recognition used in ACD software running on PC-based phone systems, another article on voice-enabled routers, another article on web-based contact centers, another on unified messaging, then yet another on fault resilient computers. And on and on with only a tenuous thread connecting them, a thread that loops around such popular buzz-words as Internet, telecom, convergence, networks, communications and solutions.

Extreme customer care, unified messaging and communications, voice processing on the LAN and WAN, wireless technologies, fax over IP, packet telephony, the primacy of voice communications – all of these things are "computer telephony." It is the successful convergence of computers and communications systems to create new, innovative and valued solutions for businesses, prospective customers of a business, and individuals.

Indeed, the computer telephony industry is such a nebulous, dynamic entity that it was high time that somebody gave it legitimacy by making it the subject of an encyclopedia. That somebody turned out to be Yours Truly.

This industry has been very good to me – far better in fact than anything I could possibly have imagined when I first strode into Harry Newton's office on that fateful day in 1994. And if you read this book and can grasp the essence of CT technology and how it can help you or your business, then it will be very good to you too.

– *Richard Grigonis*
New York, New York
June 2000

Even as this edition goes to press, the second edition is being planned. The author welcomes corrections, changes, additions, comments and suggestions. For organizational purposes a separate email address has been created. Please send all correspondence to ctencyclopedia@cmp.com

A & B (Robbed) Bit Signaling

Many T-1 connections use an in-band signaling method called "robbed bit." In-band signaling is also known as Channel Associated Signaling (CAS), because line signaling information is carried within each channel's data stream. Robbed bit signaling literally "robs" one bit from the data path on each of the 24 DS-0 channels to indicate a hook condition. In the case of a T-1 using the D4 Super Frame (D4 SF) framing technique (based on 12 frame increments), every 6th and 12th frames have the Least Significant Bit (LSB) from every 8-bit byte overwritten to indicate the hook state. Because such bits are no longer available for the actual channels, the DS-0's are now voice grade lines and only transmit at 56Kbps, as compared to the full 64Kbps in out-of-band or clear channel signaling. The bit robbed from the 6th frame is called the "A-bit" and the bit robbed from the 12th frame is called the "B-bit." Note that in the case of T-1s using Extended Super Framing (ESF), which is based on 24 frames, the LSB of frames 6, 12, 18 and 24 are used for signaling purposes.

For straight voice transmissions, robbed bit signaling doesn't compromise quality very much. But it does limit the amount of "call information" that can be buried in the channels of a T-1 circuit. By robbing bits, a circuit can only indicate ringing, hang up, wink and pulse digit dialing. This does not translate to very much "advanced intelligence" of which the Advanced Intelligent Network (AIN) is supposedly capable, which is why many developers have turned to ISDN PRI (Primary Rate Interface) for more advanced applications.

ISDN PRI turns a T-1 circuit into 23 Bearer channels and 1 Data (signaling) channel (or 23B + 1D). And while you do lose one channel for transmission purposes, you can more than make up for it with the more varied signaling information you get on the D channel, such as better, faster Automatic Number Identification (ANI). Under T-1 robbed-bit signaling, there is a margin for error in transmission or detection of a caller's DTMF tones. ISDN PRI's D channel assures flawless ANI. The D channel also lets you group together channels into flexible pipes. Standards for doing this include Bandwidth ON Demand INteroperability Groups (BONDING) and Multi-Rate ISDN.

A/D Converter

Analog to Digital Converter. A device that converts an analog signal to a digital signal before being processed by digital circuitry. The most familiar device that does analog to digital (A/D) conversion is the modem (modulator-demodulator), which converts digital signals from your computer to analog and vice versa so as to send signals over the analog public network.

A/D conversion is done by taking the analog signal and sampling the voltage level (height of the waveform) at specific time intervals and assigning a value to it. Determining the time intervals for the periodic sampling process depends upon how closely the digital signal will represent the analog signal versus how much bandwidth will be required.

Henry Nyquist's famous 1928 paper, "Certain Topics in Telegraph Transmission Theory" postulated what is now known as the Nyquist Theorem, which states that a sample of twice the highest signal frequency rate can digitally represent the waveform with suitable precision. Nyquist said essentially that the highest frequency which can be accurately represented is one-half of the sampling rate.

Of course, nothing prevents you from sampling at a higher rate – a process called oversampling – which provides better resolution but requires greater bandwidth.

To digitally encode an analog waveform and then translate the digital information back into an efficiently modulated waveform suitable for high bandwidth applications, one must first determine the sample rate, then sample the signal, then determine the resolution, which is the number of voltage levels used to represent each sample, which for k bits per sample is 2 to the power k voltage levels (for example, an 8 bit word used for video luminance needs 2 to the 8th power, or 256 discrete voltage levels).

Finally, one multiplies the sample rate by the discreet levels per sample to determine bit rate. A typical DS-0 telephone protocol samples the 4 kHz audio signal at 8,000 times a second (2 samples per cycle x 4,000 cycles) with a resolution of 8 bits (256 possible discrete values in each 8 bit sample), resulting in a 8,000 x 8 or 64,000 (64 Kbps) bit rate. So, to digitally encode a voice signal in real time over a circuit requires a 64 Kbps bandwidth.

An A/D converter may be preceded in a circuit by signal conditioning equipment for amplification or filtering. Input signals may be multiplexed (passed in turn) to the A/D converter, or there may be one A/D converter for each signal.

A/D converters are also used extensively in data acquisition applications.

Abandoned Call

A caller hangs up before any communication takes place or the call can be answered. Overly aggressive predictive dialers in call centers tend to do this, dialing too many phone numbers (to too many people sitting down to have dinner) and connecting before a sufficient number of call center agents are available to read the campaign pitch script for the next session, whereupon the system hangs up after a few seconds. Hence, you pick up the phone, no one is there, and the calling party hangs up. Crank caller? Not at all, you've just encountered an overly optimistic predictive dialer. Outbound abandoned calls are also referred to (with some justification) as "nuisance calls".

On the other hand, if you call into a call center, are put on hold for too long, get bored and hang up, that's considered an abandoned call too. If call centers note that there are too many inbound abandoned calls, they increase the number of agents so as not to lose an unacceptable percentage of business. If too many outbound abandoned calls are being made, the call center manager can simply adjust the predictive dialer to be more pessimistic when estimating the probability of agents becoming available for outbound telemarketing or collections calls.

In most states and countries where there is much outbound dialing, legislation, codes of practice, and media and customer pressure have all combined to limit maximum targets for abandoned calls to around 5%, and sometimes much lower. In Florida and Texas, a live agent must be available at all times.

Many advanced predictive dialers barely manage to achieve good results at 1% abandoned target levels, which is seen as a realistic minimum figure acceptable to call centers. In the case of power dialing, an abandoned call target of zero can be set, and this will have the effect of diverting all calls for which no logged on agent is available, to an agent in the standby pool.

There are various way of measuring abandoned calls. The two most common are as follows:

Formula 1
number of abandoned calls / live calls + abandoned calls * 100 / 1

1

Formula 2

number of abandoned calls / all call outcomes + abandoned calls * 100 / 1

In formula 2, "all call outcomes" includes not only live calls but busies, no answers, telco messages (i.e. SITs) and calls that end up connecting to answering machines, fax machines, and modems.

ACD

See Automatic Call Distribution.

ACTAS

See Alliance of Computer Telephone Application Suppliers

ACTIUS

See Association of Computer Telephone Integration Users and Suppliers

ActiveX

Microsoft's ActiveX is based upon the concept of component software, or "componentware," which is considered to be a more productive means to develop software than "reinventing the wheel." Component software allows you to simply "plug-in" functionality, extending your software development environment with new functions, with every control accessed through a common interface. Components can also be re-used by more than one application, in which case they are called component containers (as in "containers of reusable code"). Today, ActiveX controls are components (or objects) you can insert into a web page or other application to reuse the functionality a third party has already programmed.

ActiveX components fall into the highly interrelated categories of Automation Servers, Automation Controllers, Controls, COM Objects, Documents, and Containers.

Programming would never be the same again after 1991, all thanks to the appearance of Visual Basic (VB) and the great innovation of its "custom controls," software objects that you can embed in a VB program or other Windows development tool. In the "old days" (pre-Windows) you would compile your DOS program with a "library" of some precompiled subprograms and functions that would serve as reusable code. Microsoft's 16-bit Visual Basic eXtention (VBX) controls took this idea a step further and gave you what was at the time considered to be tremendous power, all within the Windows Graphical User Interface (GUI). Reminiscent of function libraries, VBXs immediately became the standard way of extending the functionality of Visual Basic.

The original Visual Basic "custom controls" were programmed by third parties and behaved identically to the controls shipped with Visual Basic: They appear in the Visual Basic toolbox, you control their behavior from your software, they generate events that your program can respond to, and they have properties that your program can change.

A VBX for telephony might have a property such as "rings" that a programmer used to specify the number of rings before a call is connected. "Pick up," "dial" and "get-digits" are other examples of telephony actions that might be performed. "Hangup" and "ring-detected" are two events that might cause a certain block of code to be executed. For example, on the ring-detected event the associated code would call the "pickup" action to connect the call and possibly the "play-file" action to speak a greeting to the caller.

There are hundreds of controls out there for Visual Basic for database management, multimedia presentations, imaging, host connectivity, etc. The ones that concern us do computer telephony functions (though anything can be leveraged, like host connectivity for IVR, etc.). In particular, setting up IVR and media-processing systems, call routing; voice recognition and text-to-speech over the phone; faxing and so on.

Whereas VB was the original recipient of control technology, today thousands of third party controls exist for everything from accessing legacy databases on mainframes to programming a Dialogic or NMS telephony resource board, all available for any Windows development environment such as Visual Basic, VBScript, Visual C++, Delphi, Visual FoxPro, and PowerBuilder. All of these packages soon abandoned Microsoft's original 16-bit VBX control technology and in the mid-1990s became collections of 32-bit Object Linking and Embedding (OLE) controls, first called OCX controls by the 32-bit VB community (which became known later as "ActiveX" technology). These worked with any 32-bit development tool supporting OLE / OCX controls, such as Visual C++, Access, Visual FoxPro, Powersoft's PowerBuilder and Borland's C++ and Delphi.

The OCX extender is still with us – just perform a search on your hard drive and you'll find scads of them in your Windows system directory. Don't be fooled though – it's all ActiveX technology.

OLE started out as a technology consisting of a set of interfaces that defined how software components could work together so one could create compound documents, which involves dynamically linking files and applications together. An "object" in this context is a combination of data and the application needed to modify that data. Such objects can be embedded in or linked to documents created with a different application. For example, OLE allows an Excel spreadsheet to be embedded within a Microsoft Word document. Double click on the embedded spreadsheet, and Excel will launch and allow you to edit the spreadsheet.

OLE didn't entirely solve the general problem of how various software components should provide services to one another.

OLE's successor, OLE2, introduced the concept of the Component Object Model (COM), a more generalized means for allowing different kinds of software to communicate and provide services to each other, based upon a programming specification for creating encapsulated, reusable and interactive objects. COM supplies the low-level object-binding mechanism allowing objects to communicate with each other, while OLE2 uses COM to provide high-level user-friendly application services involving linking and embedding so as to better integrate desktop applications and thus enable users to create compound documents.

OLE2 was much more successful with handling compound documents than OLE, but COM's capabilities developed far beyond that and became a common paradigm for component-based software design. COM was not merely OLE anymore.

Unfortunately, Microsoft liked the term "OLE" so much that they often used it instead of "COM," causing some confusion. Microsoft (and everyone else) now favors the term "ActiveX control" instead of OLE or OCX when referring to the component object. The older OLE concept is viewed as just Microsoft's original implementation of COM for the Windows Desktop, while ActiveX / OCXs essentially make the Internet an extension of the desktop, thus simplifying Internet development and integrating seamlessly the Web and the Windows operating system. With ActiveX, you can run a Visual Basic program within your browser and have it manipulate data from Microsoft applications such as Word and Excel.

When used in a network with a directory and other support functions, COM becomes the Distributed Component Object Model (DCOM). "Distributed" means that it can run clients and servers in different processes across an intranet or Internet.

DCOM works just like COM in that a client "asks" the registry the location of the server situated and, instead of providing a path to a directory on the local PC, it instead points to an IP address. Thus, COM processes run on the same machine in different address spaces, but DCOM processes are distributed across a network. DCOM is roughly equivalent to its chief competitor, the Common Object Request Broker Architecture (CORBA) another way of providing a set of distributed services. CORBA comes under the Object Management Group (OMG) at http://www.omg.org. Another ActiveX competitor is Sun's Remote Method Invocation (RMI) specification. All of these multi-tier architectures essentially provide frameworks for invoking services from middleware, so network clients don't suffer from a data traffic version of "gridlock."

For programmers in the mid-1990s, object controls with 32-bit extensions were immediately recognized as being technically superior to the old 16-bit VBXs for three main reasons:

• **Sharing**. OLE automation allowed one app component to access the functions of another. The requesting component is the client and the supplying component is the server. OLE Automation Servers – now called ActiveX Server Components – could run like a Dynamic Link Library (DLL) within the client app ("in process"), like a separate executable app ("out-of-process") or as a separate app on a different computer ("remote").

Application components can thus be reused, shared and distributed across a network, including the Internet, or a CT app can be broken down into component processes that can also be distributed and shared with other apps over a network.

A good example of this would be a touchtone-based order-entry system sharing a credit-card validation OLE server across a network. A main auto-attendant app could use OLE to pass control of calls to voice mail, fax-on-demand or IVR server components for better, flexible and more innovative integration. For application generation in general (and definitely for resource-hungry CT), this type of design architecture could be revolutionary.

• **Slicker programming.** The old 16-bit Visual Basic VBX custom controls support properties, events and actions. The newer 32-bit OLE Controls support properties, events and methods. The latter were found to be better because they eliminated the need for a developer to set properties before calling every action. For example, when using Artisoft / Stylus' old Visual Voice VBX custom control a developer would have to set the MaxDigits property before calling the action to retrieve touchtone digits. However, with the method interface of the Visual Voice OLE Control, a developer could now simply specify the maximum number of digits to be retrieved as a parameter to the GetDigits method – a much "slicker" method.

In the Visual Voice examples below you see a comparison of two Visual Basic code samples for a simple IVR order entry system. The first sample uses the old VBX control and the second example uses an OLE control. Notice that OLE Control methods and parameter passing make the code in the second example more concise.

VBX EXAMPLE

```
        Sub Voice1_RingDetected()
        On Error GoTo MainErr
        'Pick Up
        Voice1.Action = VV_PICK_UP
        'Play 'Hello. Welcome to the Order Status System.'
        Voice1.Value = "greeting.vox"
        Voice1.Action = VV_PLAY_FILE
        'Allow caller to terminate next prompt with any number
        Voice1.VTermDigits = "1234567890"
```

```
        'Prompt caller to enter order number.
        'Play 'Please enter your five digit order number.'
        Voice1.Value = "getorder.vox"
        Voice1.Action = VV_PLAY_FILE
        Voice1.MaxDigits = 5
        Voice1.DTermDigits = ""
        Voice1.Action = VV_GET_DIGITS
        OrderNum = Voice1.Digits
        'Search database
        orders.Seek "=", OrderNum
        If orders.NoMatch Then
        'Play 'I'm sorry, that order number is not in the orders database'
        Voice1.Value = "notvalid.vox"
        Voice1.Action = VV_PLAY_FILE
        Else
        "'Order number... was shipped on..."
        Voice1.Value = "ordernum.vox|FILE, " & OrderNum &
        "|CHARACTER,wasship.vox|FILE, " &
        orders("Shipped") & "|DATE"
        Voice1.Action = VV_PLAY_STRING
        End If
        'Play 'Thank you for calling. Goodbye.'
        Voice1.Value = "goodbye.vox"
        Voice1.Action = VV_PLAY_FILE
        'Hang Up
        Voice1.Action = VV_HANG_UP
        Exit Sub
        MainErr:
        Call vvProcessLineErrors(Voice1)
        Exit Sub
        End Sub
```

OLE EXAMPLE

```
        Sub Voice1_RingDetected()
        On Error GoTo MainErr
        'Pick Up
        Voice1.Pickup
        'Play 'Hello. Welcome to the Order Status System.'
        Voice1.PlayFile " greeting.vox", , ""
        'Prompt caller to enter their order number.
        'Play 'Please enter your five digit order number.'
        Voice1.PlayFile " getorder.vox", , "1234567890"
        OrderNum = Voice1.GetDigits(5, "")
        'Search database
        Orders.Seek "=", OrderNum
        If Orders.NoMatch Then
        'Play 'I'm sorry, that order number is not in the orders database'
        Voice1.PlayFile "notvalid.vox"
        Else
        "'Order number... was shipped on..."
        Voice1.PlayString "ordernum.vox|FILE, " & OrderNum &
        "|CHARACTER, wasship.vox|FILE, " &
        Orders("ShippedDate") & "|DATE"
        End If
        'Play 'Thank you for calling. Goodbye.'
        Voice1.PlayFile "goodbye.vox"
        'Hang Up
        Voice1.Hangup
        Exit Sub
```

3

```
MainErr:
Call vvProcessLineErrors(Voice1)
Exit Sub
End Sub
```

Another nicety for programmers is that the OLE Control standard is an open one. The internal VBX architecture was never documented by Microsoft, making it practically impossible for vendors to provide VBX support in their development tools.

Today, you can develop ActiveX objects with Visual Basic, Visual C++ (there is a VC++ Microsoft Foundation Class Library ActiveX Control Wizard), Visual J++, or any other OCX-type of development environment, and then you can actually use ActiveX objects and write scripts in such environments as Visual C++, the Visual Basic Scripting Edition, and the ActiveX Control Pad.

- **Stability.** One of the original objections to creating big telephony apps in 16-bit Windows was the risky cooperative multitasking scheme used by Windows 3.x. Your DLL or VBX had to be really well-behaved or you'd get hung up or crash – unacceptable with anything to do with answering phone calls. Succeeding 32-bit versions of Windows and controls were progressively more stable. With the appearance of Windows 95, for example, preemptive multitasking became available to developers, which guarantees stability so long as you are running 32-bit apps and controls. Even if something really strange does occur, 32-bit tasks can be isolated and shut down if something goes wrong, whereas problems with 16-bit apps and controls can be somewhat "contagious," bringing your entire system to its knees.

The old Windows 3.x was also notorious for running out of resources, causing all sorts of unpredictable behavior, including just not running at all. Windows 3.0 had only one 64K block for its data traffic and Windows 3.1 had two. Windows 95 had many more available resources, and Windows 98 had even more resource "breathing room." With Windows 2000, all of the old limitations will have finally vanished.

In any case, in 1996 Microsoft decided to once again relegate the term OLE to compound documents. At the Internet Professional Developers Conference (Internet PDC) in March 1996, Microsoft coined a new term, ActiveX, that was a play on the conference's slogan "Activate the Internet." At the time there was no technology that embodied the term. ActiveX would soon refer, however, to a somewhat confusing array of COM-based technologies formerly associated with OLE. Indeed, in some instances, entire documents explaining the implementation of OLE technologies were rewritten for ActiveX technologies merely by changing the word "OLE" to "ActiveX."

Like OLE2, ActiveX controls are a form of COM object. A "pure" COM object is a more generalized, abstract type of object than those tied to any particular language such as C++ or Java. Thus, client software accessing a COM object's functionality is not aware of any of its details, such as whether it exists as a DLL or as a distinct process in its own executable. Instead of forcing the developer to link to a component's functionality at design time or provide a path to the component in the source code, COM can simply get the object's location by accessing the registry, provided that the registry knows where the server object happens to be situated.

Among other things, COM allows developers to build standalone components (servers), so, today, ActiveX includes both client and server technologies. A common mistake is to confuse the broad term "ActiveX" with "ActiveX controls" which are only a type of COM object that provide a set of services to their client, or an ActiveX container, with interfaces constrained in certain specified ways when they interact with a desktop client (for example, an ActiveX control supports self-registration, and most ActiveX controls support automation and are connectable so that they can generate events for their client). The component can be modified at design time, as in the case with VB, or at run time, like in the case with ActiveX Scripting. On the other hand, "ActiveX Server Components" (which used to be called OLE Automation Servers) are the code that executes on the server, not the client, and are definitely not ActiveX controls. These server-based components can directly access machine resources, if necessary, extend functionality beyond what is available in scripting and can be distributed to another processor with DCOM.

Using Active Server Pages is a relatively painless way to communicate with server-side components via a high-level scripting language. Many of the tools used to create ActiveX controls can be used to create the automation servers that act as the server-side components in Active Server Pages (ASPs). Automation Servers are components that can be driven by the running code of another application (called "Automation Controllers" which includes the likes of Visual Basic), so they generally don't need a user interface. Such automation servers can automate access to a database, perform complex queries, and process data values used by script code, supporting an HTML stream of the desired information that is returned to the client's browser. Automation Servers can be in-process (executing in the Automation Controller's processing space), local (executing in its own space), or remote (executing in another PC's processing space).

One can "split hairs" further when discussing the ActiveX / ActiveX controls dichotomy. Business rules are generally defined not in an ActiveX control per se, but in an ActiveX Dynamic Link Library (DLL) or ActiveX EXE component. An ActiveX DLL is a code component that runs in the same process as the client application and uses the same address space as the client application that's using the component. Any components that must interact with the Microsoft Transaction Server must be ActiveX DLL components. Like Automation Servers, an ActiveX.DLL can execute as either in-process, local or remote.

Unlike an ActiveX DLL, an ActiveX.EXE runs in a separate process from the client application and has its own address space. ActiveX.EXEs must be used when a component will be running on a computer different from the computer running the client application and you don't plan to use Microsoft Transaction Server. ActiveX EXEs can be used for creating components that can also be run standalone (e.g. run by clicking on the icon).

Since ActiveX DLLs are considered easier to test than ActiveX.EXEs, you can test drive your ActiveX EXE first as an ActiveX DLL and then transform it to an ActiveX EXE using the Project Properties dialog box in Visual Basic. But in doing so remember that an ActiveX.EXE can execute only locally or remotely, not in-process.

When working with Visual Basic, "pure" ActiveX controls are different from other ActiveX components in that they normally have a visual component like a combo box or set of text boxes, making them a sort of standard user interface element that you can customize. The extension of an ActiveX control is usually OCX, but in reality it is essentially a standard windows DLL. This means that ActiveX controls cannot simply run on their own – they must first be loaded into a container of some sort that supports ActiveX controls, such as Visual Basic forms, Excel forms, or VC++, and the controls always execute in-process to the Container in which they find themselves. The world's favorite containers for running ActiveX controls are now, as you might have guessed, Web browsers, where the ActiveX controls "compete" with Java applets.

Indeed, ActiveX controls were at first promoted as a form of more efficient, "slimmed down" (less memory intensive) OLE technology with an Internet ability that enabled interactive content for the World Wide Web as well as support for both real and de-facto standards such as HTML, TCP/IP, Java, COM, etc. The evolution of component functionality from VBX to OLE / OCX to OLE2 / COM and ActiveX / DCOM has been one of keeping up with developments in network technology. Standalone machines running 16-bit VBX controls gave way to computers needing 32-bit OLE software components that could perform interactive control over LANs, all of which ultimately led to OLE being revamped for the Internet as ActiveX controls where "applet sized" objects with user-controllable functions can be embedded in a web page to provide various multimedia effects, dynamic page layouts and executable applications that can be run in real-time over the Internet.

This is all possible because the HTML <OBJECT></OBJECT> tags, approved by the World Wide Web Consortium (W3C), allows 32-bit objects to be inserted into HTML pages (<APPLET> and </APPLET> tags are for applets), and scripting languages such as VBScript, JavaScript and Java can be used to manipulate the embedded objects. Active Scripting can control the combined interactive behavior of several ActiveX controls and/or Java Applets from the browser or server. Any ActiveX-supported browser can run Java applets and integrate said applets with ActiveX controls via Microsoft's Java Virtual Machine. You can make a browser look like any other application, or you can have a separate application window open up giving your ActiveX control the guise of a full application.

Also, so-called ActiveX Documents enable users to view non-HTML documents, such as Microsoft Word or Excel files, through a Web browser. And don't forget that ActiveX functionality is not limited to browsers – ActiveX controls can be called from any other OLE-aware application.

ActiveX components encapsulate their functionality in Events, Methods, and Properties. Events are subroutines that are automatically called by your development environment when a special condition is met. Methods are built-in procedures associated with a particular control. For example, Intel / Artisoft's (Cambridge, MA – 617-354-0600, www.artisoft.com) Visual Voice control's "PickUp" method picks up a phone line and the "PlayFile" method plays a pre-recorded message over the phone line. Properties are characteristics of a control which define how it behaves. For example, the Visual Voice control's "Rings" property describes how many rings the caller will hear before your application responds. Events, methods, and properties work similarly in other development environments.

ActiveX controls also support Microsoft's Active Template Library (ATL), a set of template-based C++ classes that allow you to quickly create COM objects that are fast and small (in many cases they can run in as little as 20K or less). ATL has special support for such COM features as stock implementations of IUnknown, IClassFactory, IClassFactory2 and IDispatch; dual interfaces; standard COM enumerator interfaces; connection points; tear-off interfaces; and ActiveX controls. ATL code can be used in the production of single-threaded objects, so-called apartment-model objects, free-threaded model objects, or both free-threaded and apartment-model objects.

Although the ActiveX emphasis is obviously on development under Windows, Microsoft has made efforts to offer open, cross-platform support on Macintosh and UNIX operating systems, as well as Java and Java-enabled tools. Of course, the individual controls themselves are not cross-platform, for although in theory it doesn't matter what language an ActiveX control is written in, the control itself must still be compiled into the native machine code for every type of computer that will run it. Native compilation allows ActiveX to run faster than Java, but unlike Java it is not "recompilable," which brings us to...

The ActiveX vs Java debate. One upon a time, ActiveX had a competitor called OpenDoc, championed by Apple and IBM. OpenDoc technology never caught on, however, and today many look upon the Java language as the real competition, in particular JavaBeans, a Java-based component architecture that can be downloaded off of the Internet and run on any platform, not just under Windows. An ActiveX control is somewhat equivalent to a Java applet, ActiveX and Java are used for the same kinds of tasks and can look very similar to users. But while Java is a real programming language, ActiveX isn't – it's a "componentware" tool for creating object oriented programs, which allows you to use more sophisticated programming environments such as C++, easier programming environments such as VB, or even – Java! Thus, ActiveX can be used to extend any programming language, including Java itself, albeit Microsoft's own version of Java. In a sense, then, Java and ActiveX don't compete head-to-head. Rather, Java can be used in some way to complement ActiveX.

ActiveX can be used as a sort of "glue" – Java programmers can create and use ActiveX controls, and objects created with other programming languages can link to Java applets. In Microsoft's J++, ActiveX controls can implement methods and properties of the COM objects as if they were simply other Java classes, and can be imported by wizards. Also, Java and ActiveX can be used on the same web page.

Although Microsoft has tightly integrated ActiveX controls with its Telephony API (TAPI) spec, ActiveX will face off in the telecom arena against JTAPI, the Java Telephony API proposed by Sun Microsystems. The competition between ActiveX and JTAPI should prove to be interesting.

The Future of ActiveX. In October of 1996, Microsoft turned over control of the ActiveX spec to The Open Group (http://www.opengroup.org), a consortium that sets industry standards. It, in turn, has set up the ActiveX Working Group (http://www.activex.org) a consortium of software and systems vendors dedicated to promotion and widespread adoption of ActiveX core technologies. Functioning as an authoring group working under the auspices of The Open Group, The Active Group is directed by a steering committee of 12 vendors including Adobe Systems, Computer Associates International, DEC, Hewlett Packard, Microsoft Corporation, Powersoft-Sybase, Sheridan Systems, Siemens-Nixdorf Information Systems, Software AG, Videosoft, Visio, and Wall Data.

Although some anti-Microsoft programmers refer to ActiveX as "inactive X" (referring to their own lack of interest), more and more ActiveX controls and related developments appear every day, and the formulation of COM itself has been remarkably stable since its inception in 1993. ActiveX should be around for a long time. Most of the telephony application generators (app-gens) such as Intel / Artisoft's Visual Voice, are essentially collections of ActiveX controls. Parity's ActiveX controls for building computer telephony systems are also quite impressive. Their VoiceBocx custom controls give programmers the ability to handle Dialogic multi-line call processing cards, Dialogic DM3 IPLink cards for Internet telephony, or TAPI-compliant voice processing cards or CTI links.

OPUS Maestro from Prima (Nuns' Island, Canada – 514-768-1000, www.prima.ca) is another application generator that takes advantage of ActiveX, particularly ActiveX Data Objects (ADO), Microsoft's high-level data interface to any data store, be it a front-end database client or middle-tier business object using an application, tool, language, or even an Internet browser. ADO is becoming a major data interface for 1-to-N-tier client / server and Web-based, data-driven solution development. DCOM is the communication foundation between server modules of OPUS Maestro. Prima believes that DCOM conveys the message of many network vendors: "The

Network is the Computer," or rather, "The Network is the Server".

Fortunately, ActiveX is an open standard. Anybody can build and publish ActiveX objects that you can plug into your web site or use with a computer telephony application, and indeed there are thousands of ActiveX objects on the Internet you can download and use immediately.

Until recently, the only browser that could run ActiveX controls was Microsoft's Internet Explorer. Recently plug-ins have appeared to allow ActiveX to run on other browsers. Of course, in the case of Explorer, many useful ActiveX and VBScripts are already built-in and don't have to be downloaded to the client, such as using a button to control other elements of a web page. Before a new control is downloaded, the user is presented with a "Certificate Page" giving information about the author. The user has the option of rejecting or accepting the new control, thus continuing the download.

Since ActiveX controls sit on the client machine, they don't have the same security restrictions of Java and JavaScript. Still, at one particular JavaOne conference, Sun pulled an interesting publicity stunt by showing to an audience how to write an ActiveX control that would erase a client's disk drive when downloaded.

Problems with COM+ load balancing technology for Windows 2000 and the industry's wide acceptance of vendor independent COBRA over Microsoft's COM and DCOM appears to have spurred Microsoft to announce a new XML-based standard called the Simple Object Access Protocol (SOAP), which is meant to replace DCOM and kill off Java once and for all. SOAP will use XML formats to piggyback a distributed object protocol on top of HTTP. SOAP requests and responses will use HTTP, obviating the need for a separate server port, which could be blocked by a firewall. SOAP is in draft review with the Internet Engineering Task Force (See Simple Object Access Protocol).

Recommended Sites:

For ActiveX authoring tools and other utilities, check out http://www.ActiveX.com and
http://browserwatch.internet.com/activex.html.
Microsoft's own example of how to build a distributed application using ActiveX and DCOM can be found at
http://msdn.microsoft.com/workshop/components/activex/magic.asp.

Active Directory Services Interface (ADSI)

ADSI is ostensibly a set of Component Object Model (COM) objects acting as a proprietary API and programmability layer for querying, modifying, and publishing Microsoft Active Directory information. Besides programmers and developers, ADSI also enables administrators to automate common tasks such as adding users and groups, managing printers, and setting permissions on network resources.

Unlike vendors such as Novell who treat the directory as its own platform, ADSI is simply a Windows 2000 component. But ADSI is more than just an interface into the Active Directory. It's an abstraction layer sitting atop the APIs used by the various directory services supplied with network operating systems. The standard ADSI objects, or providers, are found within multiple "namespaces," which are typically the directory services. Thus, applications written to ADSI are essentially written to a common API, and so will work with any directory service offering an ADSI provider. For example, with ADSI, applications can access LDAP, NDS, Active Directory, and other directories with ADSI interfaces as long as the appropriate service providers are available.

As more than one pundit has remarked: "ADSI is to directories what ODBC (Open Database Connectivity) is to databases." However, being a high-level abstraction layer, ADSI does not allow access to a directory's full feature set as do calls made directly to a particular directory's API.

Still, in the current release, an ADSI-enabled PC can work with objects in each of four major namespaces of the following network providers:

- **WinNT** – Microsoft Windows NT Server 4.0 directory.
- **LDAP** – Lightweight Directory Access Protocol (LDAP), Version 2 and Version 3. Not only can this namespace provider be used to access any LDAP server, but it also works with the Wijndows 2000 Active Directory and Internet directory services such as Bigfoot, InfoSpace, and WhoWhere. There is also an LDAP interface to Microsoft Exchange.
- **NWCOMPAT** – NetWare 3.x bindery.
- **NDS** – NetWare 4.x Netware Directory Services (NDS).

Microsoft is striving to ensure that as many of their products as possible integrate with ADSI, such as the Active Directory, Microsoft Exchange 5.5, Microsoft Internet Information Server (IIS), and Microsoft Site Server.

ADSI is highly extensible., and since any language that can act as an OLE automation controller can interface with ADSI, network administrators can use whatever language they feel comfortable working with (C++, Perl, Visual Basic, etc.) to script the management of multiple directories.

ADSI is included in Windows 2000, but it can also be downloaded from Microsoft's Web site (at www.microsoft.com/adsi) for Windows 95, Windows 98, and Windows NT 4.0

Adaptive Differential Pulse Code Modulation (ADPCM)

ADPCM is a family of voice waveform compression and decompression algorithms that converts sound or analog information to binary information (a string of 0's and 1's) by taking frequent periodic samples of the sound and converting the "height" of the waveform at each sampled point into a binary number. But instead of coding a value directly, ADPCM is based upon calculating the difference between two consecutive speech samples in standard PCM signals, or between an actual value and its predicted value. Thus, the overall transmission bit rate can be greatly reduced, so less storage space is necessary than the PCM format of WAV and AIFF files.

One common ADPCM implementation takes 16-bit linear PCM samples and converts them to 4-bit samples, yielding a compression ratio of 4:1. ADPCM is used in applications as varied as voice mail / answering machines, sending conversations on fiber-optic long-distance lines, and audio storage on Sony's Mini Disc. G.761-compliant ADPCM transcoders can compress sixty 64 Kbps full-duplex voice channels carried by two E-1 trunks into sixty 32 Kbps full-duplex voice channels for transport by one E-1 trunk.

In computer telephony, Intel / Dialogic's VOX files are flat binary files containing digitized voice data samples. Each byte contains two samples and the encoding within each sample is done with ADPCM.

ADPCM can be subject to quantization noise and high frequency aliasing, errors that create audible noise on transients and fast changes of dynamics, such as speech or musical instrumental attacks having high frequency components. At 22kHz and lower sampling-rates, audible artifacts are even stronger. Still, ADPCM in some form remains the foundation for much of the voice compression and decompression done in the telecom and other industries.

ADC

See A/D Converter.

ADPCM

See Adaptive Differential Pulse Code Modulation

ADSI

1. See Active Directory Services Interface.
2. See Analog Display Services Interface.

ADSL

See Asymmetric Digital Subscriber Line

ADSL Lite

See Asymmetric Digital Subscriber Line

Advanced Intelligent Network (AIN)

See Intelligent Network.

Advanced Mobile Phone System (AMPS)

This was the first analog cellular public service in North America. Also used in Mexico, and Central and South America. AMPS underwent a two-year test in Chicago in 1976, then the Bell Telephone Company officially unveiled it in 1978. AMPS became a standard for North America in 1983.

AMPS allocates frequency ranges within the 800 and 900 Megahertz (MHz) spectrum to cellular phones. The bands are divided into 30 kHz FM sub-bands using Frequency Division Multiple Access (FDMA), called channels. Each service provider can use half of the 824-849 MHz bandwidth for receiving signals (the "reverse channel" from mobile cellular phones to the base station) and half the 869-894 MHz bandwidth for transmitting (the "forward channel" from the base station to the cellular phones), so a total of 50 MHz is divided between the two operators.

The B band (or block) is assigned to the local wireline carrier, and the A band is assigned to the non-wireline (mobile or wireless) carrier.

Other first-generation mobile phone systems deployed around the world during the AMS era include the Total Access Communication Service (TACS) in Europe and the Nordic Mobile Telephone (NMT) service, among others.

AMPS entered the digital age when FDMA was re-engineered so that each channel could be further subdivided using Time Division Multiple Access (TDMA), resulting in a service now called Digital AMPS (D-AMPS). D-AMPS can squeeze three channels into the space of each analog AMPS channel, thus tripling the number of calls that can be accommodated per channel.

AEB

See Analog Expansion Bus.

AIN

Advanced Intelligent Network. See Intelligent Network.

A-Law

Pulse Code Modulation (PCM) is the digitized compression and decompression standard for voice. Theoretically, PCM samples signal amplitude at 8,000 times a second, and is represented by 8 bits, yielding a 64 Kbps channel.

The U.S. and Japan use the mu-law compression standard with PCM, while in Europe and Asia PCM-encoded audio signals are compressed at variable rates – e.g., from 16 to 8 bits per sample. This compression standard is called A-law and it is based on non-linear quantization, which means that the quantization levels are not evenly distributed across the quantization range – they are more numerous near the zero level than they are at the maximum level, thus reducing the quantization noise of low-intensity signals. (See Pulse Code Modulation)

Alliance of Computer Telephone Application Suppliers (ACTAS)

An industry forum specializing in computer telephony in the US. Subgroup of MultiMedia Telecommunications Association (MMTA), an international forum that supports the convergence of communications and computing business applications as an open market development, public policy and educational forum for equipment manufacturers, software developers, distributors, network service suppliers, value-added resellers and systems integrators. Founded in 1970 as the North American Telecommunications Association, MMTA supports the development of open and global markets for the delivery of technology-based solutions to the business community. Its website is www.mmta.org. The MMTA is itself a subsidiary of the Telecommunications Industry Association (TIA) (www.tiaonline.org).

Alternate Mark Inversion (AMI)

The original line coding technique for formatting T-1 network datastreams. A T-1 uses a single pair of wires in each direction and the signals on those wires are the pulses which represent data (1s and 0s). A T-1 sends pulses using TDM or Time Division Multiplexing (See Time Division Multiplexing) which means that all equipment in a T-1 circuit must operate in synchronization because all devices must check the T-1 at precise time intervals to determine if a pulse (1) or no pulse (0) has been received at each bit time or "timeslot."

If when using AMI the T-1 devices detect no pulse during the time interval when one might otherwise be sent, then that non-event is considered to be a zero. As can be expected, a 1 is sent on an AMI T-1 by sending a pulse, as opposed to not sending a pulse. The "alternating mark rule" means successive ones (pulses) are alternately inverted – if the most recent pulse sent was of positive polarity, then the next pulse sent must be of negative polarity, and vice versa.

If an AMI T-1 device receives two pulses in a row of the same polarity, then a bipolar violation (BPV) has occurred. This acts as a sort of primitive error checking capability with a probability of 0.5 of detecting inserted, altered, or lost bits.

Now, if too many zeroes are sent at a time there will be no pulses on the T-1 at all and the clock circuitry in all of the hardware will rapidly fall out of synchronization, since the only way to maintain synchronization on a T-1 is by detecting the rate at which pulses are being received. If there are no pulses (too many zeroes in a row) there is an eventual loss of synchronization.

Thus AMI demands that a certain "ones density" be maintained, that a certain minimum of the bits over a certain period of time be guaranteed to be a "1" (pulse). To satisfy the 1s density requirement on an AMI T-1 one bit out of every eight bits must always be a 1, which means that one bit out of every eight is not available for voice or data traffic (any data pattern can be sent in the remaining bandwidth, such as seven zeros).

A T-1 runs at 1.544 Mbps. About 8K is used for T-1 framing, which leaves 1.536 Mbps The 1.536 is usually divided into 24 timeslots (DS-0's) or "channels" of 64 Kbps each. But by removing or "robbing" the one bit in 8 that is reserved to satisfy ones density requirement the user is left with 56 Kbps per timeslot.

A better method of guaranteeing 1s density is the more modern B8ZS format technique (See B8ZS) which allows each timeslot to transmit a full 64 Kbps, a technique used widely in data communications.

American National Standards Institute (ANSI)

The organization that defines standards, including network standards, for the United States.

AMIS

See Audio Messaging Interchange Specification.

AMPS

See Advanced Mobile Phone System

Analog Display Services Interface (ADSI)

The Analog Display Services Interface (ADSI) is a Bellcore (now Telcordia) protocol that specifies how voice and small bursts of data can simultaneously share the same analog phone line. An ADSI phone is a "screen phone" or "smart phone" that relies on the ADSI protocol to present menus and data to a user. ADSI devices tend to be analog phones with visual display screens and extended keyboards having keys that are labeled for specific functions, but the ADSI signaling protocol can also work with set-top boxes, PC phones and personal assistant appliances to support the analog transmission of voice and text with a host computer or network exchange switch. ADSI phones appeal to people who like to use phones but avoid PCs – even though a hidden PC-like infrastructure is built into the phone! Thus, instead of a PC-as-phone, such a device is a phone-as-PC.

Network service providers began to promote ADSI starting around 1995 to make it easier for subscribers to use enhanced call management services such as call waiting, call forwarding, and three-way calling. It was felt by many that the sequences of DTMF tones used to access enhanced services were becoming too difficult to remember and a user-friendly, visual interface had to be added to telephony and information services. The combination of ADSI and a screen phone can transform call waiting, for example, so that an ADSI-enhanced version of call waiting can display your voice mail selections (such as the names of your callers), on the phone's screen.

Network providers soon realized that ADSI phones could be used for many other services other than call management. Subscribers simply download a service "script" via frequency shift keying (FSK) at 1,200 bps, peruse the script and subscribe to the service if it looks appealing. One can then choose service options from a visual menu that is ready to guide the user through sequences of context-sensitive information.

E-mail messages can be sent and read aloud via ADSI phones using the customary built-in screen and soft keys. You can perform banking transactions, obtain train and plane schedules, and order tickets for sports events from your ADSI phone. Unified messaging solutions can be instantly programmed into ADSI devices. Call center agents can also view, use and store call information supplied by a caller's ADSI phone without asking the caller to repeat or enter any information.

All of these features and third-party information services are now referred to as Advanced Screen Telephony (AST), which is made possible through the ADSI protocol.

Following in the footsteps of the cellular phone marketing model, it was expected that network operators would subsidize or perhaps even give away ADSI screen phones to subscribers, making up the cost later from the revenues generated by ADSI-based services.

Third-party information services could easily be offered over the network since ADSI works on an end-to-end basis, so the voice network infrastructure does not need to be altered as it would in the case of an access service such as ISDN or xDSL. A provider need only build an ADSI host server from which services can be dispensed to any ADSI-enabled screen phone anywhere.

In the U.S. and Canada there are over one million ADSI-enabled phones in the residential market, with forecasts of up to 10 million by the year 2002.

Bellcore (now Telcordia) formulated ADSI and its associated protocol to support the bi-directional transmission of data and voice between a Stored Program Controlled System (SPCS) – a switching system or computer server – and compatible Customer Premises Equipment (CPE) which is the screen phone or other ADSI compliant device.

An ADSI telephone can work in either voice mode or data mode. Voice mode is for normal telephone audio communication, and data mode is for transmitting ADSI commands and controlling the telephone display (voice is muted in data mode).

Prior to transmitting data to an ADSI device, one must first establish "handshaking" between the SPCS and the CPE by sending an ADSI alert tone called a CPE Alerting Signal (CAS) to the phone. The tone is a dual frequency tone of 2,130 and 2,750 Hz played at -15 dB for 80 milliseconds. The tone's characteristics are recognized by an ADSI phone as the inquiry: "Are you an ADSI telephone or compatible device?" If that is the case, then the phone sends an acknowledgment DTMF digit of "a" (meaning "yes") that must be received by the SPCS within 500 milliseconds following the end of the alert tone.

Once handshaking is established, the ADSI telephone has now muted the voice path and data can be transmitted to it, such as Caller ID data messages that can be displayed on the CPE screen. The ADSI data is encoded and transmitted to the phone over the voice channel using a standard 1,200 baud modem specification. The voice is muted to allow data transfer. ADSI phone responses to the SPCS/server are mapped into DTMF sequences.

Aside from the raw data, ADSI also defines a protocol, syntax, special message sequences, and rules followed by both CPE and SPCS/servers when using the interface.

ADSI data is sent to the ADSI telephone in a message burst, corresponding to a single transmission that contains up to five messages, with each message consisting of one or more ADSI commands.

The ADSI alert tone causes the ADSI telephone to switch to data mode for one message burst or transmission. When the transmission is complete, the ADSI phone will revert to voice mode unless the transmission contained a message with a "Switch to Data" command.

After the data is transmitted, the ADSI telephone sends an acknowledgment consisting of a DTMF "d" plus a digit from "1" to "5" indicating the number of messages the ADSI telephone successfully received. Comparing this message count with the number of messages transmitted immediately reveals errors and the need to retransmit messages.

The protocol and interface are designed to communicate with a single abstract model of the CPE, making certain "standard" assumptions about what functionality exists at the CPE. For example, the abstract, standardized collection of logical components in an ADSI-compatible CPE includes provisions for virtual and predefined display storage of a certain size, regardless of the device used.

Soft keys on the abstract CPE for customer control of, or response to, a feature can be programmed by a download mechanism from the SPCS/server all at once or interactively. Similarly, data to be displayed can be downloaded into the CPE from the SPCS/server all at once or interactively. Generally, the SPCS/Server delivers data to the CPE via FSK modem signaling, but there is a built-in option to allow use of other protocols once a connection is secured using the standard ADSI DTMF/FSK modem based physical layer protocol. Under special circumstances, DC signaling can take the place of DTMF tones when a CPE transmits responses to an SPCS/server. The protocol supports only one active device at a time.

ADSI network-to-CPE communications operate in two modes:

- **Server Display Control (SDC) mode,** using real-time communication between the SPCS/server and CPE so that the SPCS/server can control the screen, the display of information and soft key labels / actions. Online, real-time stock quotes and pricing come under the provenance of SDC Sessions.

- **Feature Download Management (FDM) mode,** which provides semi-permanent downloads of one or more pre-defined CPE-resident scripts from the server to the local device, where they are interpreted and run. The server sends a script segmented into a burst of messages through the network to be loaded into memory resident in the CPE. An ADSI session between a SPCS/server's download administration feature and the CPE function that receives, manages, and stores a download of a CPE-resident feature script for later execution is called a Feature Download session. Once downloaded, CPE-resident scripts run independently on the CPE.

BellSouth, Bell Canada and U.S. West were early promoters of ADSI in 1995, as were some European telcos (where ADSI is called Server Display and Script Services, or SDSS), and CIDCO, Intelidata, Inventec, Lucent, Nortel, and Philips competed with each other to produce ADSI phones.

Unfortunately, during this period the Internet and its World Wide Web began to explode in popularity, which has slowed the acceptance of ADSI, as has the rapidly declining prices of PCs, which in some cases have made them less expensive than the phones themselves.

Some manufacturers, such as Philips Consumer Communications L.P. (Fremont, CA – 510-445-6000, www.pcc.philips.com) and CIDCO (Morgan Hill, CA – 408-779-1162, www.cidco.com) took the Internet into account when they web-enabled their ADSI phones to handle e-mail and browse the Web. Philips soon went far beyond ADSI technology and instead licensed Phone.com's UP.Browser microbrowser software so that Philips' GSM mobile phones could become "Internet companions" possessing web functionality such as e-mail, Internet, and access to corporate Intranet information.

Analog Expansion Bus (AEB)

An early analog voice-processing bus that was designed by Dialogic for its DTI/124, D/21D, D/4x, LSI/120 and other voice cards for connecting voice cards together across the AT PC bus. The AEB allows audio voice signals to be passed between different voice processing components via connectors mounted on the board. Other voice communication buses used on Dialogic hardware include the industry-standard H.100, H.110 (for CompactPCI), Dialogic's own Signal Computing System Architecture (SCbus), and the PCM Expansion Bus (PEB). The D/21D, D/41D, and D/41E voice boards use the AEB, the SpringBoards (8- and 12-channel voice boards) use the PEB, and the D/41ESC and the DIALOG/HD voice boards (D/240SC, D/240SC-T1, D/320SC, D/300SC-E1, D/160SC-LS) use the SCbus or the PEB.

ANI

See Automatic Number Identification

ANSI

See American National Standards Institute.

API

See Application Programming Interface

App-Gen

See Application Generator.

Application Generator

An application generator (or app-gen, app gen, or AppGen) is a software package that eases some of the programming workload when developing computer telephony applications. They range from high level languages, to scripting toolkits, to full-blown GUI flowcharts that can generate reams of source code for the developer. The objective in all cases is to allow the developer to quickly and easily specify "what"' is desired rather than spending time detailing each stop of "how" the IVR (or other system) is to work, as is the case with more conventional programming and development techniques.

By the late 1980s and early 1990s, it had become obvious that an Interactive Voice Response (IVR) system saves an organization considerable money. Repetitive, "no-brainer" tasks such as reading out a checking account balance from a database, making a reservation, ordering a product or a faxback white paper could be done by calling a computer, not an employee.

But installing an IVR was expensive. You had to buy a "turnkey" application, or else hire an expert to develop a custom application for you.

No doubt the "expert" you hired was taking far too long to develop your system. You wondered how he or she (or anyone else) could make a living by developing such systems on a production basis. The answer, of course, was that the expert charged considerable sums of money for services rendered.

It became evident that quickly cobbling together a sophisticated IVR system was nearly impossible, even for experienced programmers. What the computer telephony industry needed was an erector-set like software and hardware package that would let you quickly create customized IVR systems, either for yourself or a client, and produce them as one would produce cars on an assembly line.

For years in the AI field there had been a category known as "automatic programming" but much of that research had generally gotten lost in the world of compiler design, where the compilation process of a program written in a high level computer language causes "automatic programming" of assembler or machine code.

In the late 1980s there emerged script-based toolkits for telephony. These were essentially high-level languages that could control Dialogic or Natural MicroSystems voice and fax boards. Why did script-based packages become popular? The simple reason is that some people, usually those with programming experience, enjoy the control a script-based language can give them, even if a GUI development environment is available (and only a few OS/2 GUI app-gens were available in the early 1990s).

Script-based toolkits are directed at the people trained to think "linearly" (sequentially), rather than "visual" (multi-dimensionally). Indeed, the author suspects that some of us who have been programming too long tend to think sequentially all the time – ever see a programmer eat French fries one at a time in rapid succession?

The script-based toolkit ruled the industry until about 1994. In those days much development occurred using such tools as the Scenario Programming Language (SPL) from Expert Systems, now EASE CT Solutions (Alpharetta, GA – 404-338-2241, www.easey.com), V-Script from Enhanced Systems (Norcross, GA – 770-662-1503) (on January 1, 1999 Enhanced Systems became Vodavi-CT, a wholly owned subsidiary of Vodavi Technology, Inc., www.vodavi-ct.com/Index.htm), the FAR Voice from FAR Systems (Fort Atkinson, WI – 920-563-2221, www.farsystems.com), GranGen from Granada Systems Design (Hopewell Junction, NY – 914-221-1617); AppGEN from Hager International Voice Research / Hager Telecommunications (Hopkinton, MA – 508-435-9551); the Voice

Operating System (VOS), a language that's still going strong from Parity Software (Sausalito, CA – www.parity.com, acquired in 1999 by Dialogic, an Intel Company); the Voice Applications Language (VAL) from U.S. Telecom International (Joplin, MO – 417-781-7000, www.usti.com); and the Universal Scripting Language (USL) from a company called UniVoice Products (Norwood, MA – 781-255-5400, www.inea.com/uv/index.html).

Although it appeared "long ago" in 1989, Parity's VOS is still widely used by developers, probably because VOS telephony commands map onto the API for the popular Dialogic cards, and so a whole series of complex functions (such as sending a fax) can be replaced with a single VOS command. In this way VOS resembles a fourth generation database language more than a conventional computer language, but with special telephony extensions. For example, here is a snippet of VOS source code that waits for an incoming call, answers the call and plays a message:

```
program
        line=arg(1);                    # Assign line 1 to task
        sc_onhook(line);                # Hang up
        sc_watch(line,"+-rd",1);        # Watch for ring/hgup
events
        sc_wait(line);                  # Wait for ring event
        sc_offhook(line);               # Go offhook (answer)
        sc_play(line,"hello.vox");      # Play speech file
        restart;                        # Start over
        end
```

One can expand on this framework, building a more complicated, genuine IVR system:

```
        # Example of another VOS program.
        # This program:
        #  a) Waits for a call.
        #  b) Plays a greeting.
        #  c) Plays a main menu, press 1 for messages, 2 to quit.
        #  If the caller presses 1:
        #  d) Asks for a two-digit number from 10 to 99
        #   e) Plays a message (10 to 99) depending on digits
entered
        #  f) Repeats until caller hangs up.
        # Program uses 'goto' because easy to understand, could
also
        # have used 'for' or 'do' .. 'until' loops.

dec # Declare variables
        var line : 2;                   # Phone line number
        var digits : 2;                 # Digits dialed by caller
enddec
program                                 # Execution starts here
        line = 1;
        sc_use(line);
        sc_wait(line);                  # Wait for incoming call
        sc_offhook(line);               # Answer call
        sc_play(line, "GREETING.VOX");  # Opening greeting
MainMenu:
        sc_play(line, "MAINMENU.VOX");  # Main menu
        sc_getdigits(line, 1);          # Wait for one digit
        digits = sc_digits(line);
        if (digits < 1 or digits > 2)   # Is it a valid response?
        goto MainMenu;                  # Not valid, ask again
        endif
```

```
AskForTwo:
        sc_play(line, "ASK2DIGS.VOX");  # Ask for two digits 10 .. 99
        sc_getdigits(line, 2);          # Wait for two digits
        digits = sc_digits(line);
        if (digits < 10 or digits > 99) # Is it a valid response?
        goto AskForTwo;
        endif
        sc_play(line, "MSG" & digits);  # Play MSGnn where
                                          nn=10 .. 99
        goto MainMenu;                  # Repeat until caller hangs
up
        endprogram
onsignal# Program jumps here on hangup
        restart;
        end
```

As can be seen, even the earliest versions of VOS had all the flow-of-control features of a modern, structured programming language including "if...else", loops (for, while, and do...until) and "switch...case". There is also an innovative "jump" command which can take you back to the main program from a nested function call.

Like a conventional programming language, VOS supports arithmetic expressions including parentheses, logical conditions using and, or, not, and named variables. You can define your own functions and build up libraries of re-usable functions. You can define private variables within functions or subroutines for creating independent libraries.

VOS has support for FoxPro, Btrieve, Clipper and dBASE database files. VOS code is source and binary compatible across a range of operating systems: DOS, UNIX, Windows 3.1 / 95 / 98 / NT, and now Windows 2000.

Until the mid-1990s, the main market for toolmakers (not to mention tool-users) was still IVR – the so-called "giving data a voice" application (a phrase that was actually copyrighted by EASE CT Solutions, an app-gen maker). Everyone is familiar with it now. It's old hat: You "touch-tone" (DTMF) unique information into the a PC resource card over the phone, and the application checks a database for the specific information you need and then "speaks" or faxes it off to you.

In retrospect, it is quite remarkable what one could do with IVR functions on a PC of limited processing power running DOS in the early 1990s. One neat package that appeared was Glenn Stok's Contact Caller from Stok Software (Elmhurst, NY – 718-699-9393, www.stok.com) a voice application toolkit for those who needed to get a custom program up and running quickly. Still in production with upgrades over the years, it includes ready to run applications for order entry, database inquiry and bulk message dispatching that work with Dialogic telephony cards.

Yes, IVR was a large market and it still is. Any business that employs people to answer the phone and verbally reiterate simple database-accessed information to callers can and should use IVR. After all, paying people money to provide this information over the phone to customers is a waste of their minds and a company's money.

As we've seen, in response to this huge market most scripting languages were optimized for building IVR systems, but many new features were being added to telephony scripting toolkits every day and companies were always trying to "push the envelope" with whatever tools were available.

As early as 1994, for example, one could build a PC-based International Callback system based on the following stretch of VOS code:

```
        # Name of file: CALLBACK.VS
        # International callback skeleton app written in VOS
```

```
dec
# Variables:
        var inbound_line : 2;    # Phone line nr for DID call
        var callback_line : 2;   # Phone line nr for call-back
        var outbound_line : 2;   # Phone line nr for out-bound call
        var DID_digits : 7;      # Variable to store DID digits
        var PIN : 5;             # Variable to store PIN code
        var callers_nr : 16;     # Call-back number (caller's number)
        var nr_to_call : 16;     # Number caller wishes to reach
        var DID_database : 3;    # DID, PIN, call-back nr database
        var rec_nr : 12;         # Record number in DID database
        var rec : 3;             # Record handle
        var callback_nr : 16;    # Callback number of caller
# Named constants:
        const NR_DID_DIGITS = 7;   # Number of DID digits
        const NR_PIN_DIGITS = 5;   # Number of digits in PIN code
        const DTI = 1;             # DTI/212 board number
        const DMX = 2;             # DMX board number
        const CONNECT = 32;        # DMX command for full-duplex connect
        const DISCONNECT = "-1";   # D/121B disconnect
    enddec
program
    inbound_line = arg();
    outbound_line = inbound_line + 12;
    callback_line = inbound_line;
    DID_database = db_open("DID.DBF", 0, 0);
    sc_connect(inbound_line, inbound_line);
    sc_connect(outbound_line, outbound_line);
# Wait for incoming call, get DID digits but don't answer call
    sc_wait(inbound_line);
    sc_wink(inbound_line);
    DID_digits = sc_getdgts(inbound_line, NR_DID_DIGITS);
# Look up call-back number in database
    rec_nr = db_first(DID_database, "CALLBACK_NR", DID_digits);
    rec = db_get(DID_database, rec_nr);
    callback_nr = db_fget(rec, "CALLBACK_NR");
# Initiate return call
    sc_offhook(callback_line);
    sleep(20);
    sc_call(callback_line, callers_nr);
# Go off-hook, gets US dial tone on out-bound T-1 channel
    sc_offhook(outbound_line);
# Connect lines through DMX switch
    DMX_route(DMX, inbound_line, outbound_line, CONNECT);
# Disconnect D/121B channels, dial-tone now passed to caller
    sc_connect(inbound_line, inbound_line, DISCONNECT);
    sc_connect(outbound_line, outbound_line, DISCONNECT);
# Wait for one party to hang up by checking every second to see
# if the T-1 "a" signalling bit on either side went OFF.
    do
         sleep(10);
    until (DTI_trans(DTI, inbound_line, "a") or
         DTI_trans(DTI, outbound_line, "a"));
# Terminate and wait for next DID call
    restart;
endprogram
```

In 1996, PCS Telecom (Tequesta, FL – 561-745-1688, www.telephonyworld.com/platform/pcstele/index.htm) spent three weeks writing VOS code to create a ground breaking PC based telephone switching platform for long distance resellers and debit-card service providers, having 240 ports (10 Dialogic T-1 cards) in a single-chassis running under MS-DOS.

This high density (made possible by the remarkable stability of VOS) allowed PCS customers to resell millions of minutes a month of international long distance, call-back, or debit card service. The voice resources and network interfaces were provided by four Dialogic D/240SC-T1s and six DTI/240SCs. Interconnection and timeslots switching were accomplished over the then-exotic Dialogic SCbus. A few years later PCS Telecom would achieve 384 ports per chassis and would go on to become a premier provider of turnkey PC-based enhanced service platforms.

PCT Telecom proved that, with a robust app-gen or scripting tool, a "do-it-yourself" PC-based system could compete with products from the larger switch manufacturers. Indeed, PCS could offer custom features and control that could not yet be provided by the larger switch vendors, while also offering a comparable number of ports at a fraction of the cost of a big switch.

As David Dragon, then PCS Telecom's VP of Development said at the time, "We have replaced a million-dollar switch and we can still sleep at night."

But there had to be a faster, easier way than scripting to create call processing telephony applications, something that would be accessible to those individuals having less than expert programming expertise.

As it turned out, Microsoft's Basic pseudocode interpreter would go through some elaborate twists and turns and ultimately be transformed into Visual Basic (VB) running under Windows in 1991.

Programming would never be the same again, all thanks to the great innovation of VB's "custom controls," software objects that you embedded in a Visual Basic or other similar Windows development tool.

One of the great telephony-related innovators in this area came from a then-fledgling company called Stylus Innovations in Cambridge, MA. Stylus and its custom-control-based product Visual Voice started out when two students entered a business plan in MIT's Best New Company of the Year Competition in 1991 and won the $10,000 first prize. Stylus incorporated in 1992 and was bought by Artisoft in 1996 for $12.8 million. Intel later bought the rights to Visual Voice from Artisoft in November of 1999.

What set Stylus apart from other app-gen makers (and shook them all up rather severely) was its abandonment of the traditional policy of charging runtime fees. This is the standard way of doing things in the Visual Basic applications world, but was completely earth-shattering to the voice / fax processing app-gen world.

Such a pricing plan from companies like Stylus helped propel the sales of their products to unprecedented levels.

But success did not result solely from innovative pricing. There was, after all, a huge audience of embedded Visual Basic programmers waiting to get their hands on exciting new programming tools that could take the mystery out of developing telephony-related applications. These telephony illiterates latched onto CT with surprising quickness, discovering that they could use the new telephony VBX custom controls they found in their toolkits to do things they couldn't do before, such as add voice and fax front-ends to existing databases, automate routine information processes (IVR, fax-on-demand) and drive various CT resource and media processing cards for automated attendants, voice / fax mail, etc. Suddenly it was all simple.

Aside from telephony functions, there were all kinds of other customized VBX controls available for leverage, including database access and connectivity tools. The descendants of these early third-party add-ons are still a major selling point for development in all of the Microsoft

"Visual" languages.

Visual Basic CT development exploded. Dataquest soon ranked Stylus as the number one supplier of an IVR product shipped in North America.

Although technically a scripting app-gen, Visual Voice offered a more graphical development environment than its competitors and was the stepping stone to such purely GUI Visual Basic app-gens as VBVoice from Pronexus, a product that actually appeared just prior to Visual Voice in 1993. Even Parity Software, long a champion of its own proprietary telephony language (VOS), entered the fray with a Visual Basic product called VoiceBocx.

The collision of Visual Basic and telephony would propel the computer telephony industry toward a more Microsoft-styled model of affordable, mass-marketed software.

In many ways, it worked. Today, Visual Basic CT programming has become a large and mature app-gen world. With the right tools, it's simple and generally effective and now supports most CT resources (voice / fax boards, telephony network interfaces of all shapes and sizes, ASR, text-to-speech, etc.). It also continues to improve overall within computer telephony confines.

This is not to say that the great migration to Visual Basic (and later other GUI visual environments) was a smooth one.

The Visual Basic programming environment (indeed, the whole concept of Windows GUI interfaces) is founded upon an "event driven" metaphor – events signal important user actions, as when a user performs a mouse click on a form element such as a button, and some associated code is then executed.

Script-based app-gens such as VOS, however, are procedural. You input a series of commands and branching instructions sequentially. Many of the processes found in telephony are procedural (i.e. wait for call, upon ring answer, play greeting, hang up) so script based development software lends itself to both conventional programming techniques by experienced programmers and telephony itself.

In fact, another early criticism about Visual Basic development concerned its ability – or rather inability – to handle so-called high-density (i.e. many ports) CT applications.

Generally there are three programming techniques for managing multiple lines: A state machine, multiple executables and multiple threads.

A state machine is a single executable (an executable is herein defined as an executing instance of an EXE or equivalent file running on a single thread). The state machine technique can be used on any operating system. What happens is that an API call such as PlayVoxFile returns immediately to the application rather than just waiting around for the process to finish. Later, the CT resource board driver sends a message to the application when the play finishes signaling that the playing of the voice prompt has been successfully accomplished. This state machine technique allows a single executable to control more than one phone line and is the only model which works on MS-DOS, since DOS only allows one executable to run at a time.

State machine code is very difficult to write. You can't arbitrarily use the conventional commands found in a programming language (such as IF, THEN, and ELSE statements) to determine an application's flow of control. Instead, you must keep track of the current state of each phone line in a state table, a process which is described in detail in Bob Edgar's book, *PC Telephony*.

As it happens, Parity's VOS has its own internal state machine multitasking engine which enables one EXE to handle all telephony traffic and activity. Running VOS on Unix, NT or Windows 2000 creates a multitasking OS environment within the existing OS. Running VOS on

DOS, a single tasking OS, also conveniently provides multitasking for computer telephony applications.

Running multiple executables (a technique that's also called the synchronous model) is as simple as starting a different program (or a separate copy of one program) for each phone line serving the application. Sometimes a developer may want to better isolate voice processing channels by deploying computer telephony applications on a one process per voice channel basis, but for this to work it is necessary that the operating system used be capable of multitasking

Until Microsoft's release of Service Pack 2 for Visual Basic 5 in August of 1997, Visual Basic needed one EXE file for each independent phone line. If you had a T-1 line and a system having 24 ports, you needed to spawn 24 EXEs, which rapidly consumed memory and processing power.

Multi-threading is found in Windows NT, Windows 2000 and some UNIX flavors which allows a single program to have more than one execution path at the same time. A thread is a light-weight version of a process or task: It shares code and data with other threads in the same program. A thread is created and maintained for each user or process wanting access to an application program. The thread allows a program to know which user is being served as the program alternately gets re-entered for different users.

Multithreading is often confused with multitasking. Multithreading is the management of multiple concurrent uses of the same program, while multitasking allows multiple tasks to run concurrently, taking turns using the resources of the computer. Thus, an operating system such as Windows NT or Windows 2000 multitasks between all active multiple threads.

A typical use for a thread is to recalculate, search or print in the background while a user edits a word processing file or a spreadsheet in the foreground. Since threads within one program share code and data, the operating system should be able to switch between them faster than between separate executables with their separate address spaces, so threads demand less system overhead. Modern operating systems such as Windows also allow multithreading within program processes so that the system is saved the overhead of creating a new process for each thread. For high-density applications, there is a significant gain in efficiency by running multiple voice processing threads within a single process. The use of native Windows 95 and NT voice card drivers eliminates the overhead of DOS, improving system performance.

At the time (1997), VB aficionados admitted that Visual Basic was limited by the one EXE per line rule, but in the same breath they would charge that one could not build data structures of any complexity in VOS, which provides support for integers, strings, and arrays, but not objects. In VB, of course, one can easily create complex data structures (i.e. computer.price=3000, computer.drivetype="SCSI", etc.) which are a very compact and efficient way to store and process complex data sets. One cannot do this in VOS unless you declare separate variables for each sub-type.

Initially, Artisoft and Parity attempted to tackle the multitasking problem during the summer of 1997 by converting their controls into ActiveX EXE components, which developers access by selecting the appropriate Template when they start Visual Basic. They would then add the same type of code they've always written when developing the old-style single-threaded Visual Basic applications. Using this technique the code would not have to be changed to accommodate multitasking.

Pronexus, one of the true GUI Visual Basic app-gens, managed to avoid the whole multithreading controversy entirely. Although the original versions of VBVoice were based on state machines, the 32-bit version (which had debuted in February 1996) introduced a unique architecture that gave applications support for multi-threading. It did this by calling on a special control – a non-VB "engine" developed by Pronexus.

With this engine, VBVoice could run a single executable managing up to 240 lines in a single application. Each channel was streamed into a separate thread within the control. Each thread communicated with the main VB thread using messages. The engine also provided multi-threaded ODBC support.

A "VBVFrame" control sat on top of the Visual Basic form and it served as the "container" for the remaining VBVoice controls. The "LineGroup" control answered or placed a call and specified which channels belonged to a given application. It administered and managed the threads for each channel.

When the VBVoice system started, the runtime system checked all of the channels in each LineGroup control and dropped a separate processing thread for each channel. Call processing was conducted in the current control for each channel, guided by the properties of the control and the events fired for that control. On termination, the processing exited through the appropriate output node and was passed to the input node of the connecting control.

The entire process is not unlike the telephone network itself. Think of a signal arriving at a switching box to which the destination address is processed and the packets routed to the appropriate next node in the network. With a VBVoice app, the various user-defined parameters are interpreted by the runtime, which then selects the appropriate branch of the map and moves processing to another node, all along the same thread.

Even then, VBVoice supported the concept of dynamic application deployment in one computer. For example, modifications to a running application could be made off-line then started alongside the old version, taking control of inbound calls without missing a beat. Multitasking also allowed for the deployment of unrelated applications on a single computer, a situation often encountered in service bureaus operations.

Remember, this was all being done prior to Visual Basic 5.0. When Service Pack 2 for VB 5.0 became available for download from the Microsoft website in August, 1997, it suddenly became possible to place VB events into a free-threaded mode. Better yet: You could choose to use the existing serialized events or to fire events from a separate channel-specific thread.

With the existing serial event firing, all events were now generated in sequence in the VB main thread, thus ensuring freedom from thread safety issues such as those surrounding database access using Database Access Objects (DAO). The free-threading option delivered complete channel independence for the events.

Again, the choice was the main factor here: Each control could select free-threaded or serialized events, giving programmers more flexibility in their design approach.

All of this meant good news: If you were using Visual Basic to build multi-threaded ActiveX applications, you could now no longer "hang yourself," as opposed to Visual C++ and Delphi developers who have complete and total control over threads and thus responsibility for keeping them out of trouble. This is because Visual Basic applications can do threading only by playing by the rules of Microsoft's Component Object Model (COM), a basic set of rules about the creation and use of software objects or components.

One of the great things about COM is when its rules are followed, all COM objects behave properly when they access each other (even from different threads). Visual Basic was always considered as the "golden boy" of COM, so even if Visual Basic happens to access some kind of COM component that's not thread safe, COM will act like a traffic cop, step in and prevent two different threads from accessing the component simultaneously.

If two threads do try to access a thread-unsafe component at the same time, COM will force one thread to wait until the other thread is finished. All components are assumed to be thread-unsafe unless specifically coded otherwise.

So it was at that moment, in August 1997, that the Visual Basic appgen had come of age as a major CT development tool.

And with Windows 98 and NT 4.0 then looming on the horizon, it was clear even to hard-core traditional CT app-tool vendors that the up and coming 32-bit Windows environment was not going to remain merely a desktop GUI application launching pad for old DOS programs, but the chief development and deployment environment for most CT application software, which it promised to shield from background hardware with a carefully crafted API family: WOSA for database connectivity, MAPI for messaging, TAPI for telephony, and SAPI for speech technology.

Windows 32-bit operating systems coupled with the new 32-bit appgens now available gave Computer Telephony developers the flexibility to create and deploy any type of application imaginable. This included light-duty desktop applications all the way up to high-density, mission-critical applications. The multi-tasking capabilities of Windows NT and now Windows 2000 offer improved stability for mission-critical applications.

As time went on, companies demanded more from the existing development tools. Voice mail and automated attendants soon had to be added to any development environment's bag of tricks. And the applications grew more ever more sophisticated as entirely new technologies were suddenly thrust upon the scene.

The Internet became an overnight sensation, and with it web-enabled call centers. In 1994 PCs became powerful enough to handle voice recognition and speaker identification over the phone, as well as text-to-speech and secure encryption algorithms. Also, companies began to scale up their applications to higher and higher densities – to around 384 ports, which was just about the limit of character-based DOS toolkits running in a single PC chassis.

Today, board-level DSP resources can now do unbelievable things in a single slot (120 ports of fax, for example), capabilities that just a few years ago were impossible. And the H.100, H.110, S.100 and TAPI 3.0 initiatives continue to mature, adding standardized interoperability and advanced switching on digital T-1 / E-1 and PRI circuits, high-density digital line counts, multimedia (voice, fax, text-to-speech, voice rec) processing, multi-chassis platforms, etc. to our development world.

Even manufacturers of staid old PBXs and PABXs began to slowly open up to the LAN and PC-based CT servers controlled by app-gens, which are only too glad to incorporate first and third party telephony control and messaging mediums to satisfy their clients, despite the complexity of adding an old phone system as a switching component to an advanced CT system. After all, out of the complexities of CT comes new opportunities.

Still, whatever their size or complexity, CT applications are ultimately based upon two underlying technologies: Media processing (such as handling data, responding to a DTMF tone by playing an appropriate voice prompt or sending a fax) and call control (placing, answering, transferring and holding calls).

The construction of an application is also divided into two independent parts: Building the application, then running and maintaining it. Even if an application can be built within a few hours, one will be managing, updating and expanding the system for years to come. Thus, consideration of the underlying architecture of a CT system is extremely important.

While creating and maintaining such CT applications, people tend to think either visually (multi-dimensionally) or else linearly (sequentially). Some professional programmers still work with script-based toolkits, simply because they have to deal with many pages of source code all day long and have a "sequential" mindset.

But most human beings – and less experienced programmers in particular – tend to think "visually" and this fact ultimately led to the rise of the Graphical User Interface (GUI) and a whole new approach to developing applications emerged – drag and drop.

Once GUI-based operating systems became established, the appearance of the GUI app-gen was inevitable. GUI app-gens are software packages running under various flavors of Windows and UNIX that try as much as possible to shield non-hacker programmers from the minutiae of controlling voice and fax cards, queues, OS calls and other excruciating technical details. They allow in-house programmers, VARs, system integrators and application developers to quickly and inexpensively create and run CT and Web-enabled CT applications on a fast production scale with far less risk than conventional programming methodologies.

Rather than coding the application's callflow in assembler, C, or C++ and writing interfaces to the telephony and speech hardware as well as code accessing external databases, CT application development with GUI app-gens is now often done by connecting some icons together, filling out some variable values, and making sure that you have "thought out" your project logically enough for your application to work the way you intend it to.

Some telephony languages, such as Parity's VOS, were married to a graphical front end so as to become user friendly to nonprogrammers, while other development packages, such as VBVoice, Envox, OmniVox, Show N Tel and OPUS Maestro, were designed "from scratch" with the GUI paradigm in mind.

The latest and most advanced development environments now provide an easy-to-use graphic, tabular or tree interface for representing an application as a diagram or visual representation explicitly on your PC's monitor. Such app-gens as Envox, Prima's Maestro and Mediasoft's IVS have become advanced software systems, not just mere app-gens. Such development environments let you build the most sophisticated custom CT applications without having to memorize the Dialogic catalog, ponder CT design methodology, or even actually see the final source code.

Many computer telephony developers found that an app-gen with a GUI that clearly reveals the underlying structure of an application can bring structure to a computer telephony project. If two developers each write a complex program in "C", they will write it differently. By using an app-gen, the code becomes relatively structured and the work of various engineers becomes accessible. This means high-end CT apps can be developed and enhanced quickly, adapting to the international telecom and business climate.

As an aside, I have always found it strange that the one platform that could have served as a major GUI telephony development environment in the 1980s – the Apple Macintosh – pretty much fell by the wayside. The poor Mac was not really designed to handle more than one line at a time, and there was no major development effort by Apple to remedy the situation. Even with this limitation, Cypress Software scored a major success with their PhonePro product ($199.95), which is now sold by Bing Software (Mission Viejo, CA – 949-837-0070, www.bingsoftware.com).

PhonePro telephony development software uses an intuitive icon-based scripting interface. Each one of PhonePro's icons represents an action or function, such as picking up the phone, dialing a number, recording or playing a sound. Users simply drag the icons from their palettes onto a work space and arrange them in a flow chart fashion to create full function telephony applications. PhonePro 3.2 ships with prepackaged scripts that users can run out of the box or customize to their specifications. PhonePro 3.2 is also available in a Runtime version, which executes scripts developed using the full version of PhonePro 3.2. Developers can use PhonePro Runtime 3.2 to create custom double-clickable applications.

Macintosh programmers nevertheless continued to have difficulty

This is an example of a very basic answering machine built with the PhonePro 3.2 app-gen for the Macintosh. Icons are connected in flowchart form.

developing multi-line systems. One Macintosh product managed to get around this limitation. Duet by Magnum Software Corporation (Chatsworth, CA, – 818-701-5051, www.magnumcorp.com) is, as its name implies, a hybrid of two IVR products, TFLX for Macs and DAX for DOS PCs.

Back in the Dark Ages of Mac computing (the mid 1980s) there came the first GUI IVR app generator, Teleflex. Teleflex was a winner in the first annual 1988/89 Media Dimensions, Inc. Awards for Most Innovative Voice Applications. The name was shortened to TFLX the following year.

Even in 1988, the program had startling similarities to the most advanced app generators of today, giving users the ability to create incoming or outgoing voice or touchtone apps such as voice messaging, voice processing, IVR, text-to-speech and audiotex. The user could record his or her own messages or use the digitized supply provided. Apps were – and still are – programmed with Magnum's proprietary Picture Programming Language (PPL) which involves linking Task Icons in a flowchart, a process similar to other, later app generators.

The problem with this scenario was that the Mac was limited to a single line. Although one could install five or six individual TFLX lines. This requires a Macintosh, a site license and a TFLX II hardware unit per line. If you needed lines capable of sharing a database, they could be networked together.

In 1994, RAM Research in Concord, CA came to the rescue with their PC DAX voice processing software for multiple lines that works with an extraordinary number of voice / fax boards and modems from such companies as Brooktrout / Rhetorex, Intel / Dialogic, Bicom, Eletech, Pika, Music Telecom, JTDual, and Talking Technologies. Like the V'ger in the first Star Trek movie, the resulting software collision between DAX and TFLX led to Duet.

One now uses Duet as follows: You create your voice mail, order entry, fax on demand, or IVR app on a Mac with the TFLX GUI Picture Programming Language. You can even record your sound files and save them in PC voicecard-compatible file formats on the Mac.

You then generate a runtime version (a "script") and transfer it and any voice files to an MS-DOS computer using something like Apple's PC Exchange. The runtime controls the DAX software on a PC running DOS, which can handle from 2 to 128 phone lines, depending upon its configuration (the minimum configuration is a 33MHz 386DX with voice cards or voice/fax card).

The Mac can be used for other applications until you want to make revisions or additions to the program. Of course, if you only have a Mac and don't need more than a few lines, you can purchase TFLX separately (packages complete with cards range from $495). But if you want to develop PC apps too, you're going to have to buy a PC and a Mac. Although it all works, one would feel more comfortable if Magnum could port TFLX's GUI app generator over to DAX.

This is not to bemoan the TFLX-half of Duet's abilities. TFLX itself can do just about anything any other app generator mentioned in this article can, and then some: TFLX was not only the first IVR and fax-on-demand system for the Mac, it was also the first to work with videophones – it can accept caller input as voice, touchtone, video picture or fax. Still, if you don't mind spending the money on keeping an extra platform around, Duet will do the job.

Returning now to the PC world, we find that today there really are three types of app-gens. At one extreme is the truly "pure" GUI or object-oriented approach, such as VBVoice from Pronexus (Kanata, Ontario, Canada – 613-271-8989, www.pronexus.com) or MasterVox from Mastermind (Washington, D.C. – 202-298-8500, www.mastermind-tech.com), where you can drag and drop telephony features into your app and connect them into what looks like a flowchart. At the other end of the spectrum you have specialized programming languages, in the manner of Parity Software's VOS. In the gray area between you find packages that, although not strictly GUI, have one or more strong "visual" components, such as state-machine tree structures displayed onscreen.

Just because a program is pure GUI doesn't mean that it's *ipso facto* superior to a non-GUI. You can take the world's worst program, put a GUI on it and you'll still have a stinker, only now with an attractive interface.

Indeed, when a company named Cascade once gave their customers a peek at an upcoming pure GUI for their Voice Tools product, nearly everyone responded that the program worked fine as it was, and Cascade should wait on the GUI if it was going to slow down deployment of the program's new version. They were interested more in functionality than the interface.

Still, application generators are getting "GUI-er" all the time.

A system built with a GUI app-gen may be of no concern to a caller, but it certainly matters to a developer who cares how long it takes to build an application. In Visual Basic, for example, a developer can exploit a graphical environment by having multiple windows open at once – one for program execution, one for inspecting variables and one for tracing the executing code.

Other benefits of a graphical development environment include point-and-click screen design and smart editors that parse code in real-time. Anyone who remembers days of intensive programming in a character-based DOS environment should agree.

Also, the idea that a non-programmer can create at least simple IVR, fax and Web-based applications with a powerful code generator dwelling beneath an attractive, easy-to-understand call flow interface is now becoming a real possibility.

In order for an app-gen to be so incredibly user-friendly, tremendous programming effort must go into their development, since they must essentially act as a cross between an extremely high level language (so high a level that an app can be represented visually by rebus-like icons connected by lines) and an automatic programmer, usually in the form of "wizard" software that guides you through the app-building process.

Of course, friendliness has its limits. Many vendors are tempted to claim that their app-gen packages are so simple that "you don't need to know programming." Don't believe it.

Although one does not have to be a top-notch programmer to use such a user-friendly package as a CT app-gen, one should be familiar with the basics of syntax and logic, as well as how all of the pieces of your application will interact. If you're going to construct a large, complicated debit card, web-enabled call center or Internet gateway, it's definitely recommended that you have some programming experience, particularly telephony programming experience. Developing a really top-notch, complex, money-making CT application on which you will stake your livelihood is a situation demanding that you better have some idea of what you are doing, or at least hire someone with more than a modicum of programming skill. As good as they are, app-gens are not yet able to ask you what you really want or read your mind.

As more computer telephony standards have appeared and as CT systems have become more stable, app-gens have taken advantage of whatever hardware and software is out there: Fax, speech recognition, text-to-speech, and now the myriad possibilities afforded by IP technology and web-enabled call centers.

The call center is truly becoming an all-inclusive "customer contact center" designed to improve customer loyalty and wallet share. App-gens are accelerating this process, allowing developers to crank out sophisticated, top-flight systems on a near mass production basis. Even proprietary call center software packages have some kind of app-gen like module that lets you quickly put together a callflow script and provision telephony resources.

The "first wave" of interactive CT applications at many call centers involved IVR systems, but the "second wave" of self-service systems are evolving into voice-centric e-commerce platforms. This results in two new requirements for app-gens: Advanced speech recognition with speaker verification and customer relationship management capabilities. Web page interaction processing also fits neatly into the customer "self-help first" e-commerce scenario.

So, to re-cap, modern app-gens evolved from simple IVR scripting tools into something that allows you do call center-oriented applications involving speech recognition, text-to-speech, faxback functions, and now web-related processing such as website callback buttons, chat windows, and allowing a customer service representative (CSR) to surf a company website in sync with a prospective customer. VoIP is even showing up – enabling you to build apps that permit voice communications over the 'Net, and a few companies we know of are even working on video-over-IP.

Many app-gen makers such as Envox and Elix feel that their products are no longer mere app-gens. They're too flexible. They transcend what a good old IVR app-gen was supposed to do. Perhaps the industry really does need a new category above and beyond "app-gen".

But while a "super app-gen" is thrilling to ponder as a giant abstract piece of software, much equally sophisticated PC-based hardware is necessary to accomplish anything with it – indeed, the software ultimately becomes just another integral component of what are called Enterprise Communication Servers, CT Servers, or Next Generation Communications Servers, such as those from Aspect, Buffalo International, Altitude Software, Genesys Telecom Labs and Interactive Intelligence.

The reseller channel also needs to get a handle on deploying large complicated systems – yes, you could build a huge phone system and call center package with an app-gen, but would you really want to? The industry has begun to move toward flexible turnkey systems with hardware and software ready to go, rather than reinventing the wheel with each customer, no matter how quickly it can be done.

For example, Interactive Intelligence's (Indianapolis, IN – 317-872-3000, www.inter-intelli.com) Enterprise Interaction Center (EIC) has built-in software "handlers" to function as a development system for things like IVR scripting, but the software does much more than plain IVR. The "handlers" determine how a particular event is to be handled, whether it's a phone call, fax, e-mail or web interaction, but such handlers aren't really separable from the system in which they can perform all of their marvelous tricks.

Talkie from Info Systems (Kew Gardens, NY – 416-665-7638, www.infotalkie.com) started out as an IVR app-gen and then kept evolving as the hardware it ran on became more and more sophisticated. Now Info Systems offers things such as the new heavy-duty Talkie IG2 Switch, based on the Summa Four switching platform. Talkie IG2 has the ability to deliver calls via the Internet, through regular carriers or through lease lines, all within the same platform.

Super app-gens (or whatever one wishes to call them) may get a bit "lost in the machinery" of the Enterprise Communications Server, but as a team they provide an extraordinary range of services that generally center on Customer Relationship Management (CRM).

Still, if you do plan to take the "do-it-yourself" road to computer telephony application development with your faithful app-gen in tow, be prepared for a few annoyances. The first nuisance you'll encounter in your role as CT developer probably won't be the app-gen itself, but adding telephony boards to your computer and loading the drivers if you happen to be integrating the hardware yourself. Even here, however, one finds that Dialogic's latest NT drivers, for example, are much easier to install than previous ones. In the past you had to go into the NT Control Panel and fiddle with various arcane parameters and settings. Now you can just pop in a CD and the drivers will install and an automatic test locates the board and its status.

Then there are dongles, those nondescript little devices – invented by a fellow named Dongle, of course – which you must plug into your parallel port to unlock the vast capabilities of your software. Both Apex's OmniVox and Mediasoft Telecoms' IVS use these little contraptions.

Another, and perhaps the biggest nuisance, is recording and keeping track of voice prompts. Some app-gens expect your voice prompts to follow a certain kind of filename progressive naming convention (such as 1001.VOX, 1002.VOX, etc.). Others do not accept filename extenders. Others deviate from the standard 8 KHz sampling of VOX files so prompts imported from another system may sound more like a chorus of bullfrogs than a sultry female voice.

Brooktrout (who sells Show N Tel) makes it easy to add "voice sentences" to any application. Their Voice Sentence Builder lets developers create sentences that seamlessly mix pre-recorded speech and spoken variables. Also, those who have worked with Brooktrout's cards over the years will be familiar with their Prompt Development Tool Kit that can couple a Sound Blaster 16 sound board with Brooktrout's conversion software, so you can record and edit professional quality WAV speech files and then convert them into formats supported by TR Series boards such as ADPCM (24 and 32 kbps), CVSD (24 and 32 kbps) and PCM u-law (64 Kbps).

But even if your app-gen package is the rare bird that has no recording or editing capabilities, inexpensive third-party products let you record and edit voice prompts yourself and then copy them to a specified directory on the PC. We've used the ubiquitous VFEdit and Sonic Foundry's Sound Forge at CT Magazine for years, and I recently downloaded a shareware voice prompt package and produced some voice prompts for a new portal website that's based on Macromedia's Flash 3.0. It wasn't until I had finished working with it that I realized how easy the whole process had become.

Be forewarned, if many voice prompts are involved, voice prompt management is an aggravation. Some advice: Record or generate your prompts with whatever voice prompt editing tool comes with the app-gen or is recommended by the vendor who sells you the app-gen.

No doubt the reader is now ready to make his or her fortune by buying a ton of voice and fax cards, a rack of PCs, an app-gen, and proceed to mass produce CT applications.

Things never turn out to be all that simple, of course. There are a number of important things to keep in mind for the proper selection, care and feeding of your app-gen and its resulting generated application. Before you get in over your head, read these tips:

1. Plan Ahead. How many voice or fax ports will the IVR application need?

Here's a tried-and-true table for determining the number of voice ports for an interactive computer telephony system. First, determine the total minutes per day occupied by the phone calls by multiplying the number of calls by the average length of each call, then look up the number of ports needed to service the system in the table below.

Total Minutes Per Day	Est. # of Ports
300	2
600	4
1,200	6
2,100	8
3,600	12
5,100	16
6,600	20
8,400	24

As for fax ports, calculate the total number of fax pages sent per day by multiplying the number of total faxes per day by the average number of fax pages per day, and look up that figure in the table below:

Fax Pages per day	Est. # of Fax Ports
650	1
1,300	2
2,600	4
5,200	8
7,800	12
10,400	16

Here's another formula to calculate the number of phone lines needed. It's from *Computer Telephony* magazine's Chris Bajorek, otherwise known as the columnist "Dr. CT" and the founder of CT Labs (Roseville, CA – 916-784-7870, www.ct-labs.com).

Bajorek's reasoning goes like this: If you add enough lines so a caller never experiences a busy signal, the system is probably now oversized (and thus too expensive). In this case, many of the lines may be idle a good percentage of the time. To properly determine how many lines are enough, you must decide on a blocking factor and know the average length of each call. This blocking factor is the probability that a caller will experience a busy signal. A typical blocking factor to use is 1% (that is, one out of 100 calls will experience a busy signal), and will usually adequately handle peak hour traffic.

Briefly, here is how to estimate how many lines your system should have for a 1% blocking factor. First, determine the average length of each application call and then estimate the number of calls expected during the peak hour (sometimes you have to make an educated guess). The following equation will give an approximate idea of how many lines are needed so that very few callers will hear busy signals:

lines = Calls/hr (Peak) * Avg call (in MINS) * 0.0238

where:

Lines = total telephone lines required
Calls/hr (Peak) = number of calls per hour during a peak hour
Avg call (in MINS) = length of an average call in minutes
0.0238 = constant value

For example, if 250 calls were expected during the peak hour and the average call was 3 minutes long, the number of lines needed would be:

250 * 3 * 0.0238 = 17.8
or about 18 phone lines.

Once you have calculated how many lines you will need for adequate call handling, you can purchase your speech cards and related hardware.

2. Get a complete set of tools. Nothing could be more "business-critical" than your phone and data networks. Because you can't rely on independent client libraries that haven't been tried and tested in combination on large server systems for tasks like database retrieval, the two watchwords for enterprise app-gens are comprehensiveness and integration.

The app-gen should come with all possible functions to ensure that all the pieces work in concert. These include IVR, fax, database access, prompt recording / editing, hardware (phone and computer) control, and programming tools – compiler, debugger, application manager, and management / diagnostic reports.

If the system(s) you're going to build and the app-gen you're going to build it with are not "spec'd" for high tech trimmings such as speech recognition, computer-based switching, dynamic hardware resource allocation, T-1 / E-1, ISDN or text-to-speech, you can be sure that you will have to retrofit these capabilities sooner or later.

Generally, systems integrators want these different advanced technologies packaged into a neat "granular" client-server architecture, where each module is an autonomous server providing a separate technology such as database access, network interface, speech and fax resources, etc. to apps acting as clients. With granularity you won't have to rewrite apps as new technologies appear.

Because you'll want these types of "interactive" switching / information systems to handle callers based on automatic inputs like ANI and DNIS – besides traditional voice response where callers are prompted for DTMF touchtones – they need to work with digital PC network interfaces as well.

The most flexible app-gens let developers create systems on completely open, non-proprietary PC platforms. There are a lot of end users out there stuck with maxed-out proprietary switching "mainframes." To update them with flexible, programmable applications would require extensive programming and integration expertise. CT app-gen software, whose developers have had years of experience driving PC voice-pro-

cessing hardware, makes it possible to revitalize these old "clunkers."

To construct high density systems with sophisticated switching the software should have specific modules and / or script-level routines to access the APIs of the latest in PC switching resources, including cards from Amtelco, Aculab, Bicom, Intel / Dialogic, Natural MicroSystems, Pika, Music Telecom, and Brooktrout / Rhetorex.

Although the Ring! company in France has been developing a Modula-2 based app-gen that will ultimately run with any operating system and any card, at the moment no app-gen supports all hardware. App-gen makers tend to focus on the popular voice boards from the "Big Three" companies – Dialogic, Natural MicroSystems and Brooktrout / Rhetorex. I expect Amtelco, Bicom, Music Telecom and Pika to join that exclusive club one day. As for fax, Brooktrout cards are outstanding, as are the Intel / Dialogic Gammalink cards.

3. Choose the right operating system. Multiple operating systems support is not as important as it once was: When app-gens first appeared years ago, a lot of them were initially written for OS/2, such as Show N Tel from Brooktrout (Southborough, MA – 508-229-7777, www.brooksoft.com), The Electronic Workforce, from Edify Corporation (Santa Clara, CA – 408-982-2000, www.edify.com) and MasterVox from Mastermind Technologies (Washington, D.C. – 202-298-8500, www.mastermind-tech.com), since OS/2 was the "up-and-coming" GUI OS of the time. Everything else was written either for UNIX, such as OmniVox from Apex Voice Communications (Sherman Oaks, CA – 818-379-8400, www.apexvoice.com), IVS from Mediasoft Telecom / Elix (Montreal, Quebec, Canada – 514-731-3838, www.mediasoft.com) – or even protected mode DOS, such as VOS from Parity Software.

Ironically, MS-DOS remains a widely used OS among computer telephony developers, particularly for simple IVR systems that don't demand much processing power. DOS is at an advantage in that it needs no processing or memory overhead unless an explicit request is made by an application to do something. Hand-tuned code on DOS should therefore have the lowest overhead of any alternative platform running on the same processor.

However, Windows 2000 is the future's ultimate strategic platform. It's a robust, preemptively multi-tasked, fully 32-bit operating system with the world's favorite graphical user interface. UNIX, particularly Solaris, may still be the most robust OS for extremely high density systems, but Windows 2000 will overtake it, and just about every app-gen maker has already or is developing a Windows 2000 version of their products.

When Windows NT began to gain in popularity, some app-gen vendors hurriedly did a bit of "cosmetic surgery," porting their wares from the single-user, single-tasking DOS environment straight up to Windows NT. Such toolkits probably are not taking advantage of NT's multitasking abilities, unless the programmer has done a total rewrite of the code.

Although Windows 2000 may dominate in the future, one should ask about the support for multiple operating systems. Can you run apps seamlessly together as part of the same distributed CT server? Can an app be developed in UNIX and run on NT or Windows 2000? Don't marry an OS – the CT industry moves too fast. But if you have to, Windows 2000 is it.

4. Look for a Graphical User Interface. Today's enterprise app-gens have evolved into high-level object-oriented interfaces over many generations. A well-implemented GUI can shave months of development time from the efforts of even the most seasoned text-based programmer. Just because a GUI is easy to use doesn't mean it shouldn't support all third-party libraries and controls for custom and front-end application integration. Complete workgroups should be able to build re-usable object

libraries to share code from project to project. Plus, developers should have access to underlying code for unique application needs.

5. Not all GUIs are friendly. You can take the world's worst app-gen, put a sizzling GUI "front-end" on it having sexy colors, neat tabbed boxes and glitzy icons designed by a professional artist – but you'll still have the world's worst app-gen, as unfriendly as ever. You might want to take the app-gen for a test drive, to see if you're both suited for a long-term relationship. By actually testing the app-gen you'll be able to determine such things as the following. . .

6. Is the GUI for novices or power users? Some GUI-based app-gens are simple "connect-the-dots" kind of toolkits that are strictly for beginners only. Once you become a power user, they seem restrictive and cumbersome to work with. Other, more flexible packages have hundreds of different kinds of icons, menus and variables, leaving beginners completely clueless. No single approach can satisfy everyone. That's why the best app-gens offer more than one approach for creating an application: Beginners can start off with a GUI or menu technique, while power users can graduate to writing and editing an application's call flow script directly sans GUI.

7. If you're a dyed-in-the-wool programmer, you may gravitate toward using a script-based app-gen. Although I just emphasized that GUI app-gens are easy and quick to use whether or not you've done much programming, experienced programmers intent on building a giant system might become confused dealing with a GUI where the application call flow is defined by connecting icons together. A really big system built in this way can sometimes look like a GUI version of "spaghetti code." This is why some conventional programmers who tend to think "linguistically" swear by script-based, high-level telephony languages such as MediaSoft Telecom's Blabla, or Parity's VOS.

8. Allow for those nitty-gritty "real" programmers, too. Many "true GUI" tools limit developers to dragging and dropping "power boxes" or "cells" or "action icons" the dialog boxes of which are opened up, parameters set, then the icons are linked together in whatever call flow you've planned for building the application. Similarly, a script-based app-gen assumes that all of the "verbs" (functions) of the script language are enough for you to build any application.

Of course, app-gen designers couldn't possibly anticipate everything you could do with their tools. Veteran developers will tell you customers always have special requests – some crazy terminal emulation for getting data out of a mainframe database, for example. This can mean dropping down not just to script code but to the Visual Basic or "C++" routine level or perhaps even the assembler level. Make sure your app-gen allows developers and hacker folk to program at the lowest system level possible if need be.

9. Are building and running apps separate processes? Although you want to be able to modify a program at runtime, you also don't want to slow down runtime servers while developing a totally new application. You should be able to build apps off line. This approach permits several programmers to share app design duties.

10. Can the platform do access control? If you're using several programmers and you've got a tight schedule, each developer will likely be working on a different project. This situation calls for access control and system permissions (such as highly secure Rate 2 NSA access control) for developers working on the server. Your developers may grumble about having to log on to the system with a user ID and password, but this procedure helps to avoid unauthorized access, delays, and coding errors.

11. What are the app-gen's testing and debugging tools? And speaking of coding errors, does your app-gen have a runtime analyzer? Can

you "single-step" through the program a line at a time? Can you put a trace on the system to observe the value of selected "watch" variables, database search results or host access retrievals? Can you put a breakpoint in the code to stop the program at a particular point? You should be able to generate both a dynamic display of the information or capture it in an ASCII file for later analysis.

12. A game of SIM-IVR system, anyone? If you're an independent developer, you'll want a simulation tool to sell a client on your brilliant idea for an IVR system. Clients will want to "kick the tires" and hear the app's voice prompts and check every step of the call flow.

13. Make sure the app-gen supports open telephony and network standards: It should support telephony standards like TAPI, TSAPI, CSTA, S.100, H.100, H.110, MVIP, and SCSA. Avoid being locked in to a single supplier of proprietary hardware or software. Choose the most popular names with the highest market growth. What is the most common data communication protocol? TCP/IP. How about a popular database query language? SQL. Database interface? ODBC. You get the idea. Every system component should be interchangeable with at least one "second source" part.

14. Look for phone system independence. As stated above, telephony standards compatibility (H.100, S.100, TAPI, TSAPI, CSTA) is important. But phone systems (like their computer counterparts a few years ago) are just now making the transition to open systems. That means any high-end app-gen must be able to communicate with the millions of legacy phone systems using in-band and out-of-band proprietary protocols for years to come.

Besides, writing to a telephony API such as TAPI or TSAPI is like programming a "generic" telephony system. To take advantage of any advanced unique (and proprietary) functions built into modern phone systems, you've really got to be able to write to the specific API of your target phone system.

15. Get integrated messaging and call center utilities. These applications have broad appeal for any organization adopting computer telephony technologies. On the messaging side, an app-gen should support integration of voice, fax, and e-mail into a single message box, with the parameters definable at the client-level.

Ideally, the tools will allow integration using the protocol of the organization's existing e-mail system, to take advantage of already familiar features. For call center and ACD functions, support for third-party call control APIs are necessary enabling tools. The app-gen should also include the proper network controls for delivering data in screen-pop applications.

16. Get scaleable high-density port support: Can you and your system grow old together? Even if your application's initial hardware needs are small (anywhere from 2 to 24 lines in a single PC), scalability is a necessity because telephony systems can grow rapidly. As a result, the app-gen must support a minimum of 48 (preferably 96) lines per chassis, and networked clients to accommodate phone network growth into the hundreds and thousands of lines. Single chassis systems max out at between 240 and 384 lines. After that you must network them together.

Moreover, although it's possible to plug 240 lines into a bank of Dialogic or NMS cards, those cards in turn will have to be plugged into a heavy-duty rackmounted PC having a well-grounded passive backplane and massive power supplies. To go beyond a single chassis your app-gen should "understand" H.100 (or at least it's predecessors, SCSA and MVIP), and things get even more complicated if you need to centrally monitor and control a network of physically distributed servers running different call processing resources and applications.

The term "seamless scalability" is often bandied about when describing app-gens, often without regard to what it really means. If your system grows rapidly, you'll find out soon enough just how "seamless" the scalability really is.

17. Look for multitasking and multithreading. Nearly all app-gens now support it. Linking phone and fax hardware is a complex art, particularly when thousands of lines are being driven, and the systems run in perpetuity. A true multi-tasking operating system such as Windows NT, Windows 2000 and Unix / Linux isolates each process, so a breakdown in one line or in a single task won't crash the entire application. Desktop, mono-tasking operating systems like DOS have slow file systems, since disk drives can't operate concurrently with the CPU (unless they're SCSI drives with RISC chips), drastically impeding performance of multi-line systems.

18. Look for support for many databases and hosts. One important reason CT app-gens are used to build specialized call-handling systems is because of their inherent expertise in "going where the data is" without changing embedded database applications.

This is not a trivial task. Today's corporate networks have many data storehouses distributed across multiple operating system platforms. Ideally, an app-gen must deal with virtually any database and sync / async host terminals to retrieve data. And it must be able to handle as many simultaneous transactions as the back-end database will allow. In fact, the better – and usually more expensive – "enterprise-strength" packages are indeed able to integrate with almost any conceivable computer system, convincing the platform (whether mainframe, mini or, most importantly, LAN-based) that they are just another compute or terminal looking for information and then extract data seamlessly from the software. Just remember that even similar MIS plants can be configured differently. Things can get pretty weird. But a good IVR package makes it all a lot saner.

19. Beware of operating system driver compatibility: Make sure all the cards you plan to use are supported under your PC's operating system. Get the latest drivers unless your software demands a specific version.

20. Does the app-gen support Plug and Play? Installing and configuring the various components of a call processing system is usually a nightmare. How many jumpers do you set on a voice card? How many interrupts does a fax card take up, and which ones are they? What in-band signaling tones does your app-gen have to produce to control your PBX? Some app-gens, such as EASE from EASE CT Solutions (Alpharetta, GA – 404-338-2241, www.easey.com), can connect to just about any phone system with, well, ease. Look for an app-gen that can detect what kind of call processing boards you've installed and automatically generate any configuration files needed by the CT server.

21. If you're doing your own hardware integration, get a hefty PC power supply, one big enough to power all of the cards needed for your system. 500 watts and more is common, and now we're beginning to see 750 and 1,000 watts.

22. Keep it cool. Make sure the PC can keep the cards within their recommended operating temperatures. Any card that runs hot will fail in the field – it's just a matter of when. Make sure the CPU cooling fan does not interfere with any telephony-designated card slots.

23. Shore up system weak points, or get a fault resilient or fault tolerant computer. Design enough redundancy into your system so it won't suffer too much at any particular point of failure. What happens if you lose a power supply? A disk drive? A T-1 card?

If your budget permits, look into buying an industrial grade, fault resilient computer such as those from APPRO, Compaq, Crystal Group, Diversified, Force, HP, IBM, I-Bus, ICS Advent, Motorola, RAAC, Radisys,

SBS Technologies, U.S. Logic, Ziatech, etc. You can order them with hot-swappable power supplies, alarming boards (which can notify you by phone e-mail or pager of system problems), RAID Level 5 (or higher) redundant drives and in some cases hot-swappable cards, such as those found on CompactPCI (cPCI) systems.

24. Ensure high hard disk performance. Make sure your system's redundant hard disk drives or RAID is state of the art can keep up with the maximum loading of your system. Opt for Ultra3 SCSI drives (also called Ultra 160 SCSI) if a great deal of data traffic is to be handled. Some caching controllers such as those from Distributed Processing Technology (Maitland, FL – 407-830-5522, www.dpt.com) can help if the same files are loaded over and over.

25. Get runtime management and "hot scalability". When running multiple corporate telephony applications over many lines, a robust environment to manage it all is a necessity. The ability to "hot-load" applications and record and play prompts concurrently, without interrupting on-line programs, is also important, since customers hate interruptions of service for an upgrade (make sure your solutions can run non-stop, 24 hours a day, 7 days a week).

A high-end runtime manager inherently provides mechanisms for management statistics and diagnostic reporting, both integrated and compatible with third-party report generators.

Indeed, the gaping difference between a mere "app-gen" and a true IVR "platform", "system", or "solution" is how the package can handle the runtime environment. A good app-gen runtime system will come with tools – local apps or networked client applications – that permit load testing, monitoring and maintenance of running programs.

Every app-gen has some basic host or database access. Not everyone has a robust runtime environment. We've had some difficulty explaining to people what we mean by a "runtime environment," or "runtime functionality." Could you imagine reconfiguring your PBX and voice mailboxes by completely shutting down your phone system? Of course not. A phone system can be altered while it's running.

Similarly, the better app-gens can handle the runtime configuration of parameters and preferences that are voice / fax card related or application defined. Administrative security, start / stop of application groups, start / stop of individual lines, mapping of applications to lines, report generation, status display, vocabulary maintenance, database maintenance, file management functions, busy out lines, "graceful" shutdown, and debugging tools, should all be manipulable during runtime by supervisors.

Runtime functionality means the difference between a system that can be administered by a mere supervisor and a system that must be administered by a developer / programmer.

26. How about load stress tests? Bachir Halimi, former president of MediaSoft Telecom (now Elix) says: "You need to know how an app will behave under heavy call volume before putting it into service. You should be able to build samples of user inputs. Ask if you can select the number of simultaneous virtual callers, the call duration and time between calls. You should be able to 'click' to start the test and let it run for as long as you want it to. This should be done in software, before committing to a single piece of telephone hardware."

27. Check for runtime royalty fees. Some app-gens charge, others don't. Artisoft's Visual Voice gained much notoriety years ago when they were the only telephony toolkit not to charge runtime fees or royalties.

28. Is there auto-documentation? Since you must produce at least one complete set of documentation for your app, look for an app-gen that will do "automatic documentation." It should be able to generate and print the app's latest call flow, script and description of prompts.

29. Be careful, it's a new frontier. Before buying, get real-world application examples from the app-gen vendor and its developers. Ask about special developer seminars. The better vendors hold them for these advanced apps.

30. If you're still uncomfortable about what app-gen to buy, get some evaluation software from the makers. Parity Software will sell you a CD for just $89 containing evaluation software of everything they make. Of course, Parity Software was acquired by Intel because of its Topaz Project, a development effort that was meant to be a unified API for all of Parity's products. Intel has had the bright idea of taking the Topaz Project, renaming it the "object infrastructure" and then persuading other app-gen and tookit makers to follow what at first glance appears to be a universal CT API. It's actually much more than that, and will run above Dialogic's CT Media (another Intel acquisition). Intel has done more for app-gens – and computer telephony in general – in the last six months of 1999 than any other company did during all of the 1990s.

Computer Telephony magazine's fax columnist Maury Kauffman, president of The Kaufmann Group (Voorhees, NJ – 609-651-1651, www.kauffmangroup.com) and author of the book Computer-based Fax Processing, gives these tips when shopping for an app-gen that will be used to develop an application dealing with fax technology:

1. Does the tool support fax products from multiple vendors? Some fax cards are best suited for one-call fax applications. Others are better suited for two-call.

2. Does the tool do text-to-fax conversions without making calls to external utilities or C libraries?

3. Does the tool have document processing features (e.g. the ability to merge data from an IVR application into a free-form template to create customized documents)?

4. Does the tool support multiple bitmap formats or just raw T.4 fax coding? Support for PCX, DCX and TIFF/F insures maximum compatibility with document preparation and desktop publishing utilities.

5. Does the tool support the use of fax boards as a shared resource? Many fax-on-demand apps need only four ports of fax from a 24 port system. The rest may be used as an outbound fax server. Make sure you choose a tool that lets you use fax devices as a system shared resource.

6. Does the tool support digital network interfaces to fax products? On larger systems, you will want to locate fax devices behind a digital trunk (e.g. T-1, E-1, ISDN). Many app gens support only analog front-ends to fax peripherals.

7. Does the tool provide you with source code for modules which control fax transactions? Without this capability, a simple feature such as stripping (combining text and bit mapped graphics on a single page) will be impossible to implement.

8. Does the tool come with example programs that illustrate how to integrate fax into telephony apps? Without examples, you will spend a lot of time experimenting.

9. Does the tool come with bundled technical support or is that extra? Some fax products are difficult to configure, especially in high-density configurations.

10. Does the app-gen or toolkit come in an evaluation version that allows you to build and test applications? Some tools are good for some projects, but wrong for others. An evaluation system lets you figure out whether a product will work for you – before you spend money on development tools.

A final admonition would be: Don't be penny-wise. Shop features and performance first, price second. A good development tool will save you money by enabling you to get your product to market quickly. A cheap tool will be more expensive in the long run. A Hyundai does not run like a Mercedes.

Application Programming Interface (API)

A defined way for a programmer to gain access and use a series of functions or services exposed by an application or operating system. This is generally done in the programmer's source code by calling one or more procedures residing in a library of routines or functions embedded a the programming language.

ASR

See Automatic Speech Recognition

Association of Computer Telephone Integration Users and Suppliers (ACTIUS)

A prestigious industry forum for CT in Europe. It's part of the Telecommunications Managers Association (TMA) in the UK. ACTIUS provides an open industry forum for interchange and discussion on Computer Telephony between its members. A key aim is to explain the benefits of CT applications to the broadest possible range of users. ACTIUS also represents its members' interests on relevant regulatory and standards issues, both in the UK and the EU. ACTIUS defines CTI as "the functional integration of business application software with telephone based communications." Website: www.actius.org.uk

Asymmetric Digital Subscriber Line (ADSL)

Promoted as "the successor to ISDN" and the most popular form of xDSL, ADSL is a high speed digital compression, multiplexing and transmission method that works over existing copper twisted pair phone wires.

Unlike, say, Symmetrical DSL, ADSL offers a higher "downlink" or "downstream" bandwidth from the carrier to the end user than an "uplink" or "upstream" bandwidth from the customer to the carrier.

This "unevenness" was done partly to eliminate Near-End Crosstalk (NEXT) at the central office (CO), and partly to more efficiently apportion bandwidth, since, aside from videoconferencing and similar real-time high bandwidth communications, most Internet and computer telephony applications move more data in one direction than another. Think about it. You download more multimedia clips and movies than you produce; you receive more e-mails, voice mails, and faxes than you send; and you browse the Internet – it doesn't browse you (at least not yet).

Indeed, unlike HDSL that was designed to replace T-1 / E-1 repeater systems, ADSL was originally conceived as a vehicle for consumers to receive video-on-demand (VoD), and was equipped with special features such as interleaving and error protection to ensure few transmitted errors so that high video compression rates could be used. As a result, ADSL video at high bit rates will yield better picture quality than some cable TV providers. Error correction introduces about 20 msec of delay, which is tolerable by business LAN and IP-based wide area data communications applications.

ADSL was also originally conceived as a temporary stepping stone on an upgrade path leading from analog to ISDN to ADSL to ATM, but ATM use has been restricted to the network backbone and the "one wire wonder" ATM LAN phone systems, such as Sphericall, from Sphere Communications (Lake Bluff, IL – 847-247-8200, www.spherecom.com). Also, unlike ATM, which is a Layer 2 (L2) communications protocol, there has been a tremendous surge in Internet Protocol (IP) use, which is a Layer 3 (L3) protocol, all because of the immense popularity of the Internet and IP-based corporate LANs, WANs and Virtual Private Networks (VPNs).

ATM is still used on the backbone, however and ADSL modems do accommodate ATM transport with variable rates and compensation for ATM overhead, as well as IP protocols. Also, ATM layer 2 multiplexing is used by COs to deploy ADSL networks instead of the older method of using Ethernet as a multiplexing protocol.

At the time of its original formulation in 1989, the downstream rate for ADSL was 1.5 Mbps and the upstream rate was from 16 to 64 Kbps. Today, full-rate ADSL offers three basic channels: The high-speed downlink of between 1.5 Mbps and 6.14 Mbps (though equipment exists to deliver up to 9.6 Mbps), a full-duplex data channel of up to 640 Kbps, and an ordinary 4 KHz POTS channel.

To bring an ADSL circuit into a home or business, incoming lines must be provisioned correctly and the ADSL service provider's designated data termination equipment must be in place. This consists of two "modems" that must be installed, one at each end of a twisted-pair telephone line. DSL "modems" aren't exactly like conventional modems, but the term continues to be popular. One ADSL modem is situated at the subscriber's premises. Technically, it's known as an ADSL Termination Unit – Remote (ATU-R), though the term has been usurped by the more generic (and popular) Customer Premise Equipment (CPE). The other modem (usually part of a rack of modems with line cards) is located at the Local Exchange Carrier's (LEC's) central office. It's known as an ADSL Termination Unit – Central (ATU-C), though this term has been supplanted by the term Digital Subscriber Line Access Multiplexer (DSLAM).

In early installations, each CO ADSL modem was connected via an Ethernet link to a switch or hub, whereas in the modern CO's DSLAM the CO modems connect directly to a shared high-speed backplane (See xDSL).

Whereas conventional analog modems use the same 300 Hz to 3.4 KHz bandwidth as the phone system, precluding concurrent voice and data transmissions, ADSL equipment takes advantage of the fact that the physical connection (copper wires) between your home / business and the LEC can carry much greater bandwidth – about 1.1 MHz. By using a passband modulation technique, an ADSL transceiver can handle the frequencies far above those used by a conventional phone call (from 30 KHz up to 1.104 MHz).

ADSL takes this bandwidth of the copper loop and, in the manner of a frequency division multiplexed system, splits the whole bandwidth into three information channels: one high-speed downstream channel (to the user), one medium-speed duplex (upstream / downstream) channel, and one conventional voice channel. The ADSL modems or "transceivers" convert analog signals into digital streams and vice versa.

Whereas standard modems (e.g. 56k) send data streams from the sender's local loop though the public phone systems' digital switches and then to the receiver's local loop, ADSL modems do not pass signals directly into the public phone systems, so they can use much higher frequencies that the PSTN can tolerate.

In the case of full-rate ADSL, both ends of the local loop also have "splitters" (acting as passband filters) that take the low frequency voice channel and either separate it from or combine it with the higher frequency data subbands, depending on the direction of transmission.

At the subscriber's premises, the splitter is attached to the phone line near where it enters the premise. The splitter forks the phone line: one branch hooks up to the original house telephone wiring and the other branch heads to the ADSL modem.

The ADSL modem should have an RJ-45 socket into which you can plug an Ethernet LAN cable to your computer, which should have installed a Network Interface Card (NIC). Many computers (such as the iMac) already have NICs built into them. Some ADSL providers sell a combined router / hub / modem unit that will let you set up a home LAN that shares an ADSL line to the Internet.

Meanwhile, in the carrier's CO, the passband filters split off the voice channel and send it into a traditional POTS switch where it is routed onto the conventional public phone network. A DSLAM or DSL "modem" at the CO terminates the data channel and routes the high-speed data stream onto ATM, frame relay, T-1, T-3, Ethernet, or some kind of serial network, to a business office or perhaps an ISP where it may ultimately find its way to the Internet.

All this means is that the available bandwidth is divided up in such as way that the ADSL portion can peacefully coexist with the analog phone system on the same copper pair.

Indeed, ADSL's three-channel architecture allows you to send an e-mail, download a multimedia file, and make a duplex phone call over your existing phone service simultaneously. You can telecommute via a VPN to your corporate LAN, and still be videoconferencing with a customer. ADSL is also "on all the time" – there is no wait as one dials up an ISP and connects as is the case with conventional modems.

Also, most ADSL modems are designed with "passive POTS splitters" that do not require external electrical power, so that your existing "lifeline" POTS service will continue operating even if your ADSL modem or ADSL services cease to function in an emergency situation. You can still dial 911 at any time.

One big problem with full-rate ADSL are the splitters that must be installed at the customer premise. Splitter installation may require altering phone wiring connection or even installing new wiring to the DSL modem. To avoid this, various alternatives have been proposed.

In 1997, Rockwell Semiconductor Systems (San Diego, CA – 858-653-6709, www.rockwell.com) proposed a "splitterless" ADSL called Consumer DSL (CDSL), which had an 18,000 foot reach, but only had 1 Mbps download and 128 Kbps upload speeds. CDSL's main advantage was that it could be plugged into phone outlets just as conventional modems. CDSL modems could be purchased off retail shelves rather than supplied and installed by a phone company.

Like conventional ADSL, CDSL technology would still require provisioning from the CO but could maintain a continuous "CDSL all the time" connection between the PC and the phone company. CDSL also used some innovative techniques to rapidly adjust data rates in response to varying line impairments caused by concurrent operation of POTS over the same line. In this respect it was similar to Rate Adaptive DSL (RADSL), which monitors line quality and adjusts the data transmission rate accordingly, but which can reach much higher download speeds of more than 8 Mbps.

Another solution, used by equipment made by Netspeed, Inc. (Austin, TX – 512-249-8055, www.netspeed.com) is to use what they call "microfilters." Netspeed's blanket term for this technology is Easy Digital Subscriber Line (EZ-DSL). A microfilter is essentially a customer-installable low-pass filter with an RJ-11 jack on either end. A microfilter is placed between each telecom device and the wall jack it plugs into, except for the DSL modem.

There are two versions of the EZ-DSL Microfilter: An in-line version and a wall-mount version.

The in-line Netspeed Microfilter is a 2.50" x 1.00" x 1.03" plastic enclosure that houses a printed circuit board (PCB) assembly and an RJ-11 female connector at either end. The top-level assembly includes a 6' RJ-11 to RJ-11 pigtail for connection to the wall outlet. The wall-mount version is a 4.50" x 2.75" plastic plate used in conjunction with wall-mount telephones. The wall-mount Microfilter is installed in place of the standard telephone jack outlet where a wall-mount telephone is used.

The real "winner" in the splitterless ADSL race, however, appears to be ADSL Lite – a slower version of ADSL, an extension to the current ANSI standard T1.413 "full rate" ADSL that does not need a voice/data POTS splitter at the subscriber's home or office, which normally is installed at a premise during a visit by a skilled phone company technician – an event called a "truck roll".

ADSL Lite is also known as Universal ADSL (UADSL), DSL Lite or DSL-Lite. It was previously named G.Lite because the International Telecommunication Union's (ITU) (www.itu.ch) original Study Group 15 / Question 4 working group that examined a number of DSL modem standards needed a designation to differentiate among the options. This sector of the ITU standards were designated with a "G". So for working group meetings G.Lite was designated as the placeholder name for a "light" version of ADSL. There also exists a g.dmt (the ITU Standard G.992.1. International "full rate" ADSL standard for data rates up to 8 Mbps), g.hs (handshake), and g.test.

In June 1999, the ITU formally ratified the ADSL Lite G.992.2 standard. ADSL Lite technology delivers data transmission speeds of up to 1.5 Mbps downstream and up to 512 Kbps upstream at distances of up to 18,000 feet (3.4 miles or 5.5 km.) from the CO.

ADSL Lite will be used mostly by residential consumers, but perhaps even SOHOs and medium sized businesses for bandwidth-hungry applications such as high-speed Internet access, corporate network access, video conferencing, whiteboarding and data conferencing, Voice over IP (VoIP) and simultaneous use of a single line for data and phone calls.

A major effort is under way to make ADSL Lite ubiquitous so that one can buy an ADSL Lite modem from a retailer anywhere in the world, bring it home and use it immediately. ADSL Lite is backed by such enormous and influential companies as Microsoft, Intel, Compaq, and some of the regional Bell operating companies, who founded the Universal ADSL Working Group (UAWG) (www.uawg.org) in January of 1998 to gain consensus in the industry on a consumer-friendly version of ADSL.

There is also an effort to ensure that ADSL Lite will be interoperable with its full-rate ADSL cousin, ADSL T1.413 so that LECs can inexpensively install a single access termination system for both full-rate ADSL and the Lite splitterless version. This also allows ADSL Lite to provide an upward migration path to full-rate ADSL.

Fortunately, the ITU-approved versions of ADSL and ADSL Lite both employ ANSI standard Discrete Multitone (DMT) multicarrier modulation techniques (See xDSL) so it appears that ADSL Lite can easily be made interoperable with standard-compliant ADSL central office equipment.

Multicarrier encoding techniques such as DMT divide the bandwidth up into multiple, parallel subbands, also called subcarriers or subchannels. Each subband is encoded using one of the single-carrier modulation techniques, with all subbands ultimately bonded back together at the receiver. Standard (ANSI) ADSL uses 256 frequency channels for the downstream data and 32 channels for the upstream. All the channels have bandwidth of 4.3125 KHz and the frequency difference between two successive channels is also 4.3125 KHz.

ADSL Lite cuts the maximum number of subbands from 256 to 128, and it cuts the number of bits per symbol time from 15 to eight. Multicarrier modulation techniques used by ADSL and ADSL Lite require so much digital signal processing that they were not commer-

cially feasible until inexpensive, mass produced IC technology caught up with engineers' dreams. The lower bit rates used by ADSL Lite means that 60% fewer fast Fourier operations to maintain orthogonality between all of the subbands are needed as well as less demanding equalization and error-correction algorithms, and so an ADSL Lite modem can use Digital Signal Processors (DSPs) that are less powerful (fewer MIPS) and less expensive than those used in full rate ADSL modems.

Indeed, the Scalable ADSL Modem (SAM) Chipset from Integrated Telecom Express (Santa Clara, CA – 408-980-8689, www.itexinc.com) dispenses with DSPs entirely, harnessing instead the processing power of a host PC processor. SAM is designed to be housed in a PCI card and used as an internal modem.

The mass marketing of ADSL Lite will determine whether the technology performs well on interior residential wiring that supports phones ranging from expensive handsets to inexpensive bargain basement specials incorporating marginal components. Also, unterminated phone jacks can act as bridged taps, causing line interference and reduced transmission speeds. It may be that some users will have to install some type of low-cost microfilter to prevent the phones' voice service from interfering with data service and vice versa, or else rewire the premises, use an in-house LAN, or install more expensive wireless technologies.

In Europe, ADSL has enjoyed enthusiastic acceptance, while in the U.S. both ADSL and ADSL Lite have been deployed slowly and face stiff competition from cable modems.

Recommended websites:
The ADSL Forum: www.adsl.com
DSL Reports: www.dslreports.com
The DSL Center: www.dslcenter.com

Asynchronous Transfer Mode (ATM)

ATM is a LAN and WAN network transfer protocol technology that can transport realtime voice and data in the form of fixed-length packets called cells, each of which is 53 bytes long, containing 48 payload bytes and five header bytes. Information to be transmitted is buffered and placed in a cell. When each cell is "full" it is sent through the network to the destination specified within the cell's header.

The most significant benefit of ATM is its uniform handling of services, allowing one network to meet the needs of many broadband services.

ATM originated with the Consultative Committee on International Telephone & Telegraph (CCITT) International standards organization, an ancestor of the International Telecommunications Union or ITU.

When digital communications took off in the mid-1980s, the ITU began debating a standard for a Broadband Integrated Services Digital Network (B-ISDN) that would replace the entire public network and allow all voice and data traffic to be carried over one digital network fabric. In 1988 the ITU decided to adopt ATM as the transport technology for B-ISDN (the terms are often used interchangeably).

ATM resulted as a combination of the concepts of cell relay switching technology (supporting both variable and constant bit rates) and the Synchronous Digital Hierarchy (SDH) over optical fiber transmission technology (supporting large amounts of data to be transmitted over a network efficiently). Thus, ATM's cell-switching technology combines the best advantages of both circuit-switching (for constant bit rate services such as voice and image) and packet-switching (for variable bit rate services such as data and full motion video) technologies. The result is the bandwidth guarantee of circuit switching combined with the high efficiency of packet switching.

ATM is similar to packet-switched networks, but different in that ATM does provide Quality of Service (QoS) by ensuring "cell sequence integrity" – cells arrive at their destination in the same order as they left the source. This is because ATM topology is connection-oriented, establishing a temporary, virtual, "logical" circuit through a network of switches from end to end when a cell needs to be sent so that cells sharing the same source and destination are guaranteed to travel over the same route. This is unlike a network such as the Internet, where each succeeding packet can take a different route to a particular shared destination.

But unlike telephone switches that dedicate circuits end to end, the ATM connection is not dedicated to one conversation. When the ATM channel does not use the reserved bandwidth of the connection path, information from other channels can use this spare capacity. If the ATM network is idle, the logical circuits' unused bandwidth can be used to transport unassigned cells. A lull during a videoconference can be exploited by the system by sending data cells or some other cells, the system making sure that there is no "space" or "gaps" between cells.

Thus, ATM is "asynchronous" in the sense that although cells are relayed synchronously, data need not be sent at regular intervals. ATM frames are synchronous but the circuits are not allocated specific time slots within the ATM frame and the slots may vary in bandwidth. Cells from multiple sources and multiple destinations are asynchronously multiplexed between multiple packet switches. This sharing, or statistical multiplexing, can handle high bandwidth bursty traffic such as compressed video which can have a peak bit-rate 10 or more times its average bit rate.

ATM is also highly scalable, supporting transmission speeds as slow as 9.6 Kbps between ships at sea or it can be in increments of 1.5 Mbps, 25 Mbps, 100 Mbps, 155 Mbps, and 622 Mbps (OC-12) full duplex. ATM can also run at 1.224 Gbps (OC-24) and 2,488 Mbps (OC-48) services, and higher OC-n (n x 51.84 Mbps), yet ATM is not tied to any particular physical medium.

Since the cell is defined independently of speeds, framing or physical media, LANs, WANs and public networks can use the same cell format. A 45 Mbps DS3 (or T-3) line in a CO can receive cells generated by a multimedia application on a 155 Mbps LAN (either shielded twisted pair or fiber) and switch them on to a WAN-based OC-12 SONET system. ATM devices such as network cards are simply routing cells, which means that ATM technology blurs the distinction between the local and public networks since routers and bridges aren't needed, just switches residing on a fast, optical backbone.

The incredible bandwidth possibilities of ATM lead many to believe that computer telephony could now do real-time high-definition video conferencing, high speed transfer of x-ray or other medical imaging, and interactive multimedia applications of such high quality that users could truly find themselves in a realistic (or completely fantastic) virtual world – cyberspace dwellers at last. ATM did not quite have the future that its proponents had hoped.

By the early 1990s commercial equipment was developed for large backbone telco applications, and private network vendors also became interested in the high-bandwidth possibilities of ATM, so in 1991 the ATM Forum was founded, which has replaced the ITU as the specifications setting body for ATM.

ATM Specifications

The ATM specifications consist of the length and format of the ATM cell, adaptation layer functions, and signaling.

Using large or variable length cells such as the kind used by X.25 and Frame Relay systems, gives a better payload to overhead ratio, but at the expense of longer, more variable delays. In order to maintain QoS for time-sensitive voice and video communications having a variable bit rate (VBR), packetization delay variance (the time it takes to fill a cell with data) had to be reduced as much as possible, which can accomplished by specifying small cells. A small ATM cell ensures that voice and video can be inserted into the stream at a high rate of periodicity, enough to support realtime transmissions. Since smaller cells reduce the amount of information carried per cell and larger sized packets are required by data networks to increase bandwidth efficiency, ATM's 53 byte cell format was something of a compromise.

Still, the fact that ATM cells are fixed-length allows for the construction of very fast switches, since any processing associated with variable-length packets is eliminated – the system doesn't need to be looking for the end of a frame, for example.

ATM has its own protocol reference model independent of the standard seven-layer OSI model (See OSI Model), consisting of a control plane, user plane and management plane. Still, the ATM model does have some components resembling equivalent ones in the OSI layered model. The User plane (for information transfer) and Control plane (for call control) are structured in three main layers – Physical Layer, ATM Layer and ATM Adaptation Layer (AAL), with the Physical and AAL Layers further divided into sublayers.

The Physical layer (or "PHY," corresponding to OSI Physical) is usually taken to be SONET/SDH (which itself consists of four layers.) but can involve other underlying transport media.

Above the Physical Layer sits AAL and the ATM Layer.

The AAL formats the 48 byte cell payload so as to adapt ATM's cell switching abilities to the attributes of many different higher layer protocols, each having varying traffic characteristics. This service-dependent layer roughly corresponds to the OSI data link layer (data error control above the Physical layer), though some argue that it's really similar to OSI transport, as it involves end-to-end connections.

Of the original four types of AAL, two of these were combined. The ATM adaptation sublayers are now defined as follows:
- **AAL1** – Standardized in both the ITU-T and ANSI since 1993 and still used by large, traditional telephone networks for placing Voice-over-ATM, for uncompressed video and other isochronous traffic, AAL1 is used for connection-oriented services that demand Constant Bit Rate (CBR) traffic as well as specific delay and timing requirements, such as DS-1 and DS-3. AAL1 has been for many years the only standardized way to provide timing recovery, optional forward error correction, and minimization of the effects of cell loss in the network.

AAL1 figures into two of the three standard methods used to transport voice traffic over ATM networks: AAL1 Unstructured Circuit Emulation, AAL1 Structured Circuit Emulation, and AAL2 Variable Bit Rate Voice-over-ATM.

By extending AAL1 to allow replacement of 64 Kbps traditional digital voice circuits, ATM could be used to convey voice on ATM backbones instead of TDM infrastructures.

AAL1 Unstructured Circuit Emulation allows the user to establish an AAL1 ATM connection allowing ATM to replace Time Division Multiplexing (TDM) circuits to carry a plain DS-1 at fixed rates (such as a full T-1 or E-1) end-to-end over the ATM backbone, with bits in equaling bits out (no robbing of bits), and with clocking information carried within the cells.

AAL1 Structured Circuit Emulation is for "loop timed" N x 64 Kbps services, such as a fractional T-1 or E-1 where the ATM network switches the N x 64 circuits. over the ATM backbone.

In either case, emulating a circuit inside an ATM network tends to consume more bandwidth than necessary. Because of the overhead in the ATM cell and in the AAL1 layer, the ATM connection needs an overhead of 12% more bandwidth than the circuit it is carrying. A 1.536 Mbps DS-1 circuit carried across an ATM backbone requires 1.73 Mbps of ATM bandwidth. AAL1 thus tends to be a "brute force" approach to sending anything over ATM because of the permanently allocated bandwidth that is poorly utilized and inefficient, making AAL1 a wasteful solution for Voice over ATM. This should not be too surprising, since AAL1 became a de facto standard in the absence of a real, optimized specification that could handle Voice-over-ATM.

- **AAL2** – Previously known as Composite ATM or AAL-CU, AAL2 is a relatively new connection-oriented ATM Adaptation Layer, specified in ITU-T Recommendations I.363.2 (1997), I.366.1 (1998), and I.336.2 (1999), carries the specific mandate to provide highly efficient means of mapping voice and other bursty real-time into ATM cells.

Originally AAL2 was targeted for video, but AAL5 seems now to occupy that niche. It was found that it was not necessary to link a cell's bit format (specified by the adaptation layer) to its priority requirements. Instead, AAL2 uses a scheme called Variable Bit Rate Real-Time (VBR RT) wherein AAL2 sends voice-oriented minicells of variable size (up to 64 bytes in length) packed into the normal fixed size cells, allowing for multiplexing within the cell and cutting delays. Its structure also provides for the packing of short length packets into one (or more) ATM cells. The variable size packets have an end-to-end timing relationship as can be found in private enterprise or internal public trunking. This leads to a major improvement in bandwidth efficiency over either the structured or unstructured circuit emulation of AAL1, since AAL2's variable bit rate makes use of the more statistically multiplexible variable bit rate ATM traffic classes, allowing users to utilize more available bandwidth, which can be further increased using voice compression, silence suppression, and idle channel removal. Access connections using AAL2's new techniques can transport voice circuits over the same facilities as data circuits.

AAL2 also enables multiple user channels on a single ATM virtual circuit and varying traffic conditions for each individual user or channel.

- **AAL3/4** – This AAL is a merger of what was originally two distinct adaptation layers, AAL3 (connection oriented VBR) and AAL4 (connectionless VBR) both of which are subject to considerable overhead. AAL3/4 places four bytes of overhead in each cell, plus an additional minimum of eight bytes of overall overhead for each stream or "datagram."

- **AAL5** – Perhaps the most commonly used adaptation layer, AAL5 is data-oriented, supports connection-oriented variable bit rate data services but not connectionless ones, which would have allowed channel sharing.

AAL5's low and good error detection have made it popular for the efficient transport of TCP/IP over LANs, classical IP-over-ATM and frame relay traffic.

Unlike AAL3/4, AAL5 puts no overhead in each cell aside from changing a bit in the cell header. AAL5 uses variable length Protocol Data Units (PDUs) consisting of user data that can be from one to 655,635 bytes long. A Segmentation and Reassembly (SAR) sublayer of AAL5 divides the PDU into 48-byte chunks for cell transport over the network. During this process a PDU may be padded so to always be a multiple of 48 bytes. Finally, AAL5 adds an eight byte trailer to the end of the PDU. The PDU is closed with a 32-bit Cyclic Redundancy C=Check (CRC) performed over all of the PDU's contents.

Furthermore, AAL5 is more efficient to process than AAL3/4 (one need only look at a bit in the header in each cell, rather than examine the data contents) and has better error detection properties.

So, AAL5 is "leaner" and needs less overhead than AAL3/4, but this comes at the expense of error recovery and built-in retransmission rules demanding that, unless the packet is fully received, the application cannot receive any part of the packet and must ignore the segment of the packet already correctly received.

AAL5 is suitable for sending a constant data stream from one location to another, but it tends to have problems unless some form of synchronization pattern for interoperability can be embedded into or otherwise sent along with the data. If a service lacking timing clock signals tries to make end-to-end connections between synchronous narrowband equipment such as PBXs or devices with H.261 codecs, frame slips can occur since the equipment is designed to be used on networks delivering very little packet jitter. One or two frame slips is interpreted as a fault and will cause the equipment to go careening out of service.

One might wonder, then, why the ATM Forum long ago endorsed using the real-time AAL5 variable bit rate class of service for real-time video, even though most MPEG streams use fixed packets running at a constant bit rate. The answer is that real-time VBR was chosen over CBR since MPEG already has its own time base, the Program Clock Reference, in its transport stream and didn't need the time stamp found in AAL1's CBR class of service. Using real-time VBR, each video connection's jitter, latency, cell-loss and cell-error rates can be specified to satisfactory levels in each user contract. While video is sensitive to jitter and latency, the decoding process can compensate for any picture frames lost as a result of cell loss and errors.

The ATM Layer. Moving on from the adaptation layer, perhaps the most important layer is the ATM Layer that is responsible for creating cells, formatting the cell header (five bytes) and actually transporting information across an ATM network. Some claim that this too corresponds to OSI Physical, since it deals with bit transport, while others maintain it's really the OSI Data Link layer, since it involves such things as addressing, formatting and flow control.

A cell's five byte header consists of six fields containing information used in the correct routing and QoS of the cell's 48 byte data payload.

1. Generic Flow Control (GFC) This four-bit field has only local significance, since it's value is not preserved from end-to-end and it is overwritten in ATM switches. The GFC is used locally to give the User-to-Network Interface (UNI) a way to negotiate with the shared access networks about how to multiplex the shared network among the cells of the various ATM connections, thus controlling the traffic flow on ATM connections from a terminal to a network. The GFC is also used to reduce cell jitters in Constant Bit Rate (CBR) services.

2. Virtual Path Identifier (VPI) (eight bit field) and,

3. Virtual Channel Identifier (VCI), a 16 bit field. The VPI and VCI address fields store cell routing information concerning the network path.

ATM is connection-oriented only in that, prior to sending data between two endpoints, a virtual / logical connection has to be secured between the two ports. The virtual connections ATM uses for information transport are the Virtual Path (VP), also called the Virtual Path Connection (VPC) and the Virtual Channel (VC), also called the Virtual Circuit Connection (VCC). Both the VPC and the VCC make cell multiplexing possible.

The VCC is the actual connection between the source and destination endpoints in an ATM network. It consists of a series of Virtual Circuit (VC) links extending between VC switches. These links of the VCC comprise the path along which cells for a particular call are transmitted between the two endpoints. The VCC can be identified in the cell by the value of the VCI.

Communicating ports can have multiple VCCs, but only one VPC. The VCI holds the address of the VCC, while the VPI stores the address of the VPC.

As cells reach each switching point at the end of each link during the course of their journey in the network, their VCI and VPI fields are examined, their values being used to determine where the cell should best be forwarded next towards the destination.

VCIs are not centrally managed, so it is usually impossible for the VCI to rely upon the same numeric value to guide the cell through every link in its journey. Instead, a unique number is used for each leg of the journey and the VCI value changes at each link along the way. At the end of each link a cell enters a new VC ATM switch where it encounters routing translation tables that alter the cell's VCI (and perhaps VPI) to the correct value for the next leg of the cell's journey. The ATM switch then builds the new cell header having the new VCI value and sends it along the next link.

The Virtual Path (VP) is essentially a bunch of virtual circuit links that follow the same route and have the same endpoints. As with VCs, virtual path links can be strung together to form the Virtual Path Connection (VPC). VPs provide logical direct routes between switching nodes via intermediate cross-connect nodes, and can establish logical links between switches not directly physically connected. Such flexibility can be used to reconfigure the logical network structure when the network's data traffic characteristics change. Indeed, VPs can be managed together and some central switches need only be capable of switching whole virtual paths, not individual virtual channels.

As with VCs, VPs are identified in the cell header with an identifier, the VPI, which is usually either eight bits long for UNI or 12 bits for Network-to-Network Interfaces (NNI). The VPI field identifies multiple circuits destined for the same endpoint, drastically reducing the number of translation table entries of each intermediate switch. The combination of the VPI and VCI fields uniquely identifies each of the possible 2 to the (12+16) or 268 million channels that can theoretically be asynchronously transmitted across a shared link.

Every circuit on each link of an ATM network is thus identified by the unique integer fields of the VPI and the VCI. Both of these are used by the switch to ensure that circuits and paths are routed correctly and enable an ATM switch to distinguish the type of connection. As it is, ATM switches have quite a bit to do, since they are responsible for switching cells between ports, buffering cells, translating VPIs and VCIs, guaranteeing QoS, and performing both connection set-up and tear-down.

4. Payload Type (PT). This three bit descriptor field indicates what kind of data is contained in the cell, such as user data, OAM data (network internal, signaling data). It also gives information about the ATM cell that was affected by traffic congestion.

5. Cell Loss Priority (CLP). This one bit field is used to show whether a byte of information should be discarded during network congestion. If the CLP bit of an ATM cell is 0, the cell has high priority compared with a CLP bit set to 1. As a result, when congestion occurs, cells with the CLP bit set to 1 are dropped before a cell that has a CLP bit set to 0.

6. Header Error Control (HEC). The HEC field does a CRC check of the cell header contents to correct single bit errors occurring in the cell header and to detect multi-bit errors.

Where is ATM "hanging out" these days?

- Giant backbone switches for the biggest telcos and interexchange carriers.
- Medium to large switches for big private networks like banks and brokerage houses, or for hubs into public networks.

- Smaller switches that are the centers of local workgroups and which function like routers and hubs.
- Local access devices such as network adapter cards.
- ATM test equipment.

ATM continues to be deployed for mission critical network backbones by ISPs and larger enterprises, but until recently ATM wasn't cost effective for the desktop. If you wanted more bandwidth on your Ethernet 10BaseT network, you just installed 100BaseT cards. But the price of ATM NICs and hubs are dropping rapidly, and a certain "critical mass" has now been reached. Unfortunately, Ethernet NICs have dropped in price too – even faster than ATM has – and gigabit Ethernet is now being deployed, which may entirely overshadow ATM on the LAN.

Still, ATM switching in LANs is increasing although the ATM25 Alliance that promoted a slower 25.6 Mbps ATM for LANs was disbanded in 1996. ATM 25 at the desktop did provide a clean multi-service solution, and has been used by such "one-wire wonder" phone systems as Sphericall (See One Wire Wonder). But 100BaseT Ethernet is now installed just about everywhere.

For office systems and call centers another big ATM "breakthrough" came in 1998 when the ATM Forum announced the release of its "Voice and Telephony Over ATM (VTOA) to the Desktop Specification", offering support for 64 Kbps, G.711 PCM-encoded switched voice services to ATM terminals. The 64 Kbps format dovetails nicely with today's PBX technology that involves digital switching of 64 Kbps channels with services and features (i.e. call transfer, call conference, etc.), also provided from the PBX.

Existing voice over ATM equipment such as Nortel's Magellan switches and Lucent Technologies' Definity PBX were using proprietary technology and were therefore not interoperable with ATM devices. That situation has slowly changed thanks to the standards produced by the VTOA Working Group, which has focused on supporting additional PBX-like features for ATM, such as call transfer, multimedia conferencing, voice over IP connectivity and compressed voice for mobility applications.

The second major problem with using ATM for a converged telephony business solution is the question – do you incorporate your prexisting phone system or scratch everything and use ATM / PC-based switching?

To deploy ATM you will need to replace Ethernet switches and hubs with ATM switches, replace Ethernet NIC cards with voice/data ATM NIC cards, be sure to buy interoperable NICs, phone switches and ATM switches.

But with IP you can use your existing LAN and data communications infrastructure and deploy new services (such as voice and video) incrementally.

Investing heavily in ATM network cards and software are fine if you want an advanced PBX-less system, but some type of server-based call control system is still needed, such as the one used in Sphericall from Sphere Communications (Lake Bluff, IL – 847-247-8200, www.spherecom.com).

But Sphere's Sphericall is quite flexible – you can start out with a basic system incorporating your present phone system and Ethernet but ultimately install a 100% pure ATM network.

Complicated interfaces between ATM and Ethernet networks began to be perfected when the ATM Forum defined the LAN Emulation (LANE) specification to allow legacy LAN users to take advantage of ATM's benefits without requiring modifications to end-station hardware or software (LANE can emulate IEEE 802.3 Ethernet on top of an ATM network so client devices such as routers, ATM workstations, and LAN switches can use LANE server functions to emulate a LAN across ATM).

You would in any case want to run Sphericall in such a "mixed" architecture environment, not just to be compatible with legacy Ethernet equipment but to add some CT capability to the basic Sphericall product without disturbing your present LAN architecture. You can have a mixture of ATM and existing equipment, all served by the same Sphericall client / server software. Windows-based client call control is available over the existing Ethernet or token ring LAN via the client / server capabilities of TAPI. And Sphere's PhoneHub phones can serve as standalone phones (not associated to computers) with standard features accessible by touchtones. You can configure a system so that one segment of a LAN – or just a single desktop – can use ATM connections to the desktop for delivery of voice and data, while other segments remain on Ethernet and existing phone wiring serviced by the Sphere PhoneHub. In this way you can gradually convert your LAN to a high-bandwidth ATM system while still benefiting from a client / server PBX.

The Call Center Professional (CCPRO) from CellIT, Inc. (Miami, FL – 305-436-2300, www.cellit.com) can also use both ATM and Ethernet networks to deliver an enterprise-wide, multimedia system for the high call volume call center / service bureau industry. The high-end form of CCPRO is the ATM-based Broadband CCPRO (B-CCPRO) call center product that eliminates the usual PBX, telephones and associated wiring infrastructure and offers additional savings thanks to the intrinsic performance enhancements inherent in a broadband implementation (i.e., enhanced networking capabilities and the benefits of a universal interface supporting voice, video and data at the call center agent's station).

Still, a total ATM network really makes sense only if the phone and data network installation is new or if a prexisting ATM network exists in your company, since TCP/IP is now entrenched as the world's most popular network protocol.

Router and Voice over IP enthusiasts claim that good QoS can already be delivered over IP-routed networks instead of ATM, but ATM, unlike IP, actually can guarantee quality transmissions of voice and video. The probability of an ATM switch losing a cell (called the cell loss probability) is kept below one in a 100,000,000 and delay values with the range of 10^{-8} to 10^{-10}.

Still, while it's true that TCP/IP doesn't yet provide good quality of service, the slow but steady transition to the IP6 protocol will allow the current version to support much larger address spaces, encryption, and quality of service using such protocols as the RSVP. In a few years the distinction in quality between Voice over IP and Voice over ATM networks will be negligible.

Ironically, at one time ATM was in fact the only way to do high-speed IP (IP over ATM), but it's becoming more difficult to build ATM interfaces on routers that can handle the fantastic bandwidth demands imposed by the ever-expanding Internet. There are an increasing number of economic and technical reasons why IP over SONET offers a better scenario than the more complex IP over ATM.

All of the latest backbone IP routers connect directly to the fiber network without the need for, or the expense of, an ATM layer. New backbone carriers, like Level 3, have clearly stated that they plan to build pure IP backbones – without ATM.

One might think that ATM took too long to capture a market that has already been staked out by IP and the Internet – that ATM is the "betamax" to IP's "VHS".

I'll leave the final word to Brough Turner, the Senior VP, CTO and co-founder of Natural MicroSystems., who says that: "Silicon switching concepts, developed for ATM, are now being applied to IP. The first proposal came from Ipsilon (now part of Cabletron – www.cabletron.com). Then Cisco offered "tag switching" as a protocol for fast IP that can leverage ATM technology. The most important market for ATM equipment

vendors is in access circuits that must multiplex a variety of legacy services. While there is an inexorable trend to move everything onto IP, legacy services never die. It is safe to predict that, 15 years from now, there will still be 9600 baud leased data circuits, X.25 service and a wide variety of private leased lines. The most efficient way to provision these legacy services is to use ATM to multiplex them all onto fiber. This opportunity will provide market growth for ATM equipment vendors for five to ten years."

Audio Messaging Interchange Specification (AMIS)

Audio Messaging Interchange Specification. The forerunner of the Voice Profile for Internet Messaging (VPIM), AMIS is a protocol that allows different voice messaging systems to exchange messages, thus a user can record a message on one voice mail system and send it via the PSTN to one or more users of a competing voice mail system that is nevertheless AMIS-compliant.

In the early 1990s it became apparent that unified messaging on the LAN was about to become a real technology (it was the subject of *Computer Telephony* magazine's first "test issue" in the fall of 1993). Developers, however, were even then speculating about unified messaging on the WAN. Why not pass unified-styled "voice" missives across the public network from server to server, just like e-mails and faxes?

While there are no doubt incompatibilities between different e-mail systems and platforms, the basic functions of e-mailers are relatively straightforward – to store and forward ASCII text and binary attachments. The problem of interoperability between incompatible e-mail systems could be addressed with gateways and messaging middleware, and toward the middle of the 1990s it became apparent that the Internet was going to become the ultimate transport mechanism upon which all e-mail traffic journeyed. In the Internet world there appeared a perfectly good standard called Multimedia Internet Mail Extensions (MIME) which specifies how one formats multimedia messages, which was followed by a good transport protocol called the Simple Mail Transfer Protocol (SMTP), and a TCP/IP related protocol governing electronic mail transmissions and receptions.

Voice messaging, on the other hand, involves digitized analog voice data, recorded in proprietary formats at different sampling rates and bit-resolutions. Voice boards perform the analog-to-digital conversion that digitizes voice data for storage and then re-converts it to analog signals for later playback. All of this prevents the easy exchange of voicemail message files between different systems.

As a first attempt to solve this problem, the voicemail industry adopted AMIS for exchanging messages between different voicemail systems (Version 1 of the AMIS specification was issued in February 1990). Many voice messaging systems still support it.

There are actually two AMIS standards, one for transferring message files digitally and the other for analog message transfer.

AMIS Digital (AMIS-D) has some of the features of (but is incompatible with) X.400 messaging, with recommended transport over X.25. AMIS Digital is reminiscent of e-mail messaging standards, specifying message addressing and transport mechanisms, such as MHS.

The digital AMIS specification provides formats for messaging functions, such as send and receive, reply, forward, message attributes, delivery and receipt status, message recording format, etc. It also supports features such as inclusion of a message originator's spoken name, message addressing options such as delivery notification, confidential message, and future delivery. Message transfer functions, encompassing connection establishment, security, bi-directional message flow, and so on, are also included in the specification.

The analog AMIS protocol is simpler and less sophisticated than the digital version, but it is easier for developers to work with. Analog AMIS defines a messaging standard where a voice-message system dials a second system and then plays back the message to be delivered. An AMIS-compatible message contains a standard header that includes address information such as the dial-in number of the addressee's mail system, the addressee's mailbox number, etc.

To deliver an AMIS message, the sending system dials the receiving system, then plays back DTMF codes from the message header that identifies the target mailbox, followed by analog playback of the actual voice message itself. In this manner, any two AMIS-compliant messaging systems can exchange voice messages. The received message is recorded and stored in the format native to the receiving system and the issue of incompatible message file formats is thus avoided.

To deliver an AMIS message, the sending voice mail system dials the receiving voice mail system, then plays back DTMF codes from the message header that identifies the target mailbox, followed by analog playback of the actual voice mail message itself. In this manner, any two AMIS-compliant voice mail systems can exchange messages. The received message is recorded and stored in the format native to the receiving system and the issue of incompatible message file formats is thus avoided.

Although still built into messaging platforms by such companies as Active Voice, AVT, Fujitsu, Nortel and Siemens, AMIS has not become a runaway success. In order to devise a common, universal set of interoperability functions among different voice messaging systems, the AMIS specification had to be "diluted" somewhat, reducing the feature set to the point where customers no longer found AMIS to be attractive. All hopes are now pinned not on AMIS but its successor, VPIM (See Voice Profile for Internet Messaging).

Automatic Call Distribution (ACD)

Also known as Automated Call Distribution, Automated Call Distributor, and Automatic Call Dispatcher. ACD systems or subsystems are designed to answer, queue and route the many incoming calls to call center agents. They are literally an "intelligent switchboard" that lies at the core of every call center, routing more calls faster than any human operator.

Whereas a conventional phone system (PBX) enables a large group of people to share a small, economical number of CO lines, an ACD system for a call center makes the opposite assumption, placing as many agents as possible on the phone and talking to customers, which requires deployment of at least as many trunks (actual or virtual) as you have agents.

Standalone ACDs or computer telephony systems with ACD functionality are smart switches that basically answer a call, play a message to the caller letting them know how valuable their call is and how special they are, queues the call, hands the call off to cooperating automated transaction processing equipment (IVR and speech recognition systems), call-abandonment-prevention equipment (Music On Hold and audiotext) to keep the poor caller from getting bored while in the queue, or perhaps to voicemail systems if it's a holiday and no agents are available.

After being subjected to this succession of automated front-end adjuncts, the call is finally distributed to the next available call center agent in a "line hunt" group.

If the number of active incoming calls is less than the number of available agent terminals, then the call will be immediately routed to an agent or customer representative on the basis of some kind of logic as found in a special database of routing rules.

An ACD is supposed to maximize agent time spent on the phone, monitoring agent status and sending calls at an appropriate pace. Agents

are the most expensive components of a call center, so one must carefully manage them to get their highest productivity.

Early systems were very simple, routing the oldest call in queue to the next available agent – a strategy called Next Available Agent (NAA) queuing. NAA works best if traffic is very uniform and agents are equally efficient in handling calls, otherwise this strategy tends to overburden the most efficient call center personnel.

An improvement on NAA is to send the oldest call in queue to the longest-idle agent terminal, a strategy that helps balance call load.

As systems became larger and more complicated, it became necessary to establish groups of agents with a common purpose or skill-set, and setup subqueues leading to each group from an automated attendant, IVR or routing scheme based on Caller ID or DNIS. Callers could still get caught waiting for long periods in grouping system subqueues, even when other agents serving other subqueues were available. More modern systems solved this problem by providing several levels of increasingly-general fallback routing. A call is placed for a certain time interval in queue for a particular group, then, with succeeding time intervals, the call is "promoted" so as to be accessible by larger and larger agent pools.

After systems achieved optimum staffing in terms of sheer numbers, however, it became evident that increased call center productivity gains could not achieved by queuing a call to more agents but rather to the best agent for a particular call. Modern ACD systems thus try to determine the nature of the call and then route it to the agent with the skill set best able to handle the call, a technique called skills-based routing. Skills-based routing becomes a literal necessity with large complex call centers where a broad range of services are provided, or in international call centers where callers may speak any one of several languages.

The first computer-controlled, all digital, Automatic Call Distribution system in the call center industry was developed by Howard Walrath who led a team at Rockwell in the early 1970s. After the 1973 merger of Rockwell with the Collins Radio Company, engineers used Collins technology, originally developed for "front-ending" large IBM reservation systems, to develop the Collins Galaxy ACD.

By 1976 10 major airlines were using Collins Galaxy ACDs, along with many car rental companies, hotels and credit card authorization centers. The immediate success of the ACD concept led to the formation of Rockwell Switching Systems Division, which later became Rockwell Electronic Commerce (Wood Dale, IL – 630-227-8000, www.ec.rockwell.com).

In 1985 Walrath retired from Rockwell, which ceased manufacturing the Galaxy ACD in December 1997, though existing service contracts will be supported until November 1, 2007.

Early ACD systems were both amusing and exasperating to deal with as they were large, ungainly units that were expensive and difficult to install and integrate with the rest of a company's telecom and data processing systems. Today many digital PBXs, "unPBXs", PC-PBXs, and CT servers have ACD capability. Indeed, tracing the historical development of computer telephony is very much equivalent to tracing the increasing sophistication of a telephony system's ability to route voice calls and various forms of messaging.

Back in the 1970s and well into the 1980s, primitive standalone ACDs serviced the first call centers, all of which were not so jokingly referred to as telephony-enabled sweatshops. In a large and impersonal environment the ACD functioned merely as a specialized switch, queuing and routing calls to the next available agent in a pool of agents or Customer Service Representatives (CSRs).

In fact, by modern standards such a device wasn't really an ACD at all, but a far-less intelligent technological cousin called a Uniform Call Distributor (UCD). A UCD distributes incoming calls uniformly amongst a group of agents according to a pre-determined and pre-wired logic, such as "top-down" or the ever popular "round-robin" order. The UCD will not attempt to monitor or interpret the real-time voice traffic load status, or which agent is the most industrious or idle longest.

The ancient UCD / ACD environment was impersonal for the caller as well, who could expect the inevitable archetypal recording: "Thank you for calling. Please stay on the line. All of our agents are busy at this time. An agent will be with you shortly..."

This bland, restrictive situation existed because intelligent call-routing had not yet appeared. Audiotex, intelligent hold processing, IVR, caller-directed routing, and other automated facilities were still in the future. ACD "intelligence" was limited to statistical reporting centering on raw productivity (e.g., calls per agent/per hour). There was no attempt to do content analysis, data mining or place call center operations in a larger, integrated business context, since there was little or no computer integration and LANs had not yet appeared to connect different departmental voice and data systems.

Today, ACD functionality strives to keep pace as it serves the modern call center's many new personas. A call center (or "contact center" as the Gartner Group calls it) is any formal, informal, permanent, temporary or virtual facility where a sufficient volume of business-related voice traffic occurs to justify applying machine intelligence to the call-handling process. Call centers can have hundreds or thousands of agents in one facility, or apportioned into several regional centers, or spread across the country with agents telecommuting from their homes.

Today's call centers do customer care, perform telesales, handle order processing, bully people into paying their overdue bills, act as a help desk, and more. Increasingly, the formerly unskilled agents who worked in call centers have been replaced with trained specialists. They can help you by giving you medical advice, or help your computer by giving you Windows 2000 installation advice. They can place a stock trade for you, coordinate disaster-relief services, and sell you a ticket to a stage performance or sports event.

To facilitate the myriad goings on in the modern call center, a new generation of ACDs and software-hardware computer telephony systems with ACD functionality have been developed that exploit advances in switch engineering, LAN-based voice processing, and IP and computer telephony to increase personnel efficiency, improve quality of service, and (paradoxically) bring a more "human" touch to call center interactions to keep both customers and call center staff happy and the call center profitable.

Such contemporary ACDs fall into five basic categories:

1. Customer premise standalone systems, which range from about $1,000 to $10,000 per station are specialized switch "turnkey" solutions. These are optimized for call distribution, are laden with features and can handle a huge volume of calls.

For example, Rockwell's flagship Spectrum Integrated Call Center System is a classic ACD that is also "open" system in that it connects to corporate LANs with standard Ethernet and TCP/IP data link protocols that permit call centers to implement third party applications for CT, data-directed inbound call routing, and automatic customer contact systems for outbound calling.

The Spectrum has a distributed processing architecture, which means that instead of forcing a central processor to perform every function, system logic and processing power for specific system functions are distributed to subsystem cards. This "offloading" approach makes the Spectrum system highly modular, easily customizable and scalable as the call center staff expands. The Spectrum can handle call centers of between 25 and 2,400 agents, processing call volumes up to 25 calls per second (90,000 calls per hour).

These calls can involve Voice Response Units (VRUs), host database interrogation, call transfers, etc. Multiple units can be networked to provide automatic backup and load balancing.

2. Customer premise PBX/ACDs that are software or software/hardware add-ins or add-ons to a standard switch. They're less expensive than standalones (running between about $500 to $4,000 per agent), but often less powerful with fewer features.

If you already own a relatively new business phone system from a major manufacturer, it's possible that an ACD can be obtained from the vendor in the form of a new CPU card or switch software upgrade package. These in-switch ACDs, from makers like Telrad, Vodavi, and others, let you designate a set of extensions to run as a call center. Agents are prompted about caller status directly on station-set LED displays. Managers get a more comprehensive display of agent and group status. Options are available to let you plug in readerboards and other adjuncts.

The next step up is to buy a PC-based ACD call control system and plug it into your switch – an "adjunct ACD". Low-end examples of these integrate to the switch the same way voicemail systems do, attaching to analog or T-1 ports on the station side of the phone system. They can receive, queue, and transfer calls as well as monitor agents, using in-band, tone-based signaling protocols. More advanced systems add an out-of-band signaling channel – either proprietary (via your PBX's OAI port) or standard (via CSTA/TSAPI or TAPI), enabling smoother call-transfers, faster port handling, and overall better reliability.

Comdial's (Charlottesville, VA – 804-978-2200, www.comdial.com) add-on ACD solution for their Impact FX PBX is QuickQ, costing about $2,150 plus $200 to $550 per agent. QuickQ has a Windows client/server architecture that allows for such niceties as a Microsoft Access-based reporting system.

One major advantage of a PC-based adjunct ACD is that many vendors build audiotext, IVR, and/or voicemail into the same box that handles the call switching, eliminating the need to provide these facilities in separate boxes. Also, the management PC will commonly provide an informative display for the call center manager, and may be used to run adjunct applications, such as readerboard drivers. In the most advanced systems, the head-end call switching server links to your LAN to access databases and to deliver to PC agent stations such niceties as screen pops. Actual, hands-on management and processing of ACD data is performed on a separate PC, also linked to the LAN.

CallLink Agent Package from Teledata Solutions (now Apropos, Oakbrook Terrace, IL – 630-472-9600, www.apropos.com) lets you add powerful integration without the need to make changes to your operating system, switch or database. The system works with your existing switch or Centrex on a LAN. The Call Services Processor is connected to the switch and since it runs on standard PCs it's scalable and easy to implement. The system includes voice response, call distribution and CTI. Agents receive caller information on their PCs after it's collected from the VRU or through ANI. The Q-View Intelligent Call Distributor lets agents view caller info before taking the call.

3. Customer premise server-based ACDs are simply CT Servers or unPBXs having ACD capability. Rockwell's (Wood Dale, IL – 800-416-8199, www.ec.rockwell.com) Transcend, their first attempt to capture the small to medium call center market, is a perfect example of a PC-based ACD / call center system running Windows NT that conforms to Dialogic's "open" hardware-independent ECTF S.100-compliant CT-Media middleware. The famous end-to-end call activity reports and call-flow design telescripting that made the Spectrum so impressive have been ported over to the Transcend, which is designed to handle from 10 to 80 agents.

Thanks to compliance with such standards and call control conventions as TAPI, third-party software add-on products can be plugged into Transcend and their PC-based brethren, conveniently providing additional call-center functionality and application layers to Transcend, resources which are automatically shared throughout the common platform.

Rockwell's design relies on the strengths of the Windows NT operating system, in particular a simplified graphical user interface for management and supervision. Transcend also embraces the desktop telephony concept: There are no proprietary handsets and existing LANs integrate with ease.

Although Transcend is mostly a software-based ACD system, it can also avoid the need for having a PBX or key system, since it supports auto attendant, basic PBX call control and unified messaging.

4. IP-based or IP-integrated ACDs. Callback buttons on websites are often used by prospective customers to signal to a web-based call center that they're interested in buying a product but would like to have some questions answered by a live agent before a sale can take place. The simplest and most prevalent are the "call me now" button systems (pioneered by companies such as SpanLink) that let Web surfers request a callback from an agent over the PSTN. There are several ways to engineer these, with the most sophisticated systems establishing an IP connection, through a CGI script, to an ACD's outbound predictive-dialing queue, which initiates a call and connects an available agent with the customer. Less complex systems, which need not be co-located with the ACD, receive an IP message with the surfer's phone information, dial into the call center, find an agent, then place the outbound call on a conferenced line.

With additional programming effort, these Web-enabled IVR applications can perform an HTML "web page pop" at the agent's desktop, reminiscent of a conventional screen pop, which lets the agent know what Web page (URL) the customer is viewing, and can display the same page in the agent's browser.

Callback buttons normally prompt the customer to input a phone number, but if the customer has only one access line to the Internet then he or she has to disconnect to receive a call from the call center. This spurred development of more net-centric systems employing Internet telephony gateways. Yes, now that Voice over IP (VoIP) is taking hold, fanatical Internet aficionados no longer have to disconnect to communicate with the agent. New web-enabled call centers allow the agent to return the surfer's call along the same data channel used to access the Web site. One of the more efficient of these (the eFusion system) establishes a direct, circuit-switched data-link with the caller for the duration of the transaction, eliminating the latency problems that normally plague Internet telephony.

The disruptive "callback" button has thus become a more efficient "connect me" button that can quickly establish voice links from the agent to the customer over the Internet, so long as the customer has a Windows-based PC, a web browser, Internet access and an H.323 Internet phone or equivalent software. This means, however, that the ACD must monitor the IP addresses of every agent in the call center to direct the call to the next open agent. A security risk occurs if these addresses are outside the firewall.

This problem was solved by the PNX ACD 2.0 from PakNetX (Salem, NH – 603-890-6616, www.paknetx.com) one of the industry's first ACDs for Internet telephony applications, with an integrated H.323/T.120 firewall proxy that provides increased security so that call center agents can be placed behind the corporate firewall, an important security measure not possible with non-switch based Internet call center solutions.

The PNX ACD integrates a Windows NT software-based IP switch with an enhanced ACD, call management and administrative functions for Internet telephony and Internet multimedia call centers. It can be deployed as a standalone call center, allowing call center agents to handle and manage Internet telephone calls using traditional call center switch call control functions such as hold, retrieve, transfer and conference.

PNX ACD system software provides Internet telephony / PSTN gateway support for mixed network topologies, integrated firewall functions, and CT interfaces for integration with legacy phone systems.

The PNX ACD system software also supports video calls between agents and customers over the Internet. Once simply adds a desktop camera and a standard Internet phone. Callers can connect to an agent by clicking on a "Connect Me" button on the company web site, by calling the company call center with a standard Internet phone or by calling with a conventional POTS phone connected to an Internet telephony gateway. (Not to be outdone, Genesys Telecommunications Laboratories' (San Francisco, CA – 888-GENESYS, www.genesyslab.com) VideoACD can, via software, intelligently route inbound video calls through multimedia call centers and organizations. However, VideoACD only runs on POTS or ISDN lines.)

Whatever the type of call connection, the software queues and delivers the incoming call to the next available agent who can then provide the customer with video, web browser sharing, text chat, file transfer, and data collaboration services using H.323/T.120-compliant tools. The PNX ACD package is priced starting at $3,500 per seat.

5. Network-based ACDs that tie together multiple small call centers into a seamless operation, allowing ACD agent groups, at different locations (nodes), service calls over the network independently of where the call first entered the network. Such distributed ACDs rely on ISDN signaling – either ISDN PRI and SS7 signaling or a combination of conventional TIE trunks with an out-of-band ISDN D channel carrying call data between call center nodes. When calls start to overflow from an overly busy center, a Network ACD can search over the ISDN data pathway for alternative call destinations.

Networking costs can be kept low because the bandwidth for a full voice connection is needed unless a diverted call can actually be answered. When an agent becomes available, the ISDN data channel reserves that agent to prevent another call from arriving while the voice path is established.

The least complicated and least expensive (less than $1,000 per agent) form of these are work-at-home voice/data transceivers from companies such as MCK Telecommunications (Calgary, Canada – 800-661-2625, www.mck.com). These plug into digital station ports on a standard ACD, attach to an ISDN or analog phone line, and transceive full voicepath and signaling to a station set at the agent's home or a remote office.

An extreme form of a Network ACD is one that can be installed in a Central Office (thus the alternate term, CO-ACD, which also happens to be the name of a service offered by U.S. West) where the ACD functionality is sold over the network as an enhanced service to customers by network providers. Intecom (Dallas, TX – 800-468-3945, www.intecom.com) has pursued this route, as has Nortel with their Meridian Networking Automatic Call Distribution (NACD) systems used to build both private enterprise distributed ACDs and larger network provider systems who sell the ACD services under a private label, providing subscribers anywhere with advanced ACD and call center functionality.

With Network ACDs subscribers can "pay as they go" for world-class call center functionality without huge capital outlays for expensive premise equipment. As with any leased or outsourced service, upgrading obsolete equipment or software is no longer a problem (unless your service decides not to or goes bankrupt). Also, the best network ACD services still allow call center managers complete control of their day-to-day operations, even though the decentralized nature of a network-based call center allows agents to be situated anywhere. This encourages the deployment of what otherwise would be extremely expensive decentralized call centers where calls can be distributed and managed across multiple geographically dispersed sites with the same efficiencies as a single site – the so-called "virtual call centers" (See Virtual Call Center).

The virtualization of the call center environment, thanks to the Network ACDs, may yet turn out to have the greatest impact of any of the computer telephony innovations in the modern call center.

The most comprehensive "virtual ACD" products such as Distributed Call Center (DCC) from Teloquent (Billerica, MA – 978-671-5111, www.teloquent.com), provide ACD intelligence to allow virtual ACD groups to switch calls over both public and private networks to an optimal ACD location to achieve previously unreachable levels of service and efficiency. Since many manual technical and administrative changes normally required to re-balance incoming calls and agent staffing the call center staff are now a thing of the past, call center agents can now be anywhere (or at least anywhere having POTS and/or ISDN), which means that there is immediately achieved maximum flexibility in scheduling agents among sites (e.g. overstaffing to handle peak periods is obviated by overflow channels to other sites) and thus a minimization of staff and operational costs, all of which ultimately results in the greatest possible agent and site productivity.

In the case of Nortel, several multi-site virtual networking ACD solutions exist for the PSTN that bring the functionality of Nortel's private network-based ACD solutions to the masses: Meridian Network ACD (NACD) that actually connects multiple ACD location in a seamless and cost-effective call center; Virtual Networking Services (VNS) that offers applications similar to those for NACD but more economically routes calls to remote agents over the public switched network rather than a private one; Network Level Routing (NLR) which actually handles the call-by-call switched network routing, and Intelligent Call Allocation.

Here are some of the features to look for in ACDs and ACD software:

- **Routing.** An intelligent call router tells the carrier how to route the call before it reaches the ACD and then sends calls out based on customer ID numbers, screen pops, DNIS/DID, ANI, agent skills, availability, location, cost of the phone call or whatever criteria you can program.

For example, Connex by Xantel (Phoenix, AZ – 602-437-6400, www.xantel.com) is a Windows NT server that intercepts incoming calls before they reach your PBX, providing various options for handling the calls from your PC screen and can treat groups of people working together to handle customer calls as a call center work group.

A common scenario: Three calls are simultaneously ringing into your direct dial phone number. With Connex running you can observe on your PC screen the identities of the callers thanks to Caller ID or the customer number the callers entered when queried by Connex.

You answer the first caller, play a personalized announcement (you've recorded previously) to the second and the third is routed to an assistant. Thus, three calls have been taken care of in a few seconds just by executing a few keystrokes on your PC keyboard or with some mouse clicks.

You can assign high priority numbers to your best customers in your customer list so their calls will be handled ahead of others.

The same phone number can be used for voice and fax calls. The Connex recognizes the fax CNG tones and routes the call to a fax machine or PC with a fax modem.

You can even intelligently route calls to the appropriate people in your organization without relying on an ACD system having skills-based routing capability. Everyone in a workgroup can see the status of every-

one else in the group including whether people are on a call, to whom they're speaking or whether they're available.

- **Call Blending.** The most sophisticated ACDs can combine inbound call processing with outbound predictive dialing – a practice known as "call blending." With a predictive dialer (See Predictive Dialer), phone numbers are extracted from your database and outbound calls are made when your inbound ones are slow, or vice versa. Some agents find this process a bit disconcerting, but it does keep productivity high, since the agent is always doing either inbound or outbound work.

- **Skills-based routing.** Uses artificial intelligence (AI) to find the right agent for a call. Info comes from the public switched network or caller input, such as the incoming trunk group ID, DNIS, ANI, time-of-day and day-of-week, or caller-prompted info such as customer number or menu choices.

- **Messaging.** This lets supervisors display a message on an agent's screen. For example, "Finish up the call" or "There are 10 calls waiting." This can take the place of or can be used in conjunction with a readerboard (See Readerboard).

- **Workforce Management.** All call center supervisors need figures for forecasting, scheduling employees, as well as measuring, monitoring and controlling large workforces at multiple sites. You can track the number of incoming calls and calls put on hold, call abandonment rate and queued calls during the day and compile this historical data into comprehensive reports. The supervisor now has an idea of how many agents are needed during specific times in the day. And, from this info, he or she can prepare cost-effective schedules.

Plus, workforce management software monitors agent performance often in realtime. Either by agent or call group, a supervisor can display who is on a call and for how long, who is away from his desk and who is handling the most calls.

- **Remote Agents.** During busy periods and peak requirement times remote agents can be attached to the system to ease the workload. These agents can dial in from their location – perhaps even from their homes – and the system can also call them when needed. They become part of a virtual office.

- **Voicemail.** Callers in queue can leave a message and receive a return call. For a system that handles night callers, routing a call into voicemail is a good deal but during business hours, would you like to sit around and wait for someone to call you back?

- **Auto attendants.** These send callers to a certain call group selected by the caller via IVR (e.g. Dial 1 for sales, 2 for technical support, etc.)

- **Interactive Voice Response (IVR).** Using IVR, callers never need speak with an agent. They can check a bank balance, order clothes, or get information sent by just pressing some touchtones. Best of all, the callers never need be placed in a queue.

- **Computer Telephony Integration (CTI).** CTI means that databases can be accessed so agents can see screen pops of caller information, and can handle more calls in less time. Customers calls take less time since they no longer must answer routine information about themselves.

- **Workstation integration** (integrating the phone and computer station). A subset of CTI, this allows for the most exotic of computer telephony functionality. Once the agent station has been eliminated and all information and calls travel via a LAN, video call routing becomes possible. Prior to the mania over IP, ACDs were moving toward "multimedia routing" with everyone expecting ACD platforms to migrate to video over the PSTN. As it turns out, these dreams are being realized, but in the context of the Internet, not the more cumbersome and expensive traditional circuit-switched PSTN.

If your switch or computer telephony system is built with "open" CT standards such as TAPI, TSAPI, CSTA, or CallPath, you may be able to buy software enhancements to your system at regular intervals or perhaps even development them yourself.

Finally, if the cost of ACD functionality seems too high, remember that equipment related costs represent only about 5% of the cost of running a call center. Labor, at 65%, is a much more expensive component. And your ACD is the most important tool you have to maximize your labor pool's efficiency.

Some ACD vendors and PC-based systems with ACD capability:

Artisoft (Cambridge, MA – 617-354-0600) www.artisoft.com

Aspect Telecommunications (San Jose CA – 408-325-2200) www.aspect.com

AVT (Kirkland, WA – 425-820-6000) www.avtc.com

BCS Technologies (Englewood, CO – 303-713-3000) www.bcstech-nologies.com

Buffalo International (Valhalla, NY – 914-747-8500) www.opencti.com

Cintech (Cincinnati, OH – 513-731-6000) www.cintech-cti.com

Cisco (San Jose, CA – 408-526-4000) www.cisco.com

Cortelco (Memphis, TN – 901-365-7774) www.cortelco.com

CosmoCom (Hauppauge, NY – 631-851-0100) www.cosmocom.com

ECI Telecom / Tadiran (Clarksburg, MD – 301-428-9405) www.ecitele.com

Ericsson Business Networks (Research Triangle Park, NC – 919-472-7495) www.ericsson.se/Enterprise/

Executone Information Systems (Milford, CT – 203-876-7600) www.executone.com

Fujitsu Business Communication Systems (San Diego, CA – 888-322-7284) www.fbcs.fujitsu.com

Imagine (USA) Ltd. (Norcross, GA – 770-921-5433) www.imagineusa.com

Intecom (Dallas, TX – 972-855-8000) www.intecom.com

Interactive Intelligence (Indianapolis, IN – 317-872-3000) www.Inter-intelli.com

Lucent Technologies (Murray Hill, NJ – 800-372-2447) www.lucent.com

NEC America (Irving, TX – 972-518-7000) www.cng.nec.com

Nitsuko America (Shelton, CT – 203-926-5400) www.nitsuko.com

Nokia / Telekol (Waltham, MA – 781-487-7100) www.telekol.com

Nortel Networks (Santa Clara, CA – 408-988-5550) www.nortelnetworks.com

PakNetX (Salem, NH – 603-890-6616) www.paknetx.com

PARSEC Technologies (Gurgaon Haryana, India – +91-098100-86718) www.parsec.co.in

Picazo Communications (San Jose, CA – 408-383-9300) www.picazo.com

Redcom Laboratories(Victor, NY – 716-924-6500) www.redcom.com

Rockwell Electronic Commerce (Wood Dale, IL – 630-227-8000) www.ec.rockwell.com

Siemens (Santa Clara, CA – 408-492-2000) www.siemenscom.com

Stratasoft (Houston, TX – 713-795-2670) www.stratasoft.com

Teknekron Infoswitch (Fort Worth, TX – 817-262-3100) www.teknekron.com

Teloquent Communications (Billerica, MA – 978-671-5211) www.teloquent.com

Telrad Telecommunications (Woodbury, NY – 516-921-8300)

www.telradusa.com

Vodavi (Scottsdale, AZ – 480-443-6000) www.vodavi.com

Automatic Call Sequencer (ACS)

Sometimes described as "a poor man's auto attendant", an ACS or "call sequencer" is more sophisticated than a Uniform Call Distributor (See ACD) but with fewer abilities than an ACD (See ACD). An ACS is a device attached to the phone lines before they reach a call center's key system or PBX. The ACS answers incoming calls, plays an announcement that someone will talk to them in a moment, then plays some music. Since they don't have the kind of internal switching mechanisms found in ACDs, these call sequencers depend on the phone system for call routing.

Early ACS systems could generate various lamp flashing sequences indicating the length of time a call has been on hold, thus suggesting which "oldest" line a call center agent should pick up. Modern systems provide video display terminals telling agents how many calls are in queue and what line to pick up next. Call sequencers can also store statistical information on the progress of calls. Call sequencer reports should tally the number of incoming calls, lost calls, complete calls, trunk lines and similar information.

Automatic Number Identification (ANI)

A Telco service similar to CLID (Calling Line Identification, or Caller ID) which allows a called party to identify the calling party's billing number (which is not necessarily also the calling number) and other call information of the caller before the called party is connected. Originally ANI information was sent "in-band" with the information carried as audible DFMF signals in the voice path of analog trunks or analog equivalent T-1 trunks. Today, ANI information is sent using Signaling System 7 (SS7) digital information network, but ANI, unlike Caller ID, does not need SS7 to be present everywhere in the network. When using a T-1 line, the ANI information also includes the geographic coordinates of the originating call's central office. The term ANI is becoming synonymous with Caller ID, though both are different technologies: ANI information is provided by the long distance carrier for billing purposes and cannot be blocked by the recipient, while Caller ID operates within the service area of only a particular group of central offices and can be blocked for all calls or on a per call basis.

Automatic Speech Recognition

This technology reliably recognizes a certain defined subset of human speech, such as spoken digits, or particular words or phrases. ASR is the official acronym covering the whole field of speech/voice rec. The term goes back to the earliest work in speech recognition by artificial intelligence researchers in the 1960s. Nobody uses the term ASR much, except for academics and speech recognition companies themselves and of course *Computer Telephony* magazine – it's a quick way to refer to the industry en masse.

ASR is also called "speech rec" and "voice recognition," but early researchers termed "voice recognition" as the ability of a machine to recognize a particular voice. Voice recognition systems need to be trained by the user (this is also called speaker dependent recognition).

In the late 1960s and 1970s, research centers at MIT and Stanford set out to solve the problems of speech recognition and vision for robotics. With some naivete, they thought that it would take 10 years to solve the vision problem, and just a year or so to completely figure out speech recognition.

Everybody's still working on speech.

ASR technology is driven by competing theories. During early researche in the field, two basic engines emerged: Hidden Markov

Modeling (HMM) and Dynamic Time Warping (DTW). In recent years researchers have devised other, more sophisticated models, though these two remain popular.

Single word speech recognition has been around for 25 years. Using simple template pattern matching, speech is compared to a list of word templates. The template that gets the best matching score "wins" and so the word associated with that template is surmised to be the word that was uttered. Template-based systems are still used for services such as dialing by voice from a personal rolodex and automated operator services.

In the mid-1980s it became possible for ASR systems to recognize multiple words, thanks to two developments. The first was "word spotting," wherein a spoken phrase is scanned for keyword(s) in the system's vocabulary, ignoring other sounds (words). Such system can partially interpret free-form speech. For example, if the word to spot is "black," then it doesn't matter if you say "I want the black shoes" or "how about something in black?" If you're prompted to say a digit but belch into the phone beforehand, word spotting can still pick out the digit, and without being offended.

Another advance was the development of connected word recognition using level-building algorithms that search for particular strings consisting of sets of words. Continuous speech recognition can deal with multiple word commands such as "call office" and digit strings such as account numbers and telephone numbers.

It takes considerable processor power to listen to a conversation and spot key words, such as compound nouns, on the fly. Not only does an ASR system have to figure out the start and end of each word, but it has to try to match that word in a vocabulary list.

Many of these systems are still HMM or "template" based. Your utterance is matched against a template based not upon your own voice, but "ideal" templates created from the averaged utterances of anywhere from a dozen to thousands of people.

Since the early 1990s there have been many attempts to introduce grammar-based systems that can identify every word in an utterance by constraining the recognition task with finite-state grammars and thus precisely define exactly which word combinations are allowed. Such grammars have to be tailor-made for each application, since in order to fit into a telephony system a large grammar has to be pared down to recognize whatever phrases are applicable to specific customer-system interactions.

More complicated grammars might consist of all possible expected variations on "switch $500 from checking to savings" or "reserve two business-class tickets from Newark to San Francisco on May 12th at 12 o'clock." Some vendors have released testing tools that are able to measure the success of selected grammars before they're put into production. At least one toolkit, Natural Language Speech Assistant from Unisys (www.marketplace.unisys.com/nlu), works independently of its core engine.

Statistical-based systems then appeared. Companies such as Dragon Systems discovered, for example, that the word "and" is followed by the word "the" 2.8% of the time, but it is followed by the word "with" only 0.2% of the time. This leads to accurate systems, so long as the user sticks to predefined phrases.

Lucent worked with a financial services company, USAA, to test a system that greeted callers with the announcement "How may I direct your call?" Since USAA offers a wide variety of services, callers responded with a highly diverse set of answers such as "Uh, I need to refinance my house" and "Do you offer no-load mutual funds?" Such tremendous "branching" that occurs in sentences of normal discourse are more suited to statistical-based systems than grammar based ones.

In terms of user functionally, ASR falls into two major groups:

Speaker dependent vs. independent. Speaker dependent technology (generally based on DTW algorithms) is where a system is "trained" by having the user says a word over and over – it's recorded anywhere from one to three or four times.

Then you have speaker independent systems. These are of two types:

- **Discrete.** You can do single utterances, but must pause between utterances. An utterance could be a single word like "yes," "no," "stop," "go" or a digit. (Speaker dependent systems can also be discrete.)
- **Continuous.** This was originally referred to as "continuous digits," since most IVR systems based upon this technology have you enter your account number by uttering digits instead of entering them with touchtones.

Theories aside, the type of ASR that interests computer telephony developers is one where a computer is used to recognize what is said by another party in a telephone call and respond to it or act upon it in some way.

It's generally agreed that every CT application would benefit from some form of ASR – whether it's speaker-independent, speaker-dependent, continuous (usually interpreting free-form speech through word spotting) or discrete (recognizes specific words one at a time, usually after a beep).

Indeed, the use of natural language dialog systems in many call centers and telephone information services is now virtually taken for granted. Calls can be processed, forwarded and answered efficiently. For the caller, there are no lengthy waiting periods – and none of the unnecessary charges, which this often entails.

When ASR for CT first appeared, developers immediately suggested that it would be an important tool for interacting with databases and other computer systems in a natural way. Then, as now, ASR was seen as especially beneficial when you could not see or use a dialpad or keyboard or if you are connected to an IVR system while either using a phone in a foreign country or using a phone that's equipped with only a rotary dialer – if there's no dialpad, no DTMF touchtones can be interpreted by the IVR system, making the human voice the only universal controlling CT "protocol".

Indeed, proponents of speech recognition over the phone really do believe that they can eliminate the need for users inputting touchtones entirely, making phone-based systems and services more accessible to callers. This "rotary relief" can even be felt in the U.S., where there's still only about an 85% DTMF penetration. In Europe, things are often worse. Germany and Spain have only about 30% and 12% DTMF coverage respectively while France and Scandinavia are more on a par with the U.S.

This is why ASR in Europe has undergone internal research by the local PTTs and telcos. For example, British Telecom's (New York, NY – 212-418-7814) ASR goes back all the way to 1983, according to Denis Johnston, BT's head of development for speech recognition. It was a name-dialer for cell phones, "an incredibly good value," recognizing up to 100 names back in the days when cell phones cost a mere three grand.

Now that mobility is mass-market, BT has intensified its drive for analyzing speech rec requirements in the cellular network and in the network in general, speech-enabling its VPN and call center infrastructure offerings through its Concert subsidiary..

Johnston admits that the biggest deployments they currently have are internal. "We eat our own dog food first," he says. Inside BT's own labs, 4,500 people can be dialed by name.

In Europe, Johnston says, a lot of the core speech rec preprocessing is being imbedded in the mobile phones themselves, Nokia's Aurora project being one example.

Telecom networks in Asia are not that dissimilar to those in Europe, with pockets of relatively high levels of DTMF usage sitting in a veritable sea of fairly low overall usage levels.

The Asia market in particular is practically begging for ASR technology. Rough estimates of DTMF coverage in the huge Asia / Pacific arena are as follows: Australia 50%, New Zealand 70%, Singapore 100%, Hong Kong 100%, Taiwan 90%, China 40% (cities 95%), Malaysia 65%, Indonesia 45%, Thailand 65%, India 25% and Japan 75%.

In the U.S. it's easy to take touchtones for granted and believe the unwarranted assumption that everyone who purports to be civilized has touchtone service. They don't.

As the U.S. market matures, voice equipment manufacturers search for new and high-growth foreign opportunities. They're discovering that the radically different telecom infrastructure in overseas markets and lack of DTMF usage elevates the importance of speech recognition from a "gee whiz" item to an essential technology. Along with pulse-to-tone conversion, ASR is the great emancipator, both here in North America and around the world.

But even if you don't live on the road to Mandalay and happen to own a touchtone phone, you will still want the ease of speech recognition over the phone.

Once you connect a computer to a phone you have a compelling reason for your computer to understand speech – when you call in to get information out of it you'd like to be able to ask for information using speech, not by pressing through an endless "menu tree" of touchtones.

After all, how many readers out there have ever gotten frustrated while dealing with a really large IVR system? First there's a prompt asking you to touchtone-in your lengthy account number. Then you've got to press 1 for one item or 2 for something else, etc. Then there's another prompt for another touchtone menu. Then you discover you pressed the wrong button and have to abort whatever it is you're trying to do.

Speech recognition, on the other hand, can be as easy as saying "Sell 200 shares of GM and buy Ford." The ability to utter spoken commands over the phone gives you access to database information that would be difficult and time-consuming to obtain otherwise.

Speech is natural. We speak three times faster than we type. Keyboards let us make typographical errors, forcing us to retype. Navigating menus by moving and clicking a mouse can be awkward, even with the best graphical user interfaces. Calling from a cellular or speakerphone? Speech recognition frees the hands and eyes.

In the "gee whiz" days of speech recognition, the field consisted of academicians interested only in new and interesting algorithms. The industry has matured and the conferences are focusing on applications. Telephony now drives the market to a significant extent. As a result, everybody in the industry has gone from being a laid-back, professorial academic to a deal-bent, shark-in-the-water application-specific marketeer. Research organizations are coming out of the closet, spinning off commercial firms capitalizing on their technology.

Thanks to a factor of 4,000 increase in ASR system performance that occurred from 1990 to 1995, the computational needs for real time speech recognition over the phone (for a 1,000 word vocabulary) and the computational power in high-end workstations finally matched each other, making ASR over the phone finally practical. Also, ASR error rates declined by 30% per year over the same five year period, and are presently above 95% for the best algorithms.

Companies such as SpeechWorks, VPC and VCS (both now part of Philips), Nuance, and Lernout & Hauspie are running neck-and-neck in this area. They all trade in algorithms that until recently were very good

only at speaker-independent digits and command words, affectionately known in the industry as "talk to the dog stuff," like – "go," "start," "stop," "cancel," "add," "delete" and the digits. They have been working on increasing their vocabularies and many of their systems can now do continuous speech.

And during 1998, speech recognition over wireless systems became practical. Since a spoken password can't be "shoulder-surfed" many enhanced computer telephony service-bureau apps in the cellular and long-distance calling-card markets began to use ASR for security purposes.

The progress that has been made is all the more remarkable when one considers that a system that can recognize speech over the phone has a much greater challenge than just listening to you from a microphone attached to a PC in your home or office, where the sound transmission from your mouth to a microphone is direct and stable, and 8 kHz or greater bandwidths are quite easy to achieve.

Over the phone network, however, many factors make life hard on ASR, particularly speaker-dependent recognition. There are all sorts of unstable and unpredictable characteristics arising from the network itself, as well as from cordless and mobile equipment. Also, the analog bandwidth tends to be a mere 3.5 kHz with cellular phones having a more restricted frequency response than a normal wireline phone line.

Whereas you can use a high quality microphone with a desktop computer, the phone system transmits sound from all types of phone microphones, including an old carbon button mic.

The area around your desktop is probably quiet and controlled – an incoming call can be from any uncontrolled noisy environment, replete with background music and perhaps other speakers.

Speaking speed and volume also tend to be uncontrolled on the phone network, along with the appearance of non-native speakers of unpredictable gender.

Technical problems aside, there are other considerations. Even if the speech recognition rate is 99%, is the system designed to appear "natural" to use? A very "un-dialog"-like system awkward to use can discourage callers.

One of the strange things about ASR is that users demand speech systems that can achieve close to 100% recognition, yet these same people will tolerate far less accuracy from their fellow human listeners.

So, are touchtones dead yet? No, but ASR is indeed arriving, not just "coming."

Remember when every year from 1982 to 1987 in the PC world was the Year of the LAN? And when every year up until about 1992 was the Year of the CD-ROM? The same thing is about to happen in customer premise and network services for ASR. Converging technological advances, safety considerations and market forces are going to stretch ASR past the clipped "talk to the dog" responses we've been using to communicate with these systems thus far: Yes, No, Cancel, and the numbers one through ten. Speech rec, and more natural speech rec, is going to show up in the services that LD carriers, ITSPs, CLECs, cellular and land-line providers, and one-number service bureaus deploy to lure and keep customers.

ASR is going to infiltrate both customer premise systems and even the public network because readily available, all-purpose Pentium and RISC processors have finally gotten fast enough to do the intensive digitization, probability calculation, and matching necessary to make speech rec possible.

Running many speech recognition processes on a host computer is far more economical than buying a series of expensive CT resource cards, each housing a series of Digital Signal Processors (DSPs) that offloads the work of speech recognition from the formerly overburdened main system CPU.

DSP power is increasing too, though some algorithms, such as the latest version of Lernout & Hauspie's, have outgrown DSPs and must run on host CPUs. Still, any way you look at it, ASR technology is now making possible a density (lines per DSP and/or software module) out of which developers can afford to make a low-cost product. Instead of buying one SPARCstation or Intel machine and having one recognizer per one phone line you can now buy one Pentium and have 100 or more recognizers operating. For someone building a speech recognition system in the telephony network, that's affordable.

Many ASR network services started years ago using 100MHz Pentiums, and now you can get 1 GHz processors and passive backplane cards, which can fit many more speech recognizers in a single chassis. The trend is towards increasing CPU horsepower and decreasing DSP reliance.

By the same token, relatively cheap processors and single-board-computer clusters have brought the per-port price of ASR down into the same ball park, if not the same base, as good old-fashioned touchtone IVR. As late as 1997, speech rec cost ten times the cost of DTMF per port. Now it's only 25 to 75 percent more, according to Stuart R. Patterson, CEO of SpeechWorks.

The faster processors also can take on the more challenging task of larger recognition vocabularies. Companies such as Voice Control Systems (Dallas, TX – 972-726-1200, www.voicecontrol.com), now part of Philips Speech Processing (Atlanta, GA – 770-821-2400, www.speech.be.philips.com), hope to bring to telephony the capabilities of desktop ASR now being sold retail by IBM (Via Voice) Dragon (Naturally Speaking) and Lernout & Hauspie / Kurzweil (Voice Plus and Voice Pro). These PC-based recognizers can handle vocabularies up into the 60,000-word range and higher.

As the ASR vocabulary capacities grew, the simultaneous emergence of the Web put database entry within everyone's reach. Now, if I want to tell a speech-recognizing cellular provider to retrieve a name out of a speech-navigable Rolodex for me, I don't have to train it with all 150 of my contact names and phone numbers that I want it to recognize. I don't have to sit enunciating into a phone 450 times and touch-toning 150 times. I can just browse over to my user profile on CellCo's website and enter them in as text and numbers, filling in blanks. Or I can e-mail the telco a contact file straight out of Goldmine, and let them build the speech templates for me.

When did ASR algorithms figure out how to pronounce these text fields, an ability normally found in the Text-to-Speech (TTS) field? It occurred when the PC's increasing processing ability made possible a working version of an ASR processing model called "phonetic" or "phonemic". It's easily understood if you've experienced the "hooked on phonics" course for kids.

Words are broken into phonemes – depending upon what expert you quote there are between 48 and 54 of them in English. The system concatenates those sounds and goes to a dictionary to see if it can find particular combinations (Of course, this process is actually a bit more complicated than it appears – the phonemes themselves are usually broken down into lower-level diphomes and triphomes.).

Such improvements to speech rec technology moved some ASR system architectures from whole-word templates to phonetic recognition, which can now generally be found underlying any ASR product that must recognize 10,000 or more different words.

With a phonemic system it's quite easy to create customized vocabularies. For example, if you're creating a customized system and you want it to recognize the utterance "customer service," the system would use a text-to-speech engine to create a phonetic representation of the word and then use that phonetic representation to assemble the phonemes and adds it to the vocabulary.

Phonetic recognizers can accept text input and perform educated guesses on that text's pronunciation. With that, it can build a recognition model, or acoustic template. If it guesses wrong, it can usually be manually corrected with that vendor's pronunciation tools.

For example, Lernout & Hauspie's continuous phonetic system running on the Dialogic Antares card for telephony apps uses a special program called LEXTOOL that relies on their text-to-speech engine to generate custom vocabularies for the speech recognition engine.

On top of this technology, we're seeing a lot of neural net and artifical intelligence being placed inside these systems. Researchers are creating "N best" systems, where the system generates "confidence levels" of the first "N" (three, four or five) possible utterances and then runs these results through a syntactic or semantic filter to see if the words are used correctly in context.

If some aren't used correctly in context, the system throws the words out. Through the process of elimination, the correct word matching the utterance is found, so you can initiate phone calls with a phrase like "Call 617-555-1212" or even better: "Call the Electric Company." Or "Call Pretzel Airlines Reservations, please."

Remember, however, that these kinds of high-end phonetic and AI algorithms require a lot of RAM and big MIPS, lowering your density (fewer recognizers and lines per DSP), but you get better recognition accuracy and the system becomes flexible enough for anybody to use.

For ASR-based auto attendants and directory services applications, speech scientists also have worked on building huge databases of family name pronunciations, so that even rare foreign names, when entered, are automatically assigned their acoustic templates. An early VCS (now Philips) version of this kind of product was called Namebuilder.

Finally, one of the unwritten rules in ASR is that "more data is better data." In the past few years, ASR vendors have been busy collecting data across the U.S., Europe and Asia. Recognition accuracy is dependent upon the complete coverage and analysis of dialects and accents, which can only be done if someone goes through the trouble of collecting large collections of speech samples.

For example, a 1999 pan European research project, SpeechDat II, involved the collection of speech databases in 11 official European languages and dialects. Twenty European technology organizations participated, including Lernout & Hauspie, Vocalis, Siemens, and university research centers and PTTs. Speech samples were gathered from fixed and mobile phone networks for training and testing recognizers, and a third database was collected for developing verification applications. In all, 28 databases were produced, which should give all participants a leg up in developing future recognition engines in more languages.

As for safety concerns, voice-activated dialing and other ASR services may be promoted as a remedy to car accidents that result from distracted drivers. Voice activation may even become a precondition for operating cell phones in cars. Even now in parts of Europe it's illegal to chat on the phone while driving.

Present-day ASR applications for CT include the following:

- **Voice Activated Dialing (VAD)** This has been in existence for several years. At Bell Atlantic Mobile, VAD is called EasyDial. It is supplied by Intellivoice (Atlanta, GA – 404-816-3535, www.intellivoice.com) and it's available for just an additional few dollars a month on a phone bill. A speaker-dependent form of ASR, VAD typically requires users to build a small speed-dial database by repeating a name, such as "Mom," three times and entering a phone number via DTMF. The repetition "trains" the recognizer to build a small file of whole-word speech templates that match on segments of the caller's voice.

Intellivoice provides voice-activated dialing through its EasyDial network product to several telcos. They've added error-correction and other enhancements to the original VCS engine, as well as subscriber statistics and alarming. Bell Atlantic Mobile, as noted, and upstate New York Frontier Cellular both offer it to their subscribers.

But what's so good about such a "hands-free dialing" service if you end up calling an PBX or IVR system that still requires the caller to hit a series of touchtones to get the right phone extension or menu choice? No good at all, obviously, so Intellivoice also sells EasyDial Navigator, which allows the speech recognition to also generate the kind of touchtones necessary to control the auto attendant or IVR routines you're likely to encounter after your VAD call goes through.

Other wireless service providers wanting to offer Wireless Intelligent Network (WIN) services to their customers have turned to Nortel Networks' Wireless Voice Activated Dialing (WVAD). Nortel Networks' (Santa Clara, CA – 408-986-0890, www.nortel.com) Wireless VAD is a Wireless Intelligent Network service designed for networks based on AMPS, TDMA, or CDMA technology. WVAD simplifies subscriber interaction while offering the hands-free dialing. Just two simple keystrokes can provide fast access to this subscription-based service. WVAD allows subscribers to place calls on their wireless phone by verbally speaking the name, phone number, or location (such as "office" or "home") they wish to reach. Nortel Networks WVAD confirms the request and then makes the connection.

Leveraging Nortel Networks' leadership in networks, the WVAD application exploits the IS-41C standard, giving subscribers simple access (such as # [SEND]) to their personal directories – even while they roam. Nortel Networks WVAD interfaces with a service provider's IS-41 network to support subscriber access to this service for roamers in the territory – or in areas where the service provider has roaming agreements. Moreover, a full suite of operation, administration, maintenance and provisioning (OAM&P) capabilities help efficiently manage a WVAD service.

WVAD uses SS7 ISUP to ensure quick network access and call completion, and Release Link Trunk (RLT) capabilities optimize trunk usage, significantly reducing the number of trunks needed for WVAD deployment. RLT also eliminates the need for external programmable switches and avoids costly double trunking scenarios. Finally, WVAD's integration of billing indicators with a single Call Detail Record on the switch minimizes the impact on downstream billing.

Lucent is also a major player in the wireless VAD game. The Lucent Voice Activated Dialing Service has overcome the restrictions that have limited voice-dialing services in the past: Poor recognition technology and cumbersome dialing-list management. Using top-notch speech technology from Bell Labs, Lucent's Voice Activated Dialing offers recognition accuracy rates that often exceed manual dialing on a full-size keypad. List management is also controlled by voice commands, allowing names to be added or deleted with minimum effort. The subscriber initiates Voice Activated Dialing by pressing the * and SEND keys. Once connected, the mobile user has access to a range of functions, including voice dialing by name or number, voice-controlled feature activation, and name list administration.

According to several major ASR vendors, VAD should have become immensely popular – but it hasn't so far. Stuart Patterson of Applied Language Technologies says, "VAD was a very big flurry in the cell market and has died down. A lot of users don't know they have it, other networks don't have it. But voice dialing is still a big win. The problem is

how to price it, sell it, position it. Cellular usage is more challenging because of line noise. PCS will be easier."

Trials of VAD may also have been poorly marketed, explained, or perhaps VAD alone wasn't enough to sell customers on the SR idea, according to Bill Meisel, long-time publisher of Speech Recognition Update and president of TMA Associates (Tarzana, CA – 818-708-0962, www.tmaassociates.com). A voice-activated list of ten numbers or fewer wouldn't save anyone a memory chore, since most people know their top ten phone numbers reasonably well by rote. But get into a larger Rolodex, he suggests, and you have a different class of service. "It shouldn't be confused with commodities like call waiting or call forwarding. It attaches customer to company," he says.

Local telcos/cellular providers might consider it this way: Put 150 of your subscriber's business contacts on his VAD file, and now you've got him over the same kind of barrel as the ISP that hosts his e-commerce web site. It hurts your relationship with your customers too much to move to a different service.

- **Name-oriented Directory Assistance (large scale) and automated attendants (small scale).** Yves Normadin, president of ASR vendor Locus Dialogue (Montreal, , wwwCanada – 514-954-3804, www.locusdialogue.com), told us, "For most telcos, Directory Assistance is a money-losing proposition. Nortel has tried to automate with the Automated Directory Assistance Service (ADAS) as an aid to a live operator." ADAS is an automated system that asks the caller for the city and listing. An operator only gets on the line, if at all, to deliver the number. "If it saves three seconds on each call, that's a lot," says Normadin.

In some locations, a Nortel adjunct to the switch is actually recognizing the spoken city name, narrowing the search for the operator. At parts of Bell Canada, they're trying to recognize some of the listings themselves, with "vocabularies" of only a couple thousand popular listings. "They've found they can cover a large percentage of calls that way," says Bill Meisel.

Locus Dialog (Montreal, Canada – 514-954-3804, www.locusdialog.com) is also active in speech-recognizing directory assistance, combining it with a visual interface in the form of ADSI screen phones. In a pilot for Bell Canada in Ottawa and Quebec, callers name a city and see a listing of possibly-heard matches on their ADSI screens. They can then use soft keys to pick the correct town. The caller then repeats the process for the actual listing.

Locus Dialogue is a relatively new (1996) name in speech recognition, although its core technology has a ten-year history starting at the Computer Research Institute of Montreal. The company first impressed us with a demo in 1998 of a speech-recognizing directory service prototyped for Bell Canada.

Locus Dialogue uses ADSI screen phones, asks the caller to specify city, displays the list of possible cities that it heard, and asks the caller to pick the correct one with soft keys. Then it goes through a similar audio/visual procedure to locate the right person and phone number. Normandin told *Computer Telephony* magazine that half a million Canadians in Ontario and Quebec have ADSI phones, and hope that applications like this might help sell more. It's a way to add the visual component to the PC-challenged.

Bell Canada is expanding their speech-enabled directory assistance to include yellow-pages and business white pages listings, and blue pages (government agencies). They also want to put this application to work on public phones, using Nortel's Vista 350 ADSI model.

Locus Dialogue's ASR is speaker-independent, phoneme-based, and is currently bilingual, simultaneously recognizing English and French or

English and Spanish. They sell a SoftDialogue ASR server and a SDK to qualified application developers.

Normadin also suggests that speech will eventually be used alongside touchtone to activate such features as *69 call return and *70 call blocking.

During 1998 and 1999, more ASR development and sales activity appears to be occurring down at the customer premises end than in the network, particularly with speech-controlled automated attendants.

As 1999 drew to a close, Lucent Technologies (Los Angeles, CA – 888-458-2368, www.lucent.com) added speech recognition to its Intuity Audix Multimedia messaging system in the form of a new add-on, the Lucent Voice Director. The new voice messaging feature makes life a little easier by allowing business users to address messages or dial a colleague simply by speaking the person's name. The application also enables an outside caller, such as a customer, to reach an employee without going through a switchboard attendant or navigating through a touchtone-activated menu system.

Lucent said it designed Lucent Voice Director using Bell Labs' highly-accurate speech recognition technology and flexible dictionary. Lucent Voice Director can be used with millions of Intuity Audix mailboxes that are installed today. Lucent said that Lucent Voice Director will be especially valuable for companies that are increasing their mobile workforce because the application makes it easier for employees, while out of the office, to contact and leave messages for colleagues. For example, a manager who is travelling can send an important message to several team members without having to remember their extension numbers. The system plays back the recorded name of the individual before sending a message or connecting a call, allowing users to ensure their messages reach the right people.

"I can't imagine why a company with a Lucent voice messaging system wouldn't add the Lucent Voice Director option," said Bill Meisel of TMA Associates (www), a speech industry consulting firm in Tarzana, CA. "The low price and minimal support requirements make the decision easy. It will be popular whether users are calling from within the company or from the road." Lucent Voice Director uses a 20,000 name vocabulary and is specifically tuned for a mobile environment. It is the first speech recognition product to be thoroughly integrated with a messaging server, making it easy to administer and highly accurate.

Lucent Voice Director software, which is loaded on a Windows NT workstation and integrated with the Intuity Audix server, is available in the U.S. and Canada for Lucent's new Intuity Audix Release 5.0 Multimedia Messaging System. Future releases of Lucent Voice Director will include speech commands. Pricing for Lucent Voice Director software begins at $3,000.

In 1999 Voice Request, a leading provider of speech enabled call routing systems, became a unit of the Philips Speech Processing business unit of Philips Electronics (Atlanta, GA – 770-821-2400, www.speech.be.philips.com) when Voice Request's parent company, VCS, was acquired by Philips.

Philips Voice Request makes one of the more interesting ASR automated attendants, Pure ReQuest!, which automatically routes an organization's incoming or internal phone calls by enabling callers to say the name of the person or department desired. The system greets the caller, prompts him or her for the name desired, recognizes the request and routes the call to the appropriate extension, all without the intervention of an operator.

In a similar vein, Phonetic Systems (Burlington, MA – 781-229-5873, www.phoneticsystems.com) offers its PhoneticOperator and PhoneticAttendant. PhoneticOperator is a speech-enabled, telephony-based directory search and auto-attendant product designed to search more than 1,000,000 records in a single database. PhoneticAttendant, has the same incredible speed and accuracy of PhoneticOperator, but is designed for smaller directories. These tools are used for name and department searches, call center solutions, information retrieval, call routing, and directory services applications.

Rooted firmly in the customer premises rather that the network, Registry Magic's (Boca Raton, FL – 561-367-0408, www.registrymagic.com) Virtual Operator is an automated ASR-powered receptionist that can answer up to 12 calls simultaneously. It eliminates an automatic transfer to the operator from rotary telephone callers as well as prompts such as "press one for sales, press two for accounting, press three for..."

A "barge-in" feature lets callers interrupt the Virtual Operator at any time. Repeat callers who know the name of an employee or department can interrupt the Virtual Operator's announcement greeting, say the name, and connect faster than with a touchtone menu or a live operator.

The larger Virtual Operator Enterprise Edition stores up to 10,000 words or names. It enables automated updates to its on-line phonetic directory structure, allowing for the import of current human resource (HR) tables from the company database, to add/delete employees or change extensions. For example, this can be done nightly, without human intervention.

- **General Enhanced Services for the PSTN.**

Intervoice-Brite (Dallas, TX – 972-454-8862, www.intervoice.com) is a major IVR and enhanced services platform provider to call centers and networks. One major ASR-enabled network service area Intervoice-Brite is working on is collect call automation. They've got 2,500 ports of this going for one of the big-three long distance carriers in four or five call centers, running off their own OneVoice platform as an intelligent service node.

It's a simple continuous-word and discrete-word recognition application and it's in use by all the large LD carriers, says Intervoice. In this case, a VCS / Philips speech rec engine listens for the number the collect caller wants to reach. When it reaches the called party, it listens for "yes" or "no" in answer to the "do you wish to accept the charges?" prompt.

This app has been around since the mid 1990s. The interesting thing is that British Telecom and Japan Telecom are using it for their own collect-call service, and they find it cheaper to route the calls through U.S. platforms than to maintain their own operators. It switches through for these overseas callers to U.S. and non-U.S. destinations.

Late in February 1999 Intervoice announced the sale of 30 OneVoice platforms to a major transportation company. This will automate destination schedule inquiries and reservations, all through a speech-rec interface.

Intervoice developed this particular application to carrier specs with their own InVision scripting tool, and also deployed it on their own VCD voice and signal processing boards, so there's more control over the application and the engine.

Vocalis (Cambridge, U.K. – +44-0-1223-846177; Boston, MA – 781-279-2436, www.vocalis.com) was started in 1993 by a management buy-out of Logica's Speech and Natural Language Group. They are now a public company based in the U.K. Vocalis sells its SPEECHtel speech rec intelligent peripheral largely through OEM relationships with Ericsson and its Intelligent Network switch-based operator system. The Speechtel IP is SCO Unix based, with IP, SS7 and CS-1 integration. Graeme Smith, Marketing Director, notes that a Latin American client is automating its collect call processing and operator services, using their Castillian Spanish recognizer.

Vocalis is now partially automating directory assistance with call completion for Telia TeleRespons AB, Sweden, and for Telekom Malaysia. Digital Telecommunications Philippines (DIGITEL) is also using Speechtel for speech-enabled operator services and one-number access. You can also find Speechtel in networks in Australia, Ireland, Norway and Mexico.

"Person-to-person calling is a very popular service in the Philippines and automation will speed up call handling times," notes Charles Halle, Vocalis CEO. "We let network operators select intelligent network services that are most appropriate to their own market conditions."

Another Vocalis packaged speech-enabled service allows callers to record "my number has changed" announcements that automatically redirect calls to the new number.

On the subscriber side, Speechtel supports voice-activated dialing, one-number follow-me service, and unified messaging.

For computer telephony developers, Vocalis sells SpeechWare, a set of tools and APIs for adding complex speech recognition to existing applications. SpeechWare supports a wide range of languages, enabling the roll out of applications in multiple international markets. This SpeechWare core technology, has its roots in SCO Unix and there's a version for Windows NT. Vocalis also offers Transcriba, a tool which breaks down speech into components suitable for recognition, and the Scripta speech-enabling design tool.

In March 1999, Vocalis announced at CT Expo a new product, SpeecHTML, a gateway platform and a service that takes your web site and "speech-enables" it, similar to VxML, and our *Computer Telephony* magazine's own Telephony Markup Language (TML) initiative.

Nuance Communications (Menlo Park, CA – 650-847-0000, www.nuancecom.com) is another major name in natural language speech recognition. They also supply a speaker verification product, Nuance Verifier, and a developer's toolkit.

They've made a big splash in vertical markets and established leading phonetic dictionaries for travel, dates and dollar amounts, as well as a general-purpose American English stock of over 110,000 words. Everyone in the industry knows about their impressive Charles Schwab VoiceBroker application. VoiceBroker (trademarked by Schwab) is a phone service that uses speech recognition technology to provide stock, mutual fund and market indicator information to callers and lets them do business transactions over the phone.

VoiceBroker uses Nuance's conversational transaction technology, which consists of speech recognition and language understanding software that allows a caller to speak in everyday, conversational English. The staff at *Computer Telephony* magazine finally realized that ASR had truly come of age when some Nuance officials dropped by our offices and gave us a demonstration of VoiceBroker.

VoiceBroker understands and responds to inquiries about every security listed on the New York Stock Exchange, the American Stock Exchange and NASDAQ. You can just call up and say the name (or variations of the name) of more than 13,000 stocks, mutual funds or market indicators to get the latest quote information.

This job is a perfect fit for Nuance's transaction technology, since it can recognize perhaps the largest active vocabulary for phone-based services and combines this vocabulary with language understanding. This lets it recognize continuous, natural speech with meaningful comprehension.

The product is completely speaker independent and no training on the part of the user is necessary.

We had fun asking about Natural Microsystems stock (it was up) and General Electric (it was up). I told VoiceBroker to "sell 100 shares of Intel," but the system informed me that, since I didn't own any Intel, I couldn't sell any. So I bought some, then checked my portfolio.

Schwab customers can create personal stocklists so customers can get multiple quotes automatically.

To demo Nuance's speech recognition technology, you can download software from their web site at www.nuancecom.com.

Nuance also puts out an exceptionally clear and helpful backgrounder on speech recognition in general and their Nuance product in particular. Nuance has acoustic libraries for English and Spanish, is working with Kyoto-based Omron on building a Japanese product, and is integrated with IVR platforms from Voicetek, Syntellect, CCS, and IBM's DirectTalk (now the IBM Corepoint Voice Response system).

There are several network providers interested in VAD personal dialers using Nuance's ASR engine. And, according to Marketing VP Steve Ehrlich, two customers are developing personal agents that will let you pick up the cell phone and boss around your voice mail, e-mail, scheduler and calendar. Wouldn't that be a great use of time stuck in traffic, not only hearing and sending e-mail, but scheduling appointments right into your Outlook?

"It's exceptionally complex; the dialog is difficult," says Ehrlich. "You have a store of information. Your e-mail is in an e-mail database and voice mail is in a voice-mail system, your calendar and scheduler might be in Outlook. A voice interface provides a central way of accessing all those."

Ehrlich also points out that while travel and financial fields are more aggressive, telcos are much more reticent about adopting new technologies.

Still, Nuance scored some major wins with their Nuance 6 speech rec product during 1998 and 1999.. The Webley personal assistant switched over to it, General Magic finally revealed that they'd chosen Nuance for Portico, and Motorola weighed in too, with their similar Myosphere service.

Nuance and Edify have put out joint releases to brag about their Sears, Roebuck & Co. speech rec application. You can try it – just call your local Sears. You are told store hours and then prompted for a department, or even a product. If you say "refrigerators" and "microwave" it will route your call to the large and small appliances department!

John Shea, Nuance's director of product marketing, tells me that all these calls are being transferred to a central system for the recognition. It then determines where you called from and sends your call back to the appropriate store. "Our system is architected with distributed servers to balance the load among many tasks and offer fault tolerance," he says.

When our intrepid Executive Editor, Ellen Muraskin, tried it, she said, "toys," it queued her for a human operator. As it happens, Sears has no "toy" department per se, but sells toys out of "boys" or sporting goods departments. Similarly, uttering "games" and "Furbies" send you to an operator queue. The main point is, a good percentage of callers will not be held in queue, because the IVR can process many calls simultaneously. And Sears decided from the beginning that it wasn't going to ask callers to enter DTMF touchtones. The system handles more than 120,000 information calls a day for 800 stores.

How do they update their product list? Nuance 6.2 supports "dynamic grammars," says Shea, letting customers pull text strings out of databases on the fly and load them into the recognizer at specified points of the dialog.

Nuance provides speaker verification as an option. Both word- and concept-recognition, as well as verification, can be combined in one user step by matching on entered account numbers. Enrollment can also be very subtle; the first few times a caller accesses a bank account, perhaps, she'll speak a PIN number. The PIN will identify her, but also build a voice print with successive repetitions. Finally, it can substitute the voice match itself for a PIN.

Nuance is known for its financial and travel vocabularies. Shea says that six different pilots have been completed with major financial and insurance companies, due for rollout during 2000. And Nuance also does the listening for UPS' package tracking IVR, letting callers speak up to 18-character alphanumeric codes. UPS says they shaved up to 90 cents off the cost of each call.

Nuance languages include U.S., U.K. and Australian English, German, Japanese, and Latin American Spanish, Canadian French, Brazilian Portuguese, European French and Mandarin.

Nuance runs on NT and most Unix flavors, and runs on "all major IVR platforms," says Shea. They've followed SpeechWorks' lead in packaging up their code into building blocks. Nuance's Speech Objects are Java and Active X components. They've also launched a lower-priced, entry-level recognizer for vocabularies of under 20 items. It's called Nuance Express and starts at $250 per concurrent user.

Voice Control Systems (VCS) now part of Philips Speech Processing (Atlanta, GA – 770-821-2400, www.speech.be.philips.com), claims to own the lion's share of the domestic VAD market: Bell Atlantic Mobile, AT&T Wireless and SouthWestern Bell Mobility are three notable customers, the first by virtue of the VCS technology embedded in Intellivoice's EasyDial product. VCS was long known for having the widest coverage in languages of small, discrete "talk to the dog" speech rec vocabularies: Yes, no, numbers, stop, cancel. These simple commands have been used to speech-enable common IVR applications for over 50 languages in 30 countries.

Through acquisitions and partnering arrangements with larger-vocabulary speech rec vendors, they've now broadened their line to go from the industry's basic 16 words to as high as telephony-based speech recognition goes: 60,000 words or more. Their later line was called SpeechWave, and they've been working to bring these products in under a common API for developer's one-stop shopping convenience and undying loyalty.

"We first provided discrete digits and commands. It's very easy to fit 32 of those on one Dialogic Antares board for low cost," says VCS product marketing manager Jeff Engle.

When VCS purchased PureSpeech and their ReCite! product line in the late 1990s, they increased their vocabulary capacity up to 20,000 words. They gained natural language recognition, which can recognize requests and commands that are worded in a number of ways. And they gained additional phonetic recognition – good for frequently changing vocabularies – which allows users to type in words and names to be recognized. Purespeech's algorithms ran on Dialogic Antares cards.

Prior to their acquisition by Philips, VCS had also merged with Voice Processing Corporation, whose recognizer runs on Natural MicroSystems cards. They then formed an IBM partnership, developing and cross marketing a telephony-customized version of IBM's 65,000-word Via Voice desktop recognizer.

Sometimes, Engle says, developers testing the human interface perform what's called "Wizard of Oz" testing. Developers sit behind the curtain, as it were, pretending to be a (presumably) Great and Powerful automated system while noting the replies of test callers.

"We've done quite a few deals in selling speech-enabling products to companies that provide platforms to network operators," says Engle.

Tecnomen, Finland-based developers of enhanced services platforms is integrating SpeechWave into their VAD, voice mail, fax, e-mail and paging applications. Speech recognition has a stronger impetus in

Europe as a DTMF substitute, particularly in Eastern Europe, where rotary phones are still prevalent.

Philips / VCS has also scored the endorsement of Siemens, which is integrating the ASR technology with their EWSD switch. Siemens intends to focus on call screening, operator services, directory assistance line checking, collect call, and call completion.

SpeechWorks International, formerly Applied Language Technologies (Boston, MA – 617-428-4444, www.speechworks.com) offers a natural-language speech recognition product, SpeechWorks. Their Recognition Engine can pick up more than 50K spoken words and navigates though variable grammar for voice processing. And it synchronizes with a range of hardware platforms and scales to thousands of phone lines. The engine also supports languages such as: UK and Australian English, Canadian French, German, Brazilian Portuguese, Latin American Spanish, Mandarin and Cantonese.

SpeechWorks was used in the Intelliventures division of Bell South to build a speech-driven Yellow Pages, starting with just restaurant listings in Daytona and Gainesville, Florida.

Some of us at *Computer Telephony* magazine tried it, and it was a good illustration of the way natural language can grab several pieces of information out of one randomly constructed sentence. "Tell me what you'd like," it said, to which we replied, "I'd like an Italian restaurant in Daytona Beach."

The software figures out what parts of the search criteria are missing and structures subsequent questions accordingly, as in, "How much would you like to spend?" But one could just as easily have said, "Find me an Italian restaurant in Daytona Beach with entrees starting around 18 dollars," and it would have taken that all down at once. Bell South's application found no Indian and five kosher restaurants (all branches of Larry's Subs) in Daytona Beach.

About 10,000 people use this service each month, according to Intelliventure's David Shipp. They've also added some news, weather, stock quotes and sports information to the restaurant guide. We tried that: "Compaq," "Oracle" and "Polycom" it got; "Dialogic" (too many companies sound like "Dialogic") and "Excel" gave it trouble. But the system doesn't misinform; it asks "Did you say Xylogic?" and you can just say "no."

"We want to first understand the technology. Then we'll build a business around it," says Shipp. Bell South Mobility plans to test SpeechWorks' VAD on the cellular network next. "We shopped around and we were impressed with SpeechWorks' toolkit," says Shipp. It's a continuing partnership. "They're going down a path that both of us are defining."

The Yellow Pages app runs on a Sun platform and uses Dialogic's Antares DSP board to run SpeechWorks.

Stuart Patterson, CEO of SpeechWorks, says that the speech recognition engine itself is only 25% of the problem of building speech-rec applications. "When you get down to it, tools are another 25 percent. Platform integration on a Dialogic or NMS or IBM is another 25 percent, which defines scalability. Another 25 percent is getting the customer application developed. There's a lot of product out there, but not a lot of support services. With the shortage of IT professionals we have today, it's even more severe in speech."

SpeechWorks' answer to that problem is a building-block toolkit of speech rec procedures that they call Dialog Modules. "Each module is a recognition event," says Patterson. "We have dialog modules for example, recognizing continuous numbers that you can put in the call flow of an application. We have 10 to 15 of these modules, depending on two access methods: The first is a code-level API that can be written to C or C++.

You then set the module's parameters within a sequenced data file."

The second method is integrated with the service creation platforms and app gens of Voicetek Generations (Chelmsford, MA – 978-250-9393, www) or Intervoice-Brite's NSP-5000 (Dallas, TX – 972-454-8862, www.intervoice.com). In this case, each module appears as an icon, with forms for plunking in parameters. "For example, to put in an item module, your parameters will be the 60 names of a small-company auto attendant," says Patterson.

SpeechWorks has tested their modules with additional Dialogic app gen VARS and vertical VARs, using Antares and the low-density D/41 boards. "We deliver SpeechWorks with an enormous two-to three-layer dictionary, and phonetic rules that are applied. We color-code words that are outside our dictionary," says Paterson. Users can add words to the dictionary using SpeechWorks' pronunciation tools. SpeechWorks also offers a two- to three-day training course for its toolkit.

"If you have a vocabulary of under 2,000 words without much natural language, you can do it with (Dialog modules and) very little ASR expertise. And if you need more, we have a services group."

SpeechWorks is also considering leveraging their work in Dialog Modules by partnering with other speech vendors' databases. If the challenge of getting speech rec out to market is a lack of mature tools for service creation and operation, Patterson says, "Why should we spend another one to two man years on algorithms, when the recognition itself is only 25 percent of the challenge?"

While they concede to no one in recognition accuracy – they claim a 20% to 30% gain in raw recognition accuracy every quarter and their barge-in capability is especially well tuned not to be fooled by garbage noise – Patterson's largest single emphasis has always been on the user interface.

"The accuracy can be lost on a caller if the user interface is poor," Patterson says. Furthermore, "the counter-intuitive fact about Natural Language Recognition is that people don't know how to use it." They stammer if given the freedom to phrase any request in any form. "People want to be given a context. If you don't ground them, they'll be out of vocabulary a lot. We've worked with enhanced service providers who seen they're out-of-vocabulary rate go over ten percent. They must fix their prompts!"

SpeechWorks' ready answer to this problem is twofold: Dialog Modules and Testing Tools. The Dialog Modules containerize common SpeechWorks speech rec tasks, such as dates and currencies, and are easily imbedded in CT applications, or dragged and dropped via app gens or scripting languages. They've also packaged the modules into Active X controls that are going into Artisoft's Visual Voice.

If you use SpeechWorks' Dialog Modules, you can use the stats they generate in SpeechWorks' tuning tools. The meta-level AppStats tuning tool analyzes and improves a production system by tracking responses and transaction completion rates. The Show Call tool steps through an individual call. Automatic Adaptation Tools continually refine the recognition models based on actual caller interactions and allow the system to automatically learn the correct pronunciations of words in the vocabulary, and to adapt the acoustic and language models to match the characteristics of the application and the population of users. Patterson reports that an Australian developer used SpeechWorks with American English and through the tuning tools, allowed it to improve its own performance to adapt to Aussie.

Speechworks' staff expanded from 50 to 90 employees during 1998-1999, and they opened offices and created new research alliances in Mexico and Singapore.

The most recent development was deployment of Ratefinder, an app for Fedex that asks callers the type, dimensions and weight of their pack-ages, their sending and destination zip codes, and gives them the shipping cost.. NextLink Interactive, a subsidiary of Nextlink Communications, developed it with SpeechWorks' assistance. Ellen Muraskin of *Computer Telephony* magazine tried it – it's choice number five on Fedex's self-service 800 number, the same one you dial to track packages.

A friendly female voice asks for your Fedex account number, and if you can't produce it, "she" offers to calculate your order using standard business rates. The recognition is smooth and almost perfect; she only had to ask to repeat something once. She does take a long time to look up rates, though.

Patterson says that this application was designed by a "user experience architect" who also designed Wildfire, which is probably why Fedex's automated agent has that same competent, friendly but efficient quality about her. There's also a familiar, pleasant percolating sound that signifies a recognition task in progress.

Another interesting Hewlett Packard SpeechWorks app helps their customer service call center. It lets callers know whether their HP products are Y2K compliant. Answers can be (a) yes, (b) please check our web site here's the URL, or (c) please hold for the next available agent. HP supports thousands of products in multiple call centers; here's a new area for large-vocabulary recognition to shine.

SpeechWorks' famous E-Trade app is expanding to include mutual fund quotes and trading, an agreement with Guardian Life Insurance stakes a claim in that vertical market, and Dialogic's SpeechWorks Host, a low-density board that comes packaged up with Dialog Modules and SpeechWorks for up to 250-word vocabularies, is gaining ground among developers.

SpeechWorks has also launched a nationwide seminar program to teach developers how to build speech-enabled applications.

((ROBBIE, CAPTION FOR speechworks2.tif:))

SpeechWorks' Tuning Tools show developers a speech recognition task-by-task success rate as a call proceeds through its flow; dialog modules with high error rates might indicate the need to rephrase prompts or change vocabularies.

In 1998 VCS (now part of Philips) entered into a partnership with IBM to jointly market IBM's 65,000-word Via Voice recognizer to the telecom world. That, plus acquisitions from PureSpeech (also now part of Philips), were all given a common API and inserted into their Speechwave product line.

IBM's Via Voice then grew into Speechwave Enterprise. This natural language understanding is specified with BNF grammars. It's the only recognizer whose marketers claimed, however, that it understands what pronouns refer to. In other words, if you say "I'd like to move $500 into my checking account," And then say, "I'd like to change it." Via Voice will know that "it" refers to the money.

Speechwave Recite 3.0, in product manager Tim Walsh's words, "... marries the up to 20,000-word Purespeech phonetic recognizer with VCS's world class digit recognition." It's got the phonetic entry ability to recognize almost limitless names, products, and destinations, as well as the continuous-digit recognition.

The third SpeechWave product, called Impact, combines five SR technologies: discrete and continuous digits, alphanumeric recognition, an 850-word phonetic vocabulary, and patented simultaneous speech recognition and speaker verification.

• **Extremely scalable multipurpose ASR engines for developers.**

Babel Technologies (Mons, Belgium – +32-65-37-42-75; San Francisco, CA – 415-538-3792, www.babeltech.com) offers a speech recognition engine that comes out of the Mons Institute of Technology. Its singular claim is to a greater level of accuracy made possible by the combination of two ASR technologies: the established Hidden Markov Model (HMM) and newer work in Neural Network theory.

They're a relatively small shop; two principals were away at CeBit, and as soon as they got back they took some pains to respond to my e-mail request for more information on the advantages of ANN technology. If the terms "Mean Squared Error" or "relative entropy" are the kind of things that get you excited, then by all means give them a call.

The HMM / ANN combination appears to be especially good at rejection; that is, recognizing that an utterance (or noise) does not match any possible desired word and is therefore to be rejected. Engines that are weak at rejection will often strain to come up with what they consider the likeliest match. A good example is the American Presidents ASR demo over at IBM's Via Voice site. While the system correctly identifies the name of every true American President, blurting out "Itzhak Rabin" will return an ID on "John Kennedy." In fairness, it did know enough to discard "Frank Zappa."

Right now, the Belgian broadcasting company RTBF is using a Babel-driven IVR that delivers weather information on more than 500 European cities. The system currently runs on two E-1s.

Babel's ASR technology can scale up from a few words to a few thousand words. It's available for NT and most Unix platforms. It can handle English, French, and Spanish, German and Dutch.

IBM Speech Systems (West Palm Beach, FL – 416-383-9224, priority code 6N9BB001, www.IBM.com) has taken its ViaVoice 65,000-word recognizer and moved it from the dictation world to the telephony market in several horizontal and vertical applications. In one impressive demonstration of Big Blue Via Voice Directory Dialer at CT Expo 1999, the system could locate any of IBM's own 200,000 North American employees across a network. One could speak the geographical location first (20,000 names can be active for recognition at any one time), and then the employee name, which it uses to route the call.

IBM is also preparing natural-language understanding engines for verticals, starting with travel and mutual fund trading. ViaVoice is smart and "natural" enough to ask for specific clarification if certain pieces of information are ambiguous, for example, responding to a caller who says "sell half of fund A" with, "Did you mean Class A shares or Class AA shares?"

ViaVoice Telephony Runtime, as the engine is called, comes with acoustic modeling optimized for both landline and cellular telephony. Sample grammars are included for banking, commands, dates, digits, continuous numbers, times, voice mail, voice-activated network control and yes/no. Grammar development and admin tools come with a web GUI, and testing tools are there, too. It also automatically generates TTS prompts if they're not previously recorded.

The engine also dynamically adds grammars of words and phrases as the application is used, collecting data and building recognized word combinations, analogous to the way in which data collection builds the probability scales that are used, in turn, to build word models. Dr. David Nahamoo, senior manager in IBM's Human Language Technologies research department, says that "we use a statistical data-driven approach, which allows the system to learn from the way people use it." They've already got the five major European languages; Arabic, Chinese, and Japanese are covered, with more being prototyped. Only English has natural-language understanding (NLU) tools thus far.

Lernout & Hauspie (Burlington, MA – 781-203-5000, www.lhs.com) currently offers two speech recognition engines in the telecom market, with a third host CPU-based too. ASR 1500 is a software-only recognizer for applications requiring from 500 to 1,000 words. According to L&H's Tom Morse, it's used for voice commands and form-filling; he

gives a golf-reservation IVR as a good example.

"Ninety-five percent of all telephony speech rec applications use under two thousand words," says Morse. Telephony applications aren't as concerned with vocabulary size as speaker-independence, audio garbage tolerance, and processor efficiency.

ASR 1500, like the Philips and Lucent engines, works with Unisys' Natural Language Speech Assistant grammar-building tool. "NLSA provides two major benefits," says Morse. "It's a more approachable and graphical way to specify a grammar. It works in a spreadsheet format, with columns of things to be said. The user can visualize the different paths through the vocabulary." The end product is a BNF grammar specification. L&H also sells its own speech-rec toolkit for C language developers.

L&H also sells a large-vocabulary recognizer, for over 10,000 words, which derives from BBN's august Hark! Engine. "Most of the time this engine is picking from a large list of words, most of which are common names: Cities, stocks, company names, restaurant names, people's names," says Morse. It's currently in use in a number of call center applications. L&H renamed it under its own label and is working on developing an API consistent with its smaller-vocabulary offering. The large-vocabulary recognizer runs on Sun Solaris and IBM AIX.

Philips Speech Processing (Atlanta, GA – 770-821-2400, www.speech.be.philips.com) is a pioneer (and major acquirer of pioneers) in speech recognition, natural dialog and language understanding technologies. A developer of voice enabled telephony applications, Philips has a large installed base of speech recognition and natural dialogue systems in Europe and is a major speech technology provider in North America. Its natural dialog platform speech recognition engines are used for banking, travel, auto attendants, speech portals and white and yellow pages automation.

Philips is taking the natural-language understanding concept to the call center, attempting to automate a lot of what human agents understand from customers' routine inquiries and transactions.

SpeechPearl is the Philips core recognition engine, while SpeechMania 99 is a natural-language engine and toolkit that takes scripting information and allows developers to turn a call center agent's script into the automated half of a "conversation." SpeechPearl and SpeechMania 99 run on Windows NT.

SpeechPearl can let speakers initiate conversations at any part in a dialog, unlike other systems that can only recognize specific words at specified steps through a sequence. SpeechPearl also lets users switch languages on the fly. The Mania dialog builder also has an open API that accommodates a range of text-to-speech engines; TTS would be needed to dynamically build some of a conversation's prompts or confirm to the caller what the ASR engine "heard."

SpeechMania 99, running on Windows NT and based on Philips' TrueDialog technology, has become a popular natural dialog system platform for automating telephone-based information and transaction services. SpeechMania 99 can be integrated into all telephony platforms.

Philips Speech Processing attaches considerable importance to simplified integration into existing IVR and voice mail platforms. SpeechMania 99 is supplied with a LAN-based telephony interface, permitting straightforward integration into existing third-party architectures. Also SpeechMania 99 improves administration capabilities by means of standardized error and status logging, which can be evaluated by Simple Network Management Protocol (SNMP) compatible administration platforms. The top-tier functionality for dynamic application switching simplifies the application development process of SpeechMania 99, thus speeding time to market of new automated voice services.

SpeechMania 99 also comes with the SpeechXpert Toolkit, which is based on a graphic user interface. SpeechXpert, based upon a unified graphic workflow paradigm, provides application developers with all the necessary development and evaluation tools, all of which have a uniform look and feel.

The potential uses of SpeechMania 99 include such applications as the voice-controlled selection and utilization of value added services, telephone-based access to financial services, timetable information, and weather reports – with several thousand calls daily. SpeechMania 99 allows the caller to switch between individual services, so that there is no necessity to make repeated calls. As a result, the provider of information services can use existing computer workstations in the call center much more efficiently. SpeechMania 99 should be thus able to automatically (and successfully) process a large quantity of natural language information service interactions.

But Philips main thrust appears to be automating a large part of call center interactions, since their high-definition dialog language tools can accept the wide range of human responses found in call center voice traffic.

"SpeechMania [and Philips' Dialog Description Language] allows you to take [agent] scripting information and convert it into our own dialog," says Bruce Cooperman, Philips' senior VP, speech technologies. Mania also works with TTS to automate the agent's lines of the script; i.e., prompts. "Speech recognition allows us to fill in the slots, to access a database, which is a normal call center environment."

Typically, "natural-language understanding" speech rec dialog designers use what's called BNF grammars to specify all the different ways a caller might respond to a prompt. Cooperman says that Philips doesn't take such a "closed" BNF grammar route with SpeechMania. "We've found in speech rec that if you follow BNF grammar, the engine inserts grammar rules into the understanding itself and you restrict the engine's flexibility. This works with numbers, but not with Natural Language understanding.

"In Mania, we use an open grammar approach, so that the core recognizer is not thwarted by grammar rules." The recognizer first captures the words, which are then sifted through concept models. This means that the core recognition task is isolated from the step that looks for meaning. "We send a voice graph together with our understanding to speech understanding modules," says Cooperman. "That allows us to interpret the key words in a phrase without aborting them because the grammar rules are so restrictive." This is what allows callers to speak in incomplete sentences, or with imperfect grammar, and still be understood. And Philips is looking for chances to prove it: "We challenge you to bring your call center script to us," they've told us.

Thus, with a SpeechMania 99 application, callers don't really have to adhere to a predefined dialog structure; instead, they can use their own idiosyncratic questioning technique to determine what form the "conversation" with the computer takes. Everyone can speak freely with his or her own choice of words and form of expression, without being restricted to a particular syntactic structure.

SpeechPearl, and by extension, SpeechMania, support what they call dynamic context extension, allowing callers to switch active vocabularies (the range of words understood at any given time in a dialog) and even languages on the fly.

SpeechMania 99 is available in a wide range of languages, including U.S. and U.K. English, German, Swiss German, French, Dutch, Spanish, U.S. Spanish, Italian, Norwegian, and Portuguese. Further language options are being developed.

Philips has agreed to integrate the SpeechPearl engine into Unisys's BNF-compliant, NL Speech Assistant toolkit. This will be sold through an established Unisys IVR channel and is aimed at very high-density, carrier-grade applications. SpeechPearl is also promised to be integrated into Brooktrout's (Southborough, MA – 781-433-9525, www.brooktrout.com) Show N Tel application platform, and works with the platform of Brite Voice, now Intervoice-Brite (Heathrow, FL – 407-357-1000, www.brite.com).

When SpeechMania dynamically loads a vocabulary, it is loading an acoustic model for that vocabulary. Similarly, in the area of cellular services, it can enable an application in which the CO picks up the ANI of the cell phone and determines that an acoustic model optimized for GSM airlink (or any other cellphone airlink) would work best.

In Italy, Omnitel cellular is using SpeechMania for a telephony "portal." Dial into one number, and from there ask for any of a range of services, such as sending flowers, calling a cab, or hearing a weather forecast. Actimedia (formerly TeleDirect, the Yellow Pages publishing arm of Bell Canada) is also using Philips' speech processing in its portal, now in trial with restaurants.

Philips also has airline and railroad reservations systems under contract in Germany, Switzerland and Sweden, provisioned through the telcos. "The European PTTs have taken an aggressive stand by supplying their large customers with speech-enabled capabilities on their network," says Cooperman. "That's an initiative we have yet to see in the U.S. Hopefully, it'll come about with voice messaging and unified messaging."

Other well-known service providers already using SpeechMania include the "Koerslijn" of the Dutch Postbank (to do share price inquiries), the Swiss, Dutch and Swedish Railways (timetable information), RTL Multimedia (weather information service), Lufthansa (flight information service), and Bell Canada (an electronic restaurant guide for Toronto).

Syrinx Speech Systems (North Sydney, AU – +61-2-9925-0477; Princeton, NJ – 609-452-9423, www.syrinx.com.au) makes a speech recognition system called SyCon that works on Linkon's FC-4000 boards, Dialogic's Antares 6000/50 boards, and IBM's DirecTalk boards. It is a vocally trained system that works with any language or dialects.

Founded in 1990 by Dr Clive Summerfield and Professor Trevor Cole of Sydney University's Speech Research Laboratory, Syrinx has grown into an international voice processing company, based in Australia, developing speech recognition solutions with a focus on automated customer services and call centers.

Syrinx has formed strategic partnerships with IVR hardware manufacturers across the world to deliver complete solutions to end users, making its technology platform independent. Capitalizing on Australia's multilingual population, the Syrinx' team of experts has developed core technology that is language and accent independent, to service the global market. Tools and services developed by Syrinx Applications Engineering Division enable this core technology to be tailored to match any customer's specific application and vocabulary needs.

The Syrinx speech recognition engine operates reliably in noisy environments such as digital mobile phone networks. The outstanding competitive advantages of Syrinx technology have been demonstrated many times in the number of accounts where Syrinx has outperformed U.S. and European competitors.

Syrinx has extended its range of products over the years from highly successful small vocabulary applications (with a maximum of 50 words) to large vocabulary natural language speech recognition solutions. The aim here is to enable callers to interact with the computer as if it were a human operator.

The Natural Language Speech Assistant from Unisys (Blue Bell, PA – 215-986-4011, www.marketplace.unisys.com/nlu) is an advanced speech-application development tool set. Unisys NLSA not only provides application developers with the tools for designing and creating speech applications but also enables application project management, development methodology and testing. Unlike all other natural language technology, Unisys NLSA keeps developers' applications from becoming prematurely obsolete by providing open tools for design and development across multiple platforms and speech recognizers. This allows developers to "snap" speech-recognition engines in and out of their programs based on business requirements or technology improvements. Unisys NLSA has won many industry awards for innovation, including designation as Best Industrial/Professional Product by the American Voice Input/Output Society (AVIOS) and an Editor's Choice Award from C@ll Center Solutions magazine.

Companies can increase employee productivity in answering inquiries and selling sophisticated products like mortgages thanks to NL Assistant, which can analyzes the text according to rules set by NL software owners. The new product contains over 400 grammar rules. It also supports pronouns and sentence fragments. Its dictionary has more than 112,000 language entries including nouns, verbs and question constructions. The company says that for most applications, approximately 40% to 60% requests are for routine calls the NL Assistant can handle, saving agent time and expense.

You can use NL Assistant for customer service, help desk, information access, order entry and marketing response. It also processes Internet and Intranet requests / responses. The Internet version permits a dialogue between the Web site visitor and NL Assistant, thus saving on agent time and hassle in answering e-mails. Another version, NL Mortgage Assistant, has been specifically written and equipped to handle banking/mortgage inquiries. It provides monthly payment quotes for purchasing new homes or mortgage refinancing.

NL Assistant is sold as part of IVR systems made by some of Unisys's growing list of partners. These include Brooktrout, Intel / Parity, Periphonics, Mediasoft Telecom. It is also available separately to call centers that have units from these vendors. NL Assistant comes with NL API, Engine, Resource Manager and Toolkit for $8,000, plus $500/port. You have the choice of creating the application yourself or contracting with Unisys to do it for you. 800-874-8647 (ext.610).

* Personal Assistants. These ASR services are more ambitious. They encompass not only voice-activated dialing, but follow-me service, call screening, appointment reminding, conferencing, transferring, voice mail management, paging, and more. (See individual listings, below). Their voice prompts and answers exhibit distinct personalities.

Wildfire (Lexington, MA – 781-778-1500, www.wildfire.com), the oldest and most famous of the voice-controlled personal assistants, now sells its platform to service providers and has its first telco customers happily deploying the product, while Webley (Deerfield, IL – 888-444-6400, www.webley.com), a rival with a hardened Web component, is still in the service business but has been discussing with several telcos here and abroad to collocate in COs. Webley and Wildfire providers are also long-distance resellers by virtue of their follow-me and 800-number service. General Magic is also searching for network customers for its Portico product.

One obstacle the Webleys, Wildfires and Porticos may have encountered is resistance from telcos who already deploy a voice mail service. The carriers see virtual assistants as a threat to that business.

"The customer needs to understand a simple benefit at a simple price," says Bill Meisel, President of TMA Associates. "Customers still get confused when they get all these features. They feel they're paying for things they're not getting." Lucent Technologies seems to have taken this lesson to heart, offering its new personal assistant product (see more below) as a pick-and-choose collection of services, rather than a whole Wildfire-style enchilada.

Lucent also is trying to make the enrollment process even easier by grabbing VAD entries from the callers themselves. In other words, Harry calls Sally, announces himself, and Lucent's Intelligent Assistant adds Harry's utterance of his own name, attaches it to the Caller ID number, and adds it to Sally's voice-activated Rolodex. General Magic's Portico and Webley, two virtual assistants who still do let customers build directories via web, try to streamline even that process by accepting directories straight out of Outlook, other PIMs, or even a Palm Pilot.

Rich Miner, Wildfire's CTO, expected Wildfire to draw more than half a million customers by the end of 1999. These subscribers are expected to come on board via their Network Wildfire product, sold to wireless and wireline carriers.

"The existing channel is not mass-market," says Miner. "We'll shortly be 99 percent carrier based." Wildfire started in business in 1994 with Wildfire Gold, a DSP-powered version which was originally licensed to one or two authorized service providers per area code. In its Gold incarnation, the service was expensive to provide or scale up. Miner says that they've addressed this issue in their newer host CPU-based products and the Network version.

Network Wildfire's recognition routines run on Pentiums, not special DSP-based CT resource boards (though voice capture still requires the use of telephony boards, from NMS). The automated assistant "listens in" throughout a phone call so it can be pressed into service at the mention of its name: "Wildfire."

Miner says that even though it's constantly on alert, this doesn't overly consume resources – less than one to 15% of even a lowly 200 MHz Pentium. "That recognition costs me around five dollars if a Pentium Pro is $500," he says. "We can get 70 concurrent recognizers running on a Pentium II."

If Wildfire can squeeze this much recognition ability out of even a Pentium II, it's at least partly because she performs constrained, not totally unscripted, word recognition. Users must be cleverly prompted or trained to say "Put me through" or "take a message" in exactly those words. You give Wildfire one piece of information at a time, and she queries you through each piece: Where, when, who, and so on.

To Wildfire's credit, she exudes a warm yet efficient personality. She recognizes frequent callers (through ANI) with a cheerful "Oh, Hi!" She places calls, performs one-number follow-me, whispers the names of waiting callers in your ear, takes and schedules reminders, screens callers, and captures callers' names and phone numbers so subscribers can prioritize messages and return calls with simple voice commands.

She's even been programmed with a sense of humor. Lee Erdman, president of 1-800-REACHME, New York, is an authorized service provider with licenses for area codes in New York City, Phoenix and L.A. He engineered and demoed this little-known facet of her personality in this exchange:

"Wildfire?"--"I'm here!"
"Do me a favor."--"What kind of favor?"
"I'm depressed."--"You know, you deserve a day off!"
"Wildfire?"--"I'm here!"
"Do me a favor."--"What kind of favor?"
"I'm depressed."--"Oh, That's too bad!"
"Wildfire?"--"I'm here!"
"Do me a favor."--"What kind of favor?"
"I'm depressed."--"Great. Now I'm your therapist!"

"Wildfire?"–"I'm here!"

"Do me a favor."–"What kind of favor?"

"I'm depressed."–"You're depressed. I live in a box!"

Network Wildfire is also being pitched to carriers as a modular product whose components can be sold incrementally, increasing revenue and customer loyalty. The basic package is being offered as voice dialing; voice-controlled voice mail with return-number capture; WildWhisper, intelligent call waiting; WildFind, call routing; and (in the next release) WildFax, fax forwarding.

Other Wildfire modules will be sold as extra options: A 150-name voice-driven Rolodex, conferencing, tickler file messages, a PC-based interface to contact and schedule information, and one-number service for voice, fax and e-mail. Also, there is the "Virtual Hallway" – any subscriber on the system can see who else happens to be on the same server and connect with them in the manner of an office intercom.

Most attractive to carriers, Network Wildfire automates the upsell process. Through the Wildfire Automated Marketing Manager, the system might say to a subscriber, "I've identified ten numbers that you call very often. Would you like to be able to call them by name?" Wildfire not only sells, it provisions the feature.

Wildfire has carrier customers willing to announce intention to deploy, including Pac Bell Mobile and Orange Personal Communications Ltd. in the U.K. Miner says that if Pac Bell charges around $10 a month for the service, they will recoup their investment in one year. They are also working with carriers in France and Italy, working on creating personalities and speech rec for these countries.

Upgrades to Wildfire add e-mail reading and the web interface to the service. Subscribers can reply to e-mails with spoken messages, to be e-mailed as RealAudio files. They can also retrieve messages and edit contact information through the Webfire website.

Further ahead, Miner looks at creating an inter-carrier network that will let roaming subscribers retrieve messages and phone calls across multiple Wildfire-equipped carriers. This would work along the same model as cellular carriers' home locator registers, using SS7 and IP signaling to reference subscriber information at home-base servers.

Wildfire's literature tries to assure traditional authorized service providers that they won't be left out in the cold by a carrier-based product. 1-800-REACHME currently sells a full-featured service, with all available bells and whistles, for $29 a month plus usage: 15 cents a minute or less, based on volume, for session time. A toll-free number is another 10 cents a minute, and long distance outbound another 15 cents a minute. The provider absorbs the cost of outbound local calls within the session charge.

Reachme's Lee Erdman says that a lot of his customers are people on the road more than 40% of the time, and a lot are in sports or entertainment fields. He adds that a Wildfire account makes a great and economical replacement for calling cards.

Then there's the matter of calling card replacements. If you use a calling card, you're paying up to 85 cents for the privilege of hearing the bong, and up to 80 cents a minute. That setup charge is applied for every call, even if you just hit the space key between calls. With Wildfire, you can make call after call with no additional setup charges, and you can also receive calls while you're on the line. Add up his inbound, outbound, and session time charges, and you get to forty cents a minute. You can also send alphanumeric pages by speaking each lettter one at a time to Wildfire (or use touchtones), and she sends the message out immediately.

Ultimately, Erdman and his organization would like to be your CLEC. They foresee a day when subscribers will pick up a phone to hear

"What can I do for you?" instead of a dial tone.

Webley was a later (July 1997) contender in the ASR personal services category, and a Dialogic technology-based rival to NMS-based Wildfire. It offers many of the same speech-driven follow-me, voice-dial, routing, screening, scheduler features. Webley is a product of Vail Systems, a CT custom development shop. Vail's principals' CT roots go back to 1990 and their web interface experience dates back to the heyday of the Mosaic browser.

Alex Kurganov, Webley's co-developer and VP of technology, is quick to point out differences between his product and Wildfire. Webley uses VCS / Philips' Purespeech, a natural language product. This means that commands can be recognized with different wordings; users do not have to be trained to use particular phrases. That's a more processor-intensive type of recognition, but Webley saves resources in a different way because the recognition engine is not active for the duration of a call, listening for its name. You wake up Webley by hitting a touch tone. If any key will do, this does not seem like a great inconvenience or a danger while driving.

Webley, as its name implies, is also web-enabled right now. Its web site is up and running. Users can retrieve voice, fax, e-mail, and pager messages on Webley's site as well as over the phone or through their own e-mail clients. They can also use the site to edit their contact lists, schedule and manage live conferences, and view call detail records. E-mail reading through text-to-speech is also part of the system now. So is an audio version of e-mail sending, via RealAudio attachment.

According to Kurganov, users will be able to construct their own lists of phrases to command Webley agents. Speech-activated agents, another one of the Next Big Things in speech-rec services, will go comb the web and third-party data sources to read you stock quotes, sports scores, air fares, and anything else you can think of. Banging this drum loudest was General Magic (see below), with the Portico (formerly called Serengeti) "virtual assistant."

Webley runs on a switchless UNIX platform that connects multiple voice servers over the extended SCSA bus. All resources get allocated on a contention basis, and all lines in this platform, which means to date all Webley customers, are provisioned in one huge hunt group. This means that should one PC-or, more likely, one T1-go down, customers never have to know about it.

"You realize what it takes to provision 800 numbers if they're attached to a particular box. If the PC goes down, you have to reprovision a user to some other box. It's a big deal to switch a customer from one hunt group to another, and PCs only go down every day," according to Kurganov. "We've been more reliable on Pentiums than the telco," he says. Everything is redundant with inexpensive components: the database, the RAID5 systems. The platform is housed in two redundant facilities in Chicago and Deerfield, relying on two different power grids connected on an MFS loop. Dialogic builds the boxes for Webley, with Pentium processors and Antares DSP boards running the PureSpeech recognizer. Webley gets 12 ports of simultaneous recognition out of an Antares board, but remember that speech rec is only going on for a small portion of any call. Kurganov is also considering adding Nuance speech rec, for its ease of scalability.

Another major difference in Webley is that so far, it's only been sold as a service out of the Deerfield and Chicago headquarters. They're charging $14.95 a month; 800 number service is billed at 15 cents a minute, conference calls cost an additional five dollars per conference, e-mail reading is another five dollars a month, and fax retrieval another five.

Webley is certainly entertaining collocation agreements with carriers and Internet Telephony Service Providers (ITSPs). Kurganov says they're now talking to cellular carriers and providers abroad. He proposes trial runs via tie lines from his platform to SS7 or 5ESS switches. Carriers could test run Webley for their subscribers, initially incurring no capital expense, only minutes of usage.

Finally, Webley's personality is male and British. It's supposed to make you think of a loyal and efficient Jeeves-like butler. Compared with Wildfire's friendly and knowing American female, though, some people find him a bit fussy. I feel sure that given the right contract, Alex Kurganov would recast Webley's character in a New York minute.

General Magic (Sunnyvale, CA – 408-774-4000, www.generalmagic.com) has developed its own very ambitious virtual assistant service, Portico (formerly called Serengeti), which has been speech-enabled using Nuance's recognition engine. Portico lets users access their e-mail, voice mail, calendar, address book, news, and stock quote information via any telephone or leading web browser. Portico is speaker independent. You can talk to Portico in normal, conversational English and Portico will understand – no training is necessary.

Once you sign up and your account is activated, you will be assigned a personal, toll-free number that gives you access to the system.

Portico has the intelligence to screen incoming calls and manage your accessibility at any time of day. "Follow me" call routing lets you screen your calls.

If you happen to be a "road warrior" your cellular phone can easily become the best place to access communications and information. Your voice mail, e-mail messages, news and stock quotes will be delivered in Portico's friendly voice. And you can respond to messages with your choice of communication. Portico lets you keep your eyes on the road, while your voice does the dialing.

You can talk your way through a whole week of appointments as you update your address book, calendar and "to do" list. Portico synchronizes with leading desktop information managers like PalmPilot, Microsoft Outlook, Microsoft Schedule+, Lotus Organizer, ACT!, Sidekick, plus most handheld devices using Windows CE.

The Portico demo at 800-774-2229 lets you listen in on a hypothetical user's experience with Portico. The Portico subscription based service offers plans starting at $9.95 a month.

General Magic is also pursing the likeliest adopters by more tightly focussing their marketing efforts and segmenting their MagicTalk applications. They're now supporting and educating their best performing resellers. They continue to sell through carriers and ISPs and there's a market trial underway with Qwest, says GM's Buck Krawczyk. Also, GM is concentrating on a much more specialized application with Intuit's Quicken.com. web site.

Quicken.com, with two million users, is used by customers of the (very web-integrated) Quicken 99 desktop consumer product, although you don't have to run the software to use the service. Quicken.com allows customers to maintain their own centralized accounts at the site, so that updates can be entered from any browser on any PC. It also serves as a gateway to banks, allowing web-based transactions. If you own and use Quicken 99, Quicken.com will download the updates to your hard drive for installation.

General Magic's plan is to telephony-enable all of this through a speech-recognizing MagicTalk interface. This actually sounds plausible, since "road warriors" might want to access their finances, pay bills or purchase stock from their laptop while sitting in, say, a hotel room, or from a phone.

Steve Markman, CEO of General Magic, is hoping the narrowcast application plants a seed. To quote his release statement: "General Magic offers these growing companies an ideal solution and, at the same time, we can benefit from the Web's broad reach, which could contribute to more rapid market acceptance of our voice-enabled services,"

The Lucent Speech Solutions division of Lucent Technologies (Naperville, IL – 630-979-7742, www.lucent.com) is part of Lucent's Switching and Access Systems business unit. Lucent shares a 30-year speech processing history with AT&T that diverged around 1996. The common ancestral Bell Labs ASR is the speech rec behind a lot of currently deployed directory assistance and operator services, now called the Lucent Automatic Speech Recognition (LASR) 3.0 engine.

Lucent currently supplies speech rec technology in North American Spanish (for U.S. and Mexico), Canadian French, Continental French, Italian, German, Dutch and Castillian Spanish.

Their own Lucent Intelligent Assistant (LIA), mentioned previously, is out of market trials and is offered on a service node, equipped with ASR and TTS. In addition to the call management features we've heard about before, it will offer to build name directories in a novel way, grabbing them from caller's own pronunciations of their names and ANI or Caller ID as they announce themselves. Equally important, it will offer its services as a whole or in pieces, and can be priced lower accordingly.

LIA was demoed to Ameritech and in-bound pieces of the service were subsequently packaged into something called Privacy Manager. It's designed to let you eat in peace at dinner time, fending off those pesky telemarketers. If a call comes in without CallerID, Privacy Manager asks the caller to identify himself and records the reply. It then plays the recording to the subscriber, who can decide whether to take the call or let the service turn the unwanted caller away. "Pieces of Intelligent Agent make sense for different users," says Dan Furman, President, Lucent Speech Processing Solutions. "This [Privacy Manager] makes sense for cellular companies, too, because it lets people leave their cell phones on without fear of paying for unwanted callers."

"We're trying to bring the cost of an intelligent agent down to ten dollars a month. We're not aiming this at the road warrior," adds Furman. He says that "less is more" in designing personal assistant-type services, and that psychologists wrote a lot of the code behind Lucent's assistant. "If you want to change your outbound dial list, there's only two commands – add and delete," he notes.

Lucent speech technologies have also been integrated with the Unisys Natural Language Speech Assistant grammar-building and testing tool.

On the hardware side, Lucent's new division is offering a new CompactPCI speech processing board, and ISA speech processing board with 48 MB of memory, a T-1 interface board for up to five T-1s, and an echo cancellation board. An entire Lucent Speech Server, powered with Lucent's own boards, is going to be used by Intellivoice Communications as a speech-enabled IVR platform, and was picked by MovieFone to allow natural language access to its nationwide automated move and theater database.

- **Dictation Services.** The outer limits of telephony-based ASR can be sampled at www.speechmachines.com, where you can demo a service called CyberTranscriber. Let's say I'm stuck in traffic, my deadline for this book is in three days, and a great sentence (this one) comes to mind. I whip out my cell phone, dial into SpeechMachine's 800 number, and speak the words into my phone. By the time I get to work, the text shows up in an e-mail sent to me in the form of a perfectly transcribed Word file attachment.

  This is an impressive case of speaker-independent, continuous-word recognition of unlimited vocabulary, and over the phone, to boot.

Speech Machines (Redwood City, CA – 650-568-1500, www.speech-machines.com) is using a proprietary speech recognition product to do a thorough first pass on recognizing and transcribing the caller/subscriber's digitized speech. However, they then pass the text and voice file on to human editors working on an island off the coast of Scotland, who clean it up, make it as perfect as they can, and ship it back to you as an e-mail, web link or forward it to anyone else you wish.

SpeechMachines' product has roots in the British military. "Our UK offices are behind barbed wire," says Michael von Grey, Speech Machines' president. A technology built for wire-tapping, he points out, is intrinsically optimized for challenging cellular environments.

Computer Telephony magazine's tireless columnist and Executive Editor, Ellen Muraskin, tried the service when it was first deployed, calling in to 1-888-I-SPEAK-IT and thinking out loud in her best Queens English (she's from Queens). She checked into the www.speechmachines.com web site a minute later to check the status of her dictation. It confirmed that her job was queued. About half an hour later, it showed up in her e-mail, letter perfect. Even the punctuation she had forgotten to speak in came out.

Von Grey said that her voice was going over PSTN to the UK, where it was recorded and digitized. A link-up between the recording server and their database confirmed that she's a trial subscriber, and also referenced her speaking attributes to apply the correct acoustic models.

Since Speech Machines had taken care of Muraskin's enrollment, they had chosen a default "U.S. English" to categorize her speech. When you enroll yourself on the web site, you can choose among several U.S. regional accents and Irish, Welsh, Scottish, and other U.K. variants. As soon as CyberTranscriber recognizes the user's PIN, it knows to "listen" with the correct acoustic model and personalized vocabulary.

At the processing center in Scotland, queuing software applied priorities to the job. Some medical clients can wait up to 48 hours, says von Grey. Some business clients must be satisfied in under three.

Users can also e-mail in voice files recorded over PC mics or Voice-It digital recorders.

Speech Machines' recognition software produces a first-draft transcription. Proprietary proofreading software focuses the human editor's efforts on areas of low confidence. Cleaned up, the file is returned to the subscriber via e-mail. In England and the U.S. it can also be routed to distribution lists or third parties. You can also receive a special proprietary audio file and playback program, if you like. This will let you hear your voice recording as you read your transcription.

Von Grey says that his recognizer adapts to each speaker's accent, perfecting its recognition. Users can also make personalized additions to their vocabularies, over the web site. Von Grey says that users who have a lot of unintelligible words are invited to submit these additions via e-mail, so their profiles can be augmented with the new words.

One can think up hundreds of applications for CyberTranscriber. Salespeople submitting call reports while driving from one client to the next. E-mail addicts who want to travel really light. Any business person wanting to make hay while they're stuck in traffic. Indeed, von Grey admits that the peak periods in his service center correspond exactly to drive times in the UK and the US coasts.

Speech Machines is only being offered as a service right now, and as an OEM product to be rebranded by carriers. Since it's Internet-based, "we can add the necessary resources in whatever place is feasible," notes von Grey. They're looking in India now to expand their pool of English-editing talent.

Their recording servers are "dozens" of NT machines, with 300-MHz Pentium processors, ISDN lines coming in, Acculab E-1 interfaces and Brooktrout / Rhetorex cards for voice capture. Their server database is accessed via SQL. They charge end users a one-time registration fee of $29.95, and $0.015 per word with a monthly minimum charge of $7.50, all of which compares favorably with traditional transcription services that work from tapes.

Speech Machines is not afraid of competition. Grey says that there's a "fearsome" range of requirements, in addition to good accuracy, that they've had a head start in mastering: CT integration, speech rec, e-commerce, credit-card validation, confidentiality, "Plus, we're running a call center."

There's a great web interface, too. Speech Machines has actually deployed this service for UK-based Cellnet in September of 1999, the cellular network 60% owned by British Telecom. Cellnet almost immediately acquired over 7,000 customers. In so doing, they appear to be the first carrier to be giving more than lip service to the idea of dictated e-mail. Add text-to-speech and voila, you have a truly speech-enabled messaging system, not only reading, but now sending e-mails as well.

Speech Machines is still in the process of selling North American telcos on the service, and they've teamed up with Webley, the virtual assistant, to add e-mail sending to his mix of services..

They've also launched a data-collection service, called TalkForm. Built upon CyberTranscriber, it recognizes all types of data fields that might be spoken in response to automated phone surveys or any other form-filling application. Think insurance claim forms and customer orders. They design and run the IVR, and ship the data over the Internet into a client's database.

TalkForms first announced deployment is with the Visiting Nurse Service of New York, whose staff of 2,500 field nurses, rehab therapists and social workers will dial in to dictate vital signs, clinical progress, and other data. All transcribed text, including free-form entries, is to be delivered via encrypted and password-protected e-mail, and automatically slotted into VNSNY's mainframe database.

Finally, TalkForms is attacking the transcription market where a lot of it lives: in the medical field, by offering to price their system under traditional transcription houses. They've licensed some Lernout & Hauspie technology for the medical market, as well.

- **Straight IVR replacement.** Despite the promise of speech recognition software to replace touchtones, there has been relatively slow acceptance among call centers and their clients. Rex Stringham, president of the Enterprise Integration Group (EIG, San Ramon, CA www), says that voice-driven applications have been greatly refined and improved during the last half of the 1990s, but he doesn't think that speech rec is quite ready to become a standard component of IVR systems. EIG is a professional services firm that works mostly with clients that have IVR systems, and Stringham has observed how these clients have used speech recognition with their IVR systems.

"Speech recognition technology is now robust, but companies implementing it continue to have problems with deployment," explains Stringham. "The problems are usually blamed on the technology itself, but in most cases it's the design of the IVR user interface."

EIG's research suggests that many callers encountering speech rec-enabled IVR systems still follow the same menu-driven approaches as touchtone interfaces. Although callers can make verbal selections rather than inputting touchtones, they still must first endure listening to a list of options.

Stringham asserts that applying menus to speech recognition applications prevents them from realizing their full potential. "Some engineers use this as push-button systems, but you need to take into account that it's a spoken system. Applying [touchtone] thinking to speech causes problems."

Designers need to not only transmit and interpret a spoken message from a person to an IVR system, but also take into account the person's most probable psychological state.

"You want to design an application so that when a caller calls, it can respond in a way that makes sense," contends Bruce Balentine, EIG's director of speech recognition. "That's harder to do than you might think, given how person-to-person conversations can digress."

Balentine cites colloquialisms and some callers' habits of answering questions with questions as examples of human quirks that can baffle even the best speech recognition software. His concern is that when the IVR system doesn't know what to say next, neither does the person on the other end of the phone line. The caller either hangs up or the speech recognition system refers the caller back to a series of menu options.

Yet Balentine is still optimistic in his assessment of speech recognition applications and he hints at what could be done to make them more efficient.

He's only half-joking when he says that: "application designers need to overcome their bias about making their products more intelligent, and focus more on making them well-behaved."

Although the latest speech recognition software can identify and interpret words accurately, it still needs to stay focused on the task at hand – to be not just capable of speech recognition, but of "speech acts". Whenever an application is confused by human idioms or eccentricities, it needs to keep requesting the information it needs as often as necessary. If the IVR system remains consistent in pursuing a specific objective, the caller who uses it can follow its lead and get what he wants more quickly.

UltimateRECO from Alternate Access (Raleigh, NC – 919-831-1860, www.alternateaccess.com) is a product that allows callers to interact with IBM Corepoint Voice Response (formerly DirectTalk for AIX) applications through a Voice User Interface (VUI). The product is multi-lingual and has speaker independent recognition capabilities that allows continuous numbers and hundreds of command words to be used to meet a wide range of requirements. Speech Recognition gives callers an easier, faster and more natural way to input information by telephone, and helps the company utilize its customer service staff more efficiently.

Alternate Access has also deployed Speech Recognition products in companies in banking, insurance, utilities, transportation, telecommunications and government in the U.S., Europe and other parts of the world.

Natural Speech Communication Ltd. (Rishon Lezion, Israel – +972-3-9519779, www), known as NSC, is a six-year-old company offering high-density and cost-competitive recognition solutions in the form of chip sets, PC boards, servers and stand-alone software packages. Their NCS Engine chip set implements up to eight channels of speaker-independent and dependent continuous speech rec and key word spotting, barge-in, and dynamic grammar. Their NSC Board, in ISA, PCI and embedded PC104 form factors, takes on up to four chip sets, for up to 32 channels.

NSC claims a special expertise in performing under severe noise conditions, extreme channel distortions, and cross-media inputs.

Nortel (Montreal – 514-769-OPEN, www.nortel.com) demoed its OpenSpeech recognizer Developer Kit at CT Expo in March 1999, but the kit uses an engine whose history goes all the way back to 1989, when Nortel launched automated third-party call billing and directory assistance. It contains a grammar editor to tweak prepackaged grammars and create new ones, and a phonetic (type-it-in) vocabulary builder. It also comes with performance evaluation tools for fine tuning grammars.

The OpenSpeech developer kit also helps developers optimize the use of telephony ports and speech rec channels. Most fascinating: Unspecified grammars can be dynamically added through the recognizer's "experience." In other words, an unexpected phrase can be phonetically transcribed and then recognized when said by others.

OpenSpeech comes with connected-digit recognition and prepackaged yes/no recognition. It has an open API for C, C++, ActiveX control, and popular IVR platforms, and runs on Windows NT and SCO Unix.

* Speech-controlled browsers. Wouldn't it be great if you could surf the Web simply by talking to your PC? If you wanted to go to your Home Page, you'd say "Go Home". and the page would pop up on your screen. If you wanted to take a look at your list of favorites sites, you'd say "Show Favorites." To move down a page, you'd say "Scroll Down." You'd simply say it, and your computer would do it, like Robbie the Robot or HAL from 2001. Thanks to ASR, you can do this and more.

Conversa Web 3.0 from Conversa (Redmond, WA – 425-558-7554, www.conversa.com) Conversa Web is a voice-enabled, hands-free Web browser that lets people explore any Web site using their natural speaking voice. There's no need to train the system to recognize a user's voice and no need for instruction – the user simply says the links or voiceable icons (called Saycons). Conversa's proprietary, continuous speech engine, Conversa Advanced Symbolic Speech Interpreter (CASSI), is speaker-independent and accommodates an unlimited vocabulary. It can recognize almost every English-speaking person, even those with accents.

The system works on top of Internet Explorer 4.0 and higher but it has its own user interface, so your standard IE window configuration is untouched. The system automatically keeps track of all the links in the Web page you are visiting. Most links can be activated simply by uttering the first word or two in the link.

Common commands such as "Refresh Page" or "Search the Web" are included in the Saycon toolbar. Numbered Saycons are provided for duplicate or hard-to-pronounce links. Users can add new links or sites by saying "Add to Links toolbar" or "Add to Favorites"

Complaints about previous versions of Conversa Web centered on the

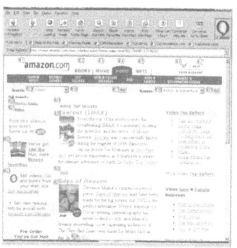

Amazon.com's Video Store gets the Conversa Web 3.0 treatment. Notice the numbered "Saycons" next to JPEG and GIF images, textboxes, dropdown menus, and hard-to-pronounce links. To activate them, you simply utter "Number 1" or whatever, instead of pronouncing the actual link.

fact that it did not provide a way to dictate into fields or dictate URL's, or keep track of buttons or drop lists in the Web pages. There were also some problems with framed pages. Version 3, which was released in May 1999, has taken care of these problems. You now use a military alphabet to voice-type URLs or search terms, and an AutoComplete feature enters previously entered addresses after the first few letters. You can also "press" buttons via voice. Images, global commands, Web addresses, drop-down menus You'll find Saycons on the menu bar, plus numbered Saycons next to JPEG and GIF images, textboxes, dropdown menus, radio buttons, and even next to hard-to-pronounce links. That way you can say "Number 3" instead of "Photomicrography." The commands "Show Favorites" and "Show history List" are also voice-activated.

Best of all, they've included a conversational ability so that you can enjoy a two-way dialog with sites that content providers have "conversationalized." This is a Web site that talks to you in an interactive fashion. Say, for example, you are preparing for a trip and want to check out the weather. You would go to a weather site that has been "conversationalized" and ask for the information. You'll hear the latest temperature and weather conditions. You can easily recognize a conversationalized site because it will be talking back to you (although your PC also talks to you when responding to your navigational requests, it is not considered a conversation). Users can converse with selected Web content at www.conversations.com,

Conversa Web does not recognize Java programs unless they employ the Java Speech API.

Conversa Web is currently available in English and German, with other languages in development. Conversa Web's retail price is $59.95, though discounts and rebates are usually available. Available at CompUSA, Fry's Electronics, Best Buy, Office Max, J&R, MicroCenter and Software City.

Conversa's competitor, the FreeSpeech Browser from Philips Speech Processing (Cambridge, MA – 617-441-0000, www. speech.philips.com) is also a speech recognition product for navigating the Web on the desktop with one's voice using Microsoft Internet Explorer 4.01 (with the latest Service Pack) or 5.0. It uses the Philips SpeechMike microphone and FreeSpeech 2000 natural language speech recognition software to provide continuous-speech recognition.

With the Philips system, if a specific link cannot be identified from your voice commands, you can still "voice click" on it by calling out a number displayed next to the link.

The Philips FreeSpeech Browser doesn't require training time and takes about 45 seconds to learn your speech patterns.

Philips' FreeSpeech Browser is as inexpensive as can be – it's a free 10MB download from the Philips website and from Internet e-commerce sites and portals.

Another "voice browser" plug-in achieves speech recognition over IP. A spin-off of Philips Electronics called Spridge, Inc. has launched a free web service called mySpeech (www.mySpeech.com) that enables users to add a speech interface to their sites, typically home pages.

After downloading and customizing the mySpeech button, publishers as well as visitors of these speech-enabled sites can navigate by speaking words starting with "my" such as "myPictures" or "myFriends", etc. Ten of these "my" words can be customized during registration. Users specify a URL for each word, so that when somebody speaks into their microphone (for example "myFamily") they are taken to the page that gives the requested information.

Also, the mySpeech button offers a distributed portal functionality. By speaking generic words such as "Books", "Travel", or "Music" users are

taken to category pages within the mySpeech domain, where they have access to additional speech-controlled applications e.g. for weather forecasts, travel information or finding concert tickets. More of these services are going to be added over time.

Currently, mySpeech works on Windows PCs having either Internet Explorer 4 and higher, or Navigator 4 and higher. It's also available for Macintosh equipment (e.g. iMac) running Navigator 4 and higher. Aside from a sound card and a microphone, no additional hardware of any kind is necessary. There is a special section within the mySpeech site, the "Audio Setup Page", that assists users in optimizing their sound system so that it works well with mySpeech.

When first-time visitors come to www.mySpeech.com, they are offered to install a small plugin (160K download) on their machine. This is all quite user-friendly – except for pressing an "OK" button, no interaction is needed. Once the plugin is installed, users now have the mySpeech button available, which serves as a push-to-talk button for speech input. mySpeech is speaker-independent and works immediately without having to go through any training procedure.

Typically the recognition is completed in less than real time because of the optimized server architecture. The plugin does some time critical speech analysis and compression, and the results are sent to the web server where a more conventional Philips speech rec algorithm does the final analysis. The result of the recognition can be used to load a new HTML page on the client browser, interact with the current page through JavaScript (or any other scripting language), or perform any type of action.

- **Toolkits for creating speech-enhanced websites that can be navigated from a PC or via a phone.** Most people find frustrating inputting the number of touchtones necessary to operate many existing IVR automated telephone services, such as those allowing users to retrieve information such as bank balances, flight schedules, and movie show times from any phone. With the amazing growth of the Internet and World Wide Web, providers of traditional IVR phone services now had to come up a way to server a new set of customers accessing information and services through the web. While in most cases customers still access automated services through the phone, providers are finding it easier to build new services that exploit the power of web technology.

Content providers today are thus faced with the dilemma of whether they should provide services only on the web, or should they also provide telephone access to their applications? There are always hardware and integration costs involved in deploying a telephone service. Typically, a provider must buy telephony hardware (CT resource boards, fault resilient computers, etc), develop an application in a proprietary application programming interface (or with an application generator), and integrate the application with existing databases. It would be great if a content provider could develop a voice application using many of the same web development tools Its programmers already are familiar with, publish the application on an existing Web server, and arrange with a service provider to handle interpretation of a special representation of the application.

In fact, there is a way to do a "special representation" of an application that provides the best of both worlds, IVR telephony and the web. There has been growing interest in the concept of using an HTML or XML-like "markup language" to define voice access to web-based applications. For several years Netphonic had a product known as Web-on-Call that used an extended HTML and software server to provide telephone access to web services; in 1998, General Magic acquired Netphonic to support Web access for phone customers. In October 1998, the World Wide Web Consortium (W3C) sponsored a workshop on Voice

Browsers. A number of leading companies, including AT&T, IBM, Lucent, Microsoft, Motorola, and Sun, participated.

Most recently, IBM has announced SpeechML, which provides a markup language for speech interfaces to Web pages; the current version provides a speech interface for desktop PC browsers.

Microsoft is also working on an HTML Telephony Engine, which is a "voice IVR" markup language.

Then there's the Wireless Application Protocol (WAP) which may actually be specifically the protocol a cell phone uses to talk to "content" (something like a server attached to the Internet) which is something other than voice content.

Some systems, such as Vocalis' SpeecHTML, use a subset of HTML, together with a fixed set of interaction policies, to provide interactive voice services. At one extreme, Vocalis (Cambridge, UK – +44-0-1223-846177, www.vocalis.com) wants to provide a service that requires the site owner to do nothing but hand them a URL: in return, he or she gets an 800 number. No changes are made to the site, they say.

Callers to this number are read aloud page contents and hyperlinks via TTS, and prompted for navigation commands by the caller speaking one of the links. A Vocalis SpeecHTML gateway does the talking and converts the HTML links, and presumably CGI scripting, into the equivalent of IVR branches and database access. It's also listening for the spoken choices.

Obviously, some sites are going to work better orally than others; art gallery sites are not going to work over the phone. (Joke heard at CT Expo 99 in Los Angeles: "Welcome to the Louvre's web site. Please say the name of the painting you wish to hear." [User says, 'Mona Lisa.'] "Thank you, you have chosen the Mona Lisa. Blue dot, red dot, red dot, brown dot...")

Stock quote sites, on the other hand, might do very well and Vocalis' SpeecHTML technology could be quite successful. ISPs will eventually buy their own Vocalis gateways, offering their web tenants both a reach into the eyes of PC owners and the ears of worldwide phone users. Since web browsing customers are not used to paying long distance tolls, this sounds like a good push for VoIP traffic, as well.

Proponents of competing markup languages such as the Voice eXtensible markup Language (VxML) described below (interestingly, Vocalis is listed among them) seem to be taking a more divergent IVR path, using a site's database and integral purpose, if not its preexisting HTML. Instead of trying to read an existing website, VxML allows a designer to add a script so that if someone is coming in through a phone portal, it can carry on a dialog with that site. It makes the phone a parallel part of e-commerce, and it makes speech a parallel way to give you customer service.

Indeed, VxML is one of the larger efforts to "voice enable" websites. VxML is actually a combination of two speech-enabled web projects: VoiceXML from AT&T Bell Labs and VoxML from Motorola, both of which are used to represent a caller's many choices and input options when interacting with an IVR or other voice automation system.

VoiceXML began as a research project called PhoneWeb at AT&T Bell Laboratories, a company since separated into AT&T Labs and Lucent Technologies' Bell Labs. Chris Ramming, David Ladd, and Ken Rehor worked on what would become the phone Web service concept. Since then, Rehor went to Lucent and Ladd to Motorola. After the AT&T / Lucent split, both companies pursued development of independent versions of a phone markup language.

The three friends and colleagues kept in touch. While AT&T was building the VoiceXML phone markup language and launching applications, Lucent worked on a similar project known as TelePortal. TelePortal's research focus veered off into service creation and natural language applications.

AT&T Labs continued to build VoiceXML and used it to construct various types of applications, ranging from call center-style services to consumer telephone services that use a visual Web site for customers to configure and administer their telephone features. AT&T's intent has been twofold. First, it wanted to forge a new way for its business clients to construct call center applications with AT&T-provided network call handling. Second, AT&T wanted a new way to build and quickly deploy advanced consumer telephone services, and in particular define new ways in which third parties could participate in the creation of new consumer services.

In the meantime, Motorola was working on it's own telephony markup approach as a way to provide mobile users with up-to-the-minute information and interactions. Continuing the association of mobile phones with productivity, Motorola's efforts focused on hands-free access, which led to an emphasis on speech recognition rather than touchtones as an input mechanism. Also, by starting later, Motorola was able to base its language on the recently-developed XML framework. This all led to Motorola's October 1998 announcement of the VoxML language.

AT&T, Lucent and Motorola finally decided to combine the best parts of their separate VoiceXML and VoxML development efforts, creating the VxML specification.

Ramming, Nils Klarlund, and Andrew Forrest, of AT&T Labs Research, worked with Lucent and Motorola to launch the VxML Forum. Seventeen other leading companies from the speech, Internet, and communications markets have agreed to support the forum and play an active role. These include 3Com Corporation, IBM, and Hewlett-Packard.

Giant IBM is among the top-billed companies on the VxML standards announcement. These days, IBM is excited about transactional access to all kinds of information with any hand-held device. That information could include mother lodes of customer data residing in IBM mainframes, suddenly accessible for customer self-service through VxML-enabled websites – if the enterprise wants it to be so available.

For the meantime, IBM has created its own speech markup language and made it available for download on www.alphaworks.ibm.com for developer use and feedback. Whatever the ultimate ML standard, IBM's hope is to implement it on its own web server platform, Websphere, containing IBM's own NLU conversational engine, the business logic and security features, and the access to (perhaps mainframe) databases that compose the other important pieces of e-commerce. Conversational engines would be integrated into the web servers, and IBM will provide tools to allow developers to build the conversational, NLU piece. These engines would be accessible to the application writer through the markup language calls and API calls, such as Java's speech APIs.

Nuance has also lent its name to the VxML Forum announcement. "There's a lot of interest in speech-in, web out," says John Shea, director of Nuance's product marketing. "It seemed like something-speech-ML was being developed every month. We were afraid that customers would hesitate. There's 65 million connected PCs and something like 1 billion phones. We don't care who wins the standards war, we just need our speech objects to be able to pull information from the web and to enable a speech command there," he says.

Philips is another member of the new VxML consortium. Says Bruce Cooperman, Philips' senior VP of speech technologies: "From the vendors' standpoint, Lucent wants to sell [voice-web server] gateways, AT&T wants to sell network services, and Motorola wants to sell fancy phones. We have a vested interest in having it all speech-enabled and to be a participant."

As currently specified, VxML is basically HTML for voice. It works like this: When a user calls a special telephone number, the call is routed to a network resource called a voice response unit (VRU), which launches a Web browser, finds and interprets a document written in VxML, and responds to the caller. Users interact with the Web either by voice or DTMF capabilities on their phone.

Also, today there are hefty hardware and integration costs to companies that want to provide automated telephone services, such as bank-by-phone systems. VxML would eliminate this. Customers don't have to buy the equipment, since it can become an enhanced network service.

Because established Web technologies are used, the integration with back-end databases can be shared with the HTML application. Because the development of the application is separated from its deployment, the content provider will have much more flexibility in deploying the application. For example, one option would be to contract with a service provider until the voice application had proven its worth, and later purchase a VxML platform to own and operate.

Consumers also can realize many advantages from a standardized voice markup language. First, the ease of deploying new voice applications with the markup approach promises to expand the range of applications accessible from the telephone. Furthermore, once a large number of services are available via the Internet, it becomes possible to interact with several unrelated services during a single phone call. In essence, individual consumers could have "Voice ISP" service in addition to, or included in, their traditional data ISP services.

VxML will include conventional telephony input, output and call control features, including: touchtone input, automatic speech recognition support, audio recording (e.g., for voice mail), the ability to play recordings (such as WAV files), speech synthesis from plain or annotated text, call transfer, conferencing, and other advanced call management features. As an XML-based definition with an HTML-like appearance, VxML will be easy to learn for experienced Web content programmers and amenable to easy processing by tools to support desktop development of VoiceXML Web applications.

AT&T and its business customers have built several examples of typical automated business applications: customer surveys, telephone e-commerce services, product promotion, recipe browsing and delivery, frequently asked question services. AT&T has also built a full consumer telephone service based on its work, which included contributions from business partners for weather, news, and stock market data. AT&T has also constructed many other prototype consumer services such as prepaid calling card and universal messaging.

Lucent has demonstrated the use of the markup language approach to create banking and other e-commerce services, a variety of information retrieval services, and interactive communications services.

Motorola has demonstrated a collection of mobile-productivity applications from three early adopters of the technology: BizTravel.com, CBS Marketwatch, and The Weather Channel. Other active areas of application development include e-commerce, consumer self-service, local events information, and corporate intranet information access.

Recommended site: www.vxmlforum.org

A marketing group from Nuance (Menlo Park, CA – 650-847-0000, www.nuance.com) came to New York on October 5th, 1999 and rented some elaborate party and theater space in the Chelsea district of Manhattan, near *Computer Telephony* magazine's neck of Silicon Alley, to make a big splash for the unveiling of their entries in the VxML / VoxML / VoiceXML, speech-enabled web sweepstakes.

The products being demoed live were Voyager – a speech-enabled "phone browser," if you will, and V-Builder, a tool that developers can use

to build voice "sites" out of preexisting screen-oriented web content. CEO Ronald Croen set the stage, introducing an enactment of a hypothetical dual-career, two-kid couple with typically hectic work schedules. Using her hotel room phone, the wife dialed Voyager, instructed it to "browse" to her address book site at excite.com, using it as her personal dialing directory to dial her office and hear her messages. Upon learning that she was urgently needed in Atlanta, she told Voyager to "browse" to travelocity.com, and booked a flight. Since her credit card billing information was already stored on Voyager's own server, she needed only to verify her identity, through Nuance's own voice verification technology, to make the purchase. She verbally hit the "back" button to return to her address book and voice-dialed her husband.

Voyager, playing what appeared to be a Webley / Wildfire-like role of virtual assistant, answered to its spoken name and obeyed commands that follow the browser metaphor: "go back" to previous sites, "bookmark" new pages, and "read it," playing e-mail messages through the mytalk.com free e-mail web site. E-mail was read using Accuvoice text-to-speech. Voyager's server, to be deployed in the carrier's or ISP's network, manages navigation and call control between locations on the "Voice Web" and enables personalization of the interface through stored user profiles and voice prints.

The demo did a convincing job of proving how visual-web options could be effectively converted into aural ones, prompting the user with, "Say 'next' when you're ready for the next step," and "I'll dial Paul unless you say cancel." It also allowed the "wife" to enter a voice site where she was able to check on the status of an order, using natural language queries. During the husband's turn, we saw him use Voyager to browse to a golf-course voice site, book a tee-time, and then get restaurants out of a voice Yellow Pages. He also got step-by-step, read-aloud driving directions out of Mapquest's voice-site equivalent. "Go next" comes in very handy there.

Then Marketing VP Steve Ehrlich took the mike to introduce V-Builder. V-Builder, a tool that empowers developers to create voice interfaces to Web content, without special skills and without modifying existing Web pages. V-Builder generates voice pages that are accessed from Voyager on the Voice Web, or as independent applications. These voice pages are optimized for phone users and created using industry standard SpeechObjects and the VoiceXML or VoxML voice mark-up languages.

During Ehrlich's demonstration, the screen showed an embedded browser opened to a preexisting travel site on the right-hand window, a voice page assembly palette on the left, and a "rendering" of the site's HTML, in hierarchical view, in the middle. The idea was to click on each HTML piece, input by input and output by output, and drag it to the voice component side, where V-Builder suggested the best Nuance SpeechObject component for each task. SpeechObjects, of course, are the customizable components (in Java and soon, ActiveX) that Nuance packages to complete discrete recognition tasks such as currencies, dates and destinations. The database access and business logic remain accessed from within the HTTP server. V-Builder, an integrated development environment, comes with prompt recording and grammar-tuning tools and can also be used to quickly develop voice interfaces for other applications for access over the phone.

Both Nuance Voyager and V-Builder take advantage of Nuance's core speech recognition and voice authentication software.

Nuance is looking to build up a network of developers, and will offer its V-Builder graphical tool free to members. Again, Nuance's target customer is an ISP or a carrier looking to supply enhanced service; or perhaps web content providers looking to reach the PC-less.

So, it seems to us, VxML proposes to do what speech-enabled IVR was always meant to do, but it leverages the enormous effort being spent on web development, in effect making an IVR developer out of your web master. And it provides a new market for the speech technology vendor, whose TTS and ASR gateway must provide the ears and voice.

Neither site provider nor browser need have the ASR hardware, with the gateway sitting in the wireline or wireless telco's central office, or the ISP's NOC.

Somewhere in the midst of these various approaches is Microsoft with its HMTL Telephony Engine, first pitched as "Web-based IVR" technology at CT Expo 1999 with about nine out of its 36 partner exhibitors. It's a run-time engine to run on top of Windows 2000, and again, the idea is to use web authoring to do IVR. Microsoft's concept is different, demanding that you to use the same site for web site and TUI access. It's not as independent of visual representation, with the web-based IVR employing HTML tags to voice-enable site components.

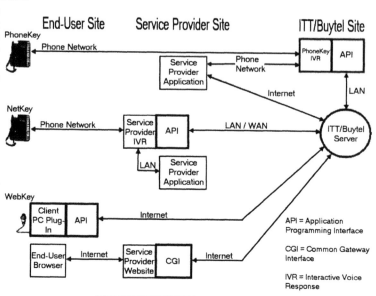

In October of 1999, Microsoft also acquired Entropic Inc., a worldwide leader of software and toolkits for speech recognition. Entropic's proven expertise in speech recognition, dialogue development and telephony integration technologies will complement Microsoft's ongoing efforts to facilitate speech-enabled applications, promote speech technologies and specifically advance server-based telephony solutions

Entropic's engineering team will help Microsoft enhance its speech application programming interface (SAPI), a reliable, open set of speech APIs for both speech engine and application developers. With SAPI, developers are able to use a rich programming model that utilizes COM and SOAP to more easily develop speech-enabled applications, while customers benefit from a broad choice of speech-recognition and synthesis engines provided by various third-party speech vendors.

Entropic is best known for its industry-leading speech R&D software toolkits, including ESPS/waves+ and HTK. "We're excited that Microsoft shares our vision for providing telephony access to speech-enable the Web and make it ubiquitous for all types of users," said John Shore, founder and vice chairman of Entropic.

The Entropic acquisition is part of Microsoft's long-term commitment to delivering speech technology in Microsoft products. Over the past several years, Microsoft has made significant investments in building world-class development teams, opening research efforts worldwide, and working with other leading companies such as Lernout & Hauspie.

- **Speaker verification / identification.** One of the hottest ASR topics these days seems to be biometrics, specifically speaker verification. The idea is that a person's voice is as unique as a fingerprint. Uses: Toll call and cellular fraud protection, financial transactions over the phone, or even ATM withdrawals. With such a system nothing transpires unless a stored speech sample matches a pattern in your spoken response. Companies will be successful that not only offer good speech recognition but good voice verification as well.

T-Netix, Inc. (Englewood, CO – 303-790-9111, www.T-NETIX.com), founded in 1997, is a provider of fraud prevention technologies. They offer the SpeakEZ Voice Print technology which determines a match between a person's spoken word and a digitally stored "voice print"

SpeakerKey Service Options. Service providers utilize SpeakerKey as a "gateway" to authenticate end users (clients or subscribers) before connecting them to information and/or services provided by the service provider's system. Elimination of "front-end" authentication by operators reduces user access time to the service provider's system and increases overall productivity of the service provider's system.

The end user starts with either a toll-free phone call or by accessing a designated Internet site. Phone calls are answered with a customized greeting. Internet connections are answered with a customized screen display.

The end user is then asked to provide a verbal input, via either a phone handset or a microphone plugged into the user's own computer terminal. The verbal input can be as simple as the user's spoken account number. This is called Account Number Verification (ANV), a user identification method perfected by ITT and Buytel. Alternatively, the verbal input can be repetition of digits or phrases prompted by SpeakerKey.

For the ANV method, Buytel's proprietary state-of-the-art enhanced speech recognition and database access software recognizes the user's spoken account number and retrieves a previously stored reference voiceprint for the account number designated. If the ANV method is not used, SpeakerKey requests the user to enter his/her account number via verbal input, phone keypad, or computer keyboard. SpeakerKey then retrieves the voiceprint for the account number entered.

Next, ITT's world-renowned SpeakerKey completes the identification of the user, by comparing the user's voice sample with the reference voiceprint for the designated account number. If there is a voice match, the service provider's system is notified and the appropriate actions are taken (e.g. the user is connected to a customer service representative (CSR); input data is taken from the user; etc.)

If the voices do not match, the user is either asked to try again or is denied access to the service provider's system. At the option of the service provider, users whose voices do not match can be switched to an operator for further action. In addition to confirming user identity, some service provider systems also use SpeakerKey to offer spoken menu selection choices, which are sent to the service provider along with confirmation of user identity.

T-Netix and Peak Network Communications (www) in turn have formed a new joint venture to develop Internet security solutions based on voice recognition technology. The new entity, Sentry Systems (Englewood, CO – 303-792-3726, www.sentry-systems.com), develops speech rec products based on T-Netix's SpeakEZ Voice Print, VeriNet WEB technologies and Peak's Web server security software. Sentry's wares will add a layer of security more specific than things such as firewalls or passwords. The first product to appear from Sentry was VoiceKey, a software product that secures enterprise level network communications and transactions by verifying individual's identities with attributes of their voice.

VoiceKey is designed to precisely verify the identity of any Internet user through the use of a personally-selected voice password – any spoken password in any language. This exact user verification will secure Internet transactions and access to network-delivered information or resources. Sentry's product uses existing hardware – the user's voice password is spoken directly into the desktop PC's standard microphone and is fully functional using the most commonly available sound cards.

Next, the SpeakerKey Services product family from ITT Industries and Buytel, Ltd. (Ft Wayne, IN – 219-451-6321, www.speakerkey.com) provide integrated speech-based user access and identification, for phone-based and Internet-based services as well as support functions, such as text-to-speech and database imaging.

SpeakerKey technolgoy allows call center operators and other service providers to do away with PINs, passwords, "mother's maiden name?" and similar archaic, time-consuming, costly, and inaccurate user identification techniques. It supports existing phone and Internet interfaces, new and legacy IVR systems, and real-time and archival reporting requirements. It eliminates capital investment requirements, development effort, and risk of obsolescence.

SpeakerKey allows service providers to pay for only the services used, with no cost for unused spare system capacity beyond immediate needs.

SpeakerKey technology is available in three basic configurations: PhoneKey (for the phone), NetKey (runs on the LAN), and WebKey (Internet-based)

The heart of SpeakerKey is the SpeakerKey server. The SpeakerKey server is a triply redundant system (additional redundancy in some areas) with an HP fiber channel RAID for voiceprint storage, multiple 8-way Xeon processors, and redundant enhanced recognition and verification software, database software, and other system software. The system is configured as multiple, completely independent, interconnected, mirrored systems, located in different physical locations for maximum operational integrity.

External communications (all locations) include multiple telephone carriers and Internet gateways, utilizing multiple, redundant 100 Mbps copper, fiber-optic, and microwave services.

Current system capacity (expandable) provides for 8 GB of main memory and storage for 64 million voiceprints (at approximately 10 KB each). There is an average of 0.5 second processing time per end-user verification.

The SpeakerKey server can accept voice information over the phone network, a local area network, or the Internet, and can respond, via either phone network or the Internet, with end-user account information, end-user identity verification decisions and other information, end-user Caller ID information, data capture (customer menu choices and/or other information, as appropriate), call forwarding of original end-user phone call (if appropriate), real-time and archival reports of transaction data and system decisions.

For PhoneKey and WebKey, the SpeakerKey servers are located at ITT/Buytel facilities. For NetKey, there is an on-site NetKey server, a smaller version of the full scale ITT/Buytel SpeakerKey server. The NetKey server may be linked, via Internet, to the SpeakerKey server at an ITT/Buytel site, to provide added redundancy and peak load overflow protection.

Keyware Technologies (Woburn, MA – 617-933-1311, www.keywareusa.com) offers an interesting approach to remote authentication with their Layered Biometric Verification (LBV) technology. They have placed their voice verification algorithms into a network server that is built to accommodate additional biometric tests (for fingerprints or the face, for example), depending on their clients' needs. The server checks previously stored voice-prints against a voice password that is sent by an LBV-compliant client application.

With Keyware's LBV technology, voice verification is a two step process. First, you enroll by repeating a passphrase three times. The passphrase should be at least two seconds in length. Good passphrase examples are "My voice is my   password" or "Keyware Technologies" because they contain strong vowel sounds. Depending on the application, the user's voice profile, called a "voice print", is recorded into a database or onto a smart card.

To gain access to data or a facility, the user's spoken password is matched against the database voice print. The verification takes place instantaneously and the user is allowed or denied access. The digitized voiceprint is then stored on an LBV server but can also be kept on a local PC, depending on which version of their software you have. Then, when an application needs to authenticate the identity of someone initiating a transaction, their software translates the voice into bits, encrypts it, and sends it to an LBV server for verification. Users are then either granted access or prevented from continuing the transaction.

Keyware sells a software development kit, VoiceGuardian, which allows developers to add ActiveX controls into apps that need voice security, including Web-based ones. Intended for applications that require more than just voice verification, the LBV server is sold separately and communicates with VoiceGuardian.

VoiceGuardian is capable of language independent voice verification using Lernout & Hauspie technology, can be text-dependent with user-selected passphrase, is available in microphone and telephone versions, has monitoring and uage reportage ability, has enhanced front end signal processing to ensure accurate recognition in case of low bandwidth signals, has small data memory requirements (3KB compressed for voice prints for standard storage, 1.8KB for low storage such as smartcards), and it can do a signal quality check, giving user feedback on signal overload, bad signal-to-noise ratio and insufficient loudness.

- **Voice Data Mining and Indexing Recorded Speech.**

First VAD was The Next Big Thing. Then voice biometric identification was going to be a bigger Next Big Thing. Then personal assistants were going to be a yet bigger Next Big Thing. Finally, VxML and its brethren were to be the biggest Next Big Thing of them all.

I thought we had just about run out of Next Big Things until the following application made its appearance in February 1999. It's not a real-time application and it's not a network service, but it uses speech rec and it sounds like the answer to a real problem: Zeroing in on the word or the phrase you need to hear that's buried in hours of recorded speech, the kind of voice traffic siphoned off from call centers and stored away on tape of disk.

That's the idea behind Dragon Systems' (Newton, MA – 617-965-5200, www.dragonsys.com) AudioMining technology. It uses Dragon's speech recognition algorithms to convert audio data into searchable text, which is in turn indexed to locations on the recorded medium. Large technical support centers, help desks, conferencing service bureaus, emergency response teams, law enforcement agencies are all natural prospects here.

ASR Dawns with the New Millenium?

For those companies not wanting to invest in customer premise ASR technology, speech rec services offered by telcos via subscription would appear to offer considerable ease of use and lack the kind of angst associated with installing complex CT systems.

Still, there are some barriers to the success of ASR on the voice network. In 1998 practically the whole staff of *Computer Telephony* magazine got caught up in the hype. We extolled the miracle of how two developments – the growth of phonetic recognition and web-based data entry – were going to set voice-activated dialing on fire. Speaker-independent ASR engines would not need subscribers to "train" dialing lists by phoning in and repeating "mom" three times before entering her phone number. Subscribers would be able to build a personal phone directory by entering phone numbers and destinations on the network operator's website, and phonetic recognizers would build word models straight from the phonemes of the entered text. By entering a personal profile via the web, you can tell such systems just what number you mean by "sister," or "summer house," or "Red Ferrari." You can enter or change your default numbers for phone, fax, cell phone, and pager. Similarly, follow-me services could allow subscribers to enter phone numbers and schedules on websites.

In 1999, however, we found that while voice-activated dialing was still alive in the network, and more feature-rich, voice-activated personal assistants 'a la Wildfire, Webley and Portico were still being offered, but they're taking longer than we expected to find large numbers of end-user customers. At least one platform provider, Accessline, appears to have temporarily given up finding such forward-looking customers. There's not much new to report with network-enabled, voice-activated dialing. "I've heard the numbers of adopters is in the one-to-two percent range," says William Meisel, publisher of *Speech Recognition Update* and president or TMA Associates (Encino, CA – 818-708-0962, www.tmaassociates.com). "Unless you enroll, you have nothing." And people who get a free month trial of VAD just don't bother to load their directories. They don't train the recognizer by speaking the names of the people they want to dial.

In 1998 the task of enrollment was being made even easier via web entry of data. But one of the companies then preparing for web entry, Accessline Communications, (Bellevue, WA – 206-621-3500, www.accessline.com) later told us that they'd aborted the effort, at least temporarily.

"One difficulty our carrier customers had was selling a complex product to someone who didn't understand the need," says Accessline's Fred Epler. They've now redirected their business away from selling VAD and follow-me platforms to carriers. As of March 1999, Accessline is marketing a three-tiered pager-notification service directly to the end user. Pager-packers are the kind of people who've already bought into the find-me concept, and should be amenable to adopting a related platform.

It seems that, as the new millenium dawns, speech recognition in the network is taking a path similar to the hand-held PDA, which wasn't accepted in a day: It took a winding road littered with rejected false starts (such as the Apple Newton) before the right price and feature set was found in the Palm Pilot, which opened up the way for competition from Compaq and HP. ASR-driven personal assistants, like CT itself,

needs a "killer app" formulation.

What else is remaining in the way of telco adoption of speech recognition? Price. Telcos must either figure out a way to get customers to pay for ASR-enabled services, or else get the costs of provisioning low enough to give it away as a loss leader. Some products that have proven to be a hard sell into enterprises, and reach better economies of scale (and require less tech support) when deployed as a service on a network.

Still, Automated Speech Recognition is fast approaching a mainstream technology. Where does it all end? The Holy Grail of speech recognition is the human-like HAL 9000 computer in the movie 2001.

According to John Oberteuffer, president of Voice Information Associates (Lexington, MA – 781-861-6680, www.asrnews.com) and publisher of *ASR News*: "The HAL 9000 is the vision that spoiled everybody's idea of what speech recognition is going to be able to do for the next 10 or 15 years. I think we're 15 or 20 years away from HAL. That doesn't mean we won't get there, but we're a long way off from that. But there will be a lot of exciting things that happen in between now and HAL."

Bob Duerr, Product Market Manager at Dialogic, says: "As people get comfortable dealing with structured person-machine speech dialogs, we'll see very sophisticated models come out, delivering what Wildfire promises to be. If the industry has a target – Wildfire – to point at exactly what they want, when they point at it, they are actually thinking of what HAL was in 2001. Wildfire will deliver it, as will a whole bunch of other applications."

Some ASR Developer Tips:

In 1999, Unisys and Microsoft got together to form the SpeechDepot Web portal (www.speechdepot.com), a single source for tools, components and technologies necessary for the design, development and deployment of speech-based applications. SpeechDepot also serves as an eForum for the speech application development community. Here developers can obtain, share and disseminate information on speech.

During his tenure as Dialogic's Product Manager for Speech Technology Products, Peter Gantchev gave us these tips for deploying large vocabulary high-end phonetic speech recognition in a computer telephony application:

1. Prompt Scripting is Important. The prompts should be friendly, concise and unambiguous. Remember that the prompts are leading the dialogue with the user and the way that these prompts are worded significantly influences how the user will respond to the system. A good prompt will limit the number of possible alternatives and minimize likelihood of an out-of-vocabulary utterance. In a banking application a good prompt might ask, "Would you like to check balances, transfer funds or review account activity?" instead of "What would you like to do?"

2. When it Comes to Vocabulary Sizes, Bigger Isn't Always Better. The larger the active vocabulary size the higher the demand on the recognition engine to distinguish between uttered words which typically results in a slower recognition time and lower accuracy. It is infinitely better therefore to bound the recognition alternatives for the task at hand in the application and to dynamically download the vocabulary sets as needed to complete the other recognition tasks.

3. When it Comes to Word And Phrase Sizes, Bigger is Better. Longer words and phrases are more easily distinguishable than short words or phrases. Austin and Boston are good examples of short words that sound almost identical. If we now add the state names: Austin, Texas and Boston, Massachusetts (or Boston, Mass) the resultant words are significantly different sounding.

4. Never Underestimate the Complexity of Capturing Yes and No

Answers. It's not just yes and no! It may also be "yup. okay, uh-hum, yes ma'am, you bet, yessir, sure, affirmative, yah,," etc. The application must take into account all of the possible ways that people say yes and no, including all regional variations.

5. Listen to How People are Interacting With the Machine. This will give you great feedback on the effectiveness of your prompts, the efficiencies of your call flow and most importantly the satisfaction with which people are using your system. Don't just listen to the responses that people give for each prompt encountered, but also listen for peoples' tones. Are they exasperated, frustrated or resigned? Any of these should be taken as a warning signal and you should reconsider your human interface strategy.

6. User Training can Make or Break Acceptance. Educate the user on the benefits of using your speech-enabled application over the use of DTMF detection or a human operator. The education process can be accomplished through bill flyers for banks, brokers and credit card companies or frequent flyer announcements for airlines. You may also want to consider offering incentives to entice prospective users: Pay customers $1 a bank transaction when they use your speech enabled application or add frequent flyer bonus miles for all reservations made automatically with speech technologies. Believe it or not: ATMs were not popular when they were first rolled out.

7. Use all of the Tools Offered by Telephony. ANI and DNIS can be used to effectively complement automatic speech recognition., particularly when it comes to automatic name and address capture for call centers. With either the ANI or DNIS the application can create specific geographic-based active vocabularies thereby streamlining the address capture process and improving recognition accuracy. The address capture application need not have the names for all of the cities, towns, streets, avenues, boulevards and lanes of the United States for a call originating in area code 212 (New York City).

8. Be Polite when the Recognizer Errs. Studies have shown that an apologetic computer and polite handling of ambiguities greatly enhances the users' experience with speech in an application. "I'm sorry, I thought I heard you say Dialogic, is that correct"?" is better than "Did you say Dialogic?". In the case of two possible responses, "I'm sorry, I thought I heard you say Dialogic, is that correct"?" is better than, "Did you say Dialogic or Dialog, Inc.?"

Jeff Hill at Dialogic says that speech recognition has become closely allied with text-to-speech (TTS) technology, since both disciplines' capabilities complement each other in CT systems. Hill gave us these tips for developers eyeing both ASR and TTS:

1. Interface Design. Pay as much attention to the design of the app's interface as the selection of the ASR vendor. A well-designed user interface coupled with a mediocre recognition engine beats a mediocre interface with the world's best ASR. Don't just take a DTMF app and replace the tone prompts with "say 1" instead of "press 1." Talk to your vendor for suggestions on application design assistance. It'll be worth every penny.

2. Never use TTS to replace a voice prompt. TTS still doesn't sound as good as a human's voice. Apps requiring commonly repeated or concatenated prompts should always use regular digitized voice (even if you're using TTS elsewhere). TTS is great for e-mail reading and reverse directories, but don't force-fit it.

3. Don't insert TTS mid-sentence. Although this might sound contradictory to tip number two, it's a good idea to use TTS to play the entire sentence where one slice of information varies greatly. Take the prompt: "The name of the city is Junction Falls, Iowa." Here, the city name varies greatly and is thus read using TTS. It makes sense to play the entire sen-

tence using TTS, allowing the listener to tune-in to the TTS cadence.

4. Think multiple technologies when selecting ASR and TTS products. "Press or say 1" apps for DTMF replacement are still popular uses of ASR. But more and more you'll see ASR and TTS mixed together. You should anticipate the need to use multiple technologies in the same system, even if it's not needed initially. Don't paint yourself into a corner. Design your base architecture to accommodate different technologies (ASR, TTS, Speaker Verification), and also different vendors. You never know what you'll need (or prefer) later.

5. Don't forget locality. Computer Telephony is used globally. Opportunities for ASR- and TTS-enabled apps flourish outside English-speaking nations. Try to pick ASR and TTS packages with support for multiple languages.

6. ASR packages are not created equally. Forget accuracy. There are many types of ASR, all good for some task or another. But each variant is not good for every application. Whole-word-based recognizers provide outstanding density, very good accuracy and a proven track record. Conversely, phonetic packages offer the flexibility to create very large vocabularies, but not the best track record. And speaker verification concentrates on biometrics, not speaker independent vocabularies. Now go back to tip number four!

Glossary for Speech Recognition (Voice Recognition)

- *ASR* – Automatic Speech Recognition.
- *barge-in* – VCS / Philips calls it "cut-through," when Dr. Anthony Bladon was at VPC he used to refer to it as "Talk-over". Others call it "interrupting prompts." Whatever you call it, this lets callers interrupt a currently playing prompt by speaking to the system. Barge-in is one of the little niceties developed by ASR companies for callers who have used an ASR IVR system for the ten thousandth time and would like to make things go faster by interrupting the speech prompts and answering them before they're finished playing. Without barge-in, callers must wait until the end of the prompt and then say their peace.
- *command and control* – using voice to issue commands to computer software, an IVR (interactive voice response system), or another piece of equipment (e.g., an automobile).
- *continuous speech* – speech recognition that can process naturally-spoken utterances that don't have pauses between every word, or at least less than 50 milliseconds of silence separating words. Long ago this kind of system was sometimes referred to as "continuous digits", since early, simple "talk to the dog" systems asked the caller for things such as account number digits. Continuous word recognition is the "best" (most comfortable) method of recognition, but also the most expensive and technologically tricky. It takes more DSP or host-based processor power to listen to and recognize phrases in a stream.
- *discrete speech input* – speech recognition that requires a speaker to pause after each word so the computer may distinguish the beginning and ending of words to demarcate what should be processed as a word to be identified. Technically, discrete speech recognition exists when more than 250 milliseconds of silence separate each word recognized. Discrete word recognition systems usually give the caller an audio beep to indicate when it's ready to recognize the next speech input. Discrete recognizers are only active when needed by the application. This makes the system less expensive because a handful of speech rec ports can be shared across a larger system.
- *keyword spotting* – a speech-recognition system that searches for specific words in what a person has said. For example, an IVR system to allow a user to say something like "Give me the loan department

please" when all it wants is the word "loan."

- *phonetic training* – a phonetic training system uses text to speech rules to build a matching template for each word. This technique reduces the amount of "training" a system needs by allowing input via phonetics rather than human speech.
- *recognition response time* – the time it takes to recognize a word after the end of the word is spoken. Most applications can tolerate a recognition response time of up to 500 milliseconds for discrete words, as long as the caller is not entering a string of digits. Discrete-word recognition with a response time less than 100 milliseconds has been referred to as connected-word recognition.
- *speaker adaptive speech recognition* – a system that cannot be used by a speaker until that speaker has provided a sample of speech to the system. Nearly all dictation technology is speaker adaptive.
- *speaker dependent speech recognition* – a system that cannot be used effectively by a speaker until that speaker has trained every word in the system.
- *speaker independent speech recognition* – can be used by people who have not first enrolled with a system or otherwise trained it to recognize their speech. Most speech-recognition systems not used for dictation are speaker independent.
- *speech to text* – generally, a synonym for speech-recognition dictation technology. It converts spoken words into printed text.
- *spoken language understanding* – technology and systems that incorporate elements of artificial intelligence (AI) to process spoken input. Sometimes this is used to refer to speech recognition systems that process "conversational speech" even if they do not use artificial intelligence.
- *voice recognition* – synonym for speech recognition, particularly speaker dependent speech recognition, the recognition of your particular voice. Voice recognition requires training. Voice recognition is sometimes used as a synonym for speaker verification.
- *word spotting* – the system looks for a particular word or phrase in the spoken stream of words and ignores anything else. For example, a system that uses word spotting would hear the phrases "I want to fly to Miami" and "When are your flights to Miami?" and return the same flight information to the caller. Systems that use word spotting often handle fewer simultaneous channels. It takes considerable DSP or host CPU-based processor power to listen to a conversation and spot key words on the fly.
- *zoning* – one interesting way to make a reliable ASR application is to use a technique called zoning. Set up a specific set of phrases that will work at each specific prompt. By limiting the number of phrases to those applicable to the prompt, it speeds up recognition and reduces the chance of having two words that sound similar.

Glossary for Speaker Verification and Identification
- *biometric-based technology* – (also biometrics) technology that verifies or identifies individuals by analyzing a facet of their physiology and/or behavior (e.g., voices, fingerprints)
- *equal error rate* – a threshold setting that results in an equal percentage of false acception errors or false reject errors.
- *false acceptance* – when a verification system allows an impostor to get in.
- *false rejection* – when a verification system rejects a valid user.
- *impostor* – a person who falsely claims to be a valid user.
- *speaker identification* – the process of finding the identity of an unknown speaker.
- *speaker recognition* – a synonym for voice biometrics. Also a synonym for speaker identification.

- *speaker verification* – the process of determining whether a person is who she / he claims to be.
- *text-dependent verification* – requires the use of a password or pass phrase.
- *text-independent verification* – can process an unconstrained utterance.
- *text-prompted verification* – asks users to repeat random numbers and/or words.
- *threshold* – the degree to which a new speech sample must match a stored voiceprint before a system will accept the claim of identity.
- *voice ID* – a generic term covering speaker verification, speaker identification, and speaker separation. A synonym of speaker recognition and voice biometrics.
- *voiceprint* – a sample of speech that has been converted to a form that a voice biometrics system can analyze.

Audiotex

Audiotext systems describe a generic class of CT systems that deliver audio information to callers based on touch-tone requests. Corporate information and dial-a-sport-score hotlines are a few examples. A very popular class of these systems are newspaper information hotlines, where callers select from hundreds of different categories to find out the latest in weather forecasts, regional sports scores, news items, and many other timely topics.

Expanded into many more capabilities. Artisoft's (Tucson, AZ – 520-670-7100, www.artisoft.com) InfoFast, however, brings the possibility of plug-and-play CT for small (2 to 8 voice lines) systems closer to reality. Coupling InfoFast with the appropriate Dialogic voice board (as small as the two-line Dialogic ProLine/2V) lets a small business get quickly started with fax-on-demand, voice bulletins, Web document-to-fax translation and voice mail with logging. Advanced end-users may be able to set up the system themselves. But it probably best serves as an entry point for VARs who want to dip their toes in computer telephony waters.

Automated Attendant

Also known as Caller Directed Call Routing (CDCR), automated attendants are one of the most recognized applications in the computer telephony industry. The system answers calls for a PBX or PC-based phone system to give callers the ability to directly connect to a specific individual or department at a business without an operator. The caller responds to prompts ("Enter the extension of the party you wish to talk to, or press 1 for a Company Director, press 2 for Sales, press 3 for Service, press 0 for the Operator or wait for the beep to leave a message") with DTMF touch-tone digits picked up by IVR, and the caller is automatically routed to the designated individual or agent group. Most voice mail products include the auto-attendant feature as part of the basic product.

There are three basic types of auto attendant:
- **Standalone auto attendants** are usually based on simple, no-moving-parts digital announcer technology and are suited for small businesses since they support few lines and menu options. They are very easy to program and install. To record prompts, you use a microphone, a handset attached to the unit, or simply download messages via an input jack.
- **Turnkey hybrids** are PC-based systems running on a variety of operating systems, including Windows, DOS, and Unix. They support dozens of lines, extensions and menu options. They usually do more than just basic auto attendant features and handle fax, IVR and voicemail. Sometimes they are PC-based but proprietary.

- **Software-based auto attendants** that come with application generators (See Application Generator) or the provisioning and management software of PC-based phone systems and CT communications servers.

Auto attendants' audiotex capabilities (see Audiotex) have a number of applications, the most popular one being call screening, where audiotex is used as soon as possible to supply answers to frequently asked questions that would normally waste a company representative's valuable time. (i.e., company hours of operation, location, etc.). Casual callers can thus get their basic questions answered via IVR without ever talking to a live person. Repeat callers or callers with more serious business will touchtone past the initial prompts to reach an operator or a second, IVR-based auto attendant menu.

More sophisticated audiotex-based call screening interactive processes can also be created. Audiotex menu items mixed with auto attendant items ("Press 1 for a company directory, 2 for Sales," etc.) can be expanded to up-sell, cross-sell or otherwise inform callers of important items before they choose to speak with company personnel.

Thanks to powerful tools such as ODBC or SQL database connectivity, client/server integration and full-blown IVR, auto attendants can now handle a fair percentage of inbound call traffic without human intervention. Customers can, for example, call in, enter an account number and hear their bank balance, or enter an order number and listen to a report on the status of their order.

The modern auto attendant has evolved to be far more than just a call router. Automatic Number Identification (ANI) and Dialed Number Identification Service (DNIS) (See ANI and DNIS) permit auto attendants to make intelligent guesses about what callers want, even before a call is answered. When a call comes in, depending on where it's coming from or what number the caller dialed, the system can route the call to the last agent to handle a call from that number, or routed to the best agent available in the group whose number has been dialed.

A more clever approach is to add IVR to the mix to gather information from callers as they're routed inwards, and coordinate such things as network access to databases to perform a screen pop of the caller's profile and history on the PC screen of the company representative with which they will end up speaking. The representative is thus given some time to be prepared to deal with the caller effectively when the call arrives at their desk.

Designing tree-structured IVR menus, however, is a tricky procedure that often results in a difficult to navigate system that frustrates callers, and frustrated callers tend to hang up. It is preferable that such systems be professionally designed, tested and implemented by experienced consultants and/or developers, since those first sequence of voice prompts are going to be the first impression your company makes on a caller.

Dispensing with IVR entirely, speech recognition is replacing the IVR function of auto attendants. Registry Magic (Boca Raton, FL – 561-367-0408, www.registrymagic.com) has concentrated all of their marketing on a single product, Virtual Operator, a speech-enabled auto attendant costing about $13,000. Phonetic Systems (Burlington, MA – 781-229-5823, www.phoneticsystems.com) offers PhoneticOperator, a speech navigable auto attendant and call routing system loaded with a comprehensive range of automated customer and employee service applications. Locus Dialogue (Montreal, Quebec 514-954-3804, www.locusdialogue.com) is another player in this space, as is Nortel (Saint John, New Brunswick – 800-466-7835, www.nortel.com) with its Business Directory System.

Philips Speech Processing (Atlanta, GA – 770-821-2400, www.speech.be.philips.com) has also entered the market with their VoiceRequest!, which works with most PBX and hybrid switches. When

a caller speaks a person's name or desired department, their call is automatically routed, whether internally or externally, to the correct person.

Many auto attendant makers have added fax functionality to their products. Fax-on-demand (FOD) subsystems, which offer menus of spec sheets, white papers, price lists, or other information that can be faxed to callers on request, were at one time very popular, though the World Wide Web has supplanted FOD to a great degree. In the auto attendant context, FOD, like audiotex, provides a "self service" function that can supply callers with information quickly and automatically, obviating the need for company personnel to deal with callers.

Some auto attendant makers are also adding fax mailboxes to their applications, to complement voicemail. A separate menu-set is provided for managing fax receipt and routing faxes to the appropriate mailbox. The recipient retrieves fax messages by calling in to the attendant from a standard fax machine. A password is required, just as with your voice mailbox.

One of the most interesting systems that combines all of these technologies is the Sound Advantage Natural Dialogue interface (SANDi) from Sound Advantage (Irvine, CA – 949-476-1400, www.soundadvantage.com) an ingenious voice-activated call processing system, an "electronic receptionist" invented by Michael Metcalf. SANDi costs about $10,000 and is designed for small and medium-sized companies.

SANDi is really a phonetic-based software application, but a complete software and hardware solution is also available. No voice programming or training is necessary with SANDi, as it uses proprietary technology developed by Sound Advantage and AT&T Watson speech recognition routines. Instead of using touchtone prompts, you simply use natural voice dialogue, or what Metcalf calls "speech-phone".

You can use SANDI as an add-on system to support your existing phone system, be it from Lucent, Nortel, Toshiba, Telrad, Panasonic, and others ranging from key systems to hybrids to PBXs. SANDi consists of a series of "erector set" modules you can put together into whatever system you like.

SANDi's main component is The Perfect Receptionist ($7,495) which is the first module one buys to build a SANDI system. It connects to your existing phone system to automatically answer calls. It can simultaneously handle multiple calls, screen incoming callers and forward urgent calls to designated numbers, such as cellular phones and home offices. The Perfect Receptionist needs a Windows NT workstation, with a least one standard four-port Dialogic board, which is connected to your existing phone system. This module can be used as a standalone system for automated attendant and call routing services, or it can be integrated with a number of add-on modules:

Next one would buy the Perfect Mail Module ($3,395) to provide voice access your voice mail. Perfect Mail enables SANDi to take phone messages, including date, time and phone numbers. You can call SANDi from anywhere at any time to check your messages and return calls, without using touchtones.

For another $1,010 you can get Perfect Dial, which maintains your internal extension list and outside Rolodex. It stores names and numbers for automatic voice dialing from any phone. Users can build personal and group address books to retrieve numbers and make phone calls.

If you have a legacy PBX system, the Perfect PBX Assistant ($1,010) interfaces with and controls your PBX so you can tell SANDi to "transfer this call" to "employee name" or "set up a conference call". The Perfect Email Module ($1,995) enables SANDi to read your e-mail to your over the phone, as well as send and receive e-mail from any phone, all by voice. Finally, the Perfect Fax Module ($1,995) stores your faxes and notifies you by e-mail when a fax arrives. You can view your fax on your PC to save paper and toner, and SANDi will let you use your voice to redirect your fax to other fax numbers or e-mail addresses

SANDi accommodates from 10 to 200 phones. The baseline SANDI system supports four concurrent sessions, equivalent of having four receptionists handling incoming and outgoing calls. SANDi can be expanded in four-port increments up to 48 ports. Sound Advantage is developing a larger enterprise version of SANDi, which will target CLECs, ISPs and other telecom providers.

A fully configured system, including hardware, is available for $9,995, or you can lease a complete SANDi system for as little as $350 per month.

Tips on How to Buy an Auto Attendant.

1. Plan your application in terms of goals. Is this a part-time auto attendant to replace a human receptionist during peak periods or while he or she is out to lunch? Or is it a full-time system, a sophisticated telephony front-end for your enterprise, one that projects a less human company image but undoubtedly increases caller self-service access to information, easing the burdens of your personnel? Whatever the goal, you must have a plan before you spend money and resources.

2. How many basic telephony functions are you currently lacking, and how many can you obtain from a new auto attendant? Need voice mail and auto attendant capabilities? Fax mail? ACD? IVR? Examine your present telephony system and determine what features are lacking, then try to find an auto attendant with those features and without at the same time presenting too much redundancy with existing systems.

3. The other side of the coin: Find out if the auto attendant you intend to buy will be interoperable with the technologies already ensconced in your company, such as voicemail, a digital announcer, etc.

4. Make sure there is an upgrade path. You will outgrow your call-processing system, so make sure your system can be incremented easily to handle increased demand.

5. Specialized programming. Look for attendants with programmable call handling procedures, voice prompts, special messages and other items that are relevant on a departmental level. Ensure as much flexibility as possible. No doubt there will be some sort of user friendly GUI app gen-like software to make changes easier to deal with.

6. Can you program the system and modify voice prompts and other operating parameters yourself (preferably, while the system is running)? Over time, your needs will change. Depending on a vendor or consultant to make even minor changes can be a costly situation.

As was stated earlier, auto attendants are usually a company's first impression with callers. A system should be designed and built around your needs and applications. Make sure that neither the prompts are so long that they bore the caller, nor that menu choices are so many and complicated as to be confusing. Your auto attendant is crucial to your business.

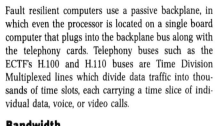

B Channel

Also called the bearer channel, is a 64 Kbps circuit switched, duplexed channel that carries voice and/or data a part of ISDN PRI and ISDN BRI. (See ISDN).

B8ZS

Binary eight-zero substitution, also called Bipolar with eight-zero substitution, was originally developed by AT&T as a line encoding option for all 24 DS-0 channels on a T-1 network (DS-1).

A "1" is sent on a T-1 by sending a pulse, a zero ("0") by not sending a pulse. The alternating mark rule demands that if the last pulse sent was of positive polarity, the next pulse sent must be negative. A T-1 device receiving two pulses in sequence of the same polarity experiences a bipolar violation (BPV).

The T-1 "1s density" rule specifies that the average number of 1s in the data should be at least 12.5% (one bit in eight) and there should be no more than 15 consecutive 0s in a T-1 data stream, lest the lack of 1s will destroy the synchronization between a sender and receiver.

B8ZS is used to prevent the loss of synchronization when long strings of zeros are transmitted. In B8ZS a specific sequence of valid pulses and intentional bipolar violations to the AMI line encoding method are used to represent a string of eight zeroes. Specifically, for any sequence of eight zeroes the following sequence is substituted: Three zeroes, a deliberate bipolar violation, a one, a zero, another deliberate bipolar violation, and another one. This pattern is recognized by configured equipment as a substitution code for eight zeros. Using this code, B8ZS ensures that a sufficient number of transitions are always on the line to retain system synchronization even when the data stream does not contain enough "1s" to do so. This means that T-1 lines can now transmit a large number of zeroes without translation problems.

Since the user no longer needs to give up one bit in eight as in AMI, B8ZS encoding allows a DS-0 timeslot to transmit a full 64 Kbps (a so-called "clear channel") instead of the conventional, reduced 56 Kbps data stream normally transmitted. B8ZS applied to all 24 DS-0s on a DS-1 increases the data throughput 14% from 1.344 Mbps to 1.536 Mbps. The advantages of B8ZS over AMI encoding will thus be readily observed in data applications.

To gain the advantages of B8ZS, your DSU / CSUs with B8ZS support are needed on both ends of the circuit, otherwise the data appears to contain bipolar violations. The most common reason that B8ZS is not seen on a line that is correctly configured to support B8ZS is that sequences of eight or more zeroes are not being generated. This often happens on T-1 lines with little or no traffic. The line must contain a minimum amount of "1" or "mark" signals. An unusually long string of "zero" or "space" signals may be interpreted by the receiver as a loss of transmission.

BABT

See British Approvals Board for Telecommunications.

Baby Bells

See Regional Bell Operating Companies

Backbone

The world's network of high speed broadband connections between switches or hubs. See Local Loop.

Backplane

PC plug-in cards and PBX line cards (circuit boards) slide in and connect to a high-speed communications line, called a backplane bus.

Fault resilient computers use a passive backplane, in which even the processor is located on a single board computer that plugs into the backplane bus along with the telephony cards. Telephony buses such as the ECTF's H.100 and H.110 buses are Time Division Multiplexed lines which divide data traffic into thousands of time slots, each carrying a time slice of individual data, voice, or video calls.

Bandwidth

In telecom, bandwidth is the total information carrying capacity of a communications channel. This is usually measured in Hertz (cycles per second) for analog transmissions and bits per second (bps) for digital communications. Early in the 20th century, Henry Nyquist noted that the difference between the highest frequency and the lowest frequency of the sine waves that compose a complex signal is the bandwidth of the signal.

Basic Rate Interface (BRI)

One of the two levels of ISDN service (the other being Primary Rate Interface – PRI), ISDN BRI allows two "B" or Bearer channels (full duplex at 64 Kbps) and one "D" or Data signaling channel at 16 Kbps. All channels are carried over one ordinary single pair of copper wires. Commonly referred to as 2B+D. Through the use of BONDING (Bandwidth on Demand) the two 64kbps channels can be combined to create up to 128 Kbps bandwidth if necessary. See ISDN.

Baseband

A baseband is most basic form of digital modulation, the pulsing of signals directly on a wire or fiber that uses the entire bandwidth capacity of the wire.

Baud

Baud is an old term that's being replaced by bits per second (bps). Baud is the number of "pieces" of information or signaling elements that can be transmitted per second. In an average digital data stream, one baud is roughly equivalent to one bit per second.

BCC

Bellcore Client Company.

Bellcore

Bell Communications Research, formed at the AT&T divestiture to provide centralized services to the seven regional Bell holding companies and their operating company subsidiaries. Now called Telcordia Technologies.

B-ISDN

Also known as BISDN, which stands for the Broadband Integrated Services Digital Network (See Asynchronous Transmission Mode).

Billing Systems

It seemingly strains one's credulity to state that Call Detail Records (CDRs) can be worth their weight in gold, but in a way they are. Not just because they're collected together and printed on the bills that result in the monthly mass-mailing of checks from subscribers to telco carriers. The "gold" referred to here goes a step beyond simply getting paid. Rather, it's the mine of usage data that helps your marketeers identify, cross-sell, retain (and possibly even win back) your most profitable customers.

These days, customers dissatisfied with their telco depart quickly, increasing a figure known as churn. To eliminate churn, telcos have offered enhanced services to customers made possible by computer telephony technology. But one often over-looked way of decreasing churn is with a good, integrated billing and customer care system having the ability to do data mining of call records, or "mine minutes," as they say.

Such a system might show one telco Customer Service Representative (CSR) all there is to know about your business dealings with your employer, or that might tell that agent, via screen pop, that you're a customer at risk for churn, one worth keeping, and, best of all, specify a nice incentive offer that gives the best probability for keeping your business.

Such a dreamboat system would produce a bill that's clear and readable (something the FCC has just begun to mandate from wireline carriers), or best yet – a website that can let you answer your own questions. A website that can even tell you about other services and products you might want to buy. A site that can even gather market intelligence. For the chance to stay out of customer service hold queues, many customers would gladly offer up mother lodes of demographic information.

The good news is that such systems have just arrived. We're beginning to see the first actual purchases and pilot programs of software modules – add-ons or included with billing packages – that specifically target churn prevention. Closely allied with that is decision-support software that helps marketers segment their market, lock in on cross-selling and new-customer targets, and even manage sales campaigns.

In the predivestiture world, billing was a back-office afterthought. AT&T, that awesome and terrifying monopoly disguised as Ma Bell, "the voice with a smile," knew that she'd get everyone's monthly payment or they'd go without phone service. Today, billing is an essential part of an aggressive marketing platform in a fiercely competitive telecom marketplace. Newer telcos, unburdened by legacy systems, have an advantage here.

"Old systems could not support new products and services in acceptable timeframes," says Avi Ofrane, president of Billing College (Teaneck, NJ – 201-833-9494), a training outfit that offers three-and-a-half-day and five-day courses on telecom billing around the world, and has included all the major IXCs, many PTTs and next-gen telcos, as well as billing vendors among its students. "Telcos can no longer afford to wait six to twelve months; they lose market share to aggressive startups – especially in the cellular world, where you have new products all the time." And where the churn rate is a stomach-churning 30% per year.

One way to prevent churn is to offer one-stop shopping: entangling customers in an almost inextricable web of products: local, long-distance, Internet access, wireless, calling card, cable TV, etc., all presented through one bill and supported through one-call customer service. That's the idea behind "convergent" billing platforms –and most billing software makers call themselves convergent now.

The other thing "convergent" systems offer is the ability to actively upsell customers, because if usage data highlights those who already pay for services A, B, and C, they can be targeted as excellent prospects for discounted D. A full-featured integrated billing and customer care (B&CC) system can pop this fact up on a CSR screen.

Of course, a savvy marketing analyst should be able to feed usage data into an Excel spreadsheet and come up with the same target prospects, but, as Saville Marketing VP Rich Aroian points out, "It's one thing to have that information available to a marketing person with Excel in a back room; it's another to make it readily available to a CSR. The key is making sure that information is actionable." Systems on the high end automate that feed into CSR screens.

Web-based Bill Presentment and Customer care.

The Yankee Group has research to show – at least in well-wired North America–that a telco's typical web-savvy customer is likely to be in a high-value group. They're more likely to use multiple services (Internet service is a given), have higher incomes, and prefer the time savings of self service over the "human" touch of phone. Electronic bill presentment and payment (EBPP) is hot, and on-line access to bills is beginning to get play in consumer advertising. The business case for carriers is equally compelling, as it reduces CSR load.

The Web presents a wonderful opportunity to present bills or just allow customer "self-care," including bill inquiry ("was my last payment received?") and even some adjustments. Up to a certain dollar amount, it's cheaper to let customers adjust their bills themselves than to involve a CSR. Of course, intelligence has to be built into this system to make sure that the right kind of customers are kept within the right crediting thresholds.

Web-based customer care also sacrifices nothing in the areas we've examined here: Churn prevention, upselling, and intelligence gathering. The web-based bill "recognizes" the logged-in customer, addresses him, and makes product offerings. Here's the almost insidious part: These offerings can be triggered by any combination of profiling factors, or by preferences tracked through previous web site clicks. Here's where all the web traffic analysis tools from the general e-tailing world come into play. (See the article "Web Site Mining Gets Granular," in the March 29, 1999 Internet Week for the latest on that front.)

Web-based bills can be presented directly from a telco's own website, or they can be aggregated through a consolidator. A consolidator could be a bank that offers a home-banking service, or a service like Intuit.com, an independent site that aggregates its users' bills. Most telcos, if they're implementing this solution, rightly prefer to present billing information through their own websites, for the opportunity to promote additional services, offload mundane CSR duty, and touch their customers.

Three interesting telco-appropriate billing solutions that are entirely or heavily directed at web presentment are: BillCast from Just In Time Solutions' (San Francisco, CA – 415-553-5505, www.justintime.com), the I-Telco version of I-Series from Blue Gill Technologies' (Ann Arbor, MI – 734-205-4100, www.bluegill.com), and the software of Novazen (Longmont, CO – 303-583-3100). As EBPP vendors, they have a great stake in proving the selling power of the web and their sites show it. Novazen's even graciously tells you all about their competitors!

The demos these companies make available for your perusal (reached via novazen.com, bluegill.com, and justintime.com), are the "future classics" or archetypes of on-line marketing systems. Try them. Just in Time's demo shows telco billing as part of a consolidated bill presented by "Bay Area Bank." Click on your LD summary item and there's call detail sorted any way you like; there's company news, upselling offers, and short, personalized e-mails from customer service: "Your line item dispute was resolved on 5.21." And "Three-way calling is now activated for your residential account." Best of all is a one-click reverse directory lookup, reminding you who you called in Phoenix last month.

Just in Time's Billcast products operate in both direct-site and consolidator scenarios. Their BillCast Presentation Server sends HTML bills – in all their drillable depth – from the biller's own site. Customers access the bill securely with industry-standard SSL and unique RSA encryption. Simple HTML tags let the biller add any element to a bill template, with no programming skill necessary.

Just in Time's OFX server works in a consolidator scenario, sending only bill summaries to consolidators through the OFX (Open Financial Exchange) open standard for exchanging billing information. This is what's known as the "thin" consolidator model. But here again, if the bill payer wants to check detail records, he or she need only click on summary information. The OFX server directs the request back to the

Presentation Server, which serves up the detailed information direct from the biller. Traffic is kept to a minimum, data security and hi-touch opportunity is kept with the biller, and the consumer can still drill down.

Just in Time's solution also integrates with existing billing and customer care apps via direct API, staging server, or print stream conversion: few carriers can scrap mailed bills entirely and none can do away with all their CSRs.

And since customer relationships with telcos are ongoing, where those with amazon.com or ebay.com might not be, the opportunity – and the instant payoff –- is there for personalized e-billing and service delivery. Just in Time's Brian Valente, marketing director, sees three core inputs here: The accumulating billing data, the segmentation data available from traditional back office applications, and the profiling information gleaned through web interaction.

BillCast's profiling engine records customer activities on the site, but the difference between general click-tracking software and BillCast is that it is immediately actionable. Answers to questions or clicks on web-presented promotions can generate dynamically generated HTML, personalized pages. "Billing begins to look less like a document, and more like an application," says Valente.

As for telecom examples, "In looking at the billing data," Valente posits, "BillCast sees that you make a tremendous amount of in-state calls. The site can send a web page that says, 'Would you like a wireless rate with no roaming charges inside California?'" Or more simply, it sees that your monthly low-price minute bucket is near empty: Click here to buy an additional 100 minutes for ten dollars.

Just in Time offers both Windows NT and Solaris versions of their BillCast family. In 1999 they announced a tremendous deal with AT&T, which makes 70 million consumers eligible for BillCast-powered online billing. The mega-carrier will start off with the consolidator model. Just in Time is also the bill-presentment piece behind Intuit.com, one of the most popular consolidator sites.

The Intuit concentrator site does not process payments directly. It goes through an e-remittance company, which in turn has agreements with banks. A company such as CheckFree, for example, itself processes electronic payments by interfaces with Integrion in Philadelphia, which is a consortium of banks. "It gets very complicated," admits Billing College's Avi Ofrane. "All those agreements have to be made on a one-to-

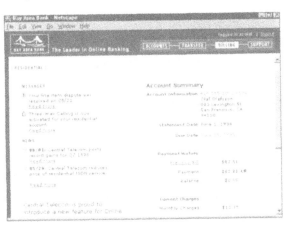

Web-based systems allow for clickable call details and embedded e-mail from a CSR, A demo of this is available at www.justintime.com

one basis. This is an issue that was resolved between cellular telcos for roaming users," but hasn't been yet in the e-bill clearinghouse arena.

The Just in Time customization for telco clients includes the BillCast software and integration. So far, web-based customer care and billing is limited to residential and small business customers. Corporate customers' bills are too large and complicated to route through IP this way, and often are presented on CD for complicated analyses. Such big customers are also less susceptible to churn, anyway.

Choosing a Billing System:

Avi Ofrane's Billing College ends one of its courses with an exercise on evaluating billing systems. "It's an eye-opener, he says, because students' RFP's for billing systems typically describe their requirements in purely technical terms: Convergent, client/server, three-tiered, and the like. But that's a technology-centric view, one that does not follow through to end goals.

"Shift the emphasis from technology to the business requirement," he says. And make the prospective vendor speak in the same terms. "I don't care if your billing package has three tiers or seventeen tiers. What you need to ask your billing vendor, if you're a telco is, how is it going to help me maximize my revenue? If you [the vendor] can't tell me that, I want nothing to do with you."

Ofrane goes further, saying that it should not be just the job of the CIO to select a billing system. "This is a revolution," he says, and one that often comes smack against departmental territorialism. "We're trying to shift the emphasis to the business unit, to the financial guys. They're the ones responsible for evaluating performance, profitability, projections."

"We promote billing selection as a joint effort. When we offer an on-site course, we always request that the group of students not be homogeneous. Not just systems analysts. Send us your marketing, sales, product marketing managers, finance people, as well as business and systems analysts and programmers. These people will have a much better appreciation for what's needed. Some of our clients have thanked us for having changed the way they think."

Convergence Billing Caveats

Carriers have to consider several issues before they plunge into "convergence." The first is consumer sticker shock: Customers who see a converged bill of $300 and more may pay more slowly or fitfully than those who see smaller bills in separate envelopes. Sticker shock may severely impact carrier cash flow. The second is customer service: If customers can resolve issues stemming from five different services with the one 800 number appearing on their bill, CSRs had better be equipped to handle all those different kinds of complaints and requests. They must have all-inclusive views of the customer, or seamless call routing intelligence that guides the caller to the appropriate agent in short order.

Most at risk are carriers looking to incorporate billing for services that they don't actually provision themselves, such as cable TV. Ofrane: "Say you're adding pay-per-view to your bill from another supplier and a customer calls to tell you 'I didn't watch that fight.' Are you taking on the pay-per-view company's call center headache?"

Outsourcing B&CC

Companies like Sprint PCS and parts of AT&T Wireless entrust all or part of their convergent customer care and billing functions to Convergys (Cincinnati, OH – 513-458-1300), a company spun off from CBIS and Matrixx. Convergys can do as little as present bills (they do this for 60 to 70% of PCS bills in the US, they say), or as much as take over first-response customer service, with more than 30 call centers and data centers. Their software, some of which they license, is also equipped for churn prevention and quick-response marketing programs. Client telcos can institute new pricing programs in minutes, through on-line GUIs to their pricing engines. These programs can be finely targeted as well, for example, "only to new retail customers in Milwaukee via XYZ marketing channel," says Convergys' Brian Henry.

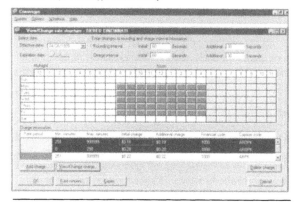

Telcos can outsource billing and customer care while still maintaining real-time control of pricing and customer retention programs, and market segmentation. This on-line interface to Convergys' pricing engine lets a client telco manager put a call center-coordinated program in place in minutes.

A Roundup of Billing Systems

We'll start by looking at these systems, which mostly run Unix and are geared for telcos serving at least half a million subscribers.

Amdocs (St Louis, MO -3124-821-3242, www.amdocs.com), is one of the big three names in the top tier of telco billing, and has even mastered black art billing for IP Telephony. But VoIP is just one module in their multi-service, convergent Ensemble package.

Ensemble is a modular but comprehensive telco solution, comprising provisioning, billing, collections, network resource management, fraud prevention, and the whole call center customer care space. They sell one system to handle the whole telco range of services, plus paging, Internet access and cable TV.

One of their optional Ensemble modules is Sales and Marketing; within this module are two closely linked software packages, Churn Management and Campaign Management. Judging by a talk *Computer Telephony* magazine's Ellen Muraskin had with their Chief Scientist of R&D, Gadi Pinkas and President of R&D, Shlomo Baleli, these tools for predicting customer behavior and tailoring interventions are about as sophisticated as this category gets.

Their Churn Management modules are sold as standalones, but as with Saville, they're integrated off the bat with Amdoc's own customer care system. This means that an Ensemble-equipped CSR in the inbound trenches knows when a caller is at high risk for churn. There is no need for a generic CRM infrastructure here, as can be found in systems from a Seibel or a Vantive.

Ensemble's CM modules help the analyst pinpoint his market, and decide best how to deploy his resources: Which business customer is worth dispatching an on-site visit, which is worth a telemarketing call, which direct mail. Campaign management, of course, is also used for targeting new customers.

As Ensemble pumps more sales and usage data into the algorithms,

the predictive models become more accurate. The models take in other, perhaps purchased market data sources as well. "If you run it frequently with new data, our system can capture changes in the dynamics of the market and produce better churn prediction," says Baleli. "You can put the carrier's incentive policy into the system, and the system will not only give you the probability of churn for a customer, but what incentive might prevent that churn."

In the words of their own literature, Amdocs employs artificial intelligence techniques, such as rule induction, statistical methods, and neural networks to discover patterns of customer behavior. To a non-statistician, that means that you can ask the system to show you patterns of those who've bought voicemail service, say, in the past, with no preconceived notions of what those patterns might be. The system mines the data warehouse, turning up correlation with zip codes, or age of customer, or age of account, minutes of usage, or percentage of overseas calls, for example. When a suitable target is spotted, the germ of a retention strategy may be planted.

The analyst can also insert behavioral variables that he has gleaned from personal experience into the models – perhaps knowledge of competing offers, or even ethnic holidays. He can perform "what-if" analyses, expand the business rules to include more or fewer subscribers, and see what results have shown in the past. The included visual presentations are maximized for the telco marketplace (see accompanying screenshots).

Amdocs runs a rather daring trial demonstration with the prospective telco customers of Churn Management. Customers give Amdocs their databases, which lists up to the last month who has churned. Based on that information, Amdocs predicts who will churn in the following month. Prospects only have to compare Amdocs' prediction with actual data of the following month to judge the accuracy of their crystal ball.

Amdocs' analysis tools have even received international recognition at a general data mining conference – the International Knowledge Discovery in Databases Conference (KDD-98), held in New York in 1998.

Ensemble runs mainly on UNIX; although some parts run on mainframes and some on NT as well. Churn management and Campaign Management can run on NT, as well as the client sides of the CRM products.

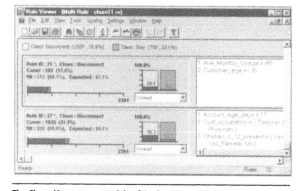

The Churn Management module of Amdocs' Ensemble billing and customer care software has come up with a behavioral segment of its customer database with an 87% likelihood of customer churn. In this screen shot, for this fictitious carrier, the segment is defined by a rule with the following conditions: a subscriber with an average monthly usage of over 491 minutes and an age under 30. The horizontal bar shows that this segment includes close to a quarter of all subscribers. In testing against sample data, 88% did in fact behave as predicted, and disconnected. Analysts can add more rules to segment the subscribers differently, and see how likely these would be to churn as well.

Amdocs prices its system based on the size of its customer, and assumes responsibility for integration and training; typically Amdocs personnel remain on-site for the implementation phases in large installations.

Say you've arrived at your rules defining segments of likely churners. You want to prevent the greatest possible number of them from churning with your limited number of telemarketers and incentives. You also want to make sure that you've chosen the best possible rules with which to find these prospective ex-customers. Some customers will be included under more than one rule. The Amdocs Churn Management module creates a model of churn behavior by applying all the rules relevant to each individual customer to generate a score reflecting the customer's overall probability of churning.

The accuracy of the model can then be checked against test data. This graph tells you that if you rank all the customers according to their modeled churn probability scores, and then access just the top 5% of customers, you will reach 15% of those about to churn. If you reach 20% of your screened subscribers, you will hit 53.7% of probable churners. The straight 45-degree line shows you that without such a predictability-scored segmentation, there's a one-to-one correlation between percent contacted and percent of churners contacted. The difference between your reach with a randomly chosen population and your reach with a predictability-scored population is the yellow area, called the "lift." A good data-mining system should give you good lift.

Apex Voice Communications (Sherman Oaks, CA – 818-379-8400, www.apexvoice.com) was originally famous for OmniVox, one of the top application generators. It then developed an equally impressive turnkey prepaid calling card system. Joining the product line is the new Apex Billing System, one of the best and most sophisticated transaction based call rating systems you'll ever see. It's for such services as long distance, debit cards, travel cards, cellular, prepaid cellular, callback, prepaid callback, and various carrier services.

The Apex Billing System can use different platforms depending on each site's specific needs. Smaller systems are based on Intel-based systems running SCO Unix. For larger systems, the preferred platform is a Bull multiprocessor Power/PC based system running AIX (similar to AS/400). Other equipment can be used so long as its been approved for running Informix or Oracle databases.

Apex Billing uses OmniVox and OmniView for call scripting (service creation and prompt recording). The standard call flows delivered with the system let you set up parameter based messaging control to play message tags, balances, and notices.

Optional billing modules are available for credit card recharge via IVR, switch interface, CDR output for post billing, and more. Apex Messaging can even be added to handle voice, fax and e-mail. Voice over IP call handling and switching control is also available within OmniVox.

Apex Billing is based on an accounting oriented database stressing flexibility. Calls may originate from any source. The billing engine allows for unlimited Multi-tiered rating, and validation is performed on the transaction format by the receiver engine.

If there isn't enough information within the raw transaction to create a valid transaction, the information is sent to a table containing raw transaction information and to an incomplete call table. If there is enough info, the call is sent to the raw transaction table. A call transaction is created in the call table, and a pointer into the call transactions table is placed in a file for the next available rating engine to acquire the call record and process the call.

Apex Billing can be set up in three different configurations. The first can is a "switch-centric" configuration where the call processing is using

a CO type switch and the billing database is using billing tickets coming from the switch. This may use Release Link Trunking (RLT) where a switch port receives the call and uses an IVR port behind the switch

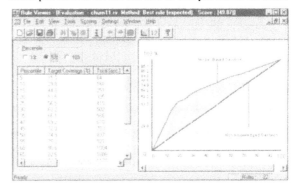

(OmniVox Intelligent Call Processor) for call setup and messaging.

The IVR port releases the call back to the switch for the duration and the call detail records or tickets coming from the switch are used to rate the call. A variation on this is to keep an IVR port camped on the call for warning messages or call control using three-way trunking.

The second kind of configuration doesn't need a CO type switch but instead uses the OmniVox Intelligent Switch for call rating. This setup may be used in a distributed configuration where multiple sites can be rated using a common database.

The third kind of configuration is for standalone billing where Apex Billing interfaces to a switch but does not provide call processing functionality. The call rating could be input from billing tickets from the CO switch, NAMS (a third-party product), or even tapes in a non-realtime mode. Apex can provide you with custom switch interface modules to interface to most major CO type switches.

Kenan Systems (Cambridge, MA – 617-225-2200, www.kenan.com), another one of the "big three" B&CC names, has its own "decision sup-

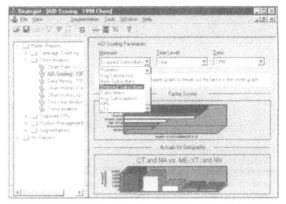

Kenan's Strategist product uses AID scoring to expose the trends and patterns that are hidden in billing and demographic data. In this example, the user wants to understand the factors contributing to churn. In the top chart, we see that the Geography bar is the longest, showing that geography is most strongly associated with churn. The bottom graph shows the actual numbers and tells us that CT and MA have particularly high rates, suggesting a service delivery problem. "Product" is the next most significant churn factor after geography, but the gender bar is not a good predictor.

port" module, which adds customer segmentation, churn/value reports, product management and campaign management to their core Arbor BP billing and customer care and Arbor OM ordering and provisioning system. The module is called Strategist,

Strategist, like other add-on tools, incorporates data from multiple sources, including its native billing system. Marketers also can analyze churn rates and customer lifetime value to target particularly valuable segments with customer loyalty programs and win-back incentives.

Customer analysis reports view the results of conducted surveys by product, geography or segment. Analysts can explore the relationship between any two factors, such as geography and gender or product and income level. The example given by Kenan: see if customers from a certain income level purchase PCS service, to see if you should target that group. Data mining here, too, reveals patterns of buying and churn behavior. Product management tools help marketers determine which products and services to bundle together, at which price and to which segments.

Strategist's server runs on Unix, clients are predictably Windows-based. The server also supports direct links to Oracle and Sybase. To this point, and alone among the big three, Kenan's billing and customer care products must run on top of a third-party call-center CRM product such as Seibel or Vantive; hooks to these are built into Arbor. David Rabkin, director of their Decision Support Group, poses this as a simple integration.

Strategist can flag CSR screens when a customer is at risk for churn. It can also generate campaign lists for outbound contact, or pipe through a script appropriate to retaining that valued customer on inbound calls.

Kenan has gone from 100 employees and $18 million in revenues to to 850 and over $175 million in three years, according to Rabkin. They've got about 100 licensees, ranging from new CLECS to France

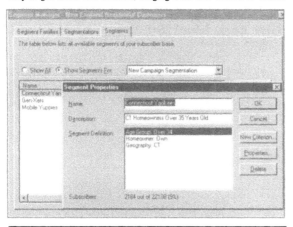

Sometimes marketers know exactly what segment they want — one that is driven by responses to a customer survey, for example. They then must be able to define their own segments in order to analyze their likely behavior. Strategist's Segment Manager lets them do that. In the example on the right, "Gen-Xers" are defined as those under 25, who rent their homes and earn less than $35,000 a year.

Telecom, for whom their system processes a billion CDRs per month. The company has been acquired by Lucent.

Rabkin explained that Strategist is configured slightly differently for each version of their billing platform; Arbor Wireline, Broadband, Mobile and Internet, because marketers look at different data and ask different questions in each business. Additional configurations can be added much

like templates, as a carrier adds more services and products. "What allows us to do this is our deep experience in all these verticals," he says. "We know what their issues are." He walked me through some of US West Long Distance's processes to illustrate.

As a startup, USW LD is prohibited by federal regulation from using the historical data of its parent, US West. So to develop a basis for segmenting their market, they purchased models of lifestyle data: such factors as "Who owns guns, what do you drive, what SIC code are you employed in. Marketing analysts use that to create their own segmentation; customers who they think are homogeneous in the marketing channel. Then they can go to town designing their programs."

They can then start communicating with those prospects by creating lists, and taking a test list to a telemarketing operation, and getting real results back. US West, in preparation for a "full-bore" rollout of long distance, used Strategist which let them look not only at purchased data, but at honest-to-God behavior, trending over time. Says Rabkin: "Who's likely to be disloyal? Who are profitable but disloyal? How can we increase loyalty? Who are not so profitable but loyal? How can we make them more profitable?"

Rabkin adds: "You can try to profile your loyal customer and use that as your basis [for marketing campaigns], and you can look at specific behaviors: Are customers ramping up the amount of long distance? Does that make them more attractive to competitors with higher usage plans? Then you want to put them on another plan."

You also want to alter campaigns based on test results, and you want results back quickly, inside of two to six weeks, says Rabkin. "If you talk to telecom marketers today, they can't do that in under six months."

While Kenan's B&CC and Strategist do not provide all the CRM plumbing, they do solve the integration of OLAP (online analytical processing), and campaign management. It's priced in scale with a carrier's size; but entry-level here, for software license only, is on the order of $300,000; and a one- to five-million dollar contract is more typical. They have relationships with several leading integration firms, including AMS.

The LHS Group (Atlanta, GA – 770-280-3100 and Frankfurt, Germany – +49 [0] 6103-482-700, www.lhsgroup.com) has over 140 BSCS telco billing and customer care systems installed in more than 60 countries. At a recent GSM Congress in Bern, Switzerland they announced their new, backward-compatible object-oriented B&CC system, Targys. It's an add-on to BSCS 5.21 and higher.

Java-based, Targys is aimed at web-based functionality for remote agents and customer self-care, and specializes in drag-and-drop customizable GUIs. Targys will be CORBA 2.0 -compliant, allowing fast bolting to third-party CORBA apps. It's got the three-tiered architecture that's beginning to be trumpeted by several makers: Layer 1 for presentation logic, Layer 3 for database, and Layer 2 in the middle for application logic.

The first component of six that were rolled out in 1999, Targys Customer Inquiry has been deployed by Swisscom AG Mobile in Bern. Order management, Customer Care, Web service, Back office, and Corporate account are to follow.

Saville Systems (Burlington, MA – 781-270-6500, www.vertexinc.com) took in $167.7 million in revenues in 1998 – the price for their packages typically runs in the multiple millions of dollars. Contract announcements have included Brazilian Netstream, Scottish Telecom and Net2000.

Saville has always had IP telephony ambitions. While convergence alone should go a long way toward preserving customers, their customer-care module, SavilleCare, addresses this issue. They're pitching it first to existing customers of their CBP Convergent Billing Platform. It will also run with other billing packages.

As a call center, sales support vehicle, SavilleCare competes with the Seibels, Vantives and Clarifys of the world, but it has the advantage, as pointed out by Richard Aroian, Saville VP marketing for the Americas, of very tight integration with CBP's billing and usage history. It also benefits from being developed from the ground up in its vertical telecom market.

It's all based on business rules. "If there's statistical information that says, for example, that the likelihood of a client leaving you goes up to x percent based on the increment of customer service calls, and the system makes that available to the CSR, then as you get near a threshold, the CSR sees that this is someone who's called us four or five times within the last 30 days. He or she needs to do something to keep that customer, such as perhaps offering a discount," explains Aroian.

By the same token, SavilleCare can also show the CSR when the loss of this customer would not be a terrible blow to the carrier. He or she would know, then, not to waste too much time or incentives on retention.

Of course, SavilleCare accepts CSR entries as they interact with customers, adding to contact history. For each call, the system can highlight potential problem areas and potential upsell opportunities. It

Here's Manuelle Fargo popping up in the CSR screen of Saville's SavilleCare system. Note the churn fields at the bottom. Based on price sensitivity, usage, time to contract end and loyalty program membership status, we can see that this valuable customer has a 100% probability of churning. So up pops a script with a retention lure: Golden Ring membership, with discounts on phone calls and family entertainment.

also lets organizations keep strict tabs on customer-specific service level agreements.

SavilleCare offers modules for integration management, for external billing, customer care, workflow, order and inventory management systems. An Inbound/Outbound call flow module includes scripting for outbound telemarketing sales campaigns and interfaces to predictive dialers. There's also scripting and screen pop functionality available for inbound calls. A workflow module is used to assign tasks to appropriate staff. A module for analysis provides out-of-box searching tools for ad hoc and regular reporting and immediate update to service reps. Web-based customer self-care is in development at Saville. A very comprehensive solution, CBP runs on UNIX and AS/400 platforms.

Mid-range Systems:

"The Holy Grail of billing systems – no vendor has one yet – is one that runs off the shelf, 'a la Windows,'" says billing expert Ofrane. The smaller packages, mostly NT based and meant for subscriber bases of under 500,000, are generally closer to that grail. They're easier to implement. But they're also much more limited functionally. They're not in the same league as the complicated products such as virtual private networks that telcos often supply to multinational corporations. They cost less.

They also generally don't come with sophisticated data mining algorithms and market analysis, and they don't pipe upselling scripts into CT-integrated screen pops. But to varying degrees, these mid-range billing products address the market-spotting and churn-prevention issue. Some get quite sophisticated, some merely allow you to output an unlimited number of reports, perhaps based on customized data fields. Some pipe their customer records into third-party data mining tools or CRM engines. Some are already offering web-based customer care.

For example, Billing Concepts (San Antonio, TX – 210-949-7020, www.billingconcepts.com) offers a B&CC package called Modular Business Applications, based on the IBM/AS400 minicomputer. It bills for LD, local phone service, wireless, Internet, cable TV and utilities.

Billing Concepts has a long history as a clearinghouse and intermediary between LEC billers and LD service providers. They still have about 65% of that market. Their acquisition of the CRM corporation bought them entree into the convergent billing software world. Their only nod to customer retention, aside from accurate bills, is the convenience and multi-relationship nature of convergent billing itself. Billing Concepts sells a site license or takes on the billing and customer care task as an outsourcer.

Daleen Technologies (Boca Raton, FL – 561-997-1612, www.daleen.com), found in 1989, offers BillPlex, which they say can produce bills for a subscriber base comprising over one million access lines.

A Unix configuration of BillPlex has been benchmarked using a 16-way Sun E6000 server, and 16 powerful 333 MHz processors, to bill 5.94 million usage records per hour. On NT, with a 4-way Xeon NT server and 12 Pentium 450 MHz processors, BillPlex billed 3.24 million usage records per hour, with capacity to support over 600,000 access lines per month. Both environments were running Oracle 8.0 RDBMS, and utilized EMC2 storage.

Daleen pitches BillPlex to optimists who may be starting with as few as 20,000 subscribers but who plan to expand quickly. The BillPlex system price scales up with customer base. Since any ratable usage data can be added, they can convergently bill across multiple services.

Table-driven, BillPlex allows for the quick introduction of new services and marketing promotions, and can also put targeted promotions and incentives on printed customer bills. Release 2.1 has an optional web interface for self-care. 2.1 also comes with four canned market packages, for CLEC, LD, Wireless and IP and DSL services. "Lock-box" payment processing debits and credits from financial institutions. Open APIs allow BillPlex to bolt onto preexisting OSSes. For data mining and churn management, it supports interfaces to third-party vendors NCR, SLP, Redbrick, and GTE. For CRM support, bolt it to Siebel, Clarify, Vantive, Versatility, GERS, and Lightbridge. Integrated with customer care, usage information to agents is near real-time.

Info Directions (Victor, NY -716-924-4110, www.infodirections.com) offers their CostGuard 2.0 billing and customer care package. It's NT-based and aimed at start-ups who want to grow with their system. Info Directions crams a great deal of functionality into their scalable billing system, their contribution to customer retention and identification is a usage analysis tool called Margin Guard. MarginGuard processes and aggregates usage records by month or year-to-date, giving views by customer, geographical region, product summary, sales/affinity group, or LATA.

CostGuard comes in three basic sizes: A six-LAN client, 500-record-per second system, CostGuard AXS, for $90,000 to $100,000, sold primarily to switchless resellers; a 200-client, 1,200 record-per-second

CostGuard Plus for around $1,000 per client, and a multi-site, 200-plus-record-per second Enterprise edition, starting at $600,000, for up to 10 million customers. The first two run on SQL server, while the low-end product gets by with Microsoft Access.

MaxBill (Ramat HaSharon, Israel – 972-3-547-2498) is a three-year startup offering an NT-based, convergent B&CC system with successful installations in England and the Continent. "Our system is rule-based, based on events that are easily user configurable," says Michah Himmelman, the president. "Our system has behind it a dynamic database that reacts to events such as a customer whose check bounces for the third time."

How about good customers? "In marketing, if a customer meets some kind of rule that validates him for specific types of discount, the system can automatically generate an offer letter or a message that will go into his next invoice." The Modular MaxBill package offers rating engine,

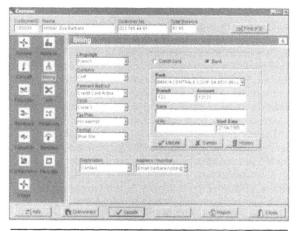

MaxBill's basic customer care screen. One merely clicks for history, follow-up, and other drill-down views.

task scheduling, customer care, general ledger, AR, and reporting pieces.

Preferred database is Microsoft SQL Server 6.5. Third-party tools can report traffic as often as hourly. A basic/minimal license fee, with customization, implementation, installation, and training is around $250,000.

Mind CTI (Yokneam, Israel – 972-4-003-7773 and Englewood Cliffs, NJ – 201-569-6967, www.mindcti.com) can do IP telephony billing with their NT-based IphonEX billing and customer care package. When the company branched out from enterprise call management to telco billing, they aimed first at VoIP, where the great majority of their customers are still found.

But IphonEX can take records from a CO switch as well as a VoIP gateway. Lior Salansky, VP of Business Development, stated that a PSTN version of IphonEX, called MindBill, has several impressive installations: One is a four-to-five-switch carrier in Italy, another is a German reseller who gets data from four carriers, and yet anther in Israel that retrieves data from cellular switches. Unlike IphonEX, MIND-Bill does not support pre-paid bills. Other functionality is the same as the new IphonEX 3.1.

Mind's web-based (or LAN-based) agent screens categorize customers graphically into groups such as debit customers with recharge-able cards, limited credit customers, and unlimited credit customers. Consider, too, that the web-based customer inquiry is also included with this mid-range product.

MindBill also lets you define (and report on) new customer fields. It also has has a CRM module – the Customer Management Journal – that

handles registration, sales, and follow-up task assignments, all color coded to specific agents. It also documents all customer interactions, be they phone, fax, e-mail, or entries to invoice, payments and adjustments tables.

Mind's built-in traffic analysis lets you see who makes the longest calls, and which destinations pull in the longest average call times. "You can do what-if analyses to see how new tariffs to those destinations might affect your revenue," says Salansky. "What if after three minutes or ten minutes I reduce the rate to that destination? That might encourage longer calls and higher revenue."

"Do a report of customers who've complained in the last month," suggests Salansky. With the CRM module, you can create a new outdialing or e-mailing campaign targeted at these at-risk customers.

MindBill systems start at $50,000. While the NT version is suitable for hundreds of thousands of customers, a Unix version will support over a million. It's also based on Oracle database. Convergent billing is on their horizon, they say.

Proxima (Montreal, Quebec – 514 875 5403, www.proxima.com), a company with roots in cable billing, sells a new telephony, cable and IP convergent billing and customer care platform called ProMedia.

ProMedia is modular, scaleable, built from scratch, and has an installation in EDS' carrier outsourcing center in Spain.

Proxima's pitch – its website and white papers – pays a lot of attention to the integration of billing with marketing. The customer profile occupies the center of the ProMedia model, its CRM screens are clear and comprehensive. Each customer category determines in part what products are available to the customer, what rates he or she pays, the terms and conditions of the service contract and tax liabilities.

ProMedia's included Marketing and Sales module helps create scripts and manage campaigns, promotions, trial periods, and purchase "loyalty" points. Reports can be generated with Crystal Reports (run-time included) or third-party tools to present, measure and interpret such variables as market penetration, customer retention rates and acquisition rates. For the agent, ProMedia detects whether customers are good targets for additional promotions, and lets her drill down to see what additional products or services might be purchased to qualify.

ProMedia also stores prospects. They offer systems intelligible in English, French, Spanish, German, Dutch, and Portuguese.

All invoicing, payment and usage history gets attached to customer profiles. Data can be stored in Oracle, Sybase, Informix and other databases, within reach of "thousands of tools" for analysis. "We let operators decide what information they want to keep on their customer profiles," says Sylvain Tétreault, Proxima's marketing manager. Optional fields might hold the number of extensions, televisions, or children at a location.

Tétreault's example: An operator designs his business process and his customer profile to note children's ages. "If I as a [convergent] operator see a high percentage of these families, I decide to offer the Disney Channel for half price for Christmas week. Right away you can go into ProMedia, create the service at half price, design a targeted mailing or a telemarketing campaign. The information is there, accessed very easily." An alert is also raised to the CSR when a qualifying customer calls in.

Win-back scenario: "When someone leaves the operator, we keep a history of that person and the reason why he churned. If the operator decides to issue a new package at a lower price, he can extract all those who left for price and contact them. Frequent-flyer scenario: "The operator wants to credit one mile per five dollar of long distance usage, or per hour of Internet connection. Again, the system allows him to configure his business process to suit."

ProMedia uses IBM's VisualAge Smalltalk tool for object-oriented development of the Windows-based GUI and application layer. GemStone's object management system is used for the management of distributed objects in the application layer and for enterprise connectivity. Although GemStone includes an object-oriented database management system, RDMS technologies, such Oracle, Sybase, Informix and DB2, may used as well.

ProMedia runs on major UNIX, as well as Windows NT server platforms. As for the cost, TÈtreault posits a carrier with 100,000 subs who spends one million dollars up front on ProMedia's installation, integration and training. If his average revenue per customer per month is $30, he says, and increases that by only $2 because he is now able to cross-sell and upsell, he's recouped his investment is six months.

Bit

The smallest amount of information that can be transmitted. Takes the values of 1 or zero (0) in binary digital communications.

Bit Error Rate (BER)

As its name implies, BER is the rate at which errors appear in datastreams. The BER may be expressed in terms of a percentage of error-free seconds or as a percentage of error-free bits.

BLEC

See Building Local Exchange Carrier.

Bong

An interactive signal prompting an originating end user to enter more information.

bps

bits per second

Bps

Bytes per second.

BRI

Basic Rate Interface. See Basic Rate Interface, ISDN.

Bridge

A internetworking device to connect LANS of a similar type (e.g. using the same protocol), but filters and selectively passes frames of data based on MAC Layer 2 information. Bridges do not perform full routing functions, but bridges can learn which addresses are associated with which network and develop a learning table so messages can be forwarded to the correct network.

British Approvals Board for Telecommunications (BABT)

This is a private, independent company and the leading telecommunications approval body in Europe. BABT certifies products and services in the fields of IT and telecom. (www.babt.co.uk).

Broadband

This is a high-capacity communications circuit or path. What "high-capacity" means, however, has been a bone of contention. Some experts consider high capacity to be a speed greater than a T-1 (1.544 Mbps), but the earliest definition specifies any channel with more bandwidth than a single standard voice grade channel.

Broadband ISDN

High speed ISDN based on ATM technology (See Asynchronous Transfer Mode).

Brouter

An amusing term referring to a bridge that also has some of the characteristics of a router (e.g. can consult a routing table to route messages between networks).

Building Local Exchange Carrier (BLEC)

A new kind of service provider that operates at the building level, a new approach to addressing the "last mile" problem in the Competitive Local Exchange Carrier (CLEC) marketplace.

For example, Gillette Global Network, Inc. (New York, NY – 212-906-0100, www.ggn.com) installs its own telecommunications Point of Presence (POP) in buildings. GGN delivers its services in commercial office buildings via their exclusive Broadband Building Infrastructure Deployment System (BBIDS). This involves "lighting" new and existing commercial buildings, which means that GGN installs a broadband, fiber-optic telecom infrastructure capable of delivering voice, data, video and Internet services to commercial tenants in those buildings.

Once GGN installs its fiber-optic infrastructure, the building is considered "lit." This indicates that GGN has become "the building phone company" and Internet Service Provider (ISP) for that building, able to provide all voice and data services to the tenants. Top-notch BLECs such as GGN offer an array of voice and data services to customers while other BLECs may only offer either a data-only solution or a voice-only solution.

Owners of buildings are increasingly interested in having their buildings "lit" because it makes the buildings more competitive in attracting and keeping tenants. By providing a "turnkey" telecom solution for building tenants, a lit building increases the building's value and allows the owner to command premium rental rates. The BLEC absorbs the costs of installing an advanced fiber-optic infrastructure in each building.

There are many other advantages related to BLECs. The benefits to a tenant in a GGN lit building, for example, include the following:
- Rapid installation and instant availability of voice and data services for new or existing tenants.
- High level of tenant support service. Since each building has a dedicated service technician, GGN becomes and "in-house" phone company and ISP for each tenant.
- Increased broadband capabilities. A fiber-optic infrastructure is capabile of offering multi-megabit per second service on demand and a true broadband experience for the tenant; and
- Cost savings – because GGN has wired the building, it controls and aggregates the voice and data traffic, enabling GGN to offer services at a more aggressive rate. The BLEC can bundle local, long distance and data services on one billing platform.
- Tenants are free to choose any provider for their voice and data services, but its usually more cost-effective and efficient to use the BLEC.

Bursty

Also called "Batchy." communications characterized by sudden high volumes of data. Ethernet and packet switched networks tend to be "bursty."

Cable Modem (CM)

Also called a Subscriber Unit (SU). Just as its analog predecessor that we have used for years to connect computers through the PSTN to ISPs and the Internet, the cable modem, instead of modulating and analog tones, transmits and receives RF carrier signals over a coaxial cable and converts them back into Internet Protocol (IP) packets, thus enabling voice and data to be delivered over a cable TV (CATV) network.

Despite this superficial similarity with analog modems, however, cable modems are a bit more complicated. They include a tuner to separate the data signal from the TV broadcast stream (integrated with a diplexer to allow for both upstream and downstream signals) as well as components from network adapters, bridges, and routers so you can connect to multiple computers. Network management software runs on cable modems so your friendly local cable company can monitor its operations, and encryption is used to prevent your data from being intercepted by someone else.

At the "headend," or central distribution point of the CATV system is where can be found the Cable Modem Termination System (CMTS), the central device for connecting the CATV network to the Internet or some other data network. At the headend, signals are combined with those from satellites and local sources, then rebroadcast through the CATV network.

One CMTS has a "downstream frequency" to cable modems in the 42 to 850 MHz range depending on plant capabilities, and can normally supply service to about 2,500 simultaneous Cable Modem users on a single TV channel. If more cable modems are placed on the network, more TV channel capacity can be added to the CMTS.

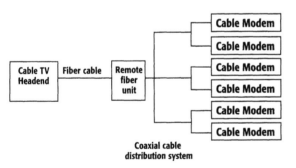

Coaxial cable distribution system

Early CATV networks consisted entirely of coaxial cable transmitting at around 330 MHz to 450 MHz to handle 60 channels of 6 MHz NTSC channels, but modern cable networks can run at 850 MHz or more, provisioned with optical fibers (to reduce signal noise) fanning out in a star topology to nodes situated in subscriber neighborhoods, whereupon the signals are transferred to a local trunk-and-branch cable system that covers the "last mile" to the subscriber premises. This optical/coaxial configuration is known as a Hybrid Fiber-Coax (HFC) network.

A number of modulation techniques have been tried with cable modem transmissions. 64-state Quadrature Amplitude Modulation (64QAM) is an efficient modulation technique used for sending data downstream. It can support up to a 36 Mbps peak transfer rate over a single 6 MHz NTSC channel. However, it is susceptible to interference when used for upstream transmissions to the headend and the Internet.

For upstream traffic, Quadrature Phase Shift Keying (QPSK) is a modulation technique that works much better, though one does not generally encounter cable systems that allow more than a few megabits per second for upstream traffic per cable modem over frequencies in the five to 42 MHz range in the U.S. and five to 65 MHz in Europe.

At the subscribers premises, many cable modems have a cable connection on one side and on the other a standard RJ-45 connector so it can connect to a 10BaseT Ethernet card installed in the subscriber's computer. The computer runs the standard TCP/IP protocol, allowing the computer to act as if it is directly connected to the Internet via an Ethernet connection. Cable is "always on" just like ISDN and xDSL, so there is nothing to "dial up". Actual cable modem configurations include:

- External cable modems, some of which use USB ports to connect to the PC.
- Internal cable modems, which are basically PCI bus plug-in cards.
- Interactive Set-Top Boxes that are really designed to provide additional TV channels on the same bandwidth via Digital Video Broadcasting (DVB). The upstream "return channel" with these devices is often through the PSTN, and enables the TV viewer to surf the web and check e-mail on the TV screen.

If cable modems can have a downstream transfer rate of 28 Mbps or more, one wonders why the data port interface is 10 Mbps Ethernet and not 100 Mbps, downstream. Indeed, in Europe the DVB/DAVIC standard does allow any type of interface, but there are various factors that keep cable modem performance equal to or below that of 10 Mbps Ethernet.

For example, there is the well-known controversy over the fact that cable modems share bandwidth to the cable headend, just as users in an office share bandwidth on a LAN. Hundreds of users sharing the bandwidth can cut individual subscriber bandwidth to a 100 Kbps or so, but things are not quite as bad as it appears.

Since the cable modem doesn't use a fixed bandwidth, if no one else on the system is using their modem (reading text from a screen, for example), then the full bandwidth is available for that time interval, even if it's just for a moment. Thus, cable modem traffic is "bursty" just like Ethernet LANs. A bottleneck of greater concern is the Internet gateway at the cable company's headend, and the fact that you can't choose your own ISP – at least not until the government grants cable systems the same "common carrier" status as the telcos.

The real problem with shared bandwidth concerns security, particularly in those systems where cable modem traffic fails to be encrypted properly.

Unlike Europe, the U.S. saw a faster roll-out of cable modems than ADSL or ADSL Lite (G.Lite) service.

An organization called Multimedia Cable Network System Partners Limited (MCNS) is developing and standardizing cable modem interface specifications in the U.S.

Some relevant cable modem websites:

Cable Modem University: www.catv.org/modem

Information on cable and xDSL modems: www.cablemodeminfo.com

The IEEE 802.14 Cable TV Protocol Working Group is the standards group developing digital cable transmission systems: http://walking-dog.com/catv/index.html

Cable Datacom News, a new on-line publication about digital cable: http://CableDatacomNews.com/

Cable Telephony

Experiments in providing voice service over cable television systems occurred as long ago as 1979, though it was not until the craze over "convergence" in the mid 1990s that the world took seriously the idea that both residential and business customers could subscribe to telephony, Internet, and entertainment services through a single cable provider.

AT&T started the industry in early 1999 when, on the heels of its merger with Tele-Communications, Inc., it announced it was forming a joint venture with Time Warner Inc. to offer cable telephony though Time Warner's cable lines in 33 states, giving AT&T access to a total of more than 40% of U.S. households. Services would include multiple phone lines per household, conference calling, call waiting, call forwarding, and individual message boxes.

Ironically, announcements such as these have frightened the Regional Bell Operating Companies (RBOCs) to the point where they are offering video services over xDSL in an attempt to nullify cable's intrusion into their territory! But that is another story.

Cable networks are a force to be reckoned with among broadband technologies, and there is little doubt that cable modems will be showing up in quite a few small offices and homes over the next few years. And while it hasn't been accompanied by the same amount of mainstream fanfare, one can argue without much hesitation that voice is going to be a key part of those bundled cable services. Cable telephony is already a reality, and quality Voice-over-IP for cable networks is not far behind.

IP technology has always been a moving target. Just as the field appears to stabilize, media pundits start yawning and the public becomes antsy over whatever new enticing technology an be viewed on the horizon.

New Yorkers have been repeatedly exposed to cable TV operator RCN's advertisements trumpeting a campaign to sell competitive phone and dialup Internet service almost as a sort of revolutionary politics. Now, of course, they've added high-speed Net access via cable modems to their mix of services. The way this type of "first-generation" cable telephone network actually works is not much different from that of a standard telco, except that the cable telephony network has Radio Frequency to Public Services Telephone Network (RF-to-PSTN) gateways.

At the "headend" in the cable provider's facility, a Host Digital Terminal (HDT) converts PSTN signals to and from a Class 5 CO switch into RF in the five to 42 MHz range. At the customer premise, a Network Interface Device (NID), which can be attached at the cable demarc (outside the house, in the basement, etc.) or built into the cable TV set-top box, pulls RF off the wire and converts it back to standard analog; attaching to conventional twisted pair copper wires, and thence to the phone equipment.

To the subscriber, this local loop process is more or less transparent: You plug phones into ordinary wall jacks and get the same calling features as the Bells offer. What's so appealing about this is that you now deal with just one communications provider for both phone and TV service. Obtaining multiple services becomes quite easy.

But even as some cable providers built out their networks to provide plain old voice using HDTs and NIDs, the Internet explosion accelerated development – and now, rapid deployment – of a different system, aimed at providing data connectivity. In such a system the "gateways" are cable modems connecting via the CATV network to Cable Modem Termination Systems (CMTSs) at the cable company's central headend.

While it's possible for the two types of gateways (HDT/NID and CMTS/CM) to work over the same strand of coax, it's redundant to maintain separate transmission mechanisms for voice and data. In fact, it's redundant to maintain a separate mechanism for video, too.

A more elegant solution eliminates RF voice in favor of IP telephony in the cable data channel. The opportunity for Multiple Service Operators (MSOs) is more or less obvious: Offer more services to subscribers with less of an investment in infrastructure and operations than would be needed in a circuit-switched environment. Also, there needn't be much fear over whether the market will be receptive or not, since cable modems provide a nice "fat pipe" to the home, satisfying consumers' ever-growing hunger for bandwidth, while lowering costs for local and long distance service and multiple phone numbers. All services can be delivered quickly and cheaply.

Cable companies are now hurriedly seeking customers in the residential market, a market where they have an existing presence and subscriber base. But as VoIP technologies continue evolving, and integrated service offerings like Virtual Private Networks (VPNs) (See Virtual Private Network) become possible, it is likely that cable will target small- to medium-sized businesses as well, putting cable modems in ever more direct competition with DSL.

We've established that VoIP makes it relatively simple to provide multiple voice channels and phone numbers, while RF approaches are limited to one or two "analog lines." But it's also not a particularly tremendous trick to make VoIP emulate T-1 (and DNIS, ANI, etc.) at the premise, affording "cable phone service subscriber" connectivity to high-density telephony equipment (key systems, IVR boxes, etc.) that were originally designed to connect with conventional digital service.

Whether or not cable operators currently offer telephony services, cable operators are going to have to give serious thought to Voice-over-IP in order to stay competitive. Several vendors, as you'll see below, have already announced combined VoIP / high-speed data access equipment for both the customer premise and the headend, and are just now beginning to appear.

Because it's still in a relatively formative stage, the range of equipment a cable operator can use to provide VoIP is somewhat more varied (particularly at the customer premise) than what's used for circuit-switched cable telephony. In a data-enabled cable network, the headend CMTS – the same headend gear used today to terminate data traffic from cable modems – may be upgraded (or replaced) to provide features for VoIP quality of service and system reliability. From the CMTS, packets can travel to an edge router and onto the IP backbone, or through a VoIP gateway and onto the PSTN.

At the subscriber's site (residence or SOHO), you have a few options. Some, like Nortel's Arris Interactive, are building a more sophisticated Network Interface Unit that doubles as a VoIP gateway and mounts on the side of the subscriber's home. Others, such as 8x8 and Nokia, provide compact add-on products that attach to cable modems inside the home, giving them voice capability. These modules typically incorporate DSPs and offer a couple of standard RJ-11 jacks to plug in regular phones or faxes. Most vendors offering an add-on voice module predict that its capabilities will be integrated into the cable modem itself as the technology develops, as is the case with Motorola's MTA and the PhoneLink adapter, which is available now.

Each of these approaches (outside-wall-mounted NIU and indoor, enhanced cable modem) has its pros and cons. NIUs have the advantage of being transparent to end users, who just plug their phones into the wall and tack a relatively unfamiliar box to the outside of their homes, as they've always done. The disadvantage, of course, is that such units can be costlier and require a more involved installation process than is perhaps necessary. While voice-enabled cable modems are cheap and easy to install (no need for additional wiring or truck rolls), they are still relatively unfamiliar to consumers. More importantly, these units can raise some significant power supply issues. Many rely on AC power, which means that in case of a power outage you could lose your phone service along with the electricity.

The question of power for CPE is, of course, only one issue out of many affecting widespread deployment of IP telephony over cable networks. Many of the other issues, as one might suspect, revolve around Quality of Service (QoS).

CableLabs (Louisville, CO – 303-661-9100, www.cablelabs.com) is a body formed to develop standards for cable voice and data equipment, governed mainly by CEOs of cable companies and made up of 250+ different vendors and manufacturers. CableLabs maintains several working groups, such as OpenCable, PacketCable, and CableModem, dedicated to defining specs for various aspects of a cable network. The CableModem initiative, begun in 1996, is well known for its development of the Data over Cable Service Interface Specification (DOCSIS). Since that time, 10 modems on the market have earned DOCSIS certification, and some headend Cable Modem Termination Systems have been qualified as DOCSIS-compliant.

In 1999, the PacketCable group issued its first specifications for IP telephony that uses a native cable form of Multimedia Gateway Control Protocol (MGCP) – which comes under CableLabs' umbrella term of Network-based Call signaling, or NCS – which were accompanied by release of the initial specs for DOCSIS 1.1.

The NCS profile of MGCP, which is known as the Network-based Call Signaling Protocol, NCS 1.x, the NCS profile, or simply NCS, has been modified from the CableLabs / PacketCable MGCP 0.1 draft in several ways – for example, the NCS protocol only aims at supporting PacketCable, the NCS protocol contains some extensions and modifications to MGCP, and the NCS protocol contains some minor simplifications from MGCP. Still, although MGCP is not NCS, and NCS is not MGCP, the names MGCP and NCS are generally used interchangeably, and "MGCP" in terms of cable telephony is taken to mean the NCS profile of MGCP.

Meanwhile, DOCSIS 1.1 addresses the problems of fragmentation and QoS for VoIP-enabled CMTSs and cable modems (voice packets are handled faster than data packets which aren't as sensitive to latency), while PacketCable seeks to define interoperability standards for any equipment used to offer packet-based multimedia services over two-way HFC plants.

The MGCP assumes a call control architecture where the call control "intelligence" is outside the gateways and handled by external call-control elements referred to as call-control elements, or Call Agents (CAs). The MGCP assumes that these CAs will synchronize with each other to send coherent commands to the gateways under their control.

The MGCP assumes a connection model where the basic constructs are endpoints and connections. A gateway contains a collection of endpoints, which are sources, or sinks, of data and could be physical or virtual. An example of a physical endpoint is an interface on a gateway that terminates an analog POTS connection to a phone, key system, PBX, etc. A gateway that terminates residential POTS lines (to phones) is called a "residential gateway" or an "embedded client." Embedded clients may optionally support video as well. An example of a virtual endpoint is an audio source in an audio-content server. Creation of physical endpoints requires hardware installation, while creation of virtual endpoints can be accomplished by software. However, the NCS profile of MGCP only addresses physical endpoints.

Connections are point-to-point. A point-to-point connection is an association between two endpoints with the purpose of transmitting data between these endpoints. Once this association is established for both endpoints, data transfer between these endpoints can take place.

NSC / MGCP call signaling specifies an instruction set used to deliver call setup and feature commands from the call controller situated at the headend to the customer premise-installed Broadband Telephony Interfaces (BTIs) and other types of distributed end-user premises gateways. The signals are transformed to and from packets in the gateways, so standard touchtone can be connected to larger IP networks via ordinary twisted-pair copper wires. MGCP also solves the problem of how the conventional Signaling System 7 (SS7) network's call setup and direction commands used by the PSTN can be channeled through IP networks.

Call Agents instruct the gateways to create connections between endpoints and to detect certain events, e.g., off-hook, and generate certain signals, e.g., ringing. It is strictly up to the Call Agent to specify how and when connections are made, between which endpoints they are made, as well as what events and signals are to be detected and generated on the endpoints. The gateway, thereby, becomes a simple device, without any call state, that receives general instructions from the Call Agent without any need to worry about or even understand the concept of calls, call states, features, or feature interactions. When new services are introduced, customer profiles changed, etc., the changes are transparent to the gateway. The Call Agents implement the changes and generate the appropriate new mix of instructions to the gateways for the changes made. Whenever the gateway reboots, it will come up in a clean state and simply carry out the Call Agent's instructions as they are received.

Call agent functions can be handled by servers at any location in the network, furnishing interfaces to both legacy Class 5 switches and the new generation of IP savvy, intelligent networking switches and devices that will enable IP enhanced services far more powerful and sophisticated than anything the could be offered via conventional PSTN technology.

The resulting new products will represent new classes of services running over cable-based packet communication networks. Such new service classes include phone calls and videoconferencing over cable networks and the Internet. These services will be delivered using the basic Internet Protocol (IP) technology that is currently used to send data via the Internet. DOCSIS 1.1 incorporates a variety of traffic engineering techniques and protocols, including those used in edge and backbone routing applications, such as DiffServ and RSVP.

The basic idea with these new standards is to create a flow of voice packets that duplicates the Constant Bit Rate (CBR) stream technology currently enjoyed by users of circuit-switched cable telephony loops. Data networks, of course, do not require a CBR and so the typical problems with doing Voice-over-IP actually involve jitter and delay issues. DOCSIS 1.1 attempts to solve these problems by building intelligence into the MAP file that the CMTS sends out to cable modems (scheduling the times granted for sending and receiving of packets), and by manipulating the Type of Service (TOS) field in the IP packet header to accommodate QoS.

DOCSIS 1.1 certified modems began to appear throughout the year 2000. PacketCable will hold additional rounds of interoperability testing on products ranging from voice-capable cable modems to call agents.

As an aside, AT&T is promoting the Distributed Open Signaling Architecture (DOSA) that demands a higher technical sophistication in user devices than MGCP and hasn't been fully developed. AT&T insists that DOSA, which is based on a new version of the Session Initiation Protocol called SIP-Plus, has better security features and supports provisioning across multiple client-server arrays more flexibly than does NCS / MGCP.

Indeed, until the necessary standards have been set and approved, it's not likely that cable VoIP is going to worry the ILECs. But, as you'll see from the roundup of companies and their products below, the vendors haven't been waiting around to follow someone else's lead. Cisco, for example, is working actively with two major Canadian cable companies to roll out telco-grade IP voice services as soon as possible. Others, like Nokia, make a strong case for MSOs to bundle second- and third-line residential service with high-speed data now, and invest in primary line and business telephone infrastructure as it starts to become more widely accepted – and when people are willing to put more trust in the abilities of their cable providers. It's a great concept for eliminating churn: Give people two additional phone lines today, with all the peripheral services they're used to such as Caller ID and call waiting; cut out the expense and waiting that traditionally had been endured by subscribers to install a phone circuit; and you're likely to hold their interest as customers, for a long time.

All the vendors we've talked to have grandiose visions for the future and surprisingly intelligent plans to back up their aspirations. But you'll also find a whole slew of products you can buy and leverage immediately – from plug-and-play end user equipment, to sophisticated end-to-end platforms. One of the exciting things about VoIP is that there isn't just one "right" way to do it. If you're an ambitious cable operator aspiring to offer multiple services to your customers, it's important to invest in equipment that lets you grow by getting the most out of the network you already have. Whatever the size or shape of that legacy cable network, the following companies can help you to transform it quickly:

VoIP Platforms for Cable:

Cisco Systems' (San Jose, CA – 800-553-NETS, www.cisco.com) uBR7246 is more than just a Cable Modem Termination System (CMTS). To be precise, it's a CMTS attached to a Cisco router, which not only gives you two products in one, but also creates an optimal platform for integrating cable with IP.

The uBR7246, a member of Cisco's universal broadband router family, is based on DOCSIS standards, to which Cisco contributed significantly in the area of voice QoS. DOCSIS 1.1 uses "unsolicited grants" to create a constant bit rate-like stream between the CMTS and the user's cable modem, reducing jitter and delay on VoIP calls travelling over HFC. In addition, the 7246's built-in routing capabilities provide support for the Resource Reservation Protocol (RSVP), which means that backbone QoS features can be extended all the way to the headend.

Included uBR7246's chassis are four slots for cable modem cards that interface with the HFC plant, two slots for port adapters, one slot for the Input / Output (I/O) controller, and one slot to hold a network processing engine. The I/O controller contains two PCMIA slots for Flash Memory cards, and an optional fast Ethernet port for 100 Mbps connection to the network. Port adapters support a variety of interfaces, ranging from 10BaseT Ethernet or 100BaseT fast Ethernet, to T1/E1, to OC-3 connections for Packet-over-SONET and ATM backbones. Cable modem cards are available with one upstream and one downstream port, or six upstream ports and one downstream. Modem cards and port adapters can be inserted and removed while the system is running. Optional redundant, hot swap power supplies are also available.

In addition to producing headend products, as well as all the backbone equipment that an MSO would need to build an HFC-based, voice/data over IP network, Cisco also offers the uBR924 cable access router. The uBR924 is a DOCSIS-compliant cable modem and router (with built-in DSPs) that offers network broadband data access and VoIP capability to SOHO environments. The unit includes a single F-connector

to interface with HFC; four RJ-45 Ethernet ports; two RJ-11 phone ports which support VoIP transmission via H.323; one RJ-11 port to provide an analog line for backup POTS service; and one RJ-45 console port to connect to an ASCII terminal or PC for local troubleshooting and reconfiguration. Using DHCP, the uBR924 dynamically assigns an IP subnet address to the cable access router each time it connects to the network.

Le Groupe Videotron ltée (Quebec, Canada – 514-380-4000, www.videotron.com) is currently leading trials using Cisco's platform to offer telco-grade VoIP service along with high-speed Internet access to customers. Videotron is testing the uBR7246 headend, together with Cisco's AS5300 VoIP gateway, 12000 series Gigabit Switch Routers, 8500 Multiservice Switch Router, Telcordia's Intelligent Gateway Call Server, and Samsung cable modems that incorporate Cisco IOS software and ASICs from Broadcom.

Lucent Technologies (Murray Hill, NJ – 888-4LUCENT, www.lucent.com) and Motorola's Multimedia Group announced an agreement to develop CableConnect, an end-to-end, integrated VoIP solution for cable operators. CableConnect is ahead of the game in terms of IP integration, and the prospects it raises for both the network and the connected residence are exciting, to say the least. Essentially, the platform combines advanced cable modems (headend and CPE) from Motorola with Lucent's PathStar Access Server and 7R/E Packet Solutions.

PathStar is everything a carrier needs to offer voice-over-IP, all in one box. Actually, two or three boxes. The system has two main components: The access shelf and the data shelf. The data shelf, which resides at the carrier's CO, replaces the functions traditionally handled by a Class 5 telephony circuit switch, digital loop carrier, VoIP gateway, and a network edge router. Using routing protocols for a variety of interfaces, the data shelf handles traffic to and from the PSTN, IP/ATM backbones, and Signal Transfer Points in an SS7 network.

One level down in the network from the data shelf sits the access shelf (up to 14 in a single PathStar system). The access shelf replaces the functions of a remote access server, and can also terminate up to 512 POTS lines, converting voice circuits into IP packets before sending them onto the data shelf. PathStar accommodates more or less any end user device (IP phone, POTS handset, PC), and it can offer almost any voice or data service (full-featured local/long distance telephony, Centrex, high-speed Internet access, VPNs). Lucent has added enhancements to PathStar for deployment in cable networks, including support for CableLabs' PacketCable call signaling model. PathStar would typically be used in smaller networks, serving around 50 to 100 homes per mile, or from 7,000 to 20,000 total subscribers.

7R/E Packet Solutions is a distributed architecture made up of multiple components that lets MSOs offer large scale (100,000 or more subscribers), primary line local and long distance service using VoIP technology. The system offers over 3,000 features and an open API. 7R/E's Call Feature Server processes calls across the network and supports services like 911, call waiting, caller ID, and Lucent's new iCentrex. The Programmable Feature Server lets you create and tear down special services such as promotional campaigns, and supports third-party development of applications. 7R/E Packet Gateways offer narrow and broadband access to virtually any type of packet network, and feature enhanced voice quality through Lucent's Echo Cancellers and other devices.

Motorola's contribution to the system is PhoneLink / MTA, a line of customer premise units that connect PCs and phones to the HFC plant, coupled with the Cable Router Cable Modem Termination System, which terminates cable modem traffic at the headend. On the backbone side,

C

the CMTS interfaces with PathStar. Currently the two communicate via ATM, but plans for OC links are being developed. The CMTS is DOCSIS-compliant in its latest release, and can serve both new DOCSIS modems as well as Motorola's legacy cable modems.

The Multiservice Terminal Adapter (MTA) is installed inside the customer premise, and can sit on a desktop aside the PC. For data, the MTA works as a standard DOCSIS compliant cable modem, providing high speed Internet access to the user's PC via an Ethernet link (though Motorola has plans to move to a USB interface in the future). With support for the PhoneLink adapter added on, however, the system can offer a customer up to four different phone lines over a single twisted copper pair. This provides a tremendous opportunity for service providers, who can now offer end users multiple lines to a subscriber's home with no extra wiring installations (in addition to the cost saving benefits inherent to VoIP). Motorola's system uses reserved bandwidth to guarantee quality of service for voice calls, as specified by DOCSIS 1.1.

CableConnect Solutions is sold by both companies, or can be purchased directly from Lucent with Lucent NetCare operational support. Kenan Systems' (Cambridge, MA – 800-77KENAN, www.kenan.com), a subsidiary of Lucent, Arbor/Broadband software can be integrated with CableConnect to perform billing, customer care, and service order management. The system has been in trials with Comcast and other cable operators.

Cable Modems and other Customer Premise Equipment:

8x8's (Santa Clara, CA – 408-727-1885,) Symphony is a four-line telephone adapter that enables broadband modems for Voice-over-IP. The compact Symphony module (about the size of a VHS cassette) attaches to DSL and cable modems via RJ-45 Ethernet connection, and connects to telephone extensions through RJ-11 jacks.

This small CPE device allows cable operators to offer four lines of full-featured local phone service over IP to residential and small business customers. Four simultaneous phone conversations can be maintained, with support for features such as call waiting, call hold, call forwarding, and caller ID. If a second call comes in to a line that is already in use, a call waiting signal notifies the user, and the call rolls over to another extension. You then have the option of flashing the switch hook to answer the call, or letting it be answered at the different extension. Symphony also features other PBX functions such as call hold, call transfer, three-way calling, call forwarding, and caller ID.

Symphony Gateways incorporate 8x8's Audacity Internet Telephony Processor, which compresses and decompresses incoming/outgoing audio signals to provide toll quality voice. Symphony supports a variety of industry standard audio codecs (G.711, G.722, G.723, and G.728), and can change codecs dynamically during a call, in order to respond to network congestion. Full-duplex acoustic echo cancellation on each line enhances voice quality. Symphony is also available in a board-level model, which integrates into the cable or xDSL modem, and also comes in two software versions – one configured for the H.323 call control protocol and one for MGCP. List price in volume is around $399.

Com21, Inc. (Milpitas, CA – 408-953-9100, www.com21.com) makes a variety of voice-enabled cable modems for the office and home. Recently, Com21 announced the Enterprise Telephony Solution (ETS), a product that offers voice connectivity and high-speed data access to telecommuters and remote call center agents. ETS consists of a two-port module, called the Telephony Application Interface Module (T-AIM 200), that plugs into an expansion port on Com21's ComPORT 1000 cable modem. T-AIM contains two RJ-11 jacks for hooking up phones or faxes. Constant Bit Rate (CBR)

traffic flow controls delay and jitter to help insure toll quality voice. Independent paths for voice and data give priority to voice traffic, and dynamic bandwidth allocation provides for efficient use of the pipe. 64 Kbps G.711 voice encoding makes the T-AIM interoperable with most voice-over-ATM and Voice-over-IP gateways. Com21 is currently developing VoATM and VoIP functionality for the device itself, and the new technology will be incorporated via software downloads as it becomes available.

Com21's even more interesting ComPORT 5000 is a combination cable modem and four-port Ethernet hub for the SOHO market. The 5000 provides transfer rates of up to 30 Mbps downstream and up to 2.56 Mbps upstream. It performs DHCP addressing, Network Address Translation, offers an enterprise-level firewall, and an optional IPSec VPN software module.

Com21's DOXport cable modems are primarily residential units designed for high speed data access. The DOCSIS-based modems feature a hardware-based Packet Accelerator Chip that enables higher throughput speeds (43 Mbps downstream, 10 Mbps up) than average (software-based) DOCSIS standard products. The unit includes a RJ-45 Ethernet connection to the user's PC, an F-type connector to the HFC plant, and an expansion slot to add on Com21's Application Interface Modules for telephony and VPN apps. LEDs provide comprehensive transmission status information. In the works are a line of DOCSIS 1.1 based products that will feature a Universal Serial Bus (USB) and built-in VoIP capabilities.

E-tech (Fremont, CA – 888-609-8885, www.e-tech.com) has developed the Integrated Telephony Cable Modem (ITCM), a CPE device that provides VoIP, video, and high-speed data to cable network subscribers. E-Tech describes their product as a conventional cable modem (such as their own CABLazer) with voice functionality added on through DSPs and other internal components.

In its first release (the product is now in beta), the ITCM will contain two RJ-11 phone jacks, an F connector for the coax cable, and a RJ-48 Ethernet interface. A standard application scenario for residential users would be primary and secondary lines for voice calls (or a voice line and a fax line) and an always-on high-speed connection to the Internet.

On its topside, the modem features six LEDs, indicating respectively: power on; active connection to HFC; Ethernet connection; message waiting (can be configured by service providers for e-mail or voice); and two displays showing whether telephone handsets (or other connected devices) are in use. The unit mounts on a wall on the interior of the subscriber's residence. A local power supply with battery backup (six to ten hours standby, about one hour talk-time) is included.

E-tech has joined Cisco Systems' NetWorks Program, licensing Cisco NetWorks software, to develop enhanced IP functions on their modems. Support for telephony features such as Caller ID, call waiting, call forwarding, and conferencing will enable cable operators to use E-Tech modems in primary line deployments of both POTS and VoIP services. All ITCM products will comply with DOCSIS/ITU standards, and will interoperate with Cisco's Universal Broadband Routers. ITCM will be branded as a Cisco NetWorks product, and be co-marketed and supported by the two companies. List price is expected to begin at around $50 over the cost of a conventional, data-only cable modem.

Nokia's (Kanata, ON – 613-591-3219, www.nokia.com) IP Shuttle is a small, slender box that can bring voice-over-IP into your home today. The IP Shuttle bears two RJ-11 jacks for connecting standard analog telephones, and a RJ-45 Ethernet port. The module contains built-in DSPs, and uses G.711 audio codecs to transmit and receive voice. IP Shuttle is not restricted to use in cable applications, but can easily plug into a cable modem, offering MSOs an easy and inexpensive avenue to provide second- and third-line voice services to their subscribers. The device can be configured and updated remotely, using Nokia's IP Client administrative server software. IP Client also features an easy-to-use browser interface that lets subscribers customize their calling preferences.

Nokia also makes IP Courier, an Ethernet-connected business telephone that includes embedded DSPs, allowing users to make VoIP calls without a PC or any other additional devices. The IP Courier can optionally be used in conjunction with a cable modem.

IP Courier and IP Shuttle form only a part of Nokia's extensive VoIP product line. In addition, the company produces back office equipment, a call processing server, and both central and branch office VoIP gateways.

Texas Instruments' (Dallas, TX – 800-336-5236, www.ti.com) two recent acquisitions have made them significant players in the cable modem industry. Libit Signal Processing (Los Altos, CA – 650-947-7800, www.ti.com), bought by TI in June of 1999, makes the core silicon technology that goes into cable modems, set-top boxes, and headend equipment. Libit's products are compliant with DOCSIS and other CableLabs specifications, and enable power efficient two-way transmission over HFC lines.

TI has also purchased Telogy Networks (Germantown, MD – 301-515-8580, www.telogy.com). Telogy develops DSP/microprocessor embedded software used to perform voice-over-IP and fax-over-IP in everything from IP phones and cable modems to VoIP gateways, cable headends, DSLAMs, and core switches and routers.

Putting two and two together, it isn't hard to imagine (and the company has confirmed it for us) that what will emerge from these two purchases is a cable modem chipset that combines Libit's data transport system with Telogy's software, embedded on TI's DSPs. The specific TI product to be employed is the TMS320 C500, a programmable DSP flexible enough to support call processing functionality and adapt to changing technology as voice-enabled cable modems begin to emerge. TI plans to remain active in the DSL market as well, and will use Telogy's software for DSL modem designs similar to those being developed for cable.

Cable Telephony Systems:

ADC Telecommunications' (Minneapolis, MN – 800-366-3891, www.adc.com) Homeworx is a modular platform that transmits video, voice, and data over HFC to subscribers' homes. A traditional cable operator can add voice and data integration to the network by way of the Homeworx Host Digital Terminal (HDT) and Integrated Service Units (ISU).

The Homeworx HDT interfaces between the HFC plant and the PSTN to offer telco-quality phone service to subscribers. Calls travel in the five to 42 Mhz upstream frequency in the local loop, until they reach the headend and are converted back into digital circuits, then passed to a Class 5 switch. The HDT supports up to 28 DS1 or 24 E-1 inputs and up to 240 DS0s over 6 MHz. TR-303, international channel associated signaling (CAS), and V5.2 switch interfaces are all available.

ADC offers a variety of ISUs for homes and businesses, in capacities ranging up to 32 POTS lines. The Home ISU is mounted on the side of a house, and serves as a network interface between the HFC and existing twisted pair wiring. The HISU comes standard with one or two POTS lines and one video connection, and you can add an expansion slot for optional Ethernet connection, ISDN service, or two additional phone lines. The Multi-dwelling ISU (MISU) ranges from 12 to 32 POTS lines, and supports coin, DDS, ISDN, DS1, E1, and Ethernet applications. Built-in cable modems offer up to 8 Mbps data access. All Homeworx ISUs can be remotely provisioned for voice and video services.

OSWorx is ADC's element management system for the Homeworx platform. OSWorx is a Telecommunications Management Network (TMN)-based system that runs on UNIX and offers a full GUI and network topology map. The system helps network managers configure and provision subscriber access, test and monitor for faults, manage performance, and insure security. In addition to OSWorx, ADC offers a number of other customized management tools for all parts of a broad/narrowband network.

ADC has plans to let MSOs migrate their existing Homeworx telephony platforms to VoIP as the technology evolves. The company is currently working to develop the MCNS-compliant gateway and routing equipment necessary to offer the same full-featured voice services currently delivered through Homeworx, while keeping equipment upgrades transparent to end users.

Arris Interactive (Suwanee, GA – 770-622-8400, www.arris-i.com), a Nortel company, lets cable operators and MSOs offer primary line phone service to customers today, with a migration path to VoIP as it matures. Arris' Cornerstone system is a carrier-grade cable telephony platform that consists in two main components.

On the CO or headend side, you have Cornerstone's Host Digital Terminal (HDT). The HDT's Modem Shelf converts incoming (upstream) signals travelling over Hybrid Fiber Coax cables into signals that can be accepted by a digital telephony switch and routed onto the PSTN. The Access Bandwidth Manager Shelf provides interfaces between the modem shelf and the digital switch, as well as interfaces for service provisioning and network management. By employing radio frequency (RF) concentration on the coaxial side, a single-bay HDT can handle services for up to 3,360 subscribers or up to 6,720 lines in optional dual configuration. On the outgoing end, each bay offers up to 28 DS-1 interfaces or 21 E1. GR-303, DMSx, and V5.2 switch interfaces are available. A Spectrum Manager component monitors the upstream spectrum and can automatically change upstream frequency to avoid noise interference in real-time.

The HDT also contains an embedded element management system. Subscriber services offered include custom calling features such as call waiting, call forward, and three-way calling; CLASS services such as caller ID, distinctive ring, and multiple number assignment; and business Centrex service. Cornerstone HDTs are NEBS compliant, with ETSI certification pending.

To communicate with the HDT from a customer premise, Arris offers a variety of products. Cornerstone Network Interface Units begin in two-line POTS models that can be installed either inside or outside a residential or SOHO site. The next step up is a four-line model that features local power supply monitoring. The Multi-Line Voice Port (MVP), designed for businesses and multi-tenant dwelling units, offers up to twelve lines for phone/fax/modem connections and supports analog Centrex services. The top of the line Universal Access Port (UAP) supports Meridian Digital Centrex service for Nortel's GR-303 Meridian Business Sets. The UAP contains two card slots, which can both accommodate Epsilon I POTS and Omega SAA (Service Adaptive Access) line cards from Nortel Networks. The SAA card provides access to services such as ISDN, Foreign Exchange, and Coin applications.

Arris has also announced plans to release a line of end-to-end IP-based products so that cable operators can provide integrated data and VoIP services. For example, the Packet Port is a DOCSIS 1.1 compliant Network Interface Device (NID) that will support up to four lines of telco-grade VoIP service, as well as a high-speed 10BaseT Ethernet connection. Also offered is the Cornerstone high capacity CMTS, a carrier-grade, fully redundant, 99.999% available system supporting up to 6,400 telephony lines. Arris is developing what they call the Cornerstone IP

Access Gateway – a voice gateway that will interface with the HDT via DS-1 or E-1 (with support for up to 672 simultaneous calls), and connect to the IP backbone via 100BaseT. Packet Port ships as part of Nortel Networks' end-to-end VoIP solution for MSOs. Arris Interactive is a Nortel Networks and ANTEC joint venture. Tellabs' (Lisle, IL – 630-378-8800,) Cablespan 2300 is a combination HDT/RSU platform that lets cable operators offer integrated voice/video/data services.

Cablespan Remote Service Units (RSU) are available in one-, two-, and four-line configurations, with a twelve-line model intended for multi-dwelling unit environments now in the works. The RSU is attached to the side of the house, and connects existing inside wiring to the local HFC plant. End users simply plug their phones and/or fax machines (as well as dial-up modems) into wall jacks as usual. For high-speed data services, the unit also offers a RJ-45 Ethernet interface. EXPRESS/path is an optional feature built onto a data card and plugged into the unit that provides a 2 Mbps upstream/downstream connection. Alternatively, a standalone cable modem can be attached to the RSU and in turn connected to the user's PC.

At the headend, the Cablespan Host Digital Terminal (HDT) transmits radio frequencies in the 350 to 800 MHz range on the downstream, and five to 42 MHz up. The HDT features "frequency agility," which allows it to respond dynamically in the case of ingress by automatically shifting frequency. On the backbone side, the HDT interfaces with a local digital switch via T1 or E1 ports and using the GR-303 or V5.2 protocol. Data is switched onto an IP router or edge device via a 10BaseT connection.

With an eye toward the emergence of Voice-over-IP, Tellabs is working on implementing the DOCSIS standard for interface between the HDT and the RSU. This will let MSOs enter the circuit telephony world with a DOCSIS-compliant platform in the near term; and, when the necessary infrastructure for VoIP is in place at the core of the network, to then migrate to IP telephony with simple software upgrades to both the HDT and RSU.

Call Center

Also referred to as a "contact center" by the Gartner Group, the call center is where computer telephony technology finds its most spectacular and innovative expression.

At one time, call centers were specific departments dedicated to centralize and manage incoming calls (internal and external help desks, technical service and problem resolution centers, reservation centers, order taking, customer service) and outgoing calls (telemarketing, collections, fundraising, polls). The calls would be handled directly in the call center or else routed to specialized departments.

Today, however, call centers are simply places where people make or take calls for a living. Entire companies can be a "call center". People telecommuting at home can be part of a virtual ACD group (See Automated Call Distribution) and thus be part of a call center. There can be "informal call centers" (a term coined by Blair Pleasant of the PELORUS Group) where there are people handling phone calls for an organization, but not in a formal location. Call centers can also function as self-sustaining businesses and can include service bureaus or outsourcers, which make or take calls on behalf of clients. (See Call Center Outsourcing.)

Indeed, with the coming of the Internet and the Company website, a "call" does not necessarily have anything to do with voice communications over the public phone network. Call centers can be web enabled (See Web-Enabled Call Centers) where agents or customer service representatives (CSRs) can respond to a web "alert" button that has been

clicked on the company website by a prospective customer. The agent response can be a phone call, a chat window, voice over IP, video over IP, or synchronized "symmetrical" surfing with the customer on the website. The call center thus becomes more of an "e-tailing" e-commerce transaction center that may never actually deal with a real voice phone call.

Estimates of the number of call centers in North America range from 60,000 to as high as 200,000. It all depends on how you define a call center. As time goes on, that definition becomes increasingly broad.

As flexible, perhaps even amorphous, as the contemporary call center has become, we can still generally group call centers into the traditional inbound-centric and outbound-centric categories:

- **Inbound Call Centers.** The majority of call centers receive toll-free calls from customers. Frequently referred to as inbound operations, these centers exist in many different types of businesses. Banks rely on them to answer customers' questions about account balances. Call centers also assist customers whose debit or credit cards were lost or stolen. Airlines, HMOs, utilities, and, most notably, catalog firms rely heavily on call centers to serve customers, often 24 hours a day and 365 days a year.

When people place calls to departments other than call centers, they usually have a certain person in mind. If they don't reach that person, they often find voice mail to be a sufficient alternative. But in call cen-

Plurimarketing in Lisbon, Portugal, is a telemarketing service bureau that handles both outbound and inbound call campaigns for clients such as TAP (Portugal's National Airline) and BRISA (Portugal's National highway company). There are over 200 agents in this call center using contact center software from Altitude Software (Milpitas, CA — 408-965-1700, www.altitudesoftware.com). Plurimarketing runs as many as 20 simultaneous campaigns for various clients during the day and even more campaigns at night. (Photo by Zippy)

ters, and especially in places such as 911 centers that receive emergency calls, callers expect to reach the first person who is available to help them. Many businesses have adopted the "911 approach" to ensure callers always have the option of speaking to live agents, even if they occasionally have to wait on hold.

Unlike phone systems in most offices, those that call centers depend on must be able to receive hundreds or thousands of calls a day and automatically route each call to the most appropriate agent. Such systems either include automatic call distributors (see ACDs) or they direct calls to agents with the help of standalone ACDs. To create some semblance of order in the way that ACDs route calls to agents, many centers divide agents into groups that relate to their respective jobs, such as answering questions or providing tech support.

Beyond establishing functional groups, call centers employ ingenious methods of determining why people are calling. Some centers create separate toll-free numbers to allow customers to call numbers that are unique to the groups of agents they intend to reach. Others use interactive voice response (IVR) systems, which callers often encounter when they first reach a call center. IVR systems usually play greeting messages to callers that prompt them to press keys on their touchtone phones that correspond to the groups of agents they want to get a hold of (See IVR).

An increasing number of centers use speech recognition systems to allow customers to request certain types of information, such as flight itineraries or store locations, without speaking to live agents and without having to listen to and navigate through IVR menus (See Automatic Speech Recognition).

Some call centers use computer telephony software and Caller ID, or ANI to identify the phone number a caller is dialing from, to ascertain whether the caller is a current customer and to specify whether the customer is placing the call from a home or business number. The computer telephony software uses the phone number from which the customer calls as the basis for looking up a customer's record in one or more data sources. If the software locates a customer's record, it finds out from the ACD which agent it is directing the customer's call to. Then it displays all or some fields from the customer's record on the agent's computer screen.

A caller waits on hold if an agent from a given group isn't immediately available to answer a call, or if agents from other groups aren't present or aren't trained to provide the particular type of assistance the caller needs. Callers have a tendency to hang up (or, in call center parlance, to abandon) if they remain on hold longer than they expect. To reduce the amount of time customers are on hold and to cut down on the percentage of callers who abandon, call centers often calculate in advance how many calls they will receive. This figure, in turn, helps them figure out how many phone lines they need and how many agents should be available to take calls at any given time.

Call centers can use workforce management software to automate the way they forecast staffing requirements and create schedules for agents. A small percentage of centers employ workforce management systems, which often require dedicated staff to manage agents' schedules.

Besides making sure agents are at their desks to answer calls, call centers measure how efficiently and effectively agents communicate with customers. Many centers set thresholds for the average number of seconds that agents take to answer each call (also known as the average speed of answer); the average number of minutes agents spend on the phone with customers; the average number of minutes callers are on hold; and the percentage of callers who abandon while they are on hold.

A measurement that call centers frequently gauge is the service level, which is the percentage of calls that agents answer within a certain period of time (usually within between 30 and 60 seconds). Service levels for call centers in some industries are under government regulation. An example is call centers in cable companies, which are required by federal law to answer 90% of calls in 30 or fewer seconds. Many workforce management tools work with ACDs to determine the number of agents a call center needs throughout the day to meet a given service level.

Many call centers pay attention to other measurements as well, although not necessarily in the same way. Nearly all call centers try to keep abandon rates below 4% and the average speed of answer below 20 seconds, but average talk times and hold times vary significantly among centers in different industries.

Centers that offer high-end services, like help desks, operate under the assumption that their customers are willing to wait on hold for as many as 20 minutes to speak to a tech support expert. In centers that generally receive a large number of calls from customers who expect immediate answers to questions, such as those about payment due dates or account balances, their goal is to keep average talk times and hold times under a few minutes. For agents in these centers, talk times above a few minutes usually indicate that agents are having difficulty with assisting customers.

Many ACDs include tools that allow call center managers to produce reports about agents' productivity. These reports provide statistics such as the number of calls agents answer, agents' average talk times and the average number of callers on hold throughout the day. To guarantee that they can respond immediately to spikes in call volumes, call center managers often look up real-time data for individual agents and for groups of agents.

If the center receives more calls than it can handle, the manager often makes the decision whether callers should hear busy signals, whether the center should immediately locate more agents or other employees to answer calls, or whether the center should play an automated greeting to let callers know that their time on hold may be unusually long. Some vendors, including ACD manufacturers, offer announcement systems that inform callers how long they can expect to wait on hold before they reach agents.

Call center managers also consult real-time and historical reports to determine how agents compare to other agents in their respective groups, as well as to find out the center's service level. Agents usually rely on electronic displays such as readerboards and TV monitors to see how many calls they have yet to answer while they're on the phone.

In keeping track of agents' performance, call center managers aren't limited to assessing agents' productivity; what the agents say during calls also matters. During each month, some centers use monitoring systems to record a random sample of agents' calls. Managers (or in some centers, dedicated trainers or quality assurance staff) listen to these calls and evaluate agents on how successfully they dealt with customers during the course of these calls. Agents are often not aware exactly when their supervisors are recording or listening to their calls; this scenario is known as silent monitoring (See Call Center Call Monitoring and Recording).

Among the more recent trends associated with inbound call centers are the proliferation of distributed call centers that operate as one center even though agents may actually be working from multiple sites. In some companies, agents can answer calls from remote offices or even from their homes.

Customer service operations are expanding their activities beyond handling calls. Agents now respond to e-mail messages, live text messages and on-line forms that customers complete from Web sites. Despite the variety of methods that people can use to reach agents, call centers continue to serve as the operational model for on-line customer service in the way they route communication from customers to agents and in the way they track agents' performance.

Many of the reporting features inherent in e-mail or text chat routing software generate statistics that are analogous to those that managers typically gather from ACDs. For example, e-mail routing products often indicate the average number of e-mail messages from customers that agents receive during a given time period, as well as the average time it takes for agents to answer these messages.

- **Outbound Call Centers.** Certain operations within an organization, such as collections or fundraising departments, rely on call centers to make outgoing calls. Telemarketers are the best-known and least-admired examples of outbound call centers, which is why telemarketing operations are on the decline throughout the U.S. Other examples include customer retention departments in businesses that experience high annual rates of churn, such as providers of wireless phone service.

The more agents an outbound center employs, the more the center benefits from predictive dialers and other types of automated dialers (See Automated Dialers) to ensure agents actually reach a high percentage of the customers or potential customers they're trying to contact.

Some call centers train agents to make outgoing calls if the volume of incoming calls is low; these centers also train agents who normally make outbound calls to answer calls if the volume of incoming calls is unusually high. This scenario, known as call blending, isn't typical in call centers, although many manufacturers of predictive dialers offer systems that work with ACDs to send incoming calls to agents.

Call Center Call Monitoring and Recording

Also known as "voice logging", the monitoring and recording phone calls between callers and call center agents formerly occurred almost exclusively in enhanced 911 (E-911) systems. Today, however, nearly all call centers have this capability, the technology having shed its old image as "dumb, expensive tape recorders" and emerged as a vital integrated management tool for call centers. Managing a growing call center is difficult, and call monitoring and recording is a tremendous aid to call center supervisors.

Although the "silent monitoring" of agents is still prevalent in call centers, when the call recording process is agent-controlled, monitoring systems no longer have to be seen as Big Brother. They can be used to set goals, coach agents and reward them in real-time, freeing supervisors to do less mundane managing. Agents can exercise some discretion over what gets recorded with on-demand recording, and event-triggered recording can keep privacy invasion to a minimum. You can eliminate discrepancies between what was said, what the supervisor remembers, and what the agent remembers.

Another former stumbling block to the acceptance of these systems – the difficulty in integrating them into call centers has vanished with the rise of integrated CT systems. From a VAR, perspective, they're an easy sale as an add-on, and some logging companies say centralized digital recording pays for itself in 18-months on people-savings alone. As customers look to get some measure of control over their call centers and boost productivity, call recording and monitoring systems enable you to find out which agents are your "top guns," and which are the duds who you'll be able to coach in real-time, using real voice samples of customer-agent conversations, not abstract flowcharts and statistical analyses. And for the paranoid supervisor, it's a dream come true to be able to actually hear how your most important customers are being handled.

With CT-enabled call recording in your call center, you can be selective, filtering out and listening to only the important calls. As for deciding what the important calls are in the first place, call center systems can help you determine this with the aid of such information such as: ANI / DNIS, IVR entered DTMF tones, customer numbers, or seeing which calls actually resulted in sales, or those that didn't. You can now review not only the call, but all the relevant data – even screen-captures from agent's PCs.

Voice logging is also a great way to confirm transactions. "Record on demand" from agent's desktops allows you to do such things as record stock trades shouted over the multiple channels of trading turret during a market crash (did Joe say "buy at 32" or "sell at 32?"). Recording technology extends to faxes – keeping a copy of important and binding legal documents never hurt. Stories of million-dollar rush orders sent to closed offices, only to be retrieved by the logging system that saved the day, may be rare but incidents like this can pay for the whole system.

Voice logging can also be important in case your business faces litigation. Bomb threats, complaints, lawsuits, etc., happen more frequently than one would prefer. Lawyers record (and sometimes deliberately don't) for obvious reasons. Medical insurance, emergency services and financial organizations record for legal reasons too, doing due diligence and self-regulation before they're forced to pay up.

If you doubt the power of call recording systems in court, just ask the 24 largest brokerage firms on the NASDAQ, whose corporate lawyers backed down from a Justice Department suit because the respective firms' traders were caught on their own loggers colluding to fix stock prices. Now they are not only forced to log, but they must also let the DOJ listen to the recordings.

On the telephony side, more PBXs and ACDs have become certified by logger manufacturers for TAPI, TSAPI, CSTA, CTConnect, etc., integration with their products. As with the rest of computer telephony technology the physical interfaces have been upgraded over time from primitive RS-232 connections to 10 Mbps and faster Ethernet. Recent software changes by PBX vendors gave digital recorders a new place to plug in besides tapping trunks or sitting on analog station ports: Service observer ports. Now instead of hassling with analog integration and conferencing a logger with a digital station, there are port groups built into switches expressly for the purpose. Trunk-side logging also got a boost from CT system vendors like NICE whose boards will allow you to do pass-through switching (take in all trunks to the logger, but pass most of them through un-logged, getting away from a 1:1 trunk-to-port ratio). High "Z" (impedence) boards ensure that calls don't get cut-off trunks when they come in.

"Free-seating" in call centers normally discourages station-side recording since one never knows which agent was sitting at what station during recording sessions. However, new systems allow for integration with agent lists and logins that will let you match the agent to the call, no matter where they were sitting.

With logger APIs released, recording can be triggered by other PC apps (on demand, or event-driven recording). With PC-based logger networking, playback became quite sophisticated. You can quickly find what you're looking for, and listen to it conveniently, even over an IP WAN. With PC, DSP, compression and storage prices dropping, and performances improving , so too has call recording systems. Basically the problems of not enough data spigots, incompatible data formats, and data systems locking each other out are beginning to solve themselves.

Technologically, call recording and monitoring is ideally suited to CT people. Anyone with experience setting up networked PCs and putting a voice card or two in them has all the major skills required for voice logging these days.

PC-based, client / server call recording systems that generate digitized voice files have pretty much replaced the unwieldy, proprietary, non-searchable, reel-to-reel tape contraptions of the past. It doesn't matter whether the logger itself is a PC-based or a proprietary (fault tolerant) box, a CT link will get PC data integration and PC GUI playback / admin functions to the logger. Thus, call monitoring and recording systems have simply become big Windows 2000 client / server databases presided over by the IT department, who consider recorded voice just another field in a comprehensive database that includes ANI, CLID, DID, DNIS and SMDR information from the switch, DTMF (Customer ID, or Agent ID) or IVR information that has passed through., screenshots of agent's networked PC data screens, and fields from other enterprise databases. Such databases can be programmed to receive data from the logger at random, on a schedule, continuously, on-demand or when a certain PC or phone event triggers it to record, then turn the logger off.

Thanks to recent higher levels of integration with CT systems, two broad cost-saving strategies in monitoring systems' deployment have emerged:

- MIS / telecom directors have started purchasing voice recording and monitoring systems as a shared and flexible resource, built to handle multiple user groups within organizations – thus making the system more affordable.
- Within organizations, call recording systems are also being deployed as centrally-managed units. SNMP alarming to WAN programming to WAN data exchange and data pooling all off-set travel costs in multisite call center organizations.

Some Call Center Monitoring and Recording vendors:

Comverse (Woodbury, NY – 516-677-7400) www.cominfosys.com

Dictaphone (Stratford, CT – 203-381-7000) www.dictaphone.com

Dynamic Instruments (San Diego, CA – 619-275-4900) www.dynamicinst.com

Altitude Software (Milpitas, CA – 408-965-1700) www.altitudesoftware.com

Envision Telephony (Seattle, WA – 206-621-9384) www.envisiontelephony.com

Eyretel (Landover, MD – 301-341-2700) www.eyretel.com

InfoBase Systems (West Palm Beach, FL – 561-681-7061) www.ctiguys.com

Mercom Systems (New York, NY – 201-507-8800) www.mercom.com

MicroCall (New York, NY – 212-759-0946)

NICE Systems (New York, NY – 800-642-3611) www.nice.com

Noble Systems (Atlanta, GA – 404-851-1331) www.noblesys.com

Racal Recorders (Herndon, VA – 703-709-7114) www.racalrecord.com

Stancil (Santa Anna, CA – 714-546-2002) www.stancil.com

TantaComm's (Omaha, NE – 800-444-8522) www.tantacomm.com

TEAC (Montebello, CA – 213-726-0383) www.teac.com

Teknekron Infoswitch (817-267-3025) www.teknekron.com

Witness Systems (Alpharetta, GA – 800-494-8637) www.witsys.com

Wygant Scientific (Porland, OR – 503-227-6901) www,wygant.com

Call Center Outsourcing

Outsourcing is the contracting out of functions, typically services, to another individual or firm. Businesses outsource when they feel the function in question can be handled at less cost in a shorter period of time and with less hassle than if they performed it themselves in-house. This leaves the outsourcing firm to focus their resources on doing what they can do best, i.e. why they are in business. This is known as "concentrating on one's core competencies".

In businesses, services such as data processing, mailroom and cleaning and security, and outbound and inbound teleservices (including e-mail / Internet handling) are outsourced. *Call Center, Computer Telephony* and *TELECONNECT* magazines outsources their customer service and subscriptions to Sunbelt Fulfillment (888-824-9793) and list rental to Rubin Response (847-619-9800, www.rubinresponse.com).

We believe that our core competencies lie in writing and editing great magazines and books, and putting on fantastic conferences and trade shows, not in quickly and efficiently answering calls about lost magazine issues, shipping magazines and marketing lists. This is best left to the experts in these fields who can do it for less than we can in-house. Outsourcing these functions therefore keeps our quality up and costs down.

In call centers there are three general types of outsourcing:

- Service Bureaus, which provide live agent contact handling, plus automated services such as IVR, e-mail response and Web self-service. Service bureaus are synonymous with outsourcers.
- Applications Hosts, which lease/rent, manage and upgrade key technologies, such as IVR (with or without speech rec) e-mail, live chat and Web self service, PBX and ACD only at their sites. Centrex is a form of applications hosting. Applications hosts do not perform live agent contact handling except under special circumstances or through partnerships with service bureaus.

  Once limited to IVR, hosts such as Call Interactive (Omaha, NE – 402-498-7000, www.callit.com), Intelogistics (Weston, FL – 877-453-5700, www.intelogistics.net) and PriceInteractive (Reston, VA – 703-620-4700, www.priceinteractive.com) are equipped with mega-multiport systems to handle consumer inquiries, surveys and disaster recovery. Applications hosting has spread to include e-mail response and outbound e-mail, web chats and PBXs. Firms such as eGain (Sunnyvale, CA – 408-737-8400, www.egain.com), Island Data (San Diego, CA – 619-487-9335, www.islanddata.com), LivePerson (New York, NY – 212-277-8950, www.liveperson.com), Vectir (Pasadena, CA – 877-7vectir, www.vectir.com) and WebEx (San Jose, CA – 408-435-7000, www.webex.com) provide e-mail / Internet app hosting. IVR app hosters have become more sophisticated, adding Automated Speech Recognition (ASR or "speech rec") to their ports. Voci International (Campbell, CA – 408-591-3783, www.vocicorp.com) is an IVR app hoster that uses Nuance's speech rec technology while Nextlink International (Bellevue, WA – 425-519-8700, www.nextlink.com) has partnered with SpeechWorks.

- **Insourcing**. This is where the outsourcer runs your call center, on facilities that you own or lease, with agents they hire/fire and pay benefits to. This can take the form of some functions, such as help desk, or an entire building. As an example Sitel (Omaha, NE – 402-963-6810, www.sitel.com), a service bureau, runs GM's Tampa, FL call center while IHS Help Desk Services (New York, NY – 800-673-2442, www.ihshelpdesk.com) hires, trains and sends representatives to work at its clients' desks; it has an after-hours help desk. The advantage of insourcing is that you have more control over both your program and the vendor because all actions are taking place on your premises. The disadvantage is that you incur the facilities costs.

Service bureaus are the largest of the three categories. Call Center magazine publishes excellent articles and buyer's guides on these companies. These firms provide business-to-consumer, business-to-business and external and internal IT outsourcing. They include well known firms such as Aegis (Dallas, TX – 214-361-9874, www.aegiscomgroup.com), Apac (Deerfield, IL – 800-688-7987, www.apacteleservices.com), ClientLogic (Buffalo, NY – 716-871-6400, www.clientlogic.com), Convergys (Cincinnati, OH – 513-723-7000, www.convergys.com), DialAmerica (Mahwah, NJ – 201-327-0200, will.dialamerica.com), ICT Group (Langhorne, PA – 215-757-0200, www.ictgroup.com), Precision Response (Miami, FL – 305-816-4600, www.prcnet.com), Sitel (Omaha, NE – 402-963-6810, www.sitel.com), Telespectrum (King of Prussia, PA – 610-878-7470, www.telespectrum.com) and TeleTech (Denver, CO – 800-835-3832, www.teletech.com), innovative up and comers such as Excell Customer Care (Tempe, AZ – 602-808-1511, www.excellcustomercare.com), MicroAge (Phoenix, AZ – 800-246-4322, www.microage.com), Sky Alland (Columbia, MD – 800-351-5055, www.skyalland.com) and TeleMark (Portland, OR – 503-652-6000, www.telemark-inc.com) and many smaller flexible highly responsive companies such as Ameridial (North Canton, OH – 800-445-7128, www.ameridial.com), CTC (DeKalb, IL – 815-748-4200) and Meyer Telemarketing, now called Meyer Associates Teleservices (St. Cloud, MN – 320-259-4000, www.callmeyer.com).

There are specialty business-to-business bureaus such as Interactive Marketing Group (Allendale, NJ – 201-327-0974, www.imgusa.com), and Techmar (Canton, MA – 781-821-8324, www.techmar.com). Others, such as eSupportNow (Boston, MA – 781-306-9797, www.esupportnow.com) and PeopleSupport (Los Angeles, CA – 310-914-5999, www.peoplesupport.com), offer primarily e-mail/Internet live and automated handling. There are specialized IT bureaus such as Cyntergy (Gaithersburg, MD – 301-926-3400, www.cyntercorp.com), DecisionOne (Frazer, PA – 888-287-9202, www.decisionone.com), 800 Support (Portland, OR – 503-684-2826, www.800support.com), PC Helps Support (Bala Cynwyd, PA – 800-HELP-412, www.pchelps.com), Stream International (Canton, MA – 888-223-8880, www.stream.com), Support Technologies (Atlanta, GA – 770-280-2630, www.supporttechnologies.com), Sykes International (Tampa, FL – 813-274-1000, www.sykes.com), Syntel (Troy, MI – 278-619-2800, www.syntelinc.com), Watts Communications (Toronto, ON, Canada – 416-255-8000, www.wattsgroup.com) and Xtrasource (Kent, OH – 330-673-3316,www.xtrasource.com).

Service bureaus provide a comprehensive suite of functions including inbound and outbound customer service, sales, lead generation and followup. They supply this with phone, and increasingly with e-mail / Internet (including web callback, collaboration, chat, and Voice over IP). Many also offer automated e-mail, IVR and web self-service. Some are preparing their operations for video – it all depends on your needs and the size of your contract.

Service bureaus can also provide many ancillary services such as database marketing, data mining and warehousing, list acquisition and analysis, data modeling and fulfillment, warehousing and shipping. Some even undertake direct mail, including creatives and lettershop. Some firms will also manage your call center and train your agents. The Centralized Marketing Company (Cordova, TN – 802-925-1974, www.cmc-max.com) developed and marketed training services and then opened a service bureau when it couldn't find suitable outsourcer partners for its clients; it used the experience of running its own center to refine its training programs.

Many bureaus are now concentrating on their own core competencies and are outsourcing services in partnership with other firms, providing one-stop shopping. For example, Centrobe (Phoenix, AZ – 602-598-9926, www.centrobe.com), a division of data processing giant EDS has partnered with IVR provider Call Interactive to hook up that firm's huge port capacity for clients.

Many leading firms are now are embracing, installing and going live with customer relationship management (CRM) software. As an example Sitel (Omaha, NE – 402-963-6810, www.sitel.com) is using Siebel's CRM package. CRM software permits outsourcers to provide complete customer contact management for you, cross-selling and upselling to callers or to called parties instead of just making/taking calls on simple outbound or inbound programs, and forwarding the leads and lists to you.

Some bureaus provide services in more than one language, most commonly Spanish and French. ICT Group, has call centers in Miami and the Canadian province of New Brunswick, Canada that have agents who are native Spanish and French speakers respectively. There are also translation services such as Language Line.

Several US-based service bureaus can support your marketing/customer service internationally. Most of the larger firms (with the striking exceptions of West) have call centers outside the US along with many mid-sized bureaus such as Affina (Peoria, IL – 800-787-7626, www.affina.com), Excell Customer Care (Tempe, AZ – 602-808-1511, www.excellcustomercare.com), Protocol Communications (Sarasota, FL – 800-435-2003, www.protocolusa.com) and TeleMark (Portland, OR – 503-652-6000, www.telemark-inc.com). There are also overseas owned firms that have operations and alliances in the U.S. such as Prestige International (New York, NY – 212-317-9333, www.prestigein.com), SNT International (Zoetermeer, Netherlands – +31-79-368-69-70, www.snt.nl) and Transcom (Carmel, IN – 888-408-7267, www.transcomusa.com).

Do not overlook locally based outsourcers, which may have just as potent capabilities as American-owned firms. These include Minacs Group (Pickering, ON, Canada – 905-837-6000, www.minacs.com) and Optima Communications (Toronto, ON, Canada – 416-581-1236, www.optima.net), Somerset Group Consulting (Upper Marlboro, MD – 301-324-0455, www.somersetgci.com), Atos (Paris, France–+33-01-49-00-90-00,www.atos-group.com) and Teleperformance International (Paris, France – +33-01-55-76-40-30, www.teleperformance.fr) and Telemarketing Japan (Tokyo, Japan – +81-3-5321-0800, www.telemarketing.co.jp), and Merchants Limited (Milton Keynes, UK – +44(0)-1908-232323, www.merchants.co.uk) and 7C Ltd. (London, UK – +44(0)-20-7505-6100, www.7c.net/index.html).

Many Canadian and Indian firms claim they can handle your U.S.-originated calls for less money than competing U.S. outsourcers, taking advantage of their countries' lower labor costs. U.S. firms that have international call centers can also offer this edge. Many American outsourcers including Affina (Peoria, IL – 800-787-7626, www.affina.com), Convergys (Cincinnati, OH – 513-723-7000, www.convergys.com), ICT Group (Langhome, PA – 215-757-0200, www.ictgroup.com), RMH (Byrn Mawr, PA – 610-520-5300, www.rhmteleservices.com), TeleMark (Portland, OR – 503-652-6000, www.telemark-inc.com), Telespectrum (King of Prussia, PA – 610-878-7470, www.telespectrum.com) and TeleTech (Denver, CO – 800-835-3832, www.teletech.com) have opened up or acquired call centers in Canada.

Outsourcing, whether by service bureaus, applications hosters or insourcers provide several key advantages of providing the same teleservices in-house. They are:

- **Tremendous flexibility.** A bureau or app hoster can handle high peak and overflow call volumes (e.g. Christmas shopping, Guy Fawkes for your family anarchist, May Day for the graying leftists from your college days) and limited-length programs (credit cards, long distance telephone service, IP telephony, contour chairs, wax sunshades, used PBXs) much easier than an in-house call center or IVR port assembly. Bureaus can hire / fire part-time and full-time agents far quicker and with less heartache than your HR department. They keep their seats and ports filled and owners/shareholders happy by seeking other programs. An outsourcer can get a program up and running in as little as a few days or a few weeks, compared with months for an in-house program.

For these reasons bureaus have traditionally been called on to perform outbound telesales and collections. IVR hosts are contacted to manage surveys and disaster recovery.

Datamonitor (New York, NY – 212-286-7400, www.datamonitor.com) identifies outbound cold calling as a function not in the core competency of most firms, hence it is the most easily outsourced of tasks. The work is part-time (typically 6 p.m. to 9 p.m at night and on weekends) and requires a different breed of agents and supervision than those found in customer service. Agents that succeed are persistent and are thick skinned; supervisors are of the same breed. Ex-Army types are ideal for such work. Don't worry about the agent's feelings. They are selected and trained to handle it, at least in most cases. A November 7th, 1999 New York Times article on cold calling to New Yorkers reported that some agents calling to that market receive "danger pay" – most companies featured in the story are outsourcers. One manager reported that New Yorkers would often angrily suggest "where the agent could place the phone," and then offer to put it there for them!

However, the same personality types that can cleverly persuade someone into buying Hello Kitty hood ornaments or extract cash from someone who was 20 minutes late in paying their bills are often not the sort who can spend five minutes in helping the caller figure out what style of garden gnomes will go with the plaster Madonna in the front yard. Or for that matter have the patience with a caller to help fix a problem they could have resolved themselves by practicing RTFM (Read The "Friendly" Manual). These inbound agents represent your company with the customer is always right attitude, no matter how idiotic the request, complaint or customer.

Still, outsourcing to bureaus is becoming more common for customer service, particularly help desks on account of the booming economy of the 1990s that has created labor shortages. If your products are mass market fire-and-forget, e.g. many consumer goods, packaged software – then you're not really building relationships (not to mention files) on every customer. Having an outsourcer take care of these contacts and forward those that need addressing (things such as product mislabeling (Windows NT for Mac) or cockroaches in the cereal), to senior management may pay off for you. Outsourced help desk firms can screen out the RTFM calls, and escalate the truly problematic ones to your in-house experts – which is your core competency.

- **Lower costs.** Outsourcer bureaus' core competencies lie in finding and training people, fast. They are not constrained by cost-burdening unions and union work rules. They can share your work with other clients, keeping the agents busy and making money for them at all times, or they can provide dedicated agents, at more cost to you. Outsourcers can bargain and haggle down the latest equipment; they need to have the latest and greatest to deliver the best services.

Outsourcers have more flexibility in locations; they set up call centers in places where they can find lower-wage/lower-turnover employees (which in turn reduces training costs) and cheap facilities. Some bureaus make/take US calls from Canada, India and Jamaica, yet the telecom quality is so good at the same time United States has become so polyethnic that you can't tell if the agent is in Delhi, NY, Delhi, Ontario, or New Delhi, India.

Bureaus have on-staff and outsourced on-call site selection firms (see Call Center Site Selection) that can find buildings and floors and have the sites set up in as little as 30 days. They can close up shop just as rapidly too, with none of the local heartache and bad press if you personally tried to close a call center. If you're an American Express executive and you had to shut down a 1,000-seat call center in Joso, Oregon, chances are your managers will face an acrid (and toxic) bonfire of AmEx cards. However if you outsourced to Zippy Teleservices and Zippy had to shut down the center because you and a couple of other clients (e.g. CT Expo) wound up their programs at the same time, you and CT Expo won't be in an excruciating situation because Zippy did its work transparently for you.

- **Less hassle.** This is where app hosting comes in very handy. Modernizing, setting up and hiring for a call center is a costly, complex, process. The new electronic technologies (as opposed to the "old electronic technolo-

gies" such as the telegraph, telephone, and Telex) make this worse. Web sites and their background interactive hardware and software take months to buy / specify, set up, program, test, debug, test again, manage and frequently upgrade. If your IT department is typical of most companies they'll whine incessantly that they don't have the time and resources to get your e-commerce site operational.

If you're doing sophisticated call handling you'll also need agents with excellent English grammar and keyboarding skills. That may be tough to find. Call center consultants bemoan the near-uselessness of today's typical U.S. high school graduate, who is as about as prepared to work in a high-tech call center as a newborn kangaroo is ready to venture out of its mother's pouch.

On the other hand, those outsourcers that have gone live with call center e-commerce functions such as e-mail / Internet contacts, have managed to do all of the hard installation work and have carefully picked and trained the right people. They are also flexible enough (depending on your contract) to open dedicated centers for you, put in technologies that meet your specs, and hire and train staff based on your requirements. All of these companies will work with you on devising agent scripts. They also promise easy system upgrades, without the budget battles from your Finance department and the "we can't do it" bellyaches from IT. It's done, it's there, it's live and it's working.

- **Easier access to overseas customers.** Contracting with an outsourcer is the best way to learn the realities of offshore markets, before making the costly and time-consuming step to servicing them with your center.

Non-U.S. centers and those U.S.-owned by locally managed outsourcers are attuned to local cultural nuances that are key to success in these markets. Countries such as Indonesia, Japan and Korea have tough-to-penetrate business and social cultures. Even Canadians, who are immersed in American culture, strongly dislike pushy outbound calls. Quebec consumers prefer to speak with someone who can communicate in their peculiar, bastardized English / French dialect. Several nations also have tough laws against outbound telemarketing, and many people in these markets are reluctant to use IVR systems. Europe's all-powerful Data Protection Directive controls what data call centers and others are allowed to keep and transmit and to whom.

Outsourcers can reach customers that you have difficulty in servicing with your own facilities. Outsourcers know the lay of the land and have the right connections, in more ways than one. You may have markets in Colombia, Eastern Europe, Indonesia, India, Mexico, Russia, South Africa (and high-priced / red-tape-choked U.S. locales like New York City) but do you really want to set up your own call center in any of those places? Even setting up in advanced countries like Japan can pose costly bureaucratic and cultural nightmares.

For pricing, service bureaus usually charge by the hour, with a minimum number of hours, plus set-up fees. Some firms, most notably DialAmerica offer performance-based pricing where instead of totally by the hour they share the risk, with or without a small hourly minimum in exchange for analyzing, sourcing and testing your lists. There are minimum sizes and prices for service bureaus, which depends on the size and complexity of your program.

The rule is – the fewer the hours the higher the cost per hour. A smaller bureau can do business with you at 500 hours a month minimum for six months, while a larger firm will consent to talk to you only if you're looking to buy 1,500 hours a month for a year. The bigger firms can charge $20 to $25 per agent hour while the smaller firm must bid at the high 20s to low 30s because of economies of scale. Some bureaus are dropping the set-up charges if you sign up for a huge, "mega-hour" multiyear contract.

There are also, however, some downsides to outsourcing. The key ones are loss of control over your agents and your customer information. Your customer contact and relationship-building skills also atrophy, yet their value depends on how important customer relationships are to your business. Some functions, such as outbound calling, have less impact on customer relationship management, where the called party is being pressed to buy, rather than inbound customer service where the caller is more engaged and wants something from you. Consultants tend to say that small / medium enterprises, and larger firms seeking a competitive advantage by stressing customer "touch" should not outsource.

Additionally, turnover tends to be greater in an outsourcer than in an in-house center, leading to higher training costs, greater error percentage and lesser quality. There is less likelihood that you will detect problems until large numbers of customers and leads have been burned.

Outsourcing may also be more costly than opening and running your own center over the long term. Part of your fee goes to the outsourcer's profits – that's money you could keep in house. Consultants estimate the tradeoff at the three year mark. If your project length exceeds that you may be better off if you opened your own center.

"The outsourcing choice makes sense if the business application is relatively simple and does not require extensive training, product knowledge, or cross or upselling skills, and when it is important to develop a call center in zero time with a very small capital investment," points out Robert Engel, principal with site selection consultancy Engel Picasso (Albuquerque, NM – 800-241-8092, www.engelpicasso.com). "The owned center choice makes the most sense where the product or service is proprietary, requires heavy agent training investment and where your development strategy employs strong cultural integration and loyalty amongst employees."

Call Center Site Selection

Site selection is the process by which you find space for your call center business, its staff, and perhaps the ACD, voice mail and other computer telephony equipment that gives it its capabilities.

To look for the right kind of call center space you need to know how much room will be needed for the functions that you want to keep on the premises – e.g. call center, software development, training, fulfillment, shipping, marketing and management. You will need space for ancillary services such as training space, reception, the "snail mail" room, supervisor offices, washrooms, computer / telecom rooms and space for amenities such as a break area, cafeteria, even bike racks, a gym, daycare, petcare and perhaps something lavish such as a barbecue pit. It all depends on the company's size, type and culture. If you're not in a downtown area or don't want to be, you'll also have to plan for on-site parking.

The shape and size of your functional areas depend on what goes in them and how you supervise your workers. If they function in teams, such as those found in advanced customer service, help desk or software development, then they need enough space and be grouped to facilitate teamwork, with a combination of cubicles and workstations laid out in such a way that so they can roll over to each other or meet en masse. When managing a help desk it's also an excellent practice to have space for the actual products so that the reps can wander over and look and play with the same machine the caller is complaining about. Supervision is less direct. If your workers function individually, but need close supervision, such as outbound telemarketing, they can be grouped into tightly spaced carrels, but with nearby supervisor bays or raised floor towers,

or placed in arenas with overlooking stations.

There are facilities consultants who can recommend optimal functional area size, layout and furniture depending on how your employees work and are supervised. They'll also tell you that in today's tight labor market, if you want to attract and keep people that you've spent a small ransom to lure and bring up to speed, then you'll have to hand over even more loot to make the place livable for your workers. If you create a workplace they like and feel comfortable in, chances are they'll be less likely to leave for the new company offering a few bucks more that just opened around the corner from the mega-mart. Yet your space still has to be functional enough for your business to make a profit off of the sweat of their brow.

When looking for a home for a white collar business such as a call center, help desk, software development or sales office, chances are you will find few if any problems in finding suitable space in most markets if you have fewer than, say, 75 to 100 employees working there. Anything bigger than that is where site selection gets really challenging, especially for call centers.

That's because call centers, unlike front office and other back-office functions, demand several unique characteristics from their surroundings that must be kept in mind when you're looking for homes for them, characteristics which can be difficult to find in many buildings. Call centers have much higher floor densities than other office uses (seven to eight workers per 1,000 square feet (sf) as opposed to four per 1,000 sf) and double the parking requirements, by the same ratio. They also place a greater drain on the HVAC and plumbing. Depending on your center you may need to be open longer hours, making excellent building and parking security more mission-critical.

Many call center employees are "butt-suckers" (cigarette butts that is, as in tobacco smokers) who need their regular fixes. This means either a pressure-ventilated indoor space (or "gas chamber") or a well shielded and frequently cleaned outdoor "puff pad". Either one should be fireproofed – what you can supply depends on the fire codes and the landlord.

Call centers also need high quality telecom / data links (such as T-1s and SONET rings) and, increasingly, backup power. The more mission-critical the function is, the more disaster-prone the area (ponder for a moment the U.S. Southland's "Hurricane Coasts", the Midwest's "Tornado Alley", the upper Northeast's "Ice Belt" and the West Coast's "Ring of Fire"), the more you need redundant power / telco. On-site power generation is a must in Latin / South America, Eastern Europe and Asia-Pacific. If your IT guru insists on having UPSs with / without a generator, your space shopping list must include room and floors strong enough to support them. Not every building floor can accommodate their weight.

Because your center is in the "people business" you may need mass transit access, which is highly recommended in the larger cities, especially those (like Atlanta, GA) that are running a foul of Federal Clean Air Act legislation. Many people don't, can't and shouldn't drive to work. Many cities are building and expanding bus and rail rapid transit lines, providing van shuttles and reverse-commute buses. You should also consider locating close to offsite amenities and services such as restaurants and shopping, and daycare and medical clinics.

If you need your center to provide fulfillment/shipping you need room for these functions and for truck access. This may or may not be part of many typical office developments.

The most important site selection determinant is labor costs and availability, overshadowing space, facilities and infrastructure. This is a people business. People account for 80% or more of a call center's costs.

The area that you call "home" for your new call center depends on how many workers are available for the optimal size of your center over its lifetime, at your expected turnover rate. Consultants say that executives often make the mistake of looking at real estate availability and costs when deciding on locations because they are tangible when it is the intangibles – labor availability and quality – that make the difference.

To illustrate this point, Susan Arledge, principal with Arledge/Power (Dallas, TX – 972-661-1100, www.arledgepower.com), a real estate consultancy, gives an object lesson by using the example of a fictional outsourcer who had a choice between two facilities in two different cities to house a center for 300 employees, with the requirement of 150 sf per worker. Each employee works 2,080 hours a year.

Building 1 offers a rental rate of $1/sf below the Building 2 rate while Building 2 offers a labor rate of $1 below the Building 1 rate. Building 1's rental advantage is $150 per employee/year ($1 rental rate x 150 sf). However Building 2 has a far greater $2,239 employee/year advantage (2,080 hours x $1 labor difference plus employment taxes).

By locating in Building 2, the outsourcer saves $2,080 per employee/year or $626,700 annually (($2,239 – $150) x 300 employees). That becomes $6.26 million over a 10-year lease. A $1 per hour labor savings offsets $14.93/sf in annual rent. Building 2's could require rental rates far above Building 1's and still offer the lower total operating cost.

"Usually when you think of site selection, you think real estate costs," says Arledge. "However site selection for the teleservices industry is really about labor, its quality and the quantity of the people who staff the site."

The common measurements of labor availability are the numbers of unemployed and the percentage of the workforce that is unemployed. This data can be obtained from the U.S. Department of Labor (DOL) and in Canada, Statistics Canada. These bodies have raw and seasonally adjusted numbers (seasonally adjusted accounts for such factors as summer employment), broken down by states / province and by major cities.

The DOL records and presents data for metropolitan statistical areas (MSAs), which are cities and metro areas above 100,000 people. Economists agree that when a given area has 4.5% unemployment rate or less it is at full employment; where rates are below that there is often rising wages resulting from competition for workers.

Yet one important labor variable often overlooked but hard to define and analyze is underemployment. Underemployment is a situation where people are working at jobs because that is what is available but they have skillsets that they may wish to utilize in other occupations, such as a customer service/sales in call centers.

One example of underemployment is where you have college-educated people working in low-paid part-time jobs, such as in retail, restaurants and private security. This workforce may be employed that way because the area has high quality of life, family and because other job markets that may use their skills are too far away and a hassle to get to. Call centers can tap into this market because they offer higher-than-average wages, better working conditions and often have flexible and more varied hours than other service sector employment such as restaurants and retail.

Yet according to consultants underemployment is very difficult to analyze because it is difficult to measure. The only way you can find out if there is underemployment is by researching the local market including talking to local economic development agencies. Also, mothers with young children, college students, retirees and the physically disabled who are not registered as unemployed are attracted to call center work.

What your people needs are depends on the skillsets your center requires, how much you're willing to pay them and how many workers you need. In the U.S. call center wages/benefits (without incentives)

ranges from $9 an hour to $15 or $16 an hour.

The more specialized the skills you need from your call center agent staff, the more challenging it is to find such agents and the more you have to pay and more wide-ranging your center staffing search must be. That in turns helps determine where your center should be.

If you're locating a help desk, you need access to technical schools or training to provide tech-thinking workers. If your center requires French and Spanish speakers, then you should locate in a community where there are native and learned speakers. Keep in mind that learned language skills may not be as acceptable to your market as mere native language skills because not even the best learned language students have a complete command of any particular language's nuance.

An Italian who bought your spit fryer that blew up in his face might really go off the handle if your Italian-speaking agent is missing out or misinterpreting some key intonations. Whereas the German who is raising sturm und drang over his kaput web browser may care more about the quality of information he's getting from the agent rather than whether the agent is communicating in north German, south German, Dutch German, Danish German, Swiss German, French German, Austrian German or Berlitz German.

If your center is going to be doing outbound telesales or collections then you need to be in a locale that has a sales culture: Workers who are pushy and thick-skinned. Not every place falls into that category. The Staten Island Advance reported in its September 13, 1999 edition that the outsourcer Unitel had to convert many of its outbound seats to inbound at its Frostburg, MD call center because its agents detested making outbound cold calls. Unitel has since opened a new outbound center in Lakeland, FL, a place which has, as they say, more aggressive "meat eaters."

Different types of call centers consume labor at varying rates. An outbound telemarketing center, because the work is very high stress, has high turnover. A help desk tends to have low turnover.

At the same time, the labor pool for minimal-skilled workers, such as outbound sales and inbound order taking, is usually larger than the labor pool for help desks, which requires people possessing a better education. Also, there are regions with sufficiently high numbers of unemployed and/or which have strong population inflows, so that your center's lifespan may be over before the pool is drained or there is enough replenishment to keep a steady flow of workers.

Too often, firms underestimate a community's population size and the labor supply with the needed skillsets from it. While firms know their turnover rate they are often unaware of its impact on the available workforce. While you might have sufficient people to open your center you may not have enough to replace them when they start leaving, let alone to fill seats when your center grows.

To ensure you have enough labor, regardless of call center type, Robert Engel, principal with Engel, Picasso (Albuquerque, NM – 800-241-8092, www.engelpicasso.com) recommends that you follow the "5% rule". This is where the total number of employees in call centers plus those that have worked in call centers does not exceed 5% of the area's total workforce.

"If the number is greater than 5% then it could become increasingly difficult to attract and keep qualified people," Engel points out.

Labor availability may force you to change your call center size. If your ideal center is a 1,000-seat barn you may not be able to entice enough agents to work for you and staff the center over its lifetime, unless your firm has deep enough pockets to out-pay and out-benefit all of the other area employers. The tremendous economies of such a massive installation are increasingly outweighed by higher wages / benefit and turnover / training costs.

Consultants say that the largest optimal center is now about 250 to 300 seats in the U.S., half that figure (125 to 150 seats) outside of the U.S., and even smaller sizes for help desks. Build a center larger than these sizes and your operating costs increase exponentially as opposed to linearly because you incur higher recruiting and training costs. Also you miss out on disaster recovery and employee promotions potential.

There are pros and cons to the various types of locations usually recommended for call centers. The most popular locations have been the suburban office belts or "bagels" off the expressways by major cities. They provide safe, easy access to the suburban / exurban labor force, are comparatively low cost and are attractive. However these locations are also desired by other call centers and back office businesses, which drive up labor and leasing costs and dry up worker supply. The lowest unemployment rates are found in such communities but the telecom / data links are generally superb.

Smaller cities are becoming an option as well as more rural areas. Costs / wages are lower in these places, the workers are more productive, turnover is less and the incentives tend to be better. Telecom services in these types of locations are improving but many out-of-the-way areas lack ISDN and xDSL (let alone T-1 service) so be careful of where you select. (A good bet is a rural area with a closed or soon-to-be-closed military base, which has fat pipes, low-rise buildings, plenty of land and parking, and a local civilian or military retiree workforce).

In such areas you also run the risk of exhausting the local labor market. If another call center moves in it might set off a destructive wage war. Faced with ticked-off call centers with staffing problems that threaten to move out, local economic development agencies are now being careful not to attract too many new centers that drain their existing employers' labor pool.

Small cities and towns "adopt" businesses, like a family member. Leaving a town is like cheating on a spouse, or leaving a "bun in the oven". If you are a brand name company and you have to shut down your call center you might be run out of town, your products chucked on a bonfire and your website hacked and trashed from infuriated local residents, long before you have time to yank out your PBX and spirit away your operation. If you're an outsourcer, however, you don't face these problems because nobody knows who you are. How many Apac or PRC dealerships are there? Is there an ICT department store?

Big cities (inside the bagel) have more labor, at less cost, and depending on the burg, may be more willing to help your call center get set up than would fully employed suburban edge cities. City locations are popular outside the U.S. If you're willing to look beyond class, ethnic and racial issues and stereotypes, accept the fact that you may indeed find a ready, willing and able workforce in the big cities, and excellent low-cost property.

City-living workers are especially ideal for outbound and inbound sales; they are tough, sharp and smart. These locales also attract immigrants, and with it native language skills. At the other end of the scale, big cities may have enough of a cool, lively downtown culture and college base to draw in tech-smart young people, who just happen to make excellent Internet sales, e-commerce and help desk reps.

To make these cities work for you, you will need to invest heavier into training and into workplace skills than in other locations. Many otherwise fine people have not had steady employment, let alone shown off any relevant job skills. You'll also have to accept mass transit, and not cars as the chief means that people who live there get around.

Locations experts recommend allocating no less than three months

and ideally more in researching and seeking sites from the moment you decide to seek a new call center to the moment when you expect to make a final decision on a given community. Selecting the space depends upon factors such as if you are leasing / renovating or building your own center, which also varies on the climate. You can open a new center for call taking / making in two to three months' time for existing property and five to 12 months for a build-to-suit – on the lower side if you are having it constructed in a non-winter climate, on the higher side in a wintry area.

Consultants estimate real estate / facilities costs for renovations and build-to-suite/lease run from $2,500/seat to $5,000/seat, assuming 100 sf per workstation. IT infrastructure (telecom, data, power) for both types run about $1,500/seat to $2,500/seat, the higher range applies for knowledge-intensive help desks. Call center hardware and software can cost between $1,000 and $2,200/seat.

Your IT costs depend on your how densely populated you want your space. If you move to a less dense environment, say 125 sf per agent, it may cost you more but you may ultimately achieve a better working environment.

Also, these costs depend on whether your center works with line of sight management or with technology, such as workforce management. Experts say that if you rely on line of sight (typically outbound) you pay more for supervisory labor. If you depend on technology, such as for help desk, you pay more for management hardware and software.

Be careful that you don't become "penny wise but pound foolish" in your site selection and property costs evaluation. Call center building costs can greatly vary between $45/sf to $120 /sf depending on building type, location and finishes. Consultants such as Bob Engel advise that the key to affordability is improved productivity, sales and customer retention.

"For example better lighting or a higher resolution terminal costs $15/month extra yet it reduces fatigue to the agent so that in the fifth or eighth hour of a shift that person is 30% more productive," Engel points out. "The return on your investment is huge. Likewise if the workstations are so tightly packed that most agents can't hear their customers for the ambient noise the overall facility costs savings will be more than spent in lost customer satisfaction."

There are several facilities types with their pros and cons. The two main options are renovations versus build-to-suit. There is a big menu of the kinds of buildings that have been renovated for call centers. These range from standard low-rise offices to funky, attractive though small floorplate old brick mills and to vacant 1950s, 1960s, or 1970s area strip malls and stores, depending on where you want your call center to be. WalMart, the world's largest retailer, even maintains a separate division that markets outmoded stores to developers; several WalMarts have been converted to call centers. There are several build-to-suit developers, such as The Alter Group (Lincolnwood, IL – 847-676-4300, www.altergroup.com) that already have floorplans designed and proven for call centers.

Both renovations and build-to-suits cost roughly the same. The chief advantage of renovation is that you can get your center set up in weeks; a build-to-suit could take months. However, with a build-to-suit you would have a center tailored to your needs, whereas compromises are almost always made when you redo an existing facility. Unless you want an everlasting monument, consultants recommend to lease rather than buy, and to ensure that you have escape and expansion clauses in your lease. The call center industry is changing too quickly for anything (especially a contract) to be set in stone.

If you are setting up a call center in another country because they have plenty of available low-cost labor (studies by The Boyd Company consistently show that Canadian locations provide better value for the money than most of their U.S. counterparts) there are unique factors to take into account. One factor is culture. Many non-Americans are not as pushy as are Americans. While aggressiveness is desirable in outbound collections and some types of sales, it is not a desired trait in inbound call handling.

Another factor is language. Some countries' consumers prefer to talk with native speakers, which, depending on the market you are targeting can significantly add to costs while others do not mind learned language speakers, which gives your center greater economies of scale by enabling your center to serve more markets. Generally the more sophisticated and knowledge-intensive a product or service is the more tolerant of learned-language agents.

Also, with the exception of the UK you will find that most European countries have tougher labor laws that make it harder to dismiss workers, yet easier to unionize. This applies more to established vertical industries such as banking / finance than in newer horizontal businesses such as outsourcing. Several countries limit the numbers and after-hours and weekend working.

Many nations' laws restrict call handling. Germany's restricts outbound telesales while India prohibits transfers from an Indian to a non-Indian call center. The European Union's Data Protection Directive controls data collected from European Union member states' residents and their transferal to non-EU countries.

Deciphering and finding out about incentives packages can be trickier outside the US. There are many more players involved, including national, provincial / state and local agencies and telcos.

When locating in Europe there are three strategies often employed: National (call center in each market), pan-European (one large center for the whole of Europe) and regional European (two to four connected centers). Each approach has its pros and cons. A national center strategy best serves each country's market and provides the greatest access to native speakers but at the highest costs of the three. The pros/cons with a pan-European strategy are reversed, especially now as the European economy improves and the call center industry grows; it is becoming difficult to get Danes and Danish-speakers to Ireland, and Italians and Italian-speakers to Denmark. The regional strategy is emerging as a doable compromise.

In the Asia-Pacific region consultants recommend a hub-and-spoke strategy. This is accomplished by having a large main center in either Australia, New Zealand or Singapore, which are affordable locales and have growing multi-ethnic populations, connected to smaller centers (and/or outsourcer partners) in other costly-to-serve and culturally tough-to-access markets, such as Japan and Korea.

There are alternative options to building your own call center. One is outsourcing (See Call Center Outsourcing). The other, but so far less popular option, is telecommuting or teleworking. Instead of having your workers take/make calls on your premises they do so on theirs. There have long been available technologies that can switch calls to agents at home, and monitor them. The telcos' discovery of high capacity copper-riding xDSL services permits one-wire voice/data and always-on Internet or Intranet access. There are companies such as Willow CSN (Miami, FL – 305-874-2000, www.willowcsn.com) that arrange for teleworked agents for your call center operations.

Teleworking does have several advantages:
- It frees up workstations and facilities, saving money, and reduces parking requirements and hassles.
- It lets you tap into potential employees who cannot easily commute,

especially if your center is a fair distance away. These workers include single mothers with very young children, seniors, the mobility-impaired and college students.
- It allows greater staff scheduling flexibility. If you need a peak evening shift filled, more workers would now be willing to do it for convenience and safety reasons (especially women) because they would not have to venture outside their homes.
- If your center is in a metropolitan area where the air quality is below Federal standards (requiring car commuting restrictions) teleworking helps you comply with them.
- It helps your agents show up on time. Commuting times and delays have become horrendous in many suburban areas in the U.S. such as Atlanta, the San Francisco Bay area, Southern California and around Boston, Chicago, Dallas-Fort Worth, New York City, Seattle-Bellevue, WA, and Washington, DC.
- It can act as an incentive to motivate agents and boost performance and retention. Many people, including agents, prefer not to commute.

Teleworking also has several disadvantages, the most important of which is lack of employee control. Unless you pay to install cameras at your employees' homes, it is difficult to see exactly what they are doing there. You risk losing agent performance and support. Your center may experience delays caused by faulty agent telecom links.

Also, with teleworking you add more technology and IT expenses and questions. Who buys and installs the computers and phones? There are also still-unresolved OSHA and liability issues over the working at home scenario. If an agent gets carpal tunnel syndrome as a result of working for you, but on their own PC and on a table and chair they own, then who is responsible for the injury?

"Teleworking can work where your agents are highly motivated and need little supervision," says Ron Cariola, vice president of Equis, a Chicago, IL-based consultancy. "These include help desk reps and high level customer service. However if from experience your call center requires high supervision, such as outbound sales, credit and collections and inbound junior level customer service and order taking then you shouldn't telework," warns Cariola.

Central Office (CO)

A building owned by a telco where the public network, control and support equipment work together to act as a nexus or main node of an inter-CO switching network. It connects local telephone subscriber's telephony equipment (phones, modems and PBXs) to the high-bandwidth switching and head-end equipment.

Centrex IP

Nortel's term for what is also known as IP Centrex (See IP Centrex).

Collaborative Communications

Also called collaborative computing, involves various forms of teleconferencing, such as the following:
- **Application Sharing** – the ability to conference with someone with an application even though the other person may not have the same application software installed on their computer. One person initiates the process by launching an application and users can view the running application's screen simultaneously. Both users can input information and control the application using keyboard, mouse, voice, etc. Files can also be shared or transferred.
- **Audioconferencing** – is the oldest form of teleconferencing, originally performed over standard circuit-switched PSTN phone lines. The first full-duplex audioconferencing bridge was introduced in

1986. Today, IP based technology employing the Internet and intranets are becoming more popular for voice teleconferencing. If more than one person is participating at one of the conferencing locations, then speakerphones or special audioconference terminal equipment (such as Polycom's) is used. If more than two locations are tied into the conference, then multipoint network bridging equipment or Internet servers and software are used.

- **Audiographics teleconferencing** – interconnects users separated by great distances using some kind of graphic display device, generally computer monitors. Participants can view the same high resolution still-frame visual at each site. Some systems allow the image to be manipulated by annotation, writing or drawing on the screen. Sophisticated systems allow collaboration on a single file or document. Audiographics is often used in distance learning since it enables students to interact with lecturers and each other (The interactive software is so students can ask questions or draw on the screen for the instructor or others to see), and the instructors can modify their presentations on the fly while a session is in progress (e.g. skip a section or elaborate on some point) or modify in between sessions (modifying subsequent lessons to accommodate student learning styles or new information).

- **Data collaboration** – is audioconferencing supplemented with some kind of graphics display. The graphics can range from slides and bitmapped computer images to documents, still-frame video, whiteboarding, faxes, etc. The conference's data traffic may or may not use the same network as the voice traffic. The most popular form of data conferencing has become web conferencing.

- **Videoconferencing (VC)** – combines audio, video, and communications networking technology for real-time interaction, often used by groups to communicate with other groups of people.

- **Web Conferencing** – users view the same website in synchonization as directed by a leader, while at the same time talking to each over an audioconferencing channel that runs either through the PSTN or via Voice-over-IP on the Internet.

- **Whiteboarding** – uses document conferencing programs so various people working from a PC or whiteboard display can work on a document in tandem without having to be co-located. You can use whiteboards to open share image files, as well as interactively edit the whiteboard canvas and mark up any images you import. Whiteboarding also allows two or more people to view and make individualized notes on the same screen over a phone line, or on a LAN. Documents can consist of text or graphics pasted from another program onto a whiteboard display or "whiteboard area" where the document can be edited. Whiteboard software generally has special tools for annotating images, allowing you to create lines, circles, rectangles, paintbrushes, and text on the image. Microsoft NetMeeting provides electronic whiteboards. Conference participants can write or draw on these whiteboards using their mouse.

Communication Server

Also called an UnPBX (Ed Margulies' term), commserver, com server, CTI server, telephony server, voice server, voice router, LAN PBX – which also includes the so-called One Wire Wonders (See One Wire Wonder), and CT Server (Dialogic / Howard Bubb's term). The term PC-PBX has also been used, though some would claim that a true PC-PBX is a PBX that has a PC chassis extension (or at least an intimate connection with a PC) in order to gain computer telephony functionality. (See PC-PBX).

Perhaps the most amusing name for these devices are "voice routers".

They aren't routers, of course, but by using the word "router" a company's IT department can claim jurisdiction over their phone system!

In any case, "commserver" appears to be the favorite nickname for the mostly PC-based devices at *Computer Telephony* magazine.

Before the development of router-based systems and solid state telephony applicances, the commserver was long thought to be the centerpiece of computer telephony technology.

A commserver generally consists of a fault resilient 19" rackmount computer running Windows NT, a WAN card for digital or analog lines, a LAN card, a voice card for VRU / IVR prompting (sometimes the functions of these three are combined into the same card), and something such as a 24-line Dialogic MSI card that let you plug station devices (phones, headsets, etc.) into the CT system.

Many commservers use phone wiring to get to the phones, so data and voice are still travelling through two different networks, and so two sets of wires go to the desktop. Others use Ethernet-enabled LAN phones or use the LAN PC workstation as a phone, even though the originating call is coming in over the conventional PSTN. Such a system is more akin to a "one-wire wonder" where IP data hubs and switches assume some of the local voice switching operations (See One Wire Wonder).

If you plug in a Fusion card from NMS or something similar into your commserver, you can also create an IP gateway and do IP telephony over the Internet.

In situations where a commserver works as a self-contained "PC-based phone system." (i.e., at smaller scales), it switches conventional (and packet) voice traffic, internally. At larger scales, the commserver may drive a larger phone system or high-capacity switch, using one of a wide variety of standard or proprietary interfaces; itself connecting to calls only to provide IVR, messaging, or similar services. Such commservers can send Caller ID or VRU data for screen pops and other data over the LAN so a call center agent can run another database related application on a PC while talking over the phone.

The most sophisticated enterprise-integrated CT Platforms (such as Altitude Software's) include a sort of virtual PBX, a high-level virtual call switching model which allows you to plug in various switching mechanisms into your system. It could be a switch from Lucent, Philips or Alcatel. It could be a PC-based system. If it can switch calls and can be connected via a LAN or directly to a PC, then you're up and running. Cisco / Geotel, with sells a highly distributed PC system that is now being reengineered to also work with Cisco IP routers, and Interactive Intelligence and Mediasoft Telecom are also broadening their switching capabilities.

PC connections to a PBX can be via an RS-232 serial cable, TCP/IP on a LAN, the Telcordia Station Message Desk Interface (SMDI), ECMA's Computer Supported Telecommunications Applications (CSTA), Lucent Technology's Adjunct Switch Application Interface or their Telephony Services API (TSAPI), Intel / Microsoft Telephony API (TAPI), IBM's Callpath Services Architecture, and Northern Telecom's Switch to Computer Application Interface (SCAI).

Commservers are easier to work with than traditional phone systems. They're easy to program – they don't have to be tweaked with DOS-like command line interfaces since most provide GUI or browser interfaces. Much of their intelligence is in software. If something breaks down, your IT people can repair the PC (try fixing a PBX sometime!). Since the hardware isn't proprietary, you can purchase inexpensive and widely available commoditized, generic components.

Most commservers offer numerous, software-based feature modules – for messaging, IVR, ACD, fax, VoIP – all of which are integrated to function seamlessly on the platform. You can build an integrated system using a common GUI application design tool that offers telephony, messaging, ACD functions, and Internet connections from a single server, all running under Windows NT or Unix. This adds enormous value to the core product; provided you like the way your commserver maker has implemented the applications in question. If you don't, then such systems can also generally be made to communicate with other PC-based client / server and desktop PC applications that have the functions you want, boosting the system's usefulness even more.

Meanwhile, enormous progress is being made by Dialogic towards promulgating an architecture under which commservers could accommodate "best of breed," third-party applications; letting you buy your phone system module from one maker and your unified messaging module from another. This architecture, called CT Media (See CT Media), is a complex software layer that isolates applications from hardware, allocates resources (DSPs, voice-recording functions, etc.) as required, and even permits applications to cooperate in providing services. For example, a properly-written voicemail application could exploit a co-resident fax server application to provide "fax on demand."

This beats buying five or six PC-like boxes to "upgrade" a proprietary PBX system to get a similar feature set. And such commservers are generally highly scalable, so you can expand your system from 10 to 120 or so ports and still be working with just one chassis.

Commservers have not generally competed with PBX at the high end of the scale, at the 300+ seat enterprise level. One exception here is the Enterprise Interaction Center (EIC) from Interactive Intelligence, which also combines call center and web center functionality.

Here are some well-known products in this area:

AltiGen's (Fremont, CA – 510-252-9712, www.altigen.com) AltiServ PC-based communications server runs on Windows NT server 4.0. Proprietary boards and AltiServ software play house in an industrial-grade computer. AltiGen has alliances with Compaq, Hewlett-Packard, Advantech, Arista, and Siliconrax.

Applications include PBX, voicemail, follow-me, call/voice processing, auto attendant, basic automatic call distribution (ACD), an e-mail server, Internet-based extension configuration, remote system administration, integrated voice-over-IP option and TI trunks. Any reasonably intelligent person can figure out AltiServ's single-point administration software. At *Computer Telephony* magazine's labs, we've done it, and without a manual.

AltiView, AltiGen's optional desktop call control application, integrates with Microsoft Outlook for screen pops, point-and-click dialing, and remote access via the Internet. Very nice is AltiConsole, the optional software operator attendant console. Big GUI buttons.

Depending on chassis configuration, you can configure 160 lines or 96 trunks (four TIS). So the ideal fit for AltiServ is any service-oriented business with eight to 150 users or multiple branch sites with 150 users per site. AltiServ is being adopted by real estate sellers, insurance, medical, commercial construction, manufacturing, ISPs, and software developers, among other vertical markets.

AltiGen's telephony boards are compliant with ISA or PCI, and MVIP standards. AltiGen's software is also compliant with industry standards TCP/IP, TAPI 2.1, DCOM, and HTTP.

The AltiServ solution supports PSTN and VoIP concurrently, letting users select the type of trunk. IP is H.323-compatible and uses G.711 and G.723.1 vocoders. You can use any analog phone with traditional phone wiring.

Depending on the technical abilities of the customer, AltiGen

resellers handle various levels of installation, support and training. Each 12-port Quantum board sells for $2,900. The AltiWare Classic Edition R2.1 is $695, while the Open Edition R2.1 is $1,295.

Meanwhile, Altitude Software's (Milpitas, CA – 877-367-3279, www.altitudesoftware.com) Easyphone software suite is a client/server solution that runs with databases from Oracle, Informix and Sybase, and supports UNIX and the Microsoft NT platforms. It offers the EASYscript intuitive application builder, which lets users without IT expertise quickly develop and customize the system for their unique business processes. Additionally, applications can be written in any ActiveX compliant tool, making the solution even easier to integrate with commercially available front- and back-office software applications.

Alitude's published application programming interfaces (APIs) provide easy integration to enterprise business applications, corporate databases, and other CTI technology, enabling a company to leverage its existing information technology infrastructure.

EASYcti is at the core of the Altitude solution. Available on Unix and NT platforms, and supporting a wide range of switches, EASYcti enables the applications and technology in your contact center to automate tasks such as placing outgoing calls and routing inbound calls. In addition, EASYcti synchronizes voice and data information, and screen-pops caller associated information to agents' screens along with the call.

As 1999 came to a close, Altitude Software unleashed Release 6.0 (working title) of their contact center software. They've completely integrated web functionality into their contact center package to the point where you now only have to deal with developing and maintaining one system. Altitude calls it "unified business." It's based upon a model they call "unified customer interaction" (uCI).

Altitude Software's Unified Customer Interaction model solves problems resulting from the integration of assisted (call center) and self-service (web-based) systems. Only one system is now in place, assuring both the assisted and the self-service interactions. Altitude calls it "unified business".

One database of customer interactions, workflow engine, customer profiles, are all common to all subsystems.

A common skills-based routing engine for all media is now used as well as a common database for customer data and profiles, a common workflow engine and a common app-gen for scripts / IVR / ecommerce web site development. This allows for the blending not just of inbound / outbound calls, but of e-mails and cooperative web sessions. Routing disciplines can be defined that assure optimal resource usage while pre-defined minimum service levels are met. The binding between the customer / agent interaction and respective data is preserved at all times, so the customer never has to repeat himself while he moves from web self-service to a contact center representative (who can synchronize agent / customer web page interactions and so can fill out web forms while the customer watches) or when the interaction is forwarded to another agent.

The ease and power afforded by Altitude's Release 6.0 in building a single platform for all possible communications commerce needs will take you to new heights (no pun intended).

Perhaps the most effective PBX client software GUI we've ever used comes from Artisoft (Cambridge, MA – 617-354-0600, www.artisoft.com). Its TeleVantage (TV) is a software PC-based phone system – NT 4.0 no less – for small to midsize companies and branch offices. It supports 48 incoming lines (analog or TI) and 144 phones. The TV uses standard telephony PC boards from Dialogic. TV servers network with clients via TCP/IP.

TeleVantage's desktop client software runs on Windows 95, 98 or NT on any desktop PC, and includes visual voicemail sporting graphical call control so you can point and click to screen calls, hold, transfer, conference, etc. There's also a call log, and a contacts and users view. And it offers personalized greetings and special call routing for identified callers.

Installation is usually done by an authorized reseller. But managing the system is quite easy: TeleVantage's Windows-based administrator lets you point and click to create new users (and auto attendants) on the system.

The newest release of TeleVantage supports IP telephony. TeleVantage uses standard IP telephony boards from Dialogic. Support will be H.323 compatible, with G.711, G.723.1, ETSI-GSM, MSGSM, and G.729a codecs.

TeleVantage software is $750 for the server software and $115 per port (where a port is a trunk, a station phone, or a desktop client). On average, a TeleVantage system ends up costing around $500 per user, including TeleVantage software, Dialogic hardware, a PC server, and phones for all users.

Sigma Systems (Denver, CO) is a 30-person software company that markets and supports full-function financial aid, loan billing and student accounts receivable software. They've set up a customer support call center using TeleVantage. The increased call completion and intelligent software-based functionality has led to a productivity increase of 30 minutes per programmer per day during the first year of use. That translates into more than $80,000 savings in personnel productivity in one year alone, not including reduced monthly phone fees and the system's competitive purchase price.

While there's now little doubt that CT Media will be an important piece in this world if (and when) we get there, another vendor and product to keep an eye on in this arena is Buffalo International (Valhalla, NY – 914-747-8500, www.opencti.com) and its Object Telephony Server NT.

Like CT Media, you can think of it as the all important middleware – accessed via an open API – in a specialized NT server that's capable of driving a standalone business telephone switch. In this case, however, OTS NT is specifically for heavy inbound/outbound call-center environments and especially for centers with predictive dialing and call-blend-ing needs, because the platform comes with Buffalo's open architecture predictive dialing engine (OAPDE).

In its latest version, it can now also support Euro ISDN trunk interfacing and extensive monitor/coach/recording functionality. Overall, we're talking a platform based on seven years of continued development and with an installed base of tens of thousands of ports.

Comdial's (Charlottesville, VA – 804-978-2200, www.comdial.com) Impact FX combines a 100 port digital Comdial DXP PBX and CT server in one cabinet (along with a set of CT applications such as an IVR system, voicemail, call control, screen pops), while the Impact NT Application Server is a stand-alone CT server for the Impact 224 and 560 and has back-end TAPI / TSAPI API support.

Also, if you have an FX PBX from Comdial, you can have a dealer add CT Voice voice- and fax-over-IP software (if you already have the right Dialogic line/DSP cards installed) in your FX's embedded PC and away you go.

After you buy a Comdial CT application, you enable it on the FX simply by having a Comdial tech call into the switch and turn it on, and perhaps load some data or some client-PC software. You can pay for CT applications by the port just by plugging in a hardware key. When we first heard of this idea we called it "Dealer-proof CT".

Comdial's ACD solution for the Impact FX PBX is QuickQ 4.0, which sells for about $2,150 plus $200 to $550 per agent. QuickQ4.0 is based on a Windows client/server architecture that enables such niceties as a Microsoft Access-based reporting system.

Comdial has also announced that its latest software release has expanded networking capability for the FX (and its more conventional DXP and DXP Plus phone systems) to support up to 3,500 end-user stations. The newest software includes one centralized Windows NT-based voicemail system, which operates across the entire network. Networking enables communications system resource consolidation including savings on long distance costs, as well as centralized system administration and maintenance. The payback period should be reduced to nearly zero from the money saved on multiple attendants.

Meanwhile, the commserver from e-Voice Communications (Sunnyvale, CA – 408-991-9988, www.evoicecomm.com) is called evoice3000. The 3000 is an NT-based switching and integrated messaging system for a small business with an Internet connection. It gives you features like caller ID, browser-based call and message control and administration, DID, voicemail, auto attendant, IVR, and e-mail integration. Evoice3000 also offers unified messaging with MS Outlook and browser-based system administration.

You need a Pentium 166, 32 MB RAM and 4 GB hard drive to run evoice3000. Mitac, Arista, and Alliance Systems are its preferred platform providers.

The evoice3000 is IP-enabled. PSTN calls can be routed real-time to another evoice3000 server over IP networks and out to PSTN destinations or internal extension phones. Its IP telephony is H.323 compatible, and it uses compression algorithms G723.1 and G729.

The IBM Personal Systems Group (Research Triangle Park, NC – 888-426-5800, www.ibm.com) sells the Computer Telephony Business Solution (CTBS) version 2.51, an NT Netfinity PC server-based telephone solution for small businesses with 16 to 96 extensions. The solution consists of an IBM Netfinity server, on-board telephony switching adapters, call control and voicemail software, and optional Nortel/IBM analog phones.

The heart of the system is really the $7,000 IBM PCI Telephony card that supports eight lines and 16 extensions, and the $5,200 PBX-1 ISA card that can connect six lines and 18 extensions. You can mix and match ISA and PCI cards in the server with the system accommodating a total of 96 extensions.

Actually, the IBM solution is mostly an OEM'd version of NetPhone hardware and software with Coresoft Technologies (Orem, UT – 801-431-0070, www.coresoft.com) add-ons running on the Netfinity server. (This may give rise to the adage that "Nobody ever got fired for buying NetPhone and CoreSoft"!).

The switching function is contained on a PCI card that can be inserted in an NT server with CTI software running on the NT server. Telephones are connected via traditional telephone wiring to punch down blocks.

The system is capable of being extended with add-on CTI applications (as mentioned, many of which are developed by Coresoft). These add-ons can be invoked from various points in the system. For example you can invoke an IVR or ACD application as a transfer option from any greeting or you can invoke any application when a voice message is left in a mailbox.

The call control client includes features such as screen pops with customer information and Caller ID, online phone directories for outbound dialing, call logging for all call activity, third-party call control for transferring, conferencing, etc, and voicemail in your e-mail inbox (unified messaging).

A no-charge SDK is also available to developers who would like to customize or develop additional add-on applications for the IBM Computer Telephony Solution.

A system with 24 trunks, 48 extensions boards, and system software (not including the IBM Netfinity server) runs about $18,000, while an 18-extension, 6-trunk configuration costs $14,500.

Interactive Intelligence's (Indianapolis, IN – 317-872-3000, www.inter-intelli.com) Enterprise Interaction Center (EIC) is a Windows NT-based "all-in-one" communications server. EIC's target markets are enterprises, call centers and service providers, with support for 200 to 300 users on a single Windows NT server. Multiple servers can be connected together to support even larger user pools, or provide feature transparency across multiple sites.

EIC is able to process all communications – phone, fax, e-mail and web interactions – via a single server. The idea is to replace proprietary devices such as PBXs, ACDs, IVRs, voicemail systems, web gateways and CTI middleware systems. So you can run EIC as a stand-alone solution, but it is capable of working with existing comm devices.

The EIC uses a graphical administrative console called Interaction Administrator. By adjusting settings in dialog boxes, administrators can implement security, add new lines, define end-user skills and create workgroups.

Interaction Client, the agent application, runs on Windows 95 or NT and turns your PC into a graphical communications console. Telephone actions such as dialing, transferring, and conferencing are drag-and-drop operations; other client functions include screen pops. The Interaction Client communicates with the EIC server via TCP/IP, which gives you work-group and queue monitoring.

A new Java version endows Java-based network computers, Java stations, and NetPCs with these same capabilities.

For IP, Interactive plans on using an H.323 gateway, built on Dialogic's DM3 IPLink platform, that will be added to the Enterprise Interaction Server. This will let EIC route interoffice calls between EIC systems over an IP network. EIC will also be enhanced with an H.323 client and gatekeeper. They'll use compression algorithms G.721, G.723, and G.729a.

Interactive offers full EIC certification for both resellers and end users. In addition, EIC comes with a built-in graphical application generator that lets many customizations be performed in-house. Cost is $1,000 to $2,000 per enterprise seat and $3,000 to $5,000 per call center seat. This includes hardware and software.

Mediasoft Telecom's (Montreal, QU, Canada – 514-731-3838, www.mediasoft.com) Office Telephony 2000 (OT 2000) is a PC-based PBX, for the small to medium-sized enterprise, featuring a suite of enterprise telephony applications that let customers and employees get information, communicate, and make transactions through public or private networks, including the Internet.

The application templates, or applets, that define OT 2000 include: IVS Front-Desk (auto-attendant), IVS HelpDesk (call center with screen-pop), IVS UniMail (unified messaging) and IVS TeleCalendar (appointment scheduling). IVS Applets are customized through a set of IVS Wizards using IVS Applet Setup (a light version of MediaSoft's IVS Studio development platform).

Designed to work with your current PBX or as a stand-alone PC-based telephone system, the OT 2000 can integrate directly with the Microsoft communications environment (Microsoft Exchange and Outlook) and MediaSoft's IVS Server.

OT 2000 supports Microsoft TAPI, the ECTF S.100 interface (CT Media), MAPI (Microsoft Exchange), the SNMP protocol and ODBC for database access. In addition, OT 2000 can run on any TAPI-compatible call processing board and can interface with any TAPI-compatible PBX. OT 2000 comes fully equipped and ready to work in your Windows NT computer.

For administrators, OTS Manager is a desktop client that manages and configures the many features of Office Telephony 2000. For users and agents, SoftPhone is a GUI that controls such call tasks as detection, caller ID, dialing, phone directories, call forwarding and conference calling. When queuing features are at work, SoftPhone generates a list of incoming calls that the agent can then review and assess status. And HelpDesk desktop client provides customer support agents with a template for follow-up communications. It provides the logged agent with a list of open issues including all related client details, and advanced search features that allow agents to quickly find related issues and resolve them on the phone using a knowledge base.

Mitel's (Kanata, ON, Canada – 613-592-2122, www.mitel.com) SX-2000 for Windows NT is based on Mitel's proven SX-2000 LIGHT PBX technology. Designed for companies with 40-120 users, it supports both standard and enhanced PBX features and interfaces, including call control, digital trunking, peripheral interfaces, auto attendant, call center routing and management, and computer telephony integration. It has GUI-based management, and messaging functionality – all pre-packaged and delivered on one server.

CTI capabilities are run through TAPI and Mitel's own MiTAI. Voice networking capability is supported through LS/GS, E&M, DID, T1, E1 and ISDN interfaces. Pioneer is manufacturing the server for Mitel's specific requirements. Data interface is Ethernet, allowing management from web browsers with several security levels.

As the system is turnkey the technician only has to program the user information. A 192 port by four T-1 link system can be programmed in a day. Mitel believes that the voice-over-IP protocols are not mature enough yet to totally replace TDM.

The Mitel SX-2000 for Windows NT is a mature product. It can be purchased today and is evolving toward being an IP solution.. The system is available from Mitel's multinational VAR channel for about $750 to $1,000 per user, depending on applications.

NEC Corporate Networks Group's (CNG) (Irving, TX – 800-832-6632) NEAX Express NT is a server-based PBX that handles all the same terminal options and features found in the NEAX2000 IVS; but is also a CTI server with TAPI 2.1 client/server applications such as voicemail, unified messaging, call log, directory dialing, screen pops, soft phone and call routing. Works with any Windows database, contact manager or PIM (Outlook, GoldMine, Lotus, Act, Day-Timer, etc.).

Made for SOHOs and branch offices, Express NT supports up to 60 ports with the embedded NT server, but you can expand to 124 ports via an expansion chassis. The NEAX Express consists of an NEC NT industrial-grade server, NEC PCIA boards that give you the PBX functionality, and third party DSP boards for voicemail. Runs on an Ethernet 10/100 network. And in the event that NT or applications lock up and a reboot is necessary the PBX continues to function without interruption, because one of the PBX boards has a traditional NEAX2000 IVS processor.

All maintenance is done via a PBX/application management GUI. Client PC software will support Widows NT, 98 or 95.

Like its other NEAX PBX products (the NEAX7400 IMX and the NEAX2000/1000 IVS), the NEAX Express will be IP-enabled on both the line (station/LAN) side and the trunk (wide area network) side. Availability is expected late this year or early next year. This IP-enabling will be implemented via an IP line card and an IP trunk card which can be easily installed into the NT Server. It will support H.373. For now MSRP estimated price will run between $15,000 to $20,000 for 20 ports.

NetPhone (West Marlborough, MA – 508-787-1000, www.netphone.com) services the small and midsized business market through its existing partnership with IBM's Computer Telephony Business Solution which is based on NetPhone technology. However, NetPhone is now also targeting the expanding enterprise market with its IP telephony solution, NetPhone Connect and NetPhone IPBX, which supports up to 1,400 users across remote offices.

NetPhone Connect and NetPhone IPBX let customers direct interoffice calls to the corporate data network, bypassing the public switched telephone network. NetPhone also offers fallback technology. All calls, regardless of the degradation or congestion on the network, are placed and routed seamlessly. No calls are lost or disconnected and users never have to redial a call. This reliability, coupled with NetPhone's open CTI applications, provides an IP telephony solution that supports both local and remote users.

Call handling/media-processing features include caller ID, dialed number translation, routing and billing functions and built-in ACD. You'll find support for TAPI, TSAPI, interfaces with Visual Basic and C++.

Price depends on system configuration, but it's around $250 per seat.

Nitsuko (Shelton, CT – 203-926-5400, www.nitsuko.com) weaves a Nitsuko-specific version of AltiGen's AltiServ system named the Nitsuko Communication Server (NCS). The new NCS will be distributed through Nitsuko America's network of more than 300 dealers and is targeted at companies requiring a system to serve up to 100 users.

With this relationship Nitsuko builds on AltiGen's proven technology while providing a unique ability to meet the needs of Nitsuko dealer channels through TAPI-based applications, VoIP and leveraging advanced research and development work.

Picazo's (San Jose, CA – 800-464-3274, www.picazo.com) next generation phone system, which they simply call Windows 2000-based Telephony Server, launched upon release of Windows 2000. The product is intended for companies with 5 to 500 users – with a switching engine that supports up to 500 lines. Soon after, look for up to 2,048 lines.

There are three key "layers" in the system – a suite of TAPI-based applications; the core switching code running under Windows 2000; and the hardware components. Each has open interfaces (TAPI, S.300, H.100),

which let other applications, switch engines, and hardware be used with the system.

A TAPI-based application suite has voicemail, automated attendant, call routing, unified messaging, ACDs, and links to websites. The client application, attendant console application and ACD management application all use either first or third-party call control. Picazo also has a line of digital display phones.

Connections to T-1 / E-1 and ISDN PRI are supported. And the product is IP-enabled for H.323 users on the LAN, supporting IP trunk lines.

List pricing for a system that includes the server, voicemail, unified messaging, line/trunk connections, digital display phones, T1/PRI and CTI clients is expected to be about $800 a user.

Telesynergy's (Santa Clara, CA – 408-260-9970, www.telesynergy.com) TelePCX is a PC-based PBX with built-in auto attendant and voicemail system housed in a single server. You get web-based unified messaging (integrated with TeleUMS), IVR, fax-on-demand, desktop call control with caller ID capability and ACD. In addition, TelePCX sports a GUI-based application generator. This software development tool kit lets you develop call flows for the IVR and fax-on-demand.

TelePCX runs on Windows NT or Windows 95. It's based on TeleSynergy's 412 PBX Board, which supports four ports by 12 extensions and two ports of voice for voice processing features. If you add the Switch Card, or the Voice Card, the TelePCX can support up to 48 ports and 96 extensions. Network compliance supports TCP/IP protocol.

With an IP telephony card, TelePCX can receive incoming calls, distribute them, and make outgoing calls through an IP network.

Operator Console is a graphical console that lets an operator answer, transfer, hold and speed dial calls. With Monitor Console, supervisors can monitor and supervise status of all telephone lines.

In a minimal, 4 x 12 configuration, with two ports of voice, all software including PBX, VM, AA, GUI-based application for IVR/FOD and web-based UMS has an MSRP of $1,550. A four-port fax board is $995.

CompactPCI (cPCI)

CompactPCI (cPCI), initiated in late 1994 by Ziatech Corporation under the auspices of the PCI Industrial Computer Manufacturers Group (PICMG), is the newest specification for PCI-based fault resilient and fault tolerant computers and defines many features that make a PC more fault resilient. The CompactPCI Specification is the result of a concerted effort of the CompactPCI subcommittee composed of the following companies: Digital Equipment, GESPAC, I-Bus, Pro-Log, Teknor, VMIC, and Ziatech.

CompactPCI offers a host of telecom features required for network applications including:

- A standard telecom bus (32 streams and 4096 time slots) for communications between cards in a chassis rack.
- Redundant chassis and board configurations for highly available resource requirements.
- A telecom form factor (3U and 6U card heights with rear panel I/O).
- Transition cards and cabling assemblies to simplify installation.
- Telecom power bus (- 48 V DC) and provision for ringing voltage.
- Hot swap capability with staged pins and system notifications with card tab release, allowing systems to be upgraded or expanded, or cards replaced without taking servers off-line.
- Software compatible with mass market PCI systems.

- Redundant power management, CPUs, disks.
- No interruption of system operation if a subsystem module fails

The principal benefits of CompactPCI include:

- Higher reliability through hot swapping of components.
- Delivery of hot-swappable telecom features required for industry (H.110, rear panel I/O, power specs).
- Compatibility of software from standard PC (PCI) systems to CompactPCI systems – existing PC software can run unchanged on CompactPCI systems.
- An open, industry-accepted specification eliminating the proprietary nature of previous high-availability systems.

But CompactPCI is not just another computer bus. It invokes passion, both good and bad.

CompactPCI always kicks up a storm of controversy whenever I publish an article on it in *Computer Telephony* magazine.

Some manufacturers warn me about the "hype" surrounding it. Many manufacturers have gone so far as to urge me to write that CompactPCI isn't yet ready for "Prime Time". Various peripheral cards haven't appeared and volume production has yet to take place, they say. They say sales of this new bus will never reach "critical mass" and that some potential buyers of fault resilient equipment are wasting their time by "fence sitting," waiting for deliverance when more CompactPCI products are available. Still other cPCI vendors demand that I write more about it. They say it's already becoming a big hit.

Indeed, CompactPCI has become perhaps the most controversial computer bus of all time. Whereas other PC buses crept into the industry quietly (no one made so much as a peep during the ISA to VL-bus to PCI transition) cPCI has brought on naysayers, doomsayers, and just plain old hostility from some fault resilient computer vendors.

In any case, all the hubbub and finger-pointing proves that CompactPCI is the hottest telephony bus going these days.

At first, you wonder why anybody would dislike the CompactPCI bus, considering all that it appears to have going for it.

Passive backplane machines are generally better built than conventional desktop computers and have done much to further the telephony industry – but they still leave a lot to be desired.

CompactPCI moves PC-based technology to a new level. It offers the same high performance of desktop and passive backplane technology, but in a much more robust form factor.

CompactPCI really sprang up at the juncture of two independent lines of developing technologies – ruggedized PCI development led by PICMG, and the Enterprise Computer Telephony Forum's (ECTF) H.100 CT Bus that was a result of the grand armistice of the MVIP-SCSA bus wars, which ultimately begat the exciting new H.110 standard and Hot Swap specification, which gives CompactPCI some of its miraculous abilities.

The conventional PCI bus has a high bandwidth but is not really "rugged". Also, you can't hot swap the components – try pulling a Dialogic or Natural MicroSystems voice card out of a running computer and you'll see what I mean. For truly mission critical telecom applications, some reformulation of the existing spec (to make it more fault resilient) was considered necessary. This became CompactPCI, which was initated in late 1994 under the auspices of PICMG.

CompactPCI can run Intel-based software because it is electrically similar to the conventional PCI bus. This was a major point because PCI has become the universal bus, tying together as it does inter-chip, intra-chip and intra-board functions. People also forget that although PCI is known as a local bus for x86-based PCs, PCI technology touches upon all modern microprocessor designs. Virtually all high performance, general purpose microprocessors use PCI: Pentiums, PowerPCs, Alphas, SPARCs and MIPS RISC processors are all built around the PCI bus. Not coincidentally, all of these processors are available on CompactPCI. In particular, the PowerPC and Compaq's Alpha chip are supported with PCI compliant chip sets and can benefit from CompactPCI's high-bandwidth that allows new high performance processors to run flat out and drive an application at maximum speed.

In fact, a number of manufacturers have placed the Sun's famous UltraSPARC 64-bit processor , the PowerPC and even DEC's 64-bit.Alpha processor on CompactPCI cards, running more exotic software such as Solaris and real-time operating systems far different than what you get in a shrink-wrapped package from Microsoft.

This matter of PCI compliance not only provides a wide choice of hardware, but also enables the free choice of operating systems from NT to UNIX to Solaris to real time. Meanwhile, the core PCI standard has added features important to CT, including Maximum Completion Time to reduce transaction latency and Message Signaled Interrupts which provides for a virtually infinite number of peer-to-peer mailbox interrupts so that voice cards can process data packets without host intervention.

But CompactPCI, though electrically similar to PCI, goes a step beyond physically. CompactPCI cards are mechanically different from standard PCI – they follow the rugged Eurocard form factor (originally popularized by the VME bus). It's an interesting, almost intimidating, style of connector. Whereas a desktop PCI card uses a card edge connector at the bottom of the board and the I/O connectors are on the side, a CompactPCI card generally uses a 3U (100 by 160 mm) or 6U (233 by 160 mm) board size with a pin-in socket connector at the bottom and the face plate and I/O connectors on the top.

The 6U cards can have up to five connectors on the rear of each card. The CompactPCI bus itself uses two of the five connectors, with the other three providing up to 315 pins for user-specified I/O connections. CompactPCI cards are firmly held in position by their connector, and there are card guides on both sides and a face plate which solidly screws into the card cage.

Thus, the high quality 2 mm metric pin-and-socket connector of the CompactPCI card (Ziatech's brainchild) is a lot more reliable than a regular PCI card edge connector. Moveover, the power and signal pins on the CompactPCI connector are staged to support the most tantalizing feature of all – hot swapping, which, along with redundancy, lies at the heart of fault tolerant and fault resilient systems, and which isn't possible on standard PCI.

CompactPCI uses a vertical passive backplane and vertical card orientation for good cooling, which is a must for today's power hungry telephony DSP cards. Cards are loaded from the front and rear, simplifying maintenance. Air is generally brought in the front bottom of the chassis, where fans direct the air up over the cards and out the rear. The cPCI bus offers many more features than other non-CT board solutions.

Unlike standard desktop PCI, CompactPCI supports twice as many PCI slots – eight vs. four. Actually, CompactPCI systems can be expanded past their eight slot limitations using PCI to PCI bridges, devices that connect to the PCI bus on one end and can drive another set of eight cPCI slots at the other.

Bridges act as a repeater, "repeating" what they see on the main bus to the bus on the other side of the bridge. You can theoretically add as many PCI bridges as you wish, building systems with scads of slots. Each bridge, however, adds a clock cycle (33 ns) to a data block transfer. If the amount of data that is moved is low (1-2 words) this additional cycle may be significant overall. Still, if a block of 200 bytes or so is moved in transfer, this added cycle should degrade performance by 1% or less. ompanies such as Radisys, however, don't like bridges circuits and have evised more sophisticated methods of bus slot expansion.

Unlike the conventional fault resilient PC, where cooling airflow ies not have regular paths and is blocked by the card's I/O bracket, disk ives and power supply unit, the CompactPCI systems are designed to ovide clear airflow paths for all plug-in cards. Cards are normally ounted vertically, allowing for natural forced air convection "chimney" oling, also reminiscent of VME. The cPCI cards can be front loaded or moved from a card cage.

Another thing it inherited from the PCI and CT world is a modified rm of the H.100 telephony bus standard, the H.110.

To understand what the H.110 standard is, one should first grasp the nplications of H.100.

Until now, the two principal CT resource buses have been Dialogic's Cbus and Natural Microsystems' MVIP-90 and H-MVIP. The wares of oth camps allow you to connect media resource cards together via a pecial cable (bypassing the host computer's regular motherboard bus)) that a single CT application can control many cards that do fax, text-)-speech and speech recognition. Also, multiple computer chassis can be ed together with these busses for huge applications.

The Enterprise Computer Telephony Forum (ECTF) then managed to nd the "CT bus wars" with a new hardware specification called "CT us" creating a "neutral" telephony bus that would work with existing Cbus and MVIP compliant products. The generic CT Bus defines a ime Division Multiplexed (TDM) single communications bus, often alled a mezzanine bus, with clock and frame system technology, across omputer chassis card slots.

CT Bus products may be developed in conjunction with a number of ardware configurations. "H.100" happens to be the PCI implementation ndeed, the first card-level definition) of the overall CT Bus spec. It provides ll the necessary info to implement a CT bus interface at the physical layer or the PCI computer chassis card slot independent of software applications.

H.100 offers some marvelous features, such as 4,096 bi-directional ime slots (permitting up to 2,048 full duplex calls) for high-volume com- munications capacity, and 128 channels per stream.

Best of all is H.100's backward compatibility and interoperability vith Dialogic's SCBus, and Natural MicroSystems' H-MVIP and MVIP- 0 telephony buses. H.100 compliant PCI products will interoperate with thers in the so-called CT Bus core mode and simultaneously with Cbus, MVIP-90 or H-MVIP (as well as ANSI VITA 6) cards using the CT us compatibility modes. It will even work with such cards plugged into SA / EISA card slots.

There is also a subset of the CT Bus specification so manufacturers an create lower-cost products.

This all helps system integrators and VARs to smoothly manage the nigration of their systems from ISA SCbus or MVIP90 to mixed ISA/PCI o 100% PCI, cleanly mixing H.100 CT Bus products with predecessor prod- icts at each step along the way. This ability to mix CT Bus and other buses illows migration of a system one card at a time, to support added features or growth, rather than having to invest in a completely new system.

H.100 supports up to 20 boards, all of which are connected with a ribbon cable that shouldn't exceed 20 inches.

So, H.100 CT Bus provides more bandwidth capacity than its prede- cessors enabling new applications and increasing the reach of existing solutions. It now makes fault resilience at this level a serious proposi- ion, making PC-based applications more robust.

And then came CompactPCI, which spurred the ECTF (who brought

you the H.100 / PCI version of the CT Bus) to become involved. They eventually announced the H.110 specification, which covers the CT Bus implementation on the cPCI form factor. The H.110 specification is elec- trically identical to the PCI standard used on desktop PCs, is functional- ly identical to H.100 and even uses the same I/O driver specification. This allows semiconductor vendors to implement a single chip design for both H.100 and H.110.

The cPCI boards with CT Bus (H.110) support can connect channels to line interfaces, to DSP resources for voice processing and signaling, or to the H.110 bus itself. These boards can run signaling protocols for line interfaces on other boards by switching these resources across the high bandwidth H.110 bus, which provides a common bridge and allows inte- gration with other board technologies and can be used to carry time- division multiplexed streams, which are divided into timeslots of equal bandwidth so as to carry up to 4,096 multiple 64 Kbps voice channels. For example, a T-1 stream consists of consecutive frames that each have 24 slots for the interleaving of samples from 24 voice channels.

Right from its inception, most members of the CompactPCI com- munity believed computer telephony would be the first big market for CompactPCI and that telephony suppliers would move quickly to it as soon as cPCI voice and fax cards were available.

Everyone believed this because the cPCI / H.110 combination sup- ported the long-awaited ability to hot-swap cards, to fix a system while it was running, without missing a beat.

Hot swap is now an official PICMG standard (a combination of hard- ware and software standards) that allows for insertion and extraction of CompactPCI peripherals in a live system without disrupting anything. Hot swap is a superset of the PCI Hot Plug specification. The difference between "Hot Plug" and "Hot Swap" is that with Hot Plug, the PCI slot must be powered down before a PCI peripheral can be inserted or removed. To do this you need an active motherboard-like backplane which is unacceptable in many high availability applications.

One of the most exciting innovations regarding CompactPCI is the hot swappability of CT Bus boards. Until now, hot-swap has been demanded by Telcos, but it couldn't be delivered by suppliers of PC- based computer telephony solutions. Now that there is a PICMG hot swap specification allowing the addition, subtraction, and replacement of boards in a running system, hot swap quickly became a key feature (not to mention a big selling point) of the high availability CT applications being developed on CompactPCI platforms

Indeed, the high availability provisions of the PICMG 2.1 Rev 1.0 stan- dard for cPCI Hot Swap means that, besides being able to remove and replace a board with the system operating, processing can automatically be switched to a second backup board already installed in the system in the event of failure in the primary processor. This allows manufacturers to create systems with "automatic failover" giving them true fault tolerance.

The ability to build CT systems on a fault tolerant cPCI foundation will give developers enough courage to create some really super-high density systems, especially since cPCI supports enough bandwidth to accommodate even the highest density CT resource board.

After all, many of these manufacturers are impatient to get to the Promised Land of 64-bit computing, a world of increased bandwidth and speed that today's top applications developers are clamoring for. Migrating from 32-bit processing to 64 may involve some unpleasant- ness, however. CompactPCI, supports 64-bit processing but the entire environment of all peripherals and software must be able to deal with 64 bit "gulps" too.

One interesting milestone in this area is Alta Technology

Corporation's Alpha processor powered CompactPCI single board computer. Alta uses a 256-bit wide memory bus and up to 1 GB of DRAM DIMM modules that allows for 60% higher bandwidth in memory performance when compared to conventional synchronous memory and a 64-bit memory bus. The higher memory bandwidth is important not only to the processor, but is a requirement for the higher I/O bandwidth of the 64-bit PCI bus.

As for software – particularly operating systems – a native Alpha version of Windows NT exists, but it's not yet fully 64-bit (Windows 2000 should be compatible). Still, as Intel's 64-bit Merced continues to take its time to market.

Then again, Linux is already fully 64-bit compliant, and has been ported to the Alpha and Sun platforms. Amazingly, the Alpha Linux port was done back in 1995! After HP has finished getting Merced out the door for Intel, Linux aficionados will port Linux to it. With its free inherent goodies such as the Parallel Virtual Machine (PVM) and Message Passing Interface (MPI), Linux becomes a real contender for real-time and distributed multiple processor applications.

Ah yes, multiprocessing. It's possible to have more than one processor board on a CompactPCI backplane, but only the main CPU can be a bus master in either a CompactPCI or PCI system. If true multiprocessing is needed (such as two processors assisting each other executing an operating system and application programs), this is usually done by coupling two or more microprocessors together using a technique called Symmetric MultiProcessing (SMP).

The coupling defines a master at the CPU board level, before encountering the PCI bus. Both CompactPCI and VME support the idea of an intelligent slave processor, although CompactPCI does a better job at makes this multiprocessor architecture transparent to the operating system and application program.

Motorola's MCPN750 board's non-transparent bridge allows the mapping of multiple processor boards' addresses onto different CompactPCI bus address spaces. A conventional system-slot processor board and multiple MCPN750 processor boards can operate in the same backplane without address conflicts. In a similar vein, Ziatech's CompactNET multiprocessing environment transforms a CPU board into a system master for up to 32 other processor boards

Even more improvements in cPCI continue to come in from all quarters. For instance, many cPCI applications in telecom require rear I/O support, simplifying service by avoiding the error-prone, front recabling process. Today, PCI mezzanine card (PMC) rear I/O, however, is expensive and time consuming since it requires the design and release of a new custom transition module for each PMC / carrier configuration. If an OEM has six PMCs and six carriers, it would need 126 transition modules to support all the possible combinations.

But a new, proposed PIM architecture from the Motorola Computer Group (MCG) would require only 12 modules-six transition and six PIMs. Called TM-PIM, it's a transition module designed to work with the proposed PMC Interface Module (PIM) architecture for modular rear I/O. This standards-based architecture allows a single transition module to replace what previously required numerous unique designs. MCG has also announced its first PIM, providing an E-1 interface for the company's MPMC860 controller.

CompactPCI single board computers (SBCs) are also progressing in leaps and bounds. For example, many now have adopted the Super Socket 7 format enabling a board to be fitted with processors from a number of different manufacturers.

And of course, unexpected developments happen with any new technology. CompactPCI systems running Windows NT and Unix seem to be fast becoming the RAS platforms of choice for OEMs targeting large enterprises, ISPs, LECs and CLECs, which should be a boon to high-density cPCI RAS solutions such as Ariel's RS2000C card.

So what's not to like about CompactPCI?

The trend in the industry has been to put "more and more eggs in one basket" by cramming more and more ports onto a card in an effort to reduce the card count in the chassis. It's all fine and dandy to say that there's nothing to worry about since everything is hot swappable, but again, with the super-high density cards one must remember that in many cases you've only got seven or eight cPCI card slots to begin with.

When a card fails, whether it's hot swappable or not, you've also lost a significant portion of your telephony network resources. It's conceivable that you could build a 96-port PC-based PBX with two boards. If you lose one of those boards, then you've lost either all of your incoming lines or all of your outgoing lines. So what is hot swap doing for you?

Another related problem is that it's going to be very typical for a particular developer to plug in a T-1 card or something similar and have the system get the network timing reference from that card, since it's the only card in the system doing the timing. Now when that card ultimately fails, you're obviously going to lose your network timing base references.

There is no easy way around this situation, except perhaps to go to lower density cards and populate the box with more cards than necessary, with each card capable of generating network timing references.

Yet another issue is that in spite of the fact that the hardware is in place to do all of this marvelous cPCI hot swapping, software is still lagging considerably behind. There are huge management issues that the people who are writing the custom code to run in a cPCI box are going to have to come to terms with, such as: "Gee a system can lose fifty percent of the outgoing analog phone lines! What now?" How do you "admin" those out? Companies such as Lucent, Nortel, Mitel and other PBX makers resolved problems like this years ago, but many developers coming from the CT industry have never encountered this scenario before on a giant big-money-at-stake scale.

So the question becomes, given the higher price of a hot swappable card, is hot swap worth it? Well, yes, but something has to be done on a management level that's up and above the simple clock and swap issues one doesn't even encounter in the everyday PC world.

Operating systems will have to be altered to take into account both hot swappability and features unique to each companies' expression of the CompactPCI architecture. Particularly much work needs to be done at the hardware device driver level. Windows NT, for example, doesn't take very kindly to having a driver added after boot time. True fault resilience or fault tolerance means that something has to tell the device driver and the operating system that, for example, something major has impacted the system and calls can't be routed the same way they were just a second ago. Let's send them along ready and waiting alternate system routes.

We'll no doubt end up with a CompactPCI "flavor" to OSes, drivers, system management, etc.

Even application generators will have to take into account the fact that their final generated executable code must be savvy about cPCI and how to re-route calls in case of any and all of the disaster scenarios that can befall a fault resilient computer.

Aside from hot swappability, that other great savior of cPCI is the idea of automatic failover, where two CPUs are working in parallel, each ready to take over the burdens of the other in case of a hardware problem. Such systems shoot for true "five nines" (99.999%) capability, with downtime of ly about five minutes or so a year. While this has been brilliantly worked at by such companies as Force, Motorola, Rittal / Texas Micro, and Ziatech, ere is the little matter of coordinating all of the other components.

Automatic failover for such things as T spans and DS3s is hard to , simply because you just can't tie two T-1 ports in parallel and run he until you decide you must suddenly run the other. You actually have have secondary switching and hot standby T-spans there to do the job. he same thing is true of something as simple as a station interface. You st don't run two source ports in parallel. There has to be something at can power a faulty device down and then bring a spare online, not mention the need for something akin to relay switching to handle the hysical connections.

Typically, rerouting physical connections is done with an external ame switch that automatically switches all the lines away from the efective Point A to the ready and waiting Point B. Such switches are otoriously expensive, unreliable, and not readily available, and one ould have to allow for some kind of output from the PC to control such switch, which is usually mounted on a wall, a spot where you'd have gather all of your phone lines, with sets of lines going in and out.

In the future all of these issues will be resolved by hardware and soft-are vendors, and even now there are measures that you can take to achieve hatever degree of reliability or high uptime that you feel your system equires or your customers will pay for. As for the moment, however, true ult tolerance in CompactPCI is a software and engineering exercise for the eveloper, not a set of clearly defined industry standards, at least not yet.

Naysayers or not, it nevertheless appears that cPCI is going to take ff. It's not a question of if, but a question of how much and how soon. ompactPCI will live and evolve.

Don't believe me? I recently had a talk with Eike Waltz, the Senior lectronics Packaging Specialist International for Rittal Corporation pringfield, OH – 937-399-0500, www.rittal-corp.com). He's also the hair of IEEE 1101.1, 1101.10 and 1101.11, which defines all of the nechanical parts that serve as the basis for CompactPCI, such as the con-ectors. (Those cPCI connectors, by the way, are becoming quite bizarre. normal 6U high cPCI card with all connectors loaded has an insertion extraction force of 118 lbs. A 9U high cPCI board needs 188 lbs of force. it has 10 rows of connectors, it takes about 340 lbs. of force!)

In any case, Waltz tells me: "Companies like ours make board acces-ories, such as the front panel for the boards, and in particular the eject injector handles, which are specified for cPCI. Rittal was the first com-any to produce these handles, in 1995, long before the secondary sup-liers came into the market in 1997 and 1998. Each handle you sell ulti-nately has a board behind it. From such a viewpoint you can judge pret-y accurately how many cPCI boards are being produced, since it's direct-y related to the sale of the handles.

Waltz continues with "Presently, we sell as many handles per month s we did during all of 1998. Yes, there's a few, say 15 percent, going into ME64X, but it's a small number in comparison to what's going into ompactPCI manufacture. So if people say that the market is not taking ff, well, it's relative. From 1995 to 1996 there was little movement, from 996 to 1997 was also relatively little movement. From 1997 to 1998 was aid to be a 160% increase, which is possible since one is dealing with uch small initial numbers.

"But this year, 1999, we are seeing a tremendous increase over last ear. The number of projects that are out there and the number of com-anies adopting cPCI, is growing continuously. Yes, most of it goes into elecom and computer telephony. A smaller number goes into things such s machine control."

After listening to Waltz, we could boldly make the prediction that the year 2000 will be "The Year of CompactPCI." Certainly sales will rise, but how much? At the moment CompactPCI is quite expensive, though there are markets, such as hospitals, that don't seem to care about the price.

One would have hoped for cPCI to have a better cost / performance ratio. But cPCI is industrial grade, it's a different world than consumer mail order PCs. It's said that a higher production volume will bring down the cost of cPCI, but in fact the production volume of CompactPCI will never be as high as its consumer cousin, PCI. You shouldn't expect a cPCI computer on your desktop anytime soon, but you'll certainly find it at the Telcos, in web-enabled call centers, at your friendly local Application Service Provider (ASP), and even on your company's own premises – yes, your Enterprise CT Server may one day be a cPCI machine.

CompactPCI looks like the logical successor to the melange of equip-ment that dominated the previous generation of fault tolerant and fault resilient systems. The H.110 bus and the hot swappability of cPCI cards will make it a big hit, even with telcos.

I expect the year 2000 to be the true "year of CompactPCI". CompactPCI's day in the sun will come after the various specifications and hardware designs are all finally "set into stone" and this year's equipment orders are finally processed, and all of the game pieces are in place for the major incursion into the fault resilient marketplace that will surely occur.

Yes, in spite of all of those naysayers, I now figure that by the end of the year 2000, cPCI will have garnered about 15% of yearly sales in the telco / CT market. Others say that cPCI will be at least a $1 billion market in three years, which, although sounding a bit optimistic, is quite possible.

And if nothing else, CompactPCI continues to engender some spirit-ed debate!

For the record, yours truly was promoting CompactPCI and writing about it in the pages of *Computer Telephony* magazine long before any-body else in the industry had even heard about it. I consider it to be the future of fault resilient and fault tolerant computer systems.

Here are major cPCI players and the products you should be familiar with.

Adax (Berkeley, CA – 510-548-7047, www.adax.com) designs, devel-ops and supports WAN software and hardware solutions for telecom and high-value datacom customers. Adax's protocol controllers, network con-trollers and software support the foundation layers (layers 1-3) of the OSI model for all the major communications protocols (HDLC, SS7, ATM, Frame Relay and X.25). Therefore, their boards make fewer demands upon CPUs, saving valuable bandwidth.

Recently, Adax migrated its SS7 solution onto the CompactPCI plat-form in the form of their two-port APC7-cPCI board and APS-SS7 soft-ware. Together they deliver very high bandwidth to SS7 applications and minimize server usage, leaving plenty of horsepower for applications like adjunct processing and intelligent switching, resulting in more calls or transactions per line (and thus more revenue).

The solution delivers complete MTP1 and MTP2 data communica-tion control conforming to the ITU, ANSI, Telcordia, and AT&T versions of SS7. Upper-level stacks, such as TCAP, ISUP, TUP, and MTP3 from any vendor can be integrated with the APC7 board through a well-defined AT&T Streams-based API to MTP2 without having to change any of the layers of SS7 software. A four-port version became available the first quarter of 2000.

Adax can deliver connections at full T-1 / E-1 and fractional T3/E3 speeds to either the Internet, virtual private network (VPN), or a private leased line through their network controller daughter card (the ANC-

TCX) that attaches to the SS7 cPCI board. The ANC-TCX provides direct connections to the WAN and is a one-slot solution. It also performs the CSU/DSU function as well as channel bank, multiplexing, and drop and insert functions.

The Adax cPCI boards operate on such open systems-based hardware platforms as Intel, Sun, IBM RS/6000, or HP Servers and all predominant UNIX and Windows NT operating systems. Customer code typically sits on top in the upper layers and communicates with the Adax lower layers through a standard AT&T UNIX Streams DLPI Interface that makes implementation of layer 2 and below transparent to the higher layers.

All of Adax's CompactPCI boards contain the Adax-designed Avalanche Chip that gives the host CPU preferential access to the board's memory and supports plug and play. The chip allows the cPCI boards to take full advantage of the 133 MBps data transfer rate over the bus and perform at up to 22 Mbps through the serial line interfaces.

Advansor is a division of American Advantech (Sunnyvale, CA – 408-245-6678, www.advantech.com). Advansor has made an auspicious debut with their Series CP7, a 7U high cPCI platform based on Intel's Pentium III and Pentium MMX(tm) processor. Its front access, slide-in and snap-on "VME type mechanics" make the CP7 highly, compact and flexible.

It comes fully equipped with memory, graphics controller, Ethernet controller, hard drive, CD-ROM drive, 250W ATX power supply and Microsoft Windows NT Workstation 4.0 (upgrades and options are available accordingly with user specifications). The CP7 has eight expansion slots for standard 6U-sized cards along with space for two 5.25" drives and one 3.5" drive. With its highly flexible modular design, different configurations can be set for individual applications.

The CP7 includes a 6U backplane that complies with the full PICMG 2.1 Hot-Swap Specification. You can build a hot-swap system by using hot-swap capable plug-in boards and software.

Standard IDE and floppy interface connectors have been placed on the backplane. Through the P3 connector, you can easily connect IDE devices and floppy drives to the backplane via the 6U single board computer modules.

A small 1U high fan module provides forced air cooling. Two 88CFM high-speed fans are mounted in a hot-swappable tray directly underneath the card slots. The fan's tachometer output is monitored by an alarm module via a 6-pin connector on the fan backplane. A protective circuit also has been designed into the fan backplane to reduce electrical spikes and noise so you can hot-swap fans.

Alta Technology Corporation (Sandy, Utah – 801-562-1010,

www.altatech.com) has coupled a 64-bit CompactPCI rugged Eurocard architecture with Compaq's (previously DEC's) Alpha microprocessor. The result is the CPCI/SBC-A500, which uses a 500 MHz 21164 microprocessor in a 6U dual slot format.

This interesting Alta hybrid design uses 64-bit, fully pipelined advanced RISC processor technology, support for up to 1 GB of ECC memory on DIMMs (there is a 256-bit pathway to memory), 2 MB of Level Three cache, and bridgeless I/O through a PCI Mezzanine Card (PMC) site.

In a CompactPCI backplane, all seven slots (or more with bridge chips) can be 64-bit peripherals. Alta offers other cPCI cards to complement the system, such as the 6U Dual PMC carrier board for both 64-bit and 32-bit peripherals, 6U 64-bit Front Panel Data Port which allows data transfer rates of up to 160 MBps, the 6U 3D Graphics Accelerator that uses 3Dlabs' GLINT and carries 40 MB EDO and 16 MB of VRAM, and the 64-bit Gigabit Ethernet PMC card that yields ten times the performance of Fast Ethernet!

As for software, for now you have to settle for the 32-bit Alpha version of Windows NT – I hope that Windows 2000 will be able to run the Alpha in a fully 64-bit environment. But remember that Linux is already fully 64-bit compliant, suggesting some interesting development opportunities.

Alta says that, with the release of even faster Alpha processors and new 64-bit peripheral chips, Alta is committed to new CompactPCI and PMC products that will continue to adopt the latest 64-bit technologies.

The XDS Infinity Series H.110 CompactPCI High Density BRI board from Amtelco (McFarland, WI – 608-838-4194, www.amtelco.com) takes the well-known Amtelco XDS boards to a new level of density and reliability, supporting both the central office and digital phones.

The XDS H.110 BRI board has 32 (2B + D) S/T ports, each configurable as either Terminal Equipment (TE) or Network Termination (NT). Protocol support is included for NI-1, NI-2, AT&T custom, EuroISDN, and INS-Net 64, and High level Data Link Control (HDLC) controllers allow transmission over the B or D channels. The board includes Terminal Endpoint Identifier (TEI) management (so that up to eight devices can be connected to one ISDN BRI line), as well as message buffering, DTMF detection and generation, and the LAPD protocol. Also, PS1 power is provided to all the ports, and line energy detection is included. Drivers are available for Windows NT and Unix.

The CPCI-CPU-1 from Analogic Corporation (Peabody, MA – 978-

Advansor's Series CP7 is a 7U high CompactPCI system that houses an eight slot backplane.

Amtelco achieves the highest ISDN density in the industry with the XDS Infinity Series H.110 CompactPCI High Density BRI board.

977-3000, www.analogic.com, now Anatel) is a 3U high 200 MHz Pentium MMX single board computer for embedded PC applications that require speed and the 80x86 architecture. The CPCI-CPU-1 includes 64 MB DRAM, a 256 KB Level 2 pipelined burst cache, an enhanced IDE hard disk controller, a floppy disk interface, COM1, COM2, LPT1, a keyboard and PS/2 mouse ports. Also, a USB is provided for distributed functions such as keyboard, modem or distributed I/O. The board also has a 10 Mbps Ethernet controller with a 10BaseT front-panel connector.

The CPCI-CPU-1 has a Plug and Play BIOS and is fully compatible with all popular PC-based operating systems such as Windows NT, 95, and MS-DOS, OS/2 and UNIX, as well as many other real-time operating systems including VxWorks, QNX, pSOS and OS-9. The CPU operates from 0°C to 55°C and is supported with an on-board, low-power fan unit and temperature sensor. There's also a real-time clock, watchdog timer and adjustments for internal and external bus and clock speeds.

Analogic's TAP-800 family of Telephony Application Processor cards began to move to the cPCI world with the TAP-810, a cPCI-based DSP resource card designed for PSTN-to-VoIP connectivity applications. Capable of handling 120 channels of G.723.1 and G.729a as well as other standard and proprietary algorithms, the board features quad T-1 / E-1 line interfaces and a 100 BaseT controller.

Integrating all these resources onto a single cPCI board allows voice and data entering the T-1 / E-1 interface to be compressed by the DSP resource and exit through the Ethernet interface without unnecessary processing by the host or bandwidth transfers over the bus. The host is free to manage call setup and teardown and to calculate and support custom calling features for packet-switched calls.

The TAP-810 is fully Hot-Swap compliant. All connections to the card are made through a rear-panel I/O transition module, simplifying cabling and maintenance. This high-density combination of inputs, outputs and DSP power allows you to build some pretty large systems – with 120 channels per board gateways, a standard 19" chassis can house up to 672 ports.

Analogic's new Network Access Card (NAC) family for CompactPCI includes DS-3 and quad and octal T-1 / E-1 cards. All three cards are 6U high, hot-swap compliant and include the H.110 protocol telephony bus.

The TAP-810 has a per-port price of $125 in OEM quantities. By the time you read this, the optional G.723 codec and H.323 stack should also be available.

The quad NAC-120 and the octal NAC-240 cards have a software-selectable T-1 / E-1 interface, Primary Rate ISDN and a passive back card. For data networking applications, they also have a 10 Mbps Ethernet interface, a Motorola MPC860 processor running VxWorks, and either 128 or 256 channels of HDLC. These boards are designed for Voice over IP trunking, Voice over IP Gateway, Remote Access Server (RAS), call center, and ISDN router applications. The NAC-120 and the NAC-240 cards are priced at $4,495 and $6,995 respectively. OEM quantity discounts are available.

The NAC-600 DS-3 card features 672 channels, all of which can be switched to any of the 4096 time slots on the H.110 interface. The prod-

uct is designed for large-scale call center, CT, and VoIP applications. The NAC-672 card with DS-3 interface is priced at $17,995; cost savings can be realized in only months, since the break-even point for switching to DS-3 is between seven and twelve T-1 lines.

The NAC-600 is a DS-3 interface controller card in the cPCI form factor. It's designed for high-density voice, data, and mixed voice and data applications. When used with Analogic's TAP-810 CompactPCI telepho-

Analogic's hot-swappable TAP-810 CompactPCI-based DSP resource card is designed for PSTN-to-VoIP connectivity applications and can handle 120 channels of G.723.1 and G.729a.

ny application processor, high-availability DS-3 scale Voice / Fax over IP Gateways, RASs, and Universal Messaging Servers (UMSs) can be built.

The NAC-600 is designed for in-band and out-of-band signaling applications, with on-board support for ISDN and robbed bit signaling. In large PBX, VoIP, and RAS applications, multiple T-1 lines are the dominant interface. The DS-3 interface can lower equipment costs and reduce monthly tariffs by integrating up to 28 T-1s (672 channels) into a single card and single service provider connection. In the U.S., the breakeven point for switching to DS-3 service is typically between 6 and 12 lines.

The NAC-600 also works in out-of-band signaling environments where SS7 and GR-303 are found. In high-density CO, ISP or Intelligent Signaling Peripherals (ITSP) installations, SS7 signaling is commonly used for call

Analogic's hot-swappable, 6U high NAC-600 DS-3 card handles 672 channels, all of which can be switchd to any of the 4,096 time slots on the cPCI H.110 bus.

control. Since SS7 is carried out of band on separate and redundant hardware, the NAC-600 provides a simple API to allow external control and switching of each channel of the DS-3.

Similarly, in internal CO applications a GR-303 signaling controller can easily manipulate the DS-3's channels. For voice and data applications, the board can switch all 672 of its DS-0 channels to any of the 4,096 H.110 computer telephony bus channels. Channels can be routed over the CT bus to other trunk interfaces, to digital signal processing (DSP) resources for voice or modem processing, and to other resources co-located in the CompactPCI chassis. An NAC-672 card with DS-3 interface costs $17,995

Ariel Corporation (Cranbury, NJ – 609-860-2900, www.ariel.com) has announced a whole family of industrial strength remote access products in the CompactPCI form factor. The first member of the family, the

RS2000C gives Windows NT, UNIX and Linux servers physical connections for remote dial-in and LAN dial-out. It combines 30 Lucent V.90 56 Kbps modems with a dual T-1 / E-1 / PRI ISDN interface to support up to 60 simultaneous remote access sessions originating from digital 64Kbps basic rate ISDN and analog 56 Kbps and 33.6 Kbps customer premises equipment on a single rack-

---

Ariel's RS2000C can support up to 60 simultaneous remote access sessions originating from digital 64 Kbps basic rate ISDN and analog 56 Kbps and 33.6 Kbps customer premises equipment on a single rackmountable cPCI card.

---

mountable cPCI card.

The RS2000C provides a single remote access call-in number that simplifies remote access installation and administration. The RS2000C automatically detects the type of customer premises equipment initiating the call (digital Basic Rate ISDN or analog modem) and routes it to the appropriate modem or HDLC controller. Support for autorate fallbacks enables the RS2000C to operate at the highest speed supported by the subscriber modem.

A standard cPCI chassis fully populated with 10 RS2000Cs can accommodate up to 20 T-1 / E-1 / PRI lines, enough to support 600 simultaneous remote access sessions. Similarly, a standard six-foot rack containing eight cPCI chassis each fully-populated with 10 RS2000Cs can handle 160 T-1 / E-1 / PRI lines, enough to support 4,800 remote access sessions.

The RS2000C provides Multi-Protocol Routing (MPR), so the board can support Windows, OS/2, Mac, and Unix remote users. Multi-link PPP (MLPPP) allows RS2000C ports to be aggregated for higher dial-in or dial-out bandwidth, and Point-to-Point Tunneling Protocol (PPTP) support, which allows the RS2000C to function as a Virtual Private Networking (VPN) device.

Ariel's cPCI products will include line cards, high-density modem pools, and integrated solutions that combine the line card and modem pool functions.

The BajaSpan CompactPCI board from Artesyn Communication Products (Madison, WI – 608-831-5500, www.artesyn.com) handles eight T-1 / E-1 spans in CT systems, wireless systems, and signaling or H.323 gateways. The board is powered by two or four 66 MHz Motorola MPC860MH / SAR PowerQUICC processors, which handle data channeled through the telecom lnks. It supports the H.110 telephony bus (delivered by a Lucent T810X chip), is hot swappable, and has a PMC expansion bus site.

Interestingly, any of the PowerQUICCs' serial ports can be routed to or from any of the H.110 or T-1 / E-1 time slots, and vice versa, so voice and data can be combined on the same T-1 or E-1 link, or 16 SS7 links can be routed on the same span.

The BajaSpan holds 16 MB of SDRAM, 8 MB Flash memory and 2 KB of EEPROM for each PowerQUICC processor. It has four serial ports and a rear-panel transition module for routing I/O. The BajaSpan costs $4,032 in lots of 1,000.

Artesyn followed the BajaSpan with the BajaSpanL ("BajaSpan Lite") a cost-reduced version of its popular Octal-T-1 / E-1 CompactPCI board. The BajaSpanL is designed for what Artesyn calls "emerging tele-

datacom" (convergence) applications such as signaling gateways, H.323 gateways, computer telephony and wireless.

Interfaces for the BajaSpanL include eight full T-1 / E-1 spans and an H.110 CTbus interface. At the heart of BajaSpanL is a time slot interchanger that allows any of the timeslots on any of the T-1 / E-1 time slots to be routed to/from any of the H.110 4096 timeslots, making the board ideal for switching or routing both voice and data within the system. Also, any timeslots from either the T-1 / E-1 spans or H.110 can be routed to / from the on-board 860 PowerQUICC processor, making BajaSpanL a nice fit for protocol processing functions.

BajaSpanL also has a 32-bit PMC expansion interface that makes it easy for designers to add custom and off-the-shelf third-party PMC cards. Support for hot swapping in standard cPCI systems and in customized fault resilient systems is also provided.

A BajaSpanL transition module provides access to the eight T-1 or E-1 spans. Rear access to the I/O via the transition module ensures a low meantime-to-repair and a convenient means of keeping the front of the system free of cables.

To reduce time to market by up to two man-years, Artesyn will offer its Portable Protocol Engine (PPE) software for the BajaSpanL. PPE offers a choice of HDLC LAPD / LAPB and / or SS7 MTP1 and MTP2 protocol stacks. PPE relieves the burden of porting low-level protocols to the hardware, reducing customers' time to market and maintenance costs.

As an alternative to the PPE protocol software package, BajaSpanL features VxWorks real-time operating system support.

Blue Wave Systems (Carrollton, TX – 972-277-4600, www.bluews.com) a leading Digital Signal Processing (DSP) solutions supplier, has announced the CPCI/C6402, a CompactPCI based DSP resource board for telecom and communications processing applications.

Incorporating eight TMS320C6202 processors, each running at 250 MHz (25% faster and smaller than the preceding TMS320C6201) and capable of a peak performance of 2,000 MIPS (alongside two Motorola MPC860T real time control and data management processors) the CPCI/C6402 board offers in one slot more than double the processing power of its predecessor, the CPCI/C6400 (which used the C6201).

The architecture of the CPCI/C6402 also means that existing software written for the C6400 platforms will port over easily, so it's possible for you to begin your new system design now using Blue Wave's CPCI/C6400 quad board, or evaluate the C6202 using the CPCI/C6202-EVM and then migrate to the CPCI/C6402 octal board.

The architecture is also optimized for resilient multi-channel processing, such as modem pools and transcoder banks that can be used in packet voice applications and cellular radio base stations. The board, which should be available by around the time you read this, can also support the more recently announced (and even more powerful) TMS320C6203 DSP.

At the moment, the technology on the C6202-equipped board can handle five T-1 spans (120 channels) of V.90 modems in a single slot. The upcoming C6203-equipped board combined with the proper software will support eight T-1 spans (192 channels) of V.90 modems in a single slot

The CPCI/C6402 board is also a core component of Blue Wave's ComStruct line of building blocks for communications processing. Other components of this ComStruct architecture include: DSP and I/O hardware, FACT (Framework Architecture for Communication Technologies)

system management software for the DSP elements of telco systems, and application specific DSP algorithms.

Prices for the CPCI/C6402 start at $8,500 in quantity.

Brooktrout Technology's (Needham, MA – 781-449-4100, www.brooktrout.com) new Netaccess CompactPCI WAN access products offer a high density T-1 / E-1 / ISDN network connectivity platform with up to eight software-selectable T-1 / E-1 ports, an H.110 CTbus, and Ethernet in a single cPCI slot, and thousands of simultaneous data or voice sessions per system.

Developers of cPCI network equipment can take advantage of hot swap and fault resilient capabilities, and support for many OSes including Solaris, QNX, LynxOS, UnixWare and Windows NT. With an intelligent on-board software stack and an HDLC packet engine, the Netaccess architecture can support such applications including internetworking equipment, wireless infrastructure, voice messaging, voice- and video-conferencing, and next-generation carrier equipment such as voice over DSL or voice over IP. The software and APIs support traffic speeds from subrates up to superchannels and data protocols including raw HDLC, LAP-B, LAP-D, LAP-F, X.25, Q.921/Q.931 (ISDN), Q.922/Q.933, Frame Relay and UDP/IP.

OEM customers achieve rapid time-to-market with worldwide homologation and switch support including Lucent 5ESS, Nortel DMS-100 and DMS-250, Euro ISDN, Ericsson, and Siemens EWSD. Partnerships with cPCI CPU/chassis manufacturers such as Motorola Computer Group and Force Computers ensures performance and time-to-market at the system level.

The DSP Mezzanine Card Option (priced from $3,000 to $3,300) for Brooktrout's CompactPCI WAN Access products includes a developer's toolkit, allowing OEM Developers to write or purchase DSP code based on the TI C54X family. Using the latest DSP Mezzanine card and Brooktrout Technology's Instant ISDN software stack, programmers can quickly integrate and run third party DSP code.

DSP Mezzanine card can support protocols like R2/MFC and DTMF to help generate and detect calls. This provides a flexible amount of DSP resources to fit each application's needs. Also, having the versatility to offer two, four or eight DSPs on the mezzanine allows a customer to configure the most cost effective solution. Developers migrating from legacy platforms such as ISA, PCI and VME to cPCI can take advantage of their high-density WAN Access cards.

By the time you read this, Brooktrout will be delivering cPCI versions of its TR1000 voice and fax messaging platform and its TR2001 IP Telephony platform, which will enable OEM product developers and service providers to build IP Telephony and messaging applications on the fault resilient cPCI platform.

Bustronic (Fremont, CA – 510-490-7388, www.bustronic.com) unveiled a CompactPCI backplane product line compatible with the kind of ATX power connectors found on all standard PC power supplies. ATX power connectors allow for an easy plug-in from the backplane to the chassis during assembly and eliminate the need to build a wire harness by using the standard PC power supply cable.

Besides the ATX connector, there are user power input options and a utility connector for power monitoring.

The latest Bustronic cPCI backplanes

come in eight, six and four slot configurations for 6U high format versions. They're fully compliant with PICMG 2.0 and 2.1 standards and take advantage of the company's computer simulated, eight layer high performance stripline design. The distributed high and low frequency decoupling capacitors and power dispersion characteristics ensure virtually zero crosstalk on Bustronic's backplanes.

Price of an eight slot cPCI backplane with ATX power connectors start at around $360 in volume.

BVM Limited (products available from Bill West Inc., Monroe, CT – 203-261-6027) manufactures a range of CompactPCI and VMEbus products based on the 680x0, PowerPC and Pentium CPUs. They also support IndustryPack and PMC mezzanine modules and carriers. To assist OEMs in developing applications, BVM's hardware is supported by lots of software ranging from full board support packs and I/O drivers to libraries and development tools.

BVM's new cPC200 is their latest 6U high SBC in CompactPCI format, available in a number of different configurations to suit particular applications.

This SBC has several things going for it: It only occupies a single slot width, allowing a left-hand system slot backplane to be used while leaving the maximum possible space for other cards in the system. The on-board PanelLink connector will drive an LCD screen up to 10 meters distant, and when used with a USB keyboard and mouse this provides a convenient user interface if you're moving around a particular area all the time. There's an S3 on-board graphics adapter, the AGP-2 based Savage 4 with 8 or 32MB of video memory.

Conforming to the Rev 2.1 Hot Swap specification, the cPC200 uses a Super Socket 7 enabling it to use variouis processors such as Intel Pentium, AMD K6-III, Cyrix and Rise, and the board can handle processor clock speeds up to 500MHz and bus speeds of 66, 75, 83 or 100MHz. The on-board power supply provides all required processor voltages. Up to 256MB of DRAM is available on-board, expandable to 512MN via a standard 144 pin DIMM socket.

10baseT and 100baseTX networking is also onboard using the Intel 82559 single chip controller, with the RJ-45 outlet on the front panel. Also on the front panel are standard PS-2 mouse and keyboard ports, and two USB ports, both supporting 12 and 1.5Mbps.

Two IDE on-board disk channels allow up to four full Ultra DMA drives, to be attached. The floppy disk controller supports all formats up to 2.88MB capacity. Two serial ports using 16550 UARTS are available via the rear I/O module.

An optional rear I/O module enables eased access to many on-board ports with the option of front or rear I/O connections. A 32 bit 33 MHz bus master PMC site allows a range of additional I/O functionality to be added still within the single slot. Extensive software support is available including Windows NT.

The 599 Series Development System from Carlo Gavazzi (Buffalo Grove, IL – 847-465-6100, www.carlogavazzi.com) is a compact 9U high, 7" wide and 14" deep. If you want to convert the system into a rackmount 19" mounting ears are available, or

Bustronic has a new line of CompactPCI backplanes in four, six and eight slot versions that use conventional ATX power connectors.

you can just sit the system on its four dainty rubber feet and use the carrying handle when you need to move it. The 599's 5U high backplane fits easily in its five slot 6U x 160mm. standard cPCI card cage with a removable 3U divider section for slots 3, 4 and 5. There is rear mounting for 80mm. transition boards for slots 1 and 2.

The unit has mounting provisions for a 3.5" floppy drive, a 3.5" card drive and a 5.25" CD ROM drive on left side of card cage. Boards, drives, and filler panels are purchased separately.

A 250 watt power supply provides +5v@25A, +3.3V@20A, -5V@0.5A, +12V@10A, -12V@1A, and +5VVSB@0.3A. A removable load resistor installed on spare drive connector regulates the power supply with light load. Cooling is provided by a 90 CFM fan.

The 599 Series Development System starts at $1,075.

Carlo Gavazzi has also just introduced a new line of cPCI Portable Systems. Designated the 585 Series and available in either five or eight slot configurations, the units are modular and allow users to easily configure them with several different backplanes (5 or 8 slot right justified CPU with H.110 support), cPCI single board computers, power supplies, and various other peripherals (hard drives, CD drives, floppy drives, CPCI peripheral cards, rear I/O cards, and mezzanine cards).

The 585 Series lightweight Portable Systems allow up to seven additional 6U high cPCI cards to be integrated into a chassis that's been environmentally toughened to surpass the normal operating specs of a standard portable CPCI System.

The 585 Series Chassis can hold 5.25" half height drive and one 3.5" half height drive (front accessible) as well as one 3.5" full height drive internally.

Continuous Computing Corporation (San Diego, CA – 619-547-8804, www.continuouscomputing.com), known as CCPU, designs and makes high availability network and application-ready computer systems for datacom and telecom equipment manufacturers. Their products are primarily SPARC-based and NEBS compliant, meeting the most stringent operational requirements of the Central Office. The company also offers OEM manufacturing services, including custom hardware and software development and product life cycle support.

CCPU recently launched the world's first Telco Starter Kit for Sun Microsystems' new UltraSPARC processor-based 19" rackmount CP 1500 compactPCI platform.

The CCPU Telco Starter Kit is an UltraSPARC-based compactPCI kit specifically for the CO, having CO-required features such as redundant dual feed power supplies and remote power cycling. The included SPARCengine 270MHz CP 1500 processor and board supports Solaris 2.X, the most popular operating system for telco apps.

This system includes all of the building blocks that will help you quickly deliver high-availability telco applications running on a NEBS compliant UltraSPARC platform.

The CCPU Telco Starter Kit for UltraSPARC compactPCI is available in simplex and duplex high availability configurations. The Kit also includes dual feed -48VDC hot-swappable power supplies, seven available cPCI slots for 6U high I/O boards providing for such functions as serial I/O, Ethernet, ATM and SS7. The system also has a 32-bit SCSI dual disk drive.

The system can take advantage of CCPU's proprietary hardware and software network, monitoring and control technology for sophisticated system maintenance.

For example, the Continuous Control Node (CCN) provides remote and local power cycling, remote and local console access, alarming, and a simplified craft interface. Via an API, these modules can be combined

with application level software to provide failover and over high-availability solutions.

A fully configured Telco Starter Kit lists at $25,000.

The Dialogic Division of Intel (Parsippany, NJ – 973-993-3000, www.dialogic.com) is enlarging its cPCI family of boards to include a full range of voice, speech, ATM, network interface, and voice/ fax over IP telephony technologies for developers.

For example, our magazine was the first to put through its paces the hot swappable cPCI version of the DM3-based QuadSpan Series. The QuadSpan platform provides up to 120 ports of both voice processing and network interface per card.

Now, Dialogic's DM3 IPLink family of IP telephony platforms is becoming available in CompactPCI. The DM3 IPLink platform allows data and telephony network interfaces and DSP resources all to operate on a single board to enable voice and real-time fax over IP.

You can plug up to 14 DM3 IPLink cPCI cards in a single chassis, allowing you to build a gateway that supports up to 840 channels in a rackmount cPCI chassis that can be accessed from the either the front or rear.

The DSP-powered DM3 IPLink uses such management tools to monitor your system as SNMP and Dialogic's BoardWatch, a runtime administration software tool that provides remote configuration, inventory, and fault management of Dialogic CT devices.

The board is H.323 compliant and complies with International Multimedia Teleconferencing Consortium (IMTC) / VoIP enhancements and refinements to H.323. The cPCI baseboard supports the H.110 CT bus spec and the board's API software is compatible with S.100 Enterprise Computer Telephony Forum (ECTF) call processing standards.

The DM3 IPLink is also available with a custom call control option that lets you run the call control on the host server with a single IP address to interface with call control agents (such as gatekeepers). This option, plus a separate porting kit (another DM3 IPLink software and documentation package) makes it possible to replace or enhance the H.323 call control module or VSR resources. The RTP/RTCP remains on the board, ensuring maximum IP packet efficiency and voice quality.

The board will support many optimized, low-bandwidth vocoder algorithms for transmitting audio over the IP network, including G.711, G.723.1, G.729a, and ETSI and MS-GSM. It also includes T.38

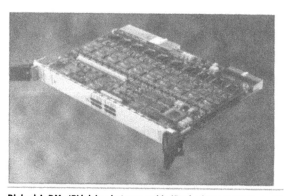

Dialogic's DM3 IPLink is a hot swappable IP telephony cPCI board with onboard DSPs, onboard Ethernet, and support for up to two T-1 or E-1 interfaces per board, making for a single-slot IP telephony gateway solution.

realtime fax over IP. Dialogic's own Voice Activity Detection (VAD) function yields maximum bandwidth utilization, and there's also embedded high-performance echo cancellation and out-of-band DTMF processing optimized for IP networks. DM3 IPLink's jitter buffers can also be fine-tuned for optimal performance.

The DM3 IPLink cPCI board can be used to quickly build a complete gateway bridging the PSTN and the IP network, since Dialogic's product conveniently provides the digital phone network interface, an Ethernet connection, and mediastream processing resources on a single board. DM3 IPLink cPCI platforms include the DM/IP241-T1c (single T-1 span in a single slot), the DM/IP301-E1c-75 or 120 (a single E-1 span in a single slot), the DM/IP481-2T1c (dual T-1s in a single slot), and the DM/IP601-2E1c-75 or 120 (dual E-1s in a single slot).

The DM3 IPLink runs under Windows NT 4 and comes with a software package containing the Dialogic NT Native Architecture, DM3 Core Software for Windows NT, Dialogic IPT Software, complete documentation, and gateway demonstration code. Standard Dialogic DM3 Direct Interface APIs are included, which gives developers flexibility for customizing and controlling the low-level application details.

The board also runs under SPARC Solaris 2.6.

Digi International (Minnetonka, MN – 612-912-3444, www.digi.com) has just introduced the compactPCI-based DataFire SYNC/2000 cPCI synchronous WAN adapter and DataFire DSP 24/48 and 30/60 access concentrators. Based on Digi's Houston Technology, they enable PC servers to perform such telecom apps as WAN routing, virtual private networking, remote access and faxing – all without relying on proprietary, "black-box" servers, or an assortment of single-purpose adapter cards.

The DataFire SYNC/2000 cPCI adapter will ship first. It supports both Windows NT and UnixWare and enables SNA connectivity when combined with the Microsoft SNA Server or IBM eNetwork Communications Server. It also includes both two- and four-port configurations enabling Internet, Frame Relay and X.25 connectivity. The adapter includes FRF.9 Frame Relay compression to boost throughput and features a main Motorola MPC860 PowerQUICC 40MHz processor. A high-performance adapter with a secondary Motorola EC603e 100MHz processor also is available.

As for Digi's dialup technology, the DataFire DSP 24/48 supports up to 48 dial-up ISDN or 56Kbps (V.90) modem connections per card via one or two T-1 or 24B+D ISDN PRI lines. The DataFire DSP 30/60 provides up to 60 ISDN channels or 56Kbps modems on one or two 30B+D ISDN PRI lines. The two products are designed for remote access, faxing and dial-out/dial-in Internet access applications, and will support all major operating systems.

Also, Digi's Driver Development Kit gives developers the ability to create drivers for their specific environment. The list price ranges from $1,195 to $2,195.

Digi has announced a multi-million dollar, multi-year agreement to embed its new cPCI-based DataFire adapters into telecom carrier switches developed and manufactured by Brazil-based Tropico Systems and Telecommunications SA. Tropico provides a third of the digital switches installed in Brazil, which together connect over five million phone lines.

The DataFire adapters will sit in a module built into the Tropico-RA central office switch, where they control all incoming Internet-bound calls to a high-speed network interconnecting Brazil's ISPs. Digi's products will be incorporated into upgrades of 400 existing switches now in use and also will be built into new switches.

Diversified Technology (Ridgeland, MS – 601-856-4121, www.dtims.com) has a huge new facility and has retooled to handle the increasing CompactPCI market. And DTI has indeed got a slew of exciting new cPCI products:

DTI's new CPC8617 is a 6U high single board computer that supports up to a 600 MHz Pentium III processor and holds up to 1GB of memory in two DIMMs (100 MHz SDRAM). The board can drive up to seven cPCI slots. Integrated on the CPC8617 is a Phoenix BIOS, PCI video with 2MB of memory, PCI UltraWide SCSI, Dual 100Mbps Ethernet ports, a solid state hard drive or flash disk, two USB ports and DTI's System Monitor. An IDE hard drive port allows the CPC8617 to support up to two IDE drives. Other expected features of the CPC8617 include floppy, parallel, keyboard, mouse and serial ports.

DTI's FTC620 is a cPCI chassis designed for mounting in standard 19" equipment racks. The system is 24.5 inches or 14U in height (add 3.5" or 2U for optional air plenums) and 14.5" in depth.

The FTC620's interesting power subsystem can support various configurations, with inputs of 110 to 220 VAC or -48 VDC. The power sub-

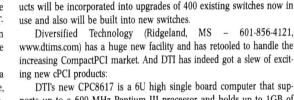

Diversified's CPC8617 6U high CompactPCI single board computer that supports up to a 600 MHz Pentium III processor and holds up to 1GB of memory in two DIMMs.

system is made up of multiple (from 3 to 7) 150 Watt modules that can be used to provide 300 to 900 Watts of power with N+1 redundancy. A failing module can be replaced without shutting down the system or generating a power glitch.

The FTC620's cooling subsystem has fans that will move 420 CFM of air. The fans are mounted in a hot swappable tray that can be pulled directly out of the chassis' front. The rotational speeds of all the fans are monitored for failure.

The FTC620 also supports many drive configurations. All drives are mounted in carriers which plug into a mid-panel mounted in the chassis. This allows for all types of drives to be serviced without having to open the system to access cables. The standard configuration has one slot which supports one 1.44MB floppy drive and one IDE CD-ROM in a single carrier. Two more slots support either two carriers with half-height IDE or SCSI devices or three carriers with Single Connect Attach (SCA) SCSI devices.

The FTC620 from Diversified Technology is a spacious14U high CompactPCI chassis.

Divesified's CPC8640 cPCI single board computer based that can support up to a 500 MHz Celeron processor and 512 MB of memory.

Four of the power supply module slots can also support additional SCA SCSI devices for a total of seven SCA devices, so you must consider the trade off between total power output and total drive support when configuring the FTC620.

The FTC620's CompactPCI card cage is slightly over twenty slots wide. The standard backplane provided with the FTC620 has slots for a processor module, two PCI-to-PCI bridge modules, and 16 cPCI expansion cards. The processor slot is two inches wide to allow for advanced CPUs with proper heat sinks without interfering with any of the expansion slots, which are all 6U high.

The backplane has cPCI Computer Telephony Specification (and therefore ECTF H.110) support at all 16 expansion slots. P3 and P5 connectors at each slot include long tails and type AB rear shrouds. Inputs are also provided for telecom power supply bus connections. There's also support of Basic and Full levels of the CompactPCI Hot Swap Specification. Systems can be upgraded in the future for support of High Availability through simply changing out one of the PCI-to-PCI bridge cards.

DTI also offers several 6U cPCI processor cards for use in the FTC620. The processor cards typically include interfaces such as 10/100 Ethernet, UltraWide SCSI, video, EIDE, floppy, parallel, and serial ports. Provisions are made for cabling to the processor card through the rear of the FTC620 via a transition card.

DTI's new CPC8640 is a cPCI single board computer that can support up to a 500 MHz Celeron processor and 512 MB of memory in two DIMMs. It has PCI video with 2 MB of memory, PCI UltraWide SCSI, a 100Mbps Ethernet port, a solid state hard drive or flash disk, a USB port and DTI's System Monitor. An IDE hard drive port allows the CPC8640 to support up to two IDE drives. There's also the usual floppy, parallel, keyboard, mouse and serial ports.

Elma Electronic (Fremont, CA – 510-656-0606, www.elma.com) has announced a new kind of automatic locking handle with improved geometry that provides a secure and dependable means of inserting and extracting hot swap boards.

What's the big deal about handles, you ask? CompactPCI and VME backplanes come with high pin density connectors, which means there are high connecting forces involved with insertion / extraction – up to 100 lbs. of force for a 6U high plug-in unit, for example. To manage these high connecting forces, Elma has developed a new injector / ejector handle that conforms to the IEEE1101.10, cPCI hot swap and VME64x standards.

They have a clean ergonomic design and a user-friendly, push button latching and locking mechanism. If you want to remove a cPCI board, the handles can be unlocked just by pressing the little red button. The button immediately activates the switch (turning it to an open position), while the button itself remains depressed. The board can then

be removed by pushing the handles outwards. When reinserting the board, the red button clicks up automatically only when the board is fully seated, thus locking the handle and activating the switch (closed position). The click of the button gives a visual and audible confirmation that the board is fully seated.

This two-step extraction process used with Elma cPCI handles also prevents the untimely removal of the board and protects the hot-swap functionality. Elma hot-swap handles serve as a drop-in replacement for most popular front panels handles and the easy unlocking mechanism of Elma cPCI handles prevents inadvertent extraction of adjacent handles.

Force Computers (San Jose, CA – 408-369-6000, www.forcecomputers.com) is another leader in CompactPCI product development, continuing to add products to the market.

Force's new Centellis CO 88520 "Cluster-in-a-Box" represents the high-end of Force's cPCI-based server platforms for communications applications. It's said to be the first NEBS-tested, SPARC / Solaris CompactPCI high availability platform. Unlike other high availability solutions targeted for use in Central Office applications, the 15U high Centellis CO 88520 supports off-the-shelf cluster software and has a fault resilient chassis tested to meet the strict disaster-resistance requirements of the Telcordia NEBS Level 3 criteria.

The CO 88520's architecture provides 2N redundant CPU and I/O cluster nodes. The ruggedized, all steel chassis construction provides extra strength to withstand Zone 4 (Richter magnitude 7.0 or higher) earthquakes.

Front accessible system components make for easy maintenance in CO frames, and the unit's front-to-back, bottom-to-top internal cooling eliminates the need for external baffles and conforms to NEBS recommended air flow.

The unit's 333 MHz UltraSPARC-IIi single board computer (with a 1 MB Level 2 cache) works in Intelligent Network applications, and drives the dual 8-slot 6U high PICMG 2.0 R2.1 cPCI passive backplane. The factory-installed Sun SPARC/Solaris 7 OS is a field-proven, off-the-shelf operating system for leveraging internal Unix software development. Solaris 7 supports an "open cluster software" topology for Sun Cluster 2.2, Legato FullTime Cluster or Veritas Cluster Server.

The system has a telecom ringing voltage bus with external input for Central Office applications that require direct connection to the Local Loop

The system has hot swap cPCI support, hot swap N+1 redundant 500 Watt power supplies, hot swap disk drives and hot swap fan trays. The independent front-to-back power supply cooling prevents power supply heat from adding to the thermal load of the card cage. The upper and lower cooling fan trays provide maximum chassis cooling for high-wattage digital signal processing applications.

The system's ingenious front-to-back cable routing provides an alternative to

Force Computers' Centellis CO 88520 "Cluster-in-a-Box" is a NEBS tested SPARC / Solaris CompactPCI fault resilient platform.

C

rear transition I/O, and the consolidated power bus bar replaces individual power connections to the backplane, increasing airflow and lowering Mean Time to Repair (MTTR).

The optional Telecom Alarm Module and alarm panel is an intelligent system monitoring device with RS-232 serial interfaces, a structured command interface language and a remote control/display panel. It has eight temperature inputs (for LM75 devices on I2C bus), seven inputs for supply voltages (including +5, +3.3, +12, -12, -48 VDC), 12 for fans (logic level or tachometer, auto sensing), four for PSU present and failure, nine watchdog timers, nine inputs for boards present, eight user defined, and Alarm Cut Off and reset inputs.

Outputs include those for telecom visual alarm relays (critical, major, minor), telecom audible alarm relays (critical, major, minor), alarm cut off relay, four telecom LEDs (critical, major, minor, ACO), PSU shut down and eight for board resets.

Each SBC can hold up to 1 GB ECC Protected EDO RAM and there's support for Fast Ethernet, WideUltra SCSI and Ultra2 (LVD) SCSI channels. The unit's media bay has space for up to six hot swap 3.5-inch SCA Ultra2 SCSI drive modules (3U high) and up to two 5.25-inch Ultra2 SCSI devices.

Force's new PowerCore CPCI-3750 is a 64-bit PowerPC 750 micro-

Gateworks' GW2600 CompactPCI SBC supports AMD K6-III processors running at speeds up to 400 MHz

processor on a 3U high cPCI single board computer. This is perhaps the industry's smallest PowerPC 750 Single Board CompactPCI Computer.

The PowerCore CPCI-3750 should find a home in such performance-driven and size-sensitive embedded applications as base station controllers, ATM and IP Layer 3 switches, and demanding command and control applications.

All PowerCore computers are designed to run a choice of embedded real-time operating systems including VxWorks/Tornado and ISI pSOS operating systems, both supported by Force. Already used by a major telecom OEM, the CPCI-3750 is based on a low power consumption 400MHz PowerPC 750 processor. The SBC includes 128 MB of high-speed synchronous DRAM, which can be expanded to 256 MB through user upgradeable memory modules and a Fast Ethernet 10/100Base TX interface. The board also contains 4 MB of programmable Flash memory to support ROM intensive applications.

The CPCI-3750 lists at $2,995.

The 6U high GW2600 CompactPCI single board computer from Gateworks Corporation (Atascadero, CA – 805-461-4000, www.gateworks.com) supports AMD K6-III processors running at speeds up to 400 MHz. The AMD K6-III processor incorporates AMD's TriLevel Cache design allowing it to outperform Intel's Pentium III processors. The TriLevel Cache design includes a full-speed 64 KB Level 1 cache and an internal full-speed backside 256 KB

Level 2 cache. Besides the processor's internal cache, the GW2600 contains an additional 512 KB of Level 3 cache.

The GW2600 is designed for telephony and routing applications which always benefit greatly from a high performance system cache.

The board also has dual 10/100 Base-TX Ethernet channels, a dual Cardbus / PCMCIA interface, up to a 32 MB SanDisk IDE Flash Drive, up to 128 MB DRAM, IDE and floppy interfaces, USB port, two serial ports, a parallel port, AT keyboard controller and PS/2 mouse port. The board supports rear-panel I/O compatible with a new line of backplanes which integrate standard PC peripheral connectors, eliminating the cost of rear I/O transition boards. The board has features for fault resilient applications such as serial port console redirection, a battery backed real-time clock and a watchdog timer. The GW2600 has been designed to run off a single 5V supply and can be used as a stand-alone SBC.

The GW2600 with a 400 MHz AMD K6-III processor, 32 MB DRAM, and 8 MB SanDisk IDE Flash disk is priced at $1,695, in OEM quantities. A low cost version with an AMD K6-2 366 MHz processor is priced at $1,425, in OEM quantities.

General Micro (Rancho Cucamonga, CA – 909-980-4863, www.gms4vme.com) has announced what is at the moment the industry's fastest CompactPCI single board computer. The C2P3 is a 550 MHz (600 or 650 MHz in the future), dual-Pentium III board that targets compute-intensive applications such as telecom, aerospace, and imaging.

For high performance at low cost, the C2P3 can accommodate two Celeron PPG370 processors currently at 466 MHz with a future expected clock speed of 666 MHz. The C2P3 Celeron version also includes 256KB of on-die cache with a one-to-one clocking which improves cache performance, and provides Pentium III performance at a fraction of the cost.

In the Desktop Slot 1 version, in order to maximize memory access performance, each of the processors is equipped with up to 512 KB Level 2 cache. The two processors also share up to 1 GB of 100 MHz SDRAM main memory. The processors, cache and memory are linked via a 100 MHz Front Side Bus (FSB) local bus, capable of supporting not only current 500 MHz processors, but also next-generation processors. A DEC 21554 Draw Bridge chip enables multiprocessing.

Aside from its raw processing performance, the C2P3 offers an array of I/O and networking options, such as dual 10/100 Mbps Ethernet interfaces, a 40 MBps UltraWide SCSI interface, and a 64-bit AGP graphics engine with four MB of video RAM optimized for 3D rendering. Also available are two UltraDMA 33 IDE interfaces, a pair of USB ports, dual Serial I/O with optional RS422 drivers and a parallel port.

The board can support hot swap I/O modules on the cPCI bus. All the I/O on the C2P3 is available via a 80 mm rear panel I/O which can also accommodate a 2.5 " 9GB IDE drive. For applications that must be

deployed without a rotating hard disk, the board also provides up to a 340 MB SanDisk 1.5" Flash IDE on the rear panel. The C2P3 runs many popular desktop and real-time operating systems, including Windows NT, Solaris x86, QNX and VxWorks. The board comes with AMI's BIOS,

General Micro's C2P3 single board computer is a screamer, capable of running two 550 MHz Pentium III processors.

onboard diagnostics software and status LEDs.

Pricing for the C2P3 starts at $1,995, less processor and memory.

GoAhead Software (Bellevue, WA – 425-453-1900, www.goahead.com) is a developer of management software for Internet appliances and devices. GoAhead has just announced a cross-platform high availability solution for the telecom and datacom industries, called GoAhead HighAvailability, that enables telcos to offer their customers "five nines" (99.999%) system uptime. GoAhead HighAvailibility greatly increases system fault detection, management and reliability, resulting in fewer dropped calls and data transmissions.

GoAhead HighAvailability is initially targeted at the CompactPCI market, allowing manufacturers of these systems to quickly incorporate a high availability solution into their products.

GoAhead HighAvailability provides dynamic configuration management, heartbeating and checkpointing for active elements within a system, including hardware, applications and operating system components. The software supports a flexible range of system configurations, including active/standby CPUs and N+1 and 2N I/O cards. Fault management capabilities provide an extensible solution for automatic detection, isolation, diagnosis and policy-based recovery for system faults. GoAhead HighAvailability provides remote, Web-based access and integrates with SNMP enterprise and network management environments.

Hybricon Corporation (Ayer, MA – 978-772-5422, www.hybricon.com) has released a new family of "CoolSlot" nine-slot portable towers that are available with Hybricon's cPCI, VME64x, or VME monolithic backplanes.

The CoolSlot towers solve overheating problems with Hybricon's patented air deflecting CoolSlot card guides which eliminate cooling "dead spots" on boards and provide more uniform airflow throughout the system. Hybricon's extensive testing (which includes thermal modeling and simulation) indicates an improvement of 50% in board areas near the front panel and backplane that are blocked by the extrusions and

Hybricon's CoolSlot is a family of nine-slot towers that can house a CompactPCI, VME64x or VME monolithic backplane.

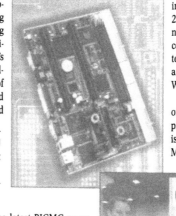

The incredibly powerful Polaris CompactPCI CPU board from I-Bus can hold dual 450 MHz Pentium II processors with MMX.

thus normally deprived of airflow.

The cPCI backplanes are compliant with the latest PICMG specs and are available with the system slot on either the left or the right side. You also have a choice of power supplies: 235 Watt embedded, 300 Watt embedded or a 350 Watt plug-in.

I-Bus (San Diego, CA – 619-974-8400, www.ibus.com), has been pushing the envelope of fault resilient computer technology for years, so it's no surprise that they would take on the CompactPCI market with gusto.

Their largest cPCI chassis is the G16, which as its name implies, is a 16-slot modular enclosure for the telecom and industrial markets. Of the 16 cPCI slots, 14 are 6U high expansion slots supporting the H.110 CT bus.. The system also supports up to three hot-swappable redundant power supplies and, with the configurable drive enclosure, up to eight drives in either a RAID or a Just a Bunch of Disks (JBOD) format.

The system has three sections: The cPCI card cage, the power supply subsystem, and an optional drive enclosure. Each of these areas has its own fan tray. Redundant fans are standard in the power supply subsystem and are an option in the drive enclosure and card cage.

The backplane is full hot-swap compatible and allows the redundant power supplies to plug directly into it, eliminating the voltage drop, noise and clutter of power cables.

I-Bus hasn't forgotten that you'll need a powerful single board computer for your G16 chassis. Their Polaris CompactPCI CPU board is based on the Intel 82443BX chipset and supports up to dual 450 MHz Pentium II processors with MMX running on an optimized 100 MHz system bus.

The board can be loaded with up to 512MB SDRAM using two DIMMS and has two RS232 serial ports, a bi-directional parallel port, a floppy disk and EIDE interface, and a real-time clock backed by replaceable battery. The front panel has ports for USB, Ethernet and video. There are also mini-DIN connectors for a PS/2 mouse, keyboard interface and a reset button. Polaris can be used alone or with an I/O companion board that has PMC interfaces and a PCI-to-PCI bridge, so you could expand your system (with a larger enclosure) by seven more backplane slots.

Inova Computers (Cotuit, MA – 508-428-1198, www.inova-computers.com) has a huge selection of CompactPCI products. For example, their ICP-MPK6 is a 3U cPCI board for multiprocessing. Powered by a Pentium-compatible AMD K6 or K6-2 running at up to 400 MHz, the board's Intel 21554 nontransparent PCI-to-PCI bridge chip enables coprocessor capabilities. Add a real-time operating system and up to seven of these coprocessors can run in a single system. The main system processor runs Windows NT.

The board holds 24MB of Flash RAM and 256KB or 512KB of Level 2 cache memory, specificially pipeline burst SRAM. 10/100Base-T or BaseTx Ethernet is included, along with a USB and IEEE 1394 port. Many other interfaces are optional, such as RS-232, CAN, DeviceNET, LONworks, Profibus, AS-i and Interbus-S.

The board supports real-time operating systems such as VenturCom's RTX extensions to Windows NT, Microware's OS-

I-Bus has created one of the larger cPCI chassis in the industry, the 16-slot G16 (formerly the Galaxy 16).

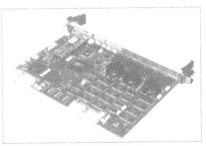

Lucent CompactPCI Speech Processing Board, a 6U high hot swappable 128 channel speech recognition board.

000 and QNX's QNX OS, and it supports interboard communications under Windows NT and Wind River Systems' VxWorks Tornado using TCP/IP protocols.

AbsolutBoot is also available, a software package that allows software booting over a network instead of a disk.

The ICP-MPK6 starts at $1,500. AbsolutBoot costs $70.

Lucent (Los Angeles, CA – 888-458-2368, www.lucent.com) now offers a CompactPCI board for speech processing, called the Lucent Compact PCI Speech Processing Board. Lucent's advanced speech applications allow you to speak naturally and in phrases, rather than merely respond to voice prompts.

Lucent's board has open, non-proprietary interfaces. It supports speech recognition, text-to-speech synthesis, and voice-coding, allowing multiple speech applications to be launched using the same board. Unlike some other systems on the market, Lucent's cPCI Speech Processing Board comes with the right to use its speech software without having to obtain individual licenses for each set of channels.

The 6U high cPCI board is hot swappable. The 32-bit bus handles a 133 MBps data bandwidth. Three RISC processors with 1 MB Level 2 caches are used (instead of DSPs) that sit on 64-bit memory busses. The board holds 192 MB of SDRAM and supports 128 simultaneous channels. You'll find available onboard 64 simultaneous channels of echo cancellation, 64 simultaneous channels of text-to-speech, 64 simultaneous channels compressed speech (16 bit linear LCCELP, ADPCM, mu-law, or a-law), 32 simultaneous channels of connected digit Automatic Speech Recognition (ASR), and 16 simultaneous channels when using a large vocabulary / continuous speech ASR.

The board can support simultaneous proportionate combinations of the above features. Operating systems you can run include UnixWare1 2.1, Solaris2 2.5, Windows 3 and NT 4.0

Lucent's Compact PCI Speech Processing Board complements Lucent's overall Speech Processing Solutions product line, which includes the Lucent Speech Server and a set of industry-standard PC cards.

MATRIX Corp. (Raleigh, NC – 919-231-8000, www.matrix.com) says that any of their VMEbus boards can be modified to have a CompactPCI interface. One example is the PENTX2, a single-slot 6U high VMEbus processor board based on the 266 MHz Pentium II Processor Mobile Module called the Embedded Module Connector 2 (EMC-2). You can load up to 128 MB SDRAM and 512 KB of Level 2 cache on the board.

Available in industrial, rugged, and conduction-cooled versions, the board supports UltraSCSI, two USB ports, 10BaseT / 100BaseTX Ethernet, one PCI Mezzanine card slot, and AGP graphics functions. ISA peripherals such as a keyboard, mouse, four serial ports, parallel port, IDE, and a floppy disk interface are supported. Also available is a Flash disk that appears to the system as an IDE disk drive.

On the conduction-cooled version, all the I/O signals pass through a 5-row DIN P2 connector and the P0 connector on the VME64 extensions backplane. OS support include Windows NT, VxWorks, and QNX.

MEN Micro (Carrollton, TX – 972-939-2675, www.menmicro.com)

offers the D2 compactPCI single board computer that holds either Slot 7 type processors or processors compliant with the Super 7 spec devised by AMD, Cyrix, IBM and IDT, so you can plug in processors such as an AMD K6 or a 266MHz Intel Tillamook . The board holds up to 128MB of DRAM and 4MB of battery-backed SRAM and 1 MB of Level 2 cache. Sporting a CompactFlash slot and 10/100MbpsEthernet interface, the board has three PC-MIP expansion-sites, along with dual EIDE; USB; ports for a keyboard, mouse and floppy-disk drive; two serial ports; a parallel port; and five PXI trigger lines for data acquisition. The board runs under Windows NT and VxWorks.

A D2 with a Tillamook processor, 32 MB of DRAM and 1MB of SRAM starts at $2,250.

The Motorola Computer Group (Tempe, AZ – 602-438-5800, www.mcg.mot.com) takes CompactPCI very seriously (they seem to announce a new product every week), and Motorola is one of the few companies building both fault resilient and fully fault tolerant "five nines" cPCI systems.

Motorola's new CPX1200 series offers central office standards compliance in a slim 3U high NEBS and ETSI compliant, single-shelf format that gives OEMs more flexibility in building telephony systems with distributed architectures, such as high-availability clustering. The high racking density possible with the CPX1200 series allows telecos to make better use of expensive floor space for both wireline and wireless services.

Besides being just 3U (5.25 inches) high and optimized for front- and rear-access equipment frames, the CPX1200 series horizontally mounts either five 6U cPCI cards (in the CPX1205 and CPX1205T), or (in the

Motorola's CPX 1200 series of compact 3U high CompactPCI systems boasts NEBS and ETSI compliance that should impress any central office.

CPX1204 and CPX1204T) four 6U cPCI cards and a flexible drive-bay module capable of supporting multiple peripherals, including floppy, hard disk, CD-ROM and tape back-up drives. Compared to many existing cPCI systems which provide eight to 16 vertically oriented 6U cards in 9U to 15U high packages, the CPX1200 series can support from 15 to 25 6U cards in the same cabinet space, nearly doubling the card capacity.

Diskless (flash-based) models are offered for real-time and network-boot applications.

The CPX1200 series also

Motorola's 16-slot CPX8216 CompactPCI chassis has been selected by Excel Switching Corporation as a component in its ONE Architecture Expandable Switching System (EXS).

Motorola now has real-time operating system support for its CPX8000 family of CompactPCI-based embedded computing platforms.

has hot-swappable redundant fan modules, a150 Watt power supply with a 300,000-hour Mean Time Between Failures (MTBF) rating, auto ranging AC and DC input versions useable in various international locations and in applications with poor national power regulations; it can also support hot swappable I/O boards per the cPCI Hot Swap Specification.

The system also supports detection and remote reporting of power, temperature and fan failure conditions.

The CPX1200 series is now available in both x86 and PowerPC CPU versions and is supported by a wide range of real-time operating systems. The list price begins at $4,265 for a network-bootable configuration.

Readers may recall that I awarded Motorola's CPX8000 family of CompactPCI chassis a 1998 Product of the Year Award. The CPX8000 was recognized for its "five-nines" (99.999%) uptime capability, which, when combined with the appropriate software, is equivalent to less than five minutes and fifteen seconds of downtime per year. This kind of reliability is essential for critical telecom applications.

As I said at the time, the telecom industry demands a system recovery time at the operating system and applications level of less than 30 seconds. The CPX8000 can do it. The elusive "five-nines" availability is possible because any active module – even system-slot CPU boards – can be swapped, fixed or upgraded while the system keeps humming.

Shortly after its debut, Motorola's CPX8216 CompactPCI chassis was selected by Excel Switching Corporation as a component in its ONE Architecture Expandable Switching System (EXS).

Also, Motorola has a deal with Sun Microsystems where Motorola will deliver IP-based CPX8216 network servers, base station controllers and base stations to create fault free wireless networks. The new architecture will allow wireless service providers and commercial developers to quickly deploy IP-based voice, data and video services. The architecture includes custom hardware from Motorola Network Solutions Sector (NSS), the PowerPC version of the CPX8216, and Sun's ChorusOS real-time and Solaris operating systems, high availability and IP services and a Java Dynamic Management Software Kit.

The latest news on the CPX8000 series is that it will be supporting real-time operating systems, which will allow telecom OEMs to standardize on a common system platform that can be used across a broad range of data and telephony applications in both carrier and enterprise markets.

MCG has licensed its high-availability technology to its real-time partners and is assisting with development and compatibility testing. The extensions to the partners' operating systems include hot swap services and management of system resources that are key to supporting a fault tolerant system with a high level of redundant capability and fault detection, correction and recovery. Partners now announcing CPX8000 support are as follows:

Enea OSE Systems AB (www.enea.com) which announced support for PowerPC configurations of the CPX8000, Lynx Real Time Systems (www.lynx.com) released LynxOS 3.1 that adds support for Pentium to the existing PowerPC configuration, Integrated Systems Inc.

(www.isi.com) released pSOSystem 3.0 for support of both Pentium and PowerPC configurations, QNX Software Systems Ltd. (www.qnx.com) announced QNX Neutrino 2.0 that supports both Pentium and PowerPC configurations; and Wind River Systems (www.wrs.com) released VxWorks version 5.4 and Tornado version 2.0 for support of both Pentium and PowerPC configurations.

Motorola also has a relationship with GoAhead Software (www.goahead.com) whereby GoAhead will provide comprehensive high-availability software (appropriately named HighAvailablity) for the CPX8000 that will help isolate a customer's application from either hardware or software faults. GoAhead HighAvailability provides dynamic configuration management, heartbeating and checkpointing for active elements within a system, including hardware, applications and operating system components.

Natural MicroSystems (Framingham, MA – 508-620-9300, www.nmss.com) is the only company other than Dialogic that makes a CompactPCI QuadSpan. If I recall correctly, NMS actually got its cPCI QuadSpan to market a hair before Dialogic did.

Computer Telephony magazine did the first public hot swap test of the Natural MicroSystems CompactPCI Alliance Generation QuadSpan in their East Coast Labs.

Natural MicroSystems' Alliance Generation 4000 Series (AG 4000) DSP-powered boards are available in both CompactPCI and PCI form factors. The cPCI versions of the AG 4000 boards can be configured as a single board or as a motherboard-daughterboard combination. The motherboard contains two or four digital T-1 or E-1 interfaces as well as 16 DSPs. The daughterboard contains 16 more DSPs.

The cPCI boards have CT Bus (H.110) support for connecting channels to line interfaces, to DSP resources for voice processing and signaling, or to the H.110 bus.

The AG 4000 Series fits into the NMS CT Access development environment (See CT Access).

The cPCI AG 4000 boards fully support the PICMG Hot Swap specification, and CT Access has API calls that enable applications to dynamically receive notification of board insertions and extractions.

The versatile Alliance Generation (AG) 4000C board is a cost-effective platform for mixed media types – including fax, voice and IP-based applications. Its single slot configuration dramatically decreases hardware costs by reducing the number of systems needed. With the PICMG-compliant hot-swap capabilities, carriers and enhanced service providers can immediately realize dramatic improvements in performance and scalability while reducing hardware costs.

As a member of NMS' Alliance Generation 4000 Series, the AG 4000C provides an impressive 120 ports of IVR/FAX and up to 60 ports of VoIP in a single cPCI slot. It consists of a base board with two or four T-1 or E-1 interfaces and 16 digital signal processors (DSPs) in a single slot, and can be supported by a daughterboard with 16 or 32 additional DSPs to minimize overhead and maximize processing time for applications. The AG

4000C supports H.323 version 2 and is an ideal platform for related emerging protocols including SGCP and MGCP, ensuring flexibility, scalability and interoperability in network deployments.

Indeed, there's support for a wide range of standard vocoding algorithms -G.723.1, G.711, G.729a, MS-GSM, ETSI GSM. Also, there's IVR and T.37 fax capabilities.

The AG 4000 Series is supported by Natural MicroSystems' CT Access development and runtime environment, a consistent set of APIs that are OS-independent to deliver true application portability. With CT Access' Natural Call Control API, programmers can quickly develop applications that run on multiple types of telecom interfaces by using a single protocol-independent API. Because of an open platform, systems that rely on existing NMS board technologies can immediately take advantage of the higher density that the AG 4000C offers, without additional development time or expense.

Among other places, you'll find the AG 4000C helping to provide carrier-class reliability in the AirCore Mobility Server Platform from TECORE of Columbia, MD, a leading supplier for the wireless switching infrastructure in the small to medium sized markets.

The NMS TX3220 board, on the other hand, is a Motorola-68060 powered board that's dedicated to delivering high-performance IP routing for VoIP applications and includes 100Base-T Ethernet support. It also provides a hardware platform for various applications involving SS7, the carrier signaling protocol used for call control and to provide Intelligent Networking (IN) and Advanced Intelligent Networking (AIN) services, and which is also used as the signaling protocol for VoIP gateways. The board can run the entire NMS SS7 signaling stack (MTP, ISUP, TUP, SCCP, TCAP) with support for up to 16 SS7 channels.

PEP Modular Computers' (Pittsburgh, PA – (412-921-3222, www.pep.com) CP602 is a Celeron-based system controller CPU for CompactPCI.

The board uses an Intel Celeron processor in a 370-pin Plastic Pin Grid Array (PPGA) package which operates at a clock speed of up to 466 MHz, and which provides a full speed 128 KB on-die Level 2 cache, yielding high performance with low power consumption.

The CP602 sports a 64-bit CompactPCI interface permitting data transfer rates of up to 264 MBps. Using the latest SDRAM technology, board provides up to 768 MB of cacheable ECC memory using three 168-pin DIMM sockets.

Besides standard PC interfaces, the new CPU offers a wide range of goodies, including a high-definition AGP/VGA graphics controller with four MB of video RAM (maximum resolution is 1,600 x 1,200 pixels), two Fast Ethernet interfaces, one UltraWide SCSI interface, and a slot for PMC mezzanine modules. Optionally, the CP602 can be equipped with an IDE-compatible hard drive disk, DiskOnChip Flash memory, and a second cPCI bus so that up to 14 slots can be addressed.

The CP602 also has rear I/O via P3, P4, and P5 connectors, and full system controller hot-swap support to allow peripheral boards

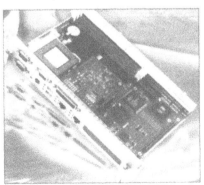

The speedy CP602 from PEP Modular Computers is a CompactPCI CPU card based on a 466MHz Celeron processor. It can hold up to 768 MB of ECC memory.

Pentair's subracks are IEEE 1101.10 compliant for VME64X and CompactPCI applications.

to be removed or added in the "power-on" condition.

The CP602 will run under Windows NT and/or real-time operating systems such as VxWorks and QNX.

PEP also offers the CP341, a 10/100 Mbps dual-independent Ethernet controller for CompactPCI applications supporting bus and star topologies. It's able to detect and automatically switch between 10Base-T and 100Base-TX data transmission, or operate at full duplex if the network operating station permits it.

Pentair Electronic Packaging (Warwick, RI – 401-732-3770, www.schroffus.com), a subsidiary of Pentair, Inc., offers enclosure solutions from over nine manufacturing locations in North America. They use Schroff brand cabinets, cases, subracks and systems; custom enclosures and consoles; high volume stamped chassis and assemblies; outdoor aluminum enclosures; and turnkey integration.

Pentair has announced in 1999 a line of 19" cPCI subrack systems, available in both 3U and 7U high backplane configurations. They are completely IEEE 1101.10 compliant for VME64X and cPCI applications.

Pentair's subrack systems include integrated ATX power supplies, EMC shielded design, 235 Watt ATX power supply, fan tray and a full range of accessories including drive bays, filler panels and front panels. The 235 Watt power supply has selectable input voltage levels of 90 to 132 VAC or 180 to 264 VAC with an input frequency of 48 to 63 Hz.

The operating temperature range is 0 degrees C to 40 degrees C and the Mean Time Before Failure (MTBF) is 100,000 hours at 25 degrees C.

Pricing for the 3U powered subrack starts at $895.

Pinnacle Data Systems (Groveport, OH – 614-748-1150, www.pinnacle.com) is a top-notch integrator that has a new NEBS compliant, CompactPCI rackmount fault tolerant solution for private label and resale for OEMs, built from redundant Sun Microsystems' SPARCengine CP 1500 single slot cPCI boards. The unit uses Fulltime's High Availability Plus (HA+) Failover software for failing over applications between processors.

The unit has two 64-bit SPARC version 9 UltraSPARC II-i processors running at either 270 or 333 MHz, featuring integrated memory and PCI interfaces, and binary compatible with SPARC application software. Mezzanine memory expansion cards are available in sizes from 64 MB up to 512 MB, yielding a maximum of 1 GB of memory per processor environment.

Power consists of four 3U high cPCI dual input -48V DC hot swappable power supplies.

DMA enabled UltraWide SCSI-2 interfaces are onboard, accessible via the rear panel. There are also autosensing dual 10/100 Mbps Ethernet channels.

The unit's Advanced System Monitoring (ASM) capabilities detect early indication of processor, power supply and cooling fan failure. The system monitoring can be configured for front panel display, GUI, or remote dial-out notification.

PDSi, can call upon its integration expertise to

configure a system to meet specific needs of the OEM. PDSi offers the OEM full life cycle support using its in-house depot repair capabilities.

PDSi offers another product, a CompactPCI Base Solution which supports Sun Microsystems' SPARCengine CP 1500 for private label and resale for OEMs. It has a single 64-bit SPARC version 9, UltraSPARC II-i processor running at 270 / 333 MHz. Up to 1 GB memory is supported. Power comes from a 350 Watt AC power supply. Like its big brother, it has Advanced System Monitoring and a three-year return to depot warranty.

Texas Micro merged into RadiSys (Houston, TX – 503-615-1100, www.radisys.com), a leading global designer and manufacturer of embedded computing solutions used by OEMs in the telecom, automation, and other industries.

This merger hasn't stopped the former Texas Micro facility from their continuing charge into CompactPCI territory. Indeed, RadiSys has joined the elite "five nines" club – building systems having an average downtime of only five minutes per year – with their new CP80-FT CompactPCI Platform. It's a fully fault tolerant NT-based system offering plug-and-play capabilities for telecom, datacom and Internet applications that demand extremely high availability. The CP80-FT is expected to be an OEM solution for companies looking for a fault tolerant platform that supports off-the-shelf software including unmodified Windows NT.

The CP80-FT was derived from RadiSys' award winning CP80 Platform, and brings together three technologies:

- RadiSys' implementation of CompactPCI technology in the CP80-FT allows OEMs to both increase Mean Time Between Failures (MTBF) and decrease Mean Time To Repair (MTTR). The new Hot-Swap technology in the CompactPCI standards allows the system to provide N+1 sparing and online repair of peripheral cards.
- The CP80-FT uses the Intelligent Platform Management Interface (IPMI), an industry standard interface which uses out-of-band signaling for management data and provides a means in the system to isolate failing components.
- Finally, rather than doing a traditional "polling" technique between two CPUs to achieve fault tolerance, the CP80-FT uses RadiSys' patented System Directed Checkpointing technology that provides transparent failure recovery of a system in less than 250 milliseconds. The failure recovery is completely transparent to the application software and much of the operating system.

Rittal Corporation (Springfield, OH – 937-399-0500, www.rittal-corp.com) is probably the biggest name in CompactPCI you may never have heard of. They work behind the scenes, providing branded subrack systems and other components to many companies. Rittal Corporation is part of Rittal International – an enormous enclosure manufacturer with offices in over 55 countries.

Recently, Rittal announced two new additions to its NEBS-compliant Ripac CompactPCI Subrack Systems line: The 9U high No. 3687302 and the 8U high No. 3687301.

Both systems have hot pluggable blower assemblies and hot swap

power supplies. The 9U No. 3687302 accepts 6Ux160mm front load boards and 6Ux80mm rear load boards and offers front load removable, hot swap RiCool blower assemblies and 350 Watt hot swappable N+1 redundant power supplies. The 8U No. 3687301 accepts 6Ux160mm front load boards only, but does offer dual RiCool blower assemblies and the 350 Watt front load removable hot swap N+1 redundant power supplies.

The standard configuration of both these units include all-aluminum construction, a maximum depth of less than 12 inches (NEBS compliant), high velocity front to rear cooling (240 CFM), acceptance of 160mm deep front boards, and two-slot to 21-slot cPCI backplane options. Options for these systems include multiple power supplies, slot keys, Type VII slim line cPCI ejector/injector handles, and drive bay chassis.

The price range for these two Ripac cPCI systems is $1,100 to $1,300.

Rittal also offers an extensive line of 3U and 6U high cPCI backplanes. Their comprehensive product line includes 32 and 64-bit versions, family related bridges, cables, termination modules, power supply backplanes and cPCI plus H.100 backplanes.

All of Rittal's modular cPCI backplanes have an extra 1/2 U for efficient power connection management on the bottom and they can be used in hot swap applications. Each module contains between two and eight slots, and operates in standalone mode with a CPU card and power supply. To build larger systems, multiple backplanes can be linked together using rear-mounting PCI bridge modules. With the modular cPCI backplane family, you're able to do things like bridge together a 3U cPCI backplane, a 6U cPCI backplane and a H.100 CT Bus backplane.

All Rittal backplanes comply with just about every standard you can think of, such as PCI 2.1, PICMG 2.0 R2.2, PICMG 2.1 R1.0, IEEE 1101.1, 11101.10 and 1101.11.

Rittal found that system designers often underestimate the need for heat dissipation, and a miscalculation of the cooling system can cause problems such as a missed product launch. So, Rittal and PIXStream have designed a system that does validation of cooling dynamics even before a particular system is built, called the RiTherm.

Rittal's RiTherm package consists of a combination of Thermal Load Boards and a Thermal system based on the cPCI architecture. RiTherm can do both sophisticated and simple evaluation and measurement of cooling in 6U CompactPCI Systems. Also, RiTherm can evaluate power supply quality on cPCI backplanes.

The RiTherm System is a 10U high cPCI Subrack equipped with two powerful hot swappable 12V RiCool blowers, a cPCI backplane, and a pluggable power supply. The system accepts up to 21 slots of front/rear mounted PCBs. Typical RiTherm System configuration is a 16 slot CPCI backplane and two 350W power supplies.

The real star of the show, however, is the RiTherm Load Board, which enables you to validate the thermal characteristics of a system before all PCBs are designed. Each load board is a 6U or 3U high cPCI instrumentation board that provides a resistive heating array of 38 configurable resis-

RadiSys / Texas Micro's amazing new CP80FT is a Windows NT-based CompactPCI system having full fault tolerance, with "five nines" (99.999%) uptime, achieved via RadiSys' patented System Directed Checkpointing technology.

tor loads and six temperature sensors (a precise airflow sensor is also available on the RP3901 load board). No sensor wires are needed.

The whole system allows you to do simulations of either I/O or SBC boards, each drawing up to 90 Watts, with support for a total of up to 21 load boards.

Rittal also has one of the best cooling solutions in the industry. Their innovative 1U RiCool blower tray can handle densely packed VME or CompactPCI systems, and is especially effective for telecom applications where rack height restrictions exist. The RiCool 19" rackmount blower tray consists of two blowers mounted side by side, together providing 220 CFM of airflow at 0" static pressure. Air is drawn up through the cPCI boards and then exhausts out the back of the subrack via powerful curved impeller blowers – while using only 1U of rack space.

At a static pressure of 0.25" of water, the RiCool blower tray provides over 180 CFM of airflow, which is at least double the airflow of typical 19" rackmount tubeaxial fan trays. For densely packed subracks, the RiCool blower tray continues to provide cooling at static pressures in excess of 1.0"of water, again far exceeding the competition. The blower tray also offers an alarm output via fan speed sensor, speed control capability, quick access, hot swap capability, and you can use it without installing any extra ducting thanks to the inherent airflow path where the impeller turns the flow direction by 90 degrees.

Rittal's has also developed a clever 350 Watt cPCI power supply that can be used in 6U cPCI computers, testing and telecom systems. The

Rittal offers a range of CompactPCI backplanes, including 32 and 64-bit versions, family related bridges, cables, termination modules, power supply backplanes and CompactPCI plus H.110 backplanes.

Rittal's No. 3687301 CompactPCI Subrack system is in the Telcorida NEBS form factor. It accepts 6Ux16omm front load boards only, but does offer dual RiCool blower assemblies and the 350 Watt front load removable hot swap N+1 redundant power supplies.

device is 8hp wide (two slots) and meets all of the PICMG cPCI spec plus N+1 redundant and hot swap application. It's high-density cooling fins are positioned directly in the airstream, allowing full power operation to 35 degrees Celsius with specified airflow. LED status indicators and a push-button inhibit switch are located on the front panel. It comes in two input voltage options and two output voltage / current options.

SBS Technologies, Inc., Embedded Computers (Raleigh, NC – 919-851-1101, www.sbs.com), formerly called VI Computer, has announced the cPRO3, a 3U high cPCI, Socket 7-based processor board that's an extension of the SBS PROTEUS family of embedded computers that are designed to ease customization for specific applications. The cPRO3 offers processor speeds up to 333 MHz (K6 or Pentium MMX), up to 128 MB of EDO DRAM, 512 KB of Level 2 cache, and up to 144 MB of Flash Disk-on-Chip.

The cPRO3 uses an Award BIOS in Flash ROM which allows future BIOS upgrades to be programmed on-site. The card has the I/O capabilities of a complete PC/AT, offering two rear panel RS-232 serial COM ports, a parallel printer port, Ethernet port, dual USB ports and front panel keyboard/mouse port. The COM-2 serial port is enhanced for software selectable RS-422 or RS-485 operation.

Pricing for the cPRO3 is $1,100 in OEM quantities.

Schroff North America (Warwick, RI – 401-732-3770, www.schroffuf.com) now offers 10-layer backplanes for CompctPCI applications that

Rittal's clever 350 Watt CompactPCI power supply can be used in 6U high cPCI computer, test and telecom systems. It takes up two slots.

meet the PICMG cPCI 2.0 R2.1 spec. These new units complete Schroff's family of VME, VME64X, cPCI and VXI backplanes.

The new 64-bit 3U, 6U and 7U high backplanes are available in four, six and eight slot versions. Schroff has incorporated stiffening rails into the design to prevent bending when units are removed and replaced.

The backplanes have a data transfer

Rittal's No. 3687302 CompactPCI Subrack system is similar to the 3687301, but accepts both 6Ux16omm front load boards and 6Ux8omm rear load boards

Rittal's innovative RiTherm system offers comprehensive evaluation and measurement of cooling in 6U high CompactPCI systems.

Rittal's hot swappable 1U RiCool blower tray provides powerful cooling of densely packed VME or cPCI subrack systems.

rate of 132 / 264 MBps, 220 pin metric connectors for hot swap applications, right or left handed controller slots, integrated connectors for power supplies, and voltage I/O can be configured for +3.3V or +5V. There are busbars on 5V, 3.3V and the Ground for optimum power distribution.

Sun Microsystems (Palo Alto, CA – 650-960-1300, www.sun.com) announced in 1999 that "in an aggressive move to secure market leadership in the growing high-availability network equipment industry," Sun has slashed prices across its entire line of 64-bit SPARCengine CompactPCI board products, including the SPARCengine 1500 and 1400 cPCI boards designed for central office use that are powered by UltraSPARC-IIi processors.

Sun sees their cPCI boards as the next growth opportunity for the OEM market. Their cPCI boards are part of a family of integrated high-availability products, which range from processors to servers for the telecom network environment, as well as Solaris Operating Environment and the ChorusOS real-time operating system (RTOS) (Sun will also let you run VxWorks, the RTOS component of the Tornado development environment from Wind River Systems).

Sun is also shooting for "five-nines availability" (99.999 percent, or five minutes and 15 seconds of downtime per year) for it's cPCI systems. Its latest cPCI boards are 425-MHz and 360-MHz versions of the SPARCengine CP1500.

The new pricing (per board, in volumes exceeding 1,000) is as follows: SPARCengine CP1500 425MHz is $2,532; 360MHz version is $1,972; 333 MHz version is $1,900; 270 MHz version is $1,850. The SPARCengine CP1400 300MHz is $1,490.

Synergy Microsystems' (San Diego, CA – 858- 452-0020, www.synergymicro.com) KGM5 is an incredibly powerful single board computer, having dual PowerPC 750 CPUs (each running up to 466 MHz) and a high-performance back-side Level 2 cache, in a single CompactPCI slot. You can also choose a single-processor version.

Interestingly, the board can function in either a system or peripheral slot with no change in configuration whatsoever. Just plug it in and it will assume the prop-

SBS Technologies, Embedded computers has announced the cPRO3, a 3U high CompactPCI, Socket 7-based processor board. It offers processor speed up to 333 MHz and up to 128 MB of EDO DRAM.

er function. This saves you from having to buy and stock two different kinds of boards for your system, simplifying things considerably.

The board can hold from 16 to 512 MB of SDRAM There's also autosensing 10/100Base-TX Ethernet, two serial ports, and an optional SCSI interface. The board also supports software such as VxWorks, LynxOS, pSOS, and Linux

The Model TL-cBOX 300 from Technoland (Sunnyvale, CA – 408-992-0888, www.technoland.com) is a 19" 3U high CompactPCI chassis with a seven slot backplane and a 230watt PS/2 hot-swap redundant power supply with an estimated MTBF of 100,000 hours.

Output voltages/currents are +5V@30A, +12V@14A 9220 Watts max.), and -5@0.5A, and -12V@0.5A with an estimated minimum MTBF of 100,000 hours. It's aluminum case keeps its gross weight down to 25.4 lbs. (12 kg.).

Technoland also has a number of single board computers. For example, the TL-cPCI 66102 is a 6U high dual or single 333 MHz Pentium II cPCI CPU module with an Intel 440LX PCI chipset that supports a fault tolerant design: In dual processor operation, the system will shut down if any CPU fails, and after the Watchdog Timer resets the system, the remaining live CPU will take over all tasks and continue operating. In single processor operation, one configures a primary and secondary CPU. The primary CPU runs all tasks and the secondary CPU is on standby mode. If the primary CPU fails, the secondary CPU will take over all tasks only after the Watchdog timer resets the system.

The board can hold up to 512 MB of EDO / SDRAM in four 168-pin DIMM sockets. There is also 256 or 512 KB cache memory. There's a dual PCI bridge (two DEC 21150 PCI bridge controllers, with each bridge able to handle seven PCI masters), four serial ports, two USB ports, a parallel port, an ATI VGA controller with 2MB of SGRAM, and an M-Systems' 72MB DiskOnChip.

Teknor Industrial Computers (Boca Raton FL – 561-470-0151, www.teknor.com) was recently acquired by Xtech Embedded Computers, a company that provides advanced embedded computer solutions for OEMs, system integrators and end-users in key technology areas including industrial single board computer design, low power/mobile computers and industrial multimedia. Their Kontron Elektronik division supplies high-quality fully integrated industrial rackmount chassis, HMI panels and portable computers.

Teknor is expanding it's pace of CompactPCI development, recently unleashing a ton of new products:

The cPCI-MXP is Teknor's hot-swappable Pentium II Low Power peripheral processor with an on-die 256 KB L2 pipelined burst cache and 440BX AGP set. It holds up to 768 MB of SDRAM with parity or ECC (for single bit error correction and dou-

Schroff's new 10-layer, 64-bit 3U, 6U and 7U high cPCI backplanes come in four, six and eight slot versions.

ble bit error detection), has interfaces for PCI Ultra DMA/33 IDE drives, 10/100Base-TX Ethernet, and a 64-bit AGP SVGA CRT controller (Intel 69000) with 2 MB video memory. Onboard you'll also find two USB ports, a parallel and two serial ports, and a floppy interface. An optional bootable CompactFlash disk interfaces to the primary EIDE channel, and is user upgradeable, master / slave.

An optional hard disk / floppy mezzanine card (cMC) is available for applications where the highest level of integration is necessary. The cPCI-MXP's ability to drive seven cPCI I/O slots can be doubled using an optional bridge mezzanine, the cMCB. More functions can be added using the onboard PCI Mezzanine Card (PMC) expansion site.

A hardware monitor keeps tabs on the voltages, temperature, fan speed, and user-defined I/Os, and there's also a software programmable two-stage watchdog timer, and power fail circuit. Price is $1,700 with a 333 MHz processor and no SDRAM. Like all other Teknor boards, it has a two-year warranty.

The CPCI-MXS is Teknor's Pentium II low power 6U high cPCI single board computer. Based on their TEK-CPCI-1003, this new SBC includes a High-Availability controller which provides three control signals for each of the seven supported cPCI I/O slots (21 signals in total). These signals are used to detect the insertion of hot swappable cPCI I/O boards ("Board Present" and "Board select" conditions), assert the board's status (such as "Healthy") and control its startup ("PCI reset").

The cPCI-MXS has the Intel 69000-powered APG Graphics engine with 2 MB of video memory integrated on-die, and two 10Base-T/100Base-TX Intel Ethernet interfaces based on Intel's 82559, which is smaller and has lower power dissipation than the 82558 silicon.

The board's processing power comes from an Intel Pentium II Low Power 333 MHz processor and includes 256 KB L2 Cache and supports up to 768 MB of ECC SDRAM on three industry standard DIMMS.

The cPCI-MXS PICMG-compliant board has an Ultra Fast/Wide SCSI 3 interface, onboard EIDE and hard disk / floppy mezzanine module, a PCI-to-PCI bridge, PMC mezzanine support, CompactFlash, four serial, one bi-directional parallel and two USB ports. The EIDE disk interface supports up to four hard disks.

To ease the task of integrating hot-swap functionality into target applications, a baseline interface to other OSs can also be provided.

The CPCI-CXS is Teknor's Celeron 6U high SBC with high availability controllers and a low cost 433 MHz Celeron processor (with

Teknor's cPCI-MXS has a 333 MHz Pentium II Low Power processor, an on-die 256 KB Level 2 cache, and holds up to 768 MB of SDRAM.

Teknor's CxP0816 is an 8U high CompactPCI system that offers four backplane options, with eight or 16 cPCI slots, PICMG-complaint and single (4 HP) or dual (8 HP) system slot. It has a hot-swap fan tray as well as a fuly rear I/O capable backplane and an ATX or AC removable or redundant power supply up to 350 Watts.

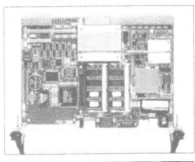
VMIC's VMICPCI-7697 is a cPCI SBC that holds a 333 MHz Pentium II.

128KB of L2 cache) that can nevertheless support up to seven cPCI slots. The cPCI-CXS SBC is pin-out compatible with all of Teknor's CompactPCI SBC family. It can hold up to 768 MB of ECC SDRAM using industry standard DIMMs, has CompactFlash support, the AGP SVGA Graphics engine with 2 MB of video memory integrated on-die, an Ultra Fast / Wide SCSI 3 interface, on-board EIDE, two 10Base-T / 100Base-TX Ethernet interfaces (based on Intel's 82559 silicon), PMC support and optional onboard hard disk and floppy disk.

Single quantity pricing of the cPCI-CXS with a 433 MHz processor and no SDRAM is $1,640.

The CxP0816 is Teknor's cPCI 19", 8U high, 16 slot Rackmount telecom enclosure. The CxP0816 offers a choice of four backplanes and three types of power supplies: a low cost ATX, hot swappable N+1 power supplies, and an upcoming DC option.

The CxP0816 integrates a 2U high, hot-swappable fan tray holding three ball bearing fans. The CxP0816 supports all Teknor CompactPCI SBCs, including the award-winning Pentium II-based TEK-CPCI-1003 and newly released cPCI-CXS Celeron processor 6U SBC. All CxP0816 enclosures support 80mm rear I/O connectors.

The selection of backplanes includes standard 8 and 15 slots, supporting Level 2 full hot-swap functionality as well as telephony 8 and 15 slot backplanes integrating the H.110 bus and supporting level 3 High-Availability hot-swap functionality. You can also have a single (4 HP) or dual (8 HP) system slot.

The CxP0816 supports up to two slot-saving storage modules (the TEK-CPCI-1085), which integrates HD and DVD drives in a 6U single slot form factor. These slim line storage modules allow you to cram up to 15 processor boards per system (one center mounted System Controller and up to 14 I/O processors). Indeed, in a 15-slot configuration, the CxP0816 offers one of the highest levels of CompactPCI integration achievable while still maintaining the capability of dual redundant, hot swappable N+1 power supplies.

The CxP0816 comes with Teknor's standard two year warranty. The single unit price of the CxP0816 with eight slots and 300 Watt ATX power supply is $2,550. OEM quantity discounts are available.

TsDesign (Atlanta, GA – 770 454 6001, www.tsdesign.com) is the U.S. office of Telesoft Design Ltd., a British company that has deployed many innovative computer telephony and SS7 solutions over the years.

Telesoft's MPAC-1022 Octal CompactPCI Line Interface Card supports up to eight T-1 or E-1 ports in a single 6U card slot. The card uses an AM186EM Lucent Ambassador chip to support connectivity to the H.110 bus. The card provides local, on-board switching which improves overall application efficiency and system scalability. Each of the interface ports is fully configurable, as either a T-1 or E-1, and port-to-port connections within any of the eight T-1 / E-1 trunks can be switched locally on the board, and do not use any of the timeslots on the H.110 bus.

Telesoft also offers their full range of industry standard T-1 and E-1 signaling interfaces – allowing the MPAC-1022 board to support both ISDN and SS7 protocols simultaneously. In fact, virtually any protocol can be selected to run on the card – you can even mix and match signaling interfaces with transmission standards. An option DSP daughter board enables simultaneous detection and generation of up to 240 circuits of DTMF and MF tone frequencies. This allows R2, CAS and V5 protocols to be supported alongside ISDN and SS#7 protocols. DTMF collection at the trunk interface can significantly reduce the overheads on dedicated voice boards by, for example, collecting notification tones in follow-on call options.

Developers can access the appropriate signaling layers to allow connectivity to switches and SS7 applications in any mobile network worldwide. This enables the MPAC-1022 card to be used for tasks such as Short Messaging Service (SMS) generation, collection and routing. The card thus can be used to build cost effective SMS Center (SMSC) platforms as well as Gateway MSC solutions written into the mobile network. GSM,

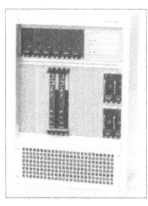

Ziatech's ZT 5083 High Availability System has 14 slots along with redundant and hot swappable components.

IS41, AIN and INAP protocols and procedures are also supported.

VMIC (Huntsville, AL – 256-880-0444, www.vmic.com) sells a Pentium II processor-based cPCI peripheral slot CPU card, the VMICPCI-7697. It complies with the high availability provisions of the PICMG 2.1 Rev 1.0 standard for cPCI Hot Swap, which means that, besides being able to remove and replace the board with the system operating, processing can automatically be switched to a second backup board already installed in the system in the event of failure in the primary processor.

The board uses an Intel embedded Pentium II module with MMX, offering speeds up to 333 MHz. An Intel 21554 embedded PCI bridge allows the onboard Pentium II to operate independently of the cPCI bus. Embedded features such as M-Systems DiskOnChip, 16 MB bootable IDE-flash, 32 KB battery-backed SRAM, three 16-bit timers, and a software-programmable watchdog timer make this board suitable for high performance realtime applications.

For Windows NT and RTX applications, the VMICPCI-7697 has a 64-bit AGP video controller with 4 MB of SGRAM and runs VMIC's VMIS-FT-9421 Ioworks Access and other IOWorks software. VXWorks support is provided using VMIC's VMISFT-7418 board support package.

Standard items include 256 MB SDRAM using two 144 pin SODIMMs, and Ultra-IDE hard drive and floppy drive controllers through the cPCI J3 I/O connector, two serial ports (one available at the front panel using an RJ11 or both ports available through the cPCI J3 connector). There's also an on-board Fast Ethernet controller supporting 10BaseT and 100BaseTX, as well as a parallel port, a USB port on the front panel (second USB port on cPCI J3), and keyboard and mouse ports on the front panel.

The price for the VMICPCI-7697 starts at $3,248,

The SuperSpan board from Voiceboard Corporation (Oxnard, CA – 93030, www.voiceboard.com) supports up to eight T-1 / E-1 interfaces (256 ports) on a single 6U high cPCI board for VoIP, V.90 modem, G3 Fax and call control applications. The T-1 interfaces also incorporate the

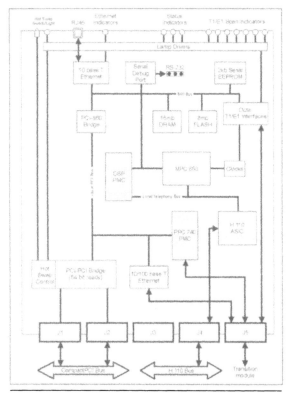

Voiceboard's SuperSpan board supports up to eight T-1 / E-1 spans on a single 6U CompactPCI board.

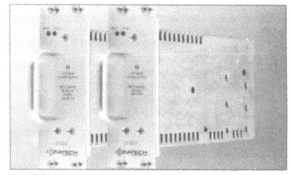

Ziatech's 150 Watt power supplies look just like 3U cards. Pictured here are the ZT 6301 (AC output) and ZT 6311 (DC output).

CSU, reducing cost and eliminating any external equipment.

Primary Rate ISDN, Frame Relay or SS7 Layers MTP1 and MTP2 can also run native onboard. SS7 layers MTP3 and above can be achieved via an Enhanced Communications Processor (ECP) option which is a single PMC slot add-on card that uses a PowerPC. The ECP can also handle H.323 and TCP/IP protocol stacks for VoIP. A dual active / hot standby configuration can be used for fault tolerant SS7 configurations.

Built on Voiceboard's MediaPro II architecture, the board has 3,200 MIPS of DSP signal processing power, H.110 switching of DS0 data streams, dual 10/100 BaseT Ethernet ports and a PowerPC 740 co-processor. SuperSpan also provides such functions as Robbed bit, CAS and CCS signaling, and it can do subrate switching down to 8Kbps which multiplies channel capacities up to eight times and efficiently handles slow speed data links and compressed voice.

Ziatech (San Luis Obispo, CA – 805-541-0488, www.ziatech.com) has joined the ranks of companies building CompactPCI systems with "five nines" (99.999%) of availability with its ZT 5083 High Availability System.

Designed for telcos and other high-end applications, the Ziatech 5083 System introduces an Intel-based redundant processing architecture for CompactPCI. Indeed, the system has built-in redundancy for active system components including system slot CPU boards, power supplies, cooling, and system alarms and can be configured with an optional RAID disk subsystem.

Ziatech's 6U high ZT 5521 cPCI single board computer can hold an Intel Pentium III Processor (Slot 1) running at speeds over 500 MHz.

The dual redundant processor architecture provides true fault tolerance. The active CPU can isolate the backup CPU if it has failed. Isolation can occur from the bus or the CPU can be powered down. The ability to power down and up again allows you to run diagnostics to determine if the takeover occurred because of a hardware or software reason. If software, the CPU may again be functional and not need to be replaced. If the hardware is faulty, then it can be swapped out.

The ZT 5083 takes the form of a 15U high, 19" wide rackmount having 14 cPCI slots. Two slots (physical slots 7 through 10) are dedicated to the system CPU subsystems. Slots (1 through 6, and 11 through 16) are provided to support up to twelve 32-or 64-bit CompactPCI peripheral cards running at 3.3V or 5V.

A standard configuration includes dual redundant system master CPUs, four load sharing, four hot-swappable AC or DC input 150 Watt power supplies (N+1 redundant), redundant fan cooling, and integrated programmable system monitoring and alarming functions in a NEBS Level 3 compliant enclosure. The system provides full support for Ziatech peripheral master CPU's (ZT 5540) for high performance multiprocessing applications.

The system is PICMG 2.1 (Hot Swap) specification compliant and accommodates IEEE 1101.11 rear I/O transition

The LinuxPCI 1000 Development System for Applied Computing from Ziatech runs MontaVista Software's Hard Hat Linux, designed for embedded applications.

cards. The system is also PICMG 2.5 (Computer Telephony Specification) compliant providing a H.110 telephony bus.

The 5083 includes the kind of software features you need to run a high availability system under Windows NT, such as comprehensive BIOS diagnostics, a development utility to simulate faults and verify system operation and a Desktop Management Interface (DMI) for the monitor and control of system-level components.

Ziatech's new ZT 5521 cPCI single board computer is 6U high and can hold an Intel Pentium III Processor (Slot 1) running at speeds over 500 MHz for optimum performance in demanding uniprocessing and multiprocessing systems such as ATM network switches. The ZT 5521 can also be ordered with an Intel Pentium II instead. The ZT 5521 supports hosting hot swap peripherals in a powered system. The board occupies two CompactPCI slots (8HP) with the processor and heat sink installed.

Based on the Intel 440BX chipset, the ZT 5521 also gets some improved performance via a 100 MHz front side bus, and dual on-board 64-bit bridges that allow it to drive up to 14 cPCI peripherals slots.

Like all of Ziatech's cPCI CPU cards, the ZT 5521 offers a choice of Windows NT, QNX, or VXWorks operating system support, and can host hot swap peripherals and serve as the System Master of up to 32 processors in Ziatech's multiprocessing environment, CompactNET.

The card can hold up to 1 GB of PC-100 ECC SDRAM DIMM modules, 8 MB of flash memory, a CompactFlash socket for further flash memory expansion, dual 100 MB Ethernet, dual USB and serial ports, a two-stage watchdog timer, an optional Advanced Graphics Port (AGP) for video, and rear I/O connections.

ZT 5521 pricing starts at $4,995, for a configuration that includes 8 MB Flash, 256 MB RAM, AGP video, embedded BIOS, and a front panel.

Ziatech has also just announced a rackmount CompactPCI development platform using Linux, called the LinuxPCI 1000 Development System for Applied Computing, which runs the first version of Linux designed for embedded applications, MontaVista Software's Hard Hat Linux.

The LinuxPCI 1000 combines a standard, fan-cooled 19" rackmount enclosure with a cPCI system based on a Pentium II processor, and peripherals and I/O needed for application development, such as a floppy drive, hard drive, CD-ROM, and a 300 Watt power supply. The system's eight-slot cPCI backplane supports 32-bit or 64-bit cards, the PICMG 2.1 Hot Swap Spec, and rear I/O. Ziatech's multiprocessing environment for cPCI, CompactNET, is an option.

The LinuxPCI 1000 Development System with a pre-tested Hard Hat Linux kernel configuration with CD-ROM has a single quantity price of $5,895 and includes 90 days of service support from MontaVista Software.

Ziatech's new hot swappable 150 Watt power supplies also deserve a mention:

The ZT 6301 (AC) and ZT 6311 (DC). One could at first mistake them for 3U high cPCI cards. Both cards have four outputs capable of providing a total of 150W for +3.3VDC, +5VDC and positive or negative 12VDC with independent output regulation. Both models use a DIN power connector, support hot swap N+1 sharing and have built-in EMI filters and status LEDs.

In 1999 Ziatech announced a strategic partnership with Lynx Real Time Systems to provide enhanced computing solutions integrating Ziatech cPCI computers and the "hard real-time" capabilities of the LynxOS for high-end communications equipment involving such things as DSL, VoIP, ATM switches, etc.

ZNYX Corp. (Fremont, CA – 510-249-0800, www.znyx.com) has partnered with the Motorola Computer Group (MCG) on a PMC Rear I/O standard called TM-PIM, a transition module designed to work with the proposed PMC Interface Module (PIM) architecture for truly standardized modular rear I/O on CompactPCI and VME systems.

Alan Deikman, chief technology officer of ZNYX – which is an MCG Embedded Connections Partner and PMC supplier – says: "MCG's active participation in the PIM standard will benefit the entire industry. ZNYX has been working with MCG since the early phases of the PIM standard. As a result, we are planning to release rear I/O multi-channel Fast Ethernet LAN products that incorporate our RAINlink High Availability technologies immediately after the PIM architecture is finalized."

Currently, ZNYX's hot-swap Ethernet board offerings for cPCI include the 64-bit NetBlaster ZX440 and ZX470 series. They come in four- and eight-channel configurations, 3U and 6U sizes and front- and rear-panel I/O.

For applications demanding front-panel I/O, the ZX440 boards manage two or four Ethernet connections. Central Offices that rely on rear-panel I/O will find that the 6U high ZX470 boards handle up to eight Ethernet connections and have a passive rear-panel transition module.

The boards come with ZNYX's own embedded Rainlink software that supports redundancy and fast deterministic failover, system-to-system trunking and system-to-switch trunking. A driver suite exists for Windows NT, Solaris, VxWorks, pSOS+, LynxOS, UnixWare, QNX and Linux. The ZX440 starts at $849; and the ZX470 at $1,599.

Also, the ZNYX ZX470 NetBlaster series is available in four (ZX474) or eight (ZX478) channel models. Prices range from $2,199 to $2,699. The ZX440 NetBlaster series is available in two (ZX442) or four (ZX444) channel models. Prices range from $1,249 to $1,999. Drivers are downloadable from ZNYX's web site at www.znyx.com.

Computer Telephone Integration (CTI)

A term coined by Jim Burton while at IBM. He is currently president of CT-Link (St. Helena, CA – 707-963-9966, www.ctlink.com). In 1985 Carl Strathmeyer developed the first computer-phone system link at DEC (now Intel / Dialogic's CT Division) had already promulgated the term Computer Integrated Telephony (CIT), but IBM did not want to use a term devised by their chief competitor. Jim Burton obliged by switching some letters of the acronym around to get CTI. Ironically, Burton has joked on many occasions about how he always liked Strathmeyer's term better than his own.

Still, the public latched onto the "CTI" and it was the principal term used to describe the technology and the industry until Harry Newton and Gerry Friesen began promoting the term Computer Telephony with the test issue of a new magazine of the same name in the fall of 1993, at the same time changing the name of their "Telecom Developers" Expo to "Computer Telephony Expo".

In response, those who had grown accustomed to using the term CTI had by this time recognized that expression "computer telephone integration" was somewhat restrictive, suggesting technology relating solely to physical computer-telephone connectivity. Still, they tried to hold onto the acronym by formulating and popularizing the "hybrid" term Computer Telephony Integration. Both "CTI" and "computer telephony" are now used interchangeably – for several years *Computer Telephony* magazine's only competitor was *CTI* magazine, published by Nadji and Richard Tehrani of the Technology Marketing Corporation (Norwalk, CT – 203-852-6800, www.tmcnet.com), who also publish *C@LL CENTER Solutions* magazine, *Internet Telephony* magazine, and sponsor the biannual CTI EXPO and Internet Telephony EXPOs.

In the March 1997 issue of *Computer Telephony* magazine, Harry Newton published an article entitled "Why CTI is not Computer Telephony" in which he explained how CTI is just one small part of the total CT picture.

People must have believed him. In January 2000 *CTI* magazine was relaunched with a new name, *Communications Solutions*.

Computer Telephony (CT)

What is computer telephony? Years ago, Newton's Telecom Dictionary first defined CT something along the lines of: "the discipline of applying computer-based intelligence to the making and receiving of phone calls, fax communications and other complex messaging and transactions involving public, private and Internet networks."

This definition, of course, leaves out reverse situations where the telephone system must control some of the functionality of the computer system, a reminder of the increasingly symbiotic relationship between computer intelligence communications.

One can say flat out and up front that there is really no generally agreed-upon definition of what exactly comprises computer telephony. This is because the field keeps changing, it is a moving target for those who would like to place it neatly in a pigeon hole and move on to other matters.

In the late 1960s the then-unnamed field of CT consisted of a few IBM 360 mainframes connected to their IBM 2750 PBXs, receiving over a public network in Germany order fulfillment-related messages from bookstores. In the 1970s it was the automatic distribution of calls to agents in large call centers with Rockwell equipment, and the Delphi Corporation's Delta 1 system used by telephone answering services in San Francisco. In the 1980s it was VMX voicemail, Periphonics IVR, and predictive dialers.

In the early 1990s computer telephony was not just about big computers any more, but little ones – microcomputers and PCs that promised to bring new and "open" programmable capabilities to both a staid telecom industry and to affordable, scalable business phone systems. Industry standardization and the availability of low-cost CT links gave rise to a new generation of CT applications that linked LAN servers, public phone switches, private PBXs and desktop PCs to phone lines.

In those days the two great names in computing where Novell and Microsoft, engendering such popular adages as "Novell owns the network and Microsoft owns the desktop." Ironically, these two companies' perceived personae would carry over into their first attempts to impose a standard computer telephony API model. In 1992 AT&T and Novell deployed the Telephony Services Applications Programming Interface (TSAPI) which only ran on NetWare, which was too expensive, but provided true client/server functionality for use in third-party call control applications, meaning that the server is intelligent and can establish and oute calls independently of the user and can do such things as control distant PBX on a WAN. Intel and Microsoft followed on TSAPI's heels with the Telephony Application Programming Interface (TAPI), a free piece of telephony software bundled with Windows that could telephony-enable applications running at the desktop (first party call control), but would not have full blown third-party client/server call control until version 3.0. This did not stop several ingenious early vendors from working around this problem by creating TAPI service providers which tricked the desktop PC into seeing a TAPI "device" inside the PC when in fact there is none, so that TAPI function calls could be redirected over a LAN/WAN connection to a PBX.

In the late 1990s, computer telephony had to be redefined yet again as the traditional circuit-switched telephone system converged with packetized technologies and networks such as IP and the Internet. At this point it became evident that one could "take the computer out of computer telephony" leaving only an abstract model of call control and messaging intelligence that was threatening to disappear into the larger machinery of customer and enterprise relationship management. In an era where concept of a platform on which to build shifted from operating systems to browsers, TAPI, having just emerged victorious from the "TAPI TSAPI war" faced in 1996 a totally new kind of competitor, the Java Telephony API, or JTAPI, which has great potential but has taken a back seat to "straight" or "ordinary" Java.

Not only has computer telephony rapidly evolved during the 1990s, but it has continually enlarged in scope. After all, computer telephony was never a single application, but a collection of them: Voice mail systems, fax servers, IVR, ACDs, intelligent e-mail systems, unified messaging, etc., all of which somehow had to work together to make life easier and more profitable for those daring enough to build such systems. This melange of technologies that comprises computer telephony goes on getting larger and larger and more diffuse, with no signs of stopping, like some great puff of smoke that expands and ultimately blends into its surroundings.

But if there is a single leitmotiv that runs through CT, it is the primacy of voice communications. Early promoters of CT defined it as a confluence of computer, telecom and voice processing systems. Indeed, when I first went to work for the flamboyant Harry Newton and his then-fledgling *Computer Telephony* magazine in 1994, we were chatting in an elevator when he suddenly turned to me and said: "You know, Richard, this field used to be just a small part of voice processing before I began to promote it into bigger and better things." (At that Newton was obviously successful, as he sold his company to Miller Freeman in 1997 for more than $130 million). Even later, when *Computer Telephony* magazine at the end of 1999 decided to revamp its tagline "The Magazine for Computer and Telephone Integration" to acknowledge the convergence of the circuit-switched and packet-switched networks, we came up with "The Voice of Converging Communications." We used the word "voice" to emphasize not only the fact that CTM was the leading publication in its field, but also to recognize the almost archetypal nature and predominance of the human voice in communications.

After all, the telephone, be it connected to a wire or wireless, has been with us for over a century and remains the most widely used business and personal communications tool. One of computer telephony's many tricks is to convert telephones into virtual "terminals," expanding their capabilities so that all sorts of messages can be sent and received with them, saving time and money.

Because computer telephony embraces so many things, we tend to lose sight of it as a distinct entity. Without realizing it, we use some form of CT every day: We check our voice mail several times a day, get our bank balances by phone, call a phone number to hear the weather forecast or call 777-FILM and to hear audiotex telling us about the movies playing downtown, make phone calls with a PC, request technical and product information via fax using the phone's keypad or even by voice command, use touchtones to get information, hear e-mail read to us over the phone via text-to-speech, instruct our phone system find us wherever we happen to be, do fax broadcasts to our customers, review all of our voice, e-mail and fax messages in one list on one screen, and use the Internet for phone calls over wireline or wireless networks. Researchers and pollsters use CT to collect and analyze information. Corporate and financial information is provided to employees or stockholders via CT. Even when your phone rings while you're eating dinner and someone tries to sell you something, that call has been placed by a telemarketer using a predictive dialing system, yet another technological denizen that lurks in the computer telephony realm.

Thus, computer telephony is not a "single anything" but a technological enabling environment that brings intelligence and choices to the communications process. It solves no particular business problem unless someone builds a suitable application upon its foundations, using the many tools that it provides. And indeed some of the new tools available to computer telephony developers, such as application generators and customizable turnkey systems, allow non-technically oriented people to easily shape their communications environment to their own vision. It's a far cry for the early days when even CT's principal tool – raw computing power – was so unsophisticated and expensive that only major entities such as AT&T, IBM / ROLM, Tandem, etc. could implement such systems.

Not only is there confusion over just what computer telephony is, there's also some bafflement concerning just "where" computer telephony resides. Similar CT applications can and have been deployed both in Customer Premise Equipment (CPE) and as enhanced services originated by telcos (or originated by third parties and resold by telcos) and offered by subscription. Further confusion results from the fact that both telecommunications and computer / data communications industries have separate ideas over what computer telephony is and where it should live in the network, which has lead to perplexity over who sells, installs and maintains CT solutions. The bright side of all this is that computer telephony is fostering the convergence of voice and data, driving the merger of packet and circuit-switched technologies for personal and business communications, and will continue to do so until it leads to some future, simplified set of communications platforms.

Surely with all of the bemusement surrounding CT, there must be great benefits associated with it or no one would be driving the market forward. In fact, CT has the potential to solve many major business problems that have escalated in recent years.

Today the CT market is driven by –
- The desire to improve customer care.
- The demand to increase personal productivity.
- The desire to control costs (fewer live agents needed to handle routine calls; the inexpensive management of a single wire to the desktop and a single device on the desktop).
- The desire to slash toll costs with IP telephony.
- The desire to upsell to customers with automated systems.
- The desire to latch onto products complying with CT standards (TAPI, JTAPI, S.100).
- The demand for instant information and the revenue that can be generated by offering and charging for valuable information.
- The desire for something that will aid the new business to business (B2B) service providers.

- Global telecom deregulation.
- The Internet and e-commerce.

Payback periods can range from a few weeks to two years for CT systems. Sounds great. So where do you begin?

Slicing and Dicing Computer Telephony

One can look at CT wearing either the hat of a developer or the hat of a person who wants to get some benefits from the technology. Do you plan to develop CT applications (in which case you would be interested in application generators and systems that can easily be integrated into those you plan to build) or are you a power user who merely wants to buy a turnkey system with enough flexibility so that you can configure it yourself?

Either way, it pays to take a detailed look at the many and varied facets of CT. But be forewarned! Trying to formulate an ontology or taxonomy of CT technology can be as mind-boggling and as complex as anything the first zoologists faced when confronting the creatures inhabiting the continents of the New World.

Computer telephony starts at the electrical engineering bus, board and "chip" level. The hardware has seen greater functionality in recent years with the arrival of higher processing and bus bandwidth capacities and the software has led to more open development systems. Major improvements have resulted from:

- High Digital Signal Processors (DSPs), which vie with host CPUs over which is the real the workhorse CT systems. DSPs are essentially processors that do multiplication extremely well! They are used to provide call progress detection, speech recognition and their increasing speed leads to greater numbers of ports on each CT resource card (greater "density").
- The CT buses are the internal bus between CT resource cards that distributes signals so the various CT resources can work together under the auspices of a software application. Restricted until recently mainly to the interconnection of boards within a PC, buses such as H.100 and H.110 will in the future interface directly to all kinds of switching equipment.
- Fault resilient computers (such as CompactPCI machines) that have redundant, load sharing components which are "hot swappable" (can be replaced while a system is running).
- Industry Standards such as SCSA, MVIP, H.100, H.110, S.100, TAPI, TSAPI, CSTA, etc.
- PBXs providing a direct or OAI connection to their PBX.
- The rise of Microsoft Windows as the dominant operating system for desktop PCs. This brought about some important new universally recognized facilities for computer telephony, including:
  - Microsoft Exchange, which offers users a single view on different forms of messaging information and hence provides a unified mailbox for both EMail and Voicemail.
  - Messaging API (MAPI) that provides client messaging applications with a standard interface to different "service providers" and hence can access MAPI compliant message stores from different vendors.
  - Telephony API (TAPI) the standard interface providing desktop PCs and CT systems with first and third party call control.
- Server Computing from Unix, Windows NT, Windows 2000 and now Linux. The greatest innovation here that helped multi-line, high density CT was the development of multitasking and multithreading.
- Microsoft's Back Office strategy wherein a range of servers provide different services to an enterprise. Of these, the Exchange Server provides information sharing and messaging functions which are most important to CT applications.

On the next rung up on the ladder, computer telephony technology at its basest level (CT resource boards, drivers, low level software) provides hundreds of discrete and interesting functions that can be manipulated by developers. These functions can be divided between call control and media processing, functions such as the following:

- Read text stored in an ASCII file on a computer to a caller using synthesized speech (text to speech).
- Play digitized speech over the phone to a caller.
- Collect statistical information about incoming and outgoing calls such as the duration of the call, date and time the call was placed, called or calling number, etc.
- Detect a busy tone or special information tones, such as those that indicate an invalid number was dialed, or there was a problem completing the call.
- Detect and answer an incoming call.
- Detect and register Dual Tone Multi-Frequency (DTMF) keys ("touchtones") pressed by a caller.
- Detect that a call has been answered and a connection established
- Detect that a call has been successfully placed and that the phone is ringing, or conversely, detect that it is not ringing.
- Detect that a circuit has been established with a dial tone.
- Detect that the call has not been answered after some predetermined amount of time.
- Detect that the call was answered by either a fax machine, a human, a modem, or even an answering machine.
- Establish a conference call by connecting three or more parties on the same call.
- Hang up a call once it is completed.
- Identify callers by communicating with the telco's central office (Telco CO) using Caller-ID or Automatic Number Identification (AIN)
- Obtain a dial tone and place a call.
- Place a call automatically and transfer the call to a live operator if it is answered.
- Place calls in a queue for holding while waiting for a person or resource to become available; while in the queue, the caller can listen to music or other pre-recorded information.
- Provide information during an active call, via tones, that an individual has an incoming call.
- Receive and store fax transmissions in a computer file.
- Route an incoming call for a given extension to another phone number
- Route incoming calls using Direct Inward Dialing (DID) information provided by the Telco CO.
- Transmit information stored in a computer file using fax technology via any number of methods of initiating the transfer.

These little components are then combined into larger collections of core technologies that ultimately work together as full-blown applications. Such telephony-enabled business applications are what are often mistakenly referred to under the umbrella term of computer telephony. A few of these are as follows:

- Automated Attendants which are essentially computers that answer a call, greet the caller and can speak pre-recorded informational messages (Audiotex), or prompt the caller through a series of menu selections (in the manner of IVR) before providing information retrieval via spoken words or fax transmissions, the routing of calls, or interfacing to voice mail, etc.
- Automated Forms Processing Systems and Services, where fax forms

scanned to a company can be scanned to extract information (for membership updates, grass roots campaigns, attendee signup for conventions, and industry research).

- Call Detail Reporting (CDR) which captures information about all outbound and/or inbound calls made from an organization and produces detailed call activity reports, enabling companies to reduce unauthorized phone use and help in selecting the least expensive long distance service. They can also help professionals that bill by the hour to accurately bill their clients for phone time.

- Call Center, Help Desk applications allows more efficient use of staff by pre-processing using an automated attendant that can offer to provide answers to common questions while the caller is in a queue, or automated call distribution (ACD) that routes calls to available operators, providing the operator with "screen pops" of information regarding the caller, or call routing based on a caller's response to prompts, maintaining holding queues, providing music or information while on hold, etc.

- Fax Server technology allows employees to transmit and receive faxes right from their desktop computers, performed in a client-server configuration that is plugged into the public network. Incoming faxes can be routed from the computer telephony system to the end user's PC using various methods (DID, OCR, etc.); faxes can also be forwarded to other individuals.

- Fax-on-Demand (FOD), also allows callers to dial in to an IVR-type system and listen to a menu telling them which documents are available by fax, and which index numbers are associated with them. The caller can make their selections by pressing touch-tone keys or by speaking the document number. At some point during the session, the caller is prompted to enter the phone number of the fax machine to receive the document. A related form of fax system, Interactive Fax Response (IFR), allows individuals to automatically receive a fax in response to a transaction performed through either the telephone or a computer. For example, a customer may receive a fax showing their charges or account balance after making a purchase.

- Interactive Voice Response (IVR), in automated attendants, is the voice that says "Press 1 for Sales, 2 for Customer Service etc." and then listens for the digit you press. IVR systems are typically used for information access over the telephone and provide a very simple front end to otherwise sophisticated databases. In its latest form it can listen to a caller's voice, do speech recognition, then perform database lookups and read off the results using synthesized speech for things like bank account balance inquiries or stock quotes. IVR systems are widely used.

- International Callback Services, also associated with "telecom resale" involve reselling lower rate "outgoing" long distance line charges to give the subscriber more economical telephone access then can be had via a direct call placed from the international location. The call is re-originated by the system from a high tariff country to a low tariff country thus enabling lower cost international calls for subscribers. These systems can be configured as either callback systems where the caller rings a number in the low tariff country which detects and rings back the caller or, in call-through where the system acts as a telecom gateway to connect callers from the high tariff country to the low where a call is then placed.

- Paging systems typically have voice prompts that guide callers through leaving their phone number and/or message, then pass this information to a radio transmission system to contact a person's pager.

- Telemarketing applications place calls from a database of numbers,

determines whether or not the call successfully completed, and then switches the call to an agent only if the call is answered by a person and not an answering machine. This aspect of the automation is called predictive dialing and automated dialing.

- Unified Messaging allows users to view and access messages of different formats and media, usually the favorite triumvirate of e-mail, voice mail, and faxes, all from a single location and a single interface (usually a computer running GUI-based software, though PDA devices are becoming more and more telephony and Internet-enabled). If the user is away from their desk, they can still call into the the system and have their messages read to them using text-to-speech. A subset of this technology is a system then converts e-mail to speech using text-to-speech technology and reads it to the user over the phone.

- Voice and Videoconferencing systems – These now use CT Resource boards, and computer telephony systems can be used to connect the conferencing parties.

- Voice Mail – The first and most popular form of CT, these started out as adjuncts to PBXs and are now often combined with IVR and other CT systems to provide telephone users with the ability to record, store, and manipulate spoken messages.

- Web-enabled call center applications, where call center agents can synchronize their web browser with that of the prospective customer so that while the agent talks or communicates in some way with the customer in real time, both can surf a company's website together. This is also known as "pushing pages" or "symmetrical surfing." The agent can even fill out a web form in conjunction with the customer.

Of course, you may prefer on vendor's fax server and another's voice mail system. How do you get them all to work together in a single system? This is where Intel / Dialogic's CT Media comes in, a middleware program for driving ECTF S.100-based applications and the SCSA multi-resource platform. (See CT Media, S.100, SCSA, H.100)

At *Computer Telephony* magazine and CT Expo, we've found that these applications and technologies naturally fall into six higher-level functional groupings, which we use to promote the industry:

1. Messaging. Voice, fax and electronic mail, fax blasters, fax servers and fax routers, paging and unified messaging (also called integrated messaging) and Internet Web-vectored phones, fax and video messaging.

2. Real-time Connectivity. Inbound and outbound call handling, "predictive" and "preview" dialing, automated attendants, LAN / screen-based call routing, one number calling / "follow me" numbers, video, audio and text-based conferencing, "PBX in a PC," collaborative computing.

3. Transaction Processing and Information Access via the Phone. Interactive voice response, audiotex, customer access to enterprise data, "giving data a voice," fax on demand and shopping on the World Wide Web, the inside-out corporate store, self-navigation over the phone.

4. Adding Intelligence to Phone Calls. Screen pops of customer records coincident with inbound and outbound phone calls, mirrored web page "pops," smart agents, skills-based call routing, virtual (geographically distributed) call centers, the "intelligent" call center, computer telephony groupware, smart help desks and AIN network-based computer telephony services.

5. Core Technologies. Voice recognition, text-to-speech, digital signal processing, application generators (of all varieties – GUI to forms-based to script-based), VoiceView, DSVD, computer-based fax routing /binary file transfer, USB (Universal Serial Bus), GeoPort, video and audio compression, call progress, dial pulse recognition, Caller ID and ANI, digital network interfaces (T-1, E-1, ISDN BRI and PRI, SS7, frame relay and ATM), voice modems, client-server telephony, logical modem interfaces,

multi-PC telephony synchronization and coordination software, the communicating PC, the Internet, the Web and the "Intranet." (IP, RSVP, and a multitude of new protocols)

6. Core Standards. The ITU-T's T.120 (document conferencing) and H.320 (video conferencing), Microsoft's TAPI – an integral part of Windows 98, NT and Windows 2000, Novell's TSAPI – a phone switch control NLM under NetWare and Sun's JTAPI. Intel's USB and InstantON. Natural MicroSystems / Mitel's MVIP and H-MVIP, Dialogic's SCSA, DM3 and the ECTF S.100 API.

Above these levels one finds Customer Relationship Management (CRM) and Enterprise Relationship Management (ERM) systems that deal with supply-chains and extremely high-level architectonic systems, where even Computer Telephony finally blends in and gets lost in the big picture.

Industry Challenges for VARs, Developers and End-Users.

For a CT developer, this technology only yields to people who understand computers, telecom and networking, people who are rarities. Just testing these applications for the international market alone is a huge nut to crack (See International CT Development).

And for the unfortunate souls who are stuck in the "channel" and who must promote, sell, resell, and otherwise work with computer telephony products and services, CT is still something of a puzzlement. The technology is complex (but somehow manages to change rapidly), expertise is low, choices are many, and confusion abounds.

Concerns thus run rampant. These are concerns that one finds in any industry – only more so in the case of computer telephony:

• Picking the right technology.

• Re-training.

• Competing standards.

• Partnering with right vendor.

• Managing channel conflicts.

• Putting convergence to work.

• Delivering better customer care through technology.

• Understanding market trends and how they will change.

• Integrating voice/data technologies and the corporate organization.

• Overcoming confusion on product/standards choices – Some of the confusion that comes out of the channel is a result of people using the terms for the technology standards, the capabilities and the products interchangeably. When sales people ask befuddled customers whether they are interested in TAPI, TSAPI or CTSA systems, one gets a lot of blank looks and apprehension. Solving business problems should take precedence over industry buzz words, trade names and acronyms.

Some of the more amusing phone calls we always get at *Computer Telephony* magazine are from people who have heard the good news about CT but don't know "where to buy it." In the old days many people thought they could build a CT system themselves – they could just buy an app-gen, a PC, a couple of Dialogic or NMS boards, and run a T-1 into their business. This "do-it-yourself" method usually leads to phone calls screaming for help. These days the tools are more powerful, but deploying large systems requires outside help.

The question then arises – who actually "sells" computer telephony? Since we've established that CT is an abstract term encompassing a number of things, it's not too surprising various parts of the channel are vying to get your business:

• **Switch Vendors.** Companies such as Nortel and Comdial realized that CT was not just a cheap competitor to their PBXs as touted in the press ("the death of the PBX!") but a value-add they could sell their customers who would naturally be interested in a product emblazoned with a brand name they know and trust. Nortel, for example, could now tack a PC box called Call Pilot onto their Meridian product line and pick up an easy $20,000 or so. Comdial even tried a "one stop shopping" approach to buying and maintaining CT-enhanced phone systems, and has built some interesting PBX-PC hybrids.

• **Switch and Computer Distributors.** Many companies don't' buy directly from vendors, but from distributors. Some of the larger ones have integration departments that tackle CT, or can refer you to somebody who can do a particular job for you.

• **Interconnect Firms.** Interconnect firms take more of a total system approach, and often deal with customers who are interested in more than just buying a PBX. Some interconnects have plunged wholeheartedly into CT.

• **LAN Integrators.** Until recently these companies knew everything about computer data networks, but not much about telecom. Now that CT is moving toward packet switched technologies such as IP, LAN integrators are taking CT more seriously and have even made efforts to understand the circuit-switched side of the CT equation.

• **Telecommunications Carriers.** These companies had the opposite problem of the LAN integrators – they know everything about telecom but are only just now becoming acquainted with computers and data networking. Some of the long distance carriers offer integration services, but some are simply offering CT functionality as an enhanced service (See Enhanced Services).

• **Consultants.** Always get a referral! With the rise of CT, many computer and telecom consultants suddenly changed their business cards to grab new business without quite realizing what they were getting into.

The Future of Computer Telephony.

Paradigm shifts can be excruciating. An infusion of IP technology into computer telephony has caused an uproar. Vendors are going to have to reinvent themselves or disappear. There are any number of directions in which the field can develop, and at the moment some large companies with deep pockets are following all directions simultaneously!

Solid state telephony appliances will gain much ground, become more powerful, moving up from their current SOHO deployments. Even KSU-less phone systems, such as the Casio PhoneMate, already have voice mail and auto attendant capability built into each phone! Add a small router-like box to serve as an IP gateway, and you're off and running with very little expense and no PC to worry about.

In the midst of this brouhaha, there's a wonderful new product category that's appearing, but none of us at *Computer Telephony* magazine know what to call it. Super app-gen? Enterprise communications server? CT development environment? The intelligent internal cloud? The "device"? (nicknamed with apologies to the Manhattan Project).

This brand new class of product forming out there is an amalgam of app-gens that got too sophisticated to be called app-gens anymore, PC-based phone systems for businesses and call centers, and IP gateways. Some of these products are software-centric, others hardware-centric, but many are right smack in the middle – turnkey PC-based systems that nevertheless have great flexibility for developers and those who would customize the system for a particular business.

They can be sold as standalone devices or used in conjunction with a legacy PBX, and some kind of app-gen software is used to configure and / or develop applications on the platform. The more adventurous developers may prefer to deal with a software-centric super app-gen and formulate their own underlying hardware solution, but the end result is usually the same.

As for the software side of the equation, the more sophisticated products from such companies as Envox, Mediasoft and Prima haven't been considered as mere "IVR application generators" in ages. True, they incorporate a range of developer tools, as well as APIs. But the underlying platform – the modules, interfaces, and facilities – are no longer limited to conventional voice-response, but provide comprehensive mechanisms for switch management and call processing, messaging, call distribution, outbound predictive dialing, electronic commerce, and other applications.

Whereas an IVR application generator of the conventional sort might be used to create and deploy a single, relatively simple, transaction-processing application such as a debit-card application, until recently you'd never use such a platform to develop (for example) an entire call center. In theory, perhaps, the facilities are there – but the developer had no assurance that they could be deployed in concert, at usable scales, with acceptable reliability.

At least not until Intel entered onto the scene. In 1999 Intel startled the industry by first buying Dialogic, then Parity Software, then making investments in Mediasoft Telecom and Prima.

What's going on here?

People look upon Intel as a chip business. Certainly Intel has reached the top of the heap in that industry, and they've searched for the next mountain to climb. What they found is an increasing form of networking "connectedness" involving both the voice and data networks, and so they decided to take charge of the situation. They want to tie PCs together with more common functionally in a manner similar to the way the phone system works.

Springing into action, Intel immediately did a number of things to make known that they are the premier component supplier to the industry. For example, they bought Dialogic and two days later they announced an entirely new business unit called the Communications Products Group. Dialogic is now part of this substantial-sized group along with a number of other pre-existing business units within Intel that had been exploring various aspects of networking, "connectedness," IP, Internet, and what-not.

A key part of Intel's strategy is Dialogic's role in the Communications Product Group. Obviously there are a lot of boundaries between traditional telephony and data networking, and Dialogic's expertise has the most impact at that boundary.

But why is Intel collecting app-gens, you ask?

The real purpose wasn't to simply acquire application generators. But one of the things that Intel and Dialogic have concluded is that the CT marketplace growth is being held back because there's no consistent view of CT resources. Now, that's not the same as saying there are too many APIs.

If you think about how database technology evolved, for example, what really got the database marketplace going was that everybody agreed on a common model for relational databases. Databases have tables, tables have rows, rows have data elements, you can do things such as "joins," etc. The now-recognized model for the database resources is what really allowed the marketplace to take off. It wasn't so much the fact that Microsoft had invented ODBC, it was the fact that when you do use ODBC there is a common conception behind it no matter if it's Oracle or Sybase or Microsoft SQL Server.

Now, we don't have anything like this kind of model for CT resources yet. Ah, you might ask, what about CT Media? CT Media is the platform or foundation upon which you can start to implement what we're looking for, but you really have to go a level of abstraction above CT media to see what the application sees. Consider the huge number of application generators out there and the way the resulting applications make requests to, say, an ActiveX control. That's happening at a different level than what CT Media can handle.

Plug any particular ActiveX control into your application to take advantage of its functionality and you ask yourself "what can this app do now?" as you check the control's specifications. The application may now be able to play messages, record messages, listen for DTMF, etc., based upon whatever the functions are in the ActiveX control, which is of course very helpful to the programmer. But the vendors who created each of these different app-gen / toolkits had a different idea of the structure of the resources being manipulated by the code written by the application programmer. They're all "kind of" the same, but they're also different enough so that they cause problems if you shift from one development environment to another.

So what Intel / Dialogic essentially has now said is: Okay, the world needs a lot of toolkits just like the world needs a lot of ways of looking into databases. Think of the database query engines that are out there, everything from complex packages all the way down to Microsoft Access. But although there are many query engines, there's only one database model, and one SQL language. Thus, Intel / Dialogic is attempting to bring the same consensus to the application generator business. Despite the fact that Intel / Dialogic bought Parity Software, licensed Visual Voice from Artisoft, and has investments in Mediasoft Telecom and Prima, they don't really want to corner the market in app-gens. They don't want to put everybody out of business and have one Intel / Dialogic app-gen, because there are too many reasons to have different flavors and user interfaces for different markets. But what they do want is for all of these app-gens to be based on a common model of computer telephony resources so that, just as database managers can look at the same database with three different query engines, so CT developers should be able to look at the same CT resources with three different app-gens. It's what Intel / Dialogic calls an "object infrastructure".

What Intel / Dialogic wishes to do is influence a whole range of app-gens and toolkits so they adopt a common object infrastructure definition of what these resources look and feel and act like. In order to persuade all of these different app-gen makers to move in that one direction, they had to use some different commercial enticements. Intel / Dialogic acquired Parity because Parity had already been doing a lot of technical work on exactly the kind of an object infrastructure they were looking for, which Parity had called their Topaz Project. Parity had developed it because they were trying to get the various versions of their own products (VOS 7, CallSuite 7, etc.) to converge on a single architecture. Topaz would offer "API transparency" to developers using all of Parity's tools. Thus, Topaz was to shield VOS / CallSuite developers from some sticky technical issues and questions and the underlying programming interfaces, including Dialogic's SR4, ECTF S.100 (including Dialogic's CT Media), TAPI, the native DM3 API (MNTI) and more. In other words, the ultimate realization of the Topaz concept was that a VOS application created today on Dialogic's SR4 API would run tomorrow on TAPI or CT Media.

Parity had done so much work on this Topaz concept that Intel / Dialogic decided to make it part of what they were doing – not necessary the Parity toolkits themselves, but the work that they had done to pull those things together architecturally, something in which Intel /

Dialogic saw great value. So Intel / Dialogic acquired Parity to get their intellectual property, skills and a staff that knew what they were doing in terms of creating a common object infrastructure.

Intel / Dialogic then looked at Artisoft, and struck a rather complex commercial deal with them, the principal element of which ensured that the Visual Voice app-gen adopted the Parity-originated object infrastructure.

As for cajoling fellow app-gen makers Mediasoft Telecom and Prima to get on the bandwagon, it was a simple arrangement where Intel/ Dialogic was able to make some equity investments in them and as part of those investments they agreed to do certain things, such as move their kits over onto platforms such as CT Media and to adopt the object infrastructure architecture that's being worked out by Intel / Dialogic.

For example, "Blabla" the scripting language underlying Mediasoft's IVS app-gen, will now operate on top of CT Media, so that a programmer using Blabla will be manipulating objects that look and feel like the same objects being manipulated by another programmer using a different app-gen such as Visual Voice or Prima's Maestro.

Whether or not taking advantage of this object infrastructure will demand that application programmers compile their programs with a common library is as yet unknown, but the goal is for programmers using Mediasoft's Blabla or Artisoft's Visual Voice or Parity's toolkits to all work with the same basic repertoire of objects and actions. Again, it's just like working on an SQL database – no matter what query language you use, you feel as if you're using the same database, objects and actions. It's a more pervasive, all-encompassing idea than just "let's all agree on an API."

This all sounds like a media-centric version of what TAPI does, but it actually deals with the meaning of the resource – the semantics rather than the syntax.

In fact, in promulgating this new object infrastructure, Intel / Dialogic demonstrated that it had learned its lessons from supporting TAPI, which was a case where Microsoft had published an elegant API, but they had forgotten to define the behavior of the resources behind the API. TAPI would let you get at and program some of the functionality of many telephony switches through the same TAPI interface, but when you specified in your application a command such as "transfer a call" you might not get the same result from each switch.

As Intel / Dialogic's Carl Strathmeyer says: "Just because you see the word 'hamburger' on a McDonald's menu and 'hamburger' on a Ritz-Carlton menu doesn't mean that you're going to get the same meal when you send in your request for a hamburger. Just agreeing on the actual word 'hamburger' doesn't do the trick. This was the problem with TAPI – it tried to address the syntax, the specific representation, without defining the words. With TAPI you could construct sentences with the telephony equivalent of the word hamburger, but no one knew what kind of hamburger they were talking about. First you have to agree as to what goes on in the kitchen! And once you do that, APIs aren't a problem anymore. I can go up to Quebec and ask for a McChicken sandwich or a McPoulet se vou plait, and I get the same sandwich. That's what we want. APIs aren't the enemy, the object model is the enemy."

So it looks as if the era of "McTelephony" is upon us – billions of Voice Bricks sold! (just kidding).

Incidentally, the Intel / Dialogic investments in Mediasoft and Prima came from the Intel Communications Fund, which is a venture capital fund set up in the summer of 1999 to invest in privately held companies that develop products and are doing important things in areas where Intel wishes to unify and steer the market. Mediasoft and Prima are not the only companies that have been talking to the Communications Fund

or who have made proposals to the fund. The fund is still entertaining more proposals from several app-gen makers.

When I heard about this "object infrastructure" plan for the first time I became fearful that the Intel / Dialogic object infrastructure might have a backlash on the CT Resource board manufacturers, who would feel it necessary to re-engineer their products so as to be more efficient when working with the new universal CT model. This would eventually lead to a situation where everyone would be making look-alike DM3 boards!

Intel / Dialogic, however, assured me that the object infrastructure won't have a major impact on hardware. Instead, there should be more of an impact on the drivers and the middleware / software levels, since the "objects" we're talking about here are "higher level" functional objects that an applications sees, such as "play a voice prompt". Most CT hardware is already capable of doing this, the more important question is how you ask for this event in your application. Intel / Dialogic wants to change not the function itself, but the way programmers "ask" for a function. They want to create a common model of "asking".

The object infrastructure concept will come in handy as we move to the next generation of CT software and hardware. As time goes on, CT systems and the well-behaved app-gens that control them will be made increasingly attractive to the less skilled staffs of small and medium-sized businesses, or perhaps even not-so-skilled would-be developers. As a result, you'll be seeing more and more super installation and configuration wizards that will "size up" what kind of hardware configuration you own (the whole network, not just the server) and enable certain features of the software that dovetail with certain hardware (a particular CT resource board or an IP gateway, for example). The GUI app-gen interface pioneered years ago will become standard for these systems and their operating environment will move over to the browser, since many systems even now can be administered via the web.

Indeed, many CT applications will be running on web servers and browsers, keeping in accordance with some of the doctrines of "thin client" techniques.

When, in December 1999, Dialogic announced version 4.0 of CT Connect (their CT integration software that allows developers to create applications that can be used, unmodified, on any of the CT Connect supported telephone systems) it was immediately apparent that the overall architecture of CT Connect 4.0 was similar to 3.0, but Java support had been added. Not the Java Telephony API (JTAPI), but straight Java. Dialogic took the native CT Connect API (which would be normally called from C or C++ or Visual Basic) and constructed a native Java equivalent of that same range of functions. So rather than inventing a new API, they just translated the existing CT Connect native API into the Java world in a native way by putting the Sun Java Virtual Machine (JVM) onto their CT Server,

Because of the way Java itself is constructed, just the fact that that JVM resource is sitting on the server means that you can get to it from any Java client. In fact, you don't need to install anything on the client end, because the client end is all standard Java. You can connect to the server via Java's built-in client /server mechanism, RMI. Thus, a way has been provided for developers wanting to write revolutionary platform-independent computer telephony applications that run from a web browser to now have a simple, direct, "native-feeling" way to get the CT features they need from CT Connect.

It all sounds quite splendid, but I can't help but think that the computer telephony industry we all know and love will evolve into something unrecognizable by today's sensibilties.

Once upon a time, back in the days of the purely circuit-switched world, it was a great achievement just to get a computer telephony system working at all. Does this fax server work with this CT server? Can you integrate so-and-so's intelligent e-mail package with voice mail? And on and on. Things limped along this way until the coming of IP, which made unified messaging and other long dreamt-of applications smoother and more reliable.

The next logical progression is that computer telephony blends into larger, more sophisticated customer relationship management systems and incredibly sophisticated e-commerce / supply chain systems. We're just beginning to see this happening on a large scale, and, ultimately, such a flexible and open communications system, almost as an afterthought, represents the truly "converged" voice and IP environment long foretold by the ancients (or at least Harry Newton).

The external "cloud" seeps into the office and becomes a sort of intelligent "internal cloud" that interacts over the "network cloud" with other "intelligent clouds" – consumers and businesses of various sizes. These interactions will be semi-automated (particularly in the case of B-to-B) and will rely heavily upon XML tags processing, even if it occurs in parallel with voice interactions or in the background.

Developing applications – indeed, whole systems-within-systems – in such a brave new world requires tools of tremendous flexibility and power. We marvel at the variety of these products as they now begin to appear. But we don't know exactly what to call them, since in order to define something you have to demarcate its working boundaries, and it's not readily apparent where you can draw the line over seamless, transparent technology that desperately seeks to blend in and get lost in the telecom machinery. How do you draw a line across a cloud?

So, any way you look at it, the future of CT is – dare I say it – kind of cloudy.

In the meantime, however, let's take a look at some of the more advanced products in this area that continue to gain momentum...

TeleVantage from Artisoft (Cambridge, MA – 617 354 0600, www.artisoft.com) is very close to being a classic example of the new category of product we've been about, particularly since we get the feeling that Artisoft may one day replace TeleVantage's existing GUI configuration software with a version of Visual Voice Pro, their famous applications generator. Even today you can use Artisoft's Visual Voice Pro to develop full-blown IVR applications and dedicate some TeleVantage hardware resources to run your IVR applications, thus integrating them with your TeleVantage server.

As it is, TeleVantage is impressive enough, a PC-based phone system for small and medium sized businesses that can work with legacy PBXs and includes an ACD along with capabilities for DID / DNIS, personal contact manager software integration, connectivity to pagers, least cost routing, intelligent multiline call control, personalized greetings and third-party call handling.

When building call center applications, you can choose how calls are distributed to agents, including choices to ring all agent phones simultaneously, or distribute calls via round robin or top-down patterns. When an agent becomes available, their phone automatically rings with the longest waiting call. You can even select a group of phones to ring simultaneously, a feature that had been requested from key system users who wanted to have multiple people handling phones as they ring.

A TeleVantage chassis can scale up to 48 trunk lines and 144 extensions. It's a flexible, upgradable software solution that operates under Windows NT and uses off-the-shelf computer, wiring and Dialogic telephony hardware, allowing it to support two digital T-1 lines as well as analog lines. When working with networked desktop PCs, TeleVantage

lets you see and hear voice and other messages, graphically manage your calls and integrate phone and e-mail messaging.

The OTS NT (Object Telephony Server) from Buffalo International (Valhalla, NY – 914-747-8500, www.opencti.com) is a modular, digital telephony platform housed in a fault resilient PC containing industry standard Dialogic voice / call processing boards. This Windows NT-based platform supports thousands of ports across multiple hot swappable hardware modules. It plugs right into your existing network, and there's no need to replace your data management software or agent workstations.

At the heart of the system is a software layer that directly controls the telephony hardware and communications with computer network applications. This layer gives you predictive dialing, inbound/outbound blending, ACD, IVR, ICR, monitoring, coaching and recording, all on one PC platform.

Buffalo's OTS NT differs from competing commservers in the sense that OTS is open and designed specifically for easy integration with any third party application. There is no proprietary user interface needed. The open architecture enables you to build or integrate call center applications using any programming environment. A variety of turnkey solutions are also available. OTS NT can standalone or work with your existing switch. As a standalone, OTS NT can support from four workstations up to 144 workstations (432 ports) on one server. OTS units can be networked through IP telephony.

New features include supporting multiple vendor apps on one system, and IP telephony and distributed call centers. Buffalo's systems are used in telesales, collections, order entry, appointment scheduling, help desks and skill based routing.

OTS NT can be controlled via a command set accessible from any computer platform via TCP/IP. The control app which performs all third party management and call control activities, is not part of OTS NT but is necessary to give commands to and await messages from OTS NT. Each station runs an agent application telling OTS NT to perform immediate call handling functions such as picking up a call, holding, conferencing and hanging up. The app usually displays information regarding the party on the phone or general campaign (call center group project) information. By allowing you to setup any control and agent application, you can reach a level of flexibility not found in proprietary solutions.

The system's Intelligent Call Transfers allows relevant data to be transferred along with the actual call. This data is transferred to the receiving user's application to do context sensitive screen pops or for other intelligent call processing purposes. The "data area" for each call is 255 bytes and can be modified at any time. Information such as which IVR options have been selected, or the caller's account number, can be added to the call data and used at later steps in the call. By default, outbound calls include dialing/data records supplied by the control application and inbound calls contain DNIS and ANI, a feature that comes in handy if you have multiple applications running on different computing platforms or if incompatible databases are used.

An average system costs about $1,000 per seat and up.

OTS lets developers add sophisticated call handling routines to your OTS NT system by using Visual Basic (or other compatible programming language) to create an ActiveX control. Each routine you develop or license is called an OTS Extension. Extensions may be invoked via a call to Buffalo's API or can be initiated by each other.

To create your own client-side software (particularly for call center applications involving Buffalo's predictive dialer), use Buffalo's Developer's Kit. The kit provides a simulated calling environment (outbound only), and comes with full documentation of Buffalo's communication protocol, as well as source code examples and communication tools to reduce development time for various programming environments.

Since the program only simulates outbound campaigns, to test the IVR and ACD functions, Buffalo provides a full-featured license for a small (four stations and lines) runtime system. At least one D41ESC card is needed for IVR and one MSI-SC card for ACD/Call Blending.

The Developer's Kit license is $299. Once you've finished creating your application, you can still use the kit, this time as a training tool as you continue to run your application within the simulated environment.

The DSP2000 is a Private Communications Exchange (PCX) product from Cortelco Systems (Memphis, TN – 901-365-7774, www.cortelco.com). Running Linux, the DSP2000 provides advanced PBX functionality, ACD capabilities, has an open architecture, and is highly scalable, from 48 to 2000 ports, providing both circuit-switched and VoIP capabilities. Also, since the enterprise comm server is a mission critical component of any organization, redundancy is a recommended option.

The system is based primarily on two components: The card chassis and the controller. Information is exchanged between these components via a high-speed, 100MB/s, TCP/IP Ethernet LAN. The card chassis can hold high-density line cards, which will scale up as needed. For example, dual T-1 boards provide 48 ports of T-1 connectivity in a single slot.

The Controller is an industrial-grade PC with a native 10/100 BaseT Ethernet connection to the outside world. This Ethernet connection provides support for networked items such as printers, monitors, system administration tools, IVR, voice mail / unified messaging, and CT links like TAPI, TSAPI, Call Path, and CSTA. The DSP2000 is powered by an Intel-based 333MHz processor and 64MB RAM to provide the "horsepower" needed in high volume locations for simultaneous internal call processing and external CT.

The system's optimized Linux kernel supports voice processing, as opposed to NT-based systems whose source is general and multi-purpose.

System Administrators can use the DSP2000 GUI tool to configure the system. This tool is web-based, so any administrator with access to a web browser and the appropriate security clearances can perform routine system administration from anywhere on the Web. This includes basic moves, adds, changes, modification of call routing plans, and programming of a user's digital phone. System hardware can even be configured remotely.

So that enterprise users won't miss their sophisticated phones, the DSP2000 supports Cortelco Systems digital phones, including 6, 12, 18, 30, and 38 button models, as well as analog phones. They even have plans for future support for IP phones. A PC-based softphone is available to provide telephony features such as speed dialing and conferencing from a user's computer.

The DSP2000's call routing capability is dynamic and can be based on real-time call conditions. The system supports up to 32,000 call routing plans, each with up to 64 steps. Other call routing capabilities such as time of day / day of week, holiday, after hours, emergency, routing across networked DSP2000s, and skills-based routing are supported. The call routing capability is independent of physical call type and can include traditional and VoIP calls. Numerous ACD options are available, including simultaneous queuing, agent call back, and support for home agents.

Future planned enhancements to the product include the develop-

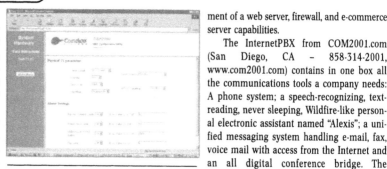

Cortelco's DSP2000 supports up to 32,000 call routing plans.

ment of a web server, firewall, and e-commerce server capabilities.

The InternetPBX from COM2001.com (San Diego, CA – 858-314-2001, www.com2001.com) contains in one box all the communications tools a company needs: A phone system; a speech-recognizing, text-reading, never sleeping, Wildfire-like personal electronic assistant named "Alexis"; a unified messaging system handling e-mail, fax, voice mail with access from the Internet and an all digital conference bridge. The InternetPBX can serve an entire company, or it can sit behind a legacy PBX and serve a key department inside a corporation, such as sales.

The InternetPBX comes in two different base configurations: The ECS-720/D (32 lines x 40 extensions) and the ECS-1440/D (56 lines x 88 extensions). Both boxes are expandable to 144 ports.

And highly fault resilient boxes they are. A Dell PowerEdge High Availability Telephony Server serves as the foundation for the InternetPBX system, which holds four 500 MHz Pentium IIIs and 1 GB of EDO RAM (expandable to 4 GB). RAID Level 1 is standard, Level 5 is optional.

Arranging a conference call is simplified, with the InternetPBX Toolbar displaying all the participants and their contact information from Outlook. For example, up to 64 people calling through the InternetPBX's digitally-balanced conference bridge can listen and talk to someone who's delivering a PowerPoint presentation using Microsoft's standards-based NetMeeting IP telephony app.

For travelers, the InternetPBX should be a joy. When you arrive at your temporary office (a client's office or home, the airport, etc.) you either dial into or connect by broadband to the Internet. The instant you log on to your corporate InternetPBX, you see an Outlook screen that shows all your voice mail, your e-mail and your fax mail. You can also view your calendar, your appointments, your contacts and your tasks to do. It's all done without any special software, just using a standard browser.

And when your phone rings in your empty, distant office, the InternetPBX pops a message across the Internet onto your PC, asking you if you want the call transferred to you, sent to voice mail, or transferred to someone else. You make your choice, and a landline or cell phone next to you rings. You're never out of touch, and your best customer never gets dumped into voice mail jail again.

The InternetPBX is built on all of the open telephony standards from Microsoft, Intel, the ITU, the Internet Engineering Taskforce (IETF) and others. This allows others (such as resellers) to build on top of COM2001.com's work and add their own value, resulting in the kind of custom-tailored phone system previously restricted to a handful of very large companies.

The InternetPBX GUI interface lets you quickly make all sorts of administrative changes (such as adding a new employee or a new mail box for a customer) from anywhere, via the Internet.

The InternetPBX Auto Attendant is a flexible IVR System, which needs no programming or IVR toolkit. In minutes an administrator can compile a script which includes state-of-the-art speech recognition (as an alternative to DTMF entry) and text-to-speech (as an alternative to recorded sound prompts). The Advanced Tab allows for automatic loading of scripts based on a Time of Day-Day of Week-Day of Month / Year method for all incoming lines or a selective group of incoming lines.

After years of research and employing an object-oriented menu tree design for unlimited levels of IVR navigation, the InternetPBX delivers a friendly and fully configurable IVR system that can be modified remotely via RAS or the Internet.

A 19" rackmount version of the InternetPBX is being manufactured and assembled for COM2001.com by Dell Computer (which now owns a minority stake in COM2001.com). This larger system can now accommodate 10 InternetPBX ECS-1440/D Servers that can handle a total of 1,440 ports with up to 20- PRI circuits (10 dual spans) and 960 extensions per rack. COM2001.com will ship systems using Pika technology, but plans to roll out a system using CTMedia version 2.1 and Intel / Dialogic's new PCI boards.

Easyphone from Altitude Software (Milpitas, CA – 408-965-1700, www.altitudesoftware.com), the archetypal future phone system for call centers (Altitude calls them contact centers) is going through a time of transition. They're Web-enhancing their Easyphone product (including Voice over IP) and expanding its capabilities, getting ready for an IPO, and are changing the name of the company to Altitude Software, all over the next several months.

Easyphone is known as a product for those wanting to build call centers. The version that will be announced around the time you read this will be even more flexible, allowing for web-enabled call centers and distributed business phone systems.

Easyphone has decided that their GUI app gen-like configuration and scripting software should be able to work with switches as well as

PC-based phone systems and routers. In order for such long range planning to be successful, extreme product flexibility has to be maintained. The best way to do this turned out to be the creation of an abstract, a call processing model that treats everything as a "call," such as fax, e-mail, web communications as well as conventional voice calls. The switching model has been abstracted to the point where the system can be adapted to any switching mechanism, be it a PBX, commserver, CT Media, CT server, IP PBX, or whatever.

At a layer above mere call control, you find that you can integrate Easyphone with such third-party customer relationship packages such as those from Siebel/Scopus, Vantive, Clarify, etc. Developers venturing up yet another layer will find that an Easyphone system can be plugged into giant systems running sophisticated Enterprise Resource Planning (ERP) software packages, which are essentially business management software systems that integrate all facets of your business, including planning, manufacturing, sales, and marketing. This includes packages from SAP, PeopleSoft and BAAN.

Essentially, the Easyphone suite of software modules comprise a client/server solution that connects to databases from Oracle, Informix and Sybase, and supports UNIX and the Microsoft NT platforms. You program it with the EASYscript application builder, designed for those without any IT experience to quickly develop a system tailored to the workings of your own business. Also, applications can be written in any ActiveX (COM / OLE) compliant tool, making the solution even easier to integrate with commercially available front- and back-office software applications.

Easyphone's published APIs allow you to integrate existing enterprise business applications, corporate databases, and other CT technology into whatever solution you'd like to build.

Easyphone comes in both Unix and Windows NT versions. It supports a range of switches, enabling applications in your contact center to automate tasks such as placing outgoing calls and routing inbound calls. Easyphone can synchronize voice and data information, and send screen-pop caller associated information to agents' screens along with the call.

Easyphone is a single integrated family of software applications. You can select the complete, integrated solution or choose only certain modules and add on additional capabilities, with minimal system integration problems as you business needs evolve.

Easyphone can deliver the full range of its integrated functions for five-agent centers as well as 1,000 agent centers, with support for seamless CT integration, intelligent call routing, IVR, automatic dialing, and call recording.

The Envox CT Studio / Script Editor from Envox (Naples, FL – 941-793-0863, www.envox.com) is the classic case of an app-gen that has become so powerful that it transcends the term. I've had occasion to see what the Envox group has been doing, and it looks as if Envox has indeed reached that certain "critical mass" of functionality that separates the pedestrian GUI scripting from more sophisticated development concepts. The term "app gen" is just too boring to describe what Envox does. The level of flexibility, the number of ready-to-go customizable applications, and the appearance of highly skilled software wizards suggests a totally new category of software development tool.

Envox is a native Windows 32-bit software development environment that lets you develop any kind of PC based telephony system that uses Intel / Dialogic technology such as Global Call, the Dialogic MSI board and the Dialogic DM3 IPLink IP telephony board. Envox runs under Windows 95, NT 3.51 or 4.0.

The kind of applications you can build with Envox is practically limitless, ranging across such things as an unPBX, contact center, IVR system, IP telephony system, fax-on-demand, or a unified messaging platform

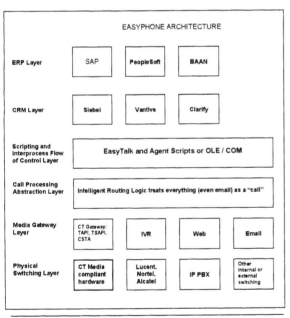

Altitude Software's Easyphone NT and UNIX contact center solution manages to stride all of our categories, thanks to its unique layered architecture. At the bottom, it can be made to work with any switching resource (PBXs, CT Media, IP PBXs, etc.). At the high end it can plug into customer relationship packages and even more gargantuan ERP software such as SAP's and PeopleSoft's.

allowing for the sending and retrieval of voice, fax, e-mail (you send mail with SMTP and retrieve it with POP3), and text-to-speech (TTS) retrieval of e-mail over the phone (any SAPI-compliant TTS engine can be used).

You can create Envox Flowchart scripts with the supplied modular software building blocks for: telephony / fax functions, speaker verification, HTML-to-fax, DHTML, ODBC database access, switching, the Microsoft Message Queue (MSMQ, which ensures all messages are prioritized and eventually reach their destination), e-mail, screen-pops, speaker verification, voice recognition (Lernout & Hauspie and VCS currently supported), text-to-speech and encryption.

Software building blocks are easiest to work with in the GUI scripting environment where you can create advanced applications without writing a single line of code. Scripts are created from graphical logical flowcharts with powerful building blocks, but functions can be called from other development environments. Programming languages such as Visual Basic, C and C++ can be incorporated into scripts using the CallDLL blocks.

For example, the connections to the phone network can be analog or digital and are handled by software building blocks such as "WaitCall" that brings the line off hook when someone calls, "ReleaseCall" that brings the line on hook when either party hangs up, and "PlayMenu," used for playing Auto Attendant messages to the caller.

Envox CT Studio now supports encryption and the provision of digital signatures using all standard cryptography engines. Envox CT studio also supports the T-Netix Speaker Verification engine that ensures secure identification using voice prints.

Just listing everything that Envox does would take an entry in itself.

Envox products are sold throughout the U.S. and Europe, but you can download the Envox Script Editor and all tools necessary to create a complete PC based switch at www.envox.com/products/download.asp.

There can be up to an unlimited number of users per system; with the actual number of ports depending on server hardware used. The scripting package is sold separately.

Icon-o-voice from Computer Telephony Systems (Melbourne, FL – 407-953-3600,) is a versatile computer telephony development environment that lets you build a wide range of applications. The advanced GUI-based application builder is based upon drawing a flow chart of the application (there are over 70 different drag and drop icons) you wish to use, or you can add on to a standard application that comes with the package.

Applications constructed using the builder include an ACD call center, an office voicemail / paging / auto-attendant system, a debit / callback system, an audiotex bulletin board system, and an Internet triggered international callback system, a conferencing platform, a mass callout system for emergency centers and special organizations, and an online text adventure game.

Icon-o-voice uses Intel / Dialogic cards, runs under Windows NT and Windows 95. Interfaces supported include analog / loop, digital T-1 / E-1 and ISDN, DID fax boards, and Dialogic integrated voice / fax boards. The system can also handle PBX integration or switching / conferencing ATM and multinode digital networks.

The system is meant for customers who want a turnkey solution with the ability to modify or add functions to their computer telephony applications. It can be used by companies of any size, including Fortune 500 companies and it's applicable to all vertical markets.

The Econ-O-Voice calling card module gives you an entire specialized system to build upon.

Icon-O-Voice has a text-to-speech capability that takes written text and reads it back to the listener. This option, together with OCR could

allow users to read faxes or e-mail, read messages to the blind, and convey database information to the user (e.g. weather forecasts, GPS positioning, phone directories, etc.) over the phone. Speech recognition is available, too.

Icon-O-Voice also gives you several fax options. You can do faxback / fax-on-demand or you can allow the caller to leave a fax in a mailbox (similar to voice mail) for retrieval at a later time. This allows the person receiving the fax to forward it to a convenient location like a hotel desk, home, office, etc. The Fax-Broadcast function lets you do a fax blast to multiple numbers.

Since most T-1 service providers will not route international calls to U.S. 800 numbers, the system can be upgraded to automatically route any U.S. 800 number to analog telephone lines.

There's even an X.25 capability, so you can tie in hundreds of callers without use of a third party or operator. The system can be used to tie in large numbers of users together from different cities or countries. Credit card interfacing permits calls to go through a packet switched data network. Significantly speeds up calling times.

The Enterprise Interaction Center (EIC) Version 1.3 from Interactive Intelligence (Indianapolis, IN – 317-872-3000, www.inter-intelli.com) is perhaps the most famous Windows NT-based "all-in-one" communications server for enterprises and call centers that can tie together multiple locations with its facilities for remote administration, network management, supervision, and reporting.

EIC's unified messaging lets you retrieve e-mail messages, voice mail messages and faxes from your desktop e-mail applications, over the Internet, or from any touchtone phone.

EIC is essentially a software-based solution running on a fault resilient Pentium server. It neatly replaces a whole slew of proprietary communication devices such as PBXs (although it can interface with legacy PBXs for departmental installations), ACDs, IVRs, voice mail systems, fax servers, Web gateways, Web callback, and CT middleware systems.

Add-on products for EIC include Interaction Web (Web chat, Web callback, and Voice over Net calls), Interaction Recorder (voice logging), Interaction Director (pre-call routing), and Interaction Dialer (predictive dialing). Two-node cluster support offers complete fault tolerance – the EIC servers are deployed in identical pairs that monitor each other. If one server fails, all operations are immediately and transparently routed to the second server. When the failed server is brought back on line, it automatically resynchronizes with the other server.

EIC supports various CO interfaces including analog, T-1, ISDN PRI, EuroISDN, and E-1.

A graphical application generator lets you customize and/or create IVR scripts, ACD rules, fax-on-demand applications, and more. EIC provides sophisticated skills-based routing, call monitoring and recording, and screen pop to applications from leading vendors including Clarify, Onyx, Silknet, Vantive, Remedy, and many others.

EIC also provides true unified messaging via Microsoft Exchange or Lotus Notes. Fax services include fax broadcasting, fax on-demand, inbound routing, OCR, and desktop faxing. You can access and manage e-mail, voice mail, and faxes from your desktop PC, a Web browser, or any touchtone phone.

EIC 1.3 has been optimized to reduce network traffic and server processing overhead. This should allow a given processor configuration (e.g. 200 MHz Dual Pentium Pro) to handle a higher overall call volume than previous versions of EIC. In particular, an optimization in EIC's use of Java has resulted in a 30-40% reduction in the CPU utilization of the Java process.

Other new features in EIC 1.3 include support for unified messaging via Lotus Notes (this is in addition to EIC's existing Microsoft Exchange support), new remote access capabilities to enable off-site call center agents and others to use EIC as fully as if they were on-site, enhancements to EIC's end-to-end reporting capabilities and built-in call recording functionality, optional Nuance speech recogntion, new built-in OCR support for fax, and support for the new Dialogic DCB conference

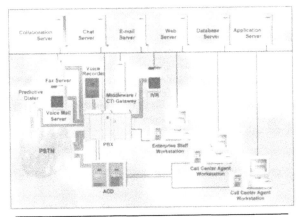

Before installing EIC by Interactive Intelligence, your enterprise's telecom facilities are a complicated place, with various proprietary systems having to get along with each other.

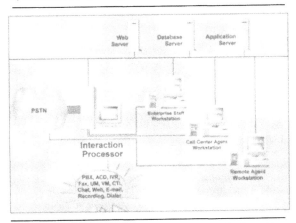

After installing EIC, the enterprise has been streamlined considerably, with call control and unified messaging brought under the software control of a Windows NT-based system.

boards that support conference calls with up to 96 participants per board and 32 participants per single call.

Interactive Intelligence is also making available two new add-on products, Interaction Director and Interaction Dialer. Interaction Director provides pre- and post-call routing for multi-site call centers. A server running Interaction Director can communicate in real-time with any number of EIC servers over a TCP / IP WAN to keep track of current queue depths, estimated wait times, numbers of available agents, skill profiles, and other statistics. It can use this information to decide where to route incoming calls.

Interaction Dialer provides distributed campaign management and predictive dialing capabilities that integrate tightly with the EIC call distribution function for outbound and blended call centers. New HTML-based client-side scripting tools ease the creation of scripts for inbound and outbound call processing.

Pricing for EIC is as follows: $1,000 to $2,000 per enterprise seat and $3,000 to $5,000 per call center seat.

InVADE from Invade International (Almondsbury, Bristol, UK – 011-44-171-575-0045, www.invade.net) is open, standard-based CT server middleware software, enabling applications from different vendors to work together on a single server, and provides basic switching functions to multiple client applications, allowing them to share hardware resources.

With InVADE, hardware of different types and from different vendors can be managed in the same server through standard interfaces. Any individual InVADE server's features are specified by choosing from a set of InVADE components and supported hardware, so InVADE based servers can perform PBX, automated call distribution, media processing, legacy PBX integration, and VoIP gateway functions in any combination and at any scale.

For example, a CT server developed around InVADE could combine PBX switching and routing services from one vendor, with applications like voice messaging, IVR, and fax-on-demand from others. Software developers can also design intelligent PBX features such as skills based routing, or combine media resources such as speech recognizers to offer voice dialing. Third party resellers can integrate whatever components a customer wants.

The system is built on Microsoft Windows NT Server and a TAPI 2.1 client-server implementation that allows "thin-client" applications to access telephony via the data network. InVADE is a service provider for digital network interfaces, IP telephony, fax and voice processing. It currently supports server based PRI ISDN telephony and WAN access cards, modular resource based SP cards, and analog station interface cards.

Each server can service hundreds of simultaneous customer interactions, and Windows NT also makes it possible to build scaleable, distributed systems. You're limited only by the speed of your LAN.

The InVADE platform consists of a configurable set of virtual line devices, all of which have a TAPi interface. InVADE has an object oriented approach to line devices and functions. A developer / administrator simply builds a line device type of choice from a large set of capabilities available to all lines. Thus, device types can be built that aren't found on traditional PBXs or even some commservers. Line devices can be associated with a physical resource at configuration time or at run time, or they can function as virtual devices, with direct physical association, such as call queues and route points.

By this method an inVADE-based IVR application controlling a voice processing resource on a given TDM channel acquires full call control capability on that channel via a TAPI interface.

Aside from TAPI compliance, inVADE ActiveX Controls further simplify the development process for programmers of various skill levels. Development tools compatible with InVADE include Visual Basic, Visual C++, etc. InVADE doesn't provide its own development tools, but integrates with familiar products through standard interfaces, so experienced PC developers can use the telephony environment without needing specialized knowledge.

For years I've tracked IVS Studio (Builder) and IVS Server from MediaSoft Telecom (Montreal, Canada – 514-731-3838, www.mediasoft.com) a Dialogic-based "soup-to-nuts" app-gen that runs under Windows NT, Unix, and Solaris, and supports up to 96 ports (for Intel platforms) and higher (for Unix). IVS can handle either analog lines or T-1s and E-1s. It's CSTA compliant and work is being done on TAPI 3.0

and CT Media compliance.

Not a company to rest on its laurels, MediaSoft then ventured well beyond the well-mapped world of app-gens, expending significant marketing and R&D resources (not to mention product development) to bring forth its new Office Telephony (OT) 2000, an intelligent PC-PBX that brings to the Small and Medium Enterprise (SME) computer telephony functions previously only available to large enterprises.

OT 2000 can run as a standalone PC-based system or with your current PBX. It yields greater functionality than traditional PBXs for less money and provides capabilities such as IVR, ACD and unified messaging, to boot.

OT 2000 provides an integrated suite of enterprise application templates, or applets, standards-based generic applications that allow customers and employees to get information, communicate and make transactions through public or private networks (including the Internet). The applets that define Office Telephony 2000 include IVS FrontDesk (auto-attendant), IVS HelpDesk (Call Center with screenpop), IVS UniMail (unified messaging) and IVS TeleCalendar (appointment scheduling). IVS Applets can be quickly customized through a set of IVS Wizards using IVS Applet Setup, a light version of MediaSoft's IVS Studio development platform, the successor to IVS Builder.

OT 2000 has a client-server architecture that resides on PCs running Windows NT and can integrate directly with Microsoft Exchange, Outlook and MediaSoft IVS Server.

Nortel Networks (Santa Clara – 408-986-0890, www.nortel.com) Enterprise Edge is a curious hybrid platform that, technologically, sits between a conventional Norstar phone system and Nortel's IP-based Internet enabled communications system, called INCA.

Enterprise Edge is a server-based communications system that offers customers three ways to communicate: Packet switching, circuit switching, or both. There's also a selection of desktop options; and a complement of software applications optimized for Windows NT server platforms.

Enterprise Edge also supports a communications model where a customer can use IP devices at the desktop communicating over the LAN to the server, and IP over the WAN to service providers and/or to other internal private network locations via a frame relay network.

Enterprise Edge has PCI slots, a processor and a disk drive like a PC, but Norstar technology has also been shrunk down to a card and is plugged into its own bus (and proprietary OS) that exists in the box, too, along with an IP gateway and Ethernet connectivity. One of the four PCI slots is taken up by a card connecting to the Norstar technology, but the other slots can be used for anything else, such as ADSL access to the PSTN. The PC part can run third party applications, and "the Norstar part" exists to make sure that there's compatibility with the 11 million Norstar phones ready and waiting out there in the market.

Using industry standard platforms in this way enables a faster "to-market" time for developers and a more cost effective model for incremental capabilities as we move into a standards based world of multimedia and networked applications.

Other features include various desktop options such as Norstar digital telephones, Internet telephones, and pre-installed software applications for Microsoft Windows NT server platforms that can be unlocked and put into action by purchasing a key code.

Enterprise Edge can also connect externally to multiple locations, customers or suppliers over the Internet or any managed IP network. Customers also get voicemail, unified messaging and an attendant console.

Pricing starts at $500 per station for the basic commserver capabilities and $615 per station with routing and VoIP capabilities.

OPUS Maestro from Prima (Nuns' Island, Canada – 514-768-1000,

www.prima.ca) has also reached escape velocity from the world of app gens and entered the realm of "integration platforms" to integrate and CT-enable any kind of telecom system.

OPUS Maestro is heavily into ActiveX technology, such as ActiveX controls having native integration to both Genesys T-Server and Dialogic CT Connect and CT Media for projects ranging from IVR systems to call centers, DCOM ("the network is the computer"), and ActiveX Data Objects (ADO) that can connect to data stores as varied as a front-end database client or middle-tier business object using an application, tool, language environment, or even an Internet browser, where it becomes a data interface for 1-to-N-tier client / server and Web-based, data-driven solution development.

OPUS Maestro runs under Windows NT and uses Intel / Dialogic cards (H.100, H.110, SCSA and PEB telephony buses) to support up to 120 ports per chassis and a total of 5,000 lines of analog, T-1s, E-1s, or ISDN PRI.

OPUS Maestro is also compliant with CSTA and IBM's CallPath. It can integrate all sorts of technologies relating to call centers, such as Siebel's customer interaction software, the Internet and other CT-related technologies.

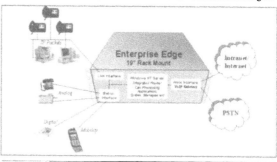

A server, a phone system, and an IP gateway, Nortel Networks' Enterprise Edge is a fascinating hybrid communication solution for small and branch offices.

The OPUS Composer module is the most "GUI app-gen part" of OPUS Maestro. But instead of a flowchart of boxes connected with spaghetti, OPUS Maestro uses a hierarchical tabbed tree control that resembles Windows Explorer. Like a GUI app-gen, you build an app by selecting "actions" from pallets to drag and drop over the call flow control. You can also drag and drop ActiveX controls, so you can add huge swaths of functionality without coding.

Of course, there's automated attendant and voice mail capability. But OPUS Maestro also lets you create software for web-enabled call centers, web-based systems, and web callback buttons. It can do web page screen scraping. Database access is via ODBC. The fax support is robust and ranges from fax-on-demand to store-and-forward to fax broadcast.

OPUS Maestro's entry level price is $3,000 (a four-port starter kit). There is a runtime fee of $400 per port (MSRP).

It also provides enhanced support for Dialogic CT Media providing hardware and switch fabric independence and application-to-application hand-off, remote deployment of application packages, and comprehensive debugging with a centralized Log Viewer.

VBVoice 4.0 and VBFax from Pronexus (Kanata, Ontario, Canada – 613-271-8989, www.pronexus.com) now lets you create reusable ActiveX components which can be inserted into the VBVoice call flow. Composite controls can encapsulate subsystems built from VBVoice controls and Visual Basic code into one component. You can build subsystems, package them up into a composite control, and connect the control to and from any other VBVoice control.

VBVoice runs under Windows 95, 98, and NT. Compatible voice/fax cards include SCSA and MVIP compliant cards from Dialogic, Brooktrout and Aculab, as well as TAPI-compliant hardware.

VBVoice incorporates Dialogic's Global Call which assists access to various global signaling protocols. VBVoice 4.0 international protocol supports shared or dedicated voice resources, multiple concurrent languages, Lucent Speech Solutions' multilingual Automatic Speech Recognition and Text To Speech (ASR / TTS) engines and other SAPI-compliant ASR / TTS engines. Pronexus has also developed new ActiveX telephony components that let you build PC-based PBXs, small call centers, and other applications.

VBVoice 4.0 has over 30 sample applications with screen pops, station interface, voice mail, messaging, call center, faxing, etc., that can be embedded into any application.

VBVoice supports up to 240 ports per chassis, with an unlimited number of users using networked PCs. The entry level price is $495. There's no runtime fee.

Call Center in a Box from Saleslan Limited (Broomfield, CO – 303 464 8838, www.saleslan.com) is an interesting modular CT call center software solution that can also be used as a PC phone system. It can handle both inbound and outbound calls simultaneously, structure call campaigns to ensure that call agents are armed with relevant on screen information.

Saleslan's client-server architecture is based on Windows NT 4.0 server and clients. Scalability is achieved by distributing the server processes over many servers so that the number of agents can be increased if needed. The server code is written in C++ using object orientated design methodology. The server uses an Oracle relational database to store all call center events, telephony, agent and application access.

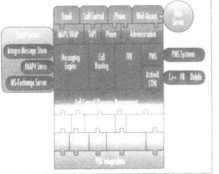

Telekol's IntegraX is based on open standards and an open architecture, running under Windows NT. The scalable design allows your business to grow by adding a wide range of software application modules.

The server connects to the PSTN via Dialogic cards, but the system's design allows you to change to cards from different vendors if necessary. Analog and ISDN connectivity is supported and the DPNSS inter-PBX protocol is also supported, as well as ISDN PRI Q-Sig voice signaling. Tight switch integration is provided by a CT module that also supplies the IVR, ACD and dialer functions, in addition to database and application integration. A TAPI interface will also be provided so that third-parties may access the telephony functions.

An application generator, Internet integration (Web page and e-mail) and a complete fulfillment system are also offered.

The interesting IntegraX Intelligent Communications Server from Telekol Corporation (Waltham, MA – 781-487-7100, www.telekol.com) is a multimedia, multi-application communications server based on an open architecture, open standards, Windows NT and Intel / Dialogic technology. Telekol's solutions that run on the IntegraX are speech-enabled, Internet-integrated, and can leverage the convergence of IP/data networks. Telekol markets its products worldwide through a network of over 400 dealers, VARs, and OEMs.

You can program your own call processing applications with the system's own API, called the IntegraCom API.

IntegraX comes in two server platform sizes: IntegraEnterprise and

IntegraBusiness. Both offer a full suite of advanced call processing and multimedia messaging applications, including call routing, voice mail, unified messaging, integration with Microsoft Exchange, desktop call control, IVR, and access to messages over the Internet.

IntegraX installs quickly with preset integrations to over 100 PBXs. The system runs on LANs having Windows NT clients, integrating with existing MAPI and IMAP4 compatible e-mail readers.

The IntegraNet module provides access and management of voice messages, e-mails, and faxes stored on a Microsoft Exchange server. Mailbox owners can access and manage their messages from any Internet-connected PC, using any major Web browser. IntegraNet uses streaming audio based on standard protocols, and works in conjunction with Microsoft Outlook Web access. Leveraging the look and feel of the Microsoft Outlook or Exchange Inbox interface, Telekol has added special icons and message header information to display the message type, sender, subject, date, and the time the message was received.

Since the last time we looked at Telekol, their software engineers have been busy adding some major new features for the new 6.2 Release. For example, there's now Lotus Notes integration with their unified messaging solution (it costs about $200 per seat in a 500 user installation), L&H speech recognition support, and a new multilingual module called IntegraGlobal, which allows callers to choose from a variety of languages and mailbox owners can have e-mail read to them in multilingual text-to-speech languages.

Telekol says they're moving in the direction of what they call "intelligent communications." They feel that voice mail, audiotext, or IVR are no longer simply discrete technologies that serve a singular purpose. There is much more "intelligence" and consolidation going on in which all communications works together to reinforce a total communications experience. For example, when someone calls into an organization, there are all kinds of ways that a call can be answered offering an opportunity to add a great deal of intelligence to this interaction. The same thing is happening on the messaging side. Telekol's fervent belief is that their intelligent call processing / transaction processing and multimedia / unified messaging solutions are two core technologies that, to be effective, must work in sync with each other.

In line with that, Telekol says their support of speech recognition represents a long term strategy to speech enable all of their applications in the future with what they believe will become the preferred interface of the user. Their first step in that direction has been speech-enabling their automated attendant. Developers can even use the speech rec technology when developing custom apps with the IntegraCom API (on the IntegraX Enterprise platform it costs $4,500 per four ports)

In a similar vein, while they deliberately targeted Exchange and Lotus Notes as their first integrations because of these two companies' market share (especially in the large enterprise and medium-sized businesses – Telekol targets), their ultimate plan is to eventually support just about any e-mail server to accommodate the mixed environments of our customers.

To sum up, IntegraX supports any kind of networking need including the PSTN, wireless, IP, and the Internet. It's scalable and supports a wide range of enterprise capacity needs from the single site to multiple networked implementations. There's seamless integration of all communications media, including voice, fax, e-mail, text-to-speech, and speech recognition. And Telekol even has a Professional Services Group to support your enterprise from installation to technical support, maintenance, and custom development.

CallCenter@nywhere from Telephony@Work (San Diego -CA – 619-410-1600, www.telephonyatwork.com) bills itself as the first platform in the industry with "Call Center for Dummies" technology, which means that programmers aren't needed for any aspect of system configuration or administration, including IVR scripts, multimedia queuing and skills-based call routing. Many platforms and app-gens have aspired to be user-friendly, but none can actually promise that programmers aren't needed at all, period.

How does Telephony@work develop such a product? To ensure ease-of-use, they periodically put together focus groups of non-technical people who are given the task of configuring a system without assistance. Wow!

In Version 2.1, systems integration can be done in minutes instead of months by allowing call centers to add custom tabs to the agent interface with fields that can read and write to any ODBC database – providing the ability to create new front ends to legacy and your own third party applications in just minutes.

Although you can configure a business phone system with CallCenter@nywhere, the product really shines as a call center platform, as its name implies. For example, the platform's integrated help desk provides a one-stop "Interaction History" stored and searchable by client that includes all recorded calls (consents to transactions, etc.), copies of all incoming and outgoing e-mail sent from or to the customer, as well as all related help desk entries. Also, call center agents who don't know the answer to a question can simply click on the "search" tab of the agent interface, the "Interaction Manager," to search all help desk records by key word. As a result, the call center's knowledge base is improved with every interaction.

Sophisticated mail queuing and distribution that ensures that agents handle workgroup mail one at a time. (An agent must complete an e-mail response before another workgroup e-mail is forwarded to that agent.)

CallCenter@nywhere queues, prioritizes and distributes interactions from all the different mediums that customers use to interact with the call center, including phone, fax, e-mail, chat, voice over web and web-callback requests. Essentially, CallCenter@nywhere redefines the "call" and replaces the concept of the "Automatic Call Distributor" with the new paradigm of the "Automatic Interaction Distributor."

Unified "call control" handles all types of interactions in an integrated way. Workgroup phone calls, web calls, e-mail, chat sessions, faxes and voicemails are now queued and distributed one at a time and appear in the agent's incoming call window as "the call." If a live inter-

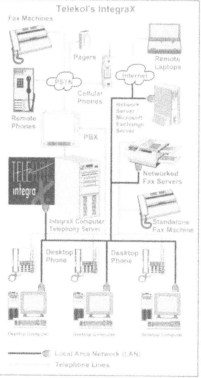

Telekol's IntegraX business communication servers can unite your phone system, LAN and the Internet, providing unified messaging and advanced call processing.

action (such as a phone call or chat session) is routed while the agent is responding to an e-mail, the agent actually puts the e-mail "on hold" before handling the live interaction, serving as a constant reminder to the agent to finish his or her e-mail and prevents them from receiving more e-mail until the current e-mail interaction is complete.

The product's integration-by-design philosophy cuts agent training costs by providing agents with a single, multi-tabbed interface instead of a confusing mess of Microsoft dialogs that some multi-vendor solutions provide. CallCenter@nywhere's integrated design also eliminates the need for many proprietary systems, including proprietary PBX's, ACD's, mail distributors, etc.

CallCenter@nywhere can also do call blending by setting thresholds about when predictive and \ or preview dial calls should be made and routed to agents based on current inbound traffic. Call centers can blend local, branch office and telecommuting agents who share all the same screen pops, visual call control and help desk capabilities.

CallCenter@nywhere gives you the choice of whether to carry voice traffic over IP or traditional phone lines or both simultaneously – while maintaining full call control for remote-agents in all cases. This enables call centers concerned about Internet sound quality to route calls over the PSTN while agents control the call with a user interface that sends commands to the PBX over IP, sort of like a remote-control for the PBX. In all cases, the data, screen pops and remote call control functions are synchronized with the call.

CallCenter@nywhere also uses a multi-threading architecture throughout the entire platform, providing a fault-tolerant architecture that prevents any individual workstation from ever "crashing" the call center.

The platform relies on open standards and doesn't use a proprietary language or compiler. The system is written entirely in C++ so that call centers and e-commerce developers can integrate with the platform quickly and easily.

CallCenter@nywhere effectively eliminates the long lead time, uncertainty and technical challenges associated with multi-vendor systems integration.

The company's largest customer is WESTAT, which runs the software in nine of its call centers and has just ordered its tenth call center platform from Telephony@Work. Interestingly, WESTAT's 2,000 agents are all local in their various facilities around the U.S., but supervision is centralized in Maryland using CallCenter@nywhere's advanced remote-supervision and remote-coaching capabilities.

CallCenter@nywhere is priced at $2,500 per port and there is no charge for seat licenses or for installing their user interfaces on local or remote agent desktops.

TelePCX 3.0 from TeleSynergy Research (Santa Clara, CA – 408-260-9970) is a PC-based PBX combined with a auto attendant / voice mail system and can be controlled by TeleSynergy's own GUI-based application generator (TeleAPG), which gives you the ability to integrate advanced features beyond that of a PBX such as unified messaging, IVR, I-phone gateway, fax servers, screen pops, call recording, etc.

Thus, you can expand your system from a PC-PBX to a call center application, and the "open platform" philosophy of TelePCX lets you add third-party software and hardware, which connects via TCP/IP.

TelePCX runs on Windows NT or 95/98. It's based on TeleSynergy's own TX-412X PBX board which supports four trunks, 12 extensions and two voice-processing ports. Through the internal switch, the two voice processing ports can connect any one of the four trunks and 12 extensions and enables the auto attendant and voice mail functions so that you won't need to buy another voice board.

Voice processing boards come in sizes of two, four, eight, 24, and 32 ports; fax boards in two or four ports, and the PBX board has four trunks and 12 extensions. There is also a dual T-1 / E-1 interface board.

When the switch control boards (Model TS – 1384A) and T-1 connectivity are added, a system can operate multiple voice (TV-2242 / 2282), fax (TF-9142), and PBX boards (TX-412X) simultaneously and support up to 48 trunks and 96 extensions, which is suitable for most small to mid-sized offices.

The PC itself can be fault resilient, fault folerant, or just an ordinary PC.

Just so you don't have to reinvent the wheel, TeleSynergy's web-based unified messaging system, TeleUMS can run on TelePCX, allowing you to send, receive and manage your voice / fax / e-mail messages from your desktop via a single mail box and friendly web based interface. Faxes can be sent through the Internet by a store-and-forward method. Remote users or road warriors can access the UMS web site and check and respond to messages.

An entry-level TelePCX 3.0 with four trunk lines, 12 extensions, auto attendant and voice mail is $1,550. A four-port voice board for IVR with TeleAPG included is $595. A four-port fax board and a four-port voice board for fax-on-demand applications including software is $1,590.

CT Access (Computer Telephony Access)

CT Access is a development environment from Natural MicroSystems (Framingham, MA – 508-620-9300, www.nmss.com) providing standard programming interfaces for telephony functions that are hardware independent. CT Access thus unifies application development across NMS AG products and provides a consistent set of OS-independent and portable APIs. It runs under Windows NT, Solaris for Intel and UnixWare. With CT Access' Natural Call Control API, you can

The Interaction Manager of CallCenter@nywhere from Telephony@work normally gets a workout when an agent doesn't know the answer to a question and uses it to search all help desk records by keyword.

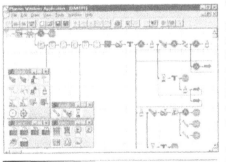

TeleSynergy's TelePCX can be reconfigured and enhanced with TeleAPG, an all-in-one GUI applications generator, which takes care of the control for voice, fax and PBX boards. With TeleAPG you can build solutions such as on-line chatting rooms, an office PBX system, IVR and fax-on-demand solutions and pack them all into just one system. TeleAPG runs on Windows NT and works with TeleSynergy voice, fax and PBX boards. Also, TeleAPG has a fully graphic voice prompt editor.

quickly develop applications that run on multiple types of telecom interfaces by using a single protocol-independent API. Natural Call Control minimizes the processing overhead on the host CPU by executing protocols on the board's control processor.

CT Bus

See H.100, H.110

CT Connect

CT Connect 4.0 is an open, standards-based computer-telephone link server from Dialogic Corporation. One of the unique features of CT Connect is its support for multiple telephony programming interfaces, including the Intel/Microsoft Telephony Application Programming Interface (TAPI), the Novell/AT&T Telephony Services Application Programming Interface (TSAPI), and the Microsoft Dynamic Data Exchange (DDE) interface. In particular, CT Connect has a unique ability to support desktop TAPI applications in server-based work-group environments where a common server approach is more cost-effective than individual desktop telephone connections. CT Connect 4.0 also supports Java and thin client development.

CTI

See Computer Telephone Integration.

CT Media

Released in 1997 by Dialogic, CT Media was the first multi-app Operating Environment middleware program for driving ECTF S.100-based applications (See S.100) and the SCSA multi-resource platform (See SCSA, H.100). Several major companies developed or announced plans for products based on CT Media after its introduction. The licensing of CT Media by Microsoft in early 1999 and the subsequent acquisition of Dialogic by Intel saw broad industry adoption and development of CT Media based products.

What are the benefits of CT Media? Just look around your office. There are probably several computer telephony media-processing applications on site working for you. And while some might be for the entire company (voice mail) and others for specific departments (fax server), they're all essentially isolated, even though they pretty much work on the same type of hardware.

This hints at a fundamental problem of computer telephony media-processing systems and apps. There's no computer telephony equivalent to a general computing server – a single central CT machine that can be accessed and manipulated by different front-end applications that intelligently share the common server's resources and the standard database information they rely on.

Computer Telephony magazine's own computer telephony installation has certainly reflected this. Over the years we've used an Active Voice Repartee voice-messaging system, a Voice Systems Research COVoice advanced auto attendant / call router, and various fax systems: A Brooktrout QuadraFax fax-on-demand system, a Copia FaxFacts fax server / blaster, and a RightFax fax server.

All of these devices work beautifully but, through no fault of their own, they all perform their tasks in isolation. They can't really communicate. They can't share data or work together. This makes our IT Manager busier that he should be. Instead of maintaining a single system, he needs to run around between five different computers, babying five different databases, learning five very different interfaces, etc.

It's also not very efficient. There's hardware duplication. Three of the machines have voice store-and-forward cards in them. Two of the machines have fax cards in them. All of them have their own telephony network interfaces in them. And, with the possible exception of the Repartee system, none of the hardware resources in them are used to anywhere near their fullest capacity.

For example, the COVoice system we used was also a full-featured voice-mail system, but we only used its excellent auto-attendant call routing for the super-tight integration it provides for the Mitel PBX we were using at the time (we have since swapped it out for a Lucent G3) and their ACD packages This means the voice boards the COVoice system contains are barely touched. Meantime, while the FOD system and the fax blaster use virtually the exact the same fax-processing hardware, they can't share resources, meaning we've got installed about twice as many fax boards than one would think necessary.

Sound familiar? We thought so, which is why we immediately realized that the Intel / Dialogic CT Media marked a new age in computer telephony system development.

CT Media represents a more open approach to building a generalized enterprise communication server environment than simply a PC-based phone system or UnPBX. While PC-PBXs claim to be open based on using open computing platforms, a standard operating system, and open hardware like trunk, station interfaces, and voice processing boards, constrain the purchaser into a limited set of applications that the supplier can develop or support. This means they are still closed at the software level and still restrict the buyer to a single source for new applications and technology upgrades. It also means that CT software developers have to worry about hardware.

In contrast, CT Media uses a completely open systems approach that lets multiple third parties develop CT servers. These servers can be purely media processing platforms that integrate with a legacy PBX, or can also include telephony switching functionality. CT Media provides developers, resellers, and system integrators the ability to mix and match technologies and software services to run on the same telephony platform, giving them the biggest "bang for the buck".

For example, a communications server developed with CT Media could combine PBX switching and routing services from one vendor, with applications like voice messaging, IVR, and fax-on-demand from others. Software developers can also design intelligent PBX features, integrating access to media resources such as speech recognizers to offer services like voice dialing. Third-party resellers can integrate these components with greater flexibility to suit specific customer needs. With the CT Media open architecture, system owners can add new services to existing systems, such as an automatic call distributor (ACD) for a telesales group, by purchasing new software from "best-of-breed" suppliers without being restricted to solutions available from the original system provider.

And for CT application developers, CT Media allows them to focus on doing a single application as superbly as possible, and presumably snap up market share for that particular application. They no longer have to worry about building CT hardware since any company deploying a CT Media application could build or buy a CT Media-based server. Then you just install the CT software modules you need and you're up and running.

CT Media is designed to be hardware independent. Any vendor's hardware that is compliant under Resource Development Kit (RDK; based upon the proposed ECTF S.300 standard) interfacing to the CT Media service provider can work under CT Media. The initial releases of CT Media were based on Windows NT while all future releases will run on the Windows 2000 operating system.

When CT Media first appeared in 1997, Rick Luhmann, then *Computer Telephony* magazine's editor-in-chief, interviewed none other than Dialogic's Director of Product Marketing for CT Media, Jim Machi, to discuss its exciting capabilities.

Machi's explained CT Media this way: "An application essentially says to a CT Media server, 'Give me a call. When you give me the call, make sure that I can play, that I can detect digits, and that I can do speech recognition.' This is different from the existing Dialogic System Release that says, 'Give me board one, channel two, speech rec board one, resource one, and fax resource number three.'

"Now, when I want to do a speech rec command, I don't have to say, 'speech rec device one, channel two, listen to timeslot three, network interface one, board two, transmit on timeslot three, speech rec one, channel two, recognize.' Instead of talking to different devices, trying to figure out which is available, trying to figure out what timeslot they're talking on, the S.100 model is: 'give me a resource, I'll call it from a Group, and then I'm going to tell that Group to recognize.' It's a simpler model to the application under the hood. We're going in and setting it all up for the app.

"Now when another app says, 'Give me a call and make sure that I can recognize,' CT Media makes sure that we grab another group of resources, another group of timeslots. Now both apps are sure that they can run in the system at the same time without stepping on each other."

"Because of our group model, we can do connections and switching much more easily. The app simply sees two groups and says, 'connect these two groups together.' So this logically could represent a call coming into a server and having and MSI card with a phone connected to it. The app is now creating a connection to it, the way it would on a PBX, but it doesn't have to know anything about how many timeslots are in the system and whether there are other apps grabbing resources at the same time, so that it would step on it. The point is that we think in terms of groups of resources and connecting them together, rather than thinking about timeslots.

"One would think that doing all of this would cost you something in overhead, but this overhead really was built into applications before. The reality is that an app wouldn't be written to do 'board one, channel two, talking over this timeslot,' but that there would be software in there that would be doing what CT Media is doing. So that one group of guys in an app company would be working on the middleware, and another group would be saying, 'For our voice mail app, we want it to answer a call, record the call, and then do DTMF detection." And their app is going to talk to the software written by the other group in the company.

"Now you've got all these app companies, all of whom have a couple of people writing this low-level middleware to abstract the resource management from the people who are doing the apps. The resource management adds no value to the total solution, so we (at Dialogic or anyone else who implements the service provider interface layer) have said we'll do it for you. Now you can spend more time on the value-added features.

"While we are adding overhead on top of our libraries, we're taking some functionality out of the app, so the net processing power of a system should still be roughly the same. The apps themselves get thinner."

CT Media server is thus genuine server software that implements, among other things, the SCSA reference specifications, providing a standard framework for app communications and standard interfaces for sharing technology resources within the server. Key features include:

- **Standard Application Programming Interfaces.** It uses the already ratified ECTF S.100 rev. 2 API specification. This means developers can build applications to run on CT Media and, since they've used an industry standard API, other servers that support S.100. It will also let their apps work with others.
- **Multiple Application Support.** CT Media handles all low-level resource management between the different hardware components installed in the computer telephony system. It will take care of resource requests from different apps and provide them with what they need to get their jobs done. This means end users can now own computer telephony systems that can be expanded with new applications and media, without buying new systems. System Integrators can also build servers with best-of-breed software that deliver precisely what a customer needs.
- **Multiple Application Integration.** Besides letting different apps work simultaneously, CT Media integrates them on an independent basis. This is provided through a feature called "call handoff," wherein one app hands off control of a call, and any associated computer telephony resources currently servicing that call, to another app.

Using call handoff, a voice-mail app from one supplier could offer the caller the option to access a FOD service after he leaves a voice message. CT Media would then hand off control of the call to a separate fax-on-demand app, which would handle the fax transaction with the caller and then either hand the call back to the voice-mail app or hang up.

- **Simplified Application Design.** CT Media does a lot of stuff that previously had to be handled by the application, including managing hardware. SCbus management and basic call control.

This means that apps no longer select specific hardware components and manage the connections between the components, but instead request sets of resources from the CT Media server. CT Media selects available and appropriate resources and maintains the low-level connections between the components on behalf of an app. For instance, when an app wants to play a prompt or voice message, the CT Media server makes the hardware component responsible for playing voice files available without the app finding and tracking what's actually processing the call.

As for call control, even the simplest media-processing app must receive calls, dial numbers and hang up. It might sound trivial, but this can get complex in the real world of telephony where different networks have different protocols for doing these tasks. CT Media removes the responsibility for managing these tasks from the application. Applications request inbound calls from the CT Media server and it takes care of it.

- **Client/Server Support.** CT Media interacts with apps using a client / server architecture. This lets system designers and end users configure media-processing applications to reside locally (on the server with the CT hardware) or remotely, where they communicate with the server over a Network. This also means that the apps and the server do not have to be in the same chassis or location. The clients (apps) can be geographically dispersed.
- **Symmetric Multi-Processing.** CT Media is capable of dynamically leveraging the increased performance of SMP capable systems without having to use system configuration tools to set process/processor affinity, although this will be retained as an option. There are three

specific user scenarios: (1) to enable CT Media to co-reside with apps and services which require significant processing performance (e.g. Host based resources, Microsoft Exchange IIS, etc.); (2) to enable a greater density of CT Media apps / channels to be supported on a single server platform; and (3) to provide a future proof platform to meet expanding customer requirements.

A CT Media system has several distinct high-level layers:

- **An Application Interface Layer** – a set of reference APIs for computer telephony, providing call control, media processing and system service interfaces. Dialogic plans to support Microsoft Speech API (SAPI), the future object-oriented ECTF S.400 interface and Microsoft ActiveX, in the future on CT Media, but for now this layer is defined, as mentioned, by the ECTF S.100 rev. 2 spec.

Think of these APIs as a programmer's control window over the CT Media Server. S.100 defines C language bindings to a suite of resource control and system management for computer telephony apps. Its own model is a simple one of configuring Groups of resources and managing them based on their features rather than their physical implementation.

The S.100 APIs supported by CT Media can be logically broken down into two types: Technology Resource APIs, which provide access to the technology offered by media-processing resources; and System Service APIs, which provide the core services needed by the Technology Resource APIs and all applications.

Technology Resources APIs include:

- Automatic Speech Recognition (ASR), which lets speech rec command and control apps;
- Runtime Facsimile for sending and receiving faxes;
- Player / Recorder for playing and recording audio data of different encoding types, text-to-speech, ADSI and TDD;
- Signal Detector / Generator for detecting and generating signals so that callers can communicate with an app; and
- Call Channel for providing a resource access to a telephone network or station set interface resource.

System Service APIs include:

- Connection / Conference Management for controlling the connection relationship between resources on the server's implied switch fabric;
- Data Management, which provides control for reading and writing S.100 data structures;
- Group Management, which creates, configures and destroys Groups;
- Session Management for creating / destroying sessions with an S.100 Server and controlling server security;
- System Call Router, which manages the switching and allocation of telephone network and station set interfaces between different resources and applications; and
- Container, which manages the storage and interchange of data.
- A Server layer – an abstract server model that includes a messaging protocol to allow language and platform independent communications between machines.
- A hardware interface layer – essentially an abstracted SCbus switch fabric model for the real-time control of streamed media data.

Within these, of course, the architecture deepens. It includes:

- **The Software Abstraction Layer – Application Interface Adapters.** The CT Media software abstraction layer consists of a set of C Language interface dynamic link libraries called Application Interface Adapters (AIAs). These AIAs convert S.100 API function calls into language-independent messages (Requests) which are dispatched to their destinations for processing. The Request messages elicit responses in the form of Response messages, which are then routed back to the AIAs. The AIAs convert Response messages into transaction events. These are queued for collection by the target application.

- **The Server Layer.** This is the core component of CT Media. It provides computer telephony services to client applications and supports associated signal-processing hardware resources. Think of it as the operating environment where resource developers must work. From their perspective, CT Media Server is a single object-oriented process composed of interacting software objects, called Service Providers.

The server layer delivers support for common services needed by all app, including client / server sessions which connect applications to the server and the system registry which tracks service providers; resource allocation and management services; switch fabric management services; media storage and management services; inter-object message-routing services; and simple call control services.

These services are provided by one or more Service Providers. The server layer is thus made up of software objects that perform specialized tasks and interact by passing messages back and forth. Individually, or in cooperating clusters, these Service Providers objects support the S.100 application programming interface.

Service Providers differ from each other by function and by characteristic. Some will be unique in the server, while others will have multiple instances running simultaneously. Some Service Providers are always present while the server is running, while others are temporary, existing, for example, for the life of a single call. Some are replaceable by third-party implementations, while others are not replaceable.

Overall, CT Media Service Providers include: the System Registry; the Signal Processing Component; the Configuration Manager; the Session Manager; the Session; the Resource Allocation Service; the Group; the Connection Manager; the SCSA System Call Router; and the Container Manager.

Rather than detail them all here, we'll touch upon each as we continue to dig throughout CT Media, specifically detailing them as they logically enter into the architecture and its application model.

These Service Providers communicate with each other using the SCSA Messaging Protocol (SMP). This is a connectionless protocol, with each Service Provider having a unique SMP address within a logical Server which may be distributed across the network.

There are three classes of SMP messages:

- Request messages, which contain commands to be executed by the target Service Provider;

- Response messages, which contain completion information and data generated by Service Providers in response to Request messages; and

- Event messages, which are unsolicited messages generated by Service Providers in response to experienced state-changes.

This leads us to our first Service Provider Profile. In this case it's the System Registry – a unique, enduring, non-replaceable Service Provider responsible for maintaining the following information: The SMP addresses of all Service Providers in its logical server, the availability of a particular Service Provider in its logical server, and useful data associated with each Service Provider in its logical server.

The System Registry has a system-standardized documented SMP address. This lets Service Providers register their SMP addresses, their availability and any other required data with the System Registry on system startup. The System Registry will automatically notify interested parties of changes in any Service Provider's operational status if required.

- **The Hardware Abstraction Layer – Resource Software.** Media processing and telephone network interface hardware are represented by software objects called Resources.

Resources, which are implemented as Service Providers, hide the implementation of a physical device from the Server and applications by providing a standardized abstract interface for each S.100-defined class of resource functionality. Resources accept Request messages from other Service Providers and pass them on to a physical resource via a driver. Conversely, Resources accept events and data from a physical device and generate Response and Event messages targeted at other Service Providers.

CT Media supports the following S.100 compliant media Resource classes through Application Interface Adapters (AIA's) and corresponding Service Provider Interfaces:

- Player – converts stored data media into an audio format for a caller;

- Recorder – converts incoming audio into a data format for storage;

- Signal Generator – generates tones and telephone network control signals;

- Signal Detector – detects and decodes tones and telephone network control signals;

- Call Channel – represents a telephone network interface for access by media resources;

- High Level Fax – sends and receives facsimile documents;

- ASR – performs recognition of spoken utterances for command or control purposes.

One particular CT Media Service Provider to be noted here is the Signal Processing Component (SPC). It's a transient, replaceable Service Provider composed of one or more Resources. There may be many SPCs per logical server.

Basically, SPC's do two things: First, when the Server is selecting resources to perform a particular media processing service, the unit of allocation is the SPC; and, second, they act as a gateway between the server and the Resource(s) they're composed of and the hardware those Resources represent.

This means it's SPCs and their component Resources which form the hardware abstraction layer in a system. In turn, this means the main task of a hardware developer who wants to play here is to create and manage the SPCs and Resources which will represent their hardware in the CT Media Server.

The way in which resources are managed by the Server depends on the architecture of its associated hardware. Some hardware architectures allow their associated resources to be stopped, started and connected to the internal switch fabric individually, while other architectures only allow "bundles" of resources to be manipulated in this way. The smallest bundle of resources that can be managed by the Server is an SPC. For example, the Dialogic D/41ESC card provides the following resources: Four Players, one per channel; four Recorders, one per channel; four Signal Generators, one per channel; four Signal Detectors, one per channel; and four Call Channels.

But because of the D/41ESC architectural design, all the resources associated with a particular D/41ESC channel must be allocated together. This means the D/41ESC supports four SPCs, with each SPC being composed of one Player, one Recorder, one Signal Generator, one Signal Detector and one Call Channel. It's also possible to design SPCs that are comprised of one resource only, for example, an ASR SPC, comprised of a single ASR resource.

- **Resource Hardware.** These are the boards and their DSP algorithms that do different media-processing and call control stuff for the server. Again, they're represented by associated Resource objects.

    Examples include telephone network and station set interface boards, audio coders and decoders, fax boards or ATM adapters. Request messages sent to a Resource are converted into driver commands by that Resource and passed down through the driver layer to the hardware where they're executed by the firmware on the board.

    Conversely, data and state-change driven events generated by the hardware are passed up through the driver layer to the Resources in the Server. Typically, the Resource would perform some pre-processing before packaging the data into an SMP message and sending it on. Events coming up through the driver will be processed by the Resource and will usually cause a state-change in the Resource and prompt the Resource to generate a Server Event message.

    CT Media is designed to be hardware independent. Any vendor's hardware that is compliant under the Resource Development Kit (RDK; based upon the proposed ECTF S.300 standard) interfacing to the CT Media service provider can work under CT Media. It presently supports many Dialogic hardware devices with more becoming available based on the usage of CT Media RDK interface.

CT Media and the S.100 Application Model

Each CT Media application will go through the same basic structure of setup, handling of calls and shutdown. Applications go through a cycle. They log into a server to create a Session with the Server. They request Groups which will automatically contain the resources they require and connect to a call. They perform a series of media processing actions on that Group. And they eventually return the Group and its resources to the Server and either request another Group or break their Session with the Server. Details:

- **Application Profile.** Since the underlying computer telephony hardware in a system is potentially shared by multiple applications, each with a distinct set of requirements, an application must make its resource requirements known to the Server to allow for the proper configuration.

    To install an application onto a CT Media-based system, an application provider supplies an Application Profile and an executable program. The Application Profile defines the combinations of features that it requires to process each call and detailed parameter settings of system resources that the application may require. CT Media then uses this information to configure groups, initialize parameters and establish run-time controls for the app.

    The Application Profile is a file external to the application's code. It contains: a Resource Block that specifies the characteristics required for resources to be used by the application; one or more Group Blocks, each containing a set of group configurations describing a collection of resources to be assembled for the app; and an Application Service Insert Block, which sets up information used by the SCSA System Call Router (SCR) Service Provider, which is a CT Media software mechanism that answers calls and hands them off to applications.

    Application Profiles can be configured to use simple default resources for relatively simple application processes; or the developer can create very detailed profiles that allow for more precise application

control of resources based on detailed characteristics.

To help keep everything coherent here, CT Media provides something called the Configuration Manager Service Provider. This is a unique, enduring, non-replaceable Service Provider that's responsible for: maintaining records of all application resource requirements specified in an Application Profiles; maintaining records of the features supported by every installed resource; and adding and removing replaceable Service Providers.

- **Applications and Session Management.** The CT Media Session Manager allows the application to establish a session with the Server by logging in. This is required whether the actual Server process is local (running in the same machine as the application) or remote (running in another machine over a LAN).

    A Session combines several concepts and services for the application. It represents: an application's view or handle to a CT Media Server and vice versa – from the Server's vantage point the Session is the source and destination of all communication with the client; a communications circuit between the application and the Server – from the client's point of view, a Session owns and controls all activity in the Server; and an event queue in the client, which is the mechanism for handling events synchronously in the application. The event queue is also the organizational structure for asynchronous event handlers, each associated with a Session / event queue.

    A Session provides a context within which CT Media objects can be created and manipulated. The session can be considered to be the owner of all objects (e.g. Groups) which are created through it. To process telephone calls, an application first must establish a Session with a specific CT Media server. The function for creating a Session identifies the application profile. With CT Media, a default Server can be designated or the application can override this with optional parameters.

    There are two CT Media Service Providers that are important at this point in the application flow:

    - The Session Manager Service Provider is a unique, enduring, non-replaceable part of a CT Media Server. It's responsible for Server-side Session management, including authenticating application log-ins and creating / deleting Server-side Session objects.
    - The Session Service Provider is transient. There may be many Sessions per logical Server. The Session is responsible for: routing commands from a client application to target Service Providers and routing responses from Service Providers to the client application; distributing a client application's resource requirements to other Service Providers; and providing a simple security firewall.

    Every application using a Server has an associated unique Session. Without one, it can't use the Server.

    An application may also establish multiple sessions with different Servers to control additional resources.

- **Applications and Resource Management.** One of the most complex problems facing computer telephony application developers is the management of scarce resources such as voice recognition or fax devices in a system. As mentioned, CT Media eases this headache. Its resource management scheme provides the following features:

    Dynamic resource allocation. Applications may retain resources between calls or use a resource for the full length of a call, a portion of the call or for simply the length of a given resource operation. Dynamic resource allocation is crucial to sharing expensive resources in a system among multiple applications, which may not need the services of that resource all the time.

- The ability to handoff ownership and responsibility of calls and associated resources between applications is one of the most important requirements of a CT Server. In a multiple application system environment, different applications may be used to handle specific transactions on a given call.

For example, one application may be responsible for answering all incoming calls and distributing them to different applications based on ANI or DNIS information. Those other applications in the system may be responsible for automated technical support inquiries, voice messaging services or document retrieval fax processing.

The key CT Media element which manages this important resource management is the Resource Allocation Service Provider, another one of those unique, enduring, non-replaceable Service Providers. It's responsible for: interpreting Application Profiles to determine what resources are required for a particular Group; locating and securing the resources it needs for its particular task; and allocating resources to Groups.

When a new resource is created, it notifies the Resource Allocation Service of its existence, registers a description of its features and any allocation constraints it may face. Once the resource is fully initialized it advises the Resource Allocation Service that it's available for allocation to a Group.

In this way the Resource Allocation Service builds a pool of resources of different classes and with different features. When a Group is being built, the Resource Allocation Service searches this pool for a resource of the correct class and with the required features to meet an application's needs.

In the same way that resources hide the individual implementations of physical resources and Groups hide the switching relationship between resources, the Resource Allocation Service hides the location and individual identity of physical resources from the application.

- **Applications and Group Management.** After a session has been established, the system's app processing resources can be fired up.

Resources are assumed to have at a minimum one input or output of circuit / packet-switched data on the internal switch fabric of the Server. Resources are shared in the sense that multiple applications may contend for access to a resource; however, once granted access, the application has exclusive ownership until it explicitly releases it.

The exact capabilities of resources may vary from vendor to vendor. For example, different vendors might implement different audio coder types. An application could be sensitive to the implemented coder type because of the format of its data. It would use standard function calls to determine the coder types supported by a particular resource.

Groups, again, provide a means for an application to access the resources it needs for processing a call. You can view them as dynamically configurable sets of resources that are used within an application to service a call. They expose their resources to the application and guarantee the existence of the resources at all times. Resources in a Group are allocated to the exclusive use of that Group until freed.

There are three Group services available to applications:

- Implicit management of raw data path relationships (for example, voice and fax) between Group members, relieving the application from having to switch between multiple resources in a Group.
- Representation of a single entity to the application (the Group ID). This saves the application the chore of maintaining separate handles or IDs for each resource in the Group and a common path for events from the resources in a Group.

- Reservation of all physical resources (CPU, memory, connections). Any resources needed to provide the application with guaranteed service for the duration of the call or beyond.

And while all members of a Group are resources, a Group has a single primary member which is distinguished only through its connection relationship to the other members of the Group. All members listen to this primary member and the primary member listens to whatever member of the Group is currently talking (transmitting), which means that only one secondary resource talks at any point in time.

For example, an incoming phone call might need a call channel resource (trunk interface), tone signaling detector, voice recognition services and an audio playback capability. The application simply creates a Group by assembling the resources according to a Group configuration in the Application Profile. When the call is complete, the application releases the resources, allowing other applications to have access to them.

On the application side, the Group Management API provides functions for obtaining characteristics and capabilities of resources within the Group, as well as changing the current value of a resource characteristic (provided the resource allows for settable characteristics).

CT Media, meanwhile, offsets this API level with its Group Service Provider, which, though transient to the runtime system, is a non-replaceable Service Provider. A Group to this software module represents a dynamically configurable set of SPCs. There may be many Groups per logical server.

From the application's point of view, these Groups are the core components for call processing. They have the responsibility to: present the resources that comprise its component SPCs as a single entity to an application; route messages between SPCs within the Group and between its member resources and other Service Providers (for example, Session); manage any persistent Runtime Control that its member Resources might be involved in; manage the streamed data switching relationships between member resources.

- **Applications and Connection Management.** While Groups provide a means for implicit (hidden from the application's view and control) connections between resources, the connection management interface provides a means for explicit resource switching management. This lets an app reserve, make and break connections between two resources through the Server switch fabric.

The actual connection to a group takes place through a virtual resource known as a switch port. The switch port is an implicit member of every Group, unless it's explicitly designated as the primary member of the Group at creation time. A switch port is an abstraction of the connection of the Group into the switch fabric of the Server.

Because the model allows the application to connect resources together rather than manipulate physical switching components, CT Media offers powerful capabilities for connecting various resources together in an abstract view from the application perspective. Creating connection or switching applications that control resources such as network interfaces, station-set interfaces, audio players and tone generators and detectors can be built independent of the underlying hardware configuration. The application does not have to manage multiple switch stages or physical addresses on a bus.

Connection Management also allows for the creation of conference bridges by connecting the switch fabric in such a way as to allow each talker to listen to all the other talkers. Each talker then has a Group associated with their leg of the call. All of these Groups are interconnected across the switch fabric by the Connection Manager.

As the conference grows, Groups are created to support the new party and attached to the existing conference bridge. In this way, the size of a conference is only limited by the number of Resources or data connections the system has available.

The Connection Manager also allows for setting up a listen-only monitoring connection across any two-party connection for attendant-supervision applications.

It has its own CT Media core component, of course. Called, simply, the Connection Manager Service Provider, it has two primary components:

The Logical Connection Manager (LCM) is an enduring, non-replaceable Service Provider that handles abstract requests to connect the switch fabric timeslots of resources together.

The Switch Domain Controller is an enduring, replaceable Service Provider that handles LCM requests by executing specific switch-fabric switching commands.

Overall, it's the responsibility of the Connection Manager to: hide the implementation of the switch fabric (or fabrics) used by the Server from the application; remove the need for applications to manage switching explicitly; manage bus bandwidth across all switch fabrics in the Server; provide streamed data paths between SPCs in a Group; and provide streamed data paths between primary SPCs in different Groups.

- **Applications and the System Call Router.** CT Media includes a System Call Router (SCR) software mechanism for handling simple call control tasks and moving resources among applications. This includes the primary method of call control within CT Media, which follows the S.100 application model, including inbound call routing to applications and outbound call routing to trunks.

- **Inbound Routing.** Today, inbound applications suffer from a special problem when sharing resources. It's not efficient to dedicate telephony resources among a large number of applications, since those applications may only be used a few times in a given day. CT Media addresses this situation by sharing lines among applications using the SCR.

The SCR performs the low-level call-control operations required to detect the arrival of a call and do whatever internal actions are necessary to make the data stream of the call available in a group. SCR answers all inbound calls and uses routing rules established by a system administrator to determine the application each call should be connected to.

Since the clients can be distributed over a LAN, there are potentially a large number of clients, each interested only in calls with specific characteristics. The SCR may be instructed to route incoming calls to different applications depending on criteria such as technology or media type, trunk ID, ANI or time of day / day of week.

Calls may be routed to a whole class of applications, instead of just one. In this case, the SCR attempts to route the call to each available application of the specified class (fax servers, for example), until it finds one that can accept the call. If no application can take the call, the SCR can hang up the call or to pass it to a default application.

- **Outbound Routing.** For shared inbound / outbound call handling, with multiple apps sharing a single Server, the SCR may be used to allocate outbound trunks. In this case, an application that wants to dial out into the network would notify the SCR. The SCR would then react by creating a Group, seizing a line, perhaps confirming dialtone to avoid glare (that is, to prevent an incoming call on the line interrupting the preparation for outbound dialing) and passing the trunk group to the waiting application.

The outbound application would then be free to dial on the outbound line. The SCR may also be instructed to perform the dialing, setting up any international tone template automatically. The SCR would then pass the group to the waiting application on connection.

The SCR also acts as means for an application to pass a call (Group) to another application. Again, this is called "call handoff."

Applications use the SCR to handoff a call to a documented application token name. Mapping of application tokens to specific installed applications may be configured by the system administrator. The receiving application must have requested a Group from SCR to receive the handed-off Group. On receiving the Group, the new application may check for Group-specific data that a previous application may have associated with the Group, such as account number or reason for handoff.

In the case of handoff failure, the Group is returned to the original application so that the caller is not lost. When handing off a group, an application may also indicate that it expects the group to be returned.

The SCR maintains an ownership stack that allows a Group (and its associated call) to be returned to the previous owner when an application is finished with that Group. When the application currently in control of the Group hands off the Group, it does so entirely, leaving the original application without any reference to the Group. When the call is terminated, an application may destroy the Group explicitly or return the Group up the ownership stack for disposal.

All of this is provided for by CT Media's SCSA Call Router Service Provider, an enduring, non-replaceable, Service Provider that: detects inbound calls and routes them to the appropriate application; makes outbound calls on the appropriate trunk based on routing rules; hands off Groups from one application Session to another; and configures Groups (using the Resource Allocation Service), so that they satisfy the needs of the application receiving the Group. The latter includes reconfiguring Groups if necessary.

- **Intelligent Switch Extensions.** CT Media also provides a special Intelligent Switch Extension (ISE) which extends the capability of the SCR by enabling the introduction of a call processing (PBX) service, which is closely coupled to and shares the same media services available to the CT Media Server.

CT Media will soon be available with an Open Standards based call processing referred to as Open ISE This will provide support for comprehensive call control using TAPI and CTI interfaces. Using these new interfaces developers will be able to build full featured switching services, supplementary services, terminal control and call detail recording services.

CT Media's Storage Features

CT Media uses Containers, Time-Varying Media and Spatial Media for data management.

- **Containers.** A Container holds one or more Data Objects. Data Objects are made up of the actual data and a set of attributes which describe the properties of the data.

In many ways, Containers and Data Objects are analogs of file systems and files in a particular operating system, except that the Container interface is independent of the actual storage strategies employed in a particular implementation. Examples of data types typically stored in CT Media Containers include: encoded audio; text; Spatial Media (for fax); and speech recognition vocabularies.

- **Open Containers.** CT Media provides an interface to enable vendor extensibility for other Data Object types. Open Container interfaces provide a method of adding different Coder types (e.g., ASCII or TIFF) and Storage Strategies that can be used to store Media Data Objects using different storage methods (e.g., MAPI as opposed to the default file / directory structure).

- **Time-Varying Media (TVM).** Containers also store this particular type of Data Object. You can think of TVM an encapsulation of an atomic piece of a time-varying medium to be recorded and played back in real-time. TVM examples include: Encoded audio data, text-to-speech data, ADSI data and TDD data.

  All of the data within a TVM Data Object is of the same format and type. The TVM encapsulates all of the information necessary for playing the data. This information is usually stored in the TVM when it is created by means of recording.

- **Spatial Media.** Spatial Media (SM) objects are Data Objects whose individual elements have geometric location information associated with them. The data portion of SM objects may be structured information such as TIFF files or may be less structured information, such as ASCII strings.

  Common examples of Spatial Media data are TIFF files, ASCII T.50 files and DCX / PCX data.

The key CT Media Service Provider here is the Container Manager. It's another enduring, non-replaceable Service Provider. Computer telephony systems typically need to store bulk data (for example during recording) and to access previously stored bulk data (for example, speech recognition vocabularies).

The Container Manager helps them take care of these tasks. It manages access to stored data and data storage facilities, provides the application with location independent to these facilities, and gives the app access to data indexed by the data's natural units (for example, pages of fax, milliseconds of speech, etc.).

CT Media Administration

CT Media provides system administration through a system administration application to assist system integrators and end users to properly configure all of its parts. The CT Media Admin application is built using the standard administration API's available in CT Media, allowing application developers to develop their own customized interfaces.

Using the CT Media admin utility, system administrators can configure resources and applications, set up a remote connection to a server, edit profiles, start and stop Service Providers, map application call handoff requests to specific applications installed on the Server, and configure SCR call-routing rules. There's also a Container Manager utility available for viewing the CT Media Container system and managing Data Objects stored in the Container system.

CT Media and TAPI

The ECTF does not have a standard call control API. When CT Media first appeared there was an attempt to compare it to (and pit it against) the Microsoft TAPI specification.

CT Media is CT Server software, which supports apps written to multiple APIs. Dialogic has plans for incorporating TAPI 3.0 call control (switching) and media processing capabilities (voice, fax, speech, etc.) into CT Media so TAPI-compliant apps can run on a CT Media-based server.

A CT Media-based CT server platform makes significant sense. You should consider it when making plans for future CT application purchases. As more ready-to-install applications become available, it will fulfill one of the visions that Dialogic had in mind years ago when they started down the long CT road.

Dialogic already has over 140 developers supporting and writing applications and resources for CT Media.

If you want more information about CT Media, I suggest you check out the Dialogic website at www.dialogic.com or go directly to.www.ctserver.com.

CT Resource Boards

Also known as telephony cards, telephony boards, fax boards, voice boards are the ISA, EISA, PCI and CompactPCI cards (and their associated drivers and development software) that you plug into PCs to help achieve the "integration" in what used to be called "Computer Telephony Integration." Certainly no self-respecting PC-based phone or IVR system would be without one.

Forgetting "pure" PC based phone systems for a moment, CT resource boards can exist "in skin" or actually embedded inside a phone system to perform some CT function, such as Telrad's plug in card to provide voice mail to their line of advanced phone systems. Indeed, many PBX makers find that the easiest way to infuse their products with CT capabilities is to extend their proprietary architecture with a PC chassis or backplane, and add industry standard CT resource boards to gain IVR, and / or analog, T-1, E-1 or IP telephony connectivity.

Since the PBX vendors are not planning to lay down and die any time soon, many CT resource board makers have found it necessary to build versions of their products that connect to a PBX rather than to a LEC interface, and some companies CT resource boards are specifically designed to work with PBXs and key systems from Lucent Technologies, Comdial, Fujitsu, Rolm, Samsung, Telrad, etc.

PCs and their resource cards have become an important part of the customer premise telephony equation because they are at the forefront of the convergence revolution, they can connect to data networks very easily and can serve as gateways between the circuit-switched world and packet-switched worlds where databases and other network resources dwell, resources that PBXs alone wouldn't have the faintest idea of how to access. Also, the cards are achieving higher and higher densities (higher port counts), more features (e.g. speech recognition, text to speech and universal port technology that enables any port to handle voice, fax or IP connectivity as needed by an application) and the cards are continually dropping in price.

By incorporating CT resource boards adhering to the Enterprise Computer Telephony Forum (ECTF) H.100 and H.110 CT Bus standards, one can mix and match boards from different manufacturers. Hot swappability is even now appearing as a feature of CT resource boards, something that PBX makers thought would never happen.

In order to effectively use a CT resource board in a CT system, you need great software development tools. Sophisticated application generators and toolkits from Intel / Parity, Artisoft, Envox, Prima, etc. are strongly tied into boards from a number on manufacturesr, Dialogic (an Intel Company), Natural Microsystems, and Brooktrout. Some developers match the board to the development tool, rather than vice versa.

While most of these boards are built for the North American market, many companies such as Dialogic, NMS and Brooktrout test their boards extensively so as to approve them for operation in 30 to 40 countries.

CT Resource boards have not as yet achieved the mass sales volume associated with say, commodity video or multimedia boards, but the company that now owns Dialogic is Intel. And Intel has a knack for mass producing technology. Perhaps one day we'll see a PC computer for the SOHO market having a motherboard sporting two or four analog ports, with onboard DTMF (touchtone) recognition and voice mail capability!

The two enormous tables that follow delineate the products of most CT resource board manufacturers. The first table consists of cards providing analog access to the PSTN, the second table lists the digital access cards (T-1, E-1, multiple spans, etc.)

Table of Analog CT Boards

Introductory Notes

Some integrated voice cards support a different number of telephony channels than their onboard DSPs can handle. These are usually designed for call completion systems such as debit cards, and international callback systems.

A standard call completion scenario has a call coming in one line and going out another, and most frequently, staying on the same card.

Computer Telephony resource boards are somewhat like the commodities market - one can never tell when a manufacturer is going to raise or lower the price of a particular model. The author asked vendors for single quantity list prices, so keep in mind that volume purchase discounts are almost always available.

For example, in the case of Dialogic, volume discount pricing is available. Contact Dialogic Sales or an authorized distributor (http://www.dialogic.com/partners/distrib.htm) for details (this applies to all pricing for Dialogic products contained within this matrix).

Also, older boards are discontinued by manufacturers and new ones with more features appear almost weekly. Therefore, the data presented in this section is not "set in stone," which means that the reader should use it only as a general buyer's guide to the market. For exact prices and details of any specific CT resource board, one should contact the manufacturer or its distributor.

Boards that have no ports are used as voice processing resource boards only, and require additional interface daughterboard cards or else they must communicate across a telephony bus (SCbus, MVIP, H.100, H.110, AEB, PEB) to be connected to analog or digital trunks

A Note on Linux: A Linux column was included in all tables, but not all manufacturers had announced Linux drivers by the time this book went to the printer. If a particular board definitely does or doesn't support Linux, that fact is indicated in the table with a "Yes" or a "No" in the Linux column. A "See Notes" entry means that you should contact the vendor, since there is a major ongoing effort in the computer telephony industry to adapt all hardware to Linux.

ANALOG TABLE LEGEND:

Vendor – Name of manufacturer

Product – Product name

List Price – Manufacturer's suggested list price in U.S. dollars (USD)

Ports – Number of telephony ports

Audio I/O – Does the board have audio jacks for recording and playback of voice prompts?

Local Port – Does it have a local POTS or headset / handset jack for a local phone or headset?

Fax – Does it support Fax onboard?

Caller ID – Can it collect inbound Caller ID?

Security – Can the serial number be queried by the application to lock the app to the board?

WAV Support – Does it offer the option to record and playback.WAV files?

Tel Bus – What telephony Bus does it support? MVIP, SCbus, H.100, H.110, AEB, PEB?

Card Length – What size is the board? Full length? Half? Three-quarters?

Form Factor – What kind of bus? ISA? PCI? CompactPCI (cPCI).

Pulse Rec – Can the board detect rotary dialed clicks?

Speech Rec – Can it do speech recognition onboard or using the Host CPU?

Conference – Can it conference two callers together?

Switching – Does it have onboard hardware to switch one channel over to another?

Pitch/Speed – Does it support pitch corrected variable speed playback?

DOS – Does the board and supporting drivers support DOS development?

Windows 3x – Does it support Windows development?

Windows 95 – Does it support Windows 95 development?

Windows NT – Does it support Windows NT development?

Unix – Does it support Unix development?

QNX – Does it support QNX real-time Unix?

OS/2 – Does it support OS/2 development?

Solaris – Does it support Solaris development?

Linux – Does it support Linux development?

TAPI – Does it support the Intel/Microsoft TAPI specification?

TSAPI – Does it support the Novel/AT&T TSAPI specification?

URL – Manufacturer's website address.

Phone – Manufacturer's phone number.

Comments – Any other clarifications or miscellaneous information about the board.

Vendor	ADAX	BICOM, Inc.	BICOM, Inc.	BICOM, Inc.
Product	APC7-PCI	SONIC C4	SONIC C2	SONIC C4 CTR 21
List Price	Call vendor	$495	$395	$625
Ports	2	4	2	4
Audio I/O	No	Yes	Yes	Yes
Local Port	No	No	No	No
Fax	No	OPT.	OPT.	OPT.
Caller ID	No	Yes	Yes	Yes
Security	No	Yes	Yes	Yes
Wav Sup	No	Yes	Yes	Yes
Tel-BUS	PMS,IAS,SBUS,ELSA,PCI,cPCI	AEB	AEB	AEB
Card Length	Half	Half	Half	3/4 length
Form Factor	PCI	ISA XT	ISA XT	ISA XT
Pulse Rec	No	No	No	No
Speech Rec	No	Opt Soft	Opt Soft	Opt Soft
Conference	No	No	No	No
Switching	No	No	No	No
Pitch/Speed	No	Yes	Yes	Yes
DOS	No	Yes	Yes	Yes
Win3x	No	Yes	Yes	Yes
Win95	No	No	No	No
Win NT	Yes	Yes	Yes	Yes
Unix	Yes	No	No	No
QNX	No	No	No	No
OS/2	No	No	No	No
Solaris	Call	No	No	No
Linux	See Notes	See Notes	See Notes	See Notes
TAPI	No	Yes	Yes	Yes
TSAPI	No	No	No	No
URL	www.adax.com	www.bicom-inc.com	www.bicom-inc.com	www.bicom-inc.com
Phone	510-548-7047	203-268-4484	203-268-4484	203-268-4484
Comments	Fractional to full T1/E1 per port with optional CSU/DSU (ANC); cPCI version available.			

Table of Analog Computer Telephony Boards

BICOM, Inc.	BICOM, Inc.	BICOM, Inc.	BICOM, Inc.	BICOM, Inc.
SONIC C2 CTR 21	SONIC C4-PTC	VM4	VMx/NT	VMx/DOS
$525	$1,595	$1,195	$1,495	$1,295
2	4	4	4/8/12	4/8/12
Yes	No	No	Yes	Yes
No	No	No	No	No
OPT.	OPT.	No	OPT.	OPT.
Yes	Yes	No	Yes	Yes
Yes	Yes	Yes	Yes	Yes
Yes	Yes	No	Yes	Yes
AEB	AEB	No	AEB	AEB
3/4	Full	N/A	Half	Half
ISA XT	ISA XT	N/A	ISA XT	ISA XT
No	Yes	No	No	No
Opt Soft	Opt Soft	No	Opt Soft	Opt Soft
No	No	No	No	No
No	No	No	No	No
Yes	Yes	Yes	Yes	Yes
Yes	Yes	Yes	Yes	Yes
Yes	No	No	Yes	Yes
No	No	No	No	No
Yes	Yes	No	Yes	Yes
No	No	No	No	No
No	No	No	No	No
No	No	No	No	No
No	No	No	No	No
See Notes	See Notes	See Notes	See Notes	See Notes
Yes	Yes	No	Yes	Yes
No	No	No	No	No
www.bicom-inc.com	www.bicom-inc.com	www.bicom-inc.com	www.bicom-inc.com	www.bicom-inc.com
203-268-4484	203-268-4484	203-268-4484	203-268-4484	203-268-4484

Vendor	Brooktrout Technology	Brooktrout Technology	Brooktrout Technology	Brooktrout Technology
Product	TruFax	TR114	RTNI-4ATI	RTNI-8ATI
List Price	starts at $799	starts at $1995	$995	$1,495
Ports	2	1,2,4	4 Trunks	8 Trunks
Audio I/O	No	No	No	No
Local Port	No	No	No	No
Fax	Yes	Yes	No	No
Caller ID	No	No	No	No
Security	No	No	Yes	Yes
Wav Sup	No	No	No	No
Tel-BUS	None	MVIP, PEB, SCbus	MVIP	MVIP
Card Length	3/4	Full	Full	Full
Form Factor	ISA XT/AT	ISA, PCI	ISA AT	ISA AT
Pulse Rec	No	No	No	No
Speech Rec	No	Yes	No	No
Conference	No	Yes	No	No
Switching	No	No	Yes	Yes
Pitch/Speed	No	Yes	No	No
DOS	No	Yes	Yes	Yes
Win3x	No	Yes	Yes	Yes
Win95	No	Yes	Yes	Yes
Win NT	Yes	Yes	Yes	Yes
Unix	No	Yes	Yes	Yes
QNX	No	Yes	Yes	Yes
OS/2	No	Yes	Yes	Yes
Solaris	No	Yes	Yes	Yes
Linux	No	Yes	See Notes	See Notes
TAPI	No	No	Yes	Yes
TSAPI	No	No	No	No
URL	www.brooktrout.com	www.brooktrout.com	www.brooktrout.com	www.brooktrout.com
Phone	781-449-4100	408-874-4116	408-874-4116	408-874-4116
Comments	Two-channel fax board for entry-level LAN fax applications.	Market leading platform for enterprise and service provider fax applications.		

Brooktrout Technology	Brooktrout Technology	Brooktrout Technology	Brooktrout Technology	Brooktrout Technology
RTNI-12ATI	RTNI-16ATI	RTNI-24ATI	RTNI-8ASI	RTNI-16ASI
$1,875	$2,245	$2,995	$1,995	$2,995
12 Trunks	16 Trunks	24 Trunks	8 Analog Stations	16 Analog Stations
No	No	No	No	No
No	No	No	No	No
No	No	No	No	No
No	No	No	No	No
Yes	Yes	Yes	Yes	Yes
No	No	No	No	No
MVIP	MVIP	MVIP	MVIP	MVIP
Full	Full	Full	Full	Full
ISA AT	ISA AT	ISA AT	ISA AT	ISA AT
No	No	No	No	No
No	No	No	No	No
No	No	No	No	No
Yes	Yes	Yes	Yes	Yes
No	No	No	No	No
Yes	Yes	Yes	Yes	Yes
Yes	Yes	Yes	Yes	Yes
Yes	Yes	Yes	Yes	Yes
Yes	Yes	Yes	Yes	Yes
Yes	Yes	Yes	Yes	Yes
Yes	Yes	Yes	Yes	Yes
Yes	Yes	Yes	Yes	Yes
Yes	Yes	Yes	Yes	Yes
See Notes	See Notes	See Notes	See Notes	See Notes
Yes	Yes	Yes	Yes	Yes
No	No	No	No	No
www.brooktrout.com	www.brooktrout.com	www.brooktrout.com	www.brooktrout.com	www.brooktrout.com
408-874-4116	408-874-4116	408-874-4116	408-874-4116	408-874-4116

Vendor	Brooktrout Technology	Brooktrout Technology	Brooktrout Technology	Brooktrout Technology
Product	RTNI-24ASI	RDSP-232	RDSP-432	Prelude Duet
List Price	$3,995	$695	$800	$495
Ports	24 Analog Stations	2 Analog	4 Analog	2 Analog
Audio I/O	No	No	No	No
Local Port	No	Yes	Yes	No
Fax	No	No	No	No
Caller ID	No	Yes	Yes	No
Security	Yes	Yes	Yes	No
Wav Sup	No	Yes	Yes	Yes
Tel-BUS	MVIP	None	None	None
Card Length	Full	Half	Half	Half
Form Factor	ISA AT	ISA AT	ISA AT	ISA AT
Pulse Rec	No	No	No	No
Speech Rec	No	No	No	No
Conference	No	No	No	No
Switching	Yes	No	No	No
Pitch/Speed	No	No	No	No
DOS	Yes	Yes	Yes	Yes
Win3x	Yes	Yes	Yes	Yes
Win95	Yes	Yes	Yes	Yes
Win NT	Yes	Yes	Yes	Yes
Unix	Yes	Yes	Yes	Yes
QNX	Yes	Yes	Yes	No
OS/2	Yes	Yes	Yes	No
Solaris	Yes	Yes	Yes	Yes
Linux	See Notes	See Notes	See Notes	See Notes
TAPI	Yes	Yes	Yes	Yes
TSAPI	No	No	No	No
URL	www.brooktrout.com	www.brooktrout.com	www.brooktrout.com	www.brooktrout.com
Phone	408-874-4116	408-874-4116	408-874-4116	408-874-4116
Comments				

Brooktrout Technology	Brooktrout Technology	Brooktrout Technology	Brooktrout Technology	Brooktrout Technology
Prelude Quartet	Vantage Volare	Vantage VPS-4	Vantage PCI-4L	Vantage PCI-8L
$595	$960	$1,495	$945	$1,692
4 Analog	4 Analog	4 Analog	4 Analog	8 Analog
No	Yes	Yes	Yes	Yes
Yes	Yes	Yes	Yes	Yes
No	No	No	No	No
No	Yes	Yes	Yes	Yes
No	Yes	Yes	Yes	Yes
Yes	Yes	Yes	Yes	Yes
None	None	MVIP	MVIP	MVIP
Half	Half	Half	Full	Full
ISA AT	ISA AT	ISA AT	PCI	PCI
No	Yes	Yes	Yes	Yes
No	No	No	No	No
No	No	No	No	No
No	No	Yes	Yes	Yes
No	Yes	Yes	Yes	Yes
Yes	Yes	Yes	No	No
Yes	Yes	Yes	No	No
Yes	Yes	Yes	No	No
Yes	Yes	Yes	Yes	Yes
Yes	Yes	Yes	No	No
No	Yes	Yes	No	No
No	Yes	Yes	No	No
Yes	Yes	Yes	No	No
See Notes	See Notes	See Notes	See Notes	See Notes
Yes	Yes	Yes	Yes	Yes
No	No	No	No	No
www.brooktrout.com	www.brooktrout.com	www.brooktrout.com	www.brooktrout.com	www.brooktrout.com
408-874-4116	408-874-4116	408-874-4116	408-874-4116	408-874-4116

Vendor	Brooktrout Technology	Brooktrout Technology	Intel/Dialogic	Intel/Dialogic
Product	Vantage PCI-4Hx	Vantage PCI-8Hx	VoiceBrick(TM)	ProLine/2V(TM)
List Price	$1,115	$1,996	$ 900 – 1100	$395
Ports	4 Analog	8 Analog	2 or 4	2
Audio I/O	Yes	Yes	No	Yes
Local Port	Yes	Yes	Yes	Yes
Fax	No	No	No	No
Caller ID	Yes	Yes	Yes	Yes
Security	Yes	Yes	Yes	Yes
Wav Sup	Yes	Yes	No	Yes
Tel-BUS	MVIP, H.100	MVIP, H.100	standalone	N/A
Card Length	Full	Full	Standalone	Half
Form Factor	PCI	PCI	Standalone	ISA AT
Pulse Rec	Yes	Yes	Yes	Yes
Speech Rec	No	No	No	No
Conference	No	No	No	No
Switching	Yes	Yes	No	No
Pitch/Speed	Yes	Yes	Yes	Yes
DOS	No	No	Yes	Yes
Win3x	No	No	No	No
Win95	No	No	No	Yes
Win NT	Yes	Yes	No	Yes
Unix	No	No	No	No
QNX	No	No	No	No
OS/2	No	No	No	No
Solaris	No	No	No	No
Linux	See Notes	See Notes	No	No
TAPI	Yes	Yes	No	Yes
TSAPI	No	No	No	No
URL	www.brooktrout.com	www.brooktrout.com	www.dialogic.com	www.dialogic.com
Phone	408-874-4116	408-874-4116	800-755-4444	800-755-4444
Comments				

Intel/Dialogic	Intel/Dialogic	Intel/Dialogic	Intel/Dialogic	Intel/Dialogic
DIALOG/4(TM)	D/21H(TM)	D/41H(TM)	D/4	D/41E(TM)
$675	$595	$995	$995	$1,195
4	2	4	4	4
No	No	No	No	No
Yes	Yes	Yes	Yes	Yes
No	No	No	No	No
No	Yes	Yes	Yes	Yes
Yes	Yes	Yes	Yes	Yes
Yes	Yes	Yes	Yes	Yes
N/A	N/A	N/A	N/A	SCbus, H.100
Half	Half	Half	Half	Full
ISA AT	ISA AT	ISA AT	PCI	PCI/ ISA AT
No	Yes	Yes	Yes	Yes
No	Yes	Yes	No	Yes
No	No	No	No	No
No	No	No	No	Yes
No	Yes	Yes	Yes	Yes
Yes	Yes	Yes	No	Yes
No	No	No	No	No
Yes	Yes	Yes	No	Yes
Yes	Yes	Yes	Yes	Yes
No	Yes	Yes	No	Yes
No	No	No	No	No
Yes	Yes	Yes	No	Yes
No	No	No	No	Yes
No	Yes	Yes	No	Yes
Yes	Yes	Yes	Yes	Yes
No	No	No	No	No
www.dialogic.com	www.dialogic.com	www.dialogic.com	www.dialogic.com	www.dialogic.com
800-755-4444	800-755-4444	800-755-4444	800-755-4444	800-755-4444

Table of Analog Computer Telephony Boards, continued

Vendor	Intel/Dialogic	Intel/Dialogic	Intel/Dialogic	Intel/Dialogic	
Product	D/41JCT-LS	D/120JCT-LS	D/160SC-LS(TM)	VFX/40ESC plus(TM)	
List Price	$1,555	$3,146	$3,995	$1,995	
Line Interface	4	12	16	4	
DS0 Channels	No	No	No	No	
DSP V ports	Yes	No	No	Yes	
Onboard DSP	Yes	Yes	No	Yes	
Tone ports	Yes	Yes	Yes	Yes	
Fax	Yes	Yes	Yes	Yes	
Security	Yes	Yes	No	Yes	
Wav Sup	SCbus	SCbus, H.100	SCbus	SCbus, H.100	
Tel-BUS	Full	Full	Full	Full	
Card Length	PCI	PCI	ISA AT	PCI/ ISA AT	
Form Factor	Yes	Yes	Yes	Yes	
Pulse Rec	Yes	Yes	No	Yes	
Speech Rec	No	No	No	No	
Conference	Yes	Yes	Yes	Yes	
Switching	Yes	Yes	Yes	Yes	
Pitch/Speed	No	No	Yes	Yes	
DOS	No	No	No	No	
Win3x	No	No	No	Yes	
Win95	Yes	Yes	Yes	Yes	
Win NT	Yes	Yes	Yes	Yes	
Unix	No	No	No	No	
QNX	No	No	No	Yes	
OS/2	Yes	Yes	Yes	No	
Solaris	Yes	Yes	Yes	Yes	
Linux	Yes	Yes	Yes	Yes	
TAPI	No	No	No	No	
URL	www.dialogic.com	www.dialogic.com	www.dialogic.com	www.dialogic.com	
Phone	800-755-4444	800-755-4444	800-755-4444	800-755-4444	
Comments					

Intel/Dialogic	Intel/Dialogic	Intel/Dialogic	Intel/Dialogic	Intel/Dialogic
CPi/100(TM)	CPi/200(TM)	CPi/400(TM)	CPD/220(TM)	DM/IP040-LSI
$660	$1,325	$2,750	$2,845	$1,520
1	2	4	4	4
No	No	No	No	No
Yes	Yes	Yes	Yes	Yes
Yes	Yes	Yes	Yes	Yes
No	No	No	No	Yes
Yes	Yes	Yes	Yes	Yes
No	No	No	No	Yes
N/A	N/A	N/A	N/A	SCbus, H.100
Half	Full	Full	Full	Full
ISA AT	PCI/ ISA AT	PCI/ ISA AT	ISA AT	PCI
No	No	No	No	No
No	No	No	No	Yes
No	No	No	No	Yes
No	No	No	No	Yes
No	No	No	No	Yes
Yes	Yes	Yes	No	No
No	No	No	No	No
No	No	No	No	No
Yes	Yes	Yes	Yes	Yes
Yes	Yes	Yes	No	No
No	No	No	No	No
Yes	Yes	Yes	No	No
No	No	No	No	No
No	No	No	No	Yes
No	No	No	No	No
No	No	No	No	No
www.dialogic.com	www.dialogic.com	www.dialogic.com	www.dialogic.com	www.dialogic.com
800-755-4444	800-755-4444	800-755-4444	800-755-4444	800-755-4448

Table of Analog Computer Telephony Boards, continued

Vendor	Intel/Dialogic	Intel/Dialogic	Intel/Dialogic	Intel/Dialogic
Product	MSI/80 GBLStation Interface	MSI/160-GBL Station Interface	MSI/240-GBL Station Interface	CT Media Starter Kit
List Price	$1,885	$2,675	$3,465	$1,495
Line Interface	8	16	24	4
DS0 Channels	Yes	Yes	Yes	No
DSP V ports	Yes	Yes	Yes	Yes
Onboard DSP	No	No	No	No
Tone ports	No	No	No	Yes
Fax	Yes	Yes	Yes	Yes
Security	No	No	No	Yes
Wav Sup	SCbus, H.100	SCbus, H.100	SCbus	SCbus, H.100
Tel-BUS	Full	Full	Full	Full
Card Length	PCI/ ISA AT	PCI/ ISA AT	ISA AT	PCI/ ISA AT
Form Factor	No	No	No	Yes
Pulse Rec	No	No	No	Yes
Speech Rec	Yes	Yes	Yes	No
Conference	Yes	Yes	Yes	Yes
Switching	No	No	No	Yes
Pitch/Speed	No	No	No	Yes
DOS	No	No	No	No
Win3x	No	No	No	Yes
Win95	Yes	Yes	Yes	Yes
Win NT	Yes	Yes	Yes	Yes
Unix	No	No	No	No
QNX	No	No	No	Yes
OS/2	No	No	No	Yes
Solaris	Yes	Yes	Yes	Yes
Linux	No	No	No	Yes
TAPI	No	No	No	No
URL	www.dialogic.com	www.dialogic.com	www.dialogic.com	www.dialogic.com
Phone	800-755-4444	800-755-4444	800-755-4444	800-755-4444
Comments	Has 32 conference resources and supports international phones.	Has 32 conference resources and supports international phones.	Has 32 conference resources and supports international phones.	Includes D/41EPCI, CT Media SDK, licenses, line simulator, and tools. Product specs same as for D/41E PCI.

Intel/Dialogic	Music Telecom	Music Telecom	Music Telecom	Music Telecom
CT Media Starter Kit PRO	NewWave 2EZ	NewWave 4EZ	NV-DSP 2HS	NV-DSP 4HS
$2,995	$395	$495	$445	$655
4	2	4	2	4
No	Yes	Yes	Yes	Yes
Yes	Yes	Yes	Yes	Yes
No	No	No	No	No
Yes	Yes	Yes	Option	Option
Yes	Yes	Yes	Yes	Yes
Yes	Yes	Yes	Yes	Yes
SCbus, H.100	AEB	AEB	AEB, MVIP	AEB, MVIP
Full	Half	Half	Half	Half
PCI/ ISA AT	ISA XT/AT	ISA XT/AT	ISA XT/AT	ISA XT/AT
Yes	No	No	No	No
Yes	Yes	Yes	Yes	Yes
No	No	No	Option	Option
Yes	No	No	Yes	Yes
Yes	Yes	Yes	Yes	Yes
Yes	Yes	Yes	Yes	Yes
No	Yes	Yes	Yes	Yes
Yes	Yes	Yes	Yes	Yes
Yes	Yes	Yes	Yes	Yes
Yes	No	No	No	No
No	No	No	No	No
Yes	No	No	No	No
Yes	No	No	No	No
Yes	See Notes	See Notes	See Notes	See Notes
Yes	Yes	Yes	Yes	Yes
No	No	No	No	No
www.dialogic.com	www.musictelecom.com	www.musictelecom.com	www.musictelecom.com	www.musictelecom.com
800-755-4444	800-648-3647	800-648-3647	800-648-3647	800-648-3647
Adds 1 week training plus 90 days technical support. Product specs same as D/41E PCI.				

Vendor	Music Telecom	Music Telecom	Music Telecom	Music Telecom
Product	Passport MTP4	Passport MTP8	Passport MTP16	Passport MTFX44
List Price	$595	$1,075	$1,795	$1,195
Line interface	4	8	16	4
DS0 Channels	Yes	Yes	Yes	Yes
DSP V ports	Yes	Yes	Yes	Yes
Onboard DSP	No	No	No	Yes 4 Pts
Tone ports	Option	Option	Option	Option
Fax	Yes	Yes	Yes	Yes
Security	Yes	Yes	Yes	Yes
Wav Sup	AEB, MVIP	AEB, MVIP	MVIP	AEB, MVIP
Tel-BUS	Full	Full	Full	Full
Card Length	ISA XT/AT	ISA XT/AT	ISA XT/AT	ISA XT/AT
Form Factor	No	No	No	No
Pulse Rec	Yes	Yes	Yes	Yes
Speech Rec	Option	Option	Option	Option
Conference	Yes	Yes	Yes	Yes
Switching	Yes	Yes	Yes	Yes
Pitch/Speed	Yes	Yes	Yes	Yes
DOS	Yes	Yes	Yes	Yes
Win3x	Yes	Yes	Yes	Yes
Win95	Yes	Yes	Yes	Yes
Win NT	No	No	No	No
Unix	No	No	No	No
QNX	No	No	No	No
OS/2	No	No	No	No
Solaris	See Notes	See Notes	See Notes	See Notes
Linux	Yes	Yes	Yes	Yes
TAPI	No	No	No	No
URL	www.musictelecom.com	www.musictelecom.com	www.musictelecom.com	www.musictelecom.com
Phone	800-648-3647	800-648-3647	800-648-3647	800-648-3647
Comments				

Table of Analog Computer Telephony Boards, continued

Music Telecom	Music Telecom	Music Telecom	Music Telecom	Music Telecom
Passport MTR16	Passport MTP2PCI	Passport MTP4PCI	Passport MTP8PCI	StationMasterMT4X12
$1,395	$425	$595	$1,075	$1,695
16	2	4	4	4
Yes	Yes	Yes	Yes	Yes
Yes	Yes	Yes	Yes	Yes
No	No	No	No	No
Option	Option	Option	Option	Option
Yes	Yes	Yes	Yes	Yes
Yes	Yes	Yes	Yes	Yes
MVIP	MVIP	MVIP	MVIP	MVIP
Full	Half	Half	Full	Full
ISA XT/AT	PCI	PCI	PCI	ISA XT/AT
No	No	No	No	No
Yes	Yes	Yes	Yes	Yes
Option	Option	Option	Option	Option
Yes	Yes	Yes	Yes	Yes
Yes	Yes	Yes	Yes	Yes
Yes	Yes	Yes	Yes	Yes
Yes	Yes	Yes	Yes	Yes
Yes	Yes	Yes	Yes	Yes
Yes	Yes	Yes	Yes	Yes
No	No	No	No	No
No	No	No	No	No
No	No	No	No	No
No	No	No	No	No
See Notes	See Notes	See Notes	See Notes	See Notes
Yes	Yes	Yes	Yes	Yes
No	No	No	No	No
www.musictelecom.com	www.musictelecom.com	www.musictelecom.com	www.musictelecom.com	www.musictelecom.com
800-648-3647	800-648-3647	800-648-3647	800-648-3647	800-648-3647

Vendor	Music Telecom	Music Telecom	Music Telecom	Music Telecom
Product	Desktop CTI MT1X1	MT2X6	MTR8PCI	Desktop PCI MT1X1PCI
List Price	$245	$1,075	$795	$245
Line Interface	1	2	8	1
DS0 Channels	Yes	Yes	Yes	Yes
DSP V ports	Yes	Yes	Yes	Yes
Onboard DSP	No	No	No	No
Tone ports	Yes	Option	Option	Yes
Fax	Yes	Yes	Yes	Yes
Security	Yes	Yes	Yes	Yes
Wav Sup	MVIP	MVIP	MVIP	MVIP
Tel-BUS	Half	Half	Full	Half
Card Length	ISA XT/AT	PCI	PCI	PCI
Form Factor	No	No	No	No
Pulse Rec	Yes	Yes	Yes	Yes
Speech Rec	Option	Option	Option	Option
Conference	Yes	Yes	Yes	Yes
Switching	Yes	Yes	Yes	Yes
Pitch/Speed	Yes	Yes	Yes	Yes
DOS	Yes	Yes	Yes	Yes
Win3x	Yes	Yes	Yes	Yes
Win95	Yes	Yes	Yes	Yes
Win NT	No	No	No	No
Unix	No	No	No	No
QNX	No	No	No	No
OS/2	No	No	No	No
Solaris	See Notes	See Notes	See Notes	See Notes
Linux	Yes	Yes	Yes	Yes
TAPI	No	No	No	No
URL	www.musictelecom.com	www.musictelecom.com	www.musictelecom.com	www.musictelecom.com
Phone	800-648-3647	800-648-3647	800-648-3647	800-648-3647
Comments				

NMS	NMS	NMS	NMS	NMS
AG 2000-8LS	AG 2000-8LSE	AG 2000/200	AG 2000-8SL	AG 2000-4LS/4SL
$2,495	$2,495	$1,895	$2,750	$2,695
8	8	0	8	8
No	No	No	No	No
No	No	No	No	No
Opt Soft	Opt Soft	Opt Soft	No	No
Yes	Yes	No	No	No
Yes	Yes	Yes	Yes	Yes
Yes	Yes	Yes	Yes	Yes
MVIP, H100	MVIP, H100	MVIP, H100	MVIP, H100	MVIP, H100
Full	Full	Full	Full	Full
PCI	PCI	PCI	PCI	PCI
No	No	No	No	No
Yes	Yes	Yes	Yes	Yes
Yes	Yes	Yes	Yes	Yes
Yes	Yes	Yes	Yes	Yes
Yes	Yes	Yes	Yes	Yes
No	No	No	No	No
No	No	No	No	No
No	No	No	No	No
Yes	Yes	Yes	Yes	Yes
Yes	Yes	Yes	Yes	Yes
Special	Special	Special	Special	Special
No	No	No	No	No
Yes	Yes	Yes	Yes	Yes
See Notes	See Notes	See Notes	See Notes	See Notes
Yes	Yes	Yes	Yes	Yes
No	No	No	No	No
www.nmss.com	www.nmss.com	www.nmss.com	www.nmss.com	www.nmss.com
800-533-6120	800-533-6120	800-533-6120	800-533-6120	800-533-6120

Vendor	NMS	NMS	NMS	NMS
Product	AG 2000-8DID	AG 2000-4DID/4LS	AG 2000-E&M,IA	AG 2000-E&M,IB
List Price	$2,995	$2,895	$2,995	$2,995
Line Interface	8	8	8	8
DS0 Channels	No	No	No	No
DSP V ports	No	No	No	No
Onboard DSP	Opt Soft	Opt Soft	No	No
Tone ports	Yes	Yes	No	No
Fax	Yes	Yes	Yes	Yes
Security	Yes	Yes	Yes	Yes
Wav Sup	MVIP, H100	MVIP, H100	MVIP, H100	MVIP, H100
Tel-BUS	Full	Full	Full	Full
Card Length	PCI	PCI	PCI	PCI
Form Factor	No	No	No	No
Pulse Rec	Yes	Yes	Yes	Yes
Speech Rec	Yes	Yes	Yes	Yes
Conference	Yes	Yes	Yes	Yes
Switching	Yes	Yes	Yes	Yes
Pitch/Speed	No	No	No	No
DOS	No	No	No	No
Win3x	No	No	No	No
Win95	Yes	Yes	Yes	Yes
Win NT	Yes	Yes	Yes	Yes
Unix	Special	Special	Special	Special
QNX	No	No	No	No
OS/2	Yes	Yes	Yes	Yes
Solaris	See Notes	See Notes	See Notes	See Notes
Linux	Yes	Yes	Yes	Yes
TAPI	No	No	No	No
URL	www.nmss.com	www.nmss.com	www.nmss.com	www.nmss.com
Phone	800-533-6120	800-533-6120	800-533-6120	800-533-6120
Comments				

Table of Analog Computer Telephony Boards, continued

NMS	NMS	NMS	NMS	NMS
AG 2000-E&M,VA	AG 2000-E&M,VB	QX 2000-4LS	QX 2000-4LSE	AG-8/80 LS
$2,995	$2,995	$995	$995	$2,495
8	8	4	4	8
No	No	No	No	No
No	No	No	No	No
No	No	Opt Soft	Opt Soft	Opt Soft
No	No	No	No	Yes
Yes	Yes	Yes	Yes	Yes
Yes	Yes	Yes	Yes	Yes
MVIP, H100	MVIP, H100	MVIP, H100	MVIP, H100	MVIP
Full	Full	Full	Full	Full
PCI	PCI	PCI	PCI	ISA AT
No	No	No	No	No
Yes	Yes	Yes	Yes	No
Yes	Yes	No	No	No
Yes	Yes	Yes	Yes	Yes
Yes	Yes	Yes	Yes	No
No	No	No	No	No
No	No	No	No	No
No	No	No	No	No
Yes	Yes	Yes	Yes	Yes
Yes	Yes	No	No	Yes
Special	Special	No	No	No
No	No	No	No	Yes legacy
Yes	Yes	No	No	Yes
See Notes	See Notes	See Notes	See Notes	See Notes
Yes	Yes	Yes	Yes	Yes
No	No	No	No	No
www.nmss.com	www.nmss.com	www.nmss.com	www.nmss.com	www.nmss.com
800-533-6120	800-533-6120	800-533-6120	800-533-6120	800-533-6120

Vendor	NMS	NMS	NMS	NMS
Product	AG-8/80 DID	AG-4 LS	CX 1000-8	CX 1000-16
List Price	$2,750	$1,695	$1,795	$2,795
Line Interface	8	4	8	16
DS0 Channels	No	No	No	No
DSP V ports	No	No	No	No
Onboard DSP	Opt Soft	Opt Soft	No	No
Tone ports	Yes	No	No	No
Fax	Yes	Yes	Yes	Yes
Security	Yes	Yes	No	No
Wav Sup	MVIP	MVIP	MVIP	MVIP
Tel-BUS	Full	Full	Full	Full
Card Length	ISA AT	ISA AT	ISA AT	ISA AT
Form Factor	No	No	No	No
Pulse Rec	No	No	No	No
Speech Rec	No	No	Yes	Yes
Conference	Yes	Yes	Yes	Yes
Switching	No	No	No	No
Pitch/Speed	No	No	No	No
DOS	No	No	No	No
Win3x	No	No	No	No
Win95	Yes	Yes	Yes	Yes
Win NT	Yes	Yes	Yes	Yes
Unix	No	No	No	No
QNX	Yes legacy	Yes legacy	Yes legacy	Yes legacy
OS/2	Yes	Yes	Yes	Yes
Solaris	See Notes	See Notes	See Notes	See Notes
Linux	Yes	Yes	Yes	Yes
TAPI	No	No	No	No
URL	www.nmss.com	www.nmss.com	www.nmss.com	www.nmss.com
Phone	800-533-6120	800-533-6120	800-533-6120	800-533-6120
Comments				

NMS	NMS	NMS	NMS	NMS
CX 1000-24	T Connect-8	T Connect-16	T Connect-24	S/T Connect
$3,595	$1,595	$1,995	$2,595	$3,295
24	8	16	24	24
No	No	No	No	No
No	No	No	No	No
No	No	No	No	No
No	No	No	No	No
Yes	Yes	Yes	Yes	Yes
No	No	No	No	No
MVIP	MVIP	MVIP	MVIP	MVIP
Full	Full	Full	Full	Full
ISA AT	ISA AT	ISA AT	ISA AT	ISA AT
No	No	No	No	No
No	No	No	No	No
Yes	Yes	Yes	Yes	Yes
Yes	Yes	Yes	Yes	Yes
No	No	No	No	No
No	No	No	No	No
No	No	No	No	No
No	No	No	No	No
Yes	Yes	Yes	Yes	Yes
Yes	Yes	Yes	Yes	Yes
No	No	No	No	No
Yes legacy	Yes legacy	Yes legacy	Yes legacy	Yes legacy
Yes	Yes	Yes	Yes	Yes
See Notes	See Notes	See Notes	See Notes	See Notes
Yes	Yes	Yes	Yes	Yes
No	No	No	No	No
www.nmss.com	www.nmss.com	www.nmss.com	www.nmss.com	www.nmss.com
800-533-6120	800-533-6120	800-533-6120	800-533-6120	800-533-6120

Table of Analog Computer Telephony Boards, continued						
Vendor	NMS	NetPhone	NetPhone	NetPhone	PacketPort.com	PacketPort.com
Product	WTI-8 LS	PBX-024	PBX-124	PBX-618	PK4004-A3	PK4004-D3
List Price	$1,500	$4,290	$7,990	$4,290	$3,725	$2,848
Line Interface	8	24 S	1T-1/24 S	6 T x 18 S	4	4
DS0 Channels	No	Yes	Yes	Yes	No	No
DSP V ports	No	No	No	No	No	No
Onboard DSP	No	No	No	No	Yes	Yes
Tone ports	No	No	Yes	Yes	Yes	Yes
Fax	Yes	Yes	Yes	Yes	No	No
Security	No	CVT	CVT	CVT	Yes	Yes
Wav Sup	MVIP	MVIP	MVIP	MVIP	MVIP, SCbus	MVIP, SCbus
Tel-BUS	Full	Full	Full	Full	Full	Full
Card Length	ISA AT	ISA AT	ISA AT	ISA AT	PCI	PCI
Form Factor	No	No	No	No	No	No
Pulse Rec	No	No	No	No	Yes	Yes
Speech Rec	Yes	Yes	Yes	Yes	Yes	Yes
Conference	Yes	Yes	Yes	Yes	Yes	Yes
Switching	No	No	No	No	No	No
Pitch/Speed	No	No	No	No	No	No
DOS	No	No	No	No	No	No
Win3x	No	No	No	No	No	No
Win95	Yes	Yes	Yes	Yes	No	No
Win NT	Yes	No	No	No	Yes	Yes
Unix	No	No	No	No	No	No
QNX	Yes legacy	No	No	No	No	No
OS/2	Yes	No	No	No	Yes	Yes
Solaris	See Notes	See Notes	See Notes	See Notes	See Notes	See Notes
Linux	Yes	Yes	Yes	Yes	No	No
TAPI	No	Yes	Yes	Yes	No	No
URL	www.nmss.com	www.netphone.com	www.netphone.com	www.netphone.com	www/packetport.com	www/packetport.com
Phone	800-533-6120	508-787-1000	508-787-1000	508-787-1000	203-831-2214	203-831-2214
Comments	Requires another DSP card.	This card is a 24 port station card used to build UnPBX or call center systems.	Not a traditional voice card, but a complete UnPBX card.	A scaleable card where multiple cards can be interconnected to build fairly large TSAPI compliant UnPBX solutions.		

PacketPort.com	PacketPort.com	PacketPort.com	PacketPort.com	PacketPort.com	PacketPort.com	PacketPort.com
PK4008-A1	PK4012-T1	PK4012-D3	PK4024-T1	PK4024-T1-BS	Powerline-I	Powerline-IV
$4,603	$4,452	$6,533	$6,295	$15,675	$195	$699
8	12	12	24	24	1	4
No	No	No	No	No	Yes	Yes
No	No	No	No	No	Yes	No
No	No	Yes	No	Yes	No	No
Yes	Yes	Yes	Yes	Yes	No	drivers
No	No	No	No	No	Yes	Yes
Yes	Yes	Yes	Yes	Yes	CVT	CVT
MVIP, SCbus	MVIP, SCbus	MVIP, SCbus	MVIP, SCbus	MVIP, SCbus	None	MVIP
Full	Full	Full	Full	Full	Half	Full
PCI	PCI	PCI	PCI	PCI	ISA XT	ISA AT
No	No	No	No	No	No	Yes
No	No	Yes	No	Yes	No	No
Yes	Yes	Yes	Yes	Yes	No	Yes
Yes	Yes	Yes	Yes	Yes	No	Yes
No	No	No	No	No	No	No
No	No	No	No	No	Yes	Yes
No	No	No	No	No	No	No
No	No	No	No	No	No	No
No	No	No	No	No	No	No
Yes	Yes	Yes	Yes	Yes	No	No
No	No	No	No	No	No	No
No	No	No	No	No	No	Soon
Yes	Yes	Yes	Yes	Yes	No	No
See Notes	See Notes	See Notes	See Notes	See Notes	See Notes	See Notes
No	No	No	No	No	No	No
No	No	No	No	No	No	No
www/packetport.com	www/packetport.com	www/packetport.com	www/packetport.com	www/packetport.com	www.tti.net	www.tti.net
203-831-2214	203-831-2214	203-831-2214	203-831-2214	203-831-2214	510-339-8275	510-339-8275

Notes on Analog Computer Telephony Resource Boards

Notes on Analog Computer Telephony Resource Boards

Table of Digital CT Boards

Introductory Notes

Some integrated voice cards support a different number of channels than their onboard DSPs can handle. These are usually designed for call completion systems such as debit cards, and international callback systems.

A standard call completion scenario has a call coming in one line and going out another, and most frequently, staying on the same card.

ISDN BRI circuits have two voice channels per circuit. US ISDN PRI loses one audio channel to out of band signaling D channel leaving only 23 audio circuits available. If more than one US ISDN PRI card is used in a system, a common signaling channel can be shared across up to 6 PRI circuits so that circuits 2-6 support 24 audio channels.

As was stated previously for the analog boards, Computer Telephony resource boards in general are somewhat like the commodities market – one can never tell when a manufacturer is going to raise or lower the price of a particular model. The author asked vendors for single quantity list prices, so keep in mind that volume purchase discounts are almost always available.

For example, in the case of Dialogic, volume discount pricing is available. Contact Dialogic Sales or an authorized distributor (http://www.dialogic.com/partners/distrib.htm) for details (this applies to all pricing for Dialogic products contained within this matrix).

Also, older boards are discontinued by manufacturers and new ones with more features appear almost weekly. Therefore, the data presented in this section is not "set in stone," which means that the reader should use it only as a general buyer's guide to the market. For exact prices and details of any specific CT resource board, one should contact the manufacturer or its distributor.

Boards that have no ports are used as voice processing resource boards only, and require additional interface daughterboard cards or else they must communicate across a telephony bus (SCbus, MVIP, H.100, H.110, AEB, PEB) to be connected to analog or digital trunks

The prices of Aculab's boards were originally provided in United Kingdom Pounds Sterling (GBP). The prices have been converted to United States Dollars (USD) using the conversion rates in effect on June 2, 2000 of 1 USD equaling approximately 0.6676681 GBP with the result rounded to the nearest whole dollar figure.

A Note on Linux: A Linux column was included in all tables, but not all manufacturers had announced Linux drivers by the time this book went to the printer. If a particular board definitely does or doesn't support Linux, that fact is indicated in the table with a "Yes" or a "No" in the Linux column. A "See Notes" entry means that you should contact the vendor, since there is a major ongoing effort in the computer telephony industry to adapt all hardware to Linux.

DIGITAL TABLE LEGEND:

Vendor – Name of manufacturer

Product – Product name

List Price – Manufacturer's suggested list price in U.S. dollars (USD)

Line Interface – Onboard POTS, LS, T1, E1, ISDN BRI or ISDN PRI?

DS0 Channels – Number of 64Kbps DS0 channels the card will support via the onboard line interfaces.

DSP V Ports – Number of voice ports supported by the card (for voice record and playback).

Onboard DSP – Does board include onboard DSP for voice or tone processing?

Tone Ports – Number of simultaneous channels the board can either generate or detect DTMF tones ("touch-tones").

Fax – Does the board offer onboard fax send / receive features?

Security – Can a developer lock their software to this particular board through the API?

Wav Support – Does the board record or playback .WAV files without conversion?

Tel-BUS – What telephony buses are supported: MVIP, HMVIP, SCBus, H.100, H.110, AEB, PEB?

Card Length – What's the length of board: Full, half, three-quarters?

Form Factor – What computer bus is used: ISA XT, ISA AT, PCI, CompactPCI (cPCI), VMEbus?

Pulse Rec – Is there onboard pulse recognition for the detection of rotary dialed numbers?

Speech Rec – Does the board have onboard automatic speech recognition?

Conference – Does the board support channel-to-channel conferencing?

Switching – Does the board support channel to channel switching?

Pitch/Speed – Does the board support pitch-adjusted variable speed playback?

DOS – Does the board and supporting drivers support DOS development?

Win3x – Does the board and supporting drivers support Windows 3.x?

Win95 – Does the board and supporting drivers support Windows 95/98?

Win NT – Does the board and supporting drivers support Windows NT

Unix – Does the board and supporting drivers support Unix?

QNX – Does the board and supporting drivers support QNX?

OS/2 – Does the board and supporting drivers support OS/2?

Solaris – Does the board and supporting drivers support Solaris?

Linux – Does the board and supporting drivers support Linux?

TAPI – Is there TAPI support for this board?

URL – Manufacturer's website address.

Phone – Manufacturer's phone number.

Comments – Any other clarifications or miscellaneous information about the board.

159

Table of Digital Computer Telephony Boards

Vendor	Aculab	Aculab	Aculab	Aculab
Product	E1 Monitor Card	Single Port E1 Card	Single Port E1 Card with DSP	Dual Port E1 Card
List Price	$3,745	$2,809	$3,483	$3,745
Line Interface	E1/T1 PRI	1 E1	1 E1	2 E1
DS0 Channels	30/24	30	30	60
DSP V ports	0	0	0	0
Onboard DSP	No	No	Yes	No
Tone ports	0	0	30	0
Fax	No	No	No	No
Security	Yes	Yes	Yes	Yes
Wav Sup	N/A	N/A	N/A	N/A
Tel-BUS	MVIP, PEB, SCbus	MVIP, PEB, SCbus	MVIP, PEB, SCbus	MVIP, PEB, SCbus
Card Length	Full	Full	Full	Full
Form Factor	ISA AT	ISA AT	ISA AT	ISA AT
Pulse Rec	No	No	Yes	No
Speech Rec	No	No	No	No
Conference	Opt	Opt	Opt	Opt
Switching	Yes	Yes	Yes	Yes
Pitch/Speed	No	No	No	No
DOS	Yes	Yes	Yes	Yes
Win3x	No	No	No	No
Win95	No	No	No	No
Win NT	Yes	Yes	Yes	Yes
Unix	Yes	Yes	Yes	Yes
QNX	No	No	No	No
OS/2	Yes	Yes	Yes	Yes
Solaris	Yes	Yes	Yes	Yes
Linux	See Notes	See Notes	See Notes	See Notes
TAPI	No	No	No	No
URL	www.aculab.com	www.aculab.com	www.aculab.com	www.aculab.com
Phone	(+44) (0)1908 273800	(+44) (0)1908 273800	(+44) (0)1908 273800	(+44) (0)1908 273800
Comments				

Aculab	Aculab	Aculab	Aculab	Aculab
Dual Port E1 Card with DSP	Single Port T1 Card	Dual Port T1 Card (Decadic & PRI)	Dual Port T1 Card (DTMF & MF R1/2)	E1/T1 Mixed
$4,419	$2,809	$3,745	$4,793	$3,745
2 E1	1 T1	2 T1/PRI	2 T1	E1/T1
60	24	48	48	54
0	0	0	0	0
Yes	No	No	Yes	No
60	0	0	48	0
No	No	No	No	No
Yes	Yes	Yes	Yes	Yes
N/A	N/A	N/A	N/A	N/A
MVIP, PEB, SCbus	MVIP, PEB, SCbus	MVIP, PEB, SCbus	MVIP, PEB, SCbus	MVIP, PEB, SCbus
Full	Full	Full	Full	Full
ISA AT	ISA AT	ISA AT	ISA AT	ISA AT
Yes	No	Yes/No	Yes	No
No	No	No	No	No
Opt	Opt	Opt	Opt	Opt
Yes	Yes	Yes	Yes	Yes
No	No	No	No	No
Yes	Yes	Yes	Yes	Yes
No	No	No	No	No
No	No	No	No	No
Yes	Yes	Yes	Yes	Yes
Yes	Yes	Yes	Yes	Yes
No	No	No	No	No
Yes	Yes	Yes	Yes	Yes
Yes	Yes	Yes	Yes	Yes
See Notes	See Notes	See Notes	See Notes	See Notes
No	No	No	No	No
www.aculab.com	www.aculab.com	www.aculab.com	www.aculab.com	www.aculab.com
(+44) (0)1908 273800	(+44) (0)1908 273800	(+44) (0)1908 273800	(+44) (0)1908 273800	(+44) (0)1908 273800

Table of Digital Computer Telephony Boards, continued

Vendor	Aculab	Aculab	Aculab	Aculab	
Product	BR4 ISDN Card	BR8 ISDN Card	MC3 Dual Port ISA Fibre Card	Prosody 1 (ISA)	
List Price	$2,809	$4,119	$3,744	$3,744	
Line Interface	4 BRI (8 Channel)	8 BRI	MC3	None	
DS0 Channels	8	16	2,423	0	
DSP V ports	0	0	0	60/30/16	
Onboard DSP	No	No	No	Yes	
Tone ports	60/30/16	60/30/16	0	60/30/16	
Fax	No	No	No	Yes	
Security	Yes	Yes	Yes	Yes	
Wav Sup	N/A	N/A	N/A	Yes	
Tel-BUS	MVIP, PEB, SCbus	MVIP, PEB, SCbus	MVIP	MVIP, SCbus	
Card Length	Full	Full	Full	Full	
Form Factor	ISA AT	ISA AT	ISA AT	ISA AT	
Pulse Rec	No	No	No	0/30/0	
Speech Rec	Yes	Yes	No	0/0/16	
Conference	Opt	Opt	No	60/30/16	
Switching	Yes	Yes	Yes	Yes	
Pitch/Speed	No	No	No	0/30/16	
DOS	Yes	Yes	No	No	
Win3x	No	No	No	No	
Win95	No	No	No	No	
Win NT	Yes	Yes	Yes	Yes	
Unix	Yes	Yes	Yes	Yes	
QNX	No	No	No	No	
OS/2	Yes	Yes	Yes	Yes	
Solaris	Yes	Yes	No	No	
Linux	See Notes	See Notes	See Notes	See Notes	
TAPI	No	No	No	No	
URL	www.aculab.com	www.aculab.com	www.aculab.com	www.aculab.com	
Phone	(+44) (0)1908 273800	(+44) (0)1908 273800	(+44) (0)1908 273800	(+44) (0)1908 273800	
Comments			Enables scaling of single PC chassis CT solution into multiple chassis	Different firmware builds give different port counts/capabilities at no extra cost	

Aculab	Aculab	Aculab	Aculab	Aculab
Prosody 2 (ISA)	Prosody 3 (ISA)	Prosody 4 (ISA)	Prosody 1 (PCI)	Prosody 2 (PCI)
$5,991	$8,237	$10,484	$4,493	$5,991
None	None	None	1, 2 or 4 E1/T1	1, 2 or 4 E1/T1
0	0	0	24 to 120	24 to 120
120/60/32	180/90/48	240/120/60	60/30/16	120/60/32
Yes	Yes	Yes	Yes	Yes
120/60/32	180/90/48	240/120/60	60/30/16	120/60/32
Yes	Yes	Yes	Yes	Yes
Yes	Yes	Yes	Yes	Yes
Yes	Yes	Yes	Yes	Yes
MVIP, SCbus	MVIP, SCbus	MVIP, SCbus	H.100, MVIP, SCbus	H.100, MVIP, SCbus
Full	Full	Full	Full	Full
ISA AT	ISA AT	ISA AT	PCI	PCI
0/60/0	0/90/0	0/120/0	Yes	Yes
0/0/32	0/0/48	0/0/64	Yes	Yes
120/60/32	180/90/48	240/120/64	Yes	Yes
Yes	Yes	Yes	Yes	Yes
0/60/32	0/90/48	0/120/64	Yes	Yes
No	No	No	No	No
No	No	No	No	No
No	No	No	No	No
Yes	Yes	Yes	Yes	Yes
Yes	Yes	Yes	Yes	Yes
No	No	No	No	No
Yes	Yes	Yes	Yes	Yes
No	No	No	Yes	Yes
See Notes	See Notes	See Notes	See Notes	See Notes
No	No	No	No	No
www.aculab.com	www.aculab.com	www.aculab.com	www.aculab.com	www.aculab.com
(+44) (0)1908 273800	(+44) (0)1908 273800	(+44) (0)1908 273800	(+44) (0)1908 273800	(+44) (0)1908 273800
Different firmware builds give different port counts/capabilities at no extra cost	Different firmware builds give different port counts/capabilities at no extra cost	Different firmware builds give different port counts/capabilities at no extra cost	Different firmware builds give different port counts/capabilities at no extra cost	Different firmware builds give different port counts/capabilities at no extra cost

Table of Digital Computer Telephony Boards, continued

Vendor	Aculab	AltiGen Communications	AltiGen Communications	AltiGen Communications
Product	T1 PCI Clipper	Standard AltiServ software with Quantum	AltiServ with Triton T1/PRI and 12 ext with 24 Voice mail ports	AltiServ with both PSTN and VoIP support
List Price	$2,397	$3,115	$5,610	$5,610
Line Interface	2 T1 Decadic	Analog LS, GS & WS and Stations	T1 and ISDN PRI	VoIP w/ G711 and G.723.1
DS0 Channels	48	12	24 (T1) / 23 (PRI)	4
DSP V ports	60/30/16	6	24	4
Onboard DSP	Yes	Yes	Yes	Yes
Tone ports	60/30/16	12	24	4
Fax	Yes	No	No	No
Security	Yes	Yes	Yes	Yes
Wav Sup	Yes	No	No	No
Tel-BUS	H.100, MVIP, SCbus	MVIP	MVIP	MVIP
Card Length	Half	Full ISA	Full PCI	Full PCI
Form Factor	PCI	ISA	PCI	PCI
Pulse Rec	Yes	No	No	No
Speech Rec	Yes	No	No	No
Conference	Yes	Yes	Yes	Yes
Switching	Yes	Yes	Yes	Yes
Pitch/Speed	Yes	Yes	Yes	Yes
DOS	No	No	No	No
Win3x	No	No	No	No
Win95	No	No	No	No
Win NT	Yes	Yes	Yes	Yes
Unix	Yes	No	No	No
QNX	No	No	No	No
OS/2	Yes	No	No	No
Solaris	tba	No	No	No
Linux	See Notes	See Notes	See Notes	See Notes
TAPI	No	Yes	Yes	Yes
URL	www.aculab.com	www.altigen.com	www.altigen.com	www.altigen.com
Phone	(+44) (0)1908 273800	888-258-4436	888-258-4436	888-258-4436
Comments				VoIP with full AltiServ feature support

Amtelco	Amtelco	Amtelco	Amtelco	Amtelco
MC-1	MC-3	SCX Multi-Channel Bridge	XDS BRI	XDS HD Conference
$1,895	$2,995	$1,795	$3,895	$1,695
None	None	None	12 BRI (24 channel)	None
N/A	N/A	N/A	24	N/A
0	0	0	12	0
No	No	No	No	Yes
0	0	0	0	256
No	No	No	No	No
OPT.	OPT.	OPT.	OPT.	OPT.
No	No	No	No	No
MVIP, SCbus	MVIP, SCbus	MVIP, SCbus	MVIP, SCbus	MVIP, SCbus
Full	Full	Full	Full	Full
ISA AT	ISA AT	ISA AT	ISA AT	ISA AT
No	No	No	No	No
No	No	No	No	No
128	128	No	No	256
Yes	Yes	Yes	Yes	Yes
No	No	No	No	No
Yes	Yes	Yes	Yes	Yes
Yes	Yes	Yes	Yes	Yes
Yes	Yes	Yes	Yes	Yes
Yes	Yes	Yes	Yes	Yes
Yes	Yes	Yes	Yes	Yes
No	No	No	No	No
Yes	Yes	Yes	Yes	Yes
No	No	No	No	No
See Notes	See Notes	See Notes	See Notes	See Notes
No	No	No	No	No
www.amtelcom.com	www.amtelcom.com	www.amtelcom.com	www.amtelcom.com	www.amtelcom.com
800-356-9224	800-356-9224	800-356-9224	800-356-9224	800-356-9224
The MC-1 card is used to bridge several telephony computers together so the voice channels are shared among more than one system.	The MC-3 card is used to bridge several telephony computers together so the voice channels are shared among more than one system.	The SCX card bridges multiple SCbus systems together so that the voice channels can be interconnected between systems.	The XDS BRI ports support either S or T interfaces so they can either drive ISDN BRI phones, or connect to ISDN trunk lines.	

Table of Digital Computer Telephony Boards, continued

Vendor	Amtelco	Amtelco	Amtelco	Amtelco
Product	XDS HD LS Interface	XDS HD Station Interface	XDS U BRI	XDS N100 MC-3
List Price	$2,895	$2,895	$2,695	$3,395
Line Interface	24 LS	24 POTS	4 U BRI (8 channel)	None
DS0 Channels	24	24	8	N/A
DSP V ports	24	24	4	N/A
Onboard DSP	No	No	No	No
Tone ports	24	24	0	N/A
Fax	No	No	No	N/A
Security	OPT.	OPT.	OPT.	OPT.
Wav Sup	No	No	No	No
Tel-BUS	MVIP, SCbus	MVIP, SCbus	MVIP, SCbus	H.100/PCI
Card Length	Full	Full	Full	Full
Form Factor	ISA AT	ISA AT	ISA AT	PCI
Pulse Rec	No	No	No	No
Speech Rec	No	No	No	No
Conference	No	No	No	128
Switching	Yes	Yes	Yes	Yes
Pitch/Speed	No	No	No	No
DOS	Yes	Yes	Yes	No
Win3x	Yes	Yes	Yes	No
Win95	Yes	Yes	Yes	No
Win NT	Yes	Yes	Yes	Yes
Unix	Yes	Yes	Yes	Yes
QNX	No	No	No	No
OS/2	Yes	Yes	Yes	No
Solaris	No	No	No	No
Linux	See Notes	See Notes	See Notes	See Notes
TAPI	No	No	No	No
URL	www.amtelcom.com	www.amtelcom.com	www.amtelcom.com	www.amtelcom.com
Phone	800-356-9224	800-356-9224	800-356-9224	608-838-4194
Comments				The MC-3 card bridges up to 20 chassis so voice channels are shared with all chassis.

Amtelco	Amtelco	Amtelco	Anatel/Analogic	Anatel/Analogic
XDS H100 BRI	XDS H110 MC-3	XDS H110 BRI	TAP-802	TAP-804
$4,295	$3,795	$4,895	$3,300	$3,300
16 BRI (32 channel)	None	32 BRI (64 channel)	None	T1/E1/ISDN (Optional)
32	N/A	64	0	24
12	N/A	32	24	30
No	No	No	Yes	Yes
16	N/A	32	24	30
N/A	N/A	N/A	OPT.	OPT.
OPT.	OPT.	OPT.	Yes	Yes
No	No	No	N/A	Yes
H.100/PCI	H.110/cPCI	H.110/cPCI	MVIP, SCbus	MVIP, SCbus
Full	Full	Full	Full	Full
PCI	cPCI	cPCI	ISA AT	PCI
No	No	No	N/A	N/A
No	No	No	N/A	N/A
No	128	No	N/A	N/A
Yes	Yes	Yes	Yes	Yes
No	No	No	N/A	N/A
No	No	No	No	No
No	No	No	No	No
No	No	No	No	No
Yes	Yes	Yes	Yes	Yes
Yes	No	No	Yes	Yes
No	No	No	No	No
No	No	No	No	No
No	No	No	No	Yes
See Notes	See Notes	See Notes	See Notes	See Notes
No	No	No	No	No
www.amtelcom.com	www.amtelcom.com	www.amtelcom.com	www.anatel.net	www.anatel.net
608-838-4194	608-838-4194	608-838-4194	978-977-6817	978-977-6817
	The MC-3 card bridges up to 20 chassis so voice channels are shared with all chassis.			

167

Table of Digital Computer Telephony Boards, continued

Vendor	Anatel/Analogic	Anatel/Analogic	Anatel/Analogic	Anatel/Analogic
Product	TAP-805	TAP-810	NAC-120	NAC-600
List Price	$6,000	$9,200	$4,495	$9,400
Line Interface	T1/E1/ISDN (Optional)	T1/E1/ISDN (Optional)	T1/E1/ISDN	DS-3
DS0 Channels	60	120	120	672
DSP V ports	120	120	N/A	N/A
Onboard DSP	Yes	Yes	No	No
Tone ports	60	120	N/A	N/A
Fax	OPT.	OPT.	N/A	N/A
Security	No	No	No	N/A
Wav Sup	Yes	Yes	N/A	N/A
Tel-BUS	H.100	H.110	H.110	H.110
Card Length	Full	6U	6U	6U
Form Factor	PCI	cPCI	cPCI	cPCI
Pulse Rec	N/A	N/A	N/A	N/A
Speech Rec	N/A	N/A	N/A	N/A
Conference	N/A	N/A	N/A	N/A
Switching	Yes	Yes	Yes	N/A
Pitch/Speed	N/A	N/A	N/A	N/A
DOS	No	No	No	No
Win3x	No	No	No	No
Win95	No	No	No	No
Win NT	Yes	Yes	Yes	Yes
Unix	Yes	Yes	Yes	Yes
QNX	No	No	No	No
OS/2	No	No	No	No
Solaris	Yes	Yes	Yes	Yes
Linux	See Notes	See Notes	See Notes	See Notes
TAPI	No	No	No	No
URL	www.anatel.net	www.anatel.net	www.anatel.net	www.anatel.net
Phone	978-977-6817	978-977-6817	978-977-6817	978-977-6817
Comments	Built-in 10/100 Mbit Ethernet	Built-in 10/100 Mbit Ethernet		PMC Slot available

Artesyn Communication	BICOM, Inc.	BICOM, Inc.	BICOM, Inc.	BICOM, Inc.
BajaSpan	GEMINI 2400D-2T1	GEMINI 2400D-T1	GEMINI 3000D-2E1	GEMINI 3000D-E1
$4,700	$4,500	$3,895	$5,895	$4,995
8 T1 or 8 E1	2 T1/PRI	1 T1/PRI	2 E1/PRI	1 E1/PRI
256	48	24	60	30
0	48	24	60	30
No	Yes	Yes	Yes	Yes
0	48	24	60	30
No	OPT.	OPT.	OPT.	OPT.
Ys	Yes	Yes	Yes	Yes
No	Yes	Yes	Yes	Yes
H.110	MVIP, SCbus	MVIP, SCbus	MVIP, SCbus	MVIP, SCbus
6U x 160mm	Full	Full	Full	Full
6U cPCI	ISA AT	ISA AT	ISA AT	ISA AT
No	No	No	No	No
No	OPT.	OPT.	OPT.	OPT.
Yes	OPT.	OPT.	OPT.	OPT.
Yes	Yes	Yes	Yes	Yes
No	Yes	Yes	Yes	Yes
No	Yes	Yes	Yes	Yes
No	No	No	No	No
No	No	No	No	No
Yes	Yes	Yes	Yes	Yes
Yes	No	No	No	No
No	No	No	No	No
No	No	No	No	No
Yes	No	No	No	No
See Notes	See Notes	See Notes	See Notes	See Notes
No	Yes	Yes	Yes	Yes
www.artesyn.com/cp	www.bicom-inc.com	www.bicom-inc.com	www.bicom-inc.com	www.bicom-inc.com
561-451-1000	203-268-4484	203-268-4484	203-268-4484	203-268-4484

Table of Digital Computer Telephony Boards, continued

Vendor	Blue Wave	Brooktrout Technology	Brooktrout Technology	Brooktrout Technology
Product	CPCI/C5420	TR114 Series	TR1014 Series	TR1000 Series
List Price	$12,000	starts at $1,995	starts at $14,845	starts at $9,505
Line Interface	4 T1 or 4 E1 via module	2 to 16	24	24 to 60
DS0 Channels	120	No	No	No
DSP V ports	N/A	No	No	No
Onboard DSP	Yes	Yes	Yes	Yes
Tone ports	N/A	No	No	No
Fax	N/A	Yes	Yes	Yes
Security	Yes	Yes	Yes	Yes
Wav Sup	N/A	SCbus, MVIP	SCbus, MVIP	SCbus, MVIP
Tel-BUS	H.110	Full	Full	Full
Card Length	6U	Full	Full	Full
Form Factor	CPCI	ISA, PCI	PCI, cPCI	PCI, cPCI
Pulse Rec	N/A	Yes	Yes	Yes
Speech Rec	No	Yes	Yes	Yes
Conference	No	Yes	Yes	Yes
Switching	No	No	Yes	Yes
Pitch/Speed	No	Yes	Yes	Yes
DOS	No	Yes	Yes	Yes
Win3x	No	Yes	Yes	Yes
Win95	No	Yes	Yes	Yes
Win NT	Yes	Yes	Yes	Yes
Unix	Yes	Yes	Yes	Yes
QNX	No	Yes	No	No
OS/2	No	Yes	Yes	Yes
Solaris	Yes	See Notes	No	No
Linux	See Notes	Yes	Yes	Yes
TAPI	No	No	No	No
URL	www.bluews.com	www.brooktrout.com	www.brooktrout.com	www.brooktrout.com
Phone	972-277-4600	781-449-4100	781-449-4100	781-449-4100
Comments	Specialized Voice/Fax/Data over IP board.	Market leading platform for enterprise and service provider fax applications.	Platform for high-density fax applications.	Platform for high-density voice and unified messaging. Optional onboard T1 interface.

Brooktrout Technology	Brooktrout Technology	Brooktrout Technology	Brooktrout Technology	Brooktrout Technology
TR2001 Series	RDSP-16000	RDSP-20000	RDSP-24000	VRS-16
starts at $5,380	$3,295	$4,195	$4,795	$3,595
24 to 60	None	None	None	None
No	N/A	N/A	N/A	N/A
No	16	16	24	16
Yes	Yes	Yes	Yes	Yes
No	24	24	24	24
No	No	No	No	No
Yes	Yes	Yes	Yes	Yes
SCbus, MVIP	Yes	Yes	Yes	Yes
Full	MVIP	MVIP	MVIP	MVIP
Full	Full	Full	Full	Full
PCI, cPCI	ISA AT	ISA AT	ISA AT	ISA AT
Yes	No	No	No	Yes
Yes	No	No	No	No
Yes	No	No	No	No
Yes	Yes	Yes	Yes	Yes
Yes	No	No	No	Yes
Yes	Yes	Yes	Yes	Yes
Yes	Yes	Yes	Yes	Yes
Yes	Yes	Yes	Yes	Yes
Yes	Yes	Yes	Yes	Yes
Yes	Yes	Yes	Yes	Yes
No	Yes	Yes	Yes	Yes
Yes	Yes	Yes	Yes	Yes
No	No	No	No	Yes
Yes	Yes	Yes	Yes	Yes
No	Yes	Yes	Yes	Yes
www.brooktrout.com	www.brooktrout.com	www.brooktrout.com	www.brooktrout.com	www.brooktrout.com
781-449-4100	781-449-4100	781-449-4100	781-449-4100	781-449-4100
Platform for voice/fax over IP.				The Vantage VRS series requires one or more separate network interface modules, which are connected via the MVIP bus.

Table of Digital Computer Telephony Boards, continued

Vendor	Brooktrout Technology	Brooktrout Technology	Brooktrout Technology	Brooktrout Technology
Product	VRS-24	VRS-32	RTNI-2E1	RTNI-2T1
List Price	$4,995	$6,995	$3,295	$2,995
Line Interface	None	None	2 E1	2 T1
DS0 Channels	N/A	N/A	60	48
DSP V ports	24	32	0	0
Onboard DSP	Yes	Yes	No	No
Tone ports	24	24	0	0
Fax	No	No	No	No
Security	Yes	Yes	Yes	Yes
Wav Sup	Yes	Yes	Yes	Yes
Tel-BUS	MVIP	MVIP	MVIP	MVIP
Card Length	Full	Full	Full	Full
Form Factor	ISA AT	ISA AT	ISA AT	ISA AT
Pulse Rec	Yes	Yes	N/A	N/A
Speech Rec	No	No	No	No
Conference	No	No	No	No
Switching	Yes	Yes	Yes	Yes
Pitch/Speed	Yes	Yes	No	No
DOS	Yes	Yes	Yes	Yes
Win3x	Yes	Yes	Yes	Yes
Win95	Yes	Yes	Yes	Yes
Win NT	Yes	Yes	Yes	Yes
Unix	Yes	Yes	Yes	Yes
QNX	Yes	Yes	Yes	Yes
OS/2	Yes	Yes	Yes	Yes
Solaris	Yes	Yes	Yes	Yes
Linux	Yes	Yes	Yes	Yes
TAPI	Yes	Yes	Yes	Yes
URL	www.brooktrout.com	www.brooktrout.com	www.brooktrout.com	www.brooktrout.com
Phone	781-449-4100	781-449-4100	781-449-4100	781-449-4100
Comments	The Vantage VRS series requires one or more separate network interface modules, which are connected via the MVIP bus.	The Vantage VRS series requires one or more separate network interface modules, which are connected via the MVIP bus.		

Commetrex	Intel/Dialogic	Intel/Dialogic	Intel/Dialogic	Intel/Dialogic
MSP-320	Media Processing + Network Interface Products	D/240-T1	D/240JCT-T1	D/300-E1-120
$3,995		$3,943	$4,534	$5,125
Optional Dual and Quad T1/E1/PRI Daughterboards		1 T1/PRI	1 T1/PRI	1 E1/PRI
120		24	24	30
120		24	24	30
Yes		Yes	Yes	Yes
120		24	24	30
Software		No	Yes	No
Yes		Yes	Yes	Yes
Yes		Yes	Yes	Yes
H.100		SCbus, H.100	SCbus, H.100	SCbus, H.100
Full		Full	Full	Full
PCI		PCI/ ISA AT	PCI	PCI/ ISA AT
No		Yes	Yes	Yes
Yes		No	Yes	No
Yes		No	No	No
Yes		Yes	Yes	Yes
No		Yes	Yes	Yes
No		Yes	No	Yes
No		No	No	No
No		No	No	No
Yes		Yes	Yes	Yes
No		Yes	Yes	Yes
No		No	No	No
No		Yes	No	Yes
No		Yes	Yes	Yes
Yes		Yes	Yes	Yes
No		Yes	Yes	Yes
www.commetrex.com		www.dialogic.com	www.dialogic.com	www.dialogic.com
770-449-7775		800-755-4444	800-755-4444	800-755-4444

Dial, CPA, Voice, VoIP, Fax, FoIP, Conf, Speech Rec, Data Modem software are optional sw licenses. Board supports Linux 2H00

Table of Digital Computer Telephony Boards, continued

Vendor	Intel/Dialogic	Intel/Dialogic	Intel/Dialogic	Intel/Dialogic	
Product	D/300-E1-75	D/300JCT-E1-120	D/300JCT-E1-75	D/240SC-2T1(TM)	
List Price	$5,125	$5,893	$5,893	$5,995	
Line Interface	1 E1/PRI	1 E1/PRI	1 E1/PRI	2 T1/PRI	
DS0 Channels	30	30	30	48	
DSP V ports	30	30	30	24	
Onboard DSP	Yes	Yes	Yes	Yes	
Tone ports	30	30	30	48	
Fax	No	Yes	Yes	No	
Security	Yes	Yes	Yes	Yes	
Wav Sup	Yes	Yes	Yes	Yes	
Tel-BUS	SCbus, H.100	SCbus, H.100	SCbus, H.100	SCbus	
Card Length	Full	Full	Full	Full	
Form Factor	PCI	PCI	PCI	ISA AT	
Pulse Rec	Yes	Yes	Yes	Yes	
Speech Rec	No	Yes	Yes	No	
Conference	No	No	No	No	
Switching	Yes	Yes	Yes	Yes	
Pitch/Speed	Yes	Yes	Yes	Yes	
DOS	Yes	No	No	Yes	
Win3x	No	No	No	No	
Win95	No	No	No	No	
Win NT	Yes	Yes	Yes	Yes	
Unix	Yes	Yes	Yes	Yes	
QNX	No	No	No	No	
OS/2	Yes	No	No	Yes	
Solaris	Yes	Yes	Yes	Yes	
Linux	Yes	Yes	Yes	Yes	
TAPI	Yes	Yes	Yes	Yes	
URL	www.dialogic.com	www.dialogic.com	www.dialogic.com	www.dialogic.com	
Phone	800-755-4444	800-755-4444	800-755-4444	800-755-4444	
Comments					

174

Intel/Dialogic	Intel/Dialogic	Intel/Dialogic	Intel/Dialogic	Intel/Dialogic
D/300SC-2E1(TM)	D/480SC-2T1(TM)	D/480JCT-2T1	DM/V480-2T1	D/600SC-2E1(TM)
$7,495	$7,097	$7,097	$8,161	$7,995
2 E1/PRI	2 T1/PRI	2 T1/PRI	2 T1/PRI	2 E1/PRI
60	48	48	48	60
30	48	48	48	60
Yes	Yes	Yes	Yes	Yes
60	48	48	48	60
No	No	Yes	Yes	No
Yes	Yes	Yes	Yes	Yes
Yes	Yes	Yes	Yes	Yes
SCbus	SCbus	H.100	SCbus, H.100	SCbus
Full	Full	Full	Full	Full
ISA AT	ISA AT	PCI	PCI	ISA AT
Yes	Yes	Yes	Yes	Yes
No	No	Yes	Yes	No
No	No	No	60	No
Yes	Yes	Yes	Yes	Yes
Yes	Yes	Yes	Yes	Yes
Yes	Yes	No	No	Yes
No	No	No	No	No
No	No	No	No	No
Yes	Yes	Yes	Yes	Yes
Yes	Yes	Yes	Yes	Yes
No	No	No	No	No
Yes	Yes	No	No	Yes
Yes	Yes	Yes	Yes	Yes
Yes	Yes	Yes	Yes	Yes
Yes	Yes	No	No	Yes
www.dialogic.com	www.dialogic.com	www.dialogic.com	www.dialogic.com	www.dialogic.com
800-755-4444	800-755-4444	800-755-4444	800-755-4444	800-755-4444

Table of Digital Computer Telephony Boards, continued

Vendor	Intel/Dialogic	Intel/Dialogic	Intel/Dialogic	Intel/Dialogic	
Product	D/600JCT-2E1	DM/V600-2E1	DM/V480-4T1	DM/V600-4E1	
List Price	$7,995	$9,195	$9,440	$10,970	
Line Interface	2 E1/PRI	2 E1/PRI	4 T1/PRI	4 E1/PRI	
DS0 Channels	60	60	96	120	
DSP V ports	60	60	48	60	
Onboard DSP	Yes	Yes	Yes	Yes	
Tone ports	60	60	96	120	
Fax	Yes	Yes	No	No	
Security	Yes	Yes	Yes	Yes	
Wav Sup	Yes	Yes	Yes	Yes	
Tel-BUS	H.100	SCbus, H.100	SCbus, H.100	SCbus, H.100	
Card Length	Full	Full	Full	Full	
Form Factor	PCI	PCI	PCI,cPCI	PCI,cPCI	
Pulse Rec	Yes	Yes	Yes	Yes	
Speech Rec	Yes	Yes	Yes	Yes	
Conference	No	60	No	No	
Switching	Yes	Yes	Yes	Yes	
Pitch/Speed	Yes	Yes	Yes	Yes	
DOS	No	No	No	No	
Win3x	No	No	No	No	
Win95	No	No	No	No	
Win NT	Yes	Yes	Yes	Yes	
Unix	Yes	Yes	Yes	Yes	
QNX	No	No	No	No	
OS/2	No	No	No	No	
Solaris	Yes	Yes	Yes	Yes	
Linux	Yes	Yes	Yes	Yes	
TAPI	No	No	No	No	
URL	www.dialogic.com	www.dialogic.com	www.dialogic.com	www.dialogic.com	
Phone	800-755-4444	800-755-4444	800-755-4444	800-755-4444	
Comments					

Intel/Dialogic	Intel/Dialogic	Intel/Dialogic	Intel/Dialogic	Intel/Dialogic
DM/V960-4T1	DM/V1200-4E1	DM/F240-T1(TM)	DM/F300-E1(TM)	DM/VF240-T1
$10,725	$12,620	contact Sales	contact Sales	contact Sales
4 T1/PRI	4 E1/PRI	1 T1/PRI	1 E1/PRI	1 T1/PRI
96	120	24	30	24
96	120	0	0	24
Yes	Yes	Yes	Yes	Yes
96	120	24	30	24
Yes	Yes	Yes	Yes	Yes
Yes	Yes	Yes	Yes	Yes
Yes	Yes	No	No	Yes
SCbus, H.100	SCbus, H.100	SCbus, H.100	SCbus, H.100	SCbus, H.100
Full	Full	Full	Full	Full
PCI, cPCI	PCI, cPCI	PCI, cPCI	PCI, cPCI	PCI
Yes	Yes	Yes	Yes	Yes
Yes	Yes	No	No	Yes
Yes	Yes	No	No	No
Yes	Yes	No	No	Yes
Yes	Yes	Yes	Yes	Yes
No	No	No	No	No
No	No	No	No	No
No	No	No	No	No
Yes	Yes	Yes	Yes	Yes
Yes	Yes	Yes	Yes	Yes
No	No	No	No	No
No	No	No	No	No
Yes	Yes	Yes	Yes	Yes
Yes	Yes	Yes	Yes	Yes
No	No	No	No	No
www.dialogic.com	www.dialogic.com	www.dialogic.com	www.dialogic.com	www.dialogic.com
800-755-4444	800-755-4444	800-755-4444	800-755-4444	800-755-4448

Table of Digital Computer Telephony Boards, continued

Vendor	Intel/Dialogic	Intel/Dialogic	Intel/Dialogic	Intel/Dialogic
Product	DM/VF300-E1	BRI/2VFD	CPi/200 BRI(TM)	CPi/400 BRI(TM)
List Price	contact Sales	$1,795	$2,370	$3,700
Line Interface	1 E1/PRI	BRI	BRI	BRI
DS0 Channels	30	4	2	4
DSP V ports	30	4	0	0
Onboard DSP	Yes	Yes	No	No
Tone ports	30	4	2	4
Fax	Yes	Yes	Yes	Yes
Security	Yes	Yes	Yes	Yes
Wav Sup	Yes	Yes	No	No
Tel-BUS	SCbus, H.100	H.100	N/A	N/A
Card Length	Full	Full	Full	Full
Form Factor	PCI	PCI	ISA	ISA
Pulse Rec	Yes	No	No	No
Speech Rec	Yes	No	No	No
Conference	No	No	No	No
Switching	Yes	No	No	No
Pitch/Speed	Yes	No	No	No
DOS	No	No	No	No
Win3x	No	No	No	No
Win95	No	No	No	No
Win NT	Yes	Yes	Yes	Yes
Unix	Yes	No	No	No
QNX	No	No	No	No
OS/2	No	No	No	No
Solaris	Yes	No	No	No
Linux	Yes	No	No	No
TAPI	No	Yes	No	No
URL	www.dialogic.com	www.dialogic.com	www.dialogic.com	www.dialogic.com
Phone	800-755-4448	800-755-4444	800-755-4444	800-755-4444
Comments				

Intel/Dialogic	Intel/Dialogic	Intel/Dialogic	Intel/Dialogic	Intel/Dialogic
CPi/2400CT-T1	CPi/3000CT-E1	Network Interface Products	DTI/240SC(TM)	DTI/241SC(TM)
$15,510	$18,330		$1,695	$3,180
1 T1/PRI	1 E1/PRI		1T1/PRI	1T1/PRI
24	30		24	24
0	0		0	0
Yes	Yes		No	Yes
24	30		0	24
Yes	Yes		No	No
Yes	Yes		Yes	Yes
No	No		No	No
SCbus, H.100	SCbus, H.100		PEB, SCbus	PEB, SCbus
Full	Full		Full	Full
PCI	PCI		ISA AT	ISA AT
No	No		No	No
No	No		No	No
No	No		No	No
No	No		Yes	Yes
No	No		No	No
No	No		Yes	Yes
No	No		No	No
No	No		No	No
Yes	Yes		Yes	Yes
No	No		Yes	Yes
No	No		No	No
No	No		Yes	Yes
No	No		Yes	Yes
No	No		Yes	Yes
No	No		Yes	Yes
www.dialogic.com	www.dialogic.com		www.dialogic.com	www.dialogic.com
800-755-4444	800-755-4444		800-755-4444	800-755-4444

Table of Digital Computer Telephony Boards, continued

Vendor	Intel/Dialogic	Intel/Dialogic	Intel/Dialogic	Intel/Dialogic
Product	DTI/300SC-75	DTI/300SC-120	DTI/301SC-75	DTI/301SC-120
List Price	$2,495	$2,495	$3,350	$3,350
Line Interface	1 E1/PRI	1 E1/PRI	1 E1/PRI	1 E1/PRI
DS0 Channels	30	30	30	30
DSP V ports	0	0	0	0
Onboard DSP	No	No	Yes	Yes
Tone ports	0	0	30	30
Fax	No	No	No	No
Security	Yes	Yes	Yes	Yes
Wav Sup	No	No	No	No
Tel-BUS	PEB, SCbus	PEB, SCbus	PEB, SCbus	PEB, SCbus
Card Length	Full	Full	Full	Full
Form Factor	ISA AT	ISA AT	ISA AT	ISA AT
Pulse Rec	No	No	No	No
Speech Rec	No	No	No	No
Conference	No	No	No	No
Switching	Yes	Yes	Yes	Yes
Pitch/Speed	No	No	No	No
DOS	Yes	Yes	Yes	Yes
Win3x	No	No	No	No
Win95	No	No	No	No
Win NT	Yes	Yes	Yes	Yes
Unix	Yes	Yes	Yes	Yes
QNX	No	No	No	No
OS/2	Yes	Yes	Yes	Yes
Solaris	Yes	Yes	Yes	Yes
Linux	Yes	Yes	Yes	Yes
TAPI	Yes	Yes	Yes	Yes
URL	www.dialogic.com	www.dialogic.com	www.dialogic.com	www.dialogic.com
Phone	800-755-4444	800-755-4444	800-755-4444	800-755-4444
Comments				

Intel/Dialogic	Intel/Dialogic	Intel/Dialogic	Intel/Dialogic	Intel/Dialogic
DTI/480SC(TM)	DTI/481SC(TM)	DTI/600SC-75	DTI/600SC-120	DTI/601SC-75
$2,695	$5,475	$3,495	$3,495	$6,295
2 T1/PRI	2 T1/PRI	2 E1/PRI	2 E1/PRI	2 E1/PRI
48	48	60	60	60
0	0	0	0	0
No	Yes	No	No	Yes
0	48	0	0	60
No	No	No	No	No
Yes	Yes	Yes	Yes	Yes
No	No	No	No	No
SCbus	SCbus	SCbus	SCbus	SCbus
Full	Full	Full	Full	Full
ISA AT	ISA AT	ISA AT	ISA AT	ISA AT
No	No	No	No	No
No	No	No	No	No
No	No	No	No	No
Yes	Yes	Yes	Yes	Yes
No	No	No	No	No
Yes	Yes	Yes	Yes	Yes
No	No	No	No	No
No	No	No	No	No
Yes	Yes	Yes	Yes	Yes
Yes	Yes	Yes	Yes	Yes
No	No	No	No	No
Yes	Yes	Yes	Yes	Yes
Yes	Yes	Yes	Yes	Yes
Yes	Yes	Yes	Yes	Yes
Yes	Yes	Yes	Yes	Yes
www.dialogic.com	www.dialogic.com	www.dialogic.com	www.dialogic.com	www.dialogic.com
800-755-4444	800-755-4444	800-755-4444	800-755-4444	800-755-4444

Vendor	Intel/Dialogic	Intel/Dialogic	Intel/Dialogic	Intel/Dialogic
Product	DTI/601SC-120	DM/N960-4T1	DM/N1200-4E1	DM/T960-4T1
List Price	$6,295	$4,945	$5,945	$8,140
Line Interface	2 E1/PRI	4 T1/PRI	4 E1/PRI	4 T1/PRI
DS0 Channels	60	96	120	96
DSP V ports	0	0	0	0
Onboard DSP	Yes	No	No	Yes
Tone ports	60	0	0	96
Fax	No	No	No	No
Security	Yes	Yes	Yes	Yes
Wav Sup	No	No	No	No
Tel-BUS	SCbus	SCbus, H.100, H.110	SCbus, H.100, H.110	SCbus, H.100, H.110
Card Length	Full	Full	Full	Full
Form Factor	ISA AT	PCI, cPCI	PCI, cPCI	PCI, cPCI
Pulse Rec	No	No	No	No
Speech Rec	No	No	No	No
Conference	No	No	No	No
Switching	Yes	Yes	Yes	Yes
Pitch/Speed	No	No	No	No
DOS	Yes	No	No	No
Win3x	No	No	No	No
Win95	No	No	No	No
Win NT	Yes	Yes	Yes	Yes
Unix	Yes	Yes	Yes	Yes
QNX	No	No	No	No
OS/2	Yes	No	No	No
Solaris	Yes	Yes	Yes	Yes
Linux	Yes	Yes	Yes	Yes
TAPI	Yes	No	No	No
URL	www.dialogic.com	www.dialogic.com	www.dialogic.com	www.dialogic.com
Phone	800-755-4444	800-755-4444	800-755-4444	800-755-4444
Comments				

182

Intel/Dialogic	Intel/Dialogic	Intel/Dialogic	Intel/Dialogic	Intel/Dialogic
DM/T1200-4E1	BRI/80	BRI/160	ATM/25Mbps	ATM/155Mbps
$8,575	$3,016	$3,595	contact Sales	contact Sales
4 E1/PRI	BRI	BRI	25Mbps UTP	OC-3
120	16	32	256	1024
0	0	0	0	0
Yes	Yes	Yes	No	No
120	8	16	0	0
No	No	No	No	No
Yes	Yes	Yes	Yes	Yes
No	No	No	No	No
SCbus, H.100, H.110	SCbus, H.100	SCbus, H.100	SCbus, H.100	SCbus, H.100, H.110
Full	Full	Full	Full	Full
PCI, cPCI	PCI/ ISA AT	PCI/ ISA AT	ISA, PCI	ISA, PCI, cPCI
No	No	No	No	No
No	No	No	No	No
No	No	No	No	No
Yes	Yes	Yes	Yes	Yes
No	No	No	No	No
No	No	No	No	No
No	No	No	No	No
No	No	No	No	No
Yes	Yes	Yes	Yes	Yes
Yes	No	No	Yes	Yes
No	No	No	No	No
No	No	No	No	No
Yes	No	No	Yes	Yes
Yes	No	No	Yes	Yes
No	No	No	No	No
www.dialogic.com	www.dialogic.com	www.dialogic.com	www.dialogic.com	www.dialogic.com
800-755-4444	800-755-4444	800-755-4444	800-755-4444	800-755-4444
			Multiple configurations available	Multiple configurations available

Table of Digital Computer Telephony Boards, continued

Vendor	Intel/Dialogic	Intel/Dialogic	Intel/Dialogic	Intel/Dialogic
Product	ATM Ring Card	DTI/DS-3	SS7 Network Interface Cards	SS7 Protocol Modules
List Price	contact Sales	$17,500	contact Sales	contact Sales
Line Interface	155Mbps UTP	DS-3	up to 8 T1 or E1	N/A
DS0 Channels	512	672	up to 240	N/A
DSP V ports	0	0	0	0
Onboard DSP	No	No	No	No
Tone ports	0	0	0	0
Fax	No	No	No	No
Security	Yes	Yes	Yes	Yes
Wav Sup	No	No	No	No
Tel-BUS	H.100	H.110	SCbus, H.100, H.110	N/A
Card Length	Full	Full	Full	N/A
Form Factor	PCI	cPCI	ISA, PCI, cPCI	N/A
Pulse Rec	No	No	No	No
Speech Rec	No	No	No	No
Conference	No	No	No	No
Switching	Yes	Yes	Yes	Yes
Pitch/Speed	No	No	No	No
DOS	No	No	Yes	Yes
Win3x	No	No	No	No
Win95	No	No	No	No
Win NT	Yes	Yes	Yes	Yes
Unix	Yes	Yes	Yes	Yes
QNX	No	No	Yes	Yes
OS/2	No	No	No	No
Solaris	Yes	Yes	Yes	Yes
Linux	Yes	Yes	Yes	Yes
TAPI	No	No	No	No
URL	www.dialogic.com	www.dialogic.com	www.dialogic.com	www.dialogic.com
Phone	800-755-4444	800-755-4444	800-755-4444	800-755-4444
Comments			Supports up to 4 SS7 Signaling links	MTP, ISUP, French & Chinese TUP, TCAP, SCCP, IS41, GSM-MAP, INAP

Intel/Dialogic	Intel/Dialogic	Intel/Dialogic	Intel/Dialogic	Intel/Dialogic
SS7 Signaling Interface Unit (SIU)	Media Processing Products	D/80	D/160SC(TM)	D/160SC-HS(TM)
contact Sales		$1,525	$2,725	$2,000
up to 20 T1 or E1		N/A	N/A	N/A
up to 600		N/A	N/A	N/A
0		8	16	16
No		Yes	Yes	Yes
0		8	16	16
No		No	No	No
Yes		Yes	Yes	Yes
No		Yes	Yes	Yes
Ethernet		SCbus, H.100	SCbus	SCbus
N/A		Full	Full	Half
N/A		PCI/ ISA AT	ISA AT	ISA AT
No		Yes	Yes	Yes
No		No	No	No
No		No	No	No
Yes		Yes	Yes	Yes
No		Yes	Yes	Yes
Yes		Yes	Yes	Yes
No		No	No	No
No		No	No	No
Yes		Yes	Yes	Yes
Yes		Yes	Yes	Yes
Yes		No	No	No
No		Yes	Yes	Yes
Yes		Yes	Yes	Yes
Yes		Yes	Yes	No
No		Yes	Yes	Yes
www.dialogic.com		www.dialogic.com	www.dialogic.com	www.dialogic.com
800-755-4444		800-755-4444	800-755-4444	800-755-4444

Supports up to 30 SS7 Signaling
links

Table of Digital Computer Telephony Boards, continued

Vendor	Intel/Dialogic	Intel/Dialogic	Intel/Dialogic	Intel/Dialogic	
Product	D/160JCT	D/240SC(TM)	D/320	D/320JCT	
List Price	$3,134	$3,900	$4,495	$5,169	
Line Interface	N/A	N/A	N/A	N/A	
DS0 Channels	N/A	N/A	N/A	N/A	
DSP V ports	16	24	32	32	
Onboard DSP	Yes	Yes	Yes	Yes	
Tone ports	16	24	32	32	
Fax	No	No	No	No	
Security	Yes	Yes	Yes	Yes	
Wav Sup	Yes	Yes	Yes	Yes	
Tel-BUS	H.100	SCbus	SCbus, H.100	H.100	
Card Length	Full	Full	Full	Full	
Form Factor	PCI	ISA AT	PCI/ ISA AT	PCI	
Pulse Rec	Yes	Yes	Yes	Yes	
Speech Rec	Yes	No	No	Yes	
Conference	No	No	No	No	
Switching	Yes	Yes	Yes	Yes	
Pitch/Speed	Yes	Yes	Yes	Yes	
DOS	No	Yes	Yes	No	
Win3x	No	No	No	No	
Win95	No	No	No	No	
Win NT	Yes	Yes	Yes	Yes	
Unix	Yes	Yes	Yes	Yes	
QNX	No	No	No	No	
OS/2	No	Yes	Yes	No	
Solaris	Yes	Yes	Yes	Yes	
Linux	Yes	Yes	Yes	Yes	
TAPI	No	Yes	Yes	No	
URL	www.dialogic.com	www.dialogic.com	www.dialogic.com	www.dialogic.com	
Phone	800-755-4444	800-755-4444	800-755-4444	800-755-4444	
Comments					

Intel/Dialogic	Intel/Dialogic	Intel/Dialogic	Intel/Dialogic	Intel/Dialogic
D/640SC(TM)	DM/V1200	DM/F240(TM)	DM/F300(TM)	CP6/SC(TM)
$6,595	$7,800	$6,768	$8,366	$3,795
N/A	N/A	N/A	N/A	N/A
N/A	N/A	N/A	N/A	N/A
64	120	0	0	0
Yes	Yes	Yes	Yes	No
64	120	0	0	6
No	No	24	30	6
Yes	Yes	Yes	Yes	Yes
Yes	Yes	No	No	No
SCbus	H.100	SCbus, H.100	SCbus, H.100	SCbus
Full	Full	Full	Full	Full
ISA AT	PCI	PCI, cPCI	PCI, cPCI	ISA
Yes	Yes	Yes	Yes	No
No	Yes	No	No	No
No	Yes	No	No	No
Yes	Yes	No	No	No
Yes	Yes	Yes	Yes	No
Yes	No	No	No	Yes
No	No	No	No	No
No	No	No	No	No
Yes	Yes	Yes	Yes	Yes
Yes	Yes	Yes	Yes	Yes
No	No	No	No	No
Yes	No	No	No	No
Yes	Yes	Yes	Yes	No
Yes	Yes	Yes	Yes	No
Yes	No	No	No	No
www.dialogic.com	www.dialogic.com	www.dialogic.com	www.dialogic.com	www.dialogic.com
800-755-4444	800-755-4444	800-755-4444	800-755-4444	800-755-4444

Table of Digital Computer Telephony Boards, continued

endor	Intel/Dialogic	Intel/Dialogic	Intel/Dialogic	Intel/Dialogic
Product	CP12/SC(TM)	CPi/2400CT	CPi/3000CT	DCB/320SC(TM)
List Price	$7,120	$12,690	$15,510	$1,095
Line Interface	N/A	N/A	N/A	N/A
DS0 Channels	N/A	N/A	N/A	N/A
DSP V ports	0	0	0	0
Onboard DSP	No	Yes	Yes	Yes
Tone ports	12	24	30	32
Fax	12	24	30	No
Security	Yes	Yes	Yes	Yes
Wav Sup	No	No	No	No
Tel-BUS	SCbus	SCbus, H.100	SCbus, H.100	SCbus
Card Length	Full	Full	Full	Full
Form Factor	ISA	PCI	PCI	ISA AT
Pulse Rec	No	No	No	No
Speech Rec	No	No	No	No
Conference	No	No	No	Yes
Switching	No	No	No	Yes
Pitch/Speed	No	No	No	No
DOS	Yes	No	No	Yes
Win3x	No	No	No	No
Win95	No	No	No	No
Win NT	Yes	Yes	Yes	Yes
Unix	Yes	No	No	Yes
QNX	No	No	No	No
OS/2	No	No	No	No
Solaris	No	No	No	Yes
Linux	No	No	No	Yes
TAPI	No	No	No	No
URL	www.dialogic.com	www.dialogic.com	www.dialogic.com	www.dialogic.com
Phone	800-755-4444	800-755-4444	800-755-4444	800-755-4444
Comments				

Intel/Dialogic	Intel/Dialogic	Intel/Dialogic	Intel/Dialogic	Intel/Dialogic
DCB/640SC(TM)	DCB/960SC(TM)	IP Telephony Products	DM/IP0821A-T1	DM/IP0821A-E1
$1,650	$2,195		$5,185	$5,185
N/A	N/A		1 T1/PRI	1 E1/PRI
N/A	N/A		8	8
0	0		8	8
Yes	Yes		Yes	Yes
64	96		8	8
No	No		Yes	Yes
Yes	Yes		Yes	Yes
No	No		Yes	Yes
SCbus	SCbus		SCbus, H.100	SCbus, H.100
Full	Full		Full	Full
ISA AT	ISA AT		PCI	PCI
No	No		No	No
No	No		Yes	Yes
Yes	Yes		Yes	Yes
Yes	Yes		Yes	Yes
No	No		Yes	Yes
Yes	Yes		No	No
No	No		No	No
No	No		No	No
Yes	Yes		Yes	Yes
Yes	Yes		Yes	Yes
No	No		No	No
No	No		No	No
Yes	Yes		Yes	Yes
Yes	Yes		No	No
No	No		No	No
www.dialogic.com	www.dialogic.com		www.dialogic.com	www.dialogic.com
800-755-4444	800-755-4444		800-755-4444	800-755-4444
			Single board solution for IP-enabled media server and enhanced services	Single board solution for IP-enabled media server and enhanced services

Table of Digital Computer Telephony Boards, continued

Vendor	Intel/Dialogic	Intel/Dialogic	Intel/Dialogic	Intel/Dialogic	
Product	DM/IP241-1T1	DM/IP301-1E1	DM/IP241-1T1c	DM/IP301-1E1c	
List Price	$8,078	$10,097	$8,078	$10,097	
Line Interface	1 T1/PRI	1 E1/PRI	1 T1/PRI	1 E1/PRI	
DS0 Channels	24	30	24	30	
DSP V ports	24	30	24	30	
Onboard DSP	Yes	Yes	Yes	Yes	
Tone ports	24	30	24	30	
Fax	Yes	Yes	Yes	Yes	
Security	Yes	Yes	Yes	Yes	
Wav Sup	Yes	Yes	Yes	Yes	
Tel-BUS	SCbus, H.100	SCbus, H.100	SCbus, H.110	SCbus, H.110	
Card Length	Full	Full	Full	Full	
Form Factor	PCI	PCI	cPCI	cPCI	
Pulse Rec	No	No	No	No	
Speech Rec	Yes	Yes	Yes	Yes	
Conference	Yes	Yes	Yes	Yes	
Switching	Yes	Yes	Yes	Yes	
Pitch/Speed	Yes	Yes	Yes	Yes	
DOS	No	No	No	No	
Win3x	No	No	No	No	
Win95	No	No	No	No	
Win NT	Yes	Yes	Yes	Yes	
Unix	Yes	Yes	Yes	Yes	
QNX	No	No	No	No	
OS/2	No	No	No	No	
Solaris	Yes	Yes	Yes	Yes	
Linux	Yes	Yes	No	No	
TAPI	No	No	No	No	
URL	www.dialogic.com	www.dialogic.com	www.dialogic.com	www.dialogic.com	
Phone	800-755-4444	800-755-4444	800-755-4444	800-755-4444	
Comments	Single board solution for IP-enabled media server and enhanced services	Single board solution for IP-enabled media server and enhanced services	Single board solution for IP-enabled media server and enhanced services	Single board solution for IP-enabled media server and enhanced services	

190

Intel/Dialogic	Intel/Dialogic	Intel/Dialogic	Intel/Dialogic	Intel/Dialogic
DM/IP481-2T1c	DM/IP601-2E1c	PBX Integration Products	D82JCT-U	D/42NE2(TM)
$11,309	$14,136		$2,995	$1,895
2 T1/PRI	2 E1/PRI		PBX - Lucent, Nortel, Mitel, Siemens, NEC	PBX- NEC KTS & PBX
48	60		8	4
48	60		8	4
Yes	Yes		Yes	Yes
48	60		8	4
Yes	Yes		Yes	No
Yes	Yes		Yes	Yes
Yes	Yes		Yes	Yes
SCbus, H.110	SCbus, H.110		SCbus, H.100	SCbus
Full	Full		Full	Full
cPCI	cPCI		PCI	PCI
No	No		No	No
Yes	Yes		Yes	No
Yes	Yes		No	No
Yes	Yes		Yes	No
Yes	Yes		Yes	Yes
No	No		No	No
No	No		No	No
No	No		No	No
Yes	Yes		Yes	Yes
Yes	Yes		Yes	No
No	No		No	No
No	No		No	Yes
Yes	Yes		No	No
No	No		Yes	No
No	No		Yes	Yes
www.dialogic.com	www.dialogic.com		www.dialogic.com	www.dialogic.com
800-755-4444	800-755-4444		800-755-4444	800-755-4444
Single board solution for IP-enabled media server and enhanced services	Single board solution for IP-enabled media server and enhanced services			

Table of Digital Computer Telephony Boards, continued

Vendor	Intel/Dialogic	Intel/Dialogic	Intel/Dialogic	Intel/Dialogic
Product	D/42-NS(TM)	Speech Products	Antares 2000/50(TM)	Antares 3000/50(TM)
List Price	$1,895		$7,950	$7,950
Line Interface	PBX-Nortel Norstar		N/A	N/A
DS0 Channels	4		N/A	N/A
DSP V ports	4		0	0
Onboard DSP	Yes		Yes	Yes
Tone ports	4		N/A	N/A
Fax	No		No	No
Security	Yes		Yes	Yes
Wav Sup	Yes		No	No
Tel-BUS	None		SCbus, PEB	SCbus, PEB
Card Length	Full		Full	Full
Form Factor	ISA AT		ISA AT	ISA AT
Pulse Rec	No		N/A	N/A
Speech Rec	No		Yes	Yes
Conference	No		No	No
Switching	No		N/A	N/A
Pitch/Speed	Yes		N/A	N/A
DOS	Yes		Yes	Yes
Win3x	No		No	No
Win95	No		No	No
Win NT	Yes		Yes	Yes
Unix	Yes		Yes	Yes
QNX	No		No	No
OS/2	No		Yes	Yes
Solaris	No		Yes	Yes
Linux	No		No	No
TAPI	Yes		No	No
URL	www.dialogic.com		www.dialogic.com	www.dialogic.com
Phone	800-755-4444		800-755-4444	800-755-4444
Comments				

Intel/Dialogic	Intel/Dialogic	Epicom Inc	Epicom Inc	NMS
Antares 6000/50(TM)	Antares 3000/60(TM)	E6720MS-T3	T3PO/PCI(TM)	AG 4000-T1
$7,950	$7,950	$7,995	$8,995	$3,495
N/A	N/A	1 T3/DS3 with M13	1 T3/DS3 with M13	T1/PRI
N/A	N/A	672	672/630(G.747)	24
0	0	N/A	N/A	24
Yes	Yes	No	YES	Yes
N/A	N/A	N/A	32	24
No	No	N/A	N/A	No
Yes	Yes	Yes	Yes	Yes
No	No	N/A	N/A	Yes
SCbus, PEB	SCbus, H.100	MVIP, SCbus	H.100	MVIP, H.100
Full	Full	Full	Full	Full
ISA AT	PCI	ISA AT	PCI	PCI
N/A	N/A	No	No	No
Yes	Yes	No	No	No
No	No	No	No	Yes
N/A	N/A	Yes	Yes	Yes
N/A	N/A	No	No	Yes
Yes	No	No	No	No
No	No	No	No	No
No	No	No	No	No
Yes	Yes	Yes	Yes	Yes
Yes	Yes	Yes	Yes	Yes
No	No	No	No	Special
Yes	No	No	No	No
Yes	Yes	No	No	Yes
No	Yes	See Notes	See Notes	See Notes
No	No	No	No	Yes
www.dialogic.com	www.dialogic.com	www.epicom-inc.com	www.epicom-inc.com	www.nmss.com
800-755-4444	800-755-4444	510-887-7070	510-887-7070	508-620-9300
		A T-3 card.		

Vendor	NMS	NMS	NMS	NMS
Product	AG 4000-E1-75	AG 4000-E1-120	AG 4000-2T1	AG 4000-2E1-75
List Price	$4,195	$4,195	$5,595	$6,595
Line Interface	E1/PRI	E1/PRI	2 T1/PRI	2 E1/PRI
DS0 Channels	30	30	48	60
DSP V ports	30	30	48	60
Onboard DSP	Yes	Yes	Yes	Yes
Tone ports	30	30	48	60
Fax	No	No	Option	Option
Security	Yes	Yes	Yes	Yes
Wav Sup	Yes	Yes	Yes	Yes
Tel-BUS	MVIP, H.100	MVIP, H.100	MVIP, H.100	MVIP, H.100
Card Length	Full	Full	Full	Full
Form Factor	PCI	PCI	PCI	PCI
Pulse Rec	No	No	No	No
Speech Rec	No	No	No	No
Conference	Yes	Yes	Yes	Yes
Switching	Yes	Yes	Yes	Yes
Pitch/Speed	Yes	Yes	Yes	Yes
DOS	No	No	No	No
Win3x	No	No	No	No
Win95	No	No	No	No
Win NT	Yes	Yes	Yes	Yes
Unix	Yes	Yes	Yes	Yes
QNX	Special	Special	Special	Special
OS/2	No	No	No	No
Solaris	Yes	Yes	Yes	Yes
Linux	See Notes	See Notes	See Notes	See Notes
TAPI	Yes	Yes	Yes	Yes
URL	www.nmss.com	www.nmss.com	www.nmss.com	www.nmss.com
Phone	508-620-9300	508-620-9300	508-620-9300	508-620-9300
Comments				

NMS	NMS	NMS	NMS	NMS
AG 4000-2E1-120	AG 4000-4T1	AG 4000-4E1-75	AG 4000-4E1-120	AG 4000C/1600-2T1 Rear
$6,595	$8,895	$10,395	$10,395	$7,945
2 E1/PRI	4 T1/PRI	4 E1/PRI	4 E1/PRI	2 T1/PRI
60	96	120	120	48
60	96	120	120	48
Yes	Yes	Yes	Yes	Yes
60	96	120	120	48
Option	Option	Option	Option	Option
Yes	Yes	Yes	Yes	Yes
Yes	Yes	Yes	Yes	Yes
MVIP, H.100	MVIP, H.100	MVIP, H.100	MVIP, H.100	H.110
Full	Full	Full	Full	6U
PCI	PCI	PCI	PCI	cPCI
No	No	No	No	No
No	No	No	No	No
Yes	Yes	Yes	Yes	Yes
Yes	Yes	Yes	Yes	Yes
Yes	Yes	Yes	Yes	Yes
No	No	No	No	No
No	No	No	No	No
No	No	No	No	No
Yes	Yes	Yes	Yes	Yes
Yes	Yes	Yes	Yes	Yes
Special	Special	Special	Special	Special
No	No	No	No	No
Yes	Yes	Yes	Yes	Yes
See Notes	See Notes	See Notes	See Notes	See Notes
Yes	Yes	Yes	Yes	Yes
www.nmss.com	www.nmss.com	www.nmss.com	www.nmss.com	www.nmss.com
508-620-9300	508-620-9300	508-620-9300	508-620-9300	508-620-9300

Vendor	NMS	NMS	NMS	NMS	
Product	AG 4000C/1600-2E1-75 Rear	AG 4000C/1600-2E1-120 Rear	AG 4000C-4T1 Front	AG 4000C-4T1 Rear	
List Price	$9,445	$9,445	$8,895	$9,045	
Line Interface	2 E1/PRI	2 E1/PRI	4 T1/PRI	4 T1/PRI	
DS0 Channels	60	60	96	96	
DSP V ports	60	60	96	96	
Onboard DSP	Yes	Yes	Yes	Yes	
Tone ports	60	60	96	96	
Fax	Option	Option	Option	Option	
Security	Yes	Yes	Yes	Yes	
Wav Sup	Yes	Yes	Yes	Yes	
Tel-BUS	H.110	H.110	H.110	H.110	
Card Length	6U	6U	6U	6U	
Form Factor	cPCI	cPCI	cPCI	cPCI	
Pulse Rec	No	No	No	No	
Speech Rec	No	No	No	No	
Conference	Yes	Yes	Yes	Yes	
Switching	Yes	Yes	Yes	Yes	
Pitch/Speed	Yes	Yes	Yes	Yes	
DOS	No	No	No	No	
Win3x	No	No	No	No	
Win95	No	No	No	No	
Win NT	Yes	Yes	Yes	Yes	
Unix	Yes	Yes	Yes	Yes	
QNX	Special	Special	Special	Special	
OS/2	No	No	No	No	
Solaris	Yes	Yes	Yes	Yes	
Linux	See Notes	See Notes	See Notes	See Notes	
TAPI	Yes	Yes	Yes	Yes	
URL	www.nmss.com	www.nmss.com	www.nmss.com	www.nmss.com	
Phone	508-620-9300	508-620-9300	508-620-9300	508-620-9300	
Comments					

NMS	NMS	NMS	NMS	NMS
AG 4000C-4E1-75 Front	AG 4000C-4E1-75 Rear	AG 4000C-4E1-120 Front	AG 4000C-4E1-120 Rear	AG-T1
$10,395	$10,545	$10,395	$10,545	$3,495
4 E1/PRI	4 E1/PRI	4 E1/PRI	4 E1/PRI	1 T1/PRI
120	120	120	120	24
120	120	120	120	24
Yes	Yes	Yes	Yes	Yes
120	120	120	120	24
Option	Option	Option	Option	Option
Yes	Yes	Yes	Yes	Yes
Yes	Yes	Yes	Yes	Yes
H.110	H.110	H.110	H.110	MVIP
6U	6U	6U	6U	Full
cPCI	cPCI	cPCI	cPCI	ISA AT
No	No	No	No	No
No	No	No	No	No
Yes	Yes	Yes	Yes	No
Yes	Yes	Yes	Yes	Yes
Yes	Yes	Yes	Yes	Yes
No	No	No	No	No
No	No	No	No	No
No	No	No	No	No
Yes	Yes	Yes	Yes	Yes
Yes	Yes	Yes	Yes	Yes
Special	Special	Special	Special	No
No	No	No	No	Yes legacy
Yes	Yes	Yes	Yes	No
See Notes	See Notes	See Notes	See Notes	See Notes
Yes	Yes	Yes	Yes	Yes
www.nmss.com	www.nmss.com	www.nmss.com	www.nmss.com	www.nmss.com
508-620-9300	508-620-9300	508-620-9300	508-620-9300	508-620-9300

Table of Digital Computer Telephony Boards, continued

Vendor	NMS	NMS	NMS	NMS
Product	AG-E1-75	AG-E1-120	AG Dual T	AG Dual E-75
List Price	$4,195	$4,195	$5,595	$6,595
Line Interface	1 E1/PRI	1 E1/PRI	2 T1/PRI	2 E1/PRI
DS0 Channels	30	30	48	60
DSP V ports	30	30	48	60
Onboard DSP	Yes	Yes	Yes	Yes
Tone ports	30	30	48	60
Fax	Option	Option	No	No
Security	Yes	Yes	Yes	Yes
Wav Sup	Yes	Yes	Yes	Yes
Tel-BUS	MVIP	MVIP	H.100	H.100
Card Length	Full	Full	Full	Full
Form Factor	ISA AT	ISA AT	PCI	PCI
Pulse Rec	No	No	No	No
Speech Rec	No	No	No	No
Conference	No	No	No	No
Switching	Yes	Yes	Yes	Yes
Pitch/Speed	Yes	Yes	Yes	Yes
DOS	No	No	No	No
Win3x	No	No	No	No
Win95	No	No	No	No
Win NT	Yes	Yes	Yes	Yes
Unix	Yes	Yes	Yes	Yes
QNX	No	No	No	No
OS/2	Yes legacy	Yes legacy	No	No
Solaris	No	No	Yes	Yes
Linux	See Notes	See Notes	See Notes	See Notes
TAPI	Yes	Yes	Yes	Yes
URL	www.nmss.com	www.nmss.com	www.nmss.com	www.nmss.com
Phone	508-620-9300	508-620-9300	508-620-9300	508-620-9300
Comments				

NMS	NMS	NMS	NMS	NMS
AG Dual E-120	AG Quad T	AG Quad E-75	AG Quad E-120	CompactPCI AG Quad T Front
$6,595	$7,095	$8,095	$8,095	$7,095
2 E1/PRI	4 T1/PRI	4 E1/PRI	4 E1/PRI	4 T1/PRI
60	96	120	120	96
60	48	60	60	48
Yes	Yes	Yes	Yes	Yes
60	96	120	120	96
No	No	No	No	No
Yes	Yes	Yes	Yes	Yes
Yes	Yes	Yes	Yes	Yes
H.100	H.100	H.100	H.100	H.110
Full	Full	Full	Full	6U
PCI	PCI	PCI	PCI	cPCI
No	No	No	No	No
No	No	No	No	No
No	No	No	No	No
Yes	Yes	Yes	Yes	Yes
Yes	Yes	Yes	Yes	Yes
No	No	No	No	No
No	No	No	No	No
No	No	No	No	No
Yes	Yes	Yes	Yes	Yes
Yes	Yes	Yes	Yes	Yes
No	No	No	No	No
No	No	No	No	No
Yes	Yes	Yes	Yes	Yes
See Notes	See Notes	See Notes	See Notes	See Notes
Yes	Yes	Yes	Yes	Yes
www.nmss.com	www.nmss.com	www.nmss.com	www.nmss.com	www.nmss.com
508-620-9300	508-620-9300	508-620-9300	508-620-9300	508-620-9300

Vendor	NMS	NMS	NMS	NMS	NMS	NMS
Product	CompactPCI AG Quad T Rear	CompactPCI AG Quad E-75 Front	CompactPCI AG Quad E-75 Rear	CompactPCI AG Quad E-120 Front	CompactPCI AG Quad E-120 Rear	AG-DSP/80
List Price	$7,095	$8,095	$8,095	$8,095	$8,095	$1,600
Line Interface	4 E1/PRI	4 E1/PRI	4 E1/PRI	4 E1/PRI	4 E1/PRI	None
DS0 Channels	120	120	120	120	120	N/A
DSP V ports	60	60	60	60	60	8
Onboard DSP	Yes	Yes	Yes	Yes	Yes	Yes
Tone ports	120	120	120	120	120	8
Fax	No	No	No	No	No	Option
Security	Yes	Yes	Yes	Yes	Yes	Yes
Wav Sup	Yes	Yes	Yes	Yes	Yes	Yes
Tel-BUS	H.110	H.110	H.110	H.110	H.110	MVIP
Card Length	6U	6U	6U	6U	6U	Full
Form Factor	cPCI	cPCI	cPCI	cPCI	cPCI	ISA AT
Pulse Rec	No	No	No	No	No	No
Speech Rec	No	No	No	No	No	No
Conference	No	No	No	No	No	No
Switching	Yes	Yes	Yes	Yes	Yes	Yes
Pitch/Speed	Yes	Yes	Yes	Yes	Yes	Yes
DOS	No	No	No	No	No	No
Win3x	No	No	No	No	No	No
Win95	No	No	No	No	No	No
Win NT	Yes	Yes	Yes	Yes	Yes	Yes
Unix	Yes	Yes	Yes	Yes	Yes	Yes
QNX	No	No	No	No	No	No
OS/2	No	No	No	No	No	Yes legacy
Solaris	Yes	Yes	Yes	Yes	Yes	No
Linux	See Notes	See Notes	See Notes	See Notes	See Notes	See Notes
TAPI	Yes	Yes	Yes	Yes	Yes	Yes
URL	www.nmss.com	www.nmss.com	www.nmss.com	www.nmss.com	www.nmss.com	www.nmss.com
Phone	508-620-9300	508-620-9300	508-620-9300	508-620-9300	508-620-9300	508-620-9300
Comments						

NMS	NMS	NMS	NMS	Odin TeleSystems Inc.	Odin TeleSystems Inc.
AG-24+	AG-30	AG-48	AG-60	Thor-2-PCI	Thor-8-PCI
$3,195	$3,895	$4,795	$5,795	$2,495	$5,495
None	None	None	None	2 x T1/E1/PRI	8 x T1/E1/PRI
N/A	N/A	N/A	N/A	60	240
24	30	48	60	60	60
Yes	Yes	Yes	Yes	Daughter Module	Daughter Module
24	30	48	60	60	60
Option	Option	Option	Option	No	No
Yes	Yes	Yes	Yes	Yes	Yes
Yes	Yes	Yes	Yes	No	No
MVIP	MVIP	MVIP	MVIP	H.100	H.100
Full	Full	Full	Full	Half	Full
ISA AT	ISA AT	ISA AT	ISA AT	PCI	PCI
No	No	No	No	No	No
No	No	No	No	No	No
No	No	No	No	No	No
Yes	Yes	Yes	Yes	Yes	Yes
Yes	Yes	Yes	Yes	No	No
No	No	No	No	No	No
No	No	No	No	No	No
No	No	No	No	Yes	Yes
Yes	Yes	Yes	Yes	Yes	Yes
Yes	Yes	Yes	Yes	Yes	Yes
No	No	No	No	N	N
Yes legacy	Yes legacy	Yes legacy	Yes legacy	No	No
No	No	No	No	No	No
See Notes	See Notes	See Notes	See Notes	See Notes	See Notes
Yes	Yes	Yes	Yes	Yes	Yes
www.nmss.com	www.nmss.com	www.nmss.com	www.nmss.com	www.OdinTS.com	www.OdinTS.com
508-620-9300	508-620-9300	508-620-9300	508-620-9300	972-664-0100	972-664-0100

Vendor	Pika	Pika	Pika	Pika	Pika	Pika	
Product	InLine 2GTb	InLine 4GTb	InLine 2 GT	InLine 4 GT	InLine ST	ISA Daytona L12	
List Price	$540	$590	$580	$630	$610	$2,720	
Line Interface	2 LS	4LS	2LS	4LS	4POTS	12 LS	
DS0 Channels	2	4	2	4	4	12	
DSP V ports	12	12	12	12	0	12	
Onboard DSP	Yes	Yes	Yes	Yes	No	Yes	
Tone ports	12	12	12	12	0	12	
Fax	OPT. Soft	OPT. Soft	OPT. Soft	OPT. Soft	No	OPT. Soft	
Security	Yes	Yes	Yes	Yes	No	Yes	
Wav Sup	Yes	Yes	Yes	Yes	N/A	Yes	
Tel-BUS	mini-mvip	mini-mvip	mini-mvip	mini-mvip	mini-mvip	MVIP	
Card Length	Half	Half	Half	Half	Half	Full	
Form Factor	ISA	ISA	ISA	ISA	ISA	ISA	
Pulse Rec	OPT. Soft	OPT. Soft	OPT. Soft	OPT. Soft	No	OPT. Soft	
Speech Rec	OPT. Soft	OPT. Soft	OPT. Soft	OPT. Soft	No	OPT. Soft	
Conference	OPT. Soft	OPT. Soft	OPT. Soft	OPT. Soft	No	OPT. Soft	
Switching	Yes	Yes	Yes	Yes	No	Yes	
Pitch/Speed	Yes	Yes	Yes	Yes	No	Yes	
DOS	Yes	Yes	Yes	Yes	Yes	Yes	
Win3x	Yes	Yes	Yes	Yes	Yes	Yes	
Win95	Yes	Yes	Yes	Yes	Yes	Yes	
Win NT	Yes	Yes	Yes	Yes	Yes	Yes	
Unix	No	No	No	No	No	No	
QNX	No	No	No	No	No	No	
OS/2	No	No	No	No	No	No	
Solaris	No	No	No	No	No	No	
Linux	Yes	Yes	Yes	Yes	Yes	Yes	
TAPI	No	No	No	No	No	No	
URL	www.pikatech.com	www.pikatech.com	www.pikatech.com	www.pikatech.com	www.pikatech.com	www.pikatech.com	
Phone	613-591-1555	613-591-1555	613-591-1555	613-591-1555	613-591-1555	613-591-1555	
Comments	Board supports Linux	Board supports Linux	Board supports Linux	Board supports Linux	Board supports Linux	Board supports Linux	

Table of Digital Computer Telephony Boards, continued

Pika	Pika	Pika	Pika	Pika	Pika
ISA Daytona L24	ISA Daytona L4P8	ISA Daytona L8P16	ISA Daytona P12	ISA Daytona P24	PCI Daytona L12
$4,760	$2,650	$4,420	$2,550	$4,220	$3,360
24 LS	4 LS & POTS	8 LS 16 POTS	12 POTS	24 POTS	12 LS
24	12	24	12	24	12
24	12	24	12	24	12
Yes	Yes	Yes	Yes	Yes	Yes
24	12	24	12	24	12
OPT. Soft	OPT. Soft	OPT. Soft	OPT. Soft	OPT. Soft	OPT. Soft
Yes	Yes	Yes	Yes	Yes	Yes
Yes	Yes	Yes	Yes	Yes	Yes
MVIP	MVIP	MVIP	MVIP	MVIP	MVIP
Full	Full	Full	Full	Full	Full
ISA	ISA	ISA	ISA	ISA	PCI
OPT. Soft	OPT. Soft	OPT. Soft	OPT. Soft	OPT. Soft	OPT. Soft
OPT. Soft	OPT. Soft	OPT. Soft	OPT. Soft	OPT. Soft	OPT. Soft
OPT. Soft	OPT. Soft	OPT. Soft	OPT. Soft	OPT. Soft	OPT. Soft
Yes	Yes	Yes	Yes	Yes	Yes
Yes	Yes	Yes	Yes	Yes	Yes
Yes	Yes	Yes	Yes	Yes	Yes
Yes	Yes	Yes	Yes	Yes	Yes
Yes	Yes	Yes	Yes	Yes	Yes
Yes	Yes	Yes	Yes	Yes	Yes
No	No	No	No	No	No
No	No	No	No	No	No
No	No	No	No	No	No
No	No	No	No	No	No
Yes	Yes	Yes	Yes	Yes	Yes
No	No	No	No	No	No
www.pikatech.com	www.pikatech.com	www.pikatech.com	www.pikatech.com	www.pikatech.com	www.pikatech.com
613-591-1555	613-591-1555	613-591-1555	613-591-1555	613-591-1555	613-591-1555
Board supports Linux	Board supports Linux	Board supports Linux	Board supports Linux	Board supports Linux	Board supports Linux

Table of Digital Computer Telephony Boards, continued

Vendor	Pika	Pika	Pika	Pika	Pika	Pika
Product	PCI Daytona L24	PCI Daytona L4P8	PCI Daytona L8P16	PCI Daytona P12	PCI Daytona P24	PCI Premiere 16
List Price	$5,950	$3,320	$5,530	$3,230	$5,270	$9,520
Line Interface	24 LS	4 LS 8 POTS	8 LS 16 POTS	12 POTS	24 POTS	None
DS0 Channels	24	12	24	12	24	N/A
DSP V ports	24	12	24	12	24	240
Onboard DSP	Yes	Yes	Yes	Yes	Yes	Yes
Tone ports	24	12	24	12	24	240
Fax	OPT. Soft	OPT. Soft	OPT. Soft	OPT. Soft	OPT. Soft	OPT. Soft
Security	Yes	Yes	Yes	Yes	Yes	Yes
Wav Sup	Yes	Yes	Yes	Yes	Yes	Yes
Tel-BUS	MVIP	MVIP	MVIP	MVIP	MVIP	MVIP
Card Length	Full	Full	Full	Full	Full	Full
Form Factor	PCI	PCI	PCI	PCI	PCI	PCI
Pulse Rec	OPT. Soft	OPT. Soft	OPT. Soft	OPT. Soft	OPT. Soft	OPT. Soft
Speech Rec	OPT. Soft	OPT. Soft	OPT. Soft	OPT. Soft	OPT. Soft	OPT. Soft
Conference	OPT. Soft	OPT. Soft	OPT. Soft	OPT. Soft	OPT. Soft	OPT. Soft
Switching	Yes	Yes	Yes	Yes	Yes	Yes
Pitch/Speed	Yes	Yes	Yes	Yes	Yes	Yes
DOS	Yes	Yes	Yes	Yes	Yes	Yes
Win3x	Yes	Yes	Yes	Yes	Yes	Yes
Win95	Yes	Yes	Yes	Yes	Yes	Yes
Win NT	Yes	Yes	Yes	Yes	Yes	Yes
Unix	No	No	No	No	No	No
QNX	No	No	No	No	No	No
OS/2	No	No	No	No	No	No
Solaris	No	No	No	No	No	No
Linux	Yes	Yes	Yes	Yes	Yes	Yes
TAPI	No	No	No	No	No	No
URL	www.pikatech.com	www.pikatech.com	www.pikatech.com	www.pikatech.com	www.pikatech.com	www.pikatech.com
Phone	613-591-1555	613-591-1555	613-591-1555	613-591-1555	613-591-1555	613-591-1555
Comments	Board supports Linux	Board supports Linux	Board supports Linux	Board supports Linux	Board supports Linux	Board supports Linux

Pika	Pika	Pika	Pika	Pika	Pika
PCI Premiere 4	PCI Premiere 8	PrimeNet PCI 30E	PrimeNet PCI 60E	PrimeNet PCI 24T	PrimeNet PCI 48T
$3,200	$5,760	$6,630	Call vendor	$5,310	$7,610
None	None	1 E1/PRI	2 E1/PRI	1 T1/PRI	2 T1/PRI
N/A	N/A	30	60	24	48
60	120	30	60	24	48
Yes	Yes	Yes	Yes	Yes	Yes
60	120	30	60	24	48
OPT. Soft	OPT. Soft	OPT. Soft	OPT. Soft	OPT. Soft	OPT. Soft
Yes	Yes	Yes	Yes	Yes	Yes
Yes	Yes	Yes	Yes	Yes	Yes
MVIP	MVIP	MVIP	MVIP	MVIP	MVIP
Half	Full	Full	Full	Full	Full
PCI	PCI	PCI	PCI	PCI	PCI
OPT. Soft	OPT. Soft	OPT. Soft	OPT. Soft	OPT. Soft	OPT. Soft
OPT. Soft	OPT. Soft	OPT. Soft	OPT. Soft	OPT. Soft	OPT. Soft
OPT. Soft	OPT. Soft	OPT. Soft	OPT. Soft	OPT. Soft	OPT. Soft
Yes	Yes	Yes	Yes	Yes	Yes
Yes	Yes	Yes	Yes	Yes	Yes
Yes	Yes	No	No	No	No
Yes	Yes	No	No	No	No
Yes	Yes	Yes	Yes	Yes	Yes
Yes	Yes	Yes	Yes	Yes	Yes
No	No	No	No	No	No
No	No	No	No	No	No
No	No	No	No	No	No
No	No	No	No	No	No
Yes	Yes	See Notes	See Notes	See Notes	See Notes
No	No	No	No	No	No
www.pikatech.com	www.pikatech.com	www.pikatech.com	www.pikatech.com	www.pikatech.com	www.pikatech.com
613-591-1555	613-591-1555	613-591-1555	613-591-1555	613-591-1555	613-591-1555
Board supports Linux	Board supports Linux				

Vendor	Telsoft Design	Telsoft Design	Telsoft Design	Telsoft Design	Voiceboard
Product	MPAC Dual PCI E1/T1 Interface Card	MPAC Quad PCI E1/T1 Interface Card	MPAC Octal cPCI E1/T1 Interface Card	IPAC Octal cPCI E1/T1/J1 Interface Card	CT22
List Price	$3,400	$4,200	$5,500	$5500	$3,495
Line Interface	2 E1 or T1 interfaces	4 E1 or T1 interfaces	8 E1 or T1 interfaces	8 E1, T1 or J1 interfaces	2 T1, PRI
DS0 Channels	60	120	240	240	48
DSP V ports	none	none	none	240	64
Onboard DSP	option	option	none	option	Yes
Tone ports	60	120	none	240	48
Fax	No	No	No	No	Yes
Security	Yes	Yes	Yes	Yes	Yes
Wav Sup	No	No	No	Yes	No
Tel-BUS	H.100, SCbus, MVIP	H.100, SCbus, MVIP	H.110	H.110	H.110
Card Length	Standard	Standard	Standard	Standard	6U
Form Factor	PCI	PCI	cPCI	cPCI	cPCI
Pulse Rec	No	No	No	No	Yes
Speech Rec	No	No	No	Yes	Yes
Conference	No	No	No	Yes	Yes
Switching	Yes	Yes	Yes	Yes	Yes
Pitch/Speed	No	No	No	No	Yes
DOS	No	No	No	No	No
Win3x	No	No	No	No	No
Win95	No	No	No	No	No
Win NT	Yes	Yes	Yes	Yes	Yes
Unix	Yes	Yes	Yes	Yes	Yes
QNX	No	No	No	No	No
OS/2	No	No	No	No	No
Solaris	Yes	Yes	Yes	Yes	Yes
Linux	See Notes	See Notes	See Notes	See Notes	See Notes
TAPI	Yes	Yes	Yes	Yes	No
URL	www.tduk.com	www.tduk.com	www.tduk.com	www.tduk.com	www.voiceboard.com
Phone	+44 1258 480880	+44 1258 480880	+44 1258 480880	+44 1258 480880	805-985-6200
Comments				on board10/100 Ethernet port	

Voiceboard	Voiceboard	Voiceboard	Voiceboard	Voiceboard	Voiceboard
CT24	CT8	CE22	CE24	CE28	C20DSP/16
$4,495	$6,495	$3,995	$4,995	$6,995	$9,995
4 T1, PRI	8 T1, PRI	2 E1, PRI	4 E1, PRI	8 E1, PRI	None
96	192	64	128	256	256
64	64	64	64	64	4,096
Yes	Yes	Yes	Yes	Yes	Yes
96	192	64	128	256	256
Yes	Yes	Yes	Yes	Yes	Yes
Yes	Yes	Yes	Yes	Yes	Yes
No	No	No	No	No	No
H.110	H.110	H.110	H.110	H.110	H.110
6U	6U	6U	6U	6U	6U
cPCI	cPCI	cPCI	cPCI	cPCI	cPCI
Yes	Yes	Yes	Yes	Yes	No
Yes	Yes	Yes	Yes	Yes	Yes
Yes	Yes	Yes	Yes	Yes	Yes
Yes	Yes	Yes	Yes	Yes	Yes
Yes	Yes	Yes	Yes	Yes	Yes
No	No	No	No	No	No
No	No	No	No	No	No
No	No	No	No	No	No
Yes	Yes	Yes	Yes	Yes	Yes
Yes	Yes	Yes	Yes	Yes	Yes
No	No	No	No	No	No
No	No	No	No	No	No
Yes	Yes	Yes	Yes	Yes	Yes
See Notes	See Notes	See Notes	See Notes	See Notes	See Notes
No	No	No	No	No	No
www.voiceboard.com	www.voiceboard.com	www.voiceboard.com	www.voiceboard.com	www.voiceboard.com	www.voiceboard.com
805-985-6200	805-985-6200	805-985-6200	805-985-6200	805-985-6200	805-985-6200

Notes on Digital Computer Telephony Resource Boards

Notes on Digital Computer Telephony Resource Boards

Customer Relationship Management (CRM)

CRM is the complete overall process of marketing, sales, and service within any organization, which can be broken down into discrete activities a company performs to identify, qualify, acquire, develop and retain loyal and profitable customers. CRM includes the kind of functionality previously found in sales force automation tools, but CRM adds the ability to perform post-sales support, service and customer maintenance. When done effectively this all leads to stronger customer loyalty, increased competitive advantage through service differentiation, improved profitability, and more efficient sales processes.

CRM also has a close cousin, Partner Relationship Management (PRM) that keeps a company's indirect channels happy and locked into a profitable relationship. PRM forces us to generalize further, leaving us finally with the concept of overall "relationship management."

If you form partnerships with others to provide service, such as a retailer who contract out warranty repair work, CRM software lets you track the interaction between companies (provided the vendor you choose offers the capability. This also applies if you are outsourcing part of your call and e-mail handling. If your department has more than one office, CRM keeps them connected).

"Any business which is dispersed divisionally or geographically, where there is interaction across multiple channels, and which needs to get their sales and customer service on the same page, is a good fit for CRM software," says Brian Tuller, senior product marketing and management director of Vantive (Santa Clara, CA – 408-982-5700, www.vantive.com).

As with much of customer-business interactions involving computer telephony, the call center or contact center is generally the focal point for successful implementations of CRM strategy.

The huge emphasis on CRM in business came about after it was discovered that it costs an estimated six to 12 times as much to add a new customer as it does to retain an existing one. In the same vein, the Harvard Business Review reports that companies can boost profits by nearly 100% by retaining just 5% more of their existing customers. And the Strategic Planning Institute reports that companies that get the highest marks for quality of customer service achieved twice the ROI, ROS and market share of companies rated as providing inferior customer service.

It's true that CRM software can be a very powerful tool in the right hands. It connects all customer touchpoints including your call center (voice, e-mail, the Internet and, in the near future, video), help desk and inside and outside sales with other internal users such as your department managers.

CRM software gives your firm's employees and managers access to uniformly organized information on each buyer or prospect contained in one or many linked databases. They utilize that data for purposes ranging from targeting customers with specific new offers to researching complaints to see what the problems are and find ways to fix them.

With CRM, every person in your company who is involved with sales and service is "on the same page." An inside salesperson could be dialing a client who at the very same moment is complaining to customer service about a missing shipment.

CRM software is especially valuable firms that sell to a limited customer base, where the buyer and the product/service has high value and whose sales loss would cause considerable pain to the balance sheet. This is especially important where there is limited or slow growth in the number of customers and prospects.

"Business to business marketers should invest in CRM if the cost of losing a single sale or customer exceeds the cost of the CRM soft-ware," says Kevin Nix, senior director Siebel Systems (San Mateo, CA – 650-295-5000, www.siebel.com).

Specialized B-to-B firms are not the only market types that can benefit from CRM software. Mass market business-to-business and business-to-consumer firms, that sell and contact different channels and which need to keep, update and dynamically use data gathered from customers and develop relationships with them may need CRM software. If you conduct business by either or including outbound and inbound telesales, e-mail and online, and collect customer data, CRM software can better coordinate and target your sales and customer acquisition/retention programs.

The individual product/service pricetags do not have to be high. Eben Frankenberg, senior vice president of sales and marketing for Onyx Software (Bellevue, WA – 425-451-8060, www.onyxcorp.com), points out that consumer sales data collected by agents with CRM software can go into better, more targeted direct marketing campaigns.

One of Onyx Software's customers, Community Playthings, saved $600,000 in catalog mailings yet increased sales by 10% by analyzing who their best customers were and sending targeted mailings to them.

But CRM is not for every company. The investment – which ranges from the tens of thousands of dollars to millions of dollars depending on the installation – may not pay off if your firm still has a natural monopoly, does not have multiple sales and service channels and/or low customer acquisition costs. CRM software is also not worth your while if you have little contact with end-customers or sell low-value products to a small customer base and have a tiny number of contacts.

"Companies that have walk-in customers and do not provide multiple sales and service channels, such as smaller consumer products firms, will have reduced benefit from CRM," says Greg Stack senior vice president with eLoyalty, a Chicago, IL-based technology consultancy.

Chances are you won't receive much payoff from CRM software if you do not have repeat business from customers, if maintaining long-term customer relationships is not a priority, or if your product/service market consists of customers who primarily buy based on the lowest price or the firm that has the sharpest cutting edge technology.

"If maintaining long-term customer relationships is not a priority for the company then you're wasting your time and resources by investing in CRM software," says Jeff Kaplan, a principal with Pittiglio Rabin Todd & McGrath (PRTM), a management consultant firm based out of Stamford, CT.

Your firm must also be organized across these different channels and functions to utilize CRM. The software becomes a difficult value proposition if there's no communication between each department.

"The businesses that stand to gain the most from CRM are those that are organized, without walls, to deliver consistent quality service and support across all departments to customers," says Clarify's (San Jose, CA – 408-573-3000, www.clarify.com) senior product marketing manager John Ragsdale. "Those that function as silos – serving and selling in only one department or product line won't benefit from CRM, and they will not be successful."

Maximizing Customer Value

The oft-quoted statistic is that 20% of customers account for 80% of your sales. To cater to the former and gain maximum profitability and retention from the latter, CRM software enables you to choose and assign a value to each customer and set up your contact handling to direct that person to the most effective interaction media.

CRM software lets you create business rules to prioritize customers and how to interact with them based on, for example, the media they select and on what they are buying or looking at on your Web site.

"You need to determine, and be able to adjust based on new information, the lifetime value of each customer," says Brian Tuller, senior product marketing and management director of Vantive Corp (Santa Clara, CA – 408-982-5700, www.vantive.com), a company acquired by PeopleSoft. "The cost of sending a call to a live agent can wipe out the profit you've made on that sale, or on future sales to that customer. If you sell $80 printers and your center has to take a call, you've just spent $20 on that call."

The key in helping determine customer value is knowing who that customer is and how they are interacting with your firm. CRM picks up data on your customer interactions from sources such as info your agents enter, software (such as problem resolution), IVR and Web site activities (e.g. cookies).

Data collected by CRM software complements ACD reports, which while highly useful, spew numbers that must be entered in or read in text-based report forms. This requires your supervisors to have experience in reading and analyzing spreadsheets, and forces them to hand-calculate what if models.

"You can't get information on the quality or impact of calls from ACD reports," says Siebel's Nix. "It can tell you who called and which agents received the calls but it can't tell you the value of these interactions."

Clarify's Ragsdale agrees. ACD data does not measure or account for how long it takes a rep to reach senior support staff or an outside field sales person offline, the telecom costs, or the value of each call or contact.

"ACD reports are excellent for information needed to schedule agents and to devise skills-based routing," Ragsdale points out. "It doesn't give you the information to help you see trends."

CRM software lets you serve and keep customers in a variety of imaginative ways. For example it can prioritize customer handling out of the IVR system, or from a Web site to a live agent, as soon as the product recognizes the caller or Web site visitor.

"We had one customer, a shipping firm, who took shipment-tracking calls on the IVR," recounts Chordiant's (Cupertino, CA – 408–517-6100, www.chordiant.com) marketing director Collin Bruce. "We put in a live agent override so that if the same caller dialed four times in five days about the same shipment we routed that person through to a live agent who could help by looking for delivery and shipping information."

There are few limits to how you can use such CRM-obtained data. Siebel's Nix recalls one client who used the call duration measurement feature to route callers who spoke slowly (captured from the IVR) to the newest agents. This helps these callers receive patient agents while providing these new employees with the time to develop expertise on such calls.

CRM software can also help you create prompts and scripts customized to the interaction mode and to the identified customer. Both have their place.

"If you have an agent that is trained on Products 1,3,5 and 7 they can use prompts because we've found that after the first two to three calls, as they gain confidence in their product knowledge they begin to improvise slightly," says Bruce. "Where scripting comes in handy is when you need to overflow calls on Products 2,4,6, and 8 to that agent. Scripts bring the agent right up to speed, without additional training."

CRM and the Business Process

An excellent way to think of CRM software is as an automated concierge for your firm. A good concierge will assess the value of each hotel guest and deliver the appropriate service level.

Yet before a hotel can take advantage of what a concierge can offer, it must know its business, its strengths and weaknesses and settle on what services and price points it will provide. It must also educate each point of service function manager on how to work with each other, the concierge and other hotel staff, and with the guests.

Without these pieces working together a good concierge can't do his job.

The same is true with CRM and your business. You must have an attractive, well-positioned-and-priced product/service and a functioning business process to make CRM work.

You also need measurements or metrics that support the CRM philosophy. One of the classic metrics is performance based on calls handled per hour. To beat that standard, agents try to hurry people off the phone, yet such practices fall counter to building relationships.

Not only must your product/service be of high quality, but there must be excellent communication between each part of the enterprise. They must share the same goals and the same common metrics.

And that can be a challenge. According to Kathryn Jackson, associate with Response Design Corporation (Ocean City, NJ – 800-366-4RDC, www.responsedesign.com), while it is relatively easy to train agents on new software, tearing down the institutional barriers between departments can be insurmountable. Your senior managers must design and agree on how to best handle customers, and make sure the performance measuring methods are compatible.

"Say your company is a telco and your agent has a high-value customer – a corporate executive working out of her home – with a line problem and needs to get the connection up immediately," explains Jackson. "The agent, who wants to keep that customer happy, contacts the repair department. Yet that department may have a metric that limits the number of repair calls in a day, to control costs, and has passed that limit. The repair head says 'no way, she'll wait for tomorrow,' Customer service and the repair department are not on the same page."

What you must do before buying CRM software is evaluate your business process and if there are any flaws, such as poor customer satisfaction, then you must redo and correct your process. CRM by itself won't fix it.

"Otherwise the CRM software will perpetuate or worsen a bad process," explains Vantive's Tuller. "You're better off using out of the box CRM processes than automating a bad process. "

The longer you must spend reorganizing and testing your business process the longer you may wait before you can install the software and realize its benefits. CRM software installation and integration can take as little as a few weeks to over a year depending on your system, requirements and on the package you select, including how much you need to customize, and on your IT staff's capabilities and time, which are often in short supply. Integration time is longer especially if you have mainframe systems.

"We recommend that you look at, set up, and implement your business process before installing our software, because without a business process and business rules you can not make CRM work effectively for you," recommends Chordiant's Bruce. "Much of the implementation work that we've done also entailed helping companies to create or modify the business process."

Clarify's Ragsdale agrees. While many CRM vendors have consulting partners and can help you with drawing up a business process, it is usually much better if you have already done this before seeking a CRM solution. He recommends going to a third party consultant or integrator who can evaluate your process and needs and give you a fair assessment because most firms cannot look at themselves objectively.

"There are two bad ways of implementing CRM: having no business process – what CRM packages give you are business rules but not the process by which you implement those rules – and taking flawed business process models and trying to automate it which also usually doesn't work," Ragsdale points out.

However, you don't have to wait until your entire company's business process has been re-engineered – which can take months if not longer – before installing CRM software. You can deploy it department-by-department, just as long as you commit yourself at some point to soon connect these departments to gain maximum value from your investment.

"Ideally it is always better to have the reengineering finished before installing CRM," explains Quintus' (Fremont, CA – 510-624-2800, www.quintus.com) Vice President of Marketing, Lawrence Byrd. "I don't find that realistic with businesses."

PRTM's Kaplan believes a firm can concurrently implement CRM software and business process changes, process-by-process, if they have reached consensus on the strategy and are prepared to make the required organizational changes in each function. Customer service, sales, marketing and distribution needs to communicate openly with each other in support of the same goals.

Once agreed to, companies should make initial changes to make CRM philosophy come to life. One method is setting uniform action priority codes so that the same customer has equal importance value across all departments, even if each department is not connected to the same software. That way if Customer X calls or e-mails, everyone knows how to treat Customer X.

Another method is setting up procedures to connect customer service and sales. One example would be for the customer service department to query the problem-tracking database before the field sales reps go out on calls so that they know the history with the product and the client.

"CRM software by itself is not going to do you a lot of good unless your company has the discipline to use it," Kaplan points out.

The consultant notes that in some cases buying tools prompts business process changes as firms realize what they can accomplish. This happens in less than 50% of the cases.

"There are many more cases where companies buy the software in the hopes of solving problems, without establishing a CRM strategy and the foundations for it, and the application fails to deliver results," explains Kaplan.

CRM Technology Selection

One mistake, according to eLoyalty's Stack, larger firms risk making when researching into buying CRM software is thinking that one package or package sets can do everything. This is not true. One size does not fit all.

Relying on a single package can create integration problems if your firm acquires or is acquired by another firm, which has a different package. Stack estimates that you can add as much 30%-50% to your software maintenance and business change time and costs by getting the two to work together.

"There's no silver bullet CRM package," Stack points out. "While most provide excellent sales force automation, most do not supply equally good integration with computer telephony, with the Web, or with other databases. While these features are adequate for small-medium enterprises the packages' architectures are not strong enough for the large firms because software is asked to do too much."

Stack recommends instead that you buy and install middleware to link your databases together, add data mining/data warehousing plus contact management software, on top of which you can buy/install CRM packages for sales force automation, field service, and customer service/help desk. The installation is then more open and easily upgradable. This lets you mix and match software to meet your particular needs without trying to integrate proprietary packages. Also, should your firm merge with another, the middleware accommodates other packages until you are ready to phase out one or both.

"When you look at CRM packages, evaluate them to see if they fit with your business," advises Stack. "Don't underestimate the amount of time it will take to integrate legacy systems or your existing data warehouse into CRM. Also, look for those packages, such as hybrids that has a layer of Visual Basic script coding that lets you customize it to your needs."

PRTM's Kaplan recommends that you have data modeling/mining tools so that you can use the information the CRM software collect. Yet too many companies focus on data entry without thinking how to extract and obtain value from that information.

"If you can't get the data out of the system and utilize it, why bother collecting and entering the data in the first place?" he asks.

Another issue to consider is whether or not you have or are planning to invest in enterprise resource planning (ERP) on the back end and supply chain management software. According to Kaplan, ERP software implementations have cut manufacturing and distribution costs, improved delivery times and order fulfillment accuracy. Thanks to these results, supply chain managers have considerable influence within an organization.

Some ERP vendors are entering the front end CRM field with products and packages and are investing sums to close the capabilities gap with their competitors. According to Kaplan, if a firm has an ERP investment it will drive the CRM application.

"If you have an ERP system, you should consider leveraging that investment with CRM packages that are compatible with it," advises Kaplan. "If not then you have a clean slate to select those systems that meet your needs and budget."

If you're a SME manager or executive, you should carefully review CRM packages to ensure that you buy only what you need. According to GWI Software (Vancouver, WA – 360-397-1000, www.gwi.com) founder Daren Nelson, most mid-sized businesses do not need the same software as a Fortune 1000 firm.

"CRM software is not one-size-fits-all," Nelson points out. "Many packages are aimed at large companies and consequently have features like back office integration that SMEs don't need, which adds to the cost and installation."

CRM Return On Investment (ROI)

Few companies realize the CRM benefits overnight and not all companies see payback in the same period of time.

The Return On Investment (ROI) comes from several sources, again depending on your installation, the package selected and your market. You can see payback in as little as three to four months. Or you may never see it because it is difficult to isolate and quantify some of your business's results due solely to CRM implementation.

CRM ROI and justification often comes from call center cost savings achieved by directing the least profitable customers to the least expensive media, thereby improving profit per customer and by increased up-selling and cross-selling. By presenting that customer record, including any special offers or messages up front, you also can shorten your call handling time, hence costs, over waiting for each screen to load up or having your agents hand off calls.

Both results: IVR/Web self-service and data-rich screens create greater productivity. You either use fewer agents or handle the same growth with your existing crew, which also saves associated costs such as facilities and equipment.

"The benefits come from the salespeople being able to make shorter, more effective sales calls because they have all the information at their fingertips and from faster assisted customer service and Web-enabled self-service," says Siebel's Nix.

ROI can also be derived from the smallest seemingly less significant factors. Clarify's Ragsdale cited the case of a large firm that required its employees to reset their passwords every 15 days: Not a big chore for head office people but a nuisance for those in the lower echelons. The firm did not change the policy, even though passwords resets became the top internal help desk problem. Thanks to CRM-type software, the managers collected the data and quantified the costs.

"When senior management saw this they changed the policy to every six months within 24 hours," says Ragsdale.

Yet determining ROI can be difficult, especially if the principal reason for investing in CRM is maintaining customer relationships, particularly through sales force automation. Vantive's Tuller has seen customers witness dramatic sales increases due to upselling while other customers have not had any measurable results.

"With sales force automation the ROI argument requires detailed tracking, which few organizations can do, " explains Tuller. "There are too many variables such as market prices, a competitor launching a new product to try and pin it on CRM."

Instead sales force automation demonstrates the value of customer retention. Tuller advises showing upper management how CRM provides better customer buying predictability and forecasts based on information gathered about customers and explaining that by having everyone on the same program reduces the chances of losing that customer and their revenue through miscommunication. "Chief financial officers like forecasts and predictability," says Tuller.

Company Types That Need CRM

There are certain vertical markets that can benefit greatly from CRM software, and in turn are buying and installing the packages. These include banking / financial / insurance firms, airlines and hotels, telecommunications and healthcare companies.

CRM, coupled onto middleware, can dramatically improve financial services marketing by linking the vast array of products and customers whose information often resides on different databases. Hypothetically one call to a bank or insurance company, or visit to their Web site, should open that customer to a targeted array of products.

This is not yet happening in many institutions, even in those that purchase CRM software, according to Chordiant's marketing director Collin Bruce. The problem is that too many businesses, most notoriously financial institutions, which buy the software, compartmentalize it in their own organizations, depriving themselves of their benefits, without any plans to get the next step: To connect the departments.

Bruce points out that banks and financial institutions offer upwards of 20 to 30 different products. However their customers on average use 1.2 products. If they eliminated the compartments they could obtain much better results. He urges that when companies buy CRM that they install it seamlessly across all parts of the enterprise to gain full value from it.

"The banks could make much more money and build far better relationships with their customers if when you called about your checking account they reminded you about your mortgage and offered to refinance it a lower rate, when they have you and when you're thinking banking," says Bruce, "rather than when you've come home from the office sorting through the junk mail or sitting down at dinner when the phone rings. By that time you might have gotten your mortgage refinanced or bought a new car elsewhere."

If you work for a telco or utility, CRM can help you keep your existing customer base by providing information such as usage patterns and services they need and cross-selling offers. For example, if your firm has Internet hosting you can use CRM to identify and target those residential customers that have more than one line. Chances are that they use that second line to connect their personal computer to a competing Internet service provider. Why not send direct mail or make outbound calls to your installed base with a special offer, including discounts on their monthly line charges?

"Before deregulation companies like the local telcos had a fixed customer base and they didn't have to worry about the poor level of service they delivered to them," explains Siebel's Nix. "All of a sudden since their customers can pick alternative carriers, they have to scramble to meet customers' needs. They can now offer long distance and Internet services. It is essentially to have software to coordinate offering these services, and serving and keeping fickle customers in a highly competitive environment."

High-tech firms can benefit greatly from CRM, especially because much of their processes: Bug identification, fixing, and new product development depends on customer input and response. These customer interactions must be recorded, responded to and analyzed quickly.

"CRM is extremely useful when you have different service level agreements (SLAs) with your customers," Clarify's Ragsdale points out. "We use it ourselves to track touchpoints with them. With SLAs linked by CRM we can automatically identify the customers, look up their SLAs and ensure, especially if they are in a high-paying top level plan that their cases are handled quicker, get the manager involved faster and have the developers give their bug fixes priority."

Quintus's Byrd sees general consumer products firms becoming excellent CRM software candidates as they change their business and sales organizations in reaction to the marketplace. These companies, which once one-way mass marketed to customers through brand advertisements and sold to them through retailers that also handled customer service, are now interacting directly with buyers. They now depend on customer service for marketplace differentiation.

"Consumer products call centers, which once dealt with fixing problems, now must manage a variety of inbound calls from complaints to people wanting free samples," explains Byrd. "Yet many of these call centers have not had the software to turn these inquiries into information to develop and maintain relationships."

According to Onyx's Frankenberg, health maintenance organizations (HMOs), can benefit from CRM software to retain high-valued clients and their employees. He cites Sierra Health Services, the Las Vegas, NV area's largest HMO, with over one million members. Within 72 hours of a new Medicare member's enrollment, an assigned member agent is prompted by the CRM software to call the new member to verify this and answer any questions they have about plan benefits or to help them find a physician.

"An HMO would deploy CRM to provide personalized, responsive service to employer groups, such as by automatically scheduling phone calls and visits to proactively identify and resolve any outstanding service issues," explains Frankenberg. "HMOs can also use it to give provider information to individual members covered by the plans."

The future of CRM

Although call center agents are being viewed as an asset rather than an expense (to the point where there are now measurements of revenue per agent rather than average handle time), there is a major movement to providing quality customer service without an agent (e.g. self-help websites) as part of the Extended Enterprise CRM (EE/CRM) movement.

Traditionally customers would pick up a phone and make a call when they wanted to buy something or if they needed help with a product or service. Until recently, most of CRM development has centered on streamlining and making more effective traditional voice connections over the PSTN to the customer.

But some customers are now on the web more often than they are on the phone. When they want help they may send an e-mail, go to the company website and search through a self-help knowledgebase, or look for a "call me back" button that will summon an agent to call them at home (or at least open up a chat box on the website).

According to research done by Dr. Jon Anton and R.F. Postmus of Purdue University, Web-based CRM customer service features fall into three levels of sophistication and robustness, namely:

1. The most general features offered are: product information, postal address, toll free number, search engine, FAQs, and a site map.

2. A second level of features are: Fax number, on-line purchasing, mailing list, purchase conditions, membership, domain fault repair, problem solving, product customization, and finally spare parts ordering.

3. An exceptional level of features are: Links to other sites, bulletin boards, product previews, introduction for first time users, site customization by the visitor, call back telephone button, chat groups, complaint recording, voice of IP, and finally site tours.

When Anton and Postmus compared Fortune Magazine's top 10 "most admired companies" with the bottom 10 (i.e., least admired companies), the average CRM Index of the top 10 was more than double the least admired companies. "One might conclude that the company's Web-based customer service substantially impacts image (i.e., branding, spin, and buzz)," they wrote.

The on-line business and e-commerce craze has resulted in an explosion of applications aimed at on-line customer management (OCM), but the best systems create a true merger of traditional brick-and-mortar and dot-com businesses, so companies can provide consistent customer service. Such super-integrated solutions tie the web to call centers, offering customers a choice of self-service or agent assisted service.

In 1999, for example, Aspect Telecommunications Corporation (San Jose, CA – 408-325-2200, www.aspect.com), a leading provider of CRM solutions, took the plunge and incorporated the web into a generalized CRM portal product.

When installed as the foundation of a company's CRM and e-commerce strategies, the Aspect Customer Relationship Portal can provide a consistent customer experience through one central place to connect customers with the right enterprise resource, regardless of the manner in which the customer contacts the business. The Aspect Portal's media-blending capabilities help a business create a multimedia customer contact center that leverages a flexible, open architecture to combine data from disparate contact center entry points – phone, web, e-mail and fax – providing consistent interaction, consistent integration and consistent information. In short, using Aspect's CRM portal, customers now have many options as to how they can contact a company.

Adaptec, Inc. (Milpitas, CA – 408-945-8600, www.adaptec.com), a global supplier of bandwidth management solutions and AnyTime Access, Inc. (Sacramento, CA – 800-852-4627, www.anytimeaccess.com), a premier provider of outsourced credit origination, underwriting and ful-

fillment services to the financial services and e-commerce industry, are both now using the Aspect Customer Relationship Portal.

In the case of Adaptec, it sought a CRM solution in which its customers could experience a consistent interface across multiple locations and throughout the enterprise, and which would allow Adaptec to leverage business rules from site to site. Using the Aspect CRM Portal, Adaptec can perform such advanced functions as integrating with a credit card authorization system, place calls back into a specific queue with all of the information needed to get a screen pop, and auto-escalate a call and allow a customer to check shipment status.

Next, Hewlett-Packard (Palo Alto, CA – 650-857-1501, www.hp.com) entered the fray with their Front Office, another advanced CRM solution enabling companies to provide consistent, comprehensive customer service through whatever communications channel (in-person, telephone, fax, Internet or e-mail) the customer chooses. E-Support provides interactive self help through an intelligent, Web based "virtual agent" that responds to frequently asked questions using a case based reasoning knowledgebase. When the customer prefers to speak with a live agent in the call center, E-Support quickly makes the connection.

Companies such as Genesys Telecom Labs (San Francisco, CA – 415-437-1100, www.genesyslab.com) and Altitude Software (Milpitas, CA – 408-965-1700, www.altitudesoftware.com) have also gone to great pains to seamlessly incorporate web functionality in what were previously purely PSTN call center software suites.

From CRM to ERM

In the late 1990s, many industry consolidations have occurred, where major CRM and Enterprise Resource Planning (ERP) vendors have acquired sales force automation and customer service and support software companies in order to diversify product capabilities and expand the customer base. Examples include Siebel's purchase of Scopus, Baan's total integration of Aurum, SAP's purchase of K & V Information Systems, and IBM's purchase of Software Artistry. Interestingly, IBM combined parts of Software Artistry along with components of other IBM-acquired CRM companies to create an entirely new CRM company, Corepoint, Inc.

The new amalgamations were resulting in something different than "classic" CRM. Just we had grown accustomed to CRM, a new support model now appears to be upon us. Ultimately CRM evolves into a more sophisticated and all-encompassing system called Enterprise Relationship Management, or ERM (See Enterprise Relationship Management), engineered by software vendors who will not only offer the expected front office sales, customer service, marketing and executive reporting functionality, but also integrated back office ERP software. Such companies include Baan / Aurum, SAP / K&V, Oracle, and soon Siebel.

We'll leave the last words on CRM to Jeffrey Shapiro, the world-renowned authority on computer telephony and author of the books *Computer Telephony Strategies* (IDG books, 1996) and *Windows Telephony* (Miller Freeman, 1997): "Achieving the aims and objective we have discussed here are difficult but they should be a welcome challenge. After more than a decade trying to solve some of the most complex problems in business and technology, I learned that unless you believe that what you are doing is fun, you are not going to succeed. The customer always knows whether you are smiling, even if they cannot hear or see you. And that is important to remember in your effort to build lifetime relationships."

Customer Loop

Also called the "local loop." It's the pair of copper wires connecting a customer's premises to the central office.

Data Encryption Standard (DES)

A standard that protects passwords from being read and then used again on the same a network to obtain unofficial access. It's also known as the Federal Information Processing Standard (FIPS) 46-1, or the Data Encryption Algorithm (DEA). The DEA is also defined in the ANSI standard X9.32. DES was developed in the 1970s in a joint effort by IBM, the National Bureau of Standards and the National Security Agency (NSA). Its purpose is to provide a standard method for protecting sensitive commercial and unclassified data. IBM created the first draft of the algorithm, calling it LUCIFER. DES became a Federal standard in November 1976.

For communications purposes, both sender and receiver using DES / DEA must know the same secret key, which can be used to encrypt and decrypt the message, or to generate and verify a Message Authentication Code (MAC). DES performs essentially only two operations on its input: Bit shifting and bit substitution. The key controls exactly how this process works. Performing these operations repeatedly in a non-linear fashion eventually generates a result which cannot be used to retrieve the original without the key, a "one way function" that will be familiar to those acquainted with chaos theory. Through the successive application of simple operations a system can achieve a state of near total randomness, which leads to what every crypto aficionado is attempting to achieve – a message or file appearing so random that it is immune to easy decryption.

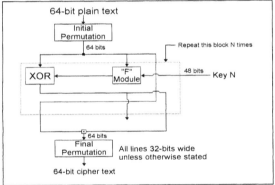

DES works with 64-bit blocks of data at a time, and uses a 56-bit key during execution (8 parity bits are stripped off from the full 64-bit key). Each 64 bits of data is iterated on from 1 to 16 times – 16 being the DES standard, since it's a symmetric 16-round Feistel cipher symmetric cryptosystem that was originally designed to be implemented directly in hardware. For each iteration a 48 bit subset of the 56 bit key is fed into the encryption block (the dashed rectangle above). Decryption is the inverse of the encryption process. The "F" module shown in the diagram is the very core of DES, which consists of several different transforms and non-linear substitutions.

NIST has recertified DES (FIPS 46-1) every five years; DES was last recertified in 1993, by default. NIST has indicated, however, it will not recertify DES again. The development of AES, the Advanced Encryption Standard is underway. AES will replace DES.

Datagram

A datagram is, to quote the Internet's Request for Comments 1594, "a self-contained, independent entity of data carrying sufficient informa-

tion to be routed from the source to the destination computer without reliance on earlier exchanges between this source and destination computer and the transporting network." The term has been generally replaced by the term packet.

DCE

Data Communications Equipment.

DCOM

Distributed Component Object Model. See ActiveX.

DDE

Dynamic Data Exchange. A Microsoft-devised software technology for computer applications to exchange data in real time

DDI

Direct Dial Inward (or Inbound), which is the same as Direct Inward Dialing (DID). See Direct Inward Dialing.

Dense Wavelength Division Multiplexing (DWDM)

Though considered by some to be the successor to standard SONET, DWDM is an even higher bandwidth version of Wave Division Multiplexing (WDM), a technique for increasing the capacity of fiber-optic transmission systems by multiplexing of multiple wavelengths (colors) of light, with each wavelength serving as a separate channel for carrying data. These channels are sometimes called virtual or logical fibers, which are all running over the actual, physical fiber.

Whereas most WDM systems allow about eight or so channels, DWDM allows more than 160 separate wavelengths or channels of data to be multiplexed onto a single optical fiber. In a network system with each channel carrying 2.488 Gbps (billion bits per second), up to about 400 billion bits can be delivered per second by the optical fiber.

DWDM came about because of the public's exponentially-increasing demand for bandwidth (about 40% per year) thanks in part to the Internet's amazing popularity.

Faced with a need to increase bandwidth on the backbone, service providers had three choices:

- Install more fiber. Digging trenches at a cost of $40,000 to $100,000 per mile is too expensive and takes too long.
- Install a faster transmission technology - a "faster pump." The technology that was originally going to save us from the bandwidth problem was none other than conventional Time-Division Multiplexing (TDM) network equipment capable of pumping 10 gigabits per second (Gbps)

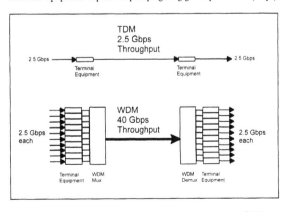

through a fiber channel. Dispersion problems (light scattering) keeps TDM transmissions below 10 Gbps, and it has become quite clear that a 10 Gbps network just isn't good enough to keep up with the rising demand for bandwidth.

- That leaves a modified form of WDM, DWDM. DWDM vastly increases the capacity of embedded fiber by assigning incoming signals to specific frequencies within a designated frequency band and then multiplexing the resulting signals as a group and transport them out over one fiber.

The technology that allows the incredible high-speed, high-volume abilities of DWDM is the optical amplifier. Optical amplifiers can boost lightwave signals, extending the distance they can reach without converting them back to electrical signals. Ultrawideband optical-fiber amplifiers have been demonstrated that can amplify lightwave signals carrying over 100 wavelengths or channels.

DWDM runs at the lowest layer of the Open Systems Interconnection (OSI) seven-layer protocol stack – the PHY or physical layer. Indeed, it actually dwells at the physical layer's lowest sub-layer – the so-called photonic layer. Since incoming signals are never terminated in the optical layer, DWDM doesn't require new electro-optical (or opto-electronic, as some prefer to call it) equipment to be at each end of the optical fiber. This also means that DWDM is oblivious to higher-layer formats, protocols or information representations, and that DWDM interfaces can thus be bit-rate and format independent, so service providers can integrate DWDM technology with legacy network equipment (of course, sometimes that integration happens to be a crude coax jumper cable between equipment, but that's another story). Such legacy hybrid scenarios generally use multiple 2.5 Gbps terminal equipment units, boosting the overall throughput far above that of a single 2.488 Gbps TDM channel.

Since a channel in a DWDM fiber is demultiplexed at the end of the transmission back into the original source, different data formats being transmitted at different data rates can be sent together. Each signal carried in an individual DWDM network channel can be travelling at rates as wildly varying as OC-3, OC-12, OC-24, etc., and each one can be in a different format – Internet (IP) data, SONET data, and ATM data can all be travelling at the same time on their own channels within the optical fiber. For example, in a 24 channel network, 12 of the 24 virtual fibers could be used to expand SONET capacity, four of the 24 could be used for SONETless ATM, two more could be used for IP over Fiber and six could be held in reserve for future use.

By using DWDM, carriers will be overjoyed that they can maintain both ATM and SONET networks, and the ATM signal doesn't have to be multiplexed up to the SONET rate to be transported over DWDM. Since no additional multiplexing is needed for the optical layer to carry signals, a new overlay network doesn't have to be deployed if a carrier wishes to quickly introduce ATM or IP services over his DWDM network.

A DWDM network with a mixture of SONET signals operating at OC-48 (2.488 Gbps) and OC-192 (9.953 Gbps) over a DWDM infrastructure can achieve capacities of over 40 Gbps. 1998 saw a surprising number of OC-48 networks deployed along with service providers planning to set up 80 wavelengths of OC-48 (199.04 Gbps in total). As the year 2000 dawned, however, 10 Gbps appeared to be the "baseline" for a modern photonic network, with the number of 40 Gbps network deployments not far behind. Carriers now began to plan for networks built with 40 wavelengths of OC-192 (398.12 Gbps total), enough bandwidth to transmit 80,000 copies of this encyclopedia in one second.

The Public Networks Group of NEC America Inc. (Herndon, VA – 800-433-2745, www.necpng.com), has under development a 160-channel version of its SpectralWave family DWDM systems, called the SpectralWave-160.

In the first few years of the 21st century, DWDM system will not only increased channel capacity but tremendous bandwidth capacity on each wavelength as well. Although some next generation terabit switch routers can support more than 100 OC-48 ports, use of a OC-192 over a DWDM system instead of four OC-48 signals "saves" three wavelengths for future use, or the three wavelengths can be rented to other carriers.

Lucent (Murray Hill, NJ – 888-458-2368, www.lucent.com) made available at the end of 2000 their all-optical WaveStar Lambda Router as a complement to its existing optical networking products, including the NX64000 packet router, the WaveStar OLS 400G DWDM products and the GX 550 core multiservice core ATM switch.

Designed for packet-based optical networks, the router is built on MicroStar technology developed by Lucent's Bell Laboratories. MicroStar consists of manipulating 256 micro-mechanical mirrors on a piece of 1 inch square silicon so as to route optical traffic to input or output mirrors 16 times faster than can electromagnetic-based systems.

Nortel Networks (–, www.nortel.com) has announced an incredible 6.4 Tbps (6.4 terabits per second) single-fiber optical transmission capability for their DWDM equipment. Nortel's system manages to cram 80 separate wavelengths on a fiber, with each channel carrying 80 Gbps of data.

"Bandwidth requirements will grow by a factor of 100 to 200 times in the next two years," says Anil Khatod, president of the Optical Networking division at Nortel.

Ciena Corp. (Linthicum Heights, MD – 410-694-5700, www.ciena.com) is another major player in this arena, as is SilkRoad Inc. (San Diego, CA – 858-457-6767, www.silkroadcorp.com).

Can a DWDM system's number of wavelengths and bit rate be upgraded? While the answer is yes for all DWDM systems, some considerable planning for this is necessary. If a service provider has constructed the DWDM infrastructure in a particular way then wishes to upgrade, one of two things must occur: Either more power must be provided or better signal-to-noise margins must be built into the system. For example, each time a provider doubles the number of channels or the bit rate, 3 dB of additional signal-to-noise margin is needed.

DWDM systems modulate optical wavelengths within what is called the "15xx" nm (nanometer) range, which appears to be optimal for transporting wavelengths, yielding the best performance and the least attenuation, and it can be optically amplified.

Because of close channel spacing, DWDM systems must be able to take into account the increasing fiber dispersion and nonlinearity encountered in the optical layer of shorter wavelengths / higher frequencies. OC-48 signals operating at 2.488 Gbps are adroitly handled by present-day DWDM systems. When one moves up to an OC-192 line running at about 10 Gbps, however, things become a bit more tricky.

The first DWDM deployments came during the period 1993 through 1994 as fiber networks providing long-distance service reached maximum capacity. Major installations started to occur during 1995 to 1996.

DWDM deployment will find its next great opportunity in the short-haul or local exchange market for campus and metropolitan areas (you too, can build the world's fastest private LAN or WAN). The greater cost of DWDM deployment is holding things up in this area, however, but some experts feels that DWDM now makes sense from a cost standpoint in short-haul applications at capacities equal to or exceeding 622 Mbps, which is equivalent to SONET/SDH OC-12/STM-4.

As for cost, companies are focussing on building flexibility into their equipment. The cost of providing a service (OC-3, etc.) is driven by the cost of providing that capacity termination outlet, or "drop," not by the cost of supporting the aggregate traffic on the ring. For example, providing an OC-3 drop on an OC-12 ring is less expensive than providing an OC-3 drop on an OC-48 ring. Optical Wavelength Add/Drop (OWAD) technology, dating from 1998, enables wavelengths to be added or dropped to or from a fiber, without using a SONET terminal. OWAD, when combined with DWDM and optical cross-connect (OXC) technology, offers a full-range of network management abilities to service providers, putting them on an even better footing than their circuit-switched network counterparts.

Chromatis Networks (Bethesda, MD – 301-657-1077, www.chromatis.com) is going after the metropolitan DWDM market with their new mutant form of DWDM called Selective WDM (SWDM) Normally, all nodes on a Sonet ring must upgrade to DWDM to be able to carry the service. Just as in the case of long-haul networks, every node in a metro ring must support DWDM even if only a few customers need the additional bandwidth. But Chromatis' SWDM can vary as a function of demand, and only the sites needing extra bandwidth will have to upgrade. High-bandwidth locations share a dedicated wavelength, while low bandwidth sites share a 1310 nanometer Sonet ring, thus essentially establishing two rings on the same fiber.

Chromatis is also combining the ATM, IP or TDM router and WDM equipment into the same unit, making it highly scalable.

Kestrel Solutions (Mountain View, CA – 650-237-7500, www.kestrelsolutions.com) is also trying to bring DWDM to the metropolitan market, this time by combining aspects of three different technologies (FDM, digital signal processing and optical modulation) into a multiplexing technology they call Optical Frequency Division Multiplexing (OFDM). Optical FDM will operate over just about any fiber and protocol.

One might ask: What about DWDM's optical predecessor, SONET and its European clone, SDH? Some say that DWDM and SONET / SDH are compatible and complementary technologies, with DWDM often being used to expand legacy SONET systems. Most believe that, over time (within ten years), DWDM will dominate and replace everything else, since the flexibility of the DWDM architecture allows carriers to support multiple topologies, protocols, and protection strategies, any one of which may be needed for whatever new and unforeseen services that will appear during the early 21st century.

After DWDM, then what?

DWDM itself is about to be challenged by yet another high bit-rate coding format called Soliton. Rather than looking upon the physical fiber's nonlinearity as a problem, Soliton signal coding actually takes advantage of fiber non-linearity to increase a signal's propagation distance. Solitons are special localized waves that exhibit particle-like behavior. In the case of optical fiber, a soliton can take the form of a narrow pulse of light that retains its shape as it travels immense distances along the fiber. The Soliton's ability to maintain its shape allows it to overcome the kind of data integrity loss brought on by chromatic dispersion.

Solitons were discovered in 1834 by Scottish engineer and boat builder John Scott Russell, who observed strange wave phenomena in water while studying the movement of canal boats on the Union Canal at Hermiston, very close to the Riccarton campus of Heriot-Watt University, Edinburgh.

As Russell described it in his "Report on Waves" published in the *Report of the fourteenth meeting of the British Association for the Advancement of Science*, York, September 1844 (London 1845), pp 311-390, Plates XLVII-LVII:

"I was observing the motion of a boat which was rapidly drawn along a narrow channel by a pair of horses, when the boat suddenly stopped – not so the mass of water in the channel which it had put in motion; it accumulated round the prow of the vessel in a state of violent agitation, then suddenly leaving it behind, rolled forward with great velocity, assuming the form of a large solitary elevation, a rounded, smooth and well-defined heap of water, which continued its course along the channel apparently without change of form or diminution of speed. I followed it on horseback, and overtook it still rolling on at a rate of some eight or nine miles an hour, preserving its original figure some thirty feet long and a foot to a foot and a half in height. Its height gradually diminished, and after a chase of one or two miles I lost it in the windings of the channel. Such, in the month of August 1834, was my first chance interview with that singular and beautiful phenomenon which I have called the Wave of Translation."

Although Russell later described the day he made his original observations as the happiest of his life, his discovery was ignored by his contemporaries, and he is known today by high school science students merely for having conducted the first experimental study of the "Doppler shift" of sound frequencies of a passing train.

By the mid-1960s, however, scientists researching nonlinear wave propagation realized that these solitary waves, or Solitons as they were now called, described a great many natural and artificial phenomena. In the 1970s, photonics researchers first proposed Solitons to counteract dispersion and nonlinearity in optical fibers, but it was not until the appearance of erbium-doped optical amplifiers in 1991 that AT&T Bell Labs could finally perform a startling demonstration of Solitons, sending them error-free over a distance of 14,000 kilometers.

Pirelli (Surrey, BC, Canada – 604-591-3311, www.pirelli.com) designed their WaveMux DWDM product in 1997 using Soliton technology, and in 1998 they held the first field trial of a commercial four wavelength OC-192 Soliton transmission product. The test was done on standard SMF-28 fiber. Practically no pulse distortion occurred. Solitons can be sent over other fiber types, such as TrueWave and dispersion-shifted fiber.

In one of those strange coincidences of history, a modern-day fiber-optic cable linking Edinburgh and Glasgow runs beneath the very tow-path from which John Scott Russell made his discovery of Solitons, and along the aqueduct now bearing his name.

DES

See Data Encryption Standard.

DHCP

See Dynamic Host Configuration Protocol.

DHTML

See Dynamic HTML.

Dialed Number Identification Service (DNIS)

A string indicating the number dialed by a caller (usually the last four, seven, or ten digits of the number), which can be used by ACDs, PBXs, or IVR systems to route the call to a specific location or be used to help generate a "screen pop." DNIS is a feature generally supported on 800 and 900 numbers. With this feature, outside callers can call various 800/900 numbers and have those call. All delivered to one set of trunks (generally digital trunks). On E&M WinkStart lines this information is provided via in-band DTMF digits.

Dialer

A dialer generates outbound calls and connects these to call center agents. There are five basic forms of Dialers:

- Preview Dialer – Preview dialing describes an automatic dialer, also called "screen dialing" or "cursor dialing" or "power dialing." Typically the prospect's account information and / or phone number appears on the PC screen before the call is made. The agent then "previews" this information and manually initiates the dialing with a mouse-click or keyboard-button push.
- Power Dialer - This is (roughly) a bulk dialer that consists of a list of numbers to be called, a series of outgoing phone lines and a group of call center agents. The dialer start to make calls as quickly as possible. It screens out busy signals and calls that ring and ring but are not answered. When it finds a "live" prospect on the line, it tries to connect the call to an agent, if no agent is available, then it abandons the call and calls again. Sometimes a recording is played to the called party asking them to be patient for a moment while the system waits to connect to an agent.

Keith Dawson, former editor of Call Center Magazine and author of the Call Center Dictionary, defines power dialing this way: "If a vendor makes a dialer that doesn't have a predictive algorithm, but they want to position it in the marketplace as more advanced than a mere preview dialer, they may tag it as a power dialer, confusing all and enlightening none. As the term with the loosest definition, it is rapidly losing all meaning."

- Progressive Dialer – A progressive dialer is similar to a power dialer, but in this case the system makes sure that at least one agent is free to take calls before the system makes a call. A progressive dialer is thus less productive than a power dialer, but it infuriates fewer callers by eliminating abandoned calls.
- Anticipatory Dialer – An anticipatory dialer uses the statistics generated by a single agent to "anticipate" when that agent will be ready for the next call. It dials the call while the agent is still talking. If that call takes too long, then you have an abandoned call, and an angry person who has been called but mysteriously hears silence on his or her phone line. It only takes only one abnormally long call to throw off the statistics, altering system performance.
- Predictive dialing – True predictive dialing should not be confused with other forms of automated dialing. A predictive dialer grabs phone numbers from databases, goes ahead and dials them and, only after finding real human prospects on the line, hands them off to call center agents. Predictive dialer software and/or hardware must therefore recognize no answers, busy signals, disconnected numbers, operator intercepts and answering machines – and keep track and report on everything.

They also use complex mathematical algorithms that consider, in real time, the number of available phone lines, the number of available operators, the length of an average conversation and the average time operators need between calls, and constantly adjust their dialing rates based on these factors. Thus, a predictive dialer is more accurate than a preview dialer because it averages talk times across entire groups of agents and is more accurate than simply examining the activities of a single agent.

You may wonder what is the real difference between these dialers in terms of how they affect the operation of a call center.

Well, manual dialing gives agents 15 to 20 minutes of average productive talking time per hour. The advent of automatic or "power dialing" extended that to 23 to 30 minutes per hour. Then came the great breakthrough of predictive dialing, which gave agents an extraordinary 40 to 57 minutes per hour.

It's not trivial technology. And it's not cheap. But if you have the need, serious outbound telemarketing, collections, etc., they will ultimately pay off in huge productivity gains. See Predictive Dialer.

Dialogic Corporation

Dialogic (Parsippany, NJ – 973-993-3000, www.dialogic.com) was founded in 1983 when Nick Zwick, one of the founders, began designing the first, open multiline voice board. Nick Zwick, Jim Shinn and company first started selling their board out of a New Jersey garage.

In 1984 Hank Magnuski at GammaLink developed the first computer-based fax board, ten years later GammaLink and its fax technology would be acquired by Dialogic. In 1985 Carl Strathmeyer at DEC would develop the first CTI computer-PBX link. Years later Strathmeyer's division would be purchased by Dialogic.

In 1986 Dialogic introduced the first telephony bus, the Analog Expansion Bus (AEB) for resource sharing, followed by and the first digital TDM bus for resource sharing the PCM Expansion Bus (PEB) in 1989. Also in 1989 Dialogic introduced the first 12 channel DSP-based voice processing board, the first T-1 interface board for voice processing.

In 1993 Dialogic led 70 companies in defining the first architecture for building CT systems based on open standards, the Signal Computing System Architecture (SCSA), a comprehensive, open architecture for providing multiple user computer telephony services in a client server environment. Dialogic's third-generation TDM telephony bus was also introduced: SCbus.

In 1994 the SCSA working groups developed the first software interoperability specification, SCSA Telephony Application Objects (TAO).

In 1995 Dialogic and other SCSA working group members formed the Enterprise Computer Telephony Forum (ECTF) to better focus on vendor interoperability issues, such as those addressed by the SCSA TAO. During that year VITA and ANSI approved the SCbus for the VME platform (ANSI/VITA 6-1994).

In 1996 the ECTF approved the S.100 API standard for software interoperability.

In 1997 the ECTF put an end to the SCSA MVIP "bus wars" by approving the H.100 CT Bus hardware standard for PCI CT resource boards, and began work on H.110 for the CompactPCI form factor.

In recent years Dialogic has re-engineered their entire line of CT resource boards to its new DM3 architecture, a scaleable and modular universal resource, "processor-neutral" architecture for voice, fax and Internet Protocol (IP) telephony applications. DM3 platforms and products today support both the VME and PCI bus as well as Windows NT, 2000 and Unix.

1998 and 1999 saw a major investment by Microsoft and a buy-out by Intel.

Though it had considerable competition over the years from some brilliant talent at Natural MicroSystems and Brooktrout, Dialogic, now an Intel company, is still very much the bellwether of the computer telephony industry and a leading manufacturer of standards-based computer telephony hardware and software platforms, with millions of ports shipped worldwide.

Every year in March, like the swallows visiting Capistrano, more than 200 beaming blue-shirted Dialogic employees scurry about the cornucopia of new technology displayed in the company's mammoth booths and those of its partners at CT Expo in California, before returning to roost back in Parsippany, NJ.

DID

See Direct Inward Dialing.

Digital Cellular

Cellular phone systems using digital speech encoding in transmission.

Digital Services (Signaling) Level 1 (DS-1)

See Circuit Switched Signaling.

Digital Signal

A discrete or discontinuous electrical signal made up of a series of discreetly timed steps of voltage level and/or phase, such as on and off pulses, signified by ones (1s) and zeroes (0s), that can be used to represent more abstract forms of information. Often for purposes of transmission, a threshold voltage reference is specified with any voltage above the threshold considered a 1, and any voltage below a 0.

Digital information must be converted to analog before it can be sent through the human analog interface (i.e., the senses).

Now that cheap microprocessing power is available to handle digital signals, digital systems are everywhere. Digital signals don't suffer from any of the drawbacks of analog communications. There's no noise or distortion accumulation, a high immunity to interference, and error checking is possible on a bit-by-bit basis. Digital signals can be compressed, encoded or encrypted in various ways. They can be multiplexed and multilayered and sent over great distances with no loss in quality. They can be copied endlessly with no degradation of data quality. Digital signals are easily stored and retrieved. Both analog and digital signals can be modulated onto a Radio Frequency (RF) or fiber-optical carrier for transport over long distances.

Digital Signal Processor (DSP)

A special microprocessor chip integral to single and multiline computer telephony resource boards that can be reprogrammed with new functions "on the fly" and is optimized (can perform multiplication and addition operations in a single execution cycle and at speeds several times that of a conventional general purpose CPU) for real-time processing of digital signal data, much the way that a math co-processor is optimized to process math data.

The DSP has been called computer telephony's miracle chip. According to Mike Coffee of Commetrex in Norcross, GA, the idea of digital signal processing actually dates back to the supercomputer array or vector processor days. These were large, expensive systems that performed repetitive arithmetic operations on data arrays (vectors). They were typically assigned to very high-value problems, such as weather forecasting and oil exploration.

In the early '80s they were scaled for microprocessors and the first commercial DSP integrated circuits (ICs) came along shortly after the more general-purpose-computing scalar microprocessors (e.g., the Intel 80xx). Though these new, very specialized digital-signal-processing microprocessors where used in all kinds of systems that needed their highly repetitive, high-performance mathematical powers (like radar and sonar), they were soon snapped up by computer telecom, where so many applications require an analog signal to be digitized (as with a codec) and then processed (as in speech processing, DTMF generation/detection, video compression/expansion and modem implementations).

In communications, Coffee therefore explains, the DSP became telecom's general-purpose "information transducer." As Coffee says, "It transforms information in one form, analog, to another, digital. That's not to say a DSP is an analog-to-digital converter (ADC). It's not. It takes the output of an ADC, for example, and derives its real information content, like the encoded ones and zeros of a data modem."

It's actually the DSP's counterparts, mixed-signal / analog products, that convert analog signals (audio, light, magnetic differences, etc.) to digital signals for processing and transmission. Analog signals from the PSTN and "the real world" are converted into a computer-readable digital format (and vice versa) with DSPs and related equipment. Wherever there's a real-time interface between the analog and digital worlds, you'll find a DSP and its associated hardware.

DSPs can be found in every modem, hard drive, CD player and digital cellular phone. They are less expensive than conventional general purpose microprocessors – just a few dollars as compared to hundreds of dollars for a Pentium III.

In computer telephony, the "signals" processed by DSPs are the Pulse Code Modulation (PCM) "media" streams. Prior to the emergence of DSPs (pre-1988), these "system-resource" functions were entirely implemented on dedicated-function, closed-architecture, multi-stream boards, interconnected by industry-standard PCM highways (MVIP, SCSA, H.100). Developers were forced to use several dedicated boards to build systems. One used a voice card to handle the voice and DTMF functions, a fax board to send and receive faxes, and sometimes switching or conferencing boards as well. DSPs changed all that.

Code for early DSP boards loaded only once during initialization, which locked down a flexible DSP platform into a fixed purpose board. More sophisticated DSP boards followed, where the configuration loaded into one or more DSP code "overlays" into the board and "told" the board how to use them. Supporting multiple overlays allows a single card to perform several functions. This in turn allowed the concept of a "universal port" to be realized in the early 1990s. Pioneered by Linkon (now PacketPort.com) "universal port" means that a port on a CT resource board can be instantly reprogrammed as needed to handle IVR, fax, ASR, IP, remote access, or what have you. This new flexibility meant that fewer boards would have to be purchased for a multi-use application. Also, offloading such telephony media-processing tasks to the DSP on CT resource boards "frees up" processing cycles that would have been otherwise wasted by the host CPU in the PC.

The "density" or number of ports that could be handled by DSP-based boards grew along with their flexibility. The first DSP-based voice-processing board supported a single port of voice. And, in an indication of things to come, it also provided a V.22bis (2,400 bps) data modem on the same chip. In 1988 there appeared the first DSP-based multiline voice boards, providing improved performance (especially in call-progress analysis) and flexibility over earlier multiline voice-processing boards, which with their old dedicated -function ICs for things like DTMF generation/detection and speech compression/expansion circuits – maxed out their capacity at four ports. Today we are achieving 256 ports.

In a modern CT resource board from Dialogic, NMS, Brooktrout, etc., the analog signal of the phone line is digitized (with a codec) to create the PCM media stream, allowing the DSP to do its work of handling speech processing, video compression / expansion, and modem implementations, etc. Most media-processing algorithms demand that the media stream data be multiplied by different constants to implement filters. For example, the current PCM sample will be multiplied by one number, the result of the previous sample's operation will be multiplied by another, the result from two samples ago by another number, and so on. DSPs are optimized for that type of operation: Perform a memory fetch, multiply by a constant, make a decision based on the result, and accumulate the result – all in one bus cycle. The newest DSPs can perform several such operations simultaneously, resulting in fantastically high MIPS ratings not to mention the ability to process more than one media stream (voice, fax, speech rec, etc.) on a single chip.

So, what we call a "voice-processing" board, or "telephony" board or a "CT resource" board, such as produced by Dialogic and NMS, is really a DSP-resource board. It only becomes a voice board when the voice-processing DSP code is downloaded to the DSP from the host PC when the system is initialized. Depending upon the code downloaded to it, a single

DSP can handle the same features which use to require many single function chips.

With a powerful enough DSP equipped with enough memory, an open-architecture DSP can just as easily support modems as it can support speech processing, including speech recognition and text-to-speech. Today, DSP-resource boards support all kinds of multiple media, DSPs have been created to sit on each port of a CT resource board and do echo cancellation, speech recognition, text-to-speech, speech / video compression, etc. During the 1990s a pioneering product in this area was MultiFax, Commetrex's independently developed fax software add-in to the NMS "voice boards". Brooktrout and Linkon (now PacketPort.com) partnered with their DSP vendors to provide support for multiple media. And Bicom and Pika jumped on the bandwagon to add new media-processing capabilities to their boards.

In the mid-1990s Intel appeared worried that the proliferation of DSPs would cut into the sales of their CPUs, so Intel attempted to "soak up unused MIPS" in Pentium processors with their "Native Signal Processing" initiative, an attempt to move less-demanding DSP applications back to a PC's host processor. Intel quickly began working on a product called IA-SPOX that would take some of the Pentium processor speed and use it for signal processing. It was essentially a ported version of SPOX that would give DSP coders an easier time getting their code running under IA-SPOX, if they already had it running under SPOX on another platform. Windows applications could connect to services through the WinSPOX API. IA-SPOX was not meant to impact on the multi-line telephony world very much, since single purpose DSPs are required for higher density multi-line hardware.

Indeed, some of the latest speech recognition code is too big to run on some DSPs (until the next generation of DSPs appears), and companies such as Wildfire have actually moved their speech recognition code to host CPUs, lowering the cost of their system sufficiently so that telcos could offer it as an enhanced service. The perpetual tug-of-war between proponents of host CPU-centric and DSP-centric systems goes on and on. But DSP camp tends to be in the lead.

After all, although eliminating a DSP makes sense for something small like a laptop, one wonders if it's cost-effective to steal all those precious (and expensive) Pentium cycles for multitasking DSP work when a $5 DSP can do the same thing without degrading overall system performance – especially when you realize that major deployments of video processing is almost upon the CT industry.

The reason a cheap DSP can beat an expensive Pentium at this game is because multiplication is an important part of the algorithms and DSPs have fabulous multiplication capabilities, as do many of the RISC processors (Lucent prefers putting RISCs on their boards as opposed to DSPs).

But Intel's work did underscore a problem that must be dealt with by all DSP vendors: Some DSP processing may be performed on the CPU while other processing may at the same time be apportioned to DSP chips. How do you coordinate processes in such a case?

And there were (and are) several problems that had to be overcome to achieve the 120 and 256 port high-density DSP-based cards we see today.

One problem has to do with board-level resource management. There is a wide range of DSP and RAM resources required, for example, to implement typical voice-processing functions and those needed to implement a data modem or speech recognition. It's possible to optimize a board for voice and eliminate any capability to process other, more resource demanding, media. So if one board can handle 120 voice streams and 60 high-speed data streams the board must include flexible resource management of MIPS, RAM, and media streams.

The Pika product line was perhaps the first (and certainly one of the few) telephony based open DSP environments where the DSP code could be moved to almost any board in the product line without software changes. The only restrictions on their product line is how much RAM each card has. Some cards have less RAM per channel than others and can't support fax and other RAM intensive DSP applications.

Since no "voice board" vendor has the resources or competencies to internally develop all media-processing technologies, there is always the tendency for vendors to develop their own proprietary, closed-architecture boards, and then going to the various media-processing technology vendors to port their code to the vendor's platform.

For example, the Dialogic Antares card can't be used in any configuration other than with another voice board. The Analogic TAP-800 is the same. All the other boards include voice and call processing. But what if you wanted to use Bicom's voice on Natural MicroSystems' AG boards? You can't do it.

So the technology vendors have their developers work on porting (technology shuffling) rather than developing newer and better technology. More effort is being spent getting current DSP software running on different boards than is being spent on developing new and better DSP algorithms. With an open hardware standard, the situation could easily reverse itself.

An elegant solution for the computer-telephony industry would be to define a board-level environment that would allow the voice technology vendors to do it just once, offering it as a board-level "application" that would run on any board which provided the industry-standard environment. Just as host-level application developers access system resources through APIs, the technology vendors would do the same, only at the board level. These APIs would be for the media stream, the external system interface (host), DSPs, and so on. And just as good host-level APIs hide the operating system (for portability) from the application developer, good board-level APIs would hide the board's operating system from the developer.

The DSP thus appears to be in the same predicament that PCs found themselves in their infancy. Anyone running DOS could only run one application at a time. Now, OS/2, Windows and other multi-tasking operating systems let us run several applications at once.

In the early days of personal computers, every computer was different. You could choose different floppies (I still have fond memories of those big eight inch floppies that held just about 80 pages of information), terminals, printers and other peripherals. Most of us didn't have to worry about hard disks, since they were not affordable then.

Software was often purchased for specific systems, since different terminal models used different control codes (even from the same manufacturer), which made software portability a big headache. Once you got the terminal configured, you then worried about your printer control codes. Some hardware vendors actually worked with software companies to re-market products preconfigured for their products so they would work better and install easier for novice users.

Then the IBM PC appeared in the market, sporting a miraculous standard Basic Input Output System (BIOS) which opened up a standard that almost anyone could code to. The IBM BIOS did have many limitations and was not very fast, but it was stable and reliable. Developers could spend more time worrying about their code rather than what it took to get their applications working on some odd terminal in Idaho.

BIOS was followed by successively higher-level operating systems such as DOS, OS/2, Windows 3.11, Windows 95, 98, NT 4.0 and now

Windows 2000. These offer a much higher level API for performing more powerful tasks on PCs. Now almost anyone can code an application for the PC and ship it out without worrying much about compatibility.

The DSP, too, should follow the same general development path as did the PC. The industry needs someone to build an open DSP platform where coders can produce code that runs on any card that conforms to the open hardware standard. Instead of buying fax code specifically for your Dialogic, NMS or Pika voice card, you should be able to select functionality for your DSP board based on price, performance and features from several vendors. In other words, we need shrink-wrapped media-processing technology.

As more manufacturers released DSPs and as DSP functionality began to get incorporated into motherboards, it became evident that what was needed was a hardware and operating system-independent "signal processing" layer, interface or API for negotiating between high-level applications and low-level DSP functions.

For a few years in the mid 1990s, we all thought that Microsoft was going to impose a DSP processing layer standard.

Microsoft was supporting SPOX from Spectron Microsystems in Goleta, CA (now part of Dialogic) which in 1994 had developed WinSPOX, a special development environment to make it easier for Windows application developers to create DSP-aware applications for multimedia applications. Microsoft had also announced a DSP Resource Manager Interface (RMI) to put all of these development environments together under the Windows umbrella.

Microsoft, seeing an opportunity to open up DSP technology to Windows 95 developers, had decided to really push RMI, which started out as a high-level API within Windows 95 allowing even small software developers with reasonable programming skill to access the DSP signal processing services of all DSP architectures in their applications.

With Microsoft behind it, RMI would be the interface of choice between the DSP management software and a series of function-specific drivers for Windows 95 APIs such as the fax API, audio recording and playback (Wave) API, MIDI API and the Telephony API (TAPI).

As development proceeded, however, RMI became a lower-level interface suitable more for OEMs, who would have to write the drivers for it to connect to the Windows 95 APIs.

RMI's "lowest common denominator" approach to standardizing access to DSPs led to some "filtering out" of some of DSPs' unique capabilities, like those of IBM's Mwave, which supports Windows 95 multimedia extensions by itself anyway.

Also, incorporating various DSP architectures into RMI was quite an involved process. Microsoft may have bitten off more than it could chew.

Furthermore, the hardwired chipset manufacturers – who would become nearly extinct if DSPs from the major suppliers become popular – sprang into action when RMI was announced, slashing prices and improving their technology. For example, many modem chipsets (like Rockwell's) are now essentially fixed-function DSPs.

At the same time they complained to Microsoft that they still dominate the multimedia market and thought that work on the RMI interface threatened the integrity of the API layer, which is called by their drivers. Microsoft didn't want to be in the middle of a programmable vs. non-programmable controversy.

Finally, many OEMs don't want to be tied into using DSPs if more cost-effective hardware comes along, which would involve the standard APIs, not RMI.

Microsoft then stunned DSP vendors in 1995 when it announced it was abandoning direct involvement with the RMI specification. Instead,

Microsoft dumped RMI development and support onto Spectron, whose SPOX DSP multitasking operating system was the genesis of RMI.

In Microsoft's version of a "Dear John" letter to DSP developers, it wrote that it would instead improve APIs for 32-bit Windows platforms "that will benefit a wider variety of hardware (editors' note – including the Pentium and successor chips) and not solely DSPs. Microsoft's efforts would include extending existing Windows APIs, introducing new APIs and extending the Windows 95 communications architecture to better support voice."

In other words, Microsoft opted out of the dedicated DSP OS business and landed itself directly in the "we support everything that looks, acts and smells like a DSP" – whether it's called a programmable DSP or just a general purpose programmable microprocessor chip acting like a DSP.

This concept of open DSP operating systems and standard processing layers was not a pipe dream, but there was no common consensus as to what should be done during the latter half of the 1990s. AT&T developed the Visible Caching Operating System (VCOS), Spectron Microsystems had SPOX (used by the DSPs in Dialogic's Antares call-processing card, which is probably why Dialogic bought Spectron). Analog Devices, which sold DSP chips to Orchid Technology and Cardinal Technologies, also had its own DSP OS called the Signal Computing Operating Environment (SCOPE), a superset of SPOX. And IBM Microelectronics (formerly the IBM Technology Products Division), promoted their Mwave processor, with over 40 independent software vendors ultimately developing to it. The popular Mwave platform is not just a chip but the whole Mwave OS consisting of a real-time scheduling kernel, the suite of algorithms (they call them tasks) to perform the actions of a modem, fax, 32-voice wave table synthesizer, digital answering machine, full duplex speakerphone, FM synthesis and Sound Blaster compatibility.

So the fact remains that DSP application software that works on one vendor's board won't run on another vendor's board. This is true even when the boards have the same chips on them, since the application interfaces vary from card to card. Most DSP software continues to be written by the board maker to support its own hardware. And while many board vendors have provided DSP developer kits or specifications to third-party software developers, they're all very different. This makes outside integration tough. In fact, it's not unusual for DSP algorithm vendors to spend as much as 50% of their time simply porting their DSP code from system to system or else limit their work to a single vendor and, usually, a single vendor board (e.g. the Dialogic Antares card).

The upshot of all this proprietary DSP juggling has slowed innovation and added expense in the development of computer telephony boards. No one, with perhaps the notable exception of CT-card market-share leaders (Dialogic, Natural MicroSystems and Brooktrout / Rhetorex) and third-party DSP software (Lernout & Hauspie, SpeechWorks, Philips, etc.), could be very happy with that.

The companies Bicom, Calibre Industries, Centigram, Cole Technical Services, Commetrex, MiBridge, Pika Technologies and QNX Software Systems were all so displeased with the realities of CT board design that they decided to change things by forming a consortium in 1997 to come up with a standard interface spec that would decouple DSP-based media-processing technologies from underlying hardware resources. The resulting spec's name was the Media Stream Processor (MSP).

As Jim Pinard, CTO of Pika, put it at the time: "Until now, a hardware resource was usually sold with proprietary media-processing software, such as voice or fax. The idea of the MSP specification is to separate the hardware from the software. We want to let voice, fax, data, text-to-speech, speech recognition and video – all produced by different companies – coexist on the same hardware resource. This philosophy has been successful in the PC world. We think it can be done at the CT resource-board level too."

Mike Coffee, MSP Evangelist and President of Commetrex (Norcross, GA – 770-449-7775, www.commetrex.com), has been one of the principal driving forces behind MSP. Commetrex has long been a very interesting DSP-based software / hardware player. Their software, for example, helped add fax-processing to NMS "voice" cards in the early 1990s. Commetrex looked long and hard at the problem of porting software to various DSPs and started development on MSP, a product "that may change the future of telephony as we see it today," as Stuart Warren (now CTO of Icon-O-Voice) used to extoll repeatedly in the pages of *Computer Telephony* magazine.

In August 1997 the MSP Consortium developed their M.100 integrated-media open-environment specification, and Commetrex MSP/CX-based card immediately complied with the spec.

The MSP Consortium's dream is for every CT developer to use an open "media neutral" CT card that could be manufactured by anyone, sold by anyone, supported by anyone, and coded on by anyone. The initial design for the board was one based on two Texas Instruments TMS320C6201 DSP chips.

Rather than limiting their system to telephony alone, the MediaStream Processor (MSP) system could be used for any task suited to a DSP. Like any good DSP-based board, each MSP can support multiple simultaneous applications such as voice, fax and speech recognition. The MSP card won't have specialized telephony interfaces for analog, T-1 or ISDN lines, but will use CT bus technology such as MVIP or SCbus (now H.100) telephony interfaces.

The prototype of the Commetrex MSP/CX was demonstrated at CT Expo 98 in Los Angeles. "We were pleased to see Commetrex running a high-performance standards-compliant media-processing system on the 'C6201 at CT Expo." said Andy Fritsch, Applications Manager, Texas Instruments. "The MSP/CX demonstrates Commetrex's innovation by being the first to offer an M.100-compliant board. The 'C6201 and its C compiler enables Commetrex to offer its customers a computer telephony board with an unprecedented 3,200 MIPs on just two DSPs and 50% reductions in media-processing software development."

Thanks to the DSPs, the MSP/CX was priced less than a four-port fax board.

Commetrex then tweaked and renamed the MSP/CX board as the Media Stream Gateway MSP-320. The MSP-320, capable of processing 256 PCM streams, now has all the resources to support the media-processing needs of a typical commserver: High-speed data and fax, Internet telephony, voice processing for messaging, video processing, text-to-speech, and speech recognition.

The MSP-320 still uses two 1,600 MIPs Texas Instruments TMS320C6201 task processors, each with 4 MB of 100MHz synchronous DRAM. The board's co-processor is a 20MHz Intel 386EX with 4 MB of 60-ns DRAM. The MSP-320 and the host PC communicate via a PCI plug-and-play interface. Telephony streams may be made available from the host PC, network interfaces on the optional daughter board, the board's 10BaseT interface, or via the H.100 industry-standard PCM CT Bus.

The MSP-320 board is sold at a fixed price regardless of the quantity purchased. OpenMedia, Commetrex's implementation of M.100, and media-processing software are licensed separately and qualify for discounts.

OpenMedia is Commetrex's implementation of the MSP Consortium's M.100 board-level environment recommendation. It is

designed to work efficiently with Open Telecommunications Framework (OTF), Commetrex's implementation of the ECTF S.100 host-level environment recommendation as well as other vendors' CT middleware.

OpenMedia comes in several versions: It can be licensed in source code for your proprietary (host- or DSP-based) media-processing products, allowing you to take advantage of the growing list of M.100-compatible products. It can be licensed as a software add-on to Commetrex's MSP Media Gateway DSP boards. And it can be licensed as a prerequisite to Open Telecommunications Framework-based media-processing products, such as PowerVOX, PowerVoIP, PowerFoIP, and PowerFAX.

As Mike Coffee told us when the board was released, "As you know, there is currently only one implementation of the S.100 specification (Dialogic's CT Media), and it pre-dated the spec's release. So Commetrex may be the first vendor to implement an S.100-conforming product since the spec's release. OTF, our implementation of S.100, is even more open than the original spec. That, coupled with our implementation of the MSP Consortium's M.100, gives Commetrex a system platform that is open in ways yet to be discovered by the market-share leaders."

Whether the MSP Consortium is triumphant remains to be seen.

Digital Subscriber Line (DSL)

Most people use this term incorrectly. DSL was actually the original term for an ISDN BRI line. As AT&T defined it: "A three-channel digital line that links the ISDN customer's terminal to the telephone company switch with four ordinary copper telephone wires. Operated at the Basic Rate Interface (with two 64-kilobit per second circuit switched channels and one 16-kilobit packet switched channel), the DSL can carry both voice and data signals at the same time, in both directions, as well as the signaling data used for call information and customer data. With the introduction of the AT&T 5E5 generic, up to eight different users can be served by a single DSL."

The entry the reader is looking for, the one describing the family of high bandwidth digital services such as ADSL, IDSL, HDSL, M-SDSL, R-ADSL, SDSL, VDSL etc., can be found under xDSL. See xDSL.

Digital Subscriber Line Access Multiplexer (DSLAM)

An xDSL line-interface device located in a telephone company Central Office that connects xDSL users to the backbone network. One side of a DSLAM connects to the customer premises xDSL Network Interface Device (NID) over the local loop. The other side interfaces with the PSTN and a wide area (ATM, Frame Relay, or IP) network system. The xDSL customer's signals are multiplexed with other customers' signals and sent over the backbone to another DSLAM that demultiplexes the signals and forwards them to the appropriate individual xDSL connections. Thus, DSLAM multiplexers enables a phone company to give customers the fastest phone line technology (xDSL) with the fastest backbone network technology (ATM, DWDM IP, etc.).

Sometimes the data traffic is so great that a DSLAM alone cannot multiplex all of the xDSL signals it receives. Just before Ascend Communications merged with Lucent Technologies (Murray Hill, NJ - 888-4-LUCENT, www.lucent.com), they released their DSL Terminator 100, a $15,000 device that aggregates and terminates up to 1,000 simultaneous incoming Permanent Virtual Circuit (PVC) or PPP sessions from multiple DSLAMs, before sending the packets on to a backbone router.

Similarly, the 6400 Universal Access Concentrator from Cisco (San Jose, CA - 800-553-NETS, www.cisco.com) is a high-density (up to 14,000 PPP subscriber connections and up to 3,500 VPN domains) DSL services platform that sits in the service provider's CO and aggregates traffic from multiple DSLAMs or from other access media (e.g. cable, wireless, or narrowband dial access). The 6400 can support on-demand, broadband networking services, so ISPs and wholesale providers can offer services such as multi-point videoconferencing, streaming video, and entertainment-driven traffic on a per-user, per-service basis.

At the low end, Ascend's DSL MAX 20 is an affordable, entry-level DSL access concentrator for the SOHO and multi-tenant unit markets. The MAX 20 offers multiple tiered service rates, ranging from 144 Kbps to 2.3 Mbps, SNMP-based network management, and Layer 2 and 3 routing capabilities. DSL MAX20 supports from eight to 32 SDSL ports, T-1 / E-1 or DS3 / OC-3 / STM-1 uplinks, and 10/100BaseT Ethernet. List price starts at $495 per port, and varies according to configuration.

Copper Mountain (Palo Alto, CA - 650-687-3300 , www.coppermountain.com) makes DSLAMs that cater particularly well to the Multi-Tenant Unit (MUT) environment. Their CopperEdge 150 is a DSL access concentrator that lets CLECs and ISPs provide DSL service to MTUs such as high-rise office or apartment buildings, hotels, and office parks and campuses. Besides high-speed Internet access (from 128 Kbps to 1.5 Mbps), CopperEdge 150 can also perform frame relay multiplexing applications, as well as Virtual Private Networks (VPNs), and DSL-based voice services. Voice features of the CopperEdge 150 include connecting remote users to the corporate PBX via DSL, and the ability to packetize voice over DSL to a packet gateway over the PSTN. Copper Mountain also offers a line card that support POTS. In addition to CLECs and ISPs, commercial property owners and managers are potential customers for the CopperEdge 150, which would allow them to provide a "DSL-ready building." CopperEdge 150 supports SNMP for local or remote management. A CopperEdge 150 starter kit, which in addition to the chassis, comes with a 24-port SDSL line card, system control module, power supply, and fan, lists for $16,995.

The CopperEdge 200 is a more robust platform that supports up to 192 SDSL and IDSL ports per chassis, and offers nine different connection speeds, ranging from 64 Kbps to 1.5 Mbps over 28,000 feet of copper. Because security is always a concern with end users connecting to the corporate LAN, the 200 model includes authentication of each user using PPP over ATM or frame relay. CopperEdge 200 is managed through CopperView, an SNMP-based interface for monitoring and managing a network of CopperEdge modules in remote locations. List price, including system control module and power supply, is $9,995.

Both CopperEdge systems are compatible with Copper Mountain's CPE, as well as with a list of companies whose CPE has been certified "Copper Compatible."

Interspeed's (North Andover, MA - 978-688-6164, www.interspeed.com) System 500 is a DSL access device for multi-tenant buildings. Unlike traditional DSLAMs, the System 500 "DSL Access Router" (DSLAR) combines the functionality of an ATM switch and an IP router, in addition to simple aggregation and multiplexing. The advantages of this are that it cuts down on equipment costs, improves efficiency through oversubscription of its uplinks, and facilitates the deployment

of multimedia services. Interspeed plans to expand their product line to support voice-over-DSL and voice-over-frame relay.

The System 500 comes with up to three 16-port, 1.168Mbps SDSL line cards, supporting up to a total of 48 ports of DSL. Its uplinks are available as either 10/100Mbps Ethernet, or OC-3 for ATM. Suggested retail price including line card and switch/router module is around $21,000.

The System 500's "big brother," the System 1000, houses 192 ports. Both have the ability to provision multiple userID/password-secured VPNs within a single chassis.

Net to Net Technologies (Dover, NH - 877-638-2638, www.nettonet-tech.com) makes an unusual IP DSLAM. Whereas most DSLAMs use ATM and frame relay for their uplink ports, Net to Net uses Ethernet framing to transport DSL over T-1s and T-3s. Coming into the box from Ethernet-connected CPE, which Net to Net also manufactures, are IP packets traveling via 10/100Mbps Ethernet, T-1, or DS3 interfaces.

Net to Net claims that the use of IP as a foundation for their system assisted in bringing the cost of the product down and in making it easier to configure and deploy. The unit includes default settings, web management, and the ability to detect live CPE ports, which make it more or less plug-and-play.

The 48 port version, IPD 4000, is specifically intended for MTUs, while the 144 line 12000 model is more of a central point of termination for a service provider's network. An unlimited number of modules can be daisy-chained together, however, making even the smaller model as scalable as is necessary. Net to Net plans to voice / VoIP-enable both their CPE products and the IP DSLAM.

Orckit's (Southbury, CT - 203-267-1000, www.orckit.com) FastInternet DSLAM uses a Rate Adaptive ADSL (RADSL) modem with the ability to download at 384 Kbps to 8.192 Mbps and upload at 128 Kbps to 768 Kbps. The system also includes ATM and IP concentrators that bridge the central office to the broadband (ATM or IP) backbone using 10 Mbps Ethernet, 100 Mbps Fast Ethernet, and 155 Mbps ATM interfaces. Orckit has worked to insure their products interoperate with G.Lite ADSL modems. Their DSLAM works with modems using Centillium Technology's G.Lite chipsets, and Fujitsu Network Communications (Richardson, TX - 800-777-FAST, www.fnc.fujitsu.com) markets SpeedPort, an ADSL DSLAM made by Orckit.

The Avidia System from Pairgain Technology (Tustin, CA – 714-832-9922, www.pairgain.com) is an integrated access concentrator housing 1,296 ports in a single, seven-foot rack. Designed for ILECs, CLECs, and ISPs, the Avidia can be configured as a DSLAM, access server, or LAN extension concentrator, and is said to be designed for telecommuting and private network applications. Avidia is based on a fully distributed ATM packet and circuit-switched architecture that offers carrier-class redundancy, scalability, and QoS support. T-1 / E-1 and frame relay are supported, as are a number of xDSL options: The Avidia System supports both full-rate DMT ADSL, G.Lite ADSL and SDSL on the same ports and uses the "handshake" protocol to automatically configure each port for the proper transmission type. A variety of chassis, line, and channel card

Full Service Network Architecture using DSLAM and ATM.

options are also available. Avidia pricing starts at $200 per port.

Promatory Communications (Fremont, CA - 510-445-3855, www.promatory.com) makes IMAS, an integrated DSL access platform and ATM switch that performs DSLAM functions along with the kind of enhanced traffic management capabilities not supported by traditional DSLAMs. IMAS can provision five different classes of service, allowing service providers to offer guaranteed quality in service level agreements. IMAS also handles the functions needed for providers to offer transparent LAN services, such as interconnecting LANs in multiple locations via DSL access links, and VPNs.

Unlike a typical DSLAM, which provisions end-to-end PVCs to establish paths between an end user's DSL modem and the ISP's backbone switch, IMAS uses Switched Virtual Circuits (SVCs), which allows user to connect to and disconnect from the network, just like they used to with dialup connections. SBCs also allow network bandwidth to be assigned dynamically. These dynamic connections allow IMAS to use statistical multiplexing, as opposed to deterministic bandwidth allocation, which means that transmissions from more subscriber ports can be crammed onto fewer DS-3 or OC-3 uplinks to the ATM backbone.

IMAS accepts line cards profiled for ADSL, SDSL, IDSL, VDSL, HDSL, and G.Lite. Promatory is working on a Voice-over-DSL platform in conjunction with with Jetstream Communications (Los Gatos, CA - 408-399-1300, www.jetstream.com), who make a voice gateway (the CPX-1000) linking DSL access networks to Class 5 switches.

Direct Dial Inward (or Inbound) (DDI)

See Direct Inward Dialing.

Direct Inward Dialing

A telephony technology that allows a caller to dial directly into a company (without dealing with an operator) via a public network number that includes extension digits.

Discrete Multi-Tone (DMT)

A technology used by ADSL to squeeze more than 6 Mbps over single pair copper wiring. The successor to CAP. See xDSL.

Distributed interNet Applications Architecture (DNA)

A methodology for three-tier application design using components that communicate via COM and DCOM. DNA is also a broad, umbrella marketing term for various Windows NT services.

DLL

See Dynamic Link Library.

DNA

See Distributed interNet Applications Architecture.

DNI

Dialed Number Identification. See Dialed Number Identification Service.

DNIS

See Dialed Number Identification Service

DNS

Can mean Domain Name System, Service or Server. Domain Name System translates a text URL (such as http://www.computertelephony.com) into the equivalent 32-bit IP address (194.73.51.228). A Domain Name Server is the computer that stores the domain addresses and routes requests to specific addresses.

DS-0

Digital Service (or Signaling) Level 0. See Circuit Switched Signaling.

DS-1

Digital Service (or Signaling) Level 1. See Circuit Switched Signaling.

DSL

See Digital Subscriber Line, xDSL.

DSLAM

See Digital Subscriber Line Access Multiplexer.

DSP

See Digital Signal Processor

DTMF

See Dual Tone Multi-Frequency.

DTMF Cut Through

The ability to input DTMF tones to a Voice Processing System while it is outputting speech. This allows the user to skip a menu or instructions.

Dual-Tone Multi-Frequency (DTMF)

Also called "touchtones " or an alternative name for Signaling System Multi-Frequency No. 4. When you press a key on a phone keypad, each digit (and * and #) is transmitted as a pair of concurrent audio frequencies. One frequency is from a high group (1136, 1209, 1477 and 1633 Hz) and one from the low group (697, 770, 852 and 941 Hz). DTMF signaling is most often used to establish connections over the PSTN, it is faster than rotary dialing (a DTMF digit takes a maximum of 0.1 second; a rotary digit takes a maximum of 1 second.). DTMF can also be used for in-band signaling CT applications, such as communicating with an IVR system during a call. DTMF is the core technology used by Interactive Voice Response (IVR) systems.

DWDM

See Dense Wavelength Division Multiplexing.

Dynamic Host Configuration Protocol (DHCP)

Every device on the Internet requires a unique IP address so as to prevent conflicts, whether it be a static address hard-wired into the client or a temporary, dynamic one assigned to a client by a server. If the latter is the case, then the client generally uses DHCP to contact the server and obtain an address. Administrators can also use DHCP to distribute IP addresses from a central point. DHCP is more advanced than its competitor, the Boostrap Protocol (BOOTP).

Dynamic HTML (DHTML)

An overall umbrella term for the ability of new HTML tags and browsers to change the contents of a Web page using script code, while it is loaded. Dynamic HTML give web documents the look and feel of GUI desktop applications of multimedia presentations.

Dynamic Link Library (DLL)

A library of functions stored as a disk file in a special Microsoft format. The code in the DLL can be called upon by application software.

E & M Signaling

A simple bi-directional signaling method developed as a standard interface connecting a phone switching system (such as a PBX) to an interoffice trunk circuit. Normally, there is a single phone number associated with a "standard" analog phone line, such as the one in your home. With Direct Inward Dial (DID) and "Ear" (or Earth) and "Mouth" (or Magnet) or E&M circuits, more than one number may be associated with one line. DID trunks usually connect customers with phone companies while E&M circuits can be used to provide internal connections between PBXs. Both DID and E&M type trunks provide similar functions but with different technology. With both DID and E&M circuits, the dialed phone number is prefixed to the call so the equipment can determine the number called.

With E&M signaling, the signaling between the trunk circuit and the associated signaling unit is done over two leads providing full-time, two-way, two-level supervision. These signaling leads are kept separate from the speech path. Originally, in the days of telegraphy, the E ("Ear") and M ("Mouth") paths occupied a single wire, with a relay contact closure or "key" on the sending side drawing current from a detector on the receiving side consisting of an electromagnet attached to a battery. These days, two wires are used per path so that the sending side draws current from itself, thus reducing induced noise in neighboring circuits.

Although E&M signaling is popular in North America, in foreign countries one comes across local technologies such as AC15 Signaling in the UK. AC15 is based on the "tone on idle" principle, which means that an idle circuit otherwise carrying no traffic is in fact carrying a permanent tone of 2.28 KHz over the line. This tone is only removed when a call is initiated and the line is "seized." In AC15, a tone "ON" is precisely similar to an un-looped E or M lead, while a tone "OFF" is the same as a looped E or M lead. AC15 uses fewer wires than E&M but requires more complicated hardware.

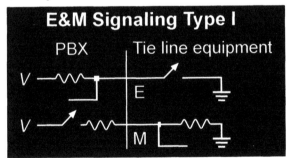

E&M Signaling Type I

The Type I interface is the most common "flavor" of E&M interface in North America. With the Type I interface the tie line equipment generates the E signal to the PBX by grounding the E lead. The PBX detects the E signal by sensing the increase in current through a resistive load (indicated in the diagram by the unconnected node branching from the E resistor's right side). Similarly, the PBX generates the M signal by establishing a current to the tie line equipment, which detects it via a resistive load. The Type I interface requires that the PBX and tie line equipment share a common signaling ground reference, normally achieved by connecting signal ground from the PBX to the RJ45 connector's SG lead (pin 8). Since there is no ground isolation between the equipment, such an arrangement may produce noise in audio circuits or may be susceptible to electrical transients. Also, the asymmetrical design of the circuit allows for only six wires to be used, but precludes using such interfaces "back to back."

E&M Signaling Types

There are five distinct physical configurations for the signaling part of the interface (Types I through V), and two distinct flavors of audio interface (two-wire or four-wire). To make matters more complicated, so-called four-wire E&M circuits can have six to eight physical wires! The difference between a two-wire and four-wire circuit is whether the audio path is full duplex on one pair or two pair of wires. The five E&M signaling standards for PBX tie line interfaces as defined by AT&T specifications, are described below.

With each signaling type, when making a call, the

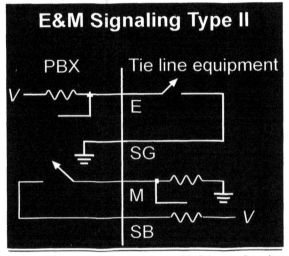

E&M Signaling Type II

The Type II interface provides complete ground isolation since it requires no common ground. Instead, each of the two signals has its own return (the circuit needs all eight wires). For the E signal, the tie line equipment permits current to flow from the PBX — the current returns to the PBX's ground reference. Similarly, the PBX closes a path for current to generate the M signal to the tie line equipment. This kind of interface is often used on Centrex lines and Nortel PBXs, and can be used "back to back" for either trunking or signaling.

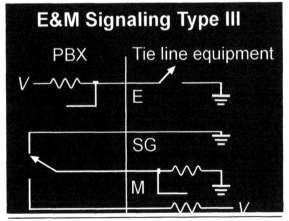

E&M Signaling Type III

Type III is a variation of Type II. Type III uses the SG lead to provide a common ground. With this configuration, the PBX drops the M signal by grounding it, rather than by opening a current loop.

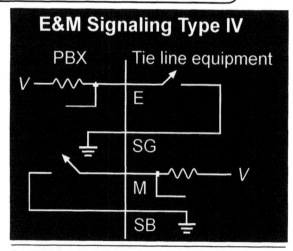

E&M Signaling Type IV

PBX — Tie line equipment

Type IV is symmetric and needs no common ground, so there is complete ground isolation. Each side closes a current loop to signal. The signal's presence is indicated by a resistive load in the flow of current.

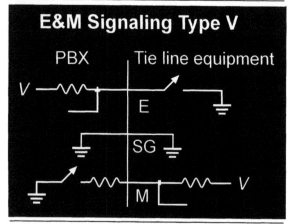

E&M Signaling Type V

PBX — Tie line equipment

The Type V interface is a simplified version of Type IV (it's also considered the easiest circuit to debug). This is also a symmetric interface (and can be used "back to back"), using only two (or six) wires. Type V does not provide ground isolation - it requires a common ground between the PBX and the tie line equipment which is provided via the SG leads.

M lead is grounded and the PBX supplies a signal, known as the M signal, which may pulse between closed and open states when the phone is dialing. When the remote site (CO) answers, the E lead gets looped to ground to show that the call has been answered and the PBX now accepts one signal, known as the E signal. This means that the tie line equipment at the CO accepts the M signal from the PBX and provides the E signal back to the PBX. The M signal accepted by the tie line equipment at one end of a tie circuit becomes the E signal output by the remote tie line interface.

When the CO wants to ring a phone, the E lead is connected to ground and the PBX indicates that it has been answered by connecting the M lead to ground.

ECTF

See Enterprise Computer Telephony Forum.

eCTI

1. See Electronic Contact Technology Integration.
2. Enterprise Computer Telephone Integration, a term devised by Genesys Telecommunications Labs to describe large-scale, distributed CTI.

Electronic Contact Technology Integration (eCTI)

Blair Pleasant, Director of CTI Research at The PELORUS Group (www.pelorus-group.com/pelorus.htm) agrees with many experts that the traditional voice-centric call center will be replaced by the "contact center" or "transaction center". These types of centers result when businesses integrate their telephone-based call-center operations with the web (See Web-Enabled Call Center). Customer contact centers allow customers to communicate with a company using their contact mode of choice, be it telephone, fax, e-mail, or the web (i.e., web chat, a Website callback request, or an Internet call using Voice-over-IP).

A report from The PELORUS Group entitled *CTI: Path To Mixed Media* reaffirms the idea that CTI is essentially plumbing, or an enabler, and that the true value is in the applications and capabilities that CTI can provide. In the new media-blended contact center, CTI, and the applications that it enables, will play a crucial role.

According to the report, eCTI is used to manage and track not only telephone calls, but multiple media or channel types, such as e-mail, Web call backs, fax, and eventually video, throughout the enterprise. Whether the customer contacts a company through the telephone, e-mail, or Web, the same fundamentals are required.

This new type of agent / caller contact is called OmniContact by Pleasant and the PELORUS Group – "omni" implying that the center can handle all types of contacts, regardless of media type, in a consistent manner based on clearly defined business rules abetting the company's service level goals.

CT is now moving from straight Computer Telephony Integration to electronic Contact Technology Integration (eCTI), which describes the integration of e-mail, web / Internet, fax, and other media in the call center, or customer contact center. eCTI may or may not include traditional CTI links. eCTI can be defined as "Using mixed-media technology to serve customers and stakeholders requiring personal assistance for information, service, or support, via the telephone or Internet."

Blair Pleasant says that: "CTI companies need to convert what they have learned in the voice-centric world, and move to the next generation, or eCTI model. Telephones and call centers won't go away, and traditional CTI will be needed for quite some time. But vendors must be ready to leverage what they know about accessing information on customers, identifying their needs, integrating data, and providing better service....We expect the trend toward Customer Relationship Management (CRM) will only help the CTI market, as CTI is an important element of a total CRM solution. eCTI and the movement toward web or Internet-based contact centers, will also help spur market growth."

Although eCTI solutions are for the moment using traditional CT integration, eventually there will be no classic "CTI" since everything will finally be fully converged into an IP platform and integrated IP / PSTN network, and a single caller queue for multiple types of incoming media.

Electronic Switching System

A generic term used to identify digital, stored-program switching systems that digitizes analog signals from the local loop, then interconnects them by multiplexing them over a time-division form of transport, assigning each signal to its own time slot. ESS switching include the old Bell System's No.1 No.2, No.3, No.4, and No.5.

Enhanced Services

Enhanced services are offered by telecommunication service providers to add value to the basic phone and network services sold to subscribers. The Enhanced Service Provider (ESP) market evolved from deregulation, particularly the Telecommunications Act of 1996, wherein service providers were allowed to offer enhanced customer services such as voice messaging, debit and credit card, single number "follow me" and other voice-based subscriber services.

At the turn of the 21st century, the ESP market entered a new phase with the convergence of voice and data. Enhanced service providers went beyond circuit switched enhanced services by "IP-enabling" their applications. This can involve simply using the IP network for inexpensive transport, with minimum change to the enhanced service application, or services can be enhanced in functionality to add value to the customer application. For example, voice messaging can be enhanced with IP to support unified messaging which combines voice mail, e-mail, fax, paging and other non-voice data services into a single, integrated application. Other examples include voice-enhanced e-commerce, where consumers at a website can speak over IP with a customer service representative, and multimedia conferencing services where one caller can "push pages" and lead the conferees through a PC-based presentation while conducting a speech or question and answer session.

Indeed, the marriage of IP and enhanced services is giving birth to a new business-quality IP service layer. Genuinely useful services are so complicated and expensive to build that they can't grow out of the enterprise network, they have to be outsourced to Service Providers (SPs) who can offer access to a new IP service layer made up of a new generation of PSTN / IP carrier-class network switches capable of handling the plethora of delivery protocols for data, voice, video, and encryption.

Some History

Trivia question: How is TV Guide like the computer telephony industry? When television sales skyrocketed in the 1950s, newspapers tried to ignore the phenomenon, hoping it would go away. The lack of pop TV coverage allowed TV Guide to step in and become one of the most-read publications in America. If the newspaper industry hadn't stuck its head in the sand, TV Guide might not have come to exist.

Similarly, computer telephony shouldn't really exist as an independent industry – AT&T and its later competitors should have been able to offer all of its functions as Centrex-type services long ago. SOHOs and even some medium-sized businesses - lacking in capital resources and expertise - are clearly adverse to installing and managing a melange of CT equipment on their premises. The opportunity is obvious.

Indeed, in theory, if telcos had been on the ball, it should have been possible to prevent corporations from building basic PBX infrastructures. One reason this didn't happen is that telcos failed to solve the user-interface issues, and give Centrex customers real control over feature configuration and provisioning. The nascent telephony CPE market did a better job, so businesses decided to take their infrastructure in-house.

Considering how long it took the unwieldy telcos just to get a simple technology such as ISDN deployed, it's no surprise that CT first appeared in customer premise equipment, and was eventually adopted by telcos only when they could easily outsource and resell it as enhanced services.

But in the "bad old days" of the telecom industry, there was little competition and considerable governmental regulation, so there was practically no incentive to develop new and enhanced services. The telecom environment consisted of local and long distance services using conventional landlines. You had one phone centrally located in your home and another at work. You commuted back and forth from one environment to the other. Ideas such as taking your office with you in virtual form via a PC or cellular phone, telecommuting, or instantly setting up a virtual call center, were relegated to science fiction.

During the 1990s, the ongoing deregulation and privatization of telecom services – in particular the deregulation of the U.S. telecommunications industry as a result of the Telecommunications Act of 1996 – made the telecom industry more interesting and competitive, introducing an era of intense intra-industry, cross-industry, and geographic competition for market share among wireline, wireless and cellular service providers.

Long distance and local exchange carriers in the U.S. now compete for local and long-distance services. In the 1990s major long-distance carriers lost customers to small and mid-sized companies, which even today continue to aggressively sell specialized services to specific niche markets, grabbing roughly a $40 billion share of a $600 billion worldwide market. "Personal selling" or "customer relationship marketing" is giving smaller companies an edge in winning market share. Domestic service providers have sometimes responded by expanding their scope and ally themselves with international telephone companies to deliver seamless global services.

Telcos also face competition from cable companies wanting to get into the telephony business, not to mention cable's high-bandwidth voice and video services.

Telcos now find that they must also compete with wireless companies for voice and data services. The introduction of digital services such as Personal Communications Services (PCS) drove down prices for mobile services, frightening the carriers.

So, thanks to a combination of deregulation and telecom-savvy consumers, a 40% to 60% customer turnover rate soon appeared among service providers. This dreaded "churn" phenomenon was quite disconcerting to carriers.

In response to this quickly evolving competitive environment, big carriers, little phone companies, and the wireless companies all scrambled to figure out how to hold on to and get new customers without lowering billing rates. The players realized that they must find new ways to differentiate themselves, such as improving existing services and rapidly introducing new value-added enhanced services offerings to not just attract new customers but to improve customer retention and increase revenue streams.

Both wireline and wireless service providers became so hungry for new and exciting services to hold onto customers and get more revenue out of them that they turned to computer telephony technology for help. One number calling, calling cards, fax, e-mail, unified messaging, speech recognition, text-to-speech, and other value-added network services all began to jazz up the formerly ho-hum telecom world.

As Andy Ory of Priority Call Management, a major enhanced services provider, once said to me: "I came of age in this industry in 1988 when I joined Boston Technology. I sat down and read all of the telecom magazines for four or five hours every day, searching for computer telephony enhanced services information. It's now a real industry. It really is. I remember displaying our first system at CT's first Expo in 1992, then called 'Telecom Developers,' which was held in a hotel underground parking lot in Dallas, Texas. Compare that with the most recent huge CT Expo in Los Angeles. The growth is unbelievable."

Computer Telephony came to the rescue of those telcos and wireless service providers desperate for new and exciting services. The ball got rolling with enhancements of existing telephone services through the addition of features such as call waiting, call forwarding, call return, caller identification, voice mail and personal 800 numbers. Later, more advanced and completely new enhanced services began to appear. Today, one number calling, fax, e-mail, unified messaging, speech recognition, text-to-speech, Internet mailboxes, and other value-added network services all are enlivening the formerly boring telecom world, enabling telecom companies to hold onto customers and extract more revenue out of them. Soon we shall all take for granted even more exotic enhanced services such as high-speed data services, data and video teleconferencing, video-on-demand, home banking and home shopping.

The Rise of the ESP

Enhanced services simply just don't appear out of thin air, of course. Somebody has to develop and market them within the existing telecom infrastructure. These are the Enhanced Service Providers or ESPs.

But before ESPs could make their presence felt in the telecom industry, the government had to develop regulations and pass legislation that would make the telecom business environment conducive to the establishment and growth of ESPs.

In the 1960s the FCC recognized that computers were becoming increasingly entangled with communications over telephone lines. In an effort to figure out both the regulatory/policy implications and what to do about such computerized communications, the FCC over a period of two decades held the *Computer I, Computer II* and *Computer III* proceedings trilogy.

FCC released Computer I in 1971. In it, the FCC conceptually separated computers used for communication from computers which performed data processing services. It defined data processing as the "use of a computer for the processing of information as distinguished from circuit or message-switching." Computers involved in the means of communications itself would be regulated as common carriers under Title II of the Communications Act of 1934, as amended. On the other hand, computers providing data processing services over the public phone network would not be regulated under Title II, though the FCC later determined that it had jurisdiction over these services under the ancillary jurisdiction of Title I. By saying this, the FCC implied that it has at least a limited jurisdiction over the Internet and over data processing services where they are transmitted over the telephone networks.

The reasoning behind *Computer I* was that large telcos would almost certainly use their economic clout to subsidize data processing services and completely dominate what should be a thriving and competitive market. This resulted in the FCC ruling that large carriers could offer data processing services only through separate subsidiaries, thus preventing any cross subsidization. The FCC believed this would foster competition, innovation and more efficient services.

However, *Computer I* was almost immediately found to be deficient, with a growing number of hybrid communication systems each forcing a case-by-case determination by the FCC over whether or not they fell under Title II. In response to this quagmire, the FCC initiated the Computer II proceeding.

Released in 1980, *Computer II* was built on the groundwork laid by *Computer I*. In addition, however, the FCC, formulated new distinctions between pure communications and data processing services by identifying the categories of "basic" and "enhanced" services. Basic service, which is regulated under Title II, includes plain old "applicationless" voice transmission: "a pure transmission capability over a communications path that is virtually transparent in terms of its interaction with customer supplied information." Enhanced services, however, are "services, offered over common carrier transmission facilities used in interstate communications, which employ computer processing applications that act on the format, content, protocol or similar aspects of the subscriber's transmitted information; provide the subscriber additional, different, or restructured information; or involve subscriber interaction with stored information. Enhanced services are not regulated under Title II of the Act." Enhanced services include data processing services under Computer I and hybrid forms of communications. Simply put, "Enhanced service is any service other than basic service". Enhanced services use the basic phone network to deliver their more advanced functionality.

In the Computer II proceeding, the FCC continued its attempts to protect a competitive market from domination by large telcos. The Computer II proceeding implemented "structural safeguards" that sustained the previous policy allowing large telcos to provide enhanced services only through separate subsidiaries. As they wrote: "In Computer II the Commission found that the exercise of ancillary jurisdiction over both enhanced services and CPE was necessary to assure wire communications services at reasonable rates. Regulation of enhanced services was deemed necessary to prevent AT&T from burdening its basic transmission service customers with part of the cost of providing competitive enhanced services".

In the early 1980s the PSTN began to diversify, with AT&T losing its century-long grip on providing a homogenous telecom infrastructure. AT&T was about to be broken up, with local and long distance services differentiating to the point where complex relationships began to appear. In the middle of all this were the enhanced service providers (ESPs) and the problem of "access charges."

"Access charges" are fees collected by the local telephone companies for the origination or termination of any interstate or foreign telecommunication. In 1983 the FCC determined that ESPs would be exempt from the access charges called "interexchange services," which normally involves local exchange carrier (LEC) telephone companies charging high per minute fees to long distance companies for providing connections to allow them to complete long distance telephone calls, which would be passed along by the local or long distance carrier to the ESP. ESPs were declared "end users" of such services and are exempt from such access charges even if the ESP uses a local telephone service to originate and terminate interstate communications. However, they do pay interstate subscriber line charges on switched access / local exchange lines and the special access surcharge on special access lines purchased from LECs. The FCC cleverly declared that ESP traffic is interstate in nature and, therefore, states may not require ESPs to pay intrastate access charges.

Incidentally, thanks to this "enhanced service provider" (ESP) exemption, Internet Service Providers (ISPs) don't have to pay these fees either – various LECs have asked the FCC to revoke the ESP exemption for the ISPs, using specious reasoning involving things such as "network congestion." The FCC's docket on this can be found on the FCC home page, at www.fcc.gov, or the special home page that follows the access fee issue: www.fcc.gov/isp.html. Although an ISP is not considered a telecommunication common carrier and is thus not regulated under Title II, an ISP such as AOL generally carries a considerable amount of "stealth voice" traffic (e.g. NetMeeting and web collaboration sessions). Many believe that the ESP exemption for ISPs will be removed by 2003. Conversely, carriers will find ways to offer their own, unregulated services.

By the mid 1980s, the definition of an ESP began to solidify. As telecom guru Harry Newton wrote in Newton's Telecom Dictionary, "an ESP is a company that provides enhanced or value-added services to end users. An ESP typically adds value to telephone lines using his own software and hardware... An ESP is an American term, unknown in Europe, where they're mostly called VANs, or Value Added Networks... Also called an IP, or Information Provider."

I define an ESP as a for-profit telecommunications entity that uses the facilities of telecommunications common carriers to establish connections with their customers and offers them one or more services to not only transmit voice and data messages but to simultaneously add value to the messages during transmission.

One popular example of an ESP is a public voice mail box provider, which involves the storage of a spoken message within the network for forwarding or retrieval at some future time. In China few people own phones, yet the appearance there of such a voice mail box enhanced service caused considerable excitement. Today many "phone-less" Chinese citizens use payphones to get their voice mail!

In the U.S., an ESP differs from a VAN provider in that an ESP offers voice as well as data services, whereas a VAN generally just supplies a data service involving the processing of data within the network and sending the results to a customer. An example of a VAN would be a database provider giving the latest airline fares. Thus, a VAN can be an IP – but so can an ESP.

Like an ESP, a VAN also employs the telecom services of commercial carriers, adding its own hardware and software to add value and allow its own services to be offered to customers.

Some VANs lease lines (e.g. T-1, 56 Kbps DDS) and allow customers to transmit their own data over a Virtual Private Network (VPN).

Whether you call them ESPs or Value-Added Networks, these are all telephone and teleprocessing services created by improving the existing infrastructure on such basic services as telephone and data interchange.

Such improvements involve the use of computers and software, as can be seen in the "official" Federal Communications Commission (FCC) definition of enhanced services as "services offered over common carrier transmission facilities used in interstate communications, which employ computer processing applications that act on the format, content, code, protocol or similar aspects of the subscriber's transmitted information; provide the subscriber additional, different or restructured information; or involve subscriber interaction with stored information."

In other words, an enhanced service is really a computer processing application that does something in some way with information transmitted over phone lines (this definition is now being expanded to include the Internet and other IP networks). This is why Transaction Services, Videotex, One Number "follow-me", Alarm Monitoring and Telemetry, Voice Mail Services and e-mail are all examples of enhanced services.

Such enhanced services usually are provided by companies with expertise in Computer Telephony (CT) or Computer Telephone Integration (CTI), since these services exist as software running on a series of networked PCs or Sun SPARCs that interface to the public network.

Although new enhanced services are offered over transmission facilities, they may be provided without filing a tariff with the Federal Communications Commission.

The FCC encourages new and existing companies to develop exciting new services and hence add value to phone lines, but in order to reach customers, all such ESPs must rely on phone lines provided by local phone companies, many of whom also like to participate in the possibly lucrative value-added telecommunications business. Indeed, local phone companies would like not just a piece of the action, but all of it - monopolistic ideas are always promulgated by the carriers, usually under the guise of desiring systems to be "homogeneous" and thus "easier to manage" for everybody's benefit.

Thus, for the FCC to allow the Bell Operating Companies (BOCs), other local phone companies and ESPs into the value-added business it had to figure out a way so that the Bell operating companies didn't grab too much market share and have an unfair advantage.

In order to encourage competition, special rules governing the provision of enhanced services were created by the FCC. AT&T, for example,

could provide enhanced services, but only through a subsidiary. After AT&T was divested in 1984, the FCC applied the same "structural separation rule" to the Bell Operating Companies (BOCs), and enhanced services continued to be provided through separate subsidiaries.

ONA and CEI

In 1986, the FCC decided that structural separation was no longer in the public interest and that there were other "nonstructural" ways of protecting fledgling enhanced service providers from the BOCs, so in fact everyone could participate in a competitive environment.

In May, 1986, in its Third Computer Inquiry, also known as the Computer III Decision, the FCC introduced the concept of an Open Network Architecture (ONA), which attempted to create through regulation a free market environment for enhanced communication services. Essentially, the FCC ordered unbundling of Local Exchange Carrier (LEC) network functionality into components that can be purchased, packaged, and sold by LEC competitors.

Comparatively Efficient Interconnection (CEI) requirements were instituted as a forerunner of ONA and were initially used for determining interim approval of enhanced services of an RBOC or AT&T before ONA was fully implemented. CEI was meant to prevent a carrier from ruining ESPs by selling them degraded or overly expensive network connections. CEI was to ensure that the same underlying services the carrier / BOC enjoyed were also made available the ESPs, both in terms of the technical aspects and at the same tariffed rate paid by the carrier.

One of the FCC's "nonstructural" safeguards in lieu of structural separation, to prevent cross-subsidization and discrimination, was the requirement that BOCs file CEI plans.

Under the Computer III regulatory regime, if a carrier got a bright idea for a new enhanced service and wanted to offer or market it to its customers, the carrier would first have to file a plan with the FCC (and obtain the Commission's approval of the plan) explaining how it could offer and sell on a nondiscriminating basis to competing ESPs the same underlying component services with technical characteristics that are equal to those that the carrier uses to create its own enhanced service. A CEI plan must also be filed if a BOC plans to act as a sales agent for an enhanced service provider or otherwise market an enhanced service, even if the BOC will not provide the enhanced service itself.

It's presumed that every enhanced service doesn't just drop out of a tree – it consists of one or more local network components, or the so-called "basic underlying elements" – also known as Basic Service Elements (BSEs) that are combined to form an actual enhanced service.

Under CEI, the carriers are obliged to unbundle the set of basic service functions that a carrier uses in an enhanced service offering and make them available to ESPs under tariff as a BSE or a set of BSEs. In essence, the carriers must provide access to their own services, and offer them to competitors on the same basis that they do for themselves.

For example, if a carrier's enhanced service is, say, a "high bandwidth follow me service" that uses calling number identification and three-way call transfer capabilities, then the Comparatively Efficient Interconnection for that service must include these basic services as a set of Basic Service Elements unbundled from other basic service offerings.

The unbundling of all the underlying services in the local access networks, or "bottleneck services," has been a focus of FCC activities for many years. The FCC's objective was to put the phone companies and the enhanced service providers on an equal footing and thus foster competition in enhanced services. ONA tried to go a bit farther in this regard than any LEC could have dreamed of, however.

How ONA goes beyond CEI

Under CEI, competing ESPs have access to the same technological services at the same price as the carriers. Amazingly, under ONA, competing ESPs – even Interexchange Carriers (IEXs) acting as ESPs – could demand access to advanced functions whether or not they were currently used by the phone companies themselves! Carriers have to accommodate the requests of their competitors for access technology, so long as it meets the FCC's criteria of demand, utility, technical feasibility, and cost feasibility, regardless of whether the carrier ever plans to offer an enhanced service.

Under ONA, the telephone companies must provide the same class of service to their own internal value-added divisions as well as to nonaffiliated (i.e. outside) valued-added companies. Thus, the carriers' architecture is to be truly "open" and everyone should be able interconnect to specific basic network functions on an unbundled and "equal access" basis.

To be more explicit, ONA includes the following:

1. Basic Serving Arrangements (BSAs). A BSA is how a customer gets access to the public network and provides for selecting Basic Service Elements (BSEs, see below). Under ONA, BSAs should be considered the fundamental connection to the network, both in terms of the switching and transport services an ESP uses to communicate with its customers through the local exchange provider's network, such basic local exchange service, private line service, and switched access. BSAs represent the minimum network functionality an ESP needs to reach customers. BSAs are access links that can be circuit switched, packet switched, or dedicated, and include signaling options and interfaces to the BSEs. Special Access is a form of BSA, but it's currently deregulated.

2. Basic service elements (BSEs). Basic Service Elements are not needed for an ESP to have a BSA, but are optional value-added unbundled features, such as Caller ID, provided by a LEC that ESPs may need in order to find useful in configuring an enhanced service on a Basic Serving Arrangement. A BSE usually consists of an access link element, a features/functions element, or a transport and usage element connecting Enhanced Service Providers to their customers.

BSEs fall into four general categories: Switching, where call routing, call management and processing are needed; Signaling, for applications like remote alarm monitoring and meter reading; Transmission, where dedicated bandwidth or bit rate is allocated to a customer application; and Network Management, where a customer is able to monitor network performance and reallocate certain capabilities. ANI, Audiotext "Dial-It" Services, and Message Waiting Notification are all examples of BSEs, which also include Central Office announcements, multiline hunt group, three-way call transfer, and uniform call distribution services. ESPs today can choose from nearly 200 unbundled BSEs in creating their own service packages for customers.

3. Complementary Network Services (CNSs). CNSs are optional unbundled basic services that an end user orders from a carrier to be able to use an enhanced service. Some examples of CNSs include: Call Forwarding Busy/Don't Answer, Call Waiting, Message waiting/indicator, Speed Calling, Three Way Calling, Virtual Dialtone, and Warm Line.

4. Ancillary Network Services. These are options available to an ESP which support and complement the provision of enhanced services. Examples of ancillary services are protocol conversion, and Direct Inward Dial (DID) with third number billing inhibited.

The FCC avoided the morass of trying to define a specific ONA standard, and instead left it up to the carriers to come up with Open Network Architectures.

Impact of the Telecommunications Act of 1996

Today, provisions for interconnection and access between telecommunications carriers found in the Telecommunications Act of 1996 are even broader than those specified by CEI and ONA.

The 1996 Act does not rely upon the FCC's earlier division of telecom into basic services and enhanced services, but instead refers to "telecommunications services" and "information services." The 1996 Act defines "telecommunications" as: the transmission, between or among points specified by the user, of information of the user's choosing, without change in the form or content of the information as sent and received. "Telecommunications service" is defined as: the offering of telecommunications for a fee directly to the public, or to such classes of users as to be effectively available directly to the public, regardless of facilities used. The 1996 Act defines "information service" as the offering of a capability for generating, acquiring, storing, transforming, processing, retrieving, utilizing, or making available information via telecommunications, and includes electronic publishing, but does not include any use of any such capability for the management, control, or operation of a telecommunications system or the management of a telecommunications service.

The Telecommunications Act of 1996 has accelerated competition, often in ways not anticipated by experts. For example, we all take for granted that universal service has always been provided as a wireline service – even with the "grand convergence" going on that will ultimately combine the public phone system and the Internet into a single packetized network, the picture we all have in our heads is of voice and data going through a wire.

In actuality, it appears that both existing and upcoming wireless services will ultimately dominate personal communications not only in the U.S. but worldwide. New and existing wireless technologies – Cellular/PCS wireless local loop, Fixed Wireless Access (FWA), Microwave Multipoint Distribution Systems (MMDS, also called "wireless cable"), Telcordia's Local Multipoint Distribution System (LMDS) for Wireless Local Loop (WLL) applications, point-to-point microwave, and satellite – will provide basic and enhanced services to customers first in difficult-to-reach low density areas, then everywhere. High-bandwidth data communications will still travel via wireline (fiber) networks, however, well into the next century.

Also, the distinction between "basic services" offered by monopolistic telcos and "enhanced services" can be somewhat blurry, since the telcos also offer certain "enhanced" services that are in fact an integral part of the telecom infrastructure – examples of such "monopolistic services" being ISDN and special services for digital wireline customers. On the other hand, the true independent services provided through the telecom infrastructure – the "competitive services" – include such items as voice mail, voice information services, e-mail services, and database information providers. Most of the services we will examine in this book are competitive services.

As convergence continues, more and more underlying packet-switched transport functions are considered to be basic services (packet-switched networks using X.25 protocols, frame relay, etc.), but Internet access itself has always been treated as an enhanced service, and ISPs have never been subject to regulation by the FCC under Title II of the Communications Act. Why? Just about every service ISPs offer (e-mail, web page hosting, chat rooms, message forums, access to various databases, etc.) are all computer related and therefore meet the criteria used to define an enhanced service.

BOCs, however, must file CEI plans when they themselves offer Internet access, so as not to compete unfairly with the ISPs.

Circuit-Switched Enhanced Services

Prior to the explosion in IP technology, the more successful circuit-switched enhanced services fell into a few categories (they still do):

1. Traditional Phone Call Resale. These money-making platforms essentially tie in and around the PSTN and wireless networks for the purpose of routing real-time voice calls.

   This would include local and long-distance resale platforms for the PSTN and cellular and PCS networks, prepaid calling-card switches and international callback boxes – mostly machines built either around industrial strength programmable switches from Excel and Summa Four or on high-density PC components and the H.100, H.110, SCbus and MVIP CT bus architectures. In general, we're talking CT-based switching mechanisms; since, after all, if you want to be a telephone company, you must be able to switch calls.

2. The Churn Stoppers and Enhanced Services Peripherals. Competition is now fierce amongst local and long-distance telcos and wireless service providers and Internet Service Providers. They're all starving for new and exciting ways to hold onto customers and generate more revenue.

   And so, computer telephony came to the rescue. One-number people finders, automated personal assistants, unified messaging, speech recognition for voice dialing, "wake me" services, Internet and basic voice /fax processing and other enhanced CT services stuff have been deployed in an effort to win and keep customers. These systems can also be used in standalone service-bureau applications.

3. Virtual Offices and Call Centers. This is a relatively new phenomenon and an extremely interesting one. The idea is to provide very sophisticated call-center CT features to small organizations for a monthly fee. What was a handful of major platform providers began to increase along with the acceptance of IP technology.

   There's a compelling notion to network-base "Virtual Office" solutions, including the fact that you can "rent" top-shelf worldwide computer telephony communication power without spending a fortune upfront and forever hassling with onsite system maintenance.

   We're not talking Centrex and base telco services here. Instead, think bigtime CT – automated call processing, multimedia messaging, follow-me switching, voice-activated call control, flexible conferencing and routing and more – all driven by open CT board-level and / or programmable switch platforms (e.g. Excel, Redcom, Summa Four), all nested neatly in the network ether, and all aimed at generic-business (rather than formal call-center) communication needs.

   The term "virtual call center" until recently has meant "distributed call center" – one still belonging to the enterprise, but capable of using the public network to route calls about as if all of the agents were sitting together in one location. GeoTel (now part of Cisco) developed and sells a large-scale system that does this. Such a "virtual call center" can also be sold as a service, however. Dare we call it a "virtually virtual call center"?

   With 80,000 call centers in the U.S. (the number grows at 20% to 30% per year) where $250 billion dollars worth of business is transacted annually, it makes sense that everybody wants a piece of the pie. Call centers with 25 to 50 agents are experiencing the most growth, but you say that you can't afford to spend the money to build one? You say your business has no room for a conventional call center – or a PBX for that matter? No problem. The virtual call center can help you.

   One definition of "Virtual Call Center" goes something like: "extending transparent call processing functionality across various kinds of technologies, and geographical locations, to agents." Virtual call centers allow for logical groupings of agents or knowledge workers who have to deal with a company's "customer contact zone," doing things such as customer services and telemarketing campaign management.

   There is another "virtual" category, which is simply a "Virtual PBX" or "Virtual Office" for business. Let's say that ten programmers are working on a new word processing program. They are organized as a company but they live at various points around the country. You call the "business" (usually an 800 number) and dial an "extension" to reach a certain person.

   Some ACD-like routing mechanism, either at the company's office or in equipment at an enhanced service provider or perhaps even built-in to the carrier's network, connects you with the person you want to talk to.

   Neither the callers nor the employees can tell whether they've been routed by an enterprise switch, by a third-party, or by a Telco Centrex service. Perhaps the agent or employee is a telecommuter working from home. It's doesn't matter so long as transactions in the "customer contact zone" are smooth and the whole operation appears as an integrated whole.

   Open, programmable computer telephony components made the whole idea of the virtual office possible, things like the CT Bus and SCbus and MVIP based media processing and switching resources and dumb programmable switches from Excel, Redcom, and Summa Four have lent the flexibility which has allowed Virtual Offices to evolve.

   If you doubt that computer telephony technology can make large-scale virtual offices possible, ask yourself what conventional Centrex, PBX or key system alone can support such advanced functions as fax mail, fax broadcast, fax-on-demand, fax-to-speech, multi party conferences from any touchtone phone, e-mail to speech, or e-mail forwarded to a fax number?

   We often extolled in the pages of *Computer Telephony* magazine the wonders of completely virtual systems (no customer premises equipment) which enable network service providers to offer a wide range of services to end-users – from basic voice mail to e-mail to fax mail, and ultimately, unified messaging.

   Those who benefit most from such unified messaging products include those running a small office / home office (SOHO), technologically advanced consumers, and mobile professionals who want the convenience of handling all forms of communications with one-number access.

   This technology offers subscribers the convenience of one place to check for all of their messages and the cost advantages of a "virtual office on the Internet," lowering office overhead costs. For example, faxes may be read and reviewed on a personal computer, eliminating fax machines, dedicated fax phone lines and recurring long distance charges.

   When retrieving messages by phone – either a telephone handset, wireless phone, or the handset of a fax machine – subscribers hear a computerized voice that provides message information. Through a PC, or any Web-connected device, subscribers can access a personal unified messaging homepage and obtain all voice mail, fax and e-mail messages.

   When the idea of virtual offices first appeared, we at first thought this technology would simply be used to virtualize specialized apps, similar to Genesys who has enterprise applications and eventually scaled up some of them so Network Service Providers can offer them as enhanced services.

   Ultimately, however, this stuff could become even more pervasive, with all of a company's switching functions and applications offered on a virtual basis, similar to the idea behind the Crosspoint Virtual PBX from Virtual PBX. com, formerly Advanced Queuing Systems (San Francisco, CA – 888-825-0800, www.virtualpbx.com). Instead of calling into an ACD at a main site and being switched out from there, many new start-up companies have no real "main site" at all, with people scattered about the country, particularly if the nature of the business is to keep employees mobile.

   Virtual offices for the masses? It can be done. States such as California are mandating that companies to set up programs that will reduce traffic and pollution. Telecommuting would be the ultimate means of complying with such state regulations.

It's known that in the case of telecommuting call center agents, the Genesys solution, for example, can keep tabs on what a home agent is doing at all times, just as if they were in an actual formal call center. The trade-off seems to be that you can better manage agents in the formal call center (management feedback is also faster) as opposed to the kind of flexibility (call overflow can be immediately farmed out to agents at home) and lower cost (no overhead or office space) that you get with using telecommuters.

Once solution is to work the agents in a rotating schedule where some days they work at home and other days they work in the office, interacting with both management and their fellow workers.

For the moment however, in larger, established companies, this kind of technology will apply mostly to connecting geographically diverse call centers, since multiple sites pooled together as a single entity allows for the creation of such things as "follow the sun" call centers, where the sun never really sets on a business.

What was really interesting was that carriers began to offer Virtual Call Centers as a Centrex service. There doesn't appear to be a particular "minimum" or "maximum" number of agents necessary to justify such a virtual office solution, and such carrier-based enhanced services allow your agent pool to go up into the thousands. Such "exotic" technology looks quite a bit more attractive when the infrastructure cost is borne by the carrier and shared amongst multiple customers.

Of course, carrier-based virtual offices means that the Telcos must build their platforms on fault resilient computers for reliability and make some hefty investments in network security (such as using RSA and SSL to secure messages) and accurate billing software. There's always a danger that the carrier is going to mess things up in a colossal way, which has always been an argument for having customer premises based systems.

IP Enhanced Services Come of Age

Deploying a wide range of CT enhanced services on a vast scale in the circuit switched world turned out to be a Herculean feat, one that was not always welcomed with open arms by subscribers. Consumers have shown a certain diffidence towards circuit-switched / phone-only enhanced services - again, mostly because of the user interface dilemma. It's difficult to manage complex services over the phone.

Meanwhile, the Internet has emerged: Forming a model for a new kind of distributed application, and solving certain persistent UI problems. The pervasiveness of the Internet, universal web browser interfaces, and e-mail have primed the public to be more receptive to IP enhanced services from carriers, CLECs, and ISPs.

Conventional, purely circuit switched-based enhanced services required expensive changes to the PSTN infrastructure and were found to not easily integrate with the kind of IP-based enterprise system where, for example, a traveler at a dial-up airport terminal in Singapore or at home in the suburbs can have direct access to everything in their company's intranet (or their PC's desktop) in New York, and their client's extranets in another country.

If you have an IP connection, you probably have people in your data center environment who are running NT servers. Thus, you already have most of the components needed to add an application atop the existing IP infrastructure, then rent the application to anyone else having an IP connection. It becomes quite an attractive proposition because all of the technological barriers have already been eliminated.

IP levels the playing field. Anybody can do IP - the ownership of IP is open so that developers can write to it, feel comfortable and don't have to worry about someone else "reverse engineering" it. Since IP is so "open"

a routing scheme, its evolution is subject to public review; reducing or eliminating the fear conventional developers have, of creating product locked to some proprietary standard that can change without notice.

If you've got the right hardware and the right sized pipe, CPE applications can be quickly scaled up and offered as an enhanced service. If a company provides a purely IP enhanced service, we tend to lump them into the Application Service Provider (ASP) category. Some new IP enhanced services, such as web conferencing, use both the IP and PSTN (for voice), giving them the appearance of more traditional enhanced services companies.

The pervasiveness of the IP standard is the secret to the triumph of IP enhanced services over those of the circuit switched world. The protocol itself is amazingly flexible and "scalable." A PDA device that clips on to your belt can have full IP interoperability with equipment up to and including the largest of routers, servers and applications. The tremendous scale to which IP has been deployed and the huge number of devices already IP compliant have given an enormous number of people the basic capabilities (not to mention web browser skills) to actually take advantage of services that leverage that infrastructure.

Also, once a would-be enhanced services or ASP developer is equipped to handle IP, they also have access to IP's "friends" - all of the other associated standard protocols such as H.323, SMTP, POP3, IMAP, and LDAP for directory provisioning and control. You have, then, a suite of protocol capabilities that combine to form a very workable, open environment on which you can begin to develop applications and services that simply couldn't have existed in a circuit-switched world, couldn't have been deployed cost effectively.

Large corporations are increasingly sensitive to costs, and the savings of outsourcing CT functions to providers offering IP enhanced services are glaringly evident.

For example, Danny Winokur, VP of Business Development of USA.net (Colorado Springs, CO - 719-265-2930, www.usa.net), tells me that "Enhanced service providers and ASPs really took off in the last six months of 1999, since big corporations are beginning to realize that an outsource relationship can significantly reduce the expense associated with providing sophisticated capabilities to their employees or customers."

Winokur continues: "If you look at the cost to a large enterprise organization of running its own e-mail or messaging environment for its employees and you compare that to the cost of a service, like our USA.net enterprise messaging offering, you have a cost that is many factors different. Some studies show that if you figure out the total cost of an advanced messaging infrastructure of an enterprise - the cost of downtime, maintenance, software upgrades, support, etc. - you find that they're paying more than $600 per year for each user mailbox. In an outsource relationship, however, a single user's annual messaging cost, depending on the exact feature configuration, drops to about $60 a year. That's an enormous difference."

All of this means that an enormous market opportunity exists. Carriers and alternative providers can't make a living anymore by simply hooking up customers to wholesale minutes on T-1s and letting them devise their own increasingly complex solutions. If you look at nex-gen carriers - and that includes both new entrants like Level 3 and Qwest and traditional telecom companies like AT&T that are, in essence, recasting themselves and their business around IP - you really do see a new strategy of presenting a wide variety of service offerings to their subscribers as part of an effort to leverage the kind of ease and flexibility that they get from the IP environment.

Telcos are finding that to really grow their revenue, they're going to have to offer a suite of services that include not just messaging functions but network traffic functionality - such as Virtual Private Networks (VPNs) – that make it easier (and cheaper) for a network manager to hook together his company's main site, branch office sites, telecommuters and client extranets all over the public IP infrastructure. And if you think about it, that's the way the public telephone network began to scale. Lots of people used the PSTN, and because of the economies of scale, the cost of transmitting voice went down (to the point where you can now shop around for the cheapest long distance service) and we all have ubiquitous voice services.

Similarly, over the next few years, service providers will roll out new services that lift the burden of deploying and managing IP network hardware and software off of the network manager's shoulders, allowing companies to use the Internet as a ubiquitous, secure, worldwide conduit for information exchange and communication. Some of these services will be at the provider's location, others will be resold by the provider.

But to forge enhanced services into a major revenue-generating machine for carriers, CLECs and ISPs, many equipment providers will struggle with complex technologies such as SS7, and look to companies such as Dialogic and NMS for assistance is creating a virtual version of the voice network's familiar functionality that will run ghostlike in the IP network. Once this new business-quality service layer of the Internet is perfected – service providers will be able to buy intelligent carrier-class equipment that can handle any and all security and delivery protocols.

Lucent is already preparing the new telecom infrastructure by providing Level 3 with its Lucent Technologies Softswitch, a Bell Labs-developed software switch that emulates circuit switching in software (hence the name "softswitch") for IP networks that combines the features customers expect from the PSTN with the inexpensive flexibility of IP technology. With the Lucent Softswitch, Level 3 will provide a full range of IP-based services indistinguishable in quality and ease of use from services on traditional circuit voice networks.

The companies have also agreed to collaborate on future softswitches and gateway products to support next-generation broadband services for business and consumers that will combine high-quality voice and video communications with Internet-style web data services.

Both Lucent and Level 3, incidentally, are founding members of the International Softswitch Consortium, (www.softswitch.org) which promotes open standards and protocols connecting distributed hardware and software that's designed to meld the traditional phone network into IP networks, and to stimulate development of value-added services for service providers and network users.

After the new paradigm takes over, instead of physically installing equipment at the customer's site, the service provider will simply "turn on" incremental services such as a VPN, firewall, web conference, voice, fax and video over IP services from a single "box" of network equipment.

Still, something resembling a GUI app-gen must be developed so that the customer has the option of doing his own customization and provisioning from his own site, otherwise companies will suffer the same kind of "lack of control" frustration that drove them to customer premises CT long ago.

For example, George S. Faigen, Senior VP at RONIN Consulting Services (Princeton, NJ – 609-452-0060, www.ronincorp.com), tells me that a wealth of companies have sprung up to offer enhanced services that were formerly used as a customer premises application, such as the wares from SAP or Peoplesoft. Faigen says: "I think this idea is a failing proposition primarily because they are trying to run only a plain vanilla application, since that's the best economy of scale they can achieve. However, most of the buyers want a tailored SAP or Peoplesoft environment. And

the degree to which these companies are able to tailor the service is minimal and is actually not a good mix compared to what people want."

It's true that IP is a wonderful, stable, neutral platform to develop enhanced services (so much so that it has allowed companies such as Cisco to disrupt the staid telecom marketplace) and has encouraged the transformation of the circuit-switched network into an IP hybrid – witness Lucent's deployment of a new class of telephony switches that are IP based – and this all suggests that it's just a matter of time before the current PSTN / IP hybrid network is replaced by an completely IP routed network.

However, all of this technological hubbub involves only the transformation of the infrastructure, which has been partly motivated by great selling opportunities made possible by the tantalizing (though ill-defined) promises always associated with new technology. If Lucent wants to sell more and more lucrative switches (and related services) but has saturated the conventional circuit-switched market, the coming of IP causes convenient confusion and disruption in the market and allows companies such as Lucent to replace a lot of infrastructure with the explosive sales of new hybrid IP switches.

But the more interesting aspect to all this is, what services will be salable once you have this new infrastructure in place? This depends upon both what enhanced services providers think they can sell (bizarre new services are being hatched all the time) and what services customers will actually buy. It's easy to forget that this is contingent on what kind of services are really needed by the public, as opposed to subscribing to a new enhanced service simply because it appears to be new and interesting. Moreover, some enhanced services fail simply because they are just not that "enhanced" to begin with, and don't live up to customer expectations. It's nice to have services such as Caller ID or voicemail, but acceptance of these services are not really a business issue, particularly when one examines the need for individual residential services.

Also, a customer doesn't necessarily have to buy all enhanced services from an ESP. Some simple enhanced services can be handled by cheap equipment on the premises. For example, you could maintain your own voicemail system on your premise in the form of one of the many small, solid-state devices now available. In the long run this can be cheaper than using an enhanced network service.

In the Internet world, where you have increasing convergence of technology, your television can be your voicemail system since it has a speaker and can in theory play back voice messages. Even General Electric microwave ovens allow you to store short voice messages to be played back by other members of your household!

Telephony service providers find the idea of providing data services quite enticing, but since no one knows exactly what to offer, a sort of "shotgun" approach has been adopted, with various trial balloons sent up in the marketplace. Technology is now available to create just about any enhanced telecom service imaginable, but there's still a lack of public interest and even a lack of awareness to overcome. For example, service providers can and do offer wireless data services over a PCS or GSM network, but while we've all had wireless phones for several years now, how many people are demanding to the ability to connect their laptop to the Internet while driving on the highway? None of this has stopped IBM from acting as technology coordinator in a series of deals with mobile operator Vodafone Airtouch, nor has it stopped Oracle from partnering with Cap Gemini, nor has it diminished financial analysts' forecasts that the market for Internet access through mobile devices will reach 500 million handsets worldwide by 2003, of a total of one billion cellular subscriptions.

On the other hand, some enhanced services are destined for immediate public acceptance. Such "no brainers" include Fax-over-IP (FoIP). For many years the process of sending faxes over a private network ("fax-over-the-backbone") has been an established technology well-known among multinational corporations, where employees send faxes to an internal corporate switch which would then transmit the fax over the backbone and then export them back to the European PSTN to fax machine closest to the destination. By renting some bandwidth on the backbone, many companies reduced their fax transmission costs to just two local phone calls, one in the U.S. and one in the destination country (or vice versa). Converting such a system to FoIP merely entails replacing the internal private backbone with an IP backbone. This is an obvious improvement since IP networks such as the Internet are far more ubiquitous and offer a greater degree of reliability than private circuit-switched networks since multiple connections can now be employed instead of just one or two.

Even with SOHOs, FoIP makes sense. If you send a fax and the receiving party's fax machine is busy or doesn't have enough memory to capture a big document, your message is stuck back in your fax machine. But a store-and-forward Internet fax server will work in the background, trying to send again and again until your fax gets through. It will also be cheaper to use than having your fax machine make a long distance call over the conventional PSTN lines.

Internet enthusiasts may argue that the point is moot since the use of faxes "should be" decreasing, being replaced by e-mail for communication. The fact is that the fax industry is quite healthy. Attorneys discovered that a signed document retains its binding legality even if it's faxed, which resulting in a fax "explosion" so large that messenger services in urban areas such as New York found themselves forced out of business. Only "fax-on-demand" services have suffered as a result of the Internet, since documentation can now be made available on websites.

Notice that there's nothing technologically startling about FoIP, yet something that is useful, easy to deploy and cost effective will always outsell something that must rely solely on the "Wow this is really cool" hype factor inherent in the promotional activities of many technology companies, particularly those steered by engineers rather than marketing people.

Thus, the short-term challenge to ESPs and SPs appears to be one of better marketing. The Service Provider must devise or resell eye-popping new services that are valued by buyers who have been sufficiently primed by advertising. The first step is for the public to develop enough awareness and interest to actually purchase or even use such services – after all, some people buy a package of services and restrict themselves to using the most basic function (such as voicemail) not realizing or perhaps being intimidated by the fact that they could be taking advantage of other, more exotic services that they're already paying for.

Before deploying a new enhanced service, a provider should ask three questions:
- Does it save money (or fulfill some other need or solve some problem)?
- Is it easy to use?
- Does it work reliably?

Ultimately, the success of IP enhanced services depends upon what the fickle public is going to do with its money, not just how cheaply a new gee-whiz technology can be deployed. No doubt there will be all kinds of unusual and repackaged services offered (e.g. VoIP voice chat rooms) and no doubt there will be some spectacular shake outs. No one knows yet who or what combination of technologies will win.

Still, in the near future, the telecom world should become quite exhilarating as new and exciting services are offered in the world's biggest free market – which consists of literally anybody with an IP connection: Global carriers and service providers, long-distance carriers, wireless companies, BOCs, PTTs, independent telcos in emerging and deregulated mar-

kets, cable companies, next-generation telcos, ISPs, CLECs, Internet portals and independent Internet-based service providers.

Here's just a smattering of what's struggling out there in the telecom jungle – pure IP services, hybrid PSTN / IP systems, flexible software switching platforms, co-located systems and services, and every combination and permutation of everything else...

Airspan Communications Ltd., (Cary, North Carolina - 919-319-0761, www.airspan.com) recently unveiled their PacketDrive technology, which delivers high-speed, low cost wireless Internet access to residential and SOHO subscribers in rural, suburban and urban environments. PacketDrive technology is offered as a seamless extension of Airspan's AS4000 System, which provides telecom SPs with Wireless Local Loop (WLL) services, including high quality voice, leased line data and ISDN. PacketDrive terminals can be added incrementally to an AS4000-based network by using Airspan's existing CDMA air interface, with no changes at the central terminal base-station, an innovation that permits smooth network evolution for existing and new Airspan-based network operators.

Allot Communications (Burlingame, CA - 650-401-2244, www.allot.com) a pioneer in policy-based bandwidth management and networking solutions and Bridgewater Systems (Kanata, Ontario, Canada - 613-591-6655, www.bridgewatersys.com) a leader in policy-based IP service provisioning and control software, have teamed up to provide a synergistic Policy Management solution at Bell South. Together they have enabled Bell South to ensure the quality of VPN, VoIP and other enhanced applications for corporate customers.

The combined Allot-Bridgewater solution will be offered to other Service Providers (SPs). It brings together the tools necessary for telcos and ISPs to offer managed carrier-grade IP services to their customers. The joint solution speeds deployment of new IP data and voice services by uniting business policies with network service policies, for all customers, across all IP services. SPs can now more readily offer value-added services to corporate customers including differentiated Quality of Service (QoS) and service prioritization, key components for mission-critical and time-sensitive business apps such as Virtual Private Networks (VPNs) and VoIP.

Allot is known for its bandwidth shaping products, the AC200 and AC300, as well as the Allot Policy Manager for administering the associated network policies, such as traffic prioritization and guaranteed bandwidth. The Allot Policy Manager solves the problem of creating scaleable, dynamic network policies by integrating with existing organizational directory information via LDAP Network policies, driven by an organization's existing databases that are dynamically updated and automatically loaded into Allot enforcement devices to match organizational goals with network resources.

Bridgewater's WideSpan solution is an integrated policy management and Authentication Authorization and Accounting (AAA) system that consolidates network-wide information about users and their service profiles and policies. It correlates business and service policies with network-level policies defined in the Allot policy management system. Together these products give Service Providers the ability to link business policies, such as user, department, organizations and domain information and priorities with network and service policies, such as QoS, and what application (e-mail, web, etc.) a user can access when.

The Bridgewater and Allot systems share user and policy information via LDAP ensuring synchronization and establishing a single view of the SP's customers and privileges. This integrated view of the users and all the services they can access eliminates the need to manage policies on a per device or per application basis.

APEX Communications (Sherman Oaks, CA - 818-379-8400,

www.apexvoice.com) gave us a preview of their new Media Gateway for Enhanced Services on what APEX calls the IP-based "Next Generation Network." The Media Gateway uses APEX's own OmniView GUI app-gen service creation tools, so Network Service Providers (NSPs) can generate their own custom services or they can use APEX's service-ready solutions such as unified messaging, prepaid VoIP and web-enabled e-mail for their IP-based network.

APEX Media Gateways are available for H.323 and TCP / RTP supporting H.245, G.711 and G.729A voice codecs.

With the combination of its Media Gateway Server for IP networks and traditional products for the existing networks, APEX now provides a tightly integrated, bridged environment for enhanced services.

Rather than selling the service directly, APEX sells the platform to SPs and spends about three months training them how to use it. I asked APEX president Ben Levy if he tends to sell his packages to telcos, ISPs, or CLECs. Levy says: "What we find is that the ISP buys these gateways. Now they have a bunch of gateways and they really don't know what to do with them or a CLEC. So what we do is to partner with the virtual switch makers because they're really important in this next generation network. So we'll bring a virtual switch in with our equipment which basically turns the IP network into a generic telephony network that can be used for much of the same things that a Time-Varying Media (TVM) resource network would be used for."

The basic gateway is a 19" rackmount holding Dialogic cards. However, simple gateways can be configured that have protocols so you can set up an enhanced services platform that doesn't really need to have anything in it other than a hard disk and Ethernet connections to the local LAN. Such a device cuts down the overall cost but it cuts down the flexibility of what the device can talk to. In this case a gateway is talking to a softswitch and cajoleing it into hooking up its datastream or RTP stream over to the enhanced media gateway. Since the DSP is already at the edged device, the gateway itself can take the media stream, do DSP encoding on it, and turn out the encoded voice call to SONET.

APEX generally avoids configuring systems below 24 lines (unless a customer wants to do some tests with the system), but using multiple machines you can scale up to about 150 ports per box. If you're working with a softswitch, you can set up routing mechanisms so you can string multiple APEX boxes together over a 100BaseT LAN.

The platform is sold as a package but the package price is based on per port pricing. APEX is opening up its billing platform interfaces to talk to these virtual switches as well, so that they can now do such things as verifying Caller IDs and things like that. There will be several different packages that APEX will be selling into these new networks.

APEX is looking for strategic partners and OEMs for its Media Gateway products.

ArelNet's (Yavne, Israel – 972-8-942-0880, www.arelnet.com) flagship product line – the i-Tone Suite – provides complete architecture with a platform of voice and fax over IP applications for carriers, ISPs and companies.

ArelNet's i-Tone is a Windows NT Server based IP gateway using Dialogic boards that can do point-to-point voice and fax messaging as well as perform as a turnkey solution for SPs wanting to offer global IP-based enhanced services such as phone-to-phone, fax-to-fax, PC-to-phone, PC-to-fax, e-mail-to-fax, fax-to-e-mail, fax broadcasting and enhanced messaging for mobile subscribers.

ArelNet's system is used by carriers, next-generation telecom companies, Internet Service Providers and value-added service providers. All are attracted to i-Tone since it can handle real-time as well as enhanced communications using a single hardware platform. The i-Tone Gateway can

act as a gateway between the PSTN and TCP/IP based data networks. There's also a Virtual Private Network (VPN) module that gives corporate subscribers many of the advantages of installing a VPN as if it were an autonomous network, but without the standalone infrastructure costs.

Call Sciences (Edison, NJ – 617-926-6665, www.callsciences.com) is now making available its advanced communications and messaging services to network operators and service providers in the Southeast U.S. and the Los Angeles area via new service bureaus.

Initially, the company will market and sell its services directly to ISPs and CLECs, allowing them to offer Call Sciences' telecom services while they focus on their own core business competencies. These enhanced telecommunication services include unified messaging, calling cards, Personal Assistant (Call Sciences' flagship single number service), and Unified Call Management, a premium call management and unified messaging application.

Call Technologies (Reston, VA – 703-995-2000, www.calltechnologies.com) provides solutions to the service provider market. Their Dialogic-based Call Courier voice mail messaging platform has been expanded to include unified voice fax and e-mail messaging, and is now called Call Courier Unify.

Call Courier Unify gives SPs a system that can take advantage of their existing POP3, IMAP4 and SMTP compliant e-mail systems, and it's even VPIM compliant for voice message exchange with other VPIM compliant voice mail systems. Subscribers can use their existing web browsers and e-mail clients to implement unified messaging in conjunction with Call Courier Unify so they can retrieve voice mail, e-mail and faxes from any phone or PC.

A user-friendly web-based provisioning tool helps subscribers set up the system to consolidate messages from e-mail and voice mail accounts. Subscribers can also provision their personal address book via the web. Subscribers can access, listen to, and reply to voice mails and e-mails using a phone with a standard DTMF interface. E-mails are accessed via an interface similar to a standard voice mail interface and are heard with a text-to-speech engine. Voice mails accessed via an e-mail client appear in the subscriber's mailbox as an e-mail with a WAV attachment.

All messages can be received from a central Call Courier Unify web server with a standard web browser, and subscribers can pick up any message from any client that has Internet access.

Centigram Communications Corporation (San Jose, CA - 408-944-0250, www.centigram.com) proclaims that they have two goals: Developing integrated enhanced services systems specifically for small to medium size wireline and wireless service providers, and enabling their customers to differentiate their offerings and increase value, thus preventing dreaded churn.

Recently Centigram announced a series of "unified communications" products to boost the popularity of personal communication services. Centigram's new product offerings include: PC-based unified messaging through a family of integrated voicemail, fax and e-mail products; one person / one number services through Smart Forwarding call forwarding; Internet and telephony integration through a family of web-based Internet call management products; Internet and wireless integration through a new generation of the company's popular Short Message Service Center (C-SMSC); and extensions to Centigram's Series 6 services platform that support multi-service integration.

You'll perhaps notice that the name of Clarent Corporation (Redwood City, CA – 650-360-7511, www.clarent.com) keeps popping up in this industry. Clarent is a major provider of scaleable, IP telephony products to carriers and ISPs. Clarent's intelligent architecture and the

Clarent Command Center enables Clarent products to route, manage, inter-connect and terminate high volumes of calls for service provider customers including the world's largest long distance telecom companies. According to a Cape Saffron report published in 1999, more minutes travel across Clarent-enabled networks worldwide than those of any other equipment supplier.

Comverse Network Systems (Wakefield, MA – 516-677-7200, www.comverse.com) is one of the largest suppliers of enhanced services platforms to wireless and wireline SPs, having swallowed up Boston Technology not too long ago. Their new InfoPeeler wireless Internet portal, part of Comverse's mobile data and wireless Internet strategy, enables wireless SPs to create information and content services, allowing subscribers to receive info from the Internet and other sources, and then have it delivered to the device of choice.

Through the InfoPeeler Internet portal, wireless operators can enable their subscribers to access information based on their personal interests and preferences. The InfoPeeler links wireless service subscribers to info from various sources, such as the Internet, news agencies, corporate Intranets, and commercial content providers via the subscriber's wireless handset. Stock quotes, weather reports and phone directories can be delivered either immediately, in "pull" mode as a response to a subscriber request, or in "push" mode according to a pre-determined schedule.

The InfoPeeler portal is device-independent, so subscribers can have information delivered to any convenient device. Services can be made available both to subscribers with Wireless Application Protocol (WAP) and non-WAP handsets. A subscriber can access services using a mobile handset and have content delivered to the wireless device, to a fax machine, to an e-mail address, to voice mail, or have the information read over the phone via text-to-speech. Moreover, when packaged with Comverse's Te@GO voice activated services module, subscribers can navigate content services via spoken commands.

Concord Technologies (Seattle, WA – 206-256-7500, www.concordfax.com) provides leading edge messaging services to clients ranging from SOHOs to Fortune 500 companies. Through the integration of the PSTN with the Internet and PCs, Concord continues to push the envelope of enhanced messaging.

Concord's unified messaging service, Concord Universal Mailbox, provides convenient access to voice, fax, e-mail and pager messages via the Web, a phone, fax machine or by using WinFax PRO software. Businesses can offer Concord Universal Mailbox to mobile or telecommuting employees to improve communications with both colleagues and customers.

Recently, Concord Technologies and Optus Software joined forces and announced the availability of Concord Internet Fax for Optus' FACSys Fax Messaging Gateway server. By combining Concord's Internet fax service with a top-selling fax server, FACSys customers can use their existing Internet connections to increase outbound fax capacity without having to buy and install fax boards and pay monthly charges for dedicated fax lines, since faxes are now transmitted via the Internet from customers' FACSys servers to Concord's global fax network, which delivers the faxes to their final destinations. Besides hardware cost savings, Concord Internet Fax for FACSys reduces fax transmission costs by up to 40 percent. The service is available to both new and current FACSys customers.

Interestingly, Concord has also added an XML-based Internet fax gateway platform to their package that incorporates Secure Sockets Layer (SSL) encryption technology to meet the security requirements of business customers. SSL encrypts the entire fax document, including the sender and recipient information, rendering the fax unreadable as it travels over the Internet to Concord's network.

Darwin Networks (Cincinnati, OH – 513-721-2300, www.darwin.net) provides high speed data communications solutions and Internet access to owners of apartment complexes, hotels, small and medium-sized businesses, office buildings and telephone companies using wireless and Digital Subscriber Line (DSL) technologies. Darwin Networks offers a full complement of marketing and support services to assist hotels, property managers, small and medium-sized businesses to market and implement the new service, including web page content development to promote high speed Internet access, and training programs.

U.S. West is offering customers Internet Call Waiting using eStream server technology from eFusion (Beaverton, OR – 503-207-6300, www.efusion.com). Internet Call Waiting lets you receive a normal telephone call while connected through a single phone line to the Internet. Incoming PSTN calls are forwarded to the eStream application server, which notifies you of the incoming call. You can accept the call, in which case the eStream server will complete the call through the Internet connection to the your multimedia PC. You and the caller can now talk while you continue browsing. You can also forward the call to a network voice mail account or to an alternate phone, or disconnect from the Internet and accept the call on your regular telephone.

Internet Direct Dial is an application that allows Internet subscribers to place outbound calls to a conventional PSTN telephone number during an Internet session. You use a PC dialpad to communicate over the Internet to the eStream server, which completes the call using the PSTN. Internet Direct Dial also supports DTMF tones, so you can navigate through IVR or mail applications while on an Internet Direct Dial call.

The eBridge Interactive Web Response (IWR) system is an internet telephony gateway system that offers eFusion's Push to Talk (PtT) capability which allows companies a Push to Talk button on their site that allows consumers to push the button and talk directly to a live customer agent to answer questions about products or services.

e~Telesales (Yarmouth, ME – 207-846-6000, www.etelesales.com), is a web-enabled call center geared for web site support that offers 24-hour, toll-free service to Internet retailers, eCommerce sites, and traditional "brick and mortar" organizations worldwide. Third-party customers can talk directly with highly trained e~Telesales agents who specialize in sales, support, and customer satisfaction.

e~Telesales and eFusion has formed a strategic alliance to implement eFusion's PtT technology within the e~Telesales call center infrastructure. Thus, e~Telesales will provide Internet retailers, eCommerce sites and traditional organizations embarking upon an Internet sales strategy with a third-party, voice-enabled customer service solution.

e:go Systems, Ltd. (New Basford, Nottingham, UK – 011-44-0-115-919-2030, www.e-gosystems.com) is a vendor independent, seamless, global enterprise messaging service that's received backing from some powerful communication companies. e:go's messaging portal evolved out of TeleConnection Europe, the UK's leading private provider of bureau based voice and fax messaging solutions.

The service is an interesting hybrid system, delivered to enterprises globally, via customer premise installed Clarent VoIP gateways, and integrated to existing telephony and data systems through e:go's local "in country" strategic partners. Customers then use the service as they would an in house messaging platform, but are routed over the Internet to a central, multi-lingual, locally time-stamped global messaging system where they can interact with other e:go customers worldwide at no extra cost and with full messaging functionality (customers can currently choose from advanced voice and /or fax messaging. and trials are currently concluding for a range of unified messaging solutions).

Here's how it works:

1. Users place calls on their telephone system.
2. Transmission goes through the current phone system – calls are routed to the VoIP gateway.
3. The voice is then digitally compressed by voice processing technology and grouped into Internet packets.
4. These packets are transported real time through the company's LAN routers to the Internet or an Intranet.
5. The packets are then sent over the Internet or Intranet network.
6. The packets are received by e:go, decompressed and presented as a full duplex call. The voice quality is guaranteed by combined voice processing and echo cancellation technologies.

e:go combines the benefits of the Internet and the PSTN to provide a global, open and seamless messaging service. Unlike other solutions providing a single messaging "in" box, e:go provides full interoperability to any other e:go customer worldwide, regardless of their telecom or ISP, without any capital investment. This allows organizations to integrate internal communication networks throughout the world with a single solution and also allows customers and suppliers to communicate on the same messaging network, providing all sorts of operational benefits and cost savings.

The AT&T Global Clearinghouse (AT&T GCH) recently began offering their members e:go IP voice messaging services as a value-added application. e:go IP voice messaging now forms part of the overall AT&T GCH service portfolio. AT&T GCH (www.ap.att.com/clearinghouse) enables ISPs to offer a range of Internet communications services, including telephone-to-telephone voice and fax services, and a variety of value-added applications such as VPNs. As a member of the AT&T Global Clearinghouse, an ISP can terminate voice traffic to over 220 countries, saving time and cost in negotiating and managing agreements with many individual ISPs.

E-Net's (Germantown, MD – 301-601-8700, www.datatelephony.com) Telecom 2000 boxes allow voice to be transmitted over data networks, including the Internet, IP, ATM, or Frame Relay. e-Net's suite of products is designed to allow Internet Service Providers, CLECs, and cable TV operators to offer competitive long distance service over their data networks. By integrating hardware, software, and billing solutions, e-Net is able to offer customers a complete VoIP solution.

e-Net and IXC Communications have agreed to jointly develop and

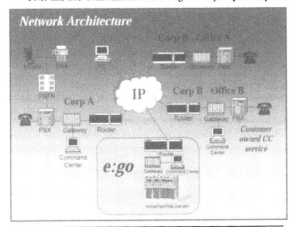

Here we see how a unified messaging service is implemented with e:go. The key ingredients are a phone call into the on-premise originating VoIP gateway, connecting to the Internet or a private packet switched network and then a terminating VOIP gateway.

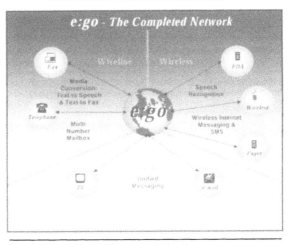

Unified Messaging, e:go style, can tell subscribers about new messages almost anywhere, via PC, PSTN telephone, mobile handset or pager. Using a PC, a mobile handset or a PSTN telephone, subscribers can reply to messages, forward messages and send them. Subscribers also can also convert messages from one media to another. They can hear e-mail messages on a telephone, and send and receive e-mail and faxes using a mobile handset.

market the Internet Telephony services of e-Net's wholly owned subsidiary, ZeroPlus.com, Inc., to consumers and businesses around the world. IXC will provide access to its 13,000-mile fiber optic Gemini2000 network, network support, co-location facilities, as well as marketing and sales support. ZeroPlus.com will provide the software and VoIP expertise to integrate PC-to-Phone calling from the current software and develop fee-based enhanced services such as call forwarding, voicemail, and unified messaging, all of which are anticipated to be revenue-generating opportunities.

IXC's Gemini2000 network is a coast- to-coast, next generation Internet backbone that carries both commercial and research community traffic. The OC-48 network was built to eliminate Internet congestion by offering communications 20 times faster than today's Internet in an effort to ensure Quality of Service (QoS) and super performance. The network's technology is supported by Cisco Systems, Newbridge Networks, and Nortel Networks.

One of the oldest names in the industry, I-Link (Draper, Utah – 801-576-5000, www.i-link.net) already delivers to over 70,000 people enhanced telephony services that are real and available today.

V-Link 3.0, is the latest offering of I-Link's communications service. I-Link has built a nationwide backbone for their Voice over IP / ATM telecom network that's independent of the Internet.

I-Link offers long distance and IP Telephony Enhanced services which includes 3.7 cents per minute long distance rates throughout the continental U.S., free business or personal 800 number service billed at less than 4 cents per minute, voice mail, call waiting "call whisper", call screening, call hold with music on hold, call transfer, do not disturb, speed calling/speed dialing, direct login, auto forwarding, personal operator, voice and fax message notification, pager notification, 12 way conference calling, one number follow me, call boomerang, voice and fax on demand, call transfer, Internet access to messages from anywhere in the world, unified messaging, i.e. integration of voice, fax and e-mail messages (allows for attaching fax or voice messages to e-mail), provides fax capability using current residential or business lines without interfering with voice calls.

I-Link also provides voice and fax broadcast features which allow

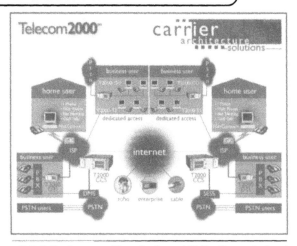

e-Net's VoIP gateways used to work only on a LAN. Now they can be found on IP networks such as IXC's Gemini2000.

you to broadcast voice and fax messages to hundreds of individuals at the same time without having to manually call or fax each person individually, which saves lots of time. Conference calling is also just a few cents per minute per line.

I-Link Inc. and Casio PhoneMate, Inc. has announced that they will manufacture and jointly market I-Link's new NetLink-IP telephone line capacity expansion device (referred to during its development as "C4"). The NetLink-IP device, developed by I-Link's wholly-owned subsidiaries, New Jersey-based MiBridge, Inc., and Israel-based ViaNet Technologies,

Integral Access' PurePacket lets carriers deliver voice and data services over a single link via uniform packet transport.

Ltd., gives a home or business simultaneous capacity of up to 24 phone lines (the equivalent of a T-1), as well as a constant high speed Internet connection on a single DSL, ISDN or cable modem line.

This cuts the number of phone lines you're currently buying from local carriers, such as U.S. West and Bell companies, etc., slashing your telecom bill. For example, 12 phone lines with your local carrier costs about $600 per month compared to $30 per month for a DSL line that yields up to 24 phone lines plus a constant high speed Internet connection, all thanks to I-Link's Netlink-IP line expansion device.

The PurePacket platform is a multi-service access platform from Integral Access (Chelmsford, MA - 978-256-8833, www.integralaccess.com) that allows competitive carriers to cost-effectively deliver multi-services (voice, tiered data, IP telephony) over a single link (DSL, T1, Fiber, etc.). PurePacket allows carriers to deploy up to 10 Mbps of fully flexible access bandwidth for integrated voice and data services to individual end users through existing copper and leased facilities.

The PurePacket family is based on Integral Access' PurePacketPower architecture, where all traffic is routed, transported and switched in a packet format, achieving bandwidth efficiencies as well as carrier savings.

The entry-level configuration for a central office PurePacket Node is less than $15,000. Customer-located equipment that delivers 24 POTS lines and 1 x 10/100BaseT over a packetized standard leased T1 costs less than $2000. Entry-level customer-located platforms are available for less than $1,000.

OmniLYNX from Intelect Network Technologies (Richardson, TX - 972-367-2100, www.intelect.com) is a protection-switched, multi-protocol, multi-point, multi-service access platform. With network management software supporting both TL-1 and SNMP, and a NEBS-compliant platform, OmniLYNX offers an huge range of protocols and services in a single shelf for both private and public networks.

The OmniLYNX architecture is designed to support SONET OC-1/OC-3/OC-12, SDH STM- 1/STM-4, xDSL, ISDN-BRI, ATM transport, IP-over-Sonet, DS1 and DS3. Services offered at the customer-premise include FXS/FXO/2W & 4W E&M, low speed data (RS-232, RS-422, RS-449, V.35), Ethernet 10/100M and Video.

Intelect has steadily extended its technology into the public network arena with a product suite that changes the business model for the Competitive Local Exchange Carrier (CLEC) industry.

"The technology we've developed for OmniLYNX is so revolutionary that we are renaming our entire product suite, with products under the OmniLYNX name specifically designed for both our traditional private network customers as well as the public network sector," said Bill Barnett, president of Intelect Network Technologies.

"We've led the private network sector for some time with innovative solutions," Barnett said. "Now we are extending that technology to the public network sector and adopting the OmniLYNX brand for all our products. We are convinced that OmniLYNX is the most flexible, cost-effective solution available for both public and private network providers." With an intelligent OmniLYNX product solution, CLECs can now seamlessly integrate voice, data and video services into a single piece of equipment, a first for the public network industry.

"OmniLYNX's unique architecture and superior bandwidth management will allow a CLEC to deliver more services with more bandwidth faster at lower cost than their competitors," said Barnett.

"CLECs depend on being able to offer bundled services at lower cost," said Barnett. "Today's customers are demanding more than the traditional voice services over T-1 and DS3 connections. Today, customers want high-speed Internet, LAN service, videoconferencing, ATM and ISDN connections too. "OmniLYNX does all that and more in one solution over existing fiber optic networks," Barnett explained. "OmniLYNX seamlessly integrates voice, data and video services into a single piece of equipment, eliminating the need for additional auxiliary equipment such as channel banks, terminal servers, video codecs and switches."

The OmniLYNX system is completely modular providing great flexibility for future expansion. Also, there's no limit on the combination of services that can be offered in each shelf or the number of shelves that can be installed in a network.

iPhonEX, from MIND CTI (Englewood Cliffs, NJ – 201-569-6967, www.mind.co.il), a Windows NT based billing and customer care system, is designed to provide ITSPs, telcos and corporations with a complete billing and analysis solution for voice and fax services over the Internet. The iPhonEX Internet Telephony Billing system offers real-time cut-off of calls when call limit is reached, creation and management of pre-paid calling cards, individualized customer rate tables and flexible fax charge options by page, duration or priority (real time or store & forward). iPhonEX allows the operator to generate invoices (both automatic and on demand), and manage payment and customer balances. iPhonEX also provides monitoring of the load on each Gateway and line and keeps track of excessive use, including fraud alarms.

Perhaps the #1 player in the carrier-class Fax over IP market, NetCentric (Bedford, MA -781-685-5200, www.netcentric.com), whose flagship product is FaxStorm, a highly-regarded package that's the main reason NetCentric has been able to secure partnerships with 12 carriers such as GTE, Singapore Telecom, Intermedia and US West.

NetCentric supports three basic services: E-mail faxing (the ability to send and receive from e-mail); enhanced fax machine (which is based on the concepts of the "never busy" fax machine and guaranteed fax delivery, so that the faxes send from a fax device are received in a network and are then delivered to a fax machine, functions such as multiple retry and automated routing will take place); and lastly volume services (things like fax broadcast and work flow faxing – where NetCentric can integrate their package right into enterprise applications). All of these solutions have a store and forward component to them.

SALIX Technologies (Gaithersburg, MD - 301-417-0017, www.salix.com) is carving out a name for themselves with their ETX5000, based on their FleXchange Architecture, that enables SPs to offer next-generation voice, fax and data services over a converged PSTN-IP network infrastructure. When SALIX launched the ETX5000, they also launched the idea of Class Independent switching, a new model for building service provider networks without the restrictions and cost associated with traditional Class 4 and Class 5 switches, made possible by its unique Service Provisioning Interface (SPI). The ETX5000 can support multiple applications from a single switch including VoIP, IP Local Loop, Internet Call Bypass, Tandem Replacement and others.

SALIX's first customer is an interesting CLEC called International Long Distance Corporation (ILDC) that will be the first provider to use a Class Independent ETX5000 switch internationally. ILDC will offer enhanced services such as Centrex and unified messaging as well as remote backup and disaster recovery to small and medium size businesses . The first deployment in Q1 2000 included a switch in the U.K.

Meanwhile, Saraide (San Mateo, CA - 650-522-1500, www.saraide.com) delivers Internet based wireless data services to carriers in North America, Europe and Asia, including Microcell, Omnitel, KPN, KDP and diAx. Wireless subscribers receive customized, time-sensitive, location-based information services such as e-mail, news updates, stock quotes, travel information, calendars, contact lists, personal banking and electronic commerce. This information can be delivered to any mobile device, over any air inter-

face, including GSM, CDMA, FLEX, GPRS, and TDMA. Saraide's range of services can turn any wireless device into a mobile information portal.

Recently Saraide has teamed up with EDS (www.eds.com) a multi-billion dollar company that's a leader in global information technology services. EDS will apply its comprehensive life-cycle services to deploy and support the operational architecture for the global delivery of Saraide's wireless data services in 50 countries.

The Sonus Open Services Partner Alliance (OSPA) from Sonus Networks (Westford, MA – 978-692-8999, www.sonusnet.com) enables carriers, ASPs, third-party independent service developers (ISDs) and telephony system software providers to rapidly develop and deliver competitive customer services and applications based on the Sonus Networks Open Services Architecture (OSA). The Alliance works closely with vendors developing value-add services, including industry leaders such as Priority Call Management, eFusion, IPeria, NetCentric and many others.

Spring Tide Networks (Boxborough, MA - 978-635-3739, www.springtidenetworks.com), makes a new kind of networking switch for telecom and Internet service providers. It makes buying advanced IP services, like VPNs or firewalls, as easy as buying ordinary phone services like call waiting or Caller ID. Spring Tide is developing the carrier-class network equipment enabling SPs to easily deploy new, value-added IP services integrating voice, data and video. Major telecos and ISPs can now offer corporate and large enterprise customers reliable, secure virtual private network (VPN) and new IP network services without high costs and complexity.

Spring Tide's new kind of carrier class, fully NEBS-compliant, 19" rackmount IP service switch has proprietary cards built for an ATM switching fabric. It's called the IP Service Switch 5000, and it complements existing Internet access and backbone equipment while providing a single service delivery point for provisioning and billing new services. Spring Tide defines IP services perhaps a little differently than their competition who just build VPN boxes that do tunneling or the IPSec security protocol or some combination of those. Spring Tide does virtual routing on a customer by customer basis. They do support IPSec and all the tunneling protocols, but they believe that the right way to define IP services is figuring out the appropriate use of many technologies.

For example, they have customers who are carriers who sell to the Wall Street community. These providers have a bunch of customers with CPE routers, all of who want to buy an IP managed service so they get an ATM pipe from a "hairbox" which comes into various virtual routers in the Spring Tide box, all of which can have overlapping address spaces grouped into private networks that shouldn't go out onto the Internet. Spring Tide's package can keep all of that straight. Spring Tide can hook them up to quality of service pipes and the ATM network on the trunk side and let the service provider bill for all that.

If the customer also has locations they'd like to hook up to their private network that also needs to be secure, then those locations can be connected to the virtual routers via IPSec. Spring Tide could also terminate thousands of users, either over PPP, LTPP, PPTP, or IPSec. So they essentially maintain a kind of connection-oriented overlay over the IP network. All of this simplifies things for MIS managers on the enterprise side, because they can go to the carrier and buy it as a service. Spring Tide's switch keeps all the routing tables per customer separate from the other customers. So the carrier gets economies of scale by having one box that can handle many different customer routers in virtual form, and all the customers have the security of not having their networks accessible to anybody else. Plus they have standard IPSec and the other security protocols.

The system can instantiate 1,000 virtual routers in one box. Spring Tide's 19" rackmount has a 14-slot chassis. The switching fabric takes up four slots, and two control processor slots leaves eight board slots ("switch modules") for handling data. The system has enough capability to handle over 100,000 IPSec tunnels simultaneously with switching between them.

The boards are bigger than the PCI form factor, more on order of VME or cPCI boards, but they aren't actually built to a standard bus spec because they plug directly into a network switching fabric. It's a sort of mid-point architecture which is hooked to a control bus so the system can sense when something is wrong and can execute a failover from one switch module to another if necessary.

Synchrony is a subscription service from an ASP called (appropriately enough) Synchrony Communications (Cincinnati, OH – 513.588.5500, www.synchronyinc.com). Synchrony is a one-stop contact center / web-based electronic relationship management (eRM) application that synchronizes all of your customers' phone, fax and Internet communications into a single view on your agent's PC. Synchrony combines advanced top-of-the-line solutions from Oracle, Cognos, Genesys, InFact, ISS, Lucent, Cisco and others. You get all the best "world class" technologies in a single package; then Synchrony manages and upgrades it for you.

By turning a call center into a one-stop contact center, Synchrony lets your customers call, fax, e-mail, Web chat, voice-over IP, even service themselves and contact an agent when they choose. Regardless of the method of contact, Synchrony routes each and every customer interaction into a single, synchronized source. It takes between 30 and 90 days maximum to get up to speed with Synchrony.

INtelligentIP and the INtelligentACD from Telecom Technologies (Richardson, TX - 972-918-0202, www.telecomtechnologies.com) are two new services that enable carriers to transfer circuit switched voice capabilities to the IP network.

IntelligentIP resides in the networks' control layer and integrates the Advanced Intelligent Network and SS7 control network with the new crop of IP-based network protocols such as SIP and H.323. It de-couples media, control and service layers for distribution of network resources where they make the most sense economically.

The IntelligentACD is for carriers looking to offer an ACD system for operator services and large network-based call center applications. It can be configured for 20 to 1,000 operator positions and includes release link trunk capability, call detail recording and skills-based agent queuing and routing.

ThinkLink (San Francisco, CA – 415-252-6200, www.thinklinkinc.com) is an Enhanced Service Provider (ESP), offering a bundled suite of IP-based messaging and voice services under the ThinkLink brand as well as on a co-branded and private label basis. The ThinkLink product is a web-based integrated messaging service that simplifies users' ability to send and receive voicemail, faxes and e-mail. The free product includes a personal local phone number, in addition to a personal 800 number, free voicemail, e-mail, fax and paging retrieval as well as integrated outbound dialing. They combine this free service with inexpensive, toll-quality calling over a private Voice over Internet Protocol (VoIP) network.

Tornado Software Development (Manhattan Beach, CA – 310-546-6319, www.tems.com) offers one of the most impressive unified messaging services, the Tornado Electronic Messaging System (TEMS), an Internet-based messaging system for sending and receiving e-mail, voice mail, faxes and pages using one unified web-based mailbox. TEMS also includes a telephone interface designed for retrieving faxes and receiving and responding to voice and e-mail.

You can retrieve TEMS voice mail via the phone, PC or Macintosh; be notified of new voice mail by fax, e-mail or pager; send or forward voice mail to other e-mail addresses via the phone or computer interface.

You can listen to your e-mail on the phone via text-to-speech, compose / send / forward mail using a touchtone phone, and receive page notifications of new e-mail. Messages can be sent to e-mail boxes, fax machines or pagers. You can even do a fax blast to multiple recipients.

As a TEMS subscriber, you get a full-service Web-accessible address book, and online member services that include an instant account balance and personal profile management

Best of all, it's just $9.95 a month – there's even a 30-day free trial. USA.NET (Colorado Springs, CO – 719-265-2930, www.usa.net) offers e-mail and advanced message outsourcing solutions for the whole gambit of users: Individuals, corporations, and service providers. With over 13 million mailboxes under their management, they have some of the largest fault resilient storage facilities around, with RAIDs that hold terabytes per box. They also offer some of the best media buys on the Net, through their Net@ddress service and through their co-branded e-mail services with Netscape and American Express – Netscape WebMail and AmExMail.

It's easy to determine which USA.NET service is right for you – just go to their website and select the appropriate subscription link. If you're an enterprise organization, select the Enterprise link and find out if USA.NET Enterprise Messaging is for you. They also offer message outsourcing services for small businesses, Web portals, service providers (such as ISPs) and telecommunications companies, as well as for VARs. They also offer classified advertisers customized e-mail programs. And for individuals, USA.NET offers a permanent and private e-mail service, Net@ddress.

Danny Winokur, VP Business Development of USA.NET tells me that: "When we offer professional messaging to the carrier we are essentially setting them up as a reseller so that they can now offer those capabilities to their own small or medium size business customers. In other words, we have built two tiers. One layer is that we're outsourcing the infrastructure directly for the carrier who may already operate his own set of dial-up accounts for his individual end-user customers.

"We're now giving them a second layer on top of that, which is the ability to actually sell a new enhanced set of services to their small and medium size business customers, which they can control and provision through an LDAP interface or through a web interface that is also available to them. And then they, in turn, sub-delegate control of the messaging 'post office' down to the small or medium size business itself so they can operate and control all of the accounts within their own domain. Another component is USA.NET enterprise messaging that functions in a fashion that's very similar to professional messaging but it adds some very important security capabilities that are of great importance to larger businesses and corporations. We offer our customers SSL encryption and support SSL for all client-server communications, including POP, IMAP, SMTP, LDAP and web (HTTP) communications."

VocalTec Communications (Fort Lee, NJ – 201-228-7000, www.vocaltec.com) and Cisco are now doing joint marketing and development of new high capacity gateways and platforms for enhance services.

VocalTec had the first PC-to-PC "Internet Phone" package in 1995, and their IP telephony was promoted with much fervor by a certain fellow named Jeff Pulver, who promptly became famous. VocalTec now actually has two major product categories: Gateways and gatekeepers. The gatekeeper is sort of the brains or intelligence behind the routing. A gatekeeper gives you the advanced features such as least cost routing, load balancing, call accounting, and authentication and authorization of both the users and the other network elements attached to gateways.

In the joint venture with Cisco, VocalTec is using their gatekeepers to add significantly more functionality to the Cisco gateways. Cisco has about 70% of the router market but they only have a basic gatekeeper that allows basic transmission. VocalTec has been developing gatekeepers far longer and the Cisco power networks can get more juice by adding VocalTec's gatekeepers. VocalTec is excited about selling to the installed base as well as new applications, since Quality of Service (QoS) is another of their gatekeeper specialities.

When it comes to the gateway between the IP world and the PSTN, that's will be a Cisco product. When it comes to the enhanced services to and all the fancy call routing and tracking capabilities, that is the responsibility and capability of the VocalTec gatekeeper.

Another advantage of using a high-powered gatekeeper is that one gatekeeper can manage and control multiple gateways. So the advantage to VocalTec's PC based platform is that you can build a VocalTec / Cisco system that scales superbly for both smaller applications and carriers.

VoiceCue Technologies (Chicago, IL – 773-481-5650, www.voicecue.com) founded in 1993, identified voice recognition as an emerging technology to deliver enhanced services for the telecommunications industry. In one year, the VoiceCue team deployed the first continuous-digit voice recognition platform on the market with its first application, Voice Dial. The next major stride taken by VoiceCue was the deployment of Directory Assistance Call Completion in 1995.

Today, VoiceCue Technologies' mission is to provide network-based voice recognition products and enhanced services to both wireless and land based carriers. These products and services can be used by the carrier to both enhance revenue growth from its current customer base and to attract new customers by offering advanced services such as voice activated dialing, directory assistance call completion, prepaid wireless, and a balance monitor.

Innovox from Voyant Technologies (Denver, CO – 888- 447-1087, www.voyanttech.com) is a platform for intelligent voice applications serving both SPs and now enterprise customers under Voyant's Service Partner Program. In addition to receiving the leading intelligent voice platform from Voyant, businesses get technical support and enhanced services from their choice of Voyant's Service Provider Partners who include Conference America, Gentner Communication Corporation and VStream.

In fact, the author had a phone briefing with Bill Ernstrom, Voyant CEO, which was set-up using Voyant's platform through one of Voyant's customers, VStream. One simply logs onto the visual portion of the conference through VStream by going to www.vstream.com, selecting "conference now" and then "participant." From there, you select "listen on the phone." You input the conference ID number, then, after logging onto the Web, you call 888-742-8686 to dial in to the voice portion of the conference over the PSTN.

Voyant faces an interesting situation as it seeks to expand into the enterprise market but without cannibalizing their service provider business. They also have released a software development kit.

WEB2PCS.COM (San Jose, CA – 408-555-1212, www.web2pcs.com) has launched Mobile Instant Messenger version 1.0, available as a free software download from Web2PCS's web site. This new service will allow a Web2PCS wireless subscriber to enjoy the same kind of instant messaging enjoyed by users of ICQ, AOL Messenger and RocketMail, and remember, the wireless subscriber need not be tied to their workstation to benefit from Instant Messaging.

Mobile Instant Messenger 1.0 is a Windows-based application that allows a Web2PCS subscriber to send text messages of up to 300 characters to any mobile phone, pager, or PDA, irrespective of the wireless carrier. The application's intuitive interface eases the process of Instant Messaging by providing four simple input fields including: From, Phone Number, Carrier and Message. Future versions of Mobile Instant Messenger are being released which incorporate new features such as a server based personal address book, the capability of sending messages to multiple recipients and a real-time wireless chat functionality.

WEB2PCS.COM also offers Email2PCS, a web-based e-mail service that channels incoming e-mail to your digital phone (you must have text messaging enabled) and a wireless Internet service, the Wireless Application Protocol (WAP) search engine, that allows subscribers using the new Internet WAP standard to access WAP-based Internet sites by keyword searches.

Enhanced Services, Conclusion.

One day, Yours Truly, the author of this weighty tome, was looking at the subscriber profile for *Computer Telephony* magazine (where I am Chief Technical Editor) and discovered that, at the time, 12.9% of our paid readership work at a telephony carrier or telephone company, internet access or service provider. That struck me as odd, since until 1997 we hadn't really published much about long-distance resale or service provisioning or billing platforms, and we write about enhanced services platforms primarily from the entrepreneurial standpoint.

Telecom executives obviously (and instantly) made the connection that we immediately didn't – that the "enhanced services" companies we were in fact writing about early on (such as Priority Call Management, Comverse and Prairie Systems) provided (and still provide) real, solid services that can attract users and can be resold profitably by carriers as they strive to meet the ever-changing needs of their markets. Carriers would love to drop their traditional reputation of being simple "bit haulers" and replace it with a new persona of "customer focused value added network and service integrators".

Of course, another explanation for our popularity with the telcos (which, being an editor, I prefer) is that *Computer Telephony* magazine is where all the new technology is being written about. It's not just the "product announcement of the week." If you want to know what the cutting edge is in telecom, *CT* is the place to start.

Historically, many of Miller Freeman/CMP's telecom magazines *(Computer Telephony, Teleconnect, and Call Center)* address the whole food chain: The people who make stuff, the people who sell it, and the people who buy it. We manage to keep things "commonsensical" enough, and focused on technologies, products and the satisfaction of needs.

Increasing coverage on enhanced services for telcos is a good thing. We're fascinated by Central Office environments, co-location, opportunities created by deregulation, etc. It's very necessary to talk to the bifurcated market of the people who are creating and selling new telecom services and the people who are buying them, because everybody eventually has to get on the same page in this new and growing area of enhanced services.

Many believe that enhanced services such as one-number or follow-me calling, web collaboration, virtual offices, virtual call centers, speech recognition, text-to-speech, fax-to-speech and wireless Internet access services will soon be used by everybody.

It couldn't happened at a better time, since computer power is now cheap enough to set up racks and racks of fault resilient PCs in the COs, ISPs and ASPs to act as application and media service distribution points, and companies are brainstorming new products and services every day.

Fortunately, while the functionality offered by "Internet Telecom" is getting more and more complicated, particularly for business, convergence ultimately implies simplification, and getting this stuff to work correctly on a large scale is less of a challenge thanks to the emergence of IP.

But another problem is that, if enhanced services live up to their promise, both individuals and organizations will use telecom devices at an unprecedented rate, and that along with the appearance of high-bandwidth services forces telecom providers to expand their systems' capacity. For example, traditional copper landline-based networks will have to be replaced and/or augmented with new fiber optic, hybrid fiber coaxial, radio frequency signal and satellite transmission networks.

Of course there is more to the problem than simply expanding hardware capacity. New network and wireless protocols are also necessary to enable the delivery of higher bandwidth and, in the case of the IP and wireless networks, address Quality of Service (QoS) issues.

Such network and wireless protocols include Asynchronous Transfer Mode ("ATM"), Switched Multimegabit Data Service ("SMDS"), Frame Relay, Synchronous Optical Network ("SONET"), Synchronous Digital Hierarchy ("SDH"), Global System for Mobile Communications ("GSM"), Cellular Digital Packet Data ("CDPD"), Code Division Multiple Access ("CDMA") and Time Division Multiple Access ("TDMA"). Advanced circuit-switched technologies, such as Common Channel Signalling System No. 7 ("SS7") and the Advanced Intelligent Network ("AIN"), are being deployed to "beef up" networks, giving them built-in intelligence so telecom service providers can develop and deploy new services quickly with a minimum of fuss.

Sometimes these efforts by telecom providers may work at cross-purposes to each other – for example, the rise of the "dumb" (meaning programmable) packet-switched IP data network that must interoperate and perhaps ultimately absorb or be fully integrated with the "intelligent" (SS7 and AIN) circuit switched voice network.

Converged enhanced services driving from the Telco end is hampered by the fact that telephone companies offer wonderful voice services, but their data services are expensive, and so out of date it can be amusing. The telcos always want "strategic partnerships," which means they want third-party enhanced services providers to do all the work for them while they enjoy as much revenue as they can get their hands on. Besides, in the past they've tended to be such clumsy partners that they end up stomping all over the ESPs and tripping themselves while they were at it. Telcos know little about moving data and nothing about distributed processing, and it is distributed processing, both in the intelligence circuit-switched and packetized world, that will be the backbone of all the new information, transaction and interactive entertainment applications.

A telco version of an enhanced service is usually a "value added service," which usually means the network gets in the way of the application doing what it really is intended to do. Telcos can be reminiscent of IBM in the 1970s.

The strategy telcos need to follow is pretty clear: Give the ESP and computer telephony industry the communication service they need so the public will stop stampeding to cable services. What the world needs is a ubiquitous, low cost, fully controllable switched digital service, some kind of xDSL.

Enterprise Computer Telephony Forum

The ECTF is an open, non-profit, mutually beneficial corporation formed to provide the leadership and direction necessary to promote an open systems environment for Computer Telephony.

The ECTF mission is to provide an open, democratic organization to promote the acceptance and implementation of Enterprise Computer Telephony based on *de facto* and *de jure* standards that will promote industry-wide interoperability.

The ECTF operates as a consensus forum led by its members. Membership is available to any and all CT vendors, developers, VARs, end-users and other interested parties. Charter membership of the ECTF included CT industry leaders Dialogic Corporation, Digital

Equipment Corporation, Ericsson Business Networks AB, Hewlett-Packard, and Nortel. The Forum's membership now includes over 70 members. A current membership list can be obtained by calling the ECTF office at 510-608-5915.

ECTF's technical efforts are divided into a number of different areas that cover all aspects of computer telephony interoperability, such as Architecture, Administrative Services (M. series, such as M.100), Application Interoperability (A. series), Call Control Interoperability (C. series, such as C.001), Computer Telephony Services Platform (S. series, such as S.100), and the Hardware Components Interoperability (H. series, such as the H.100 and H.110 specifications).

ECTF Website: http://www.ectf.org/ectf/home.htm

EPOC

A compact (fits on a ROM chip) 32-bit, multitasking operating system, application framework and application suite optimized for wireless devices, particularly portable, battery-powered, wireless information devices such as smart phones, Personal Digital Assistants (PDAs) and Wireless Application Protocol (WAP) enabled devices. Written in object-oriented style, EPOC supports a pen-based graphical user interface (GUI). The GUI layer is actually separate from the OS proper and is "owned" by Symbian, a consortium that includes Ericsson, Matsushita Communication Industrial Co. Ltd., Motorola and Nokia.

EPOC is based on an earlier OS from the British company Psion, an early maker of PDAs.

The name EPOC comes from the idea that the world is entering "a new epoch of personal convenience."

EPOC is comprised of several elements:
- A full application suite for wireless information devices, including messaging, browsing, office, PIM, and utility applications,
- Connectivity software for synchronization with data on PCs and servers,
- A flexible architecture and programming systems to support compact but powerful software on wireless information devices, and software development,
- Software development kits for wireless information device programs in C++, Java and OPL (a simple BASIC-like language), and PC-based connectivity / datasync programming in any language supporting Microsoft's COM, such as Visual C++, Visual Basic, Delphi, etc.,
- OEM tools for building and localizing wireless information devices.

EPOC has competed successfully against Microsoft's Windows CE and the PalmOS used by 3Com's PalmPilots. It is a highly efficient and incredibly stable operating system. It's said that the Psion Series 5mx (the first palmtop computer to support the EPOC runtime environment for Java) could simply be turned on once, and you'd never ever have to reset or reboot it!

ESS

See Electronic Switching System.

Extranet

A virtual network that uses authentication and passcodes to allow outside users to gain access to a company's Internet-networked web, e-mail and HTTP servers. The extranet concept in recent years has been superseded by the Virtual Private Network (VPN).

Fault Resilient Computer

A fault-resilient computer is the "computer" in "computer telephony." Both fault-resilient and fault-tolerant computer systems have "high availability" - they are more reliable than conventional PCs, allowing them to be used in "mission critical" systems (characterized by such things as high port count, multiple applications, use by service providers and Central Offices) or applications where revenue is generated on a minute-by-minute basis, such as prepaid calling card systems and enhanced services systems.

Using a conventional off-the-shelf PC for computer telephony applications can be an enormous mistake, since ordinary PCs are not designed for mission critical applications such as telephony. Standard PCs work most of the time, but most of the time is not good enough for mission critical applications.

When a single PC or LAN fails in a non-telecom environment, it may create some inconvenience to a single user or even perhaps a few departments. The failure of a mission critical CT application, however, results in a catastrophic minute-by-minute loss of revenues, or, in the case of Enhanced 911 systems, puts the public in danger. The best strategy for averting such a situation is to ensure that your platform is designed specifically for CT applications.

A fault-resilient system, for example, will remain "up" about 99.99% of the time, or will suffer downtime for only about 52 minutes a year. A fault-tolerant system will be up "five nines" or 99.999% of the time, which amounts to about five minutes per year of downtime. Some of the latest systems even have "six nines" (99.9999%) availability.

Both fault-resilient and fault-tolerant computers have redundant (and in many cases hot-swappable) storage, power and cooling components. A true fault-tolerant system will also have two CPUs, both of which work on the same application processes, but with some kind of polling process taking place to detect CPU failure.

If a CPU fails, then some form of "automatic failover" process switches control to the healthy CPU.

Stratus (acquired in August 1998 by Alameda, CA-based Ascend Communications) and Tandem (a Cupertino, CA, company now owned by Compaq) made their names developing and deploying this type of true fault tolerant technology. It's well known how to do triple module redundancy or "pair and spare" CPUs. It's also very expensive, the principal reason being that they are complex systems to build. It takes a great deal of engineering time and testing time to perfect such a system. Also, there is a limited market for these systems, so vendors immediately get into a vicious cycle of high prices to recover their investment and therefore the vendor ends up not selling very many systems. Finally, there is the software expense associated with any sort of proprietary solution.

On the other hand, a fault-resilient computer has but a single CPU, which makes it a lot less expensive than a true fault-tolerant system. However, since CPUs only tend to fail if the fans fail, and if fans are monitored by an alarming board or other monitoring subsystem, then a fault-resilient PC can give you nearly the same reliability as a similarly sized fault-tolerant system, but it'll just cost less money.

You can keep adding redundancy and other features to bring a fault resilient computer closer and closer in reliability to a fault tolerant one, but there is a point of diminishing returns where increasingly exotic technologies must do more and more (at greater and greater expense) to achieve that last little bit of ultimate availability. The goal for a business should be to build a system that's not too expensive, yet has quite a few high availability characteristics so that you can run it in such a way so as to get very good results (both reliability and profitability).

Since fault-resilient computers closely resembled fault-tolerant systems but had no single name of their own, vendors were calling their fault resilient machines "industrial", "fault tolerant", "ruggedized", "heavy-duty", "highly available" and other obtuse terms. Richard "Zippy" Grigonis, Chief Technical Editor of *Computer Telephony* magazine, settled on the term "fault resilient computer" and popularized it from 1995 onwards.

Fault-resilient and fault-tolerant computers can keep running if a component encounters trouble and can be serviced rather quickly, in most cases without even having to bring down the system. Since they tend to be 19-inch wide rackmounts, many systems can be stacked upon each other to build systems having tens of thousands of ports. Web servers, enterprise communication servers, 911 systems and many other business critical systems benefit from equipment that can "take a licking" and yet continue to run.

When shopping for a fault resilient or fault tolerant system, keep in mind that these computers should have the following well-known characteristics:

Passive backplane technology. Although there are such things as "industrial" motherboards, most fault resilient computers need more slots than afforded by a conventional motherboard. Computer telephony applications use many add-in cards for voice, fax and digital switching. Passive backplanes provide up to about 25 expansion slots and are easier to upgrade and service than motherboard-based systems.

In such systems the CPU, its memory and various I/O ports are situated on a plug-in card too, called a "single board computer" or SBC. The SBC can easily be replaced or upgraded - you just buy and plug in a new card.

The traditional motherboard itself is replaced with a passive backplane that has nothing on it other than connectors (this is why this technology is sometimes referred to as "slot cards"). The chance of a passive backplane failing is quite low.

Passive backplanes can be "segmented", so that, for example, you can have a fault tolerant system where one SBC or set of telephony resource cards can "fail over" to another set should trouble rise. With current technology you can have up to four segments in one PC enclosure.

Most backplanes and motherboards are based on the Peripheral Component Interconnect (PCI) bus, but now the CompactPCI (cPCI) bus is starting to appear, which is electrically similar to a desktop PCI bus computer, but has card connectors that are far more rugged and reliable, and which also supports "hot swappable" cards (See CompactPCI).

Redundancy. What makes fault-resilient systems genuinely fault resilient is component redundancy. For example, redundant disk drive I/O channels, fans, host controllers, and redundant power supplies that can be swapped out while the system is running ("hot swappable"). Redundant hot-swappable power supplies can also come in load sharing, (also known as "load balancing" or "current sharing") and N+1 ("need plus one") configurations.

In the case of two "load sharing" power supplies, each can be supplying 50% of the total system power plus or minus 20%. So you might have two 500 watt power supplies running in a system designed to consume only 500 watts. Each power supply is therefore subject to only a 50% load, so both hot-swappable power supplies run at cooler temperatures and their life is extended. When one of them finally does fail, the other power supply takes up the full load while you pull out the failed power supply and replace it with a new one.

N+1 architecture essentially means "more than two." In an N+1 configuration, load sharing still exists but the load is now balanced among more than two power modules.

Software can actually be a bigger nuisance when it comes to redundancy than hardware. For routing applications, telephony server redundancy is a well-known, straighforward affair, with copies of the routing application kept on each server so that it can assume the task upon failover with no interruption. But in the case of desktop screenpop middleware, things are more tricky. In order for redundancy to work here, the middleware application on the desktop must support the ability to connect to a different server when the server they are connected to fails. This is difficult to do since the DLL's supplied with the desktop actually perform the communication with the server. Many telecom products do support this redundancy to the greatest extent possible given the limitations placed on it by the DLL's but you will find that most middleware vendors do not support this ability. You will also find that some applications that are on the market today will actually crash if the server goes down. This is not considered acceptable behavior for a mission critical application. An application crashing like this often prompts all the call-center agents to call the support desk simultaneously. A little research on this subject ragarding your proposed middleware can save you and your support staff significant hardship.

RAID. Redundant data storage comes in the form of a multiple disk drive RAID subsystem. RAID means Redundant Array of Independent Disks. It's a disk subsystem architecture that in most cases writes (or "stripes") data across multiple hard disks to achieve fault tolerance, if not continuous availability. One drive in the drive array is often referred to as the "parity" drive, which contains data that can be used to recreate data on any one of the other drives, assuming that the other drives remain operational.

For example, you might have a five-drive array where one drive is designated as the parity drive. If a data drive fails and is then replaced, the drive array controller will rebuild the data on that drive using the parity drive and the other functioning drives.

There are various types of RAIDs, each indicated by a number or "level":

RAID-0 uses disk striping without parity information. RAID-0 is the fastest and most efficient array type but offers no data protection and is actually subject to the opposite of fault tolerance, since the failure of any disk will bring down the system. It's the one RAID level that's not really "RAID" at all!

RAID-1 is the array of choice for performance-critical, fault-tolerant environments. This, the most simple, secure and reliable type of RAID is also called "mirroring" or "dual copy" or "shadowing", where two hard drives are connected to the same disk controller, separate controllers, or the disk drive control is provided in software. Whenever data is written to a file, it's duplicated and written simultaneously to both disks. Thus, the data on one disk is a copy or "mirror image" of the other.

One benefit of a RAID-1 system is that if a drive goes down, your system doesn't waste time reconstructing the data. Unfortunately, mirroring is the most expensive alternative in terms of overhead (highest cost per byte), since half of the subsystem array is redundant. You always need twice as much storage as you would with a single disk drive.

RAID-2 is a disk subsystem architecture that uses disk striping across multiple disks at the bit level with parity. In RAID-2, which includes error detection and correction, an array of four disks requires three parity disks of equal size. This is seldom used since error correction codes are embedded in sectors of almost all disk drives anyway. Still, some implementations exist for supercomputer storage.

RAID-3 uses disk striping at the byte level with only one disk per array dedicated to parity information. This can be used in single-user environments and performs best when accessing long sequential records.

However, RAID-3 does not allow multiple I/O operations to be overlapped and needs synchronized drives in order to prevent performance degradation involving short records.

RAID-4 is the same as RAID 3 but stripes data in larger chunks (whole sectors or records). This allows multiple reads to be overlapped but not multiple writes. Like RAID-3, a dedicated disk stores parity information.

RAID-5 is the same as RAID 4 but data is stripped in sector-sized blocks and parity data is also striped across the disks interleaved with the data. It supports both overlapping reads and writes, but write performance is slightly degraded because of the need to update parity data.

RAID-6 is similar to RAID 5, plus additional striping so two disks can fail simultaneously, redundant controllers, fans, power supplies, etc. An array providing striping of data across multiple drives and two parity sets for increased fault tolerance. Highly reliable but suffers from slow performance.

RAID-7 is not yet an industry-standard term, but rather a product name for an RAID-like approach for multiple-host, UNIX-based environments running on various hardware platforms, including those from DEC, Silicon Graphics, Sun Microsystems, Hewlett-Packard, IBM and Sequent.

Storage Area Networks (SNAs) are independent networked-attached (or fiber channel attached) clusters of RAID-systems used for the centralized storage management of huge "server farms". Since they are independent of any particular server, reminiscent of how networked peripherals like printers are independent of any particular PC, one can perform a data back up of a SNA with special high-bandwidth hardware that doesn't interfere with the normal "conventional" network.

Pressurized forced-air cooling is also important to the reliability of PCs used in computer telephony systems, Since rackmount fault resilient computers can hold more computer telephony resource cards than conventional desktop computers, along with disk drives (if the disk drive array is internal), cooling immediately becomes a problem.

A good fault resilient PC has several fans in a plenum, circulating cooling air throughout the card cage then vented out the back of the chassis. In the case of CompactPCI designs, the system is vertical and a "chimney effect" carries the heat up and out of the unit.

Whatever the design, the fault resilient PC should have a removable, washable filter so the circulating air is free of corrosive dirt.

System Monitoring and Alarming. Even if a fault resilient computer can recover quickly from a problem, such as a blown power supply or failed disk drive, the system must somehow alert technicians that something went wrong and components must be replaced.

We once encountered a system that was running with a RAID subsystem where one drive had failed months before but there was no visual nor audio indication; no one was the wiser!

Incidents such as these indicate that a computer's monitoring / alarm subsystem is crucial to alerting maintenance technicians to system problems, enabling them to diagnose problems immediately without waiting for the system to continue "using up" its redundant components and failing completely. Such a subsystem should be able to perform related activities such as resetting unattended systems and sending distress signals over a network, modem or pager.

Alarming boards can be PCI or older ISA cards, standalone cards or rack modules with their own power connector, or built-in features on CPU cards. The best usually take the form of an intelligent card independent of the backplane with its own microprocessor and battery backup, along with user programmability, automatic switchover to a stand-by system, multichassis support, and an LED display integrated in the chassis or module. Some have the display housed in a separate package that fits into a 5.25" half-height drive bay.

NEBS. It stands for Network Equipment Building System. Some companies - large carriers - feel that their applications must be run on the most fault tolerant computers available - computers that can survive a massive earthquake, the Chicago fire, a lightning strike, or other disastrous phenomena, whether they be natural or man-made. NEBS certified equipment is designed to run applications under the most physically extreme conditions.

NEBS defines a minimum generic set of spatial and environmental /safety requirements developed by Bellcore (now called Telcordia) for use in Central Offices and other telephone buildings of the Bellcore Client Companies (BCCs). NEBS compliance is recommended only and is not required by regional BCCs. The decision to actually purchase NEBS certified equipment is totally at the discretion of each BCC.

Many companies claim that they manufacture computers that are NEBS-compliant; but how many of these machines are actually NEBS-certified? NEBS certification for a PC costs over $100,000 and takes months. Among other things, the PC chassis undergoes a "shake and bake" process that would wreck stout mechanical devices, let along an electronic computer!

Many New Roads to Fault Resilience.

Decisions, decisions. There used to be just one path to building a telecom system with high reliability: You bought a "fault resilient computer" (with a passive backplane and single board computer), made sure it had a Redundant Array of Independent Disks (RAID) storage system, plugged in your CT resource boards, and that was it.

Declining prices have spurred larger and larger purchases of fault resilient and fault tolerant PC systems, but these days there's a bewildering multitude of options for building a CT system:

Clustering involves taking groups of computers, each with its own processor, memory and disks and making the whole set of hardware and software appear as one server or logical unit to the operating system, system software, applications, and users. Clusters are loosely coupled parallel systems that decompose application processing into separate simultaneous operations. With a clustered server installation, a company can scale up the system to accommodate increased processing demands simply by adding another processor, eliminating the need to purchase larger computers.

If a processor in a cluster fails, its tasks are distributed around to other computers in the cluster and the system keeps operating. In this sense a cluster is a "processor version" of a RAID system. Perhaps we should call it "RAIP" – Redundant Array of Independent Processors.

Clusters offer greater protection than a RAID or other fault resilient systems, and it does so at a lower cost. A RAID system simply backs up data or else maintains parity data used to recompute lost data, and some fault tolerant systems force companies to buy entire dual systems that exactly mirror each other. Clusters, on the other hand, are designed under the assumption that networks have many possible points of failure and that faults may occur at the network connection, the server level or the I/O level.

Between failures, a normally operating clustered system can actually give more performance per dollar than some fault tolerant systems. Whereas a "classic" fault tolerant system has two (redundant) processors that not only duplicate each other's efforts but slow down periodically to compare their results, the loose "different-task-per-processor" parallelism inherent in a cluster can under normal conditions boost overall system performance to several times that of a single computer.

Whereas clustered computers may be in several cabinets in one or more buildings, and connected by a LAN or a WAN, Symmetrical MultiProcessing (SMP) consists of multiple processors in a single computer enclosure that share a common memory pool, bus architecture and all I/O devices. Thus, SMPs are in a sense the opposite of clusters.

SMP systems use one of the CPUs to boot up the rest of the system and load the SMP operating system that in turn boots up the other CPUs. At any one time there is only one instance of the operating system and one instance of an application in memory. The SMP operating system uses the CPUs as a pool of processing resources, all executing simultaneously, either processing data or waiting in an idle loop.

SMPs are scaleable like clusters in that you can expand them by plugging more and more processors in the chassis until you run out of room (or your system overheats). Concentrating processing power by cramming multiple CPUs into one server saves space and centralizes processor-intensive network services such as communications, client / server applications, file management, printing, etc. Also, SMP enables multiple CPUs to work concurrently on the same program thread. There are SMP capabilities built into Windows NT, Solaris and Linux.

Asymmetrical MultiProcessing (ASMP) is a form of de-coupled multiprocessing where a single master processor can coordinate the activities of multiple slave processors, each executing its tasks fairly independently out of shared memory. Like SMP, ASMP enables multiple CPUs to work concurrently on the same program, but unlike SMP, the processors are all not working on the same thread, thus making ASMP suitable for applications that need rapid real-time response.

Both SMP and ASMP are helped by segmenting the backplane - a single 16 or 20 slot backplane can be "segmented down" by keeping the same power lines across the backplane, but changing the signal routing, making sure the signal lines only go across groups of four or five slots. When the computer is fully assembled, additional resets and keyboard connectors on each group can come right off of the backplane, thus establishing separate computers in one enclosure.

Split ISA backplanes have actually been around for quite awhile. You could build a backplane and then in the manufacturing process just split it, so you could put more than one CPU and therefore more than one system on a backplane. But when you get to PCI backplanes, "splitting" the backplane now means genuinely redesigning the backplane.

Since you could now have more than one system residing on a single backplane, wouldn't it be great if you could power down one system in an enclosure if it became defective and have the other systems take up the load? You could then repair the old system and power it back up again, all without disturbing the entire configuration. Thus was born the "power-sequenced segmented backplane," ideal for applications requiring multiple servers or redundant systems.

Some technicians get a bit nervous working on one system in a 19" rackmount when there are three other systems still running in the same box. Instead of internally segmenting one 19" rackmount into four separate systems, why not just build four separate little boxes sitting side by side in the same space taken up by the original enclosure? You could channel all of your connectors into one large connector allowing you to quickly replace an entire system "on the fly" rather than open up one big box and risk disrupting the other three systems.

Crystal Group, Swemco and other companies have invaded this market and are now building what are essentially compact, "swappable computers."

Selecting Suitable Technology.

With so many choices, it's easy to lose track of what's really important with a computer telephony system - to keep it up and running and making money minute by minute, 24 hours a day.

CT applications are different than data applications because they involve voice communications, the lifeblood of any business. Your business can do without Word or Excel for a few hours, but what about losing dial-tone? Lose that and your customers will think you've gone out of business.

Naturally, it's essential that you choose the right CT platform to maximize uptime. Ordinary PCs suffer from multiple points of failure, since they are made as cheaply as possible and have no redundant let alone hot swappable components.

Fault resilient CT platforms are built so that electromechanical devices such as power supplies, cooling fans, and hard drives are built with redundancy so that if one device fails another one transparently takes its place. You want to make sure that, at a minimum, your system has redundant hot swappable power supplies. If a power supply fails, the second supply should transparently take over the system's electrical load, and a technician should be able to replace that defective supply while the system is running and with a minimum number of tools (front accessible systems are terrific in this respect). You obviously don't want to take your application out of service to correct a defect in any component.

Flexible power subsystems are a must. Whether it's AC or DC (-48 VDC). Many telecom applications, particularly in a CO, require DC power. But whatever type of power, you must make sure that your application is not going to be starved from a current standpoint. You don't want to overload the power subsystem whether it's DC or AC. If you do, you're asking for premature failure of your supplies.

Fault resilient PCs also generally rely on passive high capacity back-plane designs that can hold ten or more telephony boards. Passive back-planes handle more power than motherboards, and their time to repair will be much shorter than with non-industrial grade motherboards.

Ironically, much of the CT industry is still based upon 16-bit ISA technology. Vendors such as Dialogic and NMS still have a healthy ISA business, though this is slowly changing. Cheap off-the-shelf PCs have fewer than four full-length ISA or PCI slots. When your capacity requirements increase, such a cheap PC doesn't have enough room to allow you to expand your application. A passive backplane can hold up to 20 or more cards.

And what kind of cards are you going to plug into your CT PC? Not 56Kbps modem cards, but hefty CT resource boards. Bargain-basement PCs are designed with marginal airflow and suffer from poor cooling of critical components, since the components expected to be installed in such systems are very small light-duty peripheral devices such as network interface cards (NICs) and modems. They're not designed to deal with the extensive power and associated subsequent cooling requirements of CT resource cards for IVR, T-1 / E-1 connectivity, Fax, IP gateways, etc.

CT resource cards can do all of their wonderful tricks because they tend to be crammed full of Digital Signal Processor (DSP) chips (See Digital Signal Processor). A Dialogic 480SC or DM3 card, for example, is highly-populated with DSPs that collectively draw a considerable amount of current. Even "low density" analog four-port cards have more DSPs than a typical NIC card or modem card. The higher density the card, the more DSPs. Consequently, they draw a lot of current and generate a considerable amount of heat. CT components can easily draw more than 60 amps at 5 volts, or 600 Watts of power. New Jersey-based DAX Systems encountered a customer who had equipment that needed to draw 80 amps!

So, although you may think you've taken care of the redundancy aspects of your system, you may end up triggering the spare components sooner than you expected by loading the system with too many current-gulping and heat-generating CT networking and resource boards. Unless, or course, you've taken care of the cooling problem by buying a system with a sufficient number of fans and a well-designed pathway to direct air out of the system quickly.

The best (and most expensive) CT servers are designed from the ground up to support peripheral devices that don't draw much current and don't produce much heat energy as a by-product, or at least can channel it out of the system quickly.

One amusing problem is that whereas you can load eight fans in a system and blow heated air quickly out the back of a PC enclosure, unless provision is made to remove the heated air from the room, the heated air will simply "hang around" and be drawn back into the enclosure by the fans and then heated to an even higher temperature, thus slowly heating up the room. Your PC has now become the world's most expensive space heater. You must provide circulation or other cooling so that the ambient temperature around the PC can be maintained at normal room temperature.

Also, all of your internal chassis power, fans, temperature, and the various environmental parameters should be able to be monitored, the components being able to provide feedback to an alarming board, operating system, application, or whatever. Alarm thresholds can be used to trigger pager notifications, e-mails or SNMP traps.

No matter how reliable your PC platform is, when it does fail (and it will, somehow), the system must be quickly serviceable even though it's still on-line. Always keep additional spare components around - you can't expect Federal Express to save you. You might even opt for the small swappable PCs from Crystal Group and Swemco.

For data storage, RAID 1 (mirroring) or RAID 5 (automatic data reconstruction from a parity disk) should have individual high-quality SCSI drives in hot swap disk drive assemblies, a methodology similar to the hot swappable power supplies (IDE drives usually don't support hot swap). You should be able to replace a defective hard drive while the system is powered up, and transparent to the application.

SCSI drives have a higher mean time between failures and will give better performance in NT and UNIX installations. IDE does not typically support hot swappabilty.

Another factor affecting fault resilience is actually quite boring: Product stability. As Joe Jackson of Alliance Systems (Plano, TX - 972-633-3400, www.alliancesystems.com) says: "Something that's often overlooked when we get mesmerized by whiz-bang new technology such as redundancy, cPCI or hot swap, is that much of system downtime is usually caused by bugs in software, or in the design of the hardware. Over time, these things can be eliminated, but they're only eliminated if the basic technology is a stable platform and things don't change."

If every six months you add a whole new hardware platform to your application mix, then you will continue to uncover new problems. There is much truth to the fact that older systems tend to just run and run forever. Part of that truism is a sort of natural selection process - those computers that don't run well, don't get old. But part of it is that you can end up closely "grooving" your platform to the application - all of the defects inherent to your particular configuration and operating system and overall platform are uncovered over time, get fixed and stop causing you any more trouble. But if you decide you want a new-fangled system and you go changing out your old platform for a new one, it's almost certain that you'll soon uncover new types of problems.

Major upgrading or changes is something to think about long and hard, especially if you're going to deploy a system over a period of time. You want the best technology you can get right now, but you also want to make sure you can obtain that technology over the course of time and not have to continually change the types of components. That will be more and more of a challenge in the PC industry, which prides itself on rapid product evolution.

Care and Feeding of Fault Resilient PCs.

Periodic maintenance such as filter cleaning and internal cleanliness checks as part of periodic inspections will help your system perform longer without failure.

As for the external environment, keep the server's environment cool, clean and free of dust year round. Don't put your $100,000 fault tolerant system on a pile of sawdust, flour, or cotton mill fibers.

Always wear anti-static protection when working on internal server components. This is one of the easiest things to overlook involving platform maintenance, since few people realize that humans don't feel an Electro-Static Discharge (ESD) shock unless the voltage exceeds 3,000 volts, even though ESD events of only a few volts can cause serious permanent damage to semiconductor devices. The damage usually shows up as punctured MOSFET gates or PN junctions.

If you have onsite technicians working with any type of a hard drive, DSP-based CT resource card or CPU card, they should already know that electronic components are extremely susceptible to electrostatic discharge. Make sure that your technicians are using static straps that will enormously increase the lifespan of your equipment. The most irritating component of ESD is that it doesn't generally cause immediate catastrophic damage in most cases. Instead, ESD damage which is not immediately catastrophic leaves the component operable, but damaged, such that it will fail at a later time under normal stresses. Two weeks after your technician leaves a facility that that has never caused any problems, you may suddenly get reports of weird things happening. Very often that ESD bug is going to be the culprit.

An ESD kit is a minimal investment – only about $20. Workers who "carry" a charge into the work environment can rid themselves of that charge when they attach a wrist strap or when they step on an ESD floor mat while wearing ESD control footwear. The charge is routed to ground rather than being discharged into a sensitive component.

You should also install surge supressors on telco and data interfaces entering the server as well as filter the power going into the server. One cannot put enough emphasis on this. It's extremely important. As simple as this may appear, many people overlook putting surge suppression on telco interfaces, be it digital or analog. They'll put UPS systems on the power side of the box, but they won't protect the PSTN interface side, so very often lightning will hit the power pole outside their building and blast the CT resource boards, which are usually under a warranty that precludes lighting strikes. It can be a very expensive proposition to replace them. On the other hand, putting surge suppressors on the PSTN interfaces costs under $100.

If all else fails, extended warranties can also protect your investment to some degree. And always install a tape backup for disaster recovery. It's a time-honored technique and a very cheap way of protecting the bulk of your data.

Trends in Fault Resilience.

One big transition I've been following in this industry has been the rise of CompactPCI systems. CompactPCI is a sort of "ruggedized PCI" (it's original name) that's electrically similar to the PCI bus (successor to the ISA bus from the 1980s) but the board circuitry is enclosed in a tough Euro-card form factor. The cards are hot swappable. CompactPCI is becoming more and more popular in telecom, so take a long hard look at these systems, which are particularly suited to high-density, fault tolerant applications.

Just as there is an H.100 CT bus allowing applications to control and share the telephony resources of PCI boards, there is a similar H.110 CT bus for cPCI boards. CompactPCI systems cost more than PCI systems at the moment, but are expected to get a bit cheaper as production ramps up.

Unfortunately, the committee that did the cPCI hot swap specification work debated quite a bit on how to implement the hot swap spec. One group thought that in order to achieve the highest level of availability one would need to develop a centralized hot swap controller and have direct lines to control each card. Another group thought that, well, you don't really need all of that. You can create almost as good a solution just with some logic on the card itself and some software to control it. In the end, like any good committee, they chose both, which is a bit unfortunate, although one technique builds naturally on top of the other.

The jury is still out on whether the highest level of cPCI's so-called high availability hot swap is going to reach the level required to achieve a reasonable and acceptable hot swap scenario for most businesses and telcos, or if the next level down, called full hot swap, will be found acceptable instead, which is basically the same as the "High Availability Hot Swap" spec, but without the centralized control resource.

Another big frontier for fault resilient computers is the appearance of improved new forms of platform management. There's considerable complexity in a system when you have redundant power, redundant cooling, hot swap cards, etc., so there are many components to monitor. Predictive maintenance or looking at problems before they exist and cause downtime is certainly something you can do if you have a healthy management system running. Examples of predictive maintenance would be: If you're monitoring the RPMs of your cooling fans and you observe one fan is rotating less rapidly than normal, this could mean that a bearing is about to fail. So you can actually go to the site and replace the hot swap fan before it fails and causes a systems outage. Monitoring temperatures inside your chassis can ultimately warn you if you have impending problems such as if an air conditioning system fails or if a system's air filter is clogged and needs replacement.

The difficulty with adding platform management brings up the problem of how to maintain the portability of your software. How do you get the benefits of platform management without having to embed platform-specific code into your application?

There are at least two ways: One is to install an independent management processor, which is quite a popular solution these days. We've been calling them "alarming boards" in this entry, but the more sophisticated ones are referred to as "service processors" or "management processors." Your buy these from the platform vendor and they come integrated with the platform. Such alarming boards or management processors run completely independently of the host, sometimes with their own battery backup. They monitor system temperatures and hardware operating parameters, and they may have their own SNMP agent enabling it to alert some remote location.

The other path to platform management would be to use an industry-standard approach, something that can be borrowed from the commodity PC business. There is such a thing that's now appearing called the Intelligent Platform Management Interface (IPMI). It's being heavily promoted by Intel as well as Dell, NEC and HP, all of whom were the co-authors of the specification.

IPMI is currently under consideration by PICMG, owners of the cPCI standard, for inclusion as a standard for system management in cPCI. There's a pretty strong consensus to adopt IPMI-style management into a cPCI backplane system, which is good.

Intel's real goal is to be able to create platform management software that can plug and play across platforms. They want to achieve this by defining a standard interface for things such as power supplies and serial number information and whatever else becomes hot in the chassis during system operation, along with various electronics to be monitored.

Another important item about the IPMI standard is the Intelligent Platform Management Bus (IPMB) which is also specified by the IPMI standard. IPMB is a special bus for passing out-of-band messaging between different modules, such as fan modules, power supplies and an independent alarm processor. The routing of the IPMB into the cPCI slot architecture is what's currently the subject of standardization in PICMG subcommittees.

What this all promises is the ability for peripheral card vendors to include management capability onto their cards in a plug and play environment for cPCI systems. Because of the nature of the serial bus that's running through the system, you also have the capability to have independent third party alarm cards created that could just be plugged right into your cPCI platform, creating an instant platform supervisory environment.

CompactPCI has yet another frontier, host processor replication. Companies such as Motorola, I-Bus, Radisys and Ziatech are introducing products that are essentially fault resilient cPCI type systems having two CPU hosts in it, bringing them up to traditional fault tolerant specs.

Conclusion.

Computer telephony systems are most than just computers; they are the heart of your operation. Choose an experienced CT company to build and integrate your systems – your integrator must understand high availability technology and the consequences of failure.

Fax

The traditional (now perhaps "old fashioned") fax process goes as follows: One fax machine makes a phone call to another and sends a standard ITU-T-defined, "CNG tone" (or calling tone, which I've always called the "love warble") which is a 1,100 Hz tone emitted every three seconds. When the receiving fax machine or other device hears this tone, it knows it's an incoming fax call and the machines should "sync up". A document is then scanned, its light and dark areas converted into a series of signals, the signals are sent over a phone line to the receiving device that prints corresponding light or dark spots on a sheet of paper, resulting in a copy, or *facsimile*, of the original document.

Jules Verne describes fax machines in his 1863 novel, *Paris in the Twentieth Century*.

Fax machines were originally replaced by fax modems, the origin of which is directly traceable to computer data modems. Standard fax modem technology today is based on the four-wire, full duplex, leased-line modem originally used to connect IBM mainframe 3270 terminals over polled networks.

In the 1990s fax servers appeared (with internal fax boards providing the fax modem functionality), and now there are IP-enabled fax servers and routers with fax abilities. A fax, after all, is simply an image file, a TIF file to be exact, and so can be sent over packet networks to be read on PC screens just as easily as it can be sent over the conventional PSTN (See Fax over Packet).

In the circuit-switched world, there are various classes of fax modem devices. A fax modem "class" is a specification for communication between modems. The specification consists of a series of commands that can be used by software - or in some cases manually - to control the modem. Class specifications are numbered, and are defined by the Electronics Industry Association (EIA) in conjunction with the Telecommunications Industry Association (TIA).

There are five internationally accepted specifications for facsimile equipment. Group 1, Group 2, Group 3, Group 3 Enhanced and Group 4, there's also a Group 5 that's under development. Only 1, 2, 3 and 3 Enhanced will interoperate over conventional analog dial-up phone lines.

Group 4 is designed for digital lines running at 56/64 Kbps, and can use one of the "B" or Bearer channels of ISDN lines.

It is useful to understand that whenever anyone talks about a Class "n" modem, the reference is not just to the modulation/demodulation part, but to the entire device, including the T.30 and T.4 protocols. Generally when one refers to a Class n modem, one is referring to the whole board or external modem box that connects into the computer. The modem itself only provides the means for data to come in and go out.

Among the analog line fax machines, Group 2 is faster than Group 1, etc.

Class 1, which was approved by the EIA/TIA in 1990, was one of the first specifications for fax communication. It is a series of basic Hayes AT commands used by software to control the board. The first layer, which is addressed by Class 1 modems, operates close to the data link level. At this level, only very simple operations are performed: High level Data Link Controls (HDLCs, an ITU-TSS link layer protocol standard for point-to-point and multi-point communications), can be sent and received, an image or images can be sent and received. Besides data link operations, there are also factors such as the standard Hayes commands: dial, answer, etc.

A Class 1 device can be regarded as a small, very basic, computer, doing data link operations, such as dialing a telephone, something all Hayes-compatible modems could do. Class 1 modems also have some additional, special commands for sending and receiving HDLC information: send and receive data, check quality, etc. These are low-level, very primitive operations that the modem itself is able to do. A Class 1 modem has nothing to do with T.4 or T.30. All it can do is send or receive data. What this data means, how it is interpreted or how it is processed is passed on to the PC to perform.

Analog modems reach out to the world through an RS-232 serial port operating at a rate of 53 kbps or less. This means that every time a byte of data is sent, it must go to the serial port and there are delays inherent with that - sending 8 bits at 53 kbps takes time.

When, in 1991, Class 2 was published as a ballot standard by the TIA committee that was working on it, the ballot failed due to a number of technical and political issues. There was much disagreement over what members and industry believed the specification needed to make it work right. The TIAs committees debated many issues for over a year before agreeing on what would be in the final Class 2 specification.

Unwilling to wait for the ponderous TIA decision-making process to grind on, a number of manufacturers anxious to get into production, took the ballot as if it were a specification and went into production. The result was that after a time there were sufficient Class 2 modems out in the world to make the ballot a de facto standard, although several key areas are undefined and can vary from vendor to vendor.

After the TIA solved the issues, they published the final Class 2 specifications, which differed significantly from the Class 2 ballot. In order to distinguish the new real Class 2 specifications from the earlier ballot ones, the TIA used Class 2.0 as the new name. Thus, there is a fundamental difference between Class 2 modems based on the ballot standard and Class 2.0 modems based on the official TIA standard.

One of the most important differences deals with the packet layer protocol in the Class 2.0 specification, guaranteeing more secure throughput over the RS-232 line. Class 2 does not feature this.

Class 2 was never approved by the TIA. In fact, it is not even possible to get back copies of the original ballot, because once the committee approved the final Class 2.0 spec, all previous work was discarded.

So Class 2 is an illegitimate offspring nobody wishes to recognize and with the Class 2 installed base still out in the world, these questions and problems will not soon disappear.

In the committees' view, then, Class 2.0 and its revisions are Class 2 done right, and it has the TIAs blessing and official industry approval as a standard; that is, there is an official specification available.

As for Group 5, or G5 messaging, the G5 standard combines a superset of Internet e-mail with extensions to Group 3 fax, allowing G5 messages to travel over the Internet or be sent as faxes. G5 will allow 'point-to-point' fax transmission as well as via a third party, so that files can be sent by fax to arrive more quickly and more securely than via the Internet.

G5 messaging brings together five currently disparate messaging application areas: Image/fax, text/e-mail, voice messaging, video, and electronic commerce. It adds functionality to all these existing messaging applications. Examples include: Carrier independent delivery; end to end delivery to person, application, peripheral or department; transmission of document referencing information; electronic postmarking; and delivery confirmation.

Fax-on-Demand (FOD)

Also called Faxback, these systems enable users to call a system and request information by selecting from touchtone menus using an Interactive Voice Response (IVR) capability, and then receive the information by fax.

Fax-over-IP (FoIP)

See Fax-over-Packet (FoP).

Fax over Packet (FoP)

Fax over Packet involves technology that enables the interworking of standard fax machines with packet networks as well as the interoperability of certain classes of packet-enabled fax machines and PCs. It does this by digitizing the fax image from an analog signal and carrying it as digital data over a packet network, such as the Internet (IP), Frame Relay and ATM.

Fax is data, so it naturally belongs on packet-switched data networks – not on circuit-switched voice networks. Nailed up fax connections are wasteful, inefficient and limiting.

While most faxes still do go to plain old fax machines - faxing over packets, particularly Fax over IP, is "smarter" technology than what's been used in the past in conjunction with the circuit-switched PSTN.

How?

- You never miss faxes because your fax device is busy or out of paper. Never get a busy signal on sending.
- You now have a one fax number, accessible everywhere, even when you're mobile.
- You can store your faxes in a logical order, not in a pile.
- You can route faxes to your Inbox, automatically, or at least more efficiently.
- You can let your faxes be edited, if you want them to be.
- You can personalize your faxes with fax merges (like mail merging in Word: "Dear So-and-So...").
- You can increase your fax capacity: let any single employee blast faxes by using all the available networked ports in your enterprise
- You can efficiently, dynamically balance loads across all your fax resources. Don't wait in line to send from a fax machine, or let your fax linger in a non-networked LAN server's queue.

With all those features, fax over IP adds up to a great value-added service for people who already maintain IP networks (such as ISPs)

And if you build your corporate applications or service bureau offering using their new soon-to-be-ratified standards, no end-user will ever need to be aware of it's inner-workings.

At the low end, you can even build your own multi-line Internet fax gateway, if you're technically inclined. All you need to build your own box is some Dialogic, NMS or Brooktrout voice and fax cards, some ActiveX controls, and a copy of Visual Basic Professional Edition.

Much of the excitement of Fax over Packet centers on Fax over IP (FoIP) or "IP Fax."

A TCP/IP network (often just called an IP network) is built on a subset of the Open Systems Interconnect (OSI) model and incorporates five layers. IP stands for "Internet Protocol," and is carried in IP packets. TCP stands for "Transmission Control Protocol," which is connection-based. Packets get to where they're going in their own good time but they do get there, and in the order they were sent. Some voice/fax IP solutions use the connection-less UDP protocol (a datagram or "message in a bottle"), for rigorous timing adherence. TCP might be able to ensure packet delivery but would introduce unacceptable delays, and perhaps causing fax calls to drop. UDP, on the other hand, sacrifices guaranteed delivery in exchange for realistic speech cadence or fax time-out control.

To let the voice/fax over IP traffic through a company's firewall, sometimes both a TCP and a UDP port need to be opened at your firewall. You can open certain TCP/UDP ports on all IP addresses in your company, or just for certain IP addresses (e.g. your gateways) in your company, depending on how secure you want you firewall. Ports are usually numbered in four digits. The ports numbers are also usually unique to the particular application - one brand of servers we tested needed port 7007 open. They should tell you what port in the manual, or when you set up the software.

IP Fax and regular fax over the PSTN are different, yet have some similar characteristics:

When sending a conventional fax, your fax machine goes off-hook, dials, and the phone network completes a circuit over phone lines to the receiving fax device. You pay for the entire circuit connection. The two faxes negotiate a connection by sending some tones back and forth, synch up and exchange image data.

When you send a fax over IP, your fax machine (or PC client) transmits the image data as packets through an IP data network – the public Internet, a company Intranet, or an external Extranet - instead of the phone company's network.

Fax over IP servers need the phone network (if they need it at all) only for the local legs of calls: to deliver to regular fax machines, or receive from regular faxe machines. Thus, you save the long distance costs – and it's free if it's inside your company.

Fax over IP can happen two ways, real-time and store-and-forward.

Real-time fax over IP is similar to making a normal PSTN fax call, where the two machines "synch up" and send data over a local phone connection(s) and an IP leg between. If the fax is busy you get a busy signal. The user perceives real-time fax as one fax talking to another, without seeing all the intervening equipment.

Real-time faxing over IP depends on a well-managed IP network. Companies that provide backbone Internet services (Worldcom, SPRINT, MCI, AT&T) can provide "managed" IP networks to connect multiple locations. Connections are generally 100 milliseconds or less in overall latency, sufficient to support business-quality voice and fax. More than one to three seconds and your chances are slim.

To do real-time faxing over IP companies often take an IP telephony gateway box – one that can handle voice or fax calls on each port – and connect it directly to their PBX. They literally pull traffic from the PBX/PSTN lines and place the calls onto an IP network.

Using H.323 as IP's interoperability and call control protocol also has the important side effect of allowing voice and fax calls to hit the same ports on IP carrier's gateways.

After the call gets set up with H.323, the real-time fax can send the fax image one of two ways: Demod/Remod, or so-called "Spoofing."

Either technique appears to users as if they are making an ordinary fax call. They just don't know about the IP "trunk", or connection that sits in between the two fax devices. They dial the number and the receiving fax device on the other end hears the T.30 ITU international fax standard signals it would normally hear.

Doing real-time faxing via Demod/Remod (also called fax relay) involves de-modulating the fax modem's carrier wave signal, parsing it into binary 1s and 0s and loading them into IP packets that travel through an IP telephony gateway. The receiving fax-over-IP gateway reverses the process, re-modulating the modem carrier wave and feeding into the fax device, thus yielding the original fax. This process is very delay sensitive, however, and it's generally used only on networks with little or no IP latency, such as those used by voice gateways, since delays of even just a second can cause the session between the fax devices to time out. One T.30 session between the two fax machines must be maintained.

Spoofing, on the other hand, can be used for fax transmissions over IP networks characterized by high levels of packet latency and "jitter," or variable levels of delay of packet receipt. The original form of spoofing compensates for both long delays (up to five seconds) and the effect of jitter by using the T.30 fax protocol by padding the signal with occasional packets resembling "I'm still connected" flags to keep the session active even though IP network delay may be causing some packets to take a long time in getting to the far-end fax. The receiving fax device is thus fooled into interpreting the incoming transmission as coming from a conventional, real-time, circuit switched network. Two separate T.30 sessions, one for each fax with spoofing between, are used.

A receiving fax machine will assume that a transmission has been aborted if it has not received information after a period of time, because of it's nailed-up telephony circuit origin. Just before the receiving fax machine times out, spoofing algorithms intervene by feeding it 'fill lines'.

The fill lines produce a white streak on the fax, which on its own makes little or no difference to the readability of the fax. But it does let fax machines reliably fax over IP in real-time with almost up to three seconds one-way delay.

The downside of spoofing is call elongation. A two-minute fax might take two minutes ten seconds because of all the flags being transmitted as well as system delay, the number of pages and transitions between compression formats. A system delay of 400 ms can become a delay of about two seconds for the first page and one second for each additional page. This may become a problem for IP fax bureaus who must now do their billing based on the number of fax pages transmitted rather than the time it takes to send a fax. And with more bandwidth used, fewer free ports are available at any one time.

Real-time fax is a bit tricky, and many users will settle for store-and-forward faxing. Store-and-forward mode fax over IP works something like e-mail. You send the file to a server, that sends it to another server, and that server sends the fax. Eventually you receive confirmation as to whether the fax made it. Thus, unlike real-time faxing, you never encounter a "busy" signal, you can broadcast your fax to many devices, and you may be using fax software that can integrate with your e-mail software.

However, store-and-forward faxing eliminates the nicety of displaying to the sender an immediate confirmation message. Also, unless your store-and-forward gateway has intelligent fax boards installed, no feature negotiation will take place and the fax image and page formats transmitted will be whatever both gateways consider to be "standard."

Both the Frame Relay Forum and the International Telecommunication Union (ITU) have defined protocols for transmission of fax over a packet network. However, the principles described are equally applicable to ATM networks.

The Frame Relay Forum has defined a real-time protocol for the transmission of Fax over Frame Relay networks. Likewise, the ITU and Internet Engineering Task force (IETF) are working together to continue to evolve both the real-time Fax over IP (FoIP) network standard (T.38) that's derived from earlier standards that handled real-time fax over X.25 packet data networks, as well as the store-and-forward Fax over IP network standard (T.37), which uses Simple Mail Transfer Protocol (SMTP) addressing to send fax messages, thus allowing standardized integration into e-mail software packages.

Both T.37 and T.38 were approved by the ITU in June, 1998. Furthermore, T.38 is the fax transmission protocol selected for H.323.

Indeed, the IP voice/fax world got a big boost when Microsoft, Netscape and Intel agreed upon H.323 as the call control protocol standard (call initiation, hang up, etc). As an important side effect, H.323 allows voice and fax calls to hit the same ports on IP carrier's gateways.

And the actual process connecting fax over IP servers together is hardly more complicated than connecting your own PC to the Internet or other packet network. You don't even have to give up your regular fax machine to fax over IP. Fax can talk via autodialers to fax over IP services, or to enterprise IP fax servers. And "Internet aware" (e-mail capable) fax machines are also now available, such as Panasonic's PanaFax UF-770i.

Such Internet Fax machines send and receive documents, pictures, photos, hand-written messages and e-mail over the Internet by simply pressing a "one-touch" key. No auto-dialer is needed.

You send faxes to other Internet faxes and you can communicate directly with any other computer with an Internet connection. Panasonic's Internet Fax machine comes with G3 fax compatibility (about one 8.5 by 11 inch document per 20 seconds), as well as network scanning capabilities, so it integrates with existing network / e-mail environments.

The Panasonic machine first scans a document and compresses the binary image. It wraps the compressed image data with Tagged Image File Format (TIFF) headers and tags and then encodes the TIFF file in base64. The data is then attached to a Multipurpose Internet Mail Extension (MIME) file and passed through a predetermined STMP e-mail server. This e-mail sending is also sometimes called store-and-forward fax-over-IP. If another Internet fax is the receiving device, the original message will be printed automatically.

As the T.37 standard grows in popularity, it will be possible for Internet enabled fax machines and fax servers to integrate seamlessly. This will allow corporations to create large, integrated, enterprise fax solutions, and to leverage the cost saving of internet based routing for both paper faxes (fax machines) and computer based faxes.

One interesting company that attempted to bridge the IP network-fax machine gap early on was @Fax (Huntsville, AL – 256-721-3369, www.atfax.com), which developed the "Fax Portal", a device which can put a standard, traditional fax machine onto the Internet. This low cost, external device has a phone jack and Ethernet port, and software that allows the FaxPortal to route faxes via SMTP or even directly into a Microsoft Exchange e-mail box. This may prove to be the way fax machine traffic gets fully integrated into the computer based Internet Fax Server network.

Fax over IP is a great cost-effective way of communicating internationally. With eight to 12 time zones separating business partners, you can't always use voice. Global business transactions are waking up Americans to the fact creating e-mails using Kanji characters can be quite a nuisance.

Carriers supporting IP fax include a number of domestic United States Local Exchange Carriers (LECs) and IntereXchange Carriers (IXCs), Internet Service Providers (ISPs), and foreign postal / telephone companies (PTTs), as well as the emerging next generation providers of IP-based networks. Many fax service bureaus are adding or converting over to various kinds of fax over packet systems, along with makers of fax machines, fax servers and their respective components. IP fax-capable routers can also transmit a fax over low-delay IP networks, and can quickly fail over to the conventional circuit-switched PSTN network if packet network delays become unbearable. Such routers can provide security for fax transmissions via the IPSec (IP Security) encryption technique.

Russ Kahan (formerly with *Computer Telephony* magazine and now at Lucent Technologies) gives these seven reasons why cheap international Fax-over-IP is great for business:

1. **A signed fax sent over IP is still a legal document** – and it isn't editable. Perfect for finance, contracts, business dealings.
2. **Fax over IP is more secure.** For real security some systems support encryption, but at least two networks/protocols make intra-company faxes harder to intercept.
3. **Fax is great for complex documents and language translation.** You can hold a fax in your hand. You can take it to a translator. You can think it over, refer back to it. It's concrete.
4. **90% of faxes go through non-PCs.** Fax over IP has adapted to this, whereas e-mail is less pervasive (ditto for e-mail-to-fax services).
5. **Formatting and graphics are preserved.** Anybody who's ever gotten e-mail with odd line-breaks knows how annoying they are. Very unprofessional looking.
6. **Attachments are already "opened."** Especially tricky internationally (do you know what version of Microsoft Word they use in Taiwan?)
7. **Fax-through phone network-compatible circuits in the local country.** This makes things easy if they are available. Use ISDN fax boards in Europe. Don't trust the international telecom network to handle protocol translations converting T.30 over POTS lines to T.30 over whatever bizarre line type they favor in Uganda. You send the fax as data packets to the fax server over IP (the same everywhere), and it sends the fax like any other fax machine in that country would. Of course, you'll want this all certified by the board maker.

Chuck Hollis is a VP at NetXchange (San Francisco, CA – 415-543-7272, www.ntxc.com), one of the key players in the new and exploding corporate fax-over-Internet game. Here are some tips he shares with us:

1. **Know your objective.** It might be saving money on phone charges or simply reducing the hassle of international faxing. You might be after new capabilities, like broadcast faxing or enhanced security. Perhaps you're after better tools to understand your usage patterns and eventually create a charge-back system. Make a list. Figure out what's important now.
2. **Support extant fax machines.** Many solutions assume you will be sending all faxes from a PC device. But sending from a PC is inconvenient for hard-copy documents and forces you to abandon the fax investment you've already made.
3. **Avoid re-training.** This hidden cost can be substantial and can delay adoption of your new solution. Good solutions allow users to do precisely what they're doing today, only better. Ideally, users should be able to walk up to any fax machine and send a fax or use their favorite PC-based app.
4. **Support remote users.** More and more professionals are working from home, customer sites or the road. Your solution should be loca-

tion transparent.

5. **Know what you're spending on faxes today.** Get an estimate of how many faxes you're sending and what you're spending. Don't forget the productivity component – time spent sending the fax, checking on its status, re-sending in the event of failure and so on.
6. **Enlist a service provider.** Many fax service providers offer cost-effective fax delivery around the globe, greatly expanding your coverage and reducing both implementation and transmission costs. Consider using your own dedicated servers in high-volume locations and using a fax service provider's network for smaller locations and destinations outside your company.
7. **Keep management to a minimum.** Most companies want to spend as little time as possible managing their Internet fax network. Your solution should automatically discover new nodes as they're added to the network and transparently make changes in configurations and cost tables. A full suite of management tools should allow the solution to be managed from anywhere using Web-based interfaces. For organizations that already have well-established network management capabilities, SNMP-based monitoring tools offer an easy way to integrate fax services.
8. **Accounting and bill-back feature are key.** Experience shows accounting needs evolve over time. You want detail on every transaction and many report formats, including integration into external databases and billing packages.
9. **Security is important.** Since fax servers connect directly to the external network, your solution needs to address network security issues. Your solution should provide end-to-end encryption.
10. **Dedicated servers aren't necessary.** The ideal solution sits on top of existing network servers without requiring a dedicated system solely for fax communications. This not only saves hardware and management expense, but speeds deployment.

Israel Drori founded NetXchange in March 1995. Some call him the "father of Internet fax." Here are his tips:

1. **Use Internet Fax.** The average Fortune 500 company spends over $14 million on fax annually. Most of that money is spent on long-distance phone tariffs. Internet fax alleviates that dollar drain. The technology is also suited to the store-and-forward nature of the 'Net – unlike real-time voice communications.
2. **Make sure your solution is open.** Standards and APIs should be available throughout your Internet fax solution, as with any open product. The best solutions should be able to change, adapt and grow as your needs evolve.
3. **Support fax machines.** Many potential solutions assume all faxes will be sent from a PC. Besides being inconvenient for hard-copy documents, there's a large investment in those dreaded existing fax machines; people shouldn't have to junk them.
4. **Avoid retraining.** This hidden cost can be substantial and can delay adoption of your new solution. Users should be able to fax just as they did before Internet fax, whether that means walking up to a fax machine or using PC apps.
5. **Support remote users.** More professionals are working from home, customer sites or from the road. Solutions should provide location transparency.
6. **Security is important.** Your solution should have end-to-end encryption.

Here are some Voice over Packet vendors and services:

Alcom (Mountain View, CA – 650-694-7000) www.alcom.com
ArelNet (Yavne, Israel – +972-8-942-0880) www.arelnet.com

Array Telecom (Herndon VA · 703-787-7000) www.arraytel.com
Black Ice (Amherst, NH – 603-673-1019) www.blackice.com
Boomerang (Palo Alto, CA – 800-779-7792) www.boomerang.com
CallWare Technologies (Sandy, UT – 801-984-1100) www.callware.com
Castelle (Santa Clara, CA – 408-496-0474) www.castelle.com
Cheyenne (Islandia, NY – 516-342-5224) www.cheyenne.com
Cicso Systems (San Jose, CA – 408-526-4000) www.cisco.com
Clarent (Naperville, IL – 630-637-0448) www.clarent.com
Copia International (Naperville, IL · 630-778-8898), www.copia.com
Dialogic (Parsippany, NJ · 973-993-3000) www.dialogic.com
FACSys (Sommerset, NJ – 732-271-9568) www.facsys.com
FaxBack (Beaverton, OR · 503-645-1114) www.faxback.com
Faximum Software (West Vancouver, BC, Canada – 604-925-3600) www.faximum.com
FaxSav (Edison, NJ – 732-906-2000) www.faxsav.com
Franklin Telecom (Westlake Village, CA – 805-373-8688) www.franklintelecom.com
Faxaway (Seattle, WA – 206-301-7000) www.faxaway.com
Inter-tel (Chandler, AZ – 480-961-9000) www.inter-tel.com
MediaGate (San Jose, CA – 408-248-9495) www.mediagate.com
Micom (Simi Valley, CA · 805-853-8600) www.micom.com
Motorola Information Systems Group
(Mansfield, MA · 508-261-4583) www.mot.com
Natural MicroSystems (Natick, MA · 508-620-9300) www.nmss.com
NeTrue Communications (Fullerton, CA – 714-870-0861) www.netrue.com
NetXchange (Natick, MA – 508-653-4336) www.ntxc.com
NKO (Jacksonville, FL · 904-730-0050) www.nko.com
Omtool (Salem, NH – 603-898-8900) www.omtool.com
Open Port Technology (Chicago, IL – 312-867-5000) www.openport.com
RightFAX (Tucson, AZ · 520-320-7000) www.rightfax.com
SEPE/Fax®Star (Newport Beach, CA · 949-724-0806) www.faxstar.com
Softlinx (Westford, MA – 978-392-0001) www.softlinx.com
SoftTek/webfaxit.com (Trabuco Canyon, CA – 714-888-1181) www.webfaxit.com
Spectrafax (Naples, FL – 941) 643-8700) www.spectrafax.com
Telogy Networks (Germantown, MD – 301-515-8580) www.telogy.com
T4 Systems (Little Rock, AR – 501-227-6637) www.t4systems.com
UUNet (Fairfax, VA – 703-206-5600) www.uunet.com
Vienna Systems (Kanata, ON, Canada – 703-435-6000) www.viennasystems.com
VocalTec (Northvale, NJ – 201-768-9400) www.vocaltec.com
Voice and Data Systems (Montreal, Quebec – 514-879-8585)
http://www.lisco.com/terra_globe/commercialindex.html

Fax Server

A computer loaded with fax boards that has connections to the Local Area Network as well as the PSTN and now even IP networks. It allows users sitting at their desktop PCs to send and receive fax documents through the server.

Architecturally, most fax server or "LAN fax" products subdivide the three functions of (a) port management; (b) document assembly, rasterization and queueing; and (c) client communications at the process level, but load all three functions onto a single PC server, running NT or UNIX. In the most high-end, large-scale architectures, you can distribute functions across the LAN, filling up separate fax-card boxes to gain lots of ports; and sequencing fast, multi-CPU machines with lots of RAM and disk-space to handle raster conversion.

You don't want to skimp on board hardware; a low-grade Class 2 modem will fail during the handshaking process far too frequently. Most fax servers use intelligent fax boards (and line interface cards) from Dialogic or Brooktrout; a few use NMS hardware or Commetrex.

In one case, Castelle's FaxPress, a generic (perhaps even shared) NT server is employed for client communications, rasterization, and other functions, but this machine is slaved to a self-contained, external box, containing proprietary fax hardware and the line interface. This isn't a bad idea: the NT server doesn't have to do port management (so it can be shared with other applications), and you don't have to fool around with fax boards, which can be troublesome to install.

But you may not have to do that, in any case. Fax servers tend to be sold pretty-much turnkey, these days – so your dealer/integrator will probably assemble and configure your system. If you insist on doing it yourself, the hardest parts involve board installation and server/service configuration. Once the server is up and running, recognized across the network, with fax boards installed, feature configuration is usually not too difficult. In most cases, no more than a few hours' work (and an average of two calls to technical support) will get you up and running.

If you've ever used a faxmodem, you already have the gist of what a fax server works like from the client side. For single-destination outbound faxing, most systems arrange to have the fax server emulate a network printer, addressed by a modified standard print driver on the client machine. To fax, you hit Print, select the fax server, and go. Windows pop up to let you enter cover-sheet info, select from local fax-number lists, etc.

The big difference is that a fax server works faster. Instead of using your local CPU and hard disk for document-to-raster conversion (rasterization), and your COM port for connection management, a fax server lets you shoot your document across the LAN, where the server does the muscle-work of rasterization and port-handling. If you send a lot of multipage documents, or send the same document to multiple destinations, you'll see the productivity benefits instantly.

Once difficult to configure and impossible to use under real life conditions, fax servers are now highly scalable systems that are flexible and intelligent, that integrate with Microsoft Exchange and other e-mail systems to create unified messaging solutions. These fax servers can be administered remotely as part of a larger, Enterprise-wide, CT system.

Fax servers have had to learn new tricks to keep up with the new ways that companies use fax and other messaging technologies.

The way a company uses fax falls into four categories.

1. There are typical business applications where, for example, you want to get your invoices out to customers and you want to have it done unattended. You want your legacy IBM 3090 or AS/400 or AIX platform to now send out faxes automatically. There are also specialized fax server-based applications such as fax-on-demand, forms processing and service bureaus.
2. In the same company you could have people who don't want to integrate with e-mail, or they may have Macintosh clients. These people prefer a standalone, Winfax-Pro kind of desktop solution.
3. Unified messaging, where you send and receive faxes with Winfax Pro-like features but you do it along with your voice and e-mail while you're in Microsoft Exchange or Lotus Notes. The ability to handle multiple mail systems at the same time is becoming important too.
4. Lastly there is the move to the Internet and Intranets for faxing. Whether you've bought into the network computer concept – the "thin client" – or whether you just want to be able to do all of your faxing from a desktop browser, the Internet, for better or worse, has arrived as a fax conduit.

All these disparate methods involve sending a fax from the desktop up through the network up to the server. When you submit a fax with these packages, the fax is moved from the client to the server, using a transport such as TCP/IP or Named Pipes.

When the fax is ready for the big leap into the public network, *Least Cost Routing (LCR)* reduces phone costs by sending faxes across a Wide Area Network (WAN). If you've got a large company with an office in New York and one in London and you're sending a fax to a customer in Rome, the fax server can figure out that's it's cheaper to send it over a WAN to the Milan server then send it out as a free local call to Rome.

It's kind of like having your company's private version of an Internet hop-on hop-off server. If you can send and receive e-mail from any office, you basically can send faxes to the local office for free.

Inbound Intrigue. Whereas faxing out is a conceptually straightforward affair, faxes received on the fax server can be routed to the desktop client via various techniques. If your equipment responds with a wink signal (reversing polarity +/- on the line circuit), the CO switch can send you the digits dialed as DTMF. This is DNIS, the Dialed Number Identification Service.

You can have faxes routed to fax machines by number. Analog DNIS is called Direct Inward Dial (DID). This lets you call numbers on your company's PBX or with a three- or four-digit extension, which is associated with a particular recipient.

Computer-based fax systems can also use Optical Character Recognition OCR, besides DTMF. Using OCR, a computer-based system looks for keywords, like names or what follows "TO:" or an extension number and routes the fax that way.

All of these routing methods work, but they all have known limitations. For example, with OCR, there's no algorithm that can reliably recognize any random example of handwriting. The fax has got to be printed. And if somebody sends you a fax with a small (10-point) font size, the faxing process shrinks it a few percent more, you get aliasing (jagged letters), and other faxing "artifacts" that lower OCR's effectiveness.

Fax servers have also gotten quite cozy with the Lightweight Directory Access Protocol (LDAP) which can replicate directory information from a single database to different servers.

Everybody has adopted LDAP as the way for a client application to do a look-up in a directory: The e-mail world, voice mail vendors, Microsoft Exchange, Netscape, CCOM, and everyone who's offering directory services, all say that they're going to support LDAP as *the* protocol for an app to get information into or out of a directory.

Omtool and AVT / RightFAX for example, support LDAP. If your company is Exchange-oriented, Exchange is LDAP compatible. With inbound routing, you need to associate an extension or an individual or a delivery method. Instead of creating a new database, you can just add a new column in Exchange, call it your Inbound Routing Column, then just have all of your different DID numbers associated with a specific user's name.

Instead of creating a new database via LDAP you can use whatever database you've got, whether it be Access, Oracle, or Exchange's own database, as long as it's LDAP compatible.

Unified messaging is fast becoming a fax server's best friend. Unified messaging is starting to drive a lot of sales, particularly when Exchange integration is involved. It's relatively easy to adapt programs like Faxination and Fax Sr. to a major open platform such as Lucent's Octel Unified Messenger. Companies familiar with e-mail can install a piece of software on their Exchange Server and then every client in the company can send and receive faxes to-and-from the desktop.

In terms of e-mail integration and unified messaging, most makers offer a middleware arrangement that connects with Outlook, with or without using Exchange Server as an intermediary. Products are also available to integrate with Lotus cc:mail and Domino and Notes, but Microsoft Exchange will be the interface winner here, followed by Lotus products. The advantage of such native e-mail integration: You cut down on client software overhead and extra address directories.

Fax over IP. LAN fax technology gained in popularity almost in lockstep with the Internet. As the enterprise has become "internet enabled", it's no wonder LAN fax servers have incorporated IP technologies into their solutions.

Fortunately, there's an emerging standard for allowing IP enabled fax servers to exchange messages based on the SMTP (Simple Mail Transport Protocol) using the standard MIME format, called T.37, which was born from work in the Internet Engineering Task Force (IETF), and has been propagated to the world wide standards organization known as the ITU (International Telecommunications Union).

T.37 offers a standard mechanism for e-mail gateways to exchange fax messages as attachments. It calls out a standard image format and a format for embedding routing information into the message type.

An IP enabled fax server provides the same benefits of a traditional fax server, e-mail integration, desktop faxing, and inbound security, but with one big advantage, the ability to least cost route faxes to remote locations over public or private IP networks.

It is well known in the fax industry that the capital cost of a computer based fax solution is a fraction of the phone bill savings the fax boards will generate over time. This is one reason why intelligent fax boards are so popular in computer based fax applications. Intelligent fax boards, as the name suggests, provide "intelligence" in the hardware through advanced compression schemes which lower fax transmission times, thus lowering the overall cost of ownership.

Taking this concept a step further, IP enabled fax servers have the intelligence to route faxes over the least cost network for delivery. That network may be an alternative long distance carrier, to another LAN fax node in another city, or even over the public Internet to a global fax delivery network like Xpedite Systems.

By the way, IP-enabled fax servers still look like servers, still have fax telephony boards. Yes, a computer based fax board is even more important in an Internet-connected fax server. Many people think that fax as a document delivery mechanism is dead, and will be quickly replaced by e-mail. This could not be farther from the truth. In fact, the number of fax pages sent in the world continues to skyrocket. Some estimates say nearly 40% of all calls to and from the US and Europe are fax, and that number approaches 50% between the US and Asia.

And just because a fax server is IP enabled doesn't mean fax machines disappear. Fax machines continue to be the most common interface for fax. People continue to send paper through fax machines, and therefore, a fax modem is needed to capture the fax data into a computer. This is why computer based fax is so popular, it bridges the gap between computer / e-mail and the fax machine.

Any way you look at it, the "killer app" for fax servers is least-cost routing over the Internet or company intranet. In this vein, the most powerful mid-market product is probably AVT / RightFax, which lets you define a self-healing network of fax servers at different offices, linked by a private WAN or across the 'Net. When connectivity is functional and reliable (low latency, not too many lost packets) the servers route across the network, in store and forward mode.

Omtool's IP Fax solution posits one central IP-enabled server at headquarters and less-expensive IP-enabled fax modems -- called IP Fax Satellites – at branch offices. These can be connected through the existing corporate intranet, and can also divert office faxes from traditional stand-alone machines and ship them out over free IP.

When the 'Net is jammed up, they use regular phone lines. Other products can be configured to exploit SMTP/POP3 e-mail service for document transmission, though this method offers no immediate confirmation of successful delivery.

Maury Kauffman is managing partner of The Kauffman Group (Voorhees, NJ – 856-651-1651, www.kauffmangroup.com) an enhanced fax technology and services analysis and consulting firm, and one of the world's great fax gurus. Here are Maury's top 10 reasons to install a fax server:

1. **It's Cost Effective.** Installing a fax server reduces your total cost of faxing (hard and soft costs combined) by as much as 50-60%, compared to manual faxing.
2. **Increased Productivity.** Your fax machine no longer replaces the water cooler as "gossip central."
3. **Faster Communications.** Faxing from the desktop is even faster than e-mail.
4. **Payback in Weeks.** Sending just 30 documents per day, each two pages long, costs an organization 1,000 man hours per year in sneakerfax. How many man hours is your organization wasting?
5. **Document Confidentiality.** Afraid everyone is reading your faxes or salespeople are poaching leads, desktop faxing solves these problems.
6. **Improved Customer Service.** Sales, marketing and customer service representatives can now send and receive faxes while still talking to customers.
7. **Decreased Telecommunications Expenses.** Severs installed with intelligent fax boards, such as those manufactured by Dialogic or Brooktrout operate at considerably higher speeds than most fax machines. Faster speeds equal lower costs.
8. **Fax Management.** Now you can prioritize the bosses faxes to be sent first, those overseas to be sent after hours and schedule the marketing departments broadcasts at 3:00AM, so they don't crash your network.
9. **Ease of Installation and Education.** Hire a Fax VAR. They can have the server live and everyone faxing in no time. Plus, due to their volume discounts, often for less than it would cost you direct.
10. **Convenience.** What could be simpler than point, click, fax?

Advice from the expert: If your organization has more than ten workstations or transmits more than 50 documents per day, a LAN Fax server will save you time and money!

Here are 10 tips for choosing an IP fax server:

1. **Standards.** Be sure that your IP Fax Server is standards based, particularly supporting T.37, so your fax server can inter-operate seamlessly worldwide.
2. **Support for Intelligent Fax Hardware.** Intelligent fax hardware will save big money on your fax phone bill, often recouping the initial investment in less than 6 months.
3. **Scalability.** Be sure your solution can scale up. Most solution start small, one or two nodes, a couple of ports for fax. Be sure you can expand the nodes and ports to support your entire organization.
4. **Support for Inbound Routing.** A key feature for end users is the ability for inbound faxes to be received directly to the desktop.
5. **Least Cost Routing.** A must for all IP Fax Servers. This feature means your fax server is smart enough to find the server, anywhere in the world, that will send your fax for the lowest cost.
6. **Access to 3rd party Fax Service Bureaus.** Many fax servers today provide an IP based link to an external fax service bureau which have nodes around the world. This allows your least cost routing server to take advantage of world-wide distribution, without putting nodes in

around the world.

7. **Browser Access to Fax Mailbox.** End users, especially road warriors, need to be able to access faxes and the fax mail box over the internet, not just from your desktop.
8. **Integration with existing E-mail.** An absolute must. Your IP Fax Server needs to integrate with your current enterprise e-mail server.
9. **Directory Integration.** As the enterprise gets larger, a single, unified directory is key. As true enterprise directory services appear, be sure your IP Fax Server supports them.
10. **Security.** Sending data over the public Internet demands a secure link. Be sure your IP fax server provides some type of encryption for your faxes.

Here are some fax server vendors and their products:

AVT RightFAX Software Group

Tucson, AZ – 520-320-7000, www.rightfax.com.

Product Name: RightFAX Enterprise Suite V7.0

Hardware/OS Supported: Brooktrout Technology and Dialogic fax boards, Windows NT.

Routing options: DID, DTMF, Voice-assisted DTMF, DNIS, ANI, PBX, DNIS/DID interface, CSID, ISDN, OCR, Manual, Line/Channel, Combination Inbox/Outbox

Integrations: Microsoft Exchange, Outlook and Mail; Lotus cc:Mail and Notes, Novell GroupWise; and all SMTP/POP2 complaint Internet e-mail software. Version 7.0 introduces enhancements to the RightFAX Gateway for Lotus Notes, providing a single interface from which administrators can manage both e-mail and fax users. The gateway also features the ability to automatically synchronize user accounts in Lotus Notes with RightFAX user accounts. This means adding, removing or modifying user accounts in Lotus Notes automatically takes place in RightFAX accounts, ensuring accuracy and saving time.

IP enabled: Yes.

Pricing: Around $6,000. RightFAX Enterprise is priced by the license, not the number of users.

How best to buy: Through VARs, system integrators, distributors.

Fax Service Bureau Relationships: Sister company MediaLinq Services Group (formerly MediaTel Corporation).

According an IDC fax report, AVT RightFAX is the best-selling Windows NT fax server worldwide. In addition to LAN fax servers for workgroups and enterprises, RightFAX provides module and connector products designed to let companies create their own fax integrations, such as the RightFAX Connector for SAP/R3.

The RightFAX Enterprise Fax Manager (EFM) uses the popular Explorer-like interface to permit centralized administration of all RightFAX servers on the network. Administrators can view the status of every fax server; start and stop fax services individually or globally; determine server workload; and configure LCR rules.

RightFAX extensions to Microsoft Exchange provide a single interface from which administrators can manage both e-mail and fax users, and automatically synchronize Microsoft Exchange and RightFAX user accounts.

With Version 7.0's paging and alerting support, users can receive pages or short messages when any new fax (or one from a particular party) comes in. Screening can work with Caller Subscriber Identification (CSID) or ANI. Users can even be paged when fax sending failures occur. Users can receive faxes through RightrFAX's own user interface – FaxUtil – or via their e-mail client. On the road, they can touch tone into a nearby destination fax.

RightFAX Version 7.0 enterprise fax server software has been certified by the Baan Company Application Interface Program. The BaanConnect certification for RightFAX, AVT's highly flexible electronic document delivery software, ensures its compatibility with the Baan IV enterprise applications. Open and scalable, RightFAX seamlessly integrates with desktop, host, legacy and ERP applications to allow organizations to electronically deliver business-critical documents and better manage their overall communications. The RightFAX Connector for Baan works in conjunction with the RightFAX production system and integrates with Baan IV to enable companies to electronically deliver high-volumes of documents such as purchase orders, sales agreements and order acknowledgments automatically and unattended directly from Baan IV applications. This eliminates manual document processing, increases employee efficiency and lowers administrative costs. Version 7.0 also incorporates AVT's CommercePath Technology (see below).

AVT CommercePath Software Group
Portland, OR – 503-968-9600, http://www.appliedvoice.com
Product Name: CommercePath 2.5.
Average size of installation: 16 ports/server, out of a 2-96-port range.
Hardware/OS supported: Windows NT 4.0, SQL Server 7.0.
Routing Options: DID, DTMF, DNIS, CSID, ANI, Manual, Line/Channel, or by area code.
Inbox/Outbox Integrations: Microsoft Exchange, Lotus Notes, SMTP, MAPI, VIM.
IP enabled: Yes, with Commercepath's InternetLink Module
Pricing: $2,995 for CP Environment Software, $5,995 for Host/Production Software, $1,650 Per Channel License
How best to buy: Direct from AVT/CommercePath or through network of resellers. Distributors include Compaq, Siemens, EDS and API Wang.
Fax service bureau relationships: Sister company MediaLinq.

CommercePath, AVT's production-strength fax server, in addition to all of the standard desktop application fax/printer driver support and e-mail integration, gives companies the ability to fax EDI-format documents to their EDI-challenged vendors and customers. Their fax status screens are web-enabled, so administrators and users can browse in from anywhere to look in on their fax processes. CommercePath is also firmly on the distributed-computing bandwagon, breaking up the load across an enterprise WAN.

While the controlling, "Command" server is a full industrial-strength PC, encompassing process, notification, and routing intelligence, distributed servers can be mere fax cards and "communications" server software.

CommercePath's premade integrations include those with ERPs SAP, Oracle, and Sterling Commerce. Other integrations are offered through their own services group.

Biscom
Chelmsford, MA – 978-250-1800, www.biscom.com
Product Name: FaxCom V3.0.
Average Size of Installation: 6-8 ports per site, multiple site installations typical within the enterprise.
Hardware/OS supported: Biscom manufactures their own fax boards and hardware and works with Compaq, Appro, IBM, Dell and HP servers. FaxCom runs on Unix and Windows NT.
I/O options/Routing: Inbound: DTMF on DID and PBX phone lines and DNIS on T-1 or E-1 lines.
Inbox/Outbox integrations/: Lotus Notes/Domino, Microsoft Exchange/Outlook, Novell GroupWise and most SMTP-based applications.
IP enabled: Yes.
Pricing: A two-line, 100-user license complete with e-mail interface begins at $9,995. Thereafter, pricing is based upon users and ports.
How best to buy: Direct to end users and through channel partners

which include Ricoh, Savin, STR Software and ESD Computers.
Fax service bureau relationships: In accordance with customer's needs.
Founded in 1986, Biscom was one of the first companies to promote an enterprise fax management strategy. They offer a complete turnkey solution, as well as customized application design and support. Because FaxCom has a COM-based API, it lets users easily add fax capabilities to their existing applications. Today, that means fax-enabling workflow/imaging, transaction processing, ERP and messaging applications.

Castelle
Santa Clara, CA – 408-496-0474, www.castelle.com
Product Name: FaxPress 1500N through 5000.
Average Size of Installation: 2 ports, 75 desktops.
Hardware/OS supported: An external, self-contained "brick"-style server, FaxPress works in Windows NT or Novell networks.
I/O options: FaxPress takes analog lines and also maintains its own IP address for Internet faxing to other FaxPresses.
Routing options: Secure manual routing, line routing, T-30 sub addressing, DID, or printed on FaxPress' own self-contained print server or network queue.
Inbox/Outbox integrations: Microsoft Exchange or Outlook, Lotus Notes or cc:mail, NetWare's Groupwise and all SMTP-compliant e-mail software.
IP enabled: Yes
Pricing: One, two, four, and eight-line units for $1,495, $2,995, $5,995, and $7,495, respectively.
How best to buy: Direct from Castelle or through VARs. VARs purchase from distributors such as Ingram, Merisel or Tech Data.
Fax service bureau relationships: Fax2Net, Graphnet.

Castelle's FaxPress is a simple, gore-free fax solution that's great for companies of 25 to 75 people. An external, plug-and-play 'brick' device, it takes in analog lines in one end and network connections on the other. It comes with extensive client software and optionally interfaces with Cardiff Software for intelligent forms processing; Tally Software for reports on utilization patterns, usage anomalies, and usage-based chargeback; Goldmine for fax-enabled contact management, and other popular applications. Castelle now owns Object-Fax, an NT-based enterprise fax server from Traffic Software, and the InfoPress family of fax-on-demand and voice response solutions from Ibex Technologies.

FaxPress' Software Developer Kit (SDK) is free, allowing VARs and system integrators to bolt the fax onto any preexisting application. Castelle partners with OEMs, sells direct to VARs, and sells software enhancements and upgrades directly to end users. One "master" unit can direct up to six "slaves" to share an outbound load.

COPIA International
Naperville, IL – 630-778-8898, www.copia.com
Product Name: FaxFacts V6.146.
Average size of installation: Four lines, ten desktops. Service bureau customers average 24 lines, "and some go into the thousands."
Hardware/OS supported: Brooktrout, Commetrex, Dialogic / Gammafax boards, Windows NT.
I/O options: Analog, T-1, E-1, ISDN, PBX, e-mail-to-fax and fax-to-e-mail.
Routing options: DID, Manual, OCR, DTMF, CSID, and Line.
Inbox/Outbox integrations: Microsoft Exchange delivery of faxes, error status, broadcast reports, and voice mail.
IP enabled: only via Exchange e-mail; e-mail-to-fax gateway and fax-to-e-mail delivery.
Pricing: $365 per line, plus $50 per client.

255

How best to buy: Through resellers.

Fax Service Bureau Relationships: Faxs, Boomerang.

A full-featured scalable fax server with an included API, FaxFacts runs under an NT workstation or server that integrates with all major networks. Server features include load balancing, call routing, intelligent retries, complete call logging, LCR and status tracking. Copia's Web fax feature allows companies to keep one set of documents on a web site for both browsing and faxing, automatically converting from HTML to fax when the document is requested via FOD.

Copia's FaxFacts, makes special claim to fax-blasting finesse, claiming 600 rasterized and queued Word pages per minute on a high-end Dell server with dual processors. Desktop client software runs under Win 3.1, 95, 98 or NT.

Fenestrae

Duluth, GA – 770-622-5445, www.fenestrae.com

Product Name: Faxination 4.0.

Average Size of Installation: assuming Microsoft Exchange seats with Fenestrae Access License, 360 seats per server. Standard Edition (two lines) is for 10 to 100 users and a single Exchange Server. The Corporate Edition, for 100 to 2,000 users per server, is enterprise-enabled for least-cost routing. Corporate FELP: (Fenestrae Enterprise License Program) is for 2,000 to 50,000, and lets you pay for additional licenses as you grow.

Hardware/OS Supported: Windows NT. Supports all Brooktrout Technology boards (V.4), including NetAccess adapters. Also supports PCI bus and ISDN adapters, PRI and BRI. Supports all Gammalink (GDK. 3.1) including PCI adapters, PRI ISDN and BRI ISDN adapters. All CAPI 2.0 PRI boards.

I/O Options: T-1/E-1, ISDN, PBX.

Routing Options:. DID/DDI, Scan String, DTMF.

IP Enabled: All Corporate servers include the IP fax connector.

Inbox/Outbox Integrations: Microsoft Exchange, Mail, Lucent Technologies' Octel Unified Messenger (V.2.0), Siemens Xpressions.

Pricing: Priced by server and client access licenses. Example: Faxination for Microsoft Exchange, Standard Edition, is $1,495. 50 clients would be another $2,385.

How Best to Buy: Faxination is sold exclusively through the VAR channel. Fenestrae maintains sales and marketing offices world-wide for local VAR sales and marketing. In addition, Fenestrae has a dispersed Global Account team to ensure smooth account relations with Fortune 100 companies. Distributors include Compaq, Siemens, EDS and API Wang.

Fax Service Bureau Relationships: NetMoves.

Fenestrae was one of the first to integrate with e-mail inboxes, and tightens that further with Active Directory support in Exchange 5.5.

Fenestrae's Universal Message Routing provides intra- and extra-net Least Cost Routing (LCR), keeping tabs on both message type and message content. This allows the fax server to use Exchange infrastructure for LCR functions, while providing logical message pipes to external routing like IP fax.

Faxination can log all transmission and routing details in Exchange Public folders or MDAC sources. This data is then easily used across the enterprise in management reporting functions, cost accounting and reporting.

Fenestrae's multi-tiered, RPC-enabled architecture allows for easy scalability and "plug and play" connectivity for other host or device connections.

Lotus

Cambridge, MA – 1-800-346-1305, www.lotus.com

Product Name: Lotus Fax for Domino (FxD) 4.1.

Average Size of Installation: 4 ports, 500-1,000 desktops.

Hardware/OS supported: Brooktrout and Dialogic intelligent fax boards,Class 1, 2, and 2.0 modems, Lotus Domino Server and Microsoft NT 4.0.

Routing options: DID, DTMF, ISDN/EuroISDN, CSID, fax port, manual.

Inbox/Outbox integrations: Tightly integrated with Lotus' Notes client. Incoming faxes can be routed directly to Notes Mailboxes. Outbound faxes are composed using the standard Notes e-mail memo form. Outbound faxes can be simultaneously addressed to e-mail and fax recipients.

IP enabled: Not at present.

Pricing: Fax for Domino Enterprise: Five or more Fax Ports, unlimited number of users, $9,495. Fax for Domino Team: Up to four fax ports, unlimited users, $4,995. Fax for Domino Pro: Up to two fax ports, unlimited users, $1,995.

How best to buy: Lotus authorized resellers.

Fax service bureau relationships: None.

Lotus Fax for Domino is the perfect solution for the Notes lover in all of us. Complete integration with the Lotus Domino server means system administrators conduct all management and configuration tasks using the already-familiar Domino Directory database, locally and remotely. It's Dynamic Load Balancing feature allows multiple servers to be clustered together for maximum efficiency. Finally, inbound faxes can be routed to an individual's e-mail inbox securely in the office or on the road, using Notes' and Domino's built-in security features.

Omtool

Salem, NH - 603-898-8900, www.omtool.com.

Product Name: Fax Sr. 3.1

Hardware Supported : Brooktrout, Dialogic, NMS intelligent fax boards and Multitech external Class 2 modems.

OS Supported: Windows NT, UNIX and AS/400.

I/O options/Routing: Inbound: DID, cover page, DTMF, CSID, OCR; Outbound: global routing and least-cost routing.

Inbox/Outbox integrations: Microsoft Mail or Exchange, Lotus Notes, cc:Mail, any SMTP-type mail systems.

IP enabled: Yes.

Pricing: Fax Sr. Server with unlimited users and all clients included: $4,995. Fax Sr. Server with IP (required for each server with IP Fax functionality): $2,495. A lighter version, Fax Sr. Express, starts at $795 for a 10-user, 2-line configuration.

How best to buy: Fortune 500 companies buy direct; others via TechData, Omtool's channel partner.

Fax service bureau relationships: GTE Internetworking,, Comfax, Xpedite / Premiere Technologies, and Cable & Wireless's SureCom service.

Omtool's Fax Sr.'s is a LAN Fax server with integrated IP Fax capabilities. In addition to the IP Fax Server, the Omtool IP Fax Satellite fax-modem card, installed in office PCs on the network, gives preexisting stand-alone fax machines an on-ramp to IP, intercepting the fax and shipping it as a T.37-compliant e-mail attachment. In remote offices, the Satellite functions as an IP network node for intra-company free faxing, and as a hop-off point to local PSTN fax destinations.

Omtool's desktop clients can reside on the desktop, run thinly from within a web browser, or from within existing e-mail integrations. Omtool's solution also provides system administrators with advanced tools to manage and control fax documents and fax communications. Fax Sr. also enjoys native integration with SAP R2, R3 and R4 systems for direct, high-volume faxing sans custom programming.

Optus Software

Somerset, NJ – 732-271-9568, www.facsys.com

Product Name: FACSys Version 4.6

Average Size of Installation: 250-1,000 seats.

Hardware/OS Supported: Brooktrout, GammaLink/Dialogic, Intel, Ferrari FAX and Class 1, 2, and 2.0 modems. Windows NT and NetWare.

Routing options: DID, DTMF, CSID/ANI/DNIS, Line/Channel, Routing to e-mail, server-to-server routing via IP, network printer, forwarding to remote fax machine, manual routing to FACSys user group via subscriber ID, and combinations of the above. Also, can do OCR routing via Onset Technologies' ThruFax product.

Inbox/Outbox integrations: E-mail: SMTP/POP3, Microsoft Mail, Exchange/Outlook, Lotus Notes, Lotus cc:Mail, Novell GroupWise, Novell MHS SMF 70/71. Voice-mail: Siemens' Xpressions 470, Telekol, Lucent's Octel, and Intersis' Voixx.

IP enabled: Yes, with partners UUNET, GTE, and Concord, and server-to-server directly.

Pricing: MSRP is $45 per client license, $995 per server, $49.95 for FACSys' Exchange Connector (unlimited number of users).

How best to buy: FACSys Authorized Resellers.

Fax service bureau relationships: Premiere/Xpedite and Graphnet.

Optus Software Inc introduced FACSys in the LAN marketplace in 1990. Native integration with Exchange, being rolled out globally by Microsoft, provides FACSys with an Outlook-like appearance and support for double-byte (read: Asian-language) characters. Rich Text Formatted (RTF) messages created in Asian Exchange clients can now be rendered on the sending side, ensuring full font and format support.

FACSys' Least Cost Routing looks at a wide range of parameters – dial strings, message properties or connection costs – to automatically route faxes to other fax servers or service bureaus, via Internet, Intranet or PSTN.

Optus provides uninterrupted support programs through IBM Global Services. Optus' can also give users fax and e-mail transmission capabilities without requiring installation and distribution of the full FACSys desktop client. Finally, its Active Fax Messaging (AFM) development kit lets network administrators fax-enable all kinds of preexisting business applications with an ActiveX-like ease.

Tobit Software

Montréal, Quebec - 514-392-9220, www.tobit.com

Product Name: FaxWare 5.2,

Average Size of Installation: 4 ports, 50 desktops,

Hardware Supported : Class 2 modems, ISDN boards, intelligent fax cards and serial boards,

Routing options: DID, ISDN, CSID, line routing, manual routing,

Inbox/Outbox integrations: Any SMTP-based e-mail package and Microsoft Exchange Tobit also makes David, its own unified messaging server for integration with voice mail and e-mail.

IP enabled: Yes, with HP's JetSend technology.

Pricing: Varies.

How best to buy: Through authorized resellers and distributors.

Fax service bureau relationships: None

Tobit is a leading Novell partner, so if you're using NetWare and need a fax server, begin here. The software can fax-enable any DOS or Windows-based application and has dual-NOS applications: Netware 3.x-5 and NT 4.0. It can be upgraded to a full unified messaging server. Tobit also provides open APIs.

TopCall

Wayne, PA – 610-688-2600, www.topcall.com

Product Name: TopCall Fax and Unified Messaging 7.2x.

Average Size of Installation: 10 ports, 250-1,000 desktops.

Hardware/OS supported: Windows NT. TopCall's own proprietary RISC-based fax processors support of 2 to 60 fax lines.

Routing options: Outbound: Analog, E&M, ISDN (BRI and PRI), T-1. Inbound: Line routing, E&M DID, Bell DID, CSID.

Inbox/Outbox integrations: TopCall's own One Server voice mail system, Microsoft Exchange, Lotus Notes, Novell Groupwise, SMTP, IBM MQ Series.

Other application integrations SAP, PeopleSoft, JetForm, mainframes. Flexible APIs for intelligent handling of File Transfer and Microsoft Foundation Classes.

IP enabled: Yes.

Pricing: Hardware and software from $20,000 for entry-level systems to $500,000 for a high-volume, fault-tolerant enterprise system. Pricing based on base CPU, number of fax processors, applications to be supported and, if relevant, number of desktop "plug-in" applications.

How best to buy: TopCall sells only directly to the corporate customer, often with the involvement of system integrators.

Fax service bureau relationships: Xpedite/Premier Technologies

TopCall has entered the unified messaging server marketplace lately with voice and e-mail integration. Its modular hardware and software approach make it the perfect solution for the highest-end applications. You find support for paging, telex, and – found among production-strength servers – EDI-to-fax conversion, for automated document delivery to the EDI-unaware. TopCall also comes with high-level administration, accounting, and monitoring features, including SMNP.

The stability of the product has been proven, based on a turnkey sales and support system and the fact that the company manufacturers all hardware and software. TopCall solutions have been installed in more than 50 countries, through 20 direct sales and support offices worldwide.

Fax Service Bureau

Fax service bureaus are becoming an extremely popular outsourcing alternative to maintaining overdesigned equipment on your company's premises. Fax service bureaus also solve such problems as "Why can't I receive my faxes in my e-mail box?" "Fax this document to 3,000 people, now!" and "My international fax transmission bill was how much, last month?!"

Besides solving these problems, there are many other reasons why it makes good sense to hire a fax service bureau: for the occasional, large-scale fax broadcast; for special services (like web-based fax retrieval) offered to a few key workers (mobile execs, sellers) – even for outsourcing all your corporate faxing.

As fax guru Maury Kauffman of the Kauffman Group (Voorhees, NJ – 856-651-1651, www.kauffmangroup.com) says, "It's a question of core competency: What business are you in? Do you want to purchase fax server hardware, learn new software and locate additional ports on your PBX – all for a resource that may be underutilized? Do you want to review RFQs, make a big commitment, then discover that workers can't understand the PC-based user interface? When AC power fails or a T-1 takes a hit during your big, time-critical faxblast, do you want it to be *your* problem? Of course not. So, outsource! When you hire a bureau, there's no capital outlay. Start-up time and technical complexity are minimal. Top bureaus have state-of-the-art, fully redundant, fault-tolerant systems that guarantee little or no downtime and put no new pressures on IS/telecom managers. And if it turns out that your people don't like the user interface – find another bureau. As they used to say, the customer is always right."

Plus: A bureau can almost always provide more capacity and infrastructure than you can affordably buy. The average LAN-based fax server might have six ports. Running flat out, it'll take more than 11 hours to send the same one-page fax to 4,000 recipients, figuring one page/one port/one minute. Send a personalized fax to the same 4,000 people (so that each page sent must be rasterized individually), and you'll almost surely bring the system to its knees. By contrast, the average fax service bureau has thousands of ports (Xpedite has over 10,000), and lots of distributed CPU power: They can handle big jobs fast, without choking.

Service bureaus also have a strong financial incentive to explore new technologies for transmitting faxes at lower cost, and with greater reliability. Many have teamed up with partners to provide low-cost, IP-based fax transmission to major cities around the country, or around the world. A few have built out their own global network of nodes.

For all these reasons, some of the biggest companies – companies with the technical expertise and money to handle fax any way they like – have decided to outsource all or some of their faxing to service bureaus. You have to figure that IS / telecom management at ARCO, Cisco, CNN, Disney, Hilton, HP, IBM, ITT, Motorola, S&P, and UPS know what they're doing.

Maury Kauffman has been professionally evaluating fax bureaus since 1990 and has been fortunate to be involved in several of the world's largest bureau acquisitions. From his files of nearly 100 vendors, here's his list of America's Top Ten Enhanced Fax Service Bureaus, listed in alphabetical order:

DFI Communications (Rockford, IL – 815-398-9009, www.DFIComm.com) – Complete fax and voice bureau with many services designed for specific vertical markets. If you're one of them, look no further.

eFax.com (formerly JetFax) (Menlo Park, CA - 650-324-0600, www.eFax.com) – Provides the largest, free fax-to-e-mail service and with nearly one million users, critical mass is just around the corner.

Epigraphx (Redwood City, CA – 650-298-0100, www.Epigraphx.com) – Full-service automated and customized fax, e-mail and voice solutions. A company founded by marketers, for marketers: half of Silicon Valley uses 'em.

Fax2Net (Rockville, MD - 301-721-0400, www.fax2net.com) – 40 country, IP-based, fax messaging service available via the desktop, dialers or Philips Electronics' IP addressable fax machine (Fax2Net are partners with Philips). Interested ISPs should contact them.

Fax4Free (Marina Del Rey, CA – 1-877-Fax4Free, www.fax4free.com) – Browser-based, advertiser-sponsored service allowing anyone in the world to send faxes, within the U.S. Free.

FaxtsNow (Los Angeles, CA – 310-445-1000, www.FaxtsNow.com) – High volume fax broadcasting service for every application, with some of the most sophisticated merging capabilities available anywhere.

GTE (Burlington, MA – 617-873-2000, www.GTE-Fax.com) – Helps medium and large organizations gain control of their faxing by providing global, IP fax solutions. Unquestionably the leader among carrier-class and tier 1 ISP level providers.

MediaTel (San Francisco, CA – 415-883-4300, www.MediaTelUSA.com) – Fax broadcasts domestically, nearly a million pages every day. Call them if you need 5,000 pages sent in 5,000 seconds.

NetMoves (formerly FaxSav) (Edison, NJ – 908-906-2000, www.NetMoves.com) – Offers every IP-based fax service to all size organizations, but real strength is fax-enabling the back-end of web applications and host-based services such as faxing trade confirmations, direct deposit notifications, invoicing and shipping alerts.

Premiere Document Distribution (aka Xpedite Systems) (Eatontown, NJ - 732-389-3900, www.Xpedite.com) – by far, the world's largest fax broadcasting service bureau. If your requirements include global distribution, then bigger is better.

Fax outsourcing is fast approaching the billion dollar mark in annual sales. Services include: Fax broadcasting, Fax-on-Demand, fax to e-mail, e-mail to fax, virtual fax mailboxes, never-busy fax (inbound), guaranteed fax delivery (outbound), web-to-fax, desktop faxing, free faxing, fax-to-OCR (for survey tabulation, etc.), e-mail broadcasting, voice message broadcasting, interactive voice response (IVR), and universal messaging. The breadth of features offered by these bureaus is fantastic.

When making a selection, Maury's advice is: Don't be cheap – you will get what you pay for. Fax service bureaus know exactly what their costs are and what their competition charges. If you try play one bureau against another to shave a penny per minute, the best bureaus will turn down your business. The top bureaus are busy and smart, and refuse to play this game. Like any other vendor, look to build a mutually-beneficial relationship based on trust, not cost.

Finally, remember that hiring a fax service bureau is like calling Hertz. They pay for the car, maintenance and insurance. You only pay for "gas."

Firewall

A gateway (software or hardware) that sits between two networks, buffering and scrutinizing their data traffic so that malicious hackers cannot gain access to the system.

First Party Call Control

Gives the individual user control of the telephone call. With TAPI or TSAPI acting as the interpreter, telephone control functions are handled

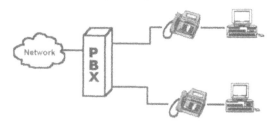

by the user's desktop PC instead of by a PBX, server, or some form of central equipment.

Here's two examples of first party call control: An incoming call is routed to a desktop telephone, which detects the Caller ID information and sends it to the desktop PC. The PC uses the information to do a "look-up" of the caller in a database, finds the record describing the caller, and displays it on the monitor screen just as the call rings the user's phone or headset. Calls can also be re-routed to a pager, another device or another location after a certain number of rings.

In an outbound scenario, the user displays a personal directory and mouse clicks on the name of a person to call, then selects "Dial". The individual's PC sends control signals to the phone to go off-hook and dial the appropriate number.

Flying Pig

The September 1995 issue of *Computer Telephony* magazine bore a cover adorned with flying pigs. The cover story was entitled, "Teaching Pigs to Fly, or PBX Makers to Listen." From that moment on, the sym-

After the Expo, an animal rights group immediately wrote to Newton to complain about the pig races, and so this would-be timeless institution ended as quickly as it had begun.

Harry last major pig promotion occurred when he had some yarmulkes manufactured with a pig emblem on top. Shortly after a big giveaway at one Expo, a group of angry rabbis confronted Newton, who proclaimed his innocence. The creature depicted on the yarmulkes was not a pig, said Newton, but a rare, featherless bird (no doubt found only in the depths of the Australian outback).

When Newton and Friesen sold Telecom Library, Flatiron Publishing and the CT Expo and Conferences to Miller Freeman (now CMP Media) the days of the flying-pig-as-corporate-symbol were over. After the sale, Newton gave friends and associates some specially built watches with – yes indeed, a pig face.

The flying pig is gone but not forgotten. It managed to achieve legendary status in the telecom industry and was one of the more inspired promotions that Harry Newton, that Master of Promotion, ever concocted.

Frame Relay

A shared-bandwidth wideband packet-based data interface standard that transmits bursts of data over WANs, often used for LAN interconnects. It's essentially a simplified, faster form of X.25 signaling and is carried over links ranging from 56 Kbps to T-1 or even T-3s. Frame-relay packets are in the form of "frames" that vary in length from 7 to 4,096 bytes. Users reserve (and pay for) a specific data rate called Committed Information Rate (CIR), but users can often burst data at higher rates. Extra frames are discarded if the carrier's network has insufficient capacity to handle them.

The Frame Relay frame structure is based on the Link Access Procedure-D (LAP-D) protocol. In the Frame Relay structure, the frame header is altered slightly to contain a 10-bit Data Link Connection Identifier (DLCI), which is the destination address of the frame, and congestion and status signal bits in place of the normal address and control fields, which the network sends to the user. The Frame Relay header is 2 bytes in length.

Frame Relay frames are transmitted to their destinations via logical paths from an originating point in the network to a destination point, called Virtual Circuits (VCs). Virtual circuits may be permanent (PVCs) or switched (SVCs). PVCs are set up administratively by the network manager for a dedicated point-to-point connection (the DLCI field represents the address of the frame and corresponds to a PVC); SVCs are set up on a call-by-call basis.

Although not generally used for voice or video, there are adherents of Voice over Frame Relay (VoFR). Frame relay is a relatively secure network infrastructure because it is managed and traffic is often monitored by the frame relay carrier, and so Virtual Private Networks (VPNs) are often created using frame relay.

Recommended website:
The Frame Relay Forum: http://www.frforum.com

bol of the flying pig, created during one of Harry Newton's more inspired moments, "took off" as they say.

Newton decided to adopt the flying pig as a sort of corporate symbol. The 1996 CT Expo in Los Angeles had a pig motif. Newton spent $10,000 on two large (12 foot) inflatable pigs. Each pig was stationed at an entrance to CT Expo at the Los Angeles Convention Center and "welcomed" visitors. The following year an alleged radical "telecom liberation" group kidnapped one of the pigs on the final day of the Expo.

The "liberators" were in fact a group of telecom workers who decided to play a trick on Newton and just have some fun. They sent in a photo of the pig (which had been damaged during the pig-napping) with a list of demands that Newton was to publish in the magazine, which he did. The group believed that Newton had a tremendous sense of humor and would appreciate the many levels of mirth engendered by these merry pranksters' outrageous act. Newton, of course, would have put them all in jail if he could have gotten his hands on them.

As things turned out, the pig was repaired and shipped back, and Newton never managed to apprehend the culprits.

The pig motif continued. Newton brought in an organization that held pig races along an oval track at the Expo. The pig races were held at three hour intervals. Visitors to the Los Angeles Convention Center during CT Expo were greeted to a booming voice periodically emerging from the Public Address System that announced such statements as: "Attention! The next pig race will be held at 2:45!" The pigs wore racing jackets that were embroidered with the names of sponsoring companies, such as Dialogic.

G.711

The ITU-T Recommendation that's become the standard for converting analog toll-quality voice lines into toll-quality digital ones. G.711 is also used extensively in newer forms of IP telephony. G.711 is a Pulse Code Modulation (PCM) scheme operating at a 8 kHz sample rate, with 8 bits per sample. Since a signal must be sampled at twice its highest frequency (as dictated by the Nyquist theorem), G.711 can thus encode frequencies between 0 and 4 kHz. The algorithm allows for the transmission and reception of A-law and u-law voice by converting linear Pulse Code Modulation (PCM) input signals (13 bits for the international A-law standard and 14 bits for u-Law) sampled at an 8 kHz sampling rate into an

8-bit compressed floating-point PCM representation. The actual technique of converting between linear PCM and G.711 PCM is known as "companding" (compressing/ expanding). The analog-speech waveform, once having been encoded as binary words is then transmitted serially, at digital bit rates of 48, 56, or 64 Kbps. ISDN channels and digital phone sets on digital PBXs use G.711. Support for this algorithm is required for ITU-T compliant videoconferencing (the H.320 / H.323 standard).

Although considered to be a popular "codec" or "vocoder," G.711 excessive bandwidth output at 64 kbps is generally considered to be pretty much uncompressed, and is thus used as a reference against which the speech quality of voice compression algorithms with high compression and lower bit rates are measured.

When the H.323 IP interoperability standard ruled the world, IP telephony relied upon vocoder algorithms with higher compression rates than G.711, such as G.723.1 or perhaps G.729a. Those are still popular and telephony equipment still uses them, but use of G.711 is now on the upswing. One reason is that bandwidth is now a lot less expensive and so using high levels of compression just isn't as important as it used to be. Instead, people want toll-quality audio.

Also the cost per port on IP telephony systems and CT resource boards has declined so much that the intellectual property costs for G.723 and G.729a are starting to appear significant. Fortunately, G.711 doesn't have any intellectual property costs, so it's not just an easier vocoder to implement and a high density vocoder, but a it's also a cheaper vocoder. Technically speaking, of course, it actually isn't a vocoder at all.

G.723 and G.723.1

These are ITU-T standardized dual rate speech coders for multimedia communications transmitting at 5.3 and 6.3 Kbps. G.723 was designed for sending compressed digital audio over ordinary analog lines as part of the H.324 videoconferencing standard.

Once there was a G.723 coder that was ultimately folded into G.726. To avoid confusion, the ITU changed the name of the currently adopted G.723 coder to G.723.1. Thus, there is no real distinction between G.723.1 and G.723 with reference to the currently adopted G.723.1 standard.

G.723.1 encodes speech or other audio signals in frames using linear predictive analysis-by-synthesis coding. Speech signal input to the G.723.1 coder are 16-bit linear PCM samples, sampled at 8 kHz. G.723.1 encodes 240 sample frames (30 ms) of 16-bit linear PCM data into either twelve 16-bit code words or twenty-four 8-bit code words for the 6.3 Kbps rate and ten 16-bit code words or twenty 8-bit code words for the 5.3 kbps rate. Total algorithmic delay is 37.5 ms (which includes 7.5 ms of look-ahead). The 6.3 Kbps rate uses as its excitation signal Multi-Pulse Maximum Likelihood Quantization (MP-MLQ) "code book" search, while the 6.3 Kbps version is based upon the Algebraic-code-excited Linear-

Prediction (ACELP) system. It's possible to switch the G.723.1 frame stream between the two rates on the fly, at any 30 ms frame boundary. There's also an option for variable rate operation using discontinuous transmission and Voice Activity Detection (VAD), which removes the bandwidth-wasting silent pauses between words prior to transmission, filling in the non-speech intervals later with low-level white noise.

The result of the algorithm is approximately 4 kHz of near-toll quality speech bandwidth under clean channel conditions. Indeed, the quality of G.723.1 measured by the standard Mean Opinion Score (MOS), normally used to rate speech codec quality, is 3.98. Since a rating of 4.0 is considered toll quality, this means that the G.723.1 vocoder has 99.5% the audio quality of analog telephony.

Since it provides high quality speech voice compression and decompression over a narrow digital band, G.723.1 is used extensively in IP telephony and it is the default voice coder for H.323. Relative to the G.729/G.729A coders, the G.723 speech coders pass DTMF "touch tones" through with less distortion, though music and sound effects suffer a bit.

Four companies collaborated closely towards the development of the G.723.1 speech technology: The Audiocodes Ltd., the DSP Group, France Telecom, and the University of Sherbrooke.

G.729 and G.729 Annex A (G.729A)

G.729 is a Conjugate-Structure Algebraic-Code-Excited Linear Prediction (CS-ACELP) speech compression algorithm (audio codec, or "vocoder") approved by ITU-T that supports 3.4 kHz speech at 8 Kbps. G.729 Annex A is a reduced complexity version of the G.729 coder.

The G.729A vocoder processes signals with 10 ms frames and has a 5 ms look-ahead resulting in a total algorithmic delay of 15 ms. The input/output of this algorithm is 16 bit linear PCM samples converted from/to a 8 kbps compressed data stream.

G.729 Annex A (G.729A) speech coder was developed for use in multimedia simultaneous voice and data applications like DSVD, where it appears in the V.70 specification.

Lucent's SX7300P is based on a coder that was submitted as a candidate for the ITU G.729A standard in 1995. Although it lost out to the slightly better performance of the University of Sherbrooke candidate, it had by far the lowest complexity of all the coders submitted and generally scored very close to the winning candidate in Mean Opinion Score (MOS) tests. Because of the short timescales involved in the G.729A selection procedure, it wasn't possible for Lucent to fine tune their candidate coder in time to win the contest. Since the ITU tests further improvements have been made in the coder design while maintaining the benefits of low complexity and delay.

Also, since the ITU contest in mid 1995 the need for a coder for Internet telephony has arisen. At the time of the work in the ITU on G.729A and G.723.1 this was not envisaged as a major application for the new coders and consequently factors such as performance under high frame erasure, delay and complexity were either not in the terms of reference for the selection procedure or were given a low weighting in the process. Although based on the design of the G.729A candidate, the Lucent's SX7300P has been optimized specifically for Internet voice applications. The coder shares with the G.729A candidate the advantage of toll quality, very low complexity and low delay characteristics while incorporating additional features such as a new and powerful frame erasure mitigation algorithm.

G.lite

Low bit-rate (1.5 Mbps) ADSL that doesn't require a line splitter. Also known as "ADSL Lite." G.lite is defined by the Universal ADSL Working Group and was adopted by the International Telecommunications Union. See Asymmetric Digital Subscriber Line, xDSL.

Gateway

An electronic signal repeater and protocol converter (hardware or software) allowing access by a computer telephony system into and out of circuit-switched or packet-switched communications networks.

At one time a gateway was used simply to connect mainframes or minicomputers to LANs or WANs. Today, a "gateway" generally involves some kind of connection between PSTN technology and IP networks, such as an H.323 gateway or an H.320 gateway, which is a server or hardware device that acts as a "bridge" between a packet-based local network and the public-switched circuit world, enabling calls between a LAN endpoint and an H.320 (ISDN) endpoint. By having an H.320 gateway on the network supporting multiple 128K ISDN (BRI) adapters, ISDN no longer needs to be installed on every desktop. All the network users can share whatever ISDN ports are available on the gateway, reducing costs.

Gatekeeper

A gatekeeper is a sort of automated software version of a network administrator. While gateways provide interfaces to other networks, handles call control and address lookup, translation, and resolution services (routing calls from gateways to terminals within a "zone" of managed terminals on a LAN) and routes calls based on available bandwidth.

With IP telephony and video conferencing, if a gatekeeper is present on the network, an H.323 endpoint (caller) must contact the gatekeeper and ask permission to make a call. In accordance with company and LAN administrator policies, the gatekeeper allows the call to go through. To make a connection, an endpoint must reference the IP address of the remote endpoint it is calling, whether on a LAN or at the other end of a gateway. The gatekeeper will resolve an IP address for all endpoints under its control, given an e-mail address or an "alias" string or an extension, such as a phone extension. This information is normally configured in the calling endpoint at setup.

Every H.323 endpoint has an IP address, either a permanent one assigned to a particular network card, or a temporary one that is assigned at network login time, via something such as the Dynamic Handshake Control Protocol (DHCP). Once the H.323 video conferencing application is invoked, it immediately looks for an H.323 gatekeeper on the network. If the gatekeeper is found, the H.323 endpoint sends its IP address, extension or alias (provided by the user in the endpoint H.323 software application) to the gatekeeper, registering itself. If a gatekeeper is present on the network, an endpoint must be registered with the gatekeeper before calls are permitted between that endpoint and another endpoint for both initiating and receiving calls. If no gatekeeper is present, the H.323 endpoint is permitted to make calls without permission from a gatekeeper.

Global System for Mobile Communications.

Also called Groupe Speciale Mobile, the standard digital wireless service in 85 countries around the world. It comes in two versions, 900 MHz and 1,800 MHz (DCS1800).

GSM

See Global System for Mobile Communications.

H.100

PCI chassis / mezzanine bus version of the CT Bus interoperability specification that is part of the Enterprise Computer Telephony Forum (ECTF) client/server architecture for scalable computer telephony systems. H.100 is a high capacity bus with provisions for redundancy and compatibility modes to enable interoperability with devices developed to existing telephony resource buses like SCbus and MVIP.

The "Bus Wars"

Computer telephony applications must access network interface and media processing resource boards plugged into a computer. CT boards must access telephony signals and sometimes must work in concert with their fellow boards. Voice, video and real-time fax have low-latency requirements and are event-driven with isochronous data streams, but PCs use asynchronous buses and interrupts lacking any guaranteed real-time processing. Normally the host CPU would handle this kind of processing, but with the proliferation of boards in high density CT systems, it became evident that the CPU and system bus would be overburdened and much of this time-critical processing of isochronous telephony traffic should be offloaded somehow.

It became evident that what was needed was some kind of internal telephony bus or "CT bus," a mezzanine bus separate from the conventional local I/O board bus, to connect the distributed-switching interface circuits on each resource board. In this way CT boards could transmit separate digitized signals (data, voice, video, fax) simultaneously over a communication medium (a backplane or a ribbon cable) by quickly interleaving a piece of each signal from each board in succession (Time Division Multiplex timeslots) and putting them on the new bus, with each board capable of sorting out the signals and putting them back together in their proper order.

Because of the increasing demand to process real-time digital signals between specialized board-level hardware, several manufacturing groups developed competing, proprietary internal PC-based Time Division Multiplex (TDM) telecom buses for large voice mail and Interactive Voice Response (IVR) platforms that required proprietary cards and backplanes.

Dialogic Corporation, which had introduced the first multiline voice board in 1985, in 1986 gave the industry a means to connect multiple analog voice resources together through a special telephony bus designed for carrying the audio and signaling information. Called the Analog Expansion Bus (AEB), this bus was similar to a bundle of two-wire analog connections, this telephony bus eliminated the need for host processing of time-critical data and made the rapid growth of the CT industry possible.

As the industry continued to grow, it became necessary to create a more powerful digital telephony bus to handle the size and complexity of newer systems. In 1989, Dialogic developed the PCM Expansion Bus (PEB), the first digital bus standard for connecting devices from multiple suppliers, featuring Pulse Code Modulation (PCM) and Digital Time Division Multiplexing (TDM) for connectivity similar to that offered by a digital T-1 or E-1 trunk line. The following year, a group of seven board vendors, led by Natural MicroSystems (NMS) and Mitel, defined another TDM bus standard, called the Multivendor Integration Protocol, or MVIP-90.

Until 1989, Natural MicroSystems had been building and selling a board costing only a few hundred dollars that you could plug into your PC, yielding an instant voice mail system. But it only worked on a single telephone line, and Natural MicroSystems was convinced they needed a multi-line system.

So, in late 1989, NMS introduced a 4-line system. Brough Turner, an NMS founder and one of the developers of the 4-port board believed (as did the rest of the management team at Natural MicroSystems) that they needed to leverage the new product technology by actively pursuing partnerships. Their goal was to get into the T-1 market as quickly as possible, because that was where they thought the opportunities for growth lay.

Turner contacted every one of the dozen companies that by 1989 had pulled a T-1 or E-1 digital telephone trunk into a PC, most of which were also interested in data related applications and had developed a multi-board set to solve their problem. Typically they used a ribbon cable to connect the cards together, and all had a different schemes for routing calls between cards. Similarly, Turner was looking for a way to route telephone data from a T-1 interface to a Digital Signal Processing (DSP) board. He also wanted to do telephone switching inside the PC, initially in order to switch calls from a voice board to a fax board, and later to build small telemarketing systems that would let you switch operators among telephone trunks dynamically. Turner and other engineers at Natural MicroSystems developed a solution that leveraged a T-1 / E-1 prototyping board made by Mitel Corporation.

Turner's research led to a Natural MicroSystems product, but not initially to MVIP. The breakthrough came about by accident (as many breakthroughs do!). While making a West Coast sales trip, Turner had an empty space in his schedule, so he filled it by paying a courtesy call on GammaLink, in Sunnyvale, California (which would eventually become part of Dialogic). The two companies had a common customer who was using both GammaLink fax boards and Natural MicroSystems voice boards.

When Turner sat down in the GammaLink conference room, he saw that the white board was covered with T-1 information, and the GammaLink engineers told him they had been instructed to figure out how to give their fax boards T-1 interfaces and, since they didn't have experience with T-1, they had hired a consultant to brief them. Turner, seeing an opportunity, explained to GammaLink that he had just spent six months working on the problem and he almost had a complete scheme of what he intended to do. He offered to give his work to GammaLink if they would agree to follow what would now become an *ad hoc* standard.

GammaLink agreed in February 1990, and that was the genesis of the MVIP standard. Turner asked the companies he had contacted earlier to see if they were interested in a common standard. About a dozen showed interest in utilizing MVIP, and seven were prepared to spend money to send a representative to New York for an initial meeting, which took place in September, 1990.

At a press conference on September 18, 1990, the seven founding companies publicly announced MVIP, their support of MVIP, and the fact that they had MVIP products available or in development. The companies were Brooktrout Technology, Inc. (Needham, MA), GammaLink (now part of Dialogic Corporation, in Sunnyvale, CA), Mitel Corporation (Kanata, Ontario, Canada), Natural MicroSystems Corporation (Natick, MA), Promptus Communications, Inc. (Portsmouth, RI), Scott Instruments (now part of Voice Control Systems, Dallas, TX), and Voice Processing Corporation (Cambridge, MA), which is now part of Philips.

Interestingly, in 1990, when the MVIP standard barely existed, two "MVIP standard" products were already in beta test at customer sites and soon to go into production. MVIP was clearly not a well-thought-out marketing strategy designed to maneuver the competition off the game board.

H

It was more the response of engineers in several different companies who recognized the fact that they needed a standard solution to a difficult problem, and that they were more likely to succeed if they worked together. One of the products already in beta test in 1990 was a Natural MicroSystems product that combined one of their DSP-based voice boards with a T-1 trunk interface made by modifying a standard Mitel product. The original engineering notes that became the MVIP 90 standard included directions for modifying the Mitel board to make the T-1 interface.

Before the end of 1990 the MVIP-90 standard was available to anyone willing to pay $295 for it. The price included the technology, documentation and unlimited rights to do anything you wanted with it. Natural MicroSystems still owned the technology and the copyrights, but they also arranged for Mitel and Nippon Telephone and Telegraph (NTT) to sell the technology.

A year later, in the fall of 1991, MVIP reached a new milestone when Rhetorex (now part of Brooktrout) adopted the standard. Rhetorex' participation was a clear signal that MVIP was not only open, but that it was not going to be controlled by any single vendor. It was on its way to becoming a legitimate industry standard.

By late 1992, more than 140 companies had licensed MVIP. The standard had been designed into numerous proprietary systems, from PBXs to enhanced service platforms. Over 30 board-level MVIP products were available from 17 different companies. Licenses were being distributed by Natural MicroSystems, Mitel, and NTT International.

1992 also saw further organization within the active MVIP community, with the formation of a Steering Committee, an Education Committee, a Technical Committee, and a Standards Committee.

At the third annual MVIP conference in 1993, there were over 200 licensees, over 55 board-level products available, and an MVIP-specific integrated circuit chip, the Flexible MVIP Interface Circuit (FMIC) was available in production quantities. The MVIP Technical Committee presented a Connection Control API standard, and reported progress on Multi-Chassis MVIP.

Also, at the 1993 MVIP conference, the MVIP Steering and Education Committees unveiled GO-MVIP, the Global Organization for MVIP, an independent body being formed to further the work of the MVIP committees and to take the MVIP protocols to formal standards. With the advent of GO-MVIP, licensing was discontinued and the standard was put in the public domain.

While both Dialogic's PEB and MVIP-90 used a similar mezzanine bus at the top of the boards for communication, the two standards were completely incompatible. Developers wanting to take advantage of the new 512 timeslot, 2 MHz MVIP-90 bus would need to abandon earlier technological investments. Because of the size of Dialogic and Natural MicroSystems, vendors began to line up in opposing camps and the CT "bus wars" began. Even more worrisome was the fact that computer telephony systems were rapidly growing larger and larger, and would need even more bandwidth and higher performance than was available with the MVIP-90 standard.

Dialogic upped the ante in 1993, when it put together an open standards-based model for building computer telephony servers that defined a more robust telephony bus, the Signal Computing System Architecture (SCSA) which was immediately supported by 70 companies.

To meet the requirements of the SCSA Hardware Model, an SCSA working group defined the SCbus, a third-generation TDM bus. In addition to offering developers up to 2,048 timeslots using an 8 MHz clock, the SCbus pioneered a distributed switching model that simplified the way applications access devices and made it easier to scale systems from

small to very high densities. By July of 1995, the American National Standards Institute (ANSI) formally recognized the power and versatility of the SCbus for high density applications by adopting it as the only telephony bus standard for the VME platform (ANSI/VITA 6-1994).

Another key area of focus for the SCSA working groups was on software interoperability. A group of 30 SCSA supporters formed working groups that were backed by over 100 companies, who acted in an advisory role, to develop the SCSA TAO Framework specification. This specification defined the components necessary for multivendor resource management, call control, and application portability. This work highlighted the need for a new industry forum that could better focus on these and other interoperability issues.

In April of 1995, an *ad hoc* SCSA steering committee consisting of Digital Equipment Corporation, Dialogic, Ericsson, Hewlett-Packard, and Nortel established the Enterprise Computer Telephony Forum (ECTF) to foster an open, competitive market for CT integration through multivendor implementations of industry standards. The forum rapidly drew new membership as it began to tackle the SCSA TAO Framework 3.5 specification, part of which formed the basis for the original 1996 version of the ECTF S.100 standard.

Towards the end of 1995, GO-MVIP announced another telephony bus, H-MVIP. Single-chassis H-MVIP is a powerful, compatible superset of MVIP-90 that extends the system to handle up to 3,072 x 64 Kbps (3,072 timeslots or 1,536 full-duplex conversations) across the bus. Also, a family of multi-chassis MVIP standards (MC-MVIP) allowed MVIP systems to scale to any size.

One of the first projects of the MVIP Technical Committee, which was formed by the MVP community in 1992, was the formulation of specifications for a multichassis system. The market was sending clear signals that there was a significant demand for larger systems. To build larger systems, it would be necessary to expand beyond the ability for CT boards to communicate with each other in a single chassis. Instead, boards in one chassis would have to be capable of communicating with boards in other chassises, allowing for the construction of computer telephony systems of immense size.

Multichassis bus communication was a big challenge. As soon as more than one CPU is involved, coordination becomes a major issue. As soon as data must be passed from one box to another, outside the computer's internal bus structure, the design must deal with issues of latency, speed on the bus or cable between computers, interference from other traffic on the bus, etc. While the multichassis standards had to be totally compatible with MVIP-90, it also had to meet a number of additional technical challenges. Even more important, from a marketing perspective, the multi-chassis MVIP standard had to give developers the ability to create a single piece of application software that could then be used across a range of hardware platforms providing different levels of performance and capacity: Scalable applications using common software. And there must be a smooth cost curve as the system grows from one, to two, to three, and to n computers.

The first documentation for the MVIP multi-chassis standard was released in 1993. The first products based on this specification, Amtelco's MC1 adapter cards, were released in 1994. In 1995, Natural MicroSystems released its MC1 adaptor. Multi-Chassis MVIP defines a common approach and common software interfaces, with alternate physical media referred to as MC1 (TDM on twisted-pair copper cable with 1,536 time-slots per cable and 3,072 time-slots in a dual cable configuration), MC2 (Fiber Distributed Data Interface II carrying up to 1,536 channels worth of 64 Kbps data between hundreds of chassis over distances of up to 60 km), MC3 (Synchronous Digital Hierarchy / Synchronous Optical Network connections operating at 155 Mbps and providing 4,800 x 64 Kbps of isochronous capacity in a dual ring configuration), and MC4 (which uses virtual circuits within an ATM network to provide a full mesh of multiple point-to-point 64 Kbps channels between chassis in a multichassis MVIP system).

Still, the industry was demanding a new telephony bus that would not only offer higher capacity and more bandwidth, but would also be a viable solution to the major compatibility issues resulting from the increasing deployment of multiple buses.

The MVIP and SCSA factions were each vying for their piece of the developers' pie, and effectively split the CT industry in half. Some board vendors "sitting on the fence" tried to support both specifications; and the entire computer telephony industry was retarded as a result.

The difficulties in either choosing one bus over the other or supporting both buses within the same computer telephony market had created industry demand for a single, universal bus standard that eliminated compatibility issues and offered the capacity to support complex multi-resource systems (eg. voice, fax, text-to-speech, speech recognition, switching, etc.)

Thankfully, the so-called "bus wars" were put to an end when the Enterprise Computer Telephony Forum (ECTF) got together technologists from both sides and officially introduced "neutral" CT Bus standards for PCI and CompactPCI (H.100 and H.110, respectively) that work with existing SCbus and MVIP compliant products.

H.100 Appears

The H.100 interoperability specification, called the "CT Bus," was developed by an ECTF working group consisting of companies such as Dialogic, IBM, Lucent, Northern Telecom, Mitel and Natural MicroSystems. The ECTF announced the official H.100 specification on May 27, 1997.

The generic ECTF CT Bus defines a TDM (See Time Division Multiplex) single communications bus, with clock and frame system technology, across computer chassis card slots.

The CT Bus' synchronous, bit-serial, TDM transport bus operates at 8.192 MHz. It's driven by two clocks, two frame sync pulses, one backup network timing reference and 32 independent bit-serial data streams. Six additional clocks are sourced by CT Bus timing masters to ensure compatibility with other TDM buses. Compatibility modes permit operation at 2.048, 4.096 and 8.192 MHz.

CT Bus master-capable interfaces inter-operating with other buses must support data rates of 2, 4, and 8 Mbps. CT bus master-capable boards must provide 128 timeslots (channels) of bus-to-bus switching at any of the three data rates. CT Bus slave cards inter-operating with other buses must support date rates of 8 Mbps and may support data rates of 2 and 4 Mbps.

A CT Bus interface must be the clock master and must provide the compatibility clocks for MVIP and SCbus.

A CT Bus may be developed in conjunction with a number of hardware configurations. "H.100" was the first card-level definition of CT Bus, the PCI bus and form factor being the lucky winner. H.110, for CompactPCI, was the second implementation (See H.110).

Thus, the H.100 specification provides all the necessary information to implement a CT Bus interface at the physical layer for the PCI computer chassis card slot independent of software applications.

H.100 offers 32 data lines that can carry 4,096 bi-directional timeslots permitting up to 2,048 full duplex calls (vs. 2,048 for SCbus or 512 for MVIP) and 128 timeslots (channels) per stream in a 125-microsecond frame. Streams needing more bandwidth (such as those for video) are able to bundle timeslots into N x 64Kbps groups, yielding 128 Kbps, 256 Kbps, etc. A digitized voice stream uses the same bytewide timeslot in each of the 8,192 frames per second to achieve a 64 Kbps constant bit rate.

There are two synchronized clock lines so that if one is lost, all devices can sync to the other signal. There are in fact two groups of redundant clocks (the telephone network reference clocks and an A and B set of core clocks for frame and bit transfers), all of which can run in primary and secondary as well as master and slave configurations.

The H.100 CT Bus functions very much like two 8 MHz SCbuses running in parallel. The CT Bus features 32 data lines and 4,096 timeslots operating at 8 MHz, while the SCbus features 16 data lines and 2,048 timeslots. Both buses can run at the same rate and maintain a distributed switching model. However, to provide compatibility modes for MVIP-90 and H-MVIP, the CT Bus must be able to handle multiple types of clock signals, selectively downgrade portions of the bus to run at slower speeds, and incorporate a centralized switching mechanism to allow data exchange with all supported buses. 16 of the 32 data lines can be selectively downgraded to 2 MHz for MVIP-90, 4 MHz for 1,024 timeslot SCbus ISA mode or 8 MHz for the 2,048 timeslot SCbus mode.

A distributed-switching Integrated Circuit (IC) switches streams to different timeslots on the CT Bus or to the local card bus. Programmable switching transfers a stream from one TDM slot to another, generally within the same frame cycle to eliminate latency. Sometimes, however, a stream being transferred from one timeslot to another can't make it in time, and so the transfer finishes during the next cycle. This can cause problems for N x 64 bundles, with sequenced data switched in an incorrect order. To handle such situations, CT Bus also has the option of constant-delay switching, which consistently delays the timeslot switch to the succeeding frame, thus keeping sequenced data in order.

H.100 switches can connect the CT Bus not only to a local bus for transfer to the resource board, but also to an external bus when needed. This allows for local connections between resources (such as DSPs) situated on the same board, CT Bus connections between devices on different boards, or connections between the CT resource boards and an external TDM bus. Once your PC chassis is full to the brim with H.100 boards, you can continue to expand your system's capacity by connecting to CT resource boards in other chassis via an external bus connecting through an H.100 master resource board.

When H.100 / CT Bus appeared, *Computer Telephony* magazine consulted with experts on both the SCbus and MVIP bus – Jeffrey Kern, senior applications engineer at Dialogic and Cliff Spencer, product marketing manager for High Availability at Natural MicroSystems, respectively – and got their tips on moving toward and mixing in the new CT Bus products. It is, as one would expect, a little tricky, but not too bad.

Moving and Mixing H.100 and SCSA Cards

To distinguish SCSA from H.100 boards, just remember that SCbus boards will typically have an SCbus ASIC (e.g. VLSI SC2K [2000] or SC4K [4000]), while H.100 boards will have an H.100 bus ASIC (e.g. OKI ML53812). These ASICs are located very close to the SCbus (or H.100 bus) connector. Some board manufacturers may have implemented the bus interface functionality into a custom ASIC. Contact your board manufacturer if you're not sure.

But just because a board has the new H.100 connector on it doesn't mean that it's an H.100 board! Dialogic is installing H.100 connectors as an update to many of its PCI-based SCbus products, which allows for easier implementation of mixed SCbus-H.100 systems. Dialogic calls these products "H.100 CT Bus interoperable," since they can interoperate with H.100 compliant products, but they still run in one of the existing legacy bus modes (SCbus).

All of the necessary SCbus signals are located on the H.100 bus, because the H.100 bus is a superset of the SCbus. Were a customer to use SCbus products that have the original SCbus connectors, he / she would have to use Dialogic's Transition Adapter to interface the board's SCbus connector to the H.100 cable's connector. The need for this adapter is eliminated by using the H.100 connector everywhere, including on the SCbus board.

Still, when you need to configure a mixed SCbus-H.100 bus system with SCbus boards that have the 26-pin SCbus connector on it, you must use Dialogic's Transition Adapter. You would also use the Transition Adapter when you need to connect H.100-connectored PCI SCbus products to SCbus-connectored ISA (or PCI) SCbus products.

As with all similar adapters, use of this device may reduce the total number of boards that can be used in a system. Also, if using the Adapter results in some unusual effect you may want to pull out some boards to determine your system's sensitivity to the Adapter. Note that use of some adapters may create a physical clearance issue with some PC chassis' side or top. Keep this in mind.

Physically, whereas the SCbus used a 26-lead flat ribbon cable to connect the cards, the new H.100 bus uses a 68-lead finer-pitch ribbon cable. The new connectors are used on both the H.100 boards and the H.100 cables. The H.100 cable can be a maximum of 20 inches in length (SCbus was 21 inches). An H.100 cable can have no more than 20 connectors (drops), typically one inch apart. (Dialogic currently offers H.100 cables in 4, 8, 12, and 16 drop configurations, though you are under no obligation to use them.)

As for the number of timeslots you now have to work with, the ISA / PCI version of the SCbus provided 64 timeslots on each of 16 streams (wires), totaling 1,024 timeslots, while the new H.100 bus has between 2,560 and 4,096 timeslots across 32 total streams, computed as follows:

- The upper 16 streams always operate at 128 timeslots / stream, providing 2,048 timeslots.
- The lower 16 streams are further divided into four smaller groups; each of which provide 32, 64, or 128 timeslots / stream on the four streams in that group. Each group of four streams can be independently configured to operate at one of the three rates just mentioned, which correspond to 2.048 MHz, 4.096 MHz, or 8.192 MHz bit rate.

Because of the possible combinations, the lower 16 streams can contribute anywhere from 32 x 16 = 512 to 128 x 16= 2,048 timeslots (H.100-only system). Typically this value will be 64 x 16 = 1,024 timeslots for mixed SCbus-H.100 systems.

All of this ultimately means that the H.100 bus will indeed operate with existing SCbus products. You can reserve the lower 16 streams of the H.100 bus to operate with the SCbus (operating at 4.096 MHz, or 4 MHz nominal). In this case the upper 16 streams are reserved for traffic between H.100 boards only.

However, whereas the SCbus didn't require bus terminations, H.100 does need terminations for four signals on the bus. Terminations are enabled on boards located at *both* ends of the bus, but never in between. Termination techniques might include physical jumper(s) and/or using a soft-jumper technique.

Soft jumpering eliminates the need for a user to install physical jumpers by using programmable registers controlled through a user interface. Some implementations of soft-jumper may have drawbacks or limitations that may or may not be important to the user. You should discuss this with your product representative.

As for placement of the boards, H.100 boards are PCI only, so you must have a free PCI slot for each H.100 board. Also, as with SCbus, you can't have more than 20 boards (total, including SCbus boards) in a system. Boards must be placed such that no more than seven inches of unused cable exists between any two H.100 boards.

One important board is the system's clock Master, designated to provide reference timing for all boards attached to the H.100 bus (and/or SCbus). This board must derive bus timing from either a digital network (e.g. "loop", if available), or as a last alternative from its own local oscillator (e.g. "independent") on the board.

The clock Master board is typically located at either end of the bus. This is much more important for H.100 than it was for SCbus – H.100 attempts to treat the bus as a transmission line, which was impractical for the SCbus.

Also, in a mixed H.100-SCbus system, the user must keep board capability and system configuration issues in mind:

The clock Master must be an H.100 (PCI) board with a network interface (e.g. T-1 or E-1). It cannot be an H.100-connectored (PCI) SCbus board nor a standard SCbus (ISA/PCI) board. H.100 boards require certain signals that can only be provided by another H.100 board and routed over the H.100 cable.

There are a lot of confusing claims about boards that "support" H.100. H.100 boards that only run in compatibility mode (SCbus) or support only the CT Bus without all of the compatibility modes (H.100-only) are not H.100-compliant.

It's important for customers to probe what a vendor means when they claim "support for H.100". Make sure that the board that you intend to use as the system's clock Master has the capability to derive timing from a network *and* that it can provide both H.100 Core Signals *and* Compatibility bus signals.

From the Dialogic perspective, there are three known "types" of PCI boards that exist today, based on telephony bus type, host bus type, and number of telephony bus connector pins:

- True H.100 boards, with H.100 connectors (68-pin); H.100/PCI/68-pins, or type HP68,
- SCbus boards, with H.100 connectors (68-pin); call this type SP68, and
- SCbus boards, with SCbus connectors (26-pin); call this type SP26.

(Note that these reference designations are not officially recognized but merely created as identifiers for this discussion.)

Locate the SP26 type boards (if any) in the PCI slots closest to the chassis' ISA slots. Next, locate the HP68 type boards in the next available PCI slots closest to the SP26 boards. Lastly, locate the SP68 boards (if any) in the next available PCI slots closest to the HP68 boards.

The HP68 board closest to the SP26 boards should use the Dialogic Transition Adapter if SP26 and/or ISA-based SCbus boards with the 26-pin SCbus connector are used in the system. Also, this same HP68 board should have termination enabled. It's highly recommended that this HP68 board also be designated as the system's clock Master. (Another HP68 board could be the clock Master, though it is best to source both SCbus and H.100 clocks from the board that uses the Transition Adapter).

The last PCI board, which should be an SP68 board, must also be terminated. If no termination provisions exist on this board, then re-arrange the boards within this group to allow you to do so. If no SP68 boards are used in the system, then you must also terminate the HP68 board at the end of the HP68 group opposite from the board with the first termination.

You may notice that your user interface (UI) looks different for H.100 bus products. Obviously, it this depends on the vendor that has developed the boards and user interface that you purchased. Realize that several new features have been put into H.100, that could have an impact on configuration. Some of these include:

Routing: The number of timeslots on the bus is now configuration-dependent. Some vendors may make handling this more intuitive than others (e.g. stream & timeslot identifiers vs. timeslot only). Fixed bit rate buses, like most implementations of SCbus, could get by with identifying a logical "timeslot" because the bit rate was fixed; a logical "timeslot" is always correlated to a specific time on a specific wire. With H.100, things are no longer fixed, but configuration-dependent.

There are also new "compatibility" modes: For working with mixed-bus systems (e.g. H.100 & SCbus).

For example, you may be required to designate the number of streams that will operate at a certain rate. Some vendors may give the user more flexibility to custom configure, with potentially more apparent complexity, and hence confusion. (e.g. Streams 0-3 bit rate? (2/4/8) Mbps; Streams 4-7 bit rate? (2/4/8) Mbps; Streams 8-11 bit rate? (2/4/8) Mbps; Streams 12-15 bit rate? (2/4/8) Mbps.)

Other vendors may choose to simplify matters for you by limiting you to an all-or-nothing concept (e.g. SCbus boards in system? If so, bus operates lower 16 streams at 4 MHz bit rate and upper 16 streams at 8 MHz. If not, then all streams operate at 8 MHz). Other vendors may happen to use a clock fallback configuration for hi-availability systems by using the redundant clocks and a backup clock Master (Fallback Master) board, or soft-jumper technology for bus terminations.

H.100 and MVIP cards

H.100 provides a total of 4,096 bidirectional 64 Kbps timeslots. The increased number of timeslots provides greater communications capacity than ever before. This increase in capacity is possible because of the introduction of a 68-pin fine-pitch ribbon cable which is physically smaller than the existing 40-pin regular-pitch MVIP cable.

H.100-base boards can be interconnected with MVIP boards via a passive transition device, commonly called a "swizzle stick," that allows the connection of the different ribbon cables. In systems comprised of a combination of boards, the master clock must be an H.100 board.

H.100 master clock circuits also include compatibility clocks for driving existing boards that operate in clock clave mode. To facilitate operatoin with MVIP boards, H.100 allows individual data lines to be programmed in groups of four to operate at 2, 4 or 8 Mbps, allowing direct connection to existing boards at their native operating speeds.

The H.100 specification incorporates technology from GO-MVIP (Global Organization for Multi-Vendor Integration Protocol) such as the programmable operting speeds technique of H-MVIP and redundant clocks from MC1 Multi-Chassis MVIP.

As mentioned, programmable operating speeds provide support for the interoperation with MVIP boards. The redundant clock eliminates a single point of failure. If any telecom board fails, including the H.100 master clock, the system will continue to operate. H.100 offers developers and integrators extensive new capabilities. It brings more capacity than any existing bus, enabling developers to deliver larger and lower-cost applications. In the future, H.100's redundant clocks will be important as they will support hot swapping of boards in CompactPCI offering customers the option of a CT system with virtually no downtime. This enables automated call centers, IP gateways and voice messaging systems to remain up and running until there is a convenient time to replace a failed board.

MVIP includes not only a specification for an internal PC-based telecom bus but also the architecture for creating hierarchical swiching systems. H.100, an interoperable superset of MVIP-90 and H-MVIP bus spec-

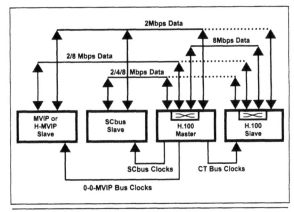

This diagram shows an H.100 "mixed" system with the possible data paths and clock distribution. Dashed lines represent optional paths.

ifications, is completely consistent with the MVIP architecture.

H.100 simply defines an alternate intra-chassis bus on which MVIP solutions can be built. The MVIP-95 device driver standard software interface directly supports H.100. And the advantages of the MVIP architecture, such as multichassis communications and unlimited scalability, are still in place.

Conclusion

With H.100 you'll finally be able to build applications that will connect to ATM 155Mb or T-3 lines. With H.100 many traditional telephony applications won't need a PBX to get the job done, since the new bus makes it possible for many cards to work with each other at high speed on the same bus and support many voice channels in the process.

There are some caveats, however:

- This SCbus / MVIP interoperability capability introduces the concept of centralized switching by so-called "master" devices), which significantly complicates routing software. Compliant master devices must also be able to drive the core CT Bus signals and all compatibility mode clock signals at 2 MHz, 4 MHz or 8 MHz.
- While it's easy to connect the new H.100 CT Bus boards to older SCbus and MVIP ISA cards, keep in mind that in such a "mixed" system the CT Bus master must handle multiple types of clock signals, selectively downgrade portions of the bus to slower speeds and incorporate centralized switching – all of which could degrade system performance where you may happen to need it most.
- There's no provision in H.100 for the hot swapping of cards, but since it's a ribbon cable bus, and the PC bus doesn't support hot swapping of cards yet, this isn't a big problem. The new Compact PCI (cPCI) format cards are hot swappable and the CT Bus for cPCI (H.110 spec) is on the way.
- Also keep in mind that the H.100 CT Bus spec is for *hardware interoperability* only. It does not address *software interoperability* for sharing those devices at the application level. However, the ECTF is addressing software interoperability through other agreements, such as the approved ECTF S.100 specification. The S.100 standard defines a standard, open API that allows developers to share a common set of telephony resources. This software standard lets developers create portable, interoperable applications that can be installed on any S.100 compliant telephony server and access the telephony resources managed by that server.

Still, vendors who were sweating away on large-scale products because they were running out of timeslots on SCbus or MVIP will love the new H.100 compatible cards. As quad T-1, T-3 and ATM cards arrive on the market, they'll be able to handle really high-density telephony solutions.

H.110

H.110 is the CompactPCI version of the H.100 standard (See H.100). The ECTF developed H.100 for PCI buses in desktop PC systems. H.110 is electrically identical to H.100, but developed for the CompactPCI form factor (See CompactPCI). It allows semiconductor vendors to implement a single chip design for both H.100 and H.110. The H.110 CT bus differs from H.100 in that it has extensions for the CompactPCI bus and uses different interfaces.

H.110 offers some advantages over H.100, such as hot-swapping ("live insertion") of boards in CompactPCI chassis from the PCI Industrial Computer Manufacturer's Group (PICMG). The prospect of building systems with virtually no downtime is attractive to the telecom industry as well as those constructing call centers, IP gateways, and unified messaging systems.

Also, instead of using a ribbon cable to connect the boards the way H.100 does, H.110 can communicate across the backplane using the CT Bus-defined P4/J4 connectors.

Additional features of the CT Bus on cPCI are: 32 data streams at 8.192 Mbps; over 260 MBps bandwidth; redundant clocks (8.192 MHz data clocks and 8 KHz Frame clocks); redundant reference clocks (8 KHz 1.544 MHz and 2.048 MHz); a maximum of 20 slots; and front panel or back panel I/O.

H.320

H.320 is a digital narrowband (N-ISDN or up to several combined ISDN "B" channels, with typical data rates ranging from 128 Kbps to 384 Kbps) standard for visual telephone systems and terminal equipment. H.320 can even interoperate with 56/64Kbps PCM phones. For conferences of more than two parties, Multipoint Control Units (MCUs) can be bought or rented.

H.320 is actually an "umbrella" standard for the ITU-T videoconferencing standards that was finalized by the International Telecommunications Union (ITU) in 1990 and appeared in manufacturers' codecs shortly thereafter. H.320 specifies H.261 for video compression, H.221, H.230, and H.242 for communications, control, and indication, G.711, G.722, and G.728 for audio signals, and several others for specialized purposes.

The related H.321 standard merely converts the H.320 standard from N-ISDN operation to B-ISDN transmission; all of the H.320 infrastructure (H.261, H.221, H.242) stayed the same to help guarantee interoperability between the two types of networks.

H.322 is used for multimedia conferencing over LANs that provide guaranteed quality of service (QoS), while H.323 was originally designed for videotelephony over Ethernet and "non QoS packet networks" and later became the principal communications interoperability standard for the Internet (See H.323).

Videoconferencing over analog telephone lines between two modems follows the H.324 standard, which combines features of both H.320 and H.323 to be efficient enough to achieve videotelephony over POTS lines.

The H.261 video compression standard is mandatory for most of the standards and recommendations listed here. H.261 has been so important that it has fostered a common misconception that H.261 compliance alone is sufficient to guarantee interoperability. H.261 supports image sizes

Common Interchange Format (CIF) (352 pixels x 288 pixels at video frame rates of 7.5, 10, 15 and 30 frames per second) and Quarter CIF or QCIF (176 pixels x 144 pixels x 30 fps). H.263 is a later ITU-standard video codec based on and compatible with H.261. It has better compression than H.261 and can transmit video at QCIF and Sub Quarter Common Intermediate Format (SQCIF) which is 128 x 96 pixels up to 30 fps.

H.320 Advantages

Since H.320 uses conventional digital lines it shares the advantages intrinsic to digital switched services: QoS and standard dialing procedures. Combining three or four low delay ISDN "B" channels yields sufficient bandwidth to provide a very realistic videotelephony experience. Prior to the rise of H.323, H.320 was the standard in videoconferencing, with much interoperability among many vendor's H.320 terminal equipment.

H.320 Disadvantages

One disadvantage is the fact that H.320 systems are inherently point-to-point systems; you need an MCU for videoconferencing more than two locations. Also, not everyone has digital lines leading into their home or business, which has prevented the widespread adoption of any kind of universal standard such as H.320. Many PBXs still don't support a comfortable bandwidth for H.320 systems or terminals, and even if they do there generally aren't any niceties such as three-way conferencing, voice mail, etc.

H.323

H.323 is an interoperability standard that specifies the modes of operation required for different audio, video, and/or data terminals to work together. It has become the dominant standard of Internet phones, audio conferencing terminals, videoconferencing and "web conferencing" (including data conferencing) technologies.

Although H.323 is associated in the public mind with Voice over IP (VoIP), H.323 started out as a videoconferencing standard for LANs. As videoconferencing technology became a cheaper proposition in the mid-1990s, it was thought that the age old dream of videotelephony for the masses would finally be realized. Unfortunately, at the time ADSL and cable telephony had not yet taken hold, and ISDN was the only digital access technology having anywhere near the bandwidth needed to support an acceptable videoconferencing user experience, which is why all videotelephony systems followed the H.320 standard.

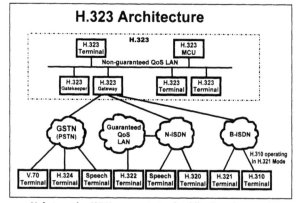

H.323 Architecture

Unfortunately, ISDN was notoriously difficult and expensive to obtain at the time (See ISDN) but analog phone and LAN connections to PCs had become quite popular, so vendors devised a different strategy and a different standard, H.323.

H.323 is an intricate, multifaceted umbrella standard, ratified by the International Telecommunications Union (ITU) back in October 1996, that, ironically uses many of the same audio, video, and data compression standards as H.320. But instead of transferring bits via a standard circuit-switched connection, the bits now move via packets over connectionless networks. "Connectionless" refers to transmissions in which no leased line or dialed channel hooks sender to receiver.

Each packet is sent and addressed independently over a packet-based network, which includes those that are IP-based (including the Internet) or Internet packet exchange (IPX)-based LANs, enterprise networks (ENs), metropolitan-area networks (MANs), and wide-area networks (WANs). H.323 can be used in an "audio only" form (for IP telephony), or both audio and video (videotelephony), or audio and data (whiteboarding), or even audio, video and data (full video and dataconferencing).

Besides the multiple codecs, call control, and channel setup specifications required to move audio, video, and data, H.323 at present has some means to attempt to lessen the effect of changing delays, long delays, and lost packets, thereby providing real-time IP telephony applications with the QoS they need.

H.323 is independent of the packet network and the transport protocols over which it runs and does not specify them. The protocols that are specified by H.323 are as follows:

- Audio CODECs
- Video CODECs
- H.225 Registration, Admission, and Status (RAS)
- H.225 call signaling
- H.245 control signaling
- Real-time Transfer Protocol (RTP)
- Real-Time Control Protocol (RTCP)

H.323 specifies four kinds of components or "entities" as they are called in the recommendation:

1. *End-point terminals.* Used for real-time bidirectional multimedia communications, an H.323 terminal can either be a PC or a standalone device, running H.323 and applications. H.323 terminal is what every user encounters when dealing with IP-telephony services, since audio support comprises the basic H.323 service. But the chief objective of H.323 is to get all connected multimedia terminals on the network to interoperate and communicate with each other. H.323 terminals are compatible with H.324 terminals on POTS and wireless networks, H.310 terminals on B-ISDN, H.320 terminals on ISDN, H.321 terminals on B-ISDN, and H.322 terminals on guaranteed QoS LANs. H.323 terminals may be used in multipoint conferences.

H.323 terminals must support the following:

- H.245 for exchanging terminal capabilities and creation of media channels.
- H.225 for call signaling and call setup.
- Registration / Admission / Status (RAS) a component for registration and other admission control with a gatekeeper.
- RTP / RTCP for sequencing audio and video packets.
- Q.931 for call signaling and call setup.

All H.323 endpoint terminals must support voice communications, and specifically the G.711 audio codec, in both A-law and M-law versions. Optionally, endpoints can support G.722, G.728, G.729, MPEG1 audio and G.723.1. A terminal should be able to send using one audio codec while receiving with another, provided it supports both.

Video and data support in H.323 are optional, such as video codecs, T.120 data-conferencing protocols, and MCU capabilities.

If video capability is provided, then H.261 Quarter Common Interchange Format (QCIF) must be supported. Optionally, a terminal may be capable of supporting H.261 Common Interchange Format (CIF) mode supporting a video image 352 x 288 pixels in size and H.263 Quarter CIF (176 x 144 pixels, or one fourth the information of a CIF image) If H.263 CIF mode is supported, then H.261 QCIF must also be supported. H.261 is also used by H.320, H.321 (for Broadband ISDN and ATM) and H.324 (PSTN analog videoconferencing). H.261 specifies the full encoding of video frames in some instances (major image changes, and encoding only the changes between a frame and the previous frame in other instances (such as the movement of objects within the same scene).

Another optional video codec is H.263, which is backward compatible with H.261, has better compression than H.261 and transmits video at QCIF and Sub-QCIF (SQCIF) rates.

2. *Gateways.* These devices act as "bridges" connecting one standard protocol (and its associated network) to another. In the past they've been used for computer data protocols, but gateways are now being designed to interconnect different telephony and videoconferencing standards. H.323 gateways convert digitized voice into addressed IP packets, translate protocols for call setup and release, map call and control signaling, map and convert media formats between different networks, and do network monitoring and administration. But most important, H.323 gateways connect to and transfer data streams between H.323 and non-H.323 networks so as to translate between H.323 conferencing endpoints and other terminal types. This function includes translation between transmission formats (i.e. H.225.0 to H.221) and between communications procedures (i.e. H.245 to H.242).

For example, a gateway can connect and provide communication between an H.323 terminal and Switched Circuit Networks (SCNs). SCNs include all circuit-switched telephony networks, e.g., Public Switched Telephone Network (PSTN). Gateways can provide interworking between such dissimilar networks as H.323 and H.320, H.324, and the PSTN (SS7,

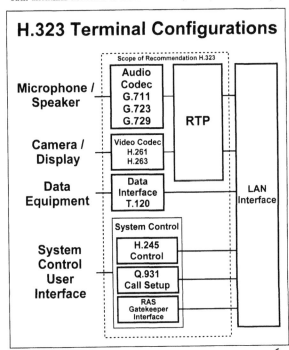

H.323 Terminal Configurations

269

ISDN, ATM). One major application of H.323 gateways is IP telephony, where an H.323 gateway must connect an IP network and switched network, such as converting H.323 on the LAN to H.320 (ISDN) on the wide area network (WAN), or H.323 on the LAN to analog voice lines on the WAN. Internet telephony caught the public's imagination with a variation on this: A gateway between H.323 on the Internet and local analog service – Hop-On-Hop-Off servers (HO-HOs) which used the Internet to transport voice, bypassing conventional long distance phone lines, but using the local phone network to get the call over "the last mile" to a user's conventional phone.

H.323-based voice-over IP systems employ hardware that converts packet-based audio into Time-Division Multiplexed (TDM) audio used by the switched circuit network. This type of gateway is often called an IP Telephony Server, offered by companies such as VocalTec, and Cisco sells devices such as their Multi-service Access Concentrator that allows for H.323-based telephony.

H.323 gateways usually sit at the boundary between a Local Area Network (LAN) and the networks of the outside world, though high-end gateway-like devices that allow switching between the IP backbone and the core of the PSTN are also appearing.

A gateway usually consists of a switched-circuit network interface (T-1 or ISDN PRI interface cards) and NICs for communication with H.323 compliant devices on the H.323 network. CT resource boards equipped with DSPs will do the voice compression, packetization, echo cancellation and DTMF regeneration. There's also a control processor that supervises all gateway functions.

Terminals send information to gateways using the H.245 and Q.931 protocols. With the appropriate transcoders, H.323 gateways can support terminals that comply with H.310, H.321, H.322, and V.70. The number of H.323 terminals that can communicate through a H.323 gateway as well as the number of circuit-switched network connections has not been standardized and is left up to the manufacturer of the gateway. Gateway designers also have tremendous leeway when planning for the number of simultaneous independent conferences supported, audio / video / data con-

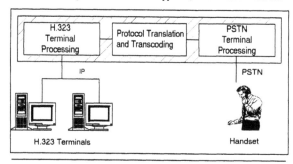

H.323 Gateway Architecture

version functions, and the inclusion or exclusion of multipoint functions.

A *proxy gateway* is a special form of gateway that maintains secured connections between H.323 sessions. For example, the Cisco Multimedia Conference Manager offers a conferencing proxy feature that provides QoS, traffic shaping, and security and policy management for H.323 traffic across secured connections.

It may come as a surprise that gateways are an optional component of an H.323 network, but an H.323 gateway is not really needed for communication between two terminals on a uniform network supporting H.323, such as a LAN.

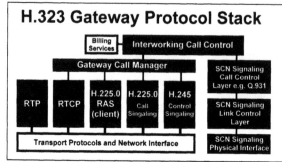

3. *Gatekeepers.* These act as virtual switches. Though not required in an H.323 system, a gatekeeper when present serves as an H.323 network's "brain" or central point for switching all calls within its domain, or "zone." A terminal must first contact the gatekeeper and ask permission to make a call. Following the LAN administrator's policies, the gatekeeper will permit the call to go through. Alternatively, a gatekeeper may instruct endpoints to connect a call signaling channel directly to one another, to bypass handling a H.225.0 signal itself. Optionally, the gatekeeper may act as defined in Q.931 to support supplementary services (e.g. call forward, call transfer). The gatekeeper can also register users and keep track of access policies and billing information.

Two important call control functions are handled by gatekeepers. The first is address translation from LAN alias addresses for terminals and gateways to IP or IPX transport addresses, as defined in the RAS specification, which involves the use of a translation table that is frequently updated by registration messages. To make a connection, a terminal endpoint must reference the IP address of the remote endpoint it is calling, whether on a LAN or at the other end of a gateway. The gatekeeper will resolve an IP address for all terminals under its control, given an e-mail address or an "alias" string or an extension, such as a phone extension. Thus, a gatekeeper earns its keep when you're trying to make an IP call but you don't know the IP address of the terminal you want to call.

The second function is bandwidth control, which involves support for Bandwidth Request, Confirm and Reject (BRQ / BCF / BRJ) messages. This may be based on bandwidth management, an extension of bandwidth control, which involves determining when there isn't any available bandwidth for a call, or if there is no more available bandwidth when a call in progress requests additional bandwidth. Some endpoints permit an upper bound for bandwidth utilization to be specified in the terminal endpoint, but, in most instances, bandwidth is controlled by an H.323 gatekeeper. For instance, if a network manager specifies a threshold for the number of simultaneous conferences on the LAN, the gatekeeper can refuse to make any more connections once the threshold of the maximum number of simultaneous conversations is reached. This limits the total conferencing bandwidth to some fraction of the total available – the remaining capacity is left for e-mail, file transfers, and other LAN protocols.

The collection of all terminals, gateways, and Multipoint Control

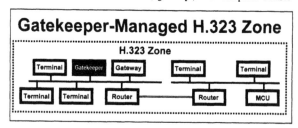

Units (MCUs) managed by a single gatekeeper is known as an *H.323 Zone.* Zones are essentially administrative conveniences similar to a Domain Name Server (DNS) domain. A zone includes at least one terminal and may include gateways or MCUs. A zone has only one gatekeeper. A zone may be independent of network topology and may be comprised of multiple network segments that are connected using routers or other devices.

A third gatekeeper task is admissions control, which uses RAS messages to authorize LAN access using ARQ / ACF / ARJ / H225.0 messages, based upon call authorization, bandwidth or other criteria set by the system manufacturer. It may also be a null function that admits all requests sans filtering.

A fourth stipulated function is zone management, which involves performing the previous three tasks for all terminals, gateways, and MCUs within its zone of jurisdiction.

Although a gatekeeper is logically separate from H.323 endpoints, designers of H.323 systems may incorporate gatekeeper functions into gateways and MCUs.

Although they are not required, gatekeepers provide important services such as addressing, authorization and authentication of terminals and gateways; bandwidth management; accounting; billing; and charging.

Gatekeepers may also provide call-routing services. Indeed, gatekeepers can be used to route H.323 calls, allowing them to be controlled more effectively and to allow for accurate billing services to be established by service providers. This service can also be used to re-route a call to another terminal if a called endpoint is unavailable. Also, a gatekeeper intelligent enough to route H.323 calls can also be programmed with routing logic to help balance and distribute call traffic among multiple gateways.

Gatekeepers can also support multipoint conferences. Normally a gatekeeper would receive H.245 control channels (but does not have to process them) from two terminals in a point-to-point conference. If the users decide to expand the conference to include more parties, the conference thus switches to multipoint mode and the Gatekeeper then redirects the H.245 control channel to a Multipoint Controller (MC).

4. *Multipoint Control Units (MCUs).* Also called Multipoint Controllers (MCs). H.323 defines two types of multipoint conferencing – centralized and decentralized.

Centralized vs. Decentralized

Decentralized multipoint is multipoint conferencing on the network done without the aid of an MCU.

Centralized multipoint conferences demand use of an MCU. MCUs mix audio and switch video between two or more endpoints, so they can do such things as conference multiple H.323 endpoints as in the case of three or more videoconferencing terminals. MCUs consist of a Multipoint Controller (MC), the endpoint that is responsible for processing call setup H.245 requests among all terminals by negotiating between terminals to determine what common audio or video codec to use (setting the conference to lowest common denominator) and managing conference resources. The MCU may also handle the media stream, though ordinarily the MCU leaves the media stream processing up to the Multipoint Processors (MPs) at the conference endpoints which can locally mix, switch and process audio, video, and data, sending the resulting streams back to the participating terminals. The MP may also provide conversion between different codecs and bit rates and may use multicast to distribute processed video. The MP can also provide video selection and audio mixing in a decentralized multipoint conference.

All terminals send audio, video, data, and control streams to the MCU point-to-point. Centralized multipoint is also commonly used with ISDN videoconferencing which uses an H.320 MCU, where all endpoints either call into, or are called from, the MCU. For H.323, the MCU sits on the network and can be in either hardware or software form. A software-based MCU running on a network server may cost little compared to a hardware-based MCU, but its performance is limited to the server's processor speed. A hardware-based MCU such as a CT resource board, however, generally supports more simultaneous users than a software-based MCU by dedicating a port with its own DSP to each user.

MC and MP capabilities can be built into a dedicated component or be part of other H.323 components. The gatekeepers, gateways, and MCUs are logically separate components of the H.323 standard but can also be combined into a single physical device.

Multicast, Unicast and Broadcast

H.323 supports multicast transport in multipoint conferences. Multicast sends a single packet to a subset of destinations on the network without replication. On the other hand, unicast sends multiple point-to-point transmissions, while broadcast sends to all destinations. In unicast or broadcast, the network may get bogged down as packets are inefficiently replicated throughout the network. Multicast transmission uses bandwidth more efficiently since all stations in the multicast group read a single data stream.

Decentralized multipoint conferences can make use of multicast technology. Participating H.323 terminals multicast audio and video to other participating terminals without sending the data to an MCU. However, the *control* of multipoint data is still centrally processed by the MCU, and H.245 control channel information is still transmitted in a point-to-point mode to an MCU.

Each endpoint supports multicast and can selectively choose what they broadcast or receive. For example, a user at a terminal may decide to listen to audio from only the endpoint currently "talking." Or the user could listen to audio from all endpoints, mixing them and producing audio from all participants. Decentralized multipoint requires terminals driven by powerful processors to perform audio mixing.

H.323 Compliance

You hear H.323-compliance claimed by many vendors, all up and down both the technological and marketing food chains, from Microsoft NetMeeting to carrier-class SS7 signaling gateways. But in reality H.323 "compliance" is no assurance of interoperability. That's because the standard leaves lots of room for interpretation. Videoconferencing platforms, by nature and necessity, have been more successful at picking compatible H.323 implementations. VoIP gateway makers are just now catching up. The interoperability problem has gotten large enough to grow its own standards initiative, administered by a new activity group within the International Multimedia Teleconferencing Consortium (IMTC) called iNOW!

Presumably, members who sign up for iNOW! will find ways of achieving and demonstrating true interoperability between different vendors' gateways and gatekeepers. The IMTC promises that its interoperability agreement will work in harmony with whatever is decided by the European Telecommunications Standards Institute (ETSI) TIPHON and ITU-T Study Group 16. For more information on iNOW! and those companies pledged to support its interoperability agreements, go to the Web at www.imtc.org.

To date, it's taken an IP wholesale carrier with the size and muscle of ITXC to make two different vendors – Vocaltec and Lucent – sync up

H

the H.323 implementations in their gatekeepers closely enough to exchange VoIP traffic. Even here, it is (to date) just for ITXC member POPs; through general availability is in progress. Lucent, Vocaltec, and ITXC, not coincidentally, were the triumvirate initially pushing iNOW! before it was handed over to the general marketplace.

While the ITU and others work on making H.323 more truly interoperable, they're also working on making it slimmer and better suited to voice traffic over the Internet.

Some of the problems arise from the fact that H.323 wasn't initially devised for a voice-carrying network. The standard came from the multimedia videoconferencing world, typically only concerned with a few highly intelligent endpoints. "That standard was hijacked by the IP telephony community, because it had everything we needed and more," says Josh Adelson, market development manager of IP telephony at Brooktrout Technology (Needham, MA – 781-449-4100, www.brooktrout.com). But it doesn't address the high densities and dumber endpoints of VoIP. "It asks too many questions about capabilities before moving media," he explains.

"The arrow thrown at [H.323] is that it doesn't scale," says Joel Hughes, product marketing manager at Natural MicroSystems (Framingham, MA – 508-620-9300, www.nmss.com). Hughes is in charge of the Fusion IP platform, which optionally incorporates market-leading H.323 protocol stack from RADVision (Mahwah, NJ – 201-529-4300, www.radvision.com).

Version 2.0 and 3.0

Version 2 of H.323, ratified in January 1998, and version 3, now in the works, have updates aimed exactly at some of the problems inherent in a standard meant for a relatively few intelligent endpoints.

Michelle Blank, vice president of global marketing at RADVision, explains some of those changes. They mainly involve enhancements to the H.323 gatekeeper, which rides herd on a "zone" of multiple gateways. First, Version 2 implements a FastConnect, two-step codec negotiation when only voice codecs must be agreed upon. It enables a media channel to open without a full H.245 media exchange (H.245 specifies the control channel governing capability exchanges and commands).

Version 2.0 added specific support for IP telephony, fax and ATM. Version 2 has extensions to non-H.323 directory services DNS and LDAP; and to authentication, authorization, and accounting servers such as RADIUS for Internet dial-up callers and Diameter for billing applications. Address translation can now occur across telephone number, URL ID, e-mail ID, as well as IP addresses. Version 2.0 also introduces the concepts of pre-granted registration request, overlapped sending of numbers for speed, including Q.931, gateway resource availability messages, and alternate gatekeepers for redundancy. H.450.1, another Version 2 enhancement, introduces a mechanism for supplementary services. To date, those include blind-call transfer and call forwarding. To come: call hold, call waiting, call park and pickup, and message waiting.

Version 3.0 (under development at the time of this publication) will add real-time fax, and gateway-to-gatekeeper and gatekeeper-to-gatekeeper support. Version 3.0 will have additional features aimed to solve other problems inherent in the previous versions of a standard designed to accommodate a relatively few intelligent terminals per session. Version 3 also begins to address large-scale production networks, bandwidth management, and more stringent quality-of-service (QoS) issues. It will also deal with SS7 integration and its role in the transfer of enhanced services to and from the PSTN.

SS7 integration is also a very important piece of the converged tele-

com scenario, because IP telephony needs to duplicate the reliability and exceed the range of services that consumers have grown to expect from PSTN such as follow-me, 800 numbers, *69 call back, *70 call block, caller ID, and voicemail. To do that, VoIP networks have to consult the great wealth of subscriber information now sent from service control points (SCPs) in SS7 networks. So we're beginning to see gateways that connect SS7 signaling to Internet-based networks. Product announcements on this front come from Lucent, Alcatel, and Nortel to name a few.

Just as SS7 call control signaling travels over a separate out-of-band network, the changes to H.323 gatekeepers and gateways suggest a similar two-connection scheme:

- Signaling sent over a signaling gateway, with an internal or external call agent for call control of trunking gateways, VoATM gateways, residential gateways, network access servers, and circuit and packet switches; and
- A separate gateway for media streams, using real-time transport protocol (RTP).

This division into two separate boxes is what's meant by a term we're going to hear more of: The "decomposed gateway." And there isn't necessarily a one-to-one relationship between the components of such a decomposed beast. One signaling gateway will be able to manage many RTP media gateways.

The idea of the decomposed gateway is to move call control intelligence out of the gateway and back into the centralized network. As Michelle Blank says: "The initial H.323 conception is peer-to-peer: intelligence in the end-points." She says that this approach tends to fall down when you get to such things as hand-held devices, where you can't store buckets of code. The tradition of the PSTN is the reverse; all the intelligence is centralized in the network, the end point is a dumb phone.

"So the issue is how to hybridize the networks," says Blank. And the result – over the past 10 months – has been to "decompose" the gateway architecture into a media gateway and a media gateway controller."

Over at CT resource board maker Natural MicroSystems, Cliff Spencer, product marketing manager of IP Telephony products says: "Perhaps H.323 is still the king, but many customers we're dealing with now, particularly those in the carrier business deploying products in networks, are looking at the Multimedia Gateway Control Protocol (MGCP) and eventually to H.248 as the gateway control protocol that they're really interested in, since it allows for control of third party media gateway hardware. We're in the process now of getting those protocols in our new products. Although we're not running yet, we will be announcing those protocols shortly."

In case you're wondering about H.248, the industry has actually made progress in the standardization of gateway control protocols. MGCP and H.248 gateway control protocols have many similarities, but significant differences in command names, parameters and syntax. H.248 also introduces the concept of a physically "decomposed" multimedia gateway, or an H.323 gateway that can be divided up into distributed, functional sub-components. This allows H.323 gateways to be both highly scalable and enables the gateways to be built with components from multiple vendors distributed across multiple physical platforms. It also encourages leverage of widely deployed network capabilities such as SS7 switches. H.248 itself defines the protocols used between the various components.

Ultimately, from a system viewpoint, there are no functional differences between a decomposed gateway, with distributed sub-components potentially on more than one physical device, and a monolithic gateway.

"By itself, H.323 was never quite homogeneous enough," says Jack Chase, market segment director of IP Telephony for Natural

MicroSystems. "Carriers found it difficult to scale H.323, gatekeepers had to go into zones, and all this complexity was hard to administer. H.323 was essentially conceived for a LAN, and that concept is collapsing under its own weight in terms of the number of variants and annexes needed to extend its functionality. And H.323 doesn't necessarily directly address a lot of the emerging needs for carrier class IP implementations."

"Many things you wanted to do on the carrier side were difficult to do with H.323," agrees Spencer. "We've seen a lot of interest in MGCP and that will eventually morph into H.248. Also, the Session Initiation Protocol (SIP) is gaining ground. Although that protocol doesn't affect the board manufacturers as much, it nevertheless looks to be a hot protocol in the industry and some of our customers are either using it or are thinking of using it in the near future."

When asked whether all of these protocols can peacefully coexist on future CT boards, Spencer thought that all these protocols are useful, each addressing its own area of expertise. "SIP to some degree will support H.323, and certainly MGCP will exist on the gateway control side. MGCP and SIP do work nicely together," he says.

"Looking at MGCP for a minute," adds Chase, "whether it's on the board or on the host processor controlling the board, essentially it offers a network API to access the functions in the board. In the case of MGCP those are gateway functions. So it's a just new interface to control the CT board, whereas something such as H.323 is more of a protocol for establishing relationships in the network. You actually specify things like 'I want to set up an endpoint and here's your what your remote site should be,' and that sort of thing."

This suite of emerging software protocols will be somewhat ubiquitous across different resource boards, so that it will eventually be expected that a product provides a full complement of standard protocols, such as MGCP interface, depending on what the product is expected to do.

As gateway decomposition has been promoted, more rival standards to H.323 have emerged and are vying for widespread adoption. As discussed previously, the Media Gateway Control Protocol (MGCP) is the decomposed contender now standing tallest among a welter of precursors from Bay Networks, Ascend, Bellcore (that's SGCP), and Level 3 (IPDC). MGCP and SGCP got their big push when CableLabs standardized on it for voice over cable. CableLabs, for its part, grew in influence after AT&T's purchase of TCI told the market that cable was a serious contender for broadband to the home, Regional Bell Operating Companies (RBOCs) be damned.

MGCP's messages are simple, transmitted in ASCII text, and are designed to allow a call agent to control a media gateway. RADVision has publicly announced that its gateway decomposition toolkits will support both MGCP and whatever is worked out jointly by the Megaco Internet Engineering Task Force (IETF) and ITU-T working groups in the closely related H.GCP/MEGACO (Media gateway control protocol). A trip to the IETF's Web site (www.ietf.org) shows five Megaco drafts; a first "stable definition" is due by early August (when this issue goes to press). Another rival from the IETF is session initiation protocol (SIP).

Brooktrout's Adelson says that MGCP, while still in a more embryonic state, has gained critical mass. "A lot of people are working on it, and it's gaining most ground in large carrier-scale implementations. We don't see it eclipsing H.323 but living alongside it."

"MGCP and H.323 will exist in the network together," echoes Jeff Lawrence, CEO of Trillium Digital Systems (Los Angeles, CA – 310-442-9222, www.trillium.com), which provides protocol stacks for both. "MGCP is a way to pull apart the pieces of the gateways and control them from somewhere else in the network. The classical model is sig-

naling gateway, media gateway, and call agent. The signaling gateway is where you try to concentrate the intelligence and conserve the [SS7] point codes, and where you spend time and effort ensuring high availability because it's going to control a lot of other equipment. The controlled media gateways, where you have the codecs residing, will have a lower price point and lots of ports. You aren't quite as concerned with reliability here because you can re-route to another box."

Now this model starts to create some interesting opportunities, says Lawrence, because you can have SS7 relayed from the public network through that call agent. You can have SIP connections terminated there, all controlled by MGCP. You have the ability, theoretically, to start providing unified services. Ultimately, you should be able to start pulling together the gatekeepers, SCPs, and call agents into one platform that straddles the boundary between the Internet and the public network.

"Over the longer term, I think you'll see that there isn't a real need to support a separate signaling network using SS7 as a transport. You'll still hold on to the application part and the user parts, ISUP and TCAP, and so on, but those will start running over IP instead of running over the lower layers of SS7. That will get us to an infrastructure where you can still support circuits, or packets. That won't happen until the service providers and carriers are confident that the IP network can provide them the low latency, availability, and security necessary. It won't happen tomorrow, but that's where I think it's heading in the longer term. I heard that the cost of running a network was 20% access and transport, and 80% was the back end – the operational provisioning, billing, and services. So integrating infrastructures, is really attractive from the carrier perspective. Charge $30 a month for cable, and $35 a month to get broadband Internet access, and for $35 and X cents a minute, get voice also," says Lawrence.

RADVision (Mahwah, NJ – 201-529-4300, www.radvision.com) began life in 1993 as a provider of audio and video gateways between IP- and ISDN-based networks. This connectivity made, and still makes, perfect sense at the juncture of the medium-bandwidth (typically 384 Kbps or three ISDN BRI lines) PSTN for business teleconferencing and the IP LANs that relay the conference from the conference-room server, over the cubicle walls, to the desktop.

Those ISDN-IP gateways predate the standard. But back in March 1996, RADVision was one of four companies invited to define an H.323 standard as part of the ITU. They were then first to market with a standards-based H.323 protocol stack, and first out with version 2. They now have about 75% to 80% of the H.323 protocol stack market, and are the only ones to provide software and hardware building blocks for all four H.323 components: terminals, gateways, gatekeepers, and enterprise-level MCUs.

RADVision has historically steered clear of its customers' turnkey VoIP gateway market, staying instead with data, voice, and video gateways, but has done two things to change that: (1) It has added a PRI interface to its LAN-to-WAN H.323-ISDN gateway, thus announcing a L2W-323P with built-in gatekeeper and support for all major CO switch and PBX protocols; and (2) it has agreed to build a voice gateway to be sold under the Siemens / 3COM name.

As far as RADVision's H.323 stack goes, developers can license either source or object code for Windows and UNIX operating systems as well as for VxWorks and pSOS real-time operating systems.

"RADVision's script language and API offer access to more than 500 functions," says Danny Levin, RADVision's vice president, engineering. All correspond to capabilities in the standard, including such basics as call make, call drop, and on-the-fly codec choice.

"We solve more problems at the stack level," says Levin when asked how one vendor's stack can differ from another. "Elemedia," for example, "forces you to solve them in the application. In our implementation, RAS can be transparent to the app, while in Elemedia you have to open sockets, you pack and unpack the messages according to ASN.1 rules and send them on the network."

There's a lot of confusion around the term "gatekeeper," admits Michelle Blank. "People don't realize that there are different kinds – carrier-class versus stand-alone for an IP manager, versus imbedded gatekeepers for imbedded components, versus gatekeepers for the service creation platform that has yet to be announced by anyone." For a good treatment of gatekeeper technologies, see RADVision's Handbook on H.323 Gatekeepers, available on the web at www.radvision.com.

"Maybe all you need is basic routing," says Blank. "If you're a carrier, you may need a gatekeeper with a great deal of flexibility for the way you do billing and provisioning. A gatekeeper here is where you plug in LDAP and RADIUS. Maybe you want something highly configurable: a gatekeeper toolkit."

Perhaps you'll want RADVision's new NGK-100 NT-based gatekeeper. It's designed to manage the company's OnLAN suite of hardware building blocks: The L2W-323P PRI Gateway, BRI Gateway, and the MCU-323.

The gatekeeper is H.323V2, and as such lets managers control bandwidth and access policies for an enterprise. It also supports 60 calls and 300 registrations on the network, and comes with built-in multimedia IVR capabilities. It keeps neighbor gatekeeper lists for optimized intergatekeeper communication, supports non-RAS endpoints, and makes use of Version 2's H.450 supplementary services for line and terminal group hunting. It comes with a toolkit, which in turn has extensions for service-creation IP environments.

RADVision licensees, including IP telco Networks Telephony (El Segundo, CA – 310-563-3900, www.networkstelephony.com), has licensed RADVision's gatekeeper with back-end suite for billing, least-cost routing, and real-time authentication. "That would take a traditional telco years and millions of dollars to implement," says Blank. The company integrated it with its customer support system, PC-to-phone, and phone-to-phone service.

RADVision's stack is also the default stack included with the NMS Fusion board, as well as the one that is bundled with the Dialogic IP Link DM3 board. NMS has followed the decomposed model, however, and hedged its bets, splitting the composite protocols between the board and the host processor.

Says NMS Fusion's Joel Hughes, "The interesting thing about MGCP and SIP is that they still rely on RTP for transport, so we've implemented the RTP stack, the real-time processing of the media stream, in Fusion on the hardware, and then let the higher levels of the stack, whether RADVision H.323 or SIP or MGCP, run on the host processor."

NMS customers have developed VoIP gateways along with call agent software. NetSpeak (Boca Raton, FL – 561-998-8700, www.netspeak.com), for example, is selling H.323, Version 2 gateways and gatekeepers to locate and control any of NetSpeak's other call management products: Route Server, Media Server, Event Management Server, WebPhone, Mini WebPhone, and Gateway Exchange. However, it is also selling SGCP-compliant call-agent software on a separate server, using SGCP to control the Fusion-based gateway. Two others in this general category of NMS customers are Telesoft (Austin, TX – 512-373-4224, www.telesoft.com) and IPVerse (Mountain View, CA – 650-919-0600, www.ipverse.com).

"They're providing gateways, but their primary business is really the call agent software itself," says Hughes.

Is this a whole new category? "It looks like it," answers Hughes. "In the H.323 world, we thought about gateways and gatekeepers. In the SIP and MGCP world, we think more about the decomposed model: media

gateways, signaling gateways, with call agents that control everything.

"Product-wise, we're seeing the call agent and the signaling gateway in the same box. The signaling gateway probably also has SS7 and SGCP/MGCP, which connects it to the public switched network for the high-level signaling. The media gateway is controlled by a call agent, and media gateways talk to each other using RTP. That's why we said for Fusion, let's definitely implement RTP in the hardware, since the high-level signaling is probably running in a separate entity, anyway. That's like the H.323 model, in that the gatekeeper has turned into the call agent.

"There's no clear winner or loser here. H.323 has a good foothold today; there are real products that work. It's very appropriate in the enterprise, clearly. H.323 has the T.120 component for data conferences, so I think it's going to have a life primarily in the enterprise.

"If you're building out a 'greenfield' (from scratch) network, you'll start with MGCP and maybe SIP because it looks like they scale better, and it's easier to find people who understand it. The IETF-driven stuff is simple. H.323 is quite complex, there aren't that many people in the world who understand it."

"The upshot is that there's still a lot of turmoil. The good news is that even if we wind up with three or four flavors of standards on the IP side, it's still not that bad. It depends on your real application: Are you an enterprise? Do you want to be multimedia? Are you concerned about bandwidth usage because you're bouncing off a satellite? Are you a new carrier using fiber? You've got to have different requirements and there's never really one size that fits all. Our assumption here is, 'Let's build platforms that definitely deliver low latency, high performance media streaming, and make sure that platforms are flexible enough to support multiple high-level protocols, H.323 or MGCP call agent.' You can write application software on top of our platform to do that."

Fusion developers can write their apps using several layers of APIs. They can write to RADVision's stack using RADVision's APIs. Below that, NMS's unified development environment, CTAccess, can be used.

Dialogic's (Parsippany, NJ – 973-993-3000, www.dialogic.com) Jim Machi, marketing manager for the IPLink board, reports that RADVision's H.323V2 will go into the next release of the IPLink DM3 board. Dialogic is now working on the split described by NMS, letting developers run MGCP or SIP "or whatever call control they want" on the host, while the packetization is performed on-board. H.323 can be run completely off the host, though, at this point.

"Our developers use IPLink for different things, so having options is important," says Machi.

As far as base-level protocol stacks are concerned, RADVision has a new competitor in Trillium Digital Systems (Los Angeles, CA – 310-442-9222, www.trillium.com), an 11-year-old company who made its name in source-code software for many international variants of SS7, ATM, ISDN, and frame relay communications.

Trillium has a great demo that deals with the industry's reservations about H.323's paunch and slowness. With its slimmed-down code size, Trillium can get its H.323 stack to perform more than 700 call setups and teardowns per second.

While that claim may or may not survive true network loads, Trillium's stack carries with it a good working relationship and a common architectural approach to its brother stacks for SS7, ATM, and ISDN. Trillium's H.323 tutorial, via its site or www.webproforum.com, is excellent.

Telogy Networks (Germantown, MD – 301-515-8580, www.telogy.com) makes Golden Gateway, a family of communication software modules that fits between the H.323 protocol stack and telephony. Telogy was ultimately acquired by Texas Instruments, which makes sense, since it loads its product on TI's DSPs for the likes of Cisco, 3Com, and Nortel.

Telogy's Tom Flanagan suggests that Telogy is to Texas Instruments as Microsoft is to Intel. "We have the equivalent of MS Office in the telephony world; compression, echo cancellation, tone detection, all those app-layer things you need to move a voice switch." The company is supplying the datacom vendors with the in-skin telephony they need to be the IP equivalent of CO or CPE switches.

Telogy has also integrated Golden Gate with its own SGCP / MGCP client signaling software. The company teamed it with NetSpeak's SGMP call agent software and Broadcom's cable modem chipset to come up with an IP-cable modem reference design, announced at VON 99. Broadcom and Telogy Networks sell to cable modem and head-end manufacturers, who in turn sell to cable TV system operators. NetSpeak markets its call agent software directly to cable system operators.

On the H.323 side, Telogy imbeds RADVision's stack and co-markets with the company.

Lucent Technologies has offerings in the H.323 and IP telephony space all up and down the techno-food chain, from a H.323 protocol stack, gateway, and gatekeeper to a carrier class SS7-integrated IPT platform. On the "elemental" end of the scale there's Elemedia (Holmdel, NJ – 732-949-2184, www.elemedia.com), an entire Lucent subsidiary dedicated to standards-based software for IP telephony.

Lucent / Bell Labs' Interoperability Lab tests H.323 interoperability between the Elemedia gatekeeper (which comes in NT, Solaris, and HP-UX versions), and gateways, and H.323 terminals of all kinds.

Gateway vendors Cisco (models AS5300 and 3640), Ascend (now part of Lucent, model MAX6000), and Lucent have all submitted models for H.323V2 testing and been shown to be basically interoperable after minimal tweaking. Part of the job of Kalpana Sheth, Interop Lab director, is to get other vendors to send their gateways in to the Labs for interoperability testing.

"We work as partners with gateway vendors," says Sheth. "In the majority of cases, it doesn't work the first time." They typically work out issues over the phone or with software downloads. They have also started to rent the lab to third parties who don't have the infrastructure or budget in place to conduct comprehensive, multivendor interoperability tests of their own H.323 products.

The setup includes a Definity ECS PBX, to bring in telephony connections for all gateways. A Lucent Cajun 115P Ethernet Switch supports the internal IP network. An internal dial plan lets any two gateways and/or H.323 terminals place calls.

Elemedia's customers are carriers. "We are targeting carrier-class gateways because we believe they're scalable and reliable," says Sheth. AT&T is their number-one customer, no surprise there, but MCI, Sonera of Finland, and Telefonica of Spain are three more licensees.

All Lucent products, Sheth assures, are now based on Elemedia's H323 stack. "That doesn't mean we won't support others," she adds. "If and other protocols emerge, we are open to supporting them."

The Lucent gateways tested include the carrier-class PacketStar IP GW1000, a trunking gateway for 28 T-1s; the PacketStar ITS-SP; and the ITS-E.

Rich Beckman, the Lab's technical specialist, demonstrated for *Computer Telephony* magazine the Elemedia NT gatekeeper that sits in the middle of all its interoperability tests. It comes with out-of-box core call control and registration functions and optional policy and management modules, and can be customized at several API levels. One exam-

ple of a low-level customization: a customer who wanted to charge IPT customers more for a richer audio codec. The low-level API allowed that customer to get into the H.245 capabilities message and reference the codec used in the call detail record for billing.

The gatekeeper comes with its own sample program that's great for debugging. Beckman added Cisco 3640 gateway endpoints, set traces, and we watched a graphical representation of the signals ricocheting back and forth against the gatekeeper: GRQ, RRQ, and so on as they registered and settled on a gatekeeper-routed, Q.931-signaled call.

Rich explained that the gatekeeper operator can make those decisions on an end point-by-end point basis. We then set up a NetMeeting 3.0 end point. Here, the gatekeeper enabled IP call waiting, because it could match the IP address with the phone number.

Over on Elemedia's Solaris gatekeeper, I saw a demo of SS7 integration, involving calling card verification and calling name delivery. The gatekeeper platform, a Sun Ultra 5, was also equipped with an SS7 interface card from DGMS, and the equivalent of a dummy SCP (SS7 switching control point). We registered a NetMeeting 2.0 end point using Elemedia's RAS helper app (this isn't needed with NetMeeting 3.0 because it registers itself). The gatekeeper's custom policy module took calling card DTMF in with its ARQ, and created an SS7 TCAP query to the internal SCP to validate the user. Similarly, Muraskin was able to consult the SS7 information to get a caller ID with name of an IP call.

The interoperability findings to date? That "out-of-box" interoperability isn't here yet, simply because engineers make differing choices in the way they locate their signals. A common example is alias information, which may be situated in one part of the RRQ message in one gateway, and sought for in a different place in the other vendor's gateway. Both H.323 "compliant," but not interoperable until both interpret the standard in the same way.

On the large-mammal end of the food chain, Lucent took aim at carriers' broadband convergence when they announced the Softswitch, also known as the PacketStar IP Services platform. This product is designed to bring the intelligent network's intelligence into packet telephony, regardless of the protocol used. Lucent offers the Softswitch as a universal signaling translator, converting SS7, MGCP, H.323, and SIP messages into its own internal protocol. Equally important, it's an open architecture and an open invitation to third-party developers. With this, Lucent also offers a service creation platform, the 7RE / Programmable Feature server. For more information on the Softswitch, go to www.lucent.com/pipsp on the web.

The first IP-based multimedia software app developed on the Softswitch was called MyNetWorks, a browser-based, universal messaging offering. MyNetWorks lets subscribers set follow-me schedules, filter phone calls, divert to voicemail, and set up conferences. It spans voice, video, e-mail, fax, wireless telephony, and PDAs. It will serve as a communications component to Lucent's Zingo, the wireless mobility Internet portal.

IP telephony carrier Level 3 startled the industry when they announced a $250 million agreement to purchase Lucent systems, including Softswitches. "The Lucent Softswitch will let us provide the kind of services our customers want from circuit-switched networks, and it lays a foundation on which our entrepreneurial partners will build a new generation of innovative broadband services," says James Q. Crowe, Level 3 president and CEO. Initially, Level 3 will offer long distance VoIP. When local VoIP is added, the plan goes, the Softswitch will enable AIN services – such as call waiting, call forwarding, billing, and operator assistance – over either traditional or IP networks.

Another Elemedia Licensee is Brooktrout Technology. Brooktrout's TR2001 IPT platform uses the H.323V2 stack written by MyBridge

(Eatontown, NJ – 732-544-2322, www.mybridge.com) and based on Lucent's Elemedia. Josh Adelson, Brooktrout's TR2001 product marketing manager, says they consider the high-level API more to CT developers' tastes, and that an IMTC bake off showed good interoperability between Lucent / Elemedia's stack, Cisco's implementation and RADVision's. He also notes that the TR2001 board has two choices of API, one straight-to-vocoders and the other higher-level. "By going down to lower-level API, you'll be able to deploy RADVision stack or MGCP when it comes."

MGCP-Compliant Offerings

Some vendors have decided that MGCP is already "here," standardization or no. Sphere Communications (Lake Bluff, IL – 847-247-8200, www.spherecom.com), the makers of IP-PBX Sphericall, offers an MGCP-compliant media gateway for providing voice and PBX functions over the enterprise. Its VBX Media gateways convert voice into standard ATM data cells. Call control is performed by media gateway controllers using Sphere's server-based software.

As Sphere announced: "With MGCP, Sphere products can adhere to the model of multiple controllers spread throughout an enterprise environment. This allows for tremendous scalability and reliability, since no single point of failure exists. In addition, MGCP will enable Sphericall to interoperate with dissimilar voice/data systems, whether they are based on H.323, SS7 or other signaling protocols. Multiple directory services such as LDAP and active directory can be accessed for call routing."

One example of the "hybridized network," comes from Alcatel (Milpitas, CA – 408-941-1800, www.alcatel.com). It's a five-component offering for IP telcos and ILECs that uses H.323 gatekeeper stacks from RADVision, VoIP codecs from Telogy, MGCP, gateway technology from an acquisition called Assured Access, and Alcatel's expertise in IN service creation and SS7.

Highlights: the Universal Access Gateway (UAG) from subsidiary Assured Access supports dial-up IP and VoIP in one product. A remote access concentrator originally built for ISPs, the UAG accepts more than 2,000 VoIP or V.90 data ports per shelf. UAG loads V.90 or VoIP codecs on the fly. A call comes into the UAG and it can sense whether it's an IP, fax or voice call. Telogy supplied the codecs and the Texas Instrument DSPs. Alcatel value-added their billing, Least Cost Routing (LCR) of VoIP, and integrated it with their intelligent network.

Several UAGs are managed by Alcatel's preexisting 1135 Service Management Center (SMC). This is the gatekeeper for setting service-level agreements and providing the accounting, authentication, and authorization functions through an incorporated RADIUS server.

Linking gateway and gatekeeper to the SS7 network is Alcatel's Call Signaling gateway, which supports MGCP and can process from 10,000 to 50,000 TCAP messages per second. (TCAP, acronym for Transaction CAPture protocol, is the part of the SS7 stack used when the IN needs to verify something with the server application. Switches can see it as a message for number translation, for example.) The idea: seamless call set up between IP and PSTN networks. Alcatel's bet on MGCP? In their opinion, the most support can be found for this standard, but they feel there's a question as to whether the IETF will accept it. There will be more flavors before it grows into a mature protocol. For the moment, though Alcatel's bet is on MGCP because it's flexible.

Two more important pieces: Alcatel's IN 1400 Service Control Point (SCP), already a fixture in 65 (mostly non-North American) PSTN carrier installations; and the new Service Creation Environment (SCE), allowing service providers to define, design, and deploy new services on SCPs. SCE will be a modular, graphical call flow interface to the calling features customers already know and use: PIN authentication, 800 number, click-to-

talk, Internet call waiting, and pre- and post-paid calling cards. Alcatel is listing a price of $219,000 for 320 VoIP / FoIP / V.90 / ISDN ports and an SMC running on a single-processor Compaq Alpha workstation.

Alcatel stresses that ISPs or CLECs can take it step-by-step on their way into Alcatel's VoIP solution: They can simply get a UAG and SMC gatekeeper to provide VoIP with data access and authorization accounting. There's no SS7 involved here.

They can follow that with the addition of the call signaling gateway, changing the T-1 / PRIs on the gateway to SS7 link connections. SS7 trunks come much cheaper than T-1 / PRIs, and offer much more flexibility in call routing. In this scenario, the actual media stream bypasses the ILEC's voice switch; only call signaling goes through it.

The final step is to introduce the IN platform for enhanced service creation.

Nortel Networks (Brampton, ON, Canada – 905-863-0000, www.nortel.com) adds to the emerging picture that delegates H.323 to enterprise LAN-to-PSTN IP and MGCP to public networks. Nortel's Tom Taylor heads up the IETF/ITU Megaco working group for MGCP standardization. According to Gregg Astoorian, Nortel's senior manager of Internet telephony marketing, the largest H.323 networks seen to date go to only 15,000 ports; and market data from Probe Research pegs the world total service provider installed base of H.323 at 200,000 ports.

Compare that 15,000 number with Nortel's IP network built for British Telecom in Spain in 2000 that initially deployed more than 90,000 lines in 10 cities. It works on the media gateway controller model, and posits distributed application and signaling servers and IP-over-ATM transport. The initial focus will be on national phone-to-phone services, with the Internet telephony network providing a transit capability between other carriers and to directly-connected customer PBXs and other IP devices.

To satisfy the Spanish interconnect regulations, the network must behave identically to a conventional voice network, with equal voice quality and features such as local number portability (LNP), emergency services access, and SS7 transparency.

Nortel's IP solution for telcos is called IPConnect, and it too features five main components: The CVX 1800 Access Switch is its media gateway, the IPConnect Call Engine (ICE) is the media gateway controller, a Java-based call control software that runs on commercial computing platforms and controls the CVX. The universal signaling point (USP) is its SS7 signaling gateway, and the universal audio server (UAS) is its application server for IVR, audiotext, conferencing services and the like. The UAS features open APIs and SDKs to lure third-party app developers. The Integrated Network Management (INM) piece is the network's service management platform, accepting status messages from network elements, both Nortel's and, say, Cisco's routers. Those messages, in turn, are typically translated from SNMP or CORBA.

Yet another SS7-IP call agent platform for IP telephony gateways is the Nuvo 500AIN, from Mockingbird Networks (Cupertino, CA – 408-342-1067, www.mockingbirdnetworks.com), a subsidiary of Opus Systems.

The Nuvo 500AIN, used in conjunction with a media gateway, appears to the PSTN switch fabric as an SSP for tandem calls. It's based on a fully redundant Solaris system, providing reliable, carrier-class operation and an open platform for custom application development. It can be expanded to support up to 25,000 simultaneous calls per logical node.

The Nuvo 500AIN appears quite adept at providing SS7 call control in live PSTN network environments. Availability is 30 days after the order is received. Suggested retail price starts at $250 per port.

"First generation media gateways had no concept of utilizing the SS7 network," notes Craig Forney, Mockingbird CEO. "And without SS7, those gateways could not provide even basic call control, not to mention advanced OSS functions, database mining, routing, and TCAP functions such as local number portability, caller profile, toll-free and toll wireline services.

By implementing the Nuvo 500AIN call agent as an intelligent adjunct, traditional media gateways can offer SS7 functionality, which enables faster call set up times, intelligent network (IN) services and new merged voice and data services."

Tips for Measuring H.323 Performance

Usually when you measure software performance, it is enough to scope the execution time for each part of the software and then combine the results. The situation is not so straightforward or linear when trying to understand the performance and other characteristics of network protocol implementations.

A reasonable question for network managers to ask when considering implementing an H.323 solution for voice and video over IP is, "how many calls per second are supported?" A real answer to this question needs to take into account not just general software performance, but also the specific call environments and scenarios that will be deployed.

Things to keep in mind when discussing performance measures with your H.323 vendor:

1. Parts of implementation can be based on other vendors' implementations, such as TCP/IP stack and ASN.1 compiler. The performance of all those components and their integration should be considered.

2. Real-world network implementation should obviously have mechanisms to optimize performance when handling many calls simultaneously. Linear extrapolation on specific components will not reflect or test these mechanisms. Ask your vendor about predictability, which can only be derived from testing real-world deployments.

3. Ascertain the level of database population that your deployment will require. Was the performance test executed in an environment close to yours? What kinds of search engine and indexing-scheme optimizations have been designed?

4. What kind of H.323 application are you designing? Different evaluation criteria apply if you are developing an embedded gateway and end point or a gatekeeper application. Performance metrics need to be considered along with feature sets, API design, architecture (for example, can RTP streams be split for distributed processing?), migration path for expanded functionality, and support for multiple operating systems.

5. Understand how the test was defined and performed in order to assess time results (see above) and to determine how to neutralize the latency factor (for example, not taking into account the network time and other-side CPU execution). Some specific measures to ask for include:

6. Best, worst, and average time per call

7. Number of simultaneous calls

8. CPU load

9. Percent of calls successfully established and disconnected
   Also ask:

- What is the exact test topology?
- Which protocols and protocol procedures were invoked?
- Which parts of software were configured or "hard coded," as compared to the ones you are going to have in real deployment? For example, what assumptions were made about capabilities of both sides?
- What were the test tools and logic?

Bottom line: Try to do the test yourself. You may discover many interesting results apart from performance such as stability, interoperability, and manageability characteristics.

Headset

Standard equipment among call center agents. They let you make and answer more calls faster, as much as 20% more calls. If you're on the phone for hours a day, you deserve a headset. Headsets can connect to PC componentry (sound cards / multimedia modems) and / or traditional telephones that hang off such devices or both (with a PC / phone-switching device).

Headsets are either binaural (two ear receivers) or monaural (one ear receiver) and come in various styles (headband, earloop, earmold, etc.). Some are convertible, so you can change the style.

Ear pads, microphones, size, cords, connections and weight vary. Some aren't. Some come with noise-canceling microphones. Some come with omni-directional mics. Some come with volume controls. That makes them more bulky. Some don't.

In any case, comfort should be the number one consideration when selecting a headset. You must wear them and test them before you buy. Headsets are a very personal item.

Expensive headsets generally sound and work better than cheap ones, though the general quality of headsets has improved in recent years.

How will you use your headset? There are PC headsets and telephone headsets. There are corded and wireless headsets. There are adapters that allow you to use one headset for both the PC and telephone. There are headsets for handheld cell phones, which let you walk down the airport concourse holding your luggage and looking like the Secret Service.

You need some sort of amplification feature because it saves time on returned calls. You'll also need a mute switch because people have a tendency to barge in on your activity. Also, you don't want distant callers hearing your musings.

"On-the-Phone" indicators can inform people around you that you're on the phone. Lights are good, but rare on headsets.

Wireless headsets are now emerging. They can't use frequencies such as those on cordless phones, since you'd get huge interference in a room full of headset wearers. The best technology used to be infrared (like TV remotes) but we're also seeing magnetic induction and eventually Bluetooth-compliant devices.

Battery life is critical in wireless headsets. The longer the life, the heavier the headset.

Help Desk

Unlike a call center which is a telemarketing interface for consumer goods and services, their agents serving merely as a persuasive middlemen between customers and goods / services, the help desk is the centerpiece of a company's service function, the central support and information contact point for end users which serves to respond to user's questions and to solve their problems.

Help Desk is a generic name. It is also known as a Customer Support Center, Hotline, Information Center, Resource Center, Response Center, Service Desk, Solutions Center, Support Desk, Technical Support Center, User Help Desk, etc.

Internal help desks provide support (generally IT support these days) to a company's staff. *External help desks* have traditionally provided "product support" to a company's customers. Whereas a call center is concerned with completing as many callS / transactions as quickly as possible, the success of a help desk is based upon the quality of information and assistance provided to callers. A "call" in modern parlance can now mean not only an actual phone call, but a voice mail, e-mail, fax, web chat, web video connection, voice over IP call, or personal visit ("walk in"). The help desk is thus a Single Point Of Contact (SPOC) for incoming customer or user queries.

Of course, a help desk cannot simply function as a general "information please" facility. The "scope" of the help desk should be defined in terms of both services (supported vs unsupported services) and customers (supported vs unsupported users), which should be communicated to customers so that they will not be disappointed following their encounter with the help desk.

In order to create a successful help desk, one must have some idea what your customers really want. Customers always want answers to their questions *immediately*, of course, which has led to "self service" websites offering access to knowledgebases, current downloads and patches, access to FAQs, etc.

As technology becomes increasingly – well, technical – the "self service" approach will fail and a user will seek help from a real person at a help desk. At this point the importance of training becomes glaringLY evident, though, too many times, we train people about the technology *per se*, but what we really need to teach them is the optimal set of "good behaviors" needed to answer the questions that come to the help desk concerning the technology. Help desk agents don't need to know everything about a product. They may be able to use the same website knowledgebase that just frustrated a caller more adroitly than the caller did to answer their question. But above all they need to know how to best interact with the customers.

Your also need to determine your help desk business balance. How much money do you have to spend on a help desk? How many resources can be brought to bear on any particular problem? What technologies will you put in place? Which one is most important? At the moment the telephone is still the most popular communications tool, but other technologies are catching up.

Technology does help a bit. New digital call monitoring techniques allow you to do instant recording of caller / agent interactions and then use these records for coaching the Customer Service Representatives (CSRs). This tends to scare CSRs, and help desks tend to have a notoriously high churn rate, so you must take care in determining your hiring and staffing model. Train them, reward them, motivate them, devise incentive programs!

You should consider having outside sources or even internal sources come in and periodically do compliance reviews. How well are you doing? Are you doing what you're supposed to be doing? If not, fix it!

There are also psychological factors centering on the caller: The help desk must provide empathy for the caller, to make sure you understand what their issues are and you make them feel good about their experience with your company's help desk. That said, note that most research shows that the general service quality at help desks is pretty mediocre. There is immense room for improvement.

Not much improvement in help desks ever occurs, of course, because help desks are cost centers, not profit centers. Customer service costs money. A help desk is almost continually admonished to be ever more efficient and cut costs, not to think of new ways of spending money. To genuinely improve service may require that you extend the length of your calls for some customers, which obviously ruins your efficiency statistics. Heaven forbid! There have even been a number of disturbing studies demonstrating that in some cases a negative correlation can occur between customer service satisfaction and profitability!

So if you are going to look at a return on investment for developing a top-tier help desk, you need to focus on observable, measurable, customer behaviors that have a bottom line impact, that are directly related to either increased revenues or decreased costs. In other words, behaviors related to profit.

Interestingly, the Number One behavior that affects profit that you at the help desk level can affect is *turnover*. The decision not to do business with your company anymore is a measurable behavior, and the Harvard Business Review says that businesses on average lose an average of 50% of their customers every five years through turnover. So the primary function of the help desk staff should be to give such good help that you can demonstrate that they actually keep customers.

Web-Enabled Help Desk or Agent-Enabled Website?

Clearly, the biggest innovation for both help desks and call centers has been the World Wide Web. But while it's a great idea to offer on-line knowledgebases and trouble ticketing so customers can solve their own problems from your website, these options are no longer enough. Live service is becoming a viable option in conjunction with on-line support.

After all, if your car were to have a flat tire, would you be pleased with the response, "Fix it yourself"?

Most likely you wouldn't, unless it took less time to fix the flat by yourself than to wait for the highway towing service to arrive.

The above example epitomizes the recurring mantra of on-line help desks – do it yourself.

Do you have a problem with paper stuck in your printer? Fill out a trouble ticket from the manufacturer's website.

Is a company's support rep taking too long to get back to you? Good. As the customer, you should know where on the company's website so you can find the solution.

Self-service is a valid option for many common problems or questions customers have. Bringing knowledgebases on-line enables customers to locate answers all the time, even while your help desk is closed. Indeed, a number of vendors are pushing the movement of knowledgebases out of the confines of the help desk and making them available for public consumption.

There are ways help desks that primarily exist on-line can offer live service from your company's website. Some companies (discussed below) make software specifically for web-based help desks.

Multima's (East Greenwich, RI – 800-532-4862, www.netkeeper.com) NetKeeper Internet, for example, offers a Webified version of its NetKeeper Help Desk Pro line of call tracking and problem resolution software. The advantage of the Webified version is that it lets reps support customers from anywhere in the world they can use a computer. Whether or not your help desk has an international clientele, you can save your customers money in service fees and, if you have a toll-free support number, your company a lot of money by letting customers refer to your Web site rather than calling a rep.

But what if your Web site doesn't provide an answer to a question? The usual method of requesting help from a Web site has been e-mail.

One growing trend is the inclusion of e-mail, and increasingly, text chat, in software that traditionally focused on problem resolution.

Right Now Technologies' (Bozeman, MT – 406-582-9341, www.rightnowtech.com) Right Now Web 2.1, for example, now includes software for importing and publishing info from your off-line knowledgebase to your Web site. If customers can't find the answers they need from searching your knowledgebase, you can have them send support requests to a generic e-mail address like support@mycompany.com. Right Now Web offers an optional module that identifies customers from their e-mail addresses and searches for any unresolved trouble tickets associated with the customer.

On-line knowledgebases use different methods of enumerating solutions to problems. Many favor a democratic approach. With Right Now Web, for instance, customers indicate whether or not a suggestion from your knowledgebase answered their questions. The greater the number of customers who deem a solution to a particular problem to be helpful, the higher the solution ranks within a list of other suggestions for solving the same problem.

Using netDialog's (San Mateo, CA – 650-372-1200, www.netdialog.com iCare, your customers can either search through your on-line knowledgebase or click on an icon on your Web site to request a live text chat session with a help desk rep. The software lets reps monitor the Web pages customers visit on your site to help them gauge the types of questions they may need help with. This way if a customer calls or sends an e-mail, reps have a record of the Web pages the customer has already visited.

As help desk reps use iCare to communicate with customers through text chat, they can also push support documents to them, such as recent updates to the user manual. iCare keeps track of all the documents and Web pages your customers and reps have consulted in the process of resolving problems.

When customers choose to find answers from your knowledgebase by themselves, they can give pass/fail ratings to any of the suggestions iCare presents. The software offers you the option of listing solutions based on their relative ratings from customers. For example, you can display support documents on your Web site to which your customers gave the most "pass" ratings for resolving a specific problem. With iCare, you can also organize solutions based on an actual number of "pass" ratings they received from your customers. For instance, you might want to place solutions that garnered 20 or more "pass" ratings at the top of the list.

For example the software allows reps to indicate how solutions to problems appear on their screens. It lets reps sort support documents containing potential answers to customers' questions by the date they were created; the name of the person who created them and by customers' ratings. By default, the software lists support documents based on how often reps or customers referred to them for answers.

iCare also lets reps e-mail customers one or more addresses of specific Web pages on your site. The Web addresses directly link customers to documents or software on your Web site that reps believe will contain answers to your customers' questions.

Some companies offer software that uses resolutions to previous problems as the basis for solving current ones. Inference (Novato, CA) offers its k-Commerce Support suite, which lets you add IVR, e-mail and text chat among the methods of harnessing the company's case-based approach to solving problems. CasePoint 3.5 integrates with Periphonics' (Bohemia, NY – 631-468-9000, www.periphonics.com) PeriProducer, which means that you can give customers the option of using your Periphonics IVR system or your Web site to find answers to questions using the same knowledgebase.

CasePoint 3.5 relies on Kana Communications' (Palo Alto, CA – 650-325-9850, www.kana.com) e-mail response software, which offers agents suggested replies to e-mail messages from customers who haven't found solutions to their problems on their own. For text chat, CasePoint uses eShare's (Commack, NY – 516-864-4700, www.eshare.com) NetAgent.

You can refer to a list of pre-written responses that correspond to specific key words within customers' e-mail messages or text chat requests but don't look up answers from your knowledgebase.

Molloy Group's (Parsippany, NJ – 973 540-1212, www.molloy.com) Internet Knowledge Kiosk 1.3 uses an eclectic combination of neural networking and fuzzy logic to solve problems using what it calls its Cognitive Processor. On its own, it learns relationships among problems, symptoms, diagnoses and solutions contained in your knowledgebases.

The Cognitive Processor avoids the crippling practical and performance limitations of older "artificial intelligence" techniques that have been applied in managing customer support, such as rule-based expert systems and case-based reasoning. The technology is a hybrid of several self-organizing knowledge processing techniques, including neural networks and fuzzy logic, as well as conventional text searching and parsing.

The browser based product was introduced in tandem with the launch of Release 1.3 of Molloy Group's Knowledge Bridge, a client / server Knowledge Management application for customer support and sales operations. Both products are components of Molloy Group's Enterprise Knowledge Management Architecture, built around an interesting knowledgebased technology called the Cognitive Processor.

Molloy's product suite helps companies enhance customer support and increase their competitiveness through Knowledge Management (KM), harvesting valuable expertise from every interaction with the customer – at the call center, and potentially from every customer contact across the entire enterprise.

"Internet Knowledge Kiosk is designed to enable online 'e-support' to work closely with the call center, help desk or other 'live' customer interaction center," said Louis Venezia, Molloy Group's COO.

"Both draw on the same, constantly growing and changing knowledge base, which is created in real time through interactions between agents and customers. The knowledge accessed in browser-based self-support may be literally seconds old. That's radically different from the more conventional 'publishing' approach used with other self-support technologies, where the knowledgebase is developed separately, offline, and then updated sporadically."

The Internet Knowledge Kiosk software offers such functionality as follows:

- Goal – a list of potential resolutions, in order of relevance or likelihood, appears on screen at all times during the problem-solving dialog, changing dynamically as the user supplies more information. The user can skip steps in the dialog and go directly to the resolution as soon as one becomes obvious.
- Transcript and Dialog Step Locator – The system displays a continuous transcript of the problem-solving dialog, including a list of steps taken in the dialog and choices the user has made. The current step is highlighted. If the user changes his mind – e.g., to choose a different symptom or diagnosis – the user can click on the name of any step on the transcript to jump backwards or forwards to that step and make the change.
- Chat option – the Kiosk will integrate with chat systems for live interaction.

Molloy's Internet Knowledge Kiosk is designed to reduce the costs of Customer Relationship Management, by reducing the incoming call volume to the call center or help desk as customers resolve their own problems. Its dialog-based interface enables the support interaction to be friendly, immediate and personal. Applications include not only problem resolution, but pre-sale product information and guided online selling.

"The Molloy Group Knowledge Management Architecture continues to grow and mature," said Richard Koloski, the Company's Vice President, Engineering. "The Kiosk product will continue to evolve to provide feature enhancements. It was designed and implemented so as to enable us to provide robust, scaleable function now, and the underlying technology allows us ultimately to move to an enterprise Java platform, when the market clearly demands that."

The Internet Knowledge Kiosk can be accessed via any frame-capable browser, including Netscape Navigator 3.01 or higher; Microsoft Internet Explorer 3.0 or higher; or America Online 3.0 and higher.

The Database Server for the application can be an ODBC-compliant database server, such as Oracle version 7.3 or higher, Microsoft SQL Server 6.5 and up, or Sybase 11.3 and higher.

Pricing for Internet Knowledge Kiosk is $40,000 per 100 concurrent users.

Keep in mind that there is no such thing as generic software for Web-enabled help desks. Applied Innovation Management (Fremont, CA – 800-942-7754, 222.aim-helpdesk.com) has a whole range of products, from defect-tracking modules to its Online Support Center, for help desks whose main communications with their customers is over the Internet.

The following companies make software that helps you Web-enable your help desks. To find more companies that provide software for your on-line help desk, visit www.callcentermagazine.com.

3Si (formerly Kimbrough Solutions)
(800-390-0641 / 303-741-9123) www.3si.com

Applied Innovation Management
(800-942-7754 / 510-226-2727) www.aim-helpdesk.com

Blue Ocean Software (813-977-4553) www.blueocean.com

Inference (800-322-9923 / 415-893-7200) www.inference.com

Knowlix (800-733-2019 / 801-924-6200) www.knowlix.com

Molloy Group (973-540-1212) www.molloy.com

Multima (800-532-4862) www.netkeeper.com

netDialog (650-372-1200) www.netdialog.com

Net Effect Systems (818-752-6600) www.neteffect.com

NetManage (800-492-5791 / 408-973-7171) www.netmanage.com

Primus (206-292-1000) www.primus.com

Remedy (650-903-5200) www.remedy.com

Right Now Technologies
(888-322-3566 / 406-522-2900) www.rightnowtech.com

Servicesoft Technologies
(800-737-8738 / 781-449-0049) www.servicesoft.com

Silknet Software (603-625-0070) www.silknet.com

Help Desk Outsourcing

One option to setting up and running your own external or internal IT help desk is outsourcing (see Call Center Outsourcing). Help desk outsourcing has several unique advantages and opportunities compared with running your desk in-house, over and above the labor and facilities cost savings and staffing flexibility in customer service/sales outsourcing.

The ability of outsourcers to more cost-effectively handle sharp and seasonal peaks and valleys and to take after-hours contacts-and their ease in hiring and letting go-can be more pronounced in help desks. Help desk reps are usually the elite of telephone/Internet customer service; they are more highly paid and have more skills than customer service/sales agents. While this can make recruiting easier it also adds to your costs. Paying reps hanging around waiting for the phone to ring or the e-mail to bounce in-and finding and seating enough of them when

you launch a new product or install a new computer system or network is an expensive proposition.

One big advantage of outsourcing your external or internal desk is that it removes a bad-vibes external and internal corporate headache. It is usually, but not always, institutionally easier to ream out a contractor and exercise your boot-out-the-door clause in a contract than to do the same to your own employees.

Of all the product areas businesses and consumers, and employees come into contact, few are more frustrating and problematic-and come in for more terabytes of virtual flak–than first level external or internal IT support. They are known in many companies as the "helpless desk". Users have long moaned about long hold times, dropped calls and unanswered e-mails and poor advice. These experiences have been backed up by sometimes widely publicized service audits on leading companies that have shown a shocking number of their reps could not solve problems even though the answers were posted on that firm's Web site FAQ.

Another reason to consider outsourcing is that you can't convince your own IT department to cooperate with your plans to expand and improve your desks. This is especially true if you want to add useful complex-to-integrate features such as remote diagnosis/resolution, chat and voice over IP.

Bill Frezza, general partner at Adams Capital Management blasted all-too-typical IT department bureaucracy and control-freak behavior in an *InternetWeek* op-ed piece published September 6, 1999, citing support problems experienced by a bank executive friend. The friend had purchased a PalmPilot with his own money only to find the desktop synchronization package is still not on the bank's support list. Then an IT technician lent him administrative rights to add the Palm Pilot software, only to be countermanded by another IT rep who "went berserk" and stripped off the software.

Justice was done, however when the friend's division decided to outsource all IT functions.

"When IT gets too centralized and insular, financed by an independent budget, it removes beyond the reach of any particular department," writes Frezza. "Requests have to be moved to the top before they have any impact. While [outsourcing] will undoubtedly take money from [the company's] bottom line at least people will be able to get their jobs done. And, best of all, they'll be treated like valued customers instead of hapless supplicants."

Yet another reason to consider outsourcing your help desk is that it frees up your high-priced geeks to solve the fun stuff: network crashes, software bugs and the like that lets them get into the bits and bytes and come up with fixes. Programmers and designers love a challenge; they get bored easily on the simple stuff, the so-called "read the friendly manual" (RTFM) questions. And you're not paying them high five-figure salaries and pouring goodies into their pockets to hold some user's hand.

Besides, there are not many techies that have people skills. Many of them are as alien to the human species as those creatures that wrote the indecipherable manuals that led to the RTFM calls in the first place. One IT outsourcer, IHS Help Desk Services (212-239-7800, www.ihshelpdesk.com), screens its people for customer service experience, and who can handle the unintelligible yells (except for the curses) of someone who can't get their [expletive deleted!] program to run-and they're on deadline – such as flight attendants and salespeople, rather than on their technical knowledge.

Still another reason to contract out at least part of your internal help desk is that they may be more familiar with the kinds of problems you may run into, therefore their reps can solve problems quicker with bigger picture LAN / WAN systems than they are with off the shelf shrink-wrapped software packages e.g. Microsoft Word, Excel et al.

Jeff Becker, Vice President of Marketing for PC Helps Support (Bala Cynwyd, PA –- 610-668-3516, www.pchelps.com) which provides packaged software support, explains that an internal help desk staff can have so many support responsibilities, ranging from LAN to hardware, that they cannot be adequately trained on all off-the-shelf software. They may experience difficulties solving problems with these applications and claims their resolution rates on such software are low.

"Internal help desk analysts are best at solving proprietary and complex applications within their particular corporate enterprise," says Becker. "They often don't have the certification or the time it takes to solve generic software package problems. They don't use these packages with extreme frequency and at times, are not as familiar with them as the callers. They're inability or difficulty in solving these types of issues can have a very negative affect on their level of job satisfaction."

There are two types of IT help desk outsourcers: Service bureaus, which handle your call/contact volume off site and insourcers, where you hire a firm to staff and run your help desk onsite. Some service bureaus, especially those that specialize in internal help desks also offer insourcing. By the same token some insourcers offer after-hours off-premises help desks, so you can shut yours down.

There are several bureaus that provide these essential first/second level services externally and/or internally. They include CompUSA Call Center Services (800-563-9699, www.compusa.com/callcenter), Convergys (800-344-3000 / 513-458-1300, www.convergys.com), Cyntergy (800-825-5787, www.cyntercorp.com), 800 Support (800-777-9608 / www.800support.com), EnvisioNet (207-373-3200, www.envnet.com) National TechTeam (800-522-4451 / 313-277-2277, www.techteam.com), Stream International (888-223-8880, www.stream.com), Support Technologies (888-2-HELP DESK / 770-280-2630, www.supporttechnologies.com), Sykes International (800-TO-SYKES / 813-274-1000, www.sykes.com), Taima (613-739-5620, www.taima.net), TRG (800-874-0062 / 919-874-0062, www.trginc.com), Watts Communications (416-255-8000, www.wattsgroup.com) and Xtrasource (330-673-3316, www.xtrasource.com).

Some service bureaus are experts in several key industries. For example, Cyntergy handles restaurant, retail and hospitality industry IT and customer service problems. Taima has experience with Internet Service Provider customer support.

Several leading bureaus, including Convergys, Stream, Sykes, Watts and Xtrasource can help your internal or external customers outside of North America. Two of these bureaus–Taima and Watts-are based in Canada, while a few others, such as Convergys have centers in that country, which indirectly gives you access to Canada's lower cost, more loyal and better educated work force.

At the other end of the scale, there are many full-service internal IT bureaus. Their desks are adept at handling Unix and other related operating software. Depending on the firm they can fix hardware problems on larger machines, such as minis and mainframes.

Firms such as CompuCon (800-488-5266 / 972-856-3600, www.compucom.com), DecisionOne (888-287-9202 (US) 800-554-5179 (Canada),

www.decisionone.com), Eltrax (770-612-3500, www.eltrax.com), Polaris Service (800-541-5831 / 508-460-1800, www.polarisservice.com), Software Spectrum (800-624-0503, www.softwarespectrum.com) can handle a vast range of issues from problem solving to asset management and field support. Such firms also manage hardware and software installation, configuration and financing. A few of these companies, such as DecisionOne, Polaris and Software Spectrum can support your desk outside the US. Some external IT bureaus such as Stream International and TRG also offer internal services.

Insourcing taps into the floating world of techies-for-hire who take such short-term jobs to keep their skills up and income flowing in while they're figuring what to do next or waiting for a great offer. It is common for firms to contract with programmer/technical consultants who often share the same workspaces as their employees; project managers can get rid of consultants instantly, but removing a poorly functioning staff member is a hassle.

There are several experienced and qualified insourcer-only companies. These include Alternative Resources Corporation (ARC) (847-317-1000, www.arcnow.com), Carlyle Technical Services (516-271-2700, www.carlyle-csi.com) and Interim Technology (201-842-0700, www.interim.com/technology).

There are pros and cons to both approaches. Service bureaus are typically less expensive than insourcers because they can control more costs with off-site centers. They buy and install (not you directly) the expensive problem management/problem resolution, diagnostic and self-healing software that you need to fix your customers' and users' hardware/software and systems. However insourcers offer you greater control because their people are on your premises. You also don't have to worry about integrating your systems with an outsourcer's.

You may wish to place all your outsourcing-IT and non-IT-with one vendor. Some IT bureaus, such as Convergys, CompUSA Call Center Services, 800 Support and Sykes also handle non-IT customer service and telesales. Analysts such as Katrina Menzigian, International Data Corporation CRM and call center services program manager, sees firms meshing their IT and non-IT service and sales into CRM, with vendors offering one-stop shopping.

She sees non-IT bureaus entering the IT help desk market, and vice versa. She recommends that you research carefully and ask questions about their support capabilities, expertise and experience, especially about how their technical specialists are trained. Similarly she advises when you bid for CRM work and when an IT bureau responds you ask about their CRM capability and expertise.

"IT companies are merging separate customer service/sales and help desk functions into what they really are, which is customer relationship management," Menzigian points out. "When you are making that move and you've decided to outsource your initial customer contact and your first and second level support you have to make sure the vendor you select is qualified for the tasks."

Recommended websites:

Association of Support Professionals – www.asponline.com

The Help Desk Institute – www.helpdeskinst.com

Help Desk Professional Association (HDPA) – http://www.hdpa.org

H

IMCCA

See International Multimedia & Collaborative Communications Alliance.

Informal Call Center

Also called a "casual call center," a term coined by Blair Pleasant at the PELORUS Group. (Raritan, NJ – 908-707-1121, www.pelorus-group.com)

It happens in almost every business – from big bank to small publishing company. So-called "knowledge workers" are humming away (thinking, calculating, writing, programming, doodling, etc.) when their phones ring. They answer. Callers are looking for information or help. These workers (hopefully) do their best to oblige. Certainly it's not as regimented as the formal call center, where well-armed agents wait ready and willing for calls and caffeine-addled supervisors drool over real-time call-processing performance pie charts. Still, the ubiquitous "informal" or "casual" call center does represent a vast arena for computer telephony technology to flourish in. Indeed, it's one that potentially dwarfs the traditional call-center market. How far have we gone in exploiting this killer opportunity area? Well, as intuition and a new comprehensive CT report suggest, we've only just begun.

Computer telephony technology is succeeding in the formal call center. Now it must slip into something a little more comfortable and saunter into what Blair pleasant playfully tagged the Casual Call Center (CCC) or Informal Call Center (ICC).

What is a small to mid–size or small and informal call center? To begin with, the actual number of agents does not really define a small or an informal call center. In fact, small call centers and informal call centers can vary from 2 agents up to 40, sometimes even 50 or 60 agents. Also, the term agent doesn't really apply in this market space. Typically, they are no longer what we would term agents, where they are sitting in cubicles with headsets and their primary function is to answer the phone. In fact, many times they are knowledge workers where answering the phones is only a portion of their overall responsibilities.

Small and informal CCCs can be found in any type or any size of business from Fortune 500s to sole proprietorships, from health care to retail to finance. Also, the business applications abound for small call centers. Any business has the need or the varying need from anywhere from reservations to help desk to support inside sales, can really benefit from the implementation of a small call center.

The CCC is a newly emerging segment that brings together knowledge workers and SOHO professionals. While these people are not call-center agents, they do spend a good deal of their time answering the phone and responding to customer inquiries. Many times, they're really the appropriate people to handle customer interaction, since they're experienced and specialized.

Of course, just like traditional call-center agents, these workers also need the right tools to intelligently and efficiently process calls. And just like in the traditional call center, this is where CT shines... if only more companies would realize it. First thing's first: they need to understand their need. Most don't.

In one poll, call-center software vendor Versatility (now part of Oracle) asked a bunch of companies: do you have a call center? Only about 10% said they did. However, when asked if they did telesales, customer service and customer support over the phone, over 60% said they did. Understand the market now?

The Differences

Casual call centers do not have specialized ACDs; rather, they use a regular PBX or key system or carrier-based Centrex service. They likely don't have predictive dialing or extensive monitoring and reporting.

CCCs are also smaller and less structured than formal call centers. They have different needs; thus, they need different products and want to buy differently than large formal call centers.

And while this is a large market, it's less sophisticated in terms of classic terminology of CT. Versatility researched this market and discovered that casual call centers prefer to buy products from someone they've purchased through before, such as a VAR, but not from small vendors they have no experience with.

Further, none of these polled CCCs specifically mentioned CT, LAN integration, screen pop or predictive dialing as requirements or technologies they were interested in. Instead, they focused on the functional requirements of their business applications.

The biggest difference between formal and casual call centers is in the definition of an agent. Formal call-center agents generally sit at phones that are next to computer terminals and their time is dedicated to doing business over the phone.

Casual call-center workers are not considered "agents"; they're considered knowledge workers who have a particular expertise and handle customer calls when necessary. Answering incoming calls and dealing with customers / callers is not their main gig, but it's something they do in addition to their primary duties. These types of workers do not generally perform the same type of service as each other and are more autonomous.

That all said, we should also point out that, although the formal and informal target markets differ substantially, the list of benefits a CCC gains from CT integration (or inherently gleans from leveraging or swapping out to the new breed of all-in-one communication server – see the October 1999 issue of *Computer Telephony* magazine for more on these platforms – page 68) looks very much like the CT benefits found in the traditional call center.

Bottom line: call takers and makers are better prepared to service callers thanks to intelligent call routing (both on- and offsite), database access, screen popping, point-and-click first-party call control and multimedia communication. Further, their expertise can in many cases provide superior call processing.

For example, all things equal (i.e. each had the same access to the same CT intelligence and resources) who would provide better customer service to an inbound caller: a help-desk agent trained in a particular app?; or a programmer who helped to code the app?

The Bad News

There are factors that retard the incursion of computer telephony technology into the CCC.

Reasons:

- Price. Surprise. The overwhelming reason why CT at the desktop for knowledge workers and SOHO professionals has been limited is the expense. Knowledge workers and workgroup members are not willing to spend a lot of money for CT.

The chief problem here, of course, is the dreaded "I" in CTI. CCCs don't want to or can't buy expensive CTI enablers such as CallPath and

Novell Telephony Services. Although it's acceptable in the traditional call center, $300 a seat (a price, FYI, that sounds good compared to the numbers of just a few years ago) is still not going to cut it in the CCC. Figure $59-$79 per user for CT to take off in this gaping niche.

- The unknown business problem. We alluded to this already. Many businesses don't even realize they have a call center, let alone understand they have to improve it with CT technology.
- Payback is hard to see. Many of the benefits of establishing a CT-based CCC mean soft, rather than hard, dollars. It makes it difficult for businesses to understand why they're upgrading to CT.

They want money-making and money-saving solutions, not "cool" technology. For example, it's hard to see the payback from desktop telephony. People have to be convinced that CT will save them time and / or money.

- Lack of "complete" solutions. The casual call center, particularly at the desktop, needs an entire solution, including equipment, network services, the telephony piece, the "I" piece and applications. CT has not been marketed as a solution, but rather in bits and pieces that need to be put together.
- CT is still not easy enough to buy. Mostly due to lack of standards (who the heck knows what works with what?). And unlike call centers, CCCs don't have dedicated resources to research and buy and implement CT. In general, knowledge workers expect installation to be easy and want shrink-wrapped applications.
- Not enough product. Many first-party applications for the desktop are "nice toys", as one vendor explained, but they do not address the needs of knowledge workers or work groups.
- Defining the market is difficult. While we know that the knowledge-worker market is huge, it's very difficult to pin down. The knowledge worker and informal call-center markets are more nebulous than the formal call center, which is pretty easy to identify. Again, in a formal call-center environment, there's a defined information systems (IS) group working to solve the needs of the agents and supervisors. There is no such group working on solving the problems of knowledge workers.
- Too much technology focus versus customer-value focus. Customers are not impressed with the CT architecture and the specific features of TAPI and TSAPI. To date there's been too much focus on technology rather than looking at why CT is good for you and creating demand based on benefits and opportunities.

Companies such as Lucent are now spending more time educating and informing customers about CT benefits, rather than trying to sell the technology. People are confused by the technology and need to understand the real business solutions.

- Lack of CT awareness. Not many knowledge workers outside the communication industry really even know about CT. If they did, they would create "pull" demand for it.

The Good News

Fortunately, the news isn't all bad. In fact, there are significant reasons why The PELORUS Group expects CT to show strong growth over the next few years in the casual call-center market.

- Computer telephony means much better customer service. This is a big reason why CT will succeed in all businesses, which realize more and more that customer service is a huge differentiator and an important key to success.

Computer telephony helps companies provide better customer service, while saving money on 800 costs and giving workers more time to do their jobs effectively. Information associated with calls and work flow management are important tools both inside and outside of the call center.

For example, by integrating TAPI with contact managers, users can get screen pop information about who's calling. This is important not only in call centers but for anyone who deals with the public. If an incoming call can be tied to a customer's claims file, repair status and so on, it makes the worker's job easier and more effective.

In order to compete with companies that have dedicated customer service centers, small and mid-sized businesses need more sophisticated systems for call handling. Companies need to make their critical customer information available to workers who answer calls.

CT lets companies tie together this critical information with the telephone system. The result is that barriers to doing business with customers can be reduced, leading to gains in productivity and customer satisfaction.

- Casual call centers will piggyback on top of formal "CTI" setups. Many companies are investing in CT for their call centers and will be able to leverage those investments throughout the rest of the organization.

For example, if a company already has TSAPI implemented for its call center, it's not very difficult or expensive to extend those capabilities to the rest of the organization and knowledge workers. If the investment has already been made in CT for the call center, then it's much easier to extend CT integration to the rest of the enterprise.

Knowledge workers will be able to "piggyback" or get the "trickle-down effect" from the call-center applications so that the incremental cost for knowledge workers is low. While no MIS director would ask for $200,000 to give CT to all of their knowledge workers, once CT is in place in an organization, the same server and switch can be used to provide CT to others in the organization, while only paying for additional user licenses, if necessary.

- Software tools are improving. And the cost of deploying CT applications is decreasing.

We're seeing a small explosion of middleware software from vendors such as Aurora and Q.Sys, making it easier to add telephony support to applications. Middleware quickly and easily adds CT support without having to modify existing applications. As new products emerge, such as MultiCall's CallFlow, CT can more easily be used in the growing CCC market.

- The communication server revolution. Rather than simply adding functionality onto the PBX, new products are being introduced that combine ACD capabilities, preview or power dialing, call blending, call routing, screen pop and other capabilities – all on one server.

The price point for this type of solution is multitudes less than traditional systems.

Call centers are getting smaller and CT is getting to the point in terms of cost and ease of integration whereby it can be implemented in smaller and medium-sized centers. No longer is CT being implemented only in the large airlines (for reservations) and banks. Rather, over the next few year, companies can increasingly afford to install CT technology.

While the number of systems shipped will increase, the average size of the call center is decreasing. Many companies stated that their CT systems were installed in companies with over 100 agents but, over the next five years, this will decrease to an average agent size of 50-60 agents.

Crossing the Chasm

The PELORUS Group sees computer telephony broken out into two different areas in respect to the high-tech adoption life cycle (from Geoffrey Moore's infamous Crossing the Chasm). For formal call-center customers, CT has crossed the chasm to the early majority. The next step is to cross the gap from early majority to late majority.

One characteristic of the late majority is that they eventually buy products / technologies – not so much because of any real belief in them

– but because they have to keep up with their competitors and the rest of their industry. This group invests in technology when products are extremely mature, competition is driving down prices and the products can be treated as commodities. CT is not yet at this point.

In areas outside of the formal call center, CT is still in the early adopter stage. In the informal call center, it's starting to make some headway, particularly in the help-desk arena. However, for the SOHO market and individual knowledge worker, CT is still in the early adopters, if not innovators, stage.

Desktop CT, including softphone stuff, unified / integrated messaging and so on, are still primarily used by innovators who are intrigued by the technology, but not necessarily the actual business case for these technologies and subsequent products. While we expect this to quickly change over the next two years, for now CT outside of the formal call center has yet to cross the chasm.

One of the problems Moore points out regarding the transition from the early market to the mainstream market (or from innovators and early adopters to early majority) is that many technologies never get defined in terms of their benefits or compelling applications.

Is a screen pop a compelling application for the desktop knowledge worker? Maybe. Maybe not.

Also: the early market is willing to take some risks, whereas the mainstream market is not. The mainstream market wants to be ensured that the product will be supported, that there are sufficient distribution channels and that other products will work with this particular product.

This is where standards play an important role and, as we all know, there are no industry standards in the CT market yet, although there are many contenders.

The mainstream market also wants to purchase products from well-established companies that will be around to support the product. The late majority in particular likes to buy preassembled packages with everything bundled at discounted prices. They want everything to be able to work together, without having to integrate the products themselves.

Moore states that the key to success is to have thought through the "whole situation" and to provide for every element of that solution within the package. The other key is to have a low-overhead distribution channel lined up that can get this package to the target market effectively.

This is why many of the switch vendors, who are well known and have good reputations, are expanding their professional services, consulting and systems integration programs. Customers feel comfortable buying products from the large switch vendors who they've had dealings with for many years and are likely to purchase CT systems from them as well.

The mainstream market is less likely to buy from small CT software start-ups and would prefer to have the software included in the entire system that they purchase from the large, known vendor.

This market segment cares about the company they are buying from, the quality of the product they are buying, the infrastructure of supporting products and system interfaces and the reliability of the service they're going to get.

Integrated Services Digital Network (ISDN)

ISDN is a digital telephone service. During the heyday of circuit switched networks, it was felt that ISDN would help carriers provide "Integrated Services," capable of handling voice, data, and video over the same circuits. ISDN follows Signaling System 7 (SS7), the international standard for signaling between the big switches on the digital telephone network, a protocol suite roughly comparable to TCP/IP in the packet-switched world. An ISDN interface is time division multiplexed into

channels (See Time Division Multiplexing) and, control and data signals are separated onto different channels, following SS7 conventions.

So-called Broadband ISDN (B-ISDN) uses ATM instead of SS7 (See Asynchronous Transport Mode).

History

Once upon a time there was a great big analog phone system. When you dialed, it was slow to connect. When you called long distance you heard lots of snap, crackle and pop on the line.

In the 1960s the telephone network was slowly but completely converted from noisy analog to a sleek, smooth, digital switching system where sound and data are converted into small packets, each containing an address as well as codes to put it in proper order with the other packets being sent.

Ironically, the only analog Plain-Old Telephone Service (POTS) left is that final mile from the local central office to the phone in your home or business.

There is a way for you to gain all of the benefits of digital communication, plus extra "channels" and call control features lacking in ordinary POTS phone service – ISDN or Integrated Services Digital Network, a circuit switched digital phone service you can order from your local phone company. ISDN uses the same copper wiring that's already in your home. The difference is the CO sends digital signals along your line.

There are two basic types of ISDN service, the Basic Rate Interface (BRI) for the home and small office, and the Primary Rate Interface (PRI) for larger businesses.

ISDN BRI

It's a tricky business sending digital signals over conventional copper wires not originally intended for them. As a result, ISDN is so sensitive to outside interference that you must be within 18,000 feet of the phone company's equipment (although "line extension" technology is available – at a cost).

In North America 90% of existing telephone lines don't need to be refurbished or de-conditioned (removing load coils) to be used for ISDN BRI service.

A Basic Rate ISDN "U-Loop" is a pair of wires that carries two 64 Kbps bearer ("B") channels and a single 16Kbps delta ("D") channel, all of which can be used simultaneously (each B channel has its own phone number). The B channels carry voice or data, and the D channel is used for control signaling, such as setting up and tearing down the call, and X.25 packet networking.

It may sound a bit strange that a whole 16 Kbps channel is used for something like "ringing" a phone, but this does have advantages, Instead of the phone company sending a massive "in-band" voltage surge to ring the bell in your phone, an unobtrusive "out-of-band" digital signal packet on a separate channel will not disturb the established connections (more than one device can share the line) and it works very quickly. The signaling can indicate who's calling, if it's a data or voice call, and what number was dialed, information that can be used for routing purposes.

BONDING

An inverse multiplexing method called Bandwidth ON Demand INteroperability Group (BONDING) lets you' split your data transfer over sets of separate channels as if their bandwidth were combined into a single channel. Thus you can combine two 64Kbps channels to surf the Internet at 128 Kbps.

Wiring

When you order ISDN, the phone company ends the cable at ("demarc") usually in your building's basement. The "inside wiring" from the demarc to your ISDN equipment (including the wall jacks) is your responsibility. The phone company will generally install and maintain it for a fee.

A "home run" or direct wiring between the ISDN wall jack and the demarc is standard procedure, also known as a star networking topology.

As for your wall jack, you can either use a standard four-wire RJ11 phone jack or a (recommended) wider, eight-wire RJ-45.

Terminal Adapters

A terminal adapter is the last piece of equipment at the user end of the ISDN chain. A terminal adapter is part of any device that runs an application and generates traffic on an ISDN line, such as an ISDN phone, a fax machine, and an ISDN "modem."

There are two kinds of ISDN Terminal Adapters (TAs), internal and external. TAs can look like modems, computer bus cards or interface cards for PBXs or routers. Your faithful digital-to-analog modem is no longer needed in the ISDN world. Some devices (such as ISDN phones) have a TA built-in that can also connect PCs, fax machines and analog, non-ISDN devices.

External ISDN TAs are easier to install (they plug into your PCs serial port) but won't give you maximum performance. Some, like the Motorola Bitsurfr Pro or the 3COM Courier actually, look like modems and are often called "ISDN modems."

Although ISDN BRI is capable of delivering 128Kbps of data, the UART chip controlling your serial port cannot handle data transmitted faster than 115 Kbps. Compression routines can get around this somewhat, and can even be used to attain an actual throughput rate that exceeds 128 Kbps.

In the US, the TA will need some configuration information from you. You need to know what switch you are connecting to in the CO, your B channel's numbers, and the Service Profile Identifiers (SPIDs).

SPIDs

Your telco allocates SPIDs just like White Pages directory numbers. A SPID number is usually identical to your 10 digit directory number with some additional digits tacked on to the beginning and end.

The "I" (Integrated) and the "S" (Services) in "ISDN" mean that a single ISDN line can use many different applications simultaneously, such as a phone, computer, fax, etc. To keep track of all these different applications, ISDN uses a SPID to identify each application. One or more SPIDs is assigned to each application on the ISDN line. Thus, an analog Plain Old Telephone Service (POTS)line has only one directory number, but an ISDN line may have several.

SPIDs tells the switch what kind of devices are connected to the line, if there are multiple devices, what services and features the switch is to provide to the devices, and helps route calls to the correct device. SPIDs are used only in circuit switched service (as opposed to packet switched services) and only in the US and Canada.

You must configure SPIDs into your phone, computer and other ISDN terminal equipment before you can connect them to the central office switch. Once you configure a SPID, the terminal equipment performs an initialization / identification process where the device sends the SPID to the network. After confirmation from the CO, the SPID is never transmitted to the switch ever again.

U and S/T Interfaces

Once you've got ISDN service, your equipment will be expecting one of two types of ISDN interfaces. The U-interface delivers ISDN signals over a single pair of wires between the CO and your home or office. Some ISDN TAs sold in North America can connect a PC directly to a U-interface.

The other ISDN interface, called the Subscriber/Termination (S/T) interface, uses two pairs of wires to carry the digital signal from the RJ11 or RJ45 wall jack to your terminal equipment. In fact, the S/T cable and connectors are the same used by 10baseT Ethernet.

Devices designed for ISDN (can plug into the S/T-Interface) are designated Terminal Equipment 1 (TE1). Non-ISDN analog devices with ordinary POTS (or "R") interfaces (phones, fax machines, modems) are designated Terminal Equipment 2 (TE2). These later devices must connect to an ISDN S/T bus via a terminal adapter.

The NT-1

If your equipment has an S/T-Interface, you need a gizmo called a Network Termination 1 (NT-1) which converts the 2-wire U-Interface into the four-wire S/T-Interface. The NT-1 is the ISDN line terminator that acts as the bridge between the phone company wiring and the end user wiring. A single NT-1 can support multiple terminal adapters. The NT-1 has a jack to accept a wire from the wall jack's U-Interface and one or more jacks for the S/T Interface connection to the PC or other ISDN or analog devices. You'll find in many cases that the wire from the wall to the NT-1 usually will have an RJ11 connector, while the wire from the NT-1 to the ISDN adapter will usually use an RJ45.

On an S/T-Interface, the Terminal or "Point T" refers to the connection between the NT-1 device and customer supplied equipment. "Terminals" (ISDN lingo for any end-user ISDN device – such as phones, fax machines, data terminals, etc.) can connect directly to an NT-1 at Point T, or there may be a PBX or customer-owned phone system there. When a PBX is on-site, the "Subscriber" or "Point S" refers to the PBX-terminal connection. Point S is physically similar but logically different from Point T allowing the CO to distinguish the PBX from other terminal equipment ITE).

The NT-1 takes the bit encoding scheme used for the long distances between it and the telephone company and translates it to the kind of encoding used between it and your devices, an encoding scheme designed not for distance but for channel sharing.

The NT-1 has an external power supply. Unlike an analog line, an ISDN line needs power to keep it going. If you suffer from a power outage, you'll suffer from an ISDN outage, unless you use an auxiliary source of power such as an Uninterruptible Power Supply (UPS).

Outside of North America, however, a complete loss of local power activates an "emergency mode operation" where the CO delivers up to 1.2 watts to the NT-1 along two or four of the "extra" wires that make up the eight-wire S/T bus. In North America there is no provision for emergency mode operation.

Multiple Devices to ISDN

If you've got an NT-1 that supports multiple S/T Interface connections (and "Terminal Equipment arbitration") you can connect up to eight devices to share the two B channels of an ISDN BRI line. – with some Nortel and AT&T switches this can be extended to 64 devices. For example, you can have an ISDN adapter in your PC, an ISDN phone and a Group 4 ISDN fax machine on the same ISDN line

How is this possible? It's true that the U interface coming from the CO only supports full-duplex data transfer over a single pair of wires, so only a single device can be connected to a U interface, namely the NT-1.

Though the NT-1 is still a full-duplex interface, there is now one pair of wires to receive data, and another pair to transmit.

Also, although the wiring and RJ45 connectors look like those of 10BaseT Ethernet, unlike 10BaseT the S/T bus isn't limited to a star configuration and can be fitted with splitters and T connectors, so you can connect various devices to your own little ISDN "network."

By assigning each device its own Service Profile Identifier (SPID) number, each device can be "called up" on its own "phone number." Each device listens for calls and will only connect to a B channel when it identifies a message requesting a service it provides, so you don't have to worry about your fax machine answering a voice call. The only stipulation is that no two devices can connect to the same B channel simultaneously.

Besides ISDN devices, some NT-1s or ISDN adapters also support analog phones, modems and fax machines by converting analog signals into ISDN and vice versa. The Motorola Bitsurfr Pro, for example, has two analog RJ11 jacks, one for each B channel.

Many of the so-called "ISDN modems" have built-in NT-1s, which makes them small, inexpensive and easy to install, but precludes adding lots of devices. For really maximum usage of ISDN, you can connect your ISDN line to a network bridge or router (such as those from Ascend) which is in turn connected to a LAN via an Ethernet card. Every PC on the LAN will then have shared access to the ISDN line. Similar equipment can be used to connect two geographically separate LANs via ISDN.

Working the D Channel

Believe it or not, you can order ISDN service having no B channels, just the D channel (0B+D service) for about $18 a month. Why would you want just a 16 Kbps D channel? Remember that the D channel can be apportioned to carry about 6Kbps for signalling and 9.6 Kbps for X.25, a packet-switching protocol defined in the early 1970s as a worldwide telecom standard.

In X.25, information is packetized and de-packetized via a Packet Assembler / Disassembler (PAD) at the sending end and receiving ends. PADs also do error-checking and error-correction. Many ISDN phones and TAs have PADs, so your phones, PCs and data terminals can communicate with other locations on the worldwide X.25 packet network. Thus, for a few dollars a month you get a permanent, full-time link to this network.

What can you do with a D channel / packet-switching link? You can do point-of-sale and point-of-service transactions. These range from credit card authorizations and medical insurance processing to food stamp debiting and building or property security.

Travel agents make reservations over such networks and Automated Teller Machines (ATMs) use them to commuticate with bank computers. Every Wal-Mart or other large store has what's known as a "Back Office Concentrator," essentially a little LAN connecting all the cash registers to a 0B+D / X.25 connection for quick (two second) credit card verification.

For a number of years there's been an X.25 "Always Connected Always Online" movement, supported by such people as Keith Thomas of ADTRAN and *Computer Telephony* magazine's own Stuart Warren (who now works for Diversified Technology). The idea is that, for example, you don't need to build up and tear down a high-bandwidth official call to, say, a remote server merely to see if you have e-mail. A little X.25 packet can do the job. Cheap X.25 connections can be used to notify you about things without the cost of a full-blown (and expensive call). One big user has been convenience markets that use D channel packet service for credit card readers. One vendor using ISDN D-channel for point-of-sale (POS) credit card authorization is VeriFone.

ISDN PRI

Most large scale computer telephony developers long ago realized that in systems using more than eight lines, it's cheaper to dump analog POTS trunks and instead use T-1 digital trunk access. With a bandwidth of 1.544 Mbps, T-1 provides simple cabling, excellent quality, and access to Automatic Number Identification, or ANI (the number of who's calling) and Dialed Number Identification Service, or DNIS (what number did they call) information. In Europe they use a 2.048 Mbps E-1 channel.

When T-1 trunks place outbound calls or receive inbound calls, they use in-band DTMF/MF tones for signaling as well as a technique of "robbing bits" from the voice channels (you can't hear the missing stolen bits – it happens too infrequently for the human ear to discern). Although these methods work, they are slow and can cause problems if a computer telephony developer needs a clean data stream.

As it happens we can layer a form of ISDN – Primary Rate Interface (PRI) – on top of T-1. Remember how ISDN BRI eliminated in-band signaling by using the D channel? ISDN PRI also has a D channel, but it's the size of a whole 64 Kbps voice channel. This contains the signaling information for the other 23 64Kbps bearer (B) channels that fit in a standard North American T-1 trunk. In Europe and the Pacific Rim ISDN PRI comes through a larger 2.048Mbps E-1 channel, so the timeslots are configured as 30 B channels, one D channel, and one channel for framing (though sometimes its a 31B+D).

A single 64kbps D channel can even be used to support eight PRI ISDN links (8 x 23 lines) using Non-Facility Associated Signaling (NFAS). This use of a single D channel to control multiple PRI interfaces frees one B channel on each interface to carry other traffic. With much NFAS capable equipment, a backup D channel can also be configured for use when the primary NFAS D channel fails.

H channels are a way to aggregate B channels and are implemented as follows:

H0 = 384 Kbps (6 B channels)
H10 = 1472 Kbps (23 B channels)
H11 = 1536 Kbps (24 B channels)
H12 = 1920 Kbps (30 B channels) – International (E-1) only

ISDN PRI speeds call connections considerably. Outbound calls connect faster and on an inbound call the ANI and DNIS information is received via message packets, which is much faster than the several seconds needed by T-1's combination of winking and tone transmission. In fact, ANI and DNIS information, critical to call center operations isn't always offered on T-1. AT&T, many local RBOCs and Bell Canada demand that you use ISDN for access to ANI and DNIS information.

Even in large non-call center-based businesses, PRI is often used for connecting the PBX and the CO. Rather than a klunky collection of BRI lines and CO switching, ISDN PRI allows the phone system to apportion B channels as needed throughout a company.

Some of the things you can do with PRI that you can't do with a regular T-1 involve "intelligent network" functions such as the transfer and rerouting of calls from one PRI to another, which means you can transfer from one call center to another at a distant location. The caller's ANI, DNIS, and account information can also be passed over the network along with the call.

An other interesting intelligent network stuff includes AT&T's Vari-A-Bill service, where a bureau can vary the billing rate of a 900 call at any time during the call. Callers can select services from a voice-automated menu, with each service individually priced.

Non-Facility Associated Signaling (NFAS) is another money-saving service. In North America, carriers typically charge a premium rate for the D channel, as much as $400 per month over and above the T-1 span charges, plus a several thousand dollar installation fee. However, this D channel can handle more signalling than is needed by the 23 B channels it's controlling on the same span or "facility." With NFAS a single D channel can control up to eight ISDN T-1 spans! Using the D channel efficiently in this way lowers overall access costs.

As in BRI, several channels in a PRI can be combined into one high-bandwidth data stream to enable such things as 30 frame per second videoconferencing or other high-speed data transfers.

Most PBXs can forward ISDN calls to remote locations (Remote Call Forwarding – RCF) using ISDN as well as providing long distance dialtone (called Direct Inward System Access DISA). These functions can be password protected.

Intelligent Network (IN)

Also called the Advanced Intelligent Network (AIN). The Intelligent Network is a way of implementing services in the Public Switched Telecommunications Network (PSTN) by migrating software for complex services and features (called Service Logic) out of the switches and into IN based servers (generally VMEbus computers running some flavor of UNIX). These servers may be called Service Control Points (SCPs), Adjuncts, or Intelligent Peripherals (IPs) depending upon their placement and function within the network. SCPs, IPs and Adjuncts all provide a standardized Service Logic Execution Environment (SLEE). In IN, a call generated by a subscriber is routed (circuit switched) to the nearest Signal Service Point (also called a Service Switching Point, or SSP), which in turn , consults the SCPs on high speed SS7 links to get the necessary information for further routing of the calls.

Many of the services that modern telephone users expect rely upon this technology, including the following:

- 800 toll-free calling services: Unlike traditional telephone numbers that explicitly identify a telephone, an 800 telephone number is simply an identification number. To route an 800 call, the network must first locate the appropriate database (which depends on the carrier)

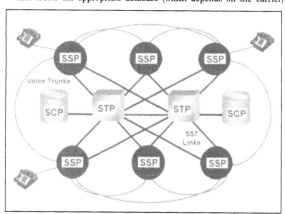

In the diagram, Signal Service Points (SSPs) are telephone switches that initiate and terminate SS7 messages. The Signal Transfer Points (STPs) are devices that route the SS7 messages within the network. The Service Control Points (SCPs) are database servers. The signaling architecture in the PSTN is extremely robust. Note that each SS7 link is duplicated, as are the STP and SCP network elements. The SS7 protocol includes comprehensive procedures to handle link and node failures, providing alternate routes and message retransmission to ensure that messages reach their desired destinations.

and then look up the terminating phone number to be used.

- Wireless roaming: When wireless users roam from their home territories, the network uses a complex series of signaling messages to enable incoming and outgoing phone calls. The subscribers' home databases are queried to determine what rights exist for roaming service in new areas. Temporary records are created in a visitor database, which is used to complete subsequent phone calls.
- Calling cards: When a calling card phone call is placed, a database is accessed to confirm the validity of the calling card and to enable proper billing to occur.
- Local Number Portability (LNP): LNP is a feature that allows phone users to change local carriers while maintaining the same phone number.

In the PSTN, IN services are controlled by SS7 protocol. The SS7 protocol is a layered protocol, with functionality isolated in specific software layers. The ISDN user part (ISUP) layer is used for setting up and tearing down phone calls. The transaction control application part (TCAP) layer is used for exchanging arbitrary information, including database queries and responses. Thus, the ability to support IN services in the PSTN requires both ISUP and TCAP support.

Development time, cost and management effort are significantly reduced by means of an IN-structured network. Also, a large range of services can be implemented in response to market demands, since the IN provides a framework to create various services in a centralized location independent of the switch: Powerful CASE tools known as Service Creation Environments (SCEs) help carriers, Enhanced Service Providers (ESPs) create their own custom services and features.

Although much fuss has been made about IP and a "dumb network," Intelligent Networks may turn out to be the "glue" that will hold technologies together. IP simply overlies networks such as ATM or frame relay or IP, and all of these "real" networks at some point have to interconnect with one another. That's where IN becomes crucial.

Interactive Voice Response (IVR)

IVR is a software application that runs in conjunction with a CT resource board capable of capturing and "understanding" touchtone (DTMF) keypad input. The touchtones are used to input numbers, to make menu selections, to answer yes or no questions, or to spell out certain words or names. Once the system understands what the user has input, it can provide appropriate responses in the form of voice prompts, faxes, callback, e-mail, etc. The whole idea is to allow users to self-navigate for information.

An IVR system gives customers 24-hour a day, 7-day a week access to a business' information and services through a simple touchtone telephone without the need for a human intermediary, such as an agent or customer service representative in a call center.

IVR was among the earliest of all computer telephony applications, making possible audiotex (777-FILM), Bank-By-Phone, Pay-Per-View, and myriad order entry applications.

Many of the old IVR companies such as InterVoice, Brite and Edify re-invented themselves as e-commerce experts, and are now using IVR (plus their expertise in back-office processing) as a "front end" to corporate websites and databases.

Jon & Jerry's Best IVR Data Access Tips

Jonathan Harwood and Jerold Francus are the principals of Alpine Solutions (Morganville, NJ – 908-536-4544, sales@alpines.com), an Interactive Voice Response (IVR) and electronic-commerce consulting firm specializing in business locator apps. Here are their tips regarding

Java

Introduced in 1995, Java is a programming language developed by Sun Microsystems that looks a bit like a simplified version of the C++ language. Differences with C++ include the fact that Java passes all arrays and objects by reference but doesn't have a pointer type, which eliminates the programming mistake of referencing a random area of memory. Java also has automatic garbage collection (just like BASIC) and C++ style exceptions are automatically generated when dereferencing a NULL pointer, accessing outside an array's bound, or running out of memory. Java's support of network protocol handlers such as HTTP, FTP, NNTP, MIME, and Sockets tip you off that it's some kind of "network programming language."

Indeed, Java is designed to be the ultimate cross platform, object-oriented programming language for creating interactive programs called "applets" that can run across networks such as the public Internet or corporate intranet. For example, Java-enabled thin-client network computing environments are perfect for distributing Java applets anywhere in a corporate environment. Also, since the client is executing the software, the server is relieved from the great processing burden of having to handle a large number of interactive web pages.

Since Java applets will run on any operating system and Java has no "extensions" unique to any operating system (except for a Windows-optimized "flavor" called Visual J++ that caused Sun to bring a lawsuit against Microsoft), its "write once, run anywhere" philosophy revives the old idea of a universal computer language. If you want to develop applications that anybody with a browser can run over the web, this is in theory the ultimate language to do it.

Java applets are written once, generally for a web page – you just include a Java application (or "applet") on your web page by inserting a special tag in your HTML code. Java application source code is compiled into an intermediate code known as Java "bytecode" which is later interpreted by a Java "virtual machine" (usually in a browser) that's tailored for the processor architecture on which the program runs. Thus, a Java applet will run on any platform having a Java interpreter – such as different kinds of computers (so long as they have a browser or some other means of executing Java bytecode), mobile phones, Internet screen phones, and other devices. The bytecode can literally run anywhere in a network on any server or client that is equipped with a Java virtual machine.

Java got an early reputation for being slow because the bytecode is interpreted a byte at a time during program execution, unlike its Microsoft competitor, ActiveX (See ActiveX). Java applets, once downloaded to the browser, contain 8-bit bytecode. ActiveX components, on the other hand, contain full 32-bit native code, another reason why Java tended to be slower than ActiveX components. To help remedy this, the Java virtual machine can be teamed with an optional Just-In-Time (JIT) compiler that dynamically compiles the 8-bit bytecode into faster 32-bit executable code during the download and execution of a Java applet.

Since the bytecodes are downloaded and executed at the client during runtime, only the latest version of an applet is ever run, which remedies the traditional problems of software distribution and upgrades. However, Java lacks "persistent" objects, unlike competing ActiveX components which are only downloaded the first time they are accessed and each time the software is updated. Since Java applets are downloaded every time they are accessed, they can add to network congestion. Some configurations exist for specific Java applets that use caching to store the applets locally, but this entails additional administration and may not work on all platforms or with all web browsers.

Also, since Java virtual machines must interpret Java byte code, during the execution of the Java applet it must also translate Java library calls into Windows operating system calls. Occasionally one will end up in a situation where there are no one-on-one mappings between Java API and Windows API calls. Java applets do not have full control of specific interface functions such as keyboard mapping. All of this can result in reliability problems, especially with host printing, user interface automation and advanced terminal emulation functions. Since ActiveX is tightly integrated with the Microsoft Windows operating system and Internet Explorer (IE) web browser, all Windows API functions are available and called directly.

Java Security

Java applets arriving at a user's browser are subject to bytecode verification. If a packet's size has changed somewhere along the way (possibly because of viruses or a Trojan horse), then the program will be instantly aborted. However, some of the original Java 1.0 virtual machine restrictions that promoted optimum security were too restrictive, such as prohibiting a Java applet from writing to the hard disk, accessing the printer and communicating with IP host addresses other than the machine it was downloaded from. For web to host applications, some of these restrictions would prevent file transfers, host printing or access to a host mainframe using other TN3270E servers. These restrictions were eliminated with the release of Java 1.1. Java's competitor, ActiveX components, use digital certificates to guarantee that the signed applet hasn't been corrupted or fiddled with by a hacker.

SSL security is also available for both ActiveX and Java web-to-host solutions. The Secure Sockets Layer (SSL) is the *de facto* Internet security standard originally developed by Netscape and provides both authentication and encryption to prevent eavesdropping and tampering of your TN3270E session(s).

Flavors of Java

Sun Microsystems has also created several subsets of Java that can be embedded in devices such as phones, smart cards, cars, TVs, pagers and more, taking Java "beyond the desktop."

Conclusion

If Microsoft completely loses the battle with Sun, then there is a high probability that Microsoft will drop support for Java, replacing their Visual J++ compiler with their experimental C++ Object Oriented Language (COOL) or some other Intermediate Language (IL) having functionality similar to Java.

Java Telephony Application Programming Interface (JTAPI)

A set of modular, abstract interface specifications of an API for use when building applications or components in Java that interface with telephony systems and must have first party or third party computer telephony call control functions. JTAPI was developed by a consortium including Dialogic, IBM, Intel, Lucent, Nortel, Novell, Siemens, and Sun Microsystems.

and interoperability, and supporting high-density telephony applications

• Support for call center, IVR, switching, voice messaging, and other functions

These standard interfaces must also be able to rely on standard responses to common functions like play, record, hold, transfer, and conference.

Look for your CT solutions vendor to supply a high-level, common, unified software abstraction layer that shields the developer from the underlying protocols and their related technology. Remember, "high-level" means using function calls, not just providing access to driver software. "Common" implies the API meets all the application's call control needs. And "unified" requires the API to be part of an overall, extensible architecture.

Be sure your network signaling API gives you everything you need to write once, and take advantage of the vast opportunity in today's global market by deploying your applications worldwide.

Top 10 Questions to Ask Your CT Partner

These tips are from Don Payette, Segment Manager, Dialogic Professional Services.

1. Is This Really a Partnership? You want a partner who acts like their success depends on your success ... because it does. Before you sign on the dotted line, make sure you can expect the level of personal attention and support you need to get to market quickly and easily with the sophisticated solutions you need to lead in your chosen markets.

2. Do You Believe in Open Systems? In the world of mainstream computing, open systems have long been the norm. But computer telephony has always been a world of separate, often proprietary systems handling functions like voice processing and computer-based fax. This has made CT systems expensive and hard to manage and stunted the growth of the CT industry. Today, the picture is changing fast. The technology is in place for next-generation CT systems based on open standards and delivered on a true CT server that allows a wide array of CT applications from different vendors to share applications and technology resources. The advantages are overwhelming: simplified development, scalability, easy manageability, shorter time to market, and lower hardware costs.

It's essential to make sure the components you use are built to the industry standards specified by key organizations like Telcordia, the American National Standards Institute (ANSI), the European Telecommunication Standards Institute (ETSI), and the International Telecommunications Union (ITU). Even better is to work with a partner active in developing industry standards through membership in organizations like the Voice over Internet Protocol (VoIP) Forum, the Internet Engineering Task Force (IETF), the Enterprise Computer Telephony Forum (ECTF), and the Intelligent Network Forum (INF).

3. How Do I Know You'll Be Around Tomorrow? Don't take chances. Find a partner with a proven track record – and even better, a long list of industry firsts. Yesterday's innovator is likely to be tomorrow's innovator. If you can, talk to some of the partners' other customers to make sure they're happy with the range of products and support they get.

4. What's Your Vision? Building a leading-edge CT application is rarely easy. And the right choices in components and technologies are rarely obvious. That's why it's essential to choose a CT partner that's in step with the industry and all the changes that are happening not only today, but tomorrow. Make sure they're providing you with a clear roadmap that's aligned with your business plan.

5. Can I Get One-Stop Shopping? The broader the partner's product line, the better. You're building a complete solution, so it's essential to deal with a single partner who can deliver all the components and options you need, and even more importantly, can help make sure they all work together. This includes the widest possible range of hardware, firmware, and software platforms and a full range of voice, fax, data, voice recognition, speech synthesis, and call center management components. Work with a partner who can provide the jump-start you need to get your application hooked up to SS7, ATM, or IP. . . and who knows the products and market to help you verify your design and approach. This kind of support will get you to market faster with less hassle – and with the right kind of product.

6. What's Your Technical Support Like? Define your expectations and needs. Make sure your CT partner is offering you experienced engineers to jump-start the design-in of their product to your application. And make sure they commit – in writing – to a guaranteed service level. You need customized, personal support you can count on. Also, make sure your partner has offices in major industrial centers around the world staffed with experienced sales professionals and knowledgeable applications engineers. Make sure they give you a choice of support plans that address your different needs as you move through your product's life cycle.

7. What's Your Marketing Support Like? Choose a partner with a strong reseller program that can add value by enhancing your market visibility and giving you market intelligence to make your sales job easier. Your partner should be able to provide you with benefits like qualified lead referrals, sales tools and regular information updates, the opportunity to purchase not-for-resale products at a discounted price, and the opportunity for cooperative participation in tradeshows, seminars, road shows, and other events.

8. Can You Help Me Go Global? If you're thinking globally, your strategy needs to include products certified to work worldwide. Make sure your partner has sales, marketing, and development in countries around the world, providing local experts and products that make it fast and affordable to build solutions for the global markets you want to enter. Know that your partner's products are approved by regulatory agencies in the countries you want to enter and that your partner knows how to work with the local certification labs. And if your CT solution includes speech, make sure your partner provides international support for speech recognition and speech synthesis, with different languages available.

9. Do You Offer Training and Certification? To help you bring solutions to market fast and build customer confidence in your expertise, it's important to work with a partner that provides an in-depth technical and reseller training program you can customize to your individual needs. Even better is a partner with a certification program that can enhance your credibility and prestige by validating you have the skills and knowledge to succeed as a CT reseller or systems integrator providing CT solutions.

10. Tell Me Why You Deserve My Business. Computer telephony is one of today's fastest changing and most sophisticated technology areas. So when you're building a CT solution, you need much more than just a component supplier. You need a real partner equipped to provide an array of technical consulting, training, certification, and support services to help speed up your application design and test cycles and support you during every phase of your product's lifecycle – from design and development through deployment and post-production maintenance. Unless your partner can offer you the complete package, they don't deserve your business. And they shouldn't get it.

Dave Krupinski's Best International CT Software Tips

Dave Krupinski is the former director of marketing and European operations for Pacific Telephony Design, a CT distributor in San Francisco that was acquired by Hello Direct.

1. Hardware makes all the difference. In most countries, the local telecom authority needs to approve any device connected to the PSTN. Check to make sure the CT hardware you're using is certified in your target markets. Dialogic and Aculab have an impressive list of worldwide approvals.
2. Digital protocols vary widely. They differ from country to country, and within the same country. For high-density digital applications select a CT hardware platform that provides a single API for all protocols. Aculab does a great job of supporting worldwide protocols with a single API. It's also the idea behind Dialogic's GlobalCall API.
3. Select the right toolkit. Your toolkit should support hardware that's approved in your target markets. Make sure the toolkit comes with (or allows you to create) multi-lingual versions of "system voice files" (the recordings and logic for speaking dates, times, prices, etc. in different languages). Ditto text-to-speech and voice rec.
4. Plan ahead. Rather than re-engineer an app for international markets, do it right the first time. For example, if the application is targeted at end-users, externalize all of the text in the user interface. This makes for easier translation.
5. Know your markets. Don't expect to succeed with an analog line app in Germany. Don't do a system in Brazil without Dial Pulse Detection. Introduce an end-user CT product into France without translating the user interface and documentation? Forget it. Don't pitch a too-intelligent app in Australia.

International Multimedia & Collaborative Communications Alliance (IMCCA)

IMCCA's mission is to foster industry growth by facilitating non-profit industry organizations to coalesce around promoting the technology and the uses of collaborative communications globally.

Their mission states that "The Interactive Multimedia & Collaborative Communications Alliance is an organization of members who share a common interest in fostering and promoting people to people communications and learning throughout the industry". Alliance members are end users, vendors, and other interested professionals involved in interactive applications.

IMCCA encourages the development of technology standards for the industry, case study success models, and broad dissemination of multimedia and collaborative communications information.

Website: www.imcca.org/cb_imcca/framesetitca.html?content

International Standards Organization (ISO)

Worldwide organization responsible for providing and defining international communications and voice standards for connectivity and interactivity. Standards include the media interface and voice/video interoperability. In particular, devoted to standardizing the Open Systems Interconnection (OSI) reference model for computer and data communications. Standards aren't published directly but must be obtained from designated third parties.

Website: www.iso.ch

International Telecommunications Union (ITU)

ITU (formerly the CCITT) is a standards body that defines compression / decompression, interconnection characteristics, and other telephony and communications related definitions relating to low level trans-

missions, typically over public networks like the PSTN. It is a resource for interoperability at the wireline level as opposed to the board level.

Website: www.itu.int

Internet

The Net. The Network of Networks. The awe-inspiring network of computers and other devices that communicate with each other via the Transmission Control Protocol / Internet Protocol (TCP/IP).

In 1960, a Harvard graduate sociology student named Theodor Holm Nelson (born 1937) came up with a computer-based hypertext idea for a term project for a computer course for the humanities. His first paper, *A File Structure for the Complex, the Changing and the Indeterminate* was published in 1965. He and his colleagues would later attempt to build a global hypertext environment called the Xanadu system. Nelson coined the term "hypertext" and "hypermedia" for non-sequential writings and branching presentations of all types. He presented a paper on "zipper lists," a key component in his Xanadu system, at a national conference of the Association for Computing Machinery in 1965. The modern World Wide Web borrows many ideas that originated with Nelson, although it is a vast oversimplification of what Nelson was trying to do.

In 1962 the U.S. Air Force commissioned Paul Baran of the RAND Corporation to perform a study about how best it could maintain command and control over its forces after a nuclear attack. The idea ultimately became one of constructing a harmless-looking military research network that could survive a nuclear strike, decentralized so sections of the country could be blown away in a nuclear war and yet the military could still secure control of their nuclear arms for a counter-attack.

Baran mulled over various ways of accomplishing this feat, then settled on a packet switched network. As he wrote:

"Packet switching is the breaking down of data into datagrams or packets that are labeled to indicate the origin and the destination of the information and the forwarding of these packets from one computer to another computer until the information arrives at its final destination computer. This was crucial to the realization of a computer network. If packets are lost at any given point, the message can be resent by the originator."

This "military research network" soon materialized in 1968 when the Advanced Research Projects Agency (ARPA, later renamed the Defense Advanced Research Projects Agency, or DARPA) awarded the contract to construct the ARPANET to Bolt Baranek and Newman (BBN). BBN first built a network based on existing Honeywell minicomputers in 1969, linking four sites or "nodes" with 50 Kbps circuits: The University of California at Los Angeles, the Stanford Research Institute (in Stanford), the University of California at Santa Barbara, and the University of Utah.

In 1972 there were 23 nodes or "hosts" and the first e-mail program was created by Ray Tomlinson of BBN.

The ARPANET originally used the Network Control Protocol (NCP) to transfer data. In 1973 a group headed by Vinton Cerf from Stanford and Bob Kahn from DARPA began development on a new and more flexible protocol, initially called the Transmission Control Protocol, and ultimately called TCP/IP, the use of which was soon mandated by the Defense Department. Cerf and Kahn make the first reference to the "Internet" in their academic paper on the new protocol. This first definition of an "Internet" was a connected set of networks, specifically those using TCP/IP. Thus, under this original definition, an "Internet" is actually a series of connected TCP/IP internets.

In 1979 the news group network known as USENET was created by Steve Bellovin, a graduate student at University of North Carolina, and programmers Tom Truscott and Jim Ellis.

In 1983, the University of Wisconsin created the Domain Name System (DNS), allowing packets to be directed to a domain name, which would be translated by the server database into the corresponding IP number.

In 1984 the ARPANET was split into two networks: MILNET (for the military) and ARPANET (for advanced research and academia). At this point MCI began replacing 56 Kbps lines with new T-1 lines (by 1991 the backbones would be T-3 links running at 45 Mbps) and IBM supplied the routers. This new network was to be called the National Science Foundation Network (NSFNET).

The Internet Engineering Task Force or IETF was established in 1986.

In 1990 the Department of Defense disbanded the ARPANET and replaced it completely by the NSFNET backbone. Around this time Tim Berners-Lee and the Organisation Européennee pour la Recherche Nucléaire (CERN) (www.cern.ch) in Geneva implemented a hypertext system on a NeXT computer called the World Wide Web (WWW) to provide efficient information access to the members of the international high-energy physics community.

In 1993 Marc Andreessen, the National Center for Supercomputing Applications (NCSA) (www.ncsa.uiuc.edu) and the University of Illinois developed a GUI client for the web called "Mosaic for X", the first browser.

Principal web protocols included the Hypertext Transport Protocol (HTTP), the Hypertext Markup Language (HTML), and the Uniform Resource Locator (URL).

In 1995, Internet telephony began to appear.

The Internet II or the "Next Generation Internet" was planned by a coalition of universities, businesses, and government. In 1996, 34 charter universities signed up and in 1997 a 622 Mbps backbone was set up with federal network interconnects (Department of Energy, NASA, etc.).

It remains to be seen whether the Internet II will ever attain anywhere near the kind of prominence enjoyed by the original Internet.

Internet Service Provider (ISP)

An ISP is a commercial organization with a permanent connection to the Internet. They lease lines from large telecommuniations companies and then resell part of their lines¥ capacity in the form of temporary Internet access connections to subscribers such as individuals and companies. The most common way to access the Internet from home is with a modem and a phone call to an Internet Service Provider (ISP). Your computer connects via modem to the ISP, which in turn is connected to the Internet with a high-speed link. ISPs also offers other related services such as website building and hosting, thus allowing a customer to put their business on-line without the cost of an in-house web server or having to train or hire staff members to maintain the server.

Edith Jacoby's "Practical Thoughts" on ISPs

Edith Jacoby is Internet Telephony Product Manager at Catalyst Telecom (formerly called the CTI Authority) (Cranford, NJ – 800-261-0247, www.ctiauthority.com) and is truly one of the most knowledgeable people on the subject. For IP Telephony Gateways using the Internet, there are many considerations in choosing your ISP. Here are her tips:

1 Choose an ISP with a large backbone and POPs near your Gateways. Check hops and delay between specific POPs.

2 Choose an ISP with worldwide presence, making it less likely you'll have to leave their backbone to reach one of your foreign gateways.

3 In places where you cannot access a large ISP directly, find out the size of the pipe from your ISP to a major backbone. It won't do you any good to have a T-1 connection to your ISP, if it then only has an ISDN connection to the rest of the world.

4 Instead of using the public Internet, you may want to consider purchasing intranet services from your ISP for predictable sound quality.

5 If you sell gateways, be sure set the customer's expectations on the voice quality.

6 If you're looking to purchase a gateway and want a demo first, beware! Almost all IP Telephony Gateway products sound great on a local area network.

7 Start with smaller analog systems, and then upgrade to scalable T-1 gateways when you've solved any problems and are ready to go full force.

Internet Telephony

Internet Telephony makes it possible to place domestic and international calls that travel at least partially over the packet switched network known as the Internet, rather than the conventional circuit switched network of the PSTN.

Internet telephony takes many forms:

- Internet telephone to analog phone via Internet and PSTN.
- Analog telephone to Internet telephone via PSTN and Internet.
- Analog telephone to analog telephone via Internet Hop-On-Hop-Off servers (HO-HOs) or similar gateways.
- Internet telephone to Internet telephone via Internet with PSTN as backup in case of network congestion and poor sound quality.
- PBX to PBX (or IP-PBX to IP-PBX) via the Internet.

When you place a call over the Internet, your voice is compressed, digitized, segmented into packets, and then transmitted using the TCP/IP protocol. Internet telephony is thus a specialized form of IP Telephony (See IP Telephony).

The whole basis of IP network-based telephony is the opportunity to make a lot of the computer telephony hardware manufacturers irrelevant. Just take a look at the Open Systems Interconnection (OSI) model. It's essentially a software model. Once you get beyond Layer 1 (where network interface cards dwell) everything else can be conceived and implemented as software. The specialty CT hardware we all know and love can be made obsolete just by developing the right kind of software.

Such homogeneous packet-based IP systems are the logical extrapolation of the "one-wire wonder" philosophy of companies such as CellIT, Selsius/Cisco, Sphere, and NBX/3COM. Indeed, NBX/3COM raised some eyebrows at Microsoft when they took TAPI 2.1 – a specification that requires an NT server somewhere in the network because of its ability to use monitors, to measure and know the status of other devices on the network – and managed to implement it on their NBX 100 one-wire system without NT whatsoever.

But before we all jump the CT ship in the face of the big bad IP monster, consider that IP telephony still has a few considerable hurdles to overcome.

DTMF tones need some kind of special encoding before they get turned into packets, then decoded at the other end. How do you address that? What's the universal standard?

Also, what happens when you route a call over a packet based network and the call route ends up returning that call back to you because the person you're calling had their calls forwarded to your voice mail? How do you resolve the kind of packet contention where you're sending and receiving and occupying needlessly excessive bandwidth? There's no logical signaling available yet that says: "Hey stop that communication out over the WAN because both endpoints are co-located at your facility!".

Although VoIP vendors adhere to the much-touted H.323 interoperability standard, H.323 is no panacea to compatibility. Company X may have H.323 products, but they may have developed a proprietary compression scheme (as in the case of Lucent and Motorola), so Company X's products can only communicate among themselves. Also, H.323 should be done purely in software – expensive gateways or specialty cards that slide into routers need not apply here.

IP telephony enthusiasts became intrigued by H.323 because it has so many aspects relating to the data communication world, such as an open interface. But what's the real functionality here? As Ed Wadbrook of NBX says: "H.323 provides communications services similar to what you can get from PBXs over nailed-up tie-line circuits. So essentially H.323 creates virtual tie-lines over packet-based networks. Beyond that the functionality is limited."

H.323 is based upon three function calls: Make Call, Conference Call, Drop Call. You can't do anything else beyond that. Various committees are working on enhancing H.323 to include Call Forwarding and Transfer, but that process is laborious and acceptance is slow.

Then of course there is the continuing problem of Quality of Service (QoS), or shall I say the lack thereof, thanks to the fact that one packet may travel directly to its destination while the next packet may get their via a route passing through China or Australia, resulting in a 10ms or greater delay that can be heard as a silent "dropout." Buffering can help here, though too much of it leads to a long irritating delay.

Many experts say the best way to get good QoS is just to throw more bandwidth at the problem. Overprovisioning. Use only 70% of your pipe.

Another solution is to use a transmission scheme that has QoS built-in, such as frame relay or ATM. When buying a frame you generally pay for a specific Committed Information Rate (CIR), the average maximum transmission speed of a user over a link to the Frame Relay network. A 64Kbps CIR, for example, will give you a guaranteed bandwidth of 64Kbps. The customer is always free to "burst" up to the maximum circuit and port, although any amount of data over the CIR can be surcharged. Excess bursts can be marked by the carrier as DE or Discard Eligible, and can be discarded if the network is congested.

Of course, many international enterprises are doing IP over ATM, not frame relay, primarily because ATM is found throughout the continental backbone. ATM also has a good QoS, but since voice is generally transmitted as Constant Bit Rate (CBR) traffic, about 20% of your bandwidth must be reserved for voice even if you're not actually sending it. If you send the voice as IP packets over ATM, then that 20% of the packet availability is lost to any other application. Still, a revised ATM spec will probably incorporate the sending of voice as Variable Bit Rate (VBR) traffic, allowing bandwidth to be allocated on an as-needed basis.

But QoS will probably receive its greatest boost from new protocols that take into account telephony. There is no single "magic bullet" IP protocol, but rather a "shotgun" splattering of families of protocols covering the whole spectrum of Internet / CT applications. These will have higher functionality and will be more scaleable than the still somewhat bloated H.323, which tends to bog down when you've got too many sites communicating with each other.

One of the new protocol contenders is Multiprotocol Label Switching (MPLS) which creates business-class IP services on ATM networks by combining Layer 3 connectivity (routing) and Layer 2 switching performance. While most carriers use ATM backbones, most enterprise networks prefer IP. MPLS integrates the two.

MPLS' QoS is achieved by assigning a label to each packet designating a priority used during transport, something like, "data, voice, video" or "silver, gold, platinum". MPLS will really shine in large enter-

prise office-to-office communications over ATM and frame-relay backbones for Virtual Private Networks (VPNs).

Another contender protocol is the Media Gateway Control Protocol (MGCP). This descended from Bellcore's circuit-based Simple Gateway Control Protocol (SGCP) and Level 3's Internet Protocol Device Control (IPDC) as a means for linking IP gateways to the PSTN. MGCP will exist at the IP / PSTN media gateways, such as those for VoIP, VoATM, as well as modem banks, cable modems and circuit cross-connects.

MGCP allows for call management software known as "media gateway controllers" to be set up throughout circuit-switched and packet-switched networks to allow for exchanges between the two networks. The media controller software or "call agents" are separate from the media gateway devices themselves, such as VoIP and VoATM gateways, as well as modem banks, cable modems, PBXs, set-top boxes, and circuit cross-connects.

Although Internet telephony technology is slowly approaching toll quality, users have found the content of multimedia, intranet communications, the cost of a single network instead of two, call completion, unified media communications and applications to be more compelling than simple "cheap minutes."

IP Centrex

Also known as Centrex IP (a Nortel term) and iCentrex (a Lucent term). Centrex-type services offered over IP networks.

With Centrex IP, Nortel lets customers connect to their carrier's Centrex service via a data IP line. Centrex IP integrates into a user's existing corporate network, providing access to any of their traditional desktop-based voice services from anywhere they can access their corporate data network. This includes all business services provided by Nortel's DMS-100 switch (Nortel is adding a gatekeeper to work in conjunction with its DMS-100 switch to offer the IP Centrex capability). The service also creates a framework for customers to build new applications such as unified messaging, the ability to access their voice mail and e-mail from their desktop computer or phone, and multimedia conferencing, which lets customers share data applications during a conference call.

Centrex IP enables users to access the full functionality of their business telephony system through a "soft client" on their PC's desktop. This allows corporate travellers to access their voice services anytime, anywhere via a simple dial-in connection from, for example home, a client office, an airport or a hotel. Centrex IP provides the unification of the delivery of voice and data over an IP connection. Additionally, delivery of a Centrex IP Ethernet-enabled feature phone allows access to the full business services portfolio at end user locations.

The standards-based offering is compliant with the H.323 Internet Protocol, G.72x, G.711, ISDN, SS7, and SNMP.

Lucent's IP-friendly Centrex package, iCentrex gives remote users access (over IP networks) to the same services they would be able to use if they were physically in their offices, such as voice mail and LAN access.

IP-PBX

Just before the dawn of this century, there appeared the idea of IP-PBXs. Not an unPBX, but IP phones and IP fax machines wired over an Ethernet connection to an IP-PBX which has a gateway built into it.

Several companies had already started along this path by developing software and hardware that could fool a PBX into seeing the Internet as a series of ordinary phone lines through which it could place and receive phone calls. Such Internet trunks behave in a manner similar to ordinary analog trunks, except that they convert voice into compressed data that's sent over a TCP/IP Ethernet interface. By putting these devices in front of a PBX, you can place and receive Internet calls the same way you make ordinary phone calls. The assumption behind such equipment is that, for the next ten years or so there will be the need for a business to hook up to either a legacy PBX or the PSTN at some point.

Lately there has been much activity and many announcements in the area of real IP-PBXs, as opposed to server based PBXs. PBXs with IP phones are appearing. There has also been a spate of IP phone announcements from Nortel, Siemens, Applio, Pentel, and others.

You'll see the enterprise in transition towards IP communications. Some of the names involved here include Cisco (which bought Selsius, makers of the first Ethernet phone), Lucent, Nortel and Ericsson. When you've got the big players reinventing themselves and introducing such products, it means that something big is going to happen. This isn't a case of a bunch of startups trying to claw their way into the PBX business with a crazy new idea. This movement signals an important shift in the telecom equipment industry. Voice can now be seen as just another application which uses the network.

IP Telephony

Simply put, IP telephony is real-time communications over a data network. It can be voice communications, or it can be any other type of communications.

The conventional PSTN uses a Switched Circuit Network (SCN) that uses a circuit-switched architecture wherein a direct connection, or circuit, is made or "nailed up" between two users across the network. The circuit provides a full-duplex connection with practically no latency, or delay, between the two endpoints, and the users have exclusive use of the circuit until they are finished and the connection is released. In the days of Alexander Graham Bell, this connection was a physical connection, a literal wire connecting the endpoints, but in the modern digital world it's a logical connection through many switches and across a variety of wiring types (twisted-pair, fiber-optic cable, etc.).

The fact that the entire PSTN was designed for circuit switching of voice calls has made it difficult to add new network services, or to increase the efficiency of traffic handling. For example, in a famous paper titled, "The Rise of the Stupid Network," David Isenberg, formerly an executive at AT&T Labs, describes an attempt to improve circuit-switched voice quality given the constraints inherent in the PSTN archtecture:

"If we had not been constrained by network architecture, the easiest way would have been to increase the sampling rate or change the coding algorithm. But to actually do this, we would have had to change every piece of the telephone network except the wires. So we had to work within the designed 64 Kbps data rate. We discovered that voice quality could be substantially improved by boosting the bass part of the signal, that part of the audio spectrum between 100 and 300 cycles per second. But as we set out to implement this conceptually simple improvement, we kept running into the problem that there were too many places in the network that had built in 'intelligent' assumptions about the voice signal; echo cancellers, conference bridges, voice messaging systems, etc. – and too many devices that depended on these acoustic assumptions for their correct operation – modems, fax machines, and a surprising number of strange devices with proprietary analog protocols. After about two years of intense effort, we made a noticeable difference."

Packet switched networks and IP telephony changes all of this.

IP telephony is not Internet telephony. IP telephony provides traditional and non-traditional telephony functions over a managed or private network that's owned and run by a single entity that can guarantee a low and predictable packet delay from one end of the network to the other.

On the other hand, telephony over the Internet refers to one application of this technology: As a replacement for long distance PSTN service, or "toll bypass" (See Internet Telephony). This is for hobbyists and amateurs. The Internet is surely IP network, but it's a patchwork of small, unmanaged networks owned and operated by disparate entities. The resulting conglomeration is a network that produces unpredictable and sometimes quite long delays in what should be real-time communications.

Still, the tantalizing idea of making "free" (or almost free) phone calls over the Internet generated a near frenzy among the more frugal of the technology-savvy consumers. This approach, identified originally as Voice On the Net (VON) was widely championed as a replacement to the PSTN. While there are millions of active Internet / IP Phone users today, VON still lacks some of the PSTN's capabilities.

The real market for "pure" IP telephony is Voice over IP (VoIP) on managed IP backbones, (e.g. Qwest, Level 3, WorldCom) and not the public Internet. Even here arbitrage is enticing, bypassing the phone tax because voice conversations are being carried over a data connection. Also, IP telephony leads to cheaper conversations not just in terms of minutes on the line, but the reduced cost of multiple calls sometimes necessary to get though to the right person at the right location. But while the public Internet may be used for a least-cost voice or non-time-sensitive data, (secondary backbone) any real-time capabilities will have to be added to it through the addition of new protocols. Internet bandwidth management, voice quality, and scaleable call control on the Internet also needs some work.

IP telephony treats voice as another form of data, which opens up all sorts of exciting new possibilities.

- More Bandwidth, More Money. Circuit switched telephony is lavish when it comes to bandwidth: 64 Kbps. Through various compression and other techniques, IP telephony takes a mere six to eight kilobits per second (even as low as 2.4). It's frugal with bandwidth, dramatically improving bandwidth efficiency for realtime voice transmission, in many cases by a factor of eight or more. The new network economics relies on the fact that networks are transmitting more data than voice, and by multiplexing the data with voice over packet based infrastructures you create a more efficient network with higher capacity than the traditional circuit based infrastructure.
- Dumb vs. Smart. Circuit-switched networks have dumb, nailed down circuits, with nothing "listening" in. You talk, that's all. Packet-switched networks send packets which have headers with space for instructions.
- New Services. IP telephony leads to the potential explosion of a thousand convenient, value-added services. These new revenue producing applications are enabled by being able to treat voice and data uniformly over the same network. It's much easier to create new classes of service in the IP world than the more rigid (and expensive) PSTN world. You can just write your own form of Custom Local Area Signaling Services (CLASS). New enhanced IP services can combine the best characteristics of real-time voice communications and data processing, such as web-enabled call centers, collaborative whiteboarding, multimedia, unified messaging, telecommuting, and distance learning. See if you can do better than the phone companies selling the next-gen version of call waiting.
- Ease of Deployment. IP telephony is not really a "new" infrastructure it is simply "more" of what already existed in terms of data communications. IP telephony can be used in conjunction with existing PSTN switches, leased and dial-up lines, PBXs and other customer premise equipment (CPE), enterprise LANs, and Internet connections. IP telephony applications can be implemented through dedicated gateways, which in turn can be based on open standards platforms for reliability and scalability.

Perhaps the most formidable challenge for the IP telephony industry is to develop networks that are not only scaleable but seamless to the subscriber in terms of ease of use and to the service provider in terms of ease of billing. This can only happen with gateways and switches that can deal with both packet networks and the PSTN.

For example, SS7 signaling was created to control PSTN voice switches and was not designed to handle IP call traffic. IP telephony's success may well hinge on how to interface IP Telephony Gateways into the SS7 network in such a way that they can be managed in a manner similar to PSTN switches. Hopefully, a standard will emerge.

Today's carriers are looking for ways to bring competitive services to market quickly and inexpensively in a hybrid network environment. Here's a look at a new kind of switch that could help navigate the divide.

Now that the dust kicked up by IP telephony's debut is beginning to settle, at least one thing has become clear: The PSTN is not just going to disappear overnight. The "legacy" class 4 and class 5 switches that carriers have used for years to move voice, along with the SS7 signaling network, are responsible for a rather extensive list of features that the average phone user simply will not give up, no matter how low the long distance charges. We aren't talking super-sexy enhanced services here, but the basics like call waiting, caller ID, and call forwarding, which are nonetheless essential in selling phone service to almost anyone.

Until recently, VoIP has been painted as a problem of network mediation, or simply changing PSTN circuits into IP packets. This is what the first generation of VoIP gateways, many of them PC-based and relatively small in scale, were designed to address. Since that time gateways have grown more robust and more capable of handling large-scale deployments in applications such as international long distance. It has also become quite clear, however, that circuit-to-packet translation is only a small part of what is necessary to make network convergence a profit-generating business that extends beyond limited niche markets.

Specifically, the missing link has been a way to mediate not only between transport protocols but between the protocols used to deliver services – taking the latter to mean both commonplace features like call waiting, and those that are more esoteric, like local number portability. "Services" has of late become something of a buzzword, repeated almost incessantly in IP telephony circles. With the opportunity for toll arbitrage having diminished almost as quickly as it appeared, it is no wonder that the industry has been drawn to IP's potential to deliver more services, more quickly and easily than the PSTN. The question, nevertheless, is how to develop these enhanced services without taking away existing ones on the one hand, and without reinventing the wheel (i.e., recreating all those esoteric SS7 and AIN applications as IP) on the other.

The answer overwhelmingly put forward in the past year is based on a new kind of switch that lets service providers, both young and old, mediate existing services and create new ones. These next-gen switches, variously termed "media gateways," "service mediation switches," and "class 4/5 replacements," have garnered a great deal of attention lately, including a feature in the *New York Times* business section (August 6, 1999) that highlighted startup Tachion Networks. Typically, a next-gen switch is architected in two parts. One holds the physical ports, line cards, and actual switching fabric; the other acts as a signaling gateway, and provides interfaces to both the SS7 network and its own databases and applications servers, where subscriber profiles are stored and new services are implemented. The latter, often referred to as a softswitch, uses the evolving Media Gateway Control Protocol (MGCP) and SIP to communicate with the core chassis.

The various names used to identify the product category can create a fair amount of confusion as to what the primary function of this type

of switch actually is, and to where exactly it would be most likely to be located in a carrier's network. The answer, of course, is multifold:

- *Internet Offload.* The first and most concrete application for a next-generation switch is to offload dialup modem traffic from existing class 5 switches. The problem with loads of people logging on to the Internet is that the traditional class 5s were not made to handle such long call holding-times in such high volumes. The resulting bottlenecks can prove very expensive to carriers if they are forced to purchase additional tandem switches simply in order to deal with the increase in traffic. By placing a significantly cheaper next-gen switch in front of the legacy tandem network, however, data traffic can be switched off either direct to the ISP via PRI lines, or to a data headend's RAS bank, where it is converted to IP and then sent to the ISP. Voice traffic travels through undisturbed to the next end office switch.

- *Class 4 Replacement.* As softswitch standards are further solidified, and signaling protocols like MGCP and SIP are approved, it will become increasingly plausible to use many of the next-gen switch offerings as class 4 tandem replacements. The primary application here is long distance VoIP. A class 4 switch is not responsible for as many features as a class 5, so as long as the signaling gateway can provide adequate SS7 interfaces, VoIP switching does not seem to present as much of a risk at this level as in the local loop.

- *Class 5 Replacement.* In certain ways, the most challenging of next-gen switching's possible applications, class 5 replacement can also be its most attractive. A big reason is simply that class 5 switches are very expensive, and present a significant barrier of entry to new CLECs and service providers. The challenge is to recreate enough of the class 5's features that subscribers won't notice a difference in service. At the same time, it is at this level in the network that the greatest opportunity for implementing enhanced services presents itself. The next-gen switch is, furthermore, particularly suited to local loop VoIP solutions, where it would accept traffic traveling over ATM or IP from an Integrated Access Device, and either switch it onto the PSTN, or keep it moving as data into the backbone. (See IP VoIP-Capable Switch Routers)

While it is easy to get excited by the promises of this new switching architecture, we have tried to maintain a healthy skepticism in approaching the topic. Even the most zealous of marketing reps won't seriously suggest that carriers are going to start ripping out their existing circuit switches and replacing them with this newer breed. More significantly, the incumbent giants – namely, Lucent and Nortel – from whom the next-gen startups are looking to steal a bit of market share, have been busy developing their own migration platforms to supplement existing 5ESS and 4ESS switches.

Lucent's 7R/E Packet Solutions and Nortel's Succession line are the embodiments of this effort. Further, there is a relatively high degree of difference in philosophy and approach even within the roughly assembled group of next-gen players itself.

The key for most of these players going forward will be for their products to remain flexible enough to reinvent themselves according to carriers' continually shifting needs – in this lies their biggest advantage over the traditionally closed, proprietary switches. In the meantime, there are a number of products hitting the streets that are well worth exploring.

Conclusion

Once all of this occurs, potential customers for IP telephony technology and applications include not only those directly involved in the Internet business but also companies in the PSTN carrier space, private networks or intranets, WANs / extranets, and enterprise networks. Just about everybody can benefit from IP telephony and, to some extent, even Internet telephony.

IP VoIP-Capable Switch Routers

High capacity switch routers are helping simplify the network's core and provide the necessary bandwidth for the not-so-distant world of data and VoIP convergence. But there's more than one way to route a packet. Backbone IP switch routers occupy a contested space in carrier networks, and the growth of voice-over-IP doesn't necessarily make things any clearer. On the one hand, as packets replace circuits and we move toward a voice and data converged universe, the importance of IP is unquestionable.

At the same time, that very convergence onto a single, common medium raises serious questions about a possible basis for differentiation: For example, how will carriers and service providers be able to tell which packets contain voice and which contain data, and then assign priority as needed to meet quality guarantees? The questions of Quality of Service (QoS), in addition to the sheer bandwidth issues that accompany faster and cheaper access to the 'Net, are central to the thoughts of those manufacturing next-gen IP routing equipment, as well as those buying it.

Until rather recently, ATM switching was the method of choice for providing converged voice, video, and data services. Though ATM cells are relatively wasteful and bandwidth inefficient, ATM's virtual circuits offered an easy solution to potential QoS problems. Using edge routers and various other products to bridge one network to another, IP traffic from LANs and gateways would be specially encapsulated to pass through ATM switches at the backbone. Though ATM has by no means been abandoned – many carriers, in fact, still trust it as the only proven way to transmit VoIP – its role in the new public network is under considerable debate. There will probably be a significant change within the coming months and years.

From a relatively early stage in the VoIP game, interest in carrier-class IP routers arose from a desire to simplify network infrastructure and to cut back on costly central office (CO) equipment. Pure IP routing would help avoid the impending bandwidth crisis, which the popularity of ATM was only accelerating, and would reduce the amount of equipment necessary to get packets to fiber at the core. Building out a routing architecture that could leverage the efficiency of IP in the data world, while providing the guaranteed quality of ATM, became the goal of a number of determined start-ups, as well as a new product market for some incumbents.

As a result, the products that have appeared, and will continue to appear, are more intelligent by leaps and bounds than are first-generation IP routers. For example, by off-loading packet forwarding functions from software onto a variety of sophisticated application specific integrated circuit (ASIC) chips, developers have found an effective way of performing table lookups without slowing down the packet flow and bottlenecking traffic. In addition, many newer routers support wirespeed forwarding regardless of packet size (crucial for voice packets, which are much smaller than data). Innovations such as those are important steps toward making the dream of a pure IP network a reality, and a convincing, if not exclusive, argument can be made for why that dream should be pursued.

In the more traditional voice-over-IP-over-ATM model, the CO can get pretty crowded. Calls travel over the PSTN locally to a class 5 switch. From there, voice circuits are handed off to a VoIP gateway and packetized before they are sent on to an edge router and finally a backbone switch. Then the packets (now encapsulated as ATM cells) must pass through multiple

SONET add/drop multiplexers before reaching dense wave division multiplexing equipment and dropped onto the fiber optic layer.

In contrast, an IP "network of the future" could be considerably simpler, though in many ways more sophisticated. Local traffic terminated on a DSLAM, a cable headend, a POTS switch, or a VoIP gateway would be passed through an access router (for example, the Cisco 7500 or newer products from Xylan, Sonus or Transmedia) at the edge, and on to a switch router at the service provider's core. The switch router interfaces directly with DWDM, and IP travels over fiber with no intervening layer of ATM.

Of course, as in many other aspects of IP telephony, we aren't quite there yet; ATM switching – in one form or another – continues to hold on, while several competing strategies duke it out in the switch router market. One of the most basic and high profile conflicts is between the high powered terabit (one trillion bits per second or higher) routers that start-ups like Avici, Nexabit, and Pluris have been busy designing, and the "mere" multi-gigabit offerings of older players such as Cisco and Nortel. OC-192 terabit products are now appearing. To give those products their due, they are built for to be scalable, and many incorporate distributed architectures that are an improvement over last generation, single-box solutions. A number of the newer entrants are also leading the way in implementing a viable standard for IP QoS.

Standards for a Difference

If one switch router and one fiber network is theoretically going to handle all kinds of traffic, from residential Web surfing to VPNs and VoIP, there obviously must be a way to distinguish between different kinds of traffic in the core. The problem with traditional IP is that it is "connectionless" and thus largely anarchical when it comes to determining routing paths and queues. Although the industry has not officially approved or accepted any single standard for classifying IP traffic, a couple of methods offer *de facto* solutions.

The differentiated services, or DiffServ, protocol is fast gaining acceptance, and effectively replacing the earlier resource reservation protocol (RSVP). DiffServ uses the "type of service" (TOS) field on IP packets, renaming it the DS byte, to carry information about an individual packet's service-level requirements. The DS byte is defined by network edge devices and read by core routers, where packets are queued accordingly and granted the necessary bandwidth. Most of the routers we'll be examining shortly support DiffServ.

Multi-Protocol Label Switching (MPLS) is another way to distinguish and classify packets, though it works on a different level from DiffServ and the two can be employed together. MPLS assigns new routing headers to individual IP packets in the form of a 32-bit label. MPLS labels determine specific routing paths for each packet and can even be used to dictate the router-to-router "hops" a packet will take along the network. MPLS is also important in IP-over-ATM applications, where information in the label can deliver each packet or flow to the appropriate type of virtual circuit (differentiating, for example, between a constant-bit-rate and a virtual-bit-rate circuit). By using MPLS for traffic engineering, the newest edge and core routers on the market are helping the move toward a possible solution to the IP/ATM conflict, or at least providing a productive compromise between existing protocols.

The switch routers covered in the following roundup illustrate the most up-to-date technologies you can find in the market. In addition, they have exciting plans and groundbreaking visions for the future of voice and data. Whatever the network of the tomorrow ends up resembling, you can be sure these companies will take a leading role in shaping it at the core.

Terabit "Super" Switches

Avici (Billerica, MA – 978-964-2000, www.avici.com) has become somewhat notorious for heralding the death of ATM. Although that claim may be slightly inflated by media hype, its Terabit Switch Router does concentrate on delivering pure IP-over-fiber (rather than over ATM) for providers of VoIP, VPNs, and e-commerce applications.

The TSR can handle a whopping 560 OC-192 or 2,240 OC-48 ports, and switches at speeds ranging from 2.5 Gbps to 5.6 Tbps. In addition to OC-192 and OC-48, TSR line cards also support OC-3 and OC-12 and can be intermixed in a single system (up to 40 cards per chassis). The TSR's switching fabric interfaces directly with dense wave division multiplexers (DWDM), reducing necessary carrier equipment and maximizing bandwidth capacity. All TSR line cards are equipped with ASICs which perform both layer 2 and layer 3 packet forwarding, and include QoS support through DiffServ, Weighted Fair Queuing (WFQ), and random early detection. The TSR uses MPLS as its main traffic protocol. Support for ATM and frame relay is also available.

The TSR chassis features redundant switching, power, and cooling components, and is NEBS compliant. List price for an initial configuration starts at around $500,000.

Nexabit Networks (Marlborough, MA – 508-460-3355, www.nexabit.com) makes the NX6400, a terabit switch router that performs high-speed IP-over-fiber functions. NX6400's port density is large – up to 192 OC-3, 96 OC-12, 64 OC-48, and 16 OC-192 channels, with a 6.4 Tbps switch capacity. And though OC-192 is only just beginning to materialize, Nexabit is already looking toward the next generation – preparing for OC-768 and eventually replacing SONET/SDH switching altogether. Built-in long-reach and short-reach lasers facilitate more successful, integrated optical internetworking.

Aside from pure horsepower, Nexabit has focused on QoS by developing an IP switch that to all appearances provides ATM-like performance. Using an internally developed protocol called IP CBR (which in effect mimics ATM's committed bit rate feature) NX6400 promises a guaranteed delay of 40 microseconds or less on any size or type of packet. Each line card supports 256 independent forwarding tables and features a programmable forwarding processor. Line cards also support a variety of packet classification rules, based on source and destination IP addresses, virtual circuit, MPLS COS bits, IP TOS, and DiffServ bytes, and layer 4 information such as protocol ID and source and destination port numbers. Since the NX6400 forwards unicast and multicast traffic at line rate, service providers can offer applications that require large packet bursts – such as streaming video, software downloads, remote medicine, and remote education – without causing significant traffic problems.

Protocols offered by the NX6400 include IP, ATM, frame relay, and MPLS, with TDM in the works. Direct connections to DWDM are also included. The NX6400 is NEBS compliant, and all major system components are redundant and hot swappable.

Nexabit was acquired by Lucent Technologies for about $900 million.

The Pluris (Cupertino, CA – 408-863-9920, www.pluris.com) 2000 Terabit Network Router is a modular system that configures from a single shelf with 90 Gbps switching capacity up to 128 shelves with 184 Terabit non-blocking switching capacity and 19.2 Tbps line capacity. Each shelf contains up to 15 line cards, and each line card holds four OC-48 ports or one OC-192 port.

In addition to its high density and port capacity, a couple of key features distinguish the TNR from other routers in its class. One is that its architecture is based on a distributed switching fabric. Unlike more traditional, "single box" solutions, which have closed backplanes and limited port capacity, the TNR features a fiber optic backplane, where multiple shelves are interconnected with fiber to form a matrix (rather than establishing connections between individual line cards). This means that as your network gets larger, you can distribute the traffic load across the TNR's various shelves and have, for example, your OC-3 ports located on one shelf while another shelf handles OC-192. Fiber interconnections also make growing the system easier and cheaper than adding multiple routers and ATM switches.

TNR line cards interface directly with DWDM or SONET/SDH-based fiber networks, eliminating costly network layers previously common to IP switching and freeing up a good deal of bandwidth in the process. Another difference from more traditional switches, crucial to VoIP implementation, is support for QoS and service-level agreement. On-board ASICs perform IP lookup and forwarding on each line card, as well as supporting the DiffServ standard for labeling and prioritizing packets.

TNR systems are NEBS compliant, and offer five nine's availability. Systems include PlurisView network management tools. Expected customers and implementations include Internet backbones, tier 1 and 2 ISP POPs, network access points, CLECs, IXCs, telco central offices, and concentration points for undersea fiber communications. TNR has a starting price of $45,000 per OC-48 port.

Gigabit IP Switch/Routers

Cisco Systems' (San Jose, CA – 800-553-NETS, www.cisco.com) 12000 series is something of a landmark in the market. While it doesn't (yet) promise the through-the-roof speeds and bandwidth of newer terabit routers, the 12000 series does offer solid support for up to 132 DS3, 44 OC-3C or STM-1C, 44 OC-12C or STM-4C, or 11 OC-48C or STM-16C interfaces, at speeds of up to 60 Gbps.

Packet forwarding functions are performed by a series of line cards plugged into a passive backplane, and routing is handled by two gigabit route processors. The line cards are interconnected to switch fabric cards via serial lines (four per slot). The 12000's switch fabric supports virtual output queues, to prevent traffic bottlenecks, and a weighted fair queuing mechanism that combines random early detection (RED), WRED, and deficit round-robin techniques. Line cards can transmit IP packets over DPT, PPP, ATM, or frame relay interfaces, and connect to other GSRs or edge routers via electrical wire or SONET/SDH fiber. Gigabit Ethernet ports provide connections to other routers in the network over fiber optic link.

The two available GSR model types, 12008 (40 Gbps) and 12012 (60 Gbps), are both NEBS and ETSI compliant. Processors, line cards, power supplies, cooling devices, and switch fabric are all redundant and hot-swap. Automatic protection switching (APS) and multiplex section protection (MPS) provide interface redundancy.

Juniper Networks (Mountain View, CA - 1-888-JUNIPER, www.juniper.net) makes the M40 Internet Backbone Router, a 32-slot chassis that holds a variety of ATM, SONET/SDH, DS3, and Gigabit Ethernet physical interface cards (PICs). In a relatively compact box, the M40 can hold eight OC-48 ports, 32 OC-12, 128 OC-3, or 128 DS3. Chassis are also rackmountable, providing for double those port densities.

Packet forwarding functions are carried out by the M40's Internet Processor, a high density ASIC with a lookup rate of 40 million packets per second and throughput speeds of up to 40 Gbps. The processor can be programmed with up to four different simultaneous layer 2 and 3 forwarding tables, and supports IPv4 and MPLS protocols, as well as virtual circuits. The M40 uses a weighted round-robin queuing scheme, with optional random early detection. Routing is performed by Juniper's JUNOS software, a modular design based on in-house protocols and an independent operating system.

A fully redundant cooling system and power supplies help insure system reliability. Several broadband service providers currently use Juniper's routers in their networks.

Lucent Technologies' (Murray Hill, NJ – 1-888-4LUCENT, www.lucent.com) multi-gigabit PacketStar 6400 IP Switches are built for carriers who want to build converged voice, video, and data IP networks. PacketStar takes aim at the traffic bottlenecks caused by over provisioning, and attempts to outsmart the old, dumb switch routers.

To this end, Bell Labs developed the concept of "classify, queue, and schedule" to establish a service hierarchy for IP traffic flow. packetstar is able to support up to 512 classification and filtering rules for layers 3 and 4 range-matching, while still performing at full wirespeed. Each slot can multiplex flows across 64,000 queues and implements multiple waited fair queuing schedulers. Those traffic management abilities are supplemented by standards-based multi-protocol label switching (MPLS), currently the most popular way for IP switches to guarantee service to end users. In addition, PacketStar's management software includes a Web-based GUI for QoS policies.

PacketStar can also be used in conjunction with PathStar – Lucent's all-in-one, combinations DSLAM, remote-access server, VoIP gateway, edge router, and telephony switch – to build a scalable carrier network from the ground up.

PacketStar supports both IP over SONET/SDH and IP over ATM. Redundant, hot-swap switching fabric, power supply, and router controller are optional. DWDM optical interfaces are included and line cards range from dual OC-3 to single OC-48. Pricing starts at around $11,000 per OC-3 port, and $27,000 per OC-12 port, with a range of different size chassis available.

Lucent's PathStar Access Server in certain ways merits its own category. Its ability to act as a class 5 switch replacement makes it possible to consider the PathStar as a next-generation switch. In addition, however, PathStar can also act as a DSLAM, an H.323-based VoIP gateway/gatekeeper, a digital loop carrier, and an IP edge router. PathStar supports between 512 and 7,000 POTS lines in a cabinet, as well as offering 10/100BaseFX Ethernet, DS1, DS3, and OC-3 interfaces. Because it can aggregate xDSL traffic, PathStar is particularly suited to voice-over-DSL applications. Although it is smaller in capacity than many of the other next-gen switches intended as class 4/5 replacements, it has the distinct advantage of being able to leverage Lucent's 5ESS technology, which makes it able to easily support as many if not more class 5 features than its competitors. PathStar also has embedded SS7 and MGCP signaling interfaces, which enable it to interwork directly with Signal Transfer Points and call agents. Pricing for PathStar starts at around $100,000.

Although Lucent has made a big deal out of IP, they recently debuted one-slot, two-port ATM uplink modules for both their Cajun P550 Gigabit Switch and the newer Cajun P880 Routing Switch, the industry's first uplinks to offer dual OC-12 (622 Mbps) bandwidth, allowing a LAN to connect to an ATM network and spew forth a total of 1.2 Gbps.

Nortel Networks' (Brampton, Ontario – 905-863-0000, www.nortel.com) purchase of Bay Networks helped solidify its position as a competitor in the IP routing market. It also added IP specific expertise to its development of the Versalar 15000 and 25000 switch routers. Each Versalar model can be used separately or together to perform both network edge and backbone IP routing for small to large service offerings. The difference in the two versions is more one of capacity than of func-

tion (both run the same software and ASICs), and each can be configured to meet varying needs. In general, however, the 25000 can be thought of as CO equipment, while the 15000 is more likely to be used as an edge device in local offices.

Though the actual voice to packet conversion takes place at a lower level of the network (via VoIP gateways and access platforms), Versalar switch routers are designed to include features specific to voice, which are not found in more traditional IP routers. For instance, because voice packets are much smaller than data packets, Versalar's processors can handle packets as small as 40 bytes and still provide wirespeed performance. To allow carriers to offer service-level agreements to customers, Versalar supports the DiffServ standard (a method of labeling packets with headers that describe their level of priority), and its processors are designed to handle multiple "filters" without slumping in performance. In addition, Versalar routers count every packet that passes through them – a feature that allows for the most sophisticated accounting and billing applications currently available.

In its first release, Versalar 25000 comes with a 240 Gbps throughput capacity, up to 10 times higher than most routers currently available, giving it the ability to handle up to 48 OC-192 ports. The 25000 also accommodates 576 DS3 ports, 96 OC-3, 24 OC-12, 24 OC-48, and 24 Gigabit Ethernet ports. The 15000 has a smaller port capacity, and supports OC-3, OC-12, and 10/100MB Ethernet. Both Versalar models are NEBS level 3 compliant, and maintain 99.999% availability, with no single point of failure and optional N+1 redundancy.

Hybrids (e.g., IP/ATM)

Cisco Systems (San Jose, CA – 800-553-NETS, www.cisco.com) has claimed its stake in next-gen switching with its recent purchase of Transmedia Communications (San Jose, CA – 408-363-8988, www.trsmedia.com). Since the acquisition, Cisco has integrated Transmedia's product, formerly known as the MMS-1600, into its multiservice access family. The product was rechristened the MGX 8260 media gateway, and made generally available in a newer version in November, 1999.

Cisco's multiservice switches have the ability to support both IP and ATM, as a migration path from circuit-switched TDM network. Dialup modem offload is expected to be the primary initial application for the MGX 8620, though Cisco's intention is to let carriers cap their investments in traditional circuit switches and use the 8620, along with other MGX products, to ultimately replace legacy class 4 and class 5 switches. Those other products include the MGX 8220 signaling gateway and the MGX 8850 multiservice IP+ATM switch. In context of Cisco's Open Packet Telephony architecture, these products can be understood as parts of an end-to-end model for IP telephony in a service provider network. Open Packet Telephony includes a service creation layer into which the MGX products integrate, and which supplies open, standards-based APIs for third-party development.

The MGX 8260 holds 170,000 voice ports in a 7' telco rack. Its line card interfaces include DS0, DS3, OC-3 to OC-48, and 100/1000 Mbps Ethernet. Pricing starts at around $50,000.

Unisphere Solutions, Inc., a Siemens company (formerly Argon Networks) (Burlington, MA – 781-313-8700, www.unispheresolutions.com) offers the GigaPacket Node (GPN), a combination ATM switch and IP router that scales from 20 Gbps to 160 Gbps. The basic theory behind Unisphere's approach is that using ATM as an underlying core for IP routing can solve IP's traffic management issues, while wave division multiplexing minimizes the effects of ATM's cell tax. Its product is targeted specifically toward providers who want to offer virtual private network (VPN) services, and know they can rely on ATM's permanent and switched virtual circuits for guaranteed QoS.

A fully configured GPN holds up to eight shelves, each with up to eight OC-48 line cards. Each switching shelf offers a 20 Gbps extension port, via which multiple shelves are interconnected. IP forwarding engines support line-rate speeds for packets as small as 40 bytes. Each line card can support a variety of protocols, including pure ATM, IP-over-ATM, IP Packet over SONET, and frame relay. GPN also supports MPLS for IP traffic management, and features a variety of classification techniques based on DiffServ standards.

GPN offers "five-nines" availability, as well as dual-redundant power supplies and hot-swap system components. List price starts at $100,000.

Unisphere acquired its "service mediation" switch from Castle Networks last year. The SMX-2100, as Unisphere calls it, has embedded SS7 links, obviating the need for a separate SS7 gateway. Having the SS7 interfaces internal to the switch chassis saves on space and cost, both of which can be precious to CLECs and startup service providers.

Recently, however, Unisphere announced that they will offer a separate softswitch, to supplement the SMX and to manage a service creation environment in addition to translating between SS7 and MGCP. Unlike most similar switch vendors, Unisphere is developing the softswitch in-house, leveraging LDAP-based directory software inherited from Siemens. The SRX-3000 softswitch can sit on any commercial hardware platform, but will use Sun servers in its first release. According to Unisphere, this hardware ambivalence is connected to the fact that fault tolerance has been incorporated into the software itself, which can be clustered for N + x redundancy, and has proven five nines' availability. The main addition to the original Siemens box is an open API, intended to support multimedia applications, such as unified messaging, IVR, voice and data VPNs, and follow-me routing, developed both by Unisphere and various third-parties.

The SMX-2100 is generally available, and has been tested for inter-operability, performing Internet traffic offload, with Lucent and Nortel class 5 switches and Alcatel's SS7 Signal Transfer Point. SRP starts at $104,000 for the chassis and channelized DS-3 trunk/access module.

NetCore Systems, now part of Tellabs (Lisle, IL – 630-378-8800, www.tellabs.com) offers the Everest Integrated Switch, a combination ATM switch and IP router on a single platform. The obvious advantage to this architecture is its two-in-one functionality, which could potentially slash a carrier's costs in half. In addition to native ATM switching and core IP routing, Everest provides integrated functions specifically targeted for carrying VoIP. Tellabs believes that QoS can, at present, be ensured only through the guaranteed bandwidth provided by ATM's virtual circuits. For that reason, Everest supports only IP-over-ATM in its first release (with the ability to select routing and switching on a per virtual circuit basis on each port, and the choice of either permanent-virtual or switched-virtual circuits). Everest also offers a variety of other features for SLAs and VPNs – including traffic classification based on DiffServ, Layer 4 protocol information, and source and destination IP addresses; weighted fair queuing and IP-aware packet discard; and provisioning of private IP routing domains.

Everest contains a modular switching fabric that scales to 2.5 terabits (in increments of 40 Gbps), with 10 to 640 Gbps interface capacity. Line-rate routing is supported for all interfaces and all packet sizes (down to 40 bytes). The upcoming release will include IP directly over fiber and MPLS support while maintaining current integrated IP/ATM features.

Everest is NEBS 3 compliant, and includes redundant, hot-swap components. Everest also features management software with a Web-based GUI, and APIs for future service provider operations. List price starts at around $15,000 per OC-3 port, $40,000 per OC-12, and $90,000 per OC-48.

Salix Technologies' (Gaithersburg, MD – 301-417-0017, www.salix.com) particular spin on the next-gen switch category is "class-independent switching." The ETX5000 holds up to 32,000 DS0s in a single chassis, and offers the choice of DS1, DS3, OC-3 to OC-48, and 100/1000BaseT Ethernet ports, all connected to one 8 GB switching fabric. On the incoming side, you have support for POTS, broadband, and PRI, and outgoing you can run any combination of TDM, IP, or ATM.

The ETX Call Agent, developed through partnerships with softswitch manufacturers, acts as an SS7 gateway, communicating with the switch via IPDC, SGCP, or MGCP. A standards-based Service Provisioning Interface (SPI) lets you program applications and deploy new services via the call agent. One point of distinction is Salix's FleXchange architecture, which separates voice and data I/O cards both from one another and from the DSP boards to create more flexible configuration options, and to let you add more cards as needed.

The recent second beta release of the product features a more compact, fully redundant chassis, with updated DSPs and higher port capacity. The long distance carrier ILDC has already announced plans to use the ETX across its international network, offering enhanced services like Centrex and unified messaging in addition to long distance voice switching. A basic configuration starts at $100,000, or around $50 per DS0 and $125 per VoIP port.

Sonus (Westford, MA – 978-692-8999, www.sonusnet.com) breaks the architecture of its switch into three parts: The GSX9000 is the core switch, and it is accompanied by both the SGX2000 SS7 gateway and the PSX6000 softswitch or Policy Server. In its first release, the product was designed mainly to act as a VoIP gateway, accepting inbound PSTN traffic and shipping it out as packets over 100BaseT fast Ethernet ports. Sonus has since added support for VoIP over ATM and Packet-over-SONET, as well as adding OC-3 and OC-12 ports for straight voice-over-ATM. The GSX9000 can handle 8,000 simultaneous calls per shelf.

The Signaling Gateway module supports ISUP variants and TCAP to let the GSX9000 interoperate with standard class 5 switches and SS7 Service Control Points. The PSX6000 softswitch is mainly intended for policy-based networking, which means that it performs multimedia call routing and implements QoS metrics based on user profiles and rules defined in LDAP-compatible databases. Open APIs and connections to IP-based application servers can be used to create new services. Communication between the policy server and the switch can take place via MGCP, IPDC, and/or SIP.

Sonus already has an active partners program, the Open Services Partner Alliance, which includes relationships for OSS integration, service provisioning, and enhanced services such as videoconferencing. 3COM recently signed on as a member of the alliance, with a package that includes Unified Messaging, IP Centrex, and fax-over-IP.

Tachion Networks' (Eatontown, NJ – 732-542-7750, www.tachion.com) Fusion 5000 is a "Collapsed Central Office" that combines class 5 switching, IP routing, frame relay switching, and SONET add/drop multiplexing into a single, 23" tall shelf. Fourteen of the shelf's sixteen available slots can hold a variety of line cards (currently available in DS-0, DS-3, OC-3, and T1/E1 models), and each line card can support a mixture of ports for frame relay, ATM, IP-over-ATM, TDM, and voice circuit applications. Line cards are connected directly to the systems two central switching cards, which are both up and running hot (one operating as primary, the other secondary) to provide full redundancy and high availability. The system currently uses embedded SS7 processors, but Tachion plans to move to an external softswitch architecture based on MGCP as the standards become accepted.

Beyond its architectural differences, the Fusion 5000 can be distinguished from others of its ilk in that it is mainly intended to act as a class 5 switch, with loop-side interfaces, as opposed to a class 4 tandem replacement. The target deployment is to have the Fusion 5000 accepting traffic from an IAD at the network edge, and providing the frequently used class 5 features such as call forwarding and automatic callback to businesses. Alternatively, the switch could be used as an adjunct to existing class 5s, to offload dial-up modem calls. At present the Fusion 5000 is mainly designed for ATM and frame relay, but a second release will include an IP routing blade.

MCI WorldCom and Comcast have both tested the product in beta. A base configuration starts at $100,000, and Tachion states that it offers everything a carrier needs to turn up service in a city for around $250,000.

Taqua Systems' (Centerville, MA – 508-778-8808, www.taqua.com) Open Compact Exchange (OCX) is specifically intended to be a next-generation class 5 switch, and comes equipped with most of the necessary subscriber services and CLASS features. The system, nevertheless, can serve in a number of different applications and points in the network, including class 5 end office, class 4 tandem, edge office gateway, and broadband local access hubs. The switch itself is fully modular, and can be distributed in more or less any fashion. The first of its key elements are a single card switch that supports up to 6,144 redundant ports in a single shelf. These can be any combination of analog, T-1 /

E-1, T3, OC-3, and ATM. Second is Taqua's Imbedded System Engine (ISE), which includes back office integration, open APIs and a service creation GUI, and dryISE, a tool which lets you simulate the switch's functions and can be used with any workstation. The API is object-oriented and extensible, and is designed to be compatible with any emergent de facto standard, such as Parlay or Softswitch. The same interface is used to access subscriber services like Centrex and to program back office functions such as billing mediation and database management. The ISEflow service creation GUI will produce State Machine Definitions (i.e. actual code) from graphically manipulable flowcharts.

Taqua developed the OCX's SS7 interfaces themselves, and use a processor-based signaling gateway to communicate with the switch via MGCP. Pricing begins at $45,000.

ISDN
See Integrated Services Digital Network

ISO
See International Standards Organization

ISP
See Internet Service Provider

ITU
See International Telecommunications Union

IVR
See Interactive Voice Response

Java

Introduced in 1995, Java is a programming language developed by Sun Microsystems that looks a bit like a simplified version of the C++ language. Differences with C++ include the fact that Java passes all arrays and objects by reference but doesn't have a pointer type, which eliminates the programming mistake of referencing a random area of memory. Java also has automatic garbage collection (just like BASIC) and C++ style exceptions are automatically generated when dereferencing a NULL pointer, accessing outside an array's bound, or running out of memory. Java's support of network protocol handlers such as HTTP, FTP, NNTP, MIME, and Sockets tip you off that it's some kind of "network programming language."

Indeed, Java is designed to be the ultimate cross platform, object-oriented programming language for creating interactive programs called "applets" that can run across networks such as the public Internet or corporate intranet. For example, Java-enabled thin-client network computing environments are perfect for distributing Java applets anywhere in a corporate environment. Also, since the client is executing the software, the server is relieved from the great processing burden of having to handle a large number of interactive web pages.

Since Java applets will run on any operating system and Java has no "extensions" unique to any operating system (except for a Windows-optimized "flavor" called Visual J++ that caused Sun to bring a lawsuit against Microsoft), its "write once, run anywhere" philosophy revives the old idea of a universal computer language. If you want to develop applications that anybody with a browser can run over the web, this is in theory the ultimate language to do it.

Java applets are written once, generally for a web page – you just include a Java application (or "applet") on your web page by inserting a special tag in your HTML code. Java application source code is compiled into an intermediate code known as Java "bytecode" which is later interpreted by a Java "virtual machine" (usually in a browser) that's tailored for the processor architecture on which the program runs. Thus, a Java applet will run on any platform having a Java interpreter – such as different kinds of computers (so long as they have a browser or some other means of executing Java bytecode), mobile phones, Internet screen phones, and other devices. The bytecode can literally run anywhere in a network on any server or client that is equipped with a Java virtual machine.

Java got an early reputation for being slow because the bytecode is interpreted a byte at a time during program execution, unlike its Microsoft competitor, ActiveX (See ActiveX). Java applets, once downloaded to the browser, contain 8-bit bytecode. ActiveX components, on the other hand, contain full 32-bit native code, another reason why Java tended to be slower than ActiveX components. To help remedy this, the Java virtual machine can be teamed with an optional Just-In-Time (JIT) compiler that dynamically compiles the 8-bit bytecode into faster 32-bit executable code during the download and execution of a Java applet.

Since the bytecodes are downloaded and executed at the client during runtime, only the latest version of an applet is ever run, which remedies the traditional problems of software distribution and upgrades. However, Java lacks "persistent" objects, unlike competing ActiveX components which are only downloaded the first time they are accessed and each time the software is updated. Since Java applets are downloaded every time they are accessed, they can add to network congestion. Some configurations exist for specific Java applets that use caching to store the applets locally, but this entails additional administration and may not work on all platforms or with all web browsers.

Also, since Java virtual machines must interpret Java byte code, during the execution of the Java applet it must also translate Java library calls into Windows operating system calls. Occasionally one will end up in a situation where there are no one-on-one mappings between Java API and Windows API calls. Java applets do not have full control of specific interface functions such as keyboard mapping. All of this can result in reliability problems, especially with host printing, user interface automation and advanced terminal emulation functions. Since ActiveX is tightly integrated with the Microsoft Windows operating system and Internet Explorer (IE) web browser, all Windows API functions are available and called directly.

Java Security

Java applets arriving at a user's browser are subject to bytecode verification. If a packet's size has changed somewhere along the way (possibly because of viruses or a Trojan horse), then the program will be instantly aborted. However, some of the original Java 1.0 virtual machine restrictions that promoted optimum security were too restrictive, such as prohibiting a Java applet from writing to the hard disk, accessing the printer and communicating with IP host addresses other than the machine it was downloaded from. For web to host applications, some of these restrictions would prevent file transfers, host printing or access to a host mainframe using other TN3270E servers. These restrictions were eliminated with the release of Java 1.1. Java's competitor, ActiveX components, use digital certificates to guarantee that the signed applet hasn't been corrupted or fiddled with by a hacker.

SSL security is also available for both ActiveX and Java web-to-host solutions. The Secure Sockets Layer (SSL) is the *de facto* Internet security standard originally developed by Netscape and provides both authentication and encryption to prevent eavesdropping and tampering of your TN3270E session(s).

Flavors of Java

Sun Microsystems has also created several subsets of Java that can be embedded in devices such as phones, smart cards, cars, TVs, pagers and more, taking Java "beyond the desktop."

Conclusion

If Microsoft completely loses the battle with Sun, then there is a high probability that Microsoft will drop support for Java, replacing their Visual J++ compiler with their experimental C++ Object Oriented Language (COOL) or some other Intermediate Language (IL) having functionality similar to Java.

Java Telephony Application Programming Interface (JTAPI)

A set of modular, abstract interface specifications of an API for use when building applications or components in Java that interface with telephony systems and must have first party or third party computer telephony call control functions. JTAPI was developed by a consortium including Dialogic, IBM, Intel, Lucent, Nortel, Novell, Siemens, and Sun Microsystems.

JTAPI version 1.1 was released January 31, 1997. JTAPI version 1.2 was released in February, 1998. JTAPI 1.3 is the latest release of the specification and was endorsed by the Enterprise Computer Telephony Forum on July 23, 1999. Sun and ECTF are now working as "partners" on JTAPI. The ECTF is now the lead organization developing JTAPI for developing JTAPI specification. An important step in solidifying the partnership was the release of the Java Community Process (JCP) by Sun in December 1998. Starting with the JTAPI 2.0 release, ECTF will follow the JCP.

JTAPI provides access to one or more of the following areas of functionality: Call Control, Telephone Physical Device Control, Media Services for Telephony, and Administrative Services for Telephony.

Telephony platforms supporting JTAPI computer telephony applications include (but are not limited to):

- Individual telephone sets (both wired and wireless)
- PBXs and CT Servers based on circuit switching technology
- Elements of IP Telephony (and other packet voice networks) including endpoints, gateways, media services servers, and call control implementations
- Computer platforms and software applications utilizing other call control (CTI) and media services APIs or protocols

JTAPI enables the creation of portable value-add Java software products that can operate on any telephony product that exposes its services through a JTAPI interface.

For example, the same JTAPI application can run on a cellular phone and a media server on a VoIP network if both expose a JTAPI interface. JTAPI makes this possible by presenting an abstraction of telephony services that is independent of a given implementation's underlying network and its relationship to that network.

JTAPI provides the following:

- *Call Control* Call Control refers to control and observation of call processing (such as address translation, call setup, call tear down, call routing, supplementary services, and access to call associated information). This includes support for wired and wireless implementations ranging in complexity from the simplest call control features to the full suite of PBX/ACD call control functionality. In its simplest implementation, JTAPI gives access to only a single telephony device (e.g. first party). But, the specification applies equally well to implementations which provide access spanning multiple devices (e.g. third party).
- *Telephone Physical Device Control.* Physical Device Control includes monitoring and control over one or more user interface elements of a telephone including: its display, buttons and lamps (real and virtual), auditory components, hook switches associated with physical telephone devices independent of call control functionality.
- *Media Services for Telephony.* Media Services includes manipulation and processing of the media streams associated with calls (including tone generation and detection, fax processing, recording and playback, text to speech, automatic speech recognition, etc.).
- *Administrative Services for Telephony.* Start-up, shut-down, and management (fault, configuration, accounting, performance and security) of telephony resources provide functionality through the JTAPI interface.

JTAPI is used to access the services of a telephone system, however not *all* the functionality required to implement a telephone system is specified by JTAPI. JTAPI does not expose individual signalling protocols (such as SS7, ISDN, GSM, etc.) and associated features that a particular telephone or telephone system implementation is likely to need internally. Nor does it include non-telephony services that Java-based telephones are likely to need (such as interfaces to draw a telephone user interface on a bitmapped display). Implementation of power management, messaging, web browsing, SNMP support, access to device-specific functions, etc., will require other complementary Java interfaces to do so.

The scope of JTAPI corresponds to the top-most layer of interfaces defined by the ECTF architectural framework and encompasses the functionality of the call control (C.xxx), media services (S.xxx), and administrative services (M.xxx) areas. For further information, refer to the ECTF architectural framework (described in documents available at http://www.ectf.org/ectf/tech/tech.htm).

JavaBean

A reusable software component, a "building block" defined by an Application Programming Interface (API) specification that can be manipulated visually in a builder tool, the visual component architecture being based on the object-oriented Java language. Any Java components conforming to the JavaBeans component model can be plugged by anyone into any other JavaBean-compliant application. Yes, Java's "write once, use everywhere" technology previously restricted to programmers can now be enjoyed by the masses. The Rapid Application Development (RAD) component software model provided by JavaBeans allows even non-programmers to build an application using a visual builder tool.

JavaBeans are thus the first truly standard component model that is platform-independent. Every Bean not only complies with the JavaBean model, but also carries with it all its properties and methods which can be easily garnered through "introspection" – a JavaBean property allowing a visual builder tool to "know" how a Bean operates.

Since JavaBeans are hostable in other component models, you can in theory use a JavaBean as a first-class ActiveX, OpenDoc, or other component. A JavaBean may be contained within an ActiveX or OpenDoc container, such as a word processor, and will behave exactly as if it were a native component.

JavaScript

Java is not the same as JavaScript, which was developed at Netscape. Java, developed by Sun Microsystems, is a platform independent, object-oriented programming language. JavaScript, on the other hand, is a scripting language, implemented as an extension of HTML. JavaScript is interpreted at a higher level, is easier to learn than Java, but lacks some of the portability of Java and is even slower than Java's bytecode.

Java Server Pages.

Server-side programming language combining HTML and Java code to generate custom servlets for dynamic web pages.

JTAPI

See Java Telephony Application Programming Interface

Key system

A key telephone system is a business phone system that lets users access several lines; make, place, transfer and conference calls; and more. Some key systems have integrated voice mail. Some have Automatic Call Distribution (ACD) capabilities. What were once electro-mechanical clunkers are now sophisticated, digital systems offering most or all of the power of large PBXs (including computer telephony integration) but in a smaller package.

By traditional definition, key systems are not switches (e.g., they lack the ability to perform functions traditionally assigned to PBXs such as selecting paths or circuits to route calls over random trunks). Traditionally, key systems have been configured "square," meaning that all outside lines (trunks) show up on every phone, and a user must manually press the button on the telephone to select a specific CO line to engage. Today, however, a good portion of products, under FCC specifications, fall into categories of both key system and PBX, hence the birth of a beast called the "hybrid" – non-square systems that serve multiple trunks and connect to a mix of digital station-sets and analog phone equipment, just like a PBX.

Other phone systems pack plenty of features, but don't have a Key Service Unit (KSU – the cabinet that contains all the switching electronics of larger systems). These so-called "KSU-less" systems are bringing big phone system power to the Small Office/Home Office (SOHO) market.

The first question a prospective buyer usually asks about a key system is: "How many lines and phones can it support?" Obviously, you know how many people need phones in your office. You probably have a rule-of-thumb (and if you don't, here's a good one) that says you need one CO line for every four to eight employees. The answer (normally expressed as a simple multiplication, e.g., "8 linesx16 stations") should tell you whether or not a product will meet your needs today, or in the foreseeable future (presuming growth).

Problem is, key system manufacturers have a difficult job when it comes to telling the market about their systems' most important characteristic, trunk and line capacity. The emergence of universal buses gives customers greater choice in configuring their phone systems but leaves the vendors somewhat frustrated in attempts to come up with a meaningful capacity designation. Life was simpler with dedicated slots for interface cards, because the familiar *small number* x *larger number* notation could carry realistic consumer data. Normally, the latter designation makes sense when the key system gives you a combo-card, a card that pairs trunk lines with station lines in some small multiple (usually around two).

For example, Vodavi can safely report that its new Triad-S flat-pack system can be expanded to 12x32: just use four of their 3x8 CKIB combo cards. So far, so good. However, most key system players have an assortment of cards in their hand, some of them higher-density, single-purpose trunk and station cards, so it's possible to report much higher station or CO counts than it is by solely using combo cards. Of course, ultimately there is a limit in the number of ports that are available. Toshiba can correctly report that its Strata DK40 has a limit of 28 station ports *and* 12 CO lines: Its base unit has room for four CO and four analog circuits and by filling the four universal slots in the expansion unit with three, eight-circuit line cards and one, eight-circuit CO card the numbers come out right.

Some vendors will give analog/digital station counts and CO line counts as a single maximum number. What's not entirely clear from examining the data is that the numbers are mutually exclusive. For example, a manufacturer may say that the maximum configuration of

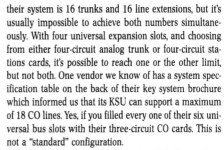

their system is 16 trunks and 16 line extensions, but it's usually impossible to achieve both numbers simultaneously. With four universal expansion slots, and choosing from either four-circuit analog trunk or four-circuit stations cards, it's possible to reach one or the other limit, but not both. One vendor we know of has a system specification table on the back of their key system brochure which informed us that its KSU can support a maximum of 18 CO lines. Yes, if you filled every one of their six universal bus slots with their three-circuit CO cards. This is not a "standard" configuration.

Here are some points to keep in mind when reviewing vendor's claims:

1. *It's the ports, stupid.* Since many manufactures have embraced universal slots, allowing interface cards to be parked anywhere, the real question that should be asked is "how many ports." Key system makers have been good about this and have used the total port number in their marketing material when expressing capacity limits. For example, NEC's Electra 192 can be grown to 24 slots and with eight ports allocated per slot, you get – you guessed it – a capacity of 192 ports. You can allocate ports between trunk and line, using combo cards and dedicated station (analog or digital) and trunk cards. There are, of course, sub-limits to the number of digital stations or analog stations that you can have, but knowing the number of ports is much more meaningful than unreachable maximum capacity limits.

2. *Voice mail counts.* Vendor's voice mail cards that are integrated into the KSU normally take up some number of available ports, leaving less ports for line interface cards. Even when the key system's maximum capacity numbers pass a sanity test, they usually do not indicate the configuration you would have if you had voice mail/auto attendant module. Although the number of voice mail ports are small, as your system grows the number of mail ports will too.

3. *Universal ports.* Universal ports are marvelous devices that allow either a single line telephone or a digital phone hang off the same port. But, normally not both at the same time. When a vendor says "with our universal port KSU, you can have a maximum of 16 digital and 16 analog lines," remember to change that "and" to an "or". Unfortunately, there is an added complexity which gives vendors the wiggle room to justify the previous claim. Some vendors will send four wires from their universal ports to a digital phone, allocating two for the digital signal and two for the analog signal. A connector in the back of these phones lets you plug a fax machine or single line phone into the digital phone and use both devices simultaneously (since they are signaling over the two spare analog wires). The analog devices are also allotted separate extensions. So, these universal port vendors can say that they have 32 ports and 64 extensions. This is true but keep in mind that these extra extensions reside in devices that live next to digital phones and can not be arbitrarily distributed anywhere in your office.

Keyphones from 3COM / NBX (Andover, MA – 978-749-0000, www.3com.com) are beautiful examples of the industrial designers craft: Any one of these curved, asymmetrical shaped sets can stand on its own as modern art. Arrayed on the right-side of the phone, the familiar line keys with BLF indicators assure us that we are looking at a key phone. However, if you felt that NBX was suggesting more than just a key phone, your intuition would be right. Trace the RJ-45 line back to its source and you're looking at an ordinary 10-baseT hub, not a KSU! Of course, extraordinary technology is at work here. With the NBX 100 Communication System, the wizards at NBX Corp have managed to turn a key phone into a device on an Ethernet network – it's got its own low-level Ethernet address – a device which transmits/receives IP packets holding digitized, compressed voice. DSPs and proprietary technology assure smooth flow of data and voice over your existing shared or switched network. Of course, normal rules of LAN segmentation apply here: a heavily used NT file server should be on its own stretch of wire.

At the top of the network, NBX 100 network server box holds a six slot chassis in which four-circuit CO cards and an auto-attendant/voice mail module sit. The NBX 100 has 148 ports which can be divided between hub ports and CO lines but no more than 48 can be allocated to the latter. An Ethernet hub module can also slide in (if you're building a LAN from scratch); otherwise, just run a wire from the box to an existing hub. The network server's primary function is to handle such standard features as conference calling, speed-dialing and call forwarding. The network server is also responsible for call setup and call break down. Once a call is connected, the network server drops out and phones are sending IP packets directly to each other. Sounds like the NBX 100 will scale up nicely.

NBX has taken full advantages of their LAN-centric approach. Want to configure the NBX 100? Just bring up a browser anywhere on the network and run Netset, an administration utility that lets everyday users customize their programmable keys. Administrators use it for privileged operations such as adding new users and services. Want to run a TAPI-compliant application on your PC? Just install NBX's own drivers – no need to add a board! – and start using your favorite PIM for call control. The wiring for this setup is simple: All NBX phones have a built-in hub repeater so your PC and phone live on the same segment from the hub.

For environments without an existing LAN infra-structure or perhaps those whose LANs carry too much data traffic, the NBX 100 may not be appropriate. However, the NBX 100 will make business sense for many. NBX says that an 8x20 configuration should run about $11,000.

BBS Telecom's (Austin, TX – 512-328-9500, www.bbstelecom.com) Plexus is a digital hybrid key system that boasts eleven universal card slots. Plexus offers plenty of interfaces to pick from: 4x8 analog combo cards, four and eight circuit analog trunk cards, and ISDN, T-1, and E-1 digital cards. With a maximum of 72 ports, you'll have the elbow-room to come up with the configuration that's just right for you. Slide a Plexus Integrated Voice Module into a spare bus slot and get 40 hours of message time and such advanced features as call record. If you don't need this much KSU, consider BBS's analog Micro-Plexus, a turn-key solution that comes configured as 4x8 but can be expanded to 12x24. Plug any single-line phone, fax machine, or a BBS keyset into its RJ-11 ports and you're ready to go. Besides the standard model, Micro-Plexus comes in two other flavors: The AA model has a built-in, four-channel, multi-level auto attendant, and the VM comes equipped with four channel voicemail. The great advantage of the Plexus series is that they can all be linked together. Use their Inter-Unit Link card to tie together two Micro-Plexus systems to get 48 ports or a Micro-Plexus and a Plexus to get 88 ports, or two Plexus systems to reach a total of 128 ports. With this kind of flexibility, Plexus should be part of your business's telecom infra-structure for years to come.

The Impact SCS from Comdial (Charlottesville, VA – 800-347-1432, www.comdial.com) lets you customize your telephones in a variety of ways. Any of the buttons can be programmed as you wish, and unlike many other systems, Impact SCS has no "fixed" feature buttons, so you can put the "hold" or "intercom" button anywhere you

want. If you add the optional DSS/BLF consoles, you can store dozens more numbers and features. Each number in the console lets you store four levels of numbers, so the 48-button console really lets you store up to 192 numbers.

The easy-to-read LCD screen makes using these telephones a breeze. You can use the interactive buttons to scroll through the system features, and you can redial or transfer a call, program speed dial, and even check your voicemail without having to remember complicated programming codes. When you're on a call, an LCD shows you the number you called, the length of the call, and even the call's cost. Each telephone in the Impact SCS system can attach to an analog device, such as a fax machine, modem, or cordless telephone.

Cortelco Kellogg's (Corinth, MS – 800-288-3132, www.cortelcokellogg.com) Aries Digital Key System, the ADKS-144, gives you nine slots in which to place an assortment of trunk and line cards. Choose from digital and analog station cards that come in eight and 16 port flavors. An eight port CO loop-start and T1 interface can also be drawn from their "deck of cards". A caller-id unit is external to the KSU, supports nine CO lines and is not required if you go with T1 and ANI. For the desktop, choose from five telephone models. Their Executive telephone set boasts a hands-free speakerphone, off hook voice announce, and a 2x16 LCD display. Their DSS/BLF unit has 50 keys from which to make connections.

Estech System's (Plano, TX – 972-422-9700, www.esi-estech.com) IVX system is a 16x34 digital system integrated with a ten port voice mail module that does not sacrifice any of the station ports. Live call screening and one touch live call recording are exceptional features of the voice mail system. Their key phones have 2x16 LCD display which can show remaining time on voice mail messages and speed dial lists. We're also fond of its audio help system.

The Medley is Executone's (Milford, CT– 800-955-9866, www.executone.com) all-in-one digital key system. Out of the box you get built-in voice mail (30 hours), auto attendant, ACD, and Caller ID. The base unit is configured as 8x16 with two ports for analog devices. By piggybacking an expansion cabinet, the Medley can grow to support 32 key phones. Attach Executone's Model 64 digital phones to the Medley and take advantage of the 64's intelligent soft keys for easy call conferencing, call transferring and call forwarding. Press the help key and audio help comes to the rescue. An LCD display lets you know who is calling.

If you were designing a KSU from scratch and wanted to bring cost-effective CT within reach of small businesses, you'd probably end up re-inventing Inter-Tel's (Chandler, AR – 602-961-9000, www.inter-tel.com) Axxent. Axxent's fully digital KSU sits in a familiar PC mid-tower and all accessories (trunk, station, and voice mail cards) slide into an eight-slot industry standard ISA bus. Five slots are dedicated for telecom use, leaving three slots for applications–for example, voice mail. This translates into a maximum configuration of 12x24. Axxent supports either ground start or loop trunks; T1 and ISDN is not available but DID is.

Inter-Tel makes the whole process of turning your PC into a virtual phone a snap. Just connect a cable from your computer's RS-232 port directly into the digital port on your Axxess Digital Terminal and you've taken care of the hardware part. Load Inter-Tel's Axxessory Connect and you're now looking at the cyber version of their top of the line Executive Digital Terminal key phone. Virtual LCD shows who is currently calling and who is pending. Click on a pending call to send it to voice mail. Click on the lookup button to view the intercom directory and search for a name. Then, click your way to a call. Program up to 20 buttons for speed dialing or special functions/features. The familiar voice mail "led" is there too. Connect can also cooperate with your favorite PIM. It sends caller id to a database or contact management program (using Microsoft's DDE protocol), triggering a screen pop-up with all the information you'll need to give great customer service.

If you still want to touch real keys, Inter-Tel's Axxess digital phones feel good and look great. The range is from Standard Digital Terminal, 23 programmable keys and 2x16 LCD display, to their Executive Terminal, which boasts a larger 6x16 display and full-duplex speakerphone. All phones have a "hot" push button keypad which can be programmed to trigger multi-key stroke functions.

Iwatsu (Carlstadt, NJ – 201-935-8580, www.iwatsu.com) key systems have long been a favorite of *Teleconnect* magazine, *Computer Telephony's* sister publication. Iwatsu's their ADIX-S key system won an Editor's Choice back in 1997 and the Omega-Voice VMI won in 1998. The ADIX-S is a digital hybrid that provides up to 52 universal ports on the station side and accommodates T-1, ISDN, DID and TIE lines on the trunk end. Their VMI is a auto-attendant/voice mail that fits into any universal slot of the Adix chassis, supporting up to 600 mailboxes and holding 300 hours of messages. What have they done lately? In late 1999, they announced the ADIX-APS , a five-cabinet tower of power that will control up to 472 ports. Their dense 16-port and 24-port interface cards make this all possible. Significant new features include external call forwarding, transfer to park and enhanced 911 service that will make the APS comply with new regulations being passed in many states. A planned TAPI 2.1 interface will make this behemoth more compatible with Windows NT.

With Release 3.0 of Lucent's (Basking Ridge, NJ – 908-582-8500, www.lucent.com) Partner Advanced Communication Systems, small business might want to take another look at this dependable, hybrid key system. The basic spec sheet looks tempting. A maximum configuration of 15 CO lines and 40 station lines. Universal station ports so you don't have to throw out your old single line phones. A phone mail system that comes in three flavors: a basic PC messaging card (PCMCIA compatible), the integrated Partner Mail VS, and an option for an external box, Partner Mail. Key phones that range from a bare bones 6 line set to a full fledged 34 line with 2x24 LCD display.

The "umbrella" 3.0 Release of Partner ACS is a generous, seasonal offering from the elves at Lucent. You'll need Release 5.0 of their Partner Mail VS to get all the features. Small business will find Lucent's cascaded out-calling to be a highly configurable voice mail function that lets demanding clients reach out and find employees. You can schedule the Mail VS to try a list of alternate numbers, pagers, or other devices, successively trying numbers if the party doesn't call back within a specified time period. Live call screening is a part of Partner Mail VS 5.0 so that answering machine-like functions are now on your desktop business phone. Law firms and accounting companies will take advantage of the Mail VS's record-a-call: Press a button and store crucial business conversations for later reference. With all these additional features, Lucent has not forgotten basics like service and reliability. All ACS configuration information, including mail greetings, are automatically backed-up to a special PCMCIA PC card. Remote administration of the KSU has also been introduced with this release. Operators can dial in and configure the KSU from a Windows GUI. Finally, for ACS release 2.0 customers, Lucent says that a quick software upgrade will bring you up to speed.

NEC's (Irving, TX – 800-TEAM-NEC, www.cng.nec.com) recently announced addition to their product line, Electra Elite 192, brings big features to small businesses. The Electra Elite's modular design grows from a standard 64 port to 192 ports, allowing a maximum of 120 digital sta-

tion lines and 64 CO lines. The Elite's architecture is also built around the industry standard ISA bus.

With Electra, NEC has really gone all out to bring digital down through the wire. They have integrated voice and data at the KSU with an eight-port phone/Ethernet hub card, creating a 10base-T LAN out of the phone lines. You plug any network-enabled PC into their VDD-U unit to bring the LAN to your desktop. A separate connector in the unit accepts a phone wire and their digital lines allow simultaneous data and voice traffic. For small businesses that want a data network and don't want to invest in new infrastructure, the 192 seems to be a very sensible and inexpensive option. Up to three card phone/Ethernet cards can be mounted in each KSU and by combining three KSUs you can network a total of 72 devices.

ISDN is also included in the Electra Elite offering. Trunk-side BRI is available now and PRI is planned for late 2nd quarter of this year.

The Elite is TAPI compliant and NEC's own ElectraCall PC attendant is a comprehensive CT application that gives operators an easy to use GUI to monitor incoming calls, email messages with attached wave file, and perform point-and-click-phone transfer. You'll need their D-term PC card, which comes with voice-recording capabilities, to operate the PC attendant. NEC also brings TAPI applications to ordinary PCs with their PC telephone adapter unit (CTA-U). NEC provides the TAPI drivers and you'll just need to plug an RS-232 cable into their adapter to run either NEC's own Phone Kits, a virtual PC that comes with the CTA, off-the-shelf telephony applications or your own. Small businesses will find this an inexpensive route to computer telephony

For the desktop, NEC's pleasing, convex-shaped phonesets vary from 16 lines–all programmable–with threex24 LCD display to a 32 line–8 fixed function–with no LCD. Their analog port adapter allows you to piggyback a single line phone (or other analog device) onto their digital phoneset. By the way, NEC (along with a select few vendors) champions the universal port concept. This means that existing single line phones may be plugged directly into the KSU, allowing small companies to go digital at their own pace.

Nitsuko's (Shelton, CT – 800-365-1928, www.nitsuko.com) 28i is the latest addition to there i-series of digital key/hybrid systems. It is a smaller version of their popular 124i and was designed with four universal slots, supporting up to a maximum of 16 trunks or 16 stations. Choose from an assortment of interface cards, including eight-circuit digital station, four-circuit analog station, four-circuit analog trunk, and 2-circuit BRI ISDN. It is software compatible with the 124i base system's release 4.06, so that any feature supported by this release is available to the 124i. This means that TAPI and Nitsuko's NVM-Series auto attendant / voice mail are within the 24i's reach.

Nortel's (Nashville, TN – 800-4-NORTEL, www.nortel.com) philosophy of giving PBX functionality at key system prices is made good in their Norstar Plus product line. From their basic Norstar Compact ICS (8x24) to their larger Modular ICS (having maximum of 242 ports), Nortel goes the extra yard to give more for less. Small businesses can start with the Compact ICS and migrate up the key system line without sacrificing their investments in cards, phone sets, and applications: It's all compatible.

Data and voice have been successfully married through out the line. At the top of the key system series, their recent Modular ICS's Integrated Data Module joins a 12 port 10baseT hub together with a T1-mux. Connected to the KSU by high speed optical fiber, the IDM dynamically allocates T-1 channels between voice and data. In the ISDN world, they have introduced station side BRI, also available on the low-end Compact ICS, so that phone and internet browsing can occur simultaneously. All that is needed is a small adapter device, which connects to the PCs RS232 port. (Trunk side PRI is also available but only on the Modular ICS). They have not forgotten about the analog world either, allocating an analog port adapter on their Compact ICS for modems and faxes.

Nortel has turned their focus to desktop applications and has yielded a range of CT-based applications. Out of the box they give you Norstar's Personal Call Manager, a TAPI-compliant application that pops-up with caller information on a Windows 95 platform, provided you're equipped to capture caller ID. Personal Call Manager lets you both answer the call from your PC and also dial out. You will need to purchase their computer telephony adapter (CTA 100), a small box with connectors for both your PC and telephone. If you've run out of desktop space and have a spare PC slot, then try their CTA 150i card. Either way, your PC has now been opened up to the CT word. For the more sophisticated, Nortel's Desktop Messaging application unifies fax, voice, and email. It integrates with the ubiquitous Microsoft Exchange, which comes standard with Windows '95 / 98 / NT / 2000. However, Norstar's Application Module (NAM), an industrial grade Pentium-powered PC is required. The NAM comes bundled with their PC Console (a GUI attendant console), Norstar Voice Mail, and Norstar Minuet, Prelude, and Cincphony ACD applications. High speed optical fiber joins the NAM to an ICS KSU. To run the NAM, you'll need at least Norstar Compact or Modular ICS Release 1.0 software or higher.

With applications and new data/voice technology, Nortel appears to have "covered all the bases."

Panasonic CSD (Secaucus, NJ – 201-348-7000, www.panasonic.com) has introduced the 7400 line of key phones which corrects some deficiencies with the 7200 workhorse and adds some new ideas. Their top of the line KX-T7436 has a 6-line display and 24 CO buttons. With their older top of the line KX-T7235 you paid for the 6-line display with only 12 CO buttons. All the 7400 phones now have "Whisper Off-Hook Call Announce" which allows a third-party to "quietly" break into a conversation without the outside party hearing. A retro jog-dial has been designed into these phones which is really a speedy way to scroll through name directories and control volume. Just tap on the dial with your finger to place a call. Of course, Panasonic's great XDP port is still standard so you can hang off a PC, fax machine or other analog devices from your key phone, without installing a daughterboard or connecting an adapter, and use both the phone and the analog device simultaneously.

At the other end of the wire Panasonic's Super Hybrid has some of the greatest flexibility in the industry in accommodating both analog and digital phones. Unlike systems that have dedicated ports, their friendly universal ports accept either single line phones or digital phones. If you go digital, you can use the XDP port found on the back of the phone to effectively double the number of stations. So, with expansion modules you can hang up to 32 digital (and 32 XDP-attached analog devices) or 32 directly connected analog phones from the 1232. With the 816 the numbers come out to 16 digital (and 16 XDP-attached analog devices) or just 16 analog.

Accompanying the 7400's is a firmware upgrade that will let you take full advantage of the phone's new features (such as the jog-dial). You'll also need the new release to operate the new KX-TD185 DID board and KD-TD171 Caller ID board. DISA capabilities are available with the KX-TD191 card.

Panasonic has also introduced ISDN support for the 1232/816, giving basic BRI support on the trunk side first with their U-interface card.

PRI support is not planned.

The Picazo VS1 from Picazo Communications (San Jose, CA – 408-383-9300, www.picaco.com) comes in two black towers, which house, separately, the Picazo Voice Server and the Picazo Port Expansion Unit (PEU). Looking a little like a metallic Stonehenge, they elicited a few oohs and aahs from passersby at trade shows. Its pure and simple design belies a very sophisticated phone system, loaded with features. Here's what a lucky Picazo customer gets, out of the box: Four-channel voicemail (software expandable to 12), three pre-configured auto attendants (20 altogether), ACD, external connectors for zone paging, and 16 ports (expandable to 192, with extra PEUs) that can be divvied out among CO lines and digital / analog station ports. Picazo also offers their Attendant98, PC-based console software; and Connect98, its CTI application. These two aren't part of the basic package, but you'll no doubt want to "splurge" for them.

With its built-in monitor and a keyboard, you can configure the VS1 directly. However, you'll want to load SiteLink management software onto a laptop (one with a modem) and configure the system from there. SiteLink's Windows software was designed to let technicians remotely log into a VS1 system. Configuration files are downloaded from the system, and the tech does MACs using an old-style, character-based application that takes up a frame within SiteLink. When he's finished the changes, the tech clicks on Upload and SiteLink sends only the changed config files back to the VS1. The great thing about SiteLink is that techs can work offline, assuming they have the original configuration files, and then upload when they're ready. One minor gripe is that if you're on-site (for example, doing a Testdrive), and the VS1 is inches away, you still have to use the modem to run SiteLink.

The VS1 comes with voicemail and an auto attendant. Don't look for voicemail configuration within SiteLink. The programming for this is strictly phone-based. The VS1 offers basic voicemail service (including message forwarding), but there are extras, as well. Pager notification, for one. Also, with a pared-down form of cascaded message notification, you can redirect voicemail messages left on an extension to a single, alternate extension. With 30 hours of voicemail storage, the VS1's voicemail system should be more than adequate for most small to midsize organizations.

The VS1 auto attendants are programmed, within SiteLink. If you navigate into the auto attendant section, you'll see three existing attendants: a basic dial-by-name attendant and two variations on the voicemail-recording attendant. Attendants can be assigned to CO lines, so the VS1 offers a very useful multi-tenant capability. With the auto attendants, VARs and dealers have plenty of opportunity to customize the VS1 for a particular business' needs.

There are a ton of other features, including Least Cost Routing and DID. And there's desktop CTI, using Picazo's Call98. It's a great phone system for configuring specialized functions, making it a great phone system for VARs, as well.

Samsung's (Miami, FL – 305-592-2900, www.samsungtelecom.com) new hybrid key system, the DCS 50si, fits snugly in the middle of the DCS product line, between their DCS Compact and their original DCS, and offers both station and trunk BRI ISDN, a feature seen on their high-end DCS 400si. The basic configuration is 0x8 and with an expansion cabinet, you can bring home six universal slots. An assortment of interface cards (2x4 combo cards and four-circuit ISDN cards), gives you a maximum of 48 B-channels to allocate between trunk and station or 18 analog trunk lines or some combination of both. By the way, the four circuits on an *individual* ISDN card can be divided between outside and inside lines.

The DCS 50si and DCS 400si are compatible with Cadence, their Auto Attendant/Voice Mail module, which was introduced back in February '98. The basic unit has four ports which can be expanded by four ports and the module fits into a slot on the DCS 50si. On the DCS 400si , Cadence grows to a maximum of 16 ports. More than just voice mail, Cadence allows one touch conversation record (a service firm favorite), answering machine emulation mode, and caller id callback. Auto-attendant completes the offering.

By connecting Samsung's Computer Telephony Module to the end of the digital wire, you open up the telephony world on your PC. Samsung's new TAPI application, Smart Control, replaces their PIM2000, offering more functionality than basic personal information management. In October of 1998, Samsung also received certification from Novell for their TSAPI drivers so now applications written for Novel servers can join the telecom revolution.

Siemens Business Communications' (Boca Raton, FL– 800-765-6123, www.siemens.com) Hicom 150E family of communications servers now fits both small and remote offices. Hicom 150E is a good fit for businesses that need to support two to 250 users. It features support for ISDN, multimedia conferencing, and video.

The Hicom 150E servers are competitively priced, easy to configure, and simple to maintain, and have been engineered to support industry-standard computer telephony interfaces such as TAPI, and CAPI.

The Optiset E telephone is completely compatible with all models of the Hicom 150E product suite. The phones come with OptiGuide, a point-and-click interface that guides you to many of the key phone features. Additionally, easy-add modular application adapters provide increased functionality and performance. The Optiset E phones support applications from basic voice to fax and modem usage, desktop video conferencing, and advanced computer telephony. The phones use ISDN (BRI) to provide advanced voice and data features at low cost.

List pricing starts at $900. Starting price for the Optiset E telephone set is less than $80. Hicom Office PhoneMail products begin at a list price of $1,350 for a complete Office PhoneMail Entry system.

Sprint's (New Century, KS – 913-791-7700, www.sprintproducts-group.com) hybrid product family, Protégé, spans the key system market. Their small CTX starts out at 6x16 and, with expansion cabinets, can be grown to 8x24. The Protégé MTX starts where the CTX ends off – weighing in at 8x24 – and can grow up to 96 ports. LTX, the eldest member of the family – it was the first Protégé Sprint introduced – can support up to 144 ports. Such advanced features as answering machine emulation and conversation record are standard, even on the CTX. Starting with the MTX, Sprint's universal slot architecture offers maximum configuration flexibility. Fit the SVP-12 voice mail card into a free slot on either the LTX or MTX and take advantage of its unlimited message capacity. You'll also get such advanced features as out calling to reach employees on the go. Full featured auto-attendant – multi-level prompts, holiday schedule, and more – comes with the voice mail.

Sprint's ProtégéDial is a great CT application that was designed specifically for the Protégé key systems. Your PC becomes a virtual phone where you can drag-and-drop your way into dialing, transferring and conferencing. Use your favorite PIM manager and connect to anyone found in your database: ProtégéDial understands Microsoft's DDE protocol so phone numbers can pass between applications.

In the easy to install and inexpensive category, one should take a look at Telematrix's (Tamarac, FL – 800-462-9446, www.telematrixusa.com) KSU-less phones. Choose from their TMX308 – up to three lines and eight stations – or the TMX508 – adds another two trunks. Wiring the sets is straightforward, and phones can be arranged in a star, loop, or combination star/loop network pattern. Whatever the pattern, if a phone fails, the rest of the network is still up.

Unlike other KSU-less phones, Telematrix doesn't force a squared configuration on you: any line can appear in any order on any phone. The sets offer such basics as speakerphone, ten speed dial keys (up to 20 speed dial number may be stored), three separate ringer volumes, and eight station keys with an in-use LED light. Little extras that the SOHO market will welcome include three way conferencing, handset volume control, and a data port with a line selector switch. At about $190 per phone, these KSU-less wonders are within the budget of even the smallest business.

Tadiran Telecommunications' (Clearwater, FL – 727-523-0000, www.tadiran.com) Coral SL communications system is designed to meet the requirements of remote offices, branch offices, and small to medium-sized businesses. The system is quite dynamic in that it now functions from as few as eight stations to as many as 200 stations.

The Coral SL supports the same family of fully featured digital telephones that are used with the larger Coral systems. The new graphical key telephone (GKT), a high-end member of the Coral Digital Key Telephone (DKT) family, is also supported by the Coral SL. The GKT maintains the entire range of DKT features, including ACD, TAPI, computer telephony, networking, and unified messaging. It also includes additional features such as multiple address books and search capabilities.

The pricing for the Coral SL is $300 to $500 per station.

Telrad's (Woodbury, NY – 516-921-8300, www.telradusa.com) key systems range from their small Synopsis (8x25) up to their mammoth Digital 400, which can grow to 144 lines and 254 stations. Telrad lets you stretch your investment dollars since boards and phone sets bought for any model in their product line are upwardly compatible. Therefore, no need to replace their top of the line Executive key set as your business grows. With speakerphone, 8x24 LCD display, 32 programmable buttons allocated for either CO lines or DSS/BLF, and a voice recognition unit, you wouldn't want to! Telrad has not been quiet in the CT area. Their TAPI compliant phones support a slew of PIM packages. ACT!, Goldmine, Lotus Organizer, and Sidekick are just some of the software that lets you point-and-click your way through call transfers, conference calls, and many other functions.

Back in July 1998, Telrad announced Release 6.6 of their software for the Digital S60, S128, and S400 systems. Existing customers will find improvements and some new features. The speed dial capacity doubled. Up to four sources of music-on-hold/background music can now be installed on a S128 system and eight on a S400. Music can be configured on a per-trunk, trunk group, DID number, or PBX station basis. A conference loop has been added and up to 30 internal users can enter a conference group in which a maximum of six member may be speaking simultaneously.

Continuing with the robust side of things, Teltronics (Sarasota, FL - 941-753-5000, www.teltronics.com) serves up two models, the VisionLS (288 ports) and the VisionMS (576 ports). They can be configured to provide all the advantages of hybrid/key, PBX, or a Centrex front-end configuration, or any combination of the three.. These configurations can be used individually, mixed in one system, or even mixed on each station, with little or no changes to system hardware. For example, you can set up a site as a PBX system, with one department at that site configured as a key system and another as a hybrid system.

Specific to the key system configuration, all the telephone lines are visible on the telephone. Consistent with traditional definition, you

K

Company	Contact	Product	Max no. lines	Max phones	List price
Cortelco Kellogg	Corinth, MS, — 800-288-3132 www.cortelcokellogg.com	Aries Digital 144 Port Key System	96	144	$3,324 $189 per station
ECI Telecom	Clearwater, FL — 727-523-0000 http://ecitele.com	Coral SL, Coral I, II, and III	6,000	6,000	$300-$500 per station
ESI (Estech Systems Inc.)	Plano, TX — 972-422-9700 www.esi-estech.com	IVX family (IVX128, IVX20)	66	84	N/A
Inter-Tel	Phoenix, AZ — 480-961-9000 www.inter-tel.com	Axxent	12	24	$200 per station
Iwatsu Telecommunications Products	Carlstadt, NJ — 800-955-8581 www.iwatsu.com	ADIX APS	200	360	$500 per station
Lucent Technologies	Basking Ridge, NJ — 800-247-7000 www.lucent.com	Merlin Magix	80	200	$1,500 $150 per station
Lucent Technologies	Basking Ridge, NJ — 800-247-7000 www.lucent.com	Partner ACS 3.0	19	40	$1,200
Nitsuko America	Shelton, CT — 203-926-5450 www.nitsuko.com	124i	52	72	$250-$450 per station
Nitsuko America	Shelton, CT — 203-926-5450 www.nitsuko.com	384i	128, 256	256	$250-$450 per station
Nitusko America	Shelton, CT — 203-926-5450 www.nitsuko.com	PORTRAIT	824	24	$200-$400 per station
NEC America Inc.	Irving, TX — 800-TEAMNEC www.cng.nec.com	Electra Elite 48 and Electra Elite 192	64	120	N/A

Features	Unique selling propositions		
	#1	#2	#3
digital, analog, SMDR, LCR, caller ID, voicemail, T-1, first party	Port architecture: 9 slots plus dedicated T-1 slot	Flexible numbering plan	Built-in modem for remote maintenance
digital, analog, SMDR, LCR, caller ID, voicemail, T-1, ISDN, first party	Multiple I/P technology	QSIG networking with I/P	Wireless operation
digital, SMDR, caller ID, voicemail, T-1, first party	Voicemail with Quick Groups Quick Moves	Comprehensive spoken and Web-based help	Seamless integration of features
digital, SMDR, LCR, caller ID, voicemail, first party	Supports Inter-Tel and third-party CTI applications	Integrated voice processing	Full-duplex speakerphone
digital, SMDR, LCR, caller ID, voicemail, T-1, ISDN, first party	e-Response Enhanced 911 Routing	Voicemail monitoring	Call storage
digital, analog, SMDR, LCR, caller ID, voicemail, T-1, ISDN, first party	Built-in call center functionality	High-speed Internet access	Application-ready for CTI, wireless, and messaging
digital, analog, SMDR, caller ID, voicemail	Remote administration back-up and restore	Simple PCMCIA card upgrades	Advanced messaging features incl. live call screen
digital, SMDR, caller ID, voicemail, T-1, ISDN, first party	Supports TIE lines	Standard and optional ACD packages	Standard hotel/motel package with optional property management package
digital, SMDR, caller ID, voicemail, T-1, ISDN, first party	Standard and optional ACD packages	Standard hotel/motel packages and optional property management packages	Intersystem networking
analog, caller ID, voicemail	2,500-type devices plugged into station ports	Directory dialing	Reverse messaging
digital, analog, SMDR, LCR, caller ID, voicemail, T-1, ISDN, first party	Conference bridge card	VoIP trunk card (2Q '00)	Internal VM/ unified messaging

K

Company	Contact	Product	Max no. lines	Max phones	List price
Panasonic Communications Systems	Seacaucus, NJ — 800-211-PANA	KX-TD1232 Digital Super Hybrid Telephone System	24	64 proprietary & 64 SLTs total 128	$365 8x16 w/all features and 16 display telephones
Samsung Telecommunications	Miami, FL — 800-876-4782 http://samsungtelecom.com	DCS Compact	10	30	$250 per station
Samsung Telecommunications	Miami, FL — 800-876-4782 http://samsungtelecom.com	Prostar 816	8	16	$225 per station
Samsung Telecommunications	Miami, FL — 800-876-4782 http://samsungtelecom.com	DCS 50si	36	40	$275 per station
Samsung Telecommunications	Miami, FL — 800-876-4782 http://samsungtelecom.com	DCS	168	172	$300 per station
Samsung Telecommunications	Miami, FL — 800-876-4782 http://samsungtelecom.com	Hotel Operator	168	172	$300 per station
Samsung Telecommunications	Miami, FL — 800-876-4782 http://samsungtelecom.com	DCS 400si	288	360	$350 per station
Sprint Products Group Inc.	New Century, KS — 913-791-7700 www.sprintproductsgroup.com	Protegé	144	144	$350-$450 per station
Telrad Telecommunications Inc.	Woodbury, NY — 800-NEW PHONE www.telradusa.com	Digital S1000	255	925	$300 per station
Teltronics Inc.	Sarasota, FL — 941-753-5000 www.teltronics.com	Vision	576	576	$450-$550 per station
Vodavi	Scottsdale, AZ — 888-287-0169 www.vodavi.com	TRIAD 3	144	252	N/A

Features	Unique Selling Propositions		
	#1	#2	#3
digital, SMDR, caller ID, ISDN	Extra Device Port (XDP)	Multiline proprietary cordless phone integration	Digital voicemail integration
digital, SMDR, LCR, caller ID, voicemail, first party	Powerful embedded caller ID features	Daughter boards allow 2 keysets using 1 extension	Available 5-year warranty
analog, SMDR, voicemail	Every line and station appears on all phones	2-year warranty	Large, easy-to-read LCD display
digital, analog, SMDR, LCR, caller ID, voicemail, ISDN, first party	ISDN trunks and stations	Cadence internal voicemail	Available 5-year warranty
digital, analog, SMDR, LCR, caller ID, voicemail, T-1, first party	Lowest cost T-1 in marketplace	Cadence internal voicemail	Available 5-year warranty
digital, analog, SMDR, LCR, caller ID, voicemail, T-1, first party	Hotel/motel system with reports, costing, and wakeup	Omni-directional property management interface	Available 5-year warranty
digital, analog, SMDR, LCR, caller ID, voicemail, T-1, ISDN, first party	Dynamic time slots and power distribution	High-density card design and compact size	Support for 12 T-1/PRI and call center software
digital, SMDR, LCR, caller ID, T-1, first party	Answering machine emulation	Station alert	Preset call forwarding
digital, analog, SMDR, LCR, caller ID, voicemail, T-1, ISDN, first party	Conference bridge	MPD redundancy	16 ACD log-ins (standard)
digital, analog, SMDR, LCR, caller ID, voicemail, T-1, ISDN	Integrated auto attendent and ACD	Automatic set relocation	Up to 16 T-1s or PRIs supported
digital, ACD, caller ID, SMDR, voicemail, T-1, ISDN	Wanderer handsets	Provides the capability for any ACD agent to log on to a primary and secondary ACD group	Improved system maintenance and user administration features

K

answer an incoming call by simply pressing the blinking outside line key. You place an outgoing call by pressing an unlit outside line key, then dial the desired number. However, the key configuration provides additional capabilities not usually found in key systems. Features such as inherent call privacy, transfer, conference, and call forward are all standard.

The hospitality industry has not been forgotten by Toshiba (Irvine, CA – 949-583-3700, www.toshiba.com) with the introduction of their new HMIS turnkey solution. Designed to hang off their modular Strata digital key systems– which ranges from the small DK 40 (maximum 12x28) to their top of the line DK424 (maximum 336 stations) – HMIS is really a 133-Mhz PC that has been bundled with a well thought out Windows application that covers the entire hotel management gamut. Although this is a PC platform, Toshiba recommends that you run only their HMIS application.

Innkeepers can quickly search different type rooms for availability. Data entered at time of reservation can be used to populate the main check-in window: the desk merely finds the pending reservation in a list-box and clicks. Of course, the system produces a guest billing statement, which can be customized to include over 15 billing item fields and any one of 14 different captions.

As you might expect, Toshiba goes the full distance on the phone side to both hotel management and their quests. Call phone records and outstanding phone balance are displayed in real-time and can be brought up in one click from the main window (data is received from a separate connection to the DK424's SMDR port). Hotel management can enter rates for long distance, international, local, and toll free calls along with tax information. Call restrictions may also be placed on rooms. For guests, a simplified voice message system is available. Each room is assigned a mail box and quests may retrieve messages (indicated by a light on the phoneset) by entering a special access code.

Toshiba has not left the DK424 alone, bringing ISDN BRI directly to the station side and supporting PRI on the trunk side. With desktop ISDN, up to two telephones can use the line simultaneously. In some areas, businesses may also find that trunk BRI is cheaper than two ana-log lines. Finally, take another look at the CT possibilities of the DK424: An optional RS-232 adapter that connects to a digital phone is an easy route to PC call control.

Vodavi's (Scottsdale, AR – 888-287-0169, www.vodavi.com) latest and smallest member of their Triad line is the Triad-S, a flat-pack system. The KSU accommodates both digital and analog stations. Their CKIB boards are designed for three CO lines and eight digital station lines; their CSIB boards take eight analog lines. Up to three CSIB boards fit into the KSU, giving 24 single line phone connections. The Triad-S is real-ly an entry-level system for small businesses but shares many advantages of the rest of the Triad family.

The Promo Package ($699) is a 6x16 KSU that comes with eight of their digital, 24-button Executive phones. Add CKIB combo cards ($225) to bring the Triad-S to its maximum configuration of 12x32. To fill those extra station ports, choose from four phone sets, all with speakerphone, including a lower-cost 12-button Executive phone ($169). LCD displays on both Executive models show caller ID. Off-hook voice announce, not always seen in smaller key systems, is available on all their phones. Attach a Triad digital phone to an external CTI module ($99), and with their Discovery software, your desktop computer becomes a virtual phone. Of course, you'll need voicemail. Vodavi's Digital Dispatch Voice Mail System (maximum eight ports) can hang off of any digital port, and comes with 70 hours of message storage. Since it's digital, you'll benefit from caller ID and quick return of calls. The two-port unit lists at $1,599.

An upgrade to four ports will run you $599. Or just spring for the four-port unit at $1,950 and save some bucks.

As your business grows and T-1 or ISDN trunks start making busi-ness sense, you'll want to move to the Triad-1, -2, or -3 Digital Key System. You can still keep all of the phones and accessories you bought for the Triad-S, since they're upwardly compatible.

Vodavi's Discovery Series brings a very flexible approach to CTI. The Discovery CT module lets users connect a PC to their 2500 series digital phones, which are engineered to be TAPI compliant. Their Discovery Windows software offers a lot of goodies: Drag-and-drop transfer, direc-tory calling, and call preview capability. If desktop real-estate is limited, why not get rid of the phone altogether? Plug Vodavi's Discovery inte-gration board into a 16-bit ISA slot, grab a headset and you're phone free. Your virtual phone displays the status of all active lines and extensions. "Keys" can be programmed for speed dialing, outside lines, conference calling or any other phone feature. Also included with Discovery is a "middleware" package called "The Link" that s lets Discovery work with ACT, Goldmine, Outlook, Maximizer, and other PIM packages

If you're just not ready to get rid of your phoneset, Vodavi's Starplus Triad digital phones gives the phoneset enthusiast tons of features. Their basic eight-button phone speakerphone has six fixed and eight programma-ble keys. An enhanced model has 24-buttons in which 12 are fixed. Their executive modes adds a 48-character LCD, which shows Caller ID informa-tion. All phones let you program the keypad to automatically dial an outside line, without having to first manually seize a trunk line. Off hook voice over, a feature not always seen in smaller key systems, is also available. Another bonus with Vodavi's phoneset is answering machine emulation mode, which lets you listen to a caller leaving a message and lets you optionally pick up.

Vodavi's Digital Dispatch is a combined auto-attendant / voice mail system that hangs off of Triad-S's digital line. Dispatch is administered by a PC (or by a phone) and can be programmed to route by name, depart-ment or extension. Voice mail can capture Caller ID and allow for speedy return of calls. Users can create up to nine personalized distribution lists and administrators can set up to 100 system-wide distribution lists. A high-end feature that service firms (law or accounting) will be sure to take advantage of is a one-button conversation record. From any phone set, a crucial phone conversation can be recorded and saved in a voice mailbox.

Although ISDN is not planned for the Triad-S, Vodavi can boast such advanced features as DISA, UCD/ACD, and LCR.

Knowledgebase

The increasingly technical nature of products and services has spurred the development of huge repositories of knowledge accessible by both help desk agents and users. These are the knowledgebases. A knowledgebase is a computer system containing a database of empirical knowledge that is accessed by special software that in many cases offers dynamic content formed according to a user's profile and company rules, and has the ability to integrate with other information sources and sys-tems. An advanced authoring tool allows for easy additions or modifica-tions to the knowledgebase.

Many companies forgo use of a knowledgebase for their intranet, substituting large sites of static HTML or online documents. As the body of information grows, of course, this becomes unwieldy.

Segmenting Your Knowledgebase

If you want to use your knowledgebase as a source for gathering answers to e-mail messages or text chat requests, here's an efficient way to do it:

Let's say you don't display your entire knowledgebase on your website. Your on-line knowledgebase contains published answers to questions; the rest of the knowledgebase can be considered a working draft. You might have a policy that states suggestions for solving problems can only be posted on your site if, based on customer feedback, they have actually worked for at least 20 different customers. In this case unpublished suggestions are visible only to support reps, who can recommend them to customers if they wish.

By creating public and private segments within your knowledgebase, you relieve yourself of the burden of having to post every possible solution to a problem on your website. The on-line knowledgebase is analogous to first-line reps, who have a good general understanding of the most common problems customers encounter. Customers who need more specialized help can tap into the rest of the knowledgebase with the assistance of a rep.

Primus' (Seattle, WA · 206-292-1000, www.primus.com) SolutionSeries 3.1 provides you with up to 127 partitions for your knowledgebase. Detailed segmentation of your knowledgebase is helpful, for example, if you want to make sure customers from competing firms don't share the same knowledgebases. The software also lets you import prepackaged knowledgebases into your own. If you want to generate and display reports about your help desk's efficacy in solving problems, you can do so using Seagate Software's on-line reporting package.

It's important to note that on-line knowledgebases are becoming essential for reps who work on corporate help desks and the employees they support. Reps can use Knowlix's (Draper, UT – 801-924-6200, www.knowlix.com) iKnow Author to create support documents to be displayed on your corporate intranet. This software helps those who create the documents catch typos and avoid generating duplicate documents.

Once the help desk posts the documents, iKnow Author keeps tabs on how often reps and employees refer to these documents, so you can determine the usefulness of the items published on your corporate support site.

Another valuable feature in internal help desk software is the ability to let reps use the software from wherever they are. For example, Remedy's (Mountain View, CA) Action Request System 4.0 internal help desk software now runs entirely within a Web browser.

There are some caveats with publishing on-line knowledgebases, whether on a public Website or a corporate intranet. You have to treat the data as though it were a physical document. Tracking all the changes to your knowledgebase can be difficult if such changes can be made in real time and, in the case of a public Website, if they can be made available to all who have Web browsers. That's why you need a strict vetting process within your help desk and a clear demarcation of who can publish solutions to problems on your site.

You also have to develop criteria for determining when support documents are ready for public viewing. Try out the solutions yourself if possible. Ask reps to do the same. Solicit customer feedback. Don't post a solution that has only worked once. When a solution does work, find out the operating system, software and peripherals the customer was using. What works for one customer may not work for another.

Whether you have dedicated staff creating your support documents, or you let reps draft them, you need to maintain quality control to make sure there is no contradiction between the resolutions you posted last week and those you posted today. As Peter Dorfman, Molloy Group's director of marketing, puts it, "There is no more volatile commodity on earth than knowledge."

Using AI and Knowledgebases

Making an investment into tools that let you search your knowledgebase should be a well-researched project. Here's some background on artificial intelligenece (AI) technologies – and some tips to consider before making a decision:

Computers that think like humans? What is all this talk about AI, and does it really work? These are logical questions any help desk manager should have when considering an expert system. Artificial intelligence studies how people think, make decisions and arrive at conclusions. An AI system learns from past experience.

The first step in determining how any (or several) of these technologies can help you is to weed through all of the AI technologies expert systems use. So here are some explanations of how these technologies work when used in the help desk.

Case-based reasoning – CBR uses past occurrences to find solutions. It will look at how a similar problem was solved and suggest the same solution if the problem seems to match. It can give you a solution that would normally require expertise to figure out, without the need to ask customers many questions. CBR needs a large number of cases in the knowledgebase to be effective.

Decision trees – ask a series of questions to which the caller's responses brings (branches) the support rep to another question. Eventually the responses can lead the rep to a likely solution to the problem. This system is most effective when used by reps with little help desk or technical training because the system leads the rep through the process.

Text retrieval – also known as key word searching. These are ways of searching the knowledgebase by typing in words that describe the problem in an attempt to come up with the correct solution.

This search tool is best used by experienced reps who are technically inclined and know how to properly describe a problem. Key word searching really doesn't use artificial intelligence. A common problem with text retrieval/key word searching is the likelihood that different people will describe a problem differently, making it difficult for the knowledgebase to suggest a solution.

Natural language processing – lets users describe a problem in their own terms to find a solution in the knowledgebase. Problems can be entered into the system informally. This technology is most effective, once again, for reps with little help desk or technical experience.

Neural network – learns patterns and relationships in data. The Help Desk Institute in Colorado Springs, Colorado, defines a neural network as a "hardware and software simulation of the human brain using artificial neurons combined in a massively parallel network."

Fuzzy logic – can draw a possible solution when there is conflicting information or no exact match for a problem description. Fuzzy Logic assigns a value of confidence based on possible solutions.

Rules-based system – uses logic rules based on the information given. A rules-based system says if x is the problem then Y must be the solution. This used to be the most common form of expert system, but has been supplanted in recent years by other forms of AI. One problem is that it requires lots of pre-programming and is only effective for a small, restricted area of knowledge that isn't likely to change often.

Now that you know what these technologies are, how can you decide what's right for you? First, you need to consider how each of these technologies will or won't benefit you. Each one can be effective in helping you match a problem with a solution. But the degree of effectiveness depends on how you operate your help desk, the size of your operation, the nature of calls and the type of people you hire to solve problems.

K

313

Dr. Lance Eliot, president of the consulting firm, Eliot & Associates, in Huntington Beach, CA, says it's important to clarify the purpose of AI. "AI strives to make the computer more intelligent, imitating the intelligence of a human being," says Eliot. "But there's a difference between the goals of artificial intelligence and what we can with it do today. The goals come no where close."

Still he thinks the technology can be very useful. "The sad thing in the help desk is that people buy an automation tool that contains an AI component but never use it because of the complexity," says Eliot. "They make the investment, but it becomes too difficult to figure out or they can't figure out how to apply it."

He goes on: "A classic problem help desk mangers face is upper management telling them to get some kind of artificial intelligence up and running, but if that person still has to field calls or is just more socially oriented, they will spend the most time at the task they're most comfortable with."

"The artificial intelligence you see today is only the first generation," says Steve Mott, president of Cognitive Systems (Stamford, CT). He says that all of these AI technologies try to meet human needs but don't have all of a human's senses.

"They're not as easy to use as some of the claims make and they can't always scale down to the problem," says Mott. He thinks the second generation of AI will be improved, and it won't be long before we start seeing the upgraded versions.

Rushing to select an expert system is a big mistake help desks make. It is vitally important that once you decide how this system can benefit you, you develop a selection process. Eliot recommends doing a cost benefit analysis figuring out the costs of following through each call without a searching a knowledgebase vs. digging up a solution once and then always being able to refer back to it.

"When evaluating tools, help desks make the mistake of checking off all the different technologies they think they'll need without really thinking about how they will use them," says Eliot.

Some help desks employ support reps with little technical experience while other help desks only use well-trained technicians to handle all calls. Other help desks use relatively non-technical support reps at the first level of support and use technicians with technical savvy to handle

problems that can't be solved at level one. The type of people you use on your help desk affects the type of product you should use.

If someone is a seasoned technician, Eliot says, it can be frustrating to be forced to key in whole sentences to take advantage of a natural language interface when trying to search the knowledgebase.

"If the system doesn't let them take the quicker route by just using key words to search the knowledgebase, then it will create frustration," says Eliot.

On the other hand, if you help desk uses mostly non-technically oriented reps it can be frustrating to them if they can only search the database using key words. A problem with simple key word search is that different people will use different key words making it difficult to search the knowledgebase if they don't hit on the right word to bring up a viable solution.

In this case a natural language interface would be more effective. The user can enter complete sentences and the computer will comprehend what they are trying to do.

You will find some products that claim to use a combination of different AI technologies. The claim that the product really offers all of these distinct techniques to search the knowledgebase may be true. But you also need to ensure how well the vendors have implemented these techniques.

"The range may be limited," says Eliot. "There may be some kind of neural network or fuzzy logic but it may be a reduced version. It's important to make sure the product's claims are bona fide. Vendors want to equal up to a help desk manager's check list with all the features they think they need, so in order to stay in the game, they may feel they have to claim their product utilizes many of these technologies."

Eliot suggests installing the product on a small scale and working with it. That's the only way to really see if it can meet your needs. And you need to have someone skilled enough to do this on staff or else hire a consultant.

"There's a real process you need to go through," says Eliot. "It's not enough to pick a product based on a simple demo or spending some time at a vendor's booth during a trade show. You need to evaluate your needs now and six months or a year down the road. I've seen help desks rush the process of selecting a problem management tool only to find they're stuck with a product that lacks the features they need after they've made a large investment."

LAN

See Local Area Network.

LCR

See Least Cost Routing.

Least Cost Routing (LCR)

Also called Automatic Route Selection (ARS). A way that your phone system automatically chooses the least expensive way of routing a voice or data call. No one phone carrier can offer the cheapest rate on every call you place. Some carriers have great rates for a specific region, but charge a premium on calls outside that region. Some carriers have excellent rates within your country, but charge excessively for international calls. Carriers subsidize their low rates with higher rates for certain types of calls, such as international calls.

With LCR, calls can be directed to different carriers based on the time of day and the destination of the call. LCR can be based upon a periodically updated software table in a PBX (or in the database of an enhanced service) to help choose the cheapest long distance provider at the moment of that the call is initiated, or it may involve using an IP network to bypass long distance services network entirely, whereupon the LCR function becomes one of finding an Internet "hop off" server that is closest to the call's destination.

LDAP

See Lightweight Directory Access Protocol.

LEC

See Local Exchange Carrier.

LEC Class 5 Office.

Sometimes referred to as a LEC's "end office." A Class 5 Office is based on a Class 5 switch and provides customer wireline access via the local loop to long distance providers, based on presubscription (by dialing "1" and the phone number) or equal access (10XXX dialed calls). The Class 5 Office has been the basic routing mechanism for carriers worldwide.

Lightweight Directory Access Protocol (LDAP)

Specification for a client / server protocol, running directly over TCP/IP, for accessing and managing online directory services.

A "directory service" is defined as something similar to a database, but tending to be more specialized and containing more descriptive, attribute-based information. Directory information is read by many users in high volume lookup scenarios, whereas changes are written to it only occasionally. Unlike a traditional Relational Database Management System (RDBMS), such directories do not need to show complex relationships between relations.

Since directories are relatively static, that can be replicated in many network locations to improve availability and reliability, while also reducing response time. Thus, a directory requires a network protocol to access the directory, and optional replication and data distribution schemes.

Directory services can be local (your PC) or global, providing service to everyone on the Internet, such as the Domain Name System (DNS), which also happens to be distributed across many cooperating computers.

Users of some network e-mail systems, enjoy the advantages of a central "white pages" directory that keeps their address book updated with the current names and addresses of local users. But "Internet standard" e-mail programs using POP or IMAP4 may not have such a function, and older Internet methods of looking up names, such as *whois, Ph,* or *finger,* are limited in their abilities and are complicated to use.

In response, major companies agreed to support a new standard, LDAP.

LDAP was originally designed for clients on PCs to access directories of X.500, the OSI directory service, but can also be used to access such things as a standalone LDAP directory service, a directory service that is "back-ended" by X.500, and any other directory system that follows the X.500 data models. Clients on the older X.500 system used the computationally burdensome Directory Access Protocol (DAP) to access the directory servers. Unlike X.500, however, the "lightweight DAP" (LDAP), also known as X.500 Lite, doesn't need the OSI stack's upper layers and is a simpler protocol to implement, particularly with respect to clients.

LDAP has become an important Internet naming standard and has been defined by and is under "change control" by the Internet Engineering Task Force (IETF) for incorporation with the Domain Name System (DNS). LDAP also permits easy NT integration with Internet and Intranet standards and applications. Imagine a simplified database where nearly all actions are simple SQL-like "SELECTs" and anyone can do these SELECT inquiries over the Internet. That's LDAP.

The LDAP standard defines: A network protocol for accessing and modifying directory information, an information model defining the form and character of the information, a namespace defining how information is referenced and organized, and an emerging distributed operation model defining how data may be distributed and referenced.

The LDAP directory service model is based on entries. An entry is a place where attributes are stored, with each attribute having a "type" (a mnemonic string such as "mail" for an e-mail address) and one or more values (such as the actual e-mail address). Collections of attributes have a Distinguished Name (DN), which is used to unambiguously refer to the entry.

In LDAP, directory entries are arranged in a hierarchical tree-like structure that reflects political, geographic and/or organizational boundaries. One or more LDAP servers contain the entry data forming the LDAP directory tree. Entries representing countries appear are at the tree's top or "root," below which are entries representing such things as follows:

- The "root" directory (the starting place or the source of the tree), which branches out to
- Countries, which branches out to –
- Organizations, branching out to –
- Organizational units (divisions, departments, etc.), branching to (including an entry for)
- Individuals (which could be people, files, and shared resources, such as printers)

X.500 and LDAP directories can be distributed among many servers called Directory System Agents (DSA). But regardless of which LDAP server a client connects to, it sees the same view of the directory.

"LDAP-aware" client software can ask LDAP servers to look up entries in various ways. LDAP servers index all the data in their entries, and "filters" may be used to select just the person or group you want, returning only the information that is pertinent to your inquiry. Directory searches are thus more like precise database queries than pattern-matching operations, such as the keyword searches you would perform with AltaVista or HotBot, which can return hundreds of extraneous data records.

Netscape includes LDAP in its Communicator suite, and Microsoft includes it as part of its Active Directory, which can be found in products

L

such as Outlook Express. Novell's NetWare Directory Services interoperates with LDAP as does Cisco's networking products.

The first implementation of LDAP was developed at the University of Michigan. Version 2 was published as RFC 1777 and RFC 1778. The current version of the protocol, LDAPv3, was completed in 1997.

The whole telecom universe has adopted LDAP as the way for a client application to do a look-up in a directory. The e-mail world, voice mail, Microsoft, Netscape, CCOM (now Lucent), and everyone who's offering directory services says that they're going to support LDAP as the standard protocol for an app to get information out of or into a directory.

Linux

Open-source UNIX-clone operating system originally formulated by Linus Torvalds and continuously developed by a distributed community of over 40,000 programmers on the Internet. It's the most exciting, "hottest" software around. Best of all, it's free. See Operating Systems for Computer Telephony.

Local Area Network (LAN)

Local Area Network. A technique for connecting (networking) a series of computers close to each other – the "local area", such as in the same building, as opposed to a Wide Area Network (WAN) in which computers in several remote locations are networked. Networking is done by installing Network Interface Cards (NICs) in the computers and running one or more common network protocols over Unshielded Twisted Pair (UTP) wiring or a even wireless 2.5 or 5 GHz radio connection.

Local Exchange Carrier (LEC)

The terms LEC and Regional Bell Operating Company (RBOC) are interchangeable. LECs made their appearance in 1984 when Judge Greene broke up "Ma Bell". Seven "baby Bells" or LECs / RBOCs were formed along with the remaining vestiges of AT&T. LECs and RBOCs handle local telephone service while AT&T and other long distance companies such as MCI and Sprint handle long distance and international calling. The Telecommunications Deregulation Act of 1996 allows both LECs and Long Distance Companies to sell both local, long distance, and international services. An Incumbent Local Exchange Carrier (ILEC) is a U.S. telephone company that was providing local service when the Telecommunications Act of 1996 was enacted.

Loop.

Wire pair (or coaxial cable) that extends from a telephone central office to a telephone instrument.

Loop Disconnect Dialing

This is another name for the pulse dialing done by rotary phones.

Loop Start

A way of signaling on trunks by seizing a phone line by giving a supervisory signal. This is done by lifting a phone's handset off hook which connects to the CO which is providing a voltage to power the line. The completed circuit allows the voltage to produce a current called the loop current.

Media Server

Originally the term "media server" meant a server involved in the sharing and broadcasting of multimedia information such as video and audio data as well as the usual HTML text as received by standard web browsers, such as Netscape Navigator and Microsoft's Internet Explorer. Any web server could be converted into a "media server" if the files and standards followed were "self contained" (didn't need special server software) such as AU, MOV, WAV and MPEG files. More sophisticated forms of such a media server involved the "streaming" standards promulgated by companies such as RealAudio and Xing Technology Corp., which enables a browser to start playing a file before it's download is finished.

At the close of the 20th century, however, the definition of a media server changed somewhat, particularly when associated with computer telephony.

All CT applications generally include a mixture of media processing and call control. Media processing is concerned with playing and recording voice files, detecting touchtones, sending and receiving faxes, and, in general, handling any type of data that may be carried on the "line" (phone line or IP-enabled wiring). Call control refers to activities such as placing, answering, transferring, holding, and conferencing calls.

Many times, applications solely providing media processing services are connected directly to the PSTN or IP network. Examples of such applications include dedicated IVR systems, fax-on-demand systems, fax servers, and fax broadcast applications. Such applications, in general, provide value-added services and do not replace nor require integration with an organization's internal phone system.

PC-based telephony servers that provide a combination of media processing and call control may also connect directly to the PSTN or IP networks. Such "unified" or "universal" commservers are becoming increasingly popular for small and growing businesses as complete replacements to closed proprietary PBX solutions.

But separating media processing and call control and placing each in a separate server does have some advantages. Indeed, companies such as Dialogic and NMS have become excited over a whole new class of product that actually goes by various names: "Media streaming gateways," "media gateways," or "media boxes." These are not strictly gateways, but media servers that perform most of their tasks in the IP world and can do such things as voicemail, fax, announcements, etc. This media streaming box will be a great new platform for enhanced services.

The modern media server is the logical extrapolation of the long tradition of CT resource board manufacturers to put different functions on different cards. This is done for several reasons. One is a physical reason. It's difficult to get eight or 16 T-1 or E-1 spans on one card as well as other functions. The other reason is that companies such as Dialogic and NMS see that, in the public network, there's eventually going to be "front-end" switches that primarily do switching services, which will be completely separate from the voice or "media" resources – which, by the way, is why you don't necessary need a one-to-one ratio of network interface ports to voice resources.

For example, on a dual span (or a QuadSpan) board today you have a one-to-one ratio of voice resources to network interfaces. But as you scale up, depending upon your application, you don't necessarily need that one-to-one ratio. One way that the public network is going to be deploying enhanced services solutions is to use powerful front-end switches. CompactPCI chassis can provide such a heavy-duty front-end

switch, which then integrates to some local fabric, such as ATM or IP. Then, hanging off that local fabric will be enhanced services or "media servers" and that's where you'll find all of the voice, fax and other CT media resource cards.

Such an adjunct "media box" can do voice, fax, and conferencing and can host multiple applications. The switching part will still be done by a switch, or the high density network interface cards (high density DTI cards, SONET, SS7, etc.). In some solutions you might integrate all voice and network functions into one box, one all-inclusive CT server, and in others you'd create a distributed application, separating the switching from the media processing.

Such media servers are finding immediate acceptance among Enhanced Services Providers (ESPs) and telephony-centric Application Services Providers (ASPs).

Enhanced services platforms were traditionally co-located with a switch or the CO but now can be deployed anywhere in an IP framework. The media server becomes just an IP endpoint in the network. You don't need a gateway to access voicemail, since the voicemail can be on a separate server in a separate location. You, sitting in New York, for example, could take your voicemail off of a server situated in California. And you could have an announcement server in yet another location. Services are no longer tied to a gateway, since a gateway has a PSTN interface, and these media servers can live in an all-IP environment. It thus makes a great enhanced services play for service providers, since it shows off the natural progression of telecom toward the IP world. The enhanced services provider can now be distributed in pieces worldwide if he or she wishes to be.

Natural MicroSystems was one of the first companies to announce a standards-based, carrier-class IP media server solution. Early in 2000 they announced the Convergence Generation (CG) family, a carrier-class, CompactPCI-based platform supporting "in-the-network" deployments. The CG 6000C, the first in this new product family had its debut demonstration at the Spring 2000 Voice On the Net conference in San Jose, California.

The CG 6000C has 240 VoIP ports per CompactPCI slot. A tightly integrated hardware/software design, the Convergence Generation supports the rapid, simplified development of IP Media Gateways, IP Media Servers and Enterprise Communications Servers, which are all critical to the creation of next-generation IP-based enhanced services and supporting infrastructures.

The CG 6000C hosts Fusion 4.0, Natural MicroSystems' VoIP software platform for the development of advanced real-time media streaming applications. The CG 6000C is also supported by Natural MicroSystems' Natural Access development and runtime environment, which provides a consistent set of operating system-independent APIs to ensure industry-leading application portability.

IP Media Server technology is a cornerstone for next-generation enhanced service offerings. The CG 6000C, a key component of Natural MicroSystems' IP Media Server architecture, was purpose-built to support low-latency media streaming of IP-based multimedia traffic. Natural MicroSystems also offers PacketMedia, a new IP Media Server technology.

The CG 6000C also includes: Universal port for IVR / Fax / Vocode / Play / Record, configurable signal processing configuration, by port, by call; T.38 IP Fax; software-selectable T-1 / E-1 interface; and MGCP v1.0 integrated with the CT Access API

Media Stream Processor (MSP)

Commetrex (Norcross GA – 770-449-7775, www.commetrex.com) is well-known for having ported its fine MultiFax media-processing software to numerous hardware platforms, including Natural MicroSystems's VBX and AG boards, BICOM's LS-E and CCS's FirstLine card.

But when the Commetrex started to do the same thing with high-speed data modems as they did with fax – make it a software-defined computer-telephony system resource – it didn't find the open-architecture hardware resources on the market they needed to do so.

Moreover, Commetrex had become convinced that – to be competitive at the system-resource level of the CT market – it had to offer a platform to OEMs which would support all media-processing needs – but on one board.

What to do? The answer was to work with other like-minded companies to produce an open environment which could attract other vendors to develop competitive boards and compatible media-processing software. From this notion evolved the MSP Consortium.

Commetrex and six other computer telephony vendors have joined forces to take the "flexible DSP" idea about as far as it can go, developing a new computer telephony value-adding interface specification they call the Media Stream Processor (MSP). This "MSP Consortium" wants to add a new software layer to computer telephony system so new "shrink-wrapped" media processing applications written in DSP code can be completely de-associated with the underlying hardware.

Just as we don't use one PC for spreadsheets, one for text editing and another for software development, we would no longer need one board for voice, another for fax, and yet another for speech recognition. The MSP effort isolates DSP firmware completely from the underlying hardware, creating an open media-processing software competitive space, where anybody can write any DSP media processing application for any MSP-compliant card.

With MSP, instead of integrating fixed-function "boards" with an industry-standard CT Bus, such as MVIP or SCSA, an OEM can integrate fax, voice, data, text-to-speech, speech recognition and video media-processing products on a single, powerful, multi-vendor DSP-resource.

The MSP specification could fundamentally change the structure of the value-adding computer telephony market: Today, developing media-processing products usually involves developing not only the media-processing software, but the hardware platform as well. But MSP creates an additional vendor-independent value-adding segment by separating the media-processing software (firmware) from the DSP-resource hardware.

With a standard definition of the media-processing system-resource platform, companies can choose to develop just MSP-compliant hardware resources; others can develop MSP-compliant media-processing products; yet others may choose to integrate MSP into different host-level software environments.

The big winners will be the industry's OEMs and end users. The ultimate mission of the MSP Consortium is to create the age of genuine off-the-shelf, plug-and-play, shrink-wrapped media-processing software.

The MSP consortium founders are soliciting the assistance of any interested company. Please visit the MSP Consortium Web site at http://www.msp.org and contact Mike Coffee, the MSP Evangelist, at mcoffee@commetrex.com or 770-449-7775.

MGCP

The Multimedia Gateway Control Protocol. (See IP Telephony, Cable Telephony)

MMTA

See Multimedia Telecommunications Association.

Mobile Phone

Called a wireless or cellular phone in the U.S. There are four varieties of mobile phones: a car phone (also called a mobile phone), transportable, portable and personal.

MSP

See Media Stream Processor

Multimedia Telecommunications Association (MMTA)

Association division of the TIA, which has a vendor-oriented focus on industrial issues surrounding telephony in general, but also a strong computer-telephony focus. This group offers a training program for both salespeople and technicians in computer telephony. Membership includes Bell Operating Companies, PBX makers, telephony board and product makers, and software vendors.

Website: www.mmta.org

Multimedia Gateway Control Protocol (MGCP)

See IP Telephony, Cable Telephony.

NEBS

See Network Equipment-Building System.

Necrophony

Receiving phone calls from dead people (!)

Perhaps the strangest episode in the history of *Computer Telephony* magazine started one day when editor in chief Rick Luhmann told the staff that he wanted something "amusing" about telephony in each issue of the magazine, starting with the April 1997 issue. In years past the magazine's staff had thought about doing special April Fool's issue including a section called "Computer Telepathy," but nothing had come of it. Indeed, the only major attempt at humor that has ever appeared in the magazine was an insert called "CT Confidential," a tabloid-style send-up of CT Expo (and the entire CT industry, for that matter). Unfortunately, "CT Confidential" apparently had infuriated as many vendors as it had amused.

In any case, Yours Truly racked his brain in an effort to come up with something new that would be amusing yet not too inflammatory to our gentle readers (an impossibility, as it turned out, but let's not ruin a good story).

Suddenly a book came to mind that had been written by D. Scott Rogo and Raymond Bayless. Entitled *Phone Calls from the Dead* (New York: Berkeley, 1979) it is a compendium of anecdotes about alleged instances where people answer telephone calls, only to briefly hear words seemingly uttered by parties known to be deceased.

Yes, what a brilliant idea, I thought. We'll do an article on phone calls from the dead. Rick loved the idea.

Freelancer and obsessive researcher, Michael J. Doherty, was all too happy to take the assignment. He even coined a new term to describe it – Necrophony. The article "Necrophony to Explode?" finally appeared in the December 1997 issue of *Computer Telephony*, Readers were greeted by the deck: "Think about it: Make or take phone calls to and from the 'living impaired.' Don't snicker. Communicating with the dead could be the hottest CT app since international callback. Here's a look at this breathtaking technological opportunity and tips on how to choose the right platform."

Much to our surprise, Doherty, a former Green Beret with a Masters degree in criminal justice who was now working as a railroad police officer in New Jersey, decided to write the article since he had experienced such a phenomenon himself!

"In 1995 I was sitting in a radio dispatch trailer," Doherty said. "It was the middle of the night. A call came in. There was a great deal of static on the line and people were having trouble hearing the caller, so I took the call. The caller, speaking in strangely archaic colloquialisms, said that an officer was in danger of being jumped by a number of people hiding in the tall grass near the scene of a reported burglary."

"His exact (and urgent) words were: 'It's the tower operator. I'm watching your bull [a term for railroad cop]. They're waiting for him in the grass! They look like they're going to roll him! I'm at the tower. I can see 'em.'"

"I quickly got on the radio and warned the on-scene officer to stop immediately. I told him he was about to be ambushed from the tall grassy area he was approaching.

"The officer halted and responded with the yell: 'Look at them all!' - as the criminals jumped up from their hiding place in the grass and fled, all of them obviously thinking that he had seen them," says Doherty

Later the railroad employees tried to thank the tower operator.

"That's when we discovered the bizarre truth," said Doherty. "The tower at that location had been torn down 10 years before. There was no other tower for miles around. What's more, after some checking, I discovered the employment position of railroad-tower operator had been eliminated decades before."

Prior to becoming a policeman, Doherty had been an Army-trained radio operator. "As for this incident, I would clearly rate the message as a so-called 'two by two' communication – barely audible, but you could still make it out."

One surprise unearthed in the article was that this literal "killer app" has a rather long history, dating well back into the 19th century (not unlike facsimile). The consensus is that the first mechanical device to receive spirit voices was built by Jonathan Koons around 1852, though the plans for the device were (and are) unfortunately lost.

The next major inventor to tackle the necrophony idea was none other than Thomas Edison, who, always on the prowl to resurrect (and steal) a good idea, tried his hand at constructing a "spirit communicator" in 1928. Though this efforts failed, an amusing follow-up to his attempts occurred in a modern-day incident involving clairvoyant Sigrun Seuterman.

Seuterman supposedly contacted Edison on "the other side" in 1967. Obviously oblivious to the ongoing Summer of Love spirit, the cantankerous and thoroughly deceasd genius carped that he was still working on the problem - indeed, he gave curt instructions on how to modify television sets to 740 mHz to facilitate paranormal communication.

A "giant leap" in spirit communication technology came with research into Electronic Voice Phenomena (EVP) - or the tape recording of normally inaudible "spirit voices" that are subsequently made audible on playback. EVP is a huge area of research that is actively practiced by investigators and enthusiasts around the world.

The first "successful" researchers to record paranormal voices were Raymond Bayless and Attila von Szalay in 1956. Their success was followed by recordings taken by Friedrich Jurgenson in 1959.

Jurgenson originally was just recording birdsongs on his magnetic tape recorder. Upon playback, Jurgenson was shocked to hear voices on the tape.

Jurgenson's biggest contribution to the field was demonstrating the technology to Dr. Koknstantine Raudive, whose name eventually became synonymous with EVP technology. Raudive compiled an exhaustively large spirit tape collection - 72,000 voice recordings in all.

But the myth and lore of the good doctor goes beyond his collection of voices. It seems that after his death, the late Dr. Raudive successfully participated in developing a communication link from "the other side" to the world of the living.

The next figure to appear, Dr. Ernst Senkowski, is credited with coining the term *transcommunication*, which has been more recently expanded to Instrumental TransCommunication (ITC), which in layman's terms is "telephoning the dead."

The goal of ITC is protracted conversation between the living and the "living impaired" without the aid of a medium or clairvoyant. Efforts have produced real devices. For example, the [[sychophone]] was developed in 1967 by Franz Seidl. Marcello Bacci and others developed and used a spirit communicator from 1971 to about 1988.

Throughout the 1970s, George Meek and William O'Neil developed the *spiricom* device along with the late Dr. George jeffries Mueller, an expired (but apparently not unreachable!) electronics expert. This device was a complex 29 mHz communications system that established "quality" two-way conversations for the first time with the Great Beyond.

The late Mr. O'Neil related through the device that experiments on his side were being conducted at 68 mHz (Note: This frequency lies between conventional FM and TV bands.)

In 1982, Meek publicized his experiments and allegedly successful replication studies were performed. Reports of contacts by tape recorder, computer, radio and television are documented.

Voice analysis performed on these recordings registers from 500 to 1,400 cycles per second and sometimes in ranges which exceed the human voicebox. Researchers have found a direct correlation between human voices and the voices of the deceased - the latter appear to be simply shifted to a higher frequency. Experts have also matched Dr. Raudive's voice with recordings made of him before and after his death.

In short, like the computer telephony industry in general, something is definitely happening here, though no one is positive exactly what. One thing's for sure though: The mystery can no longer prevent a flood of entrepreneurial necrophony initiatives.

The original article was accompanied by what we thought were some funny sidebar boxes: A review of Ed Margulies' "great new book: The UnDead PBX," A notice about Dialogic's "new Ectoplasm line of cards," and the ECTF's latest specification, called "DOA.100." The article had a bigger impact that we could ever have thought possible.

Whereas our managerial "higher ups" found the Necrophony article neither amusing nor informative, we did get a surprising amount of e-mail from scientists at major telecom companies whose "secret" hobby has been to dabble in this area. Most of them thought that the "spirit voices" were in fact those of ham radio operators, the copper wiring of a tape recorder being capable of acting as an antenna and routing the voices into the tape recording head.

Then, the Feedback column of the January, 10 1998 issue of the esteemed British scientific journal, the *New Scientist* offered "congratulations to the trade magazine Computer Telephony for an illuminating article by Michael Doherty about the technology of communicating with the dead." They journal recounted the main points of the article, ending with "Feedback foresees problems here, though. How many dead people are computer literate? And are computers on the other side IBM-compatible, or do the living impaired prefer Apple Macs?"

Moroever, we shouldn't have been too surprised to then receive the following letter from a woman whose name I'm withholding for obvious reasons:

Dear Editor:

Your true story in the December *Computer Telephony* magazine prompted me to write to you and share an experience - probably more of a spiritual nature than necrophony - but still as spine tingling.

I am telecommunications director at a large medical center. We have a good number of display-type telephones in use in all areas of the hospital. My department received a call from the secretary of the Critical Care department asking if we were doing anything new with the phones. We told her no and when questioned why she stated that the words "I Care" had just appeared on her bosses' (the director of critical care) display phone - not over to the left where everything always appears on the displays, but in the center of the display.

When we went to investigate we were told that the director, her secretary and a nurse were discussing some very intense, personal issues - specifically the director. She had been experiencing some difficult times at the medical center recently, and as is the case with most organizations, rumors had been flying fast and furious. As they were in her office discussing these rumors, the director, in frustration and some anguish, said

"Oh, I just don't care anymore - I just don't care."

The nurse, who was standing next to her consoling her, said, "Well, someone cares - look at your phone."

Then all three looked at the display on her phone and there in the middle of the display were the words, "I Care."

We investigated all the possible reasons for this phenomenon and asked all the appropriate questions. All answers lead back to the same answer: There was no "logical" explanation for this. When one considers all the various possibilities, even though none of them panned out, coupled with the timing of this message and the recipient of the message - we decided technology had nothing to do with this. Sure gave a lot of us a reason to pause and contemplate our spiritual selves!

Necrophony websites:

Man and the Unknown: Paranormal Voices - http://www.xs4all.nl/~wichm/dirvoic3.html

Visual Perception of Stochastic Resonance - http://www.ge.infm.it/~simon/srprova.html

ITC: Instrumental Contact with the Dead? - http://www.spiritweb.org/Spirit/itc-macy.html

Astrans.htm page at www.worldsite.com.br - http://www.worldsite.com.br/astrans.htm

Parasciences & transcommunication - http://parasciences.dyadel.net

Worlditc - http://www.worldic.org

La Transcommunicacion como medio de enlace con - www.ctv.es/USERS/seip/seip2.htm and seip2a.htm

IFRES - www.ifres.org

Afterlife Scientific Evidence - www.ozemail.com.au/~vwzammit

Communication Here and Beyond - www.monroeinstitute.org/voyagers/voyages/focus-1997-spring-communication-lmonroe.html

NetWare

Before the rise of Windows, NetWare from Orem, Utah-based Novell, was the major local area network operating system. Although many NetWare commands "feel" the same as those in DOS, NetWare is in fact its own operating system.

Network Equipment-Building System (NEBS)

NEBS is a subset of the family of requirements published by Bellcore known as LSSGR (LATA Switching System Generic Requirement). Bellcore published a technical reference (TR) called TR-NWT-000063, Issue 5, 1993, that outlines the NEBS "standard".

Bellcore states that the intent of this TR is to "inform the industry of Bellcore's view of the proposed minimum generic requirements that appear to be appropriate for all new telecommunications equipment systems used in central offices (CO) and other telephone buildings of a typical BCC (Bellcore Client Company)."

NEBS requirements apply to frames of equipment and *not* components or subsystems that are contained within a frame (also referred to as shelf-level components).

The requirements are broken into two major categories: spatial and environmental requirements.

Spatial requirements apply to the equipment systems' cable distributing and interconnecting frames, power equipment, operations support systems and cable entrance facilities. Compliance with these requirements is intended to improve the use of space in the CO, simplify building-equipment interfaces and help make the planning and engineering of central offices simpler and more economical. In general, the spatial requirements relate to the size and location of the equipment going into a CO, how to power it and how to cable it all together.

Environmental requirements define the conditions the equipment might be exposed to and still operate reliably and not cause any catastrophic situation such as fire spread. These environmental requirements include temperature and humidity, fire resistance (which seems to be at the top of many BCCs lists), shock and vibration (which covers transportation, office and earthquake), electrostatic discharge (susceptibility and immunity), and electromagnetic compatibility (emission and immunity). Bellcore has also defined test methods in this TR that provide procedures on how to test the equipment to ensure that it meets the requirement.

NEBS certification itself represents an extensive set of performance, quality, environmental, and safety requirements developed by Bellcore and includes earthquake, cyclic temperature, mechanical shock, and electrostatic discharge resistance tests. To gain NEBS certification, a manufactrer's gizmo must be able to –

- Sustain a Zone 4 earthquake (between 7.0 and 8.3 on the Richter Scale),
- Withstand non-operational temperatures ranging from -4 to 140 degrees Fahrenheit and operating temperatures ranging from 41 to 104 degrees Fahrenheit,
- Extinguish itself in case of fire,
- Withstand up to 8,000 volts of direct electrical contact (the equivalent

of the centerpost rackmount being struck by lightning), with no permanent loss of data or functions

Because many telecommunications companies use NEBS-compliance as a prerequisite for product deployment, the certification strengthens a products reputation. However, while many companies have NEBS compliant designs, few actually go to the trouble and expense of undergoing the rigorous process of NEBS *certification*, thus verifying that their gizmo actually meets the standard.

Testing often takes three or four months and can cost a manufacturer well over $100,000.

Newtonization

Term coined in the book, *The Digital Call Center; Gateway to Technical Intimacy* by Paul Anderson and Art Rosenberg (Doyle Publishing, www.doylepublishing.com). Defined by the authors as "the glib, topical and provocative glamorization of computer-telephony technology that is far too sophisticated and complex for the users that put it in; when plug and play is in fact plug and pray." Computer telephony has in fact come a long way since Anderson and Rosenberg wrote their scathing text, thanks in part to the coming of IP technology and the convergence phenomenon.

N

One-Wire Wonder (OWW)

This is an umbrella term that covers even the most radical of business telecom systems, business phone systems that transmit voice and signaling over LAN or WAN. In theory such systems will simplify your wiring-plant, and they'll solve huge problems with maintenance, MACs, and management, and put your business in line to reap the benefits of IP telephony.

Some of these systems also pose the thought-provoking question: "Can you take the computer out of computer telephony?" since, in it's purest form, one-wire wonders treat the network as the PBX, with the functions of the phone system absorbed by the parts of the data network, the data hubs and switches taking on the voice switching functions. At the customer premises end are boxes that act as routers and drive phones connected to the networked PCs – or the phones have Ethernet circuitry built into them and are connected to the LAN directly.

It's not too surprising that such technology would appear, since data switches and telephone switches are similar, often located in the same room, and perform similar functions. The data and voice worlds use similar connectors on cabling that can be used in either realm. The PBX after all, is a kind of voice intranet server, and connects to the PSTN, a kind of WAN. So it makes sense that both worlds would integrate.

The exact mechanism for engendering this telecom upheaval are devices that go by various names and vague descriptions. Many experts use the term "network PBX" to describe any phone system that (within the enterprise) employs a data network for phone connectivity and voice transport. It resembles very little the traditional PBX or any of its variations (including the commserver or unPBX, with which these newer systems are often grouped). Instead, at its most extreme (or arguably at its purest) stage of evolution, the newest IP-based phone systems appear very much like data network components.

As you'll find in comparing the products profiled here (or anywhere else), there is no one architecture (yet) that can be called the standard for a network phone system. When we consider that the most forward-looking products actually use the enterprise IP data LAN (and in principle, the WAN) to distribute voice and signaling as true "One Wire Wonders."

It was companies such as Cisco / Selsius and 3COM / NBX that demonstrated how the trunk and line cards from Dialogic and Natural MicroSystems found in CT systems could be replaced with IP telephony gateways, placing applications on the network rather than the PBX or PC and eliminating the need to add new boxes when increasing capacity.

An IP PBX router with an Ethernet backplane allows calls (in the form of IP packets) to come into the system from the Internet and travel right onto the LAN, hanging off of which at intervals are Ethernet-compatible devices such as Selsius phones. Since there is no PC, voice / fax / WAN / IP card or NT license, bold claims have been made that such systems will ultimately cost a mere 10% of a conventional phone system, and about half that of a PC-based phone system.

One could still hang a PC on the LAN to store voice mail, but there are "brick" like devices that are collections of flash memory chips that could easily be adapted to do the same thing. One could conceivably create a phone / messaging system with no moving parts (other than the on/off switch and keypad buttons).

As for functionality, new software and sometimes special cards can be placed in a router to enhance what it can do, just like a commserver. NBX managed to extract TAPI functionality out of NT and run it with their LAN phone system, for example.

Ironically, it was the prior appearance of the commserver (and phones can could act as a conventional or IP phone) that made corporate IP telephony a possibility and paved the way for the "pure" one-wire wonders. For example, Interactive Intelligence's commserver uses the IP network as part of it's ability to do long distance least-cost routing within the enterprise, with the server determining which network (voice or IP) is cheaper to use for any particular call.

Other products, such as Sphere Communications' (Lake Bluff, IL – 847-247-8200, www.spherecom.com) Sphericall, seek to move voice onto a data framework – but not the data framework you're using today. The Sphere solution implements an overlay LAN based on ATM – one that can deal with data services and encapsulate IP and other protocols. Sphericall uses analog phones attached to each Windows NT or 95 client in the ATM network and uses data switches to route local telephone traffic. A Windows NT 4.0 server has a card interfacing to the PSTN (not an ATM-based link to the CO).

Still other products aim at simplifying the exploitation of wide-area IP networking for inter-office communication, and at solving the basic problem of integrating network-connected apps with telephony. They're very "network-centric" phone systems; but they still deliver voice to desktop analog phones over twisted-pair wire.

A lower-end one-wire wonder hybrid is the "brick" or Network Appliance, a router with switching hub having a bank of RJ14 jacks out of which lead ordinary phone wire out to desktop handsets, as well as an RJ45 jack which leads out to the company LAN / WAN via Ethernet or ATM. Voice response and messaging services are provided by an ancillary PC server (or another type of specialized solid-state network applicance). The network hubs themselves deliver basic telephony services (i.e. dial tone, switching, etc).

Hence, what we're really discussing here is a range of loosely related products grouped under an umbrella term, a familiar problem in the computer telephony field. For many reasons, it makes sense to include all these disparate products when compiling a list for further examination. Their variations address very real lines of development and opportunities in the current marketplace: Such as the fact that most data LANs aren't yet upgraded to support multimedia QoS, not to mention the fact that most corporations have already amortized their investment in a conventional wiring plant, but still want to save money by replacing expensive proprietary stations with cheap analog phones.

But the future (we think) is in "pure play IP" one-wire wonders – network phone systems that leverage the data LAN / WAN (and potentially the Internet) and use IP for voice transport. Functionally, all one-wire wonders subdivide into similar components: a data network, a call control server, gateways (often incorporated in the central server) for PSTN connectivity, and telephones that are LAN-connected or "hung off a LAN-connected PC".

Like any other network resource, the central server (and its sometimes-separate gateway component) attach to the enterprise LAN by single wires. Multiple logical connections flow into and out of the system, over this one physical link. Phones, similarly, connect directly to the LAN (in most cases, via a daisy-chain arrangement, or through a two-port mini-hub built into the phone). In this way, an entire workstation setup (PC and telephone) can be served by a single run of cable.

Converging telephony on the IP data network doesn't eliminate (in all cases) the need to consider some LAN re-engineering. But many will

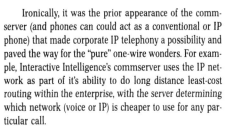

O

find they have sufficient bandwidth at all points to do so, already. Traffic on a small business LAN is generally very light. On a medium-sized LAN, it's usually pretty easy to smooth out by replacing hubs with switches, for example, and installing high-capacity backbone links between traffic zones. The payoff for such upgrades will be substantially improved data service, as well as the bandwidth to support voice.

The need for re-engineering, therefore, is mostly limited to large and heavily-trafficked LANs. In these cases, the key is that IP voice is an open, mainstream solution, heavily endorsed by major data product manufacturers. Nodal components (e.g., routers, data switches) that handle multiservice traffic are becoming increasingly common, and upgrades to "legacy" equipment increasingly available and affordable. 3Com, Cisco, and Nortel have all announced significant enhancements and additions to their multiservice switching families in the past two months, as they promote their LAN-based telephone systems more and more heavily as viable contenders for real world business deployments.

The IP LAN (and, by extension, the WAN, or even the Internet-at-large) is a connectionless cloud where resources have uniform locators (from which IP addresses can be deduced) and packets find their own routes to destinations over all sorts of transports and sub-topologies. IP-PBXs have adopted the Net's "connectionless cloud and resources" metaphors in their solutions. Most makers are moving towards embracing emerging standards such as H.323 and SIP for call control, RTP/RSVP for quality of service and insurance/traffic management, etc. Indeed, when considered in broad strokes, the most advanced IP-PBXs seem to be evolving in harmony with the softswitches, SIP proxies, MGCP servers, and other components that are coming to dominate the public-network IP telephony space.

The shift towards IP reinforces the benefits inherent to network-PBX architectures. Plug an IP-PBX server into the network at one location, plug phones into the network anywhere else, program the components, and *voilà* – everyone can negotiate the logical connections they require: to a phone in the next room, at the other end of a 1,000-mile-long WAN link; or (in principle) to a phone jacked into a dialup SOHO router at a worker's home. The complex physical topology of a conventional phone system is replaced by a logical topology that's always self-consistent; and which remains so, even if underlying physical connections change.

This logical topology is further served by making IP-PBX phones highly intelligent. Of necessity, IP-PBX phones handle internally many functions that previously depended on a permanently-wired (and powered) connection back to a central switch: They draw on battery or AC wall current to generate their own ringing and dial tones, for example, and amplify voice for speakerphone functions. IP-PBX phones also typically play a more authoritative role in system management.

In conventional PBXs, phones are interchangeable: Plug one into a physical port, and the PBX will power it up, test it, and bring it online at the port's associated extension. In IP-PBX territory, the telephone typically interrogates network resources on first connection, obtaining an IP address from a local DHCP server, and reporting its (fixed) MAC code and (dynamically-assigned) IP address to the PBX "system unit," which then maps both fragments of info to an extension map, class of service, feature set, and end-user programming specifics.

The result yields the advantages of centralized management, but the painless MACs of a more decentralized approach. Moves – to a new office, a new location, or even outside the office, to a remote site on the virtual network – can be made ad-hoc, without programming, rewiring, or adjusting PBX documentation. Adds are accomplished by programming the phone with its server's IP address, building a user account on

the central server through a web interface, and sending the phone out into the field for plug-and-go installation. Changes are processed entirely at the central site: modify the extension map, class of service, or user preferences database, and the phone works as expected.

The other benefit of intelligent IP phones – extremely significant to IP-PBX scalability – is that in most architectures (as in most network-based IP telephony models) phones transmit voice to one another via point-to-point connections (e.g., RTP over UDP) or point-to-resource connections (e.g., RTP over UDP to a gateway process, thence to a PSTN port). In many – perhaps most – cases, these connections don't involve the central server directly, except for database access during call setup. Once resources are connected, the system maintains only a loose idea of instrument/resource status required for feature support. Through this kind of unbundling, IP-PBXs can grow (in principle) much more flexibly than conventional phone systems. Need more extensions? Add them. Need more PSTN connections? Add gateway resources. The system's central intelligence can accomodate massive growth without incurring performance hits.

The point is that the IP phone system does signify a real change from a closed and centralized switching model, and does provide a real advantage by simplifying an enterprise's voice and data infrastructure into one distributed set of equipment, one network management system, and one set of professional skills necessary to manage the network.

Why Buy One?

Despite claims that a LAN-based phone system can cost as little as a tenth as much as a conventional PBX, current numbers suggest that actual cost can be as much if not more. These numbers will likely decrease over time, especially as the phones become standardized and as shipment volumes increase. There should, theoretically, be very little proprietary gear at all in such a system, so consumers and manufacturers / developers alike ought to benefit from competition in a way that was excluded by the earlier paradigm.

Initial cost savings are obviously variable; many companies already have separate voice and data networks, so abandoning the former won't save them installation costs. Greenfield installations and further buildout of existing wiring plants, however, will be somewhat cheaper and simpler. "Pure data" connectivity components (jacks, plates, cable, punchdown modules, etc.) are between 5% and 10% less expensive than mixed-mode hardware that supports Cat 5 data and conventional voice. Of course, "pure phone" non-data-capable Cat 3 hardware is even cheaper – but we don't see much demand for it, these days. We're told anecdotally that actual installation time will be about 20% less than mixed-mode, since installers don't have to implement (and then tone out, and fix mistakes on) multiple wiring plans. Tools, test equipment, and other material unique to the "phone side" can be eliminated.

But that's just the tip of the iceberg. In a one-wire system, the wiring closet is vastly simplified – the familiar 66-M150 "split blocks," wire bundles, punchdown connections, and "we added them later" modular jacks go away entirely. The entire differential cost of physically maintaining the telephony component of the wiring plant goes away. So does the need for skills unique to voice installation and maintenance.

Next, consider that in a heterogeneous wiring plant designed to support data networking and conventional telephony, you're often combining two completely-different signal carriers (one of which is carrying significant current) through the same conduit, terminating them on the same plates, then distributing to electrically-dissimilar equipment. Wiring plants in general have an unfortunate tendency to develop charge pools:

The more complex the wiring plant, the more routes spikes can travel when they discharge. And complex wiring plants are hard to protect – especially if you're looking to design the kind of "islands of protection" that most power-quality gurus recommend.

By contrast, a one-wire system carrying one kind of signal across one set of connectors is relatively easy to protect; translating directly to lower equipment costs, longer MTBF, and fewer disasters. Similarly, buyers can expect to tot up significant savings on management. They can eliminate outside charges for MACs (a big bite) and may even be able to scale back on telecom staff, or repurpose telecom-centric workers to serve data-centric functions. Certainly, any IS director of a smaller firm who's lamented at having to "manage the phone system, too" will find great joy in these devices.

Taken together, all these savings add up to some very large sums. But we're just scratching the surface. These IP-PBX systems can be deployed to take your in-house traffic off the PSTN: A multi-site enterprise will save some money by routing intra-company calls and faxes over the WAN. Smaller branch offices and remote-access-enabled home workers may be able to derive phone service from headquarters (across a WAN link or via RAS dialup), eliminating the cost of additional equipment. In principle, most of these systems will even support connections to arbitrarily-located endpoints, across the public Net, meaning that anyone can work at home or on the road. Though lack of QoS support in the greater cloud may make this impractical at the moment, that support is just around the corner.

Additional "soft savings" derive from the ability of most of these products to link together, and support a uniform dialing plan, uniform administration of higher-order feature services (like unified messaging), and other functions.

As you can see, there's a pretty compelling business case for making your next phone system a One-Wire Wonder. In the long term, what will encourage businesses to take the plunge is applications. While a lot of hype exists in the VoIP world about applications, there is also real potential waiting to be realized here, and a vendor's ability to do so could decide its future. IP telephony provides computer telephony with the opportunity to become what it has always dreamed of becoming. Part of the reason for this is that a truly converged voice/data network in some sense renders the idea of computer telephone integration (CTI) as we know it obsolete. No longer does complex and often inefficient integration between a PC and a telephone switch need to take place because, in effect, you've gotten rid of the switch. Instead, an applications server plugged directly into the LAN can integrate in a much more direct and non-hierarchical fashion with phones that, like a PC, are simply distributed endpoints on the same network. (As that network becomes mirrored by the public one as well, this model becomes even more elegant. Web-enabled call centers that use an IP infrastructure are the first example.)

The analogy between IP phones and network-connected PCs, finally, has its own important implications. The emergence of standard IP telephones means you may not always have to buy your phones from the same vendor that supplied your telephony CPE. Cost savings result. Because the architecture supports both mobility and "location arbitrariness," you may be empowered to rethink who you hire, how you accommodate them, and how your business grows. This is exciting, corporate-culture-changing stuff. It's not all about cost-savings. Some of it is about efficiency, profit, and improved competitive position.

As mentioned before, there is more than one way to build an IP phone system, and we present a number of them here. The important thing is to follow a strategy that makes sense for your business (trashing your PBX and going all IP immediately probably isn't it), and to know what you can look for and expect in a phone system, which may be a lot more than you had imagined.

The Holy Grail of IP business phone systems, the One True One-Wire Wonder, by our definition, brings IP telephony to the desktop and employs the LAN as its switching fabric. Only one set of cabling can be used to connect both phone and PC to the network. The OWW market has thus far been driven by large networking companies like Cisco, 3Com, and Ericsson, all of whom acquired smaller companies to serve as the foundation for their solutions. Start-ups like Tundo, Shoreline, and Sphere are also taking on the market with technology they've developed themselves. All use slightly different approaches, but also support open standards to a greater or lesser extent. Most are still evolving into full-scale enterprise systems, but all are shipping products today.

3Com (Santa Clara, CA – 408-326-5000, www.3com.com) got into the enterprise voice business with its purchase of NBX, whose product has since been integrated into the 3Com line as the NBX-100, a LAN-distributed system intended for small- to medium-sized businesses with up to 200 users. You have the option of installing a four-port analog line card for connection to the PSTN, or an eight-port 10BaseT hub card that lets you hook up NBX Ethernet phones. Standard features include multi-level auto attendant, voicemail, unified messaging, and hunt group capabilities. One of the biggest differentiators of the NBX-100 is that it does not use a PC server at all, but instead performs all major call control functions, and even applications such as voicemail and auto attendant, using proprietary OS and Call Processor hub.

Like others with similarly architected systems, 3Com plans to grow the NBX to fit a medium to large enterprise. An announcement made as part of 3Com's "E-Networks" initiative described plans to build the NBX's IP voice capabilities directly into 3Com's SuperStack switches, and later into the larger CoreBuilder switch. The new SuperStack II supports up to 750 users, while the CoreBuilder 9000 will take capacity to over 1,000. 3Com also offers an expanded range of interfaces on its PathBuilder WAN switches, including voice-over-DSL.

Accompanying these increases in scale, 3Com is developing (and is partnering with others to develop) next-gen applications to run on top of enterprise networks: Wireless integration between handheld PDAs and IP telephones is an example. Such apps can take advantage of emerging protocols like SIP and MGCP / MEGACO. The first of these next-gen apps will, quite logically, leverage 3Com's Palm Computing Platform to provide tighter integration between Palm Pilot and telephone.

As a broadly defined framework, E-Networks stresses the importance of building a network intelligent and discriminating enough to actually handle multiple types of traffic over the same wires without a sacrifice in quality. Conceptually, it also highlights the extent to which the next-generation phone system essentially is the data network, and the switches and routers that comprise it.

Cisco Systems (San Jose, CA – 408-526-4000, www.cisco.com) has been more in the spotlight than any other PBX vendor in history. The company's own claims for its dramatic entrance into the enterprise voice market have been so bold that everyone - particularly its competition – is anxious to see whether they will succeed.

Cisco is insistent on developing an all-IP network for enterprise voice, video, and data – a project embodied in their AVVID architecture. The voice segment of AVIDD, if it can be isolated as such, originated out of the acquisition of Selsius and focuses around the server Cisco now refers to as Call Manager.

Call Manager is a piece of software that contains the necessary rout-

O

ing tables to perform call setup and teardown. When a user picks up the handset on one of Cisco's Ethernet-connected IP phones, signals begin to change hands between the phone and the Call Manager server. First, Call Manager tells the phone to generate dial tone. Then the server analyzes digits entered by the user and performs a table lookup to determine the IP address and route of the extension dialed (an outside call is indicated by a prefix and passed out via a VoIP gateway). The server tells the phone at the opposite end to begin ringing. Once the party dialed picks up, an RTP stream is established between the two phones and Call Manager effectively drops out of the picture – though still monitoring the state of the call in order to terminate it or respond to other actions such as a transfer or conference.

Although the Call Manager software comes pre-installed on an NT server (known as the MCS 7830, Cisco ships a 19-inch rackmount box it OEMs from Compaq), don't be fooled: This is not a PC-based PBX. In fact, if you're having a hard time identifying the PBX at all in this model, you're starting to get the point. Many of the functions previously handled by a centralized switch (including ringing, dial tone, etc.) are now located in the phones themselves, which talk to each other on a direct, peer-to-peer basis.

As to the switch fabric itself, you find only the routers, hubs, and switches that already comprise an IP-based data network, all of which Cisco is eager to supply in abundance.

Although the system appears quite simple from an architectural standpoint, the features and applications it enables are rich. The Cisco IP phones use a DHCP client to automatically obtain an IP address. They report this address to the server along with their unique MAC identifiers. So they can be plugged in anywhere on the network and provide identical feature service.

The phones also have smart functions like one-button conferencing, which lets you launch a NetMeeting session instantly between two or more users. They implement quality of service based on the packet header's TOS field, and leverage QoS capabilities like weighted fair queuing and RSVP in Cisco's routers.

Other applications are being developed both by Cisco and third-party members of Cisco's Ecosystem program, using open TAPI and JTAPI interfaces. This effort includes a set of distributed contact center apps, based on Cisco's acquisitions of GeoTel and, more recently, Webline.

The MCS-7830, including the PC chassis with Call Manager software pre-installed, lists for $14,995. Cisco desktop IP phones are available in 12- and 30-button versions, priced at $385 and $485 each, respectively. A soft phone client will be available soon. Also coming soon is a newer version of the Call Manager server that can be configured for N+1 redundancy. Call Manager currently supports 200 users, though Cisco claims there is no technical reason why it cannot scale larger, and that they are currently building the service organization necessary to support large enterprise installations.

Ericsson WebCom's (Menlo Park, CA – 650-463-6000, www.ericsson-webcom.com) WebSwitch 2000 IPBX is a slightly modified version of the product it acquired when it bought TouchWave on April 12, 1999. The WebSwitch 2000 is a LAN-distributed system, targeted toward small- to medium-sized businesses and enterprise branch offices. It comes in 32- and 64-extension models that can be stacked or networked across the LAN.

The box, a self-contained network appliance, contains an embedded VoIP gateway, and supports both analog extensions and H.323-compliant client endpoints, including Symbol Technologies' Wireless LAN IP phones and software applications like Microsoft NetMeeting.

Through the embedded gateway, multi-location businesses can connect branch offices over the WAN or the Internet. Telecommuters can also gain access over the Net, and make their home phone into their business extension. Multiple locations share common dialing plans, remote dial tone (hop on/off), and centralized voicemail.

Somewhat uniquely, at least for the moment, Ericsson is also marketing the WebSwitch IPBX to service providers, who can install the box at the customer premise and have it act as a multiservice edge device. In this case, the WebSwitch would connect back to an ISP's or CLEC's POP via IP.

WebSwitch includes a long list of standard PBX features such as ACD, auto attendant, conferencing, forwarding, hold, park, dial-by-name, voicemail, and message notification. Ericsson also offers a standalone VoIP gateway as part of its WebSwitch 2000 family, for customers who wish to implement IP trunking with an existing, circuit-switched PBX.

Tundo (Cambridge, MA – 617-566-1300, www.tundo.com) is one of the few start-ups to target the medium- to large-enterprise market, with a VoIP system that architecturally resembles those of Cisco or Nortel. Tundo reports commercial deployments of more than 300 users on a single system.

At the heart of Tundo's Boundless PBX architecture is the DOT-server (Distributed Open Telephony), which handles call processing, set up, and teardown. The software runs on an NT server, and supports H.323, H.450, and other current and emerging standards.

Tundo is strong in the area of fault tolerance insofar as its servers can be configured for N+1 redundancy, while doing load balancing among all servers on a particular network. Like others that perform switching on the LAN itself, Tundo's call processing servers perform only a monitoring function once a call has been set up, and let IP phones connected via Ethernet communicate directly to one another.

While the ability to use third-party Ethernet phones is a benefit touted by most vendors of IP phone systems, Tundo is the first that we know of to actually offer their system with phones from an independent manufacturer.

So far Tundo has partnered only with Circa Communications (North Vancouver, BC, Canada – 604-990-5415, www.circa.ca) in this regard. (There are few IP phone manufacturers shipping products in volume at this point, which partially explains other vendors' hesitation.) But the company seems unquestionably committed to forming similar partnerships in the future, and has offered free licenses of its client software and its hardware reference designs to any standards-based IP telephone manufacturer.

Circa's IP feature phones sport an 80 by 160 pixel display, which shows time and date, call timer, and menus that can be scrolled through using arrow keys. The phones also include a message waiting LED, four soft keys, volume control, and full duplex speakerphone with acoustic echo cancellation.

Architecturally, they are based on a Texas Instruments / Telogy component design, and conform to TIA PN-4462, a proposed standard for VoIP phone interoperability. One of the things that PN-4462 will ensure is support of calling features across still evolving protocols like H.323, SIP, MGCP, and Megaco.

Circa's current feature set comprises a long list that includes most of what you would expect to find on a digital business phone set: transfer, forward, hold, park, calling party name, three-party conference, redial, speed dial, and more. G.711, G.723, and G.722 codecs are all supported.

Tundo supplies its own gateways for analog and digital PSTN trunk interfaces. As an alternative or supplement to the Circa IP desktop phone, you can also choose Tundo's JPhone, a Java-based soft phone, and Telport, a line-access device that lets you hook up analog phones to the DOT-server (comes in four- and 16-port configurations).

For third-party application developers Tundo offers DOT-Framework, an object-oriented software development kit that integrates existing apps like voicemail, ACD, and IVR into the system. Tundo has lso bundled various applications with its core hardware and software to arget specific environments such as call centers. In the case of a call center, the Tundo system would likely be used to integrate with existing PBX nd ACD systems, and to network multiple sites across a WAN.

Alternative Approaches

The following makers have elected to take markedly different approaches towards bringing phone service to the enterprise. While not true one-wire wonders, they're evolving in that direction. Meanwhile, they go a long way towards solving the basic network-unfriendliness of conventional PBX designs.

For example, Shoreline Teleworks (Mountain View, CA – 650-937-300, www.shoretel.com) offers an IP-PBX that scales to 360 ports, which puts Shoreline definitely at the higher end of the small- to medium-sized business range. The system itself is a combination of NT server software and IP-based voice switches that provide connections to the LAN, as well as trunk interfaces to the PSTN. These boxes, called ShoreGear, are stackable, which makes for easy expansion, and come in 12-port analog trunk and 24-channel digital T-1 flavors. Currently, Shoreline uses analog phones at the desktop, but says it will migrate to IP phones over time.

A proprietary distributed call control protocol lets you network ShoreGear voice switches across the WAN, extending the distributed call control functionality so that multiple locations can act as a single PBX. This means you get a unified dialing plan, central management, and centralized voicemail, auto attendant, and other standard applications.

The Shoreline server works in conjunction with client software that gives users one desktop interface for call control, voicemail, call history, and online company directories from the desktop. The newest release includes TAPI 2.1 support, which lets you dial out from an Outlook or Act database. A feature called AnyPhone lets you plug an analog phone into any location on the network and quickly program it (by dialing into voicemail) to become your own extension – a clever way of recreating a feature typically only available to Ethernet phones.

List price for a complete system starts at around $550 per user. The ShoreGear IP voice switch with T-1 interface costs $3,995.

Sphere Communications' Sphericall is different from other network distributed PBXs. Like its brethren, Sphericall consists of a collection of network elements, controlled by software that runs on an NT server, and uses the network as its switching fabric. The difference with Sphericall, is that this network is ATM, not IP. The justification, of course, is that for now ATM is simply a higher quality way of delivering converged voice and data.

Sphericall gives you the option of bringing ATM all the way to the desktop, but much more commonly it integrates with an Ethernet LAN and uses standard analog phones – a cheaper and typically more practical setup. The means are provided by PhoneHub, a compact box that sits between an ATM switch and the phones, acting as a gateway between 25 Mbps, 155 Mbps, or OC-3 ATM lines on one side, and analog lines on the other. PhoneHubs come in 12- and -24 port models ($4,695 and $7,195, respectively) and are rackmountable.

The other major piece of Sphericall hardware is the COHub (lists between $4895 and $5245), a trunk-side media gateway that bridges ATM to the PSTN via 24-channel, digital T-1 trunks. Sphere will be supporting the emerging MGCP protocol in all of its gateways.

While the gateways handle the patching together of various necessary but disparate wires, calls actually get set up, broken down, and managed inside of an NT server running Sphericall's server software ($1,995) and equipped with an ATM NIC card that lets it connect to the network

switches. PhoneNIC ($745) hooks the server up to a switch via 25 or 155 Mbps/OC-3 connection. The server software supports up to 250 extensions, with 96 trunk lines, and comes with a load of PBX features, including call hold, park, forwarding, caller ID, and distinctive ring. You also get auto attendant, voicemail, IVR, and unified messaging (through integration with Exchange).

An optional piece of software, Multi-Conference 2.1 ($3,995 per server), expands the standard conferencing capability from three to 12 parties on multiple simultaneous conferences. Moves, adds, and changes are managed through a Windows GUI. Optional TAPI 2.1-compliant Sphere Client software ($99 per user) extends CT features to the desktop, and uses COM objects to integrate with other applications. For one-wire at the desktop, Sphere makes the PhonePort, a small box that provides an OC-3 connection and RJ-11 jack, connecting both phone and PC to the network ($695 per unit).

PBX Vendors get into the Act

Makers of traditional PBXs are, not surprisingly, urging a rather slower migration toward IP than the pure network players. The self-interests behind each of these competing positions are not difficult to discern. What interests us, however, is that we've found most of the core group of PBX vendors to also be remarkably savvy when it comes to IP. And while they'll probably continue to move more slowly in terms of deployments than others, and may be more hesitant to support completely open standards on all of their equipment, these folks have a proven track record for delivering voice communications to businesses. Their plans for adopting IP telephony all reflect that history.

A string of acquisitions over the past year have given Alcatel (Calabasas, CA – 818-880-3500, www.alcatel.com) a wider U.S. presence, particularly in the areas of broadband access and IP telephony for carriers. Now the company – one of the largest PBX vendors in Europe – is seeking to establish itself on the enterprise level, with the introduction of its OmniPCX 4400.

OmniPCX 4400 is a UNIX-based, client/server system that ships with a variety of different handsets and lots of useful CT applications. The systems include a PWT-compliant, on-site wireless server that supports up to 256 radio base stations and over 1,000 phones, functionally identical to wired sets. One-number follow-me routing extends users' mobility off premise, as well. Client software capabilities include unified messaging, PIM and groupware integration, LDAP-compatible directories, and third-party integration through compliance with TAPI, MAPI, OLE, and other standard interfaces. The OmniPCX also has a relatively wide range of call center features that include ACD, outbound campaign management, and third-party integration with voice-logging and front-office applications. Alcatel's recent purchase of Genesys Telecommunications Labs (San Francisco, CA – 415-437-1100, www.genesyslabs.com) has further strengthened its call center offering, particularly in the area of web-based contact management and e-commerce.

Though its first release will offer TDM switching only, Alcatel plans to reshape the OmniPCX 4400 as a fully IP-capable hybrid phone system within a relatively short time frame. This will mean introducing a LAN-connected version of its existing digital phone set, as well as a soft-client version of its desktop CT app that will let remote users connect to the switch from any PC. The first Alcatel IP phone will be based on a plug-in interface card, which packetizes voice and implements QoS mechanisms such as RSVP before sending the packets onto an Ethernet cable. WAN access is provided by a gateway card installed on the call server itself. Because it can offer both IP and PSTN trunks in the same system, OmniPCX can monitor a VoIP call for packet loss and delay and switch over calls to a circuit-switched line if quality degrades beyond a certain point.

The OmniPCX 4400 starts at $40,000 for a 100-extension configuration. Price per user ranges between $350 and $900, depending on particular applications and phones.

Comdial (Charlottesville, VA – 804-978-2200, www.comdial.com) makes Impact FX, an integrated PBX and CT server. The FX line actually includes a number of different instantiations, one of which is a traditional Comdial digital PBX packaged together with the server in one cabinet, and another is the Impact FXS, which is a fully PC-based PBX. Impact FXS runs on Windows NT, and serves anywhere from 16 to 400 users. The CT server, also NT based, provides applications live voicemail, IVR, and screen pops. An optional standalone applications server offers back-end TAPI/TSAPI support.

Comdial has made advances toward voice-over-IP in recent months that show it will continue to evolve its existing platforms. Currently, Comdial offers CT Voice, a piece of VoIP/FoIP software that integrates with the systems' Dialogic cards to act as an IP gateway. Comdial intends to expand its IP support beyond simple gateway functionality, and to embrace inter-switch networking over IP, as well as doing VoIP on the line-side of the system.

These capabilities are the focus of a new board that directly installs on the FX platform. The board lets multi-site enterprises network switches between locations, running VoIP over T-1 lines. The board has much switch-like intelligence, such as the ability to terminate DID and light message waiting lamps, that will make the network of switches appear to act as one. Comdial is also adding support for Ethernet connected endpoints – phones and other multimedia devices – in the local area. The core switch software for its platform, however, will remain the same – providing users with access to all the same Comdial CT apps they presently enjoy.

Lucent Technologies (Murray Hill, NJ – 888-4-LUCENT, www.lucent.com) has a number of products that add IP capabilities to its existing PBXs (particularly the Definity One commserver). The IP ExchangeComm (IPEC), first announced in October 1998, seemed a bold statement of the vendor's commitment to following an IP path, but has since been subject to revision and a certain degree of hesitation. In December 1999, Lucent announced it would begin shipping the first iteration of the product through select VARs.

The IPEC is presently based around Lucent's IP Exchange CallManager software, which runs on a Compaq Proliant server and uses Windows NT. The server handles call control functions, including features like hold, transfer, and forward, and comes with a built-in auto attendant and voicemail system. A single server can support up to 96 extensions, though Lucent plans to increase scalability over time. Lucent also announced in December the IP Exchange Systems Development Partner Program for third-party software developers, and has already signed up Tapestry, Active Touch / WebEx, MIND CTI, and SoundLogic as applications partners for the system.

Though Lucent eventually plans to release an Ethernet-connected phone, it is currently using analog phones or Lucent digital Partner phones, which hang off a two-, four-, or eight-port adapter. The adapter (IP ExchangeAdapter) works like a gateway, encoding the voice as IP packets and sending it over the LAN to the IPEC server to be routed across the network or sent out over the PSTN. In the case of a small branch office, you could locate an ExchangeAdapter at the remote site and have the IPEC server at the main site route calls over the WAN to those extensions, obviating the need for a separate PBX or key system at

the remote office. To access PSTN trunks, the server uses a gateway card supplied by Natural Microsystems.

List price for the IP ExchangeComm is between $500 and $900 per user, depending on configuration.

Mitel's (Kanata, ON, Canada – 613-592-2122, www.mitel.com) SX2000 PBX for Windows NT server has been shipping in standard PSTN form. In 1999, Mitel announced a three-phase plan for migrating this system to put voice on the LAN and bring IP telephony to the desktop. The first phase involves the introduction of IP phones. Call control will still be handled in software on the NT server, while an internal gateway will let you connect IP phones via Ethernet. The new phones are based on Mitel's existing digital phone set designs, and include much of the same functionality. The SX2000 server connects to a peripheral node, which houses trunk and line cards. Both digital (T-1, PRI) and analog trunks are supported. In its initial release, the IP-enabled SX2000 supports around 100 IP or mixed extensions. Phase 2 of this strategy calls for Mitel to increase this scale to cater to larger enterprise customers.

In its first release, Mitel has devoted significant attention to insuring that the IP-enabled system will interoperate with, and provide the features of, its traditional PBXs. This includes support for a networking protocol that lets you link multiple systems over a wide area. Not until the third phase of its strategy will Mitel begin advocating moving users from their existing systems over to an IP-based platform. In the meantime, the system is intended for branch offices and workgroups, or standalone applications.

Mitel has also created an IP investment protection plan that guarantees smooth upgrade and transition strategies for existing and new customers. They've also committed to developing new applications based on emerging protocols including H.323, SIP, and MGCP, as well as WML and Bluetooth. An initial example may emerge out of a recently announced partnership with speech rec developer Nuance.

NEC America (Dallas, TX – 800-TEAM-NEC, www.cng.nec.com) recently began shipping NEAX Express NT, a server-based version of its traditional NEAX PBX. NEAX Express uses a form of the NEAX 2000 IVS switching fabric that has been adapted for a Windows NT environment. It also comes with a rich set of applications, including IVR, voicemail, auto attendant, and unified messaging. You can integrate these TAPI 2.1-compliant features with a Windows database, or a contact manager or PIM.

Though it performs only conventional TDM switching in its first release, NEC is now proceeding along a migration path to IP telephony. The migration involves the release of both line cards and trunk cards, to be installed directly into the NEAX Express chassis. Trunk-side connections let you network servers across the WAN via frame relay, ATM, or IP. Line-side connections actually plug the server into the LAN, and allow it to support Ethernet-connected phones. NEC's IP phone is based on the design of its traditional Dterm station sets, but supports H.323 and contains a two-port mini hub for connecting a PC through to the network. NEC also plans to offer a line of IP gateways, designed to interoperate with existing PBXs and support signaling protocols like Qsig.

The NEAX Express NT comes with up to 60 ports on a single server, or up to 124 ports with an optional expansion chassis.

From an architectural standpoint, Nortel Networks (Richardson, TX – 972-684-1000, www.nortelnetworks.com) is perhaps the most aggressive about delivering a completely decentralized, IP-based network phone system of any of the traditional PBX vendors. Nortel has released various products intended to migrate a legacy PBX or key system toward VoIP, but the essential component of its Inca (Internet Communications Architecture) is a wholly new model for carrying voice over a LAN and connecting Ethernet phones at the desktop.

Nortel's Unified Connection Manager (UCM) is software that runs on an NT server and connects to an Ethernet switch. The UCM maintains a dial plan (which associates extensions with either static or dynamically obtained IP addresses) and performs call setup and teardown.

As with Cisco's architecture, the other major element of Nortel's Inca is the phone itself. The Nortel I2004 desktop phone connects to a 10/100 Ethernet LAN using Cat 5 cables, and contains a built-in mini-hub or 2-port switch for direct connection to a PC, as well. The phones come programmed with a unique MAC address that allows them to be recognized by the server at any point in the network. They can obtain IP addresses dynamically by using DHCP and DNS. The phones themselves handle voice compression, ring and dial tone generation, and talk to one another on a peer-to-peer basis via RTP.

A separate NT server (or server cluster) connected to the network and thus linked to the Connection Manager, runs applications such as Unified Messaging and call center software, and provides links to third-party hardware (like a conference bridge) and TAPI-compliant software.

Nortel is also shipping the first in a series of trunking gateways that bridge the enterprise IP network to the PSTN. These will include stand-alone models, as well as a card that plugs into an existing Meridian 1 PBX. All Nortel enterprise gear will support H.323 and MGCP as far as they are developed, and Nortel plans to remain fully standards compliant in the future.

Siemens (Santa Clara, CA – 408-492-2000, www.icn.siemens.com) impressed us with their level-headed approach to IP telephony. They are rolling out their strategy in stages, the first of which is a line of IP phones that are shipping now. The phone system itself is the HiNet RC 3000, also shipping, but Siemens is not yet marketing it as a replacement for traditional PBXs.

HiNet falls into the distributed client-server category of next-gen phone systems. Unlike the PC-PBX or key system replacement, HiNet uses the Ethernet LAN as its switching fabric; so essentially, the product itself is software only (with the exception of a gateway for connecting to the PSTN). That software, which sits on an NT server, handles call control and management functions, communicating with a variety of clients that include Ethernet phones and soft phones.

What differentiates Siemens from others in this category is the attention it has given to integrating HiNet with its legacy HiCom systems, and protecting customers' investments in legacy PBXs. At present, Siemens is selling the HiNet product mainly to cater to specific application needs, such as branch offices, mobile sales forces, or campus expansions. These typically involve using the IP system to complement an existing PBX, which Siemens facilitates by integrating HiNet with their own existing unified messaging, voicemail, and call center applications. Siemens is also making CorNet, its PBX networking protocol, available in an IP environment.

HiNet includes its own line of standards-based Ethernet Phones, the LP 5100, which has a 24-character LCD display and the same OptiGuide interface as Siemens' traditional digital phone sets. Plans for future models include a more basic, lower-cost release, as well as a version containing a built-in hub that lets you connect PCs to the LAN through the desktop phone. Already released is an analog terminal adapter, the TA 1100, which lets you connect analog phones and fax machines to the HiNet RC 3000.

The HiNet LP 5100 phone is priced at $425 and the TA 1100 terminal adapter costs $475. A complete HiNet system costs around $750 per port.

Net-Enabled PC-PBXs

While they don't fit into the model of a distributed LAN telephony system, PC-PBX pioneers Altigen and Artisoft have both made moves toward voice-over-IP that are worth noting. These server-based systems have their niche in the small- to medium-sized business market, and offer many useful CT features to their customers. Both have already integrated IP connections on the trunk side of the network, via plug-in gateway cards that send voice out as packets across the WAN, and both can network servers across the LAN using similar technology. The next logical step is an IP telephone, which each of the two companies has said it will deliver over time. Added to their already rich client software packages, this tighter integration between phone and PC at the desktop could create a very powerful combination.

Altigen's (Fremont, CA, – 510-252-9712 www.altigen.com) AltiServ has been on the market since about 1996. With the release of version 3.5 of its AltiWare server software, Altigen is also reinventing itself as "the phone system for the Internet era." What this means is defined in a plan called "Intersect," and begins with Altigen's introduction of an embedded VoIP gateway (part of the Triton DSP board) that lets you network together two or more servers across the LAN or WAN. For the time being, the system will continue to use analog telephones (with up to 160 extensions), though Intersect defines plans to move to IP phones over time. Besides VoIP trunking, version 3.5 also supports for digital circuits, including channelized T-1 and ISDN PRI.

The other major aspect of Altigen's plan is to more explicitly focus on the call center market. To complement existing features such as auto attendant, ACD and work groups, screen pops and more, Altigen plans to introduce tighter web integration, detailed CDRs, multiple attendants, and other call center and e-commerce-specific applications. Over a slightly longer time frame, the company has also charted database and CRM integration through partnerships with other vendors. As the definition of a call center begins to blur, and more businesses are incorporating e-commerce, such features are likely to be important to a wide variety of small- to medium-sized businesses.

The vast majority of AltiServ's rich feature set is available in the VoIP version of the product as well as in the PSTN models. This includes one-number follow me routing, voicemail with up to 1,000 virutual extensions, e-mail integration, multi-location conferencing, call accounting, and the unique Zoomerang – a feature that lets you automatically return a call from voice-mail based on ANI, and then returns you directly to the next message in your voice-mail box without hanging up.

Artisoft's (Cambridge, MA – 617-354-0600, www.artisoft.com) TeleVantage uses an NT server and Dialogic voice boards, and comes standard with a big set of features like voicemail, e-mail integration, "follow-me" routing, ACD, auto attendant and least cost routing, as well as an excellent client GUI. You can add to and customize that list significantly by using Artisoft's Visual Voice, a set of Active X controls that lets you build IVR, fax, and text-to-speech apps. The most recent release of TeleVantage, version 3.0 uses Dialogic's IP Link as a built-in VoIP gateway, through which calls can be routed across the LAN or WAN. At present, Artisoft is recommending its VoIP capabilities for applications such as virtual tie lines between branch offices and hop on/off functionality (where one branch office has the option of receiving local dial tone from another for individual calls). In the future, Artisoft expects it will evolve these capabilities to bring VoIP to the desktop and to network multiple TeleVantage systems.

TeleVantage uses analog phones and a desktop Windows client for call control, messaging, and other functions. A browser-based client interface lets you check voicemail and access personal settings over the web. Integration with Microsoft Exchange gives you a unified inbox for e-mail and voicemail. TeleVantage currently scales to 264 extensions (up from its previous 144), and supports up to 96 analog, or digital T-1/E-1 trunks, including ISDN PRI.

TeleVantage is sold through resellers as a complete package, including client and server software, Dialogic boards, and analog phone sets. List price for an average configuration is around $500 per user. VoIP requires an additional Dialogic board (available in eight or 24 ports) and costs around $300 per user for the software licenses.

CellIt (Miami, FL – 305-436-2300, www.cellit.com) makes call-center software that soups up your existing PBX for futuristic call management. The Call Center Professional (CCPRO) is a client / server platform that blends inbound / outbound call center operations in narrow- and broadband environments. Basically, it's an ATM communications controller for call centers.

CellIT's CCPRO NT-based switch can communicate with a legacy PBX via 28 T-1s. Working in tandem with their CenterCord system coordinator server and database interface, the pair serves as the call center's ACD, voicemail, IVR, predictive dialer/call blender, consolidated reporting engine, and agent screen-pop engine. CCPRO can scale from zero-agent, or automated, centers to ones with thousands of agents.

The NexPath (Santa Clara, CA – 408-235-8916, www.nexpath.com) NTS Communications Server has been selling since around 1997. It combines a switching system (PBX type functions), a voicemail system, an auto attendant, a small business ACD, and network-based computer telephony in one package. Its "sweet spot" is around 80.

All Nexpath applications run on Windows 95 and Windows NT 4.0 platforms. You can use your web browser or native Windows tools to place calls, listen to voicemail, and access those hard to remember functions such as conferencing and consultation transfer.

For operators, NexDirector is NexPath's console program that controls calls and displays the status of each extension, park orbit, or outside line. NexDirector shows the caller ID and connection status of each extension. You can transfer calls to another extension, park orbit or voicemail with a click of the mouse.

The NexPath system is designed to be installed and maintained by the end user/customer. NexPath systems range in price from under $4,000 for a 5-extension system to $23,000 for an 85-extension system. The most common system is the NTS/24, which is 9 x 15 and costs $5,795.

Praxon's (Campbell, CA – 408-871-1600, www.praxon.com) Phone Data exchange (PDX) is a one-box solution integrating voice, data, networking and high-speed Internet access.

A T-1 module supports up to 24 voice/data channels. On the user side, capacity is expandable from eight to 64 users. Everything in terms of configuration and installation is done through web browser interface. The product can interface to any TAPI-based personal information manager such as Outlook or Act. Its built-in e-mail server is SMTP / POP3-compatible. The product is available through distribution channels such as TechData and Ingram Micro, and Praxon has an authorized reseller program. The Phone Data exchange has a combined solution price ranging from $300 to $425 per user.

Vertical Networks' (Sunnyvale, CA – 408-523-9700, www.verticalnetworks.com) InstantOffice system is an integrated box that includes PBX, voicemail with automated attendant, computer-telephony applications, multiprotocol routing (thank to Cisco IOS Technologies), CSU/DSU, LAN hub (Ethernet), remote access server and multiservice WAN modules. A good fit are branch offices and small businesses with under 100 users.

All features are remotely manageable and support SNMP. The platform is based on NT and uses purpose-built hardware that includes a multiple-bus design. The switching complex has a time division multiplexing (TDM) component that ensures no latency for voice services. The InstantOffice gives you up to 84 analog phone ports and up to 84 10Base-T or 10/100Base-T Ethernet hub ports.

Vertical Networks has delivered two call control applications for the InstantOffice system: OfficeAttendant and OfficeCommunicator. OfficeAttendant runs on the receptionist's PC and takes the place of the consoles used for PBXs for call handling. OfficeCommunicator, designed for the knowledge worker, is the client application that gives call control from the desktop.

Vertical Networks plans to add an H.323 gateway and an MGCP media gateway to the InstantOffice system. Support for IP telephones is also planned. Vertical also plans to support G.711, G.723.1, and G.729A.

InstantOffice has been shipping since October 1998. Price of the system for 84 users ranges from $250 per user for a base system to $475 per user for all of the features.

Open Services Gateway initiative (OSGi)

Cisco, Electrolux, Enikia, Sony and Sprint are among the leading technology companies that formed in March 1999 a consortium called the Open Services Gateway initiative (www.osgi.org). Since its formation, OSGi has since formalized its by-laws and articles of incorporation, organizing itself as a non-profit corporation in the United States. OSGi considers itself to be an open industry effort that is developing the next generation of smart consumer and small business appliances connected to commercial Internet services. These products and services are based on the Open Services Gateway specification, which provides a common foundation for ISPs, network operators and equipment manufacturers to deliver a range of e-services to gateway servers running in the home, small business or remote / branch office. The OSGi specification will create an open standard for a services gateway that functions as the platform for many communications-based services via a "services gateway" that can enable, consolidate, and manage voice, data, Internet and multimedia communications to and from the home, office and other locations.

By the fourth quarter of 1999, several technology industry leaders had become members of OSGi, including Cisco Systems (http://www.cisco.com), Electrolux (http://www.electrolux.com), Enikia (http://www.enikia.com), Invensys Controls (http://www.invensyscontrols.com), Sonera Corporation (http://www.sonera.com), Sony Corporation (http://www.sony.com), Sprint (http://www.sprint.com) and Yello Strom (http://www.yellostrom.de). With these additions, the OSGi membership now comprises 40 companies. A complete listing of OSGi member companies is available on the OSGi website.

Several OSGi member companies, such as Cisco Systems, Coactive Networks, Compaq, Echelon, GTE, IBM, Oracle, Sun Microsystems, Whirlpool and others have demonstrable OSGi-related technology.

Open System Interconnect (OSI)

An ISO publication that defines the seven generally accepted independent layers of communication protocols. See Open Systems Interconnection (OSI) Model.

Open Systems Interconnection (OSI) Model

The OSI is a standard description or reference model for how information from one point in a network (which for computer telephony is a software application in one computer) is transmitted through a telecommunications network to another endpoint (another software application in another computer).

OSI was the first worldwide effort to standardize the entire field of computer communications, or data networking, in the form of a networking framework for implementing hardware and protocols. The OSI model was developed by the International Organization for Standardization (ISO). OSI was originally supposed to be a detailed specification of computer internetworking interfaces formulated by representatives of major computer and telecommunication companies during committee meetings held beginning in 1983. Instead, the committee decided to establish a common reference model for which others could develop interfaces, that in turn could become standards.

The OSI model was completed in 1984, and it is still considered the chief architectural model for intercomputer communications. OSI continues to be administered by the ISO, so any new standard that seeks validation as an ISO standard for computer communications must be compatible with the OSI reference model. Also, the model can be used to guide product developers so that their products will consistently interoperate with other communications products. Finally, the OSI model has value as a recognized, single view of communications that gives everyone a common reference point for education and discussion about communications, since the OSI reference model is a common point of reference for categorizing and describing network devices, protocols, and issues.

The OSI reference model is purely a conceptual model, it does not do any "communicating" itself. The model is composed of an architecture or framework of seven layers, each specifying particular network functions. Everything from a cable to a web browser fits into this layered framework.

The tasks that move information between networked computers are divided into seven task groups, with each task or group of tasks then assigned to each of the seven OSI layers. Each layer has its own function and is basically self-contained, so that the tasks assigned to each layer can be implemented independently. One layer's functionality can thus be updated without affecting adjacent layers. Although manufacturers and telecom / datacom product developers do not always strictly adhere to OSI in terms of keeping related functions together in a well-defined layer, practically all communications products are described in relation to the OSI model. Different network devices are designed to operate at certain protocol levels, and each network protocol can be mapped to the OSI reference model.

OSI divides telecommunication into seven layers. The layers are in two categories: The upper layers, sometimes called the application layers, and the lower layers, or data transport layers. The upper three or four layers are used whenever a message passes from or to a user and are generally implemented only in software. The lower three layers (up to the network layer) handle data transport issues and are used when messages travel through the host computer. handle data transport issues. The bottom two layers, the physical layer and data-link layer, are implemented in hardware and software, though with IP networks only the bottom layer (the physical layer) need actually be hardware, since it is closest to the physical network medium and is responsible for actually putting information on it.

The OSI model assumes that each communicating user is at a computer equipped with hardware and software adhering to these seven functional layers. When one person sends a message to another, the data at the sender's end will pass down through each layer in that computer to the bottom layer, then over the channel, and at the other end, when the message arrives, data will flow back up through the layer hierarchy in the receiving computer and through the CT application to the end user.

The seven layers are (from the top, downward):

Layer 7. *The Application Layer.* This highest of layers in the OSI architecture ultimately leads the outside world (e.g. the user). However, the application layer is not itself the user application but only provides the system independent interface to the actual data communications application and its own user interface (though applications may indeed perform application layer functions).

Although the highest and seemingly the most abstract layer, the application layer actually consists of a complex set of standards and protocols. Application Layer Protocols are classified into Common Application Specific Elements (CASE) and Specific Application Specific Elements (SASE). Layer 7 is where communication partners are identified, quality of service is identified, and user authentication and privacy are determined. This layer contains functions for applications services, such as file transfer, remote file access and virtual terminals.

Layer 6. *The Presentation Layer.* Sometimes called the Syntax Layer, this layer is usually that part of an operating system that is concerned with the representation (syntax) of messages' data associated with an application during the transfer between two application processes. Applications routing data are simply routing binary streams, which has no meaning without a definition as to how it is to be formatted. A raw binary representation alone isn't good enough, since two computers communicating with each other may have totally different configurations: One could be using a 16 bit word size, the other 32; a PC could be using an ASCII character set, while an IBM mainframe could be using EBCDIC, etc.

The presentation layer must therefore do its part to provide transparent communications services by masking the differences of varying data formats between dissimilar systems and in general converting incoming and outgoing data from one presentation format to another (for example, from a text stream into a popup window).

The presentation layer is also concerned with methods of data encryption and data security, and compression algorithms that may have also changed the data format.

Here's how the presentation layer works: The presentation layer in one computer will attempt to establish a "transfer syntax" with the other presentation layer in the other computer by negotiating a common syntax that both applications can use. Failure to do so results in a non-connection. A widely used standards for the presentation layer are ISO 8824 and 8825

Layer 5. *The Session Layer.* This is a sort of interface between the hardware (the bottom four layers) and the software (the top two layers). This layer is the dialog manager. The transport layer handles how a data stream is directed, but it is the Session layer that says when data can flow. It negotiates and creates a connection between two presentation layers when requested, then controls the data flow. More technically, it establishes, maintains and otherwise controls the use of the basic communications channels provided by the transport layer, by setting up, coordinating, and terminating conversations, exchanges, and dialogs between the applications at each end. It handles session and connection coordination.

This layer is often combined with the transport layer.

Layer 4. *The Transport Layer.* This layer defines the rules for information exchange and manages the end-to-end flow control and delivery (for example, determining whether all packets have arrived) and error-checking / recovery within and between networks. It ensures complete data transfer.

This is the the highest of the lower layer protocols in the OSI protocol stack. It provides the means to establish, maintain, and release transport connections on behalf of session entities. It provides a connection-oriented or connectionless service. In a connection-oriented session, a circuit is established through which packets flow to the destination.

The transport layer is the first of the peer to peer layers, which means that once the lower layers are implemented, a transport layer can transparently communicate directly with the transport layer of another

O

data entity. It provides an idealized full duplex bit pipe to the upper layers in which the binary stream sent from one end, makes it, in order, to the other end. The result is that the upper layers get an idealized bit pipe and need only be concerned with what the data is, not how it arrived.

Layer 3. *The Network Layer.* This handles how data is routed from the source computer to the destination computer, sending it in the right direction over the correct intermediate nodes to the right destination, and receiving incoming transmissions at the packet level. The network layer does routing and forwarding within and between individual networks and can provide a Connection Oriented Network Service (CONS) or a Connectionless Network Service (CLNS). No matter what the route of the actual connection, the data arrives at the destination as if the two data entities were directly connected.

Layer 2. *The Data Link Layer.* This layer concerns itself with the procedures and protocols for operating the communications channels (transmission protocol knowledge and management), and provides error detection / correction and synchronization for the physical level. Computers identify themselves via MAC addresses and Network Interface Cards (NICs), and the data link layer touches upon on this level, while at the same time organizing information from the higher layers into "blocks of bits", called frames (such as "Synchronous Data Link Control" frame), for orderly transfer and error control. Sometimes this involves bit-stuffing for strings of 1's in excess of 5.

Layer 1. *The Physical Layer.* This bottom layer is responsible for activating, maintaining, and deactivating the physical connection for bit transfers between "Data Link Entities." It provides the hardware means of sending and receiving a bit stream (a "bit pipe") through the network

Application Layers		
	7. Application Layer	E-mail, Newsgroups, Web apps, Directory Services, etc.
	6. Presentation Layer	POP/SMTP, Usenet, HTTP, FTP, Telnet, DNS, SNMP, NFS
	5. Session Layer	POP/25, Port 80, RPC Portmapper
Data Transport Layers	4. Transport Layer	Transmission Control Protocol (TCP), User Datagram Protocol (UDP)
	3. Network Layer	Internet Protocol Versions 4 and 6
	2. Data Link Layer	Ethernet II, 802.2 SNAP, SLIP, PPP
	1. Physical Layer	Coaxial Cables, CAT 1-7, FDDI, ADSL, ATM, ISDN, RS-X

at the most basic electrical, mechanical, functional and procedural levels. An example of a physical layer ISO standard is the RS-232 interface.

Operating Systems for CT

Originally, the first computer operating system associated with telecommunications was Unix. It is still the dominant operating system for carriers. Linux, which is gaining rapidly, is simply a "streamlined" form of Unix. In the early 1990s, OS/2 started to appear in CT systems, but ultimately disappeared because of OS/2's failure to compete in the marketplace.

Windows 3.11, 95 and 98 have been used in smaller-scale computer telephony applications, but it is Windows NT and now Windows 2000 that has been determined to be the "CT operating system of the future." It took about six years of development and debugging on the part of Microsoft and its partners to bring NT up to the point where it was a solid, viable computer telephony platform. Nevertheless, Windows will still face some stiff competition from the various entrenched flavors of Unix and the mania over Linux.

Windows NT / 2000 eases the development and deployment of ASPs since, if you have an IP connection, you probably already have NT servers running in your organization. So your legacy equipment can be used to add an application atop the existing IP infrastructure, then you can rent the application to others having IP connections.

You may have noticed that operating systems are often "associated" with the underlying hardware platform. For example, there's been an ongoing Windows / PCI vs. UNIX / VME debate.

Demands for reliability and ruggedness in large-scale, carrier-based telecom narrows the choices to CompactPCI and VME. CompactPCI vendors have targeted telecom, which is now the biggest bus and boards market. CompactPCI runs Windows and has hot swappable boards, which is becoming attractive to carriers and service bureaus wanting to run applications found only on NT / 2000 platforms.

The PCI Industrial Computer Manufacturers Group (PICMG), the association that drives the Compact PCI standard, worked with the ECTF (See ECTF) to design a version of the H.100 telephony bus spec for CompactPCI, called H.110 (See H.110).

So, up until recently your choice of bus architecture locked you into your OS environment. Choosing CompactPCI meant going with Windows NT or 2000, while choosing VME took you to UNIX and real-time OSs. But now things have become somewhat blurred, since cPCI single board computers now carry a range of CPUs (SPARC, PowerPC, etc.), while Pentium/x86platforms on VME that can handle Windows have been deployed as well.

In programmable CO switches, the routing logic is generally controlled by high-availability, redundant UNIX clusters. More clustering software is available today from UNIX software vendors than from Microsoft software vendors.

The Java language is something of a wild card when it comes to computer telephony. Being a sort of downloadable "meta operating system," and universal application envrionment, it can reduce costs relating to operations, admininstration, and maintenance (OA&M). However, it consumes lots of CPU cycles, and it has not gained wide acceptance. It is beginning to appear in thin clients for call centers, however.

Unix is still the most important operating system for really heavy-duty computer telephony. Here are reasons why, gleaned from various companies:

1. *High Port Count.* With similar hardware configurations running, UNIX does higher line densities than other OS. The UNIX platform supports many concurrent processes and voice channels. It's ideal for large, mission-critical IVR apps. Its multi-processing environment ensures uptime if one process or app fails.

2. *Multiuser muscle.* It's a multi-user O/S; multiple users (more than four) can connect to the server.

3. *Performance.* UNIX supports high-speed call processing and reporting through memory management that supports large (greater than 16MB), shared, in-memory data caching. This allows multiple concurrent tasks to share data without unnecessary disk access. UNIX systems have higher throughput than those using other operating systems.

332

With hardware configurations running the same software, UNIX has consistently higher performance (i.e. handle more ports and transactions). To do the same in NT, you need more memory and a faster CPU. UNIX uses resources more efficiently, so a less expensive platform is required or a larger system can be built.

It's difficult to determine exactly why this is so, primarily because NT internals are not accessible. One can think of two explanations.

First, the UNIX kernel is more efficient. The NT operating system was designed around a large number of clearly defined modules built on top of a microkernel. While architecturally sound, this approach is inefficient because of the overhead of coordinating processes among various modules and going through the microkernel layer to the hardware. The significant performance improvement in NT 4.0 was partly due to reducing the modularity. UNIX, on the other hand is based on a monolithic kernel closer to the hardware.

Second, UNIX drivers are more efficient. For example, in years past, Dialogic's NT drivers were ported UNIX drivers, not native drivers.

4. *Performance Monitoring.* Built-in UNIX tools, and help, are standard for monitoring the OS and overall system performance (ipcs, sar, ps).

5. *True Multi-user.* UNIX provides a multi-user environment where every process has its own owner and multiple processes (users) can access your telephony app simultaneously. This is important if you deploy systems that will be remotely managed. Administration and control of a remote system without taking all the users off-line, is critical.

6. *Team Enabling.* UNIX's multi-user support fits in with today's multi-tasking, team-oriented workplace. It's cost-effective to support multiple users for development.

7. *Born to be client-server.* Client-server is inherent in the UNIX operating system for all services, not just file services. Take Telnet, FTP, NIS, Command Shells, etc. The TCP/IP socket based protocol provides the communication foundation for client servers in computer telephony app development.

8. *Proven Interprocess Communications.* More than any other OS, UNIX has matured in the area of interprocess communications and server to server communications. With UNIX, multiple processes easily communicate and synchronize. Some common programming techniques used by many developers involve message queues, which NT doesn't have. UNIX configures at the OS kernel level of interprocess communication facilities (shared memory, message queues). These facilities are well-suited to an architecture that's typical for event-driven computer telephony applications.

9. *Special, High Integrity File Systems.* Embedded UNIX from SCO (UnixWare / VenturCom) has read-only and journalizing file systems for high disk integrity in the event of power interruptions. The bane of high performance operating systems is disk caching, the process of keeping some portion of the file system in memory to increase throughput, since that data can be put at risk.

10. *Strong Conventional File System.* UNIX supports file systems up to one terabyte and provides support for RAID 0, 1 and 5.

11. *Beefy Databases.* The UNIX platform hosts the largest selection of industrial-strength database systems. Database vendors (Oracle, Sybase, Informix etc.) generally deploy UNIX releases first. The ability to incorporate database systems in a computer telephony app has become key.

12. *UNIX is stable.* UNIX is a very stable OS suitable for 7/24 type work that stays up for months on end in unattended environments. This includes the fact that the OS performs automated maintenance to prevent overflowing log files. UNIX has always beat NT in terms of stability. UNIX is the OS with lots more "power-on" operating hours and is simply more reliable. Even die-hard NT developers complain about NT's touchiness. At first glance, Windows 2000 appears to be a drastic improvement over NT, but we'll see.

13. *Fault Tolerance.* Going beyond simple stability, remember that there's no greater mission critical function in a small business than the telephone system. "This phone system has committed an illegal instruction and will be shut down" is not an acceptable message at any time under any circumstances. NT crashes with "blue screens of death" at odd times. UNIX does not. There are many stories of Unix systems going without a reboot for over a year. You won't find that with NT. That' why customers whose business depends on mission critical appplications demand UNIX. Mission-critical apps dictate the utmost need for fault-tolerance and system uptime. Enter Unix, with its advanced clustering capabilities. UNIX systems have a proven track record with the largest mission-critical airline reservations, banking and retail systems. Over 70% of all database apps run on UNIX.

With over 25 years in the market, this is a mature, proven OS integrated into all major computer systems on the market. UNIX is a requirement on most government bids and large corporation's RFPs. It is the system of choice for "bet your business" apps.

14. *Wide Availability.* UNIX variants can be found for almost any hardware platform. This lets developers select the best hardware platform for their projects. You can be fickle.

15. *Powerful Options.* Every hardware manufacturer caters to UNIX. The latest and greatest hardware, such as 64-bit CPUs, symmetrical multi-processing (SMP), UltraSCSI fast and wide disk drives, are all available for UNIX first. If you need app power you get it with UNIX.

16. *Multiple Flavors.* UNIX comes in more than one flavor. NT is similar to the sales pitch for the Model T, you can get it in any color you want as long as it's black. With UNIX one has a list of alternatives from Linux and FreeBSD, to the mainframe-class offerings from HP, IBM, and Sun. Platforms range from the commodity PCs to n-way multiprocessors and true fault-tolerant systems. With UNIX, you can select the vendor with the desired price-point, support, and product. UNIX spreads the wealth. UNIX is democratic.

17. *Server Choices.* UNIX is available on a wide range of servers. You can use servers from Data General, Hewlett-Packard, IBM, Intel, Sun, etc. This means savings in training, software upgrades and support contracts. Companies save money by selecting the server size best for their IVR apps. Server power ranges from economical Pentium processors to powerful RISC servers.

18. *Other Third Party Support.* UNIX supports serial I/O and other third party hardware.

19. *It's Configurable.* Compared to DOS and Windows, adding hardware to UNIX is simple. Remember the IRQ setting battles you'd have when installing multiple modem cards into a Windows system? And "plug and play" for Win95 is worse. With UNIX you simply install an S-Bus card with serial ports, plug in your modems and you're ready!

20. *UNIX is a Polyglot.* UNIX "talks to all the animals", it looks like a Novell server to IPX/SPX clients, like an NT server to Netbios clients, a UNIX server to TCP/IP clients, and at the same time provides dial-up access to a large number of different terminals (over 100 types). All transparent to your app. UNIX's support of standard protocols gives you "off-the-shelf" connectivity. This makes integrating stuff from multiple venders more doable. Open connectivity provides investment protection. Customers retain their existing databases, ACDs, PBXs, etc., as these all integrate with UNIX. Since UNIX provides SMB and IPX/SPX networking and interoperates with Microsoft and Novell networks, you can choose

the best platform for the job knowing UNIX works with existing systems.

21. *Dumb terminal support.* A single app can provide a character, a graphical, and a browser interface to users for connecting through "dumb" terminals (a PC or Mac), browsers, or whatever.

22. *Other Connectivity.* UNIX is renowned for open connectivity which is based upon its open system philosophy. IBM's AIX offers integrated aid certified support for traditional and emerging I/O including ISDN, ATM, SSA Storage. AIX Connections provide connectivity to PC's for client/server environments. AIX also supports the Distributed Computing Environment (DCE).

23. *Portability.* UNIX runs on PCs, minicomputers and mainframes. This makes for a manageable growth path. Not everything in a high-density app has to do with the CT boards. Take billing, customer database apps running on a mainframe. If all the apps on different machines are running the same OS, it's easier.

24. *It's the Portiest.* Apps written in an Intel / SCO environment can transport to an IBM / AIX or Hewlett-Packard / HP-UX servers. Portable apps provide companies with investment protection for future changes.

25. *Scalability.* Add more memory, disk, CPUs and the telephony app doesn't have to change. UNIX systems invented the word "scalable" and with the modularity of hardware systems today you can "plug in" more horsepower, from low-end 386 footprints to large-scale multiprocessor systems. One of the features about UNIX (Sun Solaris in particular) is the ability to remotely mount entire file systems across a network, providing scalability through another system and its resources. In all, Unix has better scalability over processors and servers than other operating systems. Also, Symmetric Multi-processing (SMP) is more effective and robust on UNIX than other OSs. SCO UnixWare, for example, scales to 32 processors in an SMP configuration while NT 4.0 scores two.

26. *Centralized Control.* UNIX provides remote access tools for system management, administration and maintenance support, such as FTP, Telnet, etc. This makes centralized control a snap with UNIX.

27. *Remote Access and Multi-user Support.* From the beginning, UNIX targeted remote access environments and multiple users. Today, it's still ahead in these areas. Remote control and administration is critical for many high end, sophisticated apps, especially enhanced service and ASP systems deployed around the world. Diagnosing and fixing problems remotely saves costs and aids customer satisfaction. It's advantageous for multiple users to access and troubleshoot and/or administer a system simultaneously. There are no monitors in the central office switch. Here's where you need to access the system through a serial port.

28. *Remoteness.* UNIX natively allows you to do remote monitoring, testing and development. The idea is to run multiple apps and allow multiple users to be working simultaneously on the system, doing such things as development, monitoring and maintenance tasks. The ability to do these things with NT required a lot of workarounds. Things should be different in Windows 2000.

UNIX really shines when it allows users to perform remote maintenance on systems. It's possible to log into the system over a modem using features built into the operating system. You can debug problems remotely. UNIX lets you do this without interfering with running apps.

29. *UNIX is Manageable.* Tools simplifying management are built into SCO systems. Tools include: System monitoring, proactive system maintenance, and integration with enterprise environments such as HP OpenView and IBM NetView.

This is a big plus for any telephony app you deploy. Just the ability to telnet to any problem system to review logs and do troubleshooting is a Godsend in UNIX. You can isolate problems quickly without corrupting other running programs and memory.

30. *Standardization.* Unlike many other operating systems, UNIX content and algorithms are standardized by third party committees and organizations. This commonality adds another level of portability between different UNIX platforms not shared by proprietary and single-source operating systems.

31. *UNIX is Mature.* It's a stable OS. It has been hammered on for over two decades and thousands of man months of development focused on its improvement. NT / 2000 has come a long way in development and reliability, but there are still those who think it's not ready as a mission critical OS. The telcos have always preferred UNIX, but are patiently waiting for some kind of CompactPCI Windows 2000 combination to serve as a building block for tomorrow's telecom infrastructure.

Still, UNIX has a proven track record. All phone networks are based on UNIX - NT hasn't penetrated and IT managers in large phone companies generally don't want to hear NT. Windows 2000, however, will probably start to get some real market share in telecom.

32. *It's Robust.* Technology UNIX has been around for over 25 years and is a proven and reliable operating system. It is a requirement in most government, telephone company and higher-education bids and large business RFPs.

33. *It's Open.* When compared with most other operating systems, there's very little of the UNIX kernel which cannot be accessed by a careful and persistent programmer. This is because UNIX programmers have access to "all the buttons and levers." Extensive materials on the internals of UNIX are readily available to those interested in learning more. Indeed, UNIX is the most open system available for large scale apps. Many vendors offer operating systems of equal / compatible functionality (i.e. Solaris, HPUX, AIX, OSF, SCO UNIX , LINUX, UnixWare, Ultrix, etc.). They use open industry standards such as NFS, XDR, RPC, NIS, DNS and X11. The adventuresome can get Linux and the source code to the entire OS. Companies such as NexPath choose vendors such as SCO and VenturCom for support and on-going, continued development. This doesn't reduce controlling the system. It's vital for a turnkey, zero administration system. NT is proprietary, and "handles" many of the details for the user, which often conflicts with the goals of the manufacturer.

34. *Tunability.* Since UNIX is an open system with complete programmatic access to the internals, UNIX systems can be tuned while NT is a "black box."

35. *UNIX is Low Cost.* Embedded SCO UnixWare, with real-time and file system extensions by companies such as VenturCom, is cheaper than NT for embedded apps. The UNIX OS costs less then NT Server, with similar or better capabilities.

36. *UNIX has Predictable Costs.* For a manufacturer that provides a turnkey solution, the OS cost is part of the cost of goods and affects the gross margin. Unpredictability in the cost of goods is a nightmare in making a business profitable. As NT gained ground, Microsoft adjusted the price and tinkered with their business model. Microsoft has made many adjustments on the capabilities, cost, number of users, etc. in NT. We hope that nothing similar happens with Windows 2000. A version of UNIX such as SCO UnixWare is unlimited in users, multi-threaded, multi-processor, and not limited by license for simultaneous user accesses. This gives you a cost model you can depend on. Translation: an affordable price for your product.

37. *Innovation.* Windows NT and 2000 is based on a single company's R&D. UNIX's more interesting features have resulted from the combined creative talent and collective innovation of many programmers from companies such as Data General, HP, IBM, Sun Microsystems, Novell and SCO.

38. *On-Going Evolution.* Aside from innovations, UNIX continues to steadily evolve. Vendors push the performance envelope every day. There's new security (better encryption), parallel processing (multi-CPU and threads) and connectivity (networking and the Internet). This means a shorter lab-to-market cycle than other platforms.

39. *Straightforward APIs.* The core programming API for UNIX can be documented in a single manual. The related commands, utilities, and libraries in a couple more. Another two cover the graphical user interface. In contrast, the documentation for NT ships on two CD-ROMs.

40. *Consistent APIs.* In most UNIX variants, there is a fairly substantial common set of low-level operating system "C" language functions which can be called from within a program.

41. *Can you C Why?* UNIX works well with the C language (it should, it's written in C).

42. *Stable APIs.* The API interfaces for UNIX and the device driver architecture change little between releases. This means developers can concentrate on apps and not tracking changes made by OS vendor. Contrast this stability with tracking the Windows 3.1, 95, NT, 2000 evolution. In fact, not only do Windows APIs change rapidly, they sometimes disappear. How many developers were sucked into MAW (Microsoft at Work) and MAW FAX only have Microsoft abandon the project?

43. *Textual Simplicity.* UNIX configuration files are all text files and do not require any special programs to manipulate.

44. *Development Advantages.* UNIX wins hands-down in development. Its multi-server development uses X and/or Character based tools and R commands (rsh, rlogin, rcp, etc.). There's a large selection of Command shells and scripting tools, such as csh, sh, tsh, ksh, Perl, tcl, expect, nawk, awk, grep, sed etc. UNIX uses centralized source code control systems (SCCS, RCS etc.) with file locking mechanisms. There are shell script programming features which provide ways to develop functions.

45. *Nannites Unite.* Apps can be built as an array of small cooperating executables, each with it's own specific function, rather than as a single mammoth app.

46. *Availability of Excellent Software.* The UNIX community is chock full of software at little or no cost. The GNU C++ compiler is superior to virtually all similar commercial products. The Perl language, which is interpreted and object oriented, is used by almost all web sites as the CGI programming language of choice. Also consider the Apache Web server (used by nearly half of all web sites in the U.S.), and Hylafax facsimile server software, sendmail, SMTP, daemon, etc. The list goes on and on, with royalty-free apps by some of the greatest talent in the software business. Although many of these packages can be compiled for NT and 2000, they are almost all originally developed and work best with UNIX.

47. *Internet First Helpings.* New Internet software is usually available on UNIX first. The Internet started on UNIX computers and most new software for evolving Internet protocols appear on UNIX first. Computer telephony and the Internet are growing closer. UNIX makes it possible. UNIX systems come with a certain "core" set of utilities or apps that provide access to the Internet. Take SMTP, FTP and telnet along with their associated security "wrappers". These tools provide the elements to develop and manage Internet telephony apps such as web-based call centers and Internet faxing. UNIX and its built-in Internet interoperability allows access to all platforms. SCO Advanced File and Print Server and SCO VisionFS integrate with MS Windows and NT. TCP/IP protocols were developed in UNIX. Today, the majority of Internet servers are UNIX-based.

48. *Easy TCP/IP.* TCP/IP is easy to use and configure in UNIX. Telnet, ftp, rcp, and many other network tools are available.

49. *Proven network protocols.* UNIX rules in network protocols. For example, the NexPath Telephony Server is controlled, administered, diagnosed, and upgraded via the TCP/IP network. The entire hard disk, including all app code, can be reloaded in 15 minutes from a network CDROM via a Win95 or WinNT system. Java, JavaScript, JTAPI, and TAPI control all is done over the network. In such a network centric system, it's vital that the OS be bulletproof. NT has really been playing catch-up in the network protocol area. UNIX has had all of the protocols for years, such as ftp, rlogin, rsh, httpd, sendmail, etc.

50. *Distributed Processing.* With UNIX you can launch programs from a remote machine or cascade processes from one machine to the other. This makes for a distributed runtime environment. With their NT offering, MediaSoft Telecom had to install a separate communications server and software to do this. It's cleaner with UNIX because it's built in.

51. *Modular Expansion.* UNIX's scaleable platform allows for modular expansion. As app needs increase, system processing power and capacity can expand with the addition of new, or upgraded, server(s). UNIX will run efficiently with 8MB of RAM on a 486 or on a 32 CPU SMP machine with hundreds of MB of RAM. This means your time investment and money is protected from constant re-engineering as demands increase.

52. *Building Blocks.* UNIX comes with a rich set of building blocks. The building block paradigm permeates the UNIX philosophy. An extremely powerful scripting language and a rich vocabulary of utilities make building ad-hoc utilities quick and easy. With UNIX building a simple utility to process or convert a file can be done without resorting to a programming language or purchasing additional software. UNIX is Svelte. A "Hello world" program is one line long, much faster to develop and test.

53. *Toolbox City.* UNIX comes complete with many useful utilities, including full Internet server support and an e-mail post office / server.

54. *Atomic Philosophy.* Another prevailing philosophy in the UNIX community is one of reuse and recycle. Consequently, UNIX commands are generally very "atomic." That is, they have very specific and limited capabilities. When necessary, these "atoms" can be combined into very powerful utilities. As a result, many very useful utilities have been created over the decades by users. Most are either available at little or no charge or have been incorporated into later versions of UNIX. By restricting the scope of these "atomic" units, they get debugged quicker.

55. *Flexible Shell.* One can handle esoteric customer needs without custom code in the app itself by implementing strategic sections of the app as UNIX shell scripts. This is possible because the UNIX shell not only allows you to issue commands to the OS, but it's also a programming language itself.

56. *Built-in Daytimer.* A program can execute in the background and/or scheduled intervals without specialized coding. The scheduler and background executions are a standard part of the O/S.

57. *GUI User Interfaces.* UNIX is perceived as difficult to work with. Initially, UNIX didn't have a "user-friendly" interface (a "GUI"). The user needed to be technically proficient to make the most of UNIX. Like DOS, it started as a command line operating system and evolved to GUIs. There was no desktop, no icons, no mouse-clicking. But most UNIX implementations today have a desktop GUI facility for apps, the Xwindow interface (UNIX GUI). Most users use XWindow on a separate machine for app gens. APEX's OmniVox for UNIX is an example of an app gen with an Xwindow GUI.

The "duality" of the UNIX interface provides programmer and user efficiencies. During development of a telephony app you want to use the GUI and tools that make development easy. But once the app is finished

O

and in production (i.e. answering calls or doing voice response), such actual production systems demand efficiency rather than usability. In such cases the GUI is left off to gain more memory and CPU power for the apps, and a more efficient character-based app and menu can be run instead.

So, today, a combination of the two is popular - running the GUI on a development machine and porting over to the production machine using something like APEX's service provisioning tools.

58. *Ease of Use*. UNIX has made great progress in recent years in ease of use. IBM's AIX offers an advanced operator interface (SMIT) with graphic enhancements for ease of use.

59. *CT Board Driver Support*. The top board makers have written native UNIX drivers for their boards. They are robust and reliable. Many NT drivers are simply ports from UNIX. The Santa Cruz Operation (SCO) version of UNIX has a longer history of support for UNIX on Intel / ISA platforms than NT.

60. *You can strip-down UNIX*. UNIX has versions for dedicated or embedded apps. Unlike Windows NT which comes in essentially two or three versions (For servers and workstations), UNIX comes in a myriad of versions from systems controlling large mainframe-scale multiprocessors to real-time versions that are burnt into ROMs for embedded apps. All share the same API and permit development across UNIX systems.

61. *Embedded UNIX vs. Embedded NT*. Embedded Unix from SCO UnixWare / VenturCom is modular and configurable. You can ship your product with only the necessities. If no keyboard or monitor is required, then no graphics software need be installed. Likewise other unnecessary functions need not be installed. This saves disk space for more voicemail. And it lowers the cost for your customers. The SCO UnixWare / VenturCom OS kernel, for example, can also be configured for apps to save memory at run-time. A big, heavy, one-size-fits-all OS such as NT-Server isn't necessary for most applications, so Microsoft has announced an Embedded NT product / initiative which appears to be moving slowly.

62. *Security*. SCO OSs exceed U.S. Government Orange Book C2 security requirements. An OS based on SCO technology is available with B1 security level.

63. *Virus-Be-Gone*. UNIX is resistant to viruses. It has user authentication, security levels and memory protection.

64. *Great Support*. UNIX also has a supportability advantage. App development tools and utilities are equally effective after app deployment for remote and on-site trouble shooting. Core files can be used on-site or uploaded to debug source non-intrusively to end users. UNIX uses real time operating system traces (truss, trace etc.) and network snoop capabilities. UNIX also has performance tuning utilities (iostat, vmstat, netstat) and process CPU usage utilities to identify I/O bottlenecks, memory leaks and other runtime issues.

65. *Simple licensing*. Embedded SCO UnixWare has simpler licensing arrangements than NT. NT has a 10 user/connection limit for NT Workstation. Ditto a more complicated restriction in the NT Server license (you need certain network connections for a separate Win95 or NT license). Embedded SCO UnixWare has no restriction on the number, type or additional licensing for network connections.

66. *Originally Written by AT&T*. It's an OS developed to support telephony.

67. *Positioned for the Future*. UNIX is open and flexible in supporting emerging trends like parallel computing and network computing.

68. *It works*. Talk to anyone who has used NT for development and ask how many days it has been since their NT machine has frozen and required a reboot. Numbers greater than ten deserve applause (if you can find them). Talk to anyone who has used UNIX and ask if they can remember when their system last froze. Faximum has a server running SCO UNIX on a 386/20 from 1992 and is still runs reliably. Find someone on NT who can make that claim.

69. *40 million IT Managers Can't be Wrong*. Nine out of Ten IT managers recommend it and have logged significant UNIX time.

70. *The Whole Enchilada*. Other operating systems may each provide some of the above capabilities. But UNIX incorporates them all. The call center industry relies heavily on mission critical computer telephony apps, this is THE reason for choosing UNIX.

Unix Tips & Tricks

1. Twice the Swap and None. A good rule of thumb is to ensure you have twice the swap space as you do system memory. However, try not to allow UNIX to swap! Instead, make sure there's a much RAM as possible. For real-time development, an insufficient amount of RAM means a dramatic drop in performance. If your UNIX uses paging, make sure no paging occurs during real-time production.

2. Use Cache. For CPUs that can increase the Level 2 (L2) cache, use as much L2 cache as possible. If you are using an Intel Pentium, you can usually add 256 KB of L2 cache to the 256 KB already built-in. By doubling the amount, you improve overall performance with better processing of app loop sequences.

3. Tune 'er Up. For best performance, you should individually tune the UNIX kernel. SCO Doctor has built-in kernel analysis tools to indicate which UNIX parameters need tuning. Some examples of critical parameter areas are:

- NPROC represents the maximum number of processes UNIX can support at once. Set this number to no less that 10%, but no greater than 50% larger than the total number of processes that you need in the system. If you don't have enough, you'll get an error stating, "Can't fork process." But if NPROC is too large, it will decrease efficiency (it increases the time UNIX manages the unused task slots).
- MAXUP represents the maximum number of processes that UNIX will allow one user account to run at once. If one user account is running at a time, MAXUP should be 10% greater that the maximum number of app processes used. MAXUP should be 90% of NPREC, but not equal to NPRC. If a user account attempts to exceed MAXUP, you will want enough process slots available to log in as root (the super user), kill off the app and increase NPROC and MAXUP.
- NBUF, NHBUF represents the Kernel disk buffers critical to overall UNIX performance. SCO Doctor can tune this and SCO OSs. AIX tunes these parameters dynamically.
- NSTREAM is an important TCP/IP tuning parameter. If your system does not tune these automatically, tune the NBLK parameters as well. After these tweaks, you still need a fast I/O path to your drive. Use a PCI SCSI-3 Ultra Wide adapter with a wide hard drive. This limits the amount of time BDFLUSH flushes "dirty" blocks in the kernel disk buffer to the hard drive. This buys you increased efficiency and safety. BDFLUSH stops all apps while working, so the faster the I/O to your hard drive, the faster BDFLUSH finishes. You can also increase your I/O throughput, while increasing the reliability and availability of your hard disk, by adding RAID-5 (data stripping) to your system.

4. Forks Down. Avoid indiscriminately using "fork" or "exec" in apps. If multiple processes use "fork" or "exec" simultaneously, excessive I/O will reduce productivity.

5. Conserve Your Flushes. Don't "flush" your apps too frequently. This causes file data to write to the disk and hinders other through-put causing system delays.

6. Shell Game. Shell symbols are local to your process. If another process

needs to use a new or changed symbol, that process must be created by the initial process or by running the shell script that defines the symbol.

7. To Sum Up. When transferring files between systems, use the checksum utility, "sum" to ensure file integrity. Simply checking the file size is not a guarantee for error-free transfer.

8. Iostat. A very useful tool for determining disk I/O and network bottlenecks.

9. Lint is good. If available on your specific UNIX platform, the use of Lint and a memory access checker will significantly improve supportability.

10. Take advantage of Internet USENET newsgroups: comp.UNIX.programmer; comp.UNIX.sco.programmer; comp.UNIX.sco.misc; comp.UNIX.sco.announce.

11. Use UNIX's built-in tools for monitoring OS performance. There are tools available (Olympus Tune-Up) which do this automatically.

12. Which UNIX "flavor" should you use? Any UNIX implementation provided by a major hardware vendor (Sun, HP, IBM) is an excellent implementation. You can't go wrong picking a vendor's hardware solutions on which to build your telephony apps. Besides, each one of the "major vendors" is dying to get into the computer telephony market and would love to have your business.

A personal preference is Sun and their Solaris platform. Built on the BSD model of UNIX, Sun is now evolved to the more widely accepted System V model with their Solaris 2.x releases. We like Solaris for a number of reasons: Rich set of development tools; lots of friendly Users Group support; and major apps support Solaris (Oracle, Netscape, etc.).

Optical Network

The first communications across optical fiber in an outside metropolitan fiber application occurred in the spring of 1978 in Fort Wayne, Indiana, USA. It is certain that only optical fiber technology can supply the tremendous bandwidth future telecom applications will require, and so the electronics at the endpoints will have to be changed to become optical too.

The major developments in endpoint electronics have been in the creation of the older North American Synchronous Optical NETwork (SONET) and the European Synchronous Digital Hierarchy (SDH) and the newer Wavelength Division Multiplexing (WDM) and Dense Wavelength Division Multiplexing (DWDM) technologies. See Dense Wavelength Division Multiplexing, Solutions.

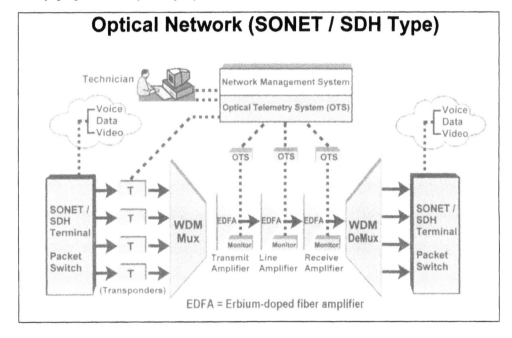

Optical Network (SONET / SDH Type)

PABX

Private Automatic Branch Exchange. A European name for a Private Branch Exchange (PBX). See Private Branch Exchange.

Packetized Telecom

The public switched telephone network, or PSTN, is a circuit-switched network. This means that a telephone call reserves a 64 Kbps physical circuit between the endpoints (caller and callee, or, for data, origin and destination). The full bandwidth is dedicated to the circuit even during periods of the call when no voice or data is transmitted. Obviously, this is a very inefficient way to make a phone call.

Packet switched networks, such as the Internet and corporate LANs divide the message into many small packets. Each packet has a header containing information that enables the switching equipment along the way to route the packet to its final destination (at the destination the header can be used to put the packet in the proper sequence with all of the other packets). Any telecom "pipe" can run flat out at maximum bandwidth, as a multitude of packets from various messages completely (and therefore efficiently) fill the pipe.

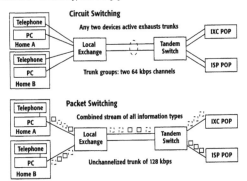

The advantage of packet-switched networks, then, is their highly adept and efficient utilization of network resources. The disadvantage of packet-switched networks, in real-time communications such as voice and visual telephony, has been their variable Quality of Service (QoS), a consequence of network congestion. Telephone network designers have found that callers can tolerate up to about 250 milliseconds (ms) delay before the interval becomes bothersome. Outside of the U.S. (particularly South America) callers appear to tolerate up to about 500ms to 750ms since, until recently, even their circuit switched phone systems haven't been particularly stable.

Besides the variable delays resulting from network congestion, there are also unavoidable fixed delays. For example, the digital signal processing necessary to digitize and compress analog voice takes up to 45ms, depending on the compression algorithm. Decompression at the receiving end takes about 10ms or less, regardless of the algorithm. When one router hands off a packet to another, 10ms is lost at each endpoint. Transmission delays are generally dependent on link speeds, the higher the speed, the less the delay. A "jitter buffer" or "smoothing buffer" adds additional delay but this is invisible since it reduces the network's variable delay by arranging the packets in a stable sequence (and regenerating missing packets if necessary) thus ensuring circuit switched quality voice at the destination.

Early low-bandwidth packet networks were based on X.25 or pro-prietary statistical multiplexing protocols that ran over modem or low-speed digital circuits, and were not capable of real-time communications.

X.25's successor in packet telephony, frame relay, became popular in the early 1990s for data communications, then voice. Frame relay's higher bandwidth was a result of the new digital circuitry then becoming available.

As the Internet's popularity exploded in the mid 1990s, any network that could carry the protocol saw increased usage. Fortunately, the Internet Protocol can travel, if need be, over all sorts of networks: Frame relay, the PSTN (ATM), etc. Pure IP networks, and IP transmissions themselves, of course, are packet-based. Since IP was originally mean to be a LAN protocol, bandwidth efficiency was not a prime consideration with its design – IP packet headers, for example, are often in the range of 24 to 48 bytes each (sometimes more), as compared to just four bytes for frame relay packet headers.

Another form of packetized telecom is the Asynchronous Transfer Mode (ATM) which divides messages into equal sized packets called "cells." ATM's use was once confined to the carrier backone, but because of its built-in Quality of Service (QoS), it is being used increasingly as a toll-quality "last mile" connection from a home or business to an optical IP backbone infrastructure. (See Asynchronous Transfer Mode.)

In most cases, organizations will want the cost savings of packet networks while conserving the known functionality and existing investments in their legacy PBXs. To do this, gateway routers need to support common channel signaling methods such as ISDN Primary Rate signaling and Q.SIG, a PBX-to-PBX peer to peer signaling standard that allows interoperability between value-added PBX functions such as conference calling and call-forwarding.

To accommodate standard voice signaling, Voice signaling is used to set up a call and monitor its end-to-end status. Voice signaling requirements vary depending on the type of interface and network locations that need to be reached. For example, Foreign Exchange Office (FXO) and Foreign Exchange Station (FXS) ports are relatively easy to support, because they don't change much (if at all) from country to country or PBX to PBX. The off-hook type is used in applications that require an audio path but have no requirements for signaling. E&M (Ear & Mouth) on the other hand, is more difficult to support because there are five major (and several minor) versions of E&M signaling protocols in use. Transparent Voice Signaling (TVS) for E&M ports is a preferable way to pass the signaling states transparently from near-end user port to far-end user port.

Here's an example of how TVS works in Motorola's Vanguard product. For each signaling transition detected at the voice interface, the Digital Signal Processing Module (DSPM) will forward a signaling packet to the far-end DSPM, informing it of the change in signaling state. This signaling packet includes a sequence number and timing information to allow the far-end DSPM to reconstruct the signaling information correctly. Because TVS converts the signal into a common digital format, it has the ability to translate signaling states between the different types of interfaces to allow for different interface types at each end: digital T1, E1 and PRI, or analog E&M. With TVS, it's possible to have a digital T1 (or E1) at one end and an analog E&M at the other end or a T1 at one end and E1 at the other.

Adding routers and packetized telephony to an existing phone system results in some interesting "converged" (or perhaps "commverged") business communications.

P

One sterling example we came across was built by Warren Prince of Prince Law Offices, P.C. (Bechtelsville, PA, – 610-323-4432, www.princelaw.com) who is interesting since he's both a lawyer and a super-techie. Self taught in telephony and programming since 1981, at one point he wrote some FoxPro apps that he demoed at a CT Expo at the request of Phonetastic (remember them?). He was even doing programming involving TSAPI back in the mid 1990s.

Prince did such a good job building a computer telephony system for his own law office that he and his colleagues now act as a VAR to other firms. His new company is called UnConundrum (Bechtelsville, PA - 610-845-0047, www.unconundrum.com). It does wide area networking, custom software development, Web enabled applications (they did part of the Federal Express Web page) and computer telephony integration.

Prince is a master at building distributed communications systems that connect attorneys in remote offices together. His own Prince Law Offices consist of headquarters in Bechtelsville, PA, and offices in Bethlehem, Camp Hill, Lebanon, North Wales, Pottstown, Reading, Scranton, and Stroudsburg.

"We're kind of unique," says Prince, speaking from his Bechtelsville headquarters. "We have many remote offices. At this and another location we're using a Comdial FXS; you know, the one that rolls around on casters. There's also an old Comdial DXP PBX at one of the remote offices. We also have WideOpen Office running here. About five of the support staff use the local FXS system."

The voice network consists of both frame relay and IP connections. The gateways consist of Motorola 6520s at five locations, and a 6560 at a sixth location. Voice over Frame Relay (VoFR) connects four of the sites together within the LATA, while two other sites are connected by jumping across the LATA over the Internet using Voice over IP (VoIP). (The LATA, or Local Access and Transport Area, is a geographical area in which the local telephone company is allowed to carry not only local calls, but long distance toll calls too. Jumping across the LATA with VoIP saves on toll calls). The connections between the routers are dedicated.

"We got all of this equipment to talk to each other," says Prince. "If you had called in here and I wasn't around, but you had known the correct extension number, then you could have reached me in our Pottstown office through our switch and the frame relay line. Now, if I were in the Lebanon or Camp Hill offices, that extension number would have routed your call instead over VoIP. The whole process remains invisible to the caller."

"No special connecting software had to be written," continues Prince. "You have to program the Comdial switch, of course, since it has a least cost routing table. There's an LCR table in the routers, too. There are two levels of routing. If I'm in a location that has a switch such as the DXP or the FXS, the switch may send the call to a router, and the LCR table there figures out which router is the next hop to send the call to."

The voice mail system used in conjunction with the headquartered Comdial system is from Key Voice Technologies (Sarasota, FL - 941-922-3800, www.keyvoice.com). No matter what office an attorney calls in from, they can access their voice mail with a toll free call. Not via an 800 number, but over the private VoIP or VoFR networks. Key Voice uses a four port analog Dialogic board, since it must connect to the IRST ports on the Comdial switch.

Some of the switching scenarios possible with this system are quite sophisticated.

Let's assume that "Mary Kay" works in the Lebanon office. If she wants to check whether she has voice mail waiting for her in the Bechtelsville office, she first picks up her phone, which is physically connected to a port in the Motorola 6520 router (that would be an FXS port in the router – not to be confused with the Comdial FXS switch!) She dials the extension for voice mail (it happens to be 2001) and those digits are passed to the router. The router looks them up in the least cost routing table, which indicates that she's making a call to Bechtelsville, and it signals that an IP connection must be used to get over to the next hop toward Bechtelsville, which is the router in the Reading office.

In Reading, the router there also sees that the packets are heading for Bechtelsville. Its LCR table indicates that the call must be routed over this last hop via frame relay, not IP. The call finally reaches the Bechtelsville router, which connects through a standard "Ear and Mouth" (E&M) port to the Bechtelsville Comdial FXS. The Comdial FXS knows that extension 2001 is a voice mail port, so it sends the call to the Key Voice voice mail system out of an IRST port. The voice mail system then picks up, Mary presses the # key and her extension, 1028. This sends the voice mail to her mail box, asks her for her password, and then plays her voice mail to her back down the chain.

"We can also use Key Voice's VCM software from the remote locations," says Prince. "That's GUI-based PC software, so that graphically you can pick up a list of voice mails and retrieve them as a series of .WAV files downloaded to your computer and listen to them through your audio card and speakers. Sometimes that seems a little slower. But it makes it easier to check voice mail if you have to RAS in from remote locations. I always use VCM here at my desk."

From a voice standpoint, the system can go one step further. "We can also seize lines at our remote locations and get dialtone from there," says Prince. "So if I'm sitting here in Bechtelsville and I need to make a call to Harrisburg, I dial a number. The table in the switch is consulted, it will send the call out through the E&M port to a router's FXS port, then via frame relay to Reading where it'll be converted to voice over IP, jump over the LATA, and get picked up in Lebanon, then routed to Camp Hill, where the router there also has an FXS port which is connected to a POTS line. It'll seize that line and dial a local number in Harrisburg without incurring a long distance toll charge."

Prince prefers PBXs and routers to fully PC-based telephony systems.

"Comdial makes the FXS look like a PC server, but it's really two machines in one," says Prince. "They're not fully integrated with one another. The PC-based Altigens are neat products, but they haven't priced themselves to be really competitive. They've priced themselves the same as many PBXs. Many people prefer to stick with the PBX that you know works and that you've been comfortable with – you don't have to learn new technology."

Prince thinks that "Motorola routers are great. I can plug a phone into one and I can dial a number just as if I'm connected to the PSTN. It will strip numbers, pad numbers, it'll do anything I want to do to that number."

"We use technology to move forward, as opposed to just throwing people all over the place," says Prince. "No one has their own particular secretary, and we do a lot through technology. For example, every piece of postal mail that comes in here gets scanned into TIF files that we store on our servers. Then we e-mail pointers to those TIFs to the attorneys. We have a firewall and a Virtual Private Network (VPN), so the attorneys can log in from their offices or homes and read all their mail as e-mail. Between the offices e-mail travels over frame relay or IP, or if the attorneys are accessing it from home they're using a plain old dial-up connection to an ISP."

So, it appears that, where the goal is an advanced packet-enabled PBX-class network, implementors are best off using the economical Voice over Frame Relay (VoFR) wherever possible, only going onto IP at sites not connected to the frame relay network. However, where the goal is to introduce new applications, implementors can move more aggressively into Voice over IP (VoIP) as long as they design their intranets with enough bandwidth to guarantee good voice quality.

PBX

See Private Branch Exchange.

PCM

See Pulse Code Modulation.

PC-PBX

This term can be applied to a whole spectrum of telecom systems:

- *PBXs Enhanced with CT Features*. PBX manufacturers make a great deal of money selling their equipment and, rather than stop making them, they would rather add new and exciting (and perhaps a bit more "open") features to their boxes to pacify their customer base. IP capability, for example, has been added to PBX architectures by fitting an existing PBX with an adjunct box that channels traffic to gateways though the intranet / Internet. Prior to its conversion to an enhanced services company, Lucent was the premier PBX maker who took IP seriously, transforming their equipment by adding built-in IP telephony gateways.
- *PBX-Embedded PCs*. Rather than setting up a full PC next to a PBX to enhance it, why not put a PC on a card and plug it into PBX to provide something such as inexpensive voice mail? That's exactly what Tadiran did.

Tadiran (Clearwater, FL – 813-523-0000, www.tadiran.com) has a product they call the CTI Bridge which adds next-gen applications to an their existing PBX. The CTI Bridge sits between your PBX and IVR, voicemail, ACD, and other applications. Calls are tracked and reported to an API. The API creates a new call/reroute command and redirects calls to the correct network, whether PSTN, FR, ATM, or IP. Trunk/line capacity is 600 ports. They use an H.323 gateway. Custom quote, but generally cost is $300 to $700 per seat.

Tadiran has also been experimenting with building PCs into their Coral phone systems, as hot-swappable add-in boards. Their $5,000, 24-port card – developed by ABS Talx – incorporates a 486 or Pentium PC running Windows 98 or NT, a full complement of hard drive and RAM storage, and draws power from the Tadiran Coral backplane - and it's hot-swappable. Direct access to TDM signal-streams on the backplane eliminates need for voice cards: instead, the card contains firmware that emulates the Dialogic Springboard API. Third-party voicemail or unified messaging software resides on the card, and acts like it's living in a conventional PC with Dialogic boards.

So, for $5,000, a 24 port card can provide voice mail for many users. It's far more economical than buying a separate PC. Bicom also puts hard drives on a card, and NEC is developing in this area too. What's interesting here is that PBX makers are including a standard PC and emulative standard voice hardware (usually emulating Dialogic boards) in their system.

Performance Measurement for Contact Centers

Measuring the performance of contact (call) centers would at first appear to be a straightforward affair, but there are many ambiguities involved. Do you measure individual activity, group activity for a particular function, or (in the case of outbound telemarketing centers) overall campaign activity? Also, different things are measured in different kinds of call centers. General Motors will be measuring different things than Continental Airlines.

Why bother measuring call center performance? Obviously, you want to see how well you're doing, to see whether you're meeting whatever the latest goals for a call center are. There exist many types of measures as to how well a center is functioning. Your company will have control over some measurements, and the Customer Service Representatives (CSRs) or agents have control over others.

Once upon a time, the call center chief measurement was that of individual call activity. This type of measurement has largely fallen by the wayside, since a CSR may encounter 30 or more types of calls, which means that there are 30 or more "ideal" call lengths. In a modern, flexible, blended center, calls can be help desk inquiries, customer service calls, course registration, etc. On the outbound side, one could be called on to serve more than one campaign, or within a particular one could be doing collections, telemarketing, teleresearch. If a center does both inbound and outbound, there should be separate sets of stats for each as well as individual stats for any blended groups. Then there are the relevant stats from front-end IVR system reports per application too!

This means that in order to get a reliable statistic, you've got to at least calculate a weighted average for the various mix of calls a day, the calls a particular agent has in a particular hour.

An agent might handle five ten minute calls in an hour, but some types of calls may take twenty minutes. Also, an agent handling five four-minute calls in an hour doesn't necessarily mean that the agent is lazy, since the actual handle time for the call is composed of two parts: The talk time plus the work time done after the call. Many call center agents spend extraordinary amounts of time doing such after-call work, but Automatic Call Distributor (ACD) reports often don't differentiate these two statistics very cleanly. The most inefficient type of after-call work is for an agent to get up and walk away to send a fax.

To determine overall agent working time, one should look at the total time an agent is logged onto the system, minus lunch, trips to the rest room, etc. which is known as schedule adherence.

Perhaps the most difficult types of measurements relate to CSR / agent quality. Nearly all call center software vendors are developing quality evaluation systems. Most of these involve sampling (recording) calls and playing them back later for scrutiny, or having a supervisor put on a pair of headsets and do live, side-by-side monitoring.

The *service level* is the percentage of the calls that get answered within, let's say, 30 seconds or 60 seconds, or whatever the target happens to be. Someone may brag that they answered 80% of their calls within 20 seconds or whatever. The Average Speed of Answer (ASA) is an ACD statistic that indicates how long the average caller waits on hold before his or her call gets to a live agent. Essentially, the ASA is 50% of the calls within some number of seconds. ASA has become the most important measure of service quality in some call centers.

The Abandon rate and Average Delay to Abandon (ADA) are closely related call center statistics. An abandoned call is a call that's answered by your ACD, put in queue but never reaches an agent. Your abandon rate is really the percentage of your callers who hang up before an agent ever gets a chance to talk to them. We must distinguish here between an abandoned call and a blocked call. A blocked call can't even get to the call center because there aren't enough trunks leading into the center. It's like having a "bandwidth problem" in the packet-switched world. The abandonment rate, however, is obviously how many callers hear the fateful words "All our agents are busy – please hold," wait in the queue a while, then hang up.

If too many callers are hanging up, then something is irritating them. Perhaps they've been waiting too long in the queue. Here's where the ADA statistic can tell you, on average, how long callers are waiting before they hang up.

P

Of course, "waiting too long" is a relative idea. Teenagers waiting on hold for tickets to see their favorite rock star will wait forever. Your Truly waits for about 15 seconds (under any circumstances) before pushing the button for an operator or somebody's extension, a trick I learned from Harry Newton (See Newton, Harry). So, the abandon rate indicates how many callers hung up while the ADA tells you how long they waited before they hung up.

Another significant measure related to ADA is the length of time that the oldest call has been in queue. If it's outrageously long, then there's something in the system that's causing calls to back up in the queue. Let's suppose that all calls have the same priority, and think of the queue as a toy train set where you're connecting box cars onto the train at the back end and giving the whole train a push, while someone at the front of the train is grabbing each box car as it arrives and throwing it into a box. The length of box cars connected at any one time is the queue. Now if you have perfectly balanced your call center, the calls are going to be arriving at the same rate they are being concluded. But if somebody walks away from their desk to send a fax or whatever, for a half hour or even for a few minutes, then the queue is going to start getting longer and longer. Although the agent ultimately returns to their station and the center is balanced again, the increased length of the queue persists unless and until some extra effort is made to reduce the queue (more agents, faster techniques to resolve a problem, or stop going to the rest room!).

All of this means that one tardy lapse on the part of a single agent can distort all throughput and quality measures of the contact center, and add expensive minutes to calls made to an 800 number. If many customers make calls to the center during their lunch hour, and you've scheduled all of your agents to take their lunches at a different time so as to handle the traffic, but they all have growling stomachs and leave for lunch at Noon anyway, then they've botched up your center's performance. Incidentally, the measure of an agents' ability to discipline themselves to have a break and go to lunch when they're supposed to according to the schedule that was put together for breaks and lunches is called *schedule compliance*.

The next measure to consider is *time available* which is the period of time that an agent is ready to take a call, but there aren't any calls to take. Too many idle agents could mean that the center is overstaffed, or that your company has funded the world's worst advertising campaign.

You can also do some measuring of center performance as a whole by using a good workforce management system. (See Workforce Management)

Customer satisfaction surveys can also be useful. They can be done via the mail or an IVR system after the call. You'll want someone who is independent of your organization to formulate the survey questions, so as not to bias the survey by concocting the questions yourself.

Surveys bring to mind sampling and theories of sampling and sampling error. Do you sample only the people who have problems? How about sampling everyone who called the center? Or perhaps just a randomly selected percentage of the calls? There are different techniques for sampling, and obviously it depends on what you are trying to measure.

Finally, the most important part of the measurement process is to communicate the results and meaning of the measurement to the entity that's controlling it, either to the agent or to the company or whatever, so that they clearly understand it.

Predictive Dialer

Computer hardware or software residing on proprietary or open telephony devices or servers that dials calls to numbers loaded from a database, and hooks the agent to the called party only when someone picks up the receiver. Dialers "throw" calls to agents at pre-set speeds, sets call length, agent breathing space between calls and how much finish-up time allocated to the agents. Dialers can also let agents leave messages on answering machines; those that don't have *homo sapiens* live on the calling end is strictly regulated by the federal government and is banned in many states.

With this level of automation, predictive dialers greatly speed up call delivery, completion and performance, hence agent productivity, compared with other dialing forms, such as manual and *automated or preview dialing. Automated dialing* pops the number up in front of the agent before the call is made and then the agent makes the calls, usually by hitting or not hitting a key.

In contrast, predictive dialers don't give the agent the choice; they simply give the caller to the agent. No thinking or fiddling about. Here's the called party, here's the name, file and script and the agent is live on the air. Go for it. Go sell'em cowboy! Squeeze that 80-year old widow for all she's worth!

Think of a treadmill whose speed is not set by you but by your trainer who only wants you on the thing for X amount of virtual miles. That gives you the dialer philosophy. The quicker the agents meet the objectives – regardless if they risk having coronaries or walking off the dang thing and yelling at their boss to put it where the sun don't shine (and we're not talking about summer in Seattle) - the more money the dialer owner makes by more sales per hour or by more completed calls per hour.

Thanks to such technological whip-cracking, dialer users can realize productivity gains of at least 100%. That means businesses can make more calls with the same number of agents, or the same amount of calls with fewer agents. Either way it's win-win, especially in a boom economy where it is harder to find someone who wants to sit all day cord-chained in a cube to make outbound calls for the usual minimum-wage-scraping-pay than it is to locate a New York City yellow cab in a rainstorm.

The dialers' cold-blooded hearts are complex mathematical algorithms that check which telephone lines are free, which agents are free and the probability of not putting the call through. They detect, by sending signals, busy signals, operator intercepts or answering devices (e.g. Centrex-based call answering or end-user answering machines). Dialers take into account all these variables. Some systems automatically adjust the dialing rate by monitoring changes in them.

The kinds of enterprises that benefit the most from predictive dialers are those who make a lot of continuous outbound calls. These include telemarketers (business-to-consumer and business-to-business sales, lead generation, appointment scheduling), charities and other non-profits, market research, public relations and collections. In-house and outsourced service bureaus alike use these devices, bureaus especially so because their bottom line depends on how efficient they are.

Like anything else predictive dialers aren't perfect. The dialer balances the benefit of keeping agents busy against the risk of reaching live prospects, only to discover there's no available agent. When the pacing works right, the dialer reaches the called party and switches the call to the agent before the called person even finishes saying hello; some dialers will wait for the "Hello" or acceptable substitute before connecting the call.

However when it doesn't, which does happen (no dialer is 100% all the time), the called party picks up the receiver, hears a click and swears, and slams down the receiver. If the call came at dinnertime or any other inconvenient moment, they may make their own outbound calls: to their legislators demanding they put these "damned telemarketers" out of business.

Which is starting to happen. The federal government requires firms to maintain Do Not Call lists; a growing number of states have DNC lists that telemarketers must scrub theirs against and calling hours restrictions more severe than called for in federal law and telemarketer registration. There has been federal legislation proposed to cut back the calling period to 5pm from 9pm. Answering devices, like those that automatically tells the agent to put the called party on the DNC lists, have grown in popularity. Some telemarketing consultants are advising the industry to move out of outbound calling to drive inbound calling by direct mail, direct response and e-mail.

The best managers who still need to make outbound calls can do is to pick an abandonment rate high enough to be viable and low enough not to annoy customers and prospects. To help telesales businesses keep in business the Direct Marketing Association (New York, NY – 212-790-1500, www.the-dma.org) recommends a rate of not more than five percent. Dialer company EIS (Herndon, VA 800-274-5676/703-326-6400), www.eisi.com recommends three percent. The American Teleservices Association (ATA, North Hollywood, CA – 800-441-3335 / 818-766-5324) advises that users keep the abandons at the lowest rate: close to zero.

Recommended abandon rates are part of the ATA and DMA dialer ethical use guidelines prepared and distributed to their members. Other worth while sections include:

- Let the dialer ring for four times or at least 12 seconds before disconnecting (to give the called party a chance to answer the phone) and release the line within two seconds if the agent is not available;
- Do not abandon the same number more than twice during the same month of a campaign and ensure that any future calls to that number are connected to an agent;
- Do not knowingly call anyone with an unpublished number and screen numbers against DNC and the DMA's Telephone Preference Service lists. (This is good advice if a business wishes to retain their staff-the resulting verbal abuse could show the lie behind the childhood rhyme 'sticks and stones may break my bones but words will never hurt me'.)

The dialers must handle DNC request handling. They must also permit users to set and stay within federal and state calling hours, adjusted by time zones. What matters is what time it is at the number the dialer is ringing up, not at the originating business or call center.

You can also set your dialer too fast. That not only burns out the agent but it ticks off the called party. That accounts for the 20-somethings who speak like chipmunks on amphetamines about some product or trade show straight from scripts, without giving the called party a chance to ask an intelligent question or find out more about what the caller is pushing. When that happens there is usually either a "don't interrupt me you jerk because I've got to meet my Hello Kitty hood ornaments sales quota" sigh from a telemarketer or an "oh @$#%^ you've got me in trouble because I'm spending longer on the call that I'm supposed to" sigh from a researcher or flack.

When complaints come in about telemarketers, market research firms and P.R. agencies it may not be the dialer per se to blame: it can be the call center's management. They're responsible for the badly targeted lists, poor scripting, hiring and keeping stupid, pushy and obnoxious agents and reps and supervisors, and their flouting of laws and common sense that are endangering the telemarketing industry and giving market research and p.r. bad names.

Yet despite their outbound heritage, dialers can also help business better handle their inbound calls and agent productivity through what's known as call blending. Outbound centers have traditionally used their dialers as inbound switches for those relatively few incoming calls. The dialer can feed outbound calls to inbound agents when their volumes are light: inbound call volumes are notoriously hard to predict. This helps eliminate the need for extra agents and may double a call center's productivity. To avoid wear-and-tear on your agents and possible schizophrenia, blending is done in time blocks, not on a call-by-call basis.

This is being made possible by changes in call center culture. The growing consumer resistance to outbound calls is occurring at the same time more people want to buy and get customer service without going into a place of business. Why wait to schlep to the mall when you can order that PC at 2am in your pajamas? Over 90% of the so-called "mail order" catalog business is actually done by phone, and increasingly over the Internet.

Also, in the old days outbound was outbound, and inbound was inbound. The two calling types, and the personalities best suited for them, never mixed. Outbound calls were made by tough, aggressive thick-skinned call-first and ask-questions-later types, their dialers firing off numbers like machine guns, the scripts read off like an M-16 on full auto. Inbound calls were answered by gentle, sweet, how can I help you and let me note that type, who calmed down the annoyed customer.

Now, thanks to the customer relationship management revolution (CRM), where every sales call is a customer contact, and every inbound customer contact is potentially a sales call, the two personality and call types are merging, with the exceptions of pure outbound sales, collections, surveys and p.r. That outbound soldier is now trained to give directions; that inbound nurse is asked to pitch when the caller is the most vulnerable.

Some dialers offer same-agent and scheduled callbacks. This is a handy CRM-building feature that allows an inbound calling customer to be called back by the same agent who took her call by tagging outgoing numbers to particular agents. A business may also want agents to "own" whole lists or subsets.

Many new dialers can be integrated with a center's preexisting telesales scripts to save agent training time. They also offer ways for businesses to design new scripts for new campaigns, with complex branching logic and outbound screen pop for live prospects.

Predictive dialers can also offer other useful features. One of them is Multi-Number Calling. These days a growing number of people have more than one number: e.g. home, work and home-at-work phones, cellphone(s) and pagers. If a business is trying to try to reach this individual, such as for collections, this lets the agent reach that individual, letting them know that "they can run but they can't hide".

Another feature is dynamic list management, which allows numbers to be added to a dial list while the list is being used, as in call-me numbers input through Web sites. Business may also be able to make real-time changes to scripts or list filters. Still another is job- or campaign-chaining. As soon as one campaign of calls is done, it assigns an agent to another one.

A dialer should allow you to switch to automated or preview dialing. These should be used for preferred called parties, or to those the predictive function called and hung up on, so that the agent is most definitely there to take the call. The dialer should permit progressive dialing, which dials one number per available agent. This limits nuisance calls, handy in campaigns where almost all calls can be expected to be picked up, such as in business-to-business.

Another good feature to have is agent-override. This permits an agent to take longer than programmed to finish a call, which can result in a sale or promise to pay that may not have occurred if the call is cut off. This feature could let agents answer a question or solve a problem the called party may have. This builds relationships between the company and the customers.

As with other technologies predictive dialers are becoming CT-enabled. Many PBXes and comm servers have predictive dialing. Stand-alone dialers are being succeeded by systems that integrate with existing switches, or incorporate their own NT-based switching platforms, but operate in a network that accepts Windows-based list management and scripting tools. Many dialers are now software rather than hardware.

Lucent's (Murray Hill, NJ 800-247-7000 www.lucent.com) ownership of dialer maker Mosiax lets the monster PBX/technology firm sell the acquired firm's dialing software. Interactive Intelligence's (Indianapolis, IN 317-872-3000 www.inter-intelli.com) groundbreaking EIC comm server incorporates predictive dialing and inbound on the NT-based switch. eShare's (formerly Melita) (Norcross, GA – 800-635-4821 / 770-239-4561, www.eshare.com) Xchange commserver has long been used as a dialer by many small-medium-sized service bureaus. Cellit (Miami, FL – 800-256-6420 / 305-436-2300, www.cellit.com) also offers dialing on its comm servers. Buffalo International's (Valhalla, NY – 914-747-8500, www.buffalo-intl.com) Object Telephony Server dialer sits on an open basic platform also for call routing and blending.

Predictive dialing is also a part of some CT and call/business management packages. Genesys Telecom Labs' (San Francisco, CA – 888-GENESYS / 415-437-1100, www.genesyslabs.com) Campaign Manager works with the firm's T-Server CT middleware. IMA (Shelton, CT – 800-776-0462 / 203-925-6800, www.imaedge.com) SoftDial Plug-in plugs into the firm's Edge software.

Some IVR vendors such as InterVoice-Brite (Dallas, TX – 972-454-8000, www.intervoice.com) have now integrated dialing into their platforms. Firms that make standalone dialers are improving their products. Noble Systems (Atlanta, GA – 888-866-2538 / 404-851-1331, www.noblesys.com) features scripting packages that permits owners to customize them for each name on the calling list, with information such as purchase histories. This enables better marketing: and less resistance when the called party picks up.

Some dialers, such as Eshare/Melita's Enterprise Explorer and Digisoft Computers' (New York, NY – 212-687-1810, www.digisoft.com) Telescript have mixed media servers. This is also an increasingly useful feature, with the rapid growth of e-mail as a communications means. Agents can reach customers/prospects by e-mail and fax as well as by phone: however the customer/prospect has told your firm they prefer.

Other dialer hardware and software makers include American Computer and Telephone (AMCAT) (Oklahoma City, OK – 800-364-5518 / 405-842-7744, www.amcat.com), Castel (Malden, MA – 781-324-0140, www.castelhq.com), Davox (Westford, MA – 800-480-2299 / 978-952-0200, www.davox.com) Electronic Technical Services (Davenport, IA – 800-747-0093 / 319-322-3562, www.ets-edge.com), Information Access Technology (Salt Lake City, UT – 800-574-8801 / 801-265-8800, www.iat-cti.com), Professional Teledata (Manchester, NH – 800-344-9944, www.proteledata.com), Stratasoft (Houston, TX – 800-390-1157 / 713-795-2670, www.stratasoft.com) SunDial Technlologies (Fort Lauderdale, FL – 800-634-2537 / 954-463-6499), www.sundialer.com and Voxco (formerly Info Zero Un), (Montreal, QC, Canada – 514-861-9255 www.voxco.com).

As with nearly every technology type, predictive dialers change and develop. Perhaps we'll see features such as those that automatically deduct wages or deliver a mild electric shock (like those "humane" jolts delivered to dogs that cross electronic fences) if the agents don't speak to a delivered call in X amount of time. Vendors mutate and morph rapidly, adding new features faster than the animal and plant life that ooze and squirm next to the New Jersey Turnpike. Dialer makers and their parents gobble up each other faster than an old Pac Man game, just like their customers.

To stay on top of who is offering what read *Call Center Magazine* (www.callcentermagazine.com), *Teleconnect* (www.teleconnect.com) and *Computer Telephony* (www.computertelephony.com) to get the details on how these things work and to sell them to end users. Also attend the various trade shows e.g. Computer Telephony Spring and Fall, Call Center Demo and Conference and CRM Demo and Conference to see these products and ask questions.

Here is a loadable time-proven list of 20 buying tips that can help your businesses, and the outbound practice to stay legal. First published in 1996 they still hold true today.

1. Call abandonment may close you down. Many lower priced predictive dialers hang up on 20% to 40% of prospects, turning your call center into a public nuisance.

People with Caller ID will get the phone number of your call center and complain. Phone companies can track the source of abandoned calls. List owners who receive complaints may stop renting you their lists. Moral: If possible, buy a system capable of a zero abandonment rate, a necessary precaution in case future legislation demands this capability.

2. Slow call transfers can cost you sales. The prospect shouldn't have to say "Hello" three or four times before the agent hears it. Your system should transfer the call so quickly that the agent hears part of the first "Hello."

3. Lower conversion means less overall sales. Having an unlimited supply of leads, with very little wait time between calls, causes the agent not to try as hard as they do when they have limited leads.

Companies using predictive dialing therefore often have a lower sale conversion rate (for example 10%) than with manual dialing (11%) or automated "power" dialing (14%). You need a contact control management capability that finds the optimal contacts per hour and control the contacts per hour to produce the most sales at the lowest cost.

4. Make sure you have the unlimited right to sell your hardware or software to a third party. Some predictive dialing manufacturers have a clause in their purchase contract that specifies that you cannot transfer the software license to another party without the manufacturer's permission.

One company who wanted to sell their used system found that the manufacturer wanted $3,000 per seat to transfer the software license, making the sale impractical.

5. Don't be forced to install extra telephone lines. Predictive dialers simultaneously dial multiple phone lines to improve connect rates. One of the most expensive systems requires three phone lines for every agent on the phone, most require two phone lines per agent, and only a handful can run okay at 1.5 lines per agent.

If a 16-station system needs three phone lines per agent, that's 48 lines or two costly T-1 spans. Even at two lines per agent, you need 32 phone lines or almost one and a half T-1 spans. Since T-1 spans come in multiples of 24 lines, you would actually have to install two T-1s. However, if the system only requires 1.5 lines per agent, you would only need 24 phone lines or one T-1 span.

6. Beware of hidden equipment costs. Besides a T-1 span or two, some predictive dialing systems, especially older secondhand models still use analog technology, requiring a channel bank to be connected to the phone lines.

Renting one from the phone company may not be cheap. On the other hand, a predictive dialer with a digital network interface can be connected directly to phone lines without a channel bank. If your system requires two or three lines per agent along with a channel bank, that's an extra cost that you may do without.

7. Don't become a captive customer. Some manufacturers boast that their system is better because it uses all proprietary technology and parts. While that may sound like an advantage during the buying process, it may quickly become a nightmare during the calling process if your manufacturer goes out of business.

Also you'll lose telemarketing revenue if you have to wait for a special repair person to travel to your location. The manufacturer may even elect to double or triple your maintenance fee after a few years or overcharge for new feature upgrades. Perhaps there never will be an upgrade! You're at their mercy. Solution: Use PCs, open APIs and cards by Dialogic, NMS, etc.

8. Does the system allow the agent to enter changed phone numbers and instantaneously redial? Alternatively, can the system detect a disconnected phone number and put it into a queue and not pass it to the agent? This seems like a minor function, yet people with changed phone numbers typically produce higher sales conversion rates, since they are being called less frequently.

9. You need call blending. The system you finally select should have the ability for every agent to handle both outbound and inbound calls from the same terminal.

This leads to higher list penetration and higher sales. The outbound telemarketer can leave a return phone number with a message explaining what the call is about. The person who returns a call is more likely to end in a successful call result. Likewise, if you need inbound call handling, your agent can automatically be passed outbound calls during slow inbound call periods.

10. Use digitized recordings. Buy a system that can simultaneously record all agents, for an entire day, and can quickly rewind, fast forward and retrieve calls for playback.

It's now possible to digitally record on a 1" x 2" computer cartridge everything that all your agents say on the phone, during an entire day, along with all the prospects' responses. You can now hear how a specific agent is opening or closing calls, what complaints are called in, and you can let trainees listen to your experienced agents or let your experienced agents critique trainees.

11. Get call tracking and custom list loading. Your system should keep a record of call attempts to be certain that each attempt is at a different time during the day and on different days. Some weaker systems have no call tracking capabilities at all, and every time you load the dialers the same people are always called first.

Some systems are so poor that on large lists you may have tried part of the list seven or eight times, while leaving as much as 20% or 30% of the list with no call attempts. To get higher list penetration during the day, the system must be able to do schedule callbacks and have list analysis capabilities, so that home rates can be determined for different zip codes or different days of the week and at different time intervals.

12. Make sure transition times are short. Predictive dialing should reduce the amount of time between when your agent hangs up the phone and when they are back on the phone talking to another person, at a given low abandonment rate, say, 2%.

Only a few predictive dialers can achieve list penetration level of 50% with daytime calling, while maintaining an average wait time between calls of nine to 20 seconds at a 2% abandonment rate.

13. Get full branch scripting. The best scripting systems allow data and calculations to be performed in the script and the screen routing is based on the answers given – not only on the current call, but on previous calls as well.

Also when an agent hits a key to change screens, it must be instantaneous, no matter how many people are on the system. A multiple-second wait time is unacceptable. The system should also provide function keys for access to questions most likely to come up at each specific place in the script.

14. Get fulfillment capabilities. An agent should be able to easily generate a personalized letter, mailing labels, invoices, sales cards or sales sheets and complete an actual order by laser printer.

15. Get security. The system should have full user security at the system level, the project level and even individual screens should be protected by User ID.

16. Get a message bulletin board. There should be a message bulletin board for individual agents, groups of agents or all agents to inform people of vital news before shifts start and after breaks end.

17. Get voice mail. This should be incorporated into the system as well as message on hold while a person is waiting.

18. Try faxing. The system should be able to do fax-on-demand.

19. Ensure campaign flexibility. There should be no limit to the number of campaigns or number of projects that can be called. Every agent should be able to call a different campaign at the same time.

20. Make sure it's expandable. What if your company becomes a big success? Plan for future growth.

Private Automatic Branch Exchange (PABX)

See Private Branch Exchange.

Private Branch Exchange (PBX)

Also known as a Private Automatic Branch Exchange (PABX), a PBX is a telephone system within a comapany's premises that switches calls between company users on local lines while allowing all users to share a certain number of external phone lines. Users of the PBX share a certain number of outside lines for making telephone calls external to the PBX.

The PBX is owned and operated by the enterprise rather than the telephone company. Private branch exchanges used analog technology originally, but today many PBXs use digital technology, with the digital signals being converted to analog for outside calls on the local loop using POTS.

A PBX includes: Telephone trunk (multiple phone) lines that terminate at the PBX, a computer with memory that manages the switching of the calls within the PBX and in and out of it, the network of lines within the PBX, and a console or switchboard for a human operator. Advanced, full-featured PBXs have some of the characteristics of audiotex devices, Automatic Call Distributors, and IVR systems. More sophisticated computer telephony functions require CTI connections to a LAN or WAN-based server. There is usually an "Open Application Interface" (OAI) to help with this integration process, though the "open" OAI is inevitably a proprietary API.

Most PBX station sets have an LCD screen that can display ASCII text. The type of information that is displayed varies with the features and programming capabilities of the switch. The information can include calling and called party ID from within the switch, ANI digits from the telephone company, hook state, time and length of call, name assigned to the extension, and message waiting notification. With a PBX integration board, this information can be easily passed "unprocessed" to a computer telephony application.

By capturing the same display messages that a phone set gets, your application can "see" and "record" the ASCII information. Sophisticated computer telephony applications can grab ANI digits from the PBX (passed along from the phone company) and may search the text string for digits and use them to access databases for customer information.

P

345

Also, many PBXs have a Station Message Detail Recording (SMDR) electrical plug, usually an RS232-C receptacle, into which one plugs a printer or call accounting system. The telephone system sends information on each call made from the system through the SMDR port. That information includes who made the call, where it went, what time of day, etc. This information can be printed or captured by the call accounting system on disk and later incorporated into various performance and management reports.

Programmable "Dumb" Switch

This type of switch is "dumb" because it doesn't have much application software. It's called "programmable" because you must add your own software.

This type of switch has four basic elements – a network interface unit (a card to connect to T-1, E-1, etc.), a switching matrix card, a bus to carry the calls between the cards and some software to control the whole shebang and to control the interface to your computer that's controlling things from the outside.

Pulse Code Modulation (PCM)

A sampling technique for digitizing analog signals, usually audio signals, though the PCM waveform coding algorithms may be used to convert any sort of bandlimited analog signal to a digital coded stream. PCM codecs are the simplest form of waveform codecs. Speech is sampled at about 8,000 times per second and then quantized. If linear quantization is used then about 12 bits per sample are needed, giving a bit rate of about 96 Kbps. It works by defining a range of input voltages, splitting this range into "bins" linearly, and assigning a binary code representing a number, in which the lowest number corresponds to the lowest input voltage, and the largest number corresponds to the highest input voltage. This technique's quality has been shown to be approximately linearly proportional to the bits allocated per sample, by the equation $S/N \sim 6N - gamma$, where S/N is the signal-to-noise ratio in dB, N is the number of bits per sample, and gamma is a constant that depends on the companding used.

For coding speech, non-linear quantization enables just 8 bits per sample to be used to deliver speech quality almost indistinguishable from the original, giving a bit rate of 64 Kbps. Two such non-linear PCM codecs have dominated since the 1960s. In the U.S. and Japan mu-law (u-law) coding is the standard, while in Europe A-law compression is used, PCM is the standard used for digital telephony in the Public Switched Telephone Network (PSTN).

Because of their simplicity, excellent quality and low delay both these codecs are still in wide use. For example, the .au audio files used to send sounds over the web are PCM files. Code to implement the G.711 A-law and u-law codes has been placed in the public domain by Sun Microsystems.

Q.931

Just as the "D" or "Delta" channel of Basic Rate Interface ISDN (ISDN BRI) is used for signaling, so Q.931 is the message-oriented signaling protocol found in the "D" channel of BRI's bigger brother, Primary Rate ISDN (ISDN PRI). It's the ITU's ISDN user-network interface Layer 3 specification for basic call control. Q.921 handles Layer 2 signaling.

Quality Assurance

A company may desire to expand its hours of service by using an outsourcer, but it's concerned how the outsourcer will treat the valued customers it has nurtured over the years. Can the outsourcer satisfy callers as well as the company's own Customer Service Representatives (CSRs)? Can you really place your customers in the hands of perfect strangers?

Fortunately there's a way to ensure that your customers are always treated exactly according to your wishes – quality assurance.

After you've selected an outsourcer, and after the outsourcer has hired and trained agents for your campaign, quality assurance is the ongoing process of ensuring that agreed-upon levels of quality is maintained for each call into and out of the call center.

If you are on a hunt for an outsourcer and are worried about quality assurance, one thing you can do is to look for a service bureau that's using state-of-the-art technologies and specializes your company's vertical market.

For example, Ameridial's (North Canton, OH – 330-497-4888, www.ameridial.com) outbound and inbound services include lead generation, appointment setting, order processing and literature fulfillment. The service bureau employs more than 700 representatives who handle 50 million calls a year.

Quality assurance has become a top priority in today's market, whether or not outsourcing is involved. About 70% of customers defect because they've encountered an attitude of indifference when dealing with customer service departments, a phenomenon that seems more pronounced because it's occurring over a phone call between strangers. But if you know exactly how agents interact with your callers each day, it's easier to pinpoint where agents need further training and catch certain types of problem calls. Once you know where the glitches are, you can fix and review, then move on and improve customer service. It's a basic learning process.

Recording and Monitoring

There's a wide range of recording devices available for call centers, including voice loggers, recording devices that perform random recordings and synchronized voice and data recording systems. Call monitoring systems let you record full conversations or parts of them. Some systems let you monitor through a web browser from any machine on your network, for added convenience. Whatever method of communication you choose, it is your responsibility to provide the means for agents to hear and/or see what happened during a customer call (See Call Center Call Monitoring and Recording).

For example, TeleRep (Glen Burnie, MD – 410-761-2424, www.telerep.com), a division of ATS Call Centers, records all calls for monitoring quality control and will allow your company to listen to random calls, thus earning your trust.

Screen monitoring is also a "hot" call center application these days because call centers are paying more attention to how call length is affected by the kind of software agents use. It might be discovered, for example, it's time-consuming to maneuver through multiple screens to get to the right information from certain types of customer information software packages.

Monitoring technology is generally seen as a good investment. If you give agents regular, fair and timely feedback, you motivate them to do a better job. They learn how to improve their performance by examining each call you share with them as good examples. The monitoring process also helps agents feel more appreciated; since they need to know what they're doing has a purpose. This is not only good for the agent, but also good for you, your center, your outsourcer (if you're using one), and your company's overall corporate image.

That's quality assurance in a nutshell, but, if you're using an outsourcer, you'll want to know as much as possible as to how your outsourcer will guarantee that every call handled by the company meets the highest standards - and perhaps you'll discover that an outsourcer can deliver even higher standards than you normally achieve internally!

How? Well, traditionally, supervisors monitor rep performance. But Assessment Solutions Incorporated (ASI; New York, NY – 212-319-8400, www.asisolutions.com) found several drawbacks with this approach. Sometimes, for example, supervisors must neglect other duties in order to find time to supervise, and are unable to monitor during peak periods – the times when reps are most pressured and more likely to rush customers. Supervisors also sometimes lack sufficient objectivity when they evaluate a person they know and, in some cases, the people they hired or trained.

ASI (and other top-notch companies like it) addresses these problems by implementing an objective, third-party monitoring and evaluation program. Its professional assessment teams remotely monitor how well your agents treat customers, then prepare detailed evaluations based on objectives and proven criteria you can use.

Indeed, a good third-party quality assurance program can teach many call center managers a lesson in understanding the call flow process, says Deanna Matsumoto, a partner with The Genesis Group (Kinnelon, NJ – 973-492-8300), a management consultant firm.

"Managers need to look at the behavior an agent exhibits from the beginning of a call to the end," she says. "Sometimes supervisors just do a tone-check – they walk the floor, hear the opening or greeting, and evaluate on this basis only. But what they have failed to do is listen to the entire conversation and analyze it from beginning to end. In the past, companies did not spend the kind of time that is necessary to coach people successfully. Now there are separate quality teams that monitor and feed the information back to the agent. With the technology today, you can take calls, play them back, make comments on-line and have reps listen to them."

The rewards you reap from monitoring shouldn't only come from using a product. A monitoring system should complement desk-side monitoring. "You should always implement desk-side monitoring as well," says Matsumoto. "This way you can see if agents really understand how to use the system, or if are they are having trouble using their screens."

In one call center Matsumoto worked with, for example, the supervisor said the calls were lasting too long. "We sat desk-side," she says, "and saw the CSR couldn't type well and had a problem with spelling. Things like this are not manifested when listening to a tape recording."

"We have advocated desk-side monitoring with our clients. We see more companies doing it, but there is reluctance because supervisors don't always have the time. But if you don't do it, it's really difficult to bring performance to the highest level."

Matsumoto issues a word of caution regarding call recording: "It's great that you can tape and listen, but make certain it doesn't alleviate the supervisor's time spent with the agent."

Still, whatever system or technique you find appropriate for your center, you'll discover that monitoring will help you pinpoint call center problems and is one of the key tools used in the quality assurance process.

Let's now take a look at Sprint Telecenters' (Orlando, FL – 800-767-2858, www.sprintbiz.com/bizpark/telecenters/welcome.html) excellent quality assurance program.

At Sprint Telecenters, quality assurance begins before a single call is made. Randy Jurecka, Director of Strategic Services, performed a benchmark study of Fortune 500 companies in several key industries, including finance, telecommunications, reservation and hospitality. From this study, Sprint Telecenters recognized 27 key measurements and criteria that it developed into its quality assurance program.

Each Teleconsultant (or "TC" as they are called at Sprint Telecenters) is monitored twice each shift by a quality assurance analyst. There is one QA analyst per 24 TCs. These QA analysts go through a formalized training program, says Jurecka. The analysts are recertified every six months. Periodic calibration sessions assure that all the analysts rate TCs the same way.

Each of the key measurements the TCs are judged on are weighted with values from ten points to two points. Vital requirements, such as conforming to the Telephone Consumer Protection Act (TCPA), have the highest value and are worth 10 points. More cosmetic voice skills, such as using pauses effectively weigh in at two points.

Some of important criteria TCs must meet are:
- All of the TCPA requirements, including requests not to be called;
- Product knowledge;
- Script adherence;
- Voice quality;
- Listening and communications skills.

Sprint Telecenters' automated system instantly creates a score, which can be sliced and diced by day, by agent, team or by campaign for analysis by management. Scores in the quality assurance program are used to recognize TCs for excellent work and to select candidates for advancement.

The system also generates reports hourly, daily, weekly and monthly so management is always on top of trends and can take advantage of opportunities. Records are archived, so if questions come up later (on TCPA compliance, for example) the data is readily available.

At The Connection (Burnsville, MN – 800-883-5777, www.the-connection.com), an inbound telemarketing specialist, customer service reps are monitored on eight calls a week by a coach. Each week they meet with that coach to evaluate and discuss their performance. The company encourages you to participate in these weekly call monitoring and coaching sessions.

The coach and CSR discuss general call control, professionalism and specific call procedures. The CSR receives bonuses and raises based on his or her overall performance and the quality of monitored calls.

The Precision Response Corporation, known as PRC (Miami, FL – 305-626-4600, www.prcnet.com) prides itself in losing its own company identity to take on that of yours. When it comes to quality assurance, you can check on PRC's performance from your office, using real-time desktop monitoring. The company's on-line reports give you the trends quickly so you always know how your campaign is going.

Reps are monitored several times each day by a member of the qual-

ity assurance team. When script changes are made, the QA personnel go into high gear, monitoring the reps more frequently.

Procedures such as these assure that your callers are always getting the highest quality care from your outsourcer. A rigorous and well-defined quality assurance program gives you peace of mind, not only when choosing your outsourcer, but for years to come.

You'll find that you play a vital role in your outsourcer's quality assurance program. Roger H. Nunley, managing director of the Customer Care Institute (Atlanta, GA – 404-352-9291, www.customercare.com), makes these suggestions in a booklet on outsourcing he wrote for the Society of Consumer Affairs Professionals (SOCAP):
- Conduct remote monitoring weekly. This is just a spot check of service quality.
- Hold a conference call with your account manager, quality assurance manager and a supervisor weekly. Occasionally include a CSR on the call. Have your outsourcer fax you quality stats before the call.
- Visit your outsourcer monthly. Meet with managers and CSRs.

Nunley's booklet is $75 for members, $150 for non-members from SOCAP (you can order it by calling 703-519-3700).

A Quality Assurance Checklist from Sprint:

Ask your outsourcer these questions about its quality assurance program before you sign on the dotted line:
- What criteria is used to evaluate the performance of agents and floor supervisors?
- What tools are used to ensure sufficient agents are on hand to deal with call volumes?
- What are your quality assurance processes and standards?
- How many calls per agent are evaluated per shift? Monthly?
- Do you offer remote monitoring? How does it work?
- Does your company have a formal quality program? Describe.
- What is your quality assurance staff to agent ratio?
- Do you have a manual or automated environment?
- What are your qualifications for quality assurance analysis?
- How do you measure caller satisfaction? Customer satisfaction?

e-talk's Quality Assurance Tips:

If you'd like to implement a successful monitoring program in your call center, e-talk Corp., formerly known as Teknekron Infoswitch (Irving, TX – 817-267-3025, www.e-talkcorp.com) offers these handy tips to keep in mind.
1. Review your business strategy, mission, goals and objectives. You need buy-in from many areas of the company. To develop and implement the program, create a diverse team that represents different job functions and levels.
2. Identify performance expectations for the center, the team and the individual agent. Collect data and obtain samples from "best in class" peers. Have the team audit calls from your center to determine success factors and relevant behaviors. Solicit input from the related areas of your company and from the team as a whole.
3. Determine observation criteria and desired behaviors. Design your evaluation form and establish a scoring system based on the data and team input. Gather as many quality form samples as possible from friends, other organizations and peers. Common question types include yes/no, point systems by scale or range, and "alpha" question types with levels based on behaviors or competencies rather than numerical values. Set up a field trial to test your system. Solicit feedback from your test group and address concerns. (You may want to create an ini-

tial draft of a quality form with notes, for explanations of form content and function based on feedback you receive.) Review and revise the monitoring form to meet your objectives.

4. Establish real standards, protocols and measurements. After you roll out the program, track trends and analyze them. Be sure to calibrate scoring between groups and centers to be sure that quality assurance monitors are grading fairly and consistently. Post results on a "wall of fame" or in newsletters. Recognize achievement in creative ways (e.g., gift certificates, "company bucks"), as well as with promotions and cash incentives.

5. Train and coach. Communicate where improvements can be made. Coach and train, so agents see the system as a development tool for their successes. Use your service observation program to provide trainers with detailed information on individual rep training needs.

6. Before implementing a service observation program, develop codes to identify the types of calls that your center or group receives. Incorporate these codes into your observation process. Put together a team to identify the reasons why customers call. Then identify your "big" and "little" categories. (e.g., Billing may be a "big" category, with balance inquiry and bill request copy as "little" categories. Assign main and sub-category type codes (e.g., B0: Billing [general info]; B1: Balance Inquiry; B2: Copy of bill request; B3: Explanation of first bill.)

7. Reports. When completing service observations, assign codes to each call based on call type. Reports can then be created to determine which areas individual reps need additional training, or who should attend a specific training class.

Quality of Service (QoS)

QoS or "kwoss" has lately been associated with the idea that circuit and particularly packet-based network characteristics can be measured, improved, and guaranteed in advance. Ideally, QoS solutions can guarantee transmission characteristics end-to-end such as available bandwidth, maximum end-to-end delay, maximum end-to-end delay variation (jitter), and maximum packet/cell loss, in order to provide continuous data streams suitable for real-time phone calls and video conferencing. In particular, packet latency must be kept low and with a low variance throughout the many switching and routing hops the packets may encounter, despite a high volume of packet traffic.

While one can compare, ponder and discuss detailed measurements relating to network bandwidth, delay and jitter, QoS encompasses a myriad of additional underlying technologies and services as well: Operating systems (real-time scheduling, threads), communications protocols, data networks, scheduling and traffic management issues, etc. A buggy application or weak endpoint system resources (too little memory, slow CPU, faulty network interface card) can ruin QoS for a user, no matter how tremendous the network bandwidth or clear the signaling. Ultimately, then, QoS is simply when a user subjectively judges the workings of an overall system to be satisfactory.

While voice and video quality is certainly in the user's interest, it also provides surprising benefits for service providers. Studies done on cellular networks show that as voice quality increases, the time users spend on the network also increases, which means billable hours – and thus revenues – increase too.

But to achieve quality on a network not originally designed for it, one needs to impose a Quality of Service (QoS) architecture.

For a QoS architecture to be successful, there must be some level of assurance that the traffic and service requirements of each network component (e.g. an application, host or router) can be satisfied, even though each component must interface with other components, each having their own set of traffic and service requirements! A true QoS must encompass the workings of all seven layers of the Open Systems Interconnection (OSI) Model (See Open Systems Interconnection Model), as well as every network element from end-to-end. Any QoS "guarantees" are only as good as the weakest link in the transmission "chain" between a source of voice / data traffic and its destination. As applications become more distributed, there is a greater need for the support of QoS at the lower network layers. This can be quite a challenge since, even on a single host, typical operating systems don't support real-time delivery.

Moreover, QoS is also something of a relative term depending on what application the user is working with. Different applications require different levels or types of QoS, and are sensitive to the values of differing parameters related to providing QoS. For example, no one would care if file transfers took a second longer, independent of the total transfer time (delay)? Similarly, who would care if, on a particular file transfer, 80% of the data was sent during the first half of the transfer and 20% in the last half (a case of delay variance or jitter)? However, Voice over IP is sensitive to latency but a few packets can be completely lost per second during a voice transmissions with no objections, whereas losing "a few bits" while wiring a billion dollars or sending an executable program could lead to disaster. Clearly, the capacity of an application to tolerate unpredictability relates to the level of QoS they require.

Most of our discussion of QoS relates to voice and data networks. The idea of applying Quality of Service to a data network first occurred in the 1970s when the X.25 network was the first recipient of limited QoS technology in the form of differentiation between X.25 data streams by assigning priority status to some packets instead of others. However, the old X.25 packet system does error checking at every step of the data's node-to-node journey, which can introduce a transmission delay of a few hundred milliseconds from source to destination. Fast packet systems such as frame relay and ATM, on the other hand, are concerned only as to where to route data traffic, not the integrity of the transmission, so the delay from sender to receiver is thus at least two orders of magnitude less than X.25.

The best known high-bandwidth packet-based transport mechanism with a built-in QoS scheme has been ATM (See Asynchronous Transmission Protocol). Because of its complexity and cost, ATM has been pretty much confined to use on the backbone or for high-end services. It is appearing more and more in "the last mile" because of its QoS but it is expected to ultimately be superseded by IP.

QoS has become an endless topic of discussion in the Internet telecom world, since the dynamic, rapidly expanding public network gives rise to traffic congestion and thus a degradation in QoS. The public network's traffic doubles every three months, while aggregate capacity has lagged. This is partly because of new Internet users and the fact that Internet-based multimedia and videoconferencing applications are finally becoming popular and rely on the public network to reach customers. QoS is under further scrutiny since businesses need a public network with a consistently high QoS as more marketing and transactions occur over it.

The public network's data traffic volume has long surpassed that of traditional Time Division Multiplexing (TDM) (See Time Division Multiplexing). Old-fashioned circuit-switched voice traffic has fallen to about 10% of data network traffic.

IP services are more successful at grabbing market share than the old PSTN-based enhanced services, but this shift from TDM-based circuit-switching to IP packet-based traffic puts great demands on even the largest routers. Moreover, additional changes in network architecture must take place since carriers must migrate the existing TDM-based

Q

access network to one optimized for delivering packet-based IP services but still able to deliver TDM-class voice services. The public network's current internetworking paradigm, based upon Interior Gateway Protocols (IGPs) used in TCP/IP networks to exchange routing information between collaborating routers in the Internet (such as RIP/RIP2, ART, OSPF or IGRP/ EIGR) have had "a jolly good run" but have reached their limits (the term "gateway" is historical, "router" is currently the preferred term for OSI Layer 3 (network) gateways). Moreover, adding TDM and ATM-based access infrastructures to all of this simply adds unnecessary and extremely burdensome overlay networks and are inefficient for an all-IP, packet-based universe.

The problem with the Internet Protocol (IP), and the architecture of the Internet and packet networks in general, is the fundamental underlying concept that datagrams (packets) carrying their own source and destination addresses can traverse a network of packet routers independently, without the aid of their sender or receiver. Thus we are dealing with so-called "dumb" networks that follow the "end-to-end argument" which stipulates that any complexity (the "smarts") should be confined to the network's end-hosts (the sender and receiver). Dumb networks have a simple architecture because they don't provide services of any complexity or sophistication. IP routers scattered throughout the network merely have to look up a packet's destination IP address in a forwarding table to determine the "next hop" toward the destination (receiver).

Typical network traffic is bursty rather than continuous, so the datagram-based (packetized) nature of IP allows it to use available bandwidth efficiently, by sharing what is available as needed. This also allows IP to adapt more flexibly to applications with varying bandwidth requirements. Routers can fragment datagrams and reassemble them at the receiving endpoint, such IP packet segmentation minimizes delays, since long data packets will no longer add to the overall delay by causing routers to pause until the entire packet is ready for forwarding–it simply chops them up into a maximum packet size that depends on the speed of the links. For example, a 64 Kbps link forces a maximum packet size of 256 bytes, while a T-1 link (1,544 Kbps) can handle 6,144 byte packets with the same delay. Ethernet-sourced packets can't exceed 1,536 bytes, which is okay since packets don't require explicit segmentation when running at Ethernet speeds.

All of these techniques allow IP to run across virtually any kind of network transmission media and between just about any set of system platforms (one is reminded of the old T-shirt reading "IP over everything") but such flexibility doesn't provide reliable data delivery. If the queue for the next hop is very long, the packets may be delayed. A full or unavailable packet can cause a router to discard one of these datagrams without it having to make this known to either the sender or receiver.

The only so-called reliability mechanisms of IP are based on upper-level transports (e.g. TCP) that can keep track of datagrams, and retransmit them, thus assuring data delivery, but these same high-level protocols cannot guarantee either the timely delivery of packets or minimum levels of data throughput, two of the characteristics that qualify QoS. This is because unlike "pure virtual circuit" technologies such as ATM and frame relay, IP does not make hard allocations of network resources.

For best-effort service, the network delivers data if it can, without any assurance of reliability, delay bounds, or throughput. IP thus provides a "best effort" single service model that can make no guarantees about when data will arrive, or how much it can deliver. An application sends data whenever it must, in any quantity, and without requesting permission or first informing the network. Every router feels at home in such a best effort environment. For example, the Cisco IOS QoS feature that implements best-effort service is First-In, First-Out (FIFO) queueing.

Best effort packet networks scale up well (witness the astounding growth of the Internet), but as service demands exceed capacity the service degrades gracefully (at least initially), as the unpredictable delivery delays (jitter) and packet data loss become more pronounced, but generally the degradation in service is not enough to noticeably affect the Internet's delay-tolerant (i.e. "elastic") data applications of e-mail, file transfer and the World Wide Web. Only when it comes to mission-critical telephony, video conferencing / streaming, multimedia and other applications with high data throughput capacity (bandwidth), low latency and real-time (and sometimes duplex) delivery demands do the failings of a "best effort" service reveal the fact that a "best effort" isn't always a good enough effort.

Because of the packetized nature of the Internet, mission critical data traffic gets mixed up with and can be delayed by contending for bandwidth with inconsequential traffic. There is no top priority rating that can be given to important messages. Following in the best of hacker traditions, the Internet is just too egalitarian!

The "real" question that arises out of all of this is: How can we increase the service level of IP (and other packet networks, if possible) and the network infrastructure, without adding undue complexity? And, one might add, being able to measure and report on service quality. That's the "Quality of Service Problem." Various companies and the Internet Engineering Task Force (IETF) have a number of promising, though competing, technologies we shall now examine.

Although bandwidth continues to expand on the Internet as many companies bury "dark" or unused optical fiber beneath railroad tracks (indeed, some companies are just burying empty conduits, waiting for new types of higher-capacity fiber to be developed), traffic continues to expand as all of the world's technologically-enabled citizens slowly come online. This leads to congestion, and much of QoS targets providing a predictable share of available bandwidth capacity during periods of congestion so that customer applications requirements may be satisfied.

QoS Parameters

So far we've touched upon just a few main characteristics that qualify QoS: Minimizing delivery delay, minimizing delay variations (jitter), and providing consistent data throughput capacity (bandwidth). Let's take a closer look at the most important parameters that help define QoS:

• *Bandwidth.* A measure of the "raw" data transmission capacity of a network, or how many bits of data can be moved from one host to another in a unit of time (almost always seconds) under ideal conditions. Bandwidth is thus usually expressed in kilobits per second (Kbps), megabits per second (Mbps), gigabits per second (Gbps), or terabits per second (Tbps). You can visualize bandwidth as a "pipe." The greater the bandwidth, the more data can travel through the pipe. There is always a certain large bandwidth that represents any particular connection's unreachable theoretical maximum capacity – unreachable because various forms of transient congestion and physical phenomena (such as crosstalk or attenuation) will deteriorate signal quality or keep the transmission below ideal levels. Nearly all networks (with the exception of some advanced fiber optic networks) are far from ideal.

The Internet and other networks tend to be a network of networks, a lacework or entangled mesh of diverse transmission media, each with its own bandwidth capacity, latency, and other diminishing factors. We've also seen how network application traffic is bursty, causing transient network congestion that results in packet delivery delays or loss of packets altogether.

Without any real Quality of Service, the only obvious way to deal with traffic congestion is to over-provision the network so that all traffic will have more bandwidth than it needs, the surplus bandwidth capacity ready and waiting for sudden peak data rates during high-demand periods. Excessively increasing resources in this way is reminiscent of the building of U.S. Interstate highways in the 1950s and 1960s, which only led to more car traffic than ever before. Also, at the rate data traffic is increasing on the Internet, "over-provisioning" is the most temporary of solutions, and the least economically viable. Furthermore, increasing bandwidth to accommodate real-time applications is still not sufficient to avoid jitter during traffic bursts. Irritating "drop outs" in voice and video applications can be experienced even over an IP network sustaining light traffic, and it's rarely possible to predict when and where in a network peak data rates are going to occur anyway.

To achieve some kind of QoS, IP services must be augmented with protocols that can be used to distinguish real-time data traffic with strict timing requirements from traffic that can tolerate packet delay, jitter and loss. QoS does not magically create additional bandwidth, but adroitly manages existing bandwidth on the basis of what an application needs. Hence, any kind of QoS offering a guaranteed service level must be able to allocate resources to individual data streams. This bandwidth flexibility must not be so extreme so that QoS-enabled (high-priority) applications don't dramatically reserve so much bandwidth that they starve low-priority Internet applications. After all, the goal of QoS is to provide some level of predictability and control beyond IP "best-effort" service.

The different voice compression / decompression algorithms (codecs) used in Voice over IP (VoIP) / IP telephony have differing bandwidth requirements. Such audio codecs range from 5.3 Kbps (G.723.1 at 3.4 kHz) to 64 Kbps (G.722 at 7 kHz):

G.711 – Pulse Code Modulation – 3.4 kHz at 56 or 64 Kbps

G.722 – audio codec – 7 kHz at anywhere between 48, 56 or 64 Kbps

G.722.1 – wideband – 7 kHz at 24 or 32 Kbps

G.723.1 – 3.4 kHz at 5.3 or 6.4 Kbps

G.726, G.727 – Adaptive Differential Pulse Code Modulation (ADPCM) speech codec – 4 KHz at 16, 24, 32 or 40 Kbps.

G.728 – audio codec – 3 kHz at 4 or 16 Kbps

G.729 – speech codec – 3.4 kHz at 8 Kbps

- *Latency.* In a network, latency, or delay is how much time it takes for a packet of data to get from one designated point to another. This can be the time between a node sending a message and receipt of the message by another node, or the average length of time for a packet to traverse a network. In rough evaluations of IP telephony systems, latency is typically measured in the milliseconds (ms) from the instant a speaker utters a word until the listener hears it. This one-way latency is known as "mouth-to-ear" latency. In many networks latency is measured differently. This is because many widely used network protocols (such as IP and IPX) require the receiving computer to periodically send acknowledgment packets to the sending PC. Since data is held by the sender until an acknowledgment is received, network performance metrics are dependent not upon simple latency, but on the "round-trip" time, which is the time between the first byte of data being sent, and the time the last byte of the acknowledgment packet for that data has been received.

The true measure of a network's throughput, then, can be calculated by dividing the amount of data transferred for a given transfer / acknowledgment cycle time. Simple latency is thus only a part of the overall equation of network throughput. In fact, in order to maximize network throughput, all modern network protocols send multiple packets of data before requiring an acknowledgment packet. Even with the circuit-switched PSTN, latency is usually measured as a round-trip latency that is the sum of two one-way latency measurements made during a telephone call. With traditional domestic analog systems, round-trip latency rarely exceeds 150 ms, which is generally unnoticeable.

In a router, latency is the "propagation delay" or the time interval between when a data packet is received and when it's retransmitted (the fundamental constant of the speed of light is a small but unavoidable form of propagation delay). A store-and-forward device such as a router must wait until an entire packet has been received before it begins to resend that packet. Because the receive time is roughly a function of packet length, latency will also vary. The "storage latency" for even a very fast store-and-forward device will be the receive time of a given packet, plus the time to calculate which port should forward the packet. In order for a router or other store-and-forward device to run at "wire speed" when receiving back-to-back minimum size packets, it must be able to determine the forwarding port calculation for the first packet during the time the next packet is being received.

On the other hand, "cut-through" devices such as traditional LAN switches begin transmitting a packet on the sending port even before it has been completely received. This technique allows a significant decrease in latency. However, even a cut-through device must wait until it has received a packet's destination address before it can retransmit that packet. So, the latency for these devices is the receive time for the destination address, plus the time it takes to calculate the forwarding port.

Even in ATM switches, "frames" consisting of several ATM cells may arrive with many gaps between the cells. These gaps in the input stream should not be counted towards the switch's contribution to the frame delay metric. A proposed solution called Message-In Message-Out (MIMO) latency improves upon First-In Last-Out (FILO) latency commonly used for continuous frame technologies such as frame relay. MIMO latency is defined as the difference between FILO latency through the switch and that through an ideal switch or any network of switches as well.

Other contributing factors to latency include "transmission delays" resulting from the dielectric constant or refractive index of the transport medium, or even the fact that large packets take longer for a round trip than smaller ones; and "computer" and/or "storage" delays where a packet on the move momentarily finds itself in the storage systems of backbone switches and bridges.

Thus, latency can encompass both delay in a transmission path or in a device within a transmission path.

Latency can become an issue even when you're performing a non-realtime activity such as surfing a website and your browser finds that it has to request ten or so images to reproduce a web page on your PC's screen. Each picture request needs to get a response from a server. If the ISP's routers don't respond quickly enough because of network congestion, then this form of latency adds seconds onto each page retrieval.

Studies have shown that humans start to get annoyed when end-to-end audio delays approach 400 ms to 500 ms, and a 700 ms to 800 ms delay reaches the limits of tolerance. The ITU-TG.114 (One-Way Transmission Time) recommendation suggests 150 ms as the maximum desired one-way latency that can occur in a conversation over a network and still be considered of high quality. An ideal delay level worthy of the best QoS architecture is probably around 45 ms or less, while video transmissions can have up to a 95 ms delay, plus variance, before frame-freezing occurs.

IP telephony is subject to a whole host of latency problems because of its reliance on a multitude of digital signal processing events. Digital Signal Processor (DSP) chips and associated software algorithms compress and decompress speech, detect signaling and touch-tones, detect silence, gener-

Q

ate tones, generate a background "comfort noise," and do echo cancellation. These processes are known collectively as voice coding or "vocoding."

Vocoding becomes a relatively efficient proposition for high density (many port) IP telephony applications only when entire frames (or batches) of data are processed at one time, allowing DSPs to use specialized, efficient, microcode instructions. However, none of the data can be processed until the frame is completely full. This leads to "accumulation delay," or "algorithmic delay." Digitized audio typically arrives from the PSTN at a fixed rate of 8,000 samples per second, so the frame size used to hold and process the data will directly affect the degree of latency. A 100 sample frame takes 12.5 ms to fill, while a 1,000 sample frame takes 125 ms to fill. The trade-off is: Larger frames, greater DSP efficiency but greater latency too.

Each of the standard vocoder methods uses its own standard frame size, so the maximum latency incurred by the framing process is directly dependent on the vocoder used by your IP telephony system.

For example, the G.723.1 vocoder has a frame duration time of 30 ms, G729a has a frame time of 10 ms, SX7300 is 15 ms, SX9600 is 15 ms and the increasingly popular G.711 isn't strictly a vocoder per se, but its data streams have considerable flexibility when specifying frame sizes.

After each frame is filled with data, the DSP algorithms must process it, which also takes time and adds to overall latency. This processing time varies but obviously never exceeds the frame collection time, or else the DSP would never finish processing one frame before the appearance of the next data-filled frame in the sequence.

Each DSP in a high-end IP telephony gateway systems processes multiple voice channels, with each DSP processing a frame from the first channel, then one from the second, then one from the third, one after the other in sequence. So while the first frame for the first channel will be completed long before the last channel, the processing of the last channel occurs just before more data for the first channel begins to arrive! Naturally, figuring out the total DSP processing contribution to latency under such circumstances is a complicated procedure, since you can't simply use the number of milliseconds it takes for a single DSP to vocode any single channel. Instead, the latency incurred due to processing is usually specified as the frame size in milliseconds. From this one can see that the total latency from framing and processing cannot exceed twice the frame size.

After the DSPs and vocoder software do their work, the resulting compressed voice data frames are often buffered before they encounter the WAN, which also adds to overall latency. Buffering reduces the number of times the DSP needs to communicate to the main CPU in the gateway and makes the vocoders' results fit into one common frame duration. For example, if a system is running G.723.1 on one channel and G.729a on another, the frame sizes are different. A system could then collect three G.729a frames into one buffer for every G.723.1 frame, allowing the system to transfer one buffer every 30 milliseconds, regardless of whatever coding algorithm is active at the time.

Also, we're dealing with a packet network, the coded voice must be assembled into packets, which also adds to total latency. Packetization can be done by the TCP/IP protocol stack, using the User Datagram Protocol (UDP) and Real Time Protocol (RTP), protocols which actually eliminate some of the overhead normally accrued from transmission acknowledgments and retries, but add 40 bytes to the packet in the IP header.

After the IP header, one or more frames of coded voice data follows. Once can reduce the inefficiency of IP packet overhead by putting more than one coded voice frame into each IP packet, but this adds additional frame periods of latency. This trade-off is another compromise that needs to be considered when deploying an IP Telephony system.

One trick that can reduce the overhead, but not increase latency, is

to let voice frames from other channels "piggyback" inside the same packet. When voice from more than one channel in the source gateway are heading for the same destination gateway, the data can be combined into a single packet.

All in all, Brooktrout (as an example) gives the following "latency budget" that lists sources and expected values for one-way latencies in an IP telephony system (as technology improves, each of these latency figures will decline).

Source	Latency (in milliseconds)
Network Interface	1 ms (1.54 Mbps T-1)
Framing	30 ms (G.723.1)
Processing Time	10 ms (worst case)
Buffering	0 ms (no additional buffering)
Packetization	30 ms (two frames per packet)
Media Access Delay	10 ms (5 - 2msec hops)
Routing	50 ms (router dependent)
Jitter Buffering	30 ms (one buffer)
Total One-Way Latency:	161 msec

The major cause of added delay can be found at the network's edge, particularly if you're working with a low-bandwidth link and low-end router. As we shall see, the first step to achieve QoS is to get delay (and jitter) under control with priority queuing instituted at the network's edges. Nearly all routers now have ways to do this, and there are external devices such as Packeteer which presumably can prioritize voice as well.

- *Jitter.* Jitter is not the same thing as delay or latency. In circuit-switched networks jitter occurs when the signal varies from its original reference timing, causing a distortion that is propagated with the signal throughout the network. In data (packet-switched) networks, jitter refers to packet jitter, not bit jitter. In packet-switched networks, jitter is a distortion of the interpacket arrival times (the interval between packet arrivals) compared to the interpacket times of the original transmission. Jitter is thus the variability of delay, the irregular arrival of consecutive packets that's measured in milliseconds. Latency is like an average delay, jitter is like the standard deviation of the delay.

Jitter is usually caused by the network a packet traverses. When a packet stream travels over the Internet there's no assurance that each packet in the sequence will travel over the same path – circuit switched networks can do that. Routing each packet on a different path is a sure-fire way to induce jitter. Other forms of jitter include host-induced jitter, which occurs when the computer transmitting packets is overwhelmed with processing duties or experiences some other momentary delays such as those associated with voice digitization, compression and packetization (high compression / low bitrate audio and video codecs require more CPU cycles and increase the likelihood of latency and jitter). More common is router-induced jitter, which appears when the rate of flow of packets out of a router is less than the rate of packet flow into the router. This can happen when a router's output link bandwidth capacity is constricted because of downstream network congestion or link problems, a problem that can expand to the point where it creates queuing problems at the network's entry or exit points.

Truly excessive jitter can lead to packet "sequence errors" where severe congestion in packet switched networks causes packets to take such different routes to reach the same receiver that packets may actually arrive out of order, which also garbles speech and video.

Normal jitter has little effect on a data-oriented applications but is perceived by users as particularly irksome when it comes to Voice over IP (VoIP), videoconferencing and other multimedia network applications. Too much packet jitter disrupts such real-time applications, impeding the ability of the receiving endpoint to smoothly regenerate voice and video – since both are inherently continuous waveforms, large varying gaps between the regenerated packets will cause reconstructed voice to sound garbled and video to have a jittery or shaky quality. Indeed, two-way or multi-way conversations and conferences demand that both jitter and delay be minimal so as to provide unperturbed interaction among the participants; even sub-second pauses can be disruptive in such situations.

The precise way to measure jitter is defined differently by different experts. Some consider jitter to be the difference between the longest and the shortest delay in some period of time. Others define jitter as the maximum delay difference between two consecutive packets during some period of time. Dan Frankowski at the University of Minnesota suggests setting up a "window size" for calculating the standard deviation and maintaining both a running sum and squared sum by subtracting the previous packet interarrival time from each, then adding the new interarrival time. The standard deviation is then equal to:

the squared sum / w · (sum / w) squared, where w is the window size

The International Telecommunications Union's (ITU's) recommended measurement of jitter involves the injection of packets into a transmission stream at regular intervals and the measurement of the variability as each one is received. The Internet Engineering Task Force (IETF) has it's own one-way Instantaneous Packet Delay Variation (IPDV) technique for determining one-way packet loss and one-way packet delay. Both the ITU and IETF techniques are "active measurements" in that they are made by injecting packet traffic into the network and measuring its properties. Other, passive measurements observe properties of data traffic by means that don't interfere with the traffic itself.

In the ATM world, the Cell Interarrival Variation (CIV) metric reveals how consistently ATM cells arrive at the receiving end-station. ATM has a built-in form of QoS and is considered to be highly stable and almost free of jitter. As for frame relay, it's "bursty" nature may result in variable delays between consecutive packets, but frame relay can set up a Permanent Virtual Circuit (PVC) that is used for an entire call. The PVC is a single path through a frame relay network which guarantees that all the packets in the call will have the same delay and won't get out of order. Earlier networks based on X.25 switching have too much delay for voice traffic but frame relay, even when using the IP protocol, can deliver very good voice quality. (Interestingly, the majority of packet switches used in telephony are frame relay or are doing VoIP over non-Internet circuits. Some companies are doing VoIP over frame relay rather than traditional Voice over Frame Relay (VoFR), probably because they realize that some of their voice calls are ultimately routed over the Internet at locations that are out of their control, which opens up the possibility of signal distortions resulting from latency, jitter and packet loss.)

The International Telecommunications Union (ITU) and The Internet Protocol Harmonization Over Networks (TIPHON) standards bodies define a one-way jitter of less than 75 ms as providing "good" Internet telephony. From 75 ms to 125 ms is considered "medium" quality and anything from 125 ms up to 225 ms is rated as "poor." The fact that packet-delay variations can be as high as 70 ms to 100 ms in some frame relay and IP networks has spurred the development of QoS techniques.

Attempts at removing jitter involves creating special receive-side "jitter buffers" in the voice gateways which collect packets and hold them long enough to allow the slowest packets to arrive, then funnel the speech packets back into a stream with the proper time interval so they can be played in the correct sequence. This causes additional delay, but does smooth packet delivery to produce a more even flow that reduces or even eliminates jitter. There is no universally accepted optimal size for

a jitter buffer since networks differ and different sessions on the same network at different times encounter differing levels of network congestion, so the buffer size will vary. A jitter buffer's size affects both jitter and latency (delay). The greater the jitter, the larger the buffer has to be to allocate more room to collect and play back previous packets at acceptable levels while waiting for new ones to arrive. Too large a buffer, however, can increase latency to where it becomes annoying to users. A typical Internet-related jitter buffer delay is 20 ms, but often reaches 80 ms to 250 ms and higher. In the case of VoIP using stable company LANs, however, the typical time to receive a packet can drop to 50 ms, in which case the jitter buffer need only hold about 10 packets to make the system acceptable to users.

So, small jitter buffers can't cope with late arriving packets, but large buffers introduce objectionable latency.

These conflicting goals of minimizing delay and removing jitter have given rise to various techniques to dynamically adapt the jitter buffer size so that it's just large enough to handle the current level of jitter on the network. For example, the buffer size could be determined by taking the standard deviation and multiplying it by some constant or derived value. The idea is to minimize the size and delay of the jitter buffer, while at the same time preventing jitter-induced buffer underflow.

Currently there are two basic approaches to adapting the jitter buffer size. The first is to measure the variation of packet levels in the jitter buffer over a time interval and incrementally adapt the buffer size to match the calculated jitter. This approach works best with networks that display a consistent jitter, such as ATM networks.

The second approach is to count the number of packets that arrive late and calculate a ratio of these packets to the number of packets that make it to the receiving application. This number is then used to adjust the jitter buffer size to target an allowable late-packet ratio. This technique works best with networks having highly variable packet-interarrival intervals, such as IP networks.

• *Packet Loss.* Lost packets also cause annoying signal degradation, but they're always ignored in the calculation of jitter. Lost packets can be conceived of as packets with infinite delay, but that would lead to spurious calculations of latency. Because standard IP networks don't guarantee service, they tend to have a higher rate of packet loss than ATM or frame relay networks.

As Chris Bajorek of CT Labs (Roseville, CA – 916-784-7870, www.ct-labs.com) says: "VoIP gateways send speech in 'frames' of 15 ms to 20 ms in length per IP packet. Since each IP packet carries with it a great amount of overhead bytes, the more speech frames you can pack into a single IP packet, the better the bandwidth utilization. So you usually can select a packing factor of one, two, or more voice frames per IP packet in the gateway configuration options. Some numbers to consider: sending a 9,600 bps vocoder-based call by packing a single voice into an IP packet yields an effective IP bit rate of 26,100 bps. Ouch – that's overhead. But packing five voice frames into a single IP packet yields an effective IP bit rate of 13,730 bps. Why not always set this value to the highest packing factor possible? Because the more frames you pack into a single IP packet, the more speech is lost when a network error causes a lost IP packet. Increasing the number of frames per IP packet also increases the end-to-end delay. So this packing adjustment becomes a balancing act based on your knowledge of data network loss factors and the desired level of speech quality."

IP networks treat voice like data, and under severe data traffic loads and congestion, the voice frames will be dropped along with data frames (a 20 ms or greater gap between VoIP packet arrival times put adjacent

Q

packets in imminent danger of loss). Whereas lost data can be retransmitted, voice packets must be delivered in real-time, since some systems use audio packets that can contain up to 40 or even 80 ms of speech information, which matches the duration of most phonemes, perhaps the most important units of speech comprehension.

Some say that a 10% random packet loss can yield almost toll quality, though ITU / TIPHON defines a loss of up to 3% as being "good" for Internet telephony. Most IP telephony experts would say that audio problems start when packet loss rises to somewhere between 2.5% and 5% of the packets lost over a one second time interval.

There are several approaches to address the problem of lost voice and video packets:

1. Small instances of packet loss (or extremely late or corrupted packets) can be compensated with "packet replay" where the receiver replays the last packet received during the interval when the lost packet was supposed to be played. This doesn't really interpolate the lost packet, it just "smoothes over the gap" by repeating the previous packet until the next packet arrives. Packet replay works well when packet loss is not severe, since speech and video don't change much from moment to moment. If a whole series of packets are lost, however, repeated packet replay becomes glaringly evident. Fortunately, well-managed corporate networks tend to have very low packet loss.

Interestingly, a packet replay technique is sometimes used by hackers to break into and disrupt computer systems on a network. Packets describing authentication sequence messages to gain access to a system can be recorded in the network and retransmitted, giving an intruder system access. Packet replay is frequently undetectable, but can be prevented by using an anti-replay service where each packet passing within the secure association is time stamped and/or tagged with a sequence number for packet sequence counting purposes. At the receiver, each packet's time stamp or sequence number is examined to see if it falls within a specified range. If it doesn't, then the packet is blocked.

2. By wasting bandwidth, one can send redundant information, such as transmitting the nth packet of voice information along with the (n+1)th packet. This corrects for the lost packet but also causes additional delay. We know from communications theory that by encoding k binary digits into one of $M = 2$ to the k power orthogonal signals one can reduce the probability of bit error. The bandwidth increases proportionally with M, or exponentially with k. Because of the increasing bandwidth, more noise and other problems may occur with the receiver and the improvement in probability of error does not increase as fast as the increase in M.

3. One can try a hybrid approach that still involves sending redundant information in the (n+1)th packet but decreases the bandwidth needed by using a low bandwidth voice coder in the first place. This still fails to resolve the delay problem.

4. When dealing with the Internet, simple packet replay and bandwidth-wasting redundant approaches are impossible or impractical and are ignored in favor of more robust techniques such as Forward Error Correction (FEC), where the receiving device is able to detect and correct any character or code block that contains fewer than a predetermined number of symbols in error. FEC methods fall into two basic categories: Intrapacket and extrapacket. Intrapacket FEC adds redundant, parity, or "check bits" to a packet or block of bits (hence the term "block codes") that allows the receiver to detect dropped bits and reconstruct them based solely on the other contents of the packet or block. Extrapacket FEC uses information from preceding packets to reconstruct the probable contents of a corrupted or lost packet. The

FEC strategy is usually implemented in the physical layer and is transparent to upper layers of the protocol.

Both methods are limited by the constraints of communications theory. The Hamming distance (d) is the minimum number of bit positions by which codewords for a particular code are different. The following two equations describe the error correcting and detecting capability of a code:

– Number of errors that can be corrected: $<= (d-1) / 2$
– Number of errors that can be detected: $<= d$.

Increasing the number of parity check bits does not lead to a linear improvement in the number of errors that can be corrected. As more and more parity bits are added, the necessary transmission bandwidth increases too, since more bits now occupy the same time interval. Increased bandwidth increases the probability of noise and additional errors. So a sort of law of diminishing returns sets in. The trick is to find the right tradeoff, selecting an optimal number of parity bits to correct as many errors at possible (thus approaching the (d-1) / 2 error correction limit) while at the same time maintaining reasonable communications throughput.

So-called cyclic codes are block codes where succeeding codewords inserted into the packet stream are simple lateral shifts of each other. Cyclic codes can be used to correct larger blocks of errors than non-cyclic block codes and are often used for both error correction and error detection. The related "convolutional codes" operate on a sliding sequence of data bits (or symbol groups) to generate the coded stream of bits. The best way to decode and reconstruct a bit stream from a convolutional coder that has picked up some noise from a network is by using the Viterbi maximum-likelihood decoding algorithm. Viterbi's algorithm is well known in clarifying the transmissions from satellites and space communications and is used in Qualcomm's Q1900 Viterbi Decoder.

Finally, a subset of the so-called BCH (Bose-Chaudhuri-Hocquenghem) cyclic codes known as Reed Solomon codes are nonbinary, multisymbol codes designed to provide multiple-error correction and are often used with IP packet routing equipment.

5. Whereas FEC strategies rely on error correction alone, Automatic Repeat Request (ARQ) uses error detection combined with feedback of information to the transmitter about good and bad packets / frames via positive or negative acknowledgments that results in the retransmission of corrupted data. Upon receiving an erred packet, the receiver requests a re-transmission of the packet from the transmitter. Generally the transmitter retransmits the message until it is either correctly received error-free in the buffer (whereupon the packets are all ordered in the buffer and then released to the system) or the error persists beyond a predetermined number of retransmissions, in which case they are considered truly lost and must be algorithmically reconstructed.

With ARQ, the number of overhead bits needed to implement an error detection scheme is much less then the number of bits needed to correct the same error, so the number of redundant bits used in FEC is much larger then in ARQ. For example, of the 40 Mbps bandwidth available per satellite transponder, only 23 Mbps transports the audio and video signals while the remaining 17 Mbps is used for forward error correction coding.

Algorithms used for ARQ include those relying on serial parity codes, modulo checksum codes and Cyclical Redundancy Check (CRC) codes (though efficient, CRC's can detect errors only after an entire block of data has been received and processed).

Hybrid ARQ / FEC algorithms have also been formulated wherein data blocks are encoded for partial error correction at the receiver and additional, uncorrected errors are retransmitted, such as Scoped Hybrid Automatic Repeat reQuest with Forward Error Correction (SHARQFEC).

Hybrid ARQs can be made more reliable than pure FEC algorithms affording increased throughput for same delay as FEC. However, the efficiency of both ARQ and hybrid ARQ schemes depend upon the packet size in both the forward channel and the reverse channel acknowledgements.

ARQ error control has been found to furnish poor throughput in the case of satellite multicasting unless one uses a combination of a terrestrial network parallel to the satellite network and a specially modified ARQ protocol. In this hybrid scheme, retransmitted ARQ frames are sent terrestrially, achieving a higher throughput than would a pure-satellite network.

- *Making the Subjective Objective.* Given all of the factors affecting voice and video quality, including latency, jitter, packet size, and packet loss - many people are confused as to how voice quality can be objectively measured, as opposed to the user actually working with the system and making a possibly arbitrary judgment based on subjective thought processes.

Most technical QoS issues center on network performance rather than end-user satisfaction. Although one can increase the intelligibility of a transmission by remedying audio packet problems, overall perceived speech *quality* is not necessarily equally improved.

A number of possible measures of sound quality for packet-based telephony have appeared. Dan Frankowski, while researching the transmission of sound on computer networks at the University of Minnesota for his thesis, came up with some measurements based on the statistical distribution of the number and lengths of the gaps in speech:

1. Gross number of silent events per conversation. This is dependent on conversation length.
2. Percent of packet arrivals that caused silent events. This factors out the length of the conversation, but for this measure what the user perceives depends on packet size. If there are ten packets per second, then 10% produces one silent event per second. If there are twenty, then 10% produces two events per second.
3. Number of silent events per second. This is a reasonable user-level parameter.
4. Total silence introduced into the conversation. This is another simple measure that also depends on the length of the conversation.
5. Percent of sound played that was forced silence. This factors out the length of the conversation. There are many more possible metrics: mean length of a silent event, median length, or even second-order statistics such as variation in the number of silent events per second, and so on. These can be considered to be too complicated without further justification.

Standards bodies such as the ITU have tried tackling the issue of voice quality and have promulgated two well-known recommendations: ITU-T P.800 (formerly known as P.80) and ITU-T P.861. P.800 defines a method to derive a Mean Opinion Score (MOS) of voice quality. The test involves recording several preselected voice samples (recorded according to another ITU recommendation, P.50) over the desired transmission media and then playing them back to a group of naive, inexperienced test subjects consisting of a mixed group of men and women under controlled conditions. The group's (hopefully) repeatable and quantifiable scores are then weighted to give a single MOS score that ranges along what's popularly known as the quality scale, a scale that beings at 1 (rating of "bad" or "worst") and ends at 5 (rating of "excellent"). An MOS of four is a rating of "toll-quality" voice.

For devices that rely on high-order compression technologies such as VoIP and videoconferencing, common carriers and major, multinational corporations usually demand MOS product performance verification tests prior to commitment, carried out by a highly-regarded independent agency such as Halhed Enterprises (Ottawa, Canada – 613-832-0451, www.hei.ca).

These subjective tests and ratings all sound very official and clear-cut, but the fact is that even the standards bodies' recommendations such as ITU-T (subjective assessment of speech transmission over phone networks) and ITU-R (subject assessment of image quality over video systems) are ultimately based upon relatively primitive five-point rating scales:

For assessing speech quality, the ITU recommended rating scale for both listening-only and conversation tests is the five-point MOS category scale, which ranges as follows: 1 (bad), 2 (poor), 3 (fair), 4 (good), 5 (excellent)

Listening-only tests can also be assessed via the listening effort scale which supposedly measures the effort required to understand the meaning of sentences. It progresses like this: 1 (no meaning understood with any feasible effort), 2 (considerable effort required), 3 (moderate effort required), 4 (attention necessary; no appreciable effort required), 5 (complete relaxation possible; no effort required)

In conversation tests, a simple binary difficulty test / scale follows the (connection) quality scale in the form of a question: "Did you or your partner have any difficulty in talking or hearing over the connection?" The answer can be 1 (yes) or 2 (no).

For rating general image quality the scale goes as follows: 1 (bad), 2 (poor) 3 (fair) 4 (good) 5 (excellent)

A special image impairment scale also has five increments: 1 (very annoying), 2 (annoying), 3 (slightly annoying), 4 (perceptible, but not annoying), 5 (imperceptible).

As for assessing image quality, conventional "single stimulus" methods are rated using the MOS or impairment scale, while the so-called Comparison Category Rating (CCR) is a subjective test where the signal from a device or service being tested is compared to "reference signals" of known-quality and then rated using a five-point, MOS-like Double-Stimulus Continuous Quality Scale (DSCQS) or a "double stimulus impairment" scale. Degradation Category Rating (DCR) is similar to CCR but is based on the test subject always listening to the reference signal which is followed by the signal passed through the device or service being tested. Finally, Degradation Mean Opinion Score (DMOS) is a listening test where the test subject compares first the reference signal, then the signal passed through the device or service under test.

P.861, also known as Perceptual Speech Quality Measurement (PSQM), was developed in Holland by KPN Research and adopted by the ITU in February 1998. P.861 attempts to automate the subjective processes going on in P.800 by defining a computer algorithm to do the quality testing. P.861 generates scores having a close correlation to the human generated MOS scores taking place under the P.800 recommendation. The unit of PSQM measurement is called the CSone (or "Compressed Sone"). PSQM measurements can be quite accurate but unfortunately PSQM specializes in the objective comparative quality measurement of telephone band (300 Hz to 3,400 Hz) speech codecs running over circuit-switched networks, not packetized voice networks (such as VoFR, VoIP and ATM) with their multitude of degradation factors that include jitter, packet / frame loss, bit errors and "clipping" caused by bad voice detection devices.

The limitations of P.861 / PSQM's automated measurement paradigm has instigated much research into other objective means of measuring voice quality. Even the ITU itself began to explore ways of replacing P.861 with a successor speech quality algorithm that best correlates with quality measurements gained subjectively with listening tests using real people.

One proposal is the Perceptual Analysis Measurement System (PAMS) which began as a project at British Telecom (BT) in 1992 to develop an engineering tool for the prediction of speech quality over a

Q

network connection, without the monetary and time costs associated with subjective testing. Under the PAMS regime of testing, a test is performed by putting a speech-like test signal, also developed by BT, into one end of a connection and recording the degraded output signal. The two signals are then compared mathematically, and then related to the subjective perception of quality as perceived by real users. The algorithms used have been developed using the data and experience acquired through years of subjective testing by BT.

At the core of PAMS is a model of human hearing that transforms the waveform representation of speech into a more relevant 'audible' representation. The speech signal is transformed into a time, frequency and amplitude depiction reminiscent of a spectrogram, the difference being that following the transformation the only parts of the signal that remain are those that would be heard by a person. This allows more relevant analysis of the difference between the original and degraded signals. Instead of simply sending bursts of monotonous waveforms, PAMS uses a linguistic approach where redundant phonemes are removed while maintaining a signal fully mathematically representative of speech. The signal, consisting of male and female voices, is generated for 30 seconds' duration. It sounds like a sequence of random nonsense words.

The accuracy of PAMS has been verified by comparing the results for a variety of speech carrier technologies against a database of subjective tests. PAMS is now a mature quality measurement technique and has been commercial available since 1998. PAMS licenses have been granted to manufacturers of network monitoring and test equipment to produce devices that assess speech quality in such applications as commissioning and monitoring, determining service level agreements, new equipment selection, equipment optimization, product design and troubleshooting network problems. More importantly, PAMS was the world's first objective speech quality assessment tool capable of assessing VoIP systems.

Still, research continues into the mysteries of just what constitutes the quality of sound and how it can be measured. Research derived from varied fields such as high fidelity audio engineering and hearing aid research have decomposed the phenomenon of "speech quality" into multiple "dimensions" of measurable variables such as intelligibility, loudness, naturalness, listening effort, pleasantness of tone etc. In their paper, "Quantifying the relationship between speech quality and speech intelligibility," (*Journal of Speech and Hearing Research*, 1995, vol. 38, pp. 714-725), J.E. Preminger and D.J. Van Tasell note that "Although a multidimensional view of speech quality has not been disputed, many researchers have taken a unidimensional approach to its investigation... When speech quality is treated as a unidimensional phenomenon, speech quality measurements are essentially judgments, and one or several of the individual quality dimensions may influence the listener's preference." The simple auditory and visual evaluations of speech or image quality currently in vogue only capture a small feature set of the great, complex and subtle world of perception. These primitive forms of evaluation prevent us from determining the degree to which the many perceived (though unmeasured) factors constituting "quality" are important, or even what their critical boundaries (minimum and maximum thresholds) are for various communications tasks. A more sophisticated level of speech and video analysis has not yet reached all related disciplines, particularly packet network telephony.

Types of Services for QoS

As we've seen, "best effort" service is always in danger of providing unacceptable service, and occasionally it can be completely swamped with network congestion or other problems, in which case it sometimes doesn't provide any usable service at all.

Thus, much of the focus of QoS concerns providing predictable service (service herein defined as the share of available capacity) even during periods of network congestion. Because each IP packet contains its own destination address/routing information and therefore needs to be handled individually by a router, the router can't discriminate "important packets" (carrying real-time voice and video) from packets carrying traffic that is not-so-important (faxes, e-mail, etc.). Because of this inability to separate out and speed important, real-time packet traffic on its way, periods of congestion usually cause such packets to arrive at their destination with delay, jitter, etc.

A simplistic way to achieve quality of service is to add more bandwidth. Since we can't immediately start digging beneath railroad tracks to string more fiber around the country, one alternative to simply adding raw bandwidth is creating a QoS-enabled network that allows for dynamic reallocations of existing shared resources (bandwidth), "reserving" router resources as needed just before a communication begins. This technique is associated with the so-called "integrated services," as we shall see.

Another major scheme doesn't reserve resources by signaling the router in advance. Instead, the priority level of the message is encoded into the packet along with the routing information. The routers and the network must therefore achieve the desired QoS "on the fly" by acting upon the service based on the QoS specified by each packet. This technique is associated with the so-called "differentiated services."

These two types of QoS are applied to two different types of packet streams, *flows* and *aggregates*:

A "flow" is a single, unidirectional data stream between sending and receiving applications. A flow is uniquely identified by five parameters: the transport protocol, source address, source port number, destination address, and destination port number. Two or more flows with some parameters in common make up an "aggregate".

Network topology, applications, and user preferences determine what kind of QoS is most appropriate for individual flows or aggregates.

Thus, Quality of Service (QoS) is really an umbrella term for a collection of technologies which allow network-aware applications to request and receive predictable service levels in terms of data throughput capacity (bandwidth), latency variations (jitter) and propagation latency from QoS-enabled IP networks which can respond to requests from critical applications for either resource allocations or differentiated levels of service among shared resources.

Being able to define, measure and report on service quality are also important attributes of QoS solutions, since some kind of verification of quality will become necessary so that service providers not only specify, contract and deliver QoS, but accurately bill their customers for QoS too. Businesses will pay for a high level of QoS, provided they can be confident that they're paying for a genuinely high level of QoS. A more general term for these capabilities is Class of Service (COS), which has come to mean the ability of switches and routers to prioritize traffic into different queues or classes

Interestingly, IP version 4 already provides an eight-bit Type of Service (TOS) field in the IP header (IP version 6 has an equivalent Traffic Class byte). The first three bits are used to indicate one of eight possible levels of packet precedence. This allows an IP node to designate specific datagrams as having higher priority than others, but the TOS is normally excluded from processing because some routers are known to change the value of this field, even though the IP specification does not consider TOS to be a mutable header field. Still, many router vendors map IP's eight different priority levels directly onto the underlying Ethernet LAN. Such "packet marking" can be done for any arbitrary QoS treatment in the network (e.g., high / low loss / delay priority).

Thus, thanks to TOS and QoS-enabled routers, COS can be experienced under controlled conditions on private managed IP networks. That's why VoIP / IP telephony sounds much better over a managed IP network than the Internet – the private managed network has low packet jitter because all of the routers on the network can be programmed to look for voice IP packets and prioritize them, putting them ahead of any data packets sitting in the router's transmit queue (such QoS functionality isn't yet used over the Internet since it requires end-to-end support). In this way, any long strings of outgoing data packets won't contribute to the jitter of voice packets. Also, large packets lead to another QoS concern over delay, since a very long video packet can delay a voice packet from exiting the router in a timely manner, a situation that can be remedied with "IP packet segmentation" where the router is programmed to segment all outbound data packets based on the WAN access bandwidth (packet segmentation and prioritization is particularly important at low bandwidth WAN access rates of 56 Kbps to 512 Kbps).

If we could permit high-priority traffic to cut ahead of other packets in Internet routers we could lower the overall latency (delay) for these special voice and video packets on the Internet too. The guarantee of low latency and jitter (the promise of COS and QoS) is critical for voice, but it's also important for video, which can use much larger packet sizes and higher transmission rates. However, there are still ATM networks comprising the backbone as well as IP networks, and a QoS solution must encompass all networks and equipment that the packets travers. ATM could provide different forms of traffic control and QoS but current higher-layer protocols cannot request ATM QoS classes specifically, since no mapping is done from the higher-level protocols to the underlying ATM network.

Moreover, despite appearances and hype to the contrary, COS doesn't actually "guarantee" anything, even in the pristine world of private managed networks. Deep down, COS is still basically a "best effort" service concept, although now the class of packets with the highest priority is designated as having the highest probability (gets the best chance) of achieving smooth, consistent throughput as desired. Even mighty ATM, with its virtual "circuit emulation" network model having the strongest (and most complicated) QoS signaling architecture that allows you to pre-select a level of quality, is ultimately measured and guaranteed in terms of the probability of cell loss, the average delay at a gateway, the variation in delay in a group of cells, actual cell losses, and the transmission error rate. All of this just doesn't seem to offer a truly rock-solid guarantee of quality to an application or a user.

One of the best expositions on the shortcomings of packet prioritization I've ever read comes from a newsgroup posting by James Carlson of IronBridge Networks (Lexington MA, – 781-372-8000, www.ironbridgenetworks.com): "(The problem with the concept of high-priority traffic is more easily illustrated with an analogy. If I want a guaranteed seat on an airline, I have to pay extra *and* I have to specify where I want to go. This works for large numbers of people, because we can, at least in principle, do global scheduling in advance. If we had a system where I could say, 'here's more money, but I won't tell you where I want to go until I show up,' this would still work as long as there weren't very many people trying this – in other words, if the price were high enough. The problem with both Switched Virtual Circuits (SVCs) in ATM and Quality of Service (QoS) in IP is that people are asking not to reveal where they want to travel, and they're apparently presuming that we can do this for large numbers of customers. This just isn't true. There is no solution to this problem

except channelization, which is, to stretch that analogy, being able to pay for a Lear jet.) (Or, another way to look at it: If everyone hits that porn site at 3 A.M., QoS goes right out the window. 20 Mbps of high-priority traffic just doesn't fit in a 10 Mbps pipe. QoS is, at best, a probabilistic statement. And as long as we're going with probabilities, we can do better and simpler than signaling by measuring and overprovisioning.)"

All of the above can surely be taken as a hindrance to those who wish a priority-enabled version of the current Internet Protocol Suite (IPS) to quickly evolve into a real-time and quality controlled Integrated Services Packet Switched Network (ISPSN). Prioritization is in fact only one of several possible implementation mechanisms, not a complete service model in itself. As soon as there are too many real-time voice packet streams, or "flows" competing for high priority status, then every stream will degrade gracefully, just like a "best effort" service. What users really want, of course, is for initial streams requesting high priority service to achieve it while other, later requests for high priority service should receive a "busy signal."

Even taking into account such criticisms of QoS models being too probabilistic or statistical in nature, the fact remains that to provide any sort of service "guarantees" at all – some reasonable level of quantifiable reliability and predictable quality over a user-determined time interval – the IP packet infrastructure layer needs to be supplemented with some kind of QoS functionality, which brings us back to the need for the network to differentiate traffic and manipulate the allocation of resources to enable different relative service levels for users and their varied applications. This same modified packet infrastructure layer must also transport voice control packets (e.g., call-control protocols such the Media Gateway Control Protocol – MGCP), as well as the voice packets themselves, and with the same quality.

Fortunately, the two major types of QoS now available, integrated service and differentiated service, are actually complementary, and can be used in combination for different network contexts. Let's examine them in more detail:

Integrated Service – IntServ

The IEFT-sponsored Internet Integrated Services (IIS) architecture expands the Internet service model so that its packet switching protocols can support Integrated Services (IntServ), which as its name implies, is the transport of both real-time and data traffic within a single network infrastructure. The architecture consists of four parts: a packet classifier in the host that determines the packet's route, an admission controller (so everybody can't grab high quality service at the same time), a packet scheduler in the host that makes the forwarding decisions (such as determining which packet is to be transmitted next wherever the queuing of packets may occur) to achieve the desired QoS, and the signaling process itself.

IntScrv is really a multiple service model handling multiple QoS requirements. In the IntServ model an application, prior to sending any data, is able to select from among multiple, specific levels of packet delivery service offered by the network. The application explicitly signals the network, apprising it of its data traffic profile and requesting a particular kind of service that can encompass its bandwidth and delay requirements. A number of these IntServ QoS traffic control classes have waxed and waned in popularity, such as "predicated service," "controlled load" that allows packets to be assigned priority so that they are not kept waiting in router queues as they cross the network (as described in the IETF's Request For Comments (RFC) document RFC 2211), and "guaranteed rate" or "guaranteed delay" (RFC 2212) where a specific amount of bandwidth is reserved and packets will not be dumped so long as traffic does-

Q

n't exceed the reserved capacity. Whatever technique used, all of the IP routers and subnets along the path followed by an application's data packets must support QoS control mechanisms.

The application sends data only after getting a confirmation from the network. The application is also expected to "keep up its part of the bargain" by sending data that stays within the data traffic profile it has described to the network.

Also to support this capability, there must exist a way for the application to signal its needs to the individual network elements along the path and to convey QoS management information between these network elements and the application. While there are any number of ways to do this, IntServ proponents have been enthusiastic over resource reservation setup protocols such as the Resource ReSerVation Protocol (RSVP) (RFC 2205). RSVP is thus not a transport or routing protocol, but rather an Internet control protocol. It doesn't carry data, but is designed to operate with existing routing protocols (such as version 4 and 6 of IP) and works in parallel with TCP or UDP data "flows." An RSVP packet is very flexible; it can vary in size and in the number of data types and objects.

The fundamental design of RSVP is based upon research performed during the period 1992 to 1993 by a collaboration that included Lixia Zhang (of Xerox PARC), Deborah Estrin (USC/ISI), Scott Shenker (Xerox PARC), Sugih Jamin (USC/Xerox PARC), and Daniel Zappala (USC). Sugih Jamin developed the first prototype implementation of RSVP and successfully demonstrated it in May 1993.

Beginning in 1993, the University of California / Information Sciences Institute (ISI) in Marina Del Rey, California, was funded by the U.S. Defense Research Projects Agency (DARPA) to perform research, development, and technology transfer on RSVP. The general thrust of this project was to move RSVP forward from a set of research ideas into Internet practice.

During the initial RSVP project (April 1993 to June 1995), ISI developed a research prototype RSVP daemon, wrote successive drafts of an RSVP protocol spec, and introduced RSVP into the IETF standardization process. The project's final technical report "The Design of the RSVP Protocol" summarizes the RSVP protocol as of July 1995. Version 1 of RSVP is specified in RFC 2205 and several ancillary RFC's documents, and the Internet Engineering Task Force (IETF) has reached consensus on the RSVP specification documents as Internet Proposed Standards.

Since July 1995, ISI has been continuing RSVP development and technology transfer under a follow-on contract. This contract, which originally carried the project title "RSVP2", was later renamed to "RSVPTNT" when it was moved underneath the "TNT" umbrella contract at DARPA's request. The primary objectives for this project are: maturing the base RSVP protocol specification and working through the IETF to make it a Proposed Standard, creating a reference implementation that also includes experimental protocol features, and performing research on the interactions with routing, on access control and accounting, and on aggregation of state.

Important pieces of the RSVP protocol and of the reference implementation distributed by ISI have been developed or are under development by other organizations.

Although ordinarily used on a per-flow basis, RSVP is also used to reserve resources for aggregates. Senders first specify their outgoing traffic in terms of the preferred upper and lower bounds of bandwidth, delay, and jitter. The network's part of the deal is to maintain a per-flow state and then perform packet classification, policing, and intelligent queuing based on that state. The main component of the RSVP architecture that achieves this is a flow specification, or "flowspec," which describes both the source's traffic stream and the application's service requirements. Flowspecs actually specify the desired QoS, and are employed to set node's packet scheduler parameters.

At this point the two fundamental RSVP message types come into play: PATH and RESV. PATH messages are emanated from the sender and contain the sender's traffic specification (TSpec) information that defines the data flow and the IP address of the previous hop. Each RSVP-enabled router along the downstream path registers the previous hop and replaces it with its own address. Each router thus establishes a "path-state" that includes the previous source address of the PATH message (which is the next hop "upstream" towards the sender). The PATH messages are ultimately sent to the destination address of the unicast or multicast receiver(s).

To make an actual resource reservation, the receivers send a RESV (reservation request) message upstream to the local source of the PATH message (the previous hop). The RESV message also includes the TSpec and an RSpec that contains the session specification's actual desired QoS type (such as controlled load or guaranteed delay), and a "filter spec" that characterizes the packets for which the reservation is being made (e.g. the transport protocol and port number). The filter spec is used to identify the sender(s). Together, the RSpec and filter-spec represent a "flow-descriptor" that routers use to identify reservations. Fields in packet headers can be scrutinized and the data packets can be filtered by determining the set of data packets to which the flowspec-defined QoS will apply. Packets meant for a particular session but not matching any of the filter specs defined for that session are handled as slower, best-effort data traffic instead of high priority.

When an RSVP router receives an RESV QoS request message, reservation setup occurs and two decision processes come into play: Policy control and admission control.

Policy control determines whether the user has authorization to make a reservation. Policy control is an amalgam of control, user authentication and accounting information. A policy server may tell a router or LAN switch that you can't do web collaboration or videoconferencing during the busiest time of the day.

Once it's decided that the user and the application are in good standing and can begin communicating across the network with another end-point, admission control determines whether sufficient available resources can be allocated to allow the requested "connection" to occur at the preferred QoS. If resources at not available then the router returns an error back to the receiver. If everything checks out okay, then the router sends an RESV message upstream to the next router. The final router in the series (which is either near the sender or at a reservation merge point for multicast flows) receives the RESV message, accepts the request and sends a confirmation message back to the receiver.

RSVP's major innovative attributes include:

1. The use of "soft state" or the "connectionless" approach in routers, servers and hosts for both unicast (one source to one destination), multicast (one source to many destinations) and many sources-to-one-destination transmissions.

"State" in regards to network nodes refers to information about network conditions stored in the router nodes by networking protocols. The network state is stored in a distributed fashion across various nodes and various protocols. For example, a teleconference or multicast application might run on top of multiple protocols. In the host, the Internet Group Management Protocol (IGMP) stores information about the multicast groups in which the host participates. Depending upon the membership information multicast routing protocols such as the Distance Vector Multicast Routing Protocol (DVMRP), Multicast Open Shortest Path First

(MOSPF), Protocol-Independent Multicast (PIM) or Core Based Tree (CBT) can create a multicast forwarding state in the routers. Reservation protocols such as RSVP or STream Protocol Version 2 (ST-II) may reserve network resources for the teleconferencing session along the multicast tree.

So that the state can be modified to reflect the changes in network conditions, the network nodes communicate with each other to exchange information. Based on these control messages the network nodes modify their stored state. For instance, a change in network topology triggers the routers into exchanging messages and perhaps modifying the multicast forwarding state if necessary.

These network node states can be categorized as "hard" or "soft" states. A hard state is one that is installed when the node receives a set-up message and is removed only when the node receives an explicit tear-down message. The reservation protocol ST-II and the multicast routing protocol CBT are based upon hard state principles. Hard state architectures use explicitly acknowledged, reliable message transport that necessitates increased protocol complexity.

"Soft state" protocols such as the Protocol Independent Multicast (PIM) and RSVP, on the other hand, use periodic refresh messages to keep the network state alive while adapting to changing network conditions. The refreshes are sent periodically after one "refresh period." A change in the route causes these refresh messages to automatically install the new state in the network components along the new route. In this way, RSVP caches the reservation state in the network nodes and uses periodic refreshes by "PATH" and "RESV" messages to maintain the soft state in the network nodes. The state is discarded if no refresh messages arrive before the end of a "cleanup timeout" interval that's usually a small multiple of the refresh period, the size of the multiplying factor used being dependent on the degree of robustness desired. The state can also be deleted by an explicit "teardown" message.

At the expiration of each timeout period or after a state change, RSVP scans its state to build and forward PATH and RESV refresh messages to succeeding hops along the network.

When RSVP and the network is in steady state, the state is refreshed hop-by-hop to enable "merging" (multicast reservations are "merged" at traffic replication points while traveling upstream, which involves complex algorithms). When a received state is different from the stored state, the stored state is updated and new refresh messages are generated and immediately propagated end-to-end throughout the network. Propagation of a change ceases when it reaches a point where merging no longer results in a state change. This minimizes change-generated RSVP control traffic and is considered essential for scaling to large multicast groups. Still, some concerns have been raised over the scaleability of soft state protocols like RSVP. With RSVP, the number of refresh packets scale linearly with the number of sessions and sending many packets results in a high overhead. Some experts feel that soft state protocols can be made scaleable only if control messages can be constrained independent of the amount of state in the network.

During quiescent, steady network conditions, hard state protocols generate less control traffic since they don't send the kind of periodic refresh / control messages that soft state protocols do. Soft state protocols can adapt faster to changes in network conditions, but at the expense of the bandwidth needed for sending the periodic refresh messages. However, when the network is dynamic (drastic and continuing traffic and/or topological changes), hard state control messages must also be generated anyway for the same adaptation purposes, which means that under such circumstances the extra bandwidth taken up by control messages for soft state and hard state paradigms would be of similar magnitude.

2. Receiver-controlled reservation requests, based on a dynamic multicast heterogeneous receiver model. Normally, Quality of Service requests on a network brings to mind a user or application sending a request for a desired level of service. One is reminded of how ATM sessions negotiate QoS parameters while setting up a session. By contrast, RSVP is rooted in the IP multicast paradigm where stations must register first before participating in a session. With RSVP, it is the receiver of a data flow, not the sender which initiates and maintains the resource reservation used for that flow or aggregate. This is because a source-initiated reservation could never deal with many different kinds of receivers participating in a multicast. One receiver may enjoy a high bandwidth connection, another a low bandwidth connection. A source-based reservation system could not know in advance about the capabilities of all of the receivers in the network (indeed, the sender knows neither the identity of the receivers nor even whether anyone is even listening), and so it might over- or underprovision resources for the transmission. Using a receiver based model, however, each RSVP receiver can initiate its own reservation of a desired level of resources, keeping it active as long as it or the user wishes. This scheme acts as a sort of distributed solution for the problem of resource reservation.

3. Flexible control over sharing reservations and forwarding of subflows.

4. Data distribution via IP multicast (RSVP receivers can establish and maintain – independently of its creation – resource reservations over a "distribution tree" of participating users). As packets from the multicast's source arrive from the Internet at a gateway host, they are classified and scheduled out using a set of queues and timers.

Interestingly, flowspecs provide no explicit loss metric for the bandwidth, and jitter is assumed to be resolved at the endpoint systems. Even though the receiver originates RSVP reservation requests, it is at the sender where the actual QoS control takes place. The resource reservation requests are passed on the reverse flow datapath from the receiver to all senders but RSVP sends its messages as IP datagrams with no enhancement of reliability. The periodic refresh messages from hosts and routers are expected to handle lost RSVP messages. If the effective cleanup timeout is set to K times the refresh timeout interval, then RSVP can tolerate K-1 successive RSVP packet losses before mistakenly deleting a state.

When RSVP packets encounter gateways that don't support RSVP, they can be "tunneled" through as ordinary packets. RSVP traffic traversing non-RSVP routers in this way creates a weak link in the QoS chain, causing the service to fall back to "best effort" with no resource allocation occurring across these links.

Furthermore, although a receiver can ask for an acknowledgment of the QoS it has requested (called "one pass service" where the receiver includes a confirmation-request object containing its IP address in a RESV message), this acknowledgment is not a true confirmation of QoS, but just an indication that the request has been accepted with a high probability of success. This is because the request jumps from node to node until it is either rejected or it encounters a node with a preexisting reservation that's equal or greater in priority to itself, at which point it stops moving through the network. And then the last node which processed the reservation sends a confirming "ResvConf" message back to the receiver.

An enhanced version of one pass service is called One Pass With Advertising (OPWA) where a receiver can predict the end-to-end QoS available by discovering the path's characteristics before making any resource reservations. It does this by sending a PATH message that includes a package of OPWA "advertising" information known as the

Q

Adspec parameter. When Adspec is received by a router it's passed to the local traffic control mechanism, which returns an updated Adspec. In a way similar to incremental route calculation, OPWA permits incremental accumulation of the delay for a reservation. The original receiver can use these Adspec "advertisements" to evaluate end-to-end delay prior to adjusting or constructing new resource reservation requests. This greatly improves the odds of a successful reservation and can make adjustments for path characteristics that change during the reservation process, such as the Guaranteed Service queuing delay.

The source of some of the complications surrounding RSVP centers are the fact that RSVP doesn't do double duty as a routing protocol. Real routing protocols such as Router Information Protocol (RIP) and Open Shortest Path First (OSPF) are able to identify the fastest paths in a network, but they have no facility to ascertain which connection has enough bandwidth to serve RSVP requests. This is a far cry from ATM's Private Network to Network Interface (PNNI) that has been in use since April 1996 and which defines both the signaling and routing protocols for use between ATM switches, enabling ATM to not only find a path, but one with a designated bandwidth capacity.

Still, RSVP implementations are available in many commercial products. See the Buyer's Guide Marketplace at www.ipmulticast.com and www.qosforum.com.

For example, the Cisco Internetwork Operating System (Cisco IOS) QoS includes RSVP along with various intelligent queuing mechanisms used to provide Guaranteed Rate and Controlled Load services.

Differentiated Service — DiffServ

Differentiated service or the "DiffServ" architecture is another IETF initiative to provide a scaleable end-to-end QoS. It's a multiple service QoS model but, unlike integrated services, an application using differentiated service does not first tell the router what QoS it requires before sending data. This was to be avoided since the goal for DiffServ is to define a scalable service discrimination model without the need to maintain the state of each flow and signaling at every hop.

Instead, we start off with what at first looks like a conventional best effort service network, but each packet is now "marked" to assign priority and thus QoS. This can be done quite simply by using the normally unused Type of Service (TOS) field in IPv4 or the Traffic Class Field in IPv6, in which case the field is now referred to as the Differentiated Services Code Point" (DSCP). One can also use the packet's source and destination addresses to "mark" the packet's traffic classification.

Each DiffServ flow is policed and marked according to the service profile at the edge of the network and policy decisions and implementations are left to local "trust" domains. Network core devices read only packet markings, not flows, so per-flow state is required inside the network.

Whatever the encoding settled upon, routers at DiffServ nodes look for the special bits and attempt to deliver a particular kind of forwarding treatment, or per-hop behavior, at each network node based on the QoS specified by the bits in each packet. DiffServ reserves enough bits per packet to define up to 64 classes of service.

Diffserv is a network layer protocol (Layer 3 of the OSI Model) and it doesn't deal with or rely upon underlying layers. DiffServ provides a "softer" form of QoS with its "coarse" and simple way to categorize and prioritize network traffic, which is one reason why it's considered ideal for aggregate flows. DiffServ traffic shaping occurs at the network's boundary nodes where the "marked" packets are aggregated into flows according to their classifiers. Indeed, all of DiffServ's most intensive data traffic classification and conditioning (metering, marking, shaping and

policing) occurs at the network's edge.

Cisco IOS QoS supports the differentiated service model with such features as Committed Access Rate (CAR), which performs packet classification through IP Precedence and QoS group settings. CAR performs metering and policing of traffic, providing bandwidth management. Cisco IOS also has intelligent queueing schemes that can be used with CAR to deliver differentiated services.

The Multi Protocol Labeling Switching (MPLS), yet another IETF initiative, borrows ideas from both DiffServ and the circuit emulation abilities of ATM (in fact, it can leverage existing ATM hardware and can even compute routes through a network based on constraints in exactly the same manner as ATM using a distributed topology database). MPLS evolved from Cisco's "Tag Switching," IBM's "ARIS," and Toshiba's "Cell-Switched Router." MPLS looks to be more of a "traffic engineering" protocol than a QoS protocol since MPLS can be used to bring connection-oriented mechanisms to connectionless network layer protocols, such as "fixed bandwidth pipes" similar to ATM virtual circuits

In an MPLS network, incoming packets at the ingress (entry) point are assigned a "label"or "identifier" by a Label Edge Router (LER). The LER then encapsulates the packet with an MPLS header containing the label information, and the packet is forwarded to the next hop along a Label Switch Path (LSP) where each subsequent Label Switch Router (LSR) makes forwarding decisions based upon the label information from the header. At each hop, the LSR strips off the existing label and swaps it out for a new label which tells the next hop where to forward the packet. The routers use the label value as an index into a table that specifies the next hop and a new label. The LSPs behave like ATM virtual circuits but have even greater resource and performance efficiency.

MPLS is akin to DiffServ in that it also marks and unmarks packet traffic network entry and exit points. But unlike DiffServ, which uses the marking to determine priority within a router, the 20-bit labels of MPLS markings are primarily designed to determine the next router hop. MPLS exists only on routers. There is no end-host protocol component, nor can MPLS be controlled by an application, since no MPLS APIs exist.

The original selling point for Label-based switching methods was to allow routers to make forwarding decisions based on the contents of a simple label, rather than by performing a complex route lookup based on destination IP address. This initial justification for technologies such as MPLS is no longer considered to be the main benefit, since Layer 3 switches (ASIC-based routers) can now perform route table lookups at tremendous speeds. However, MPLS conveys many other benefits to IP-based networks, including the following:

Elimination of Multiple Layers – Most carrier networks employ an overlay model where ATM or frame relay is used at Layer 2 and a logical IP routed topology runs over it on Layer 3. The Layer 2 switches provide the high-speed connectivity, while the IP routers at the edge provide the intelligence to forward IP datagrams through virtual Layer 2 circuits. Multilayer switching integrates OSI Layer 2 switching and Layer 3 routing so that Layer 3 can directly enjoy some of the speed advantages found in Layer 2. Using MPLS, carriers can migrate many of the functions of the ATM control plane to Layer 3, thereby simplifying network management and network complexity. Eventually, carrier networks will migrate away from ATM completely, which will eliminate losses incurred by dicing up IP traffic and fitting it into ATM cells (amusingly referred to as ATM's "cell-tax")

Traffic Engineering – Since Layer 2 information about the network links is now integrated into Layer 3 of an IP network, there is now both the ability to set the route packet traffic will take through the network, as well as the ability to set performance characteristics for a class of traffic. Also, network operators now have flexibility in terms of diverting and routing traffic around link failures, congestion and bottlenecks.

VPNs – using MPLS, service providers can create IP tunnels throughout their network, without the need for encryption or end-user applications.

Protocol Independence – MPLS is "multi-protocol" it can be used with network protocols other than IP, such as ATM, frame relay, IPX, PPP, etc. or directly over a data-link layer such as the MPLS architecture itself.

Thus, MPLS and IP can combine to offer the best of IP and the best of circuit switching technologies.

Many experts feel that DiffServ complements MPLS for enabling end-to-end QoS for IP traffic. An MPLS network can obtain a packet's QoS information from the DS byte, and DiffServ's large set of traffic management services can be used by MPLS at the network layers where they are needed most. DiffServ works at the edge where customer traffic meets a service provider network, and because it specifies QoS at Layer 3 it can run without modification on any Layer 2 infrastructure that supports IP. MPLS shines when it adds QoS-aware IP routing functionality to the network core. Such a MPLS / DiffServ combination might allow network operators to completely prioritize packet traffic and provision network bandwidth to deliver toll-quality voice calls and genuinely differentiated data services even over the Internet.

Recommended Websites:

Quality of Service Forum – www.qosforum.com

QoS Forum IP QoS Glossary – www.qosforum.com/docs/glossary

Business Communications Review QoS Glossary – www.bcr.com/bcrmag/04/499p25s2.htm

Data Communications QoS Glossary – www.data.com/issue/990507/policy_watch.html

HP/Agilent QoS Definitions – www.tm.agilent.com/tmo/iia/comms/hub/issues/qos_index.html

IP Highway QoS Glossary – www.iphighway.com/glossary.html

Orchestream QoS Glossary – www.orchestream.com/supp_06.html

Q

R.100

The Enterprise Computer Telephony Forum (ECTF) (Fremont, CA – 510-608-5915, www.ectf.org) nurtured development of R.100, an interoperability agreement that addresses the need for easily gathering, consolidating and reporting call center data. It was released in December of 1999.

The R.100 standards consist of a call/thread model and employs open application programming interfaces such as TAPI and JTAPI, and sits above the ECTF call-control architecture to define how to derive data from both circuit- and packet-switched voice calls.

Today's call centers contain a proliferation of systems that enhance the effectiveness of the center, but each one collects separate information on individual calls. All of the data collected by these different systems is needed to intelligently modify and fine-tune call center activities, but this very proliferation of systems presents a barrier to the correlation of data for an individual call. R.100 presents standardized techniques for gathering this data in a consistent manner from all the devices in a call center.

R.100 defines three types of records:
- call segment – basically a piece of data representing a call transition (similar to a state change)
- a call record – for Call Detail Recording (CDR) purposes.
- a "thread" - that ties all sub-operations together.

These techniques make the reporting process much easier and more effective. R.100 is available now, and can be downloaded from the ECTF web site at http://www.ectf.org.

"The availability of R.100 represents an important milestone for making call centers an ever more potent element of successful business strategies," says Ed Verney, ECTF chairman and president. "Today's call centers are at the very hub of effective customer relationship management strategies, but as business conditions change they require constant fine tuning to extend their effectiveness. Finally, R.100 will provide the techniques necessary to efficiently gather the right data to intelligently implement these needed changes."

"R.100's techniques for providing in-depth, actionable data on all aspects of call center performance is a boon to developers as well as to call center managers because it provides a beaten path for a host of new product opportunities," said Roger Huffadine, technical director of call center software and training specialist at Callscan in the UK. "The reporting techniques detailed in R.100 make it much easier for developers to enhance their product lines with data-reporting solutions that will be well received by their customers."

R.100 defines common terminology for elements within the call center environment, minimum data sets to be available from call center functions, a common tracking method from call initiation to conclusion regardless of number of segments, to eliminate multiple counting, a method for clock synchronization, and data structures.

In developing R.100, the ECTF took pains to ensure that it would encompass current and future computer telephony (CT) technologies including new switching fabric and media processing resource implementations. The model presented by R.100 is applicable in environments with intelligent networks, virtual call centers, networked call centers, circuit switched networks, IP telephony networks, computer telephony integration (including ECTF C.001, TAPI, TSAPI, JTAPI, and ECMA CSTA); and/or multimedia systems and calls.

RBOC

See Regional Bell Operating Company.

Readerboard

Also called an Electronic Displayboard, Electronic Wall Display or Message display Unit (MDU). These can still be found in many call centers.

A traditional readerboard is a large sign, usually placed on the wall of a call center. It displays useful information agents can view at a glance to make informed decisions on how to improve call processing. The main data displayed is most often up-to-the-minute ACD statistics. These stats are updated frequently, usually every 30 seconds or so. Common stats displayed include the number of calls in queue, the amount of time the call that has been holding the longest has been on hold, the average speed of answer and the service level.

There are many options to choose from in displays. Some models deliver a lot of different colors, sizes and lines of text. Some vendors let you add TV monitors to your existing configuration. The software driving your readerboards must be compatible with your existing networks and be able to read the data pumped out by your switch. A system is set up so it pulls data out of your ACD, formats it, then displays it (on either readerboards or PC displays) so agents can make informed decisions.

In some call centers, additional info is added to readerboards along with the standard ACD information. For example, you may want to plug in data about specials or incentives, birthdays, congratulations, news or even network info ... things you feel agents should know about or will get them pumped up and ready to take on calls.

The use of colors is a big readerboard feature. Colors indicate when things are going okay, getting tricky, and when the call center is in big trouble. Most displays use the standard red, yellow and green. Audible alarms are other helpful features, sounding to alert agents to a crisis.

One of the latest trends in displays is taking data to the agent desktop. The same information shown on full-sized readerboards may be shown, in mini form on PC displays. Most PC displays can be made to look like tiny readerboards, complete with color-coded thresholds and alarms. Agents using this type of display in theory don't need to look up from their desk because data is conveniently visible in the corner of their screen.

PC displays also solve the problem of agents who can't see the readerboard from where they are sitting. Your call center may be laid out with high-walled cubicles, or have pillars or some other obstructive object preventing a clear view of the screen.

An electronic display yields improved call statistics and improved agent morale. Readerboards let your agents know when call volume is low. They can then take action – moving to outbound to make a few calls. And when a surge of calls come in, they see it and can flip back to handle the queue. All this takes a lot of pressure off both you and your agents.

Regional Bell Operating Company (RBOC).

The RBOCs are the remnants of the dismantled AT&T / Bell System. They were created by antitrust Judge Harold Greene at the conclusion of the government's divestiture of the Bell operating companies from AT&T in 1983. There were originally seven RBOCs, each of which owned two or more Bell Operating Companies (BOCs). The original RBOCs: Ameritech, Bell Atlantic, BellSouth, Nynex, Pacific Telesis, Southwestern Bell and U.S. West.

Router

A router is usually conceived of as a device sitting at the juncture of the Internet and a LAN but it can also be software running in a computer. A router determines to which network node a packet should be forwarded toward an ultimate destination. A router sits at the juncture of at least two networks or gateways. Each network normally has a local (default) router connected to the local LAN and also to another ("remote") LAN. A router maintains and consults with a sort of "least cost routing" table containing available routes, and distance and cost algorithms to figure out the best route for a given packet, which usually involves encountering more routers. A so-called edge router interfaces with an Asynchronous Transfer Mode (ATM) network or similar core / backbone network. A brouter is a network bridge combined with a router.

To send packets from your computer on your local LAN to computers which belong to other IP networks that are not directly connected to your LAN requires the use of an intermediate system (i.e. an IP router).

One particularly controversial topic is whether to use a dedicated external router or an internal serial interface card, all of which go by the misnomer of "router cards". Computer makers will tell you that external routers are largely unnecessary devices for most scenarios if you are running one or more UNIX-like hosts already, and are less functional choices even if you're not. See www.etinc.com/routers.htm.

RSVP

See Quality of Service.

S Series

The ECTF (See Enterprise Computer Telephony Forum) specifies software interfaces used throughout their vision of a computer telephony architectural model based upon open, shared resource servers. The foundation for the ECTF's software standards is the S.100 media- and switching-services API that allows resources to be shared in CT servers. For client/server systems, the ECTF's S.200 media-services transport-protocol interface provides client inter-communication with the server. If a manufacturer happens to supply a product that encompasses both a client and server, it is not restricted to using S.200 but can instead use its own proprietary protocol. The ECTF's S.300 service-provider interface permits the telephony server to interchangeably use resources from different vendors as they work together in sharing CT resources.. The S.900 administrative-services interface includes functionality for managing the server's configuration, faults, performance, security, and accounting.

The S Series software standards are said to complement, rather than replace, existing call control API standards, such as Microsoft's Telephony Application Programming Interface (TAPI) and Novell's Telephony Services Application Programming Interface (TSAPI), the Computer Supported Telephony Application (CSTA), IBM's CallPath or the Javatel Applications Programming Interface (JTAPI).

S.100

S.100 is the software foundation for the ECTF's (See Enterprise Computer Telephony Forum) "Framework" for creating a CT computer that processes audio, fax, image and other "media" transmitted over phone calls and makes those media usefully accessible (interactive voice / fax response) and manageable (voice mail, fax servers, unified messaging, etc.).

It consists of openly defined, modular hardware-*and*-software elements – including multi-vendor / multi-resource application components (voice, fax, text-to-speech (TTS), automatic speech recognition (ASR), network interfaces, etc.), a specialized Time Division Multiplexed (TDM) bus, high-level APIs, lower-level hardware Service Provider Interfaces (SPIs), and an implied software neural network to control everything.

These pieces come together to form "servers" that make CT media-processing an open "enterprise service" – a peer alongside SQL database, directory, messaging, file-and-print and other common back-office network services.

In theory, these principles all sound good but, when it comes to computer telephony standards, principles have a habit of remaining annoyingly abstract.

It's taken quite a while for the ECTF's "standardized server" to live up to its advanced billing. After investigating the most current real-life development initiatives and available product offerings revolving in and around the Forum's Framework, one can point to several reasons why:

• Great Expectations: Frankly, there was too much hoopla too soon, and the Framework was pushed in the Press harder than it should have. Unfortunately, we pundits had no business immediately tossing around buzzwords like "shrinkwrapped" before there was anything to put in the box.

The fact is that something as ambitious as the Framework needs to be examined cautiously and step by step. Remember, when you get down to it, the major goal here is to make CT software work much like typical Local Area Network (LAN) software, and to do it from all (end-user, VAR and developer) angles. To achieve this will be quite an accomplishment.

• Multi-Vendor CT Is Not Trivial. One of the big promises of the Framework is that app developers armed with a standard high-level family of hardware-shielding APIs will be able to focus their expertise on specific app niches and specific parts of an overall media-processing solution, rather than continuing to write generic one-size-fits-all CT packages for all sorts of enterprises.

Once more, like LAN software in general, the idea is that Vendor X's stuff will work along side other vendors' stuff to form complete, more tailored solutions.

Again, sounds great. But after talking with people who are actually developing within the Framework, it's apparent that the act of passing multi-vendor software control back and forth over living telephone calls – a/k/a call handoff – has been a real stumbling block in the labs. This has kept all current system development far removed from the shrink-wrapped ideal.

• TAPI Looms. Even before Microsoft released its first glimpses into TAPI 3.0, many people who could've pounced on the ECTF media-processing server Framework didn't. Instead, they indicated they were waiting for the word out of Redmond.

Without opening up a whole other TAPI vs. Framework discussion and how the two may or may not interoperate, it's clear that the former now directly addresses (or at least promises to address) media processing and therefore does step on the latter's turf.

Of course, people can (and do) say that the application-level API of the ECTF Framework (S.100) is done and it always takes Microsoft forever to revise TAPI? (it true that Microsoft's Windows Telephony delivery track record has been sketchy at best). Still, there's no question a potential conflict is muddying the development picture here and retarding Framework development.

• Not Everyone Likes Standards. Some manufacturers and developers who could be helping with the ECTF Framework aren't, and for practical reasons.

First of all, they're not terribly upset that making the magic of CT happen is often hard. They've already spent a lot of time and energy learning the inner workings of a plethora of CT resource cards *and* the arcane telephony and networking protocols they routinely employ. They know how to generate application code that's absolutely responsible for making it all work – from interfacing the system to the telephone network, playing and recording voice messages, sending and receiving faxes, performing ASR, etc.

They also know how to charge premium fees and margins for their hard-earned expertise.

Naturally, they're not overly enthused with the idea of standardized high-level software development and certainly aren't hopping through hoops to make it happen.

• A Vital Missing Piece. To further taint the utopian scene of multi-vendor apps and hardware all working in harmony under the ECTF Framework, it's important to note the time it's taken the ECTF to finalize its crucial S.300 SPI layer.

The theory behind S.300 is that CT technology developers will have a hardware-independent software interface for their device drivers and firmware / hardware products. In other words, this is the software interface that will allow the required server *middleware* to take care of low-level CT resource management between the different hardware components installed in the server and the S.100 requests for those resources.

Without this piece, you can't expect anything but vendor-hardware-specific implementations.

• Dialogic and CT Media Backlash. Despite the fact that the ECTF includes every major CT vendor you can shake a stick at (including all of Dialogic's competitors), some companies (even those with people sitting directly on ECTF Framework technical committees) still paint the initiative as Parsippany propaganda.

Some of this is prejudice left over from the ECTF Framework's considerable roots in Dialogic's Signal Computing System Architecture TAO (Telephony Application Objects) Framework. Some of this is legitimate protestations over a specification that is relatively young and not even entirely finished.

But some of it is surely sour grapes, since, at least as far as you can talk about the Framework as it exists today, Intel's Dialogic subsidiary is closely aligned with almost all tangible news in this area. This is because Dialogic was the first major vendor to supply the required server *middleware* product for application software written to the already ratified ECTF S.100 portion of the Framework specification.

We're talking of course about their much ballyhooed CT Media platform, which is still the only Framework server software in town (and the real reason why Intel bought Dialogic), though Brooktrout is on record as saying they will provide an S.100 compliant middleware engine within their BOSTON (Brooktrout Open System Telephony) architecture.

In either case, unless S.300 is fully embraced by the industry, we're still looking mostly at systems that work with a particular vendor's server software and hardware.

Still, Rome wasn't built in a day, we're talking about the beginnings of a major industry undertaking. What's more, there are some very intriguing products emerging here, particularly in the area of adjunct servers for traditional telephone systems (PBXs / ACDs and enhanced services nodes).

Secondary Market

The "stigma" of buying used equipment for call centers has pretty much disappeared in recent years. In fact, the secondary marketplace offers several benefits that are hard to resist.

It's possible to save about 30% to 75% off the list price when buying used equipment. The age of the product, the level of demand and the degree to which it has been refurbished are all factors that affect the price. And in the call center market, keeping up with the latest systems isn't easy. Upgrades and new releases are put out constantly; most products you may end up finding will be perhaps 5% to 10% less than the list price.

Secondary vendors rarely charge to expedite delivery of parts or merchandise, and many will track things down for you and offer technical support on products they sell – free of charge – so you'll be saving money in these areas. There are other great benefits to working with a good secondary vendor:

- Refurbishing. Some of these shops take used equipment and "snazz" it up on both the inside and out to like-new condition. When it comes to refurbishing equipment, CMS Communications' (Chesterfield, MO – 800-830-1788, www.cmsc.com) depot repair program involves a 14-step process that has been refined since the company started up in 1986. The program includes component testing, 24- to 48-hour burn-in (with load simulators to create in-service conditions) and a thorough cosmetic enhancement. CMS says they make each instrument look and function like new.
- Advance Replacement. Like many other secondary vendors, CMS manages an instant repair program. CMS stocks an inventory for a vendor, so as soon as it gets a call for a replacement, it sends a new part out immediately – even before the defective part has arrived back at the company. If you're looking for a secondary marketer, make sure they

offer this service plan.
- Hard-to-find parts/extensive inventory management. A lot of the market for secondary systems comes from companies looking to get rid of their own telecom inventory. Besides running a full-service repair facility, Telrepco Services (Wallingford, CT – 203-269-5151, www.telrepco.svc.com) also resells and purchases telecom and computer products, and specializes in hard-to-find parts that support both new and secondary market platforms.

Telrepco also provides a complete repair/refurbishing service for manufacturers, third-party maintenance companies and end-users, on equipment including key systems, PBXs, PCs, networking products, terminals and other related peripherals. To support its varied customer base, other products and services (e.g., Parts Support, Assembly Load Test, Warranty Services, Depot Repair, Asset Management and Reconfiguration programs) are available.

A big secondary market player since 1971, Source (Dallas, TX – 800-608-6514, www.source.com) manages and processes over 60,000 parts, components and systems each month. The company provides communications equipment, manufacturer-neutral product support and repair and consulting services. Products are warehoused at Source's facility, where they're inventoried using a serial bar coding system. The bar coding system is critical to product tracking based on make, model and value, as well as warranty, tracking, repair and performance measures.

- Manufacturer specialization. Atlantic Telecom (East Hanover, NJ – 800-418-9199, www.atlantictelecom.com), for example, resells both new and refurbished Lucent Technologies equipment. Its product inventory ranges from the smallest peripheral to the latest technology Lucent offers. Atlantic is distinguished for its product quality, customer service and competitive pricing.
- More personalized attention. The longer you work with a secondary vendor, the closer your relationship becomes with them, the more quickly they can respond to your needs; they will know what equipment you have and what parts you need in a jam. KLF (Indianapolis, IN – 317-872-8888, www.klf.net) says its business is founded on understanding and getting to know customers over the long term.

Though many of KLF's (a division of KLF Business Communications Systems, itself now a part of Communications Systems USA) customer relationships begin in used equipment, they tend to evolve into new equipment sales and more complex forms of service. KLF specializes in Siemens/Rolm, Nortel, and PBX peripheral products and accessories.

- Maintenance/technical support of discontinued systems. One example of a secondary marketer providing these services is Quality Teleservices (QTI) (San Dimas, CA – 909-599-3863, www.qualityteleservices.com). They are authorized to provide maintenance and technical support of all manufactured-discontinued Telcom Technologies ACD systems.

Quality Teleservices developed Solution 2000, which adds new features to Telcom Technologies' used and/or existing ECD-4000EX ACD systems. Refurbished ECD-4000 EX systems equipped with this upgrade are also available for call centers looking for secondary equipment with all of the capabilities of new ACD systems at a fraction of the price.

The system's communications server provides a platform that expands the system's existing capabilities, and can be configured as your needs change. Because the PC hardware can be easily replaced and/or upgraded, the ACD is not made obsolete in the near future by new technologies. And when you use the Solution 2000 with your existing ECD-4000EX, further savings are possible, because you'll be able to reuse some of your existing equipment.

When you purchase used equipment, you're no longer dealing with the manufacturer. Your secondary market dealer becomes your new supplier of back-up parts. They should be concerned about your system and how it can help you increase productivity.

There are many companies with sterling reputations in the secondary market. But, since only you know what your specific needs and preferences are (e.g., inventory, small versus large company), it's up to you to make decisions intelligently. Just because a shop offers attractive deals doesn't mean much, if they can't help you find the right products, fix them up and keep them running long enough for you to make some money with them.

So do some research. Get a credit report on the company you are interested in. Ask the potential provider for a list of names of its customers, then call them to see what they say. Check out the stability factor, by asking how long it has been in business.

"Get a personal referral," says Don Barrett, Division Manager of KLF Remarketing, a company that has been refurbishing equipment since 1986, and been in business for the past 30 years. "You are much better off if you know who you are working with. Some of these businesses [in the secondary market] say they have a price that's too good to be true, and it is. One of our biggest strengths is that we know we might mess up, but we jump to make sure it's taken care of quickly. We back up the part."

Another good place to double-check a company is with The NATD (National Association of Telecommunications Dealers; Delray Beach, FL). This association sets industry standards for the secondary market. If you have a question, problem or concern about a dealer, you can call them at 561-266-9440.

Where to Find Secondary Equipment

Here is a list of numbers you can call for info including used equipment that's as good as new, tips on how to best finance your investment in used equipment and industry associations.

A-1 Teletronics (800-736-4397/813-576-5001) www.A1_Teletronics.com

ACD Systems International (610-792-1600) www.acdsystems.com

Advanta Business Services (800-255-0022) www.advantalease.com

Alliance Systems (972-633-3400) www.alliancesystems.com

Allied Resources (800-843-5432/707-648-0905) www.alliedresources.com

Amdev Communications (209-962-5900) www.amdevcomm.com

Answer Point (800-569-8840/702-785-8840) www.answerpointinc.com

ARC Communications (800-220-0044/732-981-9600) www.arccom.com

Atlantic Telecom (800-418-9199/973-515-5252) www.atlantictelecom.com

Call Center Technologies (800-996-1996/203-740-3530) www.callcentertech.com

Call Processing Solutions (800-472-2291) www.callprocess.com

CMS Communications (800-830-2286/314-530-2771) www.cmsc.com

Comfort Telecommunications (800-399-3224/941-945-3224) www.comfortel.com

Cornerstone (800-562-2552/408-435-8900) www.corimage.com

CT Marketplace (800-977-4005) www.ctlink.com

D&S Communications (800-879-2020) www.dscomm.com

Girsberger Office Seating (800-849-0545/919-934-0545) www.girsberger.com

KLF Remarketing (800-225-8891/317-876-6586) www.klf.net

Lucent Technologies (800-222-6200/908-582-8500) www.lucent.com

Main Resource (800-644-5089/207-797-8410) www.mainresource.com

NATD (561-266-9440) www.natd.com

Native Sun Communications (800-432-3254/562-424-0072) www.nativesunone.com

Nicom Technologies (800-285-2502/315-431-4853) www.nicomtech.com

Norstan Resale Services (612-971-6261) www.norstan.com

North American Communications Resource (NACR) (800-431-1333/612-851-9300) www.nacr.com

Pinnacle Marketing (800-642-3539/612-937-6006) www.pinnacle.com

Phillips Communications (800-999-4123/804-985-3600) www.phillipscom.com

Phonextra (800-3-PHONEX) www.phonextra.com

Quality Teleservices (909-599-3863) www.qualityteleservices.com

Resale Systems (800-777-3725/610-647-9111) www.resalesys.com

Resource Systems (800-873-3729/770-497-1283) www.resourcesys.com

ReTele Communications (800-346-1211/781-455-9900) www.retele.com

Rhyne Communications (800-634-6770/973-227-0606) www.rhynecom.com

Rockwell Electronic Commerce (800-814-8149) www.ec.rockwell.com

Rolm Resale Systems (612-416-3401) www.rolmresale.com

Search Equipment Exchange (800-252-5969) www.apchou.com

Sitel Communications (510-745-0898) www.sitel.com

Source (800-647-2205/972-450-2600) www.source.com

Specialized Resources (888-774-2400/972-664-6600) www.sritelecom.com

Supply Technology (800-767-7292/714-646-7292) www.supply.com

Technology Renaissance (404-876-4440) www.techrencorp.com

Telemarket Resources (800-793-1213/402-572-0350) www.tmkt-resources.com

Telrepco Services (203-269-5151) www.telrepcoservices.com

Ventura Telephone Equipment (520-888-7888) www.venturatel.com

Session Initiation Protocol (SIP)

A simple signaling / application-layer control protocol for Internet multimedia conferencing and telephony. SIP was developed within the IETF Multiparty Multimedia Session Control (IETF MMUSIC) working group.

SIP provides the necessary protocol mechanisms so that end systems and proxy servers can provide services such as the following:

- call forwarding, including
  - the equivalent of 700-, 800- and 900- type calls;
  - call-forwarding no answer;
  - call-forwarding busy;

- call-forwarding unconditional;
- other address-translation services;
- callee and calling "number" delivery, where numbers can be any (preferably unique) naming scheme;
- personal mobility, i.e., the ability to reach a called party under a single, location-independent address even when the user changes terminals;
- terminal-type negotiation and selection: a caller can be given a choice how to reach the party, e.g., via Internet telephony,
- mobile phone, an answering service, etc.;
- terminal capability negotiation;
- caller and callee authentication;
- blind and supervised call transfer;
- invitations to multicast conferences.

Extensions of SIP are available to allow third-party signaling for such things as click-to-dial services, fully meshed conferences and connections to multipoint control units (MCUs), as well as mixed modes and the transition between those.

SIP does not prescribe how a conference is to be managed, but it can invite users to conferences by conveying the information necessary. SIP does not allocate multicast addresses either, this is done by SAP.

SIP addresses users by an e-mail-like address and re-uses some of the infrastructure of electronic mail delivery such as DNS MX records or uses SMTP EXPN for address expansion. SIP addresses (URLs) can also be embedded in web pages. SIP is addressing-neutral, addresses can be expressed as URLs of various types.

SIP can also be used for signaling Internet real-time fax delivery. This requires no major changes. Fax might be carried via RTP, TCP (e.g., the protocols discussed in the Internet fax working group) or other mechanisms.

SIP is independent of the packet layer and only requires an unreliable datagram service, as it provides its own reliability mechanism. While SIP typically is used over UDP or TCP, it could, without technical changes, be run over IPX, frame relay, ATM AAL5 or X.25, in rough order of desirability.

Why another Internet protocol?

Even as IP telephony gains rapidly in acceptance, its proponents still make reassuring noises about legacy infrastructure. "Don't worry," they say. "Our gateways will plug right into the back of your old PBX – you won't even know they're there. Relax. IP telephony doesn't change everything overnight."

In fact, it does change everything, and more rapidly than any of us could have anticipated. Or at least it would, if legacy telecom weren't hanging around VoIP's neck like an enormous practical and conceptual boat-anchor. To achieve critical mass, today's IP telephony services and products must integrate with the Public Services Telephone Network (PSTN) and fit more or less neatly into legacy niches at the CO and customer premise. The result, perhaps inevitably, is that function tends to follow form: because these new networks, services and devices interoperate with the older architecture, they tend willy-nilly to recapitulate its dominant themes.

Dominant themes of the old network include concentration, increased local complexity, verticality, determinism. Everywhere in the converging communications space, we see evidence of these legacy tendencies at work – often fusing with next-gen principles in strange hybrids of old and new. Right now, for example, VoIP network ASPs like Telera are working hard to concentrate gateway, call processing and IVR functionality in boxes at the network edge. Considered one way, this is quite "next gen" - Telera is putting all their legacy stuff (gateway com-

ponents, conventional IVR) in one box, and pushing it out of the IP core. In another sense, it's quite "old network." In Telera's model, the gateway – conceptually a fairly simple device – becomes more complicated, more important, less generic.

Another example: The much touted IP PBXs are (naturally enough) being packaged as drop-in replacements for conventional phone systems. So products must either incorporate trunk- and station-side gateways, or use peripheral gateways for conventional connectivity. The gateways are conceptually disposable – most architectures handle call-direction and feature service entirely in software, on the IP side. But they add real complexity to what would otherwise be elegant, simple systems; becoming likely points of critical failure; and increasing cost. And until everything in the world goes IP (or at least until gateway service moves up into the network, where it belongs) they won't go away.

Centralized, deterministic, box-bound "telephony thinking" inflects most VoIP protocols, as well - but in a topsy-turvy way. When telecom-heads confront IP transport, the basic old telecom model ("drive dumb endpoints with concentrated, vertically-integrated intelligence") gets turned sideways. More intelligence is concentrated at the endpoints – both to manage what's assumed to be an extremely-fallible network, and to handle facilities negotiations between more- and less-complex devices, all residing somewhere on the (presumed) vertical continuum from voice, to video, to data, to all-of-the-above. But this distribution of intelligence to endpoints doesn't make things simpler at the network core.

H.323 – derived from the wireline videoconferencing protocol, H.320 – is an obvious case in point: Complex, deterministic, vertical. The protocol – spread across at least six major documents (not counting optional addenda and semi-official commentary) – defines every component of a voice / video / data conferencing network: terminals, gateways, gatekeepers, MCUs and other feature servers (See H.323). H.323 uses ISDN-style Q.931 signaling for call setup, plus other protocols – RAS and H.245 – for terminal / gatekeeper negotiations and codec / facilities handshaking. All these protocols – dozens of back-and-forth messages – must be managed to set up a simple, point to point voice call.

H.323 is a fairly-stable standard – you can go to a range of third parties and buy stack components for host deployment, or terminal implementations. Interoperability tests are proceeding. Scaleability concerns are being addressed. H.323 gateway networks are deployed. H.323 PC clients are widely available – NetMeeting, for one, representing a kind of low-end, de-facto standard for client-side functionality. The first, relatively low-cost H.323 telephones are in the pipeline. There's no question that H.323 works.

But anyone who looks at the standard should have questions. The New Network isn't going to be nearly as fallible as H.323 presupposes. (Nor is H.323 especially robust - the numerous messages required to set up calls mean plenty of targets for line-hits. Call setup failure due to packet loss is one of the things Chris Bajorek's CT Labs measures (See Bajorek, Chris), and when they test H.323 gateways, the performance of some systems is truly dismal.) There's going to be plenty of bandwidth – the idea that IP telephony is going to happen across 4.8 Kbps compressed connections is pretty-well outmoded, as is the idea that most IP connections will have to negotiate bandwidth shifts, mid-call. Do we really need a facilities-negotiation sub-protocol (H.245) that not only manages mid-call codec changes, but is actually capable of changing H.245 revision-levels, on the fly? And why would anyone want to use ISDN-style signaling to set up calls across an IP network?

Yes, H.323 works. But many experts feel that there's something fundamentally wrong-headed about it. It's all about concentration and control – dynamics diametrically opposed to the simple, open, horizontal, multipurpose philosophy of pure Internet technologies like e-mail and the Web.

This – in combination with other "legacy" tendencies influencing the development of IP telephony networks and CPE – entails a fundamental risk to the converging communications economy. The IP telephony revolution could hang fire – or actually fail – if persistent legacy characteristics obscure real "killer app" opportunities, or hamper IP telephony's ability to elide with the 'Net's most powerful technologies.

SIP, the Better Way?

SIP – the Session Interface Protocol – may offer a better way to do telephony in an IP environment. SIP comes at the challenge of converged communications from a horizontal, 'Net-head perspective. The result is a simple protocol with profound implications.

Like the web – originally designed as a document-sharing system for academics – SIP originated with a simple, practical brief. In the mid-90s, Henning Schulzrinne – now Associate Professor in the Departments of Computer Science and Electrical Engineering at New York's Columbia University; Jonathan Rosenberg – now Chief Scientist at SIP software maker DynamicSoft; and several others began work on a signaling protocol that defines call setup and teardown functions as simple text commands. IP telephony – as we conceive it today – wasn't yet on their (or anybody else's) radar-screen.

"At that stage," Professor Schulzrinne remarks, "IP telephony as a term probably didn't even exist, at least not in my community. Initially, SIP was intended to create a mechanism for inviting people to large-scale, multipoint conferences. After a short while, it became clear that technology-wise, it was not a significant jump from where we were to setting up point-to-point conferences – essentially 'phone calls.' And once 'IP telephony' became the thing to do, then people started looking primarily at using the protocol for voice applications. But the emphasis of SIP has always been to remain as independent as possible of the media it underlies."

Jonathan Rosenberg takes this sentiment a step further: "It's actually not even just media that SIP abstracts. The protocol makes a total separation between what it means to be a session, and what it means to establish one. SIP talks about establishing or modifying or terminating a session, but that particular session could just as easily be a multiplayer Doom game as it could be a voice channel or a videoconference."

The decision to format SIP messages as text was a profound one: Text is human-friendly and robust. More to the point, text processing lies at the heart of the 'Net's true killer apps: e-mail and the web. Extensible tagging systems, document identification, data-type declaration, parsing methods – all these have been worked over and normalized by Net-heads in the process of bringing the modern 'Net online. Schulzrinne and his colleagues understood this, and made a second leap: they decided SIP text messages would be composed in standard ISO UTF-8 (ASCII Unicode) characters, using HTTP 1.1 syntax.

"In H.323," explains Schulzrinne, "there is very much a vertical integration notion present. It specifies everything from the codec for the media down to how you carry the packets in RTP, because part of the specification is to describe the content of the data stream. In the IETF (the standards body promoting SIP), we've taken much more of a Lego-like approach, much more horizontal. What we've tried to provide are building blocks, which fit together with a number of different Internet protocols, so that we can use a common URL for naming, we can use MIME for describing content, etc."

A SIP message looks like the first five or six lines of source behind a well-formed Web page (the part that says: "Content-type: etc., etc."). SIP messages look this way because that's exactly what they are – an appli-

cation of the 'Net's simplest, most widely-implemented, most general-purpose system for document-type declaration. SIP also adopts the conventional URL format for addressing: your SIP "phone number" is "yourname@yourhost.com," with an optional port number. The URL is translated to an IP address (fixed, dynamic, temporary, etc.) through DNS, the generic nameserver system.

Rosenberg emphasizes the point: "We didn't go and define our own type of addresses because we saw that the Internet already had address formats, URLs, and we figured that people are probably going to want to throw together URLs of different types, as they have elsewhere on the Internet. And without even really considering the implications of that decision, the service possibilities it has enabled have been huge. For example, with SIP, it's just as easy to transfer someone to another phone as it is to transfer them to a web page or any other application that accepts URLs – even ones like instant messaging, which didn't exist when we wrote the spec."

The mechanism for doing this magic doesn't even belong to SIP, *per se* – it just falls out of the decision to use standard DTDs and URLs to manage telephony. The SIP client – whether a browser plug-in or a physical LAN phone – doesn't have to know all about web pages or IRC clients to make it work. It just needs to recognize when a returned destination involves a content-type it can't manage (e.g., a Web page) and refer the destination URL to an appropriate application. In a converging communications universe in which more and more apps (UM, CRM, e-commerce, etc.) are Web-mediated, SIP offers means to telephony-enable these apps with trivial effort.

SIP protocol, itself, is very light, employing a minimum set of messages to accomplish its goals. By comparison with H.323, setting up a call in SIP is a no-brainer. The calling terminal sends an INVITE message, containing codec preferences; the answering terminal sends an OKAY back (with its own codec preferences) and rings. About five discrete steps (2.5 round-trips) serve to set up, manage, and tear down a call, in the simplest, point to point, case.

To support mobility and higher-order applications, SIP defines several "useful entities" (read: simple pieces of software that sit on a well-known port) that help manage calls in different ways: registrars, which maintain a map of "what IP address a given user is at, right now"; proxies, which can act as transcoders, auto-responders, or forwarding agents; and redirect servers, which perform a subset of forwarding functions. These helpers, roughly analogous to H.323 gatekeepers and other functionaries – but much more scaleable – can be set up to provide support for mobility and all sorts of higher-order applications. For example, an ACD can be implemented as an application controlling a SIP proxy server – the application maps available agents, and employs the proxy to send an appropriate redirect (or "queued") message to each calling client.

Even in what it can't or doesn't do, SIP provides positive benefits. The protocol, for example, doesn't specify a quality of service mechanism. It does, however, provide a way for two or more endpoints, engaged in a peer-to-peer communication, to dynamically compare their resources and use what is available. "SIP intentionally doesn't reserve resources for QoS," comments Schulzrinne, "because the world has enough non-deployed QoS protocols as it is. Where SIP comes in as a value proposition is that it establishes associations – sessions – between entities, and one of the things those entities can discuss is which protocols we have in common, which we're willing to share, and on which we'll just agree to disagree."

Why SIP May Predominate

"I think, fundamentally, the success of the Internet is all about taking vertical pieces and breaking them into horizontal components," says

Rosenberg. "The reason why the Internet succeeded in a lot of cases, where BBSs and other online services had failed, is because the Internet immediately separated out transport from services, while the others tried to integrate access, transport, and services."

"Now we're starting to see Internet telephony following a similar evolution – we've already broken up the telephony gateway, for example, into a softswitch and a media gateway. But at this point, it's still a vertically integrated market. For the evolution to continue, we're going to have to break up the model into even more pieces, so that one user's services can reside in any number of different places in the network, depending on what they are."

On the Internet, this idea is a given. "An ISP doesn't build or own all the web services its users access. It lets other people build web services that are customized for a particular group of people, because that's some other person's expertise," Rosenberg adds. For telecom, however, this proposition involves some major paradigm shifting.

"There used to be a kind of black art," Schulzrinne quips, "where you had to undergo rituals and have your head shaved appropriately before you were allowed to program an SS7 service. It's not something you could just learn in school. But what we're doing is making it possible people who have a similar skill set to web page designers, who know some standard scripting languages, to develop services that are either customized for their own organization, or target some vertical market."

As Rosenberg points out, "The Internet is all about access to tools. It was because some random yahoo could sit down and say, "gee, this is a neat idea," and then whip up the service ... that there was so much innovation and so much growth in commerce, all at once. In the voice world, though, I'd have to wait for my telco to go through the three-year cycle of adding a new service, and it would still not be a true vertical-market service. We've never seen vertical-market, specialized voice services, never. But we've seen tons of vertical-market web services deployed in the past few years, and that's where a huge source of value has been. So our mission is to create a horizontal platform that lets anybody create vertical-market services incorporating voice."

A virtual Web of decentralized services. Disparate endpoints communicating with one another through nothing more than their own embedded software. Yahoo! deploying voice applications once controlled by AT&T ... Are we living in a fantasy world? Yet every possible indication that we've seen from the industry suggests that SIP, and its fundamental implications for communication, are about to make a major impact on Internet communications.

Part of the proof lies in the wide range of products that are already incorporating SIP at different levels in the network. In terms of infrastructure, Dynamicsoft is leading the way by providing SIP proxies and location services as the basis for a horizontally integrated applications platform. The softswitch community has largely embraced SIP as a way of communicating between softswitches and is strongly considering the protocol as a means for tying together softswitches and application servers. At the edges, companies like AudioTalk (now HearMe) and NetSpeak are basing software VoIP clients on SIP, and proving the technology to be usable today, for anyone with a multimedia PC and browser. And PingTel is taking the next logical step at the edge by building the first truly intelligent SIP telephones, and using a fully integrated Java environment that demonstrates the extent of the protocol's proximity to the Internet.

Perhaps most significantly, SIP is garnering a high degree of support from carriers. Level 3 has made recent announcements that describe widespread use of SIP throughout its network. And, at the most recent VON show, MCI WorldCom demonstrated a public test network that incorporated SIP-based products from at least seven different vendors, all interoperating with one another.

The scope of SIP-based products and services is likely to grow immensely in the near future. As it does, however, we expect to see the protocol less emphasized rather than more – just as one doesn't necessarily emphasize the use of HTTP in an Internet application. Already, a growing group of players are approaching it from the right direction. "What we've found," says Rosenberg, "is that all sorts of vendors and service providers have particular applications that they want to get done. When they try to figure out how, the find they have a choice of protocols, none of which has already defined the necessary feature set, but which can serve as a platform to build upon. Increasingly, we're finding they want to build on SIP. As a result, we're seeing a lot of vendors defining extensions, and doing things that we hadn't originally conceived of SIP to do; but, then again, that was sort of the whole idea, now, wasn't it?"

Some people would like to see more severe modifications to SIP. To better incorporate calls with termination points in the PSTN, SIP+, a new draft protocol, has appeared. SIP+ encapsulates SS7 ISUP, Q.931 ISDN or CAS signals into a SIP message. Because SIP is basically just an e-mail message, SIP+ embeds the binary signals as a MIME attachment to the SIP message. The MIME encoding lets the signals be tunneled between Media Gateway Controllers (MCGs), and the SUBTYPE conveys to the MGCs what type of message (ISUP, Q.931 or CAS) has been encoded.

It's beginning to look like SIP with its call control speciality will hang out at the Customer Premises, while H.323 will still be lurking deeper in the network.

Here are some companies that already are producing SIP-related products:

3Com (Santa Clara, CA – 408-326-5000, www.3com.com) has come out strong in support of SIP, and has been active in promoting its acceptance for quite some time. Initial implementations have largely centered around using SIP as a way of interfacing between a Palm Pilot and a SIP-based telephone, which the company has demonstrated at trade shows. The Palm integration, however, while holding undeniable sex appeal, as well as the potential for some truly useful applications (dialing directly out of your Palm address book, as a basic example), is really a way of directing attention toward a larger 3Com project – namely the development of an IP Centrex solution.

Essentially, the system consists of a series of SPARC servers that run Centrex-like features, and communicate with client devices over IP, through a SIP server. In addition to the call control and applications software, 3Com has developed the SIP server itself, as well as a line of SIP phones to reside at the customer premise. Nevertheless, the company plans to use fully open interfaces at every level, so that any component of the system – phones, servers, applications – could be replaced or complemented by a standards-based product from a third-party. This open architecture differs from 3Com's enterprise LAN-PBX offering, the NBX-100, which uses a proprietary protocol over Ethernet to interface with phones. Although 3Com eventually plans to migrate the NBX to an open IP protocol, they are first and foremost looking at SIP as a wide area protocol, and are in this respect in line with the thinking of many other vendors. 3Com's Ikhlaq Suhu points to SIP's inherent scalability, reliability, and simplicity – all of which are related to the fact that the protocol defines a peer-to-peer, "stateless" call model – as its main advantages for use in networks that extend beyond the local area.

In addition to participating in and hosting SIP bake-offs, 3Com is currently in an IETF proposal to specify an open standard for service provisioning and authentication in SIP-based application server architectures like its own. The company has also submitted a draft to the IETF that defines a standard method for passing SIP messages between a Palm device and a phone. By standardizing its own de facto method for this type of application, 3Com plans to enable third-party developers of Palm apps, as well as vendors of other Palm OS-based products and of other SIP phones to achieve the same type of integration.

Broadsoft (Gaithersburg, MD - 301-977-9440, www.broadsoft.com) makes an application server and service creation environment. BroadWorks, as the system is known, replaces class 5 features in a converged network, and adds enhanced services like messaging, auto attendant, and conferencing. The scope of applications can even extend to intelligent call routing, as well as personal communications services like find-me / follow-me. The system as a whole includes an integrated media server (for functions like IVR and conference setup), along with the software framework for creating and delivering applications. Also, it incorporates a web server that can be used to give subscribers browser-based access to their service profiles and accounts.

Architecturally, BroadWorks could sit behind a series of network gateways interfacing to the PSTN and SS7, or behind a softswitch and media gateway connecting to the packet network. The idea is that users (most likely small- to medium-sized businesses) would connect to the data network either with an IP phone or behind a gateway at the customer premise, and access network services hosted by their CLEC. While all of BroadWorks services can be delivered in a "legacy" environment as well as a packet network, client endpoints that are connected through IP can provide a much more elegant way of delivering web-integrated services, particularly if they can communicate using SIP.

Beyond supporting SIP as a call setup interface to end devices, however, Broadsoft is also pursuing the possibility of using the protocol as a standard for interfacing between a softswitch and an application server. To this end, the company is working on the SIP-TSI initiative with IPeria, as well as participating in SIP bake-offs to promote interoperability. So far, Broadsoft reports they've found interoperability a much more readily obtainable goal with SIP than with H.323.

The CMG/CSS 2000 Softswitch from ComGates, Ltd. a technological spin-off of Arbinet Communications, Inc. (Herzeliya, Israel – +011-972 995-0-0404, www.comgates.com) is a real-time multi-protocol carrier-level PSTN / IP Switch / Gateway that can adapt automatically to all standard signaling protocols.

Indeed, the CMG/CSS 2000 is said to be the industry's first softSwitch to provide on-the-fly multi-protocol capability by employing a unique, proprietary process that can, in real time, automatically switch traffic between all existing and emerging standards. The CMG/CSS 2000 enables telephony networks and carriers deploying various versions and releases of H.323, SIP, MGCP and MEGACO to immediately begin having interoperability so they can exchange traffic with one another as the standards develop, there-by eliminating the concern about what protocol is on the other end or to be troubled over the deployment of new, changes or emerging standards between peered networks.

ComGates' CMG/CSS 2000 Softswitch adapts instantly – without presetting – to the protocol being used by the connecting gateway. The CMG/CSS 2000 Softswitch can also simultaneously run multiple transmission protocols on the same operating platform, eliminating the need for users to commit to one protocol over another and enabling the integration of new technologies as quickly as they are developed.

The new CMG/CSS 2000 Softswitch transparently bridges and integrates ordinary public telephone networks with data communications networks. This enables carriers, Internet Service Providers (ISPs), Internet Telephony Service Providers (ITSPs), and telephone companies to seamlessly integrate into their service offerings Voice Over IP (VoIP) and Fax over IP (FoIP).

"With the fight for protocol dominance being continually waged the established PTT's like NTT speak in the flavor of H.323, while new emerging IP Telephony backbone Data network providers such as Level3 use MGCP/SIP. Integration of the CMG/CSS 2000 Softswitch lets the networks transfer and handoff traffic without having to be concerned about the native protocol that the traffic from the peered network," says Jacob Tirosh, President and CEO of ComGates, Ltd.

Designed to meet the stringent needs of the carrier market, the CMG/CSS 2000 Softswitch is Unix-based for maximum reliability. The scaleable CMG/CSS 2000 Softswitch supports up to 10,000 ports, providing full transparency between the PSTN, IP/Data networks and soon Wireless networks. The open architecture design supports all open standard signaling protocols including, CAS ISDN, SS7/C7, H.323, MGCP, SIP in any mix.

"In the current VoIP environment there are no consistent standards and no real indication as to which of the conflicting protocols will become the dominant choice in the future," observes Tirosh. "The architecture of the CMG/CSS 2000 Softswitch relieves decision makers from committing to a particular protocol and leaves the door open for incorporating new protocols as they become popular. Our unique on-the-fly adaptability of the Softswitch provides carriers and networks with an enabling technology that ensures they can deploy systems in a timely manner with the confidence that their underlying infrastructure will remain up-to-date."

The flexible, open architecture and standard independence of the CMG/CSS 2000 Softswitch particularly benefits Telephony Service Providers (TSPs) and Competitive Local Exchanges (CLECs) moving into providing Internet service, and Internet Service Providers (ISPs) moving into Internet telephony services.

"The full potential for the convergence of conventional telephony, data networks, and the Internet is yet to be determined," added Tirosh. "The business models for VoIP and related merged services are still being developed."

There's no questioning Dynamicsoft's (East Hanover, NJ – 973-736-6580, www.dynamicsoft.com) commitment to, or innovation with, SIP. Jonathan Rosenberg, one of the protocol's co-authors, serves as the company's Chief Scientist, and Eric Sumner, a former CTO and former colleague of Rosenberg at Lucent, is President and CEO of Dynamicsoft.

Although the company is young, it's moving forward quickly with developing a line of SIP-based network servers, forming multiple strategic partnerships, and even securing its first customers. Dynamicsoft's product line, grouped under the name eConvergence, includes a SIP proxy server, location server, user agent, and application server. The latter, of course, can be seen as the most valuable, least generic element in the mix, though at the moment Dynamicsoft's other products are equally if not more crucial in defining a next-gen IP architecture.

The proxy server is essentially responsible for call routing and session management. It also performs redirect functions, routing to media gateways, and user authentication. While there seems to be some functional overlap between a SIP proxy server and a softswitch, Dynamicsoft takes care to distinguish between the two, and sees the softswitch as necessary for interfacing to the PSTN, while its own product is firmly rooted in IP. Before routing a session to its addressee, the proxy must query either a domain name server or a SIP location server. The purpose of the latter is to maintain user profiles and provide subscriber registration. The

S

registration function is one of the more unique aspects of the SIP model. It effectively provides a mechanism whereby a user registers with the network each time she comes online, and can access individual profiles (contained within the same server) that specify information for routing based on a number of different criteria.

On the other side of the proxy server lies the SIP user agent, which is actually a piece of software that gets embedded in a range of other boxes – from a softswitch to an IP-PBX to an IP phone to a gateway. The user agent initiates and manages the basic connection between two endpoints. It is available as either a C++ or Java program, and incorporates an extensive API and development toolkit. The user agent also supports SIP BCP-T (for softswitch-to-softswitch communication), as well as the SIP Control Processing Language (CPL) and the Common Gateway Interface (CGI).

The application server provides a horizontal platform for interfacing to and creating new applications to run on the SIP based network. Like the user agent, the application server will support SIP-specific interfaces in addition to general Internet protocols, languages, and APIs.

Dynamicsoft has been chosen to build out an end-to-end SIP network for Level 3 Communications, and they work with smaller providers like I-Link and eStara to SIP-enabled their networks. The company is also partnering with Aravox to develop a SIP-controlled firewall for the carrier market.

IPeria (Burlington, MA – 781-993-3500, www.iperia.com) makes an enhanced services platform, IPeria Service Node, that combines elements of a media server and an IP applications server, but remains independent of network transport. While IPeria's product, as even the company's name would indicate, is clearly aimed at converging carriers, it can deliver its core services, like unified messaging and voice-activated email, over the PSTN as well as over a packet network. The guiding idea is to let users access any type of communication (voice, fax, email) from any device (phone, wireless, web, Palm, etc.).

One result of IPeria's hybrid approach is a proposed application of SIP that uses the protocol in interesting ways to merge legacy and next-gen infrastructures. Specifically, IPeria, under the direction of CTO Greg Girard, is drafting a spec called SIP-Telephony Service Interface (SIP-TSI) that seeks to define a standard way of communicating between a softswitch and an application server. The main purpose of SIP-TSI is to allow the application server, through the mediation of the softswitch, access core switching and DSP resources located in a media gateway, in order that these resources do not have to replicated locally in the app server itself (as would generally be the case in a PSTN/IN implementation). DSPs located in the media gateway are used to transmit and receive fax, and for signal detection and signal transformation of voice and DTMF tones. While SIP itself does not define commands for mid-stream control of media transmissions (as is necessary, e.g., for collecting DTMF tones), an adjunct called the SIP "INFO Method" lets mid-session control messages be passed between two SIP user agents (which in this case would be contained respectively within the softswitch and the application server). SIP-TSI, which is being submitted by IPeria and other vendors to the International Softswitch Consortium for approval, is one example of how the protocol can be extended in the network, and how elements such as a "user agent" are not necessarily bound to specific hardware or software devices.

IPeria's Service Node platform is now generally available, along with unified messaging and voice-activated email (incorporating speech rec and text-to-speech) as the first two apps. Other personal communications services will be released over time.

ipVerse (Sunnyvale, CA – 408-830-3200, www.ipverse.com) makes a softswitch that leverages SIP in a number of ways. The product, ControlSwitch, has a relatively clearly defined role in the network: It resides between the layers of media gateway / trunking equipment on one end, supplying call control and addressing services to these devices, and service delivery platforms, providing APIs and open interfaces, on the other.

One thing that ipVerse prides itself on is its openness in communicating with other systems, whether on the level of signaling or of APIs. Not only does the control switch support the full range of next-gen and legacy signaling protocols – SIP, MGCP, H.323, IPDC, Q.931, SS7 – but it boasts the ability to interface between any of them seamlessly, allowing communication between devices which on their own would not talk to one another. (This includes communication between the ControlSwitch and other vendors' softswitches via SIP-T.) In terms of APIs, JAIN, Parlay, XML-based interfaces and a host of others make it easier to connect the control-switch into multi-vendor application or policy servers.

Another key differentiator for ipVerse is its scalability: The ControlSwitch has been shown to handle up to one million simultaneous calls.

Beyond the intrinsic values of such openness and scalability, ipVerse's emphasis on these attributes raises an important point – namely, that a softswitch on the market today should be designed with a mutability of function in mind. Because network architectures are currently so much in a state of transition and flux, a softswitch should be capable of easily adapting to the somewhat disparate roles of a signaling gateway in cases of simple modem offload, an enabler of enhanced service delivery in a centralized next-gen network, or a proxy/address server in a network where intelligence is fully distributed to the edges (as may well be one of the affects of SIP). The ipVerse platform takes all of these possible scenarios into account.

Lucent Technologies (Murray Hill, NJ – 888-4-LUCENT, www.lucent.com) is using SIP on two distinct but interrelated fronts. The first concerns Lucent's Communications Software Group, a division that focuses on IP-enabling Intelligent Network applications. The most recent releases to come out of this group are the PacketIN application server and a complementary Java / XML-based service creation environment. PacketIN works in both a circuit switched environment, where it would use SS7 and TCAP to deliver services to the switch, as well as in next-gen or hybrid architectures (which are its primary focus), where it would build on a softswitch (which Lucent also supplies) and a media gateway. In the softswitch scenario, Lucent has developed some PacketIN applications that leverage SIP in interesting ways.

One example is Lucent's Online Communications Center (unofficially referred to as Internet Call Waiting on Steroids), a piece of software which Lucent had previously released but has now enabled for use with the softswitch. Besides standard Internet Call Waiting (which gives you a screen pop with caller ID info), OCC lets you accept incoming PSTN calls over the Internet, dynamically re-route calls to another phone number, and gives you other desktop call management capabilities.

What's particularly interesting, however, is that this service could be fully integrated with circuit-switched IN: Using a TCP/IP interface it developed for its own Service Control Points (SCP), Lucent can trigger a SIP message directly from the SS7 network to the PacketIN server, which in turn responds with instructions about what to do with the call.

While this level of fairly involved internetworking may seem a bit complex for an Internet Call Waiting app, it raises potentially intriguing possibilities for things like network-based contact center routing – another service Lucent is offering in conjunction with PacketIN. (Of course, running these services in a fully next-gen environment, i.e. over IP, makes for a much more elegant scenario, and lets you program everything using standard APIs – JAIN, Parlay, etc.).

The second major implementation of SIP within Lucent comes through its elemedia (Holmdel, NJ – 888-elemedia, www.elemedia.com) subsidiary, which developed a SIP Server on the basis of the Lucent softswitch. The server performs authentication, registration, and redirect, in addition to some of the same type of features that could run on a softswitch (personal mobility, time-of-day routing, etc.). The SIP server is a newer addition to elemedia's product line, which also includes an H.323 protocol stack and gatekeeper software. The project was begun under the direction of Jonathan Rosenberg (co-author of SIP) and Eric Sumner, both of whom have since moved to dynamicsoft.

Netergy, formerly 8x8, (Santa Clara, CA – 408-727-1885, www.8x8.com) offers its Netergy Advanced Telephony system, a hosted IP-PBX for service providers that makes interesting use of both MGCP and SIP. Architecturally, the product centers on a piece of Java-based software, the iPBX, that runs in a distributed fashion across a cluster of Sun Netra servers. The iPBX software manages call setup and teardown, and implements standard features like transfer, forward, and hold. At the customer premise, 8x8 offers a number of options. Currently, Netergy would use 8x8's Symphony Media Hub, a two- or four-port gateway that lets you hook up analog phones and fax machines, and connects to a router via Ethernet. In this case, MGCP is used to communicate between the hub and the "switch" (which really should be thought of more like a media gateway controller). 8x8 also offers a larger, 24-port gateway, and will interoperate with similar CPE products from Cisco. A much sexier solution, of course, would involve putting IP phones at the customer premise, and 8x8 has approached this on a couple of fronts. One is to let customers use IP phones from third-party vendors, and interoperability trials with Cisco and PingTel are underway. Another is using IP phones built around 8x8's Audacity T2 Processor (controlled by MGCP). 8x8 also offers a browser-based soft client that performs third-party call control.

On the back-end, the iPBX gives you a number of options in terms of network interfaces. The most conventional would be to an H.323 gateway/gatekeeper, and out to the PSTN. But you could also connect the servers to a SIP-compliant softswitch and use IP throughout the network.

While a lot of the interfaces for both end devices and network elements depend on products that are still being developed by others, 8x8 is remaining very open and supportive of most available protocols.

After a long period of quiet buzz, Pingtel (Woburn, MA – 781-938-5306, www.pingtel.com) unveiled its SIP-based IP phones to the public in 1999. Having now seen and heard about the product, called xpressa, we think it could very well be the first to deliver on the intrinsic, but as yet elusive, promises of IP telephony.

Simply put, Pingtel built its product from an Internet mindset. CEO Jay Batson is very clear about what this means: Intelligence in the endpoints. Batson explains that in an IP architecture, "The phone should be an open, extensible platform for delivering new software features. Ostensibly it's a piece of hardware, but really our product is a software platform for running Java applications." And when Pingtel talks about openness, extensibility, and Java applications, they aren't just paying lip service. The phones have their own, embedded JVM, a Sun developers' toolkit, JTAPI interfaces, and Pingtel is going a long way to provide resources like style guides and online communities to any potential Java developers.

The architectural model behind these initiatives is basically that some core telephony features (generally transparent to the user) will remain located at central points in the carrier's network, while enhanced services will be thoroughly distributed, and accessible through the phone

client as through a web browser. In fact, because these "services" will themselves be Java applets or HTTP-based content, any logical distinction between the web and an IP-based network of intelligent phone clients begins to break down altogether. This, of course, is a completely different way of thinking of "enhanced services" than what we know today. And the difference lies partly in the fact that the "service provider" as such (ISP, CLEC, etc.) is no longer the sole supplier of applications to a user (a proposition about which carriers, of course, have some mixed feelings). Rather, new software could now be download to a phone exactly as one downloads a new app to a Palm Pilot.

In addition to leveraging a Java framework and significant internal processing power, a key enabler for Pingtel has been its early and extensive incorporation of SIP. According to Batson, the company was attracted to SIP from the beginning for its natural extensibility and for the power it puts in the hands of intelligent endpoints (i.e., SIP's peer-to-peer negotiation model for setting up a session). Batson takes care to note that, like HTTP, "SIP as a protocol is simply a vehicle for delivering applications": That is to say, the protocol itself should not necessarily become involved in defining those applications. In this he echoes SIP's authors, and goes on to explain that with xpressa, any number of Internet protocols (http, LDAP, MIME, etc.) might get involved in the execution of phone applications. SIP's inherent ability to integrate with these other protocols only makes it a more natural choice for the platform.

Pingtel expects its phones will be primarily distributed through service providers, in the context of IP Centrex or virtual PBX applications. They also plan to form OEM relationships with IP-PBX manufacturers.

Siemens (Santa Clara, CA – 408-492-2000, www.siemens.com) has been doing some interesting and impressive work with SIP, mainly in the area of IP telephones. The LP5100, Siemens' first IP phone design, was also one of the first to market to use a standards-based architecture (with H.323 in its original iteration). Now Siemens is offering the same phone model, using all the same hardware as in the original version, but with the option of a SIP protocol stack incorporated instead of H.323. While the company will continue to offer both protocols as software options (and is looking at adding MGCP), they do report finding certain advantages to SIP. According to Siemens' Joan Vandermate, SIP has shown better performance than H.323, with faster call setup time and increased resource efficiency, using the same CPU and other hardware components. The direct implication of this, Vandermate points out, is that a richer feature set could be added to a SIP phone without necessarily requiring any hardware upgrade. And this is in line with Siemens' overall approach to the IP phone market, a crucial tenet of which is keeping the cost of the device down as far as possible without sacrificing quality or availability of features.

Siemens plans to market its SIP-based version of its IP phone as an endpoint for IP Centrex applications, and will aggressively pursue service providers and carriers as potential customers for the product in the future. The phones offer all the basic functionality necessary to do Centrex features, but Siemens remains open to developing enhanced feature support as it is called for by the market. Likely what will emerge will be a diverse line of IP phones (Siemens is touting a scaled down, less expensive version called the LP2100) that addresses everyone from consumers to high-end business users. Having an already robust manufacturing and distribution plant in place (through which Siemens' Gigaset wireless phones are produced in mass quantities) will provide a strong foundation for entering the market.

While softswitch development to this point seems to have followed a progression from infrastructure (signaling gateway, call management functions) to enhanced services, Syndeo (Cupertino, CA – 408-861-1000,

S

www.syndeocorp.com) is taking a slightly more accelerated approach with a system intended to provide class 5 features from the outset. Nevertheless, the Syndeo Broadband Services System does much more than simply replicate class 5 functionality.

While the product contains a component that calls itself a softswitch, it recognizes that the softswitch function is very loosely defined, and highly dependent on the environment in which it is deployed. The Syndeo Softswitch, therefore, is built to be able to act as a call agent in an MGCP scenario, a proxy server in a SIP network, a gatekeeper in H.323, or an SSP in SS7. Essentially what this means is that the product supports all of the above as embedded interfaces and can employ them as necessary.

The other unique aspect of the Syndeo system is a web-based management and control offering, Communication Manager, that subscribers can use to access their service profiles. Not only does this let users configure features and options via the web, but it can also be adapted to run as a persistent Java applet to serve mobility services. For instance, a user could run the applet on her laptop, and use it to assign her local profile (phone number, features) to a remote phone. Communication Managers also integrate directly with PDA and contact management applications.

With regard to SIP, Syndeo has already formed partnerships with 3Com, Pingtel, and AudioTalk, to support their SIP-based clients. For service creation on top of its platforms, Syndeo supports the JAIN API, along with other open standards.

Telephony Experts (Los Angeles, CA – 310-445-1822, www.telephony-experts.com) and Nuera Communications (San Diego, CA – 858-625-2400, www.nuera.com) have partnered on an offering that joins the former's Talking IP product with the latter's VoIP gateways in a SIP-based environment. Telephony Experts' Talking IP is a software-only enhanced services and billing platform. Nuera makes high-volume trunking gateways for service provider networks. Both companies are supporting SIP in the latest versions of their products. The way the combined scenario works begins with a call entering the Nuera media gateway and being passed to a media gateway controller or softswitch (which Nuera also produces) to receive routing instructions (here the means of communication is MGCP). If the MGC detects a request for enhanced services (e.g. prepaid calling card, messaging, etc.), a SIP session is established (via a re-invite command) between the media gateway controller and the Talking IP server.

Talking IP uses a SQL database, and includes its own service creation environment. Nuera's products support multiple standards, including SIP-T for softswitch-to-softswitch communication. The companies' relationship is non-exclusive.

Texas Instruments (Dallas, TX – 800-336-5236, www.ti.com) subsidiary Telogy Networks (Germantown, MD – 301-515-8580, www.telogy.com) offers SIP support in a number of implementations of its Golden Gateway software. Golden Gateway is voice-over-packet processing and conversion software that runs on TI's programmable DSPs and RISC / CISC microprocessors. At the DSP level, Telogy's software performs tone detection, echo cancellation, and a number of other functions related to voice / fax / modem signal processing. This software is also responsible for processing telephony signals, such as FXO / FXS, BRI / PRI, and QSig. Once a call has been packetized and stamped with RTP headers (all of which takes place in the DSP), the packets are sent to the microprocessor (typically a part of the same chipset), which runs a separate but relate part of the Telogy software to implement an appropriate packet signaling protocol. Telogy's first deployments used H.323 only for instances of voice-over-IP, but the company has since added SIP and MGCP to its repertoire.

One notable instantiation of SIP is a reference design for a voice-enabled cable modem. Telogy has also incorporated SIP into its integrated chipset for IP phone manufacturers – a product that combines TI DSPs, a RISC processor, an embedded Ethernet switch, Real-Time Operating System, and Golden Gateway software. According to Nancy Goguen, Telogy's VP of Marketing, implementing SIP on Telogy's existing product was a relatively easy project, and one that could be carried out in-house on the basis of the published specs, as opposed to its original implementation of H.323, for which Telogy chose to license the RADVision protocol stack. Other benefits of SIP which Goguen cites include its decentralized nature, its scalability, and its backward compatibility. Still, she notes, uptake from customers is still mixed ("it's about 50/50 in the IP phone world," Goguen notes), and she stresses the importance of multi-protocol support from TI's standpoint for at least the foreseeable future.

Trillium Digital Systems (Los Angeles, CA – 310-442-9222, www.trillium.com) writes source code for a variety of protocol stacks within IP, ATM, SS7, and others. Trillium software is typically embedded within a microprocessor, but the company emphasizes what it refers to as "portability" – being able to abstract the core software itself from the individual processor, OS, or compiler, and thus give equipment manufacturers a greater degree of flexibility. A set of common interfaces are essentially all that needs to be tweaked in order to move a piece of Trillium software across different underlying systems.

Trillium released its first H.323 stack in 1998 and has since licensed the product to gateway manufacturers and client software developers. More recently, with regard to VoIP, the company has been focusing on MGCP, as well as Multi-Protocol Label Switching (MPLS), for which it plans to release source code. SIP software is still being developed. While individual protocols are being added incrementally, Trillium's CEO Jeff Lawrence makes it clear that the company has a long-term vision of how the new network architecture is shaping up. Although Trillium will continue to support a wide range of (sometimes competing) protocols, Lawrence is pretty confident that "SIP will eventually become the signaling protocol that replaces SS7." Of course, Lawrence does not expect that this shift will occur overnight, but he does feel that SIP's simplicity, scalability, and recent momentum in terms of industry support will propel it forward as the successor to H.323.

Lawrence also makes the important point that SIP will need to interwork with other protocols in the network to supply functions like quality of service, security, transport, and gateway control, and points to MPLS, IP Sec, SCTP, and MGCP / MEGACO as crucial parts of the network architecture. Although most of these protocols are only now emerging, and commercially available software to support them is still being developed, Trillium says it plans to supply customers with frameworks that allow for interworking as standards are adopted.

Ubiquity Software Corporation (Kanata, Ontario, Canada – 613-271-2027, www.ubiquity.net) makes an application server and service creation environment they call Helmsman. Ubiquity's position is essentially as an enabler of applications, rather than a developer of them. While the company will offer certain core features on top of its platform, and can do customization for individual service providers, the idea is basically to make it as easy as possible for a third-party (either an outside developer or even the service provider himself) to deploy whatever applications are desired. In this way, Ubiquity's product is like a series of connectors or hooks that interface, on the one hand, to endpoint devices (phones, PCs, PDAs, etc.) and on the other to network elements (SIP servers, gatekeepers, call agents, etc.). In the case of the latter, Ubiquity is supplying its own SIP server, to perform proxy, redirect, location, and registration services. Nevertheless, the company still considers the "service enabling" layer as its sweet spot, and will partner with other vendors of network equipment to supply carriers with the components necessary for call control and packet transmission.

One significant customer win for Ubiquity has been I-Link (Draper, UT – 801-576-5000, www.i-link.com), a next-gen service provider building out a network to host distributed communications applications. I-Link is using a combination of the Helmsman server and a Helmsman Desktop client that Ubiquity designed to integrate with Windows PCs. The client app essentially lends call control and VoIP capabilities to software like Outlook or Goldmine, and can also be integrated with web servers and used in e-commerce applications. Helmsman Desktop incorporates a SIP user agent to communicate with elements in the network like a softswitch. In addition to SIP, Helmsman supports H.323, MGCP/MEGACO, and TAPI.

Unisphere Solutions (Chelmsford, MA – 978-848-0300, www.unispheresolutions .com), who offers a line of routers, switches, and gateways to the carrier market, has been devoting its most recent energy to developing its high-density softswitch, the SRX-3000, and its service creation environment (a software platform to create and deploy applications in the network). At this point, Unisphere says that its main goals – and most pressing concerns in terms of securing customers – are to ensure interoperability and any-to-any switching, regardless of the underlying protocol. Nevertheless, the company has found SIP to be an asset for performing particular tasks like softswitch-to-softswitch communication (using SIP BCP-T, a more formalized version of SIP+). Unisphere, like others, has also been looking at SIP as a way of interfacing to its service creation nodes, which would likely speed the deployment of features. They like the fact that SIP is lightweight, relatively easy to implement, and is more of a native protocol for public networks than H.323.

Unisphere's "any-to-any" switching philosophy is carried out by way of a mechanism within the softswitch that normalizes every incoming interface and implements common call handling procedures within the server. Because this was a goal from the beginning, they've constructed the system in such a way that it's easy to add a new protocol or variant without fundamentally altering the call control scripts. In its initial release, Unisphere offers interworking between SS7, SIP, MGCP, and H.323, along with limited class 5 features.

Signaling System 7 (SS7)

SS7 is the underlying signaling structure of the Intelligent Network (IN). In previous centuries, the Public Switched Telephone Network (PSTN) was made up of very large switches that could provide only those limited services that were "hardwired" into them. Introducing new services was a ridiculously lengthy, expensive and difficult process. New services were subjected to regression testing and often took up to five years, in many cases as long as the actual development cycle.

Intelligent Networking was designed to enable modularization of the network into nodes that communicate using a standard open interface, and so Intelligent Networking Application Part (INAP) was born. These nodes each perform one of the functions required in these services:

- Service Switching Point (SSP) – Provides the circuit switching for the call and is generally where a call originates.
- Service Control Point (SCP) – Provides all the service control logic that turns a call into a service.
- Service Data Point (SDP) – Stores all the data needed by the SCP to correctly control the service.

- Intelligent Peripheral (IP) – Provides tones, announcements and speech recognition. It also prompts the subscriber for information and collects it.

Each node may be supplied by a different vendor, and compliance to the standard allows the node to be replaced or updated without a need for changing other network components. This modularity allows carriers and service providers to quickly and cheaply deploy exciting new services in the network without replacing all of the legacy hardware.

Thanks to the continuing miniaturization of electronic components, a suite of SS7 telephony protocols can run on a modern computer telephony resource board plugged into a PC, giving a CT application full connectivity to networks around the world.

Simple Network Management Protocol (SNMP)

A protocol developed by the Internet Engineering Task Force (IETF) and defined in their RFC 1157 document, SNMP was originally designed to monitor and troubleshoot routers and bridges on the Internet. It then became a standard for managing TCP/IP-based networks and multi-vendor products used in those networks that could be made SNMP compliant, and rapidly gained favor among network administrators when they realized that SNMP enabled them to see network statistics and change device configurations from a central location. Although a part of the TCP/IP protocol suite and still used for managing IP gateways and network devices, SNMP is no longer restricted to TCP/IP networks.

SNMP and SNMP-based applications gives network managers the ability to monitor and communicate status information between such items as:

- Computers running Windows NT
- LAN Manager servers
- Minicomputers or mainframe computers
- Routers or gateways
- Terminal servers
- Wiring hubs

SNMP also gives one proactive abilities, allowing one to do such things as follows:

- Assign priorities for communication
- Install software on the network
- Manage databases
- Manage power on the network
- Monitor printer queues
- Set up addresses for devices

SNMP uses a distributed client/server architecture that is based on three basic interoperating components: A manager, an agent, and a database. The SNMP "manager" is usually a software program (such as HP OpenView) that transforms a computer into a Network Management Station (NMS). An NMS can control many agents.

The "agents" can be software or firmware in managed network devices or "nodes," such as a bridge, router, or host. Each agent stores management data and responds to SNMP manager queries for data. Agents can be grouped into "communities" that are similar to Windows NT or UNIX domains. SNMP messages can be sent to "community names" representing a community / domain, and entire community names can be monitored.

Finally, the "database" is known as the Management Information Base (MIB). The MIB serves as a map of the hierarchical order of all managed objects (the collection of databases for all managed SNMP-supported devices) and how they can be accessed. Any device running an SNMP agent maintains MIB variables and has its own map / database of relevant objects for which it collects statistics and performs other functions.

S

The SNMP network management process itself is also based on three components:

- The Management Information Base (MIB)
- The Structure of Management Information (SMI) which are rules specifying the format used to define objects managed on the network that the SNMP protocol accesses.
- The SNMP Protocol itself, which defines the format of messages exchanged by management systems and agents. It allow one to specify, for example, "Trap" operations or messages that alert the network management station of critical events, such as when an Uninterruptible Power Supply switches over to a battery.

Everything in an SNMP network is a node. Nodes can be categorized as "managed nodes" running an agent process that services management node requests, "management nodes" which is a workstation where the network administrator is running HP Overview or some other management / monitoring software (the administrator can also manage his own management node), and "unmanagemable nodes" that are not SNMP compliant but may be manageable by SNMP via a proxy agent running on a different device.

Because of SNMPs "lightweight" quality and ease of implementation, it has become very popular and is used on a wide range of hardware and software platforms.

Simple Object Access Protocol (SOAP)

Lets' say you had one program running on a Windows NT PC in New York and another Linux machine in San Francisco is running a different program. The program in New York needs to communicate and pass information to the program in San Francisco. Normally one imagines accessing the PSTN, setting up a call, etc., translating data formats, etc. But let's say that both machines are always connected to the Internet. Wouldn't it be wonderful if the web's ubiquitous Hypertext Transfer Protocol (HTTP) and the trendy Extensible Markup Language (XML) could also serve as a communications mechanism between the two programs, even though they are running under different operating systems?

As it happens, Microsoft, DevelopMentor, and Userland Software saw an opportunity here and jointly developed SOAP, which has been proposed as a standard interface to the Internet Engineering Task Force (IETF). In the SOAP architecture, HTTP serves as the base transport while XML documents are used for encoding of the invocation requests and responses. SOAP is thus an XML / HTTP-based protocol that can also access services, objects and servers in a platform-independent manner.

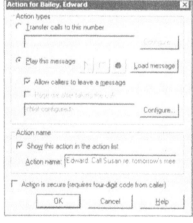

SOAP's competitors include the Internet Inter-ORB Protocol (IIOP), a protocol that is part of the Common Object Request Broker Architecture (CORBA) and Sun Microsystems' Remote Method Invocation (RMI) a Java-based client/server interprogram protocol.

SIP

See Session Initiation Protocol

Small Office Home Office (SOHO)

During the 1990s, various companies tried to bring computer telephony to SOHOs, generally with disastrous results. Many products came and went such as Creative Labs' PhoneBlaster, the Syncro board, the Phoenix software suite, a multitude of voice modems, etc.

Companies continue to try to bring CT technology to the little guy. With the improved technology and standards of the 21st century, the list of emerging telecom products and services that can assist small businesses is large and growing, ranging from full-featured NT-based PBXs to low-cost stand-alone key systems; mobile phones having the Wireless

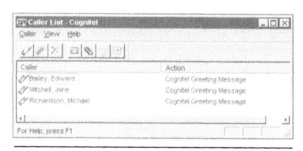

Step 1. Steve Jones is a Cognitel user. He has configured the Cognitel Caller List with three of his most common callers, all of whom currently receive the standard Cognitel greeting message.

Application Protocol (WAP) and built-in micro-browsers; programmable auto-attendants and voice-mail systems; web-based unified messaging, digital pagers; wireless PDA; voice-over-IP; bundled voice/data services from competitive interconnect carriers, etc.

An entire web site is devoted to telecom solutions for SOHOs, entitled, appropriately enough, the Soho Telecom Phone Systems web site, which can be found at www.sohotelecom.com. It is here that small businesses come together to learn (articles, newsletters, reviews), discuss (forums, opinion polls), buy and sell (auctions, classifieds and online shopping) the full range of advanced voice-enabled solutions.

One fascinating example of SOHO computer telephony we ran across is Cognitel, a speech-enabled telephony SOHO voice mail and autoattendant application. It's by NovCom (Nashua, NH – 603-886-1684, www.novcom.com) and NovCom President Eric Almeida demonstrated it in our

Step 2. Steve selects Edward Bailey, for whom he would like to leave a special message — namely that Edward should make arrangements with Susan for the big meeting tomorrow. Steve double-clicks to edit the action for Edward, then records the message using his computer's microphone. He gives it a name so he can remember what the message is about.

East Coast Lab. Cognitel software consists of a a remarkable little $49.95 software package running under Windows 98.

Cognitel not only integrates voice mail with the Microsoft Exchange Inbox to give you unified messaging, but it gives you voice-activated remote access to check your messages.

Cognitel is for you if you're a small office or home user who is frequently absent but you don't use a receptionist or call answering service, or if you want to give our callers individualized treatment, or if you just want to consolidate all of your telecom activities to a single phone number.

It works like this: You keep Cognitel running in the background. A call comes in and a screen pop appears, either generated by Caller ID or speech verification. If you've been away from your computer, activity logs tell you who's called, what time, and what action was taken by Cognitel.

It's the only SOHO product on the market to use speech recognition to identify callers.

Because proper nouns lack regularity in phonemic transcription and pronunciation, phoneme-based recognition, the kind of speech recognition technology most commonly used today, isn't well-suited to the task of spoken name recognition. Consequently Cognitel employs a pattern-based recognition approach instead of a phoneme-based approach.

There are two "models" for recognizing each caller via speech verification. One model is when you initially train the system by microphone, a manual training technique that's used just before the people in your Caller List

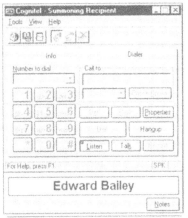

Step 3. Later on, while Steve is out, Cognitel takes the call.

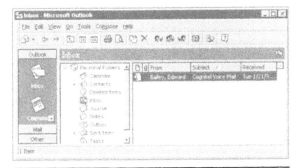

Step 4. When Steve returns home, he checks in the Inbox to find a message from Cognitel waiting for him.

directory call-in for the first-time.

Almeida and I entered a few names (Cognitel is not language specific, so I could have spoken foreign names). The name I spoke ("Harry Newton") is analyzed and a rough representation is placed in the system. When Harry finally calls into the system, he says "Harry Newton" and that sound is matched with the set of names models (patterns) by using original modifying Dynamic Time Warping (DTW) algorithms. Cognitel initiates recognition with name models trained by me via the microphone.

The recognition accuracy in such cases is about two out of three, which is not that bad, since the system must compare my utterance of a person's name with the person's own utterance of their name!

Later, when Cognitel has received at least one sample of a given caller's name spoken by the caller him / herself to the auto-attendant, an

auto-training procedure starts (the second, "call specific" model).

The more times a caller calls and announces their name, Cognitel creates an ever-more-accurate model of the caller and can reach 95% to 100% accuracy in identifying callers.

Cognitel can perform name recognition in real time with up to 50 callers entered on your Caller List. With more than 50 callers there may be a delay of several seconds between the time a caller says his / her name and the instant at which Cognitel establishes the his/her identity. However for optimal recognition results NovCom recommends that you begin with 20 callers or fewer when you first begin to train the system by microphone.

As calls are received from callers in this initial group, Cognitel's name recognition ability becomes robust surprising quickly.

If you've got Caller ID and somebody calls, Cognitel can do a screen pop without resorting to speech verification. In either case, the internal list is consulted and if the caller is on the list, then the name will appear on the pop up screen.

Still, speech verification is potentially more accurate than Caller ID, especially if the caller is calling from a PBX-based phone system, a cell phone, or a public phone. If the user has previously entered their name in the system, Cognitel can react to specific callers with custom greetings and messages, special call transfers, or paging notifications.

The real value of all this speech recognition functionality is that you can configure the system to generate one of three special actions for certain callers:

- Either a custom message greeting ("Oh Harry I'm not in but I was expecting your call – I want to confirm our meeting for 10 a.m. tomorrow.").
- People with conferencing ability (such as the Bell Atlantic service) can have certain callers forwarded to another number such as a cell phone.
- If you have a paging service, calls from certain recognized callers can be used to prompt paging actions. It works with any paging system that uses touchtone entry. The sophisticated paging systems have some limited messaging capability, such as the name of the caller being broadcast to a page – NovCom is working on making Cognitel compatible with the most sophisticated paging systems.

In normal operation you can use a phone or just your PC's microphone and speakers to place and take calls. If you pick up the phone before Cognitel answers an incoming call, then Cognitel won't answer, since it thinks either you're going to answer or else you're going to make an outbound call.

Cognitel has a neat little voice mail system. It also has the ability to identify incoming faxes and if there's a fax application open in the background

Step 5. Steve double-clicks on the message to open the Cognitel message player. He plays back the message and knows that everything will go according to plan tomorrow. Steve adds a subject and some notes to the message for future reference.

S

Cognitel will pass the fax off to the appropriate app. The same thing holds true for e-mail.

Contained in the HP computer we used was a new LT Winmodem from Lucent. NovCom has decided that TAPI-compliant devices, particularly the new Winmodems, were the way to go.

Best of all, you can download a free, 30-day demonstration version of Cognitel from www.cognitel.com. You can also buy the full version online for only $49.95.

Cognitel In Action

Smartphone

A smart phone is a wireless digital phone that's also capable of accessing the Internet whereupon it can send and receive e-mail and fax information, and act as a Personal Data Assistant (PDA).

Smartphones include such devices as the Nokia 9000i Communicator, the Motorola i1000, and the Neopoint 1000.

The Strategis Group of Washington, D.C., has issued a report claiming that annual Smartphone sales will reach 60.3 million units ($7.8 billion in sales) by 2005. More and more Smartphones have been Wireless Access Protocol (WAP)-enabled by manufacturers in anticipation of user interest in accessing the Internet from a mobile phone, creating a unique supply-and-demand situation where new users will receive handsets with wireless data capabilities whether they plan to use them or not.

In a survey of current wireless phone users, more than one-third said they would be interested in a wireless phone which can access the Internet. Even among non-users of wireless phones, The Strategis Group survey showed that 20% would be interested in a wireless phone with the ability to access the Internet.

The "WAP phone" version of the smartphone appears to be an interim solution for the public's desire for a wireless Internet.

SNMP

See Simple Network Management Protocol.

SOAP

See Simple Object Access Protocol.

Softswitch

Softswitches are open, extensible and intelligent devices installed between legacy Time Division Multiplexed (TDM) and ATM backbones and newer IP packet data networks, and are capable of routing phone calls between the two kinds of networks.

Softswitches are characterized by:
* A secure broadband call model
* An open, directory-enabled platform
* Bulletproof reliability
* Easy and quick service creation
* Scalable intelligence and call processing
* Straightforward integration with legacy networks
* Support for class-independent services over any media.

Often clustering software that can spread the softswitch intelligence over multiple servers is used to bring availability into the five and six nines (99.9999%) range at a reasonable cost.

SOHO

See Small Office Home Office.

SONET

See Synchronous Optical NETwork

Start Method

The signal that an interface gives to the PBX when it's ready to receive digits. "WINK START" means the interface will wink its control lead when it is ready for digits. "IMMEDIATE START" means that the interface will be ready to accept digits whenever the PBX goes off hook. "WINK START" is the most common mode of interface.

Synchronous Digital Hierarchy (SDH)

The ITU-T version of the Synchronous Optical NETwork (SONET) (See Synchronous Optical NETwork).

Synchronous Optical NETwork (SONET)

Born 15 years ago as sketches on a napkin in an airport restaurant, SONET is a fiber optic transport standard developed by the American National Standard Institute (ANSI) to bring some order to the early days of fiber optics, when every long-distance carrier had its own proprietary optical Time Division Multiplexed (TDM) system. Carriers had to interoperate somehow.

In 1985, Bellcore (now Telcordia) began working on what ultimately became SONET. Bellcore's Yau Ching and Rodney Boehm developed Sonet while working on a project for MCI involving a new interface to connect interoffice networks from Local Exchange Carriers (LECs) to interexchange carriers. Later, CCITT joined the effort, which resulted in a SONET standard (the first official SONET equipment hit the market in 1989) and a set of parallel CCITT recommendations (G.707, G.708, and G.709) appeared in 1989 called the Synchronous Digital Hierarchy (SDH). SDH is only slightly different from SONET. Practically all the long-distance telephone traffic in the U.S. and Europe uses trunks running SONET in the physical layer.

A SONET system has a familiar feel to it, with switches, multiplexers, and repeaters, except for the fact that all these devices are connected by optical fiber, not copper.

In the SONET nomenclature, a pristine, contiguous section of fiber stretching directly from one device to another is called (believe it or not) a section. A run of fiber between two multiplexers is called a line. The entire end-to-end connection from source to destination is called a path.

The SONET topology can be a mesh, but is often a dual ring.

The basic building block of SONET is the Synchronous Transport Signal (STS) with a transport rate of 51.84 Mbps. SONET is really a souped-up version of a traditional TDM system, with the fiber's entire bandwidth devoted to one channel containing time slots for various subchannels. Being a synchronous system, bits sent over SONET are controlled by a precise master clock with an accuracy of about 1 part in 109. The basic SONET frame is an 810 byte block sent every 125 gs.

SONET has the advantages of providing high-capacity fiber optic transport, defines a system of synchronous signal levels, includes a high-level of OAM&P (Operations, Administration, Maintenance, and Provisioning) capability, and supports automatic protection switching, and allows a high-degree of interoperability between different vendor platforms, allowing different types of formats to be transmitted on one line.

Despite its tremendous success and proliferation in global telecommunications, SONET is looking a bit long in the tooth when compared to recent advances of pure optical networking. Many wonder if Sonet will be supplanted buy other technologies or it if will somehow continue to evolve.

T-1

T-1 (also spelled T1), means "Tier 1." It represents the original transmission rate (1.544 Mbps) of a high speed, copper-wire based, Pulse Code Modulation (PCM) and Time Division Multiplexing (TDM) digital network for voice transmission called the T-Carrier system (T stands for trunk). It was developed by AT&T starting in 1955, was demonstrated (somewhat disastrously) in 1958 and successful trials finally took place in 1961. Commercial T-1 service debuted on July 25, 1961. The Bell System then began the arduous task of converting their Frequency Division Multiplexing (FDM) trunks to TDM trunks. The digital T-1 and the T-carrier system improved the signal/noise ratio on multi-line phone trunks and eventually completely replaced the backbone of what for nearly 100 years had been a fully analog telephone system. T-1 was used internally by the phone companies to interconnect Central Offices (COs) until it was made available to public individual subscribers as a tariffed point-to-point dedicated ("on all the time") service in 1983, thanks in part to the break up of AT&T also occurring at that time. AT&T first offered T-1 service under the brand called the High-Capacity Terrestrial Service, then renamed it ACCUNET around 1985 and offered additional functions such as network cross-connect reconfiguration capability. Today you can order T-1 service under the AT&T's trademark of "ACCUNET," Southern Bell's "MEGALINK," and other brand name telco services.

The original transmission rate (1.544 Mbps) in the T-1 span is still perhaps the most common form of digital transmission found in the local loop and is used extensively in the U.S., Canada and Japan (where it's called a J1). T-1 is used primarily for trunks connecting points within and between major metropolitan areas, but it's also used for nearly two million local lines to central offices (COs) and remote switches. T-1 is so common that it has become more of a utility service than a backbone service.

T-1 operates over special low capacitance shielded twisted pair cabling, although any ordinary standard twisted pair cabling can be used if precautions are taken to eliminate crosstalk, such as lowering the terminating impedance to around 100 ohms. A typical T-1 line between two locations consists of four wires in the form of two twisted pair lines, one pair for transmit and one pair for receive, running from the customer site to the CO. These pairs are usually carefully selected pairs within standard multi-pair cables already installed, and "repeaters" (signal regenerators) are placed along the cable at 6,000 foot (1.83 km) intervals and are powered by a simplex current of 60 milliamps (ma) between the two pairs. At the CO, the T-1 enters and is usually multiplexed up into a T-3 or higher for long-haul transport. The T-3 operates at 44.736 Mbps and the T-4 at 274.176 Mbps. These are known as "supergroups." At the other end, the serving central office demultiplexes the T-1 out of the higher bandwidth service and routes it to a twisted pair cable to the customer premises.

The T-1 is the major transport of Internet traffic. For example, a small Internet Service Provider (ISP) may connect to the Internet as a point-of-presence (POP) over a T-1 span owned by a large carrier. Or you, sitting at your computer at your company, may be connecting to the Internet over a shared T-1. The T-1 connects the backbone provider to the ISP provider via a telecommunications provider (telco). The T-1's digital signal comes into whatever the company uses as a "channel bank," which is a T-1 voice / data multiplexer that terminates the T-1 connection and separates or "multiplexes" it into 24 individual channels, perhaps even converting them to analog so that they can be plugged into an old phone system. A classic channel bank consists of an integrated Channel Service

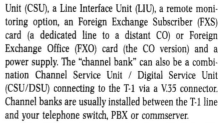

Unit (CSU), a Line Interface Unit (LIU), a remote monitoring option, an Foreign Exchange Subscriber (FXS) card (a dedicated line to a distant CO) or Foreign Exchange Office (FXO) card (the CO version) and a power supply. The "channel bank" can also be a combination Channel Service Unit / Digital Service Unit (CSU/DSU) connecting to the T-1 via a V.35 connector. Channel banks are usually installed between the T-1 line and your telephone switch, PBX or commserver.

The CSU part recovers the channelized (multiplexed) data and offers such diagnostic and protective functions as lightning protection and loopback for diagnostic testing. The loopback signal is the control signal defined in the T-1 specification. This signal enables the operator at a telco control center to command the remote equipment to loop its receive signals back onto its own transmit path. Complete end-to-end testing can be done from the control center in this way. While the link is in loopback mode, the operator in control can insert test equipment onto the line.

Don't confuse the CSU's loopback signaling with LOOP timing, which refers to when the CSU (or CSU/DSU) doesn't provide the clock from an internal oscillator but recovers it from the T-1 network. In such a case, the clock can be provided by the telco directly or by another piece of equipment such as the CSU/DSU at the other end of the circuit that is set for internal timing.

Inputs and outputs to and from a CSU are raw T-1 signals traveling over two twisted pairs (using a terminal strip, old screw terminals, a DB-15 (15 pin) connector or an RJ-48C for connection to the telco's lines). The actual wires bringing the T-1 service onto the customer premises may also be carrying a DC current source (60 or 140 mA) for powering the CSU supplied by your friendly local telco. Such a T-1 line is said to be "wet." T-1 lines not providing "span power" are called "dry." Wet T-1's ensure that even if there's a power failure at your site, the CSU will continue to function, preserving the integrity of the T-1 signal.

The repeater nearest to the CSU is guaranteed to be within 3,000 feet, half the distance of the maximum distance normally found between repeaters. If the repeater is too close or too far, the CSU may have a built-in switchable attenuator known as a Line Build-Out (LBO) to adjust the line. Automatic Line Build-Out (ALBO) equalizers can be found in repeaters in the public network, and the CSU does in some respects resemble a repeater, though it has far more diagnostic functionality.

For data applications, the CSU is integrated with the DSU, which strips off the encoding used to package the digital data, converting T-1 voltages and encoding to data for a router and vice versa. It usually connects to the router via a V.35 interface. A CSU/DSU is thus basically a kind of modem for leased-lines.

The signal then goes to the router (or the T-1 line may just go to a Dialogic T-1 board in a communications server and then across the backplane to a LAN NIC card).

From the router the signal may travel over to a master name server and may be routed to other servers such as a gateway that enables you to connect to the Internet.

Non-digital PBX's lacking T-1 connectivity can be connected to a T-1 via a full channel bank, which means that you'll also be adding some additional cards to your phone system.

The original T-Carrier spans were built using AT&T's T-1 standard design, which has a bit error rate of about 1x10E-7 (one bit wrong in 10,000,000 transmitted) over a span of 50 signal repeaters (signal regenerators), with up to 6,000 feet between each repeater. They later developed a better system for transmitting over longer distances between cities called T-1 OUTSTATE that has the same bit error rate of 1x10E-7 even after traveling through 250 repeaters. Those telcos recognizing that most of the future circuit traffic would be data began to follow the T-1 OUTSTATE specifications and have built T-Carrier networks displaying excellent error performance.

The T-Carrier system converts the normal analog signal of a subscriber pair into a Digital Signal or DS. The digitization technique used is Pulse Code Modulation (PCM), and it was felt that human speech could be represented by taking 8,000 samples per second of the voice's analog waveform and quantizing it to 8 bit precision with an analog to digital (A/D) converter. When the bits are ordered in a 64 Kbps serial stream, this signal is called a "DS0" or "Digital Signal (or Service) Level Zero" by the phone company.

The 64Kbps DS0 bandwidth for one voice channel also happens to be the fundamental building block of the so-called Digital Signal X series of standard digital transmission rates or levels based on the DS0. Both the North American T-carrier system and the European E-carrier systems use the DS series as a base multiple. The Digital Signal Level One (DS1) signal used in the T-1 carrier, is 24 DS0s (1.544 Mbps). DS2 consists of four multiplexed DS1 signals yielding a rate of 6.312 Mbps. DS3, the signal in the T-3 carrier, carries a multiple of 28 DS1 signals or 672 DS0s or 44.736 Mbps.

The Digital Signal X scheme is based on the ANSI T-1.107 guidelines. The ITU's guidelines differ a bit from these.

Also, "DS" does not mean the same thing as "T." The first two layers of the International Standards Organization (ISO) Open System Interconnect (OSI) model are the Physical and Logical layers. The Physical layer focuses on electrical characteristics such as voltage levels, waveform shape, etc., while the logical layer deals data formatting and how data can be separated from low-level protocols. The designation "Digital Signal" or "DS" thus describes the physical layer and should not be confused with the "T" designation which refers to the type of carrier used.

The T-1 is thus the digital carrier facility used to transmit the DS1 (24 64Kbps DS0 channels) using TDM (See Time Division Multiplexing) at an overall rate of 1.544 megabits per second.

You may have noticed that 24 64Kbps channels gives us a total bandwidth of 1.536 Mbps, not 1.544 Mbps. Why are more bits transmitted than are apparently necessary?

The answer lies in our use of TDM to format the bit stream. Using TDM means that we "multiplex" or sample a byte (8 bits) from each of the 24 DS0 channels in succession, creating a 24 byte "frame" of bits. Each channel thus has a recurring "timeslot" in each frame. However, when looking at a data stream, one bit pretty much looks like any other. If a terminal receiver can't distinguish between a frame's beginning and end, then the transmitter and receiver will be "out of frame sync" yielding poor performance.

Ironically, prior to 1924 synchronization of signals in the public telecom network was not even considered. There was no encoding or decoding circuity, just a wire that carried a pure analog baseband signal representing voice at its normal frequency range from one place to another, one channel per line. It was about as sophisticated as two plastic cups with a string stretched between them. After 1924, however, the Type C carrier line appeared which enabled three two-way (duplex) voice channels to travel over the same wire thanks to shifting up the frequencies of the channels. From then on, things definitely got a lot more complicated!

Indeed, these days, when a PCM bit stream is transmitted over a telecom link such as a DS1 channel flowing through a T-1 carrier, the complex processes of synchronizing signals and executing timing algorithms on a huge public network are far more daunting than for anything similar done on a LAN. Synchronization of signals must be attended to at three different levels: Bit, timeslot, and frame.

Bit synchronization refers to the need for the receiver / decoder to be able to accept bits at the same rate or frequency at which the transmitter / coder is sending them. If they get out of sync, a "slip" will occur and one or more bits will be lost. At the central office a Digital Access and Cross-connect System (DACS) is a digital switching device that can route T-1 lines and can split out and recombine DS0 channels into different T-1 circuits. These T-1s must of course run at the same rate, which means that somewhere there must be a "Master clock." Indeed, for many years a central 1.544-MHz frequency source has been housed near the geographic center of the U.S. in Hillsboro, Missouri. This was later augmented by using signals from the Global Positioning System (GPS) to synchronize clocks at some network nodes. These nodes in turn send the clock signal over dedicated links to those nodes lacking GPS sync ability.

Frame synchronization is usually attained by periodically inserting some additional bits as a marker in the bit stream. And if frame synchronization has been achieved, then timeslot synchronization should already have fallen into line.

The big challenge with T-1 circuits is the framing. If the receiver goes out of phase with the sender, data could be plugged into the wrong channels.

How do we keep all of the T-1 equipment in sync with the frames? Including more than one DS0 channel in a TDM bit stream only works if we add what are called "framing bits," that serve as boundary markers and a time reference so that the individual channels can be synchronized, properly routed and ultimately discriminated and decoded at the receiving end.

AT&T / Bell Labs looked at two possible approaches to using framing bits: Forward-acting and backward-acting synchronization. The forward-acting synchronization method starts each frame with a unique code. This allows for extremely fast resynchronization, since it occurs right at the start of each frame, so each frame can be resynchronized anew. If the bit stream gets out of sync with the equipment then only one frame is lost, unless an error occurs when receiving the framing pattern, in which case another frame can also be lost. To make forward-acting synchronization work, circuitry must be developed to produce a unique framing code for each frame, and lots of extra bits must be sent with each frame. The technology of 1961 was not up to the task.

In the case of backward-acting synchronization, a single kind of framing code is transmitted with each frame instead of a unique one for each frame. Of course, a longer, "as unique as possible" framing pattern is better for resynchronization so the system is not deceived by bits in the data stream that coincidentally resemble the framing pattern. Instead of resynchronizing every frame as in the case of forward-acting synchronization, a backward-acting receiver just verifies framing integrity. If a misframe has occurred, it will take several frames to resynchronize since the receiving equipment must methodically examine each bit in the frame to find the start of the framing pattern. Backward-acting synchronization can also allow for one or two transmission errors before declaring a misframe (AT&T wanted the synchronization to allow for single transmission errors).

To keep the cost of resynchronization inexpensive and relatively quick, AT&T chose the backward-acting synchronization technique for the T-Carrier system.

It is generally the formatter, or the CSU, that inserts the framing bits into the bit stream (in a PC-based system, of course, all of this circuitry fits on a single T-1 PC board from such companies as Dialogic, Natural MicroSystems or Brooktrout). After a byte (8 bits) of data from each channel is multiplexed (sampled in succession) and sent (24 channels * 8 bits = 192 bits) this "frame" of 192 bits is followed by an extra frame bit. 193 bits can be transmitted in 125 microseconds, so the total bit rate of a DS1 is as follows:

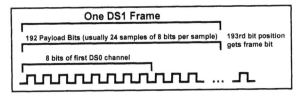

One DS1 Frame

192 Payload Bits (usually 24 samples of 8 bits per sample)   193rd bit position gets frame bit

8 bits of first DS0 channel

Total bit rate = 8,000 samples per sec * ((8 bits per sample * 24 samples) + 1 frame bit) = 1.544 Mbps

Which is what we know as a T-1 and what AT&T engineers in the 1960s and 1970s knew as a DS1, though T-1 should really refer to the raw "carrier" data rate and the copper transmission system while DS1 refers to the bit format and framed rate.

In any case, you're probably thinking that you can now run off and use all of the 1.536 Mbps bandwidth available after you subtract the bandwidth taken by the frame bits.

Well, not quite.

Line Coding

In the original 1961 standard, it was known that the signal repeaters and channel banks of the day were not quite as sophisticated as they are now, and needed all the help they could get in terms of getting the correct synchronization of pulses representing the ones and zeroes and to allow for as much distance as possible between the repeaters. To this end, AT&T began investigating possible advantageous line code techniques.

Up until that time, Return to Zero (RZ) coding seemed to make sense. RZ coding represents a logical "1" with a positive voltage pulse (+V) followed by no voltage (0V), and a logical "0" is represented by a negative voltage pulse (-V) followed again by a 0V. Every pulse can be used to provide clock information to synchronize the system.

It was soon realized that RZ line coding wasn't good enough to be used with the T-1 carrier, since RZ doesn't do a particularly good job of balancing the DC voltage on the line, and extreme DC level bias can disrupt the workings of a T-1 circuit. This is because a typical T-1 configuration includes balanced transformer couples on the sending and receiving ends to isolate the line and terminals from any DC components on the long links. The transformers act as high-pass filters that allow only high-frequency components to be transmitted and blocking low-level DC fluctuations, and excess positive or negative DC voltage will saturate the induction field of these transformers, thus distorting the signal. So, as it turns out, the "return to zero" technique wasn't living up to it's name, and didn't return the DC line voltage to zero fast enough.

Instead of RZ, a different form of line encoding was developed called Alternate Mark Inversion (AMI), which is sometimes called binary coded alternate mark inversion (See Alternate Mark Inversion). The AMI coding scheme uses bipolar but "non return to zero" pulses to represent logical 1's and 0's. The binary value of "1" is represented by a square wave (a pulse of about 3 volts), while a "0" is represented by a flat straight line for that timeslot (no pulse). The signals have a "bipolar" format: Every time a "1" pulse is sent, it is opposite in polarity to that of the previous

pulse. A pulse consisting of +3 volts would be followed by a pulse of -3 volts, then +3 volts again, and so forth. Remember that the reason pulses of alternating polarity are sent is to ensure that the signal spends an equal amount of time at each polarity and so the average direct current voltage on the line always remains zero. Keeping the overall voltage zero actually doubles the distance that a DS1 signal can travel over a pair of copper wires. The bipolar format also makes error detection easy: Two consecutive pulses of the same polarity is called a Bipolar Violation (BPV). Too many consecutive BPVs suggest that defective equipment, lines, or electrical storms are affecting the transmission.

Of course there's a problem here – what if you must send a signal with a lot of zeroes, which is just a flat line (no voltage). A long idle interval on a T-1 trunk is a disaster for the normally synchronous operation of the T-1 and the ability of the CSU's and repeaters to distinguish valid input from line noise. To correctly recognize the DS1 stream, the repeater must know when to sample the bipolar signal to determine whether a one or a zero is actually being transmitted at any given instant. To get the timing for the sampling correct, all repeaters use a clocking method based on binary pulses - the 1's – to maintain synchronization with whatever equipment is transmitting the DS1 signal.

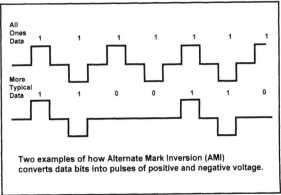

All Ones Data   1   1   1   1   1   1   1

More Typical Data   1   1   0   0   1   1   0

Two examples of how Alternate Mark Inversion (AMI) converts data bits into pulses of positive and negative voltage.

In other words, to keep things in sync, a certain minimum number of 1's have to be transmitted over a certain time interval, since the signal itself is the clocking mechanism. This requirement became known as the "ones density" or "1's density" requirement.

There are several stipulations to the ones density requirement. All of these conditions had to be met in the original T-1 facilities, and they are still relevant:

- On average, there must be 12.5% transmitted, or a "1" for every eight "0's." This is also called the "pulse density."
- One in 15 bits must be a "1" which means that there cannot be more than 15 zeroes in succession, since this would confuse the equipment waiting for the synchronizing bits (the ones density was not maintained independently of the data stream)
- In any group of 24 bits, at least three bits of those bits must be a "1".

Certainly the complicated waveforms representing human speech, when digitized, leads to bit patterns that meet all the criteria of ones density, but one could never be absolutely sure, and, in any case, data was beginning to travel over T-1's too, and data streams can contain a long string of zeroes. Early T-1 analog repeater equipment was not very sophisticated in the way bits were tabulated, and so to ensure the ratio of a 1 for every eight 0's, a primitive technique called "bit stuffing" was used, where every eighth bit (considered to be the Least Significant Bit (LSB) of each DS0) was "stuffed" or set to be a "1". At the receiving end, the stuffed bits would be discarded and the signal extracted.

Using seven bits instead of eight to convey data unfortunately reduces each DS0's bandwidth from 64 Kbps to 56 Kbps and the usable data rate for the whole T-1 dropped to 1.344 Mbps (56,000 bps * 24 channels).

Voice quality was also affected slightly, since when the original analog waveform of speech was sampled and digitized 8,000 times a second, there were now only 128 possible combinations of seven bits, and thus 128 possible digital levels of representation for the height of the waveform at each moment of sampling, yielding a dynamic sound range of about 48 dB. Since there is only a small possibility that the exact height of the waveform at the moment of sampling will exactly equal one of the 128 digital numbers, the difference between the actual value of the height of the waveform being sampled and the value of the closest digital level of representation is a sampling error known as the sampling quantization error. This type of error introduces about 1% distortion at full output during a conversation over a T-1. If eight bits are used, however, 256 levels can be brought to bear on the sampling / quantization process, which gives us a dynamic range of about 48 dB and about 0.5% distortion at full output. Each bit added to the sampling / quantization process doubles the number of levels, increases the dynamic range by 6 dB, and decreases the distortion by a factor of two.

As more and more data began to travel over T-1 lines, the 7-bit word / 56 Kbps DS0 rate became increasingly inadequate. Unlike voice, computer data could rarely conform to the ones density standards. Moreover, while a 7-bit word can be used to pass the ASCII character set, most computers use an 8-bit word, which involves elaborate reconstruction routines at the receiving end.

The need for a new line coding technique friendly to data communications led to the development of what is now called Binary (or Bipolar) with 8-Zero Substitution (B8ZS). Unlike AMI, B8ZS coding can move eight-bit words by passing all eight bits but substituting any occurrences of eight consecutive zeros with a special special predetermined bit sequence or "code word" consisting of four "1's" and four "0's" arranged in such a manner so that two intentional bipolar violations (BPVs) will occur. Whereas with AMI BPV's are considered an indicator of defective equipment or bad wiring, in B8ZS they are used as a flag to alert the equipment to perform the eight zeroes replacement. The special bit sequence can be detected at the receiving end and the eight zeroes restored.

More specifically, when the equipment replaces a string of eight zeros it does so with the pattern 00011011 having two intentional BPV's at positions four and seven. The first BPV replaces the fourth zero with a pulse having the same polarity as the last pulse in the previous word. The second BPV replaces the fifth and seventh zeros. So, if the pulse preceding the inserted code is transmitted as a positive pulse (+), the inserted code is 000+-0-+ (BPV's in positions four and seven). If the pulse preceding the inserted code had been a negative pulse (-), then the inserted code would be 000-+0+- (BPV's still appearing in positions four and seven).

In one master stroke, B8ZS allows us to regain our 64 Kbps bandwidth per DS0 channel while at the same time guaranteeing the required ones density independently of the data stream – the ones density can be met without stuffing pulses (and more possible errors) into the data stream. Also note that the average DC voltage still remains zero.

The restored, full 64 Kbps DS0 channel is called "Clear Channel Capability." AT&T references it in Publication 62411 in Appendix B as CB144. It's also part of the ANSI T-1.403-1989 standard.

Since intentional bipolar violations are used to signal that it's time to make the eight zeroes replacement, all equipment on the T-1 line must be configured for the newer B8ZS coding, otherwise the bipolar viola-

tions will be interpreted as equipment errors (CSU's for example, must not automatically remove the BPV's). If any single repeater or other network component in the path between the two T-1 endpoints is set up for AMI when a line was specified B8ZS, the line will not work and your CSU will give you error messages such as "excessive error rate" or "bipolar violations" (since B8ZS was meant for data communications, its handling of errors is a lot more stringent than AMI, which was designed for voice transmissions). Voice and data traffic will not flow. Fortunately, nearly all T-1 equipment now uses B8ZS line coding, and it's also used for Primary Rate ISDN and optical SONET transmissions.

The B8ZS encoding and decoding of a T-1 data stream is generally done by the CSUs at both ends of the line.

B8ZS line coding has a competitor: Zero Byte Time Slot Interchange (ZBTSI) can also be used to achieve a "clear" 64 kbps channel. Originally based on a method developed by Verilink, ZBTSI became a Bell standard and an ANSI standard. ZBTSI works by buffering four frames (in 96 bytes of memory), and switches time slots to avoid 1's density violations by exchanging the time slot occupied by a byte of all zeros with another non-zero byte or "channel number" at the beginning of the frame. The position of the zero byte in the data stream is indicated by a 7 bit non-zero index value. If the eighth bit is set to one, then it means that additional zero bytes are present. Either framing bits or data bits are used to notify the equipment that the data stream has been encoded using ZBTSI. Also, a five-bit "scrambler" is added to the ZBTSI data stream to minimize error multiplication

Although ZBTSI can be installed on a network more easily than B8ZS, ZBTSI has never overtaken B8ZS in popularity, probably because of the expense of installing buffering circuitry in the network components.

Yet another method is Bipolar (or Bit) 7 with Zero Code Suppression (B7ZCS) which also sets one of the eight 0 bits to a 1. Using Mu-Law PCM, the digital sequence of "00000010" is substituted for the numerical values of "00000001" and "00000000."

In fact, there's a whole family of line coding schemes, called Binary N Zero Substitution (BNZA), that are based upon generating intentional bipolar violations. DS3's use B3ZS line coding where instances of "000" is replaced with "00-" or "+0+" if the previous pulse is "-", and with "00+" or "-0-" if the previous pulse is "+". In the case of B6ZS line coding, any "000000" is substituted with either "0-+0+-" or "0+-0-+" depending on the polarity of the previous pulse.

In Europe, where they use the E-1 (32 DS0 channels and a 2,048 Kbps total bandwidth) neither AMI nor straight B8ZS line coding is used. Instead, the rather complicated High Density Bipolar order 3 (HDB3) encoding is employed.

HDB3 line coding is similar to its brethren in that it's a bipolar signaling technique (both positive and negative pulses are transmitted). It's loosely based on AMI but is like B8ZS in that it can insert one of four possible violation codes whenever there is a run of four or more zeroes. HDB3 ensures that no more than 3 consecutive zero voltage states will be transmitted on the line. The first violation transmitted is a single pulse in bit position four. Subsequent violations always occur in bit position four, depending upon the previous violation code and the last preceding pulse polarity. To be more specific, a sequence of four consecutive "0's" are encoded using a special "violation" bit which has the same polarity as the last bit set to "1" which was sent using the AMI encoding rule. So when there are four consecutive "0" bits they are changed to the bit pattern "x00V" where V is the bipolar violation bit and x is set to either "0" or "1" so that the number of "1" bits between bipolar violations is odd. In other words, the polarity of the V bit is the same as the previous non zero voltage (opposite to a "1" bit which causes a V signal with an alternate voltage according to the previous one). For example, the bit pattern 10000110 encodes under HDB3 to +000+-+, whereas the corresponding encoding using AMI is +0000-+.

This technique solves the problem of no changes in voltage for a sequence of "0's" but a new problem appears in that the polarity of the non zero bits is the same so a non-zero DC voltage would now accumulate and persist on the line. To resolve this, the polarity of the V bit is changed so it is the opposite polarity of the previous V bit. When this occurs the bit stream is changed again to B00V where the polarity of the B bit is the same as the polarity of the V bit. When the receiver first gets the B bit it thinks that it is a "1" bit but when it receives the V bit (with the same polarity) it now interprets this to mean that B and the V bit are in fact "0." Surely, only a European committee could have come up with this algorithm!

Although the ones density requirement was concocted to keep the T-1 line repeaters and CSU's in sync and operating properly (since the data clocks are derived from examining the pulses of 1's passing through the equipment), the AT&T 62411 standard came about during a time when all repeaters were analog and quite shaky. Contemporary digital repeaters don't need so many 1's to recover their clock synchronization, and can tolerate up to 60 or more consecutive zeros. Also, in 1987 the FCC realized that the then-new repeaters were not subject to oscillation anymore, so they relaxed the ones density requirement. The FCC declared that the public network would not suffer any harm even if up to 80 zeroes in a row are transmitted, as long as the average 1's density is still a minimum of 12.5%. Still, one can never be certain if your T-1 transmission is going to travel through an old repeater somewhere, so commercial CSUs still adhere to the 62411 standard.

We've seen that every 192 bits of payload data on a T-1 (which can be live traffic or some kind of test pattern) is followed by bit #193, which is a framing bit. Since the whole 193-bit block is repeated at 8,000 times per second, this is known as a "frame rate" of 8 kHz.

You may have noticed that Yours Truly apparently neglected to specify if that special framing bit is a zero or a one. Actually, I was trying to avoid opening up a whole new can of worms involving a discussion of "framing formats." A frame bit is either one or zero depending upon what frame format we're using.

Framing Schemes

In early T-1 technology, synchronization (called phase synchronism or framing) is performed by a receiver's "framing detector" circuit. The detector compares the incoming bit stream against the framing signal generated by the receiving timing circuit.

To first synchronize a receiver with the bit stream from the transmitter, the receiver's frame detector picks a random bit, assumes that it's the frame bit, then grabs every 193rd bit in the stream beyond that for up to eight cycles (frames) and looks at this sequence of what it hopes are framing bits. If these bits form a special, distinctive, predetermined sequence or "framing pattern" (such as 0101010) that isn't likely to coincidentally occur often in the bit stream, then it has found the right pattern and the bit stream is now in sync. Remember that this is a "backward-acting" process so that every frame doesn't have its own unique code – if it did, this would complicate (and slow down) signaling, as would be the case with such a "forward-acting" process. Instead, a reasonably "rare" framing pattern is used for all frames.

If the receiver makes a wrong initial guess and the wrong bit pattern appears, then a misframe condition is declared and a frame search starts

anew: Instead of waiting for one bit position and to begin the search again with the next bit, the previous bit position's time slot is actually tried next, and every 193rd bit is again grabbed for up to eight frames, forming a sequence of these bits which the receiver expects will now match the desired framing pattern. The receiver immediately moves forward one bit position whenever the framing pattern is violated without waiting to check all eight frames. In this way, a systematic search occurs through all 192 bit positions, or until the correct framing is found. This all sounds rather laborious and time-consuming, but at 1.544 Mbps, it takes only about 0.4 to 6 milliseconds (ms) to detect an out-of-frame condition and at most 50 ms to synchronize the receiver with the transmitter, which is the highest allowed Maximum Average Reframe Time (MART).

Even when the circuit is finally running normally, "in frame" as it were, the pattern of frame bits are checked. The frame is assumed to be synchronized until a received framing bit doesn't match what it should be. If a considerable deviation from the framing sequence should occur, then a misframe is declared and another frame search starts. During normal mode some error tolerance is allowed so that the occasional random bit error doesn't trigger a misframe and an unnecessary frame search. During frame hunting mode, however, no error tolerance is permitted until the goal of timing recovery has been achieved at the receiver.

DS1 framing formats include those known as D1 / D1A, D1B, D1C, D1D, D2, D3, D4 and ESF. Currently the two most popular are D4 and ESF.

D1 (later called D1A) was the first (and least efficient) framing pattern used by AT&T / Bell Telephone for T-1 signaling starting in 1961. Slight modifications to D1A led to D1B and D1C framing. AMI line coding was in vogue at the time, and so every eighth bit could not be used for either voice or data traffic, causing a loss of 8 Kbps and lowering each DS0 channel's usable bandwidth from 64 Kbps to 56 Kbps. Someone then had the bright idea of using the eighth bit of each word for signaling information. One bit can be either a 1 or a 0 and these states represent on or off hook. Since this doesn't supply much signaling information, it was decided that so-called revertive pulses would also be used to furnish two additional signaling states. These other pulses were transmitted using bit 7, reducing the number of data bits to 6. This restricts bandwidth (and voice quality) further, but only momentarily – once the called party answers, bit 7 is again available to convey user data.

None of this ingenious signaling improved voice or data quality, of course, and it was soon realized that lavishing 8,000 bps per channel on signaling was totally unnecessary.

To improve the voice quality, a so-called "logarithmic compander" (a compressor / expander) was used. Normal, uniform quantization of an analog signal tends to cause the low amplitude values to ultimately sound distorted since quantization noise is more noticeable in low level rather than high level signals. We need to do proportionately more samples at the low end than at the high end to get good sound quality. To achieve this we can use non-linear encoding techniques or we can compand (compress/expand) the signal and then use uniform quantization. Logarithmic companding achieves the desired scenario of compressing the quantizing levels at low speech amplitudes at the transmitting end and expanding these levels at the receiving end, thus reducing the quantizing distortion. Logarithmic companding thus functions in the same manner as human hearing, discarding only the information that the brain would not process anyway, and it gives good quality results for speech and music (for example, Dolby encoding does logarithmic companding to eliminate the hissing noise added by magnetic tape technology). Although companding's compression ratio is not high, it needs little processing power.

D1 channel banks used what's called the "mu-100 law" in their log-

T

arithmic compander to increase the number of quantization steps for low level signals and decrease the number of quantization steps for high level signals. Without logarithmic companding, 11 bits would be needed to represent digitized voice with the same quality as 7 bits of companded digitized voice.

Once the DS0 channels were digitized and ready to be time division multiplexed into a DS1 stream over the T-1, two compression algorithms were used in parallel to compact the data. Two compressors were used instead of one because a single compressor would need more processing power than could be delivered by the circuitry of the early 1960s. Since each compressor was handling different sets of DS0 channels, the DS0 bytes couldn't be neatly queued in round robin order in cycles of 1 to 24 channels as you would expect. Instead, the DS0 samples were interleaved during the TDM process, with the sampled channels appearing in the following order: 1, 13, 2, 14, 3, 15, 4, 16, 5, 17, 6, 18, 7, 19, 8, 20, 9, 21, 10, 22, 11, 24, 12, and 24.

A T-1 D1A, D1B and D1C channel bank uses the 193rd bit to identify the division between frames, with a pattern from frame to frame of 010101. . . This pattern was selected since it's simple to generate and shows up as a pure 4 kHz component in the transmitted or received DS0 channels.

Unfortunately, the framing pattern, as it turned out, was so simple it actually caused problems.

Remember that the pattern formed by successive frame bits is supposed to be a highly distinctive sequence that would have a low probability of just accidentally "showing up" in the data stream. This is done because, although no data bit sequence can disrupt framing during normal T-1 operation (since only the frame bits are examined), any error in the framing sequence causes a misframe to be declared and a frame search begun, which means that data bits are examined too and data bit sequences that accidentally look like framing bit sequences can confuse the framing circuit. Random data could look like a framing pattern, but typically such a pattern shouldn't exist for very long. Besides, the alternating 1 and 0 sequence is a 4 kHz tone, and such a tone can easily be filtered out prior to digitization with a low pass filter, thus preventing a bit from alternating between 1 and 0 for a long time in bit positions other than #193.

However, in the case of D1's simple 010101 framing sequence, the framing circuit could be confused by repetitive audio *after* it was digitized, and it just so happens that sending the standard Bell test frequency of 1,000 Hz (1 kHz) over one of the DS0 channels processed by a channel bank would generate an alternating one and zero pattern every 193 bits, thus tricking network equipment into synchronizing on the wrong pattern, and messing up all 24 channels. Bell quickly adopted a 1,004 Hz test tone, then moved on to use D2 and later more advanced framing patterns.

With D1 framing, any variation from the 0101010 pattern was referred to as a "violation." During normal mode the 193rd bit of each frame is collected and checked for a framing violation. If a violation occurs, then that violation is recorded in some sort of memory device and a check is made to see if a preset threshold is exceeded. If the threshold hasn't been exceeded, then the framing circuit continues to the next frame. If enough violations occur in close succession, then the threshold has been exceeded and a misframe is declared, which results in a framing search. Search mode begins with the frame counter waiting for 1 bit to pass before counting another 193 bits and checking for a violation. If there is a violation, then the search process is reinitiated immediately (move over to the next bit position, count 193 bits, check to see if that's a framing bit). If there is no violation, then there's a chance that we've found a real framing bit and the framing is now correct. A memory device is then incremented and an inspection is made to ascertain whether enough consecutive non-violations have occurred to declare a correct frame. If not, then the next frame is examined to see if the pattern persists. If a specified number of consecutive non-violations take place, then there is a high probability that the correct framing has been detected, so a "true frame" is declared and the framing circuit is returned to the normal mode of operation.

The monitor for non-violations in the search mode is referred to as a "confidence circuit". Any violation during the search mode wipes out any existing confidence that the position is correct. The confidence circuit is necessary to insure that the probability of data looking like framing is sufficiently small so that a correct frame can be declared with certainty and a return to the normal mode of operation can be accomplished.

In the old days of D1 framing using analog circuitry, one way the system could keep track of frame bits and their bit error rate was done in the D1 channel bank by pulsing a leaky capacitor. Every time a frame violation occurred, the voltage on the capacitor was incremented by a small, fixed amount. The electrical charge eventually leaked away, but a certain number of violations over a short time interval would generate enough voltage on the capacitor so that the threshold was reached and a misframe declared. The threshold was reached when three consecutive framing bits were in violation or when about three out of five consecutive framing bits were in violation.

Next came D2 framing and channel banks, in 1969. D1 channel banks had deficiencies when it came to maintaining a good signal to noise ratio (SNR) over long distances. T-1 met the original requirement of sending DS1 signals between central offices spaced at up to 50 repeater (about 50 mile) intervals. By the end of the 1960s, however, AT&T had the bright idea of converting the entire long haul network to digital. Since the big switches on the backbone where still analog, digital signals from the T-1 had to be converted to analog, fed through the switch, then converted back into a digital T-1 stream and sent to the next hop. Converting a signal back-and-forth from digital to analog to digital caused an accumulation of quantization and thus a reduction in the SNR.

D2 channel banks could convert 96 channels into four independent 24-channel DS1's. A single coder and a new compander was used for all 96 channels. Eight bits were now available for data, so the nonlinear coding technique used for the logarithmic compander was a 15 segment approximation to the what's called the "mu-255 law."

D2 sampling was done in two stages. The 96 individual channels were arranged into eight groups of 12 channels. Each channel in a group was sampled sequentially. The samples from each group were then interleaved to form a 96 channel signal, then separated out into four DS1's, with each DS1 containing samples from two of the original eight groups.

D2 framing was also different. First, D2 framing allowed all eight bits per word to be sent in five out of every six frames. Every sixth frame, however, consisted of seven bit words. This is known as 7 5/6 coding with "robbed bit" or "stolen bit" signaling. As for the frame boundaries themselves, the D2 framing bits alternate just as in D1, but whereas a D1 channel bank uses every 193rd bit to indicate the division between frames, with a cycling pattern from frame to frame of 010101. . . and so on, the D2 channel bank uses the 193rd bit of every other frame, or every 386th bit for framing. The "skipped" framing bits in the even numbered frames are used for other things such as carrying Common Channel Interoffice Signaling (CCIS) or identifying signaling frames. The resulting sequence formed if we collect all the 193rd bit position bits looks like this: 0x1x0x1x0x1x . . . where x can be bits set to anything other than the alternating sequence. To the reframe circuit, the frame's length is 386 bits instead of 193 and every 386th bit is a framing bit following the same pattern (010101. . .) as D1. If the D2 reframe circuit is activated then the reframe time is four times as long as D1, since the time needed to do a reframe is proportional to the square of the frame length.

D2's other major difference from D1 framing came with the appearance of the Super Frame (SF) or superframe, a framing technique that has persisted right up to the present day and is closely associated with D4 framing. A basic superframe is made of 12 DS1 frames (2,316 bits). Only Frames 6 and 12 in the superframe are used for line signaling information, so in every sixth and twelfth frame the speech or data is digitized into seven bits rather than eight bits.

The eighth bits in Frame 6 are the "A bits", while the eighth bits in Frame 12 are the "B bits." The four combinations of A and B bits indicate on-hook, off-hook, etc. In order to place a call on a T-1, the device must somehow signal the central office that it wants to place a call. This information is transmitted in the A and B bits. Let us assume for clarity that the B-bit mirrors the A-bit. This is standard for a T-1 wink start line. The T-1 line must be set up for wink start and the number of touchtone (DTMF) digits (sent by the CO) matches the configuration on your device, which could be a Dialogic / Gammalink board in a PC. When a user or an application controlling the device wants to place an outbound call, the device will "pick-up" the line by asserting the A-bit. Upon seeing the A-bit transition, the CO will raise its A-bit for a short period of time, then drop it. This is called the "wink." When the device sees the A-bit drop from the CO, then the device will dial the desired numbers and perform call progress. When the remote end has answered the line, the CO will assert its A-bit to indicate a call connection. When the call has ended, both ends will drop their A-bits.

Effective throughput for both the A bit (Frame 6) and the B bit (Frame 12) is 666.66 bps each.

The remaining 10 frames of the superframe (Frames 2, 3, 4, 5, 7, 8, 9, 10, and 11) are allowed to transmit eight bits per channel – this seemingly odd arrangement of frames yields approximately a 4 dB improvement in quantization noise over D1 channel banks and is only 2 dB worse than if eight bits per channel were used in every frame.

Meanwhile, the 193rd bit in each D2 frame is used not only to mark frame boundaries but also as a control bit. If you line up all 12 frame / control bits into a sequence, they form a 12-digit "control word" (100011011100). The 12 framing bits in this superframe (one framing bit per frame) goes through the 12-bit pattern 100011011100. That is, the framing bit is a 1 in the first frame, then 0 for each of the next three frames, then a 1 for the next two frames, and so on. By checking for this specific pattern in every 193rd bit, the receiver can establish frame synchronization – and determine which 8 bits are associated with which of the 24 channels.

Every superframe has this control word of 100011011100, which can be broken into two separate bytes: the Terminal Frame (ft) and the Signal Frame (fs).

If we extract the bits from the odd numbered positions from our control word of 100011011100 (corresponding to the odd numbered frames) we get the pattern 101010 – this is terminal frame. The terminal framing bits always alternate between 1 and 0. Unlike the older D1 channel banks, the technology in 1969 was good enough so that a framing bit was not needed for every frame, just every other frame. Aside from marking boundaries, these terminal bits are used to synchronize terminal equipment.

The bits from the even numbered positions of the control word yields the pattern 001110, which is the signal frame. These bits identify frames containing signaling bits that notify the equipment of state

changes. The signal frame can identify which frames have signaling information in them by a change in pattern. Treated as a series of placeholders, the six-digit binary pattern 001110 has place holders for Frames 2, 4, 6, 8, 10 and 12. The first change in pattern of the word (the first "1" to appear after a string of 0's) occurs at the third position from the left, which is the placeholder for Frame 6. The 1's persist until we see a "0" again at the last position, which is the placeholder for Frame 12. This all means that Frames 6 and 12 contain signaling information which happens to be the A bits in Frame 6 and the B bits in Frame 12. As we've pointed out, the information is stored there via bit-robbing the Least Significant Bit (LSB) in each of the 24 DS0 channel bytes (or "words" as they are called in the telecom industry). The four possible combination of A and B bits are used to signal off-hook, on-hook, ringing and dialing for that specific DS0 channel.

When a D2 channel bank gets out of sync and must be reframed, the D2 reframe process is quite fast because it examines eight bits at a time instead of one at a time as in the case of D1. The first of the eight bits is considered as having the highest probability of being the start of the framing pattern, with the other bits considered in succession. If the first bit of each sampled byte continues to follow the proscribed framing pattern, a score in a confidence counter is incremented until a certain threshold is reached, whereupon it can be declared that the frame has been found and the system is back in sync.

During the time that the first bit candidate was being considered, however, a history of past violations of the other seven bits have been recorded too. If the candidate bit fails to match the proper framing pattern, then we don't have to start the hunt process anew with the second bit position, since we already have stored its past history and can examine it immediately. The system can check to see if any violations have occurred in position 2 since the hunt for the frame began. If the past history of bit position 2 doesn't match the framing pattern, then this second bit position is eliminated from contention and the third bit position is scrutinized in the same fashion.

This process continues until a contender is found or else all eight bits have failed the test. Whenever a contender is found, it becomes the bit position with the highest probability and the confidence counter is reset to zero and incremented again. This technique allow up to eight non-framing positions to be examined and eliminated at a time rather than having to check each bit position individually by waiting for the bits to arrive.

The other difference between D1 and D2 is the alignment of the timeslots on the digital stream.

Recall that D1 framing interleaves the DS0 channel bytes this way: 1, 13, 2, 14, 3, 15, 4, 16, 5, 17, 6, 18, 7, 19, 8, 20, 9, 21, 10, 22, 11, 23, 12, 24

D2, however, aligns the DS0 bytes in this order: 12, 13, 1, 17, 5, 21, 9, 15, 3, 19, 7, 23, 11, 14, 2, 18, 6, 22, 10, 16, 4, 20, 8, 24

D2 channel banks also allowed for the first examples of Common Channel Interoffice Signaling (CCIS). CCIS is telephone signaling technique that encodes signaling information for a group of trunks in a separate 64 kbps TDM channel allocated for signaling and network management information. This dedicated signaling channel uses all eight bits of each channel in every frame for data. In the 1970s the basis for the CCIS system was the International Consultative Committee on Telephone and Telegraph (CCITT) No. 6 international standard (CCIS6), which became the foundation for At&T's Stored Program Control (SPC) network. The CCIS6 (or "SS6") network consisted of several signaling regions, each having a pair of interconnected Signal Transfer Points (STPs). The switching systems embedded in the CCIS6 that connect to

T

STPs were called Serving Offices (SOs).

CCITT Signaling System 7 (SS7) is the current standard CCIS and is immensely popular, used by nearly all IXC and LEC carriers. Instead of an in-band signaling data trunk that carries signaling information for many channels like SS6, SS7 instead relies upon a separate packet switched network of DS0 channels to perform network control along with a Service Control Point (SCP) and a database. SS7 makes possible Custom Local Area Signaling Services (CLASS) such as Caller ID and voice mail, software defined networks, virtual private networks (VPNs), portable 800 numbers and citywide Centrex services (See Signaling System 7).

In 1972, D3 channel banks appeared, designed to provide high quality connections even over long hauls. The D3's were backwardly compatible with the D2's. The D3's combined 24 channels into a single DS1 signal. They used a single coder and mu-255 law compander (so the channels could be combined in a nice round robin sequence of 1, 2, 3, 4 etc.) as well as the superframe format.

In 1973, AT&T decided to offer a conversion kit to make D1 channel banks compatible with D2 and D3 equipment. The resulting upgraded channel bank was known as the D1D channel bank. It was quite an economical retrofit, since you only had to replace 12 of the 31 boards in the original D1 channel bank. The performance of the D1D was about the same as the D3 channel bank since they both used superframes and mu-255 law companders. The D1D was not exactly like the D3, however, since the D1D used the same somewhat jumbled sampling sequence as the regular D1 channel bank.

The D4 channel banks were developed in 1975 when AT&T found that by placing the signal repeaters at 3,000 foot intervals (instead of 6,000 feet) the usable bit rate of a T-1 could be doubled to 3.152 Mbps or 48 DS0 channels, which is a bit rate between a DS1 and a DS2. They called this a T1C (or DS1C) line. To deploy this new type of digital link, AT&T needed a way to multiplex two DS1 lines into a 48 channel signal. A multiplexer could have been built into a D3 channel bank, but AT&T opted to design and build a new type of channel bank, the D4. The D4 uses the same superframe format and mu-255 law companding that first appeared in the D2 channel banks. Since the D2, D3, and D1D channel banks all had different sampling sequences, D4's sampling sequence was never specified. In other words, there really is no D4 framing *per se*. Instead, D4's circuitry can automatically "shift gears" and alter the sampling sequence, making the D4 compatible with all of the older equipment. By 1986 use of D4 framing was mandatory on AT&T T-1 lines.

The multiplexer that makes the D4 channel bank possible is the M1C asynchronous multiplexer. Since the input signals are asynchronous, the M1C needs some way of handling signal phase variations such as "jitter" and "wander." To solve this problem, AT&T chose to use a form of bit stuffing or pulse stuffing. The M1C takes the second T-1 signal and logically inverts it (1's become 0's and vice versa), and a framing bit is stuffed in two out of three code words, resulting in 26-bit blocks of data. The channels are interleaved and then scrambled by the addition of modulo 2 of the signal with the previous bit. Finally, the bit stream is combined with a control bit sequence that permits the receiver's demultiplexer to work by preceding every 52 bits with one DS1C framing bit. A sequence of 24 such 53-bit frames forms a 1,272-bit frame known as an "M-frame."

The M1C multiplexer allows you to connect a D4 channel bank to a T1C line, or you can connect a D4 directly to two 24 DS0 channel banks of any type with two T-1 lines. You can even link together two D4 channel banks to assemble a signal for use on a T-2 line.

In 1982 the D5 channel bank was deployed. The D5 is basically like a D4, but it gives you the option for multiplexing 72 and 96 channels, so a D5 channel bank can be connected to a T-2 line. The D5 also gives you the ability to use Extended Superframe Format (ESF), the successor to the superframe format, which had been formulated by AT&T in 1979.

The superframe format was far better than the original D1 framing, but it was still difficult to detect errors and line degradations. By the early 1980s Very Large Scale Integrated (VLSI) circuits had appeared with such processing power and timing accuracy that AT&T could build a channel bank with high quality oscillators that needed fewer framing bits than earlier models, allowing these bits to be used for error correction and other chores. This was the basis for the development of the Extended Superframe Format.

The ESF, or extended superframe, is indeed an extension of the D4 format. The ESF frame is larger than a superframe: Instead of 12 DS1 frames it consists of a group of 24 consecutive 193 bit long DS1 frames. Thus, the Extended Superframe can be conceived of as two standard superframes, but the frame bit is made to be more versatile. The frame bit goes through a 24-bit cycle (using 8 Kbps of bandwidth) that involves both fixed and variable framing patterns in three categories:

Of the 24 framing bits of ESF, only six bits are actually used to create the framing pattern used to achieve bit stream synchronization with the receiver. These are the Frame Pattern Sync (FPS) bits and they come from Frames 4, 8, 12, 16, 20, and 24. These six bits follow the sequence 001011. Only 2 Kbps of the total bandwidth is used by these FPS bits for synchronization purposes.

Of the remaining 18 framing bits, the six bits culled from Frames 2, 6, 10, 14, 18 and 22 are used to store a six-bit Cyclic Redundancy Check (CRC-6), which is the basis of a method of detecting the existence of errors in a transmission using polynomial division. Aside from being used for error checking purposes, the D5 channel bank can use the CRC to track the T-1 trunk's performance. The CRC is a six-bit coded "data signature" generated from data contained in the previous 24 frames (all 4,632 bits of the superframe). The framing bits are set to "1" prior to the CRC calculation to prohibit any framing bit errors from causing CRC errors. The equipment constructing the ESF performs a mathematical calculation on the signal to be transmitted across the link (excluding the control bits). The signal is transmitted across the link to the receiving equipment on the next ESF using the 193rd bit in frames 2, 6, 10, 14, 18, and 22. The receiving equipment performs the same mathematical calculation on the information received from the link and will compare the result to the six-bit word that arrives in the six CRC bit positions in the next ESF.

It's somewhat amusing to note that the same kind of CRC algorithm once used to ensure the data integrity of files being transferred over 1,200 baud analog modems could also be used by a high bandwidth receiving CSU to track a T-1's error rate. The CRC bits take 2 Kbps of the bandwidth.

The remaining 12 framing bits (in the odd numbered frames 1, 3, 5, 7, etc.) provide a 4 Kbps supervisory data channel called the Facility Data Link (FDL). The FDL is used to transmit the CRC performance information to the telco's network monitoring and testing stations that are on the lookout for line transmission degradations. It's said that this monitoring of performance results in 70% fewer outages than superframes and a 30% reduction in downtime per outage. Indeed, thanks to the CRC code, a fully outfitted T-1 system with ESF formatting can detect 98.4% of all possible bit errors. The FDL channel can also be used for functions such as non-intrusive remote configuration and monitoring of CSUs. The ESF datalink's operation depends upon whether the AT&T TR 54016 or the ANSI T1.403 version of ESF is used by the carrier. These two standards are very similar, with the only significant difference being in the implementation of the ESF Datalink Framing Bits.

So much for the advanced features of ESF. What about conventional network signaling? Robbed bit signaling can be used in conjunction with ESF, with bits being stolen from frames 6, 12, 18, and 24. Four signaling bits can be reserved in the Extended Superframe Format for each channel, which allows us to define up to 16 signaling states. The signaling bits in each frame are customarily assigned a letter according to the following rules:

- If 16 states are available, the bits are called A, B, C, and D-Bits
- If 4 states are available, the bits are called A, B, A, and B-Bits
- If 2 states are available, then the bits are called A, A, A, and A-Bits

There is also a "transparent" signaling mode where no bits are stolen. Instead, the whole 24th channel is used for signaling which provides a full 64 Kbps of Clear Channel Signaling. This unique channel is where we find our old friend, Common Channel Interoffice Signaling (CCIS).

Extended Superframes used with the D5 channel bank are the most common form of T-Carrier transmissions. Even the Europeans use a form of it on their E-1 circuits (but because they have 32 DS0 channels, they use a 256 bit frame). D5 is also closely associated with the immensely popular B8ZS line coding.

However, if you've created your own T-1 network on your own campus over your own wires in buildings less than 6,000 feet apart, then you can run them at full 1.544 Mbps with no framing bits necessary. However, most modern equipment will add the framing bits whether you want them or not.

Ordering a T-1

Before connecting your CSU/DSU to the telco T-1 service, the following telco information must be obtained from the T-1 provider:
- Line Code: AMI or B8ZS
- Framing Format: D4 or ESF
- T-1 Timing Source: It can be from the telco (T-1 provider) or it can be the customer's responsibility, in which case it could be one of the following – internal timing to one of the CSU/DSUs, timing recovered from Drop & Insert (PBX), or timing recovered from a bridge or router.

Available Bandwidth (DS0s): It could be a full T-1 (24 DS0s) or a Fractional T-1 (1 to 23 DS0s).

Once you've settled on what you want and what's available, tape this information to the T-1 jack. You are now armed with sufficient information to set up a T-1 line!

T-1 Error Conditions

T-1 has a number of defined alarm and control signals. The alarm signals have different color designations and are used to indicate serious problems on the link. These alarm signals are generally defined as:

- Red Alarm – A local equipment alarm that indicates the incoming signal has been corrupted for several seconds. The red alarm shows up visually on the equipment that detects the failure. This equipment will then begin sending a yellow alarm as its outbound signal.
- Yellow Alarm – A failure has been detected. The yellow alarm pattern has a number of different definitions. The most common D4 definition is to set 1 bit of every channel to a zero.
- Blue Alarm – The total absence of an incoming signal. This alarm also serves to keep the circuit in synchronization by sending continuous transitions (an all 1's pattern).

A T-1 is Many Things

So when someone tells you they're running a T-1, they may mean that they have a network that's moving data at 1.544 Mbps, or they may

have a network in accord with the T-1 electrical interface specification (DSX-1), or that they have a network that moves data in conformity to one of the several framing formats (D4, ESF, etc.). Most of the time we can presume that they mean all three.

Whither T-1?

The Ariel Corporation (Cranbury, NJ – 609-860-2900, www.ariel.com) masters of both PCI- and CompactPCI-based 56K / ISDN access products, has enhanced its PowerPOP architecture and 56K/ISDN access family so they can use SS7 signaling technology.

Most smaller ISPs provide dial-up access by leasing subscriber-side T-1 or T-1 / ISDN PRI lines from the local telco at a cost of $500 to $1,000 per month. ISPs can reduce this cost by replacing expensive T-1 and PRI lines with the same Inter Machine Trunk (IMT) lines used by ILECs and CLECs, which typically cost less than half as much.

To obtain these IMT lines, however, ISPs must use the SS7 signaling network, which, until now, meant you had to buy a multi-million dollar Class-5 telephone switch. Ariel's approach to incorporating SS7 technology makes it cost effective for ISPs to take advantage of the SS7 network, enabling them to employ IMT lines without the Class-5 switch.

Deploying ISP networks using Ariel's SS7-enabled access technology will also permit ISPs to reduce POP infrastructure and management costs. Today, ISPs must deploy multiple POPs within each LATA (Local Access and Transport Area, roughly equivalent to an area code) so their customers can log on without paying long distance charges. With Ariel's technology, ISPs/CLECs will be able to service an entire LATA with as little as a single POP.

Products based on the new Ariel SS7 gateway technology is sold together with Ariel's PCI and CompactPCI-based 56K/ISDN access products to ISPs, CLECs, ILECs, and co-location service providers.

"When we created the PowerPOP architecture," said Dennis Schneider, senior vice president of worldwide marketing at Ariel, "we did so with the idea of reducing remote access cost and providing an open, intelligent platform that could be used to reduce cost and boost performance throughout an ISP's network. Our RS4200 platform was an important step in that direction, cutting the capital cost of remote access by about $100 per port. Now, we are developing an SS7 technology that, together with the RS4200, will enable ISPs to reduce their dial-up access and POP costs in a way currently not possible with most remote access concentrators."

Also, the emerging high bit-rate digital subscriber line (HDSL2) standard is a sterling example of symmetric DSL technology which can deliver symmetric 1.544-Mbps data rates (the same as a T-1) to distances of up to 12,000 feet from the switch using the American National Standards Institute's (ANSI's) T1E1.4/99-006 draft standard for single-pair T-1 transport over twisted pair. With single-pair transmission, the cost of provisioning and maintaining a leased line or high-capacity service plummets. An interoperable standard such as HDSL2 should become widely used in the local loop.

This technology combines 2B1Q line code with echo cancellation techniques to transport T-1 service. HDSL is robust. It can be deployed without special conditioning of the line or partitioning of the cable bundle. Today, if you order a T-1 or E-1 service, there's an 80% chance that the actual physical transport you end up with will be HDSL.

T.120

T.120 is a family of open standards for multipoint data collaboration, established by the International Telecommunications Union (ITU). This standard has been widely adopted by most providers of software, hardware, and telecommunications around the world. T.120 can be found in systems that perform such collaborative applications as desktop data conferencing, multi-user applications, and multi-player gaming. The T.120 infrastructure can use both unicast and multicast simultaneously, providing a flexible solution for mixed unicast and multicast networks. T.120 can work alone or in conjunction with other ITU standards, such as the H.32x family of video conferencing standards. While H.320 (the ISDN videoconferencing standard) does provide a basic means of graphics transfer, T. 120 will support the kind of higher resolutions, pointing and annotation functions found in sophisticated data sharing applications.

The T.120 standard supports a broad range of transport options, including the Public Switched Telephone Networks (PSTN or POTS), Integrated Switched Digital Networks (ISDN), Packet Switched Digital Networks (PSDN), Circuit Switched Digital Networks (CSDN), and popular local area network protocols (such as TCP/IP and IPX via reference protocol). Furthermore, these different network transports, operating at different speeds, can easily co-exist in the same multipoint conference.

Over 100 international vendors, including Apple, AT&T, British Telecom, Cisco Systems, Intel, MCI, Microsoft, and PictureTel have designed, built and sold T.120-based products and services.

TCP/IP

See Transmission Control Protocol / Internet Protocol.

TDM

See Time Division Multiplexing.

Telcordia Technologies

Telcordia is the former Bellcore, which was in turn part of the former Bell Labs of AT&T. As part of the contractual agreement related to the company's purchase in 1997 by Science Applications International Corp (SAIC), Bellcore had to change its name.

Telecommunications Industry Association (TIA)

The TIA defines telecom standards for the PBX, PSTN interfaces to the PSTN and cabling / transmission media. In 1924, a small group of suppliers to the independent telephone industry organized to plan an industry trade show. Later, that group became a committee of the United States Independent Telephone Association. In 1979, the group split off as a separate affiliated association, the United States Telecommunications Suppliers Association (USTSA), and became a major organizer of telecom exhibitions and seminars. TIA was formed in April 1988 after a merger of USTSA and the Information and Telecommunications Technologies Group of EIA. EIA began as the Radio Manufacturers Association (RMA) in 1924.

TIA is now a full-service national trade organization with membership of 1,000 large and small companies that provide communications and information technology products, materials, systems, distribution services and professional services in the United States and around the world. The association's member companies manufacture or supply virtually all of the products used in global communication networks. TIA represents the telecommunications industry with its subsidiary, the MultiMedia Telecommunications Association (MMTA), in conjunction with the Electronic Industries Alliance (EIA).

Mission:

TIA seeks to provide its members a forum for the examination of industry issues and information. The association serves as the voice of manufacturers and suppliers of communications and information tech-

nology products on public policy and international issues affecting its membership.

Website: www.tiaonline.org

TAPI

See Telephony Applications Programming Interface.

TAPI versus TSAPI War

In the mid 1990s, during the "bus wars" between Dialogic's SCSA and Natural MicroSystem's MVIP, another huge controversy erupted over which telephony API would become dominant in the computer telephony industry. IBM's CallPath, and AT&T's PassageWay had been in existence for some years, but the two principal contenders appeared to be the Intel / Microsoft TAPI and the Novel / AT&T TSAPI.

At the time, comparing TAPI and TSAPI seemed like comparing apples and oranges. However, with 20/20 hindsight, we can now make some valid observations:

TSAPI is an API which is a partial implementation of the European Computer Supported Telephony Applications (CSTA) protocol standard. The CSTA standard defines switch-to-host link status and command protocol describing the message formats by which a host computer can send requests to the PBX and by which the PBX sends replies and status back to the host computer.

ECMA explicitly refused to define a standard API for CSTA, so TSAPI's claim to being standards-based is indirect at best. Both the CSTA and TSAPI focus is on traditional call centers, providing only call control to the desktops.

TSAPI was jointly defined and promoted by Novell and the part of AT&T that became Lucent, major players to be sure. Novell's strategy of the time, called "NetWare everywhere," was reflected in the TSAPI architecture in that, by definition, it required a LAN and client-server architecture, making NetWare a part of every computer telephony solution.

AT&T obviously wanted to protect its then-current business model of requiring a PBX in the closet and a proprietary digital phone on every desktop, since the phone sets have always been the revenue-generating part of the PBX business. This is also reflected in the TSAPI architecture in that TSAPI requires PBX-connected phones to sit on every desktop – you don't have the option of turning your PC into a phone, as you can with the Intel / Microsoft TAPI.

Lucent's implementation of TSAPI actually demanded that a proprietary phone be present on the desktop. So just when you got a mechanism to control your phone from the PC and trade the expensive phone for a cheap analog phone, you can't.

Also, the TSAPI SDK included a PBX simulator provided by Lucent to help testing without having to own a PBX. Needless to say, it emulated the AT&T G3. It's said that during the mid-1990s some Europeans jokingly referred to TSAPI as a CSTA translator for AT&T equipment.

Call Control. TAPI has a single call model that abstracts or "virtualizes" the different call models of different networks. Thus, TAPI can be made to support just about any call model on a wide range of phone networks and PBXs. This is similar to the way printer drivers work under Windows. There are lots of different printer models, but applications don't have to worry about the difference because they are "virtualized."

TSAPI on the other hand, is based on one call model – the CSTA design for only PBXs and key systems. Unfortunately, real-world differences between PBXs and key systems cannot always be modeled onto the confining call model provided by CSTA. This sometimes forces developers to define PBX-specific extensions for common features such as call conferencing. Even TAPI would ultimately have to deal with handling a plethora of PBX features too, especially when third party developers neglected to build comprehensive drivers or Telephony Service Providers (TSPs).

Third-Party Call Control. TSAPI originally prided itself on its claim that it could do third-party PBX / LAN switching and TAPI couldn't. In fact, TAPI soon offered four different connection models for integrating PCs and phones, including third-party call control on a LAN.

TSAPI provided more extensive monitoring of all endpoints of a third-party call than did TAPI. Still, after a third-party call has been established, the actual manipulation of such a call by an app is practically identical to the manipulation of a first-party call – namely, by manipulating the call at one of its endpoints.

Call Routing. This was supported by both APIs. In TAPI, call routing was provided using standard call control functions (call redirection). TSAPI did it via a collection of specialized call routing API functions.

DTMF Detection/Generation. TAPI can do this, making it easy to build answering machines or IVR systems. Also, an outbound call answered by a voice response system can be controlled by a TAPI-compliant app sending DTMF tones. The TSAPI didn't have this, so you had to use your phone keypad to generate those tones yourself. That means TSAPI apps couldn't incorporate DTMF input from callers either.

Call Center Agent Support. TSAPI had extensive support for call center agent state reporting – the computer system can be informed if the agent is logged off or on, ready to receive calls, etc. TAPI had none, its supporters claiming that these are not really call control functions but rather "database" functions. When the ACD becomes simply a software app, there is no reason to maintain specialized functions for accessing a database on proprietary hardware.

Multiple Independent Apps. TAPI has a set of well-defined mechanisms governing the sharing of phone calls across apps, similar to the way Windows governs the sharing of the display, keyboard, mouse and printer across apps. In contrast, the TSAPI spec assumed that you were running a single app and didn't define the behavior that occurs when multiple apps on the same client PC are listening for incoming calls.

Multimedia Integration. TAPI can treat calls as "media devices" (like audio device, fax device, data device, video device, etc.), and the corresponding multimedia API can then be used to access and manipulate the actual media info, thus preserving the WAN aspects of phone networks. This is essential when you consider both narrowband PSTN / ISDN and broadband B-ISDN / ATM / cable. TAPI lets the PC take advantage of data transport capability in these networks, which is the only reason you would use these networks in the first place.

TSAPI originally defined no mechanism for accessing a call's information content, treating the phone network and computer network still pretty much as separate and isolated worlds.

OS Platforms Supported. TAPI is strictly a denizen of the Windows family of operating systems. However, since information exchange (i.e., protocol) over the telephone network is not part of Telephony APIs, there is no true interoperability requirement regarding the PC endpoints involved in the call. So a TAPI app running on a PC can call a Apple Telephone Manager app on a Mac just fine.

Microsoft and Intel supposedly encouraged other platforms to adopt TAPI as their telephony API, since there is very little in TAPI that is truly Windows-specific, but nothing came of this.

Although there were press announcements about porting TSAPI to other platforms, it always relied on Telephony Services and NetWare networking.

Strange Bedfellows. Banyan Vines was a once-famous WAN that competed with both NetWare and Windows NT, having a relatively small

market share that nevertheless included the "cream of the crop" - the internationally-based members of the Fortune 1000. Interestingly, Banyan obviously wasn't going to support NetWare telephonic solutions, and it looked to TAPI for computer-telephone integrations.

Running TAPI apps on NetWare/TSAPI. Before Mark Lee joined the computer telephony development effort at Microsoft, he worked at Northern Telecom (now called Nortel) where he developed a free Windows module called Tmap. This interesting little piece of software could map TAPI commands to TSAPI, letting TAPI apps run on TSAPI-based integrations from any Windows client PC. TAPI apps could run unchanged on add-in cards, add-on modules, CTI phones, TAPI client-server integrations, and TSAPI NetWare/PBX integrations.

Tmap was both a regular Service Provider (driver) to TAPI as well as a regular client app to TSAPI. Tmap therefore required no upgrade or changes to the existing TAPI.DLL or TSAPI code. Tmap literally translates between the Intel / Microsoft TAPI Service Provider Interface and the TSAPI Windows client library interface. Tmap must be installed in every Windows PC, but can coexist with and be operated at the same time as other TAPI SPs in the PC such as a modem or telephony add-in card.

Tmap and TAPI together also provide the multi-application support that is missing in "raw" TSAPI.

Indeed, TSAPI-based switch integrations could now run all TAPI apps. However, it is important to note that Tmap could not allow TSAPI apps to run on TAPI integrations due to TSAPI's closed driver architecture at the client PC.

Tmap never became a superstar of the CT industry. It was apparently somewhat inefficient and was used in only a few special circumstances.

If a suitable middleware had appeared to support both TAPI and TSAPI, developers wouldn't have needed to create proprietary telephone interfaces, and the end user of an application wouldn't need to wait for the application supplier to add telephony capabilities to it so they could finally enjoy the benefits of CT. But such middleware would probably have been "bloatware," and would have taken a performance hit under real-life conditions.

Telemarketing

The advertising, promotion and sale of products and services by telephone. Telemarketing is usually (but not always) handled by call center agents working either directly as employees of the firms which are providing or reselling the products/services or indirectly for outsourcers (see Call Center Outsourcing) which have been contracted to carry out these functions.

There are two forms of telemarketing: outbound and inbound. Outbound telemarketing is when an agent calls to promote or sell to a business, consumer or institution, with names and telephone numbers from an existing customer or cold-call prospect lists. The numbers from these lists are now almost always loaded into predictive dialers (see Predictive Dialers) which selects and dials them, connecting the agent only when someone answers the phone.

Outbound business to consumer telemarketing is the activity that most people love to hate. When you hear the word 'telemarketing' the speaker or author is almost always referring to outbound. This is the fast-talker who rings you up and tries to sell you day-glo plastic garden gnomes when you're munching your dinnertime garden salad.

Yet the reason why "they" call your family up and give them indigestion is because there are many people that will buy such tacky lawn ornaments over the phone. Equally if not more importantly the accrued cost per sale with outbound calling is much less than brick-and-mortar retail, such as flogging the little people in a garden care center.

The seller doesn't have to lay these cheery miniature humanoids out nice amongst the vibrating-seat lawn chairs and the pink flamingoes, and erect fences and wire in alarms to keep out the college kids and other crazies that want them as mascots or love objects. Or provide friendly, knowledge clerks with associate degrees in outdoor design and majors in lawn ornaments who can help you decide whether the little one with the nose ring matches your plaster Madonna birdbath. In contrast outbound agents have a short period of time to make the sale. Chances are, (if you're lucky) you'll never hear from the same person again.

Neither does the marketer have to spend extra scarce and valuable retail floor space for these beings. All they need is a big, strong building on a cheap lot in some Siberia-like office/industrial park, a stable of tattoo-bearing scanner-wielding forklift jockeys whipped by a shipping department receiving orders from the call centers, either in-house or outsourced.

Outbound telemarketing can occur as part of an outbound customer care call. This is where a firm with whom you've already talked to or have done business with makes a courtesy or followup call to keep you as a customer. For example, if you had problems with your cable company that caused you to miss "Dwarf Toss-A-Mania" on pay-per-view, if they're good they'll call you, see if everything is all right. Depending on how agitated you were the agent might offer three nights free, 10% off your next bill or send you coupons knocking 50% off a new HDTV.

Outbound business to business telemarketing is conducted somewhat different from business to consumer telemarketing. Outbound b-to-b calling is often done to qualify leads for followup by senior inside or outside salespeople in addition to direct sales.

For example, say your firm makes molds and molding equipment and you've identified the lawn ornament industry as a target market. You get an e-mail from "Nevermore Novelties" saying that they're interested in your latest line of castings. You direct your e-mail to the inside salesperson that has that territory or customer classification and they ring them up. If Nevermore wants to buy a few molds, the inside telemarketing salesperson makes the sale over the phone. If the firm qualifies as a big customer, i.e., cleans out your stock, has been a Top 10 buyer and plans a big expansion (thanks to people like your family buying their gnomes from outbound telemarketers) then your inside salesperson forwards the lead to the field sales force to pay Nevermore a visit.

Most outbound telemarketing calls are made by live agents. However, not all outbound marketing calls are made by people. There are automated dialers that deliver pre-recorded artificial or real voice messages only. These are often used by small businesses that either can't afford or don't want to hire people to make such calls to drive business. There are application hosting outsourcers that can bulk-deliver massive numbers of such messages, at very little cost. These dialers' biggest liability, besides not receiving valuable information while interacting with the called parties, is that many called parties think they should go to the devil. You can't tell them to relocate to a very warm place (we're not talking about Florida in July but close enough). They're also illegal to use them on cold-calls to consumers. The same goes for commercial faxes.

Only if you're making non-commercial calls can you cold-call with autodialers. These include soliciting for a charity, for a government agency (notifying you they're about to rip up the street outside your house because the under-indictment contractor used wooden pipes on the gas main) or for polling, such as for your local council member – who had received campaign contributions from that gassy contractor).

Also, not all outbound marketing telecommunications is made by phone. Commercial outbound e-mail, sent in-house or through remailers,

has become extremely popular with marketers because they cost less than automated dialing mostly because they use the practically free Internet. Such communications can be done from anywhere to anywhere. Unfortunately unlike circuit-switched live or automated message phone calls they are more difficult to trace. Consequently it has become the favorite medium of porno pushers and scamsters as well as legitimate marketers. Such unwanted junk e-mails are known as spam after a famous Monty Python sketch where a hapless diner said he didn't want Spam whereupon the restaurant staff berated him, backed up by a chorus of Vikings singing "Spam Spam Spam Spam Spam Spam Spam Spam...wonderful Spam...!"

Inbound telemarketing is where an individual, such as yourself, decides to call in to find out more and buy a product/service. This can be driven by many sources: print, radio, broadcast, website advertising, catalog direct mail piece, e-mail, a billboard or bus/subway ads that you're staring at to keep your nose out of your fellow human being's armpit. Or by urgent needs such as mattress springs hooking your backside, spouse whining about how you never send him flowers, and your kids screaming they want the latest exploding dolls for Christmas.

Inbound telemarketing also frequently occurs when someone calls in to find out some more information, or has a concern, complaint or problem, or is performing a service such as credit card activation, and makes a purchase. This is known as cross-selling/upselling. This is why you get a sales pitch when you activate your card: the credit card company has your undivided attention.

Inbound telemarketing differs somewhat between business to consumer and business-to-business applications. Using sophisticated customer relationship management (CRM) packages and IVR/computer telephony links, you can direct high-value callers to specialized agents for more exclusive and attentive service (such as queue-jumping) before the phone is picked up. The key difference is, depending on the product/service being marketed, and the customer contacting you, the b-to-b inbound telemarketers forward high-value leads to outside field service reps.

As with outbound telemarketing, inbound telemarketing does not necessarily require live agents or circuit-switched voice. Your customers can respond to offers, inquire availability of, purchase, and take rain or tornado checks through an IVR system, by e-mail, or through a Web site, either self-service or with an Internet-enabled agent that they can converse with via text chat, voice over IP and Web collaboration.

This "call'em and sell'em/fire-and-forget" approach in traditional outbound business to consumer telemarketing is in considerable contrast to inbound telemarketing. In inbound, customer service is key in making the sale, and capturing and retaining the customer. This is known as 'one-to-one marketing' where the object is to attract and retain that customer, finding out their likes/dislikes/buying patterns so that they can harvest their dollars throughout that buyer's lifetime. If the firm doesn't have that product/service available a good operation will try and find an acceptable substitute, hail check, snow check, etc. just to maintain that good will. The agents should be able to tell what gnomes will go well with that Madonna because that agent and that company wants you back, so that when they come out with that singing fiberglass lawn Elvis that'll be ideal for your back yard you'll buy that too.

If you're wondering how come that outbound telemarketer picked on you instead of your next door neighbor the answer is that your name (and not theirs) was on a list the marketer had. Lists are the Holy Grail of direct marketing. They make outbound and inbound telemarketing, plus broadcast fax, e-mail and plain old junk mail marketing happen. A firm calls you from a list, and uses a list to direct mail, e-mail/fax to customers and prospects to drive inbound telemarketing and e-mail marketing responses. In turn the information gathered, which goes into databases, is used, to generate and populate more lists. Companies pull names from these lists for their marketing campaigns or they rent/sell them to other firms, in non-competitive fields.

CMP Media is in the list business. If you subscribe to our magazines you are on our list (and, contrary to the Lord High Executioner's ditty in Gilbert & Sullivan's operetta The Mikado) you would most certainly be missed because we make money from you. Rubin Response markets our list to companies that want to sell you products and services. Rubin Response likes CMP Media, especially *Call Center, Computer Telephony* and *Teleconnect* magazines. They kindly displayed our publications at their booth at a recent Direct Marketing Association (DMA) trade show. We make money for them. And that's because you are a very valuable person. You buy things other people want to sell you. That's why, in turn we're picky in who reads our publications. We only want important people as readers who make buying/specifying decisions.

The same goes at home. If you were on a customer list that is because you had probably purchased something (probably that Hello Kitty hood ornament) from the same company, and they figured you had the same kind of taste in lawn furnishings. Or the marketer had purchased or rented your name directly from another company or list broker, or obtained it from in-house sources such as a direct mailing response card, a contest that you entered for that machine-gun-fitted "NRA Hunting Edition Ultimate Humvee SUV" or from their Web site.

There are two types of lists, compiled and response. The American List Council, a big Princeton, NJ-based list marketer explains the differences and their pros and cons. Compiled lists are general lists pulled from sources such as directories, phone books, public records, retail sales slips and trade show registrations, such as CT Expo and Call Center Demo. Firms can use them to reach entire markets. That's why if you came to CT Expo Fall chances are you'll get a mailing to come to CT Expo Spring.

With compiled lists marketers usually know your name (generation), family, lifestyle and neighborhood characteristics and they match them against demographic and geographic profiles of lawn ornament buyers. That's one source of the name and telephone number loaded by the gnome-pushing marketer who called you up.

Another source may be response lists. They containing consumers who buy regularly through a certain channel, such as by phone. That's the other source of your name: the telemarketer rented it from "Trailer Trash Monthly". Or from that garden center where you had bought the flock of pink flamingoes from, enticed by their free offer of the infamous John Waters flick those colorful birds are most associated with.

Compiled lists give comprehensive specific market coverage, such as if a firm is opening a new store and wants to drive new business in its catchment area. Response lists generate higher response per name. Compiled lists' selections are more demographic while response lists are more psychographic (attitudes, buying patterns). Compiled lists are also less expensive than response lists.

Marketers append information to their lists, such as phone numbers if they only have your address, and vice versa. They also append buying data, demographic and geographic changes to your file. If your spouse had a baby that's why you'll get showered with goodies and offers (such as for infant-safe soft chewy crib gnomes and pink flamingoes): such information is public knowledge. They know when you move; they obtain US Postal Service's National Change of Address lists.

The marketers' object by gaining such information is to find out what you want to buy, when and how much because you will buy something,

sometime. They want to target only those offers that you are mostly likely to accept. The smart ones don't want to waste their time and resources buying lists and paying agents (and their ancillary facilities and technology costs) trying to sell you something you don't want or need, like baby clothes if you're childless. Unless of course wearing them is your thing and your name is on a targeted list that we don't want to know about.

Telemarketing, like selling used cars and contracting for cellphone services ultimately lead to ripoffs and hassles. To counteract this there is a growing maze of federal and state laws to regulate principally inbound calling mainly to prevent frauds but increasingly to stop marketers from pestering consumers. The federal Telephone Consumer Protection Act (TCPA) and the Telemarketing Sales Rule, adopted with the Telephone and Consumer Fraud and Abuse Prevention Act, permits called consumer parties to say NO to telemarketers, bans telemarketers from calling if so told, requires them to remove names from lists and to maintain do not call (DNC) lists. The laws also limit the types of numbers that can be called and restricts outbound teleselling to between 8am and 9pm. They require marketers to give their name, on whose behalf they're calling and telephone number/address where they may be contacted and bans misrepresentation.

If you're into telemarketing and you're caught telefrauding Granny on her 65th birthday a fairly new federal law, the Telephone Fraud Prevention Act gives you three years' jail time on top of what the courts had imposed. The law also orders you to forfeit property such as your office Jacuzzi and your hot-sheet roach hotel that you had provided at $100 a night for your shift-sleeping agents.

In the same spirit as the TCPA, the Direct Marketing Association (DMA), which represents telemarketers and other direct marketers provides a set of three list removal services, the Mail Preference Service, the Telephone Preference Service and the e-Mail Preference service. You can write in to the DMA and have your name removed from marketers' lists. DMA members scrub their lists against the preference services' names.

Many states require telemarketers to register with them, usually with a laundry list of exemptions such as charities, publications and already-regulated businesses such as securities dealers. A growing number of them have created DNC lists where consumers pay to have their names removed from all telemarketing databases and which telemarketers must purchase. Many states further narrow calling hours beyond the federal 8am to 9pm limit. For example Alabama and Mississippi ban outbound calling on Sundays, leaving you to eat your Sunday dinner in peace without being asked to buy garden gnomes by phone, though your son might think they're cool and insists that you boogie down to the garden center now and buy one. Several states, such as Kansas, stop marketers from making a rebuttal if you say no, you don't want the stupid little visages cluttering up the crabgrass.

Getting away with moving operations in other countries such as Canada is becoming more difficult as these nations pass and enforce laws to counteract fraud, working with US federal and state agencies. There is even an anti-fraud call center, PhoneBusters, which is staffed mostly by volunteers under the auspices of the Ontario Provincial Police in North Bay, ON, Canada. PhoneBusters not only deals with Canadian calls with many US fraud complaints. If your firm or partnering outsourcer does telebusiness in Europe your operation falls under European Union as well as national laws and regulations. Germany for example bans outbound cold calling while the EU's Data Protection Directive controls and restricts the data you're allowed to collect. There are also US privacy laws to obey as well.

Using existing customer lists and obtaining highly targeted lists for outbound and inbound telemarketing is replacing untargeted cold-call outbound and one-to-one marketing from your existing customer base, according to consultants, because they are more effective and they don't annoy customers or prospects. The days of shotgun cold-calling and junk mailing are over. Many consumers also don't want to hear from companies they already do business with.

Consumers are flocking to state DNC lists and buying call-screening answering machines and Caller-ID systems like hotcakes; some states have passed laws banning deliberate Caller ID blocking by telemarketers. One popular lovely little device (also sold by phone) is Ameritech's/SBC's Privacy Manager. It and similar devices from other vendors identifies calls before they hit consumers' phones, asks the intended called parties if they want to accept, decline or refuse sales calls and tell the telemarketers to put them on their federally-mandated DNC list.

Experts say that consumers are becoming much more harried, with much less time on their hands and don't want the hassle of dealing with answering telemarketers' calls. One famous episode in the hit series AllyMcBeal showed starring actress Calista Flockhart getting a telemarketing call at an inappropriate time and then lifting the agent's head out of the receiver, telling off the hapless telemarketer, and slamming the phone down. Many print and broadcast media are doing more stories about telemarketer harassment.

Time-stressed New Yorkers have a particularly notorious and effective way of handling telemarketers. According to a November 7, 1999 New York Times article they use their famed acid-dripping tongues to burn strips off the hapless outbound agents, in some cases giving suggestions to these poor people where they could place their phones and offering assistance with installation. Some telemarketers and survey firms award their agents "danger pay" for calling to New York City.

"What's driving this consumer concern is the high volume of calls," says NBC investigative reporter Liz Crenshaw. "People are upset about telemarketers interfering with their precious time."

What should you do if you're telemarketing? Some consultants think improved answering device detection, to prevent that annoying ring could help. However others point out that this doesn't increase the number of consumers you're trying to reach. Also, those consumers that have such answering/Caller ID units are those who have the desired buying power; those that can't afford them are less likely to afford your products or services or become good lifetime buyers.

Other consultants recommend a multimedia approach: e.g. direct mail, e-mail, broadcast response, and letting the customer/prospect decide the response mode, at their leisure. People want to buy products like they've always done, but this is the hyperinformation era where they have the knowledge, and quick access to it and they act on it when and how they want. Marketers are no longer the repositories of product/service wisdom, unlike in less savvy, less connected and empowered times.

Says Geri Gantman, senior partner with Oetting and Company, a New York City-based marketing consultancy, and who once ran a service bureau: "Consumers have been erecting moats to keep unwanted people out. The days when you can load numbers into a dialer and call away are over. You must get consumers interested, on their own time, in your product or service with means such as direct mail and the web that lets them communicate the way they want. Once that relationship is established, then as long as it is customer-sensitive, outbound begins to open again."

Telephone

It's smaller than a breadbox, and it sends and receives voice messages and data over great distances. The telephony relies on a vast, complex network of switching systems called Central Offices

(COs) or exchanges, though telephones can be wireless as well as wireline in nature.

Telephony

Pronounced "te-lef-e-ne." The transmission of sound point-to-point (rather than broadcast) over great distances, especially by telephone. It can also mean the technology and manufacture of telephone equipment. Internet telephony uses the packet network-based Internet rather than the conventional circuit-switched telephone company infrastructure to exchange spoken or other information.

Telephony Applications Programming Interface (TAPI)

TAPI is now an evolving API engendering the convergence of both traditional PSTN telephony and IP telephony in a PC development environment. TAPI started out as an API for connecting a Windows PC to traditional circuit switched phone services and giving it first party call control from the desktop (See First Party Call Control).

Although TAPI was certainly not the first telephony API in the industry, TAPI was defined with the goal of becoming the universal telephony API.

Before TAPI, everybody and their brother had what was purported to be a "standard" Open Application Interface (OAI) which was in fact both "standard" and "open" only to whatever proprietary equipment the API was written for.

There have been a number of contenders for a truly universal CT-standard in recent years, such as the European Computer Supported Telephony Application (CSTA), IBM's CallPath Services API (CSA), AT&T's PassageWay, DEC's (now Dialogic's) telephony platform and the Novell / AT&T Telephony Services API (TSAPI). TAPI's advantage was that it was included free in every copy of the Windows operating system.

TAPI was actually born at Intel – a group led by Herman D'Hooge was the "father" and TAPI lore alleges that at that time they were working on a "PBX-on-a-card" with the secret codename of a famous steam-powered South African locomotive called the Mikado!

TAPI may have begun at Intel, but later development was done jointly by Microsoft and Intel, and the first publicly available version, TAPI 1.3 (an add-on for Windows 3.1), was released in 1994. Windows programmers found they could use TAPI to enable Windows applications to share telephony devices with each other and provide a common means of handling different media (voice, data, fax, video, etc.) on a wide range of hardware platforms.

It seems only like yesterday (actually 1995) that TAPI was engaged in a market share battle with a rival, Novell's Telephony Server Application Programming Interface (TSAPI) which had appeared shortly before TAPI and was already providing third-party call control (See Third Party Call Control), a key feature that was for years sorely lacking in TAPI.

Computer Telephony magazine first entered the fray in May of 1995, when it published Yours Truly's 40 page in-depth article about TAPI, which also consisted of a series of interviews with TAPI luminaries of that era (Toby Nixon, Guy Blair, Charles Fitzgerald) and a complete roundup of existing TAPI-compliant products. Some people think that the article gave TAPI a big boost during a critical moment in its history.

Ironically, *Computer Telephony* magazine had originally come out in favor of TAPI's rival TSAPI (see Telephony Services Application Programming Interface) which had appeared before TAPI and had some features that TAPI lacked. But TSAPI's developers, Novell and AT&T / Lucent, then made the grave mistake of not returning publisher Harry Newton's phone calls. The rest is history.

TAPI's abilities have evolved over the years from simple desktop first party call control to third-party call control supporting client/server systems on a LAN.

TAPI is actually two interfaces. The application programming interface (API) is what you use to create a computer program that uses telephone functions. The service provider interface (SPI) is used to connect to a specific telephone network, much as a printer driver lets a program interact with the printer.

A "service provider" is really nothing more than a fancy name for a driver. Thus, a TAPI service provider (TSP) is a driver that allows TAPI applications to communicate with different types of TAPI hardware. The TSP translates TAPI functions into commands that the hardware can understand, and translates events from the hardware into data that the TAPI application can understand. For example, Windows 95 and NT were the first OSes to come with a built-in TSP called Unimodem. Unimodem is a "universal" modem service provider that supports a wide range of commonly used modems.

Of course, a TSP can't work miracles. If you're using a TSP that supports CallerID, but the modem you're using doesn't support CallerID, then the application will turn up empty-handed when it does a function call to TAPI to get CallerID data from the modem.

Since different telephony hardware supports different types of functions, different TSP's support different TAPI functions. To continue our modem example, if you suddenly decide to use telephony hardware other than modems, such as PBX's, voice processing cards, etc. you will need to get the correct TSP provided by the hardware vendor.

Windows 95 integrated TAPI 1.4 directly into the operating system. TAPI 1.4 supported the development of 32-bit TAPI applications and TSP developers could develop 32-bit companion applications (using the Win32 API) to supplement a TSP in Windows 95 and Windows 3.1. TSP's in Windows 95 and Windows 3.1 were slow 16 bit DLLs and TSP programmers couldn't control exactly when the TSP was erased from memory after use. If the TSP was loaded by a 16 bit application, the TAPI.DLL erased it after the application executed a "providerShutdown" function, but could be affected by a number of factors. 32 bit code is much faster than 16 bit code, and programmers now had direct access to the Win32 API, and the application could run independently in its own context. A telephony programmer could even use 32 bit companion applications in Windows 3.1 by writing them using the Win32 subset API and then installing the Win32 environment on top of Windows 3.1

TAPI's initial impact on the computer telephony industry was to alert developers and CT resource board makers to adjust their API's and large-scale platform software to accommodate Microsoft's creation.

It was glaringly evident that TAPI 1.3 (for Windows 3.x) and TAPI 1.4 (for Windows 95) were entirely local implementations, meaning that the application requesting telephone services and the service provider ("driver") delivering those services had to be on the same machine. Note that Microsoft has never been in the position of supplying either the application or the service provider modules ("drivers"); these must come from 3rd parties. Microsoft only provides the specification of how they work together, and some software within Windows that enforces that specification and helps applications and service provider modules talk to each other.

At Dialogic, their CT-Connect software was reworked to support TAPI 1.3 and 1.4 by supplying a local service provider that ran on each desktop, received service requests from that desktop, and then forwarded the requests through Dialogic's own client/server mechanism that was Remote Procedure Call (RPC)-based to the central CT-Connect server. The central server did things like replicate events (because more than one desktop can ask about a single common phone) and arbitrate incompatible actions (like a second desktop asking to hang up a phone that is already on-hook). These kinds of things make a server-based CTI environment much more complex than a desktop-based environment. CT-Connect added much more than just client/server transport, and much more than just CTI link protocols.

TAPI 2.0 appeared with the release of Windows NT 4.0. This was the first version of TAPI to work with Windows NT.

Microsoft had actually advertised two significant advances in TAPI 2.0. First, they were going to add some API features to support more sophisticated applications. For example, TAPI 1.x didn't report a call reference number, so it was impossible to track a call which goes into an ACD queue and then pops out later somewhere else. As it turned out, TAPI 2.0 did have ACD support and other PBX specific functions.

Second, they were going to implement a Microsoft-proprietary client/server mechanism so that an application on one machine could talk to a service provider on a different machine – just like a print server mechanism, which lets an application on one machine send output to a printer on another machine transparently.

It turned out that Microsoft did the API extensions but kept TAPI following a local-only model, where the NT-based application and the NT-based service provider module must reside on the same physical Windows NT machine. This doesn't matter much for PC-board type telephony service providers, where the card is inherently desktop-based anyhow, but it makes no sense in a shared-link third-party CTI environment – you would have to run all applications on the same PC that the switch is plugged into via the CTI link. So, the initial release of TAPI 2.0 did not have much to excite developers and VARs wanting to work with shared-CTI links.

Switch vendors went ahead and wrote non-client/server TAPI drivers that could connect desktop applications to their special desktop phones. But they refrained from writing the TAPI drivers for CTI links, because doing this would only make sense if TAPI officially included support for client/server stuff. Without that, you would have a CTI link going into one system and have no ability for applications on other machines to share computer telephony events. Some vendors, such as Northern Telecom, made a big deal about how a client/server enabled Microsoft TAPI would allow them to build a single TAPI driver for their Meridian Link PBX and avoid having to use packages like Dialogic's CT-Connect. But this strategy fell apart for two reasons: First, following the Microsoft strategy whole-hog meant there was no support for non-Microsoft APIs or operating environments (TAPI's rival TSAPI couldn't be used, for example). Second, Microsoft took its time to implement the client/server feature, stranding their product strategies. This actually helped the sales of products such as CT-Connect, which already had a robust, open client/server mechanism.

Probably the most irksome quality about the TAPI 2.0 architecture for high-end telecom developers was the fact that you were restricted to developing applications only for Windows 95 and/or Windows NT systems, with the concomitant Microsoft networking. There was no support for UNIX, Win 3.x, OS/2, etc. (Not that the competition was any better: TAPI's rival TSAPI had both a required client and server platform, with a proprietary LAN protocol in between).

A leading computer telephony company such as Dialogic was thus faced with the prospect of taking their CT-Connect product and building a "super-driver" to create a client/server TAPI 2.0 at the central server that could accept requests via the TAPI mechanism but also simultaneously accept requests via non-Microsoft RPC connections – thereby supporting UNIX, OS/2, Win 3.x, etc., or they could just wait for an improved version of TAPI.

CT application programmers working with the TAPI TSPs available at the time for various CT hardware (particularly multi-line voice cards

T

and some single-line DSP products) found that TAPI didn't specify enough of the proprietary PBX integration function parameters. For example, you couldn't specify in your application that a NEC hang-up tone behaves one way and a similar tone generated by a Panasonic key system sounds differently. TAPI's syntax was tied into a single "semantic" model. TAPI's creators had to make some hard and fast architectural assumptions about the consistency of the telephone environment, and of course, there's very little that's consistent about telephony, at least when you compare it to computer technology. Dialogic engineers, for example, had a tough time mapping CTI-link events into TAPI events, because TAPI wanted all phone environments to work the same way (Hint: They don't). And TAPI wasn't flexible enough to adapt; it instead expected the proprietary phone environments to adapt to itself!

The mapping of various real-world PBXs to the TAPI model had to be done by the TSP drivers provided by third parties, such as the PBX makers themselves. PBX manufacturers, however, were suspicious of Microsoft and didn't really want to cooperate by completely divulging all the details of their hardware or revealing all of their proprietary in-band tones and coding schemes. Microsoft's programmers would have taken years longer to create an all-inclusive, fully comprehensive version of TAPI anyway. TAPI thus became a sort of "vanilla" or "generic" call processing model that could never take advantage of all of the unique and arcane features offered by the world's phone systems. This lack of cooperation on the part of the PBX makers would continue and would lead years later to the collapse of Microsoft's Valhalla project. Also, the lack of a semantic rigor beneath TAPI's elegant syntax would later spur Intel to start developing a universal API-like "object infrastructure" wherein all of the world's application generators and toolkits could be certain of the behavior of the underlying computer telephony components they were controlling.

In the middle of 1997, Microsoft released TAPI 2.1 as an update supporting both Windows 95 and NT 4.0. TAPI 2.0 service providers could be used on Windows 95 if TAPI 2.1 was installed and the service provider did not use any features specific to Windows NT.

The primary new thing that TAPI 2.1 brought to the table was client / server telephony applications support. No longer was it necessary to have hardware at the desktop, for example, to connect the PC and a phone set to enable client / server telephony applications. You could have a logical connection between, say, a PBX and a server to allow screen pops and the kind of things we associate with CT connectivity to happen in your desktop environment.

TAPI 2.1 is supported in Windows 98 and Windows NT Workstation 4.0, Windows NT Server 4.0 and Windows 95. In the case of Windows 98, TAPI 2.1 support is included in the operating system. For the other two products, Windows NT Workstation and Server 4.0, their latest service packs include the most recently refreshed versions of TAPI 2.1. For Windows 95 there's a web download available, and an SDK that's been available since June of 1997. Before moving on the TAPI 3.0, Microsoft "refreshed the bits" of TAPI 2.1 somewhat to address some bug fixes and some other things.

At the dawn of the millennium, TAPI 3.0 became the new (though somewhat delayed) Telephony API of Windows 2000. We now see a remarkable new operating system on the client side and on the server side there are a number of new built-in communications goodies.

Until the rise of the Internet and high bandwidth networking, TAPI was best characterized as essentially a call control API for traditional circuit-switched telephony.

However, the evolution of TAPI from version 2.1 to 3.0 had to take into account the "hit parade" of new technologies. The "convergence" of both traditional telephony and IP Telephony is a hot item right now, and TAPI 3.0 indeed embraces emerging technologies that enable voice, data and video collaboration over existing LANs, WANs and the Internet.

Indeed, the key goals of TAPI 3.0 are to bridge the gap between the PSTN and IP transports (extending existing PBXs while building and migrating to IP telephony services), connect call control and media control so that both the call and the media stream could be managed using Microsoft DirectShow with rendering and capture devices, and to simplify development of communication applications via COM based APIs, Directory Services integration and new TSPs are included with the package.

TAPI 3.0 offers both call control as well as media streaming support for traditional telephony and IP telephony or other types of telephony. What that means is that a developer can write an application and have it work equally well regardless of whether the media is directed across a traditional circuit-switched telephony link, or across an IP telephony path or ATM telephony path. Call control functions are provided by a TSP which is responsible for translating the protocol-independent call model of TAPI into protocol-specific call setup and teardown, on a service-by-service basis.

There are four major components to TAPI 3.0: the TAPI 3.0 COM API, the TAPI Server, Telephony Service Providers, and Media Stream Providers.

Unlike TAPI 2.1, TAPI 3.0 is expressed as a suite of Component Object Model (COM) objects. COM is a language independent component architecture, not a programming language. COM is meant to be a general purpose, object-oriented means to encapsulate commonly used functions and services. The COM architecture provides a platform independent and distributed platform for multi-threaded applications.

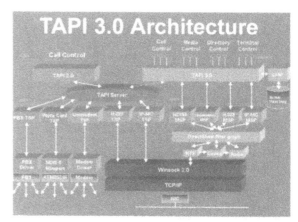

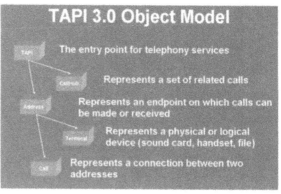

COM encompasses everything previously known as OLE Automation (Object Linking and Embedding), which was meant for letting higher level programming languages access COM "objects." An object is a set of functions collected into interfaces. Each object has data associated with it. The source of the data itself is called the data object. With COM, the transfer of the data itself is separated from the transfer protocol.

COM's object-oriented, language-neutral API and simplified interfaces allows developers to create products using not just Visual C++, but Visual Basic, Java, scripting languages, and what-not. It thus affords developers much flexibility and it opens up TAPI to a whole broader population of developers out there who may want to telephony-enable their applications or build telephony apps.

As we venture forth into the 21st century, object-oriented programming continues to be the dominant software development paradigm. TAPI 3.0's API is based upon five objects: TAPI, Address, Terminal, Call, and CallHub

The TAPI object is the application's entry point to TAPI 3.0. This object represents all telephony resources to which the local computer has access allowing an application to enumerate all local and remote addresses.

The Address object represents the origination or destination point for a call. Address capabilities can be retrieved from this object, such as media and terminal support. An application can wait for a call on an address object, or can create an outgoing call object from an address object.

A Terminal object represents the "sink" or renderer and the termination or original point of a connection. The terminal object can map to hardware used for human interaction, such as a telephone or microphone, but can also be a file or any other device capable of receiving input or creating output.

As for the Call object, think of it as a first party view of the phone call. All call control is done through the Call object, which represents an address's connection between the local address and one or more other addresses. This connection can be made directly or through a CallHub. There is a call object for each member of a CallHub.

The CallHub object represents a set of related calls. A CallHub object cannot be created directly by an application – they are created indirectly when incoming calls are received through TAPI 3.0. Using a CallHub object, a user can enumerate the other participants in a call or conference, and perhaps (because of the location-independent nature of COM) perform call control on the Call objects associated with those users, subject to sufficient permissions.

But the TAPI 3.0 COM API is only one of TAPI 3.0's four major components. The other three are: The TAPI Server, Telephony Service Providers, and Media Stream Providers.

The TAPI Server process (TAPISRV.EXE) abstracts the TAPI Service Provider Interface (TSPI) from TAPI 3.0 and TAPI 2.1, allowing TAPI 2.1 Telephony Service Providers to be used with TAPI 3.0.

The Telephony Service Providers (TSPs) translate the protocol-independent call model of TAPI into protocol specific call setup and teardown, on a service-by-service basis. TAPI 3.0 is backwardly compatible with TAPI 2.1 TSPs. Developers will no doubt be glad to hear that TAPI 3.0 drivers and applications are backwardly compatible. A TAPI 2.1 Service Provider Interface (TSPI) will be able to be used by TAPI 3.0 apps, but since none of the new media features are available in the earlier revision, obviously it would still be restricted to what's available with the TAPI 2.1 TSPI. TAPI 2.1 apps will continue to be supported as well under the 3.0 environment.

Two IP Telephony service providers (and their associated Media Stream Providers) ship by default with TAPI 3.0: the H.323 TSP and the IP Multicast Conferencing TSP, which we'll mention again later.

Finally, as for the Media Stream Providers (MSPs), these are a unified way to provide media control functions that allow an application to access the media streams and data of a call. TAPI 3.0 supports DirectShow as the primary media handler (DirectShow is Windows 2000's extensible framework for efficient unified control of streaming media via its exposed COM interfaces). TAPI MSPs implement DirectShow interfaces for particular TSPs and are needed for any telephony service that makes use of DirectShow media streaming. An MSP is required for a telephony service that uses DirectShow media streaming, with a one-to-one mapping between the TSP and the MSP. The communication channel is transparent to TAPI 3.0.

DirectShow is the mechanism of choice for use with the components Microsoft provides in the box to provide access to the media stream, but if a developer for some reason doesn't like DirectShow, he or she can alternatively use some other software mechanism.

Also under the TAPI 3.0 cover is a mechanism by which the conventional, telephony service side of TAPI 2.1 and the new component that provides access to the streams, the MSP (which is like an adjunct to the TAPI service provider) can communicate with each other via the TAPI DLL in order to provide a unified view of the stream and the call.

Microsoft defines a bunch of objects by which you manipulate the media on the stream. Besides the TAPI Call Object, they also have an associated stream object onto which you can attach terminals which are the sources of data on the stream. You can have "hard terminals" such as bound I/O devices as well as "soft terminals" such as tone detectors, speech recognizers, or whatever.

As "endpoint terminals" are projected down into the operating system through DirectShow, they could exist in User Mode, they could be in drivers running in the kernel, or they could actually be proxied onto underlying hardware. So the whole model is quite flexible.

As a result of the feedback from Windows 2000 / TAPI 3.0 Beta 2.0, over the summer of 1998 Microsoft redesigned the media service interface. The way terminals are handled, the way terminals are built and some of the interfaces with TAPI were redesigned. That design is now in beta 3.0.

TAPI 3.0 can leverage other capabilities in the operating system, mainly the Active Directory services, which is perhaps the most strategic technology to be included in Windows 2000.

Network administrators and developers should be glad that Windows NT's flat, awkward domain structure is gone, replaced in Windows 2000 with a genuine, hierarchical, distributed directory service for managing resources across an enterprise or extranet.

The Active Directory is a set of open interfaces that interoperates, abstracts and unifies the functions of various network providers' directory services – the Lightweight Directory Access Protocol (LDAP) version 3 (an Internet Engineering Task Force standard for exchanging directory information with clients and other directories), the DNS standard for naming and locating domain names on the Internet, Novell NDS or older NT-based NTDS directories (it doesn't matter so long as a service provider is available for that directory service).

The Active Directory brings these into a single view for simplifying deployment and as a mechanism to access and manage the system environment, as well as serve as a point of rendezvous to determine "who you can talk to" as it were, or how to reach that person.

A global catalog service of the Active Directory allows users to search for objects in the enterprise network by using common properties such as office location, user's last name, or e-mail address. The Active

T

Directory is also based on a sort of "delegation of authority" paradigm that permits network administrators to grant privileges to a given subset of users or objects.

The Active Directory also provides secure network access to resources via SSL, Kerberos, and / or Distributed Password Authorization (DPA). ADSI can work with LDAP servers. LDAP was defined by the Internet Engineering Task Force for integration with the Domain Name System (DNS) namespace standards. This in turn permits easy NT integration with Internet and Intranet standards and applications. Indeed, the Active Directory combines the best qualities of both DNS and X.500 / LDAP.

The Active Directory can do fault tolerant replication of directory service information via a Windows 2000 Remote Procedures Call (RPC) mechanism. An RPC is a message-passing facility that allows a distributed program to call services available on various computers in a network. Used during remote administration of computers, RPC provides a procedural view, rather than a transport-centered view, of networked operations.

The Active Directory enables a single logon for the client desktop to multiple directories (a feature that first appeared in Windows NT 4.0), makes it possible to write applications to one API but work with multiple directories, making it easier for end users to find information with rich querying of directories, and making it easier to manage multiple directories.

Interestingly, Cisco and Microsoft at one point entered into an agreement to standardize on Active Directory Services. The agreement provides tighter integration between Windows 2000 and Cisco's next generation of network products. Cisco is also partnering with Microsoft to port Active Directory to Solaris and HP-UX, and Cisco is using the Active Directory as a data store in its CNS/AD (Cisco Networking Services for Active Directory) software product. CNS/AD stores information such as hardware configuration, quality-of-service settings and policy settings for the hardware in the directory. The Cisco software runs on Cisco equipment and is targeted at service providers.

The name of Cisco always conjures up the specter of "convergence," and one of the things the convergence phenomenon is creating is the need for devices that are capable of understanding traditional telephony and / or IP telephony. Being able to enable many types of communications services on one box is one of the more interesting things about what you can do with NT Server and now Windows 2000, with it's built-in communications support and with the extensibility that's provided because of its APIs.

As for enabling pure IP Telephony, TAPI 3.0 brings call control and the control of media streams (using those MSPs again) for H.323 interoperability as well as IP Multicast, which is an extension to IP (the Internet Protocol) that allows for efficient group communication. IP Multicast arose out of a need for a lightweight, scalable conferencing solution that overcomes problems associated with real-time traffic over a datagram, "best-effort" network.

The H.323 Telephony Service Provider (and the associated H.323 Media Stream Provider) allows TAPI enabled applications to engage in multimedia conferences with any H.323 compliant terminal over the LAN.

For example, the H.323 Telephony Service Provider implements the H.323 signalling stack (Q.931, H.245, and RAS). This TSP accepts a number of different address formats, including name, machine name, and e-mail address. H.323 telephony is usually complicated by the fact that a user's network address (such as an IP address) is highly "volatile" and can't be counted on to remain unchanged between H.323 sessions.

With the TAPI H.323 Service Provider, the Active Directory Services can be used to perform user-to-IP address resolution. Specifically, user to IP mapping information is stored and continually refreshed in the

Internet Locator Service (ILS), a real-time server component of the NT Directory Service (NT DS).

The IP Multicast Conferencing TSP is mostly responsible for resolving conference names to IP multicast addresses using the IETF standard Session Description Protocol (SDP) conference descriptors stored in the ILS Conferencing Server.

It is complemented by the so-called "Rendezvous" conference controls, which are a set of COM components that abstract the concept of a conference directory, providing a mechanism to advertise new multiparty conferences and to discover existing ones. You can manipulate conferences stored on an ILS Conference Server via the Rendezvous Controls.

IP Multicast is designed to scale well as the number of participants and collaborations expand – adding one more user doesn't add a corresponding amount of bandwidth, since you're dealing with packets, not entire trunk lines per participant. IP Multicasts' transport protocol is the Real-time Transport Protocol RTP. In contrast to the ILS servers used by the H.323 TSP, there is only one ILS Conferencing Server per enterprise, since conference announcements require relatively little bandwidth.

So if you're a developer and willing to build on H.323 or provide IP multicasting (which makes efficient use of the bandwidth on a network connection for a multi-party collaborative call, for example) you won't have to really invest in reinventing the technology yourself. Developers can leverage the technology existing in Windows 2000, which should enable them to get IP solutions to market more quickly and at lower cost.

Quality of Service. In contrast to traditional data traffic, multimedia streams, such as those used in IP telephony, are extremely bandwidth and delay sensitive, making Quality of Service (QoS) more of an issue than it would be with a circuit switched network which can guarantee a minimum quality of service simply by allocating entire static circuits for every phone call.

According to Mark Lee, Product Manager for the Windows NT Server Communications Team at Microsoft, "We're going to have very comprehensive support for quality of service. That includes the Resource Reservation Protocol (RSVP) but it goes well beyond that. I think at a high level I would state that the QoS support is seen as a core customer requirement and that's why Windows 2000 has a number of QoS capabilities that are enabled throughout the operating system in different layers."

Lee continues: "That's not only true on the IP traffic side, but also very rigorous QoS support for ATM. One of the new things for Windows 2000 is native ATM support – not just ATM LAN emulation, but native ATM, with all of the associated QoS choices that are available in the ATM world."

"But back on the IP side, both Layer 2 and Layer 3 we're support different QoS capabilities," says Lee. "Our TCP/IP stack has been upgraded to enable the QoS to be supported. Besides the RSVP protocol, we also support 802.1P in the Layer 2 area, an IEEE standard. In the Layer 3 area we support the RSVP and IETF "diff-serve" or differentiated services QoS capability. To have a true QoS-enabled network infrastructure, all of the pieces along the path must support that QoS, whether it be a router or a switch or the end devices. RSVP handles the network signalling for the router.

Microsoft's Chief of Telephony Programming, Toby Nixon (celebrated as being the only Microsoft employee to wear a necktie every day to work), also tells me that "We're doing a lot of things in the OS for QoS but it kind of all stops at the wire until people upgrade their routers to the latest software loads that will allow the traffic to be carried end to end."

Nixon says that "Microsoft has a close relationship between Cisco. We're working together on the directory-enabled networking initiative, where the network administrators will be able to do the configuration of the quality of service authoritarian and provisioning all in the directory, and then have that propagated throughout the network. Once that's working it will make it much easier to administer the network than it is today."

Lee thinks that QoS concerns aren't just confined to the IP world: "We talk to our customers about why they want QoS. What are the apps that they're going to use to take advantage of QoS? Of course, IP telephony always pops up as a high priority. But frankly, IT managers tell us that some of their mission-critical applications are also leading candidates for using QoS."

Lee continues: "For example, the ability to have a guaranteed QoS when using mission-critical apps, such as an SAP database access or transactions at branch offices. These are very important things and that's we have that very broad QoS support in the OS. It goes all throughout the operating system, when you look at it. Winsock, TCP/IP stack supports, the ATM support – they all support QoS. It's going to be an important edge and yet another compelling reason why people will gravitate and use Windows 2000."

TAPI 3.0 leverages a lot of the fundamental enhancements in the operating system, including, of course, the Active Directory Service which will be useful for policy-based networking. This means that an administrator can establish policies for the types of QoS, and the types of services that an individual or a group of people within a company might be able to use. The network infrastructure will then execute that policy, with the directory service the repository for much of that policy-based information.

Whereas TAPI and its rivals (CallPath, CSTA) provide software interfaces for access to customer premise PBXs and phone systems, such telephony APIs are incapable of handling access to call setup, transfer and management services with the greater public network.

Interestingly, in late 1998, a consortium was formed to devise just such a public network API, called the Parlay API. The consortium's members include AT&T, British Telecom, Cegetel, Cisco, DGM&S Telecom (now called Ulticom), Ericsson, IBM, Lucent Technologies, Microsoft, Nortel Networks, and Siemens.

Telephony ASP

Such a company can help the Data CLECs and ISPs compete with the LECs for voice customers. The premise is that they will offer Voice service to the Data CLECs and ISPs who currently don't have that offering in their portfolio. So very quickly these providers could compete with the LECs. Voice, in turn, becomes another application.

The average small to medium sized business gets voice service from a Baby Bell and Internet Access from an ISP. What if the ISP could offer the customer both and save the customer 25% to 50% on voice service?

Such a model involves setting up "Voice Application Servers" at the ISP / DLEC site. Such technology would be true "next generation" with a network architecture based on a distributed, IP-based softswitch design. The architecture will be totally IP-based (No time division multiplexing, since Class 5 switches will be replaced with softswitches).

e-Talk Corporation (Ft. Worth, TX – 817-267-3025, www.e-talk.com), uses to term "Internet telephony application service provider" to describe a company that allows e-businesses to talk and interact with their customers online. e-Talk's E-talkNet, for example, can integrate into an e-business' website voice and data communications. Surfers click a button on the site and talk through their PC microphones and speakers. E-talk.Net's data collaboration allow web surfers and customer representatives to simultaneously view Web pages and change the data on each other's screens as they chat.

Telephony Markup Language (TML)

TML is still undergoing development. It is a form of the Extensible Markup Language (XML) devoted to telephony call control and unified messaging. TML is an attempt by *Computer Telephony* magazine to bring some order to the many different telecom-related XML's that are appearing in the industry.

First-generation computer telephony is crippled by integration headaches. Hooking voicemail to a PBX can still be an all-day undertaking (tweaking MF frequencies, fiddling around with ringback cadences, etc.). On the network side, things are only slightly easier, fruit of halfway-manageable standards such as SQL, ODBC, CORBA, Microsoft's Component Object Model (COM), Enterprise JavaBeans, and other ways of bringing data producers and consumers onto the same page.

Any time you need to get machines talking to each other, it seems an awful lot of work (not to mention a good deal of arcane knowledge) is required. That increases the cost of CT apps, complicates delivery, causes interdepartmental and political problems, and prevents some of the niftiest apps from ever seeing the light of day.

Part of the promise of convergence - second-generation CT - is that it starts making those problems go away. Or it at least puts voice and data on the same pipe, under the same protocol, making both accessible to software. The Web, too, is an incredible win for telephony, providing a ubiquitous infrastructure of readily-accessible, multi-purpose servers that transfer data on demand - both to people, and, in principle, to machines.

The people side is relatively easy to understand and implement. If you're smart enough to write a unified messaging system, it's trivial to write the HTML and CGI scripts needed to display the inbox on a Web browser. Much harder to write a client application that automatically grabs data off the web and does something useful with it - i.e., a program that can browse ComputerTelephony.com, find Richard's phone number, and stick it into GoldMine.

Yet apps like these - machine-to-machine apps - can really add intelligence to the phone call. Research agents, comparison-shopping engines, supply-chain automation, simple data sources feeding complex data sources, open-standard-based third-party telephony client applications, my messaging system talking to your messaging system ... Indeed, the whole concept of the open organization, in which contact protocols and hierarchy are "published" in machine-readable form, facilitating contact and better customer service, hangs on making machines talk to each other more fluently, with less custom programming, fiddling, debugging, and experimentation.

Ideally, you want a system in which data producers and consumers can exchange information fluently, but still evolve independently. I want to publish my dynamically-assigned IP address on my Web site so you can call me with your VoIP client. You want your app to grab my IP address and dial. I don't want to consult with you every time I change my home page. You don't want to debug your application every time my art director makes a font change. What's a body to do?

Enter XML - like HTML, it's a simplified descendant of the enormously complex Standardized General Markup Language (SGML) that's been used for high-end, highly structured publishing apps for the past 10 years. Thanks to efforts by the World Wide Web Consortium (W3C), XML has become a formal specification.

Unlike HTML, XML allows you to create your own tags (which is what "extensible" means). That's not to say XML is a replacement for HTML. It isn't. Although it's possible that a future version of HTML might be specified in XML, the two markup languages have completely different purposes. HTML tags specify the presentation of your Web information, with tags primarily defining the structure of a document, such as headers <head>, body <body>, paragraphs <P>, tables <table>, and more. Font information such as specific color or type is

397

done through attributes of the <font> tag. There are also tags for bold <b>, italic <i> and size such as <big>. On the other hand, XML tags – which can exist on the same page as HMTL - describe what the data means (e.g., first name, price, phone number, etc.) allowing you to semantically qualify information for any application on the network that's compatible with an XML parser.

Because XML specifies the semantics of data, independent from presentation, it naturally lends itself not only to business-to-consumer and business-to-business interactions but also for more intelligent search and navigation applications. Since it's extensible, XML is really a higher or "meta" level standard, a language that can define other markup "languages" to facilitate communication and transactions between consenting networked partners in any particular domain, such as e-commerce, astronomy, physics, biology, or what-not, provided that everyone agrees on what the domain's tags mean. Each domain thus has its own "vocabulary," which is a bit like an add-on library for the XML parser.

We must tread carefully here – XML allows specification of new elements and attributes, but not the ontology or semantics of these. XML essentially just defines the syntax and structure of the tags and the relationships of the elements and their attributes. There could be – and usually there is – inherent meaning in the existence of tagged elements and the containment relationships between the elements. Since the tags are meaningful words from a particular vocabulary, it is easier to associate specific semantics to them. It's like giving names to columns in a table. The meaning of the columns and how they should be used is up to the application program and the application developers. So, strictly speaking, a semantics or ontology arises out of an interpretation of the XML tags for particular domains by an application, not by XML itself.

Similar ideas have been tried before: the Common Object Request Broker Architecture (CORBA), Enterprise JavaBeans (EJBs), and Microsoft's Component Object Model (COM), all held promise but have not exactly taken the world by storm, probably since they need vast amounts of programming knowledge to make them work

The XML standard is being rapidly integrated into software products, even from longtime ruthless competitors such as Microsoft, Oracle, and IBM. Microsoft in particular has fallen in love with XML, calling it a cornerstone of its BizTalk initiative.

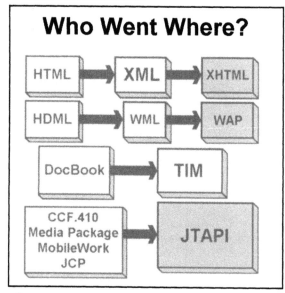

Who Went Where?

By separating structure and content from presentation, the same XML source document can be written once, then displayed or processed across multiple media by any XML compatible device: A computer monitor, cell phone display, text-to-speech device for the blind, etc. It'll continue to work on any future devices that can process XML tags, too.

Which brings us to telephony. One day it suddenly occurred to *Computer Telephony's* John Jainschigg that what the world needs is an XML vocabulary for a Telephony Markup Language (TML) to handle telephony and unified messaging. Such a language would make available similar interfaces across various communications devices and, since XML will become a standard mechanism for exchanging data as well as documents, it's possible that voice mail, video mail, faxes and e-mail messages from different vendors' repositories could be exchanged across the Internet and reviewed on any XML compatible device.

After some internal conversations, it was decided that our magazine would champion a TML vocabulary add-on to XML.

Around this time, SoloPoint's President and CEO, Arthur Chang, was in our East Coast Lab demonstrating SoloPoint's amazing new "Teleputing" network architecture for unified messaging (among other things). Chang brought up a unified messaging interface running on a laptop, then said: "At the moment this is written in Visual Basic, but in the future we'd like to do it in XML. You know, what this industry needs is something like –"

We finished the sentence for him: "– a Telephony Markup Language! You know anybody who's doing this?"

Chang didn't. And he was so taken with the idea, he said that SoloPoint would be willing to join the discussion to develop TML and would like to be the first company to incorporate it in its product line.

We had a momentary vision of becoming Internet millionaires - but then editor-in-chief John Jainschigg realized that as bona fide journalists, we had an ethical obligation to remove ourselves from direct involvement in the marketplace we report on. So it was decided that we would simply nurture TML along, report on progress, encourage companies to participate in the standard-making process, and help with coordination.

We next turned to Dr. Setrag Khoshafian at Technology Deployment International (Santa Clara, CA – 408-330-3400, www.tdiinc.com). TDI is doing massive amounts of research into XML and offers a basic service pack that consists of strategic training for XML as well as an analysis of how TDI's expertise can customize and emphasize XML technologies for your company. They also have a downloadable XML white paper at their site: www.tdiinc.com.

Indeed, TDI is a member of the W3C and its XML Schema Working Group. Currently XML documents can be either well-formed documents without a schema, or validated against a schema written as a Document Type Definition (DTD). DTDs are a set of syntax rules that tell computers what elements to expect in a document – what tags you can use in a document, what order they should appear in, which tags can appear inside other ones, which tags have attributes, etc.

Originally developed for SGML, a DTD can be part of an XML document, but it's usually a separate document. XML is a system for defining languages (or new tags or elements and their containment relationships). Each industry or scientific discipline that wants to use XML defines its own DTDs. The XML Schema working draft is intended to replace DTDs with richer constructs and simpler syntax. Currently DTD has a different syntax than XML markup. The new proposed schemas can express rules that DTDs can't, and are themselves written in XML, making it easy to integrate parts of different schemas. Still, DTDs will be around for a while yet.

In a conference call between CT magazine, Chang and Khoshafian, we did some brainstorming as to how to get our Telephony Markup Language initiative going and what the appearance of TML would mean to the industry. Here's what we came up with. . .

Originally, TDI and SoloPoint are contributing to the initiative by atomizing call and messaging functions in an effort to start formulating a list of possible telephony and messaging tags. SoloPoint is very much involved in this, trying to understand the type of messages that need to be sent and participate in the specification of the tags.

But TML will be more than just tags. There's also the organizational structure of the language that must be considered, the "base types" and the "enumeration types" that need to be listed.

For example, if you insert a type tag for an element and it has an attribute you might not want that attribute to take any distinct value; rather, you'd like the attribute to take values from a prescribed set or list of values, so you can avoid errors or you want to constrain the domain of those values (such as the range of touchtones). One typically encounters this important requirement and it must be represented in the language, so one must have the "ontology" of the tags worked out, as well as the ontology of the types and their relationships with one another.

For all our discussions of XML's handling of meaning and content, XML is ultimately syntax. What those tags mean is up to the application. That's where the domain of vocabulary standardization comes in. Many domains are dealing with this, whether it's electronic commerce or mathematics or pharmaceuticals or engineering or business. All of these XML markup languages do not actually extend the language or create a new parser, but they come with vocabularies, which set forth the ontology of what exactly those tags and their representation, containment relationships, and enumeration really mean for a particular discipline, such as telephony.

What we first need to do is to figure out the "domain knowledge". Our little group plans to hold some working sessions, and with our knowledge and expertise in the various domains, we should be able to come up with a first-order approximation of TML. Later we'll have ready version "0.01" of the tags and the domains that will be interpreted by client software.

Whatever happens, we would like to avoid the potential for conflict among multiple standards created for the same purpose. For instance, groups such as BizTalk, RosettaNet, Oasis, CommerceNet and cXML each are working on frameworks and standards that to some extent address similar needs and could potentially conflict.

But what would it mean for the industry to have a TML? You could put touchtone screens on a handheld device, and you could have universal messaging tags that would work with any device.

Arthur Chang noted that one of SoloPoint's prime directives has always been to take the Internet, the PC and all form of communications and to permeate them throughout the house without permeating multiple PCs along with them. Their goal is to keep you from running to your PC all the time, but to have the PC become a SOHO "back end server" in a sense, and that you can use SoloPoint's very inexpensive Teleporters as universal user interfaces to all the resources on the network.

That communication between those appliances, the PC and the Internet could be our proposed TML flavor of XML, but what's inside of each device may not be a browser *per se*, particularly if one is only interested in the transport of messaging and information (as you'd expect in a telephony language) although it could be. For example, on a Palm VII there might be a "microbrowser" or you might have notebook with a full-blown browser.

Fortunately, XML is up to the task, whether a browser is present in a device or not. The original markup language concept was not a case of "here's a client of a specific type" it was more along the lines of, "here's a way for a client and a server to actually talk to each other" and they will both establish in that communication what mutual capabilities they have available. HTML became "the client is the PC" by definition. Some confusion over XML and HTML also arose partly thanks to Microsoft's terrific efforts at promoting XML, which has caused to form a natural association in the public's mind between XML and HTML browser technology. Indeed, the media has described XML as "HTML on steroids."

But while XML capability can and definitely will be appearing on the browsers, one of XML's capabilities is the ability to send messages between distributed components, be they handheld devices, computers, or devices with embedded processing systems.

Actually, strictly speaking, XML cannot send messages – XML is the message content. Other applications or components or servers send messages that contain XML. Thus, you can think of XML as more of a self-describing messaging standard rather than simply some extra tags focusing on content. That's why XML can be used on the Internet, intranet, extranet, VPNs, WANs, or whatever, because XML itself is not actually sending messages. XML is standardized in the sense that when you have two processes (producer-consumer or publisher-subscriber) with a channel open between them, the two entities can speak the same language. What XML has is a structure, and there are various groups, efforts, etc. that define vocabularies or standard tags for particular application domains

We can surmise, then, that future XML "clients" won't necessarily be running on browsers most of the time, and so XML should not necessarily be perceived as merely a "browser client-like thing" talking to a "server-like thing."

One of the interesting things TML could bring to the table is for TML development to really concentrate on this pure messaging side of XML that works between servers and any type of client, be it a traditional browser-oriented client, or a PDA-like client, or a cell phone with a multiline display-like client, or even a client for the Teleporter devices that SoloPoint is developing.

TML would thus become something more fundamental than a bunch of tags for telecom devices, it would become the formulation of the underlying transport of messaging between servers and all types of different clients. The only reason you'd ever associate it with HTML would be under the circumstances when the client happens to be a PC running a browser. So what the group is trying to do with TML is to take the underlying transport of messaging and making that the focus of the language and get that permeated everywhere.

Indeed, the TML initiative may turn out to be a way to show the world how XML is really different from HTML. TML tags could be converted into any other kind of pulses, such as audio tones. Anything that has digital connections to a network could use TML.

There is some competition, of course. There are a number of standards, such as VoiceXML, sponsored by some big companies. It appears too high-level and too limited, having been defined for voice menu item management. It's used to represent a caller's many choices and input options when interacting with an IVR or other voice automation system. And yet this standard is being promoted, there's an SDK available for it, and it has a web site and everything (www.voicexml.org).

Interestingly, even the VoxML developers (some ex-Bell Labs folk) realize that "the client isn't a browser". They perceive the client as "audio over any phone" which makes sense for an IVR-related XML language. Still, the industry needs something more powerful and comprehensive, which is what it is hoped TML will be.

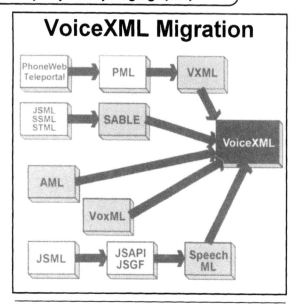

VoiceXML Migration

Here we see the transition towards VoiceXML from PML that evolved from PhoneWeb and Teleportal. SABLE evolved from JSML, SSML and STML and has been incorporated into VoiceXML as well. Motorola has also allowed VoxML to merge into VoiceXML, so we see a mixture of morphing and converging taking place as VoiceXML becomes the "favorite" for telephony markup langages. Diagram courtesy of Robert L. Pritchett.

Of course a double-edged sword appears whenever you try to be a bit different from the mainstream. The development of TML could be an opportunity to differentiate XML from HTML once and for all, or it could turn out to be the wrong approach completely because perhaps we really want to ensure its success by associating it with mainstream XML and those millions of browsers in the "front end" running on PCs.

However, there is a kind of compromise or creative twist to this: Let's say you define the TML / XML messaging standard completely. You come up with the DTD, the schemas, the tags, the whole works. What you then have is the potential of creating *virtual* devices. That is, you provide a virtual rendering on your browser so, for example, if SoloPoint's TML enabled devices are on the network, it becomes possible for a specific type of rendering a virtual representation to run on your browser.

The big challenge will be to convince everyone that TML is a language based on messaging transport which transcends whether the client is a traditional web browser or whether it's just three lines on an LCD in a cordless phone. TML is essentially a messaging protocol that allows clients and servers to communicate with each other whereas previous to this they really had no way of communicating in such a sophisticated and ubiquitous manner.

TML could be adapted to whatever new technologies or embedded devices appear. You could send signals over your home wiring or a wireless network to determine whether you left your coffeemaker on, whether the light is on in the bathroom, or what's wrong with your refrigerator. It could even be used for robotics.

Think about how people are doing messaging now in distributed systems. They use something such as COM / DCOM or any one of the very complex, non-self-describing technologies available. The nice thing about XML is that it can be used for anything, since it's based on self-describing messages.

HTML does this as well, but the tags of HTML have little or anything to do with content. If you want to extract any kind of meaningful information, such as instructions to your device, or querying of your device with specific tags, well, just about the only thing you can do with HTML is full-text search. I say "just about" because at this point HTML aficionados will quickly jump and say that there is in fact a <Meta> tag in HTML that can be used to provide keywords for searching, and some search engines use the description and content attributes of <Meta> tags for additional indexing. However, most of the time HTML searches are performed using full-text search expressions.

So, aside from <Meta> tags, normally in HTML there's no way to extract the value for a particular element and a particular attribute of this value. Now the values that you put there could be used by a device to control it or to query it or to get its status information so you can monitor it.

One big vista opening up for XML is meta-data searching. Web sites are converting from HTML because people are realizing that the "searching experience" is not so successful much of the time. HTML just doesn't cut it, whereas with XML you can now have much more meaningful search tags and provide a true search experience for the user as opposed to the full text retrieval needed for HTML. Interestingly, there is a meta-data standard called the Resource Description Framework (RDF) that provides in XML meta-data information for any resource on the network – including HTML documents. Could meta-data searching be applied in a TML environment, particularly a Wildfire-like personal assistant or "bot" that goes fetching information from directories? The possibilities are tantalizing.

During the summer of 1999, a meeting was held at TDI concerning TML. Representatives from a number of interested companies were in attendance but each company had its own completely different theory as to what TML should be like!

For example, you talk to Solopoint. They have a small appliance called the teleputer. They don't want an XML variant that uses a lot of tags and needs excessive memory and processing overhead. They can easily conceive of something that's tailored exactly to their specific appliance. Each company tends to behave like this. After talking to a room filled with company representatives, any self-respecting TML theorist has in short order a whole series of proprietary concepts on his or her hands.

We then discovered Robert L. Pritchett, who was working on a research project for a Masters Degree in Computer Science involving. . .

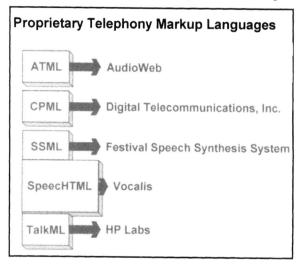

Proprietary Telephony Markup Languages

ATML	AudioWeb
CPML	Digital Telecommunications, Inc.
SSML	Festival Speech Synthesis System
SpeecHTML	Vocalis
TalkML	HP Labs

TML! Pritchett is a staunch believer in devising some kind of stable telephony markup language, lest the entire computer telephony industry will end up with high-end products that will forever be incompatible.

Shortly thereafter, we found that Telera (Campbell, CA - 408-626-6800, www.telera.com) had created a fascinating post-routing system that replaced SS7 codes with XML tags, and Pactolus Communication Software (Framingham, MA - 508-620-5519, www.pactolus.com) had developed something they called the eXtensible Telephony Markup Language (XTML).

Founded in 1999, Pactolus is now a leading provider of XML-based enhanced service platforms, applications and service creation solutions for next-generation communication networks. Pactolus is focused on delivering innovative, enhanced service solutions for the IP Telephony and Application Service Provider (ASP) markets. Their product suite, called Rapid-FLEX, is based on their XTML language, which serves as a unified service delivery framework that can combine the power of Internet-style service creation with next-generation packet networks for Internet, cable and Communications Service Providers. By leveraging the power of XTML, Rapid-FLEX enables service providers to be first to market with value-added business-to-business and residential enhanced service applications that subscribers can tailor to meet their specific needs over the Internet, using Rapid-FLEX's integrated web portal.

In a discussion with representatives from Pactolus, Telera and TDI, Pactolus told of the scope of the problem they were trying to solve with their new XTML product. They're really creating applications and services with it, particularly for Next-Gen type networks, and trying to deal with matters of call control, the message store, voice response, and database connectivity. Pactolus also built into XTML great extensibility.

As to what Telera has done with XML, it has adopted the model of a "user interface manager." It harks back to the simple, basic concept of the web, where you've got a web browser that has been instructed through HTML as to what to do when a particular user sits at his or her web browser and is visiting a website. The behavior that the user interface manager exhibits dynamically changes the website's look and feel, just as you would expect when dealing with a person browsing.

The XML language construct is meant to model the communication between an application, which is a web application in the context of what we're discussing, and a user interface manager that's sitting remote to it. This remote UIM was originally called a voice response unit (VRU) by Telera's engineers, but the concept rapidly evolved into a group of voice ports that do both voice response and call management – which means that it behaves like a remote switch.

At the first level, the voice response tags relate to what an endpoint should do when it receives a call and how it should interact with the caller. In this respect Telera's XML functionality is a bit like VoiceXML's except that Telera has brought in a lot of support for reading tones, has a better foundation of what's going on with respect to business communications today, and has made contact centers as their first targeted business space. From this perspective, the voice response application that's written for the contact center could consist of entire soft servers or it could be a simple module that's able to supervise the preliminary part of the customer self-service process and then queue the call, whereupon it will be picked up by a contact center agent. The XML part of all this was Telera's issuing of tags that could be interpreted in a linear form by the user interface manager at the network edge, and which could be extended to deal with exception handling and tone mapping for the purpose of firing different URLs in response to different tones, and so forth.

Still, when Telera saw the VoiceXML standard in action, they noted many similarities with what they were doing. Telera's approach tends to be standards oriented and so as the company moves forward it would be safe to say that they're going to migrate as much as they can to the VoiceXML standard and perhaps even influence that standard.

The one key difference between all of the XML standards that exist and what Telera has done is that, in Telera's model, the application logic is living on the web server, where there also resides programming languages of the customer's choice. The customer may use Java, C++, JavaScript, etc. to drive his site. It doesn't matter. Effectively, the logic is back there in the web server, while in the "front" is the user interface manager, which is a pure abstraction of what's going on at a very rudimentary level. You can play a prompt, you can pause, you can recognize certain tones in calls and actions can be taken as a result of them, and you can even act on exceptions that occur during the interaction with the caller. Telera has managed to construct a complete model of a caller and his or her interactions with a voice response system.

Now when applying this to contact center operation, you have an initial voice response event, and then you have the issue of queuing the call and waiting for an available agent. Telera developed additional tags in order to be able to do call routing, outbound dialing and things like that, thus treating the endpoint where the call initially arrives as a mini-switch with multiple ports on it. You can allocate new outbound ports, bridge to call, unbridge to call, wait on events, etc. All of this is somewhat modeled after CSTA messages, which are based on an abstract model of a switch and has its own call model.

The workings of Telera's model can be conceived in layers. Many of the call control primitives occur at lower layers (a bit like the OSI Reference Model), but when developing applications for the whole network one encounters the issue of providing applications that may also be shared between multiple customers with capabilities defined by policy. Therefore we can consider the whole network to be kind of distributed operating system, where an application makes requests of the system and each request is checked against policies, granted, and then lower level intelligence processes actually issue the primitives, with the privileged instructions residing within the network at a functional level equivalent to where CSTA messages would be in a conventional circuit switched system.

At a higher level (from the application's point of view), the system has high-level primitives that can be used to accomplish whatever it is we want to do with the call – how to transfer the call somewhere, for example, or how to multiparty conference, etc. This generally involves tight integration with enterprise-based call routers.

Now looking at the system's lowest level we find the messaging of the XML code itself. Down at this level "call control" has to do with being able to allocate an endpoint, issue a dialed request, and obtain all kinds of call progress and result indications.

And so, if you think about it, we've gone through all of the steps that are needed in order for a remote entity to issue a set of instructions to a dumb client (that is effectively doing low-level telephony functions) to execute instructions and issue back responses as to what it encountered during the whole communications process.

To put it all less technically, the Telera applications network is based on a model similar to that of a business communications ASP, offering the capabilities of distributed ports on the network to business customers to implement effectively whatever application they fancy (but the first market Telera itself is going after is the contact center). Other large-scale distributed applications that could easily be developed with Telera's system include unified messaging and general e-commerce applications - their component processes all conveniently map into the same types of

T

primitives that are used for the contact center applications.

Telera's development / deployment process is a bit different from establishing a classic service creation environment. For those interested in service creation environments, there already exists an XML based language called the Call Processing Language (CPL), a language that can be used to describe and control Internet telephony services and was formulated so that service providers could create a configurable network of services that they could then sell to customers.

Pactolus evaluated CPL with an eye toward using it as a development environment. While CPL does provide a service creation language, it doesn't provide call control-type primitives. CPL is pretty focused on using the SIP call control protocol, which is designed to be somewhat self-limiting for the purposes of safety. Pactolus' applications have to do a lot more than just call control, and even a lot more than SIP call control. For example, they must be able to access databases. Also, CPL doesn't want you to describe anything in a model other than a directed acyclical graph, which means that you end up with a system model having no moving constructs and other deficiencies. It also wasn't clear how you could easily extend CPL. Even VoiceXML has some rudimentary tools for extending itself. So, fundamentally, Pactolus just couldn't figure out how they could build real services that they could sell using CPL.

Most experts would agree that the basic challenge of developing in XML and the way that it's traditionally found usage within a domain of interest involves getting a large enough group of people to agree upon a shared world view that's sufficiently static so that developers have time to implement a common XML in their products, and then cajole people into buying those products. In the case of the telephony environment, the space is so broad in terms of functionality and protocols that it seemed that such shared agreements of "world view" could never be sustained long enough to really build anything useful. Pactolus realized this and so when they were building their XTML language they basically took a "divide and conquer" approach where they subdivided big problems, making it easier for everyone to agree on some of the core issues concerning how telephonic events, state information and logic could be represented, and to allow groups of people to make shared agreements on the aspects of certain protocols or device-specific processes. Achieving such a common consensus prior to developing their new variety of XML brought Pactolus into agreements on more than just XML tags; it gave new insights into communications processes.

Another not-so-obvious problem with creating a new XML standard is that everybody is "standardizing" on XML, but XML's extensibility allows us to push fundamental issues to the background: Rather than standardizing on the XML tag itself, the real problem is doing standardization on the functioning, the parameters, their types, the actions, their names, etc. If you don't do that then you'll simply become an isolated island of invention.

These matters of dealing with both the "surround stuff" and the protean nature of XML itself are in sharp contrast to the world of the old HTML standard, which is essentially a simple language shared between a producer of information and a consumer, consisting of a set of universally well-known formatting tags of limited functionality. HTML can also be conceived of as a sort of object model that's shared to a certain extent, but it's a strictly defined object model that lacks flexibility or extensibility. It's ironic how HTML, the most ubiquitous formal language of modern times is also among the least sophisticated.

Still, the HTML's simplicity is one reason that the Internet is reaching into to every nook and cranny on the planet, and XML, like HTML, relies for core transport on the well-known "web standard" –

we'll refer to it as a "standard" for if nothing else the underlying, invisible HyperText Transfer Protocol (HTTP) is used by millions of web servers and client browsers to communicate and move documents over the Internet.

XML code "rides on top" of HTTP and leverages the entire Internet infrastructure of all the known and well understood means of being able to deliver ASCII code and other data from point A to point B. The Internet is becoming as extensive as the phone network, which is good news if you want to replace some of those HTML tags whizzing through the Internet with telephony-capable XML tags. Had HTML and XML been developed in the 1960s prior to the appearance of packet networks, and if AT&T had been broken up sooner, perhaps the Internet would have consisted of data over voice channels!

Of course, XML's functionality has to do with transferring meaning. XML may be an extensible language, but it doesn't normally act the way "Turing complete" languages such as C++, or BASIC do. When Pactolus was developing XTML, they had to do what the developers of other XML-based languages such as VoiceXML had to do – reinvent the wheel. Or rather, they had to define the kind of looping, branching and other programmatic constructs (such as ways of maintaining state information) found in many of today's computer languages.

This leads to the question of why one would really want to use XML at all. Why not use an off-the-shelf programming language in popular use such as Javascript, VBScript or PERL? Why not use a defined object model that's implementable using any scripting environment we feel comfortable with?

This obvious answer is that if we did this we'd immediately lose our ability to extend the language and wouldn't be able to define new tags without reengineering the language. But even more importantly, whenever you start working with any of these languages (Java, JavaScript, C++, etc.) in a telephony context, you soon find your development efforts bound not only to the language but also to the platform of communication when it comes time for you to deal with the distribution of your application to the public.

XML, on the other hand, is independent of any transport protocol. It's self-describing, it's completely robust and completely independent in terms of its transportation through a network. You can attach XML code to an HTTP message as a message body as easily as you can make it an argument of a Java meta-invocation - it's truly independent of any language and any transport protocol.

All of the parties in *Computer Telephony* magazine's TML initiative now realize that there are several levels in telecom that need XML and need standardization of XML involving routing, voice browsing, and even perhaps higher level abstractions. These "levels" can probably be arranged in layers similar to those of the OSI Reference Model, but will have components that are more functional in nature rather than just being network-oriented.

At the time of this encyclopedia's publication, members of the TML initiative were formulating a taxonomy of the XML telephony space prior to selecting the core tags.

So, if anybody reading this would like to get involved in the the great TML crusade, give *Computer Telephony* magazine a call or send an e-mail. Like the old Steve Allen song, "this could be the start of something big."

xML Telephony Family Taxonomy

All quotes listed are from the web links shown within the body of each module below. For DTD, Schema or Specification information, check out the web links. Some modules are not "complete" because they have been superseded by other modules, as noted (thanks to Bob Pritchett for this material).

Child: Standard Generalized Markup Language (SGML)
Born: 1974
Parent: Charles Goldfarb and then the International Organization for Standardization ISO 8879
Purpose: An on-line system for integrated text processing.
Dictionary: http://www.bradley.co.uk/DICT.HTM
More: http://www.oasis-open.org/cover/sgml-xml.html

Child: HyperText Markup Language (HTML)
Born: 1990
Current Version 4.0
Parent: World Wide Web Consortium (W3C)
Morphing to XHTML
Purpose: A non-application-specific DTD developed for delivery and presentation of documents over the Web, to be composed using an HTML browser.
More: http://www.oasis-open.org/cover/sgml-xml.html

Child: Extensible Markup Language (XML)
Born: 1997
Current Version 1.0
Reborn: 10 February, 1999
Parent:W3C
Purpose: "An extremely simple dialect [or 'subset'] of SGML the goal of which is to enable generic SGML to be served, received, and processed on the Web in the way that is now possible with HTML, for which reason XML has been designed for ease of implementation, and for interoperability with both SGML and HTML."
DTD: http://www.w3.org/XML/1998/06/xmlspec-report-v20.htm
More: http://www.oasis-open.org/cover/xml.html

Child: Extensible Hyper Text Markup Language (XHTML)
Born: 22 September 1999
Current Version HTML 4.01
Parent: W3C
Purpose: It provides the basis for a family of document types that will extend and/or subset XHTML, in order to support a wide range of new devices and applications.
DTD: http://wdvl.com/Authoring/Languages/XML/XHTML/dtd.html
More: http://wdvl.com/Authoring/Languages/XML/XHTML/

Child: Audio Markup Language
Born: September 1998
Parents: T.V. Raman and David Gries, Cornell University
Licensed Parent: 4Peripherals joined VXML Forum April 1999
More: http://www.gofourth.com/news/

Child: AudioText Markup Language (ATML)
Born: August 1997
Parent: Rutgers University
Purpose: "Much as HTML tags direct a Web browser on how to display information on a computer screeen and link to other Web pages, ATML commands direct an audio brows-er on how to access, link, and process information from the Web, and read it over the phone."
Used in AudioWeb
More: http://audio-app.rutgers.edu/atml/atmlman/

Child: Broadcast HyperText markup Language (bHTML)
Born: 6 August 1998
Parent: Aninda DasGupta, ATSC T3/S17
Purpose: To be the application programming interface for a Digital Television Application Software Environment (DASE) compliant receiver.
DTD: http://toocan.philabs.research.philips.com/misc/atsc/bhtml/cover.html#s3_0
More: http://toocan.philabs.research.philips.com/misc/atsc/bhtml/cover.html

Child: Common Information Model (CIM)
Born: 9 April 1997
Current Version 2.2
Parent: Distributed Management Task Force, Inc.
Purpose: It's an approach to the management of systems and networks that applies the basic structuring and conceptualization techniques of the object-oriented paradigm that supports the cooperative development of an object-oriented schema across multiple organizations.
Specifications: http://www.dmtf.org/spec/cims.html
More: http://www.dmtf.org/index.html,
http://www.dmtf.org/spec/cim_tutorial/

Child: Call Processing Language (CPL)
Born: 26 February 1999
Parent: IP Telephony (IPTEL) working group of the Internet Engineering Task Force (IETF), Jonathan Rosenberg (Bell Labs) and Henning Schulzrinne (Columbia University)
Purpose: The CPL is powerful enough to describe a large number of services and features, but it is limited in power so that it can run safely in Internet telephony servers. The intention is to make it impossible for users to do anything more complex (and dangerous) than describing Internet telephony services. The language is not Turing-complete, and provides no way to write a loop or a function."
DTD: http://www.oasis-open.org/cover/cplDTD199902.txt
More: http://computer.org/internet/telephony/
http://www.oasis-open.org/cover/cpl.html

Child: Call Policy Markup Language (CPML)
Born: 26 July 1999
Parent: Digital Telecommunications, Inc.
Used in Extensible Service Policy Telecommunications Services Portal.
Purpose: "It opens the telecommunications network to basic IP tools that brings order to its complexity and empowers carriers, and ultimately customers, to control, customize, and extend the functionality and usefulness of the telecommunications network. It is an open language used to simply describe call processing."
More: http://www.dticorp.com
http://www.oasis-open.org/cover/cpml.html

T

Child: DocBook DTD Document Type Definition
Born: 1991
Current Version 3.1
Parent: HaL Computer Systems, O'Reilly & Associates
handed off to Davenport Group
Current Parent: Oasis DocBook Technical Committee
Used in TIM
Purpose: To provide SGML support for folks who like to write
books on-line.
DTD: http://www.oasis-open.org/docbook/docbook/index.html
More: http://www.oasis-open.org/docbook/

Child: Festival Speech Synthesis System
Born: January 1999
Version 1.3.0
Current Version 1.4.0
Parent: Centre for Speech Technology Research (CSTR)
Uses SSML
Purpose: "It offers a full text to speech system with various APIs,
as well an environment for development and research of
speech synthesis techniques."
Programming Information: http://www.cstr.ed.ac.uk/projects/festi-
val/manual/festival_toc.html
More: http://www.cstr.ed.ac.uk/projects/festival

Child: Handheld Device Markup Language (HDML)
Born: June 1997
Parent: Unwired Planet [Phone.com]
Morphed intoWML
More: http://whatis.com/hdml.htm

Child: Java Speech Application Programming Language (JSAPI)
Born: February 1998
Parent: Sun
Uses JSML
Purpose: "The JavaTM Speech API allows Java applications to
incorporate speech technology into their user interfaces. It
defines a cross-platform API to support command and con-
trol recognizers, dictation systems and speech synthesizers."
More: http://www.javasoft.com/products/java-media/
speech/forDevelopers/jsapifaq.html

Child: Java Speech or Synthesis Markup Language (JSML)
Born: 28 August 1997
Parent: Sun
Used in SpeechML, JSAPI, SABLE and VoiceXML
Purpose: "JSML is a subset of XML (Extensible Markup
Language), which is a simple dialect of SGML. By being a
subset of XML, JSML gains a standardized, extensible syn-
tax that is not tied to the Java Speech API (JSAPI)."
Specification:http://java.sun.com/products/java-media/
speech/forDevelopers/JSML/
More: http://www.oasis-open.org/cover/xml.html#xml-javaSpeech

Child: Java Telephony Application Programming Interface (JTAPI)
Born: 30 October 1997
Version 1.0

Current Version 2.0
Parent: Intel, Nortel Networks, Novell, Sun Microsystems, Inc.
More Parents: Enterprise Computer Telephony Forum (ECTF),
Dialogic, Siemens
Uses ECF S.410 JTAPI Media package, JTAPI mobile work, Java
Community Process (JCP)
Purpose: "It is an extensible API designed to scale for use in a
range of domains from first party call control in a con-
sumer device to third party call control in large distributed
call centers."
Specifications: http://www.javasoft.com/products/jtapi/#JTAPI 1.3
More: http://www.javasoft.com/products/jtapi/

Child: Phone Markup Language (PML)
Born: (before the breakup/divorce?)
Formerly PhoneWeb (AT&T) and TelePortal (Lucent)
Parent: AT&T Bell Labs, Lucent
Used in VXML
More: http://www.oasis-open.org/cover/vxml.html

Child: SABLE Consortium
Born: March 1998
Version 0.2
Current Version 1.0
Parent: AT&T Bell Labs, British Telecom, Carnegie Mellon,
Edinbugh University, Sun
Uses JSML, SSML and STML
Used in VoiceXML
Purpose: The draft SABLE specification is an initiative to establish
a standard system for marking up text input to speech syn-
thesizers.
Specification: http://www.cstr.ed.ac.uk/projects/sable_spec2.html
More: http://www.oasis-open.org/cover/sable.html

Child: Synchronized Multimedia Integration Language (SMIL)
Born: July 1998
Current Version SMIL Boston
Parent: Synchronized Multimedia Working Group
Used in RealNetworks RealSlideshow Plus
Purpose: "Define a simple XML-based language that allows
authors to write interactive multimedia presentations.
Allow reusing of SMIL syntax and semantics in other
XML-based languages, in particular those who need to
represent timing and synchronization."
Modules: http://www.w3.org/1999/08/WD-smil-boston-
19990803/Modules/smil-modules.html
More: http://www.w3.org/1999/08/WD-smil-boston-19990803/

Child: Speech HyperText Markup Language (SpeecHTML)
Born: 20 October 1998
Parent: Vocalis. Ltd.
Purpose: "The SpeecHTML service allows your customers to call
your website over the telephone."
More: http://www.speechtml.com/

Child: Speech Markup Language (SpeechML)
Born: 12 February 1999
Parent: IBM alphaWorks Laboratory

Uses JSAPI, JSGF, JSML

It has been morphed into VoiceXML.

Purpose: "It is a language for building network-based conversational applications."

More: http://www.oasis-open.org/cover/speechML.html
http://www.alphaworks.ibm.com/formula/speechml

Child: Speech Synthesis Markup Language (SSML)

Born: 1992

Parent: Paul Taylor

Version 1.0 Summer 1995

New Parents: Amy Isard, Paul Taylor

Used in Festival Speech Synthesis System and SABLE

Purpose: Converts Text to Speech.

More: http://www.oasis-open.org/cover/ssml-details.html

Child: Speech-to-Text or Spoken Text Markup Language (STML)

Born: 1996

Parent: Bell Labs, Edinburgh University

Used in SABLE

Purpose: System independent standard for marking up text for the purposes of synthesis.

DTD: http://www.bell-labs.com/project/tts/stml.dtd

More: http://www.bell-labs.com/project/tts/stml.html

Child: Talk Markup Language (TalkML)

Born: 29 June 1999

Parent: HP Labs

Purpose: Experimental language for Voice Browsers.

More: http://www.w3.org/Voice/TalkML/

Child: Telecommunications Interchange Markup (TIM) or Technical Information Markup (DTD)

Born: 1996

Parent: Telecommunications Industry Forum (TCIF)

Uses DocBook

Purpose: "It's a specification for describing the structure of telecommunications and other technical documents. TIM extends DocBook by describing additional structural constructs that are often found in technical documentation. It may now be the best general-purpose (or "interchange") DTD for technical documents in many industries."

More: http://www.atis.org/atis/tcif/ipi/dl_tim.htm
http://www.oasis-open.org/cover/swankSGML97IPI.html

Child: Telephony Markup Language (TML)

Born: August 1999

Parent: John Jainschigg & Richard "Zippy" Grigonis

Purpose: "By atomizing call and messaging functions in an effort to start formulating a list of possible telephony and messaging tags. There's also the organizational structure of the language that must be considered, the base types and the enumeration types that need to be listed. Those assist in defining constraints – such as the default values assumed by attributes when those are unspecified in a particular document."

More: http://www.oasis-open.org/cover/tml.html

Child: VoxML

Born: 30 September 1998

Parent: Motorola

Morphed into VXML

Purpose: VoxMLTM allows a developer to create a script of the conversation a user can have with an application program run by a Web server.

More: http://www.whatis.com/ Then search on VoxML

Child: Voice Extensible Markup Language Forum or now VoiceXML (VXML)

Born: 02 March 1999

Parent: AT&T Bell Labs, 4P, Lucent, Motorola (joined now by many other companies)

Uses AML, JSML, PML, SABLE, SpeechML, VoxML

Purpose: "The VXML Forum has four main objectives: (1) to develop an open VXML specification and then submit it for standardization; (2) to educate the industry about the need for a standard voice markup language; (3) to attract industry support and participation in the VXML Forum; (4) to promote industry-wide use of the resulting standard to create innovative content and service applications."

Specification: http://www.voicexml.org/specs/VoiceXML-0.9-19990817.pdf

More: http://www.oasis-open.org/cover/vxml.html
http://www.vxmlforum.org/pr19990825-1.html

Child: Wireless Application Protocol Forum (WAP)

Born: 9 August 1999

Parent: Ericson, Motorola, Nokia, Unwired Planet [renamed Phone.com] (and now, many others)

Uses HDML, WML

Purpose: Wireless Application Protocol (WAP) is a result of continuous work to define an industry wide standard for developing applications over wireless communication networks.

Specifications: http://www.wapforum.org/what/technical.htm

More: http://www.oasis-open.org/cover/wap-wml.html

Child: Wireless Markup Language (WML)

Born: 3 February 1998

Parent: Unwired Planet [Phone.com], WAP Forum

Was HDML

Used in WAP

Purpose: "WML was formed to create the global wireless protocol specification that works across differing wireless network technology types, for adoption by appropriate industry standards bodies." (see links in WAP above) "It is a language that allows the text portions of Web pages to be presented on cellular phones and personal digital assistants (PDAs) via wireless access. The Wireless Application Protocol works on top of standard data link protocols, such as GSM, CDMA, and TDMA, and provides a complete set of network communication programs comparable to and supportive of the Internet set of protocols."

More: http://www.whatis.com/wml.htm

T

Telephony Services Application Programming Interface (TSAPI)

This telephony API was developed by Novell and the part of AT&T that ultimately became Lucent Technologies. When TSAPI first appeared it was described as a "standards-based API for call control, call / device monitoring and query, call routing, device / system maintenance capabilities, and basic directory services."

TSAPI is the API for Telephony Services for NetWare, a software add-on for the Novell NetWare network operating system that basically consists of the Telephony Server NetWare Loadable Module (Telephony Server NLM). The Telephony Server NLM working in conjunction with a LAN card and a connection to a PBX allows anyone with a PC on the NetWare LAN to enjoy the functionality of third-party call control – calls and switching can be done from the server instead of from the desktop, as in the case of first party call control (See First Party Call Control and Third Party Call Control).

TSAPI / Telephony Services had a lot going for it when the whole package debuted in early 1993. It appeared before its chief competitor, the Intel / Microsoft TAPI (See Telephony Applications Programming Interface). It could do third-party call control when TAPI could only do first party call control from the desktop. It worked with PBXs from companies such as Siemens Rolm and Nortel.

However, TSAPI was a partial implementation of CSTA, which was a European telephony standard in the public domain. TSAPI was definitely not in the public domain - in fact, it cost too much. Novell apparently went through a turbulent period and suddenly decided to move their Telephony Services programmers to California, which, among other things, caused some key personnel to leave. Ultimately, AT&T / Lucent marketed TSAPI. TSAPI lost its momentum, and the rival Intel / Microsoft TAPI software soon appeared as a license-free part of the ubiquitous Windows operating system.

During the mid-1990s there was a tremendous debate over whether computer telephony programmers should use TAPI or TSAPI.

Because of TSAPI's ability to do third-party call control, it was perfect for writing applications for call centers. TSAPI greatly expanded the calling options available through an application making function calls to the Telephony Services software. For example, an application could use DNIS information to select a script for a call center agent, have a terminal automatically pick up a call, let the agent read the script to a caller, hang up a call and then signal the switch that it's ready for a new incoming call, without any direct action by the agent at the desktop.

Applications could also autodial through a PBX directly from a script, ensuring that calls could be returned at the proper time and that more important calls take priority over less important calls. It also allows one to integrate a predictive dialer with a phone switch (See Predictive Dialer).

Using TSAPI, a call center application could do "blending:" It could check the status of an ACD queue and switch agents to outbound when the number of incoming calls in the queue was below a certain specified threshold.

Text to Speech (TTS)

Text-to-speech, also called Speech Synthesis, is the process by which a computer converts any readable text into human-sounding speech output. It is the reverse of Automated Speech Recognition (ASR). A particularly interdisciplinary field, synthesized speech sits at the crossroads of linguistics, computer science, signal processing, psychology and acoustics.

TTS has lived in the shadow of its sexier speech technology sister, ASR. Perhaps that's because people are more eager to dial up or browse in and talk to computers than to listen to their admittedly robotic voices that are wearying to follow for paragraphs at a time.

Until the computer telephony revolution, the most common use of TTS was for people with vision or speech impairments. Indeed, TTS technology first evolved as the best way to give blind people a way to "read" text-based information. Audio reading devices on the market in the 1980s were the first commercial applications for these systems. Those original systems were bundled with optical character recognition (OCR) capabilities, so printed books could be "read." Those systems were expensive – a typical system cost $25,000. Ten years later there appeared less expensive talking devices such as Covox's "Speech Thing" and IBM's "Screen Reader."

Those first commercial English language synthesizers of the 1980s mispronounced many things since English language rules are so difficult to quantify. In the early, primitive days of TTS, the running joke was that all such systems sounded like "a drunken Swede." Personally, I thought the voice synthesizer on my 1986 Amiga 1000 computer sounded more like an Afghan on hashish.

By the early 1990s, TTS systems were sounding much better, benefiting from the results of research into digital audio, acoustic modeling, statistical analysis, and artificial intelligence. And today, of course, a text-to-speech synthesizer should pronounce almost every word correctly, be it in English or another language.

The original synthesis systems were large and hardware-based. The software code creating the speech was bundled with the hardware it needed to run. That hardware included a dedicated synthesis machine with interfaces to computer databases and to phone lines. Needless to say, it was quite expensive for the whole kit and kaboodle.

During the early 1990s, because text-to-speech conversion software used the same type of silicon technology used in other kinds of voice systems – namely a digital signal processor (DSP) chip – vendors found that they could move TTS technology into all sorts of embedded applications, and so TTS began to enjoy a commercial success far beyond the disabled community, finding use in educational toys (talking dictionaries) and pocket translators. General Motors was the first car company to put TTS-equipped computers (in their Orlando, FL rental cars) to help tourists navigate.

Computer telephony resource boards also use DSPs, and so it was inevitable that TTS would become a major component of digital voice storage mediums like voice mail and Interactive Voice Response (IVR) systems.

Until TTS arrived in CT systems, reading off simple scripts over the phone – bank balances, telephone numbers, small databases, etc. - was normally done by playing a recording that splices the sounds of letters or numbers together to deliver a particular type of information. These are the scripts most often used in interactive voice response (IVR) systems today.

This type of speech output is *digitized*, previously recorded human speech that's played back via a digital to analog converter. Rapidly changing information can be a problem with this method of disseminating information, since a much larger vocabulary is required to accommodate the changing data. It would take far too many pre-recorded messages to keep up with a large database.

That's where TTS systems come in. They don't use digitized speech. Instead, they employ synthesized speech generated by the computer, rather than a human user.

Indeed, TTS has actually become an integral component of many computer telephony installations, particularly those involving IVR or, to use Ease CT Solutions's trademarked phrase: "giving data a voice."

- *It's inexpensive.* One of the nice things about text-to-speech is that it eliminates the hassle and headaches of scripting and paying for studio time to record speech prompts for IVR systems, and eliminates wasted disk space – no need to hold megabytes of digitized sound files, just a compact text version of the words.

- *It's flexible.* Thanks to TTS, IVR systems can be updated quickly by just doing some text editing. You can do "search and replace" on text, which is practically impossible to do with, say, a WAV file. Data is made available to callers as soon as the database is altered.
- *It's uniform.* What happens if your voice prompts are recorded from a person who leaves your firm? Do you resort to using a patchwork of different voices? TTS is not only flexible, it provides audio uniformity.
- *It's great for unified messaging.* There are now applications for the remote reading of e-mail, fax-to-text-to-speech, speaking addresses in reverse-directory systems, etc.
- *It's going over the Internet.* Multimedia application developers are building "talking pages" on the World Wide Web. Normally, one would digitize the speech at the server, compress it using the Pulse Code Modulation (or similar) algorithm, then send it over the Internet as a file. This is too cumbersome, since you have to decompress and play the file. Ideally, you would receive the speech in real time, but servers and transmission performance on the net are as yet too variable. Text-to-speech, however, allows a small amount of text to be quickly transmitted over the Internet, which is then converted to speech locally, at the client end in your receiving PC.

Generally speaking, in terms of computer telephony, TTS is best used in voice processing applications that must rely on broad, unrelated, unpredictable vocabularies, which are too extensive (and expensive) to pre-record, such as anything involving thousands of different names and addresses.

Current examples of popular TTS applications:

- Reading e-mail (or faxmail using OCR) to those not near a modem-equipped computer or fax machine. Typically, the application reads headers or introductions of e-mailed messages to the caller, who may then choose a full reading or a faxed copy. This is a challenging application, since it must deal with the misspellings and the strange capitalization and abbreviations of e-mail correspondents.
- Providing database access: Account or inventory information. If a parts distributor, for example, wants to make a large changing inventory available to salespeople or customers, a TTS application can read any new part added straight from the ASCII database, without incurring any additional storage overhead or requiring any recording to be made.
- IVR (interactive voice response) scripts. Especially when used with a large database. Think train and flight timetables, news summaries, information response from virtual assistants (HAL from *2001, a Space Odyssey* would have employed TTS a lot), etc.
- Providing reverse directories. The caller inputs phone numbers and retrieves names and addresses.
- Reading: To the blind; to eyes-busy / hands-busy workers; to drivers (text delivered to onboard terminals by satellite or radio); to tourists in rental cars.

Like the human process it models, synthesized speech requires a lot of processing power. But while humans navigate the myriad rules of pronunciation and inflection subconsciously and effortlessly, TTS systems, before uttering any phrase, must churn through thousands of lines of rules and tables, formalized by linguists and mapped out by programmers.

At the application level, TTS can be treated as any other voice decoder: The input is ASCII data and the output is audio.

The algorithms themselves, however, do permit a wide range of customization, accepting commands that "read" affected text more slowly, insert pauses, or change voice characteristics. Several vendors of TTS systems offer a choice among male, female and even child voices, and versions for other languages as well. In addition, pitch, speed, volume, and, in some cases, prosody (the rises and falls of intonation) can be controlled through the application interface.

TTS engines are offered as Active X components, DLLs, with C programming APIs, or encapsulated components in CT application generators.

Current suppliers of TTS technology take two basic approaches to speech synthesis: One breaks down text and generates phonemes (individual speech sounds) electronically from scratch, mimicking the resonances of the human vocal tract; the other breaks down text and concatenates prerecorded fragments of actual human speech.

The first type of speech synthesis demands that we understand something about the vocal tract process that it models.

When we speak, pulses of air are sent from the lungs to the vocal chords, which vibrate at pitch frequency. These bursts of air travel on through the vocal tract (a flexible tube of varying diameters) ending at the lips. The tube/vocal tract is continually flexed as we articulate: as we move the lips, tongue, teeth and the vocal tract muscles themselves in producing speech sounds.

For each sound, then, there is an associated vocal tract shape. As the size and shape of the vocal tract changes, it selectively amplifies certain frequencies of the incoming air and dampens others. This follows the same acoustic principles as that of a flute (a tube of fixed diameter), which produces signals, if you will, of increasing frequency as its inner column of air is shortened by the flutist.

Just as each note on the flute is produced with an air column of a certain length, each voiced sound is produced with a vocal tract of a certain shape. And where a flute, being of a fixed diameter, produces a clear tone of one frequency, the vocal tract, being of many diameters, produces a sound that is a combination of many frequencies, three or four of which dominate. These dominant frequencies are called "formants," and they produce the perceived sound. It is these frequencies, or formants (and a few other parameters), that are associated with specific phonemes (or diphones or triphones, depending on the unit chosen) that are used in TTS systems.

This is why we call this first method of TTS "formant synthesis," wherein a speech waveform is generated, frame by frame. In telephony applications, these formants must work within the constraints of the usable telephone system bandwidth – 3.2 kHz.

Formant synthesis has the advantage of being adaptable to many different languages. As long as the phonemes are acoustically described and pronunciation / inflection rules added for each language, it can work in several tongues.

Japanese, for example, sounds surprisingly intelligible using phonemes described for English-speaking applications. Synthesized phonemes also take up less memory than digitized speech fragments.

The second approach to synthesized speech, concatenative synthesis, involves first recording human speech and slicing it up into phoneme segments that are stored in a table. When the time comes to generate speech, the system looks up the segments in the table phonemes, choosing those that most closely match the desired phoneme and prosody. It then glues them together and cleans up the new voice stream to match the desired output.

However, individual speech sounds vary, depending on the sounds that come immediately before or after; the "Z" sound made at the beginning of a word (zoo) may not sound (and is not produced naturally) in exactly the same way as one heard in the middle of a word such as exactly ("ig-zact-ly").

To preserve these transitions and enhance intelligibility, some TTS products store "diphones," units that attach the end of one phoneme to the beginning of the next. An even larger fragment is a "triphone," a three-phoneme string. The length of the speech fragments used vary with the vendor.

T

The latest, most advanced TTS technology uses even larger units of speech storage, such as "tetraphones" and even whole words, which are large enough to achieve an unprecedented level of natural sound quality. This is the approach taken by Lernout & Hauspie (Burlington, MA – 781-203-5000, www.lhs.com) with their RealSpeak TTS engine, and Fonix (Woburn, MA – 781-203-5311, www.fonix.com) in their Fonix AcuVoice Speech Synthesizer Software Development Kit (SDK) for Windows 95, 98, and NT 4.0.

Storing all probable combinations of phonemes for a given language occupies more storage space than individual phonemes, which can be concatenated in unlimited combinations as needed. But this is not really a problem, since CT servers have copious amounts of disk storage, speech unit databases can grow to accommodate whatever the application needs.

Like speech recognition, TTS technology has also been made better and more affordable. Advances in processor speed have made it practical to access larger and larger tables of speech units and – for low-density applications – to work without DSP resource boards through "software-only" or "native processing" implementations. A mere Pentium 120, for example, could handle up to 12 channels of L&H's TTS3000. Imagine what a 1.5 GHz processor could do. (You still need a telephony line interface board for telecom apps, though.)

Rules and Exception Dictionaries

Formant synthesized speech actually sounds better than concatenized speech if the developers have a good rules-based emulator of the language. That's the issue – how well have they modeled the language?

But whether based upon generated or concatenated speech sounds, all TTS systems are rule-based, incorporating a still-growing body of linguistics knowledge that goes back hundreds of years. To correctly read a word or a sentence, the TTS algorithm must consult thousands of such simple rules, knowing, for example, to pronounce an inner vowel long when it detects a final "e", as in the word sale.

On a much more complex level, it must be able to parse sentences (to identify subject, object, predicate) in order to know, for example, which syllables to stress in two words that may look identical: A good test sentence in this instance is "I object to that object." Also, "$100 million" should not be spoken as "one hundred dollars million." And don't forget such conundrums as "*Wind* your watch when the *wind* blows." TTS products vary in their mastery of such lexical and grammatical analysis.

Because many native and foreign words and names defy any one language's rules of pronunciation, all TTS systems consult exception dictionaries, where such words and their phonetic parameters are stored. These can be added to by the user as needed.

Generally, all text-to-speech processes are composed of these subprocesses:

Getting the Text

As its name implies, the text-to-speech process starts with a text file that may be the output of an application such as a word processor, a database, or an OCR-ed fax, for example. It may be as short as a word, keyed in via computer keyboard or telephone keypad.

Text normalization

Some initial text processing can expand and interpret certain kinds of text. A great percentage of the text we read requires our previous knowledge of special writing conventions. We have to know, for example, that Dr. is "doctor" unless we see it in an address, in which case it is read as "drive."

We know that 334-1268 is read as a phone number and not "three hundred thirty-four dash..." or that 12:15 is a time of day. The text normalization component of all TTS systems prepares text for phonetic translation by interpreting and converting abbreviations, acronyms, and deciphering numbers in context.

It breaks down all symbolic material into letters, even though at this point, these words still retain the idiosyncrasies of English (or another language's) spelling.

When the source or context of text is known, as with e-mail, a preprocessor can be used to "clean up" or normalize the text to get rid of extraneous information (like headers) and to make it more understandable.

Phonetic translation

After text is normalized, it's converted into the internationally recognized phonetic character set, first by checking words against the exception dictionary, then by applying letter-to-phoneme rules for the rule-abiding words.

This is a larger task in English than in any other written language, since (as non-native speakers can attest) the relationship between spelling and word pronunciation is least consistent here.

Analysis for Prosody

Prosody is the tune, phrasing, rhythm and emphasis that converts a group of robotic-sounding isolated words into smoothly connected spoken phrases. Here's an amusing example: "Sam struck out my friend" versus "Sam struck out, my friend."

Speech owes a lot of its perceived naturalness to its prosody. At the simplest level, this includes the lowering of pitch at the ends of declarative sentences, or the upturn at the end of a question. But prosody also clues the listener into the point of a sentence by using word emphasis, slower pronunciation, or inter-word silences.

According to Tom Morse, Director of Telecom Engineering for Lernout & Hauspie, prosody constitutes the hardest part of text-to-speech conversion, as it requires the truest "understanding" of the written text.

The prosody analysis part of the TTS process does the best it can in analyzing syntax, identifying breaks in sentences, de-emphasizing articles and prepositions, and creating an intonation pattern. Compared to the way most people read text aloud (which is more monotonic than spontaneous speech), it does reasonably well.

End users can tweak intonation patterns by inserting prosody commands directly into text. This would work in a TTS applications such as weather reporting, which could insert "Boston" and "75" into a synthesized template sentence such as "The temperature in <city name> is <degree-no> degrees."

In such cases, where the bulk of the TTS output is a known ASCII string, it is possible and practical to adjust prosody.

Phonetic Rules

In the last pass before the phonetic code is delivered to the voice generator there occurs some fine tuning and smoothing, some modification of the duration of phonemes, pitch commands, and improving the transitions between phonemes. Its output is a detailed phonetic description of an utterance, to be converted into numeric targets for the voice generator.

Combining TTS with ASR

The best TTS developers transformed their products into the front-ends for the better Automated Speech Recognition (ASR) engines.

Phonetic speech recognition systems are based on a phonetic representation of words, not the word itself. So many people want to be able to just type in a word and add it to the ASR systems' vocabulary. What they're using to do that is a text-to-speech engine.

Instead of having a text-to-speech engine output sound, the TTS program simply outputs a text file, but it's a phonetic representation of the word that gets dumped into the vocabulary of the phonetic speech recognition engine. In this way you can create a library of words for the speech rec module.

Different text-to-speech companies have focused on different marketing areas: Some companies have been tinkering with little portable devices like talking dictionaries, while Lernout & Hauspie has been working on speaking multiple languages with a single architecture.

Choosing a TTS Product

Choosing a TTS product is a more subjective process than, say, choosing speech recognition. While an ASR product either recognizes a human utterance or it doesn't, a synthesized speaker is more a matter of taste.

But not entirely: A TTS product either gets its message across to listeners or it doesn't. Test the TTS system for intelligibility by sending it text you (or a test subject) have never heard before: Do you understand it? Test its ability to normalize (to derive proper pronunciation from context) by sending it all sorts of numerical data: times of day, addresses, amounts of money. Send it information that resembles the input of your target application.

The other major measure of TTS quality, naturalness, is harder to standardize. Does the voice sound pleasant? Which product sounds more human? Friendlier?

Don't depend on your own ear, which has gotten used to the "accent" of computer speech; try it out on fresh listeners. Again, test it with unfamiliar text that typifies the input of your application.

Some TTS programs score well on naturalness but low on intelligibility. A good system achieves 95% word recognition; a careful human speaker can come as high as 97.5%, although some communication channels carrying human speech score as low as 85%.

Generally speaking, you'd favor naturalness over intelligibility when large amounts of text, like e-mail, are involved. That is because listening to a monotone for long periods can be very tiring, and also because the listener can catch more meaning from context when whole sentences or paragraphs are being spoken.

For smaller speech segments, such as phone numbers or addresses, you'd favor intelligibility; directory assistance doesn't have to sound mellifluous, but it has to be clear.

Finally, consider the platform your TTS must run on, the languages you need support for, and the preprocessor types that might be available for e-mail or other specific application areas.

In 1999 AT&T rewrote its TTS engine completely, having decided that it had taken the old Watson FlexTalk about as far as it could go. The new engine, currently called Network Watson 1.0, is already in some partners' hands.

On the business side, AT&T's ambitions for TTS are to put the technology in the network, to use it to make – not only save– money, and to make it as ubiquitous as possible. This involves partnerships with the big network hardware providers: the Lucents, Seimens, Nortels, Alcatels, Periphonics and such. Because the money in TTS is not in selling the technology itself: it's in the services that TTS makes possible, and AT&T is in the service business now.

These services, in turn, have to do with making all kinds of information retrievable through telephones, and defining "telephones" not as boring black 12-button appliances, but as anything that conveys information over distance in sound: from a cell phone with a little screen, to a PDA or a CE Windows device.

AT&T is promoting CT standards that should smooth the path of as many partners as possible. They have written interfaces to Dialogic's CT Media resource handler and to S.100.

AT&T's studies indicate that when their diphone synthesis TTS is spoken "by" a visible face (such as a Max Headroom character), they suddenly sounded much more human!

What does all this have to do with telephony? Think into the future of web-enabled call centers and pushed pages. Think of having your e-mail read to you through a screen phone or on a laptop via .WAV or .AU file, and suddenly you see where 3-D agents with synchronized lips and expressions could be used. Do you want your girlfriend's or boyfriend's face to animate your agent? Scan in the picture. Do you want to hear your e-mails in the voice of your dear departed grandfather? Give AT&T a tape of his voice and let them build a synthesizer that pronounces words his way. (Now think what happens when Disney gets hold of this.)

Looking short-term, the issues that AT&T researchers have addressed in improving synthesized speech have to do with establishing a variable scheme for choosing the size of the concatenated speech unit. The old Watson FlaxTalk had 2,700 recorded diphones. The smaller the speech unit, the lower the chance you'll need to record a new combination of phonemes. But it's also true that the smaller the speech unit, the less natural the concatenated speech sounds.

When recorded speech units are as large as words, your speech quality is excellent, but your database of possible words and way of indexing and retrieving each piece get out of hand. Unless, of course, your application is very narrow in scope, with a limited number of variables to be spoken. This is, of course, the application area we now associate with IVR.

So the work for TTS is in finding the best compromise in unit size, and in using a combination of speech unit sizes. It's also in improving the algorithms that smooth connections from one diphone to the next, and in something that appears to take a thorough engineering background in signal processing to understand, such as Harmonic Plus Noise Modeling (HNM).

To pick the human basis of its next synthesizer, AT&T auditioned hundreds of tapes of professional female voices. They whittled the contestants down and made diphone units from six of them, and then played the speech for 41 off-the-street listeners.

A new version of Network Watson appeared in 1999.

Black Ice Software (Amherst, NH - 603-673-1019, www.blackice.com), known for their toolkits and utilities in fax and imaging applications, decided to provide another piece of the unified messaging puzzle by offering their own TTS SDK.

The SpeakEasy Text-to-Speech development toolkit is available in "English and American;" a Spanish version should also be out by press time. Like all the other TTS utilities, it comes with pitch and speed control, simultaneous Multi-Channel support, and custom dictionaries of names and exception words.

SpeakEasy files include the source code for the Sample Programs in both C++ and Visual Basic Code. The SDK is compatible with Dialogic boards and Rockwell chipset-based voice modems such as Hayes and Boca. At one point, Black Ice was selling the TTS toolkit royalty-free at $1,495 for 12 ports and 1,995.00 for unlimited ports.

Elan Informatique (Toulouse, France-+33-(0)561-36-0789, www.elan.fr) offers TTS engines for US English, UK English, German, French, Spanish, Russian and Brazilian Portuguese.

T

They have three products for the telecom market. The ProVerbe Speech Platform is a hardware-based TTS resource, running on its own board, and compliant with ECTF S.100 standards. Elan Informatique also plans to conform with the S.300 standard, which is the point at which speech vendors must supply their interfaces. ProVerbe is available with drivers for Windows NT, MS-DOS, OS/2, SCO Unix, and UnixWare.

Speech Cube is Elan Informatique's host-based engine, available under Windows NT, Unix SCO, Unix Solaris and Linux. The Windows version is SAPI 4.0 compliant, and is provided with interface to Dialogic and NMS hardware. It can be used as a TTS engine or a TTS server over a network. It's available for US English, UK English, Spanish, German, and French. They also package a macro component for e-mail reading; theirs is called Dial & Play.

A SOHO or remote monitoring application might be well served by a fourth offering, a solid-state ProVerbe speech unit that takes in text over RS232 or RS485 links and reads it out over phone or loudspeaker.

All of Elan's TTS operates with PSOLA (Pitch Synchronous Overlap Add diphone synthesis) synthesis technology, making use of diphone directories collected by France telecom's CNET research lab and others.

CNET is the recognized leader in French TTS quality.

Laureate is a text-to-speech system developed at British Telecom (BT) Laboratories. It claims to be unusually natural-sounding. It also lays claim to platform independence, as it is written in C and manages its own task scheduling and memory allocation, using operating system primitives.

Laureate currently supports several accents of British and American English. They're also prototyping several other European languages, including French and Spanish.

Ask any group of application developers and platform vendors whose TTS engine they used for their project, and the majority of answers will be Lernout & Hauspie (Burlington, MA - 781-203-5000, www.lhs.com). L&H already had the telephony TTS lead years ago when they acquired competitor Berkley Speech Systems, and narrowed the field further when they purchased Centigram's TruVoice. They've also become a familiar logo to home-office consumers, via their desktop Voice Xpress dictation product, embracing the speech rec and even language-to-language machine translation fields as well. They employ more than 700 linguists world-wide, and hold over 40 patents for speech and language technology.

L&H offers the telecom developer two text-to-speech engines. TTS3000 is a software-only engine that runs on NT host-based CT platforms or on Dialogic's Antares DSP board. It comes in several languages, including US English, German, Spanish, French, Italian, Dutch and Korean. Host-based TTS also supports UK English and Japanese.

TruVoice, further developed since the core Centigram product, provides 12 channels of synthesized speech per Dialogic Antares board in English, German, Spanish and French.

L&H also offers an e-mail pre-processing plug-in as part of its SDK. This strips the unwanted header information and routing information from e-mails. Developers have obviously wasted no time putting this to use, as five or six brand new e-mail-reading services, all selling or provisioning their service through web interfaces, all responded to my call for TTS app information. Of these, four were using L&H.

L&H SDKs include four ports, starting at around $795.

When L&H's RealSpeak appeared, it was seen as a whole new approach to TTS, having a completely new TTS engine based on both concatenative technology, and equally important, on larger speech segments such as triphones, tetraphones, and even larger phoneme sequences, up to and including words.

RealSpeak TTS runs under Windows 95, 98, and NT. It requires at least a 333 MHz Pentium II, 64 MB RAM, and hard drive space of at least 1 GB.

Voice directed personal assistant services such as Webley and Wildfire use TTS to read e-mails as part of the whole personal assistant mystique. At the moment, most TTS application development centers on platforms that concentrate on reading e-mail.

These come in various business models. Most are service bureaus, but a few offer CPE platforms. All use the web for subscriber setup.

Mostly, these services either poll your existing email boxes and keep copies or retrieve e-mail headers on demand and then read whole emails to you over the phone. Web GUIs allow you to sync up your two e-mail retrieval methods as you see fit; most commonly, you'd want to leave stuff you'd only heard on the server to read later. The GUIs also allow you to filter out certain words in the header or certain senders so you don't run the meter listening to spam. Conversely, they let you choose to hear only select senders or subjects.

Third Party Call Control

When a commserver or switching device acts as a center for all calls being handled by a network of users – the network database performs a look-up necessary for the processing functions and then issues instructions to the telephone switch to enable the call to be routed to the appropriate destination.

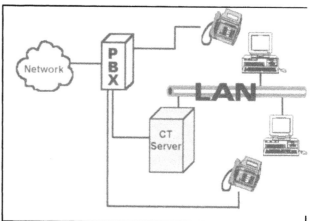

Here's an example of third party call control: An incoming call from a customer comes in accompanied by caller ID information provided by the phone company. The caller ID info goes to the network server, which looks in its database and informs the phone system where to direct the call. The server sends additional information to the appropriate PC screen or terminal to provide the user (perhaps a call center agent) with the information about the customer, what he purchased previously, what his credit is, etc. With third-party call control calls can be conferenced or transferred from desktop to desktop.

TIA

See Telecommunications Industry Association.

Time Division Multiplexing (TDM)

A digital data transmission method that takes signals from multiple channels, divides them into pieces which are then placed periodically into time slots (multiplexing), transmits them down a single path and reassembles the time slots back into multiple signals on the receiving end. Prior to TDM, analog signals were multiplexed by combining them with a carrier frequency. When there was more than one channel, this was (and is) called Frequency Division Multiplexing (FDM). FDM has basically been replaced with the digital equivalent, which is time division multiplexing. The most popular TDM system is known as Tier 1 (See T-1).

TDM's simplicity is also a source of great inefficiency. Virtual circuit / packet switch network designs such as the Internet, on the other hand, are statistical in nature and far more efficient, since bandwidth is not used until needed.

In the 1980s, computer telephony component makers were looking for ways to get the telephone data from a T-1 interface to their various application boards to build more dense and more elegant multi-resource PC-based systems that could, by necessity, handle digital telephone switching - just as the telephone company sees them - inside off-the-shelf PCs stuffed with CT processing cards.

The special ribbon cables, then and now, were the highway-like solutions for the switched 65 Kbps telephony calls to travel on. Physically, they all have wires on them for containing TDM serial data streams - the bit values (zeros or ones) represented by varying voltages on the wires. They then operate on agreed at clock speeds that put new bit-cells on each of the wires at very specific times.

To figure out what the bit value is on each of these wires, a compliant device measures the voltages every such and such nanoseconds and strictly deciphers the bit-cell cycles and when it's seeing the first of x number of these values, called timeslots.

The result is a switch fabric that allows developers to specify exactly where and how these streams (aka, phone calls) are switched and processed on the bus and within the chassis.

Such TDM buses for computer telephony boards and chassis include PEB, SCSA, MVIP, H.100, and H.110.

TML

See Telephony Markup Language.

Transmission Control Protocol / Internet Protocol (TCP/IP)

TCP/IP is the protocol suite used over the Internet. It was first proposed in 1974 by Vinton G. Cerf and Robert E. Kahn and was soon integrated into the UNIX operating system. It became the internet working technology of choice for both the government and non-government networks. It is now used for Internet access and routing.

Trunk

A trunk is a circuit that connects two switching systems (CO switches, PBXs, etc.). Technically speaking, a trunk is not exactly the same thing as a "line." For example, a line's voltage can vary, particularly in the case of a pay phone, where a line can carry +48V DC to signal keypad inhibit, +130VDC to signal coin collect, and 75VAC to ring the phone. A trunk's voltage, however, remains stable, since it is usually carrying many calls in various channels at once in a multiplexed format, making it impossible to change the voltage of an individual channel within a common wire. A digital multiplex system such as a Western Electric's SLC-96 (Subscriber Loop Carrier, version 96, or "slick") accepts 96 local subscriber lines, but only five wires may run to the CO since the signals are multiplexed. A PBX is also a switching system (albeit a small one), but its circuits to the CO are treated as lines and not trunks. To be considered as a real switch the PBX must have certain central office features. A typical PBX cannot, for example, pass long distance traffic to the rest of the public network without the CO.

TSAPI

See Telephony Services Application Programming Interface.

TTS

See Text-to-Speech.

411

Unified Messaging (UM)

Early, primitive UM systems were actually nothing more than "unified mail" systems, which had the ability to store messages of all media types (voice, fax, e-mail, paging, etc.) in one mailbox (generally Microsoft Exchange) with accessibility from either a PC or telephone. Today however, The Unified Messaging Consortium (Warrington, PA – 215-491-9966, www.unified-msg.com) defines unified messaging as "the ability to create and respond to multimedia messages with fidelity to the originator from either a telephone or PC (especially across different vendor platforms). Additionally, personal call control permits realtime control of incoming calls and call rebound with message processing."

Unified messaging should not be confused with the even more sophisticated "universal messaging" which is defined as "the ability to create any type of message and to send it to anyone without regard to the recipient's mailbox requirements." In such an ideal world you can receive or send any kind of message, any time, anywhere. It's technology that lets a user transmit voice, fax, pager and e-mail messages from a single point, and likewise to see a consolidated list on a PC or other device containing details on all types of incoming waiting messages.

UM isn't so much about the technology *per se,* as it is about using the technology to simplify your communications. So the message recipient is the key benefactor in unified messaging, since the caller doesn't know or even care if the recipient has a unified messaging mailbox.

The point of unified messaging is simplicity and ease of use. Aside from the appearance of dedicated unified-messaging appliances (and they're coming – see the entry on the Telephony Markup Language), the PC, with its powerful CPU, standard OS, and applications; its screen, mouse, keyboard, onboard multimedia facilities, and network connection, is the device best suited to handling multiple message types. As the main engine of business productivity, the PC is also where messages prompt action, and are acted upon.

Certainly few would argue that unified messaging is, in principle, a good idea. Once upon a time everyone had two phone numbers (business and home) and a mailing address. These days, corporate America has been arming employees with cellular phones, pagers, PDAs, e-mail, Internet access and other communications tools, in order to remain closer to customers and co-workers. According to a recent study, the average Fortune 1000 worker now uses an average of six communications tools.

Amazingly one of the first major UM products dates from 1984, when the late Dr. An Wang of Wang Laboratories described a product which in 1987 was brought to the marketplace. It was even given at least one "PC Product of the Year Award." Wang sought to promote the idea of unified video, voice, text, and imaging, all in one platform. The Wang phone system could tell you if you had e-mails and then play them back to you over the phone. The e-mail system could tell you that you had a voice mail but it couldn't transcribe such messages at the time because speech-to-text technology wasn't anywhere near as advanced as it is now. Faxes came into the Wang system as an image and showed up on your desktop as a computer image that could be inserted into an e-mail. Companies that used the Wang system thought it was interesting but it never became well known nor saw widespread adoption.

Unified messaging is now available in a broad range of architectures. The most familiar are extensions of conventional CPE voicemail. They live in a server that attaches to a PBX (via analog or digital station connections, and optionally via an OAI interface for call control), and talk across the LAN to other messaging servers and to clients (notably Microsoft Outlook, via

Exchange integration, but also through Lotus Notes and Domino and other popular platforms).

Every major voicemail maker has a unified messaging product or upgrade available or in the pipeline. Of those, the ones currently enjoying most-rapid acceptance are those that graft Exchange and Outlook integration onto a popular voicemail base, because they eliminate the cost of a forklift upgrade while delivering significant bang for the buck in software. Users who want on-screen review of messages, client-based account configuration, and other perks can have them without discommoding those who just want to retrieve voicemail messages and change greetings over the phone.

However, while "unified communications" technology for both multimedia messaging and Internet telephony is developing quickly, the market for such technology is just evolving and is still a bit confused. Clear definitions of the markets and real messaging needs of the users are only just now beginning to appear.

The siren's song of unified messaging has kept the whole staff at *Computer Telephony* magazine enthralled since our first issue in the Fall of 1993. The possibility of retrieving your voice mail, e-mail, and faxes from one interface has always seemed to be a proverbial Promised Land of CT, something too good to be true.

To a certain extent, it was. Like ISDN, it's a technology that's been "just around the corner" since we first cut our teeth in this business.

In fact, the technical problems of implementing UM (tying together legacy systems, for example), hazy marketing strategies, and just plain high cost have prevented unified messaging from assuming the throne of Ultimate Killer App in the CT kingdom.

But take heart, gentle reader – things are really, truly, positively, genuinely changing. The nearly decade-long gestation period of unified messaging is now over.

Deconstructing Unified Messaging

So much semantic confusion surrounds "unified messaging" that many companies are struggling to get a handle on just what it's all about. In the late 1990s The Institute for International Research (IIR) organized a famous conference in San Diego devoted to unified messaging services that was attended by over 100 delegates and speakers.

The conference was hosted by Lucent's Octel Messaging Division, itself a leading provider, and was co-chaired by Art Rosenberg, principal of the Unified Messaging Consortium, and Phyllis Huster, author of book entitled, *The Complete Messaging Handbook.*

The idea that a single, multimedia mailbox accessible from a PC or phone is all that a user needs to do unified messaging is *too* simplistic, since users still have to deal with multiple mailboxes within the contexts of where they are and what they happen to be doing when they receive or send a message. They could be at the office in front of a multimedia PC, or in a car with a pager and phone, in a hotel, or a plane, etc.

David A. Zimmer of the Unified Messaging Consortium has pointed out that the "interface" of unified messaging systems will have to be incredibly flexible based upon the characteristics of whatever access device you're using. As the devices evolve, additional changes to the interface will be made.

As Zimmer explains: "Let's say that I want to retrieve messages via a two-way pager. The screen size will probably remain small so that it can be easily carried on my belt, pocket, or purse. Unless we develop lightweight, folding screens, a 20 character display will be about the size of the screen. For reading text, it is cumbersome if the text is too lengthy.

U

Thus, the mailbox/network will have to work with the devices' capabilities to provide the message in the best format."

Zimmer takes the predicament further: "What about new message media such as video? What about new devices such as PDAs? How about the hybrid devices, pioneered by the Nokia 9000 which handles like a phone but has a screen and keyboard – what should happen to the mix of messages and how should they be handled?"

"Add to that the concept of call control (or communication control) management," continues Zimmer. "Messages/calls come into the mailbox. Rather than simply being stored in the mailbox, some calls/messages based upon my criteria would be forwarded directly to me and the device I currently have with me. The message should be 'formatted' appropriately for the device. That simple statement implies the network knows my location, the device I have, its characteristics and how to format the message. Intensive stuff."

Zimmerman concludes that, "the definition of unified messaging does not include the concept of 'one interface'. It includes 'access to one location' (actual back-end implementation of several mailboxes is possible, but to the user, it appears as one) for all message types with the proper handling of the message to meet the characteristics of the access device."

So much attention has been paid to the single "mailbox" storage issue, that everyone has ignored the user's functional needs. A better definition of UM, especially from a service perspective, is given by the Unified Messaging Consortium as, "enabling a message originator to easily create and send a message of any type, without regard to the recipient's messaging facilities."

Zimmerman himself believes that unified messaging really consists of five important elements: The mailbox, the post office, the network, the directories and the enhanced service functions (the recipient might use a follow me service where several locations are tried before forcing the caller to the mailbox).

Multimedia vs. Cross-media messaging. Another problem is that "messages" themselves are more complicated than they used to be back in the good old days when the world was run by purists and a fax was a fax and an e-mail was an e-mail.

A new-fangled "multimedia message" or "multimedia multi-part message" may contain attachments in other media: A text e-mail could have voice, data, or fax files attached, or a voice message could arrive bundled with a fax message, while "cross-media messaging" enables a message created by the originator in one medium (and mail system) to be converted to another medium for the recipient's mail facilities. Thus, an e-mail text message can be retrieved from a phone, using text-to-speech, and text from a fax cover letter can be OCRed, then similarly converted to speech.

Anyway you look at it, unified messaging is tough to do. It calls upon nearly everything in computer telephony's bag of tricks: IVR, OCR, Caller ID, voice and fax processing on the LAN, text-to-fax, text-to-speech, paging and speech recognition. The various competing UM architectures can encompass everything from mail clients to network directories, fax servers, voice mail systems, message storage servers, web servers and special administration tools.

And where does a system do this "unification" anyhow? Blair Pleasant of the PELORUS Group tells us that "There are essentially two types of UM solutions: Unified messaging and integrated messaging."

Unified messaging uses a single data location where all the different mixed-media messages are stored.

Integrated messaging integrates different independent systems, and uses multiple or separate message stores linked by a middleware-type

communications / synchronization program, employing software to integrate or tie together the different modules that capture the messages, such as the voice mail module and e-mail module, and gives the user the experience of a single mailbox.

Integrated messaging is a good interim solution for corporations that have existing investment in legacy equipment to write off but it does have some complications. Integrated systems are somewhat less flexible and more complex because of the proprietary nature of their components and in particular the requirement for synchronization, which becomes more and more complex as you add more and more different independent systems.

In particular, integrated messaging runs into trouble when confronting the issue of cross-media access and response, or when you've got different telecom devices and different kinds of communications modes and media of messaging and you need to be able to go back and forth and cross over those differences.

For example, you call in by telephone and you want to examine an e-mail. The voicemail system must now talk to the e-mail system, pull in a copy of the e-mail, do a text-to-speech conversion and read it to you. Now let's say that you listen to it and you delete it on the voicemail system – but the original e-mail is still sitting back there on the e-mail system! The voicemail system needs to communicate with the e-mail system and say "this user deleted this message, make sure it's deleted". If not, then when you come back into your e-mail you'll see a message that you thought was already deleted. That's an inconvenience, since you have to delete it again, but this time from the e-mail client.

Integrated systems also have limited extensibility – indeed, you often won't even have the kind of sophisticated notification and filtering that you get with a truly unified, standards compliant system because it must deal with individual proprietary system interfaces to each notification network that you want to use, whether it be the PSTN, a paging or a PCS network.

Also, integrated systems are sufficiently complicated so that a single expert integrator is usually called upon to help you with setting up all the integration and synchronization infrastructure, which makes you dependent upon that particular integrator.

A purely integrated solution is a compromise solution. In the case of a corporate environment where you have existing investments in old equipment, integrated messaging should be viewed as a steppingstone to a more complete unified messaging solution.

Aside from unadulterated forms of unified and integrated messaging, there are any number of hybrid unified / integrated architectures one can encounter, including:

- **Integrated desktop visual access for voice and fax messages on your PC.** No integration of voice mail and e-mail message stores or clients. Separate clients and servers for e-mail and voice mail, though voice mail and fax usually have a common store.
- **Integrated client.** Uses a single client but separate servers, with client software to integrate the media. There's no interaction between the message store servers, since the integration is done at the client level. A PC client logs into both systems and is presented with a combined interface of both stores' contents. The voice/fax server is connected to the LAN and is separate from the e-mail server, but a bridge between the clients makes it appear to the user as if there is a single mailbox or inbox.
- **Integrated server.** Consists of two separate message stores – one for voice mail and fax, and one for e-mail. You access all of your messages from either the PC or phone. The message stores are separate for voice

mail/fax and e-mail, and integration software that sits on either server is used to route and control message traffic between the two systems or message stores.

- **Synchronized messaging.** Uses synchronization between the servers (or separate message stores), as they each handle their own messages. By using a system of pointers, each message store index can be aware of the messages in the other message stores.

- **Unified messaging.** Uses a single data store for all message types and a single directory. All message types (e-mail, fax and voice mail) are actually located in the same database, or common message store. The voice mail messages are taken on the voice mail server and then moved to the common message server holding the e-mail messages. The data store holds all types of messages including those "mongrel" compound messages.

Unification at the client? Unification at the server? Should the unification be done through replication using something such as the Lightweight Directory Access Protocol (LDAP) to replicate directory information from a single database to different servers? Or perhaps the unification should be done through middleware, so the interface is simply a window that retrieves messages from various legacy systems scattered about the office?

For example, Lucent's Octel Unified Messenger uses a single server to store all message media types. That allows for simple administration with no worries about replicating or synchronizing information on different servers, and backups are easy, made right out of one directory.

Nortel's Symposium Messenger, however, uses a distributed architecture where separate servers each handle a particular type of media. The reasoning here is that herding all message types into a "tightly coupled", one server system can lead to a complete communications blackout if a hardware problem occurs. With multiple sources in a "loosely coupled" system, your voice mail could fail, but you can still get your faxes and e-mail.

But how many companies would actually pay for a complete overhaul of their voicemail, e-mail and fax systems to get a single server solution? The cost of a complete forklift upgrade to a full-blown CPE UM system may look prohibitive to those accustomed to conventional voicemail, who already have a large investment in voicemail and PBX equipment, and who are happy with their finally-stable-after-all-these-years voicemail and PBX integration. UM may be more enticing to companies if they can link together legacy e-mail and fax systems, in spite of the limitations inherent in integrated messaging systems.

So perhaps the biggest obstacle to unified messaging services are those legacy e-mail, voice mail and phone systems used both by the telephone service providers and in CPE environments.

Tying all of this stuff together requires its own Manhattan Project. And since nobody knows exactly what "unified messaging" really is, no one is going to pay an astronomical sum to achieve it under such circumstances, even though companies daring to chuck their preexisting phone system and deploy Windows NT-based communication servers find that many of them offer a terrific package of voice mail, fax, and e-mail on a single server, allowing users to have a single in-box for their messages. Some CT servers even offer unified messaging right out of the box, without needing any additional equipment.

Companies with heavy investments in legacy messaging hardware and who want a premises-based UM system will likely move towards UM only very slowly, as dictated by their current vendor's upgrade schedule, and by such unforeseen overhaul-compelling crises as NANP area code numbering changes. Startups and companies in major growth-mode

are more likely to embrace this new technology. And because the economy is seeding startups and fueling growth at an unprecedented rate, we expect lots of UM systems to "follow the money" into corporate offices over the next year.

UM Saved by the Internet?

A true unified messaging system that utilizes a single store should be a highly flexible and extensible environment. With PSTN-based unified messaging systems, there appears to be too great a cost in integrating with a variety of public network infrastructures used to give a unified messaging system its "legs" – to give it the inner connectivity of voice, paging, cellular, etc., that's needed to make it truly unified. It's very difficult to extend such a system in a standards-compliant way because you have too many different telecom interdependencies. True unification in a single system suggests that it would continue to support basic standards while new standards are ratified and seamlessly plugged into the system without encountering the many kinds of interdependencies across multiple systems that make upgrading and adding new capabilities a complicated affair.

But with the introduction of IP and the Internet, unified messaging can at last have a uniform standard foundation. Internet standards considerably help the UM crusade, such as the IMAP4 standard for e-mail retrieval. IMAP4 is a replacement for POP which adds full support for message folders, server-side searching and indexing, and partial message retrieval. Microsoft, Netscape, QualComm, CommTouch and others have IMAP support.

Another messaging transport protocol standard for voice and fax called The Voice Profile for Internet Mail (VPIM), allows voice mail and fax users to exchange messages over the Internet with users of different systems (between Lucent and Centigram systems, for example). (See Voice Profile for Internet Mail).

These and other Internet messaging standards are replacing and to some extent already have replaced proprietary system capabilities. Up until the late 1990s most e-mail addresses were hosted by proprietary systems such as Microsoft Exchange and Lotus Notes. Companies are continuing to install these systems, but now their popularity derives from the fact that such systems are now IP compliant.

The most recent generation of such systems are actually built, to some degree, on Internet standards – not completely built on them as of yet, mind you, but Internet standards are moving into the core architecture. This is significant because using such widespread standards instead of proprietary technology allows for customer premise based UM products to be replaced by services. When you rely upon standards it's no longer possible to keep technology internal and proprietary since standards engender a common body of knowledge so you now can take a product and get increased value from it in the form of a service offering.

Of course, Internet standards mean nothing in themselves unless they are actually used by everyone. Fortunately, the Internet Protocol (IP) is becoming a universal transport. The proliferation of IP in turn allows us to create for the first time a single consolidated communications network or backbone. Most of the major telecom or communications wireline infrastructures of the world (such as AT&T and MCI Worldcom) are now carrying mostly IP traffic. In fact, IP traffic overtook traditional voice circuit switched traffic on the backbone as long ago as 1996. Such networks are of course still providing voice services, various forms of data and messaging services, paging services, etc. But down at the backbone transport level many are actually running an IP infrastructure, usurping the domain formerly dominated by ATM transport.

If follows that we can expect the biggest kickstart factor for CPE UM to be voice and data convergence. As we've seen, the Internet – the substrate of convergence – offers numerous valuable facilities for messaging: The Web as a user interface, SMTP and POP3 e-mail as a reliable data-transfer mechanism for arbitrary file types, streaming media for economical delivery of voice and video to endpoints served by low-bandwidth connections, etc. But in the short term, the biggest benefit convergence brings to CPE messaging is that it lets the messaging box be a PSTN-to-data gateway .

Conventional voicemail or fax server messaging contains line interface cards laboriously integrated with analog or digital ports on a PBX. Need for this "iron" puts a hard limit on the number of simultaneous connections a server can maintain. Cost is high – ironic indeed, because installation of better messaging systems tends to increase usage, mandating higher port-counts. Five years ago, voicemail typically scaled on a ratio of one port per four employees. Today, our informants in the field tell us it's closer to 1:3, and would probably approach 1:2 if LAN-based message access wasn't reducing port/hour usage.

All this hardware and expense is mandated because a conventional messaging server has to translate between voice and data – the primal mission of all first-generation computer telephony applications. With converged communications (second generation CT) the gateway function lives outside the messaging box. It becomes a separate, centrally-managed facility, scaled directly to the enterprise's need to commute legacy voice and fax traffic on and off the PSTN. (Ultimately, we expect most PSTN-to-packet gateway functions will be taken over by telco facilities, reducing the end-user cost of convergence switchover. Everything in the local loop will be done with packets.)

Lose the gateway function, and customer premises-based UM slims way down. The messaging box becomes a simple server with DSP-based resource boards (probably running under a standard media-processing architecture, such as Intel / Dialogic's CT Media), hooked to the LAN via a high-speed interface. Scaling is no longer tied to port count – a physical manifestation locked to irreducible cost factors – but to CPU and DSP speed, chip counts, and software quality, all of which are subject to Moore's Law improvements.

And this isn't sci-fi stuff. In fact, early adopters can buy it now. For example, Active Voice's Unity unified messaging platform is compatible with Cisco's CallManager (the latest iteration of the earlier Selsius product). Chit-chat (whether from PSTN connections or across the WAN or Internet) routes through CCN directly to Unity's auto attendant, thence to extensions across the very same LAN. No integration. No media crossover. No inband signaling for supervised transfer. No fiddling around, trying to determine which MF code signifies hangup. IS and telecom managers will love this thing.

Transmedia Trends

"Everything, all the time" is a compelling idea for messaging. And today's UM architectures push the envelope, enfranchising more and more types of communicating appliance in the game. Of special interest to road warriors, and the carriers who serve them, are systems such as Bay Innovations' that fully supports the short message service (SMS) specification, enabling such nifty transmedia tricks as forwarding e-mail headers to cell phones. Most top-end UM systems now offer several types of transmedia features: for example, converting faxes to text files, using OCR, and reading them via text-to-speech over the phone. While it's easy to perceive such features as esoteric, we believe demand for them will increase rapidly as user interfaces, especially speech-driven ones, continue to improve.

Likewise, we expect user interface improvements to increase the popularity of, and demand for, other latent UM features, such as the ability to maintain multiple, outbound greetings keyed to DNIS; or have e-mails converted to faxes, and forwarded to a hotel fax machine. If those features are easy to access and modify, and if systems are designed to be fairly foolproof, people will use them, like them, and come to demand and rely upon them.

Conventional Web browsers are clearly a given for UM: it's just too easy to push messaging through a Java/ActiveX/Streaming-media-enabled browser. Indeed, it could be argued that all messaging systems are, slowly or quickly, going to converge towards the Web as a dominant means of presentation. Equally interesting are ways in which Web technology – notably XML, the "extensible markup language" – is starting to be discussed as a way of enabling such UM access via non-browser-based devices as appliances, Java-enabled PDAs, and cell phones. At the 1999 JavaOne conference in San Francisco, the must-buy item was a Java-enabled PalmPilot. Equipped with an XML compiler (a Java XML front-end is available free on the 'Net), such a device could access data from any messaging system that could present XML-tagged information according to a yet-to-be-agreed-upon industry standard.

If the industry can settle on a data-type definition (*Computer Telephony* magazine is actually participating in the work of developing such an XML interface, an initiative we call TML, for "Telephony Markup Language") (See Telephony Markup Language), the same Web page that pushes a user interface to conventional (non-XML) browsers could provide data to any other XML-capable endpoint: From a user-customized browser interface, to a third-party client app, to a PDA, cell phone, or other peripheral. Opening up messaging for access via popular client software (Outlook, Notes, etc.) was a good first step. Even more exciting is the idea of unlinking client-side from server-side development, and letting clients evolve, and grow more powerful, at their own pace.

Because convergence – and resultant gateway unbundling – offers so many powerful advantages to messaging, we expect UM sales of all kinds, and especially sales of UM CPE, to track very closely on sales of next-gen telephone equipment. UM adoption will be a little slow for the near future, but the market for this technology will go ballistic as the first decade of the 21st century gets underway.

UM Turf Wars

A related issue is "turf wars." Blair Pleasant of the PELORUS Group tells us that, "UM requires the information systems and telecom departments of end-users to work together in order to purchase, implement, and administer the UM systems. In most companies, these departments are separate, and it is often difficult to unite them... the voice mail vendors who currently market to the telecom department will have to deal instead with the MIS department. Voice mail vendors will have to find new distribution channels and methods, and will need to work with LAN resellers."

Is UM Outsourcing the Answer?

Since obstacles such as these have led to the not quite glacial-speed migration of CPE-based voice mail and e-mail systems to unified messaging, outsourcing unified messaging facilities to enterprise departmental work groups and service bureaus may become the next big business market opportunity (We can see the headline already: UM Outsourcing to Explode!).

Meanwhile, a flock of network-embedded UM products aimed at telcos, CLECs, and ISPs, are jumping into the gap, many hoping to grab the SOHO market, a subset of the road warrior population, and select businesses. Here too, the jumpstart factors are convergence and deregulation. All of a sudden, everybody's got an Internet connection and a Web browser, lots of folks have two phone lines (one for voice, one for the PC to review messages), and scads of carriers are looking for something to distinguish themselves from competitors, prevent churn, and eke out a few more bucks a month per customer. For these enterprising souls, jumping on the UM bandwagon will be like printing money.

IP technology-based UM vendors targeting service providers include –

- Traditional voice mail companies, such as Comverse, Bay Innovations (formerly Centigram), and PulsePoint Communications (formerly Digital Sound), and
- Players primarily aimed at Internet Service Providers, including Telinet, JFAX, MediaGate, Premiere, eGlobe, and MTS Advanced (MTSA). Some offer the service, while others sell the hardware and software to service providers who then sell the actual service.

UM Service Providers will include RBOCs, long distance providers, cellular and paging companies, Internet service providers, competitive local exchange carriers, broadcast and cable companies, debit and calling card companies, and others.

Even then, the real functionality of unified messaging systems are generally influenced by who's designing them, with a sort of race going on between the "e-mail" architecture camp (who treat voice and fax messages as attached files) and the "voice mail" architecture camp (who think that existing voice mail and fax systems can be modified to handle text and binary data).

The e-mail camp appears to be winning. This shouldn't come as a big surprise, since the real need for unified messaging is being propelled by the explosive growth of e-mail, and that itself is being fueled by Internet-mania. Such mass messaging activity demands effective personal message management facilities, such as message filtering, realtime message notification and delivery, and other items that fit nicely into the unified messaging world.

UM Services Markets

In terms of the markets there are, of course, always some subscribers who will immediately pay a premium for features and services when there's a clear benefit. SOHOs would be one of these categories, since it allows your home office in your extra bedroom to appear to a caller as a large business. Such people will gladly pay to have single phone number and unified messaging services.

There's another form of unified messaging which the service providers would like to popularize, since it increases their revenues or volume – the wireless markets. Almost all the wireless carriers, such as Italian Mobile, Cellular One, etc., are starting to give away a mailbox with the cellular phone, because such a service has not only become valuable to the user, but it's also a "sticky service," which means that once customers start using their mailboxes, they don't want to change their phone number or their e-mail address. This in turn sells more air time, it helps generate more tolls for them, and it helps instill customer loyalty.

Another new UM-related services market concerns advertising services, which has already come of age in the e-mail industry and is just starting to happen in the Internet telecom industry in general. Indeed, we're seeing signs (not subtle signs, mind you, but huge, flashing neon signs) that unified messaging is getting set to partake in what we've come to call "Internet Business Model Roulette" – the marvelous way in which the Internet, with its low costs of entry, negligible cost of connectivity, and audience of millions of potential consumers is making high-tech manufacturers reconsider the basic rules of money making: Namely, selling,

renting, and licensing product for a fee.

HotMail and Yahoo are classic examples of free messaging mailbox services. Indeed, all of the portal services are, deep down, basically a repository of mailboxes. Endure some advertisements and you've got your mail. Just as soap companies in the 1950s paid for the TV programming of "soap operas," so do present-day advertisers make possible free messaging services on Juno, HotMail and Yahoo – the CBS, NBC and ABC of Internet media companies.

RocketTalk (www.rockettalk.com), a 'Net-based messaging startup founded by former TTM & Associates principal Jeff Weiner, is giving Internet voicemail clients and messaging services away for nothing, riding on the assumption that he can sell advertising, because "free" draws consumers and consumers draw advertising dollars. There's a little window on the client screen, to which banner ads are pushed; OCX embedding lets users click through to the advertiser's Web-site, if they're interested. Jeff normally signs up more than 15,000 new users a week.

Meanwhile, both Yahoo! and Excite are experimenting with providing real-time VoIP clients and connectivity as a way of locking users into their portals. We figure that unified messaging will inevitably become part of this picture on a mass scale.

UM Services have an Edge

Perhaps the most savvy of all industry experts in the area of Internet-based messaging, Danny Winokaur of USA.net, has told me that "converging market and technology trends will transform today's e-mail as we know it, into a network-hosted unified messaging environment."

Winokaur notes that the first trend driving UM to the network involves an increased reliance on electronic messaging. In the early 1990s e-mail was considered something "nice to have," and corporations were half-heartedly installing their own e-mail systems. It was a convenience – if you *really* wanted to get a hold of someone, or you had information that was of any importance, you sent it by fax or you left someone a direct voicemail.

We're now getting to the point where electronic messaging is no longer a "nice to have" convenience but rather a mission-critical, essential part of a business' communication infrastructure. Indeed, some businesses rely more heavily on electronic messaging for mission critical communications than the other traditional forms of messaging such as voicemail.

Continued explosive growth in users combined with the increased reliance on messaging increases the expectations of what can be done in the messaging environment. Users now no longer tolerate the kind of "downtime" they would have years ago.

There is also an enormous need for messaging scalability as the number of users who are using Internet enabled mail or messaging platforms accelerates to massive numbers.

However, Winokaur notes that there's an increasing shortage of technical talent that's forcing corporations to outsource certain IT functions – whether it be billing systems, e-mail systems, or a variety of other technical and even customer service related functions that in past years you never thought would ever be outsourced.

Small and medium size businesses have been some of the earliest adopters of this trend, the earliest to migrate to outsourcing, since such organizations generally don't have the kind of IT resources or in-house expertise to build a dedicated messaging environment. Maintaining such an internal infrastructure requires personnel with a command of SMTP, IMAP4, POP3, Exchange, Lotus Notes, etc. Many organizations simply don't have such people and therefore have been very receptive to service

U

providers who say: "Hey, I'm going to house all the equipment for you and I'm going to give you a web interface through which you can control your mail environment, add and delete new employees, and run your system." Larger enterprises are taking longer to adopt this idea but they are starting to show signs of doing so, particularly when it comes to what one could describe as the "periphery" of the enterprise.

As its name implies, the periphery involves remote users. Messaging technology generally makes its first appearance in a company at the headquarters location where there's an adequate IT department that can handle installation, maintenance and Moves, Adds and Changes (MACs). But what happens when the company expands and a few remote offices appear scattered through different parts of the country or the world, and these remote offices are each staffed with several people, all of whom desire access to unified messaging services? It's obviously very difficult and expensive for the corporate IT people to serve those small numbers of users in geographically disparate locations, since no special messaging infrastructure was ever planned or built just for them.

Such remote users either have to suffer with a not-so-reliable, periodic "dial up" connection to the corporate message infrastructure, or they must maintain their own little remote messaging server, which can suffer from any number of problems. At this point the company will begin to seriously consider unloading their messaging problems through outsourcing, by getting their messaging from the network and not an increasingly complicated internal collection of messaging paraphernalia.

Another significant market trend that Winokaur identifies as having an impact on the future of Internet-based messaging services, is an increasingly tech savvy user population. The Internet is only now starting to mature, gaining a very sizable user population that has become familiar with technology. Prior to the Internet, of course, there was traditional networking in the form of LANs, which took hold in the corporate environment in the 1980s, so corporate users now have a good 10 to 15 years of experience using computers to interconnect with one another and to use them as a method of communicating and exchanging / sharing information.

As a result, the user population has not only grown in size but has also grown in sophistication. We now have users who understand the basics, and who now want to leverage that technology to do things that are more interesting and more sophisticated than ever before. The tantalizing possibilities for messaging relating to the Internet is driving demand for improved capabilities and upgrades which in turn continues to accelerate the development of new technologies.

Another phenomenon that has become evident is the explosion in popularity of devices such as the little phones that have screens on them and Palm devices which give you immediate access to the Internet and to other data services over a wireless network. The next generation of cellular phones and devices will have an extremely important impact on the future of messaging on the Internet because it will drive new technology development and enable a whole new breed of services that couldn't have existed until now, and most of these devices, by their very nature, are messaging-centric devices.

This next generation of handheld phone, PCS / PDA devices will be running the Wireless Application Protocol (WAP), which has gained tremendous popularity. Most new devices emanating from manufacturers such as Nokia, Ericsson, Samsung, and Motorola are going to be WAP-enabled phones that support wireless data connectivity in a standardized fashion. Such devices will therefore enable back-end service providers to have a standard method to provide services to those devices, which in turn creates a very large market given how many wireless e-mail devices are already out there. WAP will also bring commerce applications to mobile users (See Wireless Applications Protocol).

Winokaur thinks that these developments will almost completely push UM into the enhanced services or applications services arena: "Most of the messaging systems that are out there, the independent, encapsulated systems run by MIS directors at various companies, are relatively close to the edge of the network. You have a dedicated line to a company and that's where the end users are, that's the end point of the network where the messaging system resides in most cases. You're going to see those messaging systems move upstream, closer and closer to the backbone itself. The messaging capabilities and the intelligence you use when you are actually doing e-mail or doing voicemail or communicating with pagers or instant messaging, is going to move into the network infrastructure itself, onto the core backbone. All of this will enable a whole new breed of services and, of course, you then get sort of a virtuous cycle going because by moving into the infrastructure you enable new services, and new services encourage people to move their messaging into the center of the infrastructure, all of which fosters a sort of self-reinforcing cycle."

Winokaur characterizes his own company, USA.net, as a Message Service Provider, or MSP. MSPs are a breed of new companies that are getting into the business of focusing exclusively on messaging.

MSPs host mail systems on behalf of other companies and are the first step in moving the messaging infrastructure from the edge (customer premises) into the infrastructure and toward the network's core.

As Winokaur explains, "Instead of all these little independent systems, now some number of them are aggregated or consolidated together into the environment of a message service provider which can achieve a certain economy of scale. This also enables a very high degree of fault tolerance, back-up, security, and new features."

But is UM worth it?

Still, it's difficult to justify the cost of unified messaging, since there are been few if any real examples of a "time and motion study" or any kind of analysis to determine exactly how much money you save or productivity you can increase by, say, reading a fax that's just popped up on your PC screen as opposed to walking over to a fax machine and rummaging through sheets of paper.

Certainly there should be cost savings for mobile professionals, telecommuters, "occasional rovers," and even Wall Street traders who need information as fast as possible and welcome having all communications sent, received and displayed in one interface. But this "self evident" reasoning has never been stated in the kind of dollars-and-cents terms that would instantly sell a prospective customer on a system.

A study by PulsePoint Communications indicates that most direct UM savings derive from use of IP networks or even the Internet, i.e., using the Internet telephony instead of long distance for sending and retrieving voice and fax messages, and from the elimination of extra phone lines for fax or call answering voice messages while still connected to the Internet / web.

Still, productivity, responsiveness and convenience are factors too difficult to be easily quantified.

So don't expect UM to be appear way down in the consumer market yet, at least not until everyone has a cheap PC, PDA and / or TV set-top Internet access devices or new "screenphones", which could lead to a big home unified messaging market when the technology becomes goof-proof simple and dirt cheap – the man-in-the-street acceptable monthly cost range for UM service was found by a GTE study to be $10 to $15 per month; PulsePoint's study came up with an average acceptable cost of $12.37. Personally, I vote for that time-honored figure among ISPs, $19.95 per month.

Buy, Lease or use a Service?

To buy or lease a Unified Messaging System ("UMS") is a significant and expensive commitment for any business. Effectively incorporating these systems into your business is a process of selecting vendors and equipment, installing and implementing the right hardware and software for your purposes. They can also carry sizable "hidden" costs that will give pause to even the most aggressive entrepreneur. There are plenty of reasons to pass on a UM system, though some of the services appear to be reasonably priced.

But there are also plenty of reasons to continue pursuing the dream of simplified communications. Unified messaging offers such a unique opportunity for improved productivity and long-term cost savings that it's very worthwhile looking into. If you find that you or members of your staff are spending valuable time to get something off the fax, look up a phone number, retrieve e-mail, voice mail, a document, or look up a customer's history, a UMS might be just what you need. Once past the up-front costs and the initial learning curve, your productivity pay back can be both dramatic and quick.

Michael McGarry's UM Tips

Michael K. McGarry is Vice President-Operations and founder of Merlot Communications, an advanced communications systems and services company. He recommends to keep these things in mind when researching a customer premise based UM system for your business:

1. *The most critical decision you will make in the entire process is which vendor to use.* The vendor has to be sophisticated enough to evaluate your existing equipment and provide expert guidance on issues of integrating and/or replacing what you already have. A good vendor will help you discover the hidden costs of the full implementation for your organization. You have to be confident that your vendor understands your needs and priorities and appreciates the enormity of the commitment you are considering. And you should know the following things about your vendor:

2. *Longevity.* Besides knowing how long the vendor has been in business, take a look at his / her upgrade history. If you find a vendor's product has been upgraded five times in the last six months, this is probably one to avoid. Upgrades coming too frequently often indicate efforts to correct problems rather than improvements.

3. *Support Systems.* Remember, with a UMS, you essentially put all of your eggs in one basket. If the system goes down, your business can be on hold. So, check their policy on support. How much does it cost and what will you get for it? When is their service available? Do they have the ability to service you remotely? Many vendors can run remote diagnostics, troubleshoot problems from off site and even go through the Internet to send a software patch to upgrade. You want to feel confident that your system can be brought back on line and/or upgraded quickly.

4. *References.* Check references. If a vendor cannot give three to five specific business owners within your industry segment, they may not have the experience you need.

5. *Payment Options.* Does the vendor offer purchase or lease options? Can you rent equipment? Do they offer hardware and software platforms with service management for a monthly fee?

6. *Reliability issues.* Once you choose your vendor, you have chosen their hardware and software as well. Be sure they are proven and reliable.

7. *Have a good installation site for your server.* The area should be adequately open and cool, relatively dust free and secure and it should have a dedicated outlet with power surge protection.

8. *See a demonstration of the system(s) you are considering.* You want to find one that feels intuitive to you and is easy to navigate.

9. *Understand that habits change slowly.* The more you can keep of your current environment, the shorter your learning curve will be. If you introduce a radically new environment, people may be reluctant to try it. If, however, you look for a system that builds on existing processes and interfaces with your current software applications, the transition will be easier.

10. *Evaluate and plan carefully.* Think about the functions performed in your business and who performs them. Maybe you don't have to wire your entire staff. A UMS might be worthwhile only for some of your employees. And be sure to consider the needs of your "road warriors" who rely on remote access capabilities.

Remember that you always chase technology; you never catch it. So, know exactly what your needs are and what technology is best to accomplish your goals. Don't catch the technology bug. Jump in when your needs and growth require it and don't fret over each subsequent upgrade that comes along.

If you choose not to take the UMS plunge at this time, but expect to consider it again down the road, be sure the communications system you buy or have now can be upgraded later to handle UMS applications. Choose a platform now that will be a pathway towards future growth.

Active Voice Corporation (Seattle, WA – 206-441-4700, www.activevoice.com) is has more than 70,000 installations in virtually every kind of business in over 60 countries. Active Voice develops technology that helps businesses communicate better. Active Voice products are sold through a global network of independent telecommunications dealers, telephone equipment manufacturers, and computer resellers. The Company's corporate headquarters are located in Seattle, with representatives throughout the United States, as well as offices in Australia, Canada, China, France, India, The Netherlands, South Africa, Sweden and the United Kingdom.

Active Voice's Unity unified messaging server is available for Cisco Communications Network's (CCN) CallManager. Unity does unified messaging for the Microsoft Exchange Server. Using Outlook, you get single-point administration for e-mail, voicemail, and fax mail user accounts, address and distribution lists, and network configuration. You can receive, play and forward voicemail messages on the desktop.

Through standard network hardware, Unity supports direct IP connection to CCN. This means no more need for the specialty circuit voice hardware and dedicated IP-PSTN gateway ports used to connect conventional voice messaging systems to PBXs (or for that matter, to CCN, which retains the ability to link to conventional CT peripherals until such time as the whole world becomes converged).

Necessity is truly the mother of invention, particularly in a country such as Israel where international phone calls are fabulously expensive, thus spurring companies such as ArelNet (Yavne, Israel – 972-8-942-0880, www.arelnet.com) into becoming leading providers of voice and fax over IP solutions.

ArelNet's i-Tone, a Windows NT Server based IP Gateway, can do not only point-to-point voice and fax messaging, but can be a complete turnkey solution for service providers wanting to offer global IP-based enhanced services such as phone-to-phone, fax-to-fax, PC-to-phone, PC-to-fax, e-mail-to-fax, fax-to-e-mail, fax broadcasting and enhanced messaging for mobile subscribers.

Indeed, ArelNet's system is used by carriers, next-generation telecom companies, ISPs and value-added service providers. All are attracted to i-Tone since it can handle real-time as well as enhanced communications using a single hardware platform. The i-Tone Gateway can act as a gateway between the PSTN and TCP/IP based data networks.

U

There is also a Virtual Private Network (VPN) module that gives corporate subscribers many of the advantages of installing a VPN as if it were an autonomous network, but without the standalone infrastructure costs. Each company can set up its own routing and numbering plan, regardless of what other users are doing with the shared network. I-Tone's intelligent routing ensures that each subscribing company can operate its VPN with complete independence.

The i-tone Gateway PRO is housed in a fault resilient PC chassis capable of handling four E-1 / T-1s per chassis. It also has support for multiple ISDN protocols and is H.323 compliant. The Gateway consists of Dialogic boards for PSTN interfaces and an IVR module for user help and service selection. Analogic DSP-based boards are used for voice compression and Dialogic GammaFax boards are used for the enhanced fax services.

The i-Tone Gatekeeper module is the cornerstone of the i-Tone system, allowing for communication among multiple gateways along with the ability to manage one or more gateways. Normally, the main tasks of the Gatekeeper is to provide IP routing and resolving, authentication, load balancing and handling of call detail records.

The i-Tone Organization Administration and Management (OA&M) system can manage user registration as well as overall user activities, providing capabilities for Roaming and Billing, and supports different Gatekeepers with multiple zones.

The Network Management System (NMS) gives system operators real-time monitoring of all the Gateways and the status of each within the Network. With the NMS you can configure lines, monitor line activity, view processes, monitor queues and view trunk activity.

Finally, i-Tone E-Mail Gateway acts as an enhanced service center, serving as the front-end unit to the i-Tone system and handling a bi-directional exchange of messages between e-mail and voice and fax destinations.

ArelNet has joined forces for Comet Technology, a leading London-based distributor of computer products in Africa, to jointly market a complete IP messaging solution based on the i-Tone system. Comet has already installed I-Tone nodes in Europe, the Middle East and Africa, and intends to expand its network to cover the African continent, the Far East, the United States and other areas. i-Tone systems start at around $15,000.

ArelNet's ARCOM product is an integrated messaging platform that can be customized for telcos and other network operators. Based on a unified messaging module, it integrates e-mail, fax, voice, telex, telegraphy, and X.400 into a single mailbox.

ARCOM can handle millions of messages per month, and has flexible connectivity (E-1 and T-1, ISDN, X.25 and frame relay). ARCOM's modular architecture enables service providers to start with as few as one app and then expand services incrementally by adding modules as needed. Each module is designed to work with a specific service, message type, and network.

On the voice messaging side, Artisoft's (Cambridge, MA – 617-354-0600, www.artisoft.com) TeleVantage NT-based commserver gives you two options.

First, is to use the TeleVantage Client for visual access to your voicemail, identifying each caller and displaying a list of your messages on your screen. You can save, delete or forward messages, and hear them over your phone or PC's speakers. Second, you can have it send voice messages directly to your e-mail account, unifying all your messages.

Available through Artisoft's network of VARs, TeleVantage 2.1 supports T1 trunks, 48 analog trunk lines, and 144 extensions. Cost is about $750 for the server software and $115 for each trunk, station or client.

CallXpress is AVT's (formerly Applied Voice Technology, Kirkland, WA – 425-820-6000) unified messaging server. It comes in two platforms: CallXpress, which does unified messaging, network faxing, and advanced voice messaging; and the larger CallXpress Enterprise, which does UM and network faxing.

With support for up to 128 ports, CallXpress Enterprise services up to 10,000 users on a single Windows NT Server. To create a solution for a geographically diverse enterprise, you can chain additional AVT systems in an IP-based network environment to include branch offices.

For administrators, the mailbox import feature lets you add, change or delete one or more subscriber mailboxes using a single text file. This lets administrators edit en masse instead of one by one. Users can access voice, fax, and e-mail messages from a phone, desktop computer, or Web browser. Through Microsoft Outlook/Exchange or Lotus Notes you can view messages unified.

CallXpress Enterprise supports major brands of PBX telephone systems including those from NEC, Nortel, Ericsson, Lucent, Siemens, Fujitsu, and Mitel.

On all CallXpress platforms, users can access their messages by sender, by entering a subscriber mailbox number or by specifying access to messages from outside callers.

CallXpress and PhoneXpress (AVT's advanced messaging system) can be linked over analog telephone lines, and they support the Audio Messaging Interchange Specification (AMIS), an inter-industry analog protocol for connecting systems produced by different voice mail manufacturers (See Audio Messaging Interchange Specification).

An audiotext feature provides callers with commonly requested information in pre-recorded announcements, such as directions to your office, 24-hours-a-day.

Auto Directory Transfer and QuickConnect functions allow you with one key, to transfer directly to a subscriber extension from the company directory.

An Auto Message Forward, AutoCopy function let you automatically forward a copy of specified voice messages to another subscriber.

You can also receive notification of your messages at the same scheduled time and the same place each day. This notification can be to a pager as well as a telephone.

The CallXPress automated attendant is as good as any in the industry, greeting and routing callers without requiring a dedicated receptionist.

Users can choose to have message envelope information automatically announced before, after, or withheld before each message. This is configurable on a subscriber-by-subscriber basis.

An interesting "Bookmark" feature allows a user to "mark" a position in a message with a single keypad stroke, and continue to the next message. When the user goes back to the original message, the same keystroke takes them back to the marked position in the message.

Users can also maintain two personal greetings: One notifies callers that they cannot answer the phone and the other that they are on the phone.

A Call Blocking feature blocks all calls to a telephone line, putting callers directly into voice mail. This feature is available when calls are placed through the Automated Attendant.

The both CallXPress systems also allow call screening as well as Caller ID or ANI Routing. A call queuing function provides callers with the option to hold while another call is wrapped up. This feature is available when calls are placed through the autoattendant.

Taking a different approach, the Remark! Unified Messaging Assistant from Big Sky Technologies (San Diego, CA – 619-715-5000, www.bigskytech.com), turns your Lotus Notes mail database into a voice, fax and e-mail messaging system.

The Remark! server bridges your company's PBX and LAN. It includes Big Sky's Remark! Voice Server for Windows NT, unified messaging application software, text-to-speech software, and a Natural MicroSystems telephony interface board. It's available in configurations from four to 48 ports per server, and servers can be chained to build larger systems. T-1 and E-1 connections are available.

Messages are unified in the Lotus Notes mailbox, so you can access and manage them from phone or desktop. E-mail messages are read with text-to-speech technology (they use Lernout & Hauspie's TruVoice). Voice messages received from other Remark! users are played back directly, and faxes can be forwarded to any fax machine.

Black Ice Software's (Amherst, NH – 603-673-1019, www.blackice.com) NT-based Impact Voice Mail has a script editor allowing non-programmers to customize a voicemail system via a set of nodes that connect in the form of a flow chart. Each port can run a different script. The server also does voice-to-e-mail conversion in .WAV format; text-to-speech; call switching; DID inbound routing; and remote administration. The client software works with your MS or Outlook inbox, or runs with your Web browser. Impact works with both PBXs and the PSTN, and uses Dialogic, Brooktrout, NMS, or Bicom voice boards; or even voice modems with the Rockwell chipset. A four-port, 25-user version costs $2,295. Maximum size is 48 ports and about 1,000 users. Black Ice also makes fax servers and software, as well as C++ and ActiveX software development tools for building fax, voice, and imaging apps.

Meanwhile, the flagship messaging product from the Carmel Connection, Inc. (CCi) (Fremont, CA – 510-656-0222, www.cci2000.com) is CCi 2000 for Windows NT (CCi2000-NT). A voice and fax processing unit with Internet and Intranet connectivity, the idea is to use the CCi2000-NT for toll-free interoffice voice and fax messaging. And if your company subscribes to the global IP-based UniCONN Service Network, employees can send and retrieve messages for the cost of a local call from just about any location.

CCi has established a subsidiary company, United Connections Inc. (UniCONN), a global messaging service that integrates the Internet-based messaging and global telephony points of presence.

In any case, the CCi2000-NT model gives you 2 to 32 ports in two-port increments. The CCi2000-NT/NS model gives you 4 to 32 ports in four-port increments. Storage capacity is 100 hours, with additional storage optional. Mailbox capacity is 500 mailboxes, with 10,000 optional.

Carmel also offers other voice and fax processing platforms that integrate with over 50 PBXs and key systems, including those by Nortel, AT&T, Mitel, Panasonic, Ericsson, Toshiba, NEC, and Siemens.

Call Sciences (Edison, NJ – 732-494-5800, www.callsciences.com) supplies network providers with enhanced telecom and Internet networking services. Call Sciences has formed a business division, Virtel, which develops and markets enhanced communication products deliverable over global intelligent networks for corporate enterprises, wholesale distributors, and end-users. Virtel's Personal Assistant is a single number service that meshes your existing voicemail, faxes, and e-mail into a single service. All of your communications can be accessed by a single phone call or via the Web. The service can seamlessly forward calls and faxes to you even if you are out on a sales call, on a plane, or in a hotel, all within one day.

Dolphin Telecommunications, Europe's largest public mobile radio operator, has a $16 million deal to offer Call Sciences' enhanced services platform to more than 200,000 customers over its wireless TETRA network.

Playing with some leaders of the open standards-based telephony movement, CallWare Technologies (Sandy, UT – 801-984-1100, www.callware.com) developed its Callegra unified messaging app to support Intel / Dialogic's CT Media server software (CT Media is the open systems-based server software for media processing to interoperate multi-vendor apps and hardware) and Microsoft Windows NT.

Callegra lets you manage voice, e-mail, and fax messages by integrating with packages such as Outlook, Lotus Notes, and Novell GroupWise. And with all the openness inherent in the system, you get application-level features such as Microsoft Exchange server integration to unified messaging, text-to-speech and Internet telephony.

Running on Windows 95, 98, and NT-based workstations and servers, Callegra scales to 96 ports. Several digital switches and PBX/KSU vendors are supported.

Centigram's (San Jose, CA – 408-944-0250, www.centigram.com) OneView for Windows is an integrated multimedia messaging product for the desktop that runs with Centigram platforms including Series 6.3. OneView's inbox window collects all voice and fax messages in one place, so at a glance you can see sender name, the subject, the time sent, and if it's urgent. With OneView, your PC can retrieve voice, fax, and e-mail, and do text-to-speech of e-mail over the phone. Pager and cell phone notification is included, and audio PCS. The system supports 60 ports and 1,500 to 2,000 users with MAPI and VPIM e-mail systems. A four-port system with server runs about $15,000 to $18,000.

In general, the Series 6.3 platform is a comprehensive unified communications solution that enables telecom carriers and service providers to provision cutting-edge voice / data telecommunications services over wireless and wireline, and Internet-based networks. The Series 6.3 supports a wide range of software applications enabling advanced message management, administration and provisioning, call management and multimedia information management. Service providers deploying the Series 6.3 will be capable of delivering Internet-based voice, text and multimedia content to wireless and wireline phones, PCs and WAP-enabled devices.

Centigram is broadening its business efforts to focus on the burgeoning application service provider (ASP) market. Driven by two trends – the increasing sophistication of new Internet-enabled unified communications systems, and surging demand among end users for advanced messaging services – telecommunications ASPs are assuming a central role in today's Internet-based communications environment. ILECs and CLECs are finding that, in many cases, it is more cost effective to maintain unified communications offerings on an outsourced basis, and are looking to ASPs to provide the latest products and services. Centigram has close working relationships with many telecom service bureaus and ASPs, and the Series 6.3 platform enables the company to concentrate on expanding its business in this area.

Cisco (San Jose, CA – 408-526-4000, www.cisco.com) acquired the unified messaging vendor, Amteva Technologies. This will allow Cisco to offer an Outlook-style voicemail, e-mail, and fax integration on top of its IP PBX. Amteva develops middleware for enabling voicemail, fax, and e-mail messages over an IP-based network. By incorporating Amteva's middleware solution as part of its Unified Communications initiative for both enterprise and service provider customers, Cisco will make it easier for third-party developers and system integrators to add value to the open voice solution.

Prior to acquisition, Amteva marketed its Unified Messaging Plus (UM Plus), which provides such features as voice messaging over IP, fax messaging over IP, e-mail messaging over IP and single-number reach (locate a subscriber, whether mobile, in the office, or at home). The "Plus" means real-time communications; subscribers can return a call after listening to a message or create new calls from anywhere in the service.

U

Amteva's UM Plus solution is carrier class because it is built on a scalable, standards-based, interoperable, and reliable architecture designed for virtually any type of IP-based telephony app. It's also one of the few CT products that works with IP networks and leverages their ability to scale to millions of users.

Amteva could achieve this because its UM Plus solution has a distributed, object-based architecture supporting all major industry standards such as LDAP, IMAP4, SMTP/MIME, VPIM and HTTP/HTML, as well as support for centralized SNMP management and Web-based administration.

As part of a Gateserver that contains IP-based, enhanced messaging apps like those included with UM Plus, Amteva's architecture interfaces with RAS Gateways to the circuit-switched network and uses any telephone, cell phone, and fax machine to communicate with directory, media, and management services on the IP network.

Cognitel from NovCom (Nashua, NH – 603-886-1684, www.novcom.com) is a speech-enabled telephony SOHO voicemail and auto attendant Windows 98 application that costs just $49.95. Cognitel integrates voicemail with the Microsoft Exchange inbox for unified messaging and also offers voice-activated remote access to messages.

There are two models for recognizing each caller via speech verification. One model is to train the system by microphone, manually, before people in your caller list directory call in for the first time. The recognition accuracy in such cases is about two out of three – not that bad, considering the system compares your utterance of a person's name with his own utterance.

Later, when Cognitel has received at least one sample of a given caller's name spoken by the caller to the auto-attendant, an auto-training procedure cuts in. This second, "call specific," model works ever-more-accurately, each time the person calls. Ultimately, accuracy improves to between 95% to 100%.

COM2001.com (San Diego, CA – 619-314-2001, www.com2001.com) has unified messaging in its InternetPBX. But most interesting is how COM2001.com is using its newly formed partnerships to get product on the market.

COM2001.com has teamed with Sprint and Microsoft to offer complete solutions (everything) for small business customers.

The solutions, available through independent value-added providers (both Sprint partners and Microsoft certified solution providers), integrate voice and data communications services with all the hardware and software required to provide PBX, LAN, Internet, and remote access server functions. The idea is to make buying and using those solutions easy and cheap for small businesses.

"These small business solutions, enabled by Microsoft BackOffice Small Business Server and COM2001.com's InternetPBX, place advanced voice and data services within the grasp of small businesses," said Thomas Koll, vice president, Network Solutions Group, Microsoft. "Microsoft and Sprint are cooperating so small businesses can do this cost-effectively."

Comverse Network Systems (a division of Comverse Technology, Inc., Wakefield, MA – 781-246-9000, www.comversens.com) offers an Internet Call Waiting service which gives residential voice messaging and unified messaging (voice, fax, and e-mail in a single mailbox) subscribers a virtual phone line while they are online. With Internet Call Waiting, subscribers won't miss a phone call while connected to the Internet, either at home or while traveling.

Comverse's Internet Call Waiting service is activated automatically, as soon as you connect to the Internet. When a call comes in to the des-ignated phone, a message pops up on the computer screen with the name and number of the caller. You can then decide whether to accept the call, route the caller to voicemail, redirect the call to another number, play a message for the caller, or simply ignore the call.

You can speak directly with the caller over their PC, using any standards-based Internet telephony client software. You may also choose to forward the call to another number, such as a second telephone line in the house, or to a wireless phone. If you don't wish to be disturbed, the call can be diverted into your network-based voicemail box, where the caller can leave a message. You may also choose to send the caller a pre-defined message, such as, "I'll call you back in ten minutes," or even type in a custom message and have it read to the caller using text-to-speech technology.

Early in 2000, Comvers Network Systems and Siemens Information and Communication Networks demonstrated Comverse unified messaging services and a Wireless Application Protocol- (WAP) based Mobile Visual Mailbox (MVM) using Siemens' General Packet Radio Service (GPRS) Network. By leveraging WAP and GPRS, the demonstration showed how mobile users are able to instantly browse through their mailboxes and use the handset's voice capabilities to listen, reply or return a call.

Comverse's unified messaging and MVM applications enable mobile subscribers to view and manage their voice, e-mail, fax or SMS messages via GPRS. Comverse's MVM is an enhancement to traditional telephone interfaces, freeing the subscriber from the telephone's limited "serial/DTMF only" access to messages, and is available when using any WAP phone, PDA or terminal with IP connectivity.

Using MVM, voice mail attains the same superior message management qualities as e-mail. Subscribers can view the headers of multiple messages simultaneously and have full control over all the regular mailbox options (Listen, Send, Modify, Fax Forwarding, etc). Users can access a specific message without having serially to go through all the messages preceding it as with regular phones. Having selected the appropriate message, it is then either played (voice mail or e-mail with text-to-speech), shown on screen (e-mail), or transmitted (fax).

Conversa's (Redmond, WA – 425-895-1800, www.conversa.com) Conversa Messenger is a speech recognition-driven unified messaging product that manages e-mail, voice, and fax.

Conversa Messenger works with Outlook, Exchange, Microsoft Fax, and any other MAPI-compliant Windows messaging system. Messages received with Microsoft Outlook can be viewed in Conversa Messenger and vice versa. Access your system, using Messenger's conversational user interface (CUI), from a cell or regular phone to manage data and messages remotely or locally.

Supporting up to 10,000 mailboxes, Conversa Messenger runs on Windows 95 or 98. It takes up at least 50 MB of hard disk space. You also need a full duplex sound card, mic, speakers or headphone and a voice-capable Unimodem-V modem.

Coresoft Technologies' (Orem, UT – 801-431-0070, www.coresoft.com) latest CenterPoint release is the culmination of an 18-month effort to develop and deliver a uniquely extensible Windows NT-based voice and unified messaging platform. The platform is extensible through add-on apps that seamlessly integrate with the base platform to meet specific end-user needs.

Traditionally, voice and unified messaging systems have been rigid and could not be tailored to a customer's needs. With CenterPoint, customers start with a very robust base system, complete with multiple scheduled auto attendant and personal mailbox greetings, end-user configurable transfer options per greeting, and Web-based system and personal mailbox administration. The system can also be tailored to seamlessly integrate features such as IVR, ACD, and fax and messaging services.

Unified messaging modules add fax services to voice messaging, support automatic routing with DID, provide an e-mail gateway, integrate Internet-based administration, and include custom forms for Microsoft messaging. There is even a unified Call Management client that integrates with your contact manager and does screen pops. It works with Outlook, GoldMine, Act!, and PalmPilot PC apps.

Working with its own Millennium communications platforms, eOn Communications (formerly Cortelco Systems, Memphis, TN – 901-365-7774, www.eoncc.com) offers the Voice Processing System (VPS) family of voicemail, which includes four systems in port capacities from 4 to 48 ports; the Series 500, Series 1000, Series 3000 and the server-class Series 5000.

Through Outlook, Cortelco delivers unified messaging, enhancing the VPS by presenting voice, fax, and e-mail in one box. Visual call screening from the desktop, intuitive mailbox setup, and voicemail features are all available under eOn's standards-based approach.

With the VPS you also get paging options, mailbox time of day scheduling, and fax-on-demand and fax messaging and broadcast modules.

Most interesting is the Voice-Clusters digital networking module. With it you can make multiple locations run like they're on a single, virtual voicemail system. Good stuff.

Critical Path (San Francisco, CA – 415-808-8800, www.criticalpath.co.il) provides a complete range of solutions for Internet messaging and collaboration. For example, their Integrated Collaborative Applications include Messaging and Collaboration (Internet e-mail and advanced messaging for small, medium and large corporations, as well as portals and Internet service providers), Directory/Meta-Directory solutions for synchronizing and unifying resources across your corporate enterprise, web-based calendaring for corporations, service providers and portals, and resource management solutions that help corporations better schedule and manage facilities, equipment, workspace and other resources.

VoiceSupport is CTL's (Shelton, CT – 203-925-4266, www.ctlinc.com) voice processing line which includes products that satisfy the SOHO market, and small, midsize, and large enterprises. Flavors are VoiceSupport for DOS and VoiceSupport for Windows NT. Feature-wise, both are very similar.

CTL delivers unlimited mailboxes; an IVR software option; and Microsoft Exchange and Outlook integration for desktop messaging of integrated visual voice, fax, and e-mail messages. Unified messages can also be accessed by CTL's Desktop Messaging product; a GUI to be used with Novell, Lotus Notes, Banyan VINES, and the SMTP protocol.

The VoiceSupport family, both DOS and NT, includes the Momentum, with 2 to 4 ports and 2 to 4 hours of voicemail, made for 5 to 50 users; the Model 6, with 2 to 24 ports and 65 to 140 hours of voicemail, for 20 to 1,000 users; and the Model 8, with twice the port capacity of the Model 6, accommodating 50 to 4,000 users.

A 4 port 250 hour VoiceSupport NT system with 2 ports of speech recognition and 3 seats of integrated desktop messaging has a suggested retail price of $17,900.

For enterprises, organizations, ISPs and telecom providers, CyberTel (Hazlet, NJ – 732-335-0725, www.cybertelinc.com) offers its flagship product – CyberCom Server, a scalable, server-based distributed messaging platform. The general idea is to let users send, receive, forward, and broadcast all messages – fax, e-mail, voicemail or pager – from any device, to any other device, connected to any IP or PSTN network.

CyberTel's open standards "pure-play IP" platform is what supports two-way messaging. This lets you send and receive messages from all types of devices, including PCs, cellular phones, pagers, PDAs, fax machines, and the Web.

The JAVA client, called FreeCom notifies you of new messages, lists urgent messages, offers fax history, fax multicast, voice multicast, and voice e-mail.

CyberTel has formed a strategic alliance with Cybird (www.cybird.co.jp) in Japan to offer universal messaging and communications services to the Japanese wireless market in Japan (Cybird currently provides Internet content and value-added services to wireless telcos in Japan.)

With CyberTel, Cybird will let providers offer two-way Internet messaging services, tapping into Japan's 55 million wireless subscribers.

The Win Series from Digital Speech Systems (Richardson, TX – 972-235-2999, www.digitalspeech.com) can do unified messaging integration with Microsoft Outlook and Visual Fax. The system does retrieval of voice, fax, and e-mail messages. Page and cell phone notification is possible, wireless connections depend upon whether your phone systems are TAPI or TSAPI compliant. The Ethernet turnkey server supports T-1, E-1, TCP/IP and runs under Windows NT or 95. The system can expand up to 96 ports per server, with up to 65,000 users. Digital Speech will customize a system on request. Systems start at around $4,000 for a two-port turnkey system, while a four-port system with voice cards costs about $2,500.

DiRAD Technologies (Albany, NY – 518-438-6000, www.dirad.com) has found a niche by focusing on what they call interactive telephone voice and TDD (telecommunication devices for the deaf) response systems.

SoundImager, DiRAD's call processing product, provides apps, such as unified messaging, automated attendant, voicemail, audiotex, faxback, CT integration, and of course, voice recognition, and text-to-speech. UltraSilent is the patented feature on SoundImager which ensures that deaf and speech-impaired callers who use TDDs have access to the automated telephone service.

Even more helpful, DiRAD has installation and integration services, plus on-going support and maintenance. All of these turnkey solutions are made for organizations that want to improve the flow and availability of information to the people they serve, including the 29 million Americans who are deaf or speech impaired that use a TDD.

Envox Script Editor from Envox (Naples, FL – 941-793-0863, www.envox.com) lets you create a PC-based unified messaging platform giving you retrieval of voice, fax, e-mail, and text-to-speech retrieval of e-mail over the phone. The Envox product suite can be used to create IVR, UnPBX, IP telephony, contact centers, fax-on-demand, and other CT apps. It's sold throughout the United States and Europe (Envox is based in Stockholm, Sweden). Envox Flowchart scripts are created with building blocks for telephony and fax functions, speaker verification, HTML-to-fax, DHTML, ODBC database access, switching, MSMQ, e-mail, screen-pops, speaker verification, voice recognition, text-to-speech, and encryption. Programming languages such as Visual Basic, C and C++ can be incorporated into scripts using the CallDLL blocks.

There can be up to 32,000 users per system, with the number of ports depending on server hardware used. The scripting package is sold separately for $320. A four-port system for 10 clients starts at about $1,100. Additional clients cost $260 for ports, $1,760 for 50.

A result of a strategic OEM deal with AVT, OneBox is Ericsson's (Research Triangle Park, NC – 919-472-7000, www.ericsson.com) unified messaging system for enterprise customers.

OneBox lets you access all messages from either a PC, Web-based GUI or from any telephone around the world. While traveling, messages are just as easily accessible as they are in the office – via a browser or just a phone call away.

U

The OneBox telephony user interface allows a user to listen to voice-mail; you can also listen to e-mails with the integrated text-to-speech processing. The system can access e-mail from Microsoft Exchange, Lotus Notes, Microsoft MSMail, Lotus cc:Mail, Novell GroupWise or any other MAPI-compliant e-mail system.

Faxes are instantly delivered to the recipient's PC to ensure confidential material doesn't fall into the wrong hands. Outgoing faxes can also be sent directly from the users PC, eliminating the need to first print the document, then manually send it from a fax machine, and finally wait for a confirmation.

When new messages arrive, the recipient can be notified in a number of ways; by having short message service (SMS) to a mobile phone, or having OneBox call one or several pre-defined phone numbers. Notification can be activated by type of message, priority, time of day or the sender.

The OneBox GUI is incorporated as a plug-in for Microsoft Outlook, Microsoft Exchange or Lotus Notes and allows the user to manage his mailbox through point-and-click commands. The user is presented with an overview of all messages, including information on when they arrived, priority, who they're from, how long the messages are, and sometimes the topic. With this information at hand, the most important messages can be selected first, or archived for later viewing.

Esna Technologies (Richmond Hill, Ontario – 905-707-9700, www.esnatech.com) has a Telephony suite based on a client/server platform for voice processing apps. The server part is called Telephony Office-LinX (TOL) and the proprietary client software is LAN-LinX, which is what actually gives you access to visual messaging for voice and fax, visual call control with caller ID screen pops and PIM contact look ups as well as complete visual mailbox control over all single mailbox functions.

Esna's apps allow users access to solutions like desktop call control, unified messaging, network chatting, and wireless paging. The CTI server alone without any network integration will give users access to voice processing, call routing, fax on demand, fax mail, and fax server functions, multi-tenanting support for a corporate suite environment, and site networking through AMIS. An online Wizard simplifies PBX and key system integration.

Another client app offered with its TOL server is a GroupWare solution that integrates voice and fax services into any MAPI compliant SMTP/POP3 mail server. Through MAPI function calls, Telephony Office-LinX lets you access your voice, fax, and e-mail directly from your regular e-mail package (such as Outlook). Telephony Office-LinX will also tag the caller ID information to the message so you know where the voice or fax message came from.

You can forward voice and fax messages directly over the Internet just like e-mail. You can also send broadcast voice messages and fax messages over the Internet and have the messages delivered directly to a desktop PC instead of a fax machine. You can even send messages over the Internet from remote locations over the phone.

The package needs a Windows PC with a Gammafax 200 CPI or Wildcard (Puredata) fax card.

The e-voice3000 commserver from e-Voice Communications (Sunnyvale, CA – 408-991-9988, www.evoicecomm.com) rolls out the PC-based phone system that includes unified messaging, for small to mid-sized organizations.

The e-voice3000 system also delivers features such as caller ID, browser based call and message control and administration, DID, auto attendant, IVR, e-mail integration, and voice-over-IP.

Microsoft Outlook can be used to forward voicemail or faxes to any e-mail address. Plus, e-Voice does voicemail messaging over Internet, and

e-mail to voicemail via text-to-speech. The PBX board provides 48 ports and can be expanded to 144 ports.

Excel Switching's (Hyannis, MA – 508-862-3000, www.xl.com) EXS Media Gateway is a carrier-class IP telephony media gateway supporting up to 3,840 VoIP ports in a single seven-foot rack and 15,360 VoIP (EXNET system) in a NEBS-compliant, fault-tolerant architecture. With the combination of the EXS platform and Excel application development partners' enhanced services, carriers and service providers can IP-enable their telecom networks and unified messaging apps.

Figment Technologies (North York, Ontario – 416-499-1818, www.figment.net) fronts UniExchange to let you have your e-mail read to you at the office, at home, on some deserted tropical island. You just call up the server and it reads your e-mail to you.

But that's not all. You can remotely forward your messages to any fax machine; view faxes as e-mail attachments; print e-mail to any fax machine; or convert voicemail into e-mail attachments. All messages include a time stamp and optional caller ID. And Uni-Exchange's auto attendant greets callers in a variety of languages (some of which, we presume, they'll understand).

The server is compliant with Internet standards like SMTP, POP3 and IMAP4. And UniExchange handles and stores all voicemail messages using the multipurpose Internet mail extensions (MIME), meaning most of the popular Internet mail software apps will also be able to handle your voice and fax messages.

Portico from General Magic (Sunnyvale, CA – 408-774-4000, www.generalmagic.com) is a virtual assistant monthly subscription service. Portico lets users access their e-mail, voicemail, calendar, address book, news, and stock quote information via any telephone or leading Web browser. Once you sign up and your account is activated, you will be assigned a personal, toll-free number that gives you access to the system.

Sign up on www.generalmagic.com. Once you've done so, a Portico customer service representative will contact you to introduce you to your virtual assistant service and walk you through the functions, which takes approximately 20 to 30 minutes.

Portico is speaker independent (it uses algorithms from Nuance). You just talk to your virtual assistant in normal, conversational English and Portico will understand. Portico can read you e-mail to you over the phone via text-to-speech, and you can record your voice response and then forward it to the original sender as a .Wav file attachment. All the recipient needs is a RealAudio plug-in to hear your response.

You can also access your voicemail as a .WAV file, and you can access the Portico service through the Web by going to the Portico Web page (www.portico.net) and entering your personal account and passcode information.

Glenayre Electronics' (Charlotte, NC – 704-553-0038, www.glenayre.com) MVP systems let service providers offer enhanced and value-added services from a single platform; so you can stay competitive in today's crammed market.

With Constant Touch service from the MVP system, your subscribers can combine all of their phone numbers (business, cell, pager, fax, home, etc.) into one number. With the unified messaging service, they can control and manage all of their mailbox messages, no matter the format.

"Meet me" services notify a subscriber's pager that a caller is on hold. "Find me" services automatically find a subscriber by dialing different devices in any order they choose. With voice dialing, your subscribers can navigate and place a call from their mailbox using a wireline phone or mobile handset by simply speaking a pre-programmed name, telephone number or command.

At the high-end, the MVP 4240 accommodates up to 240 ports on a single node. And the MVP 4240 provides CPU and clock redundancy, along with redundancy for other critical components of the system: voice disks, system disks, power supplies, digital interfaces (T1, E1) and SCSI disk controllers. The platform supports SS7 and interfaces with all standard network configurations.

GTE Internetworking (New York, NY – 212-425-0572, www.bbn.com) turned to Telcordia to help it deliver the rollout of its nationwide IP-based unified messaging service. The carrier-grade, IP-based UM service is said to be the first nationwide service to combine voicemail, fax, and e-mail in one central repository.

The solution Telcordia helped GTE develop is a scalable, protocol-driven, standards-based architecture deployed over GTE's high-speed, high-capacity, private fiber network, the global network infrastructure (GNI). The wholesale approach frees service providers, Internet service providers (ISPs) and competitive local exchange carriers (CLECs), from the worries of installations, configurations, capacity limits, and upgrades.

For GTE's UM service, Telcordia provided and integrated Amteva's Unified Messaging Plus software. Telcordia also served as prime contractor for the hardware and messaging components of the solution, helping to integrate Software.com's directory and message servers, as well as Real Audio for audio streaming and Sun and I-Bus hardware. GTE also worked with Cisco for the network infrastructure and EMC for high reliability disc storage.

Hewlett-Packard (Palo Alto, CA – 650-857-1501, www.hp.com) has a messaging-solutions portfolio for e-mail, phone, fax, Internet, and cell communications for small to large enterprises and service providers. HP uses its own OpenMail and HP Smart Internet Messaging, with products from Microsoft, Software.com, and Nortel to enable integrated messaging services for phone, computer and fax appliances.

For service providers, HP is integrating its Smart Internet service-management technologies with Software.com's InterMail messaging application. This lets service providers monitor and manage messaging services, quantify service delivery, and mine data for growth and planning.

HP's Microsoft Exchange messaging solution for enterprises combines HP NetServer systems, HP OpenView ManageX, Windows NT, Exchange, and Smart Plug-In for Exchange to give you manageable messaging apps.

And with the HP OpenMail 6.0 messaging/collaboration solution, HP firms up its UNIX and Linux messaging offerings. OpenMail 6.0, based on Internet standards, lets enterprises implement scalable and functional corporate e-mail client support. As a matter of fact, with HP's Microsoft alliance program, OpenMail is the only non-Microsoft server supporting Outlook's collaboration functionality.

By working with third-party unified messaging companies, IBM's (Research Triangle Park, NC – 919-543-5221, www.ibm.com) Lotus Notes division is enabling the market for UM products that work with its Domino messaging platform. Lotus realizes that it's no longer simply a battle of "e-mail versus e-mail" and that companies need unification of voice, e-mail, and fax for access from the desktop or mobile devices.

For example, Big Sky Technologies works with Lotus to help define technology standards and APIs by which UM functionality can be more effectively integrated within the current Domino messaging architecture. With the newest release of Remark! Unified Messaging Assistant, Big Sky Technologies has gained a lot of attention by delivering UM solutions to both small and large Notes-based organizations to enhance their current Domino messaging environment. The Messaging Assistant, using a single data store architecture, turns the Notes mailbox into a single source

for voice, e-mail, pager, and fax messages.

A Remark! Unified Messaging Assistant bundle pricing for a four port entry level system is $9,995 which includes the Remark! Voice Server for Windows NT, Remark! Unified Messaging Assistant application, Natural MicroSystems telephony interface board, text-to-speech licenses, and one year maintenance and telephone support.

Interactive Intelligence's (Indianapolis, IN – 317-872-3000, www.inter-intelli.com) Enterprise Interaction Center (EIC) is a premier Windows NT/Dialogic card-based phone system that does UM as a sideline while it takes over the PBX, ACD, Web, and IVR functions of a call center.

Anyone on the system with a Microsoft Exchange account can now use it for voicemail and fax as well as e-mail. Workstations run Microsoft Outlook as the e-mail client. E-mails bearing voice messages show envelopes plus speaker icons; e-mails bearing fax images show an envelope with the word "fax" next to it. The Subject field shows the sending fax number as well. All messaging media are stored on the Exchange server.

You can attach a text message to a voicemail, to perhaps elaborate on the voice message or give instructions. Interactive Intelligence includes absolutely everything in the concept of UM, including paging and Web-based chats.

Intersis Technologies (Atlanta, GA – 770-980-6615, www.us.voixx.com) cultivates VoiXX, a unified messaging solution for Microsoft Exchange environments, and all related Microsoft BackOffice components. The enhancement is in the integration of the desktop phone and the PC for voice messages; and the architecture of a virtual messaging network across the LAN or the Internet via a Web browser.

Messages are viewed or heard in Outlook, Exchange or a thin client. Or you can listen to your e-mails using text-to-speech. This is where the integration of the PC and phone come in handy.

With VoiXX Web you can set up a company directory on your Web site or HTML portal for Intranet or public Internet call through. Meaning your portal is voice-enabled, so you can leave voice messages for anyone in the directory. Of course, you can hear messages left for you as well. In all, the virtual messaging network. VoiXX is a stand-alone voice messaging infrastructure, with or without a PBX. Cuts down on toll charges too.

Inter-Tel (Chandler, AZ – 602-961-9000, www.inter-tel.com) is doing something interesting with its unified messaging solution, Axxessory Talk Central. It's available in three configurations to individualize for customers.

Axxessory Talk Central Level 1 displays all messages within the customer's e-mail program, including voicemail messages, faxes and traditional e-mail.

When an Axxessory Talk mailbox receives a message, the message can be forwarded to e-mail, copied and forwarded to e-mail or remain at the fax machine or voice mailbox. Voice mail messages come across as .WAV files; fax messages as .Tif files.

Axxessory Talk Central Level 1 supports e-mail systems based on three protocols: MAPI used by Microsoft Mail or Microsoft Exchange, vendor independent messaging (VIM) used by cc:Mail or Lotus Notes, and SMTP/POP3 commonly known as Internet mail.

Axxessory Talk Central Level 2, which only supports MAPI based e-mail systems, such as Microsoft Exchange client, provides all the functionality of Level 1 and more. And Axxessory Talk Central Level 3 provides all the capabilities of Level 2 plus outbound fax.

One of the leading unified messaging services, JFax.Com (Los Angeles, CA – 888-438-5329, www.jfax.com) brings all of your voice messages and faxes to your e-mail inbox. Once in there, you can then manage them by computer or by phone.

JFax.Com gives you a personal phone number. When a fax or voice

U

message arrives, it is converted into an e-mail message and sent immediately to your inbox. Click to view your faxes or listen to your voicemail messages if around. Or, if away from your computer, dial a toll-free number to check your messages.

Cost is only a onetime service activation fee of $15 and a monthly service fee of about $12.50 depending on the scheme you choose. The toll-free number to access your e-mail inbox by phone runs $0.25 a minute. For sending faxes in the United States, they charge $0.05 per fax page.

Key Voice Technologies' (Sarasota, FL – 941-922-3800, www.keyvoice.com) Corporate Office NT and Small Office NT are Windows NT-based systems offering advanced auto attendant and voice messaging with an enhanced feature set that adds visual call Management (VCM) and unified messaging, as well as text-to-speech, e-mail reading and fax-from-the-desktop, along with 60 other standard features.

With VCM, users can view and modify their mailbox setups from their Windows 95 or Windows NT PC desktops. Users can manage mailbox messages, record, activate, and manage mailbox greetings; even set up message delivery and pager delivery numbers and time schedules, all from VCM's graphical screens. VCM also gives you complete inbound call control, through screen pops. You can decide to redirect each call, record or queue the call, screen the call, or play a certain greeting to the caller.

These upgradable systems are available in 4 to 8 port configurations with Small Office NT, and 4 to 64 port configurations with Corporate Office NT. All Key Voice NT-based systems provide users with up to either 100 or 10,000 boxes.

Lucent Technologies' (Murray Hill, NJ – 908-582-3000, www.lucent.com) www.messenger is a Web-based tool that lets voice messaging users manage and prioritize voice, fax and e-mail messages through the Web.

Users of www.messenger can access their voice, fax, and e-mail messages via the telephone or multimedia computer. The system is based on industry standards and does not need any desktop deployment. Users can quickly access this application over the Web and immediately view all the messages in their mailbox - be they voicemail, e-mail or faxes.

Lucent's www.messenger is accessible via common Web browsers, and can be used in Windows-, Macintosh-, and Sun UNIX-based computers, and can be deployed on Lucent's voice messaging servers across a wide range of PBXs.

Lucent defines unified messaging as a set of capabilities rather than a single product. As a key element in Lucent's overall unified messaging strategy, www.messenger is part of Lucent's Visual Desktop Messaging suite of unified messaging solutions, which also includes Visual Messenger, for Octel Messaging server users, and Message Manager, for INTUITY AUDIX server users.

To implement www.messenger, customers must purchase www.messenger client software licenses. Depending upon the configuration and number of seats purchased the average price per user ranges from $20 to $80. To find out more, go to www.messenger.com.

MacroVoice (Boca Raton, FL – 561-994-9781, www.macrovoice.com) sells several models of voicemail systems to fit your needs. A stand-out, the MVX 2000 Series of voice processing systems can operate as stand-alone units or can be connected to a Novell or Microsoft LAN. When connected to a LAN, the MVX 2000 offers CT capabilities and unified messaging.

Running under the Windows 95/NT platform, the MVX 2000 integrates with most key and PBX systems. As your port requirements grow, simply install additional voice cards and software.

It's capable of handling up to 64 ports and over 100 hours of message storage. Users view calls using the MVX 2000 Personal Call

Manager. Text-to-speech is available for reading of faxes and e-mails. And let's not forget speech recognition, which is dabbled in.

iPost for Edgecommander from MediaGate (San Jose, CA – 408-248-9495, www.mediagate.com) is part of MediaGate's Universal Communications Software suite. It is an Internet-based UM application for use on MediaGate's product line of Windows NT Universal Communication Servers; it lets service providers offer universal voice, fax, e-mail, and page messaging services to their customers. iPost gives users a single mailbox for all of their voice, fax, e-mail, and page messages, and makes them universally accessible via phone, fax, e-mail or Web browser. iPost uses media conversion technologies that let you retrieve messages with multiple types of devices. Notification and forwarding of messages are provided through a variety of media types and devices.

iPost comes in two versions, the iPost Universal Box and iPost Universal Courier Software. It's based on a distributed architecture enabling systems with unlimited scalability. Capacity can be increased by simply adding the appropriate MediaGate Universal Communication Servers to the network. Use of this technique allows distribution of service horizontally between multiple MediaGate systems without a single point of failure. iPost is compatible with SMTP and IMAP4 e-mail servers. For voicemail systems, iPost supports the voice profile for Internet messaging (VPIM)

A 96-port, 3,000-user server costs about $44,000.

MediaSoft Telecom's (Montreal, Quebec – 514-731-3838, www.mediasoft.com) Office Telephony 2000 has an integrated suite of application templates (applets), which include IVS FrontDesk (auto-attendant), IVS HelpDesk (call center with screenpop), IVS UniMail (unified messaging) and IVS TeleCalendar (appointment scheduling).

IVS UniMail lets you leave and read multimedia messages. Voice messages can be played on your PC from Outlook (messages are stored on Exchange server). And you get new mail notification by voice call, e-mail, or pager. MediaSoft uses multiple greetings in different languages and text-to-speech to read e-mail over the phone.

MicroCall's (Ra'anana, Israel – +972-9-760-1193, www.micro-call.com) U-Mail NT is a unified messaging system for voice, fax, and e-mail. The U-Mail Server runs 2 to 120 ports per system, and integrates voice and fax mail features with popular e-mail programs. You can retrieve, view, read, play, hear, print, compose, record, edit, reply, forward, and send all your messages, anytime from anywhere, on the road or at home.

U-Mail's remote filter module lets you set pre-defined filters and message scheduling. By defining parameters such as caller, keyword, and events; you decide which messages you want to receive, when and where, via telephone or fax.

U-Mail runs with MicroCall DSP2000 hardware boards for voice and fax, providing one integrated hardware and software solution platform.

Mirapoint (Cupertino, CA – 408-517-1300, www.mirapoint.com) offers infrastructure solutions for e-mail outsourcing, premise equipment, and co-located services, to midsize and large corporations and ISPs. Mirapoint's Internet messaging appliances are standards-based devices providing high uptime, scalability and simplicity, giving them significant advantages over general-purpose servers.

Mirapoint's M100 and M1000 are e-mail appliance servers - function-specific, standards-based, application servers for e-mail. Using embedded operating and application software combined with industry-standard hardware, these "black boxes" were built to off-load dedicated services and to co-exist with other application or network storage servers. The Mirapoint appliance server works seamlessly with any e-mail client that supports Internet-standards-based protocols: SMTP, POP3 and IMAP4, giving independence to enterprise users that may prefer one user interface to another.

Using Mirapoint's highly flexible messaging platform, service providers can incorporate additional services beyond e-mail. Mirapoint's platforms can be deployed rapidly, and managed through delegated administration services, providing unified messaging solutions for the enterprise that can be quickly integrated with edge devices such as VPIM/FPIM gateways. The Mirapoint messaging appliance can pave the way to offer a cost effective unified messaging infrastructure that blends IP-based e-mail, voice, and fax services into a single, secure communication tool.

Mitel's (Kanata, Ontario – 613-592-2122, www.mitel.com) Mediapath is a Windows NT based commserver which provides an integrated suite of telephony, unified messaging and CT services for up to about 100 users. Mitel's new OnePoint Messenger software was the first integration with the company's Centigram (now Baypoint Innovations) acquisition. OnePoint Messenger users can access voicemail, e-mail, and faxes from any telephone or Internet Web browser.

MTT (Halifax, Nova Scotia, Canada – 800-688-9611, www.mtt.ca) offers up another one of those one-number services, MTT Unified Communications, that hopes to change the way we communicate. MTT Unified Communications lets you be contacted through one number for all existing telephone and fax numbers. Retrieval of voice, fax, and e-mail messages can be managed through one central location.

Two ways to get messages – one, from any touchtone phone; two, from your personal unified communications Web page setup by MTT.

On your personal Web page, you simply log in to the MTT Unified Communications Web address, enter your personal number and your secure passcode. The quantity and type of all your messages is shown on one screen. A sound card in your computer lets you listen to your telephone messages over the PC.

Fax messages stored in your mailbox can be routed to another MTT user, sent to another fax machine or printed on your local or network printer. You can choose to have your faxes sent directly to your desktop as an e-mail attachment (.TIF image) or pick them up from your Web inbox.

NEC (Irving, TX – 972-582-6000, www.nec.com) has two unified messaging products – Unity 2.1 and the existing NEAX Mail AD-40.

The NEC Unity 2.1 is an Active Voice, Windows NT-based unified messaging solution. When used with the NEAX 2000 or NEAX 2400 phone systems, Unity 2.1 is installed on a NEC server connected to the switch. This lets users access and manage their e-mail, voicemail, and fax messages from the desktop PC, a touchtone phone or over the Internet. Messages coming into the PC are displayed in Microsoft Outlook's Inbox.

The NEAX Mail AD-40 Voice Processing System mixes four areas: Voice messaging, fax integration, e-mail integration, and call management. They say, "by giving individuals complete control over telephone, fax, and e-mail messages, NEC provides subscribers with the freedom to access information at any time."

Nitsuko America (Shelton, CT – 203-926-5400, www.nitsuko.com) offers NVM-Desktop Messaging for the NVM-2000 Voice Mail system. NVM-Desktop Messaging consolidates voicemail, fax mail, and e-mail in one integrated desktop environment and allows the mailbox subscriber to operate voicemail functions through a PC. Message functions such as recording, listening, forwarding, and saving can be performed with the click of a mouse. NVM-Desktop Messaging also provides chronological lists of new, saved, and held messages. Message information also includes the length of each as well as the date and time each was received.

Mailbox subscribers can also access e-mail via NVM-Desktop Messaging. The mailbox subscriber can read, send, reply to, and forward e-mail messages from the NVM client. Compatible e-mail packages include Microsoft (MAPI), Novell (MHS), Banyan Vines, Lotus Notes Mail and cc:Mail (VIM).

NVM-Desktop Messaging is compatible with client operating systems such as Windows 3.1/3.11, Windows 95, Windows NT and Windows for Workgroups.

Nokia Intelligent Applications (Waltham, MA – 781-487-7100, www.telekol.com), a company formerly called Telekol that was acquired in 1999 by Nokia, is committed to integrating all of its products to the Web and leveraging enhancements the Internet offers. One of the latest products to make the jump is Telekol's signature unified messaging application IntegraMail.

The Web-based version, IntegraNet provides access and management of voice messages, e-mails, and faxes stored on a Microsoft Exchange server. IntegraNet is offered on Telekol's IntegraX Enterprise platform. It uses streaming audio, based on standard protocols, and works with Microsoft Outlook Web access.

With access from the phone you listen to voice messages, hear e-mails through text-to-speech technology, and forward faxes to a fax machine convenient to the caller. With access from a PC, you listen to voice messages if your PC is sound-enabled and read e-mails and faxes.

Nortel's (Irving, TX – 972-650-9000, www.nortel.com) CallPilot is a unified messaging solution that takes the core call answering functionality of Meridian Mail, adds speech-activated messaging (speech recognition), and also throws in standards-based unified messaging that supports many e-mail programs (including MAPI and IMAP4 Web clients. Also supported are enhanced management interfaces such as the Nortel Application Builder, letting the user create voice processing apps. A Windows 95, 98 or NT GUI permits administration and reporting.

MessageASAP is a free Internet messaging service developed by startup OfficeDomain (Austin, TX – 512-499-1561, www.officedomain.com). MessageASAP gives users access to their existing e-mail, voicemail, and fax messages combined into a single Internet-accessible location.

MessageASAP is available as a free download from www.officedomain.com and from leading Internet download sites including www.download.com, www.tucows.com, and www.sba.gov.

Onebox.com (San Mateo, CA – 650-356-1200, www.onebox.com) provides a free Internet-based UM service offering voicemail, fax, e-mail, and paging for the Web and telephone. Based on patent-pending technology and a pure IP infrastructure, Onebox.com services can scale to support millions of consumers worldwide.

The company generates revenue via advertising and revenue sharing deals with sites. Onebox.com's next step will be to offer message functions to online communities, letting users set up their own message services.

Panasonic (Secaucus, NJ – 201-348-7000, www.panasonic.com) sells the TVS75, TVS100, and TVS200 Voice Processing Systems for digital integration with Panasonic KX-TD1232, KX-TD816, and KX-TD308 telephone systems.

An auto attendant answers incoming calls and routes callers to the appropriate extension. Each extension can be set up with options to screen calls, send calls directly to mailbox, and even have calls forwarded to another extension, an outside phone, a cell phone or a pager. Fax detection recognizes an incoming fax and routes it directly to the fax machine – no dedicated fax line needed. We hear, when a Panasonic TVS Voice Processing System is linked to a digital phone system your life becomes much easier.

PhoneSoft (San Diego, CA – 619-618-1900, www.phonesoft.com) was founded by the original developers of Lotus PhoneNotes, and is a

U

telephony module that sits on top of Lotus Notes, which enables you to tap into a Notes database through a touchtone phone, and lets you access, create, forward or edit Notes documents and play them over the phone through text-to-speech.

PhoneSoft's fourth generation Notes telephony solutions run on Windows NT, support all versions of Lotus Notes and Domino, and are compatible with all telephone systems and PBXs.

Their Unified MailCall application turns your existing Notes environment into a complete unified messaging system. To outside callers, Unified MailCall sounds like a standard voicemail system. But Unified MailCall can read any configuration information from any Notes Name and Address Book, and send voicemail to any Notes mailbox. Voice messages are simply Notes documents with voice recordings attached. Your mailbox can now become a "unified mailbox" containing all messages including e-mail, voicemail, faxes, and pager messages.

Orchestrate E-mail by Phone from Premiere Technologies (Atlanta, GA – 404-631-7470, www.premtec.com) is a service that lets you manage and access all of your voicemail and e-mail messages from the phone (text to speech is used for e-mails) or the Internet.

They offer flat-rate pricing of $19.95 per month with unlimited local access. No hardware or software is required and this Web-based technology runs on Premiere's own global network consisting of more than 4,500 cities in 12 nations.

For developers, Pronexus' (Carp, Ontario – 613-271-8989, www.pronexus.com) VBVoice is an ActiveX development toolkit for Visual Basic that provides an environment for creating IVR systems. These include unified messaging, voicemail, info hotlines, ACD, talking classifieds, order-entry, help desks, fax-on-demand, chat lines, telemarketing, call-back, auto-attendant, database IVR, patient reminder, surveys, and message forwarding. So as far as "enhanced" goes, you can do anything.

They say you can develop and maintain CTI apps up to 10 times faster than with any other toolkit. VBVoice 3.5 supports up to 192 channels of IVR on a single PC. And Workgroup Ready VBVoice enables custom client-server computer telephony apps that can automate user interaction and office workflow.

PulsePoint Communications' (Carpinteria, CA – 805-566-2000, www.plpt.com) next-generation enhanced services solution for telcos is a new kind of enhanced services platform called the PulsePoint Enhanced Application Platform (EAP), a carrier-grade open-system EAP based on Windows NT Server and Dialogic QuadSpan boards. It uses off-the-shelf hardware and software and the PulsePoint Middleware Tools to ensure scalability and availability and enable rapid application creation.

The system is engineered for reliability by incorporating technologies such as Microsoft Windows NT Server and Cluster Server using mirrored RAID storage. The PulsePoint Enhanced Application Platform will scale from 96 to 1920 ports in increments of 24 ports. The first central office deployment of the PulsePoint Enhanced Application Platform is being completed by GTE.

PulsePoint Messaging is the first available application for the platform. It's a NextGen messaging solution that consists of four components, targeting large enterprises, medium-sized businesses, small businesses, and the home office and consumer markets.

Unisys Corp. has declared that it will buy PulsePoint in a $100 million stock swap.

RocketTalk (Fullerton, CA – 714-449-8702, www.rockettalk.com) offers a number of major feature upgrades and other enhancements for its flagship product, RocketTalk. Of note, the newer version 1.2 includes a nice user interface, an enhanced player utility, a new Web-based mes-

sage retrieval service and sent- or received-message confirmation.

RocketTalk is free from www.rockettalk.com. It's downloadable. It's a software client that lets subscribers send voice messages to other RocketTalk users, and to e-mail accounts. Basically, anyone can send and receive messages.

The player utility, RocketPlayer, is sent with any RocketTalk e-mail message. Messages look like a single file attachment in a receiver's inbox. For marketing's sake, RocketPlayer includes a hotlink direct to the RocketTalk Web site where the product can be downloaded.

Of course, you can view and play your messages from any computer with Internet access, also thanks to RocketPlayer. You even get message histories.

StarTouch International (Snellville, GA – 402-996-8525, www.startouch.com), formerly InTouch, was originally in the long distance services business. In July 1996, they entered into the unified messaging business. Several companies claim to have electronic assistants, but StarTouch's Electronic Secretarial Administrator (ESA) is quite sophisticated. It has a switch-based installation located in Omaha that takes care of call screening and forwarding, fax-on-demand, fax and voice broadcasting, and Internet integration.

Each ESA system has its own personal toll-free number. Your callers have no PIN or code numbers to dial. All features are bundled together for you in one telephone number. Interestingly, ESA is not restricted to any number of lines. Whether you receive one call or 30, ESA handles them all. ESA sends you the calls you want, and sends the rest to voicemail. You could fax broadcast to 500 recipients while you're on an eight-way conference call and others are listening to your audio advertising.

TeleSynergy Research's (Santa Clara, CA – 408-260-9970, www.telesynergy.com) TeleUMS is a tough unified messaging server. It integrates with TeleSynergy's TelePCX phone system or on its own, delivering on TSR's "take only the piece(s) of the pie you need" philosophy.

By integrating a standard SMTP / POP3 e-mail server, TeleUMS handles all the internal and external messages with intelligent routing capability. TeleUMS supports both dial-up and leased line connections.

Incoming Mail Alert pops up on the screen to tell you messages are waiting. An inbox lets you check your mail (voice, fax, e-mail), and create folders to manage them. Check voice and fax messages remotely with a phone or fax machine. Forward and broadcast voice and fax messages.

The TeleUMS also supports server-to-server connections via Internet. This means you are able to create your own global network; so you can send a fax to branch offices through the Internet, and then send these faxes through local PSTN (public switching telephone network) to your customers. Saves money. And even voice messages can be sent through the Internet from your local TeleUMS to your branch offices.

Telephony Experts' (Los Angeles, CA – 310-445-1822, www.telephonyexperts.com) award-winning Talking NT Multimedia Smart Switch also comes in a Talking NT Enterprise SQL version that runs on Windows NT and uses Microsoft SQL Server. This architecture lets you build a distributed, global switching solution using IP networks for switch management, data replication, call processing, and real-time billing. The system also lets you build large-scale multi-chassis Talking NT switches. Call traffic can be originated and terminated through a variety of media: dedicated long-distance T1, voice-over-IP, frame relay, ATM, local T1, E1 or ISDN, Analog POTS lines, and loop-start trunks.

Applications supported include prepaid debit card, international callback, 1+ feature Group D, prepaid wireless, tandem switching, toll-free number termination, one number locator, and voicemail with paging. Talking NT Enterprise SQL can be purchased as software only or as a turn-key system.

Telephony Experts sells direct and through a global network of VARs.

If you've ever wanted to capture the content of a voicemail message – permanently – Timberline Communications (Sunnyvale, CA – 408-743-9600, www.transcribe.com) has your answer, with its voicemail transcript service. Once your voicemail system has been configured to use the service, you can selectively forward important voicemail messages to Timberline Communications, and in return receive an accurate transcription in the form of e-mail.

Siemens Information and Communication Networks (Boca Raton, FL – 800-765-6123, www.icn.siemens.com) makes Xpressions 470, a unified messaging offering that converges voicemail, fax and e-mail technologies into a unified messaging platform.

Users can access, send, and receive voice, fax, and e-mail messages anytime, anywhere, via the communications device of their choice: phone, PC or fax. This is done with technology that converts voicemail messages to voice-based e-mails, and uses a text-to-speech engine to read e-mail messages and fax headers to you over a phone connection. You can even send and receive messages that combine voice and text into a single message.

Xpressions 470 supports many common e-mail interfaces, including Microsoft Exchange and Outlook. Lotus Notes and Novell GroupWise integrations are planned for future release. For fax messaging, Xpressions 470 integrates with Fenestrae, Omtool, Optus, and TOPCALL.

The Xpressions 470 hardware and software solution comes in two models for now. Xpressions 472 supports workgroups that requires between 4 to 16 voice sessions, while the Xpressions 475 supports workgroups that require up to 48 voice sessions.

Both models come standard with a Microsoft NT-based server. Standard features include auto attendant, audio text and voice forms, as well as Web-browser interfaces.

Sphere Communications (Lake Bluff, IL – 847-247-8200, www.spherecom.com) is the developer of the unique Sphericall, an ATM-based PC-based PBX that comes with messaging. Sphericall voicemail uses Exchange for message storage, so you get unified voice and e-mail messages in your Exchange or Outlook client. Sphericall runs on Windows NT, and you get up to 120 analog ports per server. They even sport a proprietary development environment.

The TraveLink Unified Messaging Server is SphereLink's (San Jose, CA – 408-452-8300, www.spherelink.com) flagship telecommunications solution. TraveLink combines voice, fax, and e-mail messages into a single unified mailbox, which can then be accessed from any convenient standard communication device, such as phone, PC or laptop, cell phone, or fax machine. From a phone, you can listen to e-mail via text-to-speech, and you can forward fax and e-mail messages to any convenient fax machine. TraveLink can be set to alert you to incoming messages, and a call forward "follow me" option ensures that critical calls reach someone immediately.

TelePost's (Santa Cruz, CA – 831-466-8000, www.telepost.com) Message Center is its unified messaging service. It provides users with a single mailbox to retrieve and manage all voicemail, faxes, and e-mail. It is accessible through a Web browser, an Internet e-mail account and any touchtone phone from virtually anywhere in the world. Pager notification of new messages is also available.

The Message Center and TelePost's other services are offered to ISPs, telecom companies and other communications providers so TelePost partners can offer their customers an extensive business communications suite that integrates the telephone system with the Web. These services require no initial capital investment by service providers, as they are hosted entirely by TelePost.

A unified messaging system for global service providers serving mobile subscribers, Telinet Technologies' (Norcross, GA – 770-239-1000, www.telinet.com) MediaMail lets service providers customize the user interface for language and content and access messages from any continent with a phone call. It also provides management of globally deployed MediaMail servers.

Message Roaming gives subscribers a method for checking messages via a local phone call from anywhere in the world. By dialing into a local MediaMail server, the subscriber's messages are sent from home server to local server via Internet.

For multiple language voice prompts, non-U.S.-character sets are supported along with a programmable telephone user interface (TUI). And a programmable interface for multi-language text-to-speech (TTS) engines allows for the reading of e-mails in several languages.

Making good on its vision for "the convergence of voice and data networking to enable unified messaging, data access for maintenance and upgrades to the PBX, as well as the capability to create Virtual Private Networks," Telrad (Woodbury, NY – 516-921-8300, www.telrad.com) does unified messaging for the ImaGEN system.

ImaGEN's unified messaging takes the approach of integrating voice messages into a family of Microsoft Exchange products for message management. Meaning, your voice and fax mail shows up with your e-mail messages in an Exchange or Outlook client, running on a Windows 95, 98 or NT PC.

Requirements for Telrad's unified messaging are a Pentium voice server with 32 MB of RAM, 2.1 GIG Hard Drive, Windows 95 (at least), ImaGEN 6.5 and Microsoft Exchange or Outlook.

Koor and the Canadian company Nortel has signed a binding memorandum of understanding under which Nortel will sell its 20% stake in Telrad Telecommunications to Koor, which owns the remaining 80 percent, for $45 million.

The two will continue their partnership by establishing a joint company, to be called Nortel Networks Israel (NNI), that will buy some of Telrad's activities for $95 million. Koor will invest $49 million in NNI, of which part will be a loan and the rest will be in exchange for a 28 percent stake in the new firm.

As a result of the deal, Telrad will essentially be split in two, with its business equipment division and part of its public networks division being sold to NNI. The remainder will turn into a private company fully owned by Koor. The new Telrad will outsource most of its production work, and focus instead on being a holding company for start-up companies.

Tornado Development's (Manhattan Beach, CA – 310-546-6319, www.tems.com) TEMS (Tornado electronic messaging system) product was the official unified messaging system for CT Expo Fall 1999.

A grand idea, attendees can get their own TEMS e-mail address, a toll-free TEMS number, their own online message center, and in essence, become accessible on the crazy show floor.

With TEMS, your e-mail, voicemail, faxing and paging integrate into a single Web-based inbox that you can access from any Web-enabled computer or touch-tone telephone, anywhere in the world. Read, answer, filter and reroute messages. Forward faxes to remote numbers. Listen to voicemail from your computer. From any phone call your TEMS inbox and use the IVR menu to listen to your e-mail read to you over the phone. Compose, forward, and reply to e-mail.

And TEMS works on any Windows, Macintosh or other platform. A free 30-day trial is available from www.tems.com. A subscription to TEMS costs just $9.95 per month.

U

uReach (Holmdel, NJ – 732-817-0600, www.ureach.com) competes with Onebox.com and TeleBot in the free Internet-based unified messaging arena. uReach.com gives you a permanent phone number and a permanent e-mail address. You can send and retrieve voicemail, e-mail, and faxes for free. You get an 800 number and 30 free minutes per month of phone usage and unlimited access to your account on the Web. Additional phone usage will be available soon at an "attractive price."

Virtualplus (New York, NY – 212-252-9258, www.virtualplus.com) is a UM company that offers its custom Messagepoint technology to an alliance of ISPs around the world. Messagepoint lets you collect all your messages; e-mail, voicemail, faxes, and phone messages from one inbox. You can also travel light as you can collect all your messages not only from your PC, PDA or hand-held organizer, but also locally over any phone or any Internet machine. Virtualplus can store six message types in the inbox (soon to be eight). The company describes the system as the first "node-based" UM system – i.e., it works as a distributed system instead of an add-on to a PBX. A full account costs $18 to set up and $19.95 a month.

VocalData's (Richardson, TX – 972-354-2100, www.vocaldata.com) IP*Star is an IP local exchange system. IP*Star combines the functionality of Centrex with that of a distributed IP-PBX. And its architecture lets the product be configured in carrier and enterprise network apps.

IP*Star includes network-based Telephony Server software, a unified messaging package and Ethernet-IP phones. The Telephony Server software is platform-independent, meaning it runs on a pair of redundant UNIX (carrier class) or Windows NT (enterprise) servers. In carrier networks, Telephony Server software operating on a single pair of UNIX servers gives you support for a whopping total of 10,000 IP telephone users.

This lets service providers (ISPs, CLECs, ILECs) deliver enhanced services and high-speed Internet access all over a single packet-based connection (xDSL or IP on a digital link) to multiple customers. A single T1 connection between a service provider and customer site supports up to 300 VocalData IP telephones, while simultaneously providing over 1 Mbps of high-speed Internet access.

VocalTec's (Fort Lee, NJ – 201-228-7000, www.vocaltec.com) Internet Phone lets you make calls from your computer to regular phones around the world, accommodates video, has a community browser for finding people in cyberspace, and – ah ha! - has voicemail to get in touch with others who are offline. Internet Voice Mail is a software program that lets you send voicemail over the Internet for free to anyone with an e-mail address. You just enter a person's e-mail address, record a message, and send it off. You can even add text or attach files.

Internet Voice Mail has a suggested list price of $29.95 and includes a free bonus license. It's available for Windows 95, Windows NT, Windows 3.X, Windows for Workgroups 3.11, and the 68030, 68040 and Power Macintosh platforms.

Vodavi's (Scottsdale, AZ – 800-843-4863, www.vodavi.com) PathFinder Onelook Unified Messaging uses Microsoft's Exchange technology to bring your messages to the desktop. Obviously, your voice, fax, and page messages left in your PathFinder mailbox appear in your Exchange or Outlook inbox.

Onelook runs alongside other PathFinder apps, including voicemail and auto attendant. Play voice messages via speaker or phone. View voice, fax, MS-Mail, and Internet mail in one inbox.

Nuance Communications' speech recognition and speaker verification capabilities, support for Lernout & Hauspie TruVoice text-to-speech (TTS), and support up to 200 simultaneous users on a single PathFinder platform are all available.

Call'EM from Voxtron (Sint-Niklaas, Belgium – +32-3-760-4020, www.voxtron.com) makes e-mails accessible from any DTMF-capable phone. After checking your user ID and password, Call'EM downloads the e-mails from the mail server and reads the headers. You can choose to listen to the body and even send a reply which is sent back as a .Wav attachment.

The system uses Lernout and Hauspie text-to-speech technology and nine languages are available. Call'EM never deletes the original message from the ISPs mail servers. Call'EM is designed for telecom operators, Internet service providers, and corporations.

VSR (Voice Systems Research) (Rocklin, CA – 916-624-6300, www.vsrusa.com) has two voicemail and voice processing systems, made in response to demand from small businesses.

Voicentre Express, a smaller version of the original Voicentre, is wall-mountable and comes with 2 to 8 ports and 100 hours of voice storage. Features include voicemail, auto attendant, fax tone detection, distribution lists, caller queuing, screening, and it works with over 200 phone systems.

COVoice Express is an entry-level system modeled after VSR's Mitel integration. It uses the only Mitel licensed PC-based COV integration board. Features offered mirror the bigger COVoice system: voicemail, auto attendant, scheduled events, loudspeaker paging with call retrieve. It's wall mountable and is available in 2 or 4 port configurations and 70 hours of voice storage.

CommPoint is a product VSR acquired from Telephone Response Technologies (TRT) (Roseville, CA – 916-784-7777, www.trt.com). CommPoint is a Windows NT-based client/server communication platform that offers a messaging server and IVR capability as its primary strengths. WinVM is its desktop unified messaging program, a companion to the CP/Mail messaging server product.

With WinVM, you can view voice and SMTP-compatible e-mail messages in their WinVM folders, and they have graphical control over your own mailbox options and settings.

WinVM is made for Windows 95 and NT workstations with sound card and local phone support. CP/Mail includes 10 free WinVM desktop licenses, an Internet (SMTP) messaging gateway and integrated address book.

VTG (Voice Technologies Group) (Buffalo, NY – 716-689-6700, www.vtg.com) embeds voicemail systems and can readily engineer a hardware platform to fit in any KSU/PBX like a line card. This means tightly integrated voicemail platforms that work directly with the KSU/PBX backplane. This is good for dealers who no longer need to offer voicemail as an add-on outside the PBX.

Webley (Deerfield, IL – 888-444-6400, www.webley.com) is a voice activated personal assistant service. Connect to a Webley website and call Webley over the phone. If you get e-mail, Webley can read it over the phone or display it on e-mail client software. Webley doesn't store e-mail, just accesses it and reads it to you. You can also add contact names to an address book via the Web. The Web site thus becomes a unified messaging system.

Uninterruptible Power Supply (UPS)

Uninterruptible Power Supplies are devices which are connected between the mains power supply and the devices drawing power (known as "loads"). They are intended to offer protection against a variety of power supply problems, the degree and quality of protection depending on the type of UPS employed. Most are able to offer a degree of power conditioning and sufficient back-up power to allow controlled shut-down of computer systems during outages.

Some types of power anomalies you can expect from the power company include blackouts (no power), dropouts (very short blackout), brownouts (lower voltage than normal), surges (higher voltage than normal) and phase shifts. Some power problems are caused by the customer (you or your neighbor). These can be blackouts (overloading local sub-station), brownouts (starting a large motor), phase shifts (using low power factor loads that require the power company to switch in power factor correcting capacitors), transients, and high frequency noise (such as from arc welding). Mother Nature can occasionally be blamed for transients (lightning hitting a power line) and blackouts (wind blowing down a power line).

The Best Power National Power Laboratory's five-year Power Quality Study (conducted from 1990 to 1995) estimated that the average computer is subjected to 289 potentially damaging power disturbances per year, which is a little more than one "event" per business day. So you definitely need a UPS.

All UPSs consist basically of a battery charger / rectifier, battery and inverter. Manufacturers generally add other components to enhance the performance and feature set; for example, radio frequency interference filters. Modern UPSs can be operated by unskilled personnel and range in capacity from small units which can sit next to your PC at home to large rackmounted units capable of supporting the power requirements of a series of PCs.

There are three categories of UPSs: Off-line, line-interactive and on-line.

An off-line UPS is also called a standby UPS. An off-line UPS is an inexpensive system that's generally used to protect single-user workstations and other less critical applications. You plug an off-line UPS into an AC outlet. The unit consists of a basic battery / power conversion circuit and a switch that senses the occurrence of a power cut. The load is usually connected directly to the mains, and power protection occurs only when line voltage decays enough to be considered an outage, whereupon DC power is drawn from the batteries and the inverter converts the DC back into AC power that your system needs. Off-line UPSs typically offer surge suppression, too. The time it takes to transfer from your power source to the UPS's batteries can range from two and four milliseconds.

One disadvantage of off-line UPSs is that they're too quick to rely on battery power. Further up the power protection hierarchy is the line-interactive UPS, which boosts the voltage coming from your AC power supply and eliminates noise. Line-interactive UPSs don't immediately go to battery when the AC voltage decreases. Instead, they keep the voltage within a certain range, say 10 to 15%, above or below the voltage you need.

Line-interactive UPSs tolerate lower voltages than off-line UPSs before drawing from the battery. The transfer time for line-interactive UPSs, like their off-line cousins, is about two to four milliseconds.

A line-interactive UPS is basically an off-line UPS with enhancements to provide better power conditioning. Generally an interactive inverter / filter is permanently on-line and interacting with the changing state of the AC mains line and is always sharing the load and providing continuous though limited linear output voltage regulation. During normal operation, the inverter might provide a small fraction of the load, with most of the current coming from the AC mains line. When a voltage sag occurs, the inverter automatically supplys more power to the load. This is why the line-interactive type of UPS is also called a load-sharing system or a hot standby. "Buck and boost" are terms relating to voltage regulation "Buck" lowers a high voltage and "boost" raises a low voltage.

A Line-interactive UPS is cheaper than an on-line one and offers better protection than an off-line UPS. However, a line-interactive UPS is usually bulkier and heavier than an off-line UPS, has short battery run times, has no bypass circuit (which protects from high start-up current or overload), and the "break in" power on switching may damage delicate electronic equipment.

On-line UPSs are the most advanced (also the most costly) and provide the highest quality of protection and are suitable for all applications where consistent computer grade supply is required. The distinguishing feature of an on-line UPS is that the inverter is supplying all of the load all of the time. On-line UPSs don't need separate battery power because they continually recharge themselves. They continuously convert incoming AC supply to DC current, and fix any glitches in the way of noise, undervoltage or overvoltage. Then they change the DC current, which is now presumably noise-free and at the right voltage, back to AC, thereby regenerating the mains waveform and removing all disturbances. This "double conversion" process is the most effective form of power conditioning, effectively regenerating an independent AC sinewave waveform. As a result the on-line UPS fully protects the computer load from all ongoing and often transparent power problems on the utility line, something that's not provided by other types of UPS systems.

The on-line UPS's rectifier / battery charger, battery and inverter are constantly on-line and because the batteries are constantly on-line, there's no transfer switch to be activated (hence, no transfer time and so no power interruption either) upon a break in AC power when the batteries take over the load in the event of an outage.

Ideally, batteries in a UPS should be hot-swappable so you can replace the dead batteries with live ones without shutting off power. You should also find out if you need special types of plugs or wiring to connect the UPS to your equipment.

UPSs, as well as other power protection devices, should also provide a way to monitor how much power you've used. You should be able to keep track of your UPS' performance from a remote location. UPSs should let you check on them using a simple network management protocol (SNMP) adapter or even from a Web site.

The "Smart" UPS

A "smart" UPS has a built-in microprocessor and an RS-232 interface port. Many people mistakenly believe that in order to have an automatic shutdown capability, a UPS must be "smart". In fact, a UPS with a contact closure serial interface port can normally do an automatic shutdown. A "smart" UPS adds the additional functionality of being able to monitor more operating parameters, for example temperature, humidity, voltage level, etc.

During a power interruption, a decision must be made whether to gracefully shutdown the computer system. If someone is always working near the computer, this is usually done manually – when the UPS buzzes and kicks in during a power failure. However, in the case of an under-attended or unattended computer system, nobody is around to make that decision. In that case, during a power failure, the UPS will kick on and run the computer system, but only until its battery is depleted. Therefore, one should have automatic shutdown software that will gracefully shut down the computer system after a pre-configured time period of when the computer receives a low battery signal from the UPS. UPS monitoring and automatic shutdown software also can broadcast power fail messages to network workstations, keep an historical log of abnormal power conditions, and alert the system to page the administrator.

The Mystery of Power Ratings

Selecting a UPS with the right capacity for your equipment can be a bit tricky. Volt-Amps, or VA, is the amount of apparent power supplied to equipment, or volts x amps. In high school you may have been taught that Watts = Volts x Amps, so you might jump to the conclusion that VA = Watts. It isn't quite that simple. Wattage is the actual amount of the

U

power that is available to do work, and is always less than or equal to the VA rating. The ratio of watts to VA is the "power factor" of the equipment, which is a number between zero and one. Light bulbs, space heaters, toaster ovens, etc., use all of the raw energy put into them and so have a Power Factor of one. Devices based upon more delicate and complicated electronic components such as those containing transformers and capacitors, do not necessarily use all their power rating all the time and have a Power Factor of less than one, generally 0.5 or 0.6. Thus, a computer may need 160VA to run correctly but it actually only uses 80 Watts so the Power Factor is 80 / 160 = 0.5. Another way to find the VA is to remember that VA = Watts / Power Factor.

This also means that Watts = VA x Power Factor. You might be tempted to look at your computer, see the Watt rating of 80 Watts, then run to your friendly neighborhood computer store and buy an 80 Watt UPS. Don't do it! Things are not that simple, since Watts = VA x Power Factor, so 80 Watts = 160VA x 0.5 for our imaginary computer. This means that we need a 160VA unit to protect and supply power for an 80 Watt computer. Unfortunately many UPS manufacturers use a generous Power Factor of 1.0 when they advertise the ratings of their VA devices and are thus listing the rating in Watts instead of in VA capacity. They figure that "Watts" is a more familiar electrical term to non-technical people than VA and, besides, a Power Factor of 1.0 gives the largest value for the Watt rating, so 80VA x 1.0 = 80 Watts. This causes the unsuspecting consumer to buy an underpowered UPS. An "80 Watt" UPS may really be an 80VA unit that can actually only handle a 40 Watt computer! To protect a 160 Watt computer, you would have to buy a 320VA UPS.

And even if you think you've "figured out" what the right size is for your UPS, be sure to add another 20% capacity for good measure.

Tips For Selecting and Using a UPS

How do you find the right Uninterruptible Power Supply for your application or call center's needs? Here are some key features that you should look for:

- Look for monitoring and shut-down features, as well as Simple Network Management Protocol (SNMP) compatibility.
- The UPS should be able to go online with less than a 2 millisecond reverse transfer.
- Look for the ability to extend the backup time with the addition of external battery packs or systems, for up to four hours or more.
- Look for intelligent line conditioning (for standby units) that maintain the output voltage within selectable limits, without switching to battery operation. This improves the availability of the protected system while prolonging the life of the battery.
- Don't assume your UPS has a built-in power surge protector. Unless it specifically says so, assume the UPS has no surge protection and install the protection yourself. UPS systems are just as susceptible to being fried by power line surges as your other delicate electronic equipment (ACDs, computers). And doesn't it seem that the power always surges just before your electric service is interrupted? Make sure you are protected.
- The UPS you need depends on the load, or amount of AC, your system is pulling and the equipment the UPS will support. You can determine what you need by following these steps:

1. Determine all the hardware items that will be connected to the UPS.

2. Determine the amp rating for each hardware item and add the ratings together. You can find the amp ratings on the back of the equipment.

3. Multiply that total by 120 (for domestic applications) or 220 (for international applications) to get the volt-amps total. If your equipment is rated in watts, rather than amps, multiply the wattage by at least 1.4.

4. Request a UPS that exceeds your volt-amps total by at least 20%.

- What not to use with a UPS: Power loads that are highly inductive can cause serious damage to the UPS. Use your UPS to support computer or telephone system loads only for the best overall performance. Examples of equipment that should not be attached to the UPS include: Air conditioners, drills, space heaters, vacuum cleaners, buffing machines, fans, laser printers, transformers (step up / step down).
- A key non-technical question when buying UPSs is how well a vendor's warranty covers your equipment if it is damaged, for instance, by lightning.

Self-maintenance Tips

- Complete a full system run down (discharge) of the battery or batteries in your unit every three to four months.
- Don't overload the UPS, and keep inductive loads off the unit.
- Press the test button on the UPS occasionally to let the unit complete a self-diagnostic test.
- Get in early every Monday. You're not the only one who hates Mondays. If your electronic devices (telephones, ACDs, computers) are running but unattended over the weekend, expect to find glitches when you come in on Monday. Power outages, surges and sags throw your electronic gear out of whack. They accumulate in your two days "off." So get in early and be prepared to reboot.

Power Trivia You Should Know

1. Power problems are the single most frequent cause of phone and computer system failure. The average IVR system, for example, is hit with a significant power fluctuation (spike, surge, brownout) approximately 400 times a year. The problem is getting worse, as regional power grids are forced to adapt to increased consumption.

2. Power-related damage is among the most difficult types of damage to recover from. That's because it does two things: it cripples the hardware, often necessitating a costly replacement (complete with waiting time for delivery), and it wipes out data.

3. Multiple connections (to trunks, networks, peripherals, etc.) increase the number of routes through which power surges can enter and cripple an integrated call center. The more components you connect together (and we count data sources here as a component) the more vulnerable you'll be when something hits.

4. Power protection is one of the cheapest forms of insurance you can buy. The technology is solid. It's been proven to work. The added cost of protecting hardware is roughly 10% to 25% of the hardware's value. That does not count the value of the data. You could make the argument (we often do) that a call center's data is far more valuable than its hardware.

5. When you factor in the cost of potential losses, and weigh that against the likelihood of problems, the cost of protection becomes negligible.

6. With all that, why aren't more companies concerned? In a recent study commissioned by Comdisco (Rosemont, IL), it was found that 62% of the companies in the U.S. have a formal disaster recovery plan in place. Of those, only a tiny fraction (8%) have included their telemarketing operation in the plan.

7. And the really scary part? This only asks about recovery for computer operations. The companies that have an organized approach to protecting and recovering their telecom systems are far fewer. And those that are thinking about how to survive a cable cut are fewer still.

Power Glossary

AC – Alternating Current – Electrical current that continually reverses direction, with this change in direction being expressed in hertz, or cycles per second.

Amp or Ampere – Quantitative unit of measurement of electrical current. Abbreviated as A.

Blackout – A total loss of electrical power.

Brownout – A low voltage condition over an extended period of time.

Clamping Level – The voltage level above which a surge suppression device diverts energy away from the load.

Clamping Time – The response time of a surge suppression device in clamping or diverting away from the load a voltage above the claming level.

Common Mode Noise – Abnormal signals that appear between a current-carrying line and its associated ground.

Current – The flow of electricity expressed in amperes. Current refers to the quantity or intensity of electricity flow, whereas voltage refers to the pressure or force causing the electrical flow.

DC – Direct Current – electrical current which flows in one direction.

Dip – A short term voltage decrease. See also "Sag".

EMI – Electro-Magnetic Interference – or electrically induced noise or transients.

Ferroresonant Transformer – A transformer that regulates the output voltage by the principle of ferroresonance: When an iron-core inductor is part of an LC circuit and it is driven into saturation, causing its inductive reactance to increase to equal the capacitive reactance of the circuit.

Ground Fault – An undesirable path that allows current to flow from a line to ground.

Harmonic Distortion – Excessive harmonic (a frequency that is a multiple of the fundamental frequency) content that distorts the normal sinewave waveform.

Hertz or Hz – The unit of measure of the frequency of alternating current (AC). Also a well-known car rental agency.

Inverter – The part of a UPS that converts the battery's DC output into AC power.

Isolation – The degree to which a device like a UPS can separate the electrical environment of its input from its output while still allowing the desired transmission to pass through.

Joules – The amount of energy measured in watt-seconds that a surge suppression device is capable of directing away from the load in case of a surge or spike.

KVA – a Thousand VA

Load – an electrical device connected to a power source is a "load." In reference to a UPS, the load is the amount of current that is required by the attached electronic equipment. Rated loaded described in the specifications of the electronic equipment is often higher than the actual power consumption of the equipment in real world use.

Noise – An undersirable signal that is irregular and is riding on top of the desired signal.

Overvoltage – An abnormally high voltage, like a surge but lasting for a longer period of time.

Power Factor – The relationship of actual power to apparent power. In reference to a UPS, the relationship between watts and VA (volt-amperes). It is expressed as watts divided by volt-amperes (W/VA) and is usually in the range of 0.6-0.71.

Pulse Width Modulation (PWM) – Process of varying the width of a train of pulses by tying it to the characteristics of another signal.

Rectifier/Charger – That part of a UPS that converts the incoming AC utility power to DC power for driving the inverter and charging the batteries.

Sealed Lead-Acid Battery – A battery containing a liquid electrolyte that has no opening for water replenishment

Sinewave – A fundamental waveform produced by periodic, regular oscillation that expresses the sine or cosine of a linear function of time or space or both.

Single Phase – The portion of a power source that represents only a single phase of the three phases that are available.

SPS – A term referring to a stand-by or offline type UPS.

Surge – An abnormally high voltage lasting for a short period of time.

Switching Time – The amount of time it takes a stand-by or offline type UPS to switch from utility output to inverter output after the UPS senses a power interruption. Normally expressed in milliseconds. See also Transfer Time.

Three Phase – An electrical system with three different voltage lines with sinewave waveforms that are 120 degrees out of phase from one another.

Transfer Time – The amount of time it takes a stand-by or offline type UPS to sense a power interruption and switch from utility output to inverter output. Normally expressed in milliseconds. See also Switching Time.

Transformer – A device used to change the voltage of AC power or to isolate a circuit from its power source.

Transient – An unabnormal and irregular electrical event, such as a surge or sag.

Undervoltage – An abnormal low voltage lasting for a longer period of time than a sag.

VA – See Volt-Ampere

Volt – The quantitative measure describing electrical force or potential.

Volt-Ampere – The unit of measure of apparent power that is the tradiitonal unit of measure for rating UPSs. Compare this to watts, which is the unit of measure of "actual" power.

Voltage Regulator – A device that provides constant or near-constant output voltage even when input voltage fluctuates.

Watts – The unit of measure of actual power. Compare to volt-amperes (VA), which is the unit of measure of apparent power.

Waveform – The graphic form of an electrical parameter.

UPS / Power Protection Companies:

Abacus Controls Inc. (Somerville, NJ – 908-526-6010, www.abacuscontrols.com) makes power control / conditioning equipment such as UPSs.

ADS The Power Resource Inc. (Dallas, TX – 800-443-4742, www.adspower.com) makes power control / conditioning equipment such as UPSs.

American Power Conversion (West Kingston, RI – 401-789-5735, www.apcc.com) is a major manufacturer of computer accessoreis, power control / conditioning equipment and UPSes.

Best Power (Necedah, WI – 800-356-5794, www.bestpower.com) makes power control / conditioning equipment and UPSs.

Controlled Power Co (Troy, MI – 248-528-3700, www.controlledpwr.com) makes power control / conditioning and UPSs.

Fenton Technologies (Santa Ana, CA – 949-474-3800, www.fentonups.com) makes UPSs.

Gordon Kapes Inc. (Skokie, IL – 847-676-1750, www.gkinc.com) makes UPSs, as well as CT testing tools and line simulators.

Liebert Corp. (Columbus, OH – 614-888-0246, www.liebert.com) makes power control / conditioning equipment and UPSs.

MGE UPS Systems (Costa Mesa, CA – 714-557-1636, www.mge-ups.com) makes UPSs and other types of related power control / conditioning equipment.

Omni Computer Products (Carson, CA – 310-638-2500, www.rhinotek.com) offers power control / conditioning equipment, UPSs, as well as modems, NICs, and computer accessories.

ONEAC Corp. (Libertyville, IL – 800-327-8801, www.oneac.com) makes UPSs, UPS interface communications software, power control / conditioning equipment, and voice and data line protection devices.

Panamax (San Rafael, CA – 415-499-3900, www.panamax.com) makes some of the best surge suppression and other forms of power control and conditioning equipment. They no longer make UPSs.

Para Systems (Carrollton, TX – 972-446-7363, www.minutemanups.com) manufactures the MINUTEMAN line of UPSs.

Philtek Power Corporation (Blaine, WA – 360-332-7252, www.philtek.com) makes power control / conditioning equipment, and UPSs.

Powerbox USA Inc. (Broomfield, CO – 303-439-7220, www.powerbox.se) is a value-added reseller of power control / conditioning equipoment and UPSs.

Powerware Corp. (Raleigh, NC – 877-797-9273, www.powerware.com) makes UPSs and power monitoring software.

Tripp Lite (Chicago, IL – 773-869-1111, www.tripplite.com) makes power control / conditioning equipment, UPSs, surge suppression devices, power inverters, and isolation transformers.

US Logic (San Diego, CA – 619-467-1100, www.uslogic.com) makes UPSs as well as motherboards, single board computers and fault resilient computer chassis.

UPS

See Uninterruptible Power Supply.

Versa Module Europa (VME) Bus

VME bus (Versa Module Europa) is an open-ended, flexible computer backplane bus with a 32-bit wide data path, built upon the Eurocard standard (typical card sizes are 160 x 216 mm and 160 x 100 mm). The VMEbus specification was introduced by Motorola, Phillips, Thompson, and Mostek in 1981, itself having been based on an earlier "Versa bus" used by Motorola in the 1970s and a distributed arbitration scheme that was basically DEC's Unibus scheme altered to avoid DEC's patents. VME is defined by the IEEE 1014-1987 standard.

The VME bus architecture is comprised of four sub-buses: the Data Transfer Bus, the Arbitration Bus, the Priority Interrupt Bus and the Utility Bus.

Before the coming of CompactPCI technology, VME reigned supreme in the worlds of high-end telecom, military, industrial, and other real-time computing systems. VME has a huge installed base. VME uses a completely memory mapped scheme, and every device can be viewed as an address, or block of addresses. Addresses and data are not multiplexed – it's an asynchronous bus, so data can be transferred to each board at its own optimum speed. A typical transfer consists of an arbitration cycle (to gain bus control), an address cycle (to select the register) and the actual data cycle. Read, write, modify and block transfers are supported.

Since it's been around for so long, VME supports a huge number of protocols which allows newer, faster products to be added to a system while at the same time supporting older boards. At last count, VME also supports an incredible 105 realtime operating systems. It has an efficient interrupt scheme and it lets you put a full 21 slots on a backplane.

Among VME's drawbacks are: The asynchronous bus protocol, though flexible, has the kind of problems inherent in any asynchronous circuit (such as difficulty in developing clean interfaces); a slow distributed arbitration scheme, high powered signaling (which prevents it from being as low power as its PCI cousins), and about half the I/O pins of CompactPCI (though this was alleviated somewhat when the VME64X standard came along which allowed for 160 pins). One can build long busses but this tends to slow performance.

While the speed of VME-compatible processors and I/O devices steadily increased over the years, the aggregate peak bandwidth of the VMEbus was still follows the original spec of 80 MBps. The tremendous real-time I/O processing needs of some high performance systems, far beyond the capabilities of a single microprocessor, forced the development of distributed processing systems implemented as an interconnected network of multiple processors. Since the bandwidth requirements of interprocessor communication networks scale in direct proportion to their processing throughput, the 80 MBps VMEbus bandwidth limit soon became a bottleneck. This resulted in 1994 in the VMEbus International Trade Association (VITA) publishing an upgraded specification called the VME64 standard. VME64 doubles the data transfer rate by multiplexing the address and data buses. Additional enhancement of VITA members were collected together and published as the VME64X standard.

By the end of the 1990s, yet another increase in bandwidth came with the VME320 spec, which defines a new backplane that allows 512 MBps transfers between compatible cards, yet is backward compatible with slower legacy boards. Extensions to VME320 have enabled data transfer rates as high as 1 GBps.

Getting the CompactPCI and VME camps to snipe at each other is always great fun, since engineers are always subject to a testosterone-fueled "my bus is (bigger, better, more scaleable, reliable) than your bus" and, indeed, VME aficionados can bring some big guns to bear on their arguments in favor of using VME for telecom applications.

Take, for example, Benjamin Sharfi, the CEO of General Microsystems (Rancho Cucamonga, CA – 909-980-4863, www.gms4vme.com), a company that makes both VME and cPCI boards, and an amazing PCI-to-VME bridge chip called the OmniVME.

Sharfi says that "all high end multiprocessing systems are continuing to be developed under VME. All high end rugged systems are still and exclusively being done under VME. Now, everybody's been talking about VME's 'weakness' and it's partially true that until recently VME didn't support the functionality of plug and play as in cPCI. But with an implementation of VME 64X and our General Microsystems VME NT drivers, VME becomes truly plug and play under NT or Windows 2000 just like cPCI, except now you have VME's 21 slots per segment to play with instead of cPCI's eight."

Sharfi thinks that cPCI will have a role in low end systems. "I'm not talking about systems where you need multi-processing capabilities or multi-mastering, such as imaging, factory automation and military aircraft detection systems. The kind of systems where cPCI has done well are the telco datacom boxes where basically you have one 'postmaster' managing hundreds of network channels. CompactPCI has been effective in those applications."

"Now here's the interesting part that keeps getting lost – companies have not been adopting cPCI by abandoning what would have been purchases of VME equipment; instead, companies that are deploying cPCI used to build their own products. Nortel has built their own product for a long time. Cisco Systems now builds their own boards, too."

The other big lie in the industry that cPCI is completely hot swappable, says Sharfi. You cannot to date swap the Slot 1 controller, the system master. You can only replace the slave I/O cards. Also, how do you swap multiprocessors? That hasn't been worked out. "Moreover," continues Sharfi, "tell cPCI vendors you want them to put a chassis in front of you and then repeatedly plug and unplug their processor board and their rear panel I/O card into the chassis 20 times. Watch how many pins bend on the board!"

Sharfi says that VME has five distinct advantages over CompactPCI:

First, it has stable specifications. CompactPCI still has some "moving" specifications. "I'm in those cPCI committees," says Sharfi, "we've had six different specifications in two years and we're still changing things so if I built a product a year ago on cPCI, which I did, it does not now comply with today's specifications." A truly redundant system involves much complexity, and the cPCI camp still has some very serious issues to be dealt with in terms of nailing down a spec in that area. "So if you want stability in design and specifications you must use VME," says Sharfi. "Let's leave all other issues aside – from the stability of architecture and the stability of specification VME is clearly a winner."

The second thing is that PCI was developed for a processor to peripheral communication interface, *not* as a specification for a bus, "so we're making it do something that it was not intended to do and we therefore have limitations," says Sharfi.

Third, VME will have always more slot capacity than cPCI, with 21 slots per chassis, while cPCI does eight or four, depending on the bus speed.

Fourth, VME is multi-master ready, meaning that the system can designate every processor board to be a master processor with its own

I/O. While there are proprietary cPCI approaches, the cPCI spec itself doesn't yet encompass this.

Fifth is "the lie that cPCI has more I/O pins available in the rear panel boards," says Sharfi. "CompactPCI has five connectors, J1, J2, J3, J4 and J5. (J stands for 'Jack' or female connector). 6U high CompactPCI boards have 315 user defined I/O pins via three of the backplane connectors, J3, J4 and J5. J1 and J2 are for the PCI interface that goes to the PCI bus. J3 is left open for I/O. J4 and J5 are for the telephony and specialty buses that we cannot put I/O on. So J3 is the only I/O connector that's available for rear panel I/O. Now, J3 is exactly the same connector and is in the exact same mechanical position to a new connector that's been defined for VME64X called the P0/J0 connector. (P stands for "plug" or the male connector) P0/J0 is a 95 pin 2 mm connector that fits between the P1 and P2 connectors. It is defined strictly for user defined I/O."

VME64X also defines a forward compatible 160 pin connector that replaces the 96 pin P1/J1 and P2/J2 connectors. This connector adds two rows of 32 contacts per row on the outer shell. Boards using this new 160 pin connector can plug into the original 96 pin connector backplanes, and the original VME boards with 96 pin connectors will plug into VME64x backplanes with the 160 pin connector. Every other pin in the expanded P2/J2 connector's "z" row is assigned to grounds, plus two pins in the "d" row are assigned to VME hot swap control, providing 46 pins for additional user defined I/O.

So how can VME position itself to be more attractive to the telcos? Sharfi says there are three to-do's: First, stop the myth that VME is much more expensive than cPCI. Second, stop "the biggest lie of the century," that VME does not suit itself to a Pentium or a Wintel architecture because of the big and little endian issues in the processor. Third, "the industry must show that VME can be made to have a greater bandwidth across the bus than PCI," says Sharfi.

Sharfi concludes with: "If companies like mine can show those three items, I think that the VME is going to regain its popularity very quickly among many of these key projects."

Complete documentation on the VME bus can be obtained from VITA, the VME bus International Trade Association (Scottsdale, AZ - 602-951-8866, www.vita.com).

Versit

A "global initiative" (consortium) composed of computer and phone switch vendors (Apple, IBM, Novell, Lucent Technologies, Seimens Rolm) on a quest to develop a unified telephony API. Versit was officially formed on November 30, 1994. Unlike previous attempts that excluded anyone other than telco vendors, Versit tried to standardize at both the switch and the application level with their own communications signaling infrastructure for computer telephony components and other devices. Their original efforts to develop a spec led in 1995 to the The CTI Encyclopedia, which was a set of specifications including feature definitions, protocols and APIs based on TSAPI.

In the mid 1990s Versit also developed the vCard, specification. The vCard spec defines a common electronic business card format for capturing information about a person, such as their name, address, phone number, e-mail user ID, with multimedia support for photographs, sound clips, company logos, etc. Since vCard was expected to become a ubiquitous electronic business card format, it would allow an individual to consistently identify themselves without restating or rekeying the information. For example, vCards can be exchanged across the Internet using web clients, sent as attachments to e-mail messages, imported into personal address or phonebook software programs, or transmitted from tele-

phones. vCard personal information can even be carried in smart cards. Versit followed this with vCalendar, an electronic calendaring and scheduling exchange format. Beginning in December, 1996, the Internet Mail Consortium (www.imc.org) took on responsibility for the development and promotion of both vCard and VCalenar technologies.

Versit's original efforts at developing a universal CT spec were immediately recognized as having deficiencies. Then the sudden appearance of Java spurred Versit consortium members to develop and deploy a unified Java telephony API.

In late July 1995, Versit for all practical purposes merged its activities into the ECTF.

Videoconferencing (VC)

Also called "video conferencing" and "videoteleconferencing." It is a combination of video, computing, and communication technologies that converts offices, classrooms, hospitals and courtrooms into what are essentially interactive television studios so that people in different locations can to meet face-to-face and perform most of the same meeting activities they would perform if all participants were in the same room.

When I was a kid back in the 1960s, a friend of mine's father (who was a VP at AT&T) decided to install in his home his company's first-generation videophone, a device I had only seen previously at the 1964-65 World's Fair in New York City. It cost $1,200 a month, needed about three analog POTS lines, and looked pretty unimpressive.

For the next 20 years videophones, videoconferencing and data conferencing would remain something of a joke, relegated to public demonstrations at World's Fairs and cameos in science fiction movies.

In 1980 videoconferencing in public studios with heavyweight fixed units was introduced. From then until 1990, videoconferencing was an option but was still considered to be something exotic, used only by companies that could afford to splurge on bulky, outrageously expensive conferencing systems and the technicians to maintain them, or else companies that had enough advance time to plan conferences through an outsourcer.

The first corporate videoconferencing systems appeared in 1985 and were private studios designed to send and receive compressed audio and video over leased line network connections at a dedicated rate of transmission and with predictable service (point-to-point or fractional T-1, switched connections using ISDN). Proprietary ISDN systems appeared in 1990. When the ITU-T approved in March 1996 the two principal communications standards for video interoperability over ISDN (H.320 and the H.261 protocol), different manufacturers' videoconferencing systems could communicate with each other for the first time. A point-to-point connection was set up similar to a telephone call, relying on a dedicated circuit-switched ISDN connection. The big conference room and giant screen-type multipoint solutions immediately adopted and have followed the H.320 spec religiously ever since. Indeed, much of what was called "videoconferencing" in the early 1990s became associated with multiple-channel ISDN circuit switched networks and relied upon the ITU's H.261 / H.320 video encoding standards.

In fact, an H.320 based system was used in October 1995 in the first videoconference between NASA mission control and the space shuttle. NASA's system was called the Ground-To-Air Television (GATV). Its purpose was to demonstrate the capability and utility of uplinked compressed video and associated audio via the Orbiter 128Kbps Ku-band uplink channel on STS-73. The video / audio data was multiplexed into a single 128 Kbps H.320 data stream, 16Kbps of which was devoted to audio processed with the G.728 codec. The video / audio data is multiplexed into a single 128Kbps H.320 data stream, 16Kbps of which is

G.728 audio. NASA used H.320 so as to provide compatibility with overseas-originated, compressed video / audio signals. The 10.5 pound unit provided full-screen video at 352 x 288 pixels with color motion at 8 to 15 frames per second.

The video quality delivered by H.261 / H.320-compliant systems over a 384 Kbps triple BRI ISDN line (six 64 Kbps "B" channels combined, called 6B, 3xBRI, or "switched 384" with an inverse multiplexer) was "business quality" Common Intermediate Format (CIF) video consisting of 30 frames per second (fps) of a 352 x 288 pixel frame or 15 fps of CIF when a 2B bandwidth was used. The H.261 framing protocol also supports a lower resolution Quarter CIF (QCIF) format of 176 x 144 pixels. The bandwidth saved by sending a low resolution image means that QCIF can be transmitted at a higher frame rate than CIF. In fact, videoconferencing frame rates can be as high as 30 fps and as low as zero fps, depending upon available bandwidth and network congestion. Some VC products used Intel's Indeo standard instead of H.320, which has lesser image quality.

However, hardware and software developers working on Desktop Video Conferencing (DVC or DTVC, depending upon whose acronym you like) didn't really like H.320 and claimed that their own algorithms offered better video quality. Also, software manufacturers pointed out that H.320 had some problems in communication via ISDN technology.

Bigger problems with H.320 equipment were the price and the "ease" of installing ISDN. Although H.320 has been a standard since 1990, Intel spent something like $50 million developing and promoting their PC-based ProShare ISDN communications system, which didn't exactly become a household item (although ProShare would also run over IP-based networks, and Intel had devised a proprietary video compression technology called Indeo, which did not produce the kind of image quality offered by H.261, which also happens to be an ITU-T video compression standard). To home users of that era, the sales of camcorders, V.34 modems, and video capture cards continued to overshadow that of ISDN H.320 / H.261 units.

The ITU-T went back to the drawing board and worked on a number of H.320 follow-ups, making great progress with multiple document conferencing and video conferencing over POTS and LANs.

The ITU then formalized the H.324 standard, which had an improved compression algorithm yielding better audio and video quality over POTS lines. Along with it came T.120, a standard permitting document sharing during any H.32x conference, with or without video.

Around 1996, with the rise of the Internet and other packet-based networks, the H.323 interoperability standard became recognized as an important emerging ITU standard for videoconferencing over existing WANS and LANS. Early H.323 systems were not interoperable with the large installed base of ISDN / H.320 systems. This required a gateway to interconnect H.320 and H.323 systems.

Video compression and interoperability standards now include the following:

- H.320 is used for ISDN, Switched 56 and a similar H.321 used for "H.320" over ATM (See H.320)
- H.221, is a protocol that defines and maintains the frame structure for a 64 to 1920 KBps (1 to 30 64KBps Bearer channels combined) in audiovisual teleservices. Because it's a framing protocol, H.221 does not have to rely upon bitstuffing or start and stop bits for framing on a WAN link. H.221 transmits fixed sized frames of 80 bytes.
- H.261, also known as p x 64, dates back to November 1990. It specifies the video coding algorithms, the picture format, and forward error correction techniques so that different vendor's video codecs can successfully interoperate, following a uniform process for the codecs to

read incoming signals.
- H.322 used for the now nearly-defunct ISO-Ethernet LAN standard (which, however, did deliver guaranteed QoS)
- H.323 is used for packet switching without guaranteed QoS (IP) (See H.323)
- H.324 is used on analog network (PSTN) via ordinary modems. V.80 is the application interface defined in the H.324 ITU videoconferencing standard. A V.80 modem provides a standard method for H.324 applications to communicate over modems. With V.80, an application developer can design and test his video conferencing software on one modem interface, saving development time. Also, H.324 video phone applications are synchronous, sending and receiving data in synch with a timing device or "clock". Serial ports and modems are asynchronous – they accept and receive independent of any clocking device. V.80 converts the synchronous data stream of an H.324 application so that it can communicate through an asynchronous modem connection. These modems adjust to different line conditions throughout a call. Under bad conditions a modem will slow down. When conditions clear, a modem will resume at top speed. A V.80 modem alerts an H.324 video phone of its rate adjustments, allowing the application to adjust the rate at which it sends video and audio. Finally, during transmission data can be lost due to buffer overflows, phone line errors, and a number of other issues. Under these conditions, a V.80 modem communicates lost data information to the H.324 application helping it to keep real time audio and video flowing to both sides of the call.
- H.310 is used on high bandwidth ATM networks
- T.120 is for data interoperability / sharing. allowing file transfer and shared notepads – what are sometimes confusingly referred to as shared whiteboards. T.120 actually represents a suite or protocols that are based upon layers. T.120 applications expect the underlying transport to provide reliable delivery of its Protocol Data Units (PDUs) and to segment and sequence that data, and so T.120's lower level layers (T.121, T.122, T.123, T.124, and T.125) specify an application-independent mechanism for providing multipoint data communication services to any application that can use these facilities. For example, T.123 specifies transport profiles for Public Switched Telephone Networks (PSTN), Integrated Switched Digital Networks (ISDN), Circuit Switched Digital Networks (CSDN), Packet Switched Digital Networks (PSDN), Novell Netware IPX (via reference profile), and TCP/IP (via reference profile) The T.123 layer presents a uniform OSI transport interface and services (X.214/X.224) to the MCS layer above. It includes built-in error correction facilities so application developers do not have to rely on special hardware facilities to perform this function. Meanwhile, the T.120 standard's upper level layers (T.126 and T.127) define protocols for specific conferencing applications, such as shared whiteboarding and multipoint file transfer. Applications using these standardized protocols can coexist in the same conference with applications using proprietary protocols. In fact, a single application may even use a mix of standardized and non-standardized protocols. (See T.120).

There are three varieties of videoconferencing systems:

1. *Desktop systems* are for single users and consist of one or two boards inserted into PC slots. A small fixed focus video camera, a microphone, speakers, Windows-based videoconferencing software and some kind of ISDN, xDSL or LAN connection enables a user to videoconference and share data with remote sites. Since the single PC monitor is used for all display purposes, both data and video must be accommodated on the same screen. Although DeskTop Videoconferencing (DTVC) had existed in some form prior to the PC, from 1995 on the introduction

437

of cheap Intel Pentium and Motorola PowerPC CPUs brought sufficient processing power for videoconferencing to nearly all desktop and portable computers. Inexpensive yet powerful Digital Signal Processors (DSPs) included with such systems also meant that sound was now a standard part of PCs, along with high-resolution, millions-of-colors graphic subsystems. All of this coupled with the parallel development of cheap digital cameras made DTVC accessible to the average user for the first time.

2. *Rollabout systems* are complete videoconferencing packages contained in a cabinet and typically mounted on a wheeled trolley so that they can be moved from room to room. The most popular of videoconferencing systems, Rollabouts are designed for an audience of 10 or 12 people, thus filling a gap between large room systems intended for large groups and desktop systems for individuals Typically rollabouts offer high bandwidth connectivity and have one or two large monitors housed in the cabinet, along with at least one Pan, Tilt and Zoom (PTZ) camera, a desktop microphone, the control system, a software codec, and if required, a data monitor for display of PC applications.

A rollabout's audio system consists of an echo canceller, microphones, speakers and amplifiers. The control system provides the meeting participants with control over the video images, camera orientation, audio levels and other peripherals. The camera in the rollabout can be remotely controlled to select different views of the room. Presets allow you to easily switch between commonly used viewing angles. And there's also usually a graphics or document camera, which is used to share documents, charts, maps, objects and other graphics.

The codecs are designed to transmit and receive two video signals – a motion video signal, usually from the "people camera" at the front of the room, and a captured still image graphics signal, usually from the special graphics camera. A two-monitor system can display the live video on one monitor and the captured still image on another; single monitor system uses picture-in-picture to display both on one monitor. While the still graphic signal is being sent to the remote location, the motion video will briefly freeze until the transferal of the still image is finished.

3. *Room systems* are designed for large groups of up to 40 people sitting in a theater designed for a high-quality presentation. A "built-in" room system can reside on shelves behind a facade wall creating a stately, permanent look. Large format displays for both data and video are used along with sound systems capable of several inputs (desk microphones and/or radio microphones) and output via amplifier and speaker system. A programmable movable camera and a document camera typically complete such a system. Although the capabilities of these systems are similar to rollabouts, they often accept a greater variety of peripherals and are more customized to specific applications.

Basic forms of the three system types discussed above are designed for point to point conferencing. If additional parties need are to be included in the conference a multipoint bridge must be used. With a bridge, everyone participating in the conference dials in, and the bridge connects everyone to each other. Since it's impractical (but perhaps one day will be possible) to look at 30 people simultaneously in a "Hollywood Squares" view on one screen, most bridges let you see just one site at a time, employing a voice activated control that switches the image to that of whoever is speaking (so sites are usually asked to mute their microphones until they speak). Other systems designate a "director" who can control whose image is being seen thoughout the conference at any one time. If a bridge seems too expensive because you're going to use it infrequently, then you can rent one by the hour from telecom operators.

R. Ceballo and A. Lill of GartnerGroup, Inc. (Stamford, CT – 203-316-1111, http://advisor.gartnerweb.com) have published a case study indicating that desktop videoconferencing does not replace group videoconferencing, and without a specific business requirement or task, utilization of desktop VC is very low. They also report that desktop videoconferencing systems are primarily used for point-to-point meetings with one or two associates per desktop. Perhaps the most important finding was that when more than two participants are at one site, the video meetings take place via a rollabout or group / boardroom system and *not* with the desktop systems. Desktop-based videoconferencing meetings average 70% point-to-point and 30% multipoint, with an average of four sites per conference. In mixed meetings of group and desktop systems 90 minutes is about the average meeting time.

Auxiliary Equipment

There is a wide variety of ancillary equipment that can enhance VTC systems to support many different applications:

- Ancillary equipment such as additional cameras, document cameras, and extra high quality microphones for greater sound coverage enhance the system but are also more expensive.
- Document cameras can capture a still image of an object or a transparency and transmit the still graphic to the remote end.
- Dual monitor systems allow participants to share stationary graphics and other documents while still maintaining eye contact.
- Most rollabout systems come with a moveable camera that can pan, tilt, zoom and focus on individuals, pairs and/or combinations of people or objects.
- Scan converters take a VGA computer display and convert it to National Television Standards Committee (NTSC) format for display on the video monitors. Participants can have collaborative work group sessions through document sharing and PC applications.
- The VC system can have one or more cameras. Multiple cameras are generally auxiliary items for larger systems.
- Video printers are able to generate a hard copy of the television monitor screen image.
- Voice activated cameras automatically track speakers and have found wide use in the classroom.

Video Meets the Web

Because of the rise in popularity of IP and other packet transport networks, what was called "videoconferencing" (and later "multimedia conferencing") in the circuit switched network world is now being transformed and is rapidly becoming part of what's known as web collaboration and conferencing (See Web Collaboration and Conferencing).

As with virtually everything else that can be grouped under the broad heading of computer telephony, the web has woven its way into audio- and videoconferencing technology, with immediate, palpable impact. The integration of conferencing with the web means greater ubiquity, wider dissemination of information, and lower costs of operation.

More specifically, IP itself is bringing applications like videoconferencing closer into the fold of existing corporate networks, streamlining business processes, and making the technology more accessible to a bigger audience of users.

In fact, Rose Rambo, a general manager at Polycom (San Jose, CA – 408-526-9000, www.polycom.com), has told us that "The integration of conferencing and collaboration systems with servers and Internet browsers creates the ability to hold multi-site meetings and engage in simultaneous transition of computer-borne information. The widespread [use] of the Internet offers significant opportunities for additional productivity tools, as well as an alternative means to connect people any-where, any time."

You can't connect hundreds of users into a videoconference using circuit-switched equipment – the PSTN can't handle it. There aren't enough lines coming out of each switch that can be tied up simultane-ously. But, in theory, you can achieve massive multipoint connections using a packet network such as the Internet.

One example of full multipoint functionality in an IP videoconfer-encing environment is offered by Brinckmann & Associates' (Norcross, GA – 770-248-1878, www.brinckmann.com) MP-Controller family of videoconferencing interface software designed for use with White Pine's Meeting Point software and Intel's videoconferencing products.

The MP-TeamController, a fully integrated interface, provides each Intel TeamStation user with a detailed list of conference participants. Beside each name is a button indicating participation: Green is the con-ference manager, yellow is the current selected party, and the remaining participants are red. Rather than viewing only the speaking party, a user can view any party simply by clicking on his name in the sidebar.

The MP-ProController enhances the conference manager environ-ment with separate pop-up windows for the cyber conference room, remote video, and conference party list, offering the manager control and flexibility while using tools like application sharing, whiteboard, and file transfer.

Both MP-Controllers feature conferencing capabilities such as auto-adjustable video presence, audio filtering, and muting controls. The MP-Controller family features three options for displaying video: Voice Activated shows the speaking party, Time Activated moves between par-ties at the selected time interval, and Touch Activated is the aforemen-tioned point-and-click option. MP-Controller users also maintain full interoperability with users of standard H.323 clients including Microsoft NetMeeting.

A 10-user license for either the MP-Team or MP-ProController lists for $3,495, but if you buy the MeetingPoint software from Brinckmann too, a discount is available.

Based on the general trends and particular products and services that have been appearing, it looks as if conferencing applications – and especially video, the industry's long-time sleeper – will finally achieve widespread adoption.

In addition to the great enablers, IP and the web, a number of relat-ed forces will also contribute to the impending explosion in video / web conferencing applications:

- *Exploding Bandwidth.* Bandwidth is becoming faster, cheaper, and more plentiful than ever before, so that virtually everyone – including consumers and small businesses – will soon have access to a broad-band pipe. This, in turn, makes almost anyone into a potential end-point for traditionally high-bandwidth apps like video.
- *Better network infrastructure, higher QoS.* Networking vendors are preparing their core devices – routers, switches, etc. – to handle mul-tiple traffic types and high bandwidth-consuming transmissions by implementing quality of service mechanisms like TOS-bit forwarding, DiffServ, RSVP, and weighted fair queuing. This makes it easier to put voice and video on the same network as your data, and subsequently reap the benefits of IP with those applications.
- *Mobile workers and e-commerce driving demand.* As businesses become more and more distributed, both internally and in relation to their customers, the demand for access to information and for real-time communications will only increase. This trend extends to telecommuters, global knowledge workers, web-based call centers, and

business-to-business e-commerce relationships.

All of these factors, and a potentially long list of others, have made conferencing solutions more practical and more lucrative investments for many businesses. They've also given rise to a flurry of new market entrants and new product offerings from conferencing incumbents. Most major long distance carriers now offer both audio- and videoconferenc-ing services, and most (if not all of those) integrate optional web compo-nents like streaming or online presentations. A number of startups have also entered the service side of the business, and are using IP and the web to lower the inordinately high costs of traditional conference calling, and to increase the opportunities to share different types of media on an impromptu basis.

Indeed, "webconferencing" lets you share far more kinds of infor-mation than a plain audioconference and yet is not as cumbersome to set up as an old-time circuit-switched videoconference. And because even the smallest SOHO's are becoming dependent on an Internet connection to keep the business alive, more companies are offering audio, video and dataconferencing services via the web, since they know that everyone has access to it. Suddenly conferencing becomes as simple as surfing to a cer-tain website and clicking a mouse to set up your conference.

Again, most of these companies have a prerequisite demanding that you establish an account for billing purposes, but that's it. You just surf over to the conferencing website, upload your data presentation (some services, like WebSentric's Presentation.Net, provide platform-indepen-dent apps for you to create your presentation in case you're the last per-son on Earth who doesn't have PowerPoint), click on a few buttons and you're in business. Far-end participants can cruise to the site, log in with a pass code, and watch your demonstration. In some cases you can also use a whiteboard to illustrate points in real time as the conference pro-gresses, and solicit suggestions from participants.

Yes, the infusion of IP allows what would be otherwise plain "video-conferencing" products and services to perform many new and interest-ing functions, such as web marketing, software application sharing, e-learning, on-demand e-commerce customer service, and various forms of customer relationship management.

For example, joining the growing legions of videoconferencing ser-vices and/or bureaus is Visitalk.com (Phoenix, AZ -602-850-3360, www.visitalk.com), an Internet company that provides direct voice, video and information calling, plus things like voice messaging, "buddy lists," document sharing and multipoint conferencing. Visitalk uses Microsoft NetMeeting, Cu-SeeMe or any other H.323 compliant conferencing soft-ware. For a person looking for a quick "in" without much hassle or expense, this method of connectivity seems like a befitting route to the arena of video conferencing.

At the higher-end, Vialog (Andover, MA – 978-975-3700, www.via-log.com) has introduced WebConferencing.com. Just open an account at www.webconferencing.com and you can schedule conferences and pre-sentations at your convenience. You can also use WebConferencing.com for messaging services, such as voicemail broadcast, to deliver voice mes-sages to a large audience; e-mail broadcast, to distribute personalized e-mail messages to a large group; and fax broadcast, to send a fax or file to an unlimited number of recipients without tying up your fax machine.

In the same vein as airline frequent-flyer programs, V-Span (King of Prussia, PA – 888-44-VSPAN, www.vspan.com) once offered an amusing promotion for their videoconferencing service, called the Infrequent Flyer program. Every time you used V-Span's services to run a videoconference, you were awarded points; points were compiled on a quarterly basis and could be redeemed at any time for airline tickets or other prizes. You could also use your points for a company credit or to donate a gift to your favorite charity.

Finally, the cost of owning an in-house videoconferencing system has gone down as simpler products such as set-top boxes have been introduced. These devices are both easier to use and more intelligent than their predecessors; again, the web has played a role by enabling features like browser-based management, and standards like H.323 and T.120 have solidified and are starting to become more interoperable with a wider variety of equipment.

The first so-called "WebTV" turned the family room into a communications station by letting you send e-mail via your television. Now InfoView (San Jose, CA – 408-432-5400, www.innomedia.com) goes a step further and lets you videoconference right from your television with no computer, no special phone lines, and no fees. The InfoView PVP (for Personal VideoPhone) connects to any standard TV and analog phone line to become a videoconferencing system perfect for SOHOs and telecommuters, especially considering its $349.95 price tag. About half the size of a standard VCR, it contains a high-quality digital camera (with zoom, pan, and tilt capabilities) and modem that transmit and receive audio and video (adjustable up to 15 fps) right over your phone line. Just set it on top of your TV and dial. You control all the InfoView's functions with a simple GUI and your telephone keypad.

The privacy mode lets you see others without being seen, for those times you feel an obscene finger gesture coming on when the person you're speaking with says something stupid. You can also freeze-frame and take snapshots of your other party, in case you catch him making a finger gesture back at you (useful for blackmail). And if you want to share your videoconference with someone who can't be in on the call, just press Record on your VCR and send him a tape of the whole thing.

The InfoView can interface with other InfoView phones or any videophone that meets the standards for videoconferencing over regular phone lines. It supports NTSC and PAL formats, which means you can even conference with many Asian and Western European countries. And best of all, it has a $349.95 price tag.

Another example of a standalone videoconferencing system that even a novice can use is Gentner Communications' (Salt Lake City, UT – 800-945-7730, www.gentner.com) APV200-IP, which attaches to any size or brand television set, LCD, DLP projector, flat screen or any other non-intelligent output display. You can use the GUI, improved remote control, and built-in tutoring device to make videoconferencing easy the first time out; onscreen prompts guide you through the entire process.

The H.323-ready APV200-IP is also compatible with any videoconferencing system across ISDN or T-1 / E-1 lines, and it boasts a built-in Ethernet connection so all you need is an Internet / intranet to configure or service the system from wherever you happen to be. It can be used by itself or combined with Gentner's Audio Perfect products to improve the sound of your conference. Price for the APV200-IP is $7,395.

Then there's EnVision from Sorenson Vision (Logan, Utah - 435-792-12-, www.s-vision.com) a company that's been a major force in developing video compression algorithms, or codecs, for squeezing video into small, low bandwidth pipes (Apple chose their proprietary codec for their pure-software QuickTime product). Now, with EnVision, they've trickled down some of their compression research into a desktop system, one which sports an internally developed H.263-compliant hardware codec. With their software expertise burned into a chip, even owners of underpowered 90MHz Pentiums can now have a videoconferencing experience.

EnVision, lets you videoconference right from your desktop, on any PC with Windows 95, 98, or NT, 16MB RAM (32MB recommended),

15MB hard drive space, and a Pentium 90 or faster. You'll also need an IP network card and a VGA/SVGA monitor with 256 colors.

The H.263-compliant video comes through at a constant 15fps at QCIF resolutions, even at 32Kbps data rate, or you can get down to 5fps at CIF; you can choose whether to tradeoff on quality or speed, depending on which is more important. Audio supports G.711 and G.723.1 standards and includes echo cancellation for full-duplex operation, voice activity detection, and speaker/mike audio mute.

Call control features include speed dialing, call screening, call history, CallerID, and password control. And so you know you look your best before the conference call, a self-view window lets you see the image you're sending out. You can share data and applications, or transfer files with EnVision.

EnVision goes for $899 ($799 without camera) and comes with a 30-day satisfaction guarantee and unlimited tech support.

Increasingly higher levels of circuit integration eventually shrink videoconferencing devices to the point where they become embedded systems or appliances, such as the compact and affordable ViGO from VCON (Austin, TX – 512-583-7700, www.vcon.com), a videoconferencing appliance for both desktop and laptop configurations. The ViGO is purchased, installed and serviced as a high-performance appliance, improving desktop videoconferencing quality and reducing Total Cost of Ownership (TCO).

Other advances in technology eliminate the need for you to make a trek to a special conferencing room. Videoconferencing no longer has to be a static meeting between two parties at two different locations, thanks to Teleco Video Systems' (Greenville, SC – 800-800-6159, www.teleco.com) Video Response System, which lets you take the videoconference out of the conference room and down the hall, or over to your company's warehouse. A wireless camera on a mobile tripod sends live video back to the conferencing system via an RF link. Those at both the near- and far-end sites see the images that the battery-operated camera is capturing. Now instead of just trying to explain to a growling company executive at the other end of the conference line that an earthquake has left your factory in rough shape and you need new machinery, you can show him, and he doesn't even have to get out of his overstuffed chair from behind his huge mahogany desk. This system is also helpful for off-site managers to assess and solve production problems remotely, saving money and travel time.

So, thanks to exponential technological advancements and the accessibility of the web, some form of video / webconferencing is now within reach of nearly every business, regardless of their budget or technical acumen.

Because of the widespread accessibility of the Web and the growing ease of use, it seems like webconferencing is going to be around for some time, and that the videoconferencing aspect of it will become more of an everyday experience, both in the workplace and at home.

Cisco Systems' (San Jose, CA - 408-526-4000, www.cisco.com) Architecture for Voice, Video and Integrated Data (AVVID) makes the assumption that video is a part of the "convergence" communications phenomenon occurring in the corporate enterprise. AVVID compliant systems incorporate the key elements needed to deliver converged data, voice and video – multiservice router / voice gateways based on Cisco's IOS technology, application servers running core voice applications, call processing software, integrated web-based system management, and a data switching interface for seamless connectivity to recommended Cisco Catalyst QoS enabled switches.

And as the video aspect of conferencing becomes more commonplace, a whole cottage industry will develop around the psychology of videoconferencing and how to best use the intimacy of face-to-face contact during essential negotiations. And the more commonplace concerns of eye contact, body language, clothing and makeup will also come under scrutiny.

Ultimately, videoconferencing has (potentially) limitless possibilities. From friends goofing around on their home PCs with inexpensive cameras and bundled applications such as NetMeeting, to doctors saving people's lives via high-bandwidth telemedicine applications, the potential market is enormous.

The best advice to someone in the market to buy a videoconferencing system or service is to maintain a sense of proportion and balance. Yes, you can expect and demand more from conferencing applications, with lower barriers to entry. But it's also important to sift through the hype and identify those apps that actually bring benefits to your business. Talk to the major networking vendors: Most of them have developed conferencing apps or are associated in some way with companies who have, and they can all offer helpful advice for optimizing your data network to support the newest high-end systems. Keep an eye out for startups, which have appeared with exciting plans at all levels of the market. And visit those who you know as long-standing videoconferencing specialists: Many of them have released new products or enhanced existing ones in ways that could surprise you.

Finally, be realistic. You probably don't need broadcast-quality, two-way video to conduct an informal staff meeting, for instance. In fact, it may be enough to let your work-at-home team member see PowerPoint slides while talking on the phone. The point is, plenty of conferencing products out there address needs at all points of the spectrum, and you can select the one that best meets yours. The most effective will be those that integrate innovative new technologies most closely with the processes, software, and infrastructure components you already use.

Tips on VC Sound

It's easy to become so wrapped up in video technology and concepts that one can forget about how much information in a conference is conveyed via sound. For successful videoconferencing, a quiet, accessible room works best (just like a TV studio), with a carpet on the floor and preferably no windows. Any existing conference room can be adapted for use as a videoconference room by making adjustments based on the needs of video and audio equipment. A room 20 feet x 30 feet is a common size, but this will depend on the planned number of attendees. The ambient noise level in the room should be within NC-30 or NC-35, and the reverberation time should be less than 500ms (.5 sec). Inexpensive acoustical panels can be used to adjust the reverberation.

When it's your turn to speak during a videoconference, try to speak at the lowest pitch level that you find comfortable, but project your voice at your normal volume and use your regular patterns of inflection. Change the speed of speaking to emphasize content, it's okay to use silent pauses for emphasis and of course allow time when waiting for an answer.

Try not to read things verbatim. And try to avoid slovenly verbal habits, such as any distracting repetitive phrases that are magnified by a VC system, which can consist of things as simple as such "non U" utterings as "uh" or "you know" at every pause.

Navitar's Tips for a "Video Friendly" Room

Do you look like a raccoon on video? That's because your lighting is coming from ceiling lights positioned at too high an angle.

Now that everybody is starting to use videoconferencing, everyone needs to pay any attention to what they look like on screen. The "director of photography" or "lighting cameraman" is a venerable (and valuable)

profession in both the motion picture and television business, and with good reason, if you've ever seen any of the more decidedly unphotogenic Hollywood stars.

The majority of videoconferencing problems are directly linked to the source, direction and abundance of light in the room. Wall coloring and participants' clothing also effect the on-screen image in a videoconference.

The following variables are major contributors to image quality:

- Type of light being used, along with light angle and color temperature
- Background/wall colors
- Amount of white in camera view
- Furniture and accessory placement

To transmit a professional-looking image, a videoconference facility must use light at the right level, angle and color temperature. Angling the light correctly is also important. For instance, movie directors like to shoot early in the day, when the earth is tilted at a 45∞ angle to the sun. This gives them the best image depth and tone. Without enough light, the both film and video appears dark and grainy, or "noisy."

Of course, you don't have to be a Charles Rosher or a Gregg Toland to get pleasing results. Navitar (Rochester, NY - 800-828-6778, www.navitar.com) is one of the few companies that understands everything about lighting and what it takes to make a person (or even a document) look good on screen. Nivitar has become perhaps the world's leading manufacturer and supplier of superior quality optical and electrical products for the

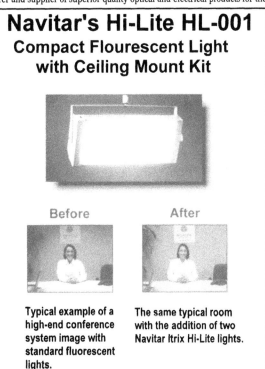

Navitar's Hi-Lite HL-001
Compact Flourescent Light with Ceiling Mount Kit

Before · After

Typical example of a high-end conference system image with standard fluorescent lights.

The same typical room with the addition of two Navitar Itrix Hi-Lite lights.

On the left, with "normal" room lighting, you'll notice that the facial features are not distinct, skin tones and colors are inconsistent and lack detail, lighter shades are "flaring out", due to lack of contrast. On the right we can see that by using Navitar's Itrix Hi-Lites the shadows have been removed and the facial featuresnow have detail, the skin tone becomes more natural and flattering, and "flaring" and "hot spots" are minimized and contrast is increased between the subjects and the background.

machine vision, audio visual industries and video-conferencing industries.

Navitar's Itrix Hi-Lite lights provide studio-quality lighting specifically designed for the the videoconferencing environment. The lights drastically improve how subjects look, making them appear warmer and more alive. The Hi-Lites manage to strike a balance between the functionality of a broadcast studio quality light and the esthetic and logistical considerations of different compressed video environments.

Here are Navitar's Tips for creating a video friendly environment:

1. Make sure that participants' faces are the lightest color that the camera sees. Although you may not realize it, there is a "white-balance" control in your videoconference room and it's critical to understanding the best way to fix a poor screen image. Your videoconferencing system averages the varying tones of color in the shot, and then compensates for a certain percentage of "gray" (around 20%). Thus, if a shot contains a high percentage of white or light tones (such as a white background wall or white clothing), the system will attempt to darken the overall shot, making the darker sections even grayer. The trouble is, faces are often electronically perceived as some of the darkest objects in view and are dimmed more. Thus, unflattering shadows will appear or deepen on the face of a person sitting in front of light-colored walls. People with darker skin tones are especially affected.

2. Remove all white background from the scene if possible. Videoconference attendees should avoid wearing white clothing. Participants should wear dark or neutral clothing, although people with darker skin tones will need to wear lighter clothing to maintain face distinction.

"Horseshoe" Set-up

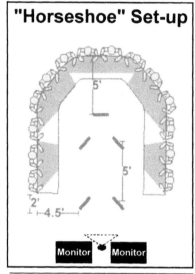

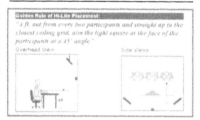

The "Horseshoe" configuration is another commonly found arrangement in executive conference rooms. Lights, such as Itrix Hi-Lites are placed in the center of the room, facing outward, to give equal and proper coverage, equally, to all participants.

Also, conference attendees should avoid wearing clothing with stripes or busy patterns. These will cause the camera's focus to oscillate, hurting picture clarity.

3. Keep your walls dark. If possible, avoid white backgrounds. Dark colors work best, so make sure your wall or backdrop is a deep shade. Royal blue or robin's egg blue will produce the most pleasing results. However, any dark shade will improve the shot. If you're conducting a training session, don't set-up a large white grease board behind the professor. This will cause the camera's electronic devices (i.e., the automatic iris) to overcompensate for the bright back ground.

4. Choose a tabletop that is non-white and non-reflective. It's best to use a darker table with a matte finish. White and/or high gloss, reflective table will create glare and cause the camera to overcompensate. Lighting is installed with regards to where participants will be sitting and facing the camera. If you change the size of your table, it will be necessary- to readjust your lighting to fit the new set-up.

5. Install lights with the proper color temperature and intensity. What many videoconferencing presenters don't know is that it's crucial to select lights with the right color temperature. Lighting at the proper color temperature is crucial for healthy skin coloring and achieving a "warm," pleasing look for the conference participants. Color temperature of light is measured in degrees Kelvin (K). If fluorescent-cent fixtures of improper color temperature are used, participants will appear either green & jaundiced, or cold and blue. For indoor video-conferencing, 3200∞K is the optimum color temperature. Occasionally, there are some instances, where 5500∞K (daylight color) bulbs may work best.

Conference Room

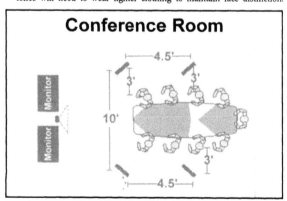

The 6-8 person oval conference room table is the most common configuration. This type of room generally has a simple set-up with a group system or "roll-about" at one end of the table. Four lights are used, at 45 degree angles, to provide coverage. Additional lights can be added for individual coverage.

Distance Learning

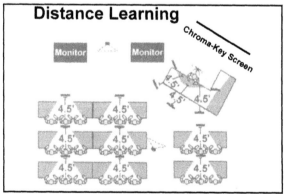

This diagram of a typical distance education configuration shows the front of a traditional lecture hall. Lights such as Navistar's Itrix Hi-Lite lights are used to give the on-camera teaching area, and the individual student desks, proper coverage.

Most major codec manufactures have a setting specifically for 3200∞K lighting. Videoconferencing managers should consult their manual and check the default settings to make sure settings are optimized for their particular room. Keep in mind that low-voltage, fluorescent fixtures are very inconsistent in color temperature out-put and produce the flickering which is detrimental to codec performance.

Note: It is important to understand that color temperature does not affect the "brightness" of the light, or the amount of light being thrown. Remember "whiter is not brighter!"

Light intensity is measured in foot-candles (fc). One foot-candle offers illumination equal to the amount of direct light cast by a single candle, one-foot away, on a square foot of surface. Higher foot-candle or light levels reduce fine-grained, static "noise" in the video signal.

Most video cameras are designed to function with only one foot-candle! However, we don't recommend trying it! Consider that the drop ceilings commonly found in most office suites are 9-feet high. With the usual fluorescent fixtures, the average foot-candle reading, 4 1/2 feet from the ceiling, will be between 12 and 30 foot-candles. For a quality video-conference, you need to provide over 40 foot-candles of soft, directional light. Optimal foot-candles for videoconferencing are between 60 to 80 foot-candles.

To provide some frame of reference:

- Starlight measures a mere 0.000005 foot-candles
- Moonlight provides about 0.01 foot-candles
- A bright office is illuminated by approximately 40 foot-candles
- TV studios are lit by about 150 foot-candles
- Sunlight, on an average day, ranges from 3,000 to 10,000 foot-candles
6. If there are windows in the room, have participants sit facing the window. Curtains or blinds should be installed to control the amount of light streaming in. One common mistake in corporate videoconferencing is setting up the conference room in a corner office with two glass walls. The exposure is wasted because heavy curtains will have to be installed to shut out some of the light.
7. Optimize the brightness and auto-iris controls of your camera to best suit your rooms needs. Don't settle for default! Once you've eliminated the lighter colors in your room, don't be afraid to adjust the brightness and contrast controls on your camera. Remember that factory defaults assume the worst of conditions. Without the danger of white creating flare in your shot, try opening up the iris one-half stop or increasing the brightness of the camera.
8. Use canned lighting or a string of track fixtures. Using canned or track lighting fixtures aimed at the wall produces a "wall-wash" effect that will provide contrast and greater three-dimensional perspective to the shot. This is an advanced option. Aim them at the walls!
9. Throw in a fake plant or tree to humanize the setting! It's important not to clutter up your videoconference room setting, but you don't want it too sterile looking either. The addition of something as simple as a plant will humanize your setting without being intrusive to your videoconference.
10. Don't forget to tilt pictures, framed degrees or awards downward to avoid glare.

Virtual call center

The first definition of "virtual call center" is a system that allows a company to extending transparent inbound and outbound call processing functionality across multiple locations, and across various kinds of technologies (such as networked ACDs from multiple vendors), to agents.

Call centers with 25 to 50 agents are experiencing the most growth in the U.S., but you say that you can't afford to spend the money to build one? You say your business has no room for a conventional call center – or a PBX for that matter? No problem. The virtual call center and virtual office have arrived.

Traditional call centers (or contact centers, as they are also termed) tend to be located in one building or even one room, with the agents supplied calls from an automatic call distribution (ACD) system. The ACD generally makes it difficult to distribute the call center agents across multiple locations or to allow for telecommuting, home-based workers.

It has been most of all the arrival of IP telephony that has enabled the modeling and provisioning of a call center as a flexible, software-driven application. Agents or Customer Service Representatives (CSRs) use VoIP client software on their desktop PCs instead of traditional "turrets" and so they can now be located anywhere (especially at home), and they'll still have full access to all call center facilities.

Such virtual call centers allow for logical groupings of agents or knowledge workers working within a company's "customer contact zone," doing things such as customer services and telemarketing campaign management. Such systems can route calls to agents anywhere, looking at agent skill sets as a whole, without segmenting them into specific sites.

In small companies, employees in different departments might handle customer inquiries, billing and complaints and can collectively be considered a virtual call center. In large companies, a virtual call center might include call centers in different cities.

There is another "virtual" category, which is simply a "Virtual PBX" or "Virtual Office" for business. Let's say that ten programmers are working on a new word processing program. They are organized as a company but they live at various points around the country. You call the "business" (usually an 800 number) and dial an "extension" to reach a certain person.

Some ACD-like routing mechanism, either at the company's office or in equipment at an enhanced service provider or perhaps even built-in to the carrier's network, connects you with the person you want to talk to.

Neither the callers nor the employees can tell whether they've been routed by an enterprise switch, by a third-party, or by a telco Centrex service. Perhaps the agent or employee is a telecommuter working from home. It doesn't matter so long as transactions in the customer contact zone are smooth and the whole operation appears as an integrated whole.

Of course, open, programmable computer telephony components made the whole idea of the virtual office possible – things like the H.100 / H.110 CT Bus based media processing and switching resources and dumb programmable switches from Excel, Redcom, and Summa Four have lent the flexibility which has allowed virtual call centers and offices to evolve.

If you doubt that computer telephony technology makes the virtual office possible, ask yourself what conventional Centrex, PBX or key system alone can support such advanced functions as fax mail, fax broadcast, fax-on-demand, fax-to-speech, multi party conferences from any touchtone phone, e-mail to speech, or e-mail forwarded to a fax number?

When the idea of virtual call centers and offices first appeared, we at first thought this technology would simply be used to virtualize specialized apps, in a manner similar to Genesys Labs (San Francisco, CA - 415-437-1100, www.genesyslab.com), who took their enterprise applications and only scaled up some of them so Network Service Providers could offer them as enhanced services.

Ultimately, however, this technology (and lifestyle) could become even more pervasive, with all of a company's switching functions and applications offered on a virtual basis, similar to the idea behind the Crosspoint Virtual PBX from VirtualPBX.Com, Inc. (San Francisco, CA - 888-825-0800, www.virtualpbx.com), formerly Advanced Queuing Systems.

The CrossPoint VirtualPBX enables companies to create a seamless virtual office environment in which calls can be routed and transferred anywhere in North America. Callers are greeted by a professional automated attendant and are given the impression that everyone is working out of one office.

With the CrossPoint VirtualPBX, there are also "virtual extensions:" calls can be routed to extensions that are connected to a cell phone, a home office phone, any touch-tone telephone. Callers call in, are greeted by a professional automated attendant, and then connect to your employees or distributed agents, wherever they are. They can also send a fax to any extension. Faxes can be retrieved at any fax machine.

CrossPoint can also distribute incoming calls to a Virtual Call Center. Your employees or agents log into the call center and then the VirtualPBX routes the calls to the agents in a first come first served basis. If all of your agents are busy taking calls, your customers are placed on hold until an agent is free. They'll never know the Sales Rep they're connecting to is not at your office! This is the ideal solution for sales and support organizations with remote, on-the-road, or distributed teams.

And because the CrossPoint VirtualPBX is a service that you can use with any phone lines or as an overlay to your existing phone system, there's no need to purchase, install or maintain new phone equipment.

When you sign up for the CrossPoint VirtualPBX service, you get your own private VirtualPBX, with your own toll-free number, auto-attendant greeting, private extensions, options, music-on-hold, ACD queues, etc. Because the system is housed and maintained by AQS, you don't have to purchase, install, or maintain expensive PBX phone equipment. You simply sign up for the CrossPoint VirtualPBX service, and begin using your VirtualPBX immediately!

Many companies are using outdated telephone systems for their businesses, or have bought used equipment from other businesses. With the CrossPoint VirtualPBX, you can upgrade your PBX without buying any new equipment.

As long as your existing PBX supports either DID (Direct Inward Dial, where each user has their own direct telephone number), or an automated attendant that a caller can dial through to an employee's desk, you can use the CrossPoint VirtualPBX as an overlay to your legacy phone system. In minutes, you can add voice and fax mail, company wide directory services, follow me roaming, call screening, new message paging, and ACD call routing to your office phone system without even touching your existing PBX.

The CrossPoint VirtualPBX is a truly affordable alternative to installing and maintaining your own expensive PBX system.

You get all the features of a full PBX system (plus many more), without the costs associated with purchasing, installing and maintaining PBX equipment onsite. With CrossPoint, there is no large up-front cost, and no ongoing monthly fee. You pay only for the time you spend on the phone!

The CrossPoint VirtualPBX is only one example of an interesting trend. Instead of calling into an ACD at a main site and being switched out from there, many new start-up companies have no real "main site" at all, with people scattered about the country, particularly if the nature of the business is to keep employees mobile.

Virtual offices for the masses? It can be done. States such as California are mandating that companies set up programs that will reduce traffic and pollution. Telecommuting would be the ultimate means of complying with such state regulations.

It's known that in the case of telecommuting call center agents, solutions from companies like Genesys Labs can keep tabs on what a home agent is up to at all times, just as if they were in an actual formal call center.

Trade-offs? You may get a slightly better managed agent in the formal call center (management feedback is also faster) but you get terrific flexibility (call overflow can be immediately farmed out to agents at home) and lower cost (no overhead or office space) with telecommuters.

For the moment however, in larger, established companies, this kind of technology will apply mostly to connecting formal call centers separated by some distance, since multiple sites pooled together as a single entity allows for the creation of such things as "follow the sun" call centers, where the sun never really sets on a national or multinational business.

Cisco Systems' (San Jose, CA · 408-526-4000, www.cisco.com) Architecture for Voice, Video and Integrated Data (AVVID) provides a way of connecting the various branches of an enterprise that, once it is in place, makes those connections seem to disappear. One example is the IP Contact Center, which Cisco has demonstrated, showing integration of GeoTel's products (another company Cisco acquired) into the AVVID architecture. AVVID virtualizes the call center, making all underlying transport agnostic to both the applications running on top of it, as well as the physical locations from which those apps are begun and terminated. The point is that this architecture does not have to be implemented in any one particular way, or even using one particular vendor's products throughout.

Essentially, AVVID pulls together Cisco's IP-based technologies for the enterprise market. The products AVVID encompasses can be differentiated into four layers. On the level of network infrastructure, Cisco offers a complete line of gateways, multilayer switches, and voice-capable routers. The Cisco CallManager offers core call control across the network, and has enough functionality to enable PBX replacement (though the entire system interoperates with existing circuit-switched equipment). On the applications layer, server-based apps like unified messaging can be distributed enterprise wide. Finally, Cisco has produced a line of client devices, including IP phones and soft phones that provide a user interface to network applications.

Such uses of VoIP / IP telephony reduces call center infrastructure costs and, through a closer coupling of voice and data services, improves overall call center efficiency and service to the customer.

Virtual Private Network (VPN)

A private data network which offers a point-to-point connection for data traffic between a client and a server or multiple servers that is established over a shared or public network (particularly the Internet) with the aid of a special form of encryption or some other security technology. The VPN connection across the Internet logically operates as a Wide Area Network (WAN) link between the sites, with data being exchanged between peers "in secret." A VPN is similar to a self-contained network. It travels over but isn't really connected to the external network (unless you want it to be) and ideally there's no exterior indication of the private network's existence, which hopefully wards off hackers and crackers. One would therefore be tempted to call a VPN a point-to-point connection over the Internet with added security.

Critical components of VPNs include:

• Security
  • Encryption for privacy and integrity of data - this is the "private" part of virtual private networking
  • Access control to maintain security of network connections
  • Authentication process for verification of user identity
• Traffic Control
  • Reliability
  • Quality of Service
• Policy Management
  • Manageability of security policy for the entire enterprise

With VPNs, organizations can extend their network over the Internet to branch offices and remote users, essentially creating what could be interpreted as a virtual LAN or private WAN via the Internet. Also, VPNs can be used by telco Service Providers (SPs) looking to provide POP outsourcing services to ISPs

For many years, the telephone companies delivered voice and data services using what they called "virtual private networks." In fact, just about all software-defined networks are considered VPNs by the phone companies. Corporations proved in the late 1980s that they liked the concept of virtual networks when they moved most of their corporate voice traffic from private leased lines onto the then-new voice VPN services offered by their public network service providers.

But the current generation of VPNs is very different than what was defined by the previous, very generic networking term. VPNs are now generally considered to be a combination of tunneling (a methodology that guarantees the safe passage of a packet of data through the Internet using encryption to protect the data payload, as well as the source and destination addresses), encryption, authentication, and access control technologies and services used to carry traffic over the Internet, a managed IP network or a provider's backbone. The traffic reaches these backbones using any combination of access technologies, including T-1, frame relay, ISDN, ATM or simple dial access.

Security is a major issue concerning VPNs. Indeed, one might say that VPNs are a by-product of network security. Using a VPN involves encrypting data before sending it through the public network and decrypting it at the receiving end. An additional level of security involves encrypting not only the data but also the originating and receiving network addresses, so that entire IP packets are encapsulated or "tunneled." The tunnels are set up between a network access point (such as Shiva Corporation's LanRover Access Switch) and a tunnel terminating device on the destination network.

IP and IPX packets are encapsulated in a tunneling protocol (see below) and these packets are, in turn, packaged by an IP packet containing the address of the corporate network, the packet's ultimate destination. It's possible that when you initiate the session the local network access server at the POP can be used to give your own IP addresses, allowing you to retain your unregistered address for working with the LAN. The encapsulated packets can be encrypted end-to-end using IPSec or its equivalent.

The Virtual Private Network Consortium (www.vpnc.org) believes that there are three major protocols that can be used to create VPNs:
• Internet Security Protocol (IPSec) is a generic structure defined by a large number of standards and recommendations that were initiated and codified by a working group of the Internet Engineering Task Force (IETF) Many expect IPSec to become the dominant way to provide security services for the Internet Protocol (IP), both IPv4 as well as IPv6, involving encryption algorithms, authentication algorithms, and management of encryption key exchange.

The University of Arizona Department of Computer Science has created an IPSec kernel module for Linux, and the National Institute of Standards and Technology (NIST) Internetworking Technologies Group also offers Cerberus, an IPSec implementation for Linux.
• The Point-to-Point Tunneling Protocol (PPTP) - a proprietary protocol originally designed as an encapsulation mechanism to allow the transport of non-TCP/IP protocols (such as IPX) over the Internet using the Generic Routing Encapsulation (GRE) protocol. PPTP was developed by and has been proposed as a standard protocol by Microsoft, Ascend, 3Com, and several other companies. The PPTP specification itself is fair-

ly generic, and allows for a variety of authentication mechanisms and encryption algorithms, as outlined in the IETF's Request For Comment (RFC) documents 1701 and 1702. Most of its security features were added later, not included in it from the beginning. Microsoft has built PPTP into its line of Windows server software. PPTP is usually installed on a company's firewall server. Novell's BorderManager also supports this protocol, as can other operating systems using PPTP software.

PPTP-Linux is C. Scott Ananian's free PPTP client for Linux that allows you to connect a Linux or other Unix computer to a PPTP server. Ports of pptp-linux to other Unix variants should be trivial, but have not yet been performed. Ananian is also working on a PPTP server version for Linux. The PPTP-Linux home page is www.pdos.lcs.mit.edu/~cananian/Projects/PPTP/.
• Layer 2 Tunneling Protocol (L2TP) is an IETF standard tunneling protocol that evolved from a combination of PPTP and the proprietary Layer 2 Forwarding (L2F) protocol found in Cisco's IOS. Whereas PPTP is geared toward ISPs and has provisions for call origination and flow control, while L2F has less overhead and is suited for managed networks. L2TP has the best features of both protocols. L2TP is implemented by ISPs to provide secure, node-to-node communications in support of multiple, simultaneous tunnels in the core of the Internet or other IP-based network. End user access to the ISP is on an insecure basis. The Internet Draft of L2TP was submitted in March 1998

Several distinct VPN applications are now popular, such as remote access to a corporate site wherein secure connectivity is provided to remote and mobile users, secure site-to-site connectivity for multiple branch offices, extranets where secure connectivity is provided to customers / suppliers / key business partners, and internal corporate usage on intranets.

When it comes to remote-access VPNs, the basic concept is to give telecommuters and mobile workers a way to get back to a corporate network over the Internet or a service provider's backbone.

In a typical remote-access VPN, telecommuters and mobile workers can connect back to a corporate network over the Internet or a service provider's backbone after the user's special client software recognizes the specified destination and negotiates a secure tunnel between the network access server and a tunnel-terminating device on the LAN. To do it, the user just dials into a Service Provider's (SP's) or Internet Service Provider's (ISP's) Point of Presence (POP), where the SP or ISP authenticates the user and establishes a tunnel either over the SP's private network or the Internet "cloud" back to the corporate LAN's edge, where the user may once again have to authenticate himself or herself to gain access to the corporate network. The encrypted packets traveling back and forth are wrapped in IP packets to tunnel their way through the Internet during the course of the encrypted VPN session. The VPN server negotiates the VPN session and decrypts the packets. The unencrypted traffic then flows as it would normally to other servers and resources.

Contrast the above procedure to the traditional "dial-in" technique whereby a user makes an expensive long distance call into a bank of modems, a remote-access server or concentrator located at headquarters. Some companies can cut telecom charges from $1,000 per month per person with dial access to less than $20 per month per person when using an ISP service that changes only a flat monthly rate.

Before switching over to a VPN, a company may have had a dedicated link to an ISP for Internet access and a 24 channel T-1 line into a remote access server to support dial-in users. Using a VPN means that you don't need the costly T-1 line for dial-in access – the traffic from these users now travels over your existing Internet access line. In one fell

V

swoop, a single VPN gateway and one "pipe" can eliminate the need for multiple network edge devices and at the same time support many different kinds of applications.

In the case of site-to-site connectivity, as in the remote-access scenario, branch offices connect to corporate headquarters through "tunnels" that transport traffic over the Internet or via a provider's backbone. Again, as in the case of remote access, a company can reduce communications costs by paying only for the access line from a branch office to the nearest service provider's (POP), rather than paying for a long distance link all the way to corporate headquarters. In this way offices in foreign countries can circumvent international tarrifs.

Also, many sites have multiple access lines: One to carry data back to headquarters and a second for Internet access. In fact, some industry studies have found that more than 70% of sites have multiple access lines. Using VPN technology for site-to-site connectivity allows a branch office with multiple links to remove the data line, moving traffic over the existing Internet access connection instead

A site-to-site VPN can be installed as quickly you can secure a high speed Internet connection (a few weeks) as opposed to installing a leased high speed data services line (several months in the U.S., even longer in foreign countries).

VPN forms of extranets are becoming popular. One can create extranets without VPN technology, of course, but VPN-based extranets give IT managers another option, especially if they already have a VPN installed for some other purpose. VPN-based extranets use the VPN's access control and authentication services to deny or grant customers, trading partners and business associates access to specific information they need to conduct business.

With a VPN-based extranet application, outside parties get to the corporate firewall by tunneling across the Internet or a service provider's network, where VPN access control services decide whether or not they can continue on through the firewall to the corporate data store.

Some IT managers use a "soft-dollars" argument for justifying a VPN-based extranet: An extranet can give different classes of customers different levels of privileges. For example, a brokerage house could segregate users into classes so that a client that spends the minimum amount is only allowed to trade electronically; a higher-spending customer gets to trade and gets a free Internet account; and a premium client gets all of this along with access to special internal stock market research reports.

A similar internal corporate use exists for VPNs. One can use the VPN's encryption, authentication and access control services to segment populations on a corporate network or intranet. For security purposes, VPN technology can control access to various classes of data for various groups. For instance, a Human Resources (HR) department could allow employees check on their accrued vacation time, but not be able to see their performance reviews. Or a national sales manager might be granted access to the sales performance records of all sales associates, while each associate only has access to his or her own records.

In the past, IT managers tried to segment user populations using Virtual LAN (VLAN) technology. With VLANs, the idea is that a manager can quickly create *ad hoc* groups of workers who appear to be on a single LAN segment. A manager can dynamically assign users to specific groups and restrict others from any one group.

Unfortunately, most VLAN approaches are proprietary and don't work in environments where there's a melange of different vendors' hubs and switches.. VPNs can cut across the mixed-equipment environment by using IP-based tunnels between a user's workstation and a server, thus creating a milieu that's similar to physically segmenting users on discrete LAN segments.

All of these various applications for VPNs are not mutually exclusive. A company could deploy a VPN to reach its branch offices, then let single remote users access the network too, and ultimately allow the network to be accessed by privileged outsiders, all done by just changing some settings in software.

The VPN products themselves can be grouped into three categories: Hardware-based systems, standalone software packages, and firewall-based systems. These days all such products support both LAN-to-LAN and remote dial-up connections.

Hardware VPN products are typically encrypting routers that store encryption keys on a chip, and are more difficult to crack than their software-based systems. Encrypting routers also have greater throughput than software based systems, and should be used whenever a large "pipe" enters the enterprise. Some hardware VPNs demand that you install a special card in each PC on the network.

Software-based VPNs are generally more flexible than hardware-based VPNs, and is used in situations where remote sites encounter a mixture of VPN and non-VPN traffic, such as web surfing. Software systems have less performance than hardware-based systems and tend to be more difficult to manage.

Firewall-based VPNs, on the other hand, can leverage the firewall's existing security mechanisms, including restriction of access to the internal network. They also do address translation and can perform strong authentication procedures. Most commercial firewalls "strip out" services that may pose security problems, thus hardening the host operating system kernel. Firewall-based VPNs are inexpensive and are great for intranet-only implementations.

VPN Telephony

Of course, what we're interested in is not just sending private data connections for VPNs, but voice traffic with a high Quality of Service (QoS) as well - VPN telephony. Once VPNs are in wide use, they'll provide the opportunity to integrate other types of communication such as multimedia and Voice over IP (VoIP). Some major telecom players are interested in IP-based VPNs. Up until late 1997, the Regional Bell operating companies (RBOCs) appeared to have little interest in IP VPNs. Following a wave of IP VPN efforts by AT&T, GTE, MCI WorldCom, Concentric Networks and others, the RBOCs suddenly played catch-up, buying up companies with VPN expertise. VPN telephony products for the customer premises is also starting to appear.

For instance, the Clarent VPN Telephony Suite from Clarent (Redwood City, CA - 650-306-7511, www.clarent.com) can run on a Celeron-based PC with 64 MB of SDRAM and equipped with telephony connections such as a digital T-1 / E-1 or analog POTS lines (RJ-11 jacks). The Clarent VPN Telephony Suite gives service providers a new way to add value and flexibility to corporate communications services by enabling voice and fax to travel over VPNs. The product is essentially a voice and fax over IP gateway for connecting telephone voice calls and fax over VPN and public IP networks.

Clarent developed their comprehensive voice and fax for VPN solution by combining their best-in-class voice and fax processing technology with an interesting routing and billing structure based on open database standards, together with a understanding of corporate telecom needs.

Each call is routed to a gateway within the VPN if one is designated. If the VPN does not cover certain cities or countries, corporate traffic can utilize public IP voice networks for call completion. Calls can access the public network from within the VPN, but public calls cannot access the VPN from outside. This yields low-cost voice and fax communications anywhere in the world, least-cost routing, and the security of a closed private network

In terms of voice quality, patent pending technology improves packet loss recovery, reduces latency and delivers the best sound quality within the lowest bandwidth even while other traffic consisting of fax and data also travels along the single, secure, flexible network. For VPN service providers, the ability to add voice and fax to the VPN leads to a total communications partnership with the customer, allowing the provider to move beyond simple bandwidth leasing to a complete one-stop shop for communications service on a global basis. Also, service providers can offer more competitive rates to customers because the cost of administration, billing, and account maintenance are consolidated into a single system.

Clarent Command Center provides billing and management for networks of multiple VPN domains, providing hundreds of thousands of simultaneous telephone calls through a single point of administration. Different inbound and outbound charges can be set for each VPN domain. Free Calls or Call Blocking options can be set for individual corporate customers. In terms of call detail recording, the system records both inbound and outbound call information including customer account, time and duration of call, destination and call termination code.

Clarent Command Center consists of a software program attached to an ODBC-compliant relational database. The Clarent Command Center package can run on a variety of high-availability Windows NT platforms connected to the Internet, from simple servers to high-speed systems with RAID drives, tape backup and hot-swappable power supplies. Multiple Clarent Command Centers running on separate machines and attached to the a single billing database can increase uptime through redundancy – keep the system running even when one Clarent Command Center is unavailable.

Aside from NT servers, the database can be housed on any high-availability server from an NT server to a mainframe, running a standard RDBMS, such as Oracle or Microsoft SQL Server. By increasing the database speed, Clarent Command Center can support additional gateways within a single network.

The Clarent Service Editor and the IVR toolbox allow flexible customization of voice prompts either on premises or remotely through a service provider office

The system can scale from four to thousands of simultaneous calls

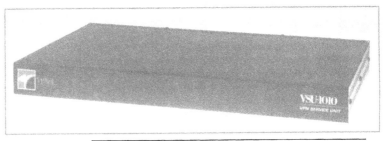

VPNet Technologies' VSU-1010E is a mid-range VPN gateway that supports up to 600 simultaneous tunnels at 10 Mbps. Typical users include mid-sized enterprises and branch / partner offices.

in each corporate site, with easy upgrades as branch offices grow in size and traffic.

VoDSL

See Voice over DSL, xDSL.

VoFR

See Voice over Frame Relay, Frame Relay.

Voice enabled DSL

See Voice over DSL (VoDSL).

Voice over DSL (VoDSL)

Also spelled "voice-over-DSL" and a specialized, proprietary form from Aware, Inc., is also known as Voice-enabled DSL (VeDSL).

Voice over DSL allows for up to 16 to 24 digital channels to come into your home or small business, also giving you Internet access and allowing conventional analog POTS service over the same wire too. VoDSL is a "back room" technology, which means that you shouldn't ever have to tweak any arcane equipment. Your carrier will provide you with a "black box" to convert one or more DSL lines into Ethernet (data) and POTS (analog voice) ports, so all of your existing data and voice equipment will seamlessly integrate to this new kind of service.

VoDSL is expected to find a home in the "middle market" defined as consisting of the Small and Medium Enterprise (SME) and Small Office Home Office (SOHO) segments.

Despite the lack of complete VoDSL standards, during 1999 and into the 21st century, just the promise of simple, cheap, high-speed Internet access over Digital Subscriber Line (DSL) technology had consumers and small businesses in excited anticipation of a new access technology cheaper than anything that had come before it, one that it was hoped would soon offer a range of new services. For many, acquiring DSL services of any sort has proved as frustrating as it is desirable. But as internal service organizations are sufficiently built out and more and more competitive elements take hold in the marketplace, DSL is finally beginning to deliver on its initial promises: lots of bandwidth, always available, at an exponentially lower cost than any previous access technology.

Nonetheless, for almost every DSL provider, incumbent and competitive alike, the push for broadband services has never just been about Internet access. As with any telco service these days, the bundle is king; and, in most cases, voice is considered to be an essential part of the value proposition, from both the carrier's and the

The VSU-1200 from VPNet Technologies, Inc. (Milpitas, CA — 408-404-1400, www.vpnet.com) is a high-availability VPN gateway with hardware redundancy that supports up to 7,500 simultaneous tunnels at 90 Mbps. Typical users include large enterprises and managed VPN data service providers.

customer's perspective.

Initially, the reasons for marrying voice and data under the rubric of DSL seemed clear: by using the high-bandwidth data pipe now available to homes and small offices, one could easily pile on multiple phone lines to the service offering without incurring the cost, time, and labor previously associated with lighting up a new voice circuit. Once the revenue potential of this idea became evident, a highly targeted and architecturally simple model quickly arose.

Essentially, this model consists of two gateways. One, a Next Generation Integrated Access Device (NG-IAD, but usually just called an IAD), sits at the customer premise and mediates between a single DSL connection and anywhere from two to around 24 analog telephones. The IAD is a clever device that eliminates the need for a DSL modem, bridge or router. Such VoDSL equipment works in a manner reminiscent of multiplexers for T-1 and frame relay circuits. Rather than provisioning individual circuits for each phone line, many calls can be multiplexed onto a single DSL line, which doubles as a high-speed data circuit for Internet access. Voice is digitized and compressed (usually 32Kbps ADPCM compression) by digital signal processors (DSPs) in the IAD ensconced at the subscriber's site and connected to the DSL circuit. The IAD serves as a circuit / packet gateway by converting outbound voice traffic to IP packets or ATM cells, and interleaving each voice channel with ordinary data traffic (multiplexing) then sending everything to a DSL Access Multiplexer (DSLAM) situated at the local central office.

The DSLAM separates both data and voice packets and routes them in two different ways: Data packets are routed to the ISP's network of choice (for example, ATM or frame relay) where it's treated as ordinary packetized traffic (which means that ATM traffic from multiple DSL links gets aggregated and multiplexed onto a common upstream link, and each cell flow is directed toward its destination by one or more ATM switches). Voice packets, on the other hand, are sent to a carrier-class voice gateway – the second major "gateway" in our VoDSL model – in the Regional Switching Center (RSC) of the Competitive Local Exchange Carrier (CLEC). The voice gateway acts essentially as a bridge between the circuit-based voice switch and the packet-based DSL access network, converting voice packets to traditional circuit-switched traffic, reconstructing the calls before sending them to the Class 5 voice switch (and thence out over the PSTN) through standard time division multiplexing (TDM) trunks using a GR-303 interface (GR-303 is a Telcordia-defined interface between a Class 5 switch and remote digital terminals that provide PSTN access to analog loops). Some gateways support T-1 trunking which means that calls can be sent directly to Class 3 or 4 long distance switches, thus bypassing reciprocal charges for Class 5 switch termination. Outside of North America, many voice gateways are produced with V5.2 interfaces.

The first VoDSL gateway systems were touted by vendors like Tollbridge, Jetstream, and CopperCom around 1998; and, after several rounds of lab testing and troubleshooting, these vendors' products are now starting to show up in actual CLEC deployments around the country.

Yet, as VoDSL becomes a reality, it is simultaneously entering into an identity crisis. At a faster pace than many expected, voice is becoming a commodity. And while this commoditization is a relatively well understood effect of convergence, its implications for carriers who have traditionally relied on phone service as their primary source of profit margin are still far from clear. Soon, the instant and inexpensive delivery of multiple phone lines, accompanied by a high-speed data connection, will be a minimum requirement for capturing any small business as a customer. Retaining that customer, however, and most importantly, making money

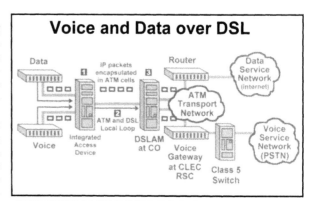

Voice and Data over DSL

1. A customer premises next-gen integrated access device encapsulates IP data packets into ATM cells for DSL transmission. It also takes care of routing and IP address management.

2. ATM is used as Layer 2 DSL transport portocol, allowing different types of traffic to travel over the same copper wire.

3. A DSL Access Multiplexer (DSLAM) routes encapsulated voice traffic to a voice gateway, which converts it back into voice signals.

from him, will take more than just a dumb pipe. Given this scenario, DSL, whether carrying data or voice, becomes simply an enabler - albeit an innovative and significant one - which must be understood within a much broader and more general technological and commercial context.

VoDSL Is (Not) Internet Telecom

In its first phases of appearance, VoDSL was conceived implicitly as its own segment of packet telephony, partaking in the overall movement of convergence, but serving a separate market and even employing a separate set of technologies from what more generally would be known as "Internet telecom." At a time when the latter was viewed, more or less accurately, as a niche market whose primary application was cheap long distance calls, this distinction had an undeniable degree of utility. To successfully offer primary line local phone service, no matter what the potential cost savings, any alternative to POTS would first have to prove itself capable of delivering the same level of quality and reliability as standard, circuit-switched telephony.

From the start of 1999 on, however, Internet telecom has undergone some significant alterations. At the most basic level, industry mindshare has shifted from cost savings to "enhanced services," and this conceptual shift is rapidly being backed up at a technical level. The emergence of the softswitch, for instance, has promised to bring about a way of bypassing the circuit-switched network altogether, as opposed to simply building bridges within it (See Softswitch). And new signaling technologies such as SIP are making it possible to understand the development, delivery, and execution of voice applications in an altogether different way than they were understood in the PSTN (See Session Initiation Protocol). In each case, the economic and underlying technical models brought to bear on telecom are fundamentally those of the Internet itself, and the two are now beginning to be seen as intertwined in ever more explicit and meaningful ways.

The initial model for VoDSL, consisting of an IAD and GR-303 gateway, is without a doubt a useful and necessary step in the evolution of packet-based telephony. Yet it seems just as certain that it is also a model which must eventually be superceded, for the simple reason that it presupposes a dumb access network. The potential to offer multiple, simultaneous voice connections over a single access line should by no means be underestimated. Provided that deployment of the solutions occurs in a relatively timely fashion, voice could very well be the first killer app to push DSL toward a truly expansive subscriber base. But it must still be recognized that the connections provided in this scenario, while higher in number and lower in cost than circuit-switched telephony, are fundamentally the same in kind. In fact, the professed aim of most currently available VoDSL equipment is to facilitate a service that essentially mirrors the PSTN - allowing all the standard Class 5 and SS7 features to pass through unhindered from the CO, and allowing end users to hang on to all the CPE they already own, including PBXs, key systems, fax machines, and analog phones. Again, this type of legacy interoperability is absolutely necessary and desirable, at least for the short term. In looking toward the broader aims of Internet telecom as a whole, however, it invokes a certain dilemma. Specifically, the drive to imitate the PSTN in terms of service delivery tends to preclude an authentically disaggregated services model, generally assumed to be the real value inherent in the Internet.

The Intelligent Edge

In most cases, the same vendors who led the way in developing the first VoDSL systems are already well aware of the somewhat precarious position they occupy. CopperCom's acquisition of softswitch maker DTI, for example, expresses poignantly that DSL-based voice services will inevitably be understood as the first part of a distributed and integrated multimedia packet network, rather than a last mile adjunct to the PSTN. At the same time, the opening of DSL networks to a more generic IP services framework would mean essentially obviating the need for DSL-specific equipment to access voice. In a softswitch-controlled model, for example, next-generation media gateways and intelligent IP phones could make a strong play for serving roles once seen as the exclusive province of VoDSL products. Perhaps more important, however, than predicting what particular form of CPE or gateway will end up being favored (the answer is almost certainly a mixture, anyway) is to understand the changes taking place today within the DSL infrastructure to bring about the possibility of a more intelligent network edge.

While VoDSL gateways have, naturally, received the most public attention, innovations that are arguably just as significant have been taking place in the area of DSL access multiplexers (DSLAMs). The DSLAM is a hardware device in the network, typically co-located at a telco central office, responsible for aggregating and terminating multiple individual subscriber lines. DSLAMs also serve as gateways to the carriers' backbone network, translating the signals they receive into standard ATM streams and handing them off to a high-capacity switch. First-generation DSLAMs were essentially simple devices, made only to serve a single function. With the advent of multiservice DSL, however, progressively greater degrees of intelligence have been added. Just to make a VoDSL application possible, the DSLAM must at least be capable of separating out voice and data streams, passing the voice to a GR-303 gateway, and the data to an ATM switch. In a typical case, this involves provisioning separate ATM permanent virtual circuits (PVCs) to and from the customer premise equipment, in order to provide a broad differentiation among media types. As more involved quality of service considerations enter the picture, however, the ATM layer 2 awareness of the DSLAM must increase significantly. For example, to make a Quality of Service (QoS) parameter such as variable bit-rate (VBR) service, which specifies

a method for packing multiple voice calls into a single PVC (to optimize bandwidth utilization), work effectively, it must be understood by the DSLAM as well as the CPE and gateways. To add this type of intelligence, while still maintaining basic (and ever-increasing) capacity and speed requirements, DSLAMs have themselves become (at least partially) ATM switches, moving functionality that was previously handled higher up in the network closer to the edge.

If ATM was the basis for the first wave of DSLAM innovation, however, the second belongs firmly in the realm of IP. And though one's immediate tendency may be to assume a dichotomy in any statement that includes both of those two terms, the latest trends in DSL actually seem to point toward a happy medium, somewhere between the two historical foes. Cisco makes perhaps the most unambiguous statement of its aims in this regard, promoting an internal development project it calls "IP+ATM," though many other companies are currently working toward a similar end - to add IP routing capabilities on top of a platform that is, in most other regards, ATM-centric.

Initially, this effort, like its predecessor, has taken the form of a quality of service initiative. And indeed, the centrality of standards such as MPLS and DiffServ to the various DSLAM vendors' product upgrades would suggest that QoS - specifically, the need to differentiate among individual packet streams within a given virtual circuit - is driving the movement toward IP. At a higher level, however, it becomes quite clear that the real significance behind introducing IP routing functionality closer to the edge of DSL networks, and accompanying it by an effective QoS mechanism, is to create a more direct interface to IP-based services. While ATM would still likely be the most prevalent transport method used in DSL lines, a blended network based on MPLS could facilitate a much more direct avenue to providing services like IP VPNs to a subscriber, irrespective of the underlying transport layer. And, most pertinent to our present considerations, such a network architecture would enable a more transparent and standards-based deployment of voice-over-IP in the local loop, across any broadband connection type, tying subscribers more tightly into Internet-based voice applications while managing QoS at every layer.

Mining the Loop

Though the possibilities raised in the sphere of IP services are undoubtedly exciting, and one could speculate endlessly on future killer apps for DSL providers based on the rapidly evolving network infrastructure, it should nonetheless be pointed out that there are in fact plenty of concrete opportunities available today for DSL providers looking to differentiate themselves from the herd. Among the most prominent we found in talking with CLECs, equipment vendors, and actual customers are:

- *Multi-Tenant Unit / Multi-Dwelling Unit.* Increasingly, access to services like broadband data and voice has become almost as integral to tenants signing a lease as heating and air conditioning, and DSL provides a particularly well-suited method for serving this short loop application. As you'll see from our roundup, vendors of DSLAMs and IADs have been awake to this opportunity for some time, and a significant portion have released MTU/MDU-optimized versions of their products.
- *Virtual Private Networking.* As distributed workforces become more and more prevalent, VPNs have become one of the most popular data service offerings, and DSL can provide a very cost effective way of building them. Almost co-extensive with this type of service, of course, is remote access for telecommuters and remote workers, and DSL's strength as a residential installation makes it a natural solution.
- *ASP Offerings/Partnerships.* The most lucrative partnership opportuni-

ty for CLECs, once they have established broadband connectivity to the home or business, will undoubtedly be ASPs, providing everything from video-on-demand to virtual call centers.

Each of these applications, along with the slightly more ethereal protocol-related considerations we've discussed, is more than adequately represented in the DSL products we are profiling here. And the sheer volume of products outlined herein is itself a kind of testament to the fervor building around DSL technology. What's perhaps most exciting is that the range of equipment now available can serve anyone from the most established RBOC to the freshest greenfield carrier. Whatever your application needs or desires, you're likely to find a platform that can help you deliver them here.

DSLAMs And More

3Com (Santa Clara, CA 800-NET-3COM - www.3com.com) now OEMs Copper Mountain's DSLAM products, providing the company with an immediate entry to the DSL market. 3Com's reasons for selecting Copper Mountain, particularly with regard to the networking company's general packet telephony strategy, provide an interesting perspective on broadband voice.

According to Todd Landry, VP of Product Management for 3Com Carrier Systems, the company was interested in Copper Mountain for two key reasons. First, Copper Mountain's major target market is CLECs serving business users; and second, Copper Mountain uses a packet-centric architecture, as opposed to most DSLAM vendors, whose products focus on ATM. "Packet networks and IP-based services are the central focus of 3Com's overall CommWorks architecture," notes Landry. "Copper Mountain is the only DSLAM we looked at that wasn't ATM at the core internally." Of course, this does not mean that 3Com believes there is no place for ATM at all in the network. Copper Mountain's product, in fact, has extensive support for ATM-based networks, and has worked to integrate closely with ATM-based voice gateways such as Jetstream and CopperCom. But while 3Com plans to offer such architectures as an option in the short term, the company firmly maintains that the long term play for multiservice broadband is in voice-over-IP.

As opposed to the more traditional gateway model for VoDSL (or at least in addition to it), 3Com believes that DSL will be used as a means for connecting IP devices at the customer premise (like a SIP-based Ethernet telephone) to a distributed, IP call signaling system in the network (like a softswitch). It is not surprising that 3Com would advocate this model, given that both of these elements, softswitch and IP phone, are part of the company's own product line. But by incorporating the access element from Copper Mountain, 3Com can provide a tight integration among these various elements that positions them strongly, particularly among highly competitive next-gen carriers. On the backend, 3Com also plans to integrate its existing VoIP gateway, the Total Control 1000, as well as its full line of routers into a complete broadband access architecture.

Accelerated Networks (Moorpark, CA - 805-553-9680, www.acceleratednetworks.com) makes a full line of DSL products, all designed to support converged voice and data. This includes IADs, DSLAMs, and a VoDSL gateway. Though supplying both a DSLAM and a gateway is somewhat unusual among DSL equipment vendors, a unified approach has the advantages of facilitating interoperability, as well as ensuring that elements such as quality of service protocols are implemented consistently throughout the system. Accelerated has also designed each of its products to be interoperable with third-party equipment, so that a service provider can choose to work with different vendors for specific parts of the network.

All of Accelerated Networks' products are ATM-based, and use both

AAL1 and AAL2 for voice, along with the standard QoS mechanisms - CBR, UBR, and VBR. In terms of hardware, the AN-3200 Voice Services Gateway and the AN-3200 Multi-Service Access Platform are designed similarly. Both offer the option of a four- or a 20-slot chassis, which means that the product can be installed in either a central office or a multi-tenant unit. The Access Platform (DSLAM) offers both SDSL and T-1 (ATM) interfaces, so that a multiservice broadband offering can be made in locations not served by DSL (T-1 is also an option on the IADs). In addition, the DSLAM includes IP routing capabilities, as well as the ability to pass Ethernet and frame relay over ATM, though its core is an ATM switching fabric.

Though its gateway could be used in a variety of different scenarios, the Accelerated Network system as a whole seems to target business subscribers, and this focus is reflected in its IADs. The AN-24 and AN-28 support, respectively, four and eight analog voice lines over a single SDSL, ADSL, or T-1 connection. Each supports enhanced data access capabilities as well, like DHCP/DNS, network address translation, and virtual private networks. One level higher in the product line is the AN-30, which offers two digital T-1 voice connections or twelve analog ports, along with all of the standard data and voice (QoS, CLASS feature pass-through) features.

AccessLan (San Jose, CA - 408-435-1380, www.accesslan.com) distinguishes its products from traditional, first-generation DSLAMs by coining the term "i-SLAM" (intelligent-DSLAM). The distinction is more than a bit of clever marketing. AccessLan is highly focussed on enabling true multiservice DSL deployments, and has laid out a plan for effectively building out an intelligent network edge based on its product evolution.

The company's two most prominent products are the PacketLoop-1000 and 2000, two DSLAMs intended respectively for multi-tenant unit and central office implementations. The PL-1000 is a stackable platform, so that extra line-card shelves can be added incrementally as a higher port capacity is required. Altogether, the system can expand to 880 SDSL ports - enough to serve a very large building or office park. The PL-2000, intended for central offices, is a somewhat larger system that uses a more conventionally architected chassis. Each access shelf handles up to 240 ports. In addition to SDSL, the 2000 also supports IDSL and the emerging HDSL2, as well as ATM and frame relay on the trunk side (DS-3, OC-3).

Like other products in their class, the PacketLoop products add intelligence to the standard multiplexing functions by employing ATM-based switching and its associated quality of service protocols. AccessLan, however, sees its implementation of QoS as part of a three-phase strategy, of which the standard ATM QoS mechanisms (CBR, UBR, VBR/AAL2) form only the first segment. The second and third stages, by contrast, are largely IP-centric, in the sense that they seek to differentiate traffic not only at the level of transport, but on the basis of the individual media streams being carried over transport circuits. In terms of protocols, this means adopting MPLS in phase two (currently underway), as a way of defining paths throughout the network for the various IP media streams carried within an ATM PVC. Phase three, still several months out, would go even further, using a protocol like DiffServ to analyze media streams on a packet-by-packet basis, and determine routing in terms of individual applications.

AccessLan's strategy with regard to quality of service, particularly its third phase, is interesting insofar as it assumes that the underlying basis for creation and delivery of applications will be IP, rather than ATM. And while this is a relatively widely held belief, AccessLan takes it a step further, to say that VoDSL must in fact be understood within the context of voice-over-IP to deliver on a real value proposition for service providers. Thus, while the company has been active for sometime in forming partnerships with suppliers of Class 5 gateways like Jetstream, Tollbridge, and CopperCom, they are also looking keenly at coupling their product to next-gen models, like the softswitch, as a way of migrating to a more purely converged Internet architecture.

Alcatel controls the lion's share of the market in ADSL DSLAMs, with its ATM Subscriber Access Multiplexer (ASAM). Because the company has remained almost exclusively focussed on ADSL (with the exception of some next-gen VDSL equipment development work), Alcatel sells mostly to carriers and ISPs focussed on the residential access market, as opposed to businesses where SDSL is a more common offering. The ASAM is a large-scale platform, intended for central office deployments, though Alcatel also offers smaller versions of the product (including 8-, 24-, 48-, and 192-line models) for deployment in MDU or Digital Loop Carrier environments. In addition to providing the line interfaces for a variety of different ADSL rates and flavors (including G.Lite), the ASAM also includes an integrated POTS splitter, for separating data transmissions from standard circuit-switched voice over the copper loop.

Alcatel makes a wide range of ADSL modems and chipsets to complement its DSLAMs. The company also sells a voice-enabled modem, the Speed Touch IAD, which offers up to four lines of voice over a single ADSL connection. All standard quality of service protocols for ADSL and ATM are supported by Speed Touch, as well as Alcatel's central office equipment.

Cisco Systems' (San Jose, CA - 408-526-4000, www.cisco.com) DSL product line centers around a multiservice access device, the 6400 Universal Access Concentrator, that has been shipping for close to two years. Though the 6400 can accept traffic from more or less any type of broadband access network (cable and wireless included), its primary deployments to date have been as an aggregator for multiple DSLAMs, used by both DSL wholesalers and CLECs/ISPs. At the core of the product is a high-speed ATM switch fabric, but it also incorporates an IP router blade, and can perform PPP termination and tunneling, which form a part of its fully featured VPN support.

The gluing together of IP and ATM capabilities at the network edge is in fact at the heart of a broad-based initiative Cisco is calling (intuitively) IP+ATM. The goal, though bold, is familiar enough - to somehow combine IP's flexibility in the creation and delivery of enhanced services with ATM's inherent traffic management benefits into a single, fluid network infrastructure. Cisco's particular spin on the project focuses on MPLS-based tag switching, which essentially defines a topological method for creating an IP routing-like traffic pattern (i.e., cheap, non-deterministic, bandwidth efficiency) over an ATM network, taking advantage of ATM-like QoS and service segmentation.

In addition to the 6400, Cisco also offers a line of next-generation DSLAMs, as well as a complete family of CPE. Like the 6400, all of these products are designed to support multiple services (voice, video, data), and include intelligent, ATM-based architectures with comprehensive quality of service mechanisms. The Cisco 6100 and 6200 DSL Access Concentrators are mainly focussed on medium- to large-scale carrier deployments (the largest scales to around 3120 ports in a system), and offer both ADSL and SDSL line support. Cisco's DSL CPE was originally data-only, and included everything from single-user ADSL modems to SDSL routers to Ethernet-to-DSL bridge/routers. The company's first voice-capable router was the 827, which lists at around $1,000. One can expect to see more voice/data integrated CPE from Cisco in the coming months. One can also expect tighter integration between the access concentrators and Cisco's developing IP- and ATM-based voice gateway technologies.

Copper Mountain Networks (Palo Alto, CA - 650-687-3300, www.coppermountain.com) is a prominent manufacturer of both DSLAMs (CopperEdge) and DSL CPE (the Copper Rocket family). The company's

first specific VoDSL product was an SDSL router that supported up to eight integrated voice lines, but the company has always supported VoDSL both through its DSLAM architecture and through its partners and interoperability programs.

The company's core products are the CopperEdge 200 and CopperEdge 150 DSL Concentrators. The main difference between these two models is capacity - the 150 comes in 24- and 48-port versions, while the 200 scales to 192 ports per chassis. Each can take SDSL, IDSL, or G.Lite line cards, and each includes the necessary protocols and QoS mechanisms to support multiservice offerings. Services includes a well-developed VPN solution, based on the aggregation and hand-off of ATM Permanent Virtual Circuits (Copper Mountain also supports PPP over ATM or frame for secure VPN connections). For VoDSL deployments, Copper Mountain supports both ATM and IP quality of service mechanisms, and has tested its equipment for interoperability with nearly all of the major gateway vendors. The IP capabilities of CopperEdge extend further than most ATM-based DSLAMs, and actually include the ability to perform layer 3 IP multiplexing, giving carriers a significant degree of flexibility in how their networks are architected, and reflecting a more general trend toward blending IP and ATM at the network edge. This blending also helps facilitate integration to converged backbone networks, through next-gen switching platforms like the Lucent PathStar, with which the CopperEdge has been proven interoperable.

At a level above the technology, Copper Mountain further differentiates itself with a unique set of customer-centric service initiatives. One example is the CopperPowered Building program, which defines a complete strategy and product package for multi-tenant unit deployments (a market which the company targets directly with CopperEdge 150). The program includes system install and maintenance (which is handled through a partnership with Lucent's NetCare Services division), NOC management, equipment financing, and even WAN backhaul services through partnerships between Copper Mountain and a number of data CLECs. The point of these outsourced services is mainly to enable start-up and non-facilities-based providers (which, in the case of MTUs, could even include a landlord or a real estate trust) to enter their target markets quickly and without a huge initial investment. Another Copper Mountain program, FastStartDSL, is conceived along the same lines. By providing market research, pricing analysis, and sales training, Copper Mountain helps competitive carriers who may not have a great deal of experience in either data or telecommunications strategically plan the rollout of services.

Interspeed (North Andover, MA - 978-688-6164, www.interspeed.com) makes a next-gen DSLAM, conceived to serve as part of an IP-based network edge, as well as simply terminating a DSL access loop. What distinguishes Interspeed's System 1000 and System 500 (its two core DSLAM platforms) is mainly the ability to do IP routing. While many DSLAM vendors are currently working on incorporating various levels of IP functionality into their products, Interspeed is among the first to incorporate the protocol into its network interfaces (T-1 / E-1, Ethernet), where most support only ATM and frame relay. One area in which this proves advantageous is the implementation of IP-based Virtual Private Networks. Interspeed has a robust VPN offering that employs PPTP, L2TP, IPSec for tunneling, and DES3 encryption. The products also support the more standard ATM-based DS-3 and OC-3 links.

In general, Interspeed's focuses on IP centers using the protocol to enable applications, while retaining ATM as a transport method for the WAN/local loop. At least in the short term, the company recognizes the need for supporting ATM, particularly with regard to VoDSL, where most gateways are ATM-based and IP quality of service issues haven't yet been ironed out. According to Interspeed's VP of Marketing Ken Osowski, however, "the real next-gen play will be voice-over-IP carried over DSL." In accordance with this belief, Interspeed is already working on things like IP-based traffic prioritization, and plans to incorporate IP signaling protocols like SIP into the product as well.

One area in which Interspeed has found favor with CLECs is in multi-tenant unit deployments. While the System 1000, which supports up to 192 ports of SDSL could be located in a central office, the System 500 (48 ports) and the Interspeed DART (8 ports) are optimized for location in a building basement.

Lucent Technologies (Murray Hill, NJ - 888-4-LUCENT, www.lucent.com) began building its DSL product line on the basis of its acquisition of Ascend, and has since reached significant market penetration with its DSL access concentrator, the Stinger. Stinger is actually just one addition to a line of DSLAMs, mostly inherited from Ascend, which includes the MAX 20, TNT, and the Terminator 100. These range from small, MTU-focused products (32 ports in the MAX 20) to a large-scale central office platform (Stinger holds up to 672 ports in a single chassis). Aside from capacity, what distinguishes Stinger is that it incorporates a true ATM switching fabric, which classifies it as a next-gen product by way of its ability to effectively handle multiple media types.

From the time of its introduction, Stinger has been positioned as an enabling component for VoDSL. At the level of QoS, Stinger implements the standard ATM protocols (like CBR, UBR, realtime VBR). The product's 1.6 Gbps switching fabric also provides the benefit of pure speed, an important component to switching through large volumes of voice and data without introducing delay. And Lucent's Navis Access System, which applies to a broad range of systems, adds the benefit of a Java-based graphical provisioning and monitoring interface. Stinger has been tested to work with the major VoDSL gateway vendors, as part of Lucent's "Wired for DSL" interoperability program.

Lucent has also made significant moves on the CPE front, such as a line of VoDSL-enabled integrated access devices, called CellPipe IAD. The purpose of the CellPipes is essentially to extend ATM-based QoS mechanisms all the way to the customer premise, where analog voice lines voice and data from the LAN is initially packed into cells. So far, the product family includes 4-port ADSL and SDSL models, an 8-port SDSL model, an 8-port SDSL DSU (to attach to existing routers), and a modular version that scales from four to 24 ports. Lucent expects the devices to be deployed in conjunction with the various GR-303 gateways available (Tollbridge, Jetstream, CopperCom, etc.), but has also planned for next-gen gateway deployments and Class 5 switch replacements, including the company's own PathStar Access Server and 7R/E. (PathStar is a combination media gateway, switch. and packet-based call control server; 7R/E is Lucent's strategy for migrating from a Class 5 to packet-switched infrastructure). To this end, Lucent supported protocols like AAL2 and MGCP in the CellPipes' initial release.

The CellPipe IADs start at a list price of $995. The products' first announced customer was broadband wireless provider Winstar Communications, who are using a combination of CellPipe to distribute voice and data traffic throughout multi-tenant buildings.

Net to Net Technologies (Portsmouth, NH - 603-740-9377, www.nettonettech.com) is a startup DSLAM vendor that takes a different approach from most. Specifically, Net to Net's products are purely IP-based, and employ Ethernet technology where a more traditional DSLAM would use ATM. Though the company's first product, the IPD 12000 (144 ports, 2.4 Gbps backplane), could serve as a central office multiplexer,

subsequent product releases have focused on scaling downward, and targeting aggressively the MTU/MDU and campus markets. The IPD4000 serves up to 48 ports of SDSL, and the Mini-DSLAM, Net to Net's newest product, is a 12-port system. The latter is rather unique in its design - an environmentally hardened, 1U high rackmount unit that's meant to be located inside a telco's Remote Terminal. Because it backhauls the DSL lines from the RT to the CO via T-1 or DS-3 trunks, this architecture provides a clever solution to the problem of extending the reach of SDSL to greater distances from the central office. (A more traditional SDSL implementation would not be able to extend across a remote terminal due to difficulty in interoperating with existing digital loop carriers).

Despite the significant technological differences, Net to Net's main appeal is a very practical, tangible one - namely, cost effectiveness. Aside from a relatively low price tag (the IPD-12000's pricing breaks down to around $995 for the chassis, $4,995 for the multiplexer module, and SDSL line cards starting at around $7,500 each; the Mini-DSLAM costs $8995), the products can be easier to provision and manage due to their use of familiar Ethernet technologies as opposed to ATM. This is important particularly in the context of Net to Net's target markets, which include not only standard multi-tenant units, but university campuses where an IT department would have a higher degree of hands-on interaction with the equipment. Basic operations are intended to be plug-and-play, and most additional management can be accessed through a web-based GUI.

Net to Net's future plans include integrating a POTS splitter to support ADSL (the products currently support SDSL and IDSL). While the products do not include full VPN capabilities, they do support 802.1Q-based VLANs, and can be coupled with third-party firewall and encryption equipment.

Nortel Networks made an aggressive entry into the DSL marketplace through its purchase of Promatory, a next-gen DSLAM vendor. Promatory's IMAS (Intelligent Multi-service Access System) provides a natural complement to Nortel's existing line of ATM Passport switches. The IMAS also forms a solid foundation for VoDSL deployment, as it was developed around a 10Gbps ATM switch fabric. The incorporation of ATM switching adds both scalability and quality of service to the platform. IMAS can support up to 1728 ports in a seven-foot rack (with multiple chassis), and includes all the major ATM QoS protocols (CBR, UBR, VBR), as well as weighted fair queuing. And to help make service provisioning (traditionally a high cost source for DSL providers) easier, IMAS uses a rather unique mechanism called Soft-PVC, which provides a software-based interface for provisioning service to subscribers. At a network level, Soft-PVC essentially automates the configuration of permanent virtual circuits across an ATM network based on the destination switch address, as opposed to requiring cross-connects between each individual switch. The concept of Soft-PVCs also includes a built-in failover mechanism, which allows for circuits to be dynamically rerouted if one network node fails.

IMAS offers optional SDSL, ADSL, IDSL, and G.Lite line cards, all of which can be mixed-and-matched within a single chassis. Support for G.Lite is particularly important to carriers targeting the residential market, and one of Nortel's more telling moves is the release of a self-configuring HomeDSL system packaged with a Creative Labs DSL modem and sold through retail outlets.

IMAS is interoperable with a wide range of CPE, and all the major GR-303 gateways. (Nortel has demonstrated VoDSL in conjunction with Jetstream at trade shows, and is working with customers on actual implementations). Nortel has also discussed plans to release its own line of CPE in the near future.

Nokia (Petaluma, CA - 707-793-7000, www.dsl.nokia.com) got into the DSL market with its acquisition of Diamond Lane. The original Diamond Lane product, a next-gen DSLAM known as SpeedLink, has since been rechristened the D50, and has undergone some interesting enhancements.

The D50's fascinating architecture is a modular hardware design that allows for extremely flexible configuration options. Specifically, the D50's line card shelf is made to be fully separable from the central control shelf, which means that a single system can be distributed geographically, as, for example, would be useful in an MTU/MDU setting. Nokia also offers an environmentally hardened, miniature version of its line card shelf, which can be co-located in a remote terminal to extend the reach of DSL in networks that employ digital loop carriers. In either case, traffic would be backhauled to the control shelf in the central office via T-1, DS-3, or OC-3 links.

The D50 was also relatively early to market with providing ATM QoS capabilities, and was engineered from the beginning with multiservice offerings, rather than just high-speed Internet access, in mind. Currently the system supports all of the standard ATM protocols for functions such as priority queuing, traffic shaping, and policing. Additional support in the area of IP-based QoS is expected to be added over time.

One of Nokia's CLEC customers, Virginia-based Picus Communications, has already rolled out a VoDSL service using the D50 DSLAMs in conjunction with VoDSL gateway equipment and CPE from CopperCom. Nokia also has the distinction of supplying equipment to Covad, one of the largest competitive wholesalers of DSL access in the country.

Separately, Nokia has been developing a fixed wireless broadband platform that will leverage aspects of the existing SpeedLink product.

VoDSL Gateways and CPE

Bizfon (Salem, NH - 800-282-6163, www.bizfon.com) has an interesting proposition for the DSL CPE market. For some time, the company has been shipping the Bizfon 680, a SOHO phone system with six lines and up to eight extensions. The product's main appeal has been a rich set of features - eight private voicemail boxes, five-party conferencing, auto attendant, automatic fax detection, call forwarding, caller ID - in a small, low priced (less than $2,000) package.

Though the 680 already packs quite a bit into a very compact box, Bizfon's newest model adds a great deal more to the device, without much altering its price or form. Based on a partnership formed with gateway/IAD vendor CopperCom, Bizfon has developed the 680xDSL, which essentially takes the eight analog extensions of the original product and turns them into VoDSL ports. All features remain the same; the only difference is that the box will now include an SDSL network interface (eventually offering the options of HDSL2, SHDSL, and ADSL), and a 10BaseT Ethernet connection. In addition, the 680xDSL will retain space for two incoming POTS lines, as well as a lifeline. All of the necessary compression, packetization, and DSL connectivity is provided inside the box by way of technology that Bizfon licensed from CopperCom (the same basic platform as that which the latter uses for its own IADs).

Until now, most DSL-based IADs have offered only basic connectivity functions, and would generally be used to front-end an existing PBX or key system. Bizfon's 680xDSL appears to be the first product to actually integrate the two devices. And the integration makes sense because Bizfon's own target market makes up a significant portion of the initial target for VoDSL.

Bizfon has already tested the original 680 for interoperability with Tollbridge Technologies' VoDSL equipment, and one can expect further gateway interoperability testing to continue. Multiple 680xDSL systems can be networked together to support up to 24 extensions.

Clarent (Redwood City, CA - 650-306-7511, www.clarent.com) made an interesting move to enter into the broadband access market by developing a line of CPE products designed to work with DSL, cable, and fixed wireless networks. Clarent's voice-over-IP gateways have traditionally been used in core networks for the purpose of long distance tandem replacement, i.e. to bypass long distance toll charges by routing phone calls as IP traffic. While this application has been a strong one for the company, particularly in providing products to international long distance carriers, Clarent's strategy going forward is to expand its reach in the network by extending its VoIP capabilities into the local access loop. Until now, this is an area in which pure play gateway makers have found it somewhat harder to compete, at least in the U.S. market. Because of Clarent's already strong international presence, selling to U.S. CLECs may not be an initial top priority anyway, though if successfully implemented, the products could provide a significantly less expensive model for delivering packetized voice over a broadband network. Because they do not require any ATM technology or Class 5 switch integration, solutions such as Clarent's could remove much of the cost associated with more traditional VoDSL systems, though the end-to-end next-generation model which they posit is likely some time away from widespread deployment.

Clarent signed MOUs with five Taiwanese vendors to manufacture the devices, and the design for the device itself is based on a combination of Clarent software and an integrated chipset and software from Texas Instruments / Telogy.

Convergent Networks (Lowell, MA - 978-323-3300, www.convergent-networks.com) makes a media gateway and softswitch designed to serve as a backend to broadband access networks. The Integrated Convergence Switch 2000 (ICS2000) collapses the functions of both a media gateway and an ATM switch into one box, and forms the basis of a programmable software platform for enhanced services and Class 5 features. Though the pragmatic role of products such as Convergent's has not yet been fully defined by the marketplace - at this point, they seem to be something more than just a gateway, and something less than a full Class 5 switch - Convergent does position itself as one step beyond the GR-303 gateway model. What ICS provides, in effect, is an architecture that can fully interoperate with the existing PSTN, but could also just as well be deployed in the context of a pure packet-based, next-gen network.

Convergent's target market is mainly aggressive CLECs, most of whom either do not wish to deploy Class 5 switches at all, or at least plan on capping their current investments in circuit-switched equipment. The fact that the ICS is ATM-centric appeals particularly to DSL providers, and its high-density (10Gbps switching fabric, OC-3 or OC-12 interfaces) supplies plenty of power to terminate a large volume of incoming voice and data traffic. When positioned as the front-end to a Class 5 switch, the ICS converts ATM to TDM, and can even interface directly to the SS7 network (interworking with ATM Q.2931 signaling via the softswitch). Alternatively, in a Class 5 replacement scenario, the Convergent switch would pass traffic directly to the ATM backbone network, receiving its signaling information from the softswitch, and more than likely still interfacing to SS7 for network-layer services like local number portability.

Convergent has formed a number of IAD partners, both for DSL and for pure ATM local access.

CopperCom's (Santa Clara, CA - 408-567-9277, www.coppercom.com) CopperComplete DSL is a VoDSL gateway architecture that includes both an IAD and a large-scale gateway intended for a central office. Though the basic architecture is similar to other VoDSL vendors, CopperCom has

V

made a significant move to break apart from the pack with its acquisition of DTI, a softswitch developer. The acquisition is interesting on a number of levels, as it merges not only two segments of the public network - "access" and "backbone" - but two distinct infrastructures - DSL/ATM and IP - as well. CopperCom explains that the marriage of local access to core switching in the public network is key to moving from simple transport to enhanced service delivery. At the same time, CopperCom is retaining a hybrid network - keeping ATM in the local loop for its QoS advantages, and doing the necessary IP integration in their gateways. In the matter of call control protocols, the company has committed itself to H.248 (MEGACO), which many feel will be the successor to MGCP. The most important aspect of the softswitch model for CopperCom is having an extensible platform in the network on which to create new services. It also means that the company can aggressively target greenfield CLECs who want to avoid Class 5 switches altogether, in addition to the RBOCs and ILECs who have traditionally been the strongest targets for GR-303-based gateway products.

Apart from the softswitch, CopperComplete encompasses three basic elements. At the customer premise, the CopperCom MXR (four or sixteen ports) accepts input from analog phones (faxes, key systems, etc), as well as from standalone PCs or a LAN, and converts it into ATM cells to be passed over a single DSL line. Once the traffic is terminated, and data is separated out, the CopperCom Gateway converts the voice back into circuits before being passed to a Class 5 switch (the scenario would, of course, be different when a softswitch is employed). The final element of the system is service management. CopperCom's software, known as CopperCommander, serves as an element management system, with enhanced features that give it many of the capabilities of an OSS. Service creation, along with monitoring and auditing, can all be handled remotely. A Windows-based GUI and embedded web server make features more accessible.

Efficient Networks (Dallas, TX - 972-852-1000, www.efficient.com) is one of the strongest players in the market for DSL customer premise equipment. After transitioning its focus a few years ago from ATM LAN equipment to DSL products, the company's first product shipments were DSL modems aimed at residential consumers and telecommuters. Since acquiring Flowpoint in December, however, Efficient has solidified its position in the small- to medium-business market as well.

All of the company's DSL-related products are grouped under the SpeedStream family name, which includes everything from routers and Ethernet bridges to one-line internal and USB-attached modems. While the majority of product deployments by Efficient's customers (CLECs, ISPs, ILECs, RBOCs) have been for standard, data-only, Internet access services, the company put voice on its roadmap early, and has already announced its first two voice-enabled products. The SpeedStream 7451, which was based on an existing Flowpoint product, is targeted at SOHO users and offers four analog ports and a single DSL connection (with integrated hub and router; lists for $995). For somewhat larger offices, Efficient subsequently introduced the SpeedStream 8600, a flexible IAD that handles up to 24 analog voice ports. In addition, the 8600 offers connections to a 10/100BaseT Ethernet LAN, and a variety of wide area interfaces, including ADSL, SDSL, T-1 (carrying ATM), and HDSL-2. Efficient has formed partnerships with most of the major vendors of VoDSL gateways, to ensure interoperability of its products. The company has also discussed plans to incorporate next-generation voice signaling protocols, such as MGCP and MEGACO, into its products. At this point, however, Efficient's most pressing concern is to provide a product that is easy to use, and can deliver a business quality voice service. "Toll-quality voice with 100% CLASS feature pass-through is an absolute necessity," notes

Peter Bourne, Efficient's Vice President and General Manager of Integrated Access.

Efficient expects to add more voice-enabled products to its offerings over time, and will continue its strategy of offering multiple, differentiated CPE products, targeted at a wide variety of customer needs.

General Bandwidth (Austin, TX - 512-681-5400, www.generalbandwidth.com) is one of the newest players to enter the VoDSL market, and certain aspects of its product reflect the company's youth in a positive way. One is that the G6, General Bandwith's core gateway product, is designed to support packetized voice over any type of local loop broadband connection, including DSL, cable, and fixed wireless. And within those broader connection types, the gateway is also built with the ability to provision, voice-over-ATM, voice-over-Frame Relay, or voice-over-IP, at the carrier's individual discretion. In the case of VoDSL deployments, where ATM is most common, the G6 implements any of the standard ATM QoS parameters (including AAL5, as well as CBR, VBR, and UBR). And although native VoIP is currently more common in hybrid fiber coax networks, the G6 is rather unique in its extensive support of IP protocols, both in terms of signaling (MGCP, H.323, SIP), and in QoS (MPLS, DiffServ, etc.). General Bandwidth has also taken a rather leading-edge stance in forming partnerships with softswitch vendors, leveraging its product's support of next-gen protocols to demonstrate its use in an end-to-end packet-based voice and data network. Despite its bent toward next-gen networks, the G6 still provides complete interoperability with Class 5 switches and the PSTN, and includes a rather unique method for delivering lifeline capabilities directly from the gateway using digital added main line (DAML) technology. Unlike others, General Bandwidth's lifeline solution can actually dynamically switch live calls from packet to circuit-switched in case of a power failure, on up to eight lines at the customer premise.

The G6 is intended to interoperate with a wide range of standards-based CPE. Up to twelve separate G6 shelves can be networked together to work as an integrated system, offering up to 96,192 total simultaneous calls.

Jetstream Communications (Los Gatos, CA - 408-399-1300, www.jetstream.com) is one of the core players in VoDSL, offering a carrier-class gateway and a family of integrated access devices to combine voice and data access. The CPX-1000 is typically located in the carrier's network, between a DSLAM and a Class 5 switch. On the access side, it accepts ATM over DS-3 or OC-3 lines, and on the switch side it interfaces via GR-303 to any major manufacturer's Class 5 (Nortel, Lucent, Siemens). In terms of line capacity, the system can scale from a minimum of 200 subscribers, to up to 7,500 with 4:1 oversubscription (typical for business-level service), or 17,000 with 9:1 oversubscription. For voice compression, the system offers dynamically selectable G.711 PCM and G.726 ADPCM. Standard G.165/168 echo cancellation is also included. CPX supports various flavors of DSL, along with T-1 and broadband wireless.

To complement its gateway, Jetstream makes two integrated access devices, an 8-port and a 24-port model. The Jetstream IAD supports pass-through of major CLASS and Centrex features, and interoperates with a wide range of DSLAMs. The 24-port model is actually a modular system that can be configured with anywhere from four to 24 ports in the same box. The IADs are made to be self-configuring by means of Jetstream's element management system, JetEMS.

JetEMS is CORBA-based management software that provides a graphical interface for functions like configuration, performance monitoring, and troubleshooting. The network administrator can view any individual subscriber connection in the network, and can remotely initiate and manage things like software upgrades. When a new user comes online, the Jetstream IAD is automatically detected by the EMS, and registered with a local gateway.

Jetstream is strong in the area of interoperability with both DSLAM and IAD vendors, and has a comprehensive testing and certification program for all types of broadband equipment. Another move aimed at speeding VoDSL deployment is Jetstream's JetStart Business Solutions Program, which offers service providers support in the areas of customer service, marketing, financial planning, and project management as they plan their roll-outs of VoDSL.

With its acquisition of General Instrument, Motorola (Austin, Tx – 512-530-5500, www.motorola.com) has made a particularly strong play in the area of cable modems and set-top boxes. The company has also been quite active with regard to DSL, however, both in developing modem and router technology, as well as through its semiconductor and chipset manufacturing lines of business.

One of the more interesting pieces of news to come out of the company is a release from its Software Products Division, which has developed a product family known as SoftDSL. Essentially, the product is a purely software-based DSL modem that can be embedded in PCs, as well as in standalone access devices or other network appliances. The main advantage of such a system, of course, is that it eliminates the need for a dedicated DSP on the modem and instead uses the existing host processor of the device on which it runs, resulting in a significantly reduced cost. In the context of DSL deployment, the product (which would likely be used primarily by residential subscribers) goes a long way toward solving the carrier's problem of installation and provisioning of CPE. The major target for SoftDSL will be PC OEMs who would package the modem with retail PCs, though the company has also discussed the possibility of partnering with service providers for distribution. Motorola already offers a similar product for dialup access.

In addition to SoftDSL, Motorola also makes a full line of DSL routers, mainly for small- to medium sized businesses, under its Vanguard Instant Access line. That entire line is served by Activator, a service management platform automates much of the provisioning and configuration process, and allows full remote management of the routers through a web-based GUI.

Polycom (Milpitas, CA - 408-526-9000, www.polycom.com), perhaps the best-known vendor of audio and video conferencing equipment, acquired Atlas Communications, adding an interesting dimension to its product line. Atlas makes broadband IADs, now sold as the Polycom NetEngine family under a newly formed Network Access division. While pure access technology has not traditionally been a part of Polycom's business, the move to acquire Atlas makes sense strategically insofar as widespread deployment of broadband connections is crucial to the adoption of applications like videoconferencing. And the fact that Atlas' products concentrate mainly on DSL is indicative of a clear shift away from ISDN (on which most videoconferencing products were originally based) in the small- and medium-sized business markets.

Before the acquisition, Atlas had been shipping a voice-enabled DSL router (largely through OEMs) for around a year, making the company a relative veteran in the VoDSL industry. In its first versions, the company focussed on quality of service, by making sure that ATM functionality was supported end-to-end, as well as interoperability and cost effectiveness. The NetEngines interoperate with all major (and many non-major) DSLAMs, as well as most available GR-303 gateways. The first product in the line to include voice support is the NetEngine 8000, a modular system that scales from four to 24 voice ports, though several other models are currently being introduced. The product is actually built on a very

flexible design, including integrated routing and bridging functionality, and can even act as an inverse multiplexer. Both SDSL and ADSL interfaces are available. The NetEngine 4000, introduced more recently, offers similar voice and data capabilities, and supports multiple WAN interfaces including SDSL, ADSL, T-1 / E-1, and ATM-25. Polycom's newest product is the NetEngine 6000 family, which offers four- and eight-port models, and targets small- to medium-sized businesses using either SDSL or T-1 / E-1. The NetEngine products are designed to support complete CLASS and Centrex feature pass-through, and include extensive bridging and routing capabilities for IP data as well. The products are priced competitively, starting under $1,000 for a four-port model.

Polycom has made a flurry of partnership announcements surrounding the NetEngine family, and Jetstream, AccessLan, and Nortel (among others) have all signed on as resellers or OEMs of the product. They have also made an investment of $5 million in InternetConnect (Marina del Rey, CA – 800-896-7467, www.internetconnect.com), DSL and enhanced services provider.

Ramp Networks (Santa Clara, CA - 408-988-5353, www.ramp.com), who early last year began offering xDSL versions of its popular WebRamp product (a shared access router originally made for the dialup market) has added voice into the mix, with the release of WebRamp 2000V.

The 2000V comes with eight analog voice ports (expandable for up to 16), one lifeline (for a single POTS line), and an Ethernet jack (for connecting to a LAN). An integrated SDSL router gives you a 2.3 Mbps connection to a single SDSL line. Eventually, Ramp intends to add other flavors of DSL, as well. Ramp's history as a data-centric manufacturer proves valuable when it comes to including robust features like built in DHCP and DNS server, along with advanced Network Address Translation, and support for IPSec and triple-DES security protocols. Other concerns which Ramp clearly took seriously in building the box were reliability and cost. The 2000V, like all WebRamps, is priced under $1000, and in the interest of cutting back on truck rolls, includes a browser-based interface for remote configuration, provisioning, and management. Upgrades are accomplished through software downloads.

Like most VoDSL systems, the Ramp box offers the choice of 64Kbps uncompressed (PCM) voice and 32Kbps compressed (ADPCM). The product interoperates with the major gateway vendors - Tollbridge, Jetstream, CopperCom - as well as major DSLAMs, like Copper Mountain.

TollBridge (Santa Clara, CA - 408-585-2100, www.tollbridgetech.com) is generally one of the first names that comes to mind in connection with VoDSL equipment. Its TollVoice system is essentially a VoDSL gateway in two-parts: An IAD that compresses and packetizes voice at the customer premises, and a box that sits between a DSLAM and a Class 5 switch, using a GR-303 interface to change voice packets back into circuits. This formula, however, is not quite so simple or straightforward as it once was, largely due to the fact that alternative network architectures are growing more and more prevalent. TollBridge responded to changes in the market with a makeover of its product.

Architecturally, TollVoice appears more or less identical. The most significant change, however, is that the system can now support packetized voice over cable networks, as well as DSL, in both the CPE and the POP equipment. The system has also been tweaked to fit in with next-gen switching infrastructures on the backend. While the TB200 would typically be deployed in conjunction with a Lucent or Nortel Class 5 switch, TollBridge has now added support for MGCP and SIP, allowing for the possibility of interfacing with a media gateway controller or softswitch.

TollBridge makes two models of its integrated access device, the TB50 and the TB55. Both support up to eight analog POTS lines, but the

V

TB55 includes a built-in SDSL router, whereas the TB50 hooks up to an external broadband modem (cable, DSL, etc.). The TB200 Local Exchange Gateway is a carrier-grade, rackmountable chassis that can switch up to 5,000 DS-0s. The line interfaces terminate DS-3 or OC-3 ATM circuits. And the difference between a DSL and cable implementation means only a difference in programming the line card. In either case, the cards feature DSP software with echo cancellation and jitter buffering, and support both G.711 and G.726 audio codecs.

The TollBridge system comes with a graphical element management system based on HP OpenView, and also includes a command line interface and a web interface.

Unisphere Solutions (Burlington, MA - 781-313-8700, www.unispheresolutions.com) is steadily gaining favor among competitive DSL carriers with its Edge Routing Switch (ERX) platforms. The ERX sits at one level deeper in the network from the access loop, typically aggregating ATM traffic from a number of DSLAMs. And while DSL is not the system's exclusive application - it could also be used to backend cable or other broadband networks - its ATM switching capabilities make sense in most DSL architectures. Once incoming sessions have been terminated, the ERX applies layer 3 routing mechanisms to send traffic directly onto either ATM or IP (packet-over-SONET) backbone networks, via high-speed OC-3 or OC-12 uplinks. This blending of ATM switching with IP routing is mainly what differentiates the ERX from a more traditional edge device, not simply because it creates a new transport model, but more importantly because of the application model it enables.

In addition to building routers and gateways, Unisphere has spent the past several months developing software platforms for service creation and delivery. These efforts include the SRX-3000, a softswitch that manages the IP-to-PSTN connection, in conjunction with Unisphere's Service Mediation Switch, as well as the UMC, a modular system that handles all aspects of subscriber and service management. One essential part of the UMC is a meta-directory that stores subscriber profiles, along with information about specific applications and network services. The Service Selection Center, perhaps the UMC's most unique component, provides a web-based interface to subscribers themselves, through which they can access on-demand applications as well as account-specific information. If a subscriber makes a change to his or her service, the information is automatically updated in the provider's database, and the requested service is automatically provisioned. Providers have control over the options available to a user, and on their own end, have a powerful graphical interface through which to define new services, billing information, or an individual subscriber's profile.

One customer, DishnetDSL, once purchased 54 of Unisphere's ERX switches, building an infrastructure to provide high-speed Net access, along with educational content, to residential subscribers in India.

Core Components

Aware (Bedford, MA - 781-276-4000, www.aware.com) develops reference designs for ADSL chipsets, modems, and transceiver modules, which it sells to a variety of OEM customers. The company has introduced a technology for VoDSL that takes a rather different approach than most other DSL solutions, in order to serve a very specific application. Aware's solution, which it calls VeDSL (Voice-enabled DSL) is intended solely to provide second-line telephone service to residential customers, combined with a broadband access connection (ADSL or the G.Lite version of ADSL). As calls are placed or received, VeDSL dynamically allocates data bandwidth to support the second-line voice call or dial-up session. When the call is complete, the bandwidth can be reallocated to maximize Internet access

speed. VeDSL will enable service providers to use a single copper wire to deliver standard POTS telephone service and high-speed Internet access, plus additional lines for voice, fax, and 56K dial-up modems.

The biggest distinction the company draws between its own technology and that of other DSL solutions in the market is that Aware does not packetize the voice, but instead uses physical layer (Layer 1) transport of the OSI Reference Model to transport the voice traffic within the ADSL data stream itself. Doing this thus avoids using either ATM or IP to carry voice and it eliminates the complexity associated with these protocols, while at the same time assuring reliable toll-quality voice. Thus, while it cannot attempt to scale beyond one extra phone line, the system does create an easy and relatively inexpensive way for service providers to enhance their offering to consumers, yet versatile enough to work with evolving packetized networks. The Aware technology is designed to work with existing central office equipment, without requiring any hardware separate from that which carriers already widely deploy. Because it does not use any traditional gateway functions, the CPE for such a model would also be intended to cost less than the majority of VoDSL solutions.

Peter LeBlanc, Aware's vice president of marketing, says: "Aware's Voice-enabled DSL technology offers the flexibility and simplicity that service providers are looking for to bundle second-line voice and data services over ADSL. Service providers will be able to use these bundled services to increase revenue streams and gain new customers, and consumers will benefit from the variety of additional services available over their existing telephone lines." Dave Burstein, editor of DSL Prime added: "U.S. service providers, including the largest, are looking for an inexpensive way to bundle high-speed Internet access and second voice line services for the residential consumer, and I believe the market potential is enormous."

Broadcom (Irvine, CA - 949-450-8700, www.broadcom.com) makes chipsets and integrated circuits for use in a variety of different broadband devices, including DSL equipment. The company also has forged an interesting partnership with Next Level Communications (Rohnert Park, CA - 707-584-6820, www.nlc.com) to develop a line of next-gen set-top boxes based on the emerging VDSL (very high-speed) standard. The box leverages VDSL's high-speed capabilities to deliver up to 52 Mbps of bandwidth over a copper loop to offer multiple VoIP lines, Internet connections, and MPEG-2 digital video applications. The system is essentially intended to serve as a residential gateway, which would link multiple devices within the home, including telephones, PCs, and televisions. Part of the solution will also likely integrate a wireless home networking technology which Broadcom also introduced, known as iLine10.

Such systems clearly indicate the direction in which most carriers hope to move, by capitalizing on the high revenue potential of interactive, digital video and other on-demand services. Broadcom is one of the first companies to offer a reference design for VDSL transceivers, and also offers a number of ADSL designs.

Other Applications

Artel (Marlborough, MA - 508-303-8200, www.artel.com) makes an ATM-based digital video system, designed to deliver broadcast quality, MPEG2 video transmissions over broadband networks. Cross Stream, as the product family is known, consists of a variety of devices that a carrier would locate in the central office and connect to a DSLAM, in order to perform demuxing and remuxing of the MPEG video streams and to implement jitter management ensuring broadcast quality. At the customer premise would be located a next-generation digital set-top box, which would essentially act as an MPEG2 decoder, as well as providing the basic network access to connected devices in the home. Artel will not develop the CPE devices themselves, but will partner with others to provide for interoperability with the central office equipment.

While systems like Artel's are emerging as a way for telcos to compete with cable TV providers and others in the area of digital video (which is likely to include applications like video-on-demand and interactive television), some significant upgrades such as the implementation of VDSL will have to take place before they can be deployed over the copper loop infrastructure.

mPhase Technologies (Norwalk, CT - 877-674-2738, www.mphasetech.com) has developed a bleeding-edge system for delivering not just voice and data, but also digital television, over a single, twisted copper pair. If the architecture mPhase is proposing takes off, it could provide telcos with a way to compete more directly with multiservice cable providers, and create a much stronger value proposition for DSL in the residential market.

So far, one of the company's most significant moves has been to create a joint venture, mPhase Television, with AlphaStar (www.mediacrossing.com), one of the original providers of direct satellite television. AlphaStar provides the necessary satellite facilities from a central location in Connecticut, while mPhase equipment located in the carrier's central office handles the local aggregation and distribution of signals.

The mPhase system, known as Traverser, consists of four distinct elements. A video distribution shelf accepts incoming satellite and local video signals. A digital headend - the key component in the system concentrates video and Internet data traffic (performing some of the basic functions of a DSLAM). A POTS splitter combines the video and data traffic with voice from the PSTN before sending it out over the same DSL line. And an IAD at the customer premises separates out each traffic stream to its appropriate end device, using an RJ-45 (10/100 Ethernet) for data, an RJ-11 for data, and a coaxial cable for television.

mPhase uses Rate Adaptive DSL, a variety of ADSL that lets the modem dynamically adapt its operating speed to available bandwidth, and is thus well-suited to bandwidth-hungry apps like video. Realtime digital video signals are delivered according to the MPEG-2 specification. Because the system works on full-rate ADSL, 2.4 Mbps of downstream bandwidth (1 Mbps upstream) is available to data traffic even after a dedicated chunk has been carved out for video. Version 2.0 of the system mPhase offers converged, packetized voice, with up to four ports hanging off the existing box.

mPhase is targeting mid-sized independent local exchange carriers as its initial customers. The company plans to operate on a revenue sharing model, between the carrier and mPhase Television. They will also assist carriers in marketing the service offering to subscribers. Trials have been conducted with a Georgia-based CLEC, Hart Telephone.

Ridgeway Systems (Austin, TX - 512-502-1775, www.ridgewaysys.com) makes a web-based application platform aimed at DSL providers looking to offer enhanced services to their subscribers, whether or not they are currently implementing VoDSL. For example, one of Ridgeway's applications, called netHance, is designed to add multimedia Internet content to regular, POTS calls. The basic idea is that an ISP could offer the application as a service through its own web portal, and allow its subscribers to add videoconferencing or data sharing capabilities, leveraging the broadband connection, to a standard telephone call. Once in place, subscribers can access the applications dynamically, with no special connection between their phone and PC, as the integration is managed on the server side.

In addition to developing the applications themselves, Ridgeway also provides the underlying hardware and middleware necessary to syn-

chronize the delivery of voice, video, and data. The VX-120 is Ridgeway's core server architecture, which processes media streams through an asynchronous packet backplane and includes an array of DSPs and associated software. The VX-120 is designed specifically to blend IP media content with circuit-switched telephone calls. Added later to the company's offerings has been a middleware platform called VX Centrex, which, by contrast, is intended for delivery of packet voice applications. VX Centrex performs policy management, enables applications like bandwidth on demand, and can even negotiate the passage of VoIP transmissions through corporate firewalls.

Viagate's (Bridgewater, NJ - 908-595-6400, www.viagate.com) 4000 Multimedia Access Switch products are designed to deliver ATM-based digital video and data over VDSL lines (27 Mbps downstream, 3 Mbps up). The system uses MPEG2 encoding to provide broadcast quality video, and is mainly intended for delivering applications like video-on-demand, interactive TV and gaming, and direct broadcast satellite (DBS) programming. Also, Viagate has demonstrated its products working in conjunction with videoconferencing systems, offering a possibly more business centric-application. The device itself can be located either in telco facilities or in a building basement, and is complemented by a line of CPE that acts as a gateway for connected end devices. ViaWay is the company's residential gateway, which is designed to attach to televisions, PCs, and telephones (voice can either be split off as POTS or carried over the DSL connection, while data and video are carried as separate ATM media streams).

ViaGate also offers a digital set-top box, called ViaPort, which is intended more exclusively for video applications, combining interactive, 2-way communication features, with more traditional VCR-like functionality using digital storage. The complete ViaGate system got its first trials in Shanghai, China, running video-on-demand and high-speed data applications over a VDSL connection.

VoDSL Without the Wait

No one can tell when VoDSL will finally becomes generally available in most parts of the U.S. But if you're just itching to get more out of that modem than pure data, a number of products are cropping up to act essentially as one-port gateways, using the Internet (or a managed backbone) to let you make free (or cheap) long distance calls. Two examples are products from DSG Technology (Walnut, CA - 909-595-8908, www.dsgtechnology.com) and IntelliSwitch (West Palm Beach, FL - 561-832-4840, www.intelliswitch.com).

DSG makes a variety of small hardware devices designed to enable VoIP calling. The simplest is InterStar, an Ethernet phone box that plugs into a DSL modem through a 10/100 Ethernet connection and lets you hook up a single analog telephone. By connecting to DSG's own network of gateways, you can call any phone number, or, you can dial another InterStar device directly by its IP address and avoid all charges. Recently, DSG introduced a complete IP-based digital phone set built around the same technology of Interstar, which also incorporates a two-port Ethernet hub to let you connect your PC to the network through the same pair of wires.

Similar to InterStar, InterPhone is a combination of InterStar and a telephone set with function keys like a traditional digital telephone set. With a built-in two-port hub it may connect to another InterPhone or PC.

A suitable Next-Gen PBX solution from DSG Technology would be composed of a DSG InterServer connected to the network together with InterPhones which play as extension. By connecting the network to a DSL modem or DSLAM, the system becomes a revitalized PBX with full functionality and built-in IVR, and with a DSL connection to the Internet to boot.

DSG Technology's high density gateways include the IP1000 and IP3000. IP1000 is a low-cost embedded gateway with capacity up to 8 ports. Their IP3000 is a carrier-class gateway which bridges and extends all Centrex features of any Class 5 switch to the IP network. Both IP1000 and IP3000 support the MGCP protocol. DSG Technology is solving the "last-mile" access problem by offering products appearing in the network right after Class 5 switches and joining devices such as the IP1000 with the IP3000.

Intelliswitch similarly makes a range of devices, extending beyond just VoIP gateways to include things like prepaid calling card platforms and back office applications. The company also offers individual lines of gateways for both the enterprise and carrier markets. Recently, Intelliswitch began producing one- and four-port Ethernet-connected devices, which let users make IP calls directly via their DSL modems. Intelliswitch is targeting both businesses and consumers with the products, and, like DSG, maintains its own gateway network allowing hop-offs to the PSTN.

Multiple Services — More Than Just Fast Modems

Rhythms (Englewood, CO - 303-476-4200, www.rhythms.com) is among the top national data CLECs specializing in DSL service. As with others in the field, it's high-speed Internet access that gets top billing in the company's ad campaigns and mainstream media coverage. But a slightly deeper look at how Rhythms' network is constructed reveals that theirs is anything but a one-dimensional play.

One of the most obvious competitive differentiators for Rhythms is speed. Unlike most, Rhythms engineered its network to support full-rate asymmetrical DSL (ADSL) on all of its lines, as opposed to a partial rate service. This means that individual line speeds ostensibly peak out at around 7 Mbps dowstream, and over 1 Mbps upstream, whereas other ADSL offerings are capped at around 1.5 Mbps down and significantly less upstream bandwidth. Catherine Hapka, Rhythms CEO and founder, asserts strongly that the reasons for this beefed-up bandwidth amount to much more than just faster Internet access. "From the beginning," Ms. Hapka notes, "we believed this business would be about putting multiple services over a single connection, not just providing a single commoditized application. In fact, we provision as much speed as we can on each circuit we install, irrespective of what the customer wants to pay for at that particular time, so that when that customer decides he does want to add video or audio streaming, or VoDSL, we're able to deliver that service immediately."

Interestingly, Ms. Hapka goes on to explain that the simple fact of having raw bandwidth available can put a new perspective on things like class of service and traffic engineering, around which a good deal of the discussion about multiservice DSL been focussed. "Our competitors are stuck with traffic shaping, and having to dictate which application - voice or data - is given priority, because they've basically maxed out their circuits with the single application of Internet access. But just by having more bandwidth, we've created much more of a carrier-class platform."

Yet bandwidth is not the only thing that sets Rhythms apart. While the integration of IP capabilities into DSL networks is just starting to garner attention in the industry, Rhythms saw it early on as another key enabler for offering differentiated services. "We are the only fully routed IP Layer 3 network of this kind," Hapka boasts, "and that means that we have a platform that can support multiple internetworking applications. It also means our network is much more efficient and much more upgradeable than the Layer 2 networks that most of our competitors maintain. And it enables us to compete in aspects of the market which simply would not be available to us otherwise, such as voice-over-IP in the local and wide area networks."

When it comes to voice, Rhythms is essentially agnostic as to how it gets delivered, and plans to deploy a combination of VoDSL and VoIP solutions, based on what is most logical for a particular application. But voice is just one of the applications the company believes its network will carry. "In the enterprise," observes Hapka, "we see the first element in the bundle being a teleworker application, providing remote connectivity to the Internet or the corporate LAN. From there, it's a relatively natural outgrowth to branch office, or even partner connectivity, and this is the segment of the market that will displace growth in frame relay. Once the a high-speed broadband connection has been made between remote locations, we think businesses will almost immediately want to use the same connection to run voice-over-IP, or extend their PBX functionality, out to teleworkers and branch offices, ripping out their RBOC private lines. And finally, we're going to see a significant growth in ASP applications, particularly customer relationship management, and perhaps videoconferencing. One segment that's already taken off for us is remote call center management, which basically builds on the teleworker and branch office applications, and we expect movement to continue in that direction."

As a company, Rhythms has a clear sense of where it is going, and an already quite mature network infrastructure has put it well on the way of getting there. A deal to become Excite@Home's exclusive provider of DSL service only brightens the company's future. But the opportunities and technologies that Rhythms is leveraging also suggest that we are only at the tip of the iceburg in terms of what DSL can deliver to its subscribers, and that the potential for creative ISPs, ASPs, and carriers to capitalize on broadband is just starting to be realized.

Voice over Frame Relay (VoFR, VOF)

Some consider frame relay to be a cheaper alternative to a T-1, while others see it as a low-risk, logical stepping stone to higher-bandwidth Asynchronous Transfer Mode (ATM). In today's deregulated communications environment, normally data-only services providers may be tempted to gain some additional revenue by also offering voice and fax on their frame relay infrastructure.

Voice over frame relay can be used for well-known voice applications such as interconnecting Private Branch Exchanges (PBXs), use of Off-Premises Extensions (OPX) and Private Line Auto Ringdown (PLAR).

VoFR is to be most often used to connect a company's regional locations with the company headquarters. Branch sites can be connected to the regional location with an analog line or sometimes with a 64Kbps digital line. A regional node supporting several remote branches might need ten or more voice ports, which would connect to a PBX via T-1 or E-1 digital interface.

At the regional node (or at truly distance branch offices) one would then install a voice-enabled Frame Relay Access Device (FRAD) or a voice-enabled router from Cisco Systems or Motorola. The FRAD connects to the PBX and router on one side, and to the frame network on the other. Voice-enabled routers are linked to the PBX and the LAN on one side and the frame network on the other. In either configuration, the PBX can then be programmed to recognize the phone numbers for the branch offices and switch those calls to the PBX port connected to the voice-enabled FRAD or router where it's compressed, packetized, and sent between sites over Permanent Virtual Circuits (PVCs), or logical point-to-point circuits established over the frame relay "cloud" for you by whoever you're buying the service from. The whole procedure is reversed in the distant office receiving the call.

For acceptable voice quality, the total round trip delay for the packets should be less than 400 milliseconds, about the same as for IP Telephony (See Quality of Service). To minimize end-to-end delay, a combination of voice compression algorithms and adaptive smoothing delay methods can be used.

Since voice packets ultimately end up traveling through a frame relay infrastructure along with data packets, some kind of prioritization is in order in an effort to somehow send packets carrying the voice ahead of the data packets. To be effective, prioritization should be combined with other methods that also speed up the flow of voice packets. For example, Motorola's 6520 MPRouter with Voice Relay gives voice packets priority over other frame relay data such as Bypass and Annex_G frames, which contain legacy and LAN data.

Prioritization also limits the size of frames from the Bypass and Annex_G stations to a size that require only 5 to 10 ms to transmit when voice is present. This limits the number of Bypass and Annex_G frames to be queued up in front of any voice frame to two. Utilizing these features, the 6520 with Voice Relay controls the delay caused by queuing to between 10 and 20 ms.

Although undoubtedly useful and relatively inexpensive, VoFR has been totally overshadowed by the public's mania over Voice over IP (VoIP).

Voice over IP (VoIP)

The term "IP telephony" appears to be supplanting "voice over IP" in popularity, though both are used extensively (See IP Telephony). The basic idea behind VoIP is that your phone call is sent via a series of data packets on a fully digital communication channel. This is done for more efficient bandwidth utilization; allowing voice and "regular data" mixed on the same infrastructure, etc.

Human speech covers a range of frequencies (including harmonics) of up to about 20 kHz most of the information that makes speech intelligible is located in the band between 300 Hz and 3,400 Hz. We can economically send just that information using very little bandwidth, and still be understood.

Once we've pared down the frequency range of human voice, we still have to convert it into a numeric (digital) representation so that it can be sent on a digital channel. We can do this in one of two ways:

Waveform codecs. These appear to waste bandwidth (such as the G.711 PCM codec that outputs 64 Kbps), since they generate higher bit rates (16 Kbps or more) but, as in the case of any tradeoff, they offer good speech quality with low latency (delay), are fairly simple, and require little CPU power.

Vocoders. These are very economical, reducing bit rates to as low as 2.4 kbps, but at the expense of greater CPU processing and lower speech quality. Instead of trying to reproduce the actual time domain waveforms like waveform codecs, these vocoders analyze each "frame" of speech (about a 20 millisecond interval) and outputs numbers that represent the sound produced by a theoretical abstract shape of the human vocal tract and vocal chord "driving function" during that frame period.

These low bit-rate vocoders can't transmit fax or modem signals with any accuracy. To send such tones, gateways using vocoder technology must have an additional feature that can detect when fax or modem calls are being made, then invoke a process to encode and transmit such signals. This is done on separate processing algorithms and software, and perhaps hardware.

Even DTMF (touchtone) signals don't make it through the low bit-rate vocoder process. Modern, sophisticated gateways deal with this by detecting (and suppressing) touchtones on the side doing the sending, encoding them in special packets for transmission, then regenerating the tones at the far end. This feature is called "DTMF relay."

After the speech is digitized, it's cut up into packets, each packet consisting of a header, trailer, and other "overhead" bytes, in addition to a data "payload" of one to several "frames" of encoded speech. Packing multiple frames of speech into a packet reduces overhead, as opposed to sending each frame in its own packet, with its own header, trailer, and other framing bytes. If a packet is lost however, then several frames are lost too, with a concomitant loss in voice quality.

Packets travel through the network independently, can arrive at their destination at different times or perhaps sometimes don't arrive at all. This all affects voice quality, increasing factors such as latency and delay (See Quality of Service).

Now, a fledgling IP communications company wanting to offer an alternative to long distance service from the major carriers like AT&T, Sprint, or MCI must create Points of Presence or POPs. You have to locate a VoIP gateway in every city, country, or hamlet where you're going to offer your phone-to-phone calling service. Why? Because you want any caller to initiate a call from the PSTN, then "hop on" to the Internet or private network, then "hop off," get back into the local PSTN, and get to the called party, all without any toll charges. That's why these devices are known as Hop-on-Hop-off (HO-HO) servers.

This sounds impossible. How can you locate gateways everywhere? Net2Phone seems to have originated the idea of making deals with regional Internet Service Providers (ISPs) to co-locate VoIP gateways at their facilities, thus eliminating expense of opening up new offices and hiring technicians to maintain the gateways and related IP network equipment. It also provides a built-in mechanism for localized billing for phone services and provides local customer support, too. The ISP's can also generate some additional revenue from the new voice service.

But to achieve reasonable channel density in an IP telephony service or product, the technical requirements of a VoIP trunk require considerably greater processing resources than a standard circuit-switched telephony trunk.

Most of the current issues will be resolved with the benefit of (a) higher levels of DSP chip integration, (b) prioritization of packets (voice will cost more to send than data) and (c) more processing power. Second generation products and services will overcome the current limitations, and be much easier to deploy as a result.

Voice over Net (VON)

There are those experts who like to split hairs. Voice over Net originally meant "voice over the Internet" but now it appears to mean "voice over packet" with an emphasis on IP telephony.

Voice Profile for Internet Messaging (VPIM)

Also known as Voice Profile for Internet Mail, it is the successor technology to the Audio Messaging Interchange Specification (AMIS), another standard for networking voicemail systems.

Based on SMTP/MIME (Simple Message Transfer Protocol / Multipurpose Internet Mail Extensions), VPIM uses the Internet to transmit voice and fax messages between voice mail systems. VPIM enables intra-company as well as inter-company networking since it's able to deliver voice and fax messages to messaging systems from different vendors, including both customer premise equipment (CPE) and service bureau environments.

Unlike the earlier analog-based AMIS technology that didn't take into account network fax routing, VPIM enables networking not just of voice messages but of fax messages, too, and over the Internet to boot. Hence, VPIM is better suited to more sophisticated applications such as unified messaging.

VPIM is supported by almost all messaging vendors: Applied Voice Technology, Lucent, Nortel, Siemens, Octel, and Centigram have committed to developing VPIM compliant networking products.

The originator (and chief promoter to get everybody to use VPIM) was Centigram Communications (San Jose, CA – 408-944-0250, www.centigram.com), which came up with VPIM's underlying concept of "Intentional voice messaging" in 1996. To explain what that is, one must first understand that, historically, voice messaging began with automatic telephone answering. You called a person, they weren't at their desk, their phone rang three times, it got answered by a voice mail machine and it took a message. That configuration was useful for awhile.

Then people said, "Sometimes I don't actually want to have a real-time live conversation with a person, I just want to leave them a voice message." There are scenarios that we've all experienced first-hand, where you're working late at night, the phone rings, you answer it, and the caller exclaims "Oh! I didn't want to talk to you, I just wanted to leave you a message."

That's intentional messaging.

Once you get into intentional messaging, there's all sorts of capabilities, where you can for example make an intentional message for a whole distribution list of people. When you get a message you can forward it on to people that need to hear it. It's essentially the whole e-mail paradigm brought to voice. It's popular with corporations and enterprises. People within a business have come to rely on using Centigram systems with these abilities.

But the problem with intentional messaging is that you're restricted to communicating with only certain people, those whose mailboxes are on your system. If you wanted to send a voice message to someone who was on a different company's system, You still had to make the phone call to them. If you wanted to give a message to five people, maybe a few people on your system, and several outsiders, you would have to make phone calls outside to those individuals and leave the message for them.

Now comes VPIM. What VPIM does is "Universal Intentional Messaging" where now you can make a voice message and fax message also and you can address it, not only to the people in your enterprise who are on your voice mail system and not only to people who maybe are networked via a private network to your system, but to anyone in the world, whose voice mail system supports the VPIM protocol.

Unlike the earlier analog AMIS standard, VPIM is fully digital, and produces high quality audio. It's based on e-mail standards like MIME and SMTP, and all the major voice mail vendors including Centigram, who supply both to corporations for CPE and to service providers, are also jumping on the VPIM protocol.

When everybody has their VPIM products rolled out, I'll be sitting there in my office in New York, working on an article on, say, VPIM, and let's say I've interviewed people from Centigram, Lucent, Nortel, and I want to make a voice message for all three people. In fact, I even want to send a copy to my Editor-in-Chief. With VPIM I can do that from my own voice mail system with one message that I make, sitting there logged into my voice mail system.

First the voice get's digitized. The VPIM companies originally standardized on G.726 – which is a 32Kbps ADPCM that sounds very good. Over time they've been adding other, more impressed encodings. G.726 was attractive to all the vendors because it wasn't proprietary to any one vendor or line card.

The digitized voice file gets encoded with MIME and sent out using

SMTP, travels over the Internet and is delivered to the various VPIM compliant voice mail systems through their respective servers, not the PSTN. The files are then decoded and placed in voice mailboxes. No toll charges. It's all "free" since it's using the Internet.

The originator of the message could be doing it from a phone at an airport, or sitting at a soundcard-equipped desktop PC if one's voice application or messaging server supports a desktop graphical interface – which Centigram's does. The nice thing about the way the message is addressed is that you just use the person's 10-digit phone number as their address. So all you need is someone's business card with their phone number on it and you have to mention that your voice box is VPIM enabled.

In principle, from the voice mail session, you just address this to the phone number, and the system figures out where on the Internet your voice mail machine is and sends it over to you.

You might say that it's the voice mail equivalent of a fax server.

As for actual faxes, the VPIM protocol also supports fax because most of the high-quality voice mail systems, including Centigram's, do support fax. So VPIM is essentially an alternative way to do a fax server.

Phone companies can roll this out to their residential customers or their SOHO customers, or anyone to whom they are providing messaging services. That means, for example, let's say you're sitting at your desk, it's late at night in California, and you want to send a message to your mother because it's too late to call her in New Hampshire. You can just send a voice mail message to her phone number and thanks to VPIM it will show up in her service provider's mail box. She'll get her "stutter dialtone" or whatever signal the provider sends customers to let them know they have a new voice message.

VPIM is called a profile, because when the software engineers created the specification they just decided to not reinvent the wheel, or go off and be like weird telephony geeks and do some bizarre new protocol that's arbitrary.

After all, there's already a perfectly good standard called Multimedia Internet Mail Extensions (MIME) which involves how you format multimedia messages. And there's already a perfectly good transport protocol called the Simple Mail Transfer Protocol (SMTP) in the Internet world, a TCP/IP related protocol governing electronic mail transmissions and receptions. So why not just take those and see how one can profile a subset of them that a voice mail server could support.

Ideally VPIM's creators wouldn't have had to invent anything. It would have been a pure profile of something that already existed. As it turned out, Centigram needed to create a new audio encoding scheme, but that wasn't too much of an ordeal, because the framework for MIME is that MIME is constantly evolving. People are constantly developing new content types which you submit to the Internet Engineering Task Force (IETF). So Centigram submitted their audio encoding and that's what was accepted. VPIM is thus a profile of the stuff that existed before Centigram started plus their own audio encoding scheme. Unlike AMIS, VPIM has nothing to do with the ITU.

VPIM is kind of like "voice mail disguised as e-mail," though Centigram would view it more as "multimedia Internet mail." Centigram specified, for example, that a VPIM compliant system doesn't have to support text messaging, where you think of a normal e-mail system or a unified mail system as having to support text. Instead, they said: "Here's the subset of MIME and SMTP that you must support to say that you're VPIM compliant."

Of course, for phone companies to support VPIM they have to install a series of servers with Internet connections. The sizing of it is just a systems engineering issue, but the bottom line is: They do need to have an

Internet or Intranet connection, and they need to support TCP/IP, SMTP and MIME. But there's nothing to say that you can only run VPIM on the public Internet. One way that various vendors want to rolling out VPIM is doing it in a private network.

With an extranet solution, one has more control over the Quality of Service (QoS) and you can install a high-bandwidth connection. You're still using all Internet protocols, but now they're no longer running over the great unwashed, unpredictable Internet.

Centrigram and Lucent were the first two companies to demonstrate VPIM by actually installing early versions of the product on each of their own corporate voicemail systems and then actually using VPIM to do intentional messaging from high-level executives at Centigram to high-level executives at Lucent.

This simple idea evolved into a more complicated scenario where instead of sending "toy" messages back and forth, both companies actually decided to use the VPIM protocol to collaborate in the writing and preparation of the press release and media alert both companies issued announcing the debut of VPIM! They worked 16 hours a day on it. Both sets of employees could write and send their messages at odd times of the day and night with no fear of disturbing anyone. They accomplished everything very rapidly over a period of two days, with about 20 iterations of the material going back and forth between Centigram and Lucent before they got it right. The whole project turned out to be quite successful, and it revealed some of the real value of VPIM.

Could VPIM be used in a Unified Messaging System? "Unified messaging" is used to refer to voice, fax and e-mail in one mailbox. "Universal messaging," which is what VPIM is offering, is the ability to message anyone anywhere, no matter what vendors' system they are on, and no matter whether it's a service provider system or a corporate system. The two are actually complementary but the nice thing about the way VPIM has been defined is that it does use e-mail standards for the protocol SMTP and MIME. It's very consistent with a unified message protocol.

So, it's possible that a unified messaging system could use VPIM. It could show up as an e-mail on a screen, you would click on it and the sound would come through your phone or speakers. It would show up as a multimedia message.

Voice Response Unit (VRU)

A sort of voice computing device. A computer has a keyboard to enter data, whereas a VRU of an IVR system accepts touchtones from telephones that have called into the system. A computer has a video display monitor to give you information, but a VRU reads back voice prompts to you.

Besides a PBX, a call center can have one or more VRUs, to provide an IVR "front end" to the call center and help callers direct themselves to the appropriate customer service representative (CSR).

A VRU can perform long, complex user interactions, such as providing multilevel menus. The VRU can call external functions, enabling it to interact with proprietary databases, allowing it to do sophisticated (though slow and somewhat unwieldy) call qualification as well as transaction processing (such as moving money from a savings account to a checking account). VRUs also can perform rudimentary software telephony services by using whatever VRU script language is included in the product, allowing the agent to transfer a call or hang up a call from his computer, rather than from his telephone set. VRUs also allow caller-specific reporting (a caller may want to raise his or her credit limit, for example).

However, a VRU does not supply a complete CTI solution. A VRU can't provide a common data repository. It can pass information to the agent, but

the agent cannot continue to access the databases with which the VRU interacted. After the VRU delivers the call to the agent, the call disappears as far as the VRU is concerned. Therefore, even if the VRU has a private data log, the agent cannot access it. Also, VRUs can't provide efficient call qualification. VRUs are optimized for announcement and touchtone collection, so doing significant data access and calculations is very expensive in terms of system resources. Adequate service in high usage call centers can require the use of several VRUs. All of this results in high system overhead, substantial physical space requirements (VRUs having a non-trivial footprint), and a significant expense. VRUs aren't able to assist in integration of heterogeneous hardware and software. Indeed, scripts must be rewritten for each kind of VRU, even if each script is to perform identical tasks. Finally, VRU's can't easily provide sophisticated software telephony services.

Voicemail (VM)

Voicemail is one of the oldest forms of computer telephony technology. If you have but one computer telephony item in your business, it is undoubtedly voicemail.

Just in case you've been participating in a suspended animation experiment for the past 40 years, here's a definition of voicemail: Whether you've successfully navigated through the automated attendant, gone through a live switchboard operator, or dialed directly to a person's desk, if that person cannot answer the call you may hear a recording of that person's voice saying, "This is Zippy at CMP Media, please leave a message and I will call you back." You may then leave a message that can be retrieved by the person you called. That's Voicemail.

Voicemail has come a long way from the days of recording messages at night when the office is closed. As it matured, it began to eliminate illegible scraps of handwritten messages that fell by the wayside. It increased the privacy of messages as well, and newer, better ways have been created to make sure that those messages reach a living, caring person (escalation and call forwarding being two features that get voicemails out of jail). The addition of DTMF, IVR, and speech recognition to voiecmail / auto attendants have made it even easier for callers to leave secure, private messages, indicating urgency and a number to call back, if they choose.

All that said, in this day and age, you don't have to buy a huge, enterprise-class voicemail / unified messaging server: Sweet and simple systems are still available. In other words, there's a voicemail for your price range and techno-sophistication level. The accompanying tables should help you decide through feature-by-feature comparison which best answers your business' needs.

But I Want Live People Answering Calls!

What are you losing by not having a voicemail system in the 21st century? If you're in a techno-industry, then there's a good possibility your company will be perceived as behind the times.

Strangely, some callers actually prefer voicemail, for reasons of privacy. When a caller leaves a voicemail message, there's no one else to overhear (and / or snicker). Voice messages also better convey the caller's tone (and anger level) to the message recipient, which can be crucial in knowing what to say when returning a call.

Not insignificant, too, is the new ability of voicemail to integrate with other types of messaging, such as fax and e-mail, which makes forwarding messages and retrieving messages, at any time of the day or night, a possibility beyond live receptionists.

Everyone Else Was Doing It

For better or for worse, the 1990s was the decade when voicemail (among thousands of other things) became commonplace. It was during those years that voicemail technology matured and became a staple in most places of business. Through information gathered from the "1999 MultiMedia Telecommunications Market Review and Forecast Report," price decreases for VM equipment and the ability of VM systems to (potentially) enhance an enterprise's productivity, while at the same time reducing personnel costs, propelled the market during the 1990s.

The average price per voice port dropped from $2,860 in 1993 to $2,400 in 1998 (that's 16.1%). Over the same time period, however, shipments of voice ports rose from 359,000 to 1,000,000 – that's a threefold gain. Spending on VM more than doubled, rising from $1 billion in 1993 to $2.4 billion in 1998.

Connections and Integrations

Enterprise voicemail typically runs on a standalone device made of voice boards and software running on a proprietary voice interface. Sometimes it's on circuit boards contained within a PBX. Increasingly often it's on a scaleable open NT platform. How do voicemail systems integrate with phone systems? Most connect using analog ports, and some connect via digital connections. A voicemail system may connect to a PBX using the same type of port that a proprietary digital telephone would use. These port connections are used to communicate with "outside" callers (those people leaving messages) and "inside" users (employees retrieving messages or reconfiguring mailbox options). Ports are also commonly used to signal the attached phone system using message-waiting lights on station sets. If the ports are dynamic, the same port may be used for any of these functions. Some older systems require you to designate specific ports for specific functions. Even if your system has dynamic ports, you might want to think about dedicating some for specific uses. For example, you might want to configure more ports for callers and fewer for system users who are retrieving their messages, changing their options, or recording their own messages. Voicemail systems are usually sold in two-, four-, or eight-port increments. The cost of adding ports to a system can vary dramatically, with four-port cards costing from $2,000 to over $10,000, depending on the system.

Expert Insight

Gioia Ambrette, president of Newcastle Communications Inc. (New York, NY – 212-780-9680, www.nccomm.com) and a 19-year veteran of the call processing game, tells about a trend she advocates, one against proprietary equipment and in favor of purchasing into an NT-based VM platform. "Why jump into bed with just one vendor," she asks, "when the only positive argument a vendor will give you for doing so is exactly that you'll just have one vendor to deal with? What about upgrades? Having only one vendor locks you down. If you need to upgrade storage, your vendor may charge you $10,000. If you are running on an open NT platform, a significant memory upgrade costs only the few hundred bucks it takes to buy another SCSI hard drive."

Ambrette also speaks of the impending convergence of media – unified messaging. NT lets you run Microsoft Exchange and even Lotus notes. A LAN can integrate with a speech recognition application to query an incoming call with a network database and provide caller ID, integratable PIMs, and even call accounting software. Sounds terrific. That said, voicemail still isn't infallible: On occasion when I've tried contacting one of my sources for an article via the speech recognition application the company used, I am sometimes left listening to Mozart's Concerto in B for bassoon for ten minutes. After hanging up and calling through the human operator, I was connected immediately. Chalk up at least one for the human beings. (Blame it on the auto attendant, or speech rec, if you will.)

We also talked to Ed Rebello and Doug Matsui, product managers at Siemans, a proprietary equipment maker. Their take on voicemail vendors is not extraordinarily different from Ambrette's. It's their contention, however, that they are responsible for inventing a lot of the technology used in open, non-proprietary equipment. However, they admit that it's harder to employ programmers limited to a single proprietary scripting "language." Therefore, says Rebello, their new-generation Expressions Platform is going to be based on a more open-standards unified messaging system.

PC Voicemail

The majority of PC-based voicemail systems can be linked to a LAN (using a standard network interface card) and equipped with optional software that has third-party capabilities such as call control and integrating applications. For example, a user has the ability to communicate directly to a PBX to make outgoing calls without (in theory) picking up a handset set. Also, a voicemail system may be able to notify users via a screen-pop when calls are transferred to their extensions, giving them the option of either answering calls or redirecting them to voicemail. More powerful clients may permit PC-based access to voicemail messages, allowing playback through the desk phone or PC sound card and speakers.

IVR

Another seemingly endless growth area in the design of voicemail is inventing a more intuitive "interface" (when searching through a system for an employee's extension), such as interactive voice response (IVR). The financial sector (e.g., banks and credit unions) finds IVR to be an excellent technology to employ while asking customers about account information, etc.

The telecom industry itself – with a sense of irony, perhaps – uses IVR quite frequently for billing inquiries and service troubleshooting.

Voicemail Should Be As Much Fun For Callers As For Users

So, voicemail should be set up to not only help employees retrieve their messages, but to help callers leave messages without going away angry. If someone calls into an extension and goes directly to a voicemail announcement, give him or her an option of pressing "0" to escape to a live person, as it has become something of a convention for people to dial "0" to reach a live operator. The call could be extremely urgent, or a quick question that another employee could answer. And a live answer is better than a late callback from an old forwarded message.

Some voicemail announcements tell the caller to press another extension number to "reach my assistant." When this happens, it is typical to get the voicemail of the assistant. Now the poor caller cannot even leave a message for the person he was calling since there are never instructions for how to go back to the called person's voicemail. Have a way to reenter the original employee's box.

Use a smart, relevant message that tells the caller something useful: Your name, your job, your extension number (the caller could have reached it by accident), and if there's a better way to reach you. How about those annoying messages that say, "I'm on the other line or away from my desk." What difference does that make to the caller? If you're on another line, in this age of CallerID, you're ignoring his call, aren't you? Why tell him that?

Some "on-the-go" companies update their outgoing messages daily – sometimes to excess. "Hello. Today is Thursday, April 8. I will be in the office at 10:00. After that I will be taking a meeting at 11:30, which may or may not run into a lunch meeting. If it does, I'll be back in the office at 1:00. From 1:30 to 1:35 I'll be in the restroom brushing my teeth and perhaps defecating, depending upon how much Pepsi I drank, which acts as both a diuretic and a laxative..."

It's a better idea to cut to the chase: "This is the voicemail of Patti Smith. Please leave a message and I'll call you back as soon as possible." If you have to leave explicit instructions on your outgoing message, preface the announcement by saying: "If at any point you want to bypass this message, hit 0."

JANE LAINO'S VOICEMAIL TIPS

Jane Laino (Jlaino@digby4.com) is a famous telecom consultant and president of DIgby 4 Group, Inc (New York, NY - 212-370-5369, www.digby4.com) DIgby 4 helps clients to manage their telecommunications resources. Support for improving call coverage is one of their services.

Here are some of Laino's suggestions for successfully using Voicemail to support call coverage:

1. Establish a company or departmental policy and procedure for using voicemail and stick to it. Some organizations decide to answer all calls live, then offer to transfer the caller to voicemail if the called person cannot be reached and the caller wishes to leave a message. Some may offer a caller the option of leaving a message with the person who answered, which is always good form if time permits. While this takes more manpower, it's considered by many to be a kinder approach than just dumping the caller into the voicemail announcement.

2. If someone calls into an extension and goes directly to a voicemail announcement, give him an option of pressing "0" to escape to a live person. Sometimes a question can be answered or a problem can be solved by another, present, person at the company. Since it has become something of a convention for people to dial "0" to reach a live person, it does not make sense to have a voicemail system that uses any other digit or series of digits and letters along with the * or # sign (the worst!). The # sign is frequently used as a calling card signal, which may not only signal your voicemail system that the caller wants to review the voicemail message he's recorded, it may signal the caller's carrier that he's ready to make another call using his calling card.

3. Some voicemail announcements instruct the caller to press another extension number to "reach my assistant." When this happens, it is typical to get the voicemail of the assistant! Now the poor caller cannot even leave a message for the person he was calling since there are never instructions for how to go back to the called person's voicemail. Having a caller escape from one person's voicemail and end up in another's is yet another form of caller abuse.

4. In terms of helping callers "escape" from voicemail, we recommend that callers pressing "O" should not be sent to a switchboard attendant, who will have no idea what is going on in the department, but rather to a departmental extension that is always answered. To avoid relying on just one person for this backup, designate a separate button on everyone's telephone that, when it rings, they will know that someone has escaped voicemail and wishes to speak to a live person. The display on the telephone will indicate whose voicemail the caller encountered, so the call can be answered appropriately. In the case where you must send voicemail escapees back to the switchboard, be sure that the switchboard attendant can identify that this caller has been in voicemail and can give the call priority.

5. Develop a standard for the scripting of individual voicemail greetings within your organization. Don't leave this to chance or callers will get an impression of inconsistency, depending upon who is called. We rec-

ommend that the voicemail greeting include the name of the company. "This is John Prescott at Comware Systems, please leave a message and I will call you back." Remember, a caller may not personally know the individual he is calling. Suppose someone calls a direct dial number and an unfamiliar name answers (this often happens when one person forwards his calls to another person for answering). If the person's greeting includes the company name, the caller will at least know he reached the right organization.

6. Avoid overused voicemail greeting phrases such as "I'm either on the phone or away from my desk." Some voicemail systems can determine whether you are on the telephone and, if so, will give the caller the option of pressing 1 to wait for you or pressing 2 to leave a message. Otherwise, what difference does it make to the caller if you're away, or talking, since you're not taking his call?

7. Decide on a company policy in terms of how long it takes to return voicemail messages. Avoid stating it in the voicemail greeting however, as it may create an expectation you cannot always live up to such as, "I will return your call within 4 business hours." Check out Vitel Software (508-831-9700). They make a product that lets you track how frequently your staff is retrieving voicemail messages.

8. In general, we suggest that you not say too much in a voicemail greeting, taking up the caller's time, nor create a greeting that needs changing very often. Invariably someone will be calling you on Wednesday and your greeting will say it's Tuesday, making you and your company appear disorganized.

9. Train your staff on how the telephone system is set up to work with the voicemail system. Most people don't know the different circumstances under which callers are sent to voicemail. Also train staff on how to transfer callers into the voice mailboxes of others. Very few people know how to do this properly, resulting in callers having to wait too long or hear an inappropriate greeting such as "Please enter your password."

10. Don't program your telephone system so that callers are only sent to voicemail if you press a button on your telephone. If someone forgets to press the button, calls may never be answered. The button can be used to send callers to voicemail without their having to wait for your telephone to ring, if you are truly not at your desk.

Voicemail and automated attendant are typically capabilities of the same system. Recognize that this is a separate system with separate care and feeding requirements. In many cases, it's not simply a capability of the telephone system. Arrange for regular maintenance and review reports on system operation to ensure that it continues to provide good call coverage support that you have put so much time and energy into planning.

10 VOICEMAIL IVR TIPS FROM VODAVI

With some form of IVR appearing on more and more voice mail servers, you can probably use some advice from the pros. Vodavi Communication Systems (Scottsdale, AZ – 888-422-2305, www.vodavi.com) obliged us with the following:

1. Before designing your audiotex menus, have your agents/operators keep a log of the types of questions asked. Any repeat questions customers keep asking can then be included during the initial menu design instead of redesigning the menu later.

2. Consistency is crucial. Always keep your "go back" and "repeat menu keys" the same. For example, if "star" takes the caller back one menu level at one point in the call flow, and "pound" repeats the information, keep it that way throughout.

3. Always provide the caller with a "zero out" to operator option.

Company Name	City, state — phone	Web site	Name of product	Physical system configuration	Maximum # of ports	Supported voice boards	Supported phone systems	Maximum voice storage (hours)	How does it communicate with switch?	Max No. Mailboxes	Types of boxes included				No. of greetings
											Group	Guest	Fax	Bulletin Boards	
ABS TALKX	Bay Shore, NY — 800-825-5944	www.abstalkx.com	Baby TALKX	Solid State Flash Based	2 & 4 expandable to 8	Proprietary	All Major brands	1.5 to 6	Single line extension	64	N	N	N		3
ABS TALKX	Bay Shore, NY — 800-825-5944	www.abstalkx.com	Simplicity for Coral and Coral SL	In-Skin Proprietary T1 voice mail card for the ECI Coral and Coral SL	4,8,12 and 24 port	Proprietary	Coral and Coral SL	200	Direct Install	unlimited	Y	Y	N	Y	6
ABS TALKX	Bay Shore, NY — 800-825-5944	www.abstalkx.com	Simplicity Plus	Turnkey PC-Based system	4,6 & 8 port	Dialogic	All Major brands	200	Analog integration	unlimited	Y	Y	N	Y	3
ABS TALKX	Bay Shore, NY — 800-825-5944	www.abstalkx.com	Starter TALKX	Turnkey PC-Based system	4,6 & 8 port	Dialogic	All Major brands	200	Analog Integration	500	Y	Y	N	Y	3
ABS TALKX	Bay Shore, NY — 800-825-5944	www.abstalkx.com	Ultra TALKX	Turnkey PC-Based system	8,12,16,20 & 24 port	Dialogic	All Major brands	200	Analog Integration, T-1 Voice Bridge, Centrex	unlimited	Y	Y	N	Y	3
ABS TALKX	Bay Shore, NY — 800-825-5944	www.abstalkx.com	Uni TALKX	Turnkey PC-Based system	unlimited	Dialogic	All Major brands	500+	Analog, serial, T-1	unlimited	Y	Y	Y	Y	3
Active Voice	Seattle, WA — 206-441-4700	www.activevoice.com	Unity 2.3	PC-Based	120					TAPI, inband, serial		Y	N	Y	N	6
Black Ice Software	Amherst, NH — 603-673-1019	www.blackice.com	Impact Voice Mail Server Deluxe	PC based	255	Dialogic, NMS, Brooktrout	Any	1000	Tone, DNIS	100,000	Y	N	Y	N	4
eOn Communications	Memphis, TN — 800-955-5080	www.eonoc.com	eOn Voice Processing System (VPS)	PC Turnkey	48	Natural Microsystems	eOn Millennium (serial); others via inband	250	Serial, Inband	3,000+	Y	Y	Y	Y	6
ESI	Plano, TX — 972-422-9700	www.esi-estech.com	VoiceWorks 18	Standalone	16	N/A	Many! (Contact us for details)	140 (optional; 70 hrs standard)	66 block	1,000	Y	Y	N	Y	3
E-Voice Communications	Sunnyvale, CA — 408-991-9988	www.evoicecomm.com	evoice2000	PC-based, Turnkey, or Board Software	32	Brooktrout	most major PBX's	determined by computer hard drive	Inband integration	1,000	N	N	N	N	
E-Voice Communications	Sunnyvale, CA — 408-991-9988	www.evoicecomm.com	evoice2000 Lite	PC-based, turnkey or board/software	32	Brooktrout	most major PBX's	determined by computer hard drive	Inband integration	1,000	Y	N	N	N	
Figment Technologies	Toronto, Ont — 416-499-1919	www.figment.net	UniExchange Comms Server	NT Server-based PC, standard or rack-mount	96	PIKA, Dialogic	Most Analog-based	60	DTMF/ normal transfer	unlimited	Y	N	Y	Y	3
HOMISCO	Melrose, MA — 781-665-1997	www.homisco.com	HVMS	Desktop PC turnkey installation	24	Natural Microsystems AG 8, VBX400	Mitel, NEC, Northern, Hitachi, Lucent	limited only by hard drive space	Inband integration or VTQ interface	3,000	Y	Y	N	Y	5
Interactive Intelligence	Indianapolis, IN — 317-715-8288	www.ININ.com	Enterprise Interaction Center (EIC)	PC-Based	currently limited by Dialogic hardware to 480 ports	Dialogic and Aculab	analog phones	limited only by available disc space on server	Built-in PBX	unlimited	Y	N	Y	N	unlimited
Inter-Tel	Milford, CT — 203-876-7600	www.inter-tel.com	Executone Infostar/DVX	Turnkey	4	embedded design	Executone IDS 42 & 84	45 (with capability to set limits per user)	Digital Integration	unlimited	Y	Y	N	Y	6
Inter-Tel	Milford, CT — 203-876-7600	www.inter-tel.com	Executone Infostar/EVX	Turnkey	12	embedded design	Executone Eclipse	100 (with capability to set limits per user)	Backplane integration	unlimited	Y	Y	N	Y	6
Inter-Tel	Milford, CT — 203-876-7600	www.inter-tel.com	Executone Infostar/VX3	PC	24	Dialogic	Executone Eclipse & Executone IDS	100 (with capability to set limits per user)	Inband (RS-232)	unlimited	Y	Y	N	Y	6
Inter-Tel	Milford, CT — 203-876-7600	www.inter-tel.com	Executone Repartee	PC	80	Dialogic	Executone Eclipse, Executone IDS, Lucent, Nortel and Rolm	200+ (with capability to set limits per user	Inband	unlimited	Y	Y	Y	Y	6
ITS Ltd	Azur, Israel — 972-355-76965	www.its-tel.com	Vocal Voice Mail and Auto attendant Sys.	PC-based features (solid state)	4	32	most well known	8	Inband DTMF/ RS-232	128	Y	Y	Y	Y	39
Key Voice Technologies	Sarasota, FL — 800-419-3800	www.keyvoice.com	Debut	Turnkey	4	Dialogic D/41H (hybrid)	Over 100-call for details	4	Inband	50	Y	Y	Y	Y	10
Key Voice Technologies	Sarasota, FL — 800-419-3800	www.keyvoice.com	Small Office Lite	PC-Based, Turnkey	4	Dialogic: DIALOG/4 Rhetorex Comdial IVPC	Over 100-call for details	60	Inband	50	Y	Y	Y	Y	10
Key Voice Technologies	Sarasota, FL — 800-419-3800	www.keyvoice.com	Small Office	PC-Based, Turnkey	4	Dialogic: DIALOG/4, ProLine/2V, D/41H, D/42D, D/42SL, D/42SX Rhetorex 232 432, Comdial IVPC	Over 100-call for details	125	Inband	100	Y	Y	Y	Y	10
Key Voice Technologies	Sarasota, FL — 800-419-3800	www.keyvoice.com	Small Office NT	PC-Based, Turnkey	8	Dialogic: D/41H D/160SCLS	Over 100-call for details	200+	Inband	100	Y	Y	Y	Y	10

Types of schedules	Record a call	Live screen	Cascade out calling levels	Forward/ rewind/ pause/ resume	Max system lists	Max per sonal lists	FIFO/ LIFO msg playing	Park and Page	Msg w/ receipt or comment	Copy msg w/ comment	Auto forward unread msg	Audiotext directory		AVR/ Q&A present	LAN connect pt chan.	Msg play track on PC?	PIM required?	Remote administration?	MSRP	Distinguishing features
day, night, temporary, outcall	N	N	N	Y/Y/Y/Y	0	0	N	N	N	Y	N	Y		Y	N	N		Via modem and telephone	$1,100	Excellent voice quality. Fast response. Music-on-hold and paging built-in.
time, day, holiday, outdial, wakeup, override	Y	N	Y/5	Y/Y/Y/Y	50+	18	Y	Y	Y	Y	Y	N		Y	Y	N		Via modem and telephone	starts at $4,000/ 4 ports	Uses standard Dialogic API calls. T-1 solution that integrates with any Coral system. No additional cards or expansion needed for standard voicemail.
time, day, holiday, outdial, wakeup, override	N	N	Y/5	Y/Y/Y/Y	50+	18	Y	Y	Y	Y	Y	N		Y	Y	N		Via modem and telephone	starts at $2,700/ 4 ports	App gen feature customization. Auto clean up and disk defragmentation. Programmable w/ either computer or keyboard/monitor package.
time, day, holiday, outdial, wakeup, override	N	N	Y/5	Y/Y/Y/Y	50+	18	Y	Y	Y	Y	Y	N		Y	Y	N		Via modem and telephone	starts at $7,700/ 4 ports	App gen feature customization. Auto clean up and disk defragmentation. Mailbox linking, options, types and class-of-service customizable.
time, day, holiday, outdial, wakeup, override	N	N	Y/5	Y/Y/Y/Y	50+	18	Y	Y	Y	Y	Y	N		Y	Y	N		Via modem and telephone	starts at $6,500/ 8 ports	App gen feature customization. Auto clean up and disk defragmentation. Mailbox linking, options, types and class-of-service customizable.
time, day, holiday, outdial, wakeup, override	Y	Y	Y/5	Y/Y/Y/Y	50+	18	Y	Y	Y	Y	Y	Y		Y		Y	Outlook	Via modem and telephone	starts at $5,750/ 4 ports	Standalone NT system utilizes Ultra's complete feature set. Auto clean up and disk defragmentation. Mailbox options, types and class-of-service customizable.
day, evening, caller ID	Y	Y	Y/4	Y/Y/Y/Y	500+	20	Y	Y	Y	Y	Y	N		N	N	Y		Web-based Internet Explorer 4.0	$9,280	Directory and message store via Exchange. Uses traditional and IP-based PBX. One point of administration via web.
day, evening, departmental	Y	N	N	N/N/Y/N			N	N	N	Y		N		Y		Y		Browser	starts at $2,000/ 4 ports	Script editor. Text to speech. Supports all major vendor hardware.
integration-dependent	Y	Y	Y/25	Y/Y/Y/Y	9,900	98	Y	Y	Y	N	Y	Y		Y	Y	Y	w/ V-card	remote communications software client	$7,500	Optional software-only fax capabilities. Advanced networking, distributed voice-mail between switches. TCP/IP-based digital networking using VoiceClusters technology.
Day, night, holiday	N	N	Y/5	Y/Y/Y/Y	0	0	Y	Y	Y	Y	Y	Y		Y	Y	N		DTMF or built-in modem and VM installer software	$7,995	Up to 16 built-in ports of voicemail. Less expensive than most systems. Very easy to install and use.
Urgent, non-urgent	N	Y	Y	Y/Y/Y/Y	un-limited	un-limited	Y	Y	Y	Y	Y	Y		Y	Y	Y	Goldmine, ACT, MS Outlook	Phone or PC		Windows and browser based admin. On-screen message control and database integration. Competitive pricing.
Urgent, non-urgent	N	Y	Y	Y/Y/Y/Y	un-limited	un-limited	Y	Y	Y	Y	Y	Y		Y	N	Y		Phone or PC	$1,499	Windows and browser based admin. Feature rich. Competitive pricing.
	N	N	N	Y/Y/N/N	un-limited	0	Y	Y	N	N	Y	N		N	N	Y	Goldmine, Outlook	Browser-based	$3,499/ 4 ports	Ultra low cost. Users can access from any phone or internet-capable machine from anywhere. Provides great value.
annual charge required	N	N	N	Y/Y/Y/Y										N	N	N		Modem	$1,500–$4,000	Combines with TCS call accounting system, designed for hotels with max of 800 extensions. Supplies 13 reports. Operates under OS/2 Warp.
day, evening	Y	Y	Y/2	Y/Y/Y/Y	un-limited	un-limited	N	Y	Y	Y	Y	Y		Y	N	Y		internet administrative tools	starts at $1,000 per seat	Software-based, fully customizable. Unified architecture. Open system, integrates easily with 3rd party applications.
day, evening, caller ID, departmental	Y	N	Y/5	Y/Y/Y/Y	50	19	Y	Y	Y	Y	N	N		Y	N	N				Low-cost, feature rich, automated attendant, outdial and integrates to a digital station port.
day, evening, caller ID, departmental	Y	N	Y/5	Y/Y/Y/Y	50	19	Y	Y	Y	Y	N	Y		Y	Y	N				Voicemail with up to 12 ports for Eclipse. Automated attendant and outdial, and remote call notification.
day, evening, caller ID, departmental	Y	N	Y/5	Y/Y/Y/Y	50	19	Y	Y	Y	Y	N	Y		Y	Y	N				Powerful system includes Remote Call Director, Emergency notification, automated attendant, Outdial, and voice activated speed dial.
day, evening, caller ID, departmental	N	Y	Y/4	Y/Y/Y/Y	9,900	9,900	Y	N	Y	N	N	Y		Y	Y	Y	Phonefax, ODBC-compliant	Via Co-sessions software		Unified messaging. PC call control.
day, evening, holiday, break	Y	Y	Y	Y/Y/Y/Y	5	128	Y	Y	Y	Y	Y	Y		N	N	Y		DTMF		Call recording. No. of messages (old/new) displayed. Dial by name, 3 languages, pager notification.
day, evening, caller ID, departmental	Y	Y	Y/5	Y/Y/Y/Y	5+	4	Y	Y	Y	Y	Y	Y		N	N	N			$1,990	Economical, over 60 standard features, outstanding technical support and dealer services.
day, evening, caller ID, departmental	Y	Y	Y/5	Y/Y/Y/Y	5+	4	Y	Y	Y	Y	Y	N		N	N	N			$3,190	Leverage to obtain NT-based voice processing, over 60 standard features, outstanding technical support and dealer services.
day, evening, caller ID, departmental	Y	Y	Y/5	Y/Y/Y/Y	20+	4	Y	Y	Y	Y	Y	Y		Y	Y	Y			$3,990	Leverage to obtain NT-based voice processing, available w/ optional features, outstanding technical support and dealer services.
day, evening, caller ID, departmental	Y	Y	Y/5	Y/Y/Y/Y	20+	4	Y	Y	Y	Y	Y	Y		Y	Y	Y	ODBC	NT-class workstation, Exchange	$6,490	Leverage hardware and software investment, value added service module, outstanding technical support and dealer services.

V

465

Company Name	City, state — phone	Web site	Name of product	Physical system configuration	Maximum # of ports	Supported voice boards	Supported phone systems	Maximum voice storage (hours)	How does it communicate with switch?	Max No Mailboxes	Group	Guest	Fax	Bulletin Boards	
Key Voice Technologies	Sarasota, FL — 800-419-3800	www.keyvoice.com	Corporate Office NT	PC-Based, Turnkey	64	Dialogic: D/41H D160SCLS	Over 100-call for details	200+	Inband	10,000	Y	Y	Y	Y	
Key Voice Technologies	Sarasota, FL — 800-419-3800	www.keyvoice.com	INTerchange	PC-Based, Turnkey	64	Dialogic: D/41H D160SCLS	Over 100-call for details	200+	Inband	10,000	Y	Y	Y	Y	1
Lucent Technologies Messaging Solutions	Milpitas, CA — 408-324-6734	www.lucent.com/octel	Insuty Audix	PC-based	512	proprietary	Many including Lucent, Nortel, Siemens	10,000	X.25, Inband, TCP/IP	73,000	Y	Y	Y	Y	
Lucent Technologies Messaging Solutions	Milpitas, CA — 408-324-6734	www.lucent.com/octel	Lucent AnyPath Messaging Platform	Proprietary	720	proprietary	All central office and mobile exchange switches supporting SS7, SMDI, or Inband Connectivity	10,000	Inband, SS7, SMDI	up to 1,000,000	Y	Y	Y	Y	
Lucent Technologies Messaging Solutions	Milpitas, CA — 408-324-6734	www.lucent.com/octel	Octel 250	Turnkey	144	proprietary	Many including Lucent, Nortel, Centrex	1350	Inband, RS-232	30,000	Y	Y	Y	Y	
Lucent Technologies Messaging Solutions	Milpitas, CA — 408-324-6734	www.lucent.com/octel	Octel 300	Proprietary, customer configurable	128	Link cards that support voice/fax recording are proprietary; some are dependent on phone system being supported	Many including Lucent, Nortel, ROLM	1,085	Inband, RS-232, set emulation	10,000	Y	Y			
Lucent Technologies Messaging Solutions	Milpitas, CA — 408-324-6734	www.lucent.com/octel	Octel 350	Turnkey	144	proprietary	Many including Lucent, Nortel, Siemens	1350	Inband	30,000	Y	Y	Y	Y	
Lucent Technologies Messaging Solutions	Milpitas, CA — 408-324-6734	www.lucent.com/octel	Unified Messenger	PC-based	240	Brooktrout	Many including Lucent, Panasonic, Siemens, Nortel	unlimited	Inband, RS-232, set emulation	unlimited	N	N	Y	Y	
Mitel Corporation	Kanata, ON — 613-391-2321	www.mitel.com	Mitel Express Messenger	In skins	8	N/A	Mitel 5x200 mitel	100	DNIC	250	Y	Y	N	N	
Mitel Corporation	Kanata, ON — 613-391-2321	www.mitel.com	Mitel Nupoint Messenger	Turnkey	240	NMS/Brooktrout	Mitel, Lucent, NEC, Fujitsu, Nortel, Siemens, Toshiba	3600	TAPI, Inband, Tone, DNIC	28,670	Y	Y	Y	Y	
Mitel Corporation	Kanata, ON — 613-391-2321	www.mitel.com	Mitel OnePoint Messenger	Turnkey	24	NMS/Brooktrout	Mitel, Lucent, NEC, Fujitsu	unlimited	Inband	unlimited	Y	Y	Y	Y	
Nitsuko America	Shelton, CT — 203-926-5498	www.nitsuko.com	Vangard	Turnkey	8	N/A	All Nitsuko Systems (i-Series, DS2000, Portrait, Onyx family, DS01)	130	Inband	200	Y	Y	N		3
Nitsuko America	Shelton, CT — 203-926-5498	www.nitsuko.com	NVM-2000	PC-based	24	Brooktrout	All Nitsuko Systems (i-Series, DS2000, Portrait, Onyx family, DS01)	270	Inband	1,000	Y	Y	Y		3
Nitsuko America	Shelton, CT — 203-926-5498	www.nitsuko.com	NVM-NT	NT Server Based	24	Dialogic D141H	Nitsuko i-Series	250	Inband	10,000	Y	Y	Y		3
Nitsuko America	Shelton, CT — 203-926-5498	www.nitsuko.com	NVM-2E	Turnkey	4	N/A	Nitsuko i-Series	3	Inband, Tone	50	Y	N	N	N	1
Panasonic	Secaucus, NJ — 201-392-6220	www.panasonic.com	Courier	Turnkey	32	Rhetorix	All that provide inband signaling	350	Inband	10,000	Y	Y	Y		10
Paraxip Electronics	Aberdeen, NJ — 732-290-1900	www.voicesaver.com	VoiceSaver NT	PC Based and Rack-mountable	256	Dialogic, Bicom and Music Telecom	Mitel, Nortel, NEC, Lucent, Wio, Nitsuko, Intertel, BPL, Alcatel, Telrad, 10Systems	14,000	T-1/E-1, SMDI, C7, SS7 and analog	30,000	Y	Y	Y		
Samsung Telecom	Miami, FL — 800-876-4782	www.samsungtele-com.com	Cadence	In skin	16	Proprietary	DCS 50si, DCS, DCS 400si	Over 100	Backplane	1,000	Y	N	Y		9
SpeechSoft	Armonk, NY — 914-273-5560	www.speechsoft.com	TopCAT	Turnkey	4	N/A	Analog, SMDI, Centrex	12	TAPI, Inband, tone, SMDI	10,000	Y	Y	Y	Y	
SpeechSoft	Armonk, NY — 914-273-5560	www.speechsoft.com	CallMASTER	Software for Windows NT	96	Analog, T-1 Dual T-1	Analog, T-1, SMDI, Centrex	Unlimited (based on drive capacity)	TAPI, Inband, tone, SMDI	1 billion	Y	Y	Y	Y	3
SpeechSoft	Armonk, NY — 914-273-5560	www.speechsoft.com	SpeechMASTER	Software for DOS or Windows 98/95 DOS shell	32	Analog, T-1, Dual T-1	Analog, T-1, SMDI, Centrex	Unlimited (based on drive capacity)	TAPI, Inband, tone, SMDI	10,000	Y	Y	Y	Y	
Sprint Products Group	New Century, KS — 913-791-7708	www.sprintproducts.com	ProtegeVoice SVP-4	PC-based	4 (expandable to 8)	Proprietary	Protegé key systems	Maximum 70	Inband	unlimited	Y	Y	N	Y	
Sprint Products Group	New Century, KS — 913-791-7708	www.sprintproducts.com	ProtegeVoice SVP-12	PC-Based	12	Proprietary	Protegé key systems	Minimum 70	Inband	unlimited	Y	Y	N	Y	
Telrad Networks	Woodbury, NY — 516-921-8300	www.telradusa.com	ImaGen	PC-based & turnkey	16	Proprietary	Telrad digital	100	digital integration	10,000	Y	Y	N	Y	5

...es of ...dules	Record a call	Live screen	Cascade out calling levels	Forward/ rewind/ pause/ resume	Max system lists	Max per serial lists	Broad cast msg paging	Park and Page	Msg w retrieval comment	Copy msg w comment	Auto forward unread msg	Auto dial directory	Custom call routing	IVR/ Q&A	LAN based pc client	Msg play-back on PC?	PIM integrate?	Remote administration?	MSRP	Distinguishing features
...evening, ...er ID, ...trmental	Y	Y	Y/5	Y/Y/Y/Y	500+	4	Y	Y	Y	Y	Y	Y	Y	Y	Y	Y	Y	DDE	NT-client workstation/ Exchange	$9,900	Leverage hardware and software investment. Value added service module. Outstanding technical support and dealer services.
...evening, ...er ID, ...trmental	Y	Y	Y/5	Y/Y/Y/Y	500+	4											Y	DDE	NT-client workstation/ Exchange		Internet standards run integrated and unified on same PC. Value added service module. Outstanding technical support and dealer services.
	N	N	N	N/N/N/N		999	Y	Y	Y	Y	N	Y	Y	Y	Y	Y		Message Manager Address Book	Terranova management software, TCP/IP		Multimedia mailbox (voice, fax, text, file attachments). Internet standards-compliant. Text to speech. Integrates telephony and IP networks
...eetings	Y	N	N	Y/Y/Y/Y	un-limited	15	Y	N	Y	Y	Y	Y	Y	Y	Y	Y			HTML browser over TCP/IP		Scalable architecture. Embedded email server and single message storage for voice, fax and email
...evening	Y	N	Y/3	Y/Y/Y/Y	300	15	Y	Y	Y	Y	Y	Y	Y	Y	Y	Y			Ethernet, TCP/IP		Reliable SCSI drives and redundant power supplies. Global message redundancy. Hot plug components such as drives and port cards permit maintenance while server is in service.
...le, day, ...liday	Y	Y	Y	Y/Y/Y/Y	250	9	Y	Y	Y	Y	Y	Y	Y	Y	Y	Y	Visual messenger		Modem		Full call processing and caller care features. Sophisticated technical support and admin. Reliability rating of 99.979% server architecture.
...selected	Y	N	Y	Y/Y/Y/Y	3,000	15	Y	N	Y	Y	Y	Y	Y	Y	Y	Y			TCP/IP		Reliable SCSI drives and redundant power supplies. Online call detail records. Hot plug components such as drives and port cards permit maintenance while server is in service.
...day, hol-...iday	N	N	N	Y/Y/Y/Y	un-limited	un-limited	Y	Y	Y	Y	Y	Y	Y	Y	Y	Y	Microsoft Outlook		MS Exchange		Voice, fax, and email in single unified mailbox. All messages available from phone or PC. Wide array of switch integration.
...evening, ...rtmental	Y	N	Y/1	Y/Y/Y/Y	5	5	Y	N	Y	Y	Y	N	Y	N	n/a	n/a		Via Built-in modem	$3,599	Great integration. Unmatched price/performance. Simplicity.
...evening, ...artmental	N	N	Y/1	Y/Y/Y/Y	99	99	Y	Y	Y	Y	N	Y	Y	Y	Y	Y		SNMP over TCP/IP	Starts at $7,500	Scalability (4-240 ports). Real-time OS. Handles calls in up to 12 languages simultaneously.
...evening ...rtmental	N	N	Y/1	Y/Y/Y/Y	un-limited	un-limited	Y	Y	Y	Y	Y	N	N	Y	Y	Y	Outlook		MS Exchange		True unified messaging with MS Exchange. Brooktrout Show N Tel Toolkit based.
...evening, ...ller ID, ...rtmental	Y	Y	Y/3	Y/Y/Y/Y	1,000	100	Y	Y	N	Y	Y	Y	Y	Y	N	N		Internal Modem		Speech recognition. 3 desktop seats. Interactive voice response.
...artmental	Y	Y	Y/3	Y/Y/Y/Y	100	100	Y	Y	Y	Y	N	Y	Y	Y	Y	Y		Internal modem		Conversation record. Park and Page. Personal answering machine evaluation.
...artmental	Y	Y	Y/3	N/N/N/N	100		Y	N	Y	Y	N	Y	Y	Y	Y	Y		Internal modem		Conversation record. Park and Page. Personal answering machine evaluation.
...artmental	Y	Y	Y/1	Y/Y/Y/Y	1	0	Y	N	Y	Y	Y	N	Y	N	N	N		Via Touchtone phone		Conversation record. Message count display. Personal answering machine evaluation.
	Y	Y	Y/4	Y/Y/Y/Y	100	100	Y	N	Y	Y	N	Y	Y	Y	Y	Y		Via Modem LAN connection		Off hook voice announce when using API integration. Remote admin through a LAN or Internet. Receive and send voicemail through PC.
...day, non-...working ...s, lunch, ...night	Y	Y	Y/10	Y/Y/Y/Y	200	10	Y	Y	Y	Y	Y	Y	Y	Y	Y	Y		Browser, pcAnywhere	$750 per port turnkey	Integrates with Winfax to provide faxback. Uses off the shelf fax/modem cards. Integrates with Quick Books for IVR. Built-in school app.
...y, Night	Y	Y	Y/100	Y/Y/Y/Y	1,000	1,000	Y	Y	Y	Y	Y	Y	Y	Y	N	N		Via modem	$1,299 4 ports	Integrates fax and voicemail in one system. Easy installation. Integrated keyset displays with multiple language support
...evening, ...artmental	Y	Y	Y/ multiple	Y/Y/Y/Y	999	999	Y	N	N	Y	N	Y	Y	Y	N	N		Modem		Cost. Features. Ease of use.
...evening, ...artmental	Y	Y	Y/ multiple	Y/Y/Y/Y	999	999	Y	N	N	Y	N	N	Y	Y	Y	N		Remote software		Cost. Features. Ease of use.
...evening	Y	Y	Y/ multiple	Y/Y/Y/Y	999	999	Y	N	N	Y	N	Y	Y	Y	N	N		Remote software		Cost. Features. Ease of use.
Day	Y	Y	Y/9	Y/Y/Y/Y	50	19	Y	Y	Y	Y	N	Y	Y	Y	N	N		Modem	$3,399	Digitally integrated. Provides two ports of voicemail for one digital link. Provides control and flexibility.
Day	Y	Y	Y/9	Y/Y/Y/Y	50	19	Y	Y	Y	Y	N	Y	Y	Y	N	N		Modem	$5,249	Digitally integrated. Provides two ports of voicemail for one digital link. Provides control and flexibility.
32	Y	Y	Y/9	Y/Y/Y/Y	50		Y	N	Y	Y	Y	Y	Y	N	Y	Y	MS Outlook	Co-session remote	$3,160 per line	Voicemail and phone system mfg by Telrad. GUI icons. ACD supervisor stations large display...

V

Company Name	City, state — phone	Web site	Name of product	Physical system configuration	Maximum # of ports	Supported voice boards	Supported phone systems	Maximum voice storage (hours)	How does it communicate with switch?	Max No. Mailboxes	Group	Guest	Fax	Bulletin Boards
Teltronics	Sarasota, FL — 941-751-7758	www.teltronics.com	VisionWorks	PC-based Software & Hardware application available in midtower & rack mounted configurations	32	Rhetorex	Teltore's Vision & most major PBX's	280	Inband, serial CTI port	unlimited	Y	Y	Y	Y
Toshiba Telecommunications	Irvine, CA — 800-222-5805	telecom.toshiba.com	Stratagy DK	Turnkey	8	n/a	Toshiba Strata	65	Inband, Tone, SMDI	unlimited	N	Y	N	Y
Toshiba Telecommunications	Irvine, CA — 800-222-5805	telecom.toshiba.com	Stratagy Flash	Turnkey	4	no additional	Toshiba	4	Inband, SMDI	unlimited	N	N	N	N
Vodavi	Norcross, GA — 770-662-1500	www.vodavi-ct.com	PathFinder/ Pathfinder IVR	PC based, turnkey or kits	30 (upgradeable to 96)	Dialogic	Vodavi, Northern Telecom, NEC 2400, Norstar, Mitel, Centrex, Inband integrations	100	Analog, digital emulation, T-1, SMDI	9,999,999	Y	Y	Y	Y
Vodavi	Norcross, GA — 770-662-1500	www.vodavi-ct.com	TalkPath	Turnkey	8	Vina	Inband support	70	Inband	9,999,999	Y	Y	Y	Y
Vodavi	Norcross, GA — 770-662-1500	www.vodavi-ct.com	Dispatch	Turnkey	8	Vina	Inband support	70	Inband	9,999,999	Y	Y	Y	Y
VoiceGate	Markham, ON — 905-513-1403	www.voicegatecorp.com	VoiceGate ICS	PC-Based, Turnkey, Windows NT	24	Dialogic, VTG	most popular	1,400	TAPI, inband, SMDI, digital phone emulation	10,000	Y	Y	Y	Y
VoiceGate	Markham, ON — 905-513-1403	www.voicegatecorp.com	VIP4000	PC-Based, Turnkey	24	Dialogic, VTG	most popular digital and analog phone systems	1,400	Inband, digital and analog phone emulation, SMDI	1,000	Y	Y	N	Y
VoiceGate	Markham, ON — 905-513-1403	www.voicegatecorp.com	VoiceGate Lite	PC-Based, Turnkey	4 upgradable to VIP4000	Dialogic	Meridian Norstar	920	Inband, digital phone emulation	60	Y	Y	N	Y
VoiceGate	Markham, ON — 905-513-1403	www.voicegatecorp.com	VoiceGate Wizard	Solid State, Flash Based	4	Dialogic Voice Brick	most popular phone systems	10	Inband, SMDI	60	Y	Y	N	Y

4. Include speech recognition alternatives to touch tone input whenever possible. Over 15% of the phones in the US are still pulse dial (rotary).

5. Keep the menu to no more than four or five options; too many options confuses most callers. Also position the menu items in descending order of popularity.

6. Analyze the calling patterns and menu usage after the first two or three months to determine if callers are finding all the features and functions in your menu tree.

7. Keep an eye on your busy hour calls and line blockage to see if you are running out of capacity in your current system. Erlang tables can help determine your current capacity.

8. If you want to have different phone numbers, go to different IVR functions (e.g., one number for repair, one number for sales, one number for support, etc.). Consider using a T-1 from the central office and passing the DNIS information to the IVR so that it can automatically determine what service the caller desires.

9. Make sure you can easily record and delete a custom front end prompt for the first menu that can be used to announce special hours or closures for holidays.

10. When confirming digit entry by a caller, try to provide a more meaningful confirmation beyond simply repeating the digits. For example, query the database and tell the caller, "You have requested to refill your prescription for Amoxicillin. Press 1 if correct, or 2 to try again."

VoIP

See Voice over IP.

VPIM

See Voice Profile for Internet Messaging.

...s of ...dules	Record a call	Live screen	Cascade out calling/ levels	Forward/ rewind/ pause/ resume	Max system lists	Max personal lists	Broad cast mes saging	Park and Page	Msg w/ receipt	Copy msg w/ comment	Auto forward unread msg	Autodial directory	Custom call routing	IVR/ Q&A present	LAN based pc client	Msg play-back on PC?	PIMs integrate?	Remote administration?	MSRP	Distinguishing features
...oting, ...r ID, ...mental	Y	Y	Y/10	Y/Y/Y/Y	un-limited	99	Y	N	Y	Y	Y	Y	Y	Y	Y	Y	MS Outlook, Goldmine, ACT	pcAnywhere	Starts at $11,425	Screen pop on station-to-station calls, direct inward dial calls, and from Vision system auto attendant. Server dialback, Vision phone display update.
...ening, ...r ID, ...mental	N	N	Y/10	Y/Y/Y/Y	7	7	Y	Y	Y	Y	Y	Y	Y	Y	N	N		Strategy admin software		Busy station ID. Night transfer alternate routing. New user tutorial.
...ening, ...ID, ...mental	N	N	Y/10	Y/Y/Y/Y	7	7	Y	Y	Y	Y	Y	Y	Y	Y	N	N		Strategy admin software		Unlimited mailboxes. New user tutorial. Supports Strategy Token Programming Language
...ontrols, ...ate, hol-...eekday, ...d, year		Y	Y/limited	Y/Y/Y/Y		10	Y	N	N	Y	Y	Y	Y	Y		Y	MS Exchange or Outlook	LAN, direct RS232 serial link, modem, RAS	$9,406-65,111	Offers IVR and Web IVR on same NT platform. Call queuing module can be added for ACD feature. Same user interface for all products
...ited	Y	Y	Y/10	Y/Y/Y/Y	un-limited	20	Y	Y	Y	Y	Y	N	Y	N	Y	Y	OneLook, Outlook	Window-based, DTMF	$1,640-3,290	
...mited	Y	Y	Y/10	Y/Y/Y/Y	un-limited	20	Y	Y	Y	Y	Y	N	Y	N	Y	Y	OneLook, Outlook	Window-based, DTMF	$1,295-2,695	
...ening	Y	N	Y/5	Y/Y/Y/Y	20	10	Y	Y	Y	Y	Y	N	Y	Y	Y	Y		Through Voicegate ICS	$3,500	Speech recognition, visual voicemail for mail.
...night	Y	N	Y/5	Y/Y/Y/Y	20	10	Y	Y	Y	Y	Y	N	Y	Y	N	N		Modem	$2,700	Trunk to trunk transfer, true tenanting. Overhead paging and call logging. NS Gateway to Program Norstar.
...vening	Y	N	Y/5	Y/Y/Y/Y	20	10	Y	Y	Y	Y	Y	N	Y	Y	N	N		Voice Voicegate Lite	$2,500	External Trunk to trunk transfer. Automated overhead paging. NS Gateway to Program Norstar.
...evening	Y	N	Y/5	Y/Y/Y/Y	1	10	Y	Y	Y	Y	Y	N	Y	N	N	N		Modem	$1,400	External Trunk-to-trunk transfer. Automated overhead paging. SMDI.

V

WAP

See Wireless Application Protocol.

Web Collaboration and Conferencing

Using a web-based system for workgroup collaboration and conferencing no matter how far apart the participants may be.

For years, conferencing technology has been out of step with its time. Visual conferencing in particular - both interactive video and data sharing - illustrates this sense of anachronism. On one hand, the promise of an application like videoconferencing could hardly be in greater demand. What business, for instance, that is to any degree a distributed organization would not benefit from some form of remote visual communication among its own members and with its customers? And yet, for all that potential efficiency and (let's be honest) just plain sexiness, the technology has decidedly failed to make it big.

Some simple, if not comprehensive, reasons for this are well known: It's too expensive, it's too difficult to use, and it's too limited in scope. No matter how developed the technology becomes, and how much the quality improves, these basic barriers will still prevent the majority of businesses from deploying products and services they would theoretically love to have.

So why, despite all this negativity, has the mood of longtime players and new entrants in the conferencing industry lately been one of excitement and creative energy? The Net, of course. Conferencing, as much if not more than any other form of communication, stands to gain from the influence of IP and the Web. Integrating conferencing apps into an IP infrastructure can solve many of the problems that have stood in the way thus far, as well as making new ones possible. And while it's most appealing to speculate on killer apps to come, we can also point to some significant ways IP has effected conferencing today.

Collaboration over the Web is already taking place. Defined broadly, this can mean sharing documents or other visual materials with some degree of further interaction among the parties involved. On the more basic end, you have the ability to share a PowerPoint presentation in sync between two Web-connected PCs. Slightly more complex are applications sharing and collaborative whiteboarding. In many instances, streaming media technologies also play a role, whether they involve audio only or audio and video - investor relations calls, press conferences, and corporate training are all using streaming to reach large audiences.

Increasingly, we find carriers offering collaboration apps on an ad hoc basis to enterprise users. Colorado-based Vstream (Louisville, CO – 303-928-2400, www.vstream.com), in fact, has built a business around it. Using a kind of next-gen conference bridge supplied by Voyant Technologies, Vstream users can set up on-the-fly conferences from a Web site, and combine multipoint audio over the PSTN with visual content shared via the Web. Further, any conference can be recorded and streamed, either in realtime or archived format, to a potentially vast number of individuals. While it may not be loaded with frills and realtime interactive features, such a service is extremely pragmatic: It is low cost, convenient, uses absolutely ubiquitous technology (phone and Web browser), and it integrates processes that are very much a part of the way most any business operates.

Polycom (Milpitas, CA – 408-526-9000, www.polycom.com) offers a somewhat more specialized type of product, though similar in concept, aimed mainly at customers with an installed base of conferencing equipment. Its StreamStation, for instance, hooks up to existing audio or video conferencing gear and captures the content of a call for unicast or multicast streaming to any users equipped with a Real Player G2 client. For presentation of documents and visual materials, Polycom makes a system that IP-enables LCD projectors, as well as an IP-based projector that can read Microsoft Office documents.

What these products and services, and several others like them, have in common is that they use the Web and IP as a complement to existing technologies that people use and trust. By making a broader range of content available to a larger group of people, they broaden the channels of communication without demanding fundamental changes in them.

While Web collaboration can be an extremely useful supplement to standard conference calls, informal workgroups, and a myriad of other environments, we are still a long way from being able to use the Web for effective realtime, interactive videoconferencing. At the same time, there has been a good deal of progress in freeing up traditional videoconferencing systems from the barriers that have slowed their deployment.

Aside from streaming media, most of the attention has centered on H.323 as a replacement for H.320. From our standpoint, there seems to be no question that IP will eventually replace ISDN as the method for transporting video traffic, and that video will become "just another app" on the converged broadband network. The protocol shift, however, is not in itself adequate to solve many of the concrete problems associated with videoconferencing. Rather, much of the progress being made in this direction is happening to some degree independently from the protocol debate, or at least on a course parallel to it.

A good deal of videoconferencing endpoint vendors have released "protocol agnostic" products in the past year, that can support both IP and ISDN. As importantly, however, many of them have introduced products which, even if they are mainly still used for ISDN conferencing, have come significantly down in cost and are much easier to use and integrate. The set-top box, for example, are a viable, inexpensive product that let even smaller companies who need videoconferencing actually start to use it. Similarly, more intuitive user interfaces and especially browser-based tools have made the process of scheduling and initiating a conference a rather less arcane task.

On the backend, and further along in the network, we are also seeing a number of positive changes. Accord Networks (Atlanta, GA – 770-641-4400, www.accordnetworks.com), for example, now offers an MCU with built-in IP/ATM/ISDN gateway functionality, as well as multi-way transcoding that tailors available bandwidth, frame rate, and compression to individual users' endpoints. And FVC.com (Santa Clara, CA – 408-567-7200, www.fvc.com) is building out managed network operations centers to let carriers integrate video as a service offering directly into their broadband networks.

To get to a point where we can have videoconferencing "dial tone," with instant connectivity, there is still a good deal of network integration and enhancement that needs to take place. With players like Cisco Systems getting involved, however, and building the necessary QoS, IP multicasting, and directory services capabilities into their routers and switches, the industry appears to be edging closer to that point of convergence.

Audio and Beyond

Cisco's (San Jose, CA - 408-526-4000, www.cisco.com) recent activity in conferencing, in fact, can serve as an interesting roadmap, or at least a rough outline, of the path we see conferencing taking in general. On one level, the company is using its IP/TV streaming product to address the particular needs of large IP broadcasting. They've also recently signed an OEM deal with RADVision from which Cisco gains an H.323 MCU and 323-to-

W

471

320 gateway. These OEM products are a necessary migratory step - they are the type of product that lets businesses get into videoconferencing now, and they have the forward looking slant of being IP-based. At the same time, these products are not positioned in isolation, but rather as one part of one layer of an enterprise architecture - what Cisco is calling AVVID (Architecture for Voice, Video, and Data). And, ironically, some of the most interesting aspects of AVVID for IP-based conferencing wouldn't necessarily be recognized as conferencing products at all.

Presently, AVVID centers around an IP PBX, Cisco's Call Manager. As it evolves, however, the architecture is becoming much less a phone system and much more a distributed collection of network elements that interrelate in a modular, rather than a strictly hierarchical, fashion. The whole point of this design is to remove the applications layer of the network from the transport and access layers. And if we accept that, in the future, "the network will be the phone system," would it not therefore be possible to imagine that the network will also be the conference bridge? Given the potential of IP multicasting, there does not seem any theoretical reason why this should not be.

Conventional multi-party / multi-location audio conferencing employs a bridge: A central server that accepts voice input on multiple ports, sums it together, applies signal processing to remove additive noise (the so-called "noise floor"), and sends it back out, so that everybody can hear what everyone else is saying.

It's an efficient system for a switched-telephony environment. The bridge offers a central, call-in destination, collects ports and required DSP resources efficiently in one place, and permits centralized conference management. But the approach has some real limitations. PSTN-derived inputs vary widely in quality, and even with very few inputs, it's hard to equalize them and suppress noise – the problem looms larger as more people/ports are conjoined. Sound quality in the back-channel is therefore frequently poor – "state of the art" teleconferences we've participated in, lately, have left at least one party unintelligible.

Adding an IP gateway lets you use a conventional conference bridge in an IP telephony or hybrid PSTN/IP environment. Technically, this should work fine. We suspect, however, that today's audioconferencing bridges will require some reworking (i.e., signal-processing upgrades) to adequately conference together a mix of "real PSTN" and "gateway-mediated VoIP" inputs. Part of the problem, here, is on the input side: signals derived from VoIP connections have different acoustic characteristics from PSTN-derived inputs, so must be processed differently before summing into a "conferenced" output signal. We're also not sure how well high-compression VoIP codecs will handle input of a signal containing multiple voices.

More troubling, however is what happens on the output side. A G.729 compression routine does a great job of squeezing one 64kbps voice signal down to 4.6 Kbps; but it's not so great at compressing a signal containing multiple voices. Where bandwidth is not an issue, a 64 Kbps G.711 codec should do much better.

In any case, as IP telephony begins to take root at the enterprise level, we think we'll start to see this hypothesis crystallize first as a tool for simplified, more efficient audioconferencing, and perhaps later as a way of blurring the line between audio and video. For the moment, however, we've rounded up a whole slew of new products that are out there today, using IP and the Internet to help you improve your workgroups and extend your presence.

Here's smattering of some players in the field:

Accord Networks' (Atlanta, GA - 770-641-4400, www.accordtelecom.com) MGC-100 addresses two key issues that have traditionally plagued MCUs: network flexibility and ease-of-use. The MGC-100 is one of the only MCUs we know of that can support ISDN (H.320), ATM (H.321), and IP (H.323) all in the same chassis. Built-in gateway functionality lets it translate between all these protocols. And, perhaps most significantly, a multi-way transcoding features ensures that each user is assigned the bandwidth, frame rate, and compression ratios most suited to his or her individual connection.

To make the system controls more intuitive, users and operators can manage conferences through a Windows-based drag-and-drop GUI, a browser-based Web application, or touch-tone commands on a standard phone. A Greet and Guide feature lets you create customized welcome screens that walk participants through the process of joining a conference.

The MGC-100 holds from 8 to 132 IP ports, or up to 96 ports of ISDN or ATM. Accord's ARENA, a newly released open API, lets you develop IP-based visual applications for ecommerce, help desk, call center, and chat room environments to work in conjunction with the MGC-100. List price for the MGC-100 starts at around $25,000.

AudioTalk Networks (Mountain View, CA - 650-988-2040, www.audiotalk.com) provides a number of Web-based VoIP applications that include click-to-talk call centers, voice-enabled chat rooms and instant messaging, and interactive distance learning. The service works with any H.323 or SIP-compliant client endpoints, and uses a combination of VoIP gateways and applications servers in the network. The distance learning app lets you do a streaming broadcast of any presentation, and offers a variety of Q&A and other interactive features. Students (or conference attendees) can also form smaller groups in which to chat with one another live, separately or as a break-out from the main conference. Integrated audioconferencing lets participants dial in from a standard phone, in addition to a VoIP client.

As we touched upon earlier, Cisco Systems has been very active in developing conferencing applications and equipment at a number of different levels of its IP telephony architecture. Cisco's OEM deal with RADVision allows them to offer an H.323-based MCU, an H.323-to-H.320 gateway, a Video Terminal Adapter, and Cisco's Multimedia Conference Manager (MCM), H.323 gatekeeper and proxy. (The MCM is a part of Cisco IOS, which runs on all of Cisco's enterprise routers, and is used to manage bandwidth and improve QOS on video conferences). Collectively, the product family is called IP/VC.

Separately, Cisco has also been developing IP/TV, a line of broadcast servers and streaming software built on Microsoft's Windows Media Technologies. IP/TV uses IP multicasting to distribute large scale events to a multitude of users - as exemplified in Cisco's famous 14 hour live broadcast of the NetAid benefit concert in conjunction with the University of Oregon.

Finally, Cisco is integrating conferencing features directly into its IP-based enterprise phone system (based on the former Selsius product, now Cisco Call Manager). The first and most concrete feature to emerge from the effort is a one-button conference feature on Cisco's IP phones. Users simply press a button while speaking to another IP phone, and NetMeeting is launched on both parties' desktops.

Cisco plans to continue its efforts as part of the developing AVVID initiative, basing its systems and software IP multicasting technology on a more converged and distributed model for multipoint voice, video, and data conferencing.

Ezenia! (Burlington, MA - 781-229-2000), formerly known as VideoServer, makes the Encounter line of IP conferencing products. Encounter includes an IP/ISDN gateway and a software-based gatekeeper. The core product in the line is Encounter NetServer, an MCU for H.323 conferencing across an enterprise. NetServer supports both POTS and H.323 connections, so that endpoints can be any combination of standard phone lines, H.323-compatible conferencing units, or multimedia PCs with a TCP/IP connection and T.120 data. Using the Encounter gateway, traditional H.320 endpoints can also be brought into a conference. You can schedule and manage conferences using a browser-based interface. The NetServer also performs audio transcoding between G.711 and G.723, to mix POTS and VoIP in the same conference. The NetServer ADX1000 is an entry-level model that comes with the same audio and data conferencing capability as the original, but supports video only with a software upgrade.

Ezenia! has added the Interactivity Server to its offerings. Incorporating the same basic design and functionality as NetServer (though without standard support for video), this model includes a set of APIs that let web developers build web-based conferencing applications like online communities and chat rooms. The primary focus of Interactivity Server is to promote interaction between a firm and its clients, or to create communities among a company's customer base.

Forum Communication Systems (Richardson, TX - 972-680-0700, www.forum-com.com) makes the Consortium Conference System, a PC-based audioconferencing bridge that runs on NT server. Consortium scales between 24 and 96 ports and connects to a PBX via T1, and to the LAN/WAN via 10BaseT Ethernet. Client software is installed on users' desktops, and lets them access configuration, scheduling, and setup options - communicating with the server via TCP/IP. Conference management and most other features can also be accessed via the Web, either though the Internet or a corporate intranet. Further integration with the enterprise LAN and WAN lets you send out e-mail notifications to participants, as well as distribute meeting documents via e-mail or the Web. Other features include: internal digital recording, dynamic breakout sessions, PIN/security features, and blast dialing. Full duplex audio and automatic gain control are standard. Consortium tailors particularly well to ad hoc conferencing: its features are easy to control, and it has special capabilities to do immediate scheduling and initiation of conferences.

FVC.com (Santa Clara, CA - 408-567-7200, www.fvc.com) is a videoconferencing infrastructure provider. Its main business is to build and maintain Video Operations Centers (VOCs), from which it offers managed network equipment to carriers who want to provide videoconferencing services without owning their own hardware. An FVC.com VOC contains MCUs, web servers, IP multicasting systems, gateways, gatekeepers, switches, and OSS databases. A broadband fiber interconnection links the VOC to a service provider's own network. As part of the facilities, FVC.com offers a Video Portal, which consists of a browser-based interface through which customers can schedule conferences, and access capabilities like multipoint bridging and video recording and streaming. While FVC.com's facilities can support ISDN (H.320) conference endpoints via gateways, it is specifically intended for delivery of video services over broadband pipes, whether they be IP or ATM based.

FVC.com also sells all of the systems it uses in its Operations Centers directly to customers.. Most recently, the company announced a strategic partnership with White Pine Software (now Cuseeme Networks, Nashua, NH - 603-886-9050, www.cuseeme.com) to jointly develop and market IP video products. As a result, FVC.com will OEM White Pine's MeetingPoint software-only conference server, and White Pine will OEM FVC.com's V-Gate 4000, an IP-to-ISDN-to-ATM gateway.

Gentner Communications (Salt Lake City, UT - 801-975-7200, www.gentner.com) recently released the APV200-IP, an H.323-compatible videoconferencing codec. The box connects to any TV, LCD projector, flat screen, PC, or laptop, and works with most any other videoconferencing system or MCU using ISDN or T1/E1 lines. The system can operate standalone, or be combined with other products in Gentner's Audio Perfect line, which supports up to 64 attached microphones, multiple cameras, and recording equipment such as VCRs and DVD players.

An embedded Web server and 10/100BaseT Ethernet port lets you access the system for configuration and management via LAN, WAN, or Internet. A remote control operated GUI helps users through set up, and a tutoring device provides self-prompting instructions to guide novice users in configuring the system. The APV200 also includes an API to support customization and enhancements.

Global Crossing (-) offers Ready Access, an on-demand teleconferencing service that utilizes the carrier's backbone fiber network. Rather than having to make reservations in advance to schedule a conference call, users are given a permanent access code and PIN with which they can dial into the network and access the service at any time. The chairperson has complete control over the conference and its participants, and can also choose to moderate a call via the Web. This includes roll call, deciding who can enter a conference and when, dial out functions, flexible billing options (e.g. billing participants for a portion of the call), recording, broadcast mode (making all or certain participants "listen only"), and break-out sessions.

Ready Access accounts are available 24/7, and are billed on a per port and per minute basis.

Latitude's (Santa Clara, CA - 408-988-7200, www.latitude.com) MeetingPlace server is a conference bridge designed for integrated voice and data communications. MeetingPlace connects to both your PBX and LAN/WAN, and lets you conference anywhere from 8 to 120 users or 60 simultaneous meetings (with up to 960 ports in 8 networked servers). Latitude also offers a version of the product, MeetingPlace IP, that works with IP phones and IP PBXs (interoperability with Cisco's Call Manager was the first to be undertaken), as well as VoIP clients such as Microsoft NetMeeting. MeetingPlace IP includes an NT-based gateway that lets POTS and VoIP users participate in the same conference.

Among MeetingPlace's advantages are easy scheduling via phone, web, or local PC; automatic e-mail notification to participants; outdialing and auto dialing; and conference recording and archiving, with file attachment and commentary options. In addition to traditional and IP audio, MeetingPlace lets you share documents, and supports Microsoft NetMeeting or other T.120-compliant endpoints as user interfaces.

MeetingPlace IP's pricing will remain the same as the standard MeetingPlace system - starting at around $100,000 for the software licenses, with the cost of hardware adding to that figure a bit.

Lucent Technologies (Murray Hill, NJ - 888-4-LUCENT, www.lucent.com) makes an MCU based on its Definitey Enterprise Communications Server, which supports audio, H.320 video, and T.120 data conferencing, and is sold by Polycom (Milpitas, CA - 408-526-9000, www.polycom.com) under the MeetingSite label. Recently, Lucent expanded its agreement with Polycom, and Lucent will now offer Polycom's full line of videoconferencing and webconferencing products, as well as working to jointly develop new IP-based audio and videoconferencing systems.

One of the first efforts of this partnership will be to add video to Lucent's Definity AnyWhere software (formerly Lucent OneMeeting). Definity AnyWhere lets remote workers share data with other members of the enterprise over the corporate WAN, while participating in an audio conference over the PSTN. Remote users also get one-button access to voicemail, as well as features like call forwarding, transfer, hold, and caller ID. List price is between $120 and $220 per user license.

Lucent has also developed a line of wideband speech coders that improve sound quality over Internet, voice, and wireless networks. Polycom has agreed to OEM the product - called ClearPresence Audio Coder. Essentially, ClearPresence is intended as a high quality, low bandwidth alternative to the G.722 coding standard, and can operate at 16, 24, or 32 Kbps (lower than G.722's 48, 56, and 64 Kbps bit rates).

MAX Internet Communications (Dallas, TX - 800-479-7146, www.maxic.com) makes a multi-function IP videoconferencing card for PCs

473

and Internet appliances. The product, MAX i.c.Live, is an interesting approach to desktop conferencing, and has more power than one might expect. With a broadband Net connection, you can do two-way, full motion videoconferencing at up to 30fps; full motion streaming at 30fps; video and audio record, edit, and playback; play MPEG-1 and DVD; hear AC3 surround sound audio; and incorporate Microsoft NetMeeting, NetShow, Media Tools, and Audio Works.

It works by using an on-board media processor that offloads video from the PC's Pentium. The processor itself performs all functions in software, eliminating the need for dedicated function chips (and replacing separate VGA, Video capture, DVD/MPEG Decoder, and sound cards). A 600Mbps internal bus handles decompressed data, to prevent system bus overload. And a proprietary operating system (MAX-OS) integrates with Windows.

The possible applications for this type of system are multifold: including corporate conferencing, distance learning, telemedicine, home entertainment, and e-commerce. Recently, MAX released a complete system that will be sold to end users, based on its core board technology. The MAX i.c.Live Video Communication Station comes in a box about the size of a DVD player, and hooks up to a broadband Internet connection and a TV or monitor, to give you access to video and audio conferencing. A camera, microphone, remote control, and on-screen menu interface are all included. List price is a reasonable $1,499.

MCI WorldCom (Clinton, MS - 800-480-3600, www.nmc.mci.com) provides a broad range of audio, video, and Web-based conferencing services. The latter include three new services that the company recently introduced. Meeting View is a feature offered with large audio conferences (e.g. investor relations calls) that lets the host and up to three other conference leaders view a list of participant names and line status using any Java-enabled browser. Conference leaders can also text chat with one another, manage question-and-answer sessions, and access polling results. Some of these features, like web-based polling, fit into MCI's broader Net Conferencing offering, which also includes document sharing and collaboration, interactive web pages, online registration, and online tech support. The Net Conferencing application uses software from PlaceWare (see below) as its foundation.

For more impromptu staff meetings, MCI offers a reservationless service called Instant Meeting. Instant Meeting gives you a toll-free or toll based number for a standard audioconference, and is available 24 hours a day.

MSHOW (Littleton, CO - 888-99MSHOW, www.mshow.com) is a Web broadcasting service intended for corporate presentations. The service uses streaming video, high quality audio, and can incorporate any combination of graphics and text. MSHOW presentations include interactive features like realtime voting and polling, leader guided Web site tours, and group profiling. You can also have integrated audioconferencing to accompany the Web-based presentation. The system can support thousands of audience members and up to seven co-presenters. 24/7 help desk support is offered to both leaders and conference attendees.

Firetalk from Multitude (East Greenwich, RI – 800-532-4862, www.netkeeper.com) is an Internet-based voice communication service that lets users talk, text chat and Web browse together-simultaneously-over the Internet. In a single, powerful application, Firetalk provides free worldwide calls, unlimited conferencing, voice and text chat at any web site, shared voice-enabled web browsing, instant messaging and voice mail.

Firetalk uses sophisticated client-server technology to provide an easy-to-use solution for communicating with the widest possible range of PC/Internet users, with the quickest setup time and lowest learning curve.

Unlike other voice products, Firetalk is a complete set of services for a broad spectrum of scenarios, from one-on-one voice and voice chat to large-scale public meeting rooms.

PictureTel's (Andover, MA - 978-292-5000, www.picturetel.com) enormously extensive product line includes several IP/Internet-based and H.323 compliant systems, ranging from desktop and conference room endpoints, to codecs, gateways, and multipoint conference servers, to interactive whiteboards and projectors. The PictureTel 330 Netconference Multipoint Server is, more or less, what its name implies - an NT based software server that supports multipoint conferencing between H.323-based clients (including Intel TeamStation and Proshare, of which PictureTel is the exclusive distributor). Netconference comes in 8 and 24 port configurations, and can host reserved or ad hoc conference "rooms." Users can collaborate with audio, video, and T.120-based data. H.320 clients can be included via a PictureTel 320/323 gateway.

The PictureTel StarCast, co-developed with Starlight Technologies, is a software-based streaming system that runs on an NT server. The server sits between a videoconferencing endpoint and a corporate LAN, and uses IP multicasting to deliver streaming content to users with multimedia PCs.

PictureTel also offers dataconferencing-only software that runs on standard PCs and works with NetMeeting. Its Interactive whiteboards and flip charts are made for conference room environments, and include a large, 77" whiteboard that lets you mark-up PowerPoint, Excel, and Word documents.

PlaceWare's (Mountain View, CA - 888-526-6170, www.placeware.com) Conference Center 3.5 is the software used by several major providers of Web-based conferencing services, as well as large enterprise customers. Conference Center offers two major meeting options - Auditorium Places and Meeting Places. An Auditorium Place conference can have up to 1,000 attendees, and presenters have control over the level of interactivity. Meeting Place is intended more for ad hoc meetings and workgroup collaboration. Other major features of the software include pre-recorded audio and video streaming; drag-and-drop PowerPoint presentations; on-the-fly polling; screen captures; whiteboarding; and live Q&A. A new feature called LiveDemo lets you run and display any application (e.g. for sales demos), in either a one-on-one, or one-to-many environment. Added to its utility is dynamic recognition of each client's connection speed, through which the system can tailor its frame rate to individual users.

Conference Center 3.5 costs around $375 per seat to license the software, or around $300 per seat per year, with support and upgrades included, to contract with PlaceWare for the service.

Polycom's (San Jose, CA - 408-526-9000, www.polycom.com) product line has lately become more IP-centric in both audio and video conferencing. ViewStation is a settop box, which supports H.320 and H.323 with a software upgrade. An embedded Web server and Ethernet hub are both included, so you can pull PowerPoint slides, for example, from the Web or LAN and integrate them into a conference. Different versions include embedded streaming capability, a multipoint model that connects up to four different endpoints without an external bridge, and an inexpensive small office system. Standard features include voice tracking camera, extended microphone, and full-duplex digital audio.

StreamStation is a box that hooks up to Polycom's ViewStation or SoundStation audioconferencing units, captures media content, and either stores it or delivers it live in IP unicast/multicast streaming format. The only requirement on the end user's side is a PC with Real Networks' Real Player G2 plug-in installed. Stored files can be up to two hours long, and are each held on the server for 14 days. Polycom also offers the WebStation, a box that lets you distribute content from an LCD projector to the Web; and ShowStation, an IP-based projector that reads Microsoft Office files.

Interestingly, Polycom has acquired Atlas Communication Engines, an OEM supplier of IADs and DSL routers. On the basis of the acquisition,

Polycom plans to develop its own line of IADs that aim at delivering voice, video and data over DSL. The acquisition signifies both the trend toward greater integration of conferencing as a converged network application, and an expansion of Polycom's business into the mainstream of network CPE.

RADVision's (Mahwah, NJ - 201-529-4300, www.radvision.com) protocol stacks, gateways, and gatekeepers are licensed and OEM'd by several major IP conferencing manufacturers. RAD also markets its own MCU-323, an IP-based, reservation-less MCU for voice and video. The system can be set up in centralized or distributed fashion - the latter involves separating the control unit from multiple processor units. One unit can handle up to 15 simultaneous multimedia calls (at up to 1.5Mbps), or 24 uncompressed, voice-only calls (64Kbps). The system is purportedly plug-n-play, and lets users initiate and join conferences by dialing into a designated number and supplying a unique conference password. Call control is provided through RADVision's NT-based NGK-100 gatekeeper, and RAD's ONLAN L2W-323 gateway is optionally available for PSTN interfaces.

Among RADVision's partnerships are Tandberg, who will integrate the MCU, gateway, and gatekeeper into its existing line of endpoints, and Cisco, who will OEM RADVision products as part of its IP/VC line. Included in the latter will be the RADVision VIU-323, a video terminal adapter. Both the VIU and MCU are built on a RISC architecture and use SNMP-based administration and management.

Sprint Conferencing (Atlanta, GA - 800-669-1235, www.sprint.com) offers the Internet Collaboration Center, a website (www.sprint.com/icc) from which users have access to a number of collaboration tools. As a result of partnering with broadcast.com, Sprint also offers streaming capabilities as part of its collaboration and conferencing line. Available services include online scheduling for audio and video conferences, browser-based text Q&A, and Sprint Online Presenter, a web-based data conferencing application. Any audio or video content can be streamed and archived for presentation to a wide audience. Streaming services are offered turnkey, and include provisioning, hosting, billing, and Sprint customer service.

Sprint's conferencing strategy is part of a wider initiative that centers around ION, the Integrated On-demand Network. ION is a next-gen, converged fiber network through which Sprint plans to provide broadband access and IP-based content to businesses and consumers. Sprint's conferencing services - audio, video, and web-based - are intended eventually to all run the ION network, and to become part of a bundled services package.

Tandberg (Reston, VA - 800-538-2884, www.tandbergvision.com) makes a variety of videoconferencing systems, in both rollabout and set-top box form. Tandberg's larger systems are all H.323-enabled. Its smaller systems support H.320 only, but feature some innovative designs, such as the Vision 600 and 770 models. These are compact units, not much bigger than a standard office phone set, have their own monitors and can plug into TV screens, as well. They support video at 15fps and 30fps, and also have integrated T.120 data capabilities.

The Vision 2500, Tandberg's latest and greatest rollabout system, uses solid state hardware as opposed to PC components. The system, however, is PC-compatible, and lets users import documents or presentations from any desktop. In addition to high quality video, it also features strong audio capabilities, with advanced noise reduction and automatic gain control.

Through its VideoTele.com business unit (VT.c), Tektronix (Wilsonville, OR – www.tektronix.com) offers broadcast-quality interactive videoconferencing systems, as well as broadband video trunking and delivery. Most recently, the VT.c introduced ConferenceMaker 7.0, a new version of its scheduling and management software for distance learning networks. Other products that ConferenceMaker operates with include

video edge devices, switches, and ATM NICs and MPEG-1/MPEG-2 codecs.

The M2C edge device, which ConferenceMaker was specifically designed to control, uses MPEG-2 to enable realtime, two-way videoconferencing at the very high quality demanded by applications like distance learning and telemedicine. MPEG-2 is a standard for digital video that can travel over any broadband protocol. To foster interoperability in this area, Tektronix serves as a charter member of the Interactive MPEG-2 Forum, led by the University of Akron.

VCON Telecommunications (Herzliya, Israel - 972-9-9590059, www.vcon.com) makes video conferencing room systems that support both H.320 and H.323, but are heavily focused on IP video applications. Unique to VCON is its PacketAssist technology, included on the latest version of VCON's MeetingPoint software, version 4.0. PacketAssist helps in bandwidth management, using an adaptive adjustment technique, and improves quality of service through IP precedence, delay adjustment, jitter correction, and (recently added through a free upgrade) support for DiffServ. Also featured in MeetingPoint is Interactive IP Multicasting, which lets you transmit one-way streaming content to multiple users with the same VCON equipment used for standard, two-way conferencing. The multicasting feature has been tested for interoperability with Cisco's IP/TV broadcasting platform.

MeetingPoint and PacketAssist are available as part of a variety of VCON desktop conferencing systems, housed on a single PCI card, as well as the MediaConnect 6000 and 8000 systems, which fit small to medium sized conference rooms and workgroups. The main difference between these two systems is that the 6000 uses a TV monitor, while the 8000 comes with a Super VGA monitor, optimizing it for applications/document sharing and collaboration. A developers' kit of 32 bit OCX controls lets you build custom apps and integrate with existing ones.

ViewCast.com (Dallas, TX - 800-250-6622, www.viewcast.com) makes codecs and peripherals that are used in all segments of the video industry. The ViewCast.com Osprey-100 is the video capture card used by both Real Networks and Microsoft as a standard for streaming software encoding. The company also offers a hardware-based encoding card that works with RealSystem G2. This card forms the basis of ViewCast's Encoding Station, a Windows NT-based system that can handle up to six video feeds from tape or DVD, and lets you broadcast streams at multiple different bandwidths simultaneously. The ViewCast Streaming Server, which comes bundles with either RealServer or Windows Media Server, can broadcast encoded streaming content to more than 1,000 simultaneous endpoints.

On the higher end of the streaming market, ViewCast.com offers the Viewpoint VBX, a server and codec array that distributes uncompressed NTSC or PAL video throughout an enterprise over existing building wiring. Because it utilizes dedicated twisted pair wiring for sending audio and video traffic, the VBX does not put an excessive strain on your LAN, using the LAN only for client-server control. Users with connected desktops can share locally the resources of equipment like Cable TV, DVD, digital satellite, and VCRs. They can also access multipoint video conferencing systems and compressed video WAN connections. An external hardware interface and client software equips any Windows-based PC for two-way communication through the VBX system. The Viewpoint VBX is available in two different server configurations. The larger, a rackmount industrial chassis running Windows NT, uses full length ISA cards and supports up to 240 one-way or 128 two-way users. The small office version is a tower PC running NT, and supports 32 one-way or 16 two-way users.

Viewcast.com has recently licensed and is developing to the RADVision protocol stack to enable its products for IP-based point-to-

W

point conferencing, and converged voice and video over IP.

VNCI's (Portsmouth, NH - 603-334-6700, www.vnci.com) VidPhone is a two-way videoconferencing and broadcast system that operates over a building's twisted pair wiring (CAT 3 or above). The video is uncompressed, so quality is high, and audio is delivered in stereo. VNCI's product is not IP-enabled, but runs over ATM or ISDN based WANs. The system itself consists of a 13-slot broadband switch and multiple endpoints, or VidPhone Stations. The endpoints are PC-based systems, equipped with camera, stereo speakers, mic, and audio/video overlay cards, and plug into standard RJ-11 phone jacks. An optional Conference Bridge card lets users initiate and control multipoint conferences.

Voyant Technologies (Westminster, CO - 888-447-1087, www.voyant-tech.com) is a spin-off from Frontier-Confertech whose product is made for Internet-based conferencing that uses PSTN voice. The idea is that while IP works well for collaboration and streaming media, it isn't yet ready to support high quality, multipoint audioconferencing. Voyant's Innovox, therefore, is a conference bridge that combines PSTN voice calls traveling over T-1s with IP-based visual applications. It can also convert PSTN voice to an RTP stream and send it via Ethernet to a streaming server for applications such as Webcasting. Sitting on top of the platform, Voyant provides an open API for developing such Web-based conferencing and collaboration apps.

Voyant is partnering with service providers, such as Vstream (see below) to make the platform available to enterprise users. The platform is priced at around $1,500 per port.

V-Span (King of Prussia, PA – 888-44V-SPAN, www.vspan.com) offers a wide variety of video, audio, and Web conferencing services. In some sense, its niche market has been created out of the shortcomings of conferencing systems: V-Span's job is to make dissimilar systems speak to one another. This includes providing multipoint bridging, gateway functionality, and network management, along with a sizable list of specialized services like event management and public room rental.

Additionally, V-Span offers Internet-centric services, including Web-enhanced audioconferencing (where visual materials can be distributed via the Web to supplement a traditional audioconference), and Internet streaming for both live events and archived content.

Vstream (Boulder, CO - 800-VSTREAM, www.vstream.com) is a Web-based conferencing service provider. Netcall is its bread-and-butter service that lets you initiate reservation-less conferences and on-the-fly webcasting from the Vstream web site. Using conference bridge hardware from Voyant Technologies, Vstream lets users share documents and graphics, collaborate, and send text messages over the Web while engaged in a traditional or Web-based audioconference. Participants can join a conference via a standard phone, or listen to a conference using RealAudio, through Vstream's webcasting option. Live conferences (broadcast as webcasts) have synchronized audio and visual content, and can be recorded and archived for playback. Vstream tracks participants on both live and recorded conferences, and authorized users have access to this information online, along with billing and account details.

Other services Vstream offers include a free utility for recording voice messages to send as e-mail attachments, voice-enhanced webcast recording, and video encoding for Web broadcast. Vstream offers up to 245Mbps of guaranteed bandwidth, with up to 750Mbps available with an hour's notice. The site is equipped to host up to 45,000 simultaneous users.

VTEL (Austin, TX - 512-437-2700, www.vtel.com) makes videoconferencing endpoints, as well as an H.323-based multimedia conference server. The VTEL Galaxy family, which can support both H.323 and H.320, comes in dual or single monitor versions, and includes four cam-

era inputs, as well as support for VCR video playback and record and recently added streaming capabilities. Integrated T.120 and support for Microsoft NetMeeting add data collaboration and application sharing features, and an Ethernet port gives you access to the LAN. Document camera with automated control and an optional interactive whiteboard are also available. Two omnidirectional mics come standard, and a 16-bit Sound Blaster is included with support for PC and DVD applications. Remote control and wireless keyboard control a Windows-based GUI that helps you easily create and manage conferences.

VTEL's conference server, the MCS/IP, does audio transcoding to mix H.323 and POTS calls, and supports T.120 data. VTEL also offers an H.323/H.320 gateway and a gatekeeper that can either be integrated or run on a separate NT server.

SmartStation is VTEL's desktop conferencing system, which has recently become available in a lower priced, IP-only model that doesn't require a separate WAN card for ISDN/DDM.

White Pine Software (now Cuseeme, Nashua, NH - 603-886-9050, www.cuseeme.com) is best known for CU-SeeMe, the popular web-based video chat software it acquired from Cornell University around a year ago. For corporate users, the company also offers MeetingPoint, a software MCU for desktop H.323 conferencing and collaboration. MeetingPoint runs on NT or Sun Solaris servers, and works with a variety of conferencing apps, including CU-SeeMe, Intel ProShare or TeamStation, Microsoft NetMeeting, and PictureTel's LiveLAN. Simple, browser-based tools for conference setup let you setup and schedule conferences, and will send out automatic e-mail invitations to participants. A built-in gatekeeper manages bandwidth on a per-conference, per-user basis, and provides authentication services. Core functions include IP multicasting, applications sharing, whiteboarding, and audio/video conferencing. A Java applet lets you choose which participants you want to see and hear during a conference. The system is compatible with RADIUS authentication, billing and tracking systems. ClassPoint is an add-on to MeetingPoint, designed for distance learning and training apps.

Zydacron's (Manchester, NH - 603-647-1000, www.zydacron.com) comCenter is a videoconferencing room system designed for intra-company workgroups. The comCenter supports both IP and ISDN, as well as T.120 data through full integration of Microsoft NetMeeting.The system also supports streaming media capabilities, and uses a browser-based interface to access files on the Internet or intranet. Other features include 27" SVGA monitor, on-screen camera control with auto-tracking, and wireless keyboard and mouse. Through an OEM deal with RADVision, Zydacron offers comCenter as part of a complete solution that includes RAD's H.323-based MCU, gateway, and gatekeeper. SRP for a standard comCenter unit (not including the RADVision products) is $12,995.

Zydacron also offers the OnWAN340 IP, a desktop IP conferencing product that delivers 30fps two-way video. OnWAN340 supports dynamic switching between ISDN and an IP-based LAN connection. The standard package includes a video/audio codec, camera, conferencing software, and Microsoft NetMeeting. Runs on Windows 98 and NT. List price is $1,195.

Web Enabled Call Centers

Instant Web callback and on-line access to a help desk's knowledge bases are two of the most popular ways companies are meshing call centers with their web sites. Web applications used by the call centers tend to fall into three categories. One is the use of web callback so customers can request a phone call from agents while visiting a Web site. Another is the creation of an on-line knowledge base so Web-browsing customers can retrieve solutions to problems from them. A third is the use of a special queue for handling live text chat sessions. Can video over IP conferencing with an agent be far behind?

All three types of Web applications are emerging as standard capabilities of Webified call centers. Perhaps the most intriguing story we know of is that of a youth hotline service, which received many responses over the Internet when it conducted a survey on teenage suicide. The participants in the survey largely perceived the Internet as a medium that is more confidential than the phone.

Public knowledge bases are convenient for customers and help desk reps. Free Internet services can attract millions of customers in a matter of months. For help desks to keep up with sudden growth, the best choice is to let the reps deal with difficult problems. Let the customers solve the simple ones themselves.

Whether it's e-mail or the phone customers typically rely on to reach agents, communication from a Web site is always a good idea.

As electronic commerce achieves mainstream acceptance as a way of doing business, Web-enabled call centers are also earning recognition for their potential to appeal to a wider range of customers than traditional call centers.

It's no wonder, then, that Internet call center products, which let Web surfers reach call center agents from a Web site, have evolved from the novelties they were two years ago to the necessities they are today.

The time is ripe for incorporating the lessons call centers have learned from providing customer service over the phone - like establishing service levels, segmenting customers and forecasting agents' schedules - and applying these lessons to Web sites. It appears, though, that consumer Web sites may first learn these lessons the hard way.

To highlight this point, the Times cited the "First Annual MCA E-Retail Sales Analysis," a study by the Westport, CT-based Marketing Corporation of America (MCA), which projected that on-line sales during the Christmas season would be as high as $5 billion. (MCA is a subsidiary of The Interpublic Group, a group of advertising and communications services firms.)

Despite high expectations for Christmas shopping on the Web, companies weren't always ready for the actual customers. As the Times' piece noted, some retailers were so overwhelmed by traffic to their Web sites that they had difficulty handling on-line orders and, in some cases, couldn't deliver their products to customers by Christmas. Order processing problems on Web sites affected companies whether they were brick-and-mortar department store chains or on-line purveyors of books and music.

Clearly, firms have to make it a priority to resolve back-end technical issues with their electronic commerce operations. Provided they get on-line order processing to work consistently, firms then need to focus their attention on the customer service they offer from their Web sites. That's where the use of software for Web-enabled call centers can have a major impact.

Like the latest voice processing products that combine IVR and predictive dialing in one system, the newest Internet call center software combines different approaches to electronic communication in one product. Such software gives you the flexibility to determine the best methods of responding to customers based on their types of requests.

For example, surfers who need answers to questions right away may prefer live text chat, and, where feasible, IP telephony. Other visitors to your Web site may be willing to wait for an agent to get back to them, and might be more comfortable with Web callback or e-mail.

An increasingly viable alternative to installing Internet call center software at your company is hiring a service bureau to handle your on-line inquiries.

Consolidated Communication

One trend emerging among Internet call center products is the integration of formerly separate methods of reaching agents on-line, such as e-mail, text chat and Web callback.

A number of products that became available in 1999 offer various ways to contact agents from a Web site.

For example, WebLine's (Burlington, MA) WebLine 2.0 software lets customers with one phone line communicate with agents using text chat or voice over IP. Because of its ability to work with a variety of computer telephony middleware products and integrate directly with leading telephone switches, including those from Aspect and Lucent Technologies, the software also lets surfers request a call back from an agent. If some of your customers have multiple lines, WebLine 2.0 lets agents push information to these customers' PCs or help them complete on-line forms while talking with them on the phone.

WebLine 2.0 does more than consolidate a variety of on-line communication methods. After you capture information about visitors to your site on your Web server, the software lets you share the data among other servers, such as database, directory, e-mail and fax servers. It also adds Web page-sharing capabilities to MCI WorldCom's (Jackson, MS) Click'N Connect IP telephony service.

Acuity's (Austin, TX – 512-425-2281, www.acuity.com) WebCenter Enterprise is among a growing number of products that combine handling of e-mail, text chat, IP calls and regular phone calls.

Also, Acuity's other product, WebCenter Express, addresses three main concerns callers have while shopping via the Web. When some Web surfers can't find what they're looking for, they may simply abandon their search, a potential loss to a company. Other prospective customers end up e-mailing the product's manufacturer, which could be a problem if the company does not have the resources to handle large volume e-mail- based inquiries. In some cases, prospective shoppers end up phoning the call center.

WebCenter Express is designed for pre-sales support at call centers. When customers start browsing the center's home page, they can click on a frequently asked questions data base. If they're unable to find an answer to their question or have a complex one, they can escalate the session by hitting the help button, which may be on another page. The advantage of such a setup is that it encourages the customer to browse and try to find answers to their questions. However, customers can immediately escalate the session by simply hitting the help button. A dialogue box then lets the customer know that an agent will be with them shortly. Once an agent is ready to attend to the customer, the agent initiates a text chat with an introduction and asks if the customer is looking for a specific item.

WebCenter Express includes an IP-based Web ACD. Customers and prospects are routed to agents in a number of ways. For example, platinum customers are sent to designated agents, while others are handled on a first-in first-out (FIFO) single queue with all customers, including telephone callers. Requests are routed to agents that have the longest idle time.

WebCenter Express automatically e-mails a transcript of the dialogue to the customer so that they have a record of the response and information. This also means that suppose the customer needs to interrupt the call in order to go somewhere, they can abandon the text chat knowing that the response to their inquiries will be e-mailed and waiting for them. Pricing starts at $8,000 for five seats.

What makes Melita's (Norcross, GA) PhoneFrame Explorer unique is its combination of predictive dialing and Web callback. WebContact, which Melita offers as part of its PhoneFrame Explorer predictive dialing system, lets customers fill out an on-line form to schedule a call from the dialer.

If you manage multiple centers, Cisco / GeoTel's (San Jose, CA – 800-

553-NETS, www.cisco.com) GeoTel Enterprise Web is an optional addition to the company's Intelligent Call Router that lets agents throughout all your locations respond to Web callback requests or IP telephony calls. Users of Melita's PhoneFrame Explorer predictive dialing system will be happy to know that GeoTel's Enterprise Web integrates with Melita's WebContact.

Mixing and matching different methods of reaching agents, whether customers contact them from a Web site or from a phone, have always been features of communications servers. Although the focus of the article isn't on these types of products, it's important to note that systems like Interactive Intelligence's (Indianapolis, IN – 317-872-3000, www.inter-intelli.com) Enterprise Interaction Center (EIC) queue and route different types of electronic communication, such as e-mail, text chat and voice over IP. They also let you manage call routing, computer telephony applications and IVR from the same system.

The Message Defines the Medium

Another trend with Internet call center software is an increasing emphasis on classifying incoming on-line communication. Whereas products that debuted in 1997 or early 1998 focused on determining which queues to place e-mail messages or text chat requests, newer software categorizes on-line messages with the goal of identifying the best method to respond to them.

For example, Business Evolution's (Princeton, NJ – 609-951-0216, www.businessevolution.com) @Once Service Center lets your Web site offer three types of communication from your Web site - live messaging, secure text chat and e-mail.

Live Messaging Channel is ideal for visitors to your Web site who want immediate answers to questions. It lets them indicate their requests by clicking on an icon to generate an on-line form.

Based on how surfers fill out the forms, and by using rules you set up to determine each request's priority, the software lets surfers communicate with agents by e-mail or through a limited form of text chat. For example, you can establish priorities based on key words within the on-line form so that a pricing question has the greatest priority. Since @Once Service Center lets you place icons for the Live Messaging Channel throughout your Web site, you can, for instance, set the highest priority for requests that come from Web pages where customers can key in credit card numbers to make on-line purchases.

If a request warrants an immediate response, Live Messaging Channel creates a small text chat box within the Web browser on the surfer's screen. The surfer selects a category for his question from a drop-down menu. The menu, which is located inside the box, typically contains a list of the most frequent requests from visitors to your site.

After choosing a category for his request from the menu, the surfer elaborates on the information he needs by typing a message within the text box. To send the message, he indicates his e-mail address (if he didn't already provide it on the on-line form). If the request can't be answered immediately, either because it's not a top priority for your center at the moment, or because agents aren't available to respond to it, @Once Service Center refers the message to an e-mail queue.

For your customers who frequently buy products from your Web site, the Chat Channel lets them click on an icon to engage in text chat sessions with agents over a secure network when they have questions that require detailed explanations. Agents use Chat Channel to push Web pages or other on-line collateral materials to surfers.

To appreciate the difference between live messaging and secure text chat, think of a Web site where all surfers have access to some Web pages and only registered users have access to the entire site. Live messaging

is analogous to the unrestricted part of the site because you allow all surfers - whether or not they're customers - to initiate communication with live agents by completing on-line forms.

Secure text chat is analogous to the special portions of your Web site that you make available only to your customers. Since the communication that occurs through Chat Channel is secure, your best customers may have to identify themselves using a password or an account number. Once they do, they can spend as much time as they need asking questions or viewing on-line materials agents push to them about the items you sell from your Web site.

Also available as part of @Once Service Center are Email Channel and Call-back Channel . Email Channel lets surfers click on icons to send e-mail messages to agents and lets you establish priorities for messages based on key words, the Web page from where the surfer originated the message or the age of the message.

For example, you can establish a maximum response time of three hours for most messages. But for important messages, such as questions from surfers who are about to make a major purchase from your Web site, you can set the maximum response time at, say, five minutes.

Call-back Channel lets surfers click on a button to request a regular phone call from an agent. It is similar to AT&T interactiveAnswers Enhanced Service, a network-wide Web callback service provided by the IP services unit of the long-distance carrier. Especially useful to surfers who have more than one phone line, Call-back Channel enables agents to push Web pages to surfers while they're speaking with them on the phone.

Sitebridge, now part of eGain (Sunnyvale, CA – 408-737-8400, www.egain.com) lets you incorporate on-line forms with its CustomerNow 2.0 text chat software. After surfers fill out these forms, CustomerNow gives you several ways to handle them. It can direct customers' responses to the on-line forms to live agents. If surfers ask questions your center frequently receives, the software can push a Web page to them or enable live agents to share onscreen material with them. CustomerNow can also send completed on-line forms to other applications, such as customer management, call tracking or e-mail software.

E-Mail Routers

E-mail routers are best known for placing e-mail messages in queues and directing them to individual agents or groups of agents. Mustang Software's (Bakersfield, CA – 661-873-2500, www.mustang.com) Internet Message Center (IMC), which we use at our editorial offices, lets agents respond to e-mail messages within their queues. The software lets you report on e-mail response times for agents, queues and your entire center.

Some e-mail systems provide knowledge bases from which agents can search for examples of appropriate responses to different types of e-mail messages. Other systems suggest electronic documents to attach to responses. Like a growing number of e-mail systems, they can reply to routine e-mail questions automatically. Such systems usually respond either with canned messages or perform lookups of databases that contain answers to customers' questions.

Although Mustang doesn't make an off-the-shelf auto-responder, it offers a development service called IMC AutoAgent that lets you automate replies to e-mail messages your center receives. The service is best for centers whose customers indicate their questions by filling out and e-mailing on-line forms.

When you use an e-mail router, it doesn't necessarily have to reside at your center. For example, eGain can host and maintain its Email Management System (EMS) for you from its offices for a monthly fee.

Although a lot of e-mail management software products are stand-alone, some of them are integrated into much larger systems. For example, since Genesys' (San Francisco, CA – 415-437-1100, www.genesyslab.com) Genesys E-mail is part of the company's Genesys Suite, version 5.1 call center software, you can use the same system to track e-mail and phone calls. The Genesys Suite also provides call routing, automated dialing and computer telephony applications.

IP Telephony

Based on numerous interviews with call centers that offer IP telephony from their consumer Web sites, we've found that Web surfers prefer text chat, at least for now.

But IP telephony is gaining ground, especially for catalog call centers that have Web sites. For example, IDT is a long-distance phone company that offers IP telephony software and services, lets visitors to the Web sites of 800-FLOWERS and Lands' End click on icons to place IP calls to live agents at regular phones.

One advantage of IP telephony over other forms of on-line communication is that it lets agents do what they do best - have conversations with callers. Coupled with text chat, IP telephony can be an attractive option for international visitors to your Web site who want live service but don't want to pay the cost of making toll calls to your company.

CosmoCom's (LaJolla, CA – 858-456-7574, www.cosmocom.com) CosmoCall lets you route IP calls, text chat requests or e-mails from surfers to agents with or without an IP telephony gateway. If your customers place IP calls from your Web site to reach agents, they can now do so without performing a special download or being concerned about a firewall getting in the way.

CosmoCall now includes a dialer so agents can make IP calls or start chat sessions to customers. It also lets agents push items to your customers' Web browsers or share Web pages with them. Like an ACD system, CosmoCall logs whether agents are on an IP call, unavailable or logged in or out.

PakNetX's (recently acquired by Aspect, San Jose, CA – 408-325-2200, www.aspect.com) PNX ACD 2.0 and CosmoCall let agents use traditional call handling capabilities, like holding, transferring and conferencing, with incoming voice over IP calls. Upcoming Web columns and our August issue will discuss more on IP telephony in call centers.

Wireless

Wireless technology is literally any communications device "without wires." They generally refer to what are known as mobile phone systems. These are based upon radio, and the origins of radio go back almost as far as the telephone, to the last decades of the 19th century, derived from experiments and demonstrations by Tesla and Marconi.

Interestingly, early in the 20th century the police forces in Europe and in the U.S. were already using radio telephony equipment. In the 1920s, police departments in Detroit, Michigan and Bayonne, New Jersey and the Connecticut State Police used in their patrol cars the technology that had improved the safety of oceangoing vessels following the Titanic disaster – radiotelephone service. The equipment was still quite bulky and unwieldy, however.

A breakthrough came in 1935, when Edwin Howard Armstrong unveiled his invention, Frequency Modulation (FM), to improve radio broadcasting, which produced a higher-quality and more static-free signal than Amplitude Modulated (AM) technology, and at the same time used smaller radio equipment.

The outbreak of World War II spurred the United States into developing mobile FM-based communications for the battlefield. Companies such as AT&T, Motorola and General Electric focused on refining mobile and portable communications. Motorola's FM Handie-Talkie and Walkie-Talkie achieved legendary status during the war years and their technology was carried over into peacetime use.

The first commercial wireless telephone service was available in 1946. Known as the Improved Mobile Telephone Service (IMTS), it used vacuum tubes and filled a car trunk. Also, frequency restrictions allowed only a few radiotelephone conversations to be held at one time in a given city. In 1949, for example, the Federal Communications Commission (FCC) of the U.S. government would allow only very small systems that could only handle 23 simultaneous calls. Each system could only have a total of about 250 customers in each city.

Later, during the 1950s and 1960s, more and more radio telephone networks became available for U.S. civilian customers. Such wide availability underscored the lack of radio frequencies in officially-designated frequency bands.

In the 1960s and 1970s, new transmission techniques such as dynamic channel allocation and cell-based networks (made possible in the 1970s by microprocessors) were developed in order to decrease radio traffic congestion. The first test of a rudimentary cellular service took place in 1962. During the 1980s, several analog cellular radio networks appeared around the world. Each country has developed along its own lines in devising and adopting network standards, some of which are compatible, but most of which aren't.

Interestingly, Europe initially pulled ahead of the U.S. in per capita wireless phone usage. Even Portugal, perhaps the poorest country in Europe, had nearly the highest per capita use of mobile phones. The U.S. has now caught up and is pulling ahead.

In general, we can divide the history of mobile systems into three generations: The first generation was simple analog transmission. This included the early cellular service called the FM-based Advanced Mobile Phone System (AMPS) that appeared in the U.S. in 1978, the Total Access Communication Service (TACS) in Europe in 1985, the Nordic Mobile Telephone (NMT) that appeared in 1981 in Scandinavia, and various other analog systems.

The second generation of mobile phone systems centered on the transition to digital modulation and transmission techniques. The immensely popular Global System for Mobile Communications (GSM) or "Group Special Mobile," swept over Europe starting in 1992, while paging services were handled by the European Radio Messaging System (ERMES). Short-range Cordless Telephone Standards (CT2, CT3) also appeared, as did the Digital Communication Service (DCS) 1800 system, and the Digital European Cordless Telephone (DECT) standard. In the U.S., AT&T provided a connection between the AMPS standard GSM via their Universal Wireless Communications Network (UWCN) that today provides seamless connection to over 46 countries located within Europe, Asia, Oceania, Africa, and the Middle East. Also, the "next generation" analog version of the AMPS system appeared, called the Narrowband Advanced Mobile Phone Service (NAMPS), which uses some digital technology that enables the network to carry three times the number of conversations as does AMPS, as well as offering some enhanced user functionality voicemail notification and the Short Message Service, or SMS (indeed, the ability of customers to use short messaging to check on the status of their bills has been a compelling feature of the PCS-type services). NAMPS phones automatically switch to the AMPS mode when the user is in an area where NAMPS systems are not used.

The third and latest generation of mobile systems involve the unifi-

W

cation of an eclectic combination of technologies. This involves such developments as the Future Public Land Mobile Telecommunication System (FPLMTS), the Universal Mobile Telecommunication System (UMTS), and International Mobile Telecommunication 2000 (IMT-2000).

Cellular Mania

In America, the "cell phone" has become both a convenient and somewhat annoying aspect of daily life. How many times have you been walking down the street only to hear a phone ring, your heart suddenly skipping a beat as you wondered whether you had brought your cell phone with you?

But whatever you do, don't confuse a cellular phone with a *cordless telephone* which is simply a phone with a very short wireless connection to a local phone outlet.

As was mentioned previously, cellular networks came about in the 1970s because of the paucity of frequencies that were allocated by organizations such as the FCC (in the U.S.) to radio telephone services.

Cellular networks get their name from the division of a geographical area covered by a network into a number of smaller areas called cells. To use the airwaves more efficiently, AT&T engineers decided to stretch the limited number of radio frequencies available for mobile service by scattering multiple low-power transmitters throughout a metropolitan area, and "handing off" or "handover" calls from transmitter to transmitter as customers moved around in their vehicles. More customers could now simultaneously access the system simultaneously, and when more capacity was needed, the area served by each transmitter could be divided again. This was the birth of the most popular form of modern wireless technology.

Still, it took 20 years to perfect the call "handoff" technology - handing off a call from cell site to cell site as the user drives his or her car around – and for the FCC to give tentative approval for cellular service to proceed. In 1973, Motorola introduced the DynaTAC mobile phone, a brick-sized radiotelephone set. In 1977, the FCC authorized two experimental licenses – to AT&T in Chicago, and to Motorola and American Radio Telephone Service, Inc. in the Baltimore/Washington, D.C. corridor.

Today, cellular systems still are based upon this underlying technology, with each designated cell area covered by a fixed radio transmitter / receiver facility (called a "cell site" or "base transceiver station") that uses short-wave analog or digital transmissions to connect to the wireless mobile devices in its own particular cell. Cells can be as small as an individual building or installation (such as an airport or sports arena) or as large as a mile or even 20 miles across. When a subscriber wants to place a call, the base station allocates a transmitting frequency which is then used between the subscriber and the base station. Also, each cellular phone is assigned a unique Electronic Serial Number (ESN), which is automatically transmitted to the cellular tower station every time a cellular call is made.

The mobile phone industry was initially confined to about 45 megahertz MHz of spectrum bandwidth, which would limit each cellular carrier to 396 frequencies or voice channels per cell. Fortunately, because of the system's low power output, each cell uses radio frequencies only within its boundaries, which means a frequency being used to carry a phone conversation in one cell can be used to carry a different conversation in a nearby cell with a low probability of interference. Thus, individual radio frequencies are used over and over again, which allows a cellular system to handle a much greater call capacity than one would encounter in radio systems such as Citizens Band (CB) in which all users must try to get airtime on the same set of channels. The way that wireless service providers

organize, or "configure," their cells is a fundamental factor achieving the highest possible levels frequency reuse and establishing an area's maximum possible calling capacity. Geographically speaking, most cells are laid out in a honeycombed pattern of equal-sized cells.

Each cellular tower station is linked to a Mobile Switching Center (MSC) or Mobile Telephone Switching Office (MTSO) which is a central switch that controls the entire operation of a cellular system and connects your wireless call to the local "wired" telephone network. Wireless carriers own MSCs. They are basically computer systems that monitor all cellular calls, tracks the location of all cellular-equipped vehicles traveling in the system, arranges handoffs, keeps track of billing information, etc.

When a subscriber using a wireless phone approaches a cell boundary, the wireless network (specifically, the MSC) detects that the signal is weakening and automatically hands off the call in progress from the current cell to the antenna in the next cell into which the caller is traveling. The hand off process ensures that the call is handled by the cell that has the strongest signal with the mobile phone (because as we drive or walk, we move further from one cell and closer to another). Sometimes a hand off occurs between radio channels in the same cell as a method to minimize co-channel interference between cells. The new base station takes over the control of the call and assigns the call to a frequency that is different from the the one used in the first cell. The original frequency used in the first cell is then released and is now available for use by other callers.

The hand off process usually means that the call has a slight outage (goes silent) and the caller is out of contact for about 250 milliseconds, not really noticeable to most callers, but it may confuse a modem device that is attempting to receive or transmit data over a wireless connection, tricking it into believing that the call has been terminated.

When subscribers travel beyond the range of the entire cell system in their home geographical area, they can still make or receive wireless calls, since the cell system can transfer them to a neighboring company's cell system without them being aware of it. For this to happen, the wireless carrier in the area where they are traveling must provide the service, which is called *roaming*. A *roaming agreement* between two network operators allows for the transfer of call charges and subscription information back and forth, as their subscribers roam into each others areas.

Unpredictable impairments to cellular service include "fading" which is the absorption and / or reflection of radio signal power by air humidity, rolling foothills, lakes, forests and buildings; co-channel interference; intermodulation distortion where two transmitters on different frequencies can manage to interfere with each other if they are too close or have too powerful an output; background or thermal noise generated by the user's receiver itself; atmospheric noise from thunder, tornadoes, etc.; and radio noise generated from high-voltage industrial or electrical equipment.

PCS

A newer wireless service similar to cellular is Personal Communications Services (PCS). PCS is sometimes called digital cellular (although there are cellular systems that are digital). Like cellular, PCS is for mobile users and uses a series of antennas to cover a geographical area. As a user moves across the landscape, his or her phone signal is picked up by the nearest antenna and then forwarded to a base station that connects to the wired network.

PCS has become an "umbrella" term encompassing a wide range of wireless mobile technologies, chiefly two-way paging and cellular-like calling services. The "personal" in PCS differentiates itself from cellular service by emphasizing that, unlike classic cellular, which was designed for use in a car, PCS is designed for greater user mobility. It uses more cell transceivers for coverage, transmits at lower power and higher frequencies than previous cellular services, and has fewer "blind" spots.

Cellular systems in the U.S. operate in the frequency bands between 824 to 849 megahertz (mHz) and 869 to 894 mHz, which is close to the band used by Ultra High Frequency (UHF) television stations. PCS services operate in the part of the electromagnetic spectrum that the FCC in September of 1993 allocated for emerging communications technologies. The spectrum was divided into three major categories of PCS services: Narrowband PCS in the 900 to 901 mHz range; 930 to 931 mHz and 940 to 941 mHz bands; broadband PCS in the 1,850 to 1,990 mHz band; and an unlicensed portion of spectrum at 1910-1930 mHz.

Several incompatible air access interfaces are used to implement analog and digital cellular networks:

• Frequency Division Multiple Access (FDMA) allows multiple users to share the same physical channel by multiplexing the transmissions so that the bandwidth of the channel is divided among the population of stations. For example, with 30 stations the frequency range of the channel is divided by 30 and each station gets its own private frequency. This guarantees that there's no interference between users, since the channels are specified by the actual used frequency, and since there are often interfrequency protection bands designated (which unfortunately wastes bandwidth). With mobile phone services several hundred duplex channels are available.

FDMA is algorithmically and technically simple, and is fairly efficient when the number of stations is small and the traffic is uniformly constant. However, it's not particularly scaleable nor can handle a varying station population. If the traffic happens to be bursty, bandwidth will be wasted. It also has no broadcasting capability. FDMA is a basic technology in the analog AMPS and the Total Access Communication System (TACS). The Digital-Advanced Mobile Phone Service (D-AMPS) also uses FDMA but adds Time Division Multiple Access (TDMA) to get three channels for each FDMA channel, tripling a channel's call capacity.

• Code Division Multiple Access (CDMA) is also known as spread spectrum transmission because the signal is "spread" across a wide bandwidth, the power of the transmitted signal now occupying a bandwidth that is much wider than the minimum bandwidth of the original signal. The distribution over the spectrum is done using a code that "marks" the spread signal. In this way multiple users can share the same channel by multiplexing the transmissions in the code space. Each transmission is encoded with its own code (key) and they all coexist on the same channel. The code thus acts as a scrambling algorithm that "randomizes" the transmission so it appears as a noise to all receivers except the intended receiver. As a part of call setup (when the SEND button on the telephone handset is pushed), the recipient phone is given the specific code used for that call, and unscrambles the transmission. All other signals encoded with different keys continue to resemble noise. The spectral power distribution (mW/Hz) of the spread signal is very small, which minimizes narrowband interference and reflection problems such as fading.

Spread spectrum technology was first conceived during World War II by an unlikely pair: The American music composer, George Antheil, and the Hollywood actress Hedy LaMarr. It was LaMarr who conceived of an idea to guide torpedoes by sending information across the multiple radio frequencies in a random pattern. The pattern would be received and decoded into commands for servo motors attached to the torpedo's control surfaces. The idea depended upon the transmitter and receiver staying in synchronization. LaMarr explained the idea to Antheil, who solved the synchronization problem. Their "Secret Communications

System" was patented in 1942 and presented free of charge to the U.S. government, which promptly forgot about it until 1962, when the idea was revived. Today, "spread spectrum" technology is used by the military of the U.S. and those of foreign countries.

CDMA technology itself used for third-generation (3G) mobile communications will come in three modes: "time division duplex" and "direct spread," both based on the wideband CDMA standard; and "multicarrier," based on the cdma2000 standard. Another 3G development path has emerged in the EDGE standard, which combines elements of the international GSM communications standard with Time Division Multiple Access technology.

• Time Division Multiple Access (TDMA), when used as a digital air interface technology, allows multiple callers to share the same voice channel by multiplexing their transmission in time. Each conversation is divided into short lengths, with each piece transmitted during it's own short consecutive periodic time slot. The channel's data rate is thus the sum of data rates of all the multiplexed transmissions. A similar technique is used in PCs for the CT Bus that allows for the sharing of computer telephony resource boards (See Time Division Multiplexing).

GSM, the Global System for Mobile Communications is a type of TDMA digital wireless network with encryption features. GSM runs at 900mHz in Europe and 1900mHz in the U.S.

In the U.S., AMPS, NAMPS, TDMA, and CDMA technologies are available access methods for the 800+ mHz bands, while TDMA, CDMA, and GSM are available for the 1,900 mHz. bands.

Cellular Growth

By 1990, construction permits had been issued for at least one cellular system in every market in the United States. By the end of 1990, before most systems had even come on-line, the wireless subscriber count topped five million. Subscribership broke the 10 million mark on November 23, 1992.

In August 1993, the FCC received authorization to auction PCS licensees when President Clinton signed the Omnibus Budget Reconciliation Act of 1993.

The wireless competition that began in 1983 with two licensees per market (one called a wireline company and the other called a nonwireline carrier) was expanded in 1995 to allow for up to nine carriers per market. By 1995 there were approximately 85 million users of cellular telephony worldwide (32 million users in the United States alone). Today, 239 million Americans can now choose between three and seven wireless service providers in their area. As of June 30, 1999, there were over 76.3 million wireless subscribers. Today, more than 90 million Americans can choose from among six or more wireless providers, and 88 million Americans can choose from among five wireless providers.

The number of digital subscribers had reached more than 28 million by mid-year 1999 – 38 percent of all wireless subscribers in the U.S. Currently, there are over 86 million wireless subscribers, with a new wireless subscriber being added every two seconds, or 42,338 new subscribers per day.

Wireless Packet-Switched Data

Conventional wireless communications systems are circuit-switched, which means that a dedicated, continuous circuit is "nailed up" and maintained between the sender and the receiver during the transmission. The radio channel remains open even when no data is being sent. Circuit-switched system connections involve significant overhead, especially when only a small amount of data may need to be transferred over a certain time period.

With packet-switched data, however, we see a transmission model reminiscent of the Internet. A computer processor connected to the cell phone sends and receives bursts, or packets, of data. A radio channel is occupied only for the duration of the data transmission instead of continuously, making packet-switched data more efficient than circuit-switched. At the switch, packet services interconnect with the Internet or directly with corporate intranets using conventional networking technologies such as frame relay.

There are a number of wireless packet-switched data technologies in the running:

- General Packet Radio Service (GPRS) is a form of packet-switched data technology being developed for GSM networks and third generation networks. Initial speeds for GSM will be up to 115Kbps. AT&T and London-based British Telecom are testing international roaming services on GPRS-based networks so that, eventually, international travelers will be able to use a single wireless device in various countries.

Other packet-switched data technologies are already in operation but are mainly used to transmit short messages to "smart phones", personal digital assistants (PDAs), handheld computers, and notebook computers.

- Cellular Digital Packet Data (CDPD) is a packet-data wireless technology developed by AT&T Wireless Services and other cellular carriers. CDPD is IP based. It connects seamlessly with the Internet. With CDPC, IP has been engineered as an overlay to analog 800 mHz AMPS networks, enabling AMPS to carry packetized data alongside voice. CDPD is primarily used to transmit brief messages for Personal Digital Assistants and "smart phones" (such as the AT&T PocketNet, with its CDPD modem and scaled down web browser) for simple messaging and credit-card transactions, but in theory you should be able to pop a CDPD modem into a laptop and access the Internet. PPP or SLIP is used between the mobile computer and the CDPD modem. CDPD uses either idle voice channels or dedicated data channels depending on the network configuration. The service is available in most major cities in the US and Canada and is provided by cellular companies including AT&T Wireless Services, Ameritech, Bell Atlantic Mobile, and GTE. In Canada, service is provided by BC TEL Mobility and TELUS Mobility.

Any IP based application will work over CDPD, but the pricing (five to 10 cents per transmitted kilobyte) and performance encourages short and bursty applications such as as brief messages. CDPD's raw throughput is 19.2 Kbps but actual throughput after protocol overhead is 10 to 12 Kbps. Round trip delays are typically less than one second. The 19.2 Kbps channel is shared by all active users in that cell or cell sector.

Wireless modem vendors include AirLink Communications, Compaq, INET, Novatel, Sierra Wireless, Tellus Technology, Uniden and Z Square Telecom. Smart phones are currently available from Samsung and Mitsubishi.

- Mobitex is a secure, reliable, open-architecture, two-way digital wireless packet switching network developed in 1984 by Eritel for the Swedish Telecommunication Administration. Mobitex now comes under the auspices of the Mobitex Operators Association (MOA), which controls its specifications, and the network infrastructure is manufactured by Ericsson Mobile Data Design AB. Mobitex networks are operated in 17 countries around the world, by network providers such as BellSouth Wireless Data in the U.S., and Cantel in Canada.

To connect to a Mobitex network, all radio modems and fixed terminals, such as hosts and gateways, must have an active Mobitex Access Number (MAN). A MAN is assigned to every user subscribing to the Mobitex network; it is analogous to a phone number on a telephone network. The MAN for a mobile user is stored in the mobile's radio modem, just as a telephone number is stored inside a cellular phone. Every network has a different range of MANs. In the U.S., MAN numbers are in the 15,000,000 to 16,999,999 range; in Canada, the MANs are in the 5,000,000 to 5,999,999 range.

In Mobitex, a packet is called an MPAK (short for "Mobitex packet"). Each MPAK can have no more than 512 bytes of data; longer messages are divided into multiple packets. MPAKs include information about the origin, destination, size, type, and sequence of data to be sent, enabling packets to be transmitted individually, in any order, as traffic permits. Individual packets may travel along different routes, in any order, without interfering with other packets sent over the same frequency by different users. At the receiving end, all packets are rounded up and reassembled into the original message. Set-up time is eliminated and the network connection is instantaneous. This type of packet-switching makes far more efficient use of channel capacity, typically allowing 10 to 50 times more users over a radio channel than a circuit-switched network.

- RAM Mobile Data (renamed as BellSouth Wireless Data, on March 18, 1998) is a wireless packet-data service, based on Mobitex technology, that's offered in most cities and towns in the U.S.

Mobitex defines both wireless-specific protocols as well as end-to-end networking protocols. The raw throughput is 8 Kbps but actual throughput after protocol overhead is about half this rate. Round trip delays usually range from four to eight seconds. The 8 Kbps channel is shared by all active users in the operating area of the base station.

Mobitex modem vendors include Ericsson and Research in Motion. One interesting new device for this service is the Inter@ctive Pager from Research in Motion. Typical service plans are around 25 to 30 cents per kilobyte.

- DataTAC is a wireless packet-data network based on technology originally developed by Motorola for IBM field service engineers using a Specialized Mobile Radio (SMR) frequency. In 1990, ARDIS was formed as a joint partnership between the two companies to commercialize this network.

A form of DataTAC called ARDIS then came under the ownership of American Mobile and can be found in many major U.S. cities and towns. DataTAC is also used for wireless networks in Europe and Asia.

In 1991, Motorola and BCE Mobile (a unit of Bell Canada) formed Bell-ARDIS - an ARDIS-compatible radio data network for Canada, which covers metropolitan centers from Newfoundland to British Columbia. In the U.S., ARDIS uses Datatac 4000. Datatac 5000 is used in most other areas, including Canada and the Asia Pacific region.

DataTAC uses trellis code modulation, interleaving, error-detection and automatic retransmission. DataTAC systems since 1993 use the Radio Data LAP (RD-LAP) wireless channel protocol, which applies 4-level FSK modulation to a 19.2 Kbps rate on 25 kHz channels, and achieves 9.6 Kbps on 12.5 kHz channels. Raw throughput drops down to 4.8 Kbps when the MDC4800 protocol is used instead of RD-LAP. Actual throughputs after protocol overhead are about half these rates. Round trip delays typically range from four to eight seconds. The 19.2 Kbps or 4.8 Kbps channel is shared by all active users in the operating area of the base station.

Key components of the DataTAC infrastructure include the Network Management Center (NMC), Area Communications Controller (ACC) and base site equipment.

The NMC is an advanced client-server based management tool

which provides all network administration, operation and maintenance functions. It acts as the central point of control in the DataTAC system.

The ACC consists of the Radio Network Gateway (RNG), Radio Network Controller (RNC), and Communications Hub. The ACC is primarily responsible for all message switching and routing functions, as well as providing the key communication link between host computers and remote base stations. Connectivity between the host computer and the infrastructure is done using industry standard X.25 or TCP/IP host link protocols. The ACC maintains all customer-specific information for each device on the system and allows users within a given geographic area to connect to host application services. It manages user device authorization, roaming control and base site control. It also collects all system usage / traffic information for accounting / billing purposes and detailed analysis of usage patterns.

DataTAC Data System Station (DSS) base site equipment is located at various remote sites in the operational area of coverage and provides the RF link between the DataTAC infrastructure and end user devices. They convert the host data messages into the RD-LAP radio channel protocol for transmission to the user devices and reverse the process on the return connection. They also allow device roaming between base site coverage areas in a seamless manner completely transparent to the end user.

DataTAC networks support an inherent service called DataTAC Messaging. DataTAC Messaging is designed to provide efficient, short message services in a peer-to-peer fashion between wireless data subscriber devices. Examples of applications well suited for this DataTAC Messaging service include two-way text messaging between wireless subscriber devices, acknowledgment paging, and store/forward e-mail (without attachments).

The communications protocol used between fixed hosts and the DataTAC network infrastructure is known as Standard Context Routing (SCR). SCR is a specialized protocol designed for communications and management control of wireless data subscriber devices.

Native Mode interface, which allows applications developed on the computing device to provide high levels of control and management of the wireless modem. This method is used to develop commercial, highly efficient wireless applications. The protocol definition of the native mode interface is known as Native Control Language (NCL).

DataTAC modem vendors include Motorola and Research in Motion. The Inter@ctive Pager from Research in Motion is also available for the ARDIS network. Typical service plans run about 30 cents per kilobyte.

- Ricochet is a packet-data network operated by Metricom (Los Gatos, CA - 408-399-8200, www.metricom.com) in the unlicensed ISM 902 MHz to 928 MHz band. Service is currently available in the San Francisco Bay Area, Seattle, Boston, and Washington, D.C. metropolitan areas; in select areas of New York City; on select corporate campuses; and at gate areas at major airports throughout the U.S. (For more information, call 1-800 Go-Wireless.) The original Ricochet worked with any Internet application that could run well at 14.4 Kbps or higher. Data throughput use to vary between 14.4 Kbps and 28.8 Kbps, but the new Ricochet service can do an impressive 128 Kbps. Modem-based applications can access a modem gateway. Ricochet uses PPP between the mobile computer and the network for IP based communications to the Internet, as well as to corporate networks over private connections. Round trip delays are typically less than one second. The radio channel is shared by all active users in the operating area of the base station.

Wired Access Points (WAPs) are the central points of connection between Ricochet local wireless networks and the local wireline and national Internet backbone networks. WAPs are integral to the Ricochet meshed frequency-hopping network architecture that incorporates shoe-box-sized micro cell radio transmitters hung on streetlight and utility poles. These intelligent self-configuring "poletop" radios allow users to send and receive data from Ricochet wireless data modems anywhere in coverage areas. Ricochet WAPs collect RF data packets from the poletop radios and convert them for transmission over the local wireline and national Internet backbone networks.

Ricochet modems are available only from Metricom. Ricochet service pricing runs about $30 per month for unlimited use.

Satellite Technologies

Somehow, the thought of satellite-based telephony services always seemed ridiculous. Now that we've seen some actual multi-billion dollar systems in operation, we can declare that, indeed, some of them really *are* ridiculous!

Normally, taking advantage of satellite communications required large antennas and support equipment to communicate with geostationary satellites hovering 36,000 km above the planet. But Low Earth Orbit (LEO) systems use many satellites whizzing by at much lower altitudes (500 to 2,000 km). Communicating with such orbiting "base stations" demands much less power that communicating with a geostationary satellite, indeed, the necessary power output can be achieved by handsets only slightly larger than today's cellphones.

In a LEO system the satellite acts much like the tower site of a cellular system, only much higher in altitude. It grabs the call from earth and usually passes it to an earth-based switch. Because of the satellite's speed, it must frequently hand off a particular call to the next satellite approaching over the horizon. This is reminiscent of the cellular system, except that in this case it's the cell site that's moving through space rather than the subscriber moving in a car.

These satellite services are just coming on line (some, such as Iridium, are going offline just as quickly, collapsing because of a lack of customer interest). Most will offer a combination of voice, data, messaging, and location services worldwide. Data services will typically range from 2.4 Kbps to 9.6 Kbps for both circuit-switched connections to the PSTN and packet-switched connections to the Internet. Like cellular data services, users generally have a serial connection between their mobile computer and the handset.

Priced from 50 cents to $5.00 per minute, satellite services will not compete directly with terrestrial systems, but will instead complement them by offering service in areas without terrestrial coverage. A number of vendors are already developing dual-mode handsets that automatically use a terrestrial cellular connection when available but fall back to a satellite connection when in remote areas.

Principal LEO systems include:

- Globalstar. A CDMA system with 48 satellites. 9.6 Kbps data service. A San Jose, Calif., subsidiary of Vodafone AirTouch plc, Globalstar was officially launched February 28, 2000. Globalstar, like Teledesic and ICO Global, uses only "bent pipe" transmissions rather than Iridium's satellite-to-satellite communications. The easier maintenance and lower cost of a bent pipe model allows Globalstar to offer calls at an average of $1.50 per minute. Iridium typically charged from $2 to nearly $5 per minute. Globalstar will ultimately offer data services such as e-mail, fax and file transfer at speeds up to 9.4K bps.

- ICO Global Communications. A TDMA system with 10 satellites. 2.4 Kbps data service. ICO Global which started as a satellite phone service before filing for Chapter 11, refocused on data services after its reorganization in 2000. ICO Global's services will appear in 2002.

W

Teledesic and ICO Global may end up merging into one company.
- Iridium. A TDMA service with 66 satellites. 2.4 Kbps data service. On March 17, 2000 Iridium officials told a New York bankruptcy court they couldn't find a buyer and were ceasing operation. Motorola Inc., Iridium's primary investor, amazingly used Iridium's last $8 million to steer the failed company's 66 satellites out of Earth's orbit so that they could burn up on re-entry. Iridium's disappearance left without service some 55,000 customers, many of them with the U.S. Department of Defense.
- Orbcomm Global LP of Herndon, VA. A Service with 28 satellites. Two-way short text messaging service that's available today.
- Teledesic LLC, of Kirkland, WA, is Craig McCaw's data satellite company. It plans to offer gigabit per second data services supported by 288 satellites by 2004. It may merge with ICO Global Communications.

Mobile Radio Technologies

Mobile radio is generally used for dispatch applications by the emergency services, transport sector and utilities. Mobile radio services were introduced to alleviate the pressure on the mobile radio frequency bands through the introduction of trunking technology. Analog mobile radio services are commonly known as Specialized Mobile Radio (SMR) services in the USA, and Public Access Mobile Radio (PAMR) in Europe and in Asia Pacific.
- SMR was introduced in the USA in 1979 and is designed for closed user group voice services with support for overlay packet data. It operates in the 800 MHz and 900 MHz spectrum and the radio equipment for the two bands is incompatible. SMR has around two million users in the USA but only 150,000 use data. There are many local and regional operators but licenses are not awarded for national networks.

PAMR services in Europe and Asia Pacific tend to use the MPT 1327 specification for trunked networks, which originated in the UK in 1985. By mid-1997 there were several hundred national and local networks operating in Europe and Asia Pacific, many offering data services. The standard has been extended to incorporate data transmitted over specially designated data channels. There are several thousand data users in Europe and in Asia Pacific.

Digital services with data services as a common feature are being introduced using either proprietary technology from the leading suppliers or the European TETRA or USA APCO25 standard. One digital ser-vice gaining ground in the USA is the Integrated Dispatch Enhanced Network (iDEN) technology developed by Motorola and most broadly deployed by Nextel. Nextel now offers both circuit-switched and packet-switched data services.

One interesting "hybrid" product in this area is the GSM Pro product family from Ericsson (Stockholm, Sweden – 01-46-8-757-00-00, www.ericsson.com) that allows wireless operators to offer an enhanced service having dispatch-like and group communication capabilities to serve the emerging market of workgroup communications for mobile professionals, sales teams, public safety (police, fire, ambulance), transit authorities, public transport (busses, railways, airports, cargo services), couriers, public works agencies (road authorities, municipalities), etc.

The product line consists of the GSM Pro Server for the network operator, a PC-based Dispatcher Console software for subscribing organizations, and special ruggedized phones for mobile end users.

GSM Pro users can make group calls over a wireless or fixed infrastructure using either mobile phones or special, rugged terminals that can operate as a mobile phone or a dispatch radio. In this way, Ericsson's Pro products are also a replacement for traditional PMR and PAMR systems.

Until recently, most PMR operators have built their own infrastructure, and PMR systems ranged from small, single-channel, hand-held radios not requiring an infrastructure to large, trunked digital systems that dispatch speech and data to thousands of users. Some large systems also include connections to a PBX and the PSTN.

The Ericsson Pro products are a "package" of services that following installation are available in the entire network, including every network with which the operator has a roaming agreement. The Pro products are based on standard globally mass-marketed technologies such as TDMA and GSM.

Thus, Ericsson Pro products can be combined with most other existing and future wireless services, such as high-speed packet data, and location-based services.

The Ericsson Pro products fall into two categories:
- TDMA Pro is geared toward mobile professionals and their support

Overview of the Ericsson *Pro* concept

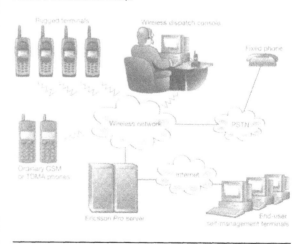

Here is the architecture of a conventional Private Mobile Radio (PMR) system. Normally, unrelated PMR systems do not interoperate with one another. Commercially operated Public Access Mobile Radio (PAMR) systems provide users with the same features as a PMR. Only a small portion of the PMR market has migrated from traditional PMR systems to public network systems, primarily because of a lack of support for PMR functions until now.

The Ericsson Pro concept brings PMR functionality to the GSM and TDMA wireless worlds by means of a rugged terminal (phone), Mobile Switching Centers (MSCs) of the wireless network, the cPCI-based Pro server and the dispatch console.

teams. Small businesses and families can now use one number for contacting fixed or wireless phones (or both). Closed user groups can be set-up for controlling end-user access and, ultimately, their wireless costs.

- GSM Pro is a replacement for PMR and PAMR systems normally used by people working in the field.

The GSM Pro concept architecture is built from four elements: A GSM network, a server that adds functionality to the network, a dispatch console (PC software unique to GSM Pro), and rugged phones with PMR-type functions.

Ericsson's experience of PMR systems in Europe suggests that 80% of calls are point-to-point and can be handled by the GSM network in typical GSM fashion. The group calling functionality, however, resides in the GSM Pro Servers, which are Windows-NT based CompactPCI machines supporting "hot swappable" cPCI boards from Natural MicroSystems. No proprietary hardware was developed for the industry-standard Pro servers, which communicate with each other via TCP/IP and connect to the network's Mobile Switching Centers (MCSs), interfacing with billing, operating and maintenance (O&M) and customer care systems.

The server-based approach means that the system can be deployed rapidly and the GSM Pro functionality is network-independent, so the server will work with multi-vendor networks.

Customers can manage group setting remotely, and customize the way the dispatch application operates, by accessing part of the Pro server via a Web browser. On a network level, Ericsson's GSM Pro solution allows 55,000 "voice groups" to be established, with up to 16 users in each group.

The GSM pro Dispatch Console is a PC-based tool that simplifies

Detailed architecture of the Ericsson Pro concept

BSC - Basestation Switching Cente
MSC - Mobile Switching Center
SN   - Switch Node

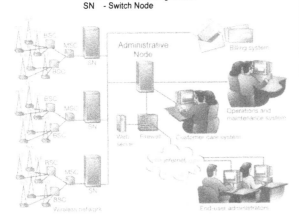

Ericsson's Pro servers add PMR functions to GSM by interfacing with MSCs and billing, operation and maintenance (O&M), and customer-care systems. The Pro server can be split — logically and physically — into several nodes, such as the Administrative Node, which serves as a central database repository, as an interface to O&M and customer care systems, and as a Web server enabling end-user administrators to manage group settings remotely. There are also one or more switch or signaling modes, which provide an interface and handle real-time traffic to MSCs. The nodes also embody switching and database hardware / software needed for Pro end-user voice functions, such as group calls. The switch nodes receive configuration information and subscriber data from the administrative node.

call dispatcher communications to GSM Pro users in the field, and enables traffic to be monitored on-screen. The dispatcher can make outgoing voice calls or transmit Short Message Service (SMS) messages to individuals or groups helped by a directory of names and phone numbers. Every call or SMS message is logged, and details of any call or message can be retrieved at any time.

The Dispatch Console can be customized, for instance to present incoming alert calls as top prority, and SMS signatures, short-cut keys, audio-visual notification details and other settings can be programmed to suit users' needs.

Finally, Ericsson has launched the world's most rugged phone, the dual band R250 PRO for GSM 900 and GSM 1800 networks. It supports both GSM phase 2+ technology and the GSM Pro system, thus combining the advantages of GSM phones with PMR functions. The R250 Pro phones operate as push-to-talk, hip-hanging radio phones as well as reg-

With GSM Pro, Ericsson has added group communication and Private Mobile Radio (PMR) functionality to the GSM system. Here we see Ericsson's ruggedized R250s PRO phone being used like a traditional PMR handset — it has a built-in loudspeaker and "push-to-talk" button. There are two modes: Push-to-talk or phone mode. Push-to-talk mode is the classical "PMR" mode, in which the built-in loudspeaker emits all received audio signals but can only transmit when the PT button is depressed. You can passively listen to calls while wearing the terminal on the hip or carrying it in a pocket. You talk into the terminal in this mode generally by holding it in front of your face, as seen here. In phone mode, however, the earpiece volume automatically drops and the terminal is operated like any other phone.

At the terminal end of the Ericsson Pro system is the incredibly rugged, dual-band GSM Pro terminal R250s PRO. To meet the stringent demands of PMR users, a new enclosure was developed for this phone that withstands weather, dropping, vibration, shock, dust and other rough treatment.

ular GSM phones. Regular GSM phones can also be included in workgroups. The system is a natural fit for any dispatching application, event-management, public safety, or for any workgroup whose members must stay within constant reach. Depending on assigned privileges, workgroup members can make one-to-all, or one-to-one calls to anyone in their assigned group or subgroup.

Wireless Application Protocol (WAP)

WAP is an open, global specification that enables mobile users with wireless devices to easily access and interact with information and services instantly. What type of devices will use WAP? Handheld digital wireless devices such as mobile phones, pagers, two-way radios, smartphones and communicators.

The WAP is aimed at turning a mass-market mobile phone into a "network-based smartphone". So, simply put, WAP is just the actual application protocol that allows a WAP-enabled Mobile phone to connect to the Internet.

With the conventional Short Messaging Service (SMS) one can send text messages to mobile phones, e.g. receiving today's news or sports results. But SMS can only give the mobile phone user a limited degree of interaction. One can compare SMS with e-mail: one can send and receive messages but it is not really possible to take any action or easily navigate to other information. WAP could be compared with an Internet browser since it supports hyperlinks in a microbrowser for quick navigation and forms that you can easily fill in, send, and receive a response. The difference between WAP and SMS could be compared with the difference between an Internet browser and e-mail.

WAP has two key components: the WAP Gateway and the microbrowser, both of which enable mobile phones to interact with the Internet. The gateway connects phones to the Internet, while the microbrowser uses an XML document format, the Wireless Mark-Up Language (WML), to display pages.

The WAP gateway acts as a proxy, interpreting requests from phone micro-browsers and retrieving content via standard HTTP requests. You won't need to change your legacy Web servers to accommodate WAP applications - though you will need to translate your HTML and redesign your pages using WML.

A WAP gateway will compile WML pages for more efficient transmission to a mobile phone. It's probably best to think of a WAP gateway as a specialized form of Internet proxy, similar to familiar Web proxies and caches.

The micro-browser lurking in the mobile handset can process WML code using the phone's standard display of three or four lines. Of course, interfaces will vary from manufacturer to manufacturer and phone to phone.

The micro-browser works in the limited memory and processor power of a mobile phone, so it isn't as sophisticated as Internet Explorer or Netscape Navigator - at least not yet!

The WAP layers are:
- Wireless Application Environment (WAE)
- Wireless Session Layer (WSL)
- Wireless Transport Layer Security (WTLS)
- Wireless Transport Layer (WTP)

The WAP was conceived by four companies: Ericsson, Motorola, Nokia, and Unwired Planet (which is now Phone.com).

WAP is designed to work with most wireless networks such as CDPD, CDMA, GSM, PDC, PHS, TDMA, FLEX, ReFLEX, iDEN, TETRA, DECT, DataTAC, and Mobitex.

Since WAP is a communications protocol and application environment, it can be built on any operating system including PalmOS, EPOC, Windows CE, FLEXOS, OS/9, JavaOS etc. It provides service interoperability even between different device families.

Recommended Websites:
The WAP Forum - www.wapforum.org
WAPNet - www.wapnet.com
WAP.NET - www.wapnet.net

Wireless Broadband

Also called Broadband Wireless, is rapidly on its way to becoming the method of choice for getting local loop access into multi-tenant constructions. The advantage of wireless access is that it avoids a number of the obstacles faced by competing technologies. These include the challenging task of revamping an older building's existing wiring, the expense of leasing local loop T1/T3 trunks from a local phone company, and the expense and functional difficulties associated with pulling fiber to the curb. Once installed, the rooftop antennas can support a range of connections to customer equipment and IADs, including Ethernet and in-building fiber. Landlords are a particularly easy sale for permitting rooftop access, as the antennas are relatively unobtrusive and inconspicuous.

A large part of the reason wireless antennas are sprouting up so quickly in places like New York, however, is due to the aggressive campaigns of companies like WinStar, Nextlink, and Teligent.

The most commonly used technology for this kind of deployment (business MTUs) is LMDS (local multipoint distribution service), which operates in the high frequency 24GHz to 40GHz bands. WinStar Communications (New York, NY - 888-WINSTAR, www.winstar.com) was one of the earliest champions of LMDS, and has been one of the most successful in rolling it out to customers. Like others, WinStar's strategy seems to be to get antennas onto the roofs of as many buildings as possible, and worry about sorting out deals with landlords and signing on customers afterward. (The latter seems to be at least a partly grassroots effort, as evidenced by the WinStar salesman periodically seen to roam the halls of our own New York offices). WinStar uses the 38GHz band, and is reportedly capable of supporting up to 4Gbps of total bandwidth per building (more than enough to offer both phone and ultra high-speed data services to every tenant).

Teligent (Vienna, VA - 888-354-4368, www.teligent.com) targets small- to medium-sized businesses across the country with its LMDS-based SmartWave technology. Teligent's pitch is to offer local and long distance phone service, along with high-speed Internet access (up to 45Mbps), billed monthly at a flat rate of around 30% less than a users current rates for these services. Signals from a number of different buildings' antennas are aggregated at a base station antenna, and in turn routed to a Teligent switching center. There, Teligent uses a combinations of ATM switches, routers, and Nortel DMS switches to separate off traffic onto public and private data networks and to the PSTN.

NextLink (Bellevue, WA - 800-900-6398, www.nextlink.com) is by no means short on resources. It is currently the largest holder of fixed wireless broadband spectrum in the U.S., and is now in the process of building out a 16,000 mile IP-centric fiber network across North America. NextLink is one of the newer entrants to broadband wireless, but the combination of its wireless and fiber infrastructure will help to quickly establish its presence as a voice and data service provider.

Wireless IP Telephony

Wireless IP telephony is mobile communications over a data-centric infrastructure that provides a lower cost network with increased capacity over the existing network infrastructure, and a new range of applications for the end user.

There's been a lot of activity in the wireless area, and much of it is geared toward increasing the data rates. With these increased rates, the wireless networks will support services other than just voice, services like fax, data and even video. As that trend continues, you'll see more and more applications of wireless that will increase the amount of data traffic over that wireless network. At some point, the data traffic will surpass the voice traffic, then the network become more efficient by transforming into a packet based infrastructure from the old circuit based infrastructure. The same drivers that are moving the wireline world towards IP telephony will move wireless there too.

What's even more interesting (and better) is that the new standards that are being developed to increase those data rates on the wireless network are packet based, making it even more natural to support a packet based infrastructure for voice and everything else, too.

Both network economics and arbitrage drive the wireless industry towards adopting IP telephony into the network. The wireless network is simply an Edge Access Technology, with a device on the edge – a communicator or a smart docking station – and all the wireless devices are connected through the ubiquitous IP network that all other IP communication devices are connected to, including landline phones and PCs.

IP and wireless systems have been emerging together, so it makes sense that "convergence" will be completed there quite rapidly. One can see it in the headlines. Motorola, a large manufacturer of wireless infrastructure equipment, makes a strategic investment in Netspeak, one of the leading gateway providers in the IP telephony space. Then there's Nortel, which made a very large acquisition of Bay Networks. Ericsson has demonstrated a GSM-based office system ("GSM on the Net") where you've got an in-building based GSM system using an IP network to connect the different elements of the wireless network together.

Nokia has demonstrated similar systems and Hughes Network Systems is also working on IP-based access to wireless data networks.

Wireless IP networks will have to provide better solutions to existing markets, whether they be cellular, wireless PBX and building wireless or wireless local loop. They must leverage the direct connection with IP telephony networks, which can be used to provide access to voice over IP equipment and networks, or to reduce back haul costs, and it could be to provide new services. The other objective of wireless IP networks is to reduce the infrastructure costs by removing proprietary equipment (though some wireless based IP gateways may be based on proprietary technology), thus making it easier for application developers to add new features and new services.

WorkFlow Management and Automation

Also called Workforce management and call center management. Software for call center management helps interpret your call center's past statistics to improve future service levels. It tells you how much staff you need and what will happen when you make changes to agent schedules.

Call center management software takes historical data from your ACD and uses it to your advantage. This data reveals important information, including how many people are holding, the longest hold time, where agents are assigned and how many of them are idle. Most of this data is shown in real-time, in a graphical, colored format that's easy to read.

If your call volume has always doubled on Tuesday, (perhaps after a major mailing promotion) it will probably double next Tuesday, too. Knowing this pattern, you'll be armed and ready with agents.

A good call center management package also helps you with scheduling. Scheduling involves making a timetable of agent hours and shifts, taking into account vacation days, breaks, training time, lengths of shifts and forecasting information. Most call center managers know how time-consuming that can be when scheduling manually or using a spreadsheet program. Management software automates the process. It can even take into account specific agent preferences for hours or days they wish to work.

Here are some more useful advantages of using call center management software:

- Detailed reports. Good call center management systems track historical info for up to a couple of years. Instead of searching back to see statistical info, your software should give you detailed reports. Reports should contain info like the number of calls, times they came in, the call length and type and log-in/log-out times. You can schedule the software to print out reports on demand or at specified times. You should also be able to compare actual performance and call conditions with your goals. Many packages can make these comparisons in a choice of formats, such as bar, pie or 3D graphs.

- Answers service level questions. At the end of the work day, you can look over reports to find out why you aren't meeting service levels. Someone may have taken a break early or come in late to work. Your department may have had more abandons or a bigger call load than usual. With individual agent stats, you'll see when and for how long each agent is logged out.

- Reduced costs. With scheduling and forecasting capabilities, you won't pay overtime for people you don't need. You'll be able to justify hiring more people to handle calls when management sees reports showing the number of agents needed to keep up with service levels.

- Multi-center coordination. If you have several call centers, look for software that lets you see real-time stats in both your center and the others. If you're managing one center, you and all other managers should be able to get the same info from one database. One software package should be enough to let each user log on through a LAN.

- Call center simulations. When making projections about call load and

W

staffing, you assume a certain level of service. Changes to the projections lead to changes in service level. Look for software that lets you play with different situations.

Remember, you want to build customer loyalty, no matter what your business is, so you are trying to give the customer the very best possible service. In terms of objectives of workforce management, one should put as the #1 objective, customer relationship management. So, for certain types of customers, you may want to have certain types of service, and may want that service to be delivered by a certain category or type of agent. To be able to do that effectively, you have to have those people in place at the right time, and all of their skills are not equal and, as we all know, all customers are not equal.

So workforce management helps to support true customer relationship management, if it's done correctly. It can also improve productivity, because if you have your skilled people in the right place at the right time, you're going to find that you're able to answer the more complex calls with more talented people who can do it more quickly, and leverage those lower volume times with things like training, soft skills and meetings, maybe non-call transactions which are less time pressing.

Hand in hand with improving productivity is decreasing cost. Improving productivity almost automatically means that you've decreased costs because you can now more efficiently take a larger number of calls with the same number of people.

xDSL

Also called "DSL." Generic acronym coined for the many flavors of digital subscriber line access services provided by telcos (ILECs, CLECs, etc.) to their customers. The "x" stands for any one of a number of implementations: A = Asymmetric, H = High-bit-rate, R = Rate Adaptive, S = Symmetric or Single-line high-bit-rate, V = Very-high-bit-rate, etc. We can thank Bellcore (now Telcordia) for nearly all of the xDSL acronyms.

Let's start by getting two pieces of trivia out of the way: First, "xDSL" refers to the type of digital modem, not a line *per se*. Homes and companies already own ordinary copper pair lines. Put a pair of ADSL modems at either end of them, and you've got an asymmetrical digital subscriber line. Put SDSL modems on the same line and you've got SDSL service instead of ADSL. Same line, different modems. So, "DSL" is a modem, not a line, and a "line" needs two modems for there to be a "service."

Trivia item number two: Originally, way back in the mists of time, the term "DSL" itself meant the modem used for Basic Rate ISDN, which was considered as futuristic in its day as ADSL and its cousins are now.

In any case xDSL can carry duplex voice and data simultaneously, as well as the signaling data used for call information and customer data. xDSL routing equipment uses digital coding techniques to more efficiently transfer data over traditional copper phone lines. Most xDSL services are "data only" circuits. xDSL does not traverse the PSTN, with the exception of the "last mile" to your home or business. At the telco's central office (CO) the xDSL signals are redirected to a high speed data backbone network just before reaching the voice switch and the switched network at the telco's central office (CO). This frees the digital service from the bandwidth limitations inherent in old-fashioned voice networks that were never engineered for high-speed data service, while at the same time allowing the user to continue using his or her analog phone equipment – there are, however, Voice over DSL (VoDSL) technologies now being deployed to replace conventional phone service too (See Voice-over-DSL).

xDSL technologies arose because of continually increasing Internet "on ramp" congestion. Whereas the public inter-Central Office (inter-CO) network consists of a vast overprovisioning of extremely high capacity ATM and optical fiber ring technology (SONET or SDH) connecting COs that consist of high capacity Digital Access and Cross-Connect (DACS) systems and T-1 / E-1 carrier transmission equipment, in most cases the "last mile" to consumers - actually a 12,000 to 18,000 foot segment from the CO's Main Distribution Frame (MDF) to the network termination equipment in your home or office – still consists of copper loops of unshielded twisted-pair telephone wires. This "last mile" is also known as the local drop, local loop, or Access Network.

Industry estimates of replacing copper wiring with optical fiber are around $1,500 for an average customer, or about $900 billion for 600 million users worldwide.

Given that the world is stuck with huge amounts of copper wiring over the "last mile" of the network, the sensible approach is to leverage the existing copper infrastructure by inventing a new high-speed transmission method that would work over the existing copper network without any need for rewiring

Attempts to squeeze more and more information through the legacy copper telecom infrastructure ultimately run up against the laws and theorems of physics and information.

Shannon's Theorem tells us that without any background noise or other impediments, even the paltry frequency band reserved for analog

phone calls is theoretically capable of carrying an infinite amount of information. Such "band limited channels" as an analog line are restricted in that the signal must lie within a certain band of frequencies W cycles per second wide (the bandwidth). The Nyquist sampling theorem states that a signal of this type can be specified by giving its values at a series of equally spaced sampling points 1/2 W apart. Once can therefore say that such a function has two W degrees of freedom, or dimensions, per second. In the case of an analog line we wish to capture a 4 kHz voice band, which means we must sample at twice the highest frequency, which means we must sample at the rate of 8,000 samples per second.

Each sample gets digitized and exists at one of a certain number of possible discrete amplitude levels. If we could dispose of the Heisenberg Uncertainty Principle and establish perfectly noiseless conditions on the channel, it would be possible to distinguish an infinite number of such discrete amplitude levels for each sample. Consequently, in principle, an infinite number of binary digits per second could be transmitted, and the capacity C (transmission speed) of the channel would be infinite. Interestingly, even if we introduce the inevitable line noise, as long as there are no limitations placed on the transmitter power P the channel's capacity can still be made infinite, since at each sample point an unlimited number of different amplitude levels could still be distinguished. Only when noise is present *and* the transmitter power is limited in some way does the capcity C become finite. This capacity depends on the statistical structure of the noise as well as the nature of the power limitation.

The simplest type of noise is resistance noise, produced in an electrical resistor by thermal effects. This type of noise is completely specified by giving its average power N.

The simplest limitation on transmitter power is the assumption that the average power N that's delivered by the transmitter is not yet greater than P. If a channel is defined by these three parameters W, P and N, the capacity C can be shown to be.

$$C = W \log_2 ((P + N) / N) \text{ bits per second}$$

As Claude Shannon himself wrote: "The implication of this formula is that it is possible, by properly choosing the signal functions, to transmit binary digits per second and to recover them at the receiving point with as small a frequency of errors as desired. It is not possible to transmit binary digits at any higher rate with an arbitrary small frequency of errors. Encoding systems in current use, pulse-code modulation and pulse-position modulation, use about four times the power predicted by the ideal formula. Unfortunately, as one attempts to approach more closely this ideal, the transmitter and receiver required become more complicated and the delays increase."

When we introduce Gaussian noise, the theoretical information carrying capacity C of a voice-band access channel can still be calculated using the Shannon theorem:

$$C = W \log_2 (1 + S/N) \text{ bits per second}$$

where W is the bandwidth (about 3,000 to 4,000 Hz for an analog line) and S/N is the signal-to-noise ratio (SNR) at the receiver, which would be about 1,000, leading to a maximum rate of between 30,000 and 40,000 bps that can be transmitted without error over a conventional analog copper line.

Plugging any realistic SNR into the equation immediately reveals that the conventional analog network of copper wires cannot handle the kind of high bandwidth one would need to achieve such things as video-conferencing or video-on-demand. Normally, one could also increase the

X

frequency of transmission to increase the amount of information that can be sent, but attenuation factors and other sobering items derived from the Maxwell / Heaviside electromagnetic field equations tell us that we simply cannot send arbitrarily high frequency transmissions through a wire – the wire will act as an antenna and radiate the transmitted energy away into space long before the message reaches its destination.

Signal loss, or attenuation, is a function of frequency. Increase the frequency, and the distance a signal travels traveled decreases by the square root of the frequency. As an example, a 40 MHz signal travels half as far as a 10 MHz signal.

Fortunately, in 1989 Bellcore (now Telcoria Technologies) thought up a radical new technology, ADSL, to use enhance the information carrying capacity of ordinary twisted-pair copper networks to a degree previously thought impossible.

ADSL's roots actually go back prior to 1989, when the telecom industry was wrestling with the problem of attenuation and devising new modulation schemes (also called line-coding techniques) to handle the copious data streams promised by the up-and-coming technologies of ISDN, T-1 and E-1. All three of these digital services, like Plain Old Telephone Service (POTS), manage to operate over conventional twisted copper wire pairs.

Various incompatible modulation schemes have appeared, but they generally fall into two classes: Single-carrier techniques and multiple-carrier techniques.

As their name implies, single carriers have a single channel that occupies all their bandwidth, and can encode information into a waveform by manipulating one or more of its three primary characteristics: Amplitude, frequency, and phase. High frequency Continuous Wave (CW) carriers can thus be modulated with Amplitude-Shift Keying (ASK), Frequency-Shift Keying (FSK), and Phase-Shift Keying (PSK). Digital services are all based upon combinations of amplitude and / or phase modulation, but not frequency modulation (FM).

"Spectral efficiency" is a measure of the number of digital bits that any given modulation technique can encode into a single cycle of a wave form. The duration of a single cycle of a wave form is referred to as the "symbol time."

T-1 and E-1 employ very simple modulation techniques devised in the 1960s that have a low spectral efficiency. T-1 uses Alternate Mark Inversion (AMI) and E-1 uses High Density Bipolar 3 (HDB3). T-1 AMI coding transmits "ones" and "zeroes" (the bits) via corresponding "peaks" and "valleys" of analog waveforms over the copper wire. Therefore, AMI coding supports 1 bit per waveform cycle (1 bit per baud) and so the 1,544,000 bps of payload and framing information in a T-1 transmission requires a copper wire that can handle a frequency of 1,544,000 Hertz or 1.544 MHz. 22-gauge wire can handle this only if repeaters are installed at 6,000 foot intervals to boost the signal (See T-1).

Simple modulation schemes have a one-to-one correspondence between the bit rate (the number of bits transmitted per second) and the baud rate (the number of signaling elements or changed states or events that can occur per Hertz per second). If only two states can exist, then the baud rate is equal to the bit rate.

To build longer loops, experiments were conducted in sending lower frequencies (reducing attenuation) and using more advanced modulation techniques to load more than one bit of information onto each signaling element (Hertz), improving the spectral efficiency and thus making up for the loss of information carrying capacity normally encountered when transmitting at lower frequencies. In such cases the bit rate is equal to the baud rate times the number of bits used to represent a line state. If

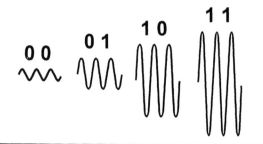

When Baud is less than Bit Rate

When a signaling element corresponds to more than one bit ("dibits" shown)

there happens to be 16 line states available and each state encodes four bits, then the bit rate is therefore four times the baud.

The appearance of Basic Rate Interface ISDN (ISDN BRI) surprised some in the telecom industry when it was announced that it could send 2 bits of information per cycle via a remarkably simple amplitude modulation form of baseband line encoding called 2 Binary, 1 Quaternary (2B1Q), which was defined in the 1988 ANSI specification T1.601. ISDN BRI's two 64Kbps bearer channels, 16 Kbps delta channel, and other overhead (all adding up to about 160 Kbps) would normally need a 160,000 Hz bandwidth for transmission, the uppermost frequencies being too high to successfully traverse an 18,000 foot long 26 gauge wire. Thanks to 2B1Q modulation, however, ISDN BRI could use half the frequency range, or 80,000 Hz, which does the trick.

2B1Q is a "multi-level line code" or "multiple modulation scheme," because it uses four levels of amplitude (voltage) to encode 2 bits. Each of the four voltage levels translates to 2 bits per Hertz (cycle), ranging over the bit set 00, 01, 10, and 11. Each voltage level is called a quaternary. There are two positive and two negative voltages, so the receiver must be able to detect both the signal's amplitude and polarity, but not the phase.

One could argue that ISDN is actually a form of xDSL, the only difference being that the "I" in ISDN means that both voice and data are digitally integrated as they are transmitted through the phone network,

Quaternary Multi-level Line Code

Bit Rate = 10   **Baud Rate = 5**

1 0   0 0   1 0   0 0   1 0

while true xDSL transmits voice and data on qualitatively and logically separate networks - analog and digital.

Encouraged by the successful application of 2B1Q to ISDN signalling, Bellcore engineers in the late 1980s thought that 2B1Q's spectral and distance characteristics would be retained even if pushed to higher speeds. They decided to replace the AMI encoding used by T-1 (1.544 Mbps) and the High Density Bipolar 3 encoding (HDB3) for E-1 (2.048 Mbps) with 2B1Q so as to provision T-1 and E-1 services without the need for repeaters and thus simplify the overall deployment of high-bandwidth networks.

The approach taken with T-1 was to place a special encoding card at the central office and another similar card at the customer premises, then split the service into four wires (two pairs), each pair running at 784,000 bits per second. This lower frequency allowed for loops up to 12,000 feet long on 24 gauge wire, or 9,000 feet on 26 gauge wire. This arrangement became known as High Speed (or High bit-rate) DSL, or HDSL. HDSL is the oldest and most heavily deployed version of xDSL. It's most recent incarnation is HDSL2, which delivers full T-1 speeds on just one copper pair and which has better interoperability that conventional HDSL technology.

When 2B1Q encoding was originally applied to the greater bandwidth of E-1 (2.048 Mbps) to create an E-1 HDSL, the service was split across six wires (three twisted pairs). Eventually the chipset technology improved to the point where E-1 HDSL could be implemented as a four wire (two pair) arrangement similar to T-1, but with each pair carrying 1.168 Mbps. Today, HDSL commonly uses two pairs to deliver symmetric T-1 or E-1 service by sending half the data on one pair, and half on the other, both operating as full-duplex echo-cancelled links, with 768 Kbps for T-1 and either 1,168 or 1,024 Kbps for E-1, depending upon the need to send any extra bits (or "quats" which are two bits long) for compatibility with SDH signal formats.

HDSL is not generally considered to be a residential or consumer access line technology, unlike ADSL. Telcos first used HDSL to replace T-1s and E-1s only on "problem" lines, but as production costs were reduced, HDSL became the preferred method of carrying T-1 and E-1 services within the telco's copper networks (and some large private campuses) and delivering these services to commercial customers for such applications as: Connecting PBXs to telco offices, connecting wireless basestations into the PSTN, and for "pairgain" applications where many voice channels must be crammed into one line.

Today, most HDSL lines are still used for T-1 / E-1 provisioning - if your business recently installed a "T-1" line running on two pairs of copper wires, it's most likely HDSL or HDSL2. Actually, one should more correctly refer to the service as DS1 – Digital Signal, Level 1 – instead of "T-1" which is associated with the older physical transport technology.

Advocates of HDSL have pointed out that HDSL has been around for so many years that the Digital Signal Processor (DSP) chips used to process and analyze the signals in HDSL equipment have gone through several iterations, and the equipment is now relatively inexpensive. Such HDSL aficionados have always hoped that it would ultimately become a common residential as well as business high-speed data access service. Unfortunately, HDSL needs two sets of twisted pair copper wires and does not allow simultaneous analog POTS and digital data transmission on those wires, unlike ADSL, as we shall see. HDSL can only carry analog voice if 64 Kbps of digital bandwidth is reserved for it.

The 2B1Q line coding scheme is relatively robust though it is somewhat susceptible to impulse noise. The 2B1Q technique of using four voltage levels 2B1Q to squeeze 2 bits of information per baud (or "symbol time") reached its technological limit with HDSL. To transmit even more bits per second, one must use more complicated modulation techniques

requiring more voltage levels to yield a higher spectral efficiency. Conventional amplitude modulation techniques have limited spectral efficiency making high bit rate transmissions extremely difficult or impossible. To encode k bits in the same symbol time you need 2 to the k power voltage levels. As the number of voltage levels is increased in an effort to achieve higher and higher data rates, it becomes increasingly difficult for the receiver to accurately discriminate among the many voltage levels. Still, even with current technology, if the signal-to-noise ratio (SNR) of the channel is 10 dB or more, up 10 bits can be transmitted for each Hertz or cycle of channel bandwidth.

A modulation technique more spectrally efficient than 2B1Q is Quadrature Amplitude Modulation (QAM), which simultaneously modulates both the phase and amplitude of a carrier signal. A signal with arbitrary phase and amplitude can be represented by a linear combination of a cosine and a sine wave. Since these two are 90 degrees out of phase, they are said to be in quadrature.

QAM has been used by modems since the 1970s. Whereas ASK, FSK and PSK send only one of two possible signals per symbol time, multi-level QAM coding greatly increases the information capacity of a band-

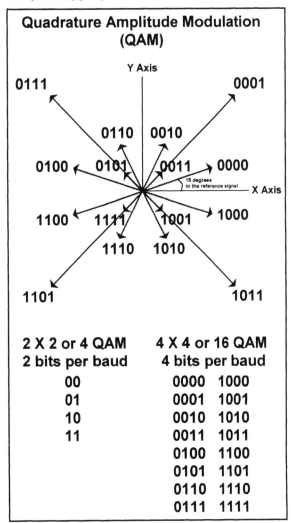

Quadrature Amplitude Modulation (QAM)

2 X 2 or 4 QAM 2 bits per baud	4 X 4 or 16 QAM 4 bits per baud	
00	0000	1000
01	0001	1001
10	0010	1010
11	0011	1011
	0100	1100
	0101	1101
	0110	1110
	0111	1111

X

width-limited channel by encoding large sets of bits within a single symbol time.

With QAM, a single carrier's phase is shifted by 0, 90, 180, or 270 degrees from a reference carrier. It's even possible to modulate the signal at 22.5 or 45 degrees to that of the reference carrier, but receivers have more difficulty discriminating among such small phase angles.

Various types of QAM are designated either in the format QAM n or n QAM, where n is an integer indicating the number of states per Hertz. The number of bits per symbol time is k, where 2 to the power of $k = n$. If, for example, 4 bits per Hertz (bHz) are encoded, the result is QAM 16 (or 16 QAM); 6 bHz produces QAM 64; 8 bHz yields QAM 256. These numbers designating the possible number of states can be represented in a two-dimensional in-phase (I) and quadrature (Q) plot, also referred to as a constellation plot. A combination of phase and amplitude establishes a constellation point for each symbol on the x-y plane. The phase is plotted like a compass, with a ray pointing from the center of the axes. The amplitude is the length of the ray. At every baud the phase and amplitude are measured and matched with a constellation point called a decision point or symbol.

The QAM illustration shows a constellation plot of a QAM 16 signal. The modulation is made up of the amplitude changes of I and/or Q in four discrete positive and negative steps. Each combination of I and Q levels is a decision point. With 12 possible phase shifts and two possible amplitudes, we find in the case of 16 QAM that each decision point can represent one of 16 possible states or 4-bit patterns per line condition (which means we can do "four bits per baud" signaling). Increasing modulation requires more decision points. A QAM 64 signal has eight I and Q amplitude levels, and QAM 256 has 16 levels.

The number of available levels of amplitude and the number of phase angles is a function of line quality.

The earliest QAM systems were used for microwave telecommunications (Indeed, QAM modulation is still used in digital microwave radios to achieve data transmission rates of up to 155 Mbps over relatively narrow bandwidths). QAM then appeared in Synchronous Digital Hierarchy (SDH) networks. Ordinary analog modems ranging in speed up to 56 Kbps and digital radio (135 Mbps) use QAM as their modulation technique, as does the Digital Video Broadcast (DVB) systems in Europe. QAM is also being used extensively in digital cable modems that use cable TV (CATV) grade coaxial cable or standard twisted-pair copper phone lines. QAM 64 and QAM 256 modulation has been specified for CATV digital video by the Society of Cable Television Engineers (SCTE)

Another, special limiting case of QAM modulation, Quadrature Phase Shift Keying (QPSK) is equivalent to QAM 4 and is often used with direct broadcast satellite transmissions, owing to its simple, robust nature. Whereas CATV may use QAM at a maximum bandwidth of 8 MHz, satellites use QPSK modulation at bandwidths up to 36 MHz.

QPSK uses only half the bandwidth of PSK for the same bit rate, but QPSK does not have twice the capacity of PSK. PSK modulation is no where near optimum in terms of bandwidth use, and so QPSK's greater bit rate has nothing to do with an increase in capacity it's just better at using bandwidth.

Another variation of amplitude and phase modulation is Carrierless Amplitude Phase modulation (CAP) developed by AT&T Paradyne. In the case of phase modulation a reference carrier is shifted a certain number of degrees to encode information. A CAP receiver is able to derive the carrier, whereas a QAM receiver must have the carrier sent from the transmitter. CAP and QAM are so similar that some CAP receivers can receive QAM transmissions.

CAP is used as a modulation scheme for ADSL.

Whereas 2B1Q, QAM, QPSK and CAP are examples of single-carrier modulation techniques, multicarrier modulation techniques such as Orthogonal Frequency Division Multiplexing (OFDM) and Discrete Multitone (DMT) are used by more advanced forms of high bandwidth service such as ADSL. Multicarriers are essentially forms of Frequency Division Multiplexing (FDM) that divide the bandwidth up into multiple, parallel subbands or channels. Each subband is encoded using one of the single-carrier modulation techniques, with all subbands ultimately bonded back together at the receiver.

Of these, Discrete Multi-Tone modulation (DMT) competed with CAP to become the standard xDSL modulation scheme. Amati Communications (now part of Texas Instruments) developed DMT, which is considerably more complicated than CAP.

In the case of ADSL, the basic difference between these two line coding methods is the way in which the optimum speed between the CO and the receiver over the single copper pair can be determined. CAP treats the entire usable frequency spectrum as a single channel and optimizes the entire data rate while DMT divides it into 256 subchannels (these should actually be called subcarriers, but subchannels now seems to be the accepted term), each having a bandwidth of 4.3125 kHz and the frequency difference between two successive channels is also 4.3125 kHz (yielding a total bandwidth of 1.104 MHz on the loop). During transmission, each 4 kHz subchannel carries only a small portion of the total data rate. Each sub-channel is optimized separately, its signal-to-noise ratio constantly monitored by the system to determine how many bits per second can be carried successfully. Any deviation in measured line conditions causes the channel to be dynamically adjusted accordingly. Indeed, if any frequency ranges in the spectrum of subchannels are found to be too noisy, the corresponding channels are deactivated entirely until line conditions improve.

All of the channels of DMT-modulated full-rate ADSL can be grouped into three segments:
- The 0 to 4 kHz range, which is used for plain old telephone service (POTS),
- the 26 kHz to 138 kHz range, used to transmit data upstream, and
- the 138 kHz to 1.1 MHz range, used to transmit data downstream.

Some DMT channels have special functions (for example, channel #65 at 276 kHz is reserved for pilot signals) and the bottom six channels (25.875 kHz) are generally not used at all, to allow room for the 4 kHz passband analog voice and a so-called guardband that separates the analog voice area from the DMT signals. Just above the analog voice area are the 32 upstream channels, followed by 250 downstream channels (of which only about 218 are typically available for user data).

CAP and DMT became fierce competitors and received all of the attention from standards bodies because they used a lower range of the frequency spectrum than earlier alternative line coding techniques and thus enjoyed reduced signal attenuation and a reasonable distance covered (the loop "reach"). However, a newer third modulation scheme known as Multiple Virtual Line (MVL), is also a possibility.

CAP appeared first and underwent longer testing than DMT, but DMT has been accepted as the standard by ETSI and ANSI (known as T1.413), since it appears to be better than CAP in terms of communications speed, bandwidth efficiency, spectral compatibility, performance, robustness, and power consumption (there are CAP enthusiasts reading this who will vehemently disagree with Yours Truly, of course!)

Ultimately, however, the issue of whether CAP or DMT is better is less important than things such as the choice of a forward error correction algorithm, how well the analog front end is protected, and other pro- prietary and semi-proprietary factors that distinguish the performance of one DSL transceiver from another.

Even with these developments, there are, however, many impedi- ments to enhancing the capacity of the copper networks with ADSL and its DSL brethren. Prior to deployment, the carrier must assess the local loop's condition to determine its suitability for DSL. In particular, there is the problem of the inhomogeneous nature of the copper infrastructure, and the effects of interference and noise on ADSL transmissions.

One reason that various forms of interference and noise foil attempts at high bandwidth transmissions over copper wire has to do with the fact that telcos generally have built high speed access systems using constant bit-rate (CBR) technology. CBR technology is predicated on providing a steady transmission on a specific frequency and, unlike the burst-mode or "bursty" packet technology as can be found in LANs, CBR technology does not automatically adapt and adjust its frequency characteristics if interference and incompatibility are encountered.

For CBR-based DSL technology to function efficiently, the transmis- sions should occur over nice, uniform, contiguous copper wire loops, the thicker the wire gauge the better, since the most important parameter of any conductor is its cross section area which determines DC-resistance and most of the conductor's current capacity. Troublesome line attenua- tion increases with line length and frequency and decreases as the wire diameter increases. Non-ferrous conductors are rated in AWG (American Wire Gauge, formerly known as B&S or Brown and Sharpe Wire Gauge). The higher the gauge number, the smaller the diameter and the thinner the wire. For xDSL and T-1 transmissions, wires are usually 24 AWG (0.0201 in. or 0.511 mm in diameter) but can sometimes get by with 26 AWG (0.0159 in. or 0.404 mm).

Other factors impacting upon xDSL deployment are the so-called Carrier Serving Area (CSA) rules that define the distribution of twisted wire pairs from Digital Loop Carrier (DLC) systems. The radius covered by CSA rules are up to 9,000 feet of 26-gauge wire and up to 12,000 feet of 24-gauge wire – what the standards body ANSI defines as a "full CSA". The CSA rules are defined as follows:

- There should be only nonloaded cable.
- Multigauge cable is restricted to two gauges.
- Total bridged tap length may not exceed 2,500 feet, and no single bridged tap may exceed 2.0 kft.
- The amount of 26 AWG cable may not exceed a total length of 9 kft, including bridged taps.
- For single gauge or multigauge cables containing only 19, 22, or 24 AWG cable, the total cable length may not exceed 12,000 feet, includ- ing bridged taps.
- The total cable length including bridged taps of a multigauge cable that contains 26-gauge wire may not exceed the results of the follow- ing equation:

12 - ((3 * total length of 26 gauge wire) / (9 - the total length of bridged taps in the cable))

The above CSA guidelines don't include any wiring in the CO nor any drop wiring and any wiring in the customer premises.

In spite of such neat rules, the copper infrastructure is unfortunate- ly something of a hodge-podge. Wires of one gauge often connect to a wire of a different gauge. It's common practice to run out from the CO around 10,000 feet of 26 gauge cabling, then change to thicker 24 AWG or even 19 AWG (0.0359 in. or 0.912 mm) for longer loops. Also, it's been estimated that the average U.S. local loop has 22 splices.

Moreover, there are often to be found in the local loop so-called bridged taps, which are unterminated wires dangling off a local loop at an intermediate point. These are sometimes created when carriers provi- sion service to a new subscriber by splitting off (or "tapping off") one of the unshielded twisted pairs in a binder cable without disconnecting the loop's unused section, or when the carrier extends a circuit beyond the subscriber's location for future connections to terminating equipment, or when consumers or field technicians attach or remove devices without completely disconnecting wires to the old device. Even the extra phone wiring in your house is essentially a combination of short bridged taps. This is why a Plain Old Telephone Service (POTS) splitter is used to iso- late the house wiring and assure a direct path for the xDSL signal to pass unimpaired to the xDSL receiver.

Bridged taps can be as long as 5,000 feet, and can cause problems even for the conventional voice passband systems, such as leakage (which causes a loss of signal strength) and signals can echo back off the unter- minated end into the main network, garbling transmissions.

After many years of service and modifications, more and more undocumented bridged taps and wire gauge changes accumulate on the local loop, causing problems for ADSL deployment. While ADSL can tol- erate some bridged taps, their existence on the line does reduce the cir- cuit's overall performance. Also, impedance-derived echoes will appear in transmissions as they travel through a series of wires of varying gauges.

ADSL also cannot work with the standard "loading" coils (or induc- tors) that are often placed at 6,000 foot intervals on very long local loops to improve voice quality by "flattening" the transmission's frequency response much in the manner of an equalizer in a consumer stereo sys- tem. Line capacity is decreased and the distance of signal transmission is increased when such inductors are equally spaced over a local-loop cir- cuit of a phone system. Also, by limiting power at high frequencies out- side the voice channel and moving the energy to lower frequencies, load- ing coils essentially act as filters, blocking high frequencies and preclud- ing the wire from transmitting digital services such as ISDN, ADSL, or a leased-line T-1. About 20% of the local loops in the U.S. have these coils, but they are slowly being removed from the access network.

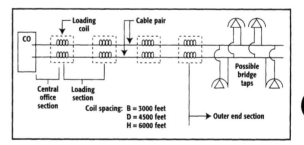

Also ADSL and its xDSL brethren (with the exception of ISDN DSL, see below) won't travel through remote terminals such as digital loop carriers.

Other transmission problems that can prohibit ADSL deployment centers on the fact that copper wires tend to be bundled together, which results in electromagnetic effects. The copper "local loop" consists not just of unshielded twisted pair wires leading to each phone. Groups of 25 or 50 such wire pairs are collected into cables called "binder groups," the name having been derived from the coil of cotton ribbon that binds the unshield- ed twisted pairs together. These binders are usually color-coded to help technicians, since some cable collections can hold over 1,000 local loops.

CBR-based high-speed transmissions through binder groups can become impeded because they are highly susceptible to crosstalk (interference entering a communications channel caused by transmitted energy leaking from one wire pair into another). Alexander Bell invented twisted pair wiring in an effort to minimize the electromagnetic radiation or capacitive coupling that causes the interference of signals from one cable to another, but signals do couple, and couple more so as frequencies and the length of line increase. Moreover, sending signals with symmetric bandwidths in many pairs within a cable significantly limits the data rate and length of line you can attain, which spurred the development of asymmetrical transmission techniques.

The various access channel and noise sources include the following:

- Near-End Crosstalk (NEXT)– the most troublesome form of crosstalk occurs at the Central Office (CO) end of the line. The large cable bundles converge and "bunch up" at the CO, which makes the transceivers at the CO end heavily susceptible to crosstalk. To be more specific, NEXT is interference that appears on another copper pair at the same end of the cable as the source of the interference, which usually turns out to be the transmitters of certain loops that can easy interfere with the input of the co-located transceivers of another loop – the transmitter signal leaks into the adjacent receiver via capacitive and inductive coupling paths. Thus, NEXT affects technologies that transmit in both directions simultaneously such as echo-canceling systems, where the upstream and the downstream signals are sent across the copper pair at the same frequencies, i.e. they overlap. NEXT limits the range of ISDN and HDSL in particular. NEXT can usually be eliminated by not transmitting in both directions in the same band at the same time, separating the two directions of transmission either into non-overlapping intervals in time or into non-overlapping frequency bands. (e.g. Frequency Division Multiplexing (FDM) Systems). Also, carriers must generally avoid interference between lines by ensuring that their various DSL services don't travel in the same copper line bundles.

This is also why many forms of xDSL are asymmetrical – at the user's end there are fewer problems with NEXT so bandwidth can be made greater from the CO to the user. High frequencies unfortunately lead to higher attenuation, so the upstream speed in ADSL must be made much less than the downstream rate, since such lower frequencies arriving at the CO reduce the attenuation at the exactly the spot where all the cable bundle endpoints are in close proximity to each other, and where each wire is most susceptible to all the other signals in the binder group due to electromagnetic induction (cross-coupling of signals). In the downstream direction, however, although the high frequencies still attenuate, they have a better chance of avoiding crosstalk at the customer end since most subscribers don't have large bundles of cables running into their premises.

Admittedly, the utopian dream of xDSL deployment for the telcos has always centered on the quest for perfect "spectral compatibility," which is the ability for any xDSL service to operate harmoniously with other services in the same bundle without NEXT or similar interactions. The term spectral compatibility, however, describes an impossible situation, since you can't really put two xDSL services in a binder without some kind of interaction occurring. Thus the real concern is more with the damage control of minimizing and managing inevitable interactions, and so the term "spectral compatibility" has been somewhat usurped by the term "spectral management."

- Far-End Crosstalk (FEXT) – arises when CO signal transmitters on other copper pairs in the same cable leak into the input of the wrong transceiver at the subscriber end. FEXT is usually less of a problem than NEXT since the far end interfering signal is attenuated as it traverses the loop.
- Resistance and the "skin effect" – as signals transmitted through metal-

lic wires at very high frequencies leads to a phenomenon called the skin effect, where electricity migrates to the wire's outer surface. Since the wire's core is no longer conducing electricity, in extreme cases it's possible to replace the core with inexpensive nonconductive material such as plastic or wood. Unfortunately, since the electricity is now confined to a smaller volume of the wire (just the "skin"), resistance increases, which weakens the signal. The skin effect prevents the deployment of services using frequencies above 1 GHz over copper wires.

- Radio Frequency Interference (RFI) - a wire not only can act as a transmitting antenna, but as a receiving one as well. See Narrowband interference, below.
- Phase Error - a signal's velocity through a wire is also a function of frequency. The higher the frequency, the slower the signal. Variations in speed can cause a phase error, which wreaks havoc on modulation techniques dependent on phase (in linear circuits the output sine wave might rise to a peak later than the input sine wave), resulting in a "phase shift" or "signal delay" that causes bit errors. The combined effects of delay and attenuation are generally lumped under the term of "signal distortion."
- Impulse noise (burst noise) - are short electrical transients caused by any number of possible sources. They occur when one wire picks up a spurious signal that lasts for only a few microseconds, but which in that time period can cause interference over a wide range of frequencies. Sources of impulse noise include other wires, electronic devices, and wire imperfections.
- Narrowband interference - is in a sense the opposite of impulse noise, as it affects a small number of frequencies over a long period of time. Narrowband interference includes AM, FM and amateur radio transmissions. Over the years the reader has probably used devices (such as headsets) incorporating wires having loose fittings, outer damage or improper isolation from the exterior environment, whereupon one will start hearing an AM radio station. This type of interference can be avoided simply by identifying the noise's frequency range and refraining from transmitting over the wire on that frequency. This method is used by the current modulation standard of ADSL.

DSLs Galore

Why are there so many different kinds of xDSL Technologies?

Different applications demand different data rates, and there's an xDSL technology to complement whatever applications "hat" your business wears. For example, HDSL and SDSL deliver services at T-1/E-1 speeds. RADSL can dynamically adapt its speed to line conditions. ADSL can achieve speeds between 1.5Mbps to 8 Mbps downstream. IDSL delivers services at 128 Kbps and MSDSL can support eight distinct rates. xDSL's chameleon-like nature becomes evident when one realizes that, since xDSL in itself is not a service so much as it is an enabler of high-speed services and applications, the list of existing industry transmission standards, formats and bit rates it can accommodate includes POTS, T-1 / E-1, Frame Relay and IP. Anyone and anything can be served, from an individual surfing the Internet in the pricacy of their home to performing T-1 replacement between COs using HDSL2.

Here are the xDSL contenders:

- ADSL - The Asymmetric Digital Subscriber Line is the most popular of all the xDSLs, taking its name from the comparatively high bandwidth coming down from the CO (up to 8 Mbps), with low bandwidth going from the user back to the CO (up to 1.5 Mbps). ADSL uses a single copper pair for transmission specifying loops up to 18,000 ft. at a wire thickness of 0.5mm (24 gauge). (See Asymmetric Digital Subscriber Line).

ADSL has a range of downstream speeds depending on distance from the DSLAM or switch:

Up to 18,000 feet – 1.544 Mbps (T-1)
16,000 feet – 2.048 Mbps (E-1)
12,000 feet – 6.312 Mbps (DS2)
9,000 feet – 8.448 Mbps

Most experts feel that 8 Mbps is sufficient bandwidth to support real-time broadcast services and pre-recorded interactive video services; and to have multiple video and data activities underway simultaneously. Indeed, the originally intended application for ADSL was video-on-demand (VOD) using between 6 Mbps and 8 Mbps downstream. VOD turned out to be more difficult to engineer than was previously thought, and the market demand for it was not clear.

Today, ADSL is seen more as an enabler for high-speed Internet and other multimedia service delivery. The deployment of ADSL will be driven by applications with such asymmetric traffic demands such as: Web Surfing, file downloads, telecommuting, and distance learning. ADSL lets you have phone conversations and transfer data simultaneously.

at the Main Distribution Frame (MDF), which in turn is connected to a central ADSL transmission unit (ATU-C). The issue of where the POTS Splitter belongs in both locations has been widely debated and is yet to be resolved to anyone's satisfaction. The ATU-C is connected to the access node, which is the aggregation point for broadband and narrowband data sources delivered from a DSL access multiplexer, or DSLAM (some would say that the term ATU-C really means the DSLAM). The DSLAM allows Internet access and a wide variety of other data types to share access to the ADSL-equipped local loop.

With such a configuration, a service provider can provision a high-bandwidth, multiservice network

ANSI has published their industry standard (known as T1.413) for full-rate ADSL in the U.S. Meanwhile, the ITU has approved a nearly identical full-rate ADSL for the entire world, known as G.992.1. The ANSI and ITU specifications call for operation rates of up to 8 Mbps downstream and up to 640 Kbps upstream when operating over telephone lines at a distance of up to 24,000 feet.

• ADSL Lite or G.Lite – A low-fat version of DSL? Not exactly. ADSL Lite (also known as Splitterless ADSL and Universal ADSL) is actually a lower bit rate version of ADSL that uses same DMT modulation

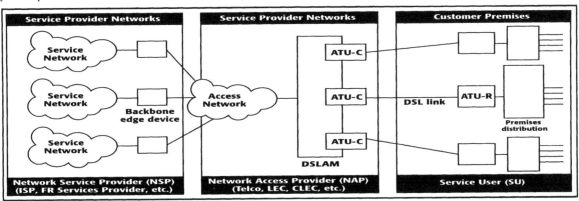

In a typical ADSL installation, at the customer premises the user service modules (which could be PC interface devices, set-top boxes or routers) attach to the Premises Distribution Network (PDN). This is a fancy name for the premises wiring scheme that interconnects the devices to the local loop.

The PDN is attached to a remote ADSL transmission unit (ATU-R), which is the equipment installed at the user's premises so they can connect to the xDSL loop. The connection is typically 10Base-T, V.35, ATM-25, or T-1 / E-1. Depending upon the type of service provisioned, ATU-R's come in a number of varying configurations. Besides supplying basic xDSL modem functionality, many ATU-Rs can do routing, bridging, TDM multiplexing or ATM multiplexing.

The ATU-R in turn connects to the local loop via a splitter. The splitter logically separates the low frequency voice from the higher frequencies that transport the various downstream and upstream channels of data traffic.

At the circuit's network side, the loop may terminate at another voice / data splitter

scheme as ADSL, but eliminates both the need for a POTS splitter or any microfilters that would normally be placed on every phone or answering machine at the customer premises. The ADSL signal is thus

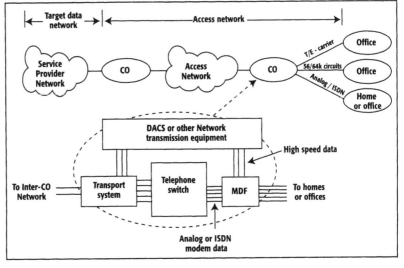

carried over the entire house phone wiring, which means that greater noise and other disruptive phenomena present forces a reduction in bandwidth (for example, picking up a telephone handset results in a change in the characteristics of the frequencies used by G.Lite, which must be dealt with). ADSL Lite's capacity is 1.544Mbps downstream and 512Kbps upstream. It has about the same reach as full-rate ADSL, or around 4 km. This form of ADSL was originally proposed as an extension to ANSI standard T1.413 by the Universal ADSL Working Group (UAWG) formed in 1997 and led by Microsoft, Intel, and Compaq. ADSL Lite is also known as G.Lite in the ITU standards committee. In October 1998, the ITU determined the G.Lite standard, which it renamed G.992.2. In June 1999, the ITU approved the G.Lite G.992.2 standard.

Like full-rate ADSL, G.lite systems generally divide a 550 kHz bandwidth on copper wire into three segments:

- The 0 to 4 kilohertz (KHz) range is used for POTS,
- The 26 kHz to 138 KHz range for upstream data, and
- The 138 KHz to 550 KHz range for downstream data.

ADSL Lite is seen as the "mass market" version of ADSL, easily deployed because no splitter or separate phone line to the DSL modem need be installed by a trained technician.

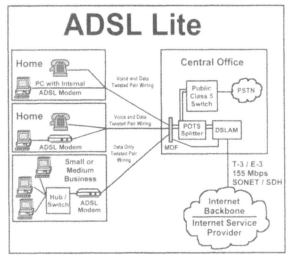

The office application shown at the bottom of the diagram is a data-only service. Many businesses may instead want both voice and data on the same line, similar to the home example. In the data-only business example shown here, several PCs are connected via an Ethernet LAN to an Ethernet switch or hub which is in turn connected to a DSL modem. At the central office (CO), the Main Distribution Frame (MDF) jumpers the subscriber line directly to the DSLAM. In many cases the DSLAM need not be located in the CO. Many competitive local exchange carriers (CLECs) have been unable to obtain co-location privileges and are providing high-speed data services by installing their DSLAMs separate from but close to the telco's CO.

Other claimed advantages for ADSL Lite include the following:

- Enables microfilterless operation, limiting them to only those phones whose voice quality appear to be impaired by the ADSL Lite signal;
- allows any phone jack in the home to be used as a high-speed ADSL Lite outlet; not just the one jack provisioned by the separate

phone line from the voice-data splitter;

- maintains the robustness of the in-home network as new devices, such as phones or answering machines, are installed after initial installation is complete;
- improves performance in changing noise and crosstalk environments;
- enables quicker connection times when turning DSL modems on;
- ADSL Lite enabled modems have the feel and "plug and play" functionality as the older, more familiar voiceband modems;
- enables PC vendors to integrate ADSL Lite capabilities into PCs, just as they do with voiceband modem technology; and
- enables faster transitions from low power modes and sleep modes.

- CDSL – The Consumer Digital Subscriber Line is a proprietary technology announced in the Fall of 1997 and trademarked by Rockwell Semiconductor Systems in Newport Beach, CA. Consumer DSL is slower than ADSL, allowing for only 1 Mbps downstream and 128 Kbps upstream in loops that have an 18,000 foot reach. Like ADSL Lite, CDSL doesn't need a splitter. Instead of using DMT or CAP modulation, however, CDSL employs its own line coding technique. And like Rate Adaptive DSL (RADSL), CDSL has circuitry that can rapidly adjust data rates in response to changes in line characteristics caused by simultaneous operation of analog voice service over the same line. In this respect it's similar to Rate Adaptive DSL (RADSL). At one point around 1998 a few Baby Bells such as BellSouth Corp. were championing CDSL, but it has not gained the same level of acceptance as ADSL.

- HDSL – As was discussed previously, High-Bit-Rate Digital Subscriber Line technology was the earliest variation of xDSL to be used on a large scale, developed in the early 1990s as a superior substitute for T-1 (1.544 Mbps) and E-1 (2.048 Mbps). The original idea behind HDSL was to apply the same line coding (modulation) technique as was used in ISDN (2B1Q), increase the bits per baud rate, reduce the frequency spectrum needed and split the service on two subscriber lines so that half of the T-1 payload (a bit rate of 784 Kbps) is carried on each wire pair – so there's four wires in the form of two twisted pairs for a T-1 (and three pairs for an E-1, though this requirement was later eliminated when better line coding techniques were used). Conventional T-1 requires repeaters every 6,000 feet to regenerate the signal strength. HDSL could achieve a longer range than T-1 without the aid of repeaters to allow transmission over distances of up to 12,000 feet (or 3.6 km., a "reach" considered equivalent to a telco's so-called CSA or Carrier Serving Area) on local loops having a wire thickness of 0.5mm (24 gauge), although a reach of 26,000 feet (5.0 miles or 7.93 km) can be achieved with heavier wire guages (22 gauge) or HDSL repeaters, which are also called "doublers" since they "double" the loop length.

Both the upstream and downstream signals occupy the same frequency band, so an echo canceler (using a technique similar to that used by V.32 and V.34 modems) is employed to separate the two directions of transmission on the subscriber line. Echo cancellation uses bandwidth more efficiently, but increases complexity and expense. Echo cancellation is also necessary because transmission systems with such overlapping signals is affected by a form of NEXT called self-near-end crosstalk (SNEXT). If there is no crosstalk in the cable, then the performance of the HDSL transceiver is limited only by the performance of the echo canceler. Although self-crosstalk is something of a concern with HDSL (as it is in SDSL and other DSLs), other physical requirements are not so strict. For example, two bridged taps are generally allowed to dangle on the line if the length of each is less than 5,000 feet (1.525 km).

Subsequently, the 2B1Q line coding was replaced in some implementations with CAP, which enabled utilization of even lower frequency ranges, resulting in less attenuation and even longer loop reach. CAP is widely used in Europe, South America and Central America for two-pair E-1 HDSL. Using these more advanced modulation techniques, HDSL transmits 1.544 Mbps or 2.048 Mbps in bandwidths ranging from 80 kHz to 240 kHz, depending upon the line coding technique, rather than the excessive 1.5 MHz demanded by the Alternate Mark Inversion (AMI) scheme used with the early T-Carrier system (See Alternate Mark Inversion, T-1). These days, HDSL can transmit T-1 or E-1 equivalent duplex signals over two twisted copper pairs for a distance of about 5 km (16,000 feet), or up to 12 km (39,000 feet) by using repeaters. HDSL standards committees of both ANSI and ETSI have endorsed these line codes in technical reports, but, ironically, HDSL never actually became a true standard and lacks the true interoperability of its successor technology, HDSL2.

The distinguishing characteristic of HDSL is that, unlike ADSL, it is symmetrical – an equal amount of bandwidth is available upstream and downstream which is the main reason why it's used as a replacement for T-1 and E-1 service. Since HDSL implementations of 12,000 feet or less are robust enough to eliminate the need for repeater equipment over normal distances and the removal of most bridged taps, it simplifies installation and maintenance. It also reduces the time, cost, and effort of isolating faults and taking corrective action when a failure does actually occur.

Typical applications for HDSL are practically the same as those for T-1. They include PBX network connections, digital loop carrier systems, cellular antenna stations, interexchange POPs, Internet servers, and private data networks.

- HDSL2-The high bit rate Digital Subscriber Line Service 2, as in current HDSL, is full-duplex symmetric, providing the same bit rate in both directions. HDSL2 also has the same bandwidth as HDSL but needs only a single twisted copper pair instead of two. Accomplishing this with a single wire pair was made possible thanks to a new line coding scheme called Overlapped Pulse Amplitude Transmission with Interlocking Spectra (OPTIS), developed by Dr. George Zimmerman of PairGain Technologies. OPTIS is based on 8-PAM (Pulse Amplitude Modulation) line code, an extension of the original 2B1Q line encoded HDSL (also known as 4-PAM) line code, with the addition of highly sophisticated trellis-coded modulation (trellis coding is a mathematical operation performed on transmitted data to improve a system's noise immunity). OPTIS for HDSL2 uses a spectrally shaped waveform that overlaps the upstream and the downstream in an interlocking manner. OPTIS has about half the latency of an equivalent Quadrature Amplitude Modulation (QAM) system. The ANSI T1/E1.4 committee adopted the OPTIS line coding scheme as the standard modulation for HDSL2. All service providers must use OPTIS to be able to claim interoperable, ANSI-compliant HDSL2 products. However, while HDSL2 transceivers that are compliant with the specifications are interoperable, the overall HDSL2 specification is not intended to provide for interoperability with prior generation HDSL transceivers.

ADC Telecommunications (Minnetonka, MN – 952-938-8080, www.adc.com) first proposed the idea of "automatic power control" (see ANSI document T1E1.4/98-131) to optimize the HDSL2 transmit signal level so as to increase performance and reduce crosstalk between lines. Rather than transmitting at peak signal levels on all loops, automatic power control reduces transmit levels on shorter loops, reducing HDSL2 crosstalk to absolute minimum levels in order to enable telecom operators to deploy different types of DSL in the same bundle of lines. HDSL2 service should get along nicely with other services co-resident in the same cable bundle, including HDSL, ADSL, IDSL, ISDN and T-1.

Of course, HDSL2 transceiver units were painstakingly designed to deal with a variety of wire pair characteristics and typical impairments (e.g., bridge taps, crosstalk and noise) so as to be spectrally compatible with other xDSL services anyway, but automatic power control allows the entire HDSL2 network to be prepared to address spectral compatibility issues regarding as yet undefined xDSL services of the future. Telcos like power control because it sustains the loop's broadband capacity without any concern over sophisticated upgrades or other matters of loop management.

Since HDSL2 has equal or better spectral compatibility / manageability than traditional HDSL, HDSL2 has a full CSA loop reach of 12,000 feet on 24 AWG wire, maintaining a signal-to-noise ratio of 5dB (which is only 1 dB less than the theoretical target of 6 dB) even while contending with worst-case crosstalk. This means that only half as many wire pairs extending from the CO are needed to deliver equivalent service to the same number of customers – such Digital Loop Carrier (DLC) "pair relief" allows service providers to deliver service to twice as many customers with the same copper feeder network by using HDSL2.

This also means that if a service provider connects HDSL2 service to a business with two twisted copper pairs in the same manner as the old HDSL, the service can achieve double the former data rate (i.e. 4 Mbps) for the same carrier serving area distance of about 12,000 feet. Moreover, by reducing the bandwidth back down to 1.544 Mbps or 2 Mbps, such a two-pair HDSL2 system (known as "reach-extended 2-pair HDSL2") can transmit for a significantly greater distance than current HDSL, well beyond the normal CSA. One would think that the loop reach should double to 24,000 feet since the HDSL2 transceivers are operating at half speed (784 Kbps). This is not the case, however, primarily because of attenuation and other physical factors. Instead, reach-extended 2-pair HDSL2 can realistically achieve about 15,000 feet with 26 AWG wire or 18,000 feet with 24 AWG wire. Furthermore, to transmit over really long-reach (beyond 18,000 feet) distances, carriers generally employ existing 2B1Q or CAP HDSL with doublers.

Like a T-1, an 8 Kbps overhead channel is also specified for activities that include Operations, Administration, Maintenance, and Provisioning (OAMP).

- IDSL – ISDN Digital Subscriber Line was originally developed by Ascend Communications (now part of Lucent Technologies). Its name "ISDN DSL" is somewhat of a misnomer since this service is really closer to ISDN than to the DSLs we've examined so far. IDSL is essentially an ISDN configuration without a switch. Unlike most xDSLs, IDSL doesn't run on POTS phone lines. Instead, ISDN signal is sent through a digital line carrier over the existing twisted copper pair between the CO and the customer premises, and connects to an ISDN-like IDSL card that plugs into a DSLAM or plugs into existing D4 channel banks or multiservice access multiplexers / concentrators at the CO. The IDSL card acts like an ISDN-enabled Class 5 CO switch when your ISDN terminal adapter or router initiates an ISDN call. Rather than switching IDSL calls to a destination (which would be quite a trick since there's no switch), the equipment establishes a dedicated unchannelized data circuit between the end user and an ISP or some other specific point. Beyond the CO, data is usually transmitted over a frame relay backbone network.

ISDN-like features of IDSL include its use of the same 2B1Q line coding used for the two 64 Kbps ISDN BRI bearer "B" channels (designated B1 and B2) and the 16 Kbps delta ("D") channel, which means that subscribers with ISDN BRI terminal adapters can use them along with their current routers and bridges for connecting to IDSL lines). Also like

X

ISDN, IDSL supports symmetric data transmission rates of 128 Kbps or 144 Kbps over loops of up to 18,000 feet when using a single copper pair made of 24 gauge wires. IDSL is so much like ISDN that it uses ISDN provisioning and testing and can easily coexist with legacy analog and ISDN services. IDSL can even operate through Digital Loop Carriers (DLCs). Other flavors of xDSL need their own special line card in the DLC, but IDSL can pass through standard DLC equipment to the switching office. DLCs serve about 20% of all U.S. phone lines.

IDSL can also make use of U-loop ISDN repeaters that increase IDSL's reach to 36,000 feet (10.9728 km), much further than other xDSL access technologies.

Any commonly used transport protocol can be used over an IDSL line such as frame relay, PPP, MP, or MP+. Using the MP+ protocol, multiple IDSL circuits bandwidths may be combined via BONDING. For example, when a 3Com IDSL Modem talks to a Copper Mountain CopperEdge DSLAM, an in-band signaling channel already exists, so there's no need for signaling on the 16 Kbps D channel. Therefore the D channel can be used for data too. The Service Provider can configure the link in the following configurations: Just the B1 Channel (at 64 Kbps), a B1 + B2 BONDed channel (at 128 Kbps), and the B1, B2, and D channels BONDed to deliver 144 Kbps.

For even greater bandwidth using IDSL, equipment such as the R-Series IDSL Routers from Netopia (Alameda, CA - 510-814-5100, www.netopia.com) offer an IDSL BONDING (IMUX) option for speeds up to 576 Kbps over four IDSL lines when used with Copper Mountain central office equipment and when the router is upgraded to DSL BONDING capability by installing a TER/31U or TER/32U single or dual IDSL WAN module along with the TER/IMUX firmware add-on.

Non-ISDN characteristics of IDSL include the fact that ISDN uses the voice network and terminates on the CO switch, setting up a circuit switched connection through the voice network for data transmission, while IDSL transmits data over a data network to the destination, totally bypassing the voice network, thereby avoiding costly upgrades to the end office switches. However, it also means that IDSL doesn't normally give you the ability to use those same lines for voice phone calls as you can with ISDN. You'll need to keep your existing POTS service for voice, unless your service is connected to something like an Ascend MAX or MAX TNT (with an IDSL line card installed) at the CO and you at the customer premises end are using an Ascend Pipeline set to the "IDSL switch type" which supports ISDN Q.931 en-bloc dialing. Only when the switch supports en-bloc dialing, can you make voice calls (En-bloc dialing reports the dialed number in the set up message sent to the CO equipment, which uses the information to route the call to the voice network). To make a call you would then obtain a trunk group number from the CO administrator. The trunk group number actually receives the call and routes it to the voice network. To dial a call, you would enter the trunk number, followed by the phone number, followed by the # key, which tells the CO equipment that you've entered the entire phone number and to initiate the call.

ISDN requires a call setup procedure while IDSL is a dedicated "on all the time" service. ISDN may involve per minute call fees after a certain number of minutes are used, while IDSL is generally billed at a flat rate.

Why should you be interested in IDSL service now when ADSL is around the corner? Here are some reasons given by Lucent:

- IDSL uses well known loop technology which can operate 18,000 feet or even longer using repeaters. This means service providers can offer IDSL as the base service, which is available to all customers. Other DSL technologies will not operate under all conditions and, hence, will have to be supplemented by IDSL. Also, other DSL technologies will not operate where DLCs are deployed. Depending upon location, up to 5% to 50% of customers are served through DLCs.
- IDSL operates at much higher speeds than modems available today but not at high enough speeds to require an immediate backbone upgrade.
- IDSL provides the only available DSL technology which has interoperability today.

- MSDSL – Also shown as M-SDSL, M/SDSL and mSDSL. The Multirate Symmetrical Digital Subscriber Line is based upon single pair SDSL technology (see SDSL below) and Carrierless Amplitude Phase modulation (CAP) line coding (CAP was selected because the bandwidth of the CAP systems is about half that of the equivalent 2B1Q line coding used in older xDSL systems; so CAP line coding causes less interference at higher frequencies with other xDSL services co-resident in the cables than would be the case if 2B1Q line coding was used). MSDSL supports integrated voice and data (like ADSL) and videoconferencing at MPEG 2 rates. Like RADSL, it has an "autorate" ability consisting of echo cancelers and adaptive equalizers to automatically and constantly adjust the data transfer rate in accordance to line conditions. MSDSL supports at least eight distinct symmetrical transfer rates and permits data transmission speeds between 160 Kbps and 2,320 Kbps specifying loops up to 29,000 ft at a wire thickness of 0.5mm (24 gauge) and 2 Mbps at 15,000 feet. Popular MSDSL bit rates are 160 Kbps (uses coded 8-CAP modulation), 272 Kbps (uses coded 16-CAP), 400 Kbps (also uses 16-CAP), 784 Kbps (64-CAP), 1,560 Kbps (64-CAP), and 2,320 Kbps (64-CAP). CAP-based systems contain a two-dimension eight-state trellis code that affords additional resistance to crosstalk.

When using 2.32 Mbps MSDSL, the bandwidth is generally split between a full E-1 payload (2.048 Mbps) with the leftover bandwidth divided into three voice channels or two ISDN channels.

Pushing the technology even further, GlobeSpan Semiconductor Inc. (Red Bank, NJ – 732-345-7500, www.globespan.net), a maker of xDSL chipsets, achieved a range for xDSL (particularly MSDSL) telecom applications of more than 30,000 feet over copper wires using high performance algorithms running on GlobeSpan's XDSL2 chipset platform. Introduced in May 1998, GlobeSpan's XDSL2 product family integrates a multi-channel digital signal processor (DSP) and framing functions into a single chip. This high integration allows two DSL channels to operate simultaneously and independently, with data rates ranging from 144 Kbps to 2.3 Mbps, using only a single crystal or oscillator. xDSL equipment using the GlobeSpan XDSL2 platform can automatically take advantage of its software downloadable capability to receive microcode upgrades.

- RADSL - Rate Adaptive Digital Subscriber Line is a form of ADSL that supports a wide range of data rates depending on the line's transmission characteristics. RADSL automatically adjusts the volume of data per second over a particular line by performing a series of tests to calculate the maximum possible transmission speed. This is particularly advantageous in situations where the line characteristics have a lower quality than expected owing to a large variance in the length and quality of the local loop, the gauge of the wire or fluctuating weather conditions, and thus a lower data rate can be tolerated (all the way down to 64 Kbps). Under optimium conditions, RADSL can transmit 8.192 Mbps over a 12,000 foot distance.

Since the downstream data rate from the CO to the subscriber is a function of line conditions and the signal to noise ratio (SNR) characteristics found on the line, the upstream rate back to the CO, which is dependent upon the downstream rate, will also vary. Upstream data rates can vary from 16 Kbps to 768 Kbps, while simultaneously providing duplex POTS, all over a single unconditioned twisted copper wire pair. The downstream and upstream channels can be split to several subchannels (up to four subchannels on the downstream and three bi-directional sub-channels) to help administer to several applications simultaneously.

Like ADSL, RADSL also allows users to have duplex phone conversations and transfer data simultaneously over a single copper pair. RADSL uses Carrierless Amplitude and Phase modulation (CAP) line coding.

The term RADSL can be applied to any rate adaptive xDSL modem – indeed, any T1.413 standard DMT modems are also technically RADSL – but RADSL originally specifically referred to a proprietary modulation standard designed by Globespan Semiconductor.

RADSL is used in the Traverser, the flagship product of mPhase Technologies (Norwalk, CT – 203-838-2741, www.mPhaseTech.com). Developed in alliance with Georgia Tech Research Institute, the mPhase Traverser Digital Video and Data Delivery System (DVDDS) works in conjunction with mPhase's Traverser Intelligent Network Interface (INI) to provide simultaneous digital television, Internet access and telephone service to homes or businesses. The Traverser enables ILECs, CLECs and RBOCs to simultaneously deliver over the existing copper loop high-speed Internet access at speeds of 6 Mbps downstream and up to 1 Mbps upstream, and up to 400 channels of broadcast quality MPEG-2 digital television programming – as compared to an average 59 channels offered by CATV and 175 channels offered by Direct Broadcast Satellite (DBS) service – as well as traditional voice services.

The Traverser DVDDS can be installed at the home and telco central office within hours as compared to the long lead times associated with laying fiber or upgrading the Hybrid Fiber Coax (HFC) architecture. The transmission of MPEG-2 digital television over copper wire can benefit areas with low cable penetration, and also provide an alternative to cable TV and DBS service.

• SDSL – Also known as S-HDSL. The Single-pair Digital Subscriber Line at first glance simply appears to be a single line (2 wire) version of the traditional two-pair HDSL, transmitting T-1 or E-1 signals over the single twisted pair, and in most cases supporting analog POTS service as well. SDSL supports bandwidth at speeds ranging from 128 Kbps to 1.544 Mbps.

Unlike HDSL, however, SDSL brings high speed digital access to customer premises where only a single phone line is practicable, and it can serve applications that demand symmetric data rates (such as servers, 768 Kbps fractional T-1, digital imaging, campus and large facility LAN-to-LAN connectivity, and remote LAN access such as telecommuting) while maintaining the existing POTS on the same local loop.

SDSL uses 2B1Q line encoding (same as T-1 or ISDN)

Still, SDSL might find use in customer premises where the phone company has refused to provision ADSL service because of crosstalk problems (NEXT and FEXT) caused by interaction with services running on other lines bundled in the same binder group – however, 2B1Q modulated systems such as SDSL, ISDN, and HDSL cannot avail themselves of any special coding or forward error correction, while the CAP systems (such as RADSL and some ADSL implementations) use a two-dimension eight-state trellis code that produces a 4-dB asymptotic coding gain, which gives them a greater resistance to NEXT interference. Most symmetric 1.544 Mbps SDSL systems don't meet the more stringent SNR margin requirements of HDSL2 in terms of crosstalk, and many SDSL systems can't maintain a full T-1 bandwidth past 11,000 feet over 24 gauge wire, a distance that's less than the reach of HDSL2 and is a distance over which

ADSL achieves rates above 6 Mbps. Still, the name SDSL has become more generic over time, and it can refer to many loosely-related symmetric services running at assorted data rates over a single loop, with some implementations even leaving out POTS support altogether, which makes for considerably differing capacity and reach limitations.

For example, Lucent / Ascend's High Performance Symmetric Digital Subscriber Line (HS) technology takes advantage of the latest SDSL technology and can achieve speeds up to 1.5 Mbps at distances up to 14,000 feet.

The Ascend DSLTNT and MAX TNT SDSL-HS line card provides 24 SDSL lines per card, and the DSLPipe-HS router has a four port Ethernet hub and multirate capabilities that provide eight different rate settings including 144, 272, 528, 784, 1,168, and 1,552 Kbps. These HS products will work in excess of 20,000 feet at reduced data rates versus 16,000 feet with previous Ascend SDSL products as the MAX TNT SDSL 16 port line card and the DSL-Pipe-S and DSLPipe-2S routers.

One of the popular 3Com SDSL Modems, on the other hand, supports these speeds and distances between the CO and the customer premises:

• 1.544 Mbps up to 2,896 m / 9,500 ft.
• 1.024 Mbps up to 3,811 m / 12,500 ft.
• 768 Kbps up to 4,542 m / 14,900 ft.
• 384 Kbps up to 5,456 m / 17,900 ft.
• 256 Kbps up to 5,791 m / 19,000 ft.
• 192 Kbps up to 6,248 m / 20,500 ft.
• 128 Kbps up to 6,940 m / 22,770 ft.

As you can see, SDSL equipment tends to be proprietary with little or no of the vendor interoperability enjoyed by, say, HDSL2 technology.
• UDSL – Unidirectional DSL is a variation of HDSL proposed by a small group of European companies. It's essentially a unidirectional version of HDSL.
• VDSL - Very High Speed Digital Subscriber Line is in essence a superfast version of ADSL. VDSL had been called "Video ADSL" (VASDL) and "Broadband DSL" (BDSL) prior to June, 1995, when the ANSI T1E1.4 Committee chose "VDSL" as the official acronym.

There have been some wildly varying suggestions by various companies over what the performance characteristics of VDSL should be. Because we are dealing with tremendously high transmission rates, attenuation forces the transmission distance along with wire to be quite short.

Maximum downstream rates, for example, are as follows:

12.96 to 13.8 Mbps over 4,500 ft.
25.92 to 27.6 Mbps over 3,000 ft.
51.84 to 55.2 Mbps over 1,000 ft.

Upstream rates fall in a range from 1.6 Mbps to 2.3 Mbps, though it's felt that 19.2 Mbps or higher will one day be possible. This is highly asymmetrical, though the reason the ANSI T1E1.4 Committee decided against the name "VADSL" for VDSL was to get rid of the "A" – they didn't want to rule out the possibility that VDSL might one day be symmetric. Indeed, VDSL can be configured to operate in a symmetric mode where a bit over 10 Mbps can travel in each direction over a 5,000 foot line.

Recently, a group of companies named the VDSL Coalition collaborated on a VDSL specification based on Single-carrier technologies. The objectives of the group are as follows:
• Investigate and resolve technical issues related to the use of single carrier technologies for VDSL,
• design a cost effective and practical integrated solution that will meet realistic system deployment requirements for VDSL,
• Create an interoperable technical specification for the resulting solution.

X

The group is open to any corporation or firm interested in the development of such a system. For further information please contact Vladimir Oksman at Bell Labs in Holmdel, NJ (voice: 732-949-1553, e-mail: oksman@lucent.com).

The VDSL Coalition's Transceiver Draft Technical Specification, Revision 0.02 (March 12, 2000) states that VDSL access may be set for both symmetric and asymmetric transmission. It is intended to deliver broadband services and can be used as a bridge to transport both the Plesiochronous Digital Hierarchy (PDH) and Synchronous Digital Hierarchy (SDH) carriers in the access network. The downstream data rates can reach up to 52 6 Mbps and the upstream data rates up to 30 Mbps. The VDSL transceiver uses Frequency Division Duplexing (FDD) as a main duplexing method.

The VDSL Coalition's document contains a complete specification of the Physical Medium Dependent (PMD) and Transmission Convergence (TC) sublayers, OAM&P functionality, link handshake and link activation procedures. It also defines the minimum set of requirements to provide both interoperability between the VDSL transceivers from different vendors and spectral compatibility between the VDSL and other xDSL services.

VDSL is a faster, though simpler form of ADSL. VDSL only deals with ATM network architectures (the only architecture with a comparable bandwidth), which means that it doesn't concern itself with channelization and packet handling, as does ADSL. VDSL will even allow you to connect multiple VDSL modems to the same line in your home or business, just as you would add an extension POTS phone to your home wiring.

To date a real standard has not yet been agreed on. But the VDSL Coalition, as well as ETSI and ANSI, the European and American standards bodies, are all mightily struggling to hammer out a standard for VDSL.

When VDSL is finally deployed (perhaps to deliver holograms or virtual reality on demand!) it will probably arrive in your home or business via an optical-fiber-to-the-curb network, with the last 1,000 feet or so being a copper tail circuit.

The Future of xDSL

Even for a medium-sized company, a T-1 (or its DSL equivalent) seems to be too small a pipe. Yet a T-3 (45 Mbps) seems to offer excessive bandwidth and is too costly. For most businesses there appears to be a magical "sweet spot" for bandwidth usage between the two, but the T-Carrier technology has left a gap at exactly this spot! Some carirers do offer fractional T-3 service, but it's usually still too expensive and difficult to provision.

With Inverse Multiplexing Over Asynchronous Transfer Mode (IMA), however, multiple DSL or T-1 channels can be bonded together (reminiscent of the way BONDING can put together multiple ISDN Bearer channels) allowing for transmissions of up to 10 Mbps in bandwidth.

Also, to help promote competition and speed up xDSL deployment, the Federal Communications Commission (FCC) in an order on line sharing and spectrum unbundling (FCC 99-355) released December 9, 1999, now requires local telcos to share existing phone lines with competitors who provide xDSL services. The FCC has ruled that the same lines can simultaneously transmit voice and high-speed xDSL data from different service providers.

In so doing, the FCC officially named Hotwire MVL (Multiple Virtual Lines) technology from Paradyne Networks (Largo, FL – 727-530-2000, www.paradyne.com) as compatible for line sharing.

In its order, the FCC states:

"Voice-Compatible Forms of xDSL. We require incumbent LECs to provide unbundled access to the high frequency portion of the loop to any carrier that seeks to deploy any version of xDSL that is presumed to be acceptable for shared-line deployment in accordance with our rules. xDSL technologies that meet this presumption include ADSL, as well as Rate-Adaptive DSL and Multiple Virtual Lines (MVL) transmission systems, all of which reserve the voiceband frequency range for non-DSL traffic. Among these, ADSL is the most widely deployed version of xDSL that is currently presumed acceptable for deployment on a shared line. Because line sharing as contemplated by this Order can occur only on lines that carry traditional analog voiceband service, lines that are not used for these services could not be shared. We conclude, therefore, that incumbent LEC arguments that we should not require unbundling of the high frequency portion of the loop because not all forms of xDSL technology are compatible with a line sharing arrangement are misplaced. Our rules ensure that xDSL technologies deployed in line sharing arrangements will not cause substantial interference to simultaneous voiceband services." (FCC 99-355, p. 35-36)

The Hotwire MVL System was designed and optimized for Network Service Providers (NSPs), Incumbent Local Exchange Carriers (ILECs), and the Competitive Local Exchange Carriers (CLECs) to reach the mass market with xDSL services. The Hotwire MVL system is impervious to bridged taps, can operate over voice lines and can reach up to 25,000 feet, thus doubling the market coverage area around a CO. Up to eight Hotwire MVL Modems can simultaneously share high-speed access and transfer files locally, sharing printers.

And as wireless technology becomes more sophisticated and pervasive, we should expect telcos in the xDSL business will free xDSL from its wire shackles and offer wireless broadband service (See Wireless Broadband).

We began to see this with the appearance of "outdoor HDSL" when some GSM operators started to install base stations for GSM "micro-cells" outdoors as well as wall-mounted or pole-mounted implementations.

Another indication of what's happening in this area can be found at Darwin Networks (Louisville, KY – 502 213-3600, www.darwinnetworks.com) which is a national xDSL provider and ISP that now has a fixed wireless delivery option. Their wireless Internet connectivity services are suited to users working at the small office or apartment complex level.

Endgate (Soquel, CA – 408-737-7300, www.endgate.com) also makes a complete line of millimeter-wave wireless equipment, from headends and antennas to plug-and-play wireless modems. Among other things, Endgate's systems can be deployed as broadband wireless local loop networks and Internet access systems in the 32-40 GHz range. Other Endgate systems can be used for high-capacity point-to-point wireless connections to link sites or buildings without a costly wired backbone.

Also, NEXTLINK Communications (McLean, VA – 703-547-2000, www.nextlink.com) and Concentric Network Corporation (San Jose, CA – 408-917-2800, www.concentric.com) has combined forces to build a broadband Internet access service. The initial service will be provided by NEXTLINK's fiber network, but broadband wireless service is scheduled to follow soon after. NEXTLINK, which already holds fixed wireless licenses blanketing the 30 largest markets in the U.S., is in a good position to become the premier wireless xDSL provider.

Recommended Websites

The ADSL Forum (www.adsl.com) is the industry association designed to promote ADSL. The site presents ongoing information about ADSL – the applications, the technology, the systems, the market, the trials, the tariffs. Orckit's Executive Vice President of Marketing and Sales, Dan Arazi, is on the Board of Directors of the ADSL Forum, and North American VP New Business Development, Nigel Cole is Chairman of the

Technical Committee of the ADSL Forum.

ADSL Experience, a Broadband Portal (www.adslexperience.com) has links to hundreds of rich media websites, and businesses related to DSL and broadband. Here's part of their list of broadband portals: www.everythingdsl.com, www.scour.net, www.searchbot.net, www.flip2it.com, www.broadcast.com, www.chello.com, www.rampt.com, www.perki.net/pages, www.onbroadband.com, broadcast.go.com, www.fastasia.com, http://speed.snap.com, http://dslnet, http://works.snap.com/, www.windowsmedia.com.

   Data Consultancy – www.thedataconsultancy.com/adsl.htm
   Digital Audio-Visual Council - www.davic.org
   DSL Digest – www.dsldigest.com
   DSL general information – www.everythingdsl.com
   DSL Life – www.dsllife.com
   DSL Links – www.dsllinks.com
   GTE's DSL Solutions Page – www.gte.com/adsl/index.html
   HDSL2 Forum – www.hdsl2.org
   One stop for DSL Related Information – www.dsl.com
   The High Bandwidth Web Page – www.specialty.com/hiband/
   VDSL Coalition - www.vsdl.org
   xDSL news – www.nwfusion.com/dsl
   xDSL News and Articles - www.xdsl.com
   xDSL Resource – www.xdslresource.com

A copy of the xDSL standards can be obtained from the American National Standards Institute at http://www.ansi.org or the European Telecommunications Standards institute at http://www.etsi.org.

   Websites of some xDSL companies:
   2Wire – www.2wire.com
   3COM – www.3com.com/xdsl/index.html
   ADC – www.adc.com
   Adtran – www.adtran.com
   Alcatel – www.alcatel.com
   Ameritech – www.ameritech.com
   Aware – www.aware.com
   Cisco – www.cisco.com/warp/public/779/servpro/solutions/dsl/

Connexant – www.connexant.com
Globespan – www.globespan.net.
Interspeed – www.interspeed.com
Kentrox – www.kentrox.com
Level One – www.level1.com
Lucent Technologies – www.lucent.com
Metalink www.metalink.com
Motorola – www.motorola.com
Nortel Networks – www.nortel.com
Orckit – www.orckit.com
PairGain – www.pairgain.com
Paradyne – www.paradyne.com
Ramp Networks – www.rampnet.com
Rockwell – www.rockwell.com
SBC – www.sbc.com
Texas Instruments – www.ti.com/sc/access
U.S. West – www.uswest.com
Westell – www.westell.com
XEL Communications – www.xel.com

XML

The eXtensible Markup Language XML is a markup language for use on the web, intranets, and elsewhere, that is based on semantic tags placed in documents containing structured information, as opposed to just inserting presentation tags as can be found in XML's well-known predecessor, HTML.

Structured information contains both content (words, images, etc.) along with some intimation of how that content behaves in different contexts. Almost all documents have some structure. "Document" refers not only to traditional web documents (lines of ASCII code) but also to other XML "data formats" such as e-commerce transactions, mathematical equations, object meta-data, etc.

In the case of telephony, XML tags used for unified messaging, IVR, "talking web pages" and call control, ultimately take on the appearance of a protocol that runs over both circuit and packet-switched networks (See Telephony Markup Language).

X

Y2K

At the moment the year 2000 rolled around, computer clocks and software the world over were supposed to go "haywire" (a quaint term we oldsters used to utter in the last millennium). They would all be victims of the "Millennium Bug," the name given to the problem caused by hardware and software which use only the last two digits of the year rather than all four. As the clock struck Midnight on December 31, 1999, essential computer systems would tick over from "99" to "00" – making some of them interpret the year as 1900 and plunging the world into chaos, or so it was thought.

As January 1, 2000 approached, apprehension increased, along with considerable confusion: Some people phoned hospitals saying that they feared they were "infected with the millennium bug"! These were no doubt the same individuals who in the mid-1960s called the U.S. Coast Guard to report "four individuals stranded in a place called Gilligan's Island" and demanded that they be rescued.

As the Moment of Doom approached, some people began hurridly stockpiling canned goods and toilet paper to get them through the impending disaster that was allegedly approaching.

Even Yours Truly was ready in the old bunker with several hundred bottles of Pepsi (which I normally keep on hand anyway), cans of Campbell's Chunky Soup, a large aging plastic canister of Brylcreem I had brought back from London in 1995 (I figured I might as well slick my hair down in case there was a shock wave), a box of chocolate truffles flown in from Switzerland (courtesy of Teuscher Chocolates on Fifth Avenue in New York City), and a bottle of "medronho" (Algarve firewater) that I picked up from the natives in the mountains of Monchique in Portugal – just to stave off the cold and the damp, of course!

But as it turned out, little if anything happened.

According to John Koskinen, Chair of the President's Council on Year 2000 Conversion, the costly effort undertaken over the past two years to deal with the Year 2000 computer problem prevented massive disruptions in systems and services during the date rollover into the new millennium.

In a January 18, 2000 interview in Washington D.C. with Paul Malamud of the Office of International Information Program, Koskinen said that the relatively problem-free date change that occurred is an indication not that the Y2K problem wasn't serious, but that the work devoted to fixing thousands of computer systems worldwide was successful.

Koskinen said the absence of serious Y2K disruptions in developing countries, where remediation efforts had lagged behind those in industrial countries, is explained by the less intense reliance in those countries on digital technology.

Yagi Antenna

A "linear end-fire" (directional) antenna. In the early 1900s two Japanese inventors devised the most popular of all directional antennas. One of the inventors was named Dr. H. Yagi (a Japanese physicist who translated into English an antenna design based on a parallel array of dipole antennas) and the other was named Uda, who may have actually originated the idea. Electrical engineers know this type of antenna as –

you guessed it, the Yagi-Uda array. Citizens Band (CB) and amateur radio enthusiasts know them as "beam antennas." If you don't have cable TV, the antenna on your roof is a cousin of the Yagi design but are called "Log-Periodic" because they serve a large number of frequencies.

Yagi antennas have tremendous "gain" and "directivity" since they can direct almost all of their signal in just one direction instead of dispersing all of the energy around a 360 degree partial sphere.

The Yagi design involves placing a slightly longer dipole behind, and a slightly shorter dipole in front of, the driven dipole to direct the radiation in the direction of the longer-to-shorter elements. To this end, the Yagi antenna consists of a "boom" (a long horizontal bar) across which are "elements" or small thin rods supported by the boom. These elements are usually vertically positioned as seen in the diagram. The front of the antenna has the shorter elements (the "director" and "driven" elements) and the rear has the longer ones (the "reflector" elements). The difference in element dimension is small compared to household Log-Periodic TV antennas. Starting from the rear, first comes the reflector element then the "driven" element then from one to any number of "directors" but usually from one to 13 or so. The shorter director elements and one of the longer reflector elements are separated by a small fraction (about 10 percent) of the wavelength of the driven element. As a resonant dipole, the driven element has a length of one-half the wavelength being used.

The signal output is off the front end of the boom. The signal "beam" is anywhere from 20 to 90 degrees wide with the longest or most elemented design having the narrowest width and highest gain or reach.

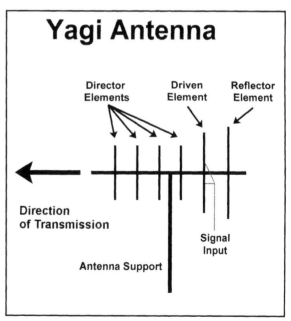

Zasu

Amusing first name, as in movie actress ZaSu Pitts (born January 3, 1898, in Parsons, KS; died June 7, 1963, in Hollywood, CA). A gifted comedienne, ZaSu was perhaps the first stereotyped funny-women to succeed in a straight, dramatic role when she portrayed the character of Trina, the obsessed miser in Erich von Stroheim's 1924 masterpiece, *Greed*. Zasu got her name from two maiden aunts, EliZA and SUsie - get it?

(Okay, The CMP Media book division keeps telling me that I should have amusing entries in my encyclopedia, "just like Harry Newton does in his Telecom Dictionary." They assure me it's "the key to his success!" (That's funny, I always thought it was his dirty jokes). So okay folks, here it is – Zippy cuts loose!

Zener Diode

A semiconductor that acts as a normal rectifier until the voltage applied to it reaches a certain critical point, called the zener voltage or the avalanche voltage, whereupon the the zener diode becomes conducting. Zener diodes make great voltage regulators.

The basic parameters of a zener diode are:
- The zener voltage must be determined. The most common range of zener voltage is 3.3 volts to 75 volts, but other voltages are possible.
- A tolerance of the specified voltage must be stated. While the most popular tolerances are 5% and 10%, more precision tolerances as low as 0.05 % are available. A test current (Iz) must be specified with the voltage and tolerance.
- The zener diode's power handling capability must be specified. Popular power ranges are: 0.25, 0.5, 1, 5, 10, and 50 Watts.

Zero Code Supression

The forced insertion of a "1" bit so that eight or more consecutive "0" bits will not be transmitted. Part of the early "ones density" requirement of early T-1 systems (See T-1).

Zippy

Nickname of Richard Grigonis, the Chief Technical Editor of *Computer Telephony* Magazine. The legend of Zippy began when *CT* Editor-in-chief Rick Luhmann found that too many people were confusing the names "Rick" with "Richard." He decided to devise a clever nickname for Richard. Did he ever. Luhmann later said that he gave Richard the name Zippy because "he zips through information and holds it like a Zip disk." Zippy himself is not quite sure about this origin of this sobriquet, having spent many accumulated hours patiently explaining to new acquaintances that he has absolutely nothing to do with the Zippy the Pinhead cartoon character.

Needless to say, Harry Newton (then publisher of the magazine) was absolutely delighted with such a deft example of character assassination, and so the name Zippy was spread far and wide, making Richard (or Zippy) into an industry icon, of sorts.

For the record, Zippy has no connection with the national Zippy the Clown party franchise, nor did he get his nickname by hanging out repeatedly at the Zippy Coin Laundry in Oakland Park, FL. Zippy doesn't use Zippy Mail, doesn't shop at Zippy's Mini-Mart in Priest River, ID, and while he is considered to be a professional photographer, he has yet to do a celebrity endorsement of the Zippy One Hour Photo Shop in Valley Village, CA. He doesn't own the Zippy Boat Works in Arden, NC nor does he have a controlling interest in Zippy Auto Glass of Lehighton, PA – though Zippy does wish he owned the lucrative national chain of Zippy Lube shops headquartered in Rexburg, ID, or the chain of Zippy Marts in the great American Southeast, or even the string of 21 Zippy's restaurants that serve mouthwatering chili to the denizens of the beautiful Haiwaiian islands.

Perhaps Zippy will one day eat Mexican in Hawaii, in which case he'll be sure to bring along some salsa from Zippy's Hot House Salsa Co. in Petersburg, AK and some buffalo wing appetizers from Zippy's Zesty Wings in Okeechobee, FL.

And if *Computer Telephony* magazine should ever fold for good, Zippy will be sure to send his resume to Zippy Typing Services in New Castle, DE. It looks like his kind of place. Maybe he'll even bump into Rick there.

Zulu Time

Coordinated Universal Time, otherwise known as Greenwich Mean Time (GMT), a borough in London located on the Prime Meridian, from which all longitude was reckoned (provided that you had an accurate chronometer).

Zzzzzzz

This damn encyclopedia is finished, now I can get some sleep!

Z

(This Appendix is not affiliated with the Marquis Who's Who Reference Work in any way, shape or form!)

Alissi, Geno

Vice President and General Manager, Computer Telephone (CT) Consulting Services Division of Dialogic (Parsippany, NJ – 973-993-3000, www.dialogic.com) an Intel company.

Geno Alissi has been the vice president and general manager of the Dialogic Computer Telephone Division since its establishment in 1995. In 1999, the CT Division was expanded to include the Dialogic Professional Services Business Unit and Parity Software Division (a 1999 acquisition).

Before taking on his current responsibilities, Alissi served as general manager of Digital Equipment's Computer Integrated Telephony (CIT) business unit, which provided computer telephone applications and solutions for the call center and customer management markets. CIT developed and introduced Digital's Computer Integrated Telephony product line, the first software-link product in the computer telephony industry, and was the first system integrator in the industry to deliver "one-stop" call center solutions for the financial services, utilities, retailing, healthcare, and telecommunications vertical markets.

Alissi holds a Master of Economics degree from the University of Hartford and a Bachelor of Economics degree from American International College.

Bajorek, Chris

Chris Bajorek is co-founder and President of CT Labs, a firm started in 1998 that specializes in CT product testing. Prior to CT Labs, Bajorek founded Telephone Response Technologies, Inc. in 1986, a company that developed a variety of innovative CT products. Before that, he worked at two CT product startup companies Integrated Office Systems and TSP, starting in 1977. Bajorek is author of the "Ask Dr. CT" column appearing monthly in *CT Magazine*. Bajorek may be reached at cbajorek@ct-labs.com.

Becker, James

Jim Becker is Amtelco's Vice President of Sales and Marketing and Director of XDS division. (McFarland, WI – 800-356-9148, www.amtelco.com).

Becker arrived at his present job of directing Amtelco's XDS CTI specialty board division by way of NASA. He started his career at Allis-Chalmers in Milwaukee, which had just scored the contract for the Apollo moonshot program, making the spacecraft's fuel cells. He says he still gets a certain thrill when he views the Saturn 5 rocket on display at the Smithsonian.

Becker is a similarly hale veteran of CT. A veteran of Amtelco since 1985, he's guided the company through the development of its two and four-wire E&M tie line boards to its PC switching resources to its mammoth conferencing boards and now on to e-business software.

Becker cleverly made sure Amtelco didn't go after the voice board marketplace – which he knew was crowded. Instead, Amtelco focused on telephone interfaces and switching that just wasn't available anywhere else. Amtelco went first from DID to E&M and dedicated station ports, the switch matrix, conferencing boards, and analog interfaces, to multichassis interconnection and BRI ISDN specifically for Central Office Applications. That's what the XDS line of boards is all about. Amtelco is also still something of a "boutique" house.

He has a Bachelor's Degree in Marketing from the University of Wyoming and has taught Direct Marketing at the University of Wisconsin.

Jim is situated in Amtelco's home office just outside Madison in McFarland, Wisconsin.

Binal, Mehmet E.

Mehmet E. Binal, Ph.D, is President and CEO of BICOM, Inc. (Monroe, CT – 203-268-4484, www.bicom-inc.com) a leading PC based voice processing board manufacturer for the telecommunication and office automation industries. BICOM boards are used in voice mail, audiotex, telemarketing, conference bridging and automated attendant systems.

Binal received his Bachelor's and Master's degrees at the Technical University of Istanbul, Turkey in 1969 and his Ph.D from the University of Aston in Birmingham, England in 1974.

Before founding BICOM in 1988, Dr. Binal was Director of Engineering at Dictaphone Corporation, in charge of development of the DX7000 Digital Express central dictation/transcription system. He worked at Marconi in England, The Turkish Scientific and Technical Research Council in Turkey, Bell Northern Research in Canada, and SHAPE Technical Center (of NATO) in the Netherlands as a researcher and systems engineer. Before joining Dictaphone in 1985 he was the Vice President of Voice and Data Systems in Canada.

Bowles, Ian

President of Pronexus (Kanata, Ontario, Canada - 613-271-8989, www.pronexus.com). Mr. Bowles is a graduate of Electrical Engineering at the University of Sussex. His career includes contributions to innovations at International Computers Ltd (Circuit Design for Mainframe computers), Nortel (Circuit board and firmware Design for DMS phone switches), BNR Computing Research Labs (Voice Response Systems) and Pika Technologies Inc. (Voice Processing Hardware Drivers).

His vision of custom components delivering sophisticated voice and fax processing functionality within the Visual Basic environment spawned Pronexus in 1993, delivering an innovative suite of rapid application development tools (featuring VBVoice) to the computer telephony developer market. A full division for custom application development services was soon added in order to complement the Pronexus product line and to better meet the demands of the marketplace.

Pronexus has grown to become one of the leading providers of tools and solutions in the industry with customers in over 70 countries around the world and strategic partnerships with Active Voice, Nortel Networks, Dialogic and Brooktrout Technologies.

Bubb, Howard G.

President of Dialogic, an Intel Company and Vice President of Intel's Communications Products Group. Intel acquired Dialogic in July of 1999 to build servers for convergent communications markets. Previously, Howard was President and CEO of Dialogic since 1991 and led their growth from $40 million through their IPO to $340 million. Under Bubb's direction Intel / Dialogic has become a leader in the burgeoning computer telephony industry providing IP telephony, voice, fax, data, networking, hardware, and software products for open systems. One-third of CT platforms worldwide containing Dialogic products.

Bubb joined Dialogic from Lexar's Telenova subsidiary where he was Senior Vice President and General Manager. As a founder of Lexar, an international provider of telecommunications systems, he was instrumental in developing the first integrated voice/data PBX. He has also held top management positions at United Technologies, Memorex, and Telex corporations. He earned a Bachelor's degree with honors in Electrical Engineering in 1976 from California Institute of Technology. He serves on the Board of Directors of the MultiMedia Telecommunications Association (MMTA), the American Electronics Association (AEA), and PairGain Technologies.

Editor's Note: One of Harry Newton's favorite stories claims he was instrumental in getting Bubb his job at Dialogic. Bubb, however, gave Newton the idea for the "Telecom Developer's" Expo that soon metamorphosed into the annual series of CT Expos that are today held worldwide by CMP.

Burton, James A.

Jim Burton (707-963-9966) is founder and CEO of C~T Link, Inc., a consulting firm specializing in voice/data integration. As a result of his 25 years of work in both the computer and telecommunications industries, Burton decided to address the need for shared integration information between the two industries by providing specialized consulting and information services.

Burton is recognized as the leading authority in the voice/data integration industry and is credited with "coining" the term computer-telephone integration (CTI). Burton has played a variety of key roles in the development and promotion of industry standards including the Microsoft and Intel Telephony Application Programming Interface (TAPI) and Universal Serial Bus (USB).

C~T Link consults with major vendors in both the computer and telephony industry regarding their CTI product development, distribution and partnership strategies. Clients include: 3 Com, AT&T, Apple, Artisoft, CallWare, Compaq, Dialogic, Ericsson, Fujitsu, Genesys, IBM, Intel, Lucent, Microsoft, NBX, Nortel, Nuance, Picazo, Siemens, Softbank, Sphere Communications, Telogy Networks, Wildfire, and ZD.

During his career, Burton founded several successful start-ups. Prior to C~T Link, Burton founded Control Key Corporation, a manufacturer of telephone accounting systems and software – a joint venture with NEC America.

Burton is a frequent keynote speaker and is regularly quoted in the trade and business press. He is an advisor to several industry associations and trade shows. He is a founding member and currently vice chairman of the board of directors of the CTI Division of the MultiMedia Telecommunications Association (MMTA) and serves on the Board of Governors of the MMTA. In addition, Burton is a contributing writer to several major industry and business publications.

Burton is Chairman of the CommUnity Program Committee and an advisor to several industry associations and trade shows including CommUnity, CT Expo, MMTA, NetWorld + InterOP, PBX 2000 and SuperComm.

Carney, Allen

Vice President of Marketing for Natural MicroSystems (Framingham, MA – 508-620-9300, www.nmss.com). Allen Carney joined Natural MicroSystems in April of 1996. In this position, he is responsible for the creation and execution of the corporate marketing strategy. Prior to joining NMS, Carney spent four years at Lotus Development Corporation of Cambridge, MA, initially responsible for international marketing, and most recently as Vice President, Applications Marketing. In this capacity, he was responsible for the creation and execution of the marketing strategy for Lotus' productivity product line, including Lotus -

1-2-3. Prior to joining Lotus, Carney spent ten years with Atex, Inc., a turnkey supplier of prepress automation systems, where he was most recently Vice President, European Operations.

Mr. Carney holds a B.A. from Yale University.

Coffee, Michael B.

President of the Commetrex Corporation (Norcross, GA – 770-449-7775, www.commetrex.com), and father of the MSP/CX Media Gateway, an open-architecture, 3,200 MIPS H.100 PCI DST board for voice, fax, data, ASR and other CT media that incorporates the industry's first implementation of the MSP Consortium's M.100 specification for vendor-independent integrated-media environments.

Dating back to 1987, Mike Coffee's tenure in the computer telephony industry nearly spans the industry's entire history. Cliff Schornak, Commetrex's co-founder, developed a four-port PC-based voice board back in 1984, giving him the right to lay claim to being a part of the industry's founding. So Commetrex's two founders can rely upon a first-hand historical perspective when viewing the CT industry. And it's helped them come up with some major innovations, not the least of which was proposing that NMS develop an inter-board PCM highway–what was to become MVIP–while Coffee was VP, Marketing, at NMS. Now Commetrex is intent on lowering the cost of media convergence in the CT industry through the open-specification efforts of the MSP Consortium, which Commetrex helped found.

But Coffee's path to founding Commetrex Corporation began by taking an EE degree at Georgia Tech in 1965. He them joined a mini-computer start-up in Fort Lauderdale, FL where he designed computers and computer systems until, in 1976, he got talked into "carrying a bag" for another South Florida computer company, beginning a 20-plus year career in marketing.

After working with Schornak in the late '80s and spending nearly two years with NMS, he and Schornak founded Commetrex to help the CT industry move beyond fixed-function media-processing resources to software-defined "any media. . . anytime" technology on ultra-powerful DSP-resource boards.

Edgar, Robert

Bob Edgar, Ph.D, founded (in 1989) and is CEO of Parity Software Development Corporation (Sausalito, CA – 415-332-5656, www.paritysoftware.com).

Dr. Edgar started his software career as a high-energy particle physicist simulating quark fields in massive mainframe calculations. Later, he became involved designing and building language translator operating systems and multi-user database systems for UNIX, XENIX, PRIMOS and MS-DOS.

Since even before the inception of his company, Edgar led the development team designing and creating the Voice Operating System (VOS), an application language for building computer telephony applications in PCs using Dialogic hardware.

In 1999, Parity announced a new telephony engine, code-named Topaz, that would offer "API transparency" to developers using Parity's VOS and CallSuite award-winning telephony tools to drive CT platforms, particularly those built on Dialogic hardware (which has traditionally been Parity's preferred app development environment).

Today's computer telephony developers working in C or C++ are confronted with a confusing and difficult decision: Which API should they use for their systems? One of Dialogic's old (but still active) drivers? S.100? TAPI? Worse, will they have to re-engineer if they decide to go another route in the future?

Topaz was meant to shield VOS and CallSuite developers from these sticky questions and the underlying programming interfaces, including Dialogic's SR4, ECTF S.100 (including Dialogic's CT Media), TAPI, the native DM3 API and more. In other words, the theory was that a VOS application created today on Dialogic's SR4 API could run tomorrow on TAPI or CT Media. The Topaz engine was to be the foundation of VOS version 7 and CallSuite version 7, which was scheduled to ship in the 3rd Quarter of 1999.

Instead, Parity was acquired by Intel Corporation in 1999, and the Topaz project suddenly became the basis for a common "object infrastructure" to be used by all of the CT industry's app-gens and toolkits. The "object infrastructure" will sit atop the Intel / Dialogic CT Media software, and will concern itself with how applications see CT capabilities at a level even higher and more general than the S.100 abstractions.

Edgar remains a respected authority on software development who is known for his pioneering work in voice processing and from his articles in *Dr. Dobb's Journal, Teleconnect, Computer Telephony* and other magazines. He is popular as a guest speaker at industry trade shows, including Computer Telephony Conference and Exposition and has presented many training seminars on PC Telephony. Edgar is also the author of a book *PC Telephony* (available from CMP Media) that has gone through several editions.

Fiszer, Max

CT visionary, consultant, current President of MarTek Consulting, Inc. (Palo Alto, CA – 650-868-7517, martek@pacbell.net) specializing in strategic marketing for firms positioning themselves in e-commerce.

Max Fiszer started his high technology career as a software engineer with the IBM Corp and progressed through a variety of management positions in software development and marketing. He left IBM to join the ROLM Corp as the Director of Technical Planning with responsibilities for planning future products and software releases for the PBX, desktop telephones, voice mail and call center product lines. With the purchase of ROLM by Siemens he became more deeply involved in computer telephony and call centers and became the Director of Global Call Centers and CTI Solutions. He left Siemens to be the Founder and CEO of MultiCall, Inc. a computer telephony software company focused on architecturally integrated solutions for mid-market companies. specializing in strategic marketing for firms positioning themselves in e-commerce.

Fiszer holds a bachelor's degree in mathematics from the City University of New York and was a Sloan Fellow at Stanford University's Graduate School of Business. He has written for a variety of industry magazines and is the recipient of a number of industry awards including "Star of the Industry" by *Computer Telephony* magazine, "Industry Pioneer" by *Call Center* magazine and "Key Contributor" by *Telemarketing* magazine

Giler, Eric

President of Brooktrout Technology (Needham MA – 781-449-4100, www.brooktrout.com). Giler has a BS in Management Science / Industrial Engineering from Carnegie Mellon University and an MBA from the Harvard Business School.

Giler's partners, David Duehren and Pat Hynes, had all worked at Teradyne in Boston before starting Brooktrout in 1984. Giler's specialty was marketing while Duehren and Hynes were design engineers. Hynes, an avid fly fisherman who "gets his most creative thoughts when he's fishing," was the one who gave Brooktrout its name.

Although known for its fine fax CT resource boards and "industrial strength" fax technology, Brooktrout under under Giler has gone to con-

siderable lengths to ensure that the public knows that Brooktrout "isn't just about fax." Even at the beginning of the company, "we knew we were in the electronic messaging business," says Giler, "and with fax, we're still in that business. So the business plan from day one has not markedly changed. Actually, the first physical product we came out with in 1984 was a voice board, to which we added fax capabilities in 1985."

Giler can be reached by e-mail at egiler@brooktrout.com

Goubau, Stéphane

Vice President, EMEA Operations for Dialogic (Parsippany, NJ – 973-993-3000, www.dialogic.com). A member of the Dialogic team since May 1992, Stéphane Goubau has been heading the company's operations in Europe, the Middle East, and Africa (EMEA) – first as Managing Director, and later, as Vice President. Under his leadership, Dialogic has become the dominant technology supplier in the European CT industry.

Previously, Goubau was Director of Information Systems with the Société Générale de Belgique, and before that, he held the positions of Manager Software Development and General Manager German Operations with Eyemetrics. Goubau holds an MS degree in electromechanical engineering from the University of Ghent, and an MBA degree from Harvard University.

Gousse, Maxime

Marketing Director of Prima (now Elix as of May, 2000).

Mr. Maxime Gousse is Director of Marketing for PRIMA (Nuns Island, Canada – 514-768-1000) since 1998 where he joined the company in 1996 as Product Manager. As Director of Marketing, he helped PRIMA launch OPUS Maestro, a highly scalable and manageable interaction portal, while receiving numerous awards from the industry for it.

Mr. Gousse has been active in computer telephony for more than 10 years. For a previous employer he was instrumental in programming a powerful voice mail application and later worked in Paris, France, as European Director in charge of distribution of its software line.

Mr. Gousse studied in New York City, where he obtained an International Baccalaureate in Mathematics. He also received a Minor in Business Management at the Montreal Economics School. This quite diverse course has given Mr. Gousse the skills required in driving the convergence of technologies and business requirements.

Halimi, Bachir

The founding President of MediaSoft Telecom Inc. (founded 1987), which merged with PRIMA in 2000 to become Elix. Halimi holds master's and bachelor's degrees in Computer Science from the Universite de Montreal.

Prior to creating MediaSoft, Halimi had founded two other companies: Alis Technologies, which specializes in enabling computer systems and peripheral equipment to communicate in languages like Hebrew, Arabic and Chinese (prior to this, much equipment only could make use of the Latin alphabet), and Bachal Telematique, which provides videotex technology platforms to telcos and service providers worldwide.

"I got into computer telephony as I was always very interested in communications and I believe CT is another better way of communicating and making transactions. I saw a huge market and lots of challenges. Telephones, computers and network management touch so many fields. I saw huge growth."

In founding MediaSoft, Halimi maintained his global orientation, hiring Israeli, Chinese, Arabic and French speakers to develop the MediaSoft IVS Builder / Server CT applications generator software for users speaking myriad languages.

For the IVS Builder Halimi oversaw the development of a custom

509

development language called "Blabla." IVS Builder is a GUI app gen that turns the graphical representation of a call flow into Blabla object code. Blabla runs on UNIX or NT servers using the IVS Server software.

Halimi is a highly acclaimed speaker at international industry conferences and has discussed the topic of web and computer telephony at many major CTI conferences, including the Computer Telephony Conference and CTI Expo in the United States and Asia, Voice Europe in England, Voice Power in Canada and CTI in the Middle East. Halimi spent several years as a member of the Management Committees of Inno-Centre, a Montreal-based organization that helps high-technology start-ups.

Heymann, Robert N.

Robert (Bob) Heymann is vice president of Asia and general manager of Dialogic S.E.A Pte Ltd., located in Singapore. He is responsible for leading the company's Asian operations and strategy, as well as managing sales and support offices throughout the Asia/Pacific region. Prior to taking this position in 1996, Heymann initiated Dialogic activities in Internet telephony as vice president of business development. He also served as vice president of product development for three years, with the responsibility for all engineering and product marketing of Dialogic voice processing and telephone interface products. He was also instrumental in the rollout of the Signal Computing System Architecture (SCSA) open systems model.

Heymann has over 15 years of experience in the telecommunications industry. Before joining Dialogic as a field applications engineer in 1988, he worked in various engineering and product management roles for ROLM and IBM. Heymann earned a bachelor's degree in engineering in 1983 from Princeton University, and a master's degree in electrical engineering administration in 1988 from Stanford University.

Horak, Ray

President of the Context Corporation (Mt. Vernon, WA – 360-336-3448, www.diginet.com/contextcorp.html), an independent consultancy that serves carriers, manufacturers, distributors and end users. Ray has published well over 100 articles in leading industry trade publications and is a member of the Advisory Board of Datapro Information Services Group. He serves on numerous Editorial Advisory Boards, and is Consulting Editor for several publications. He has over 25 years' experience in networking, having held management positions with Communications Group Inc., CONTEL and Southwestern Bell Telephone Company. Horak is internationally recognized for developing and delivering world class seminars on telecom technologies, services and management systems. He is the author of the best-selling book *Communications Systems & Networks: Voice, Data, and Broadband Technologies.* He can be reached by E-mail at ray@contextcrp.com.

House, Chuck

Executive Vice President for Research and Products for Dialogic (Parsippany, NJ – 973-993-3000, www.dialogic.com) an Intel company. Chuck House joined Dialogic in February 1998, when the company acquired Spectron MicroSystems. Spectron is a wholly-owned subsidiary located in Santa Barbara, California, that builds real-time operating systems and development tools for DSP designs. His research is focused on a Video Integrated with Voice Integrated with Data (VIVID) toolkit, and voice-enabled applications and technologies. House is also interested in issues of communications modalities, especially graphical and multimedia literacy; and societal issues raised or illuminated by computing technology, such as "reality processing," questions of public safety, and ethical behaviors. Chuck and Dialogic have been instrumental in establishing the new Center for

Information Technologies and Society at the University of California, Santa Barbara, where House serves as chairman of the advisory committee.

Chuck received a BS degree in engineering physics from the California Institute of Technology, an MSEE degree from Stanford University, an MA degree in the history of science from the University of Colorado, and an MBA degree in strategic studies from the Western Behavioral Sciences Institute (now the University of California, San Diego).

Karneef, Peter

CEO of Pika Technologies in Kanata, Ontario, Canada (www.pika.ca). Karneef has a Bachelors of Commerce and Computer Science from Carleton University and did two years of Photogrametry studies at the British Columbia Institute of Technology

Karneef founded Pika in 1987 along with Jim Pinard, a former chief engineer at Mitel. Kaneef, for his part, had worked in aerospace. He helped build the space-arm simulator for the Space Shuttle.

Tracing Pika's history takes one through the evolution of computer telephony. They've gone from making busy lamp fields for Mitel PBXs to building a vast suite of CTI integration and MVIP-compliant media-processing / call-control resources.

Kauffman, Maury

Maury Kauffman founded in 1990 The Kauffman Group Inc., an enhanced facsimile technology and services analysis and consulting firm. As managing partner, Maury Kauffman has confidentially advised more than 40 of the world's most recognized fax, voice, telecommunications and Internet-related companies.

Respected as a leading authority on enhanced fax, Kauffman is the "Life In The Fax Lane" columnist for *Computer Telephony* magazine. He has also written over 30 articles for such publications as *Information Week, VAR Business, Sales* and *Marketing Strategies* and *Voice Asia.* Kauffman has been quoted in publications ranging from *Business Week* to *Forbes,* from *Upside* to *Web Week* and from *TeleTalk* (Germany) to the *South China Morning News.* Kauffman's second book is *Internet and Computer Based Faxing.* (460 pp, $34.95, Telecom Publishing, 1-800-LIBRARY or telecom@rushorder.com.)

Kauffman is a perennial speaker at technology and communications conferences on three continents. Besides every Computer Telephony Expo, Kauffman has been a featured speaker at: ISPCon, NetWorld+Interop, Internet World US & UK, VON, FON, FaxWorld, Fax Directions and the Direct Marketing Association. He was the only US-based consultant invited to speak at the first ASIAFAX conference, held in Hong Kong. Kauffman has addressed the European Association of Newspaper and Magazine Publishers at their conference in Zurich. He was Chairman of FAXASIA '96, held in Singapore.

The Kauffman Group clients include: AT&T, AVT Corp, Brooktrout, Broadwing, Dialogic, GTE, HP, IBM, Intel, iReady, Lucent, Mail.Com, MCI/WorldCom, Pitney Bowes, Sprint, US West and Xpedite Systems.

The nature of the projects included: Strategic Alliances, Mergers and Acquisitions; Global Industry Perspectives, Positioning and Monitoring; Seminar Selling to Customers, Prospects, VARs and Partners; Product and Market Strategy Development and Implementation and Sales Training, etc.

Maury Kauffman can be contacted directly at: maury@kauffman-group.com or 856-651-1651, fax: 856-651-1652.

Kontopidis, George

George Kontopidis. Ph.D is Senior Vice President of Engineering for Natural MicroSystems (NMS) (Framingham, MA – 508-620-9300, www.nmss.com). Dr. Kontopidis joined NMS as the Vice President of

Engineering in 1989. In this position, he has contributed to the design and implementation of multiple NMS hardware and software products, and is responsible for staffing, organizing and managing the engineering group.

Before joining Natural MicroSystems, Dr. Kontopidis worked as Director of Engineering for Pacer Systems and Sea Data Corp., where he was involved with the design of autonomous underwater instruments and data loggers, all utilizing DSP technologies. Dr. Kontopidis has also commercialized software packages for scientific visualization, run his own technical service bureau, and designed a pilot hospital information system. His experience also includes teaching Computer Engineering and Electronics courses at various colleges and universities, including Harvard Summer Extension School.

Dr. Kontopidis holds an undergraduate degree in Electrical Engineering from the National Technical University of Athens, a Master's Degree in EE/Automatic Control and a Ph.D in EE/Signal Processing, both from the University of New Hampshire. He is a member of the IEEE, the ACM and the Boston Computer Society.

Laino, Jane

A consultant and President of DIgby 4 Group, Inc. (New York, NY – 212-883-1191, www.digby4.com), a company that assists clients with telecommunications planning, implementation, administration and expense management. She has a B.A. from Queens College in New York City.

Laino is the author of *The Telephony Book*, available from CMP Media.

She can be reached by e-mail at jlaino@digby4.com

Laakso, Melvin T.

As Vice President of Product Development for Dialogic (Parsippany, NJ – 973-993-3000, www.dialogic.com) an Intel company, Melvin T. Laakso is responsible for Core and DM3 Product Family Development at Dialogic. Laakso is a seasoned senior executive with over 30 years of hardware and software development and management experience. He joined Dialogic from Systemsoft Corporation, where he was Executive Vice President of the Platform Business Unit. Laakso also served Systemsoft as Vice President of Platform Engineering and Senior Director of BIOS Engineering. Previously, Laakso spent 26 years at IBM in a variety of management positions. He holds a Bachelor of Science degree in electrical engineering from the University of Washington and completed post-graduate work at Systems Research Institute (SRI) and Carnegie-Mellon University.

Landau, John

As Vice President of Strategic Marketing for Dialogic (Parsippany, NJ – 973-993-3000, www.dialogic.com) an Intel company, John Landau provides support to the company's business and technology teams in defining and developing "next generation" directions. He plays an intricate part in driving the overall strategic direction for Dialogic. One of his most important responsibilities is to cultivate partnerships with major customers and technology suppliers.

Landau also leads the company's efforts to foster the acceptance of open standards for CT systems. He spearheaded the Signal Computing System Architecture (SCSA) initiative, cofounded the Enterprise Computer Telephony Forum (ECTF), the Intelligent Network Forum (INF), and the Voice over IP (VoIP) Forum.

Before rejoining Dialogic, Landau was a founder and Vice President of Marketing at Benchmarq Microelectronics in Dallas, Texas, which specializes in integrated circuits for managing rechargeable batteries. Before joining Benchmarq, Landau was the first Remote Area Sales Manager for

Dialogic, following several years at Advanced Micro Devices where he managed the telecommunications product marketing department.

Leyland, Steve

Vice President of the Core Product Group for Dialogic (Parsippany, NJ – 973-993-3000, www.dialogic.com) an Intel company. Active in the CT industry since 1989, Steve joined Dialogic at the beginning of 1995. He manages the company's Media Business Unit, which focuses on developing and marketing a wide range of media processing products including voice, fax, and speech recognition components for the small business, enterprise, and network provider markets. During his twenty-year career in telecommunications and electronics, Steve has held engineering, marketing and general management positions. Before joining Dialogic he was an independent consultant. Steve earned a bachelor's degree in electrical engineering with honors in 1977 from the University of Nottingham, England.

Levy, Ben

Ben Levy is president and chief technical officer of APEX Voice Communications, responsible for the management of corporate and technical affairs.

Mr. Levy is the original architect of APEX's core software products. In 1989 when most call processing solutions were still proprietary, Mr. Levy saw the opportunity to break into the open architecture call processing market using UNIX-based PC systems and Dialogic boards, and co-founded APEX with Elhum Vahdat.

Under Mr. Levy's direction, APEX has become a global presence and now manufactures enhanced services and billing solutions for leading network service providers worldwide.

Prior to co-founding APEX, Mr. Levy was Manager of New Technologies at Dial Info / ACP where he worked for two years. In 1981, Mr. Levy opened After Hours at Suzanne's, a late-night eatery located in Palo Alto, which he sold in 1983. He was nominated by Dan Witter for Price Waterhouse's Entrepreneur of the Year award in 1998.

Mr. Levy received a Bachelor of Science degree in computer science from San Francisco State University in 1985 and has completed an Executive Management program at UCLA's Anderson School of Business.

Mr. Levy resides in West Hills, California with his wife and three children.

Luhmann, Rick

As editor-in-chief of *Teleconnect* magazine, Rick Luhmann literally helped forge an industry when he began to unearth, describe, name and evangelize a strange new breed of business communication products and services in the mid to late 1980s. Built primarily on and around open PC-based hardware resources, these revolutionary platforms and their applications thankfully added modern-day, computer-driven intelligence to the art and making and taking phone calls and processing the multimedia (voicemail, e-mail, faxmail) messaging monster.

During these salad days – many of which fell well before Rick's "mentor," the legendary Harry Newton, finally stumbled upon the right term to describe the "computer telephony" revolution – Mr. Luhmann was a true high-tech publishing pioneer.

Among other things:

• He was the first person ever to detail a PC-based autoattendant / voicemail software package written to a Dialogic board (September of 1986)
• He discovered the "Open Application Interface" term and its technology in a July 1987 profile of an NEC PBX.
• He virtually defined the Interactive Voice Response (IVR) market and separated its contenders and pretenders in the Fall of 1988

- He coined – with the help of EASE CT Solutions' (nee Expert Systems'). Carlton Carden – the phrase for the "international callback" application the following year.
- He was the only member of the press at the original MVIP announcement (1990) who had an inkling of what was going on.
- He was the only journalist Dialogic's Jim Shinn met with the day after the MVIP introduction (so that Jim could explain why Dialogic's PEB was better than MVIP... which even Rick didn't believe).

In short, it's safe to say that Luhmann broke more ground on computer telephony and its core technology than all other high-tech journalists wrapped together in a giant bad-suit-wearing / free-food-eating ball – a fact, of course, that made him the natural choice for the founding editor position when *Computer Telephony* magazine prepared for a test issue in the Fall of 1993.

During the magazine's first year of operation in 1994, Luhmann almost single-handedly wrote every one of the bimonthly issues himself. It was not until October of 1994 that "Zippy" Grigonis came on board and enough material could now be generated to establish a monthly version of the magazine, which started in January 1995 and persists to this day.

In April of 1999, Luhmann resigned his position as editor-in-chief of CT to become a house husband for several months. He then became editor-in-chief of content for ComputerTelephony.com – the website adjunct to the fine magazine of the same name – in November of 1999.

Says Luhmann: "Our mission is to provide top-notch original content for ComputerTelephony.com, exploiting the Web's 'instant publication' ability to cover breaking news in a way our monthly magazines can't. Our goal is to provide the same timely coverage as a wire-service feed, but to make the reader experience more nutritive by enriching raw stories with industry-insider background info, comments and interpretation. At the same time, we're exploiting the freedom the Web gives us to publish longer stories, to cover more-specialized topics, and to link in more-varied source material (industry white-papers, etc.) than conventional magazines can."

Luhmann also produces a twice-weekly email newsletter, bringing summaries of top stories to subscribers of ComputerTelephony.com, and providing links to web-based copy.

Luhmann has also been a much sought-after technical editor and communications expert over the years. One major effort on that front was the best-selling *Communications Systems & Networks: Voice, Data, and Broadband Technologies*.

Rick Luhmann received a Master's Degree in English literature from William Paterson College in 1985.

Margulies, Edwin

Currently CEO of Telephony@Work. Ed is responsible for channel development, marketing, sales, media, supply lines, alliances and production. He is a 20-year industry veteran, having specialized in telecommunications start-up companies including Dialogic and Voicetek. Margulies is also an inventor and author of nine best-selling books on telecommunications and the Internet. He is among the industry's foremost evangelists and strategists.

Margulies' most recent post was group director for CMP Media's Converging Communications Group events and conferences. He had P&L responsibility for the company's U.S.-based events including Call Center Demo, CRM Demo, Internet Telecom Expo and the world-renowned Computer Telephony Expo. Here, he created and launched five new businesses for the company.

Before running media businesses at CMP, Margulies was director,

sales and marketing for Enhanced Platforms at Dialogic Corporation (now an Intel Company). Here he launched the worldwide SCSA architecture campaign and authored numerous successful business plans for the company, leading to its IPO in 1993. Margulies was also director of Unix marketing for Unisys' Communications Industry Systems Division, and manager of national account development for Voicetek Corporation (now Aspect Communications). He worked on both regulated and non-regulated sectors as a communications consultant for CONTEL (now GTE) in the early eighties. He and two partners ran a multi-city interactive messaging service bureau, which gave him a unique perspective on the challenges facing enhanced service providers and ASPs.

Margulies' sales, marketing and product management duties have encompassed hundreds of designs for telephone company deployment and CPE (customer premises equipment). These include Network-Based Routing and Intercept, HOBIC replacement, Interactive Voice Response, Communications Controllers and Cable Pay-Per-View systems. These experiences prompted his invention of an ANI Converter system, patented in 1991 and now assigned to Intel Corp. This was the basis from which he authored books and created conference content for students of telecommunications.

Since 1993, Margulies authored nine best-selling books on the Internet, Call Centers and Computer Telephony including: *Understanding Java Telephony; Secrets of Windows Telephony; Understanding The Voice-Enabled Internet; 1001 Computer Telephony Tips, Secrets and Shortcuts; SCSA: Signal Computing System Architecture; Client Server Computer Telephony; 337 Killer Voice Processing Applications, Audio Teleconferencing – The Complete Handbook;* and *The UnPBX – The Complete Guide to the New Breed of Communications Servers.* Between 1996 and 2000, he penned the monthly "CT Periscope" column for *Computer Telephony* magazine in addition to a weekly on-line column for *VARBusiness.*

Marks, Gary

As Vice President of Marketing for Dialogic (Parsippany, NJ – 973-993-3000, www.dialogic.com) an Intel company, Gary Marks is responsible for setting the company's marketing strategy and for directing the groups that carry it out – including Channel and Partner Marketing, Web Marketing, and Corporate Communications. Marks joined Dialogic from SyQuest Technology where he was Executive Vice President of Marketing. He has also served as Vice President of Marketing for Conner Peripherals and Western Digital Corporation. Prior to that he held various management positions at AT&T/Western Electric. Marks serves on the Board of Governors for the MultiMedia Telecommunications Association (MMTA). He holds an MBA degree in finance from Fairleigh Dickinson University and a BBA degree in business management and statistical analysis from the University of Miami.

Newton, Harry

The inventor of computer telephony? At least that's what the more uninformed history books will show (the same ones that will list Al Gore as inventor of the Internet). But as Henry Ford said, "history is bunk." Here's the real story. Harry and his partner, Gerry Friesen, were running a monthly telecommunications magazine called *Teleconnect*. It's doing okay. But one area they didn't have covered (as regards advertising) was computer telephony. Someone named Marc Ostrofsky (and then Advanstar whom Ostrofsky sold his properties to) had a bunch of trade shows and a magazine entitled *Voice*. They were doing very well. Newton needed to kill off the competing Voice shows and magazines, so he coined the term "Computer Telephony" and told anyone who would listen that computer telephony was "bigger" than "Voice" and that limiting your participation to Voice would limit your sales and your profits. He didn't know whether this was true. But it was a good story. And many people, it came to pass, listened to Newton and believed him. He and Friesen put out *Computer Telephony* magazine (listed in the February, 2000 *Folio* magazine as one of the 20 hot start up magazines of the 1990s) and the Computer Telephony Conference and Exposition, and effectively killed the Advanstar's Voice shows and magazine, which no longer exist (except in England). In September, 1997, Newton and Friesen then sold their whole empire of successful magazines and trade shows – *Call Center, Computer Telephony, Imaging, Teleconnect* and Computer Telephony Conference and Exposition (CT Expo) to Miller Freeman, Inc. (now CMP) for a smidgen over $130 million. *Folio* magazine listed it as the fourth biggest magazine deal of 1997.

Newton has moved on. He now focuses his life on updating his immensely successful dictionary – *Newton's Telecom Dictionary* (published by CMP Books; www.HarryNewton.com). And, in partnership with Gerry Friesen, he has started a new magazine called *Technology Investor* (www.TechnologyInvestor.com). You can get a free subscription to his new magazine by going to his new website.

Pulver, Jeff

The Internet telephony industry's first celebrity and most vociferous proponent, Jeff Pulver (516-753-2640) is President / CEO of pulver.com, Inc., an Internet based consulting firm. Pulver.com, publishes Internet Technology related research, such as The Pulver Report, produces trade shows such as the Voice on the Net and other related conferences and workshops., Pulver.com also provides "Thought Leadership" and consulting services to the telecommunications industry.

Jeff is one of the leading experts in the field of streaming audio and video technologies and their effect on business communications. He is a pioneer in the field of Internet Telephony, having set up one of the first Vocaltec servers in his home in 1995. Jeff is the moderator of several related mailing lists including the Voice on the Net (VON) and Internet Phone mailing lists, and has developed a related resource library on the net at www.pulver.com

Jeff Pulver is also the founder of the Voice on the Net (VON) Coalition, a not-for-profit group formed in March, 1996 to help keep the Internet Telephony Industry free from regulations.

Jeff is frequently quoted in the trade and business press and is available for speaking engagements. Contact Barbara Foster at Keynote Speakers in Palo Alto, CA.

Quintas, Carlos

Founder and Chief Executive Officer of Easyphone in Portugal, now called Altitude Software (Milpitas, CA – 408-965-1700, www.altitudesoftware.com).

Carlos Quintas is a pioneer in the customer contact industry and has been recognized by *Tornado Insiders* magazine as one of Europe's top entrepreneurs. In 1993, Quintas and his team developed the industry's first integrated inbound and outbound contact center solution. Prior to Altitude Software, he was with SSF, Portugal's leading provider of business applications for the finance and credit industry. Earlier in his career, he worked at EEI, the first company in Portugal to distribute UNIX-based computers. In addition, Quintas also held a computer science teaching position at the Technical University of Lisbon. He is a recognized expert in the field of computers and telephony, and has published papers on the subjects of relational databases and speech recognition. He received his degree in electronics, with an emphasis in computer sciences and

telecommunications, from the Technical University of Lisbon.

Rainville, Francois

Co-founder in 1988 and president from 1992 to 2000 of PRIMA, a Montreal-based company which merged with MediaSoft Telecom in 2000 to become Elix. Prior to becoming PRIMA's president, Rainville held various strategic functions within the organization. His responsibilities include business development, major account management and establishment of the Toronto office.

He is a member of the Board of Directors and Executive Committee of the Conseil executif du Centre de promotion du logiciel quebecois, was named Entrepreneur of the Year by the Montreal Junior Chamber of Commerce in 1996 and "Personnalite de la semaine: by the newspaper *La Presse*. Rainville has a bachelor's degree from the Universite de Montreal in Computer Science. Prior to co-founding PRIMA in 1988, he worked as a Computer Consultant for Andersen Consulting.

Schechter, Robert

President, CEO and Chairman of Natural MicroSystems (NMS) (Framingham, MA – 508-620-9300, www.nmss.com). Bob Schechter holds a B.S. from Rensselaer Polytechnic Institute and an MBA from the Wharton School at the University of Pennsylvania.

Schechter joined Natural MicroSystems in April 1995 as President and Chief Executive Officer, and became chairman in March 1996.

Prior to joining NMS, Schechter spent eight years at Lotus Development Corporation in Cambridge, Massachusetts, first as Senior Vice President of Finance and Operations and Chief Financial Officer, and most recently as Senior Vice President of the International Business Group. In this capacity he was responsible for all sales, marketing, customer service and product development activities outside of North America as well as Lotus global consulting services. Prior to joining Lotus, Schechter was a partner with Coopers & Lybrand in Boston, where he served as Chairman of the Northeast Region High Technology Practice.

Schechter is a member of the Board of Infinium Software, Inc., the Boston Children's Museum, and the Massachusetts Telecommunications Council.

Shapiro, Jonathan

Founder and President of Alliance Systems in Plano, Texas (Plano, TX – 972-663-3400, www.alliancesystems.com). Shapiro has a BBA from Hofstra University in New York.

A pioneer and visionary in the communications industry, Shapiro is also considered to be the "father of the CT secondary market." He'd worked in telecom for many years at Rolm, IBM, VMX and Brite Voice Systems. In 1991 he saw the writing on the wall. Voice processing was by then almost entirely PC erector-set based. Previously used pieces were becoming available. Shapiro got his hands on 50 used Dialogic boards and resold them immediately. Then he bought $50,000 dollars worth of boards and sold them immediately. The rest is history.

Remember the 900 "dirty talk" industry that died when advanced boredom set in? Shapiro and company bought all their equipment and resold the parts out of Shapiro's garage. That's when another thought struck him. Not only did people need good-quality CT boards, but they needed all the other stuff too – software, cables, PCs, and help.

Every card maker sold direct and the small developer needed to contact eight to ten different vendors to buy all of the bits and pieces necessary to put a whole system together.

There was a need for One Stop Distribution of CT components – new and old. Shapiro pounced on the opportunity, building the world's largest distributor of CT technology, becoming Dialogic's first authorized distributor and creating a family of thousands of customers in 70 countries.

Shapiro then moved Alliance Systems' activities from being that of a reseller to a manufacturer of components for the Internet infrastructure. They also produce SS7 and wireless platforms.

Shapiro's company is an incredible success story, having consistently maintained profitability and experienced an average annual sales growth of 58%. The company's annual revenue run rate is currently over $65 million.

An increasingly major force in the industry, Alliance Systems is a charter member of the Intel / Dialogic CT Media Value Network and has initiated a program to certify Alliance open communication systems with CT Media applications. Indeed, in 2000 they sponsored the CT Media Olympiad Developers' Competition, with a first prize of $20,000 or a Harley-Davidson motorcycle.

Sher, Barry

President of Telephony Experts (Los Angeles, CA – 310-445-1822, www.telephonyexperts.com). Barry graduated from the University of Minnesota with a bachelor's degree in Economics. Barry is a founding member of Telephony Experts and brought to the company an extensive background in sales and marketing. Barry's experience in the computer industry began in the mid 1980s where he led in sales for the Hewlett Packard, Panasonic and Kyocera product lines with the largest office automation company in Minnesota. Barry has led Telephony Experts to a sustained 100% growth rate since the company's incorporation and is responsible for the well-managed state and continued profitability of the company.

You can reach Barry at Barry@telephonyexperts.com

Strathmeyer, Carl R.

Director of Marketing for Dialogic Corporation's Computer-Telephone Division, which was formed in 1995 when Dialogic acquired the CIT technology from Digital Equipment Corporation. In this position, he is responsible for determining and executing Dialogic's product strategy for computer-telephone integration software products and related professional services.

Strathmeyer had been with Digital 17 years in a variety of management, marketing and technical positions in the fields of computer-telephone integration, public telecommunications network applications, computer network engineering, corporate telecommunications, and business systems design.

Before joining Digital, Mr. Strathmeyer was a business systems consultant with Index Systems (now CSC Index) and Casher Associates. He holds a Bachelor of Arts degree from Dartmouth College in mathematics and computer science and is a former chairman of the computer-telephone division of the Multi-Media Telecommunications Association (MMTA), the principal trade association for the computer-telephone industry.

Strathmeyer has written numerous papers and articles on message systems and voice-related computing which have appeared in publications of the International Federation for Information Processing (IFIP), the IEEE, and the computer-telephone industry trade press. He has contributed material to several books, including *Client-Server Computer Telephony*, edited by Ed Margulies and published by CMP.

Tal, Moshe

Co-Founder, President, CEO and Director of MUSIC Telecom (Hackettstown, NJ – 908-684-1300, www.musictelecom.com). The company was founded in 1991 as Newvoice, Inc. and was subsequently acquired by MUSIC Corporation in 1996, and renamed MUSIC Telecom.

Prior to founding NewVoice, Mr. Tal was Senior Electronics Engineer for InnoVoice, a division of Extrema Systems International Corporation, where he was responsible for the design and implementation of analog and digital products using Digital Signal Processing (DSP) technology. During the 1980s, Mr. Tal was a Design Engineer with Telkoor, Ltd., Israel, and with Meeda Ltd., Israel.

Mr. Tal holds a B.S. in Electronic Engineering from Tel Aviv University.

Thomé, Henrik

Henrik is the President of Envox (Naples, FL – 941-793-0725, www.envox.com), one of the world's leading providers of server software for the development of communications applications that integrate voice, Web, database, fax and other technologies. Henrik founded his first company, an offset printing business, when he was 16 and was a regular freelance journalist for the National Swedish radio while still at college. He later moved into the film business, working at Mexfilm and Primamedia before becoming an award-winning freelance video/film producer, creating internal productions for blue chip Swedish companies such as Volvo, Vattenfall and Skandia. In 1990, Henrik founded Svensk Börsinformation, a company that provided financial information and real time stock quotes over the phone before selling this business to Bonnier, the largest publishing company in Sweden. In 1991, he founded Linewise, a system integrator and Service bureau for IVR, fax, Intranet and public Web solutions which was also recently sold.

Trumbull, Dean

As Vice President of CT Switch Products for Dialogic (Parsippany, NJ – 973-993-3000, www.dialogic.com) an Intel company, Dean Trumbull oversees the company's development of open switch platforms and products, creating hardware and software platforms for telco and enterprise switching environments. The CT Switch Products business unit delivers the Dialogic line of switching boards and the CT Media product, all of which are now being accepted by leading call center and PBX companies to build next-generation business communication platforms (BCPs). Under Trumbull's guidance, the company is accelerating CT Media enhancement to support a complete range of Dialogic products and customer solutions.

Before joining Dialogic, Trumbull spent three years at Melita International in Norcross, Georgia, serving as Vice President of Advanced Technology. Among many other accomplishments, he developed Melita's next-generation switching platform known as Mpower. Trumbull also spent 12 years in various management and strategic development positions at Intecom, Inc., in Dallas, Texas.

Turner, Brough

Senior Vice President of Technology, Chief Technology Officer of Natural MicroSystems (NMS) (Framingham, MA – 508-620-9300, www.nmss.com).

Turner was a co-founder of Natural MicroSystems and key contributor to the development of early NMS products. He also served as Vice President, Operations for the first 8 years of the company's growth. He currently serves as Senior Vice President of Technology, overseeing the evolution of product architectures and their underlying technology, and participating in new market development. In addition, he writes a monthly column in CTI Magazine and speaks widely on behalf of NMS and the Open Telecommunications industry.

In 1984, Mr. Turner created one of the industry's first public voice response demo systems which, for several years, gathered more sales leads than any other NMS marketing program. In 1989-90, he created the Multi-Vendor Integration Protocol (MVIP) a telephony bus and telephony switching architecture that, by the mid-1990s, had been adopted by hundreds of companies. Mr. Turner originated the "End the Bus Wars" initiative that eventually led to the Enterprise Computer Telephony Forum's CT Bus specification (H.100). He was an early champion of CompactPCI for Telecommunications, and in 1999 he created the Open Source for Open Telecom Initiative (www.opentelecom.org). While Mr. Turner has had far ranging telecommunications experience, digital signal processing (DSP) expertise has remained a key focus for more than 20 years.

Prior to co-founding NMS, Mr. Turner spent 13 years in engineering and engineering management at Digilab, Inc., an analytical instruments manufacturer and Block Engineering Inc., both subsidiaries of Bio-Rad Laboratories. In these positions, Mr. Turner was involved with the design of analytical and biomedical instruments, making extensive use of computers for digital signal processing, pattern recognition and scientific visualization. Rumors of a telephone hacking career during his college years are greatly exaggerated.

Mr. Turner holds a B.S. in Electrical Engineering from the Massachusetts Institute of Technology. He is a member of the IEEE and the ACM. He also participates in relevant industry and Internet bodies including the IETF, ECTF, GO-MVIP and IMTC.

Brough (rhymes with "gruff"... even though he isn't at all) Turner is a real pioneer and one of the most knowledgeable people you'll ever find in this business.

Warner, Bill

Executive Vice President, Signal Computing Products Bill Warner has over 30 year's experience in the computer industry. He is responsible for the overall strategy, marketing, and development products made by Dialogic. Before joining Dialogic, he was senior vice president of product management and development at Banyan Systems.

Previously, Bill served as vice president and general manager of the platform software business for SystemSoft. During his twenty-six-year tenure at IBM, he held a number of management and executive positions in networking hardware and software, operating systems, and middleware, leading him to the position of vice president, systems management software.

Wilson, Thomas C.

As Vice President of Customer Engineering and Services for Dialogic (Parsippany, NJ – 973-993-3000, www.dialogic.com), Tom Wilson is responsible for all aspects of technical service delivery to Dialogic customers, worldwide. Since joining Dialogic in November 1992, Wilson has overseen the continued growth of the Dialogic Applications Engineering organization, restructured the company's field trial processes, increased customer technical communications through the FirstCall InfoServer(tm) support Web site, and launched new CTI Education and Consulting Services functions. These changes resulted in the formal announcement of the Consulting Services business unit in May 1998. Wilson has also been an active participant in promoting software quality engineering methods and metrics within Dialogic, and is a member of the Americas Sales and Services organization and the Products and Services business management group.

Before joining Dialogic, Wilson spent three years as founder and principal at a software company consulting with major computing industry ISVs and government defense contractors in the area of software engineering metrics. Before that, Wilson spent nine years with Prime Computer, where his last position was Director of Service Planning. Prior to joining Prime Computer, he held the position of Staff Scientist at Nichols Research Corporation. Wilson is a voting member of the Association for Computing Machinery (ACM), the Institute of Electrical and Electronics Engineers (IEEE), and the Association of Field Service Managers International (AFSMI).

Wilson holds a BS degree in physics and an MS degree in computer science from the University of Michigan.

Woolf, Howard

As President of the Americas for Dialogic (Parsippany, NJ – 973-993-3000, www.dialogic.com) and Intel company, Howard Woolf is responsible for overseeing the entire Americas Sales and Services operation, which includes the sales teams in North America, Latin America/Caribbean, and the Global Account, Channel Sales, and Market/Product Segment units. The technical sales group, customer and credit services team, and field sales support staff also fall under his responsibility.

Woolf has over 25 years of experience in the technology industry. Before assuming his current role with Dialogic, he served as Vice President of the Communications Industry Group at Sapient Corporation. Prior to joining Sapient, he built a distinguished career at Compaq/Digital Equipment Corporation, ending his tenure there as Director of Worldwide Sales Operations. Mr. Woolf began his career with the General Electric Company in their Manufacturing Management Program.

Mr. Woolf received his MBA from Boston College and his BSCE from Clarkson University in Potsdam, New York.

535

Milton Keynes UK
Ingram Content Group UK Ltd.
UKHW051923141024
449569UK00027B/1328